Practice Problems

for the Mechanical Engineering PE Exam

A Companion to the *Mechanical Engineering Reference Manual*

Thirteenth Edition

Michael R. Lindeburg, PE

Professional Publications, Inc. • Belmont, California

Benefit by Registering This Book with PPI

- Get book updates and corrections.
- Hear the latest exam news.
- Obtain exclusive exam tips and strategies.
- Receive special discounts.

Register your book at **ppi2pass.com/register**.

Report Errors and View Corrections for This Book

PPI is grateful to every reader who notifies us of a possible error. Your feedback allows us to improve the quality and accuracy of our products. You can report errata and view corrections at **ppi2pass.com/errata**.

PRACTICE PROBLEMS FOR THE MECHANICAL ENGINEERING PE EXAM

Thirteenth Edition

Current printing of this edition: 4

Printing History

date	edition number	printing number	update
Apr 2014	13	2	Minor corrections.
Apr 2015	13	3	Minor corrections. Minor cover update.
Jul 2016	13	4	Minor corrections.

Printed in the United States of America.

PPI
1250 Fifth Avenue
Belmont, CA 94002
(650) 593-9119
ppi2pass.com

ISBN: 978-1-59126-415-6

Library of Congress Control Number: 2013938441

F E D C B A

Topics

Background and Support
Fluids
Thermodynamics
Power Cycles
Heat Transfer
HVAC
Statics
Materials
Machine Design
Dynamics and Vibrations
Control Systems
Plant Engineering
Economics
Law and Ethics

Where do I find help solving these Practice Problems?

Practice Problems for the Mechanical Engineering PE Exam presents complete, step-by-step solutions for more than 850 problems to help you prepare for the Mechanical PE exam. You can find all the background information, including charts and tables of data, that you need to solve these problems in the *Mechanical Engineering Reference Manual.*

Table of Contents

Topic XII: Plant Engineering

Topic XIII: Economics

Topic XIV: Law and Ethics

Preface and Acknowledgments

Practice Problems for the Mechanical Engineering PE Exam has humble beginnings. The first few editions were simple collections of weekly homework assignments with handwritten solutions for the students in my classes. The assignments contained targeted practice problems that I wrote to illustrate the most important concepts in my lectures, which coincided with the most important exam concepts. In those days, the number of assigned problems was a function of my perception of what an average employed engineer could accomplish in a week. I always assigned a few more problems than most people could finish, because I wanted everyone to be working to capacity. Still, the amount of work was limited, and many important engineering concepts remained unrepresented in the weekly homework.

It was a great revelation to me, as a wet-behind-the-ears educator, that the problems in my homework assignments could cover subjects that I didn't lecture about. Being able to "cover" subjects in the homework freed me up immensely. I found that I didn't have to feel guilty about spending an extra ten minutes in the classroom explaining something important or interesting that took me off of my lecture schedule. The same logic applies to synchronization between the *Mechanical Engineering Reference Manual* and *Practice Problems for the Mechanical Engineering PE Exam*. Okay, I have to admit that they're not 100% perfectly synchronized. Yes, there are problems in this book that are not represented by theory in the *Mechanical Engineering Reference Manual*. But, there aren't many of these. If you think you've discovered a problem that draws on knowledge not in the *Reference Manual*, use the index, as it is more likely that you're looking in the wrong chapter.

Someplace along the way, the purpose of this book changed from supporting a live classroom course to supporting a thorough at-home review. Each subsequent edition has added content, expanded coverage, and increased girth. This edition, in particular, greatly expands coverage, exposing you to much more of the *Mechanical Engineering Reference Manual*. Those original, targeted practice problems are still here, of course, although they have been dismembered into smaller pieces and reformulated into multiple-choice format. But, accompanying them are many new 6-minute zingers that require more than common knowledge, typical thinking, and at-hand references.

I'm sure you'll recognize when you're working one of the new zingers, as opposed to one of the original problems. Most of the new problems are stand-alone, multiple-choice problems designed to be solved in six minutes or less, just like the exam. The original problems, although also formulated into multiple-choice format, usually come in sets of interrelated parts sharing a common scenario. And, the original problems are often more complex, requiring more digging, and taking more than six minutes. As an educator and author, I rely on these more difficult problems to cover more concepts than I could with a book full of one-liners. I've written about this in other books before, but I'm rephrasing it slightly when I say that you'll learn more from five 30-minute problems than from twenty-five 6-minute problems. The 6-minute zingers usually involve a single concept, one equation, one reference book, and drill-down thinking. The 6-minute zingers can be made as difficult as desired by basing them on obscure concepts, equations, references, and data. However, no matter how difficult it is, there's not much synthesis in a drill-down zinger. A 30-minute problem requires synthesis of many concepts, equations, and references. A 6-minute zinger illustrates the exam format. The harder problems teach you engineering. You'll see.

As for improvements over the previous edition, there are many, but you are probably only interested in one change: This edition contains 301 more problems than the previous edition. As an author, I care that the content from the previous edition has been edited, improved, rewritten, made more consistent, and brought up to style. I care that there are more equations, fewer skipped steps, more references to sources, better recognition of significant digits, and strict adherence to the inclusion of units. I care that the chapters have been reorganized; some have been split up; some have been renamed. I appreciate the editing, proofreading, illustrating, and pagination that collectively constitute this new edition. However, since you don't have (don't know about, and don't care about) the previous edition, all that you need to know is that you're going to be well-prepared for the exam after working through all of the problems in this book.

Now, I'd like to introduce you to the "I couldn't have done it without you" crew. The talented team at PPI that produced this edition is the same team that produced the new edition of the *Mechanical Engineering Reference Manual*, and the team members deserve just as many accolades here as I gave them in the Acknowledgements of that book. As it turns out, producing two new books is not twice as hard as producing one new book. Considering all of the connections between the two books, producing this edition is greatly complicated by the need for consistency with and cross-references to the *Reference Manual*. Producing a set of interrelated books is not a task for the faint of heart.

Editing, typesetting, and paginating: Roger Apolinar, Tom Bergstrom, Ryan Cannon, Lisa Devoto Farrell, Kate Hayes, Tyler Hayes, Chelsea Logan, Scott Marley, Magnolia Molcan, Connor Sempek, Ian A. Walker, and Julia White

Proofreading: Bill Bergstrom, Ryan Cannon, Lisa Devoto Farrell, Tyler Hayes, Jennifer Lindeburg King, Chelsea Logan, Scott Marley, Magnolia Molcan, Connor Sempek, Bonnie Thomas, Ian A. Walker, and Julia White

Illustrating: Tom Bergstrom, Kate Hayes, and Amy Schwertman La Russa

Calculation checking: Andrew Chan, Todd Fisher, Prajesh Gongal, Scott Miller, Allen Ng, Jumphol Somsaad, Alex Valeyev, and Akira Zamudio

Cover design: Amy Schwertman La Russa

Management: Sarah Hubbard, director of product development and implementation; Cathy Schrott, production services manager; Jennifer Lindeburg King, Chelsea Logan, and Julia White, editorial project managers; Christine Eng, product development manager

For me, each new edition is an opportunity to be humbled. (Actually, it's an occasion to be humbled.) Each new edition requires a retreat from the security of pages whose kinks have supposedly been ironed out in a previous edition, and a charge into the critical, hot spotlight of a new edition. You would think that after all these years of writing problems and solutions I would know virtually all of the ways I make mistakes. I don't. But even if I did, I wouldn't be able to avoid all of the mistakes. In many cases, it takes the keen eyes of readers to catch what I didn't avoid. Here are some of the people who graciously made suggestions, recommended content, and pointed out errata during the lifetime of the previous edition. They didn't have to take the time to share their thoughts with me, but they did. And, everyone gets to benefit from their involvement.

Jonathan Brewer; Amanda Busch; Daniel Castell; Richard Davis; Alejandro Franyie; Ben Gossett; Brandee McKim; Logan McLeod; Karl Peterman; Adam Renaldo

If you've been mentioned in this Preface, I hope that you know how much I appreciate what you've done. I've been writing long enough to know that it takes a village to publish a book. Thank you for being a villager.

This edition has so much new material (and, accordingly, so many opportunities for showing my fallibility), that I'm already shaking in my boots. So, if you think you've found something questionable, or if you think there is a better way to solve a problem, I hope you'll help me out by visiting the PPI website at **ppi2pass.com/errata**. I personally respond to all comments made about my books. I like to learn new things, too. So, now it's your turn. Teach me.

Thanks, everyone!

Michael R. Lindeburg, PE

Codes Used to Prepare This Book

The documents, codes, and standards that I used to prepare this new edition were the most current available at the time. In the absence of any other specific need, that was the best strategy for this book.

Engineering practice is often constrained by law or contract to using codes and standards that have already been adopted or approved. However, newer codes and standards might be available. For example, the adoption of building codes by states and municipalities often lags publication of those codes by several years. By the time the 2013 codes are adopted, the 2015 codes have been released. Federal regulations are always published with future implementation dates. Contracts are signed with designs and specifications that were "best practice" at some time in the past. Nevertheless, the standards are referenced by edition, revision, or date. All of the work is governed by unambiguous standards.

All standards produced by ASME, ASHRAE, ANSI, ASTM, and similar organizations are identified by an edition, revision, or date. However, although NCEES lists "codes and standards" in its lists of mechanical engineering PE exam topics, unlike for the civil engineering PE exam, no editions, revisions, or dates are specified. My conclusion is that the NCEES mechanical engineering PE exam is not sensitive to changes in codes, standards, regulations, or announcements in the Federal Register. That is the reason that I referred to the most current documents available as I prepared this new edition.

How to Use This Book

This book is primarily a companion to the *Mechanical Engineering Reference Manual.* As a tool for preparing for an engineering licensing exam, there are a few, but not very many, ways to use it. And, at least one of those ways isn't very good.

For many editions, I envisioned this book being taken to work, on business trips, and (for the truly dedicated few) even on weekend getaways to the beach. I figured that a lighter book would "carry" a lot easier than the big *Mechanical Engineering Reference Manual*, so the practice problems would irresistibly call out to you every time you left the room. My vision was that you'd naturally want to bring the problems everywhere you went in order to maximize your preparation time. I never thought you'd be taking this book on a backpacking trip, but I figured you'd spend quite a few lunch breaks with the problems in your company's break area.

Now that this book has grown to behemoth size, it's less likely that it will leave your house. That's okay, as long as the problems continue to call out to you. The big issue is whether you really work the practice problems or just skim over them. Some people think they can read a problem statement, think about it for about ten seconds, read the solution, and then say "Yes, that's what I was thinking of, and that's what I would have done." Sadly, these people find out too late that the human brain doesn't learn very efficiently by observation alone. Under pressure, these people remember very little. For real learning, you have to spend some time with the stubby pencil.

There are so many ways that a problem's solution can mess with your mind. Maybe the stumble is using your calculator, like pushing log instead of ln, or forgetting to set the angle to radians instead of degrees. Maybe it's rusty math. What are $\text{erf}(x)$, $\cosh(t)$, and $\ln e(x)$, anyway? How do you complete the square or factor a polynomial? Maybe it's in finding the data needed (e.g., the specific heat of ice cream) or a unit conversion (e.g., watts to horsepower). Maybe it's trying to determine if an equation expects L to be in feet or inches, or if the volumetric flow rate is in gallons per minute or cubic feet per second. Maybe it's the definition of a strange term. Is the retardance coefficient the same as Manning's roughness constant? Getting past these stumbles takes time.

And unfortunately, most people learn by doing and have to make a mistake at least once in order not to make it again. Since making a mistake while taking the exam isn't an optimal strategy, working with the stubby pencil in your company's break area is looking more and more attractive.

Even if you do decide to get your hands dirty and actually work the problems (as opposed to skimming through them), you'll have to decide how much reliance you place on the published solutions. You'll naturally probably want to maximize the number of problems you solve by spending as little time as you can on each problem. After all, optimization is the engineering way. Are you stuck on a problem? It's tempting to turn to a solution when you get slowed down by details or stumped by the subject material. However, I want you to struggle a little bit more than that—not because I want to see you suffer, but because the "objective function" to be optimized is your exam performance. There are no prizes for minimizing your study time. When you get stuck, do your own original research as if you didn't have a detailed solution a few pages away. Start with the *Mechanical Engineering Reference Manual.* You'll be surprised what you can find in that book.

Learning something new is analogous to using a machete to cut a path through a dense jungle. By doing the work, you develop pathways that weren't there before. It's a lot different than just looking at the route on a map. You actually get nowhere by looking at a map. But cut that path once, and you're in business until the jungle overgrowth closes in again.

I chose each problem for a reason. If you skip problems, your review will be piecemeal. So, do the problems. All of them—even if you think you're not going to work in some subjects. Do them twice, once in customary U.S. units, and then, again, in SI units. Look up the references, and follow the links. Don't look at the answers until you've sweated a little. And, let's not have any whining. Please.

1 Systems of Units

PRACTICE PROBLEMS

1. Most nearly, what is 250°F converted to degrees Celsius?

(A) 115°C

(B) 121°C

(C) 124°C

(D) 420°C

2. Most nearly, what is the Stefan-Boltzmann constant (0.1713×10^{-8} Btu/hr-ft^2-°R^4) converted from English to SI units?

(A) 5.14×10^{-10} W/m^2·K^4

(B) 0.95×10^{-8} W/m^2·K^4

(C) 5.67×10^{-8} W/m^2·K^4

(D) 7.33×10^{-6} W/m^2·K^4

3. Approximately how many U.S. tons (2000 lbm per ton) of coal with a heating value of 13,000 Btu/lbm must be burned to provide as much energy as a complete nuclear conversion of 1 g of coal? (Hint: Use Einstein's equation: $E = mc^2$.)

(A) 1.7 tons

(B) 14 tons

(C) 780 tons

(D) 3300 tons

SOLUTIONS

1. The conversion to degrees Celsius is

$$\begin{aligned} T_{°C} &= \tfrac{5}{9}(T_{°F} - 32°F) \\ &= \left(\tfrac{5}{9}\right)(250°F - 32°F) \\ &= \boxed{121.1°C \quad (121°C)} \end{aligned}$$

The answer is (B).

2. The Stefan-Boltzmann constant represents a certain amount of energy (in Btu). Since it has hr-ft^2-°R^4 in the denominator, the energy is reported on a per hour basis and represents power. The energy is an areal value because it is reported per square foot. Since a square meter is larger than a square foot, the energy must be multiplied by a number larger than 1.0 in order to put it into a per square meter basis. (For example, on any given day, the energy that can be derived from a square meter of sunlight is greater than can be derived from a square foot of sunlight.) 1/0.3048 is the conversion between feet and meters, and it is larger than 1.0.

The Stefan-Boltzmann constant is reported on a per degree basis, and the same logic applies. One Kelvin (K) is a larger (longer) interval on the temperature scale than one degree Rankine. The amount of energy per Kelvin is larger than the amount of energy per degree Rankine. Therefore, the energy must be multiplied by a number greater than 1.0 to report it on a "per Kelvin" basis. This requires multiplying by $9/5$. (An incorrect conversion will result if the unit in the denominator is thought of as an actual temperature. Any given numerical value of temperature in degrees Rankine is larger than the same temperature in Kelvins. However, this problem does not require a temperature conversion. It requires a unit conversion.)

Use the following conversion factors.

$$\begin{aligned} 1~\text{Btu/hr} &= 0.2931~\text{W} \\ 1~\text{ft} &= 0.3048~\text{m} \\ 1°\text{R} &= \tfrac{5}{9}\text{K} \end{aligned}$$

Performing the conversion gives

$$\sigma = \left(0.1713 \times 10^{-8}\ \frac{\text{Btu}}{\text{hr-ft}^2\text{-}^\circ\text{R}^4}\right)\left(0.2931\ \frac{\text{W-hr}}{\text{Btu}}\right) \times \left(\frac{1\ \text{ft}}{0.3048\ \text{m}}\right)^2 \left(\frac{1^\circ\text{R}}{\frac{5}{9}\text{K}}\right)^4 = \boxed{5.67 \times 10^{-8}\ \text{W/m}^2\text{·K}^4}$$

The answer is (C).

3. The energy produced from the nuclear conversion of any quantity of mass is

$$E = mc^2$$

The speed of light, c, is 3×10^8 m/s.

For a mass of 1 g (0.001 kg),

$$\begin{aligned} E &= mc^2 \\ &= (0.001\ \text{kg})\left(3 \times 10^8\ \frac{\text{m}}{\text{s}}\right)^2 \\ &= 9 \times 10^{13}\ \text{J} \end{aligned}$$

Convert to U.S. customary units with the conversion 1 Btu = 1055.1 J.

$$\begin{aligned} E &= \frac{9 \times 10^{13}\ \text{J}}{1055.1\ \frac{\text{J}}{\text{Btu}}} \\ &= 8.53 \times 10^{10}\ \text{Btu} \end{aligned}$$

The number of tons of 13,000 Btu/lbm coal is

$$\frac{8.53 \times 10^{10}\ \text{Btu}}{\left(13{,}000\ \frac{\text{Btu}}{\text{lbm}}\right)\left(2000\ \frac{\text{lbm}}{\text{ton}}\right)} = \boxed{3281\ \text{tons}\quad (3300\ \text{tons})}$$

The answer is (D).

2 Engineering Drawing Practice

PRACTICE PROBLEMS

1. In which type of view is the building shown?

(A) isometric

(B) orthographic

(C) three-point perspective

(D) oblique

2. A $2 \times 2 \times 2$ cube is drawn two different ways. Which projections have been used?

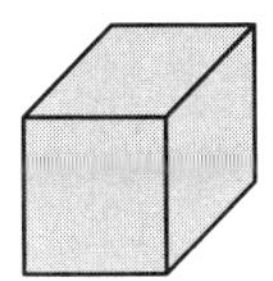

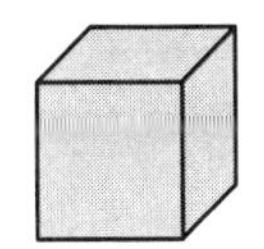

(A) oblique and isometric

(B) isometric and principal

(C) oblique and orthographic

(D) cavalier and cabinet

3. The portion of a staircase shown is presented in which type of view?

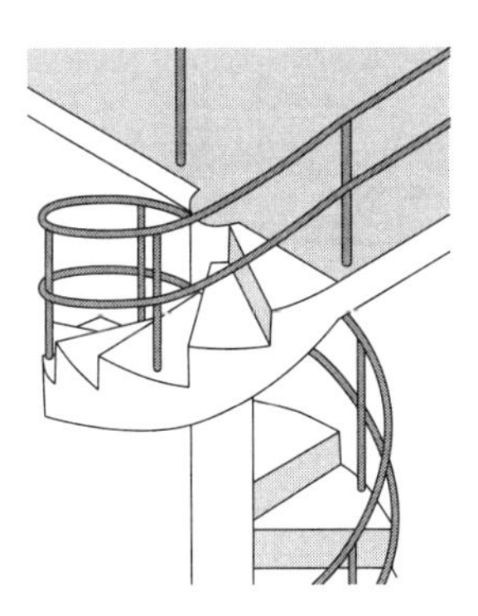

(A) isometric

(B) dimetric

(C) trimetric

(D) orthographic

4. Two views of an object are shown. Which option represents the missing third view?

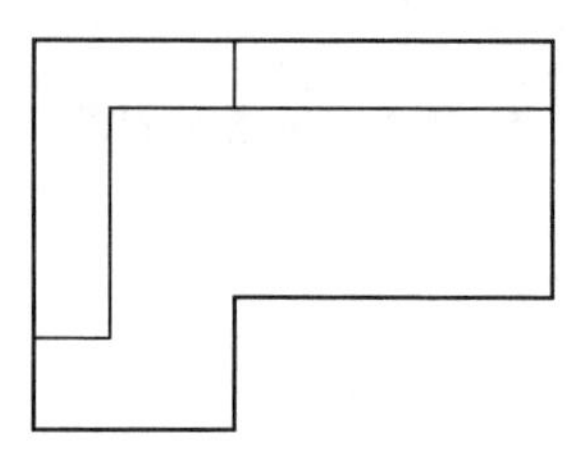

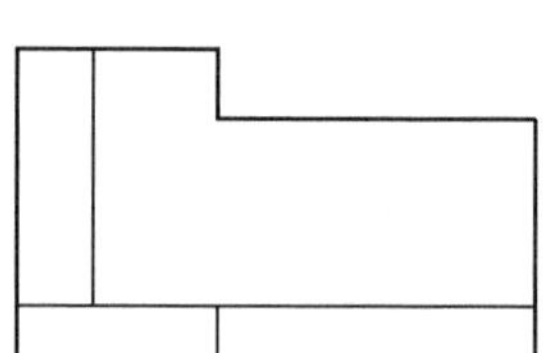

(A)

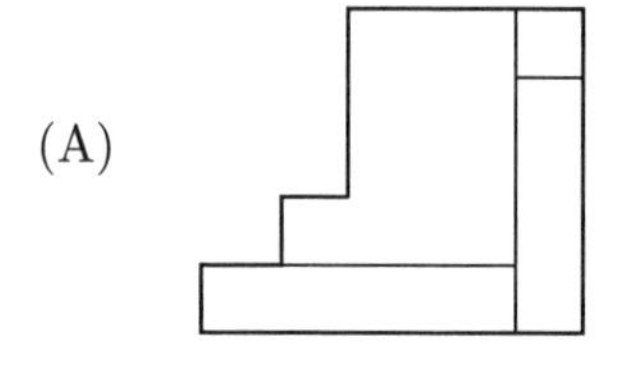

(B)

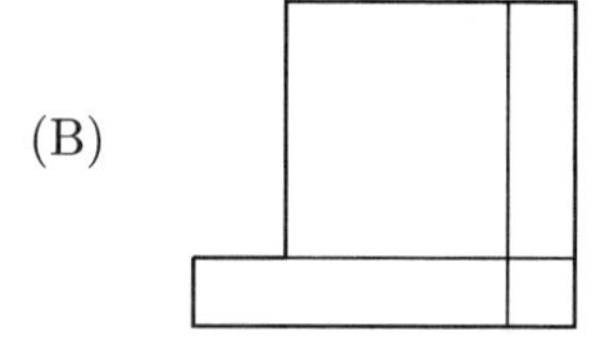

(C)

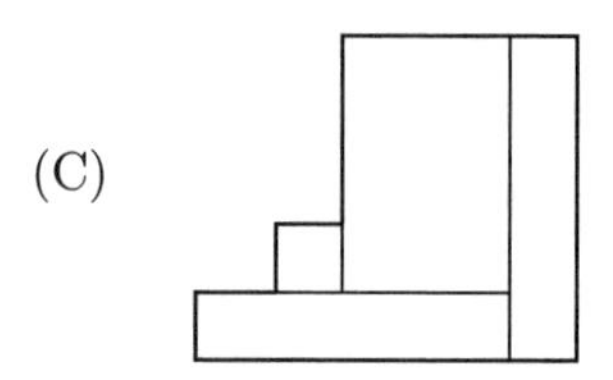

(D)

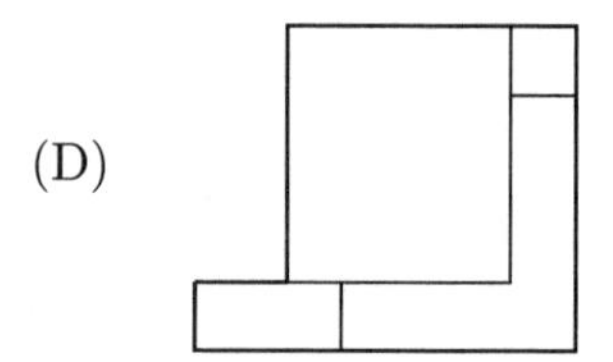

5. Two views of an object are shown. Which option represents the missing third view?

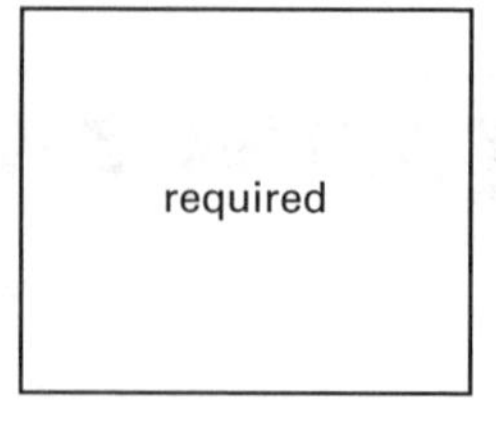

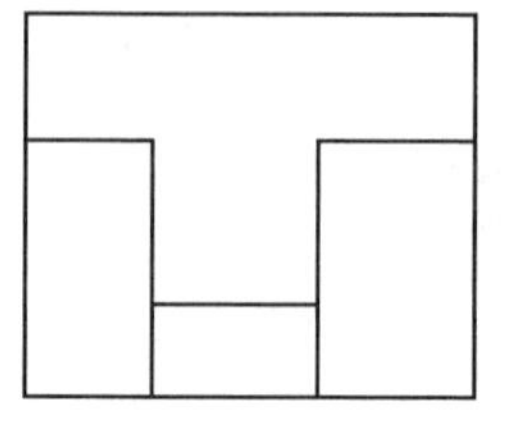

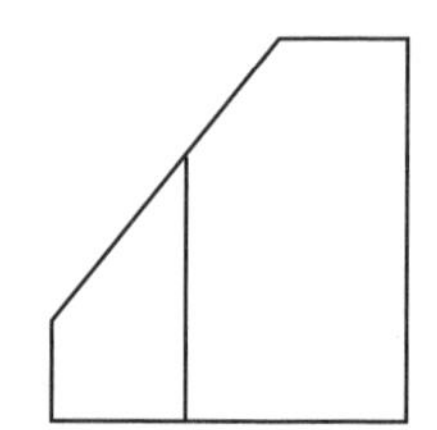

(A)

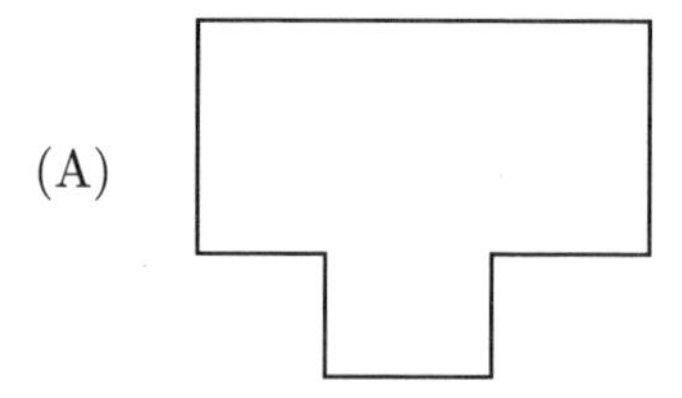

(B)

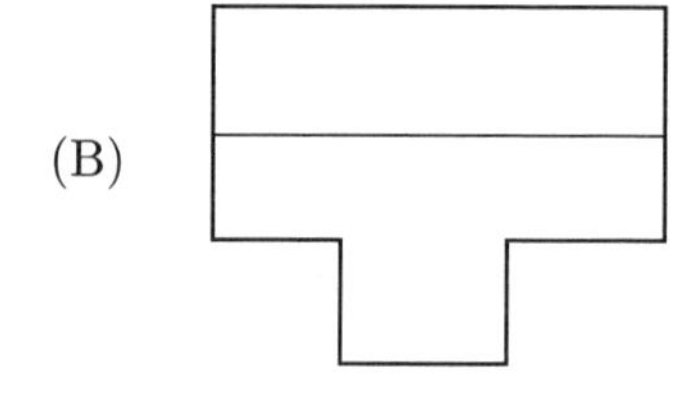

(C)

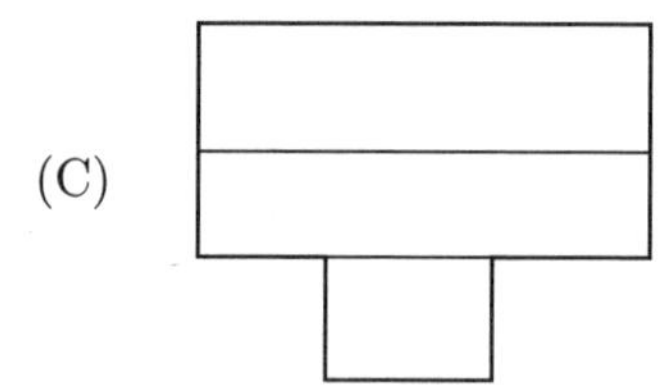

(D)

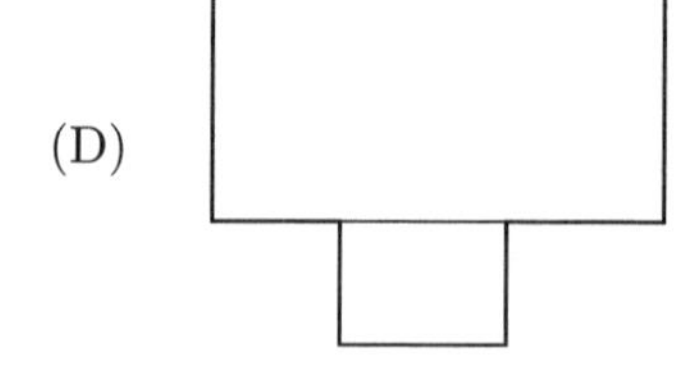

6. Two views of an object are shown. Which option represents the missing third view?

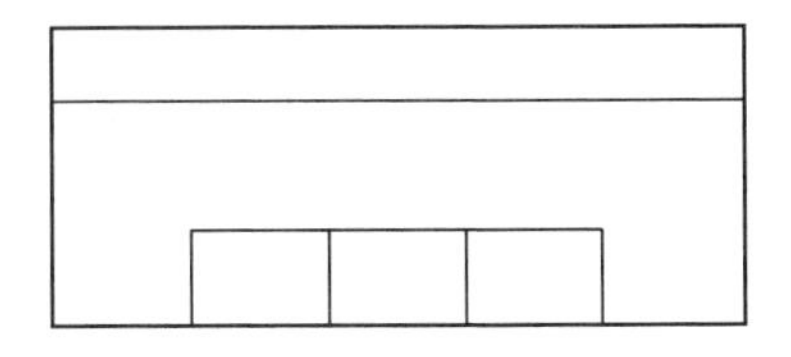

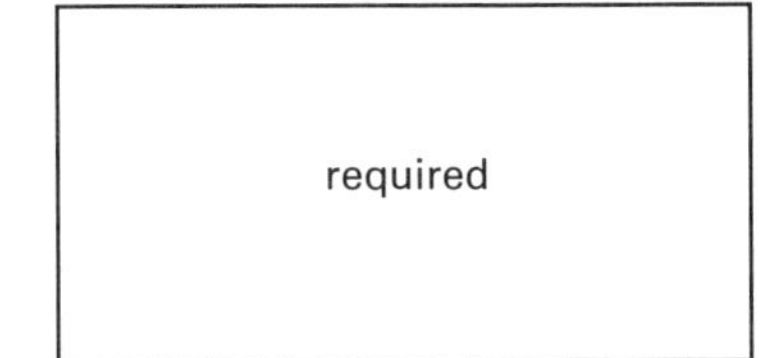

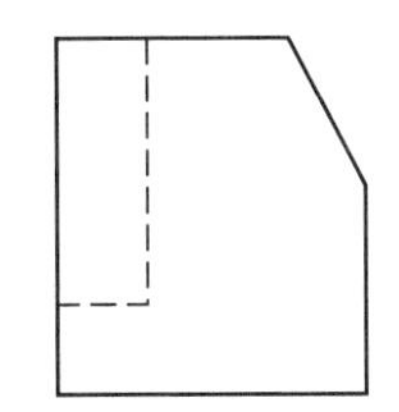

(A)

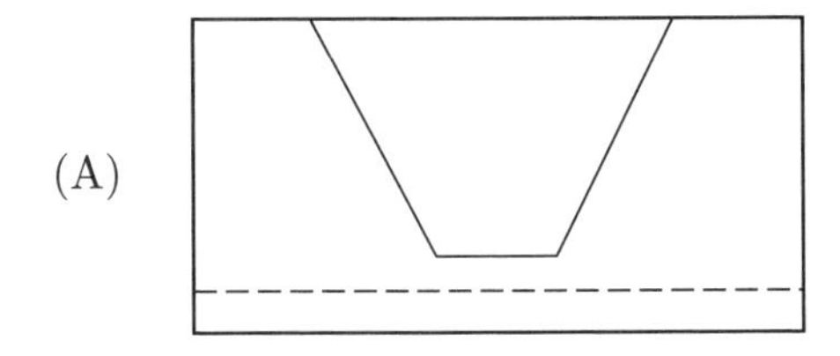

(B)

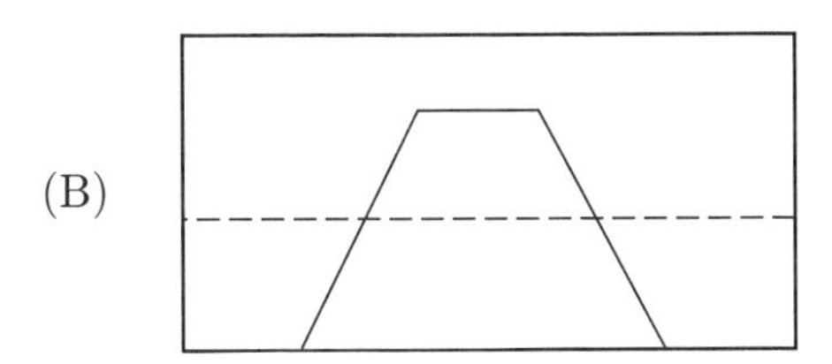

(C)

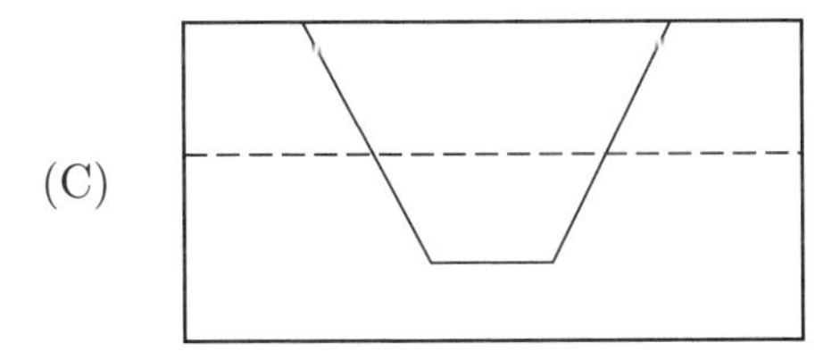

(D)

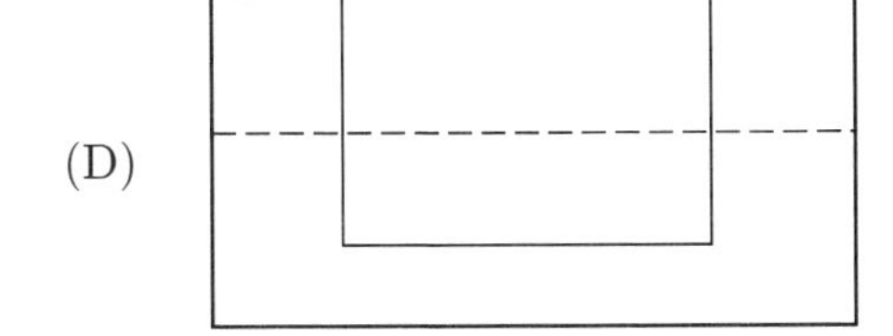

7. A pictorial sketch of an object is shown. Which option represents a three-view, orthographic drawing set?

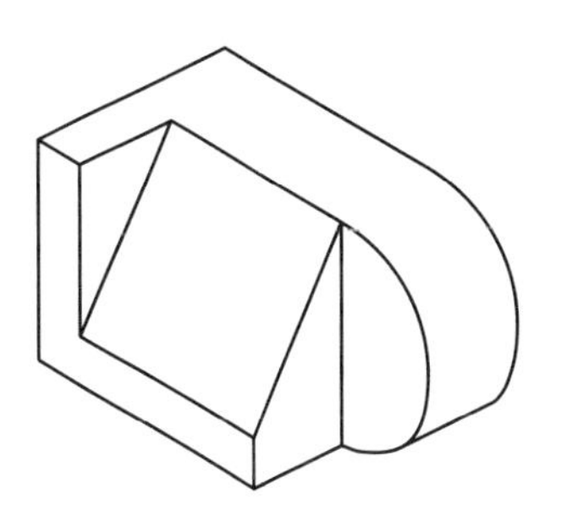

(A)

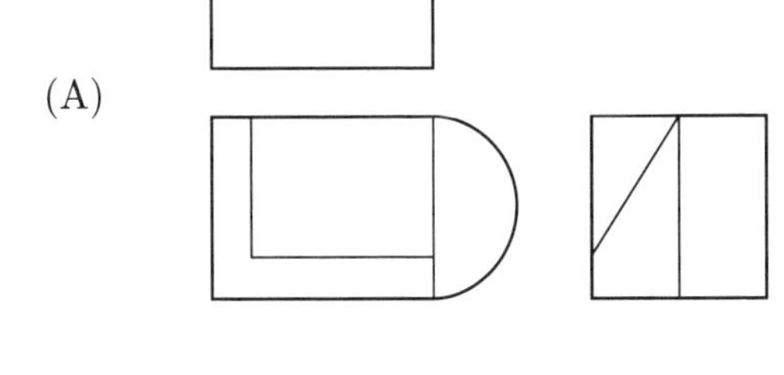

(B)

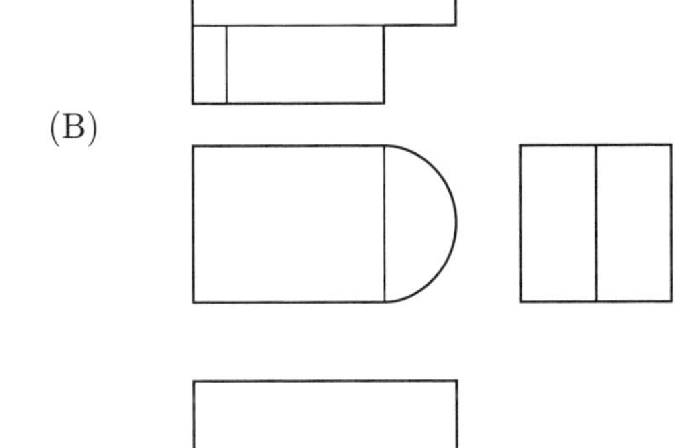

(C)

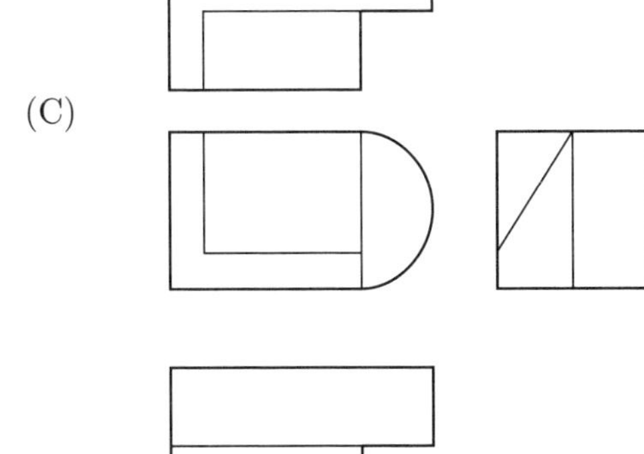

(D)

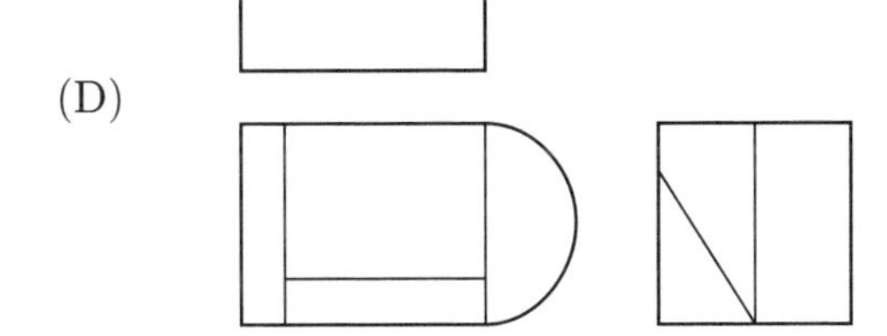

8. Two views of an object are shown. Which of the four options best represents the missing view?

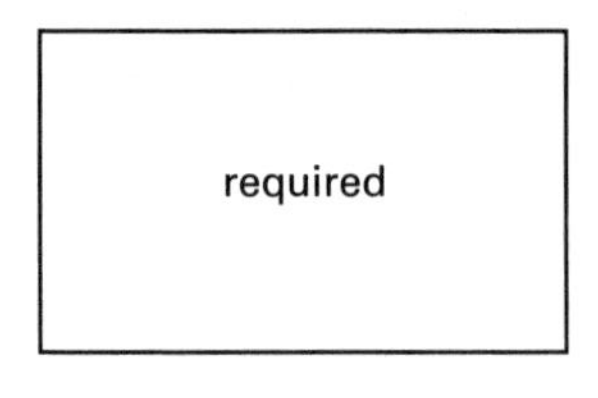

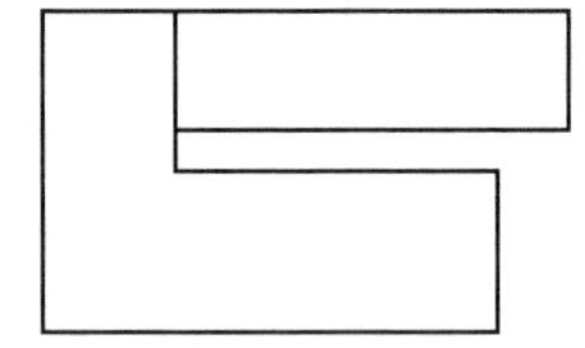

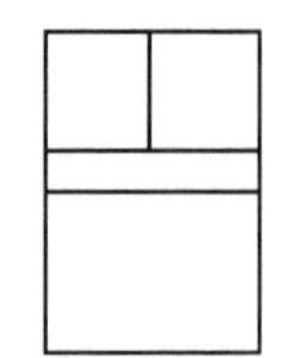

(A) 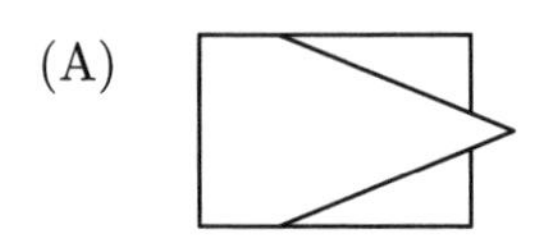

(B) 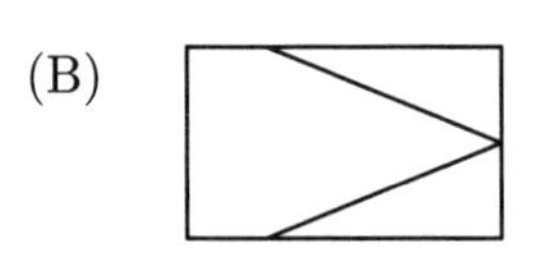

(C) 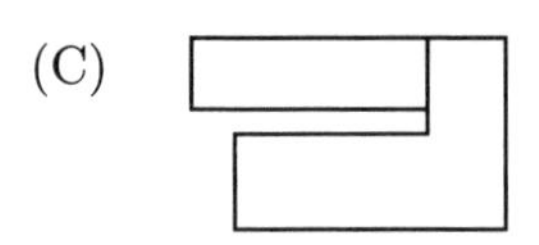

(D) 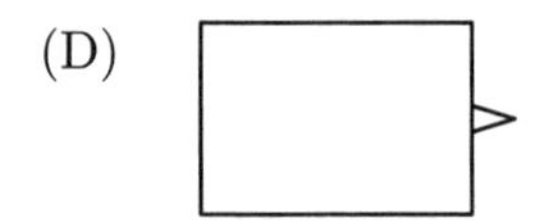

9. Two views of an object are shown. Which of the four options best represents the missing view?

required

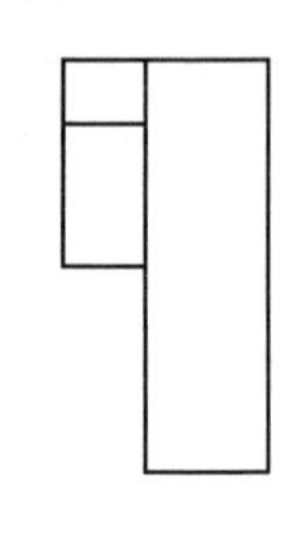

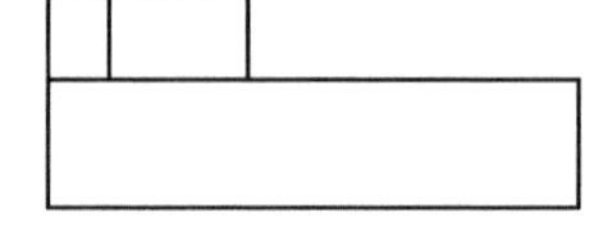

(A) 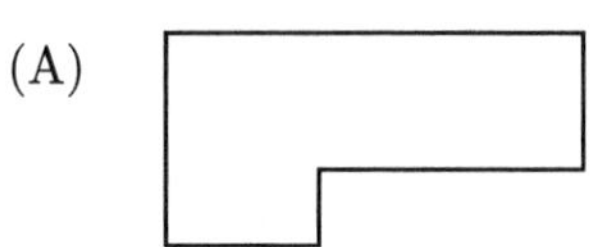

(B) 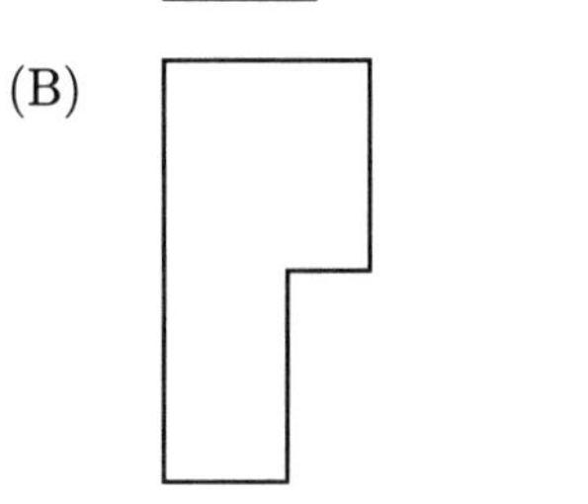

(C) 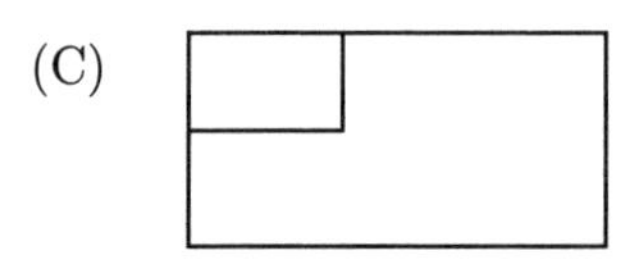

(D) 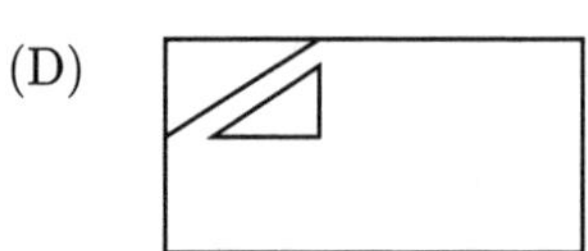

10. Elevation and plan views of an object are shown.

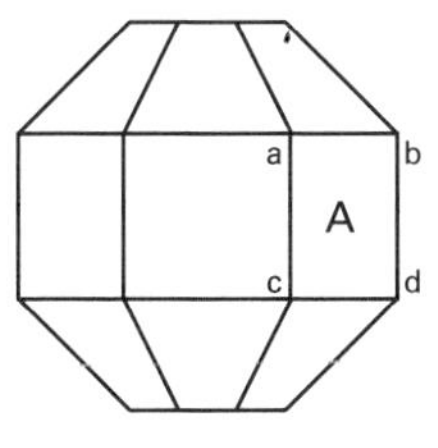

elevation view

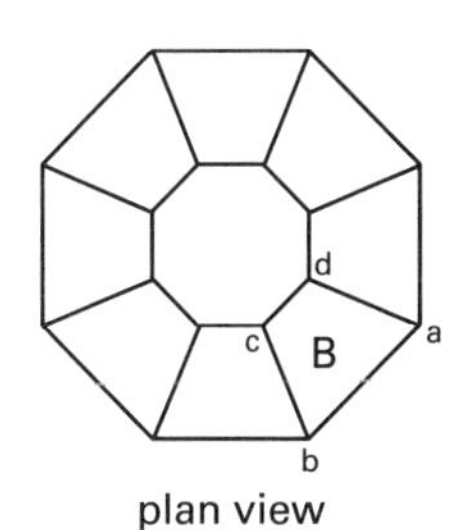

plan view

(a) Which option represents surface A in its true shape?

(A) 

square

(B) A rectangle

(C) A rectangle

(D) 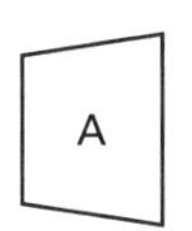

trapezoid

(b) Referring to part (a), what is this view of surface A called?

(A) auxiliary incline view

(B) auxiliary elevation view

(C) developed elevation view

(D) cross section view

(c) Which option represents surface B in its true shape?

(A) 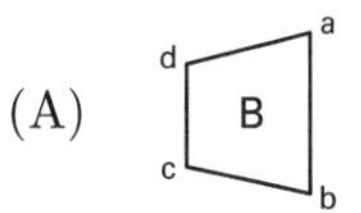

trapezoid

(B) d a B c b square

(C) 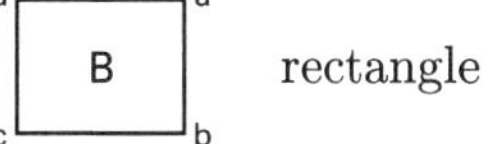

rectangle

(D) d a B c b trapezoid

(d) Referring to part (c), what is this view of surface B called?

(A) cavalier projection

(B) oblique view

(C) cabinet projection

(D) isometric view

Background and Support

11. Which option represents an isometric drawing of the object shown? Hint: Lay off horizontal and vertical (i.e., isometric) lines along the isometric axes shown.

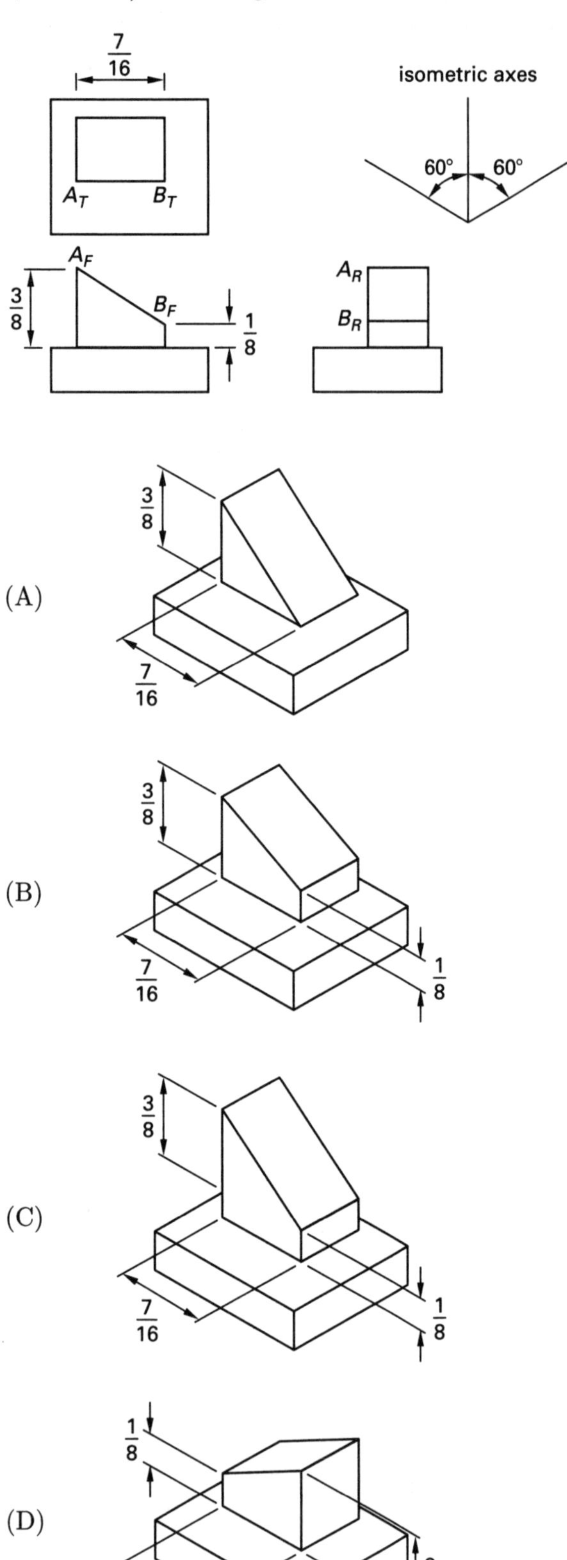

12. Two lines in an oblique position are shown. The frontal reference plane (FRP) axes are shown. Which option represents a right auxiliary normal view?

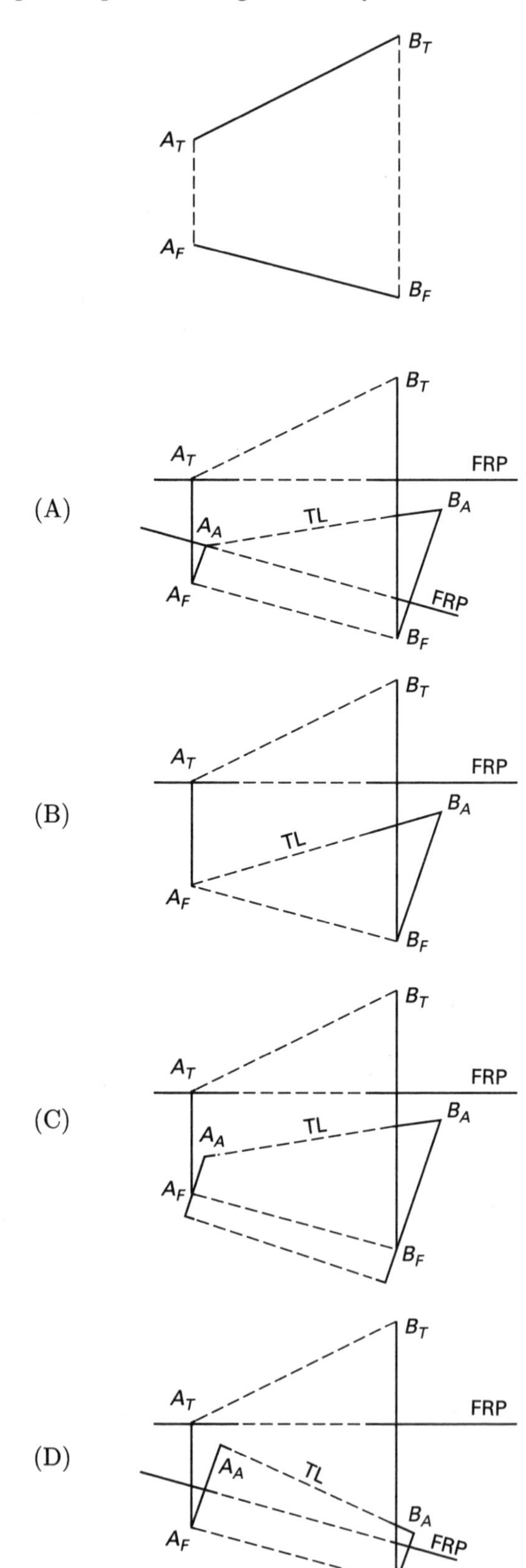

13. Two lines in an oblique position are shown. The profile reference plane (PRP) axes are shown. Which option represents a rear auxiliary normal view?

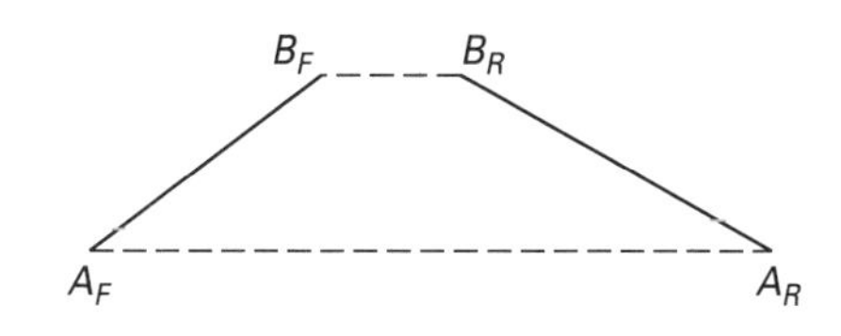

(A)

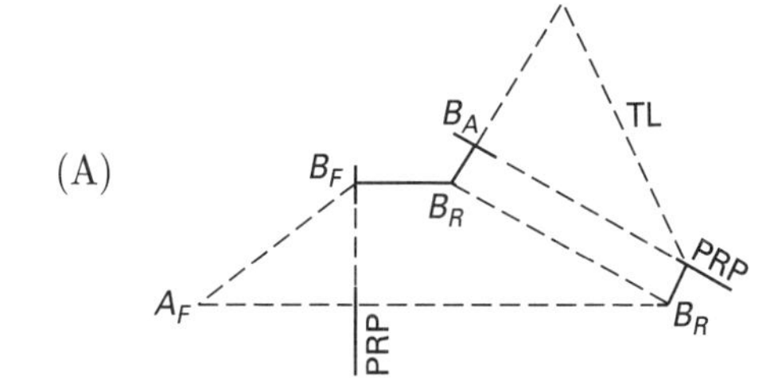

(B)

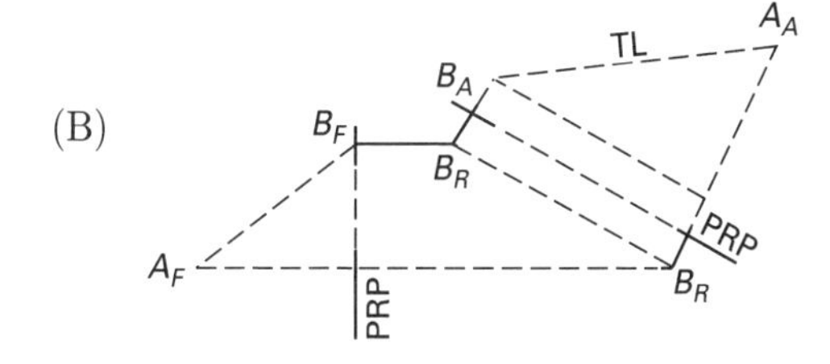

(C)

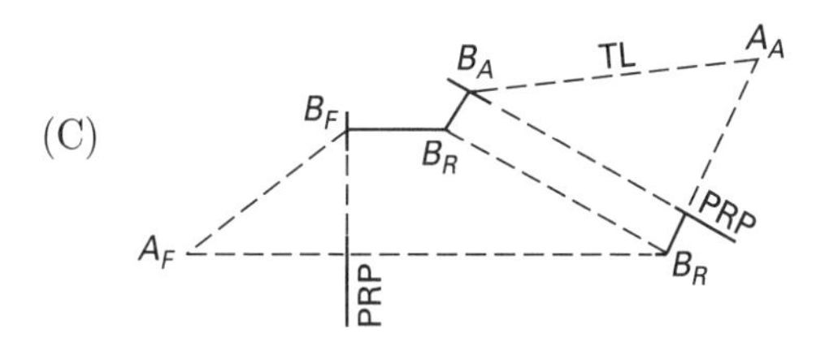

(D)

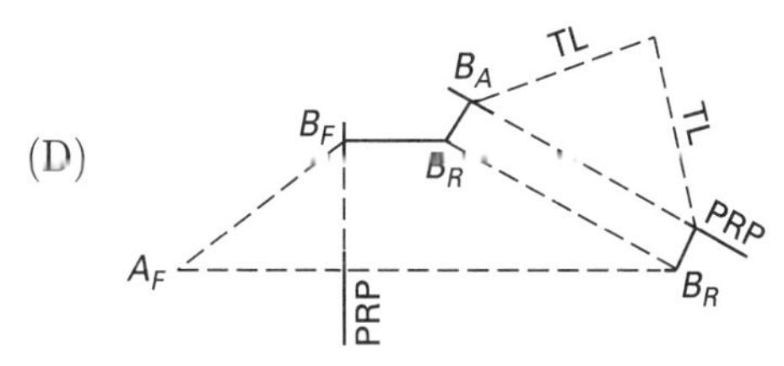

14. A parallel-scale nomograph has been prepared to solve the equation $D = 1.075\sqrt{WH}$. Using the nomograph, what is most nearly the value of D when $H = 10$ and $W = 40$?

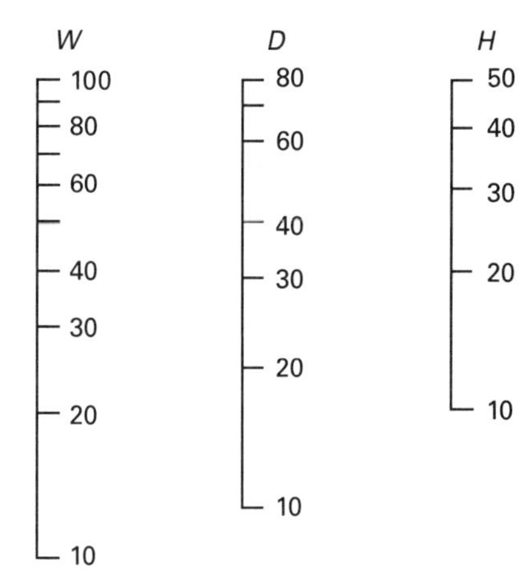

(A) 20

(B) 22

(C) 24

(D) 27

15. How would the following surface finish designation be interpreted?

0.0013

32

0.003

⊥

(A) The waviness height limit is 0.0013 μm.

(B) The lay is perpendicular.

(C) The surface roughness height limit is 0.003 μin.

(D) 32% of the surface may exceed the specification.

16. How would the following surface finish designation be interpreted?

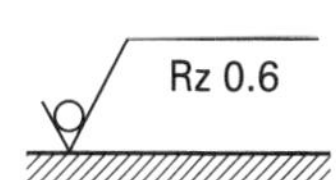

(A) The surface is rolled.

(B) The maximum roughness width is 0.6 μm.

(C) The piece is to be used exactly as it comes directly from the manufacturing process.

(D) The sample length is 0.6 m.

17. How would the following surface finish designation be interpreted?

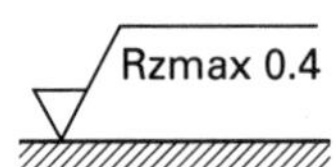

(A) The maximum waviness is 0.4 μm.

(B) The surface is to be stress relieved by bead blasting.

(C) No part of the item may exceed the specification.

(D) The maximum length that must be tested is 40% of the item's length.

18. Which of the following geometric characteristic symbols is applied using a datum reference per ASME's *Dimensioning and Tolerancing* (ASME Standard Y14.5)?

(A) straightness ⏤

(B) flatness ⏥

(C) circularity ○

(D) perpendicularity ⊥

19. The sawtooth profile of a 2 in high machined block as seen on an optical comparator is shown. What is most nearly the height of the straightness zone (actual straightness tolerance) of the profile centered about the median line?

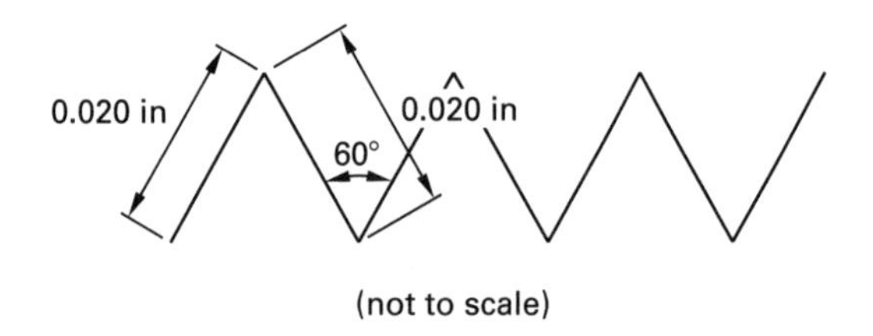

(not to scale)

(A) 0.014 in

(B) 0.017 in

(C) 2.008 in

(D) 2.013 in

20. The dimensions (in inches), locations, and tolerances of a drilled circular disc are shown on a manufacturing document. What approximate diameter pin gage (gage pin) should be used to check the location tolerance of the hole pattern from the primary reference datum?

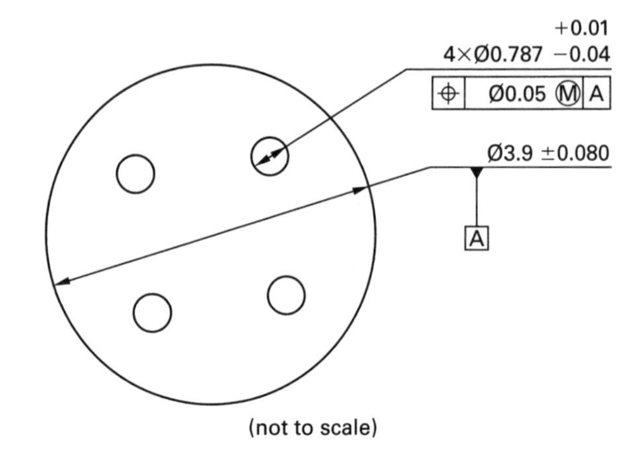

(not to scale)

(A) 0.68 in

(B) 0.70 in

(C) 0.75 in

(D) 0.80 in

21. A block with a projecting cylindrical plug is shown. All dimensions are in millimeters. How many degrees of freedom are constrained by datum B?

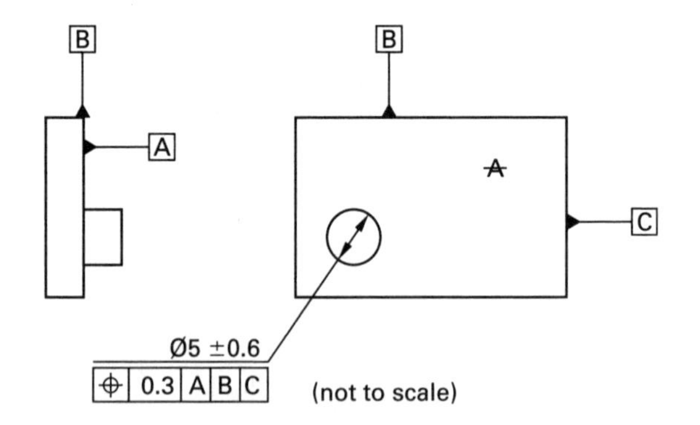

(not to scale)

(A) 1

(B) 2

(C) 3

(D) 6

22. Design specifications of a mass-produced flange are shown. All dimensions are in inches. One flange was randomly selected for inspection. The diameter of hole no. 1 was measured as 0.803 in. Relative to datum A, what total amount of positional tolerance is most nearly available for hole no. 1?

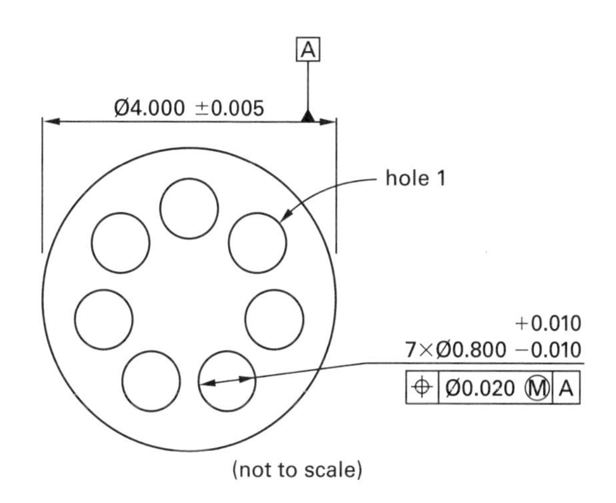

(A) 0.010 in

(B) 0.013 in

(C) 0.020 in

(D) 0.033 in

23. Design specifications for a plate with a circular hole and an accompanying circular plug are shown. All dimensions are in inches. What is the type of fit between the hole and plug at their virtual conditions?

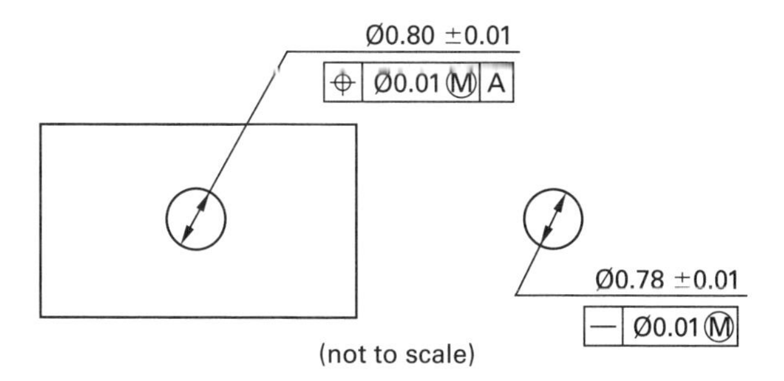

(A) clearance fit

(B) interference (press) fit

(C) transition fit

(D) line-to-line fit

24. A planimeter consists of two 14 in articulated arms (the pole arm and a tracer arm) with a 3.0 in diameter wheel centered along the length of the tracer arm. While tracing out a closed area, the planimeter wheel turns 7.65 revolutions. Most nearly, what is the area of the area traced?

(A) 210 in^2

(B) 330 in^2

(C) 670 in^2

(D) 1000 in^2

25. Measurements on a planimeter are read from a dial meter and the tracer wheel's vernier. A result is shown.

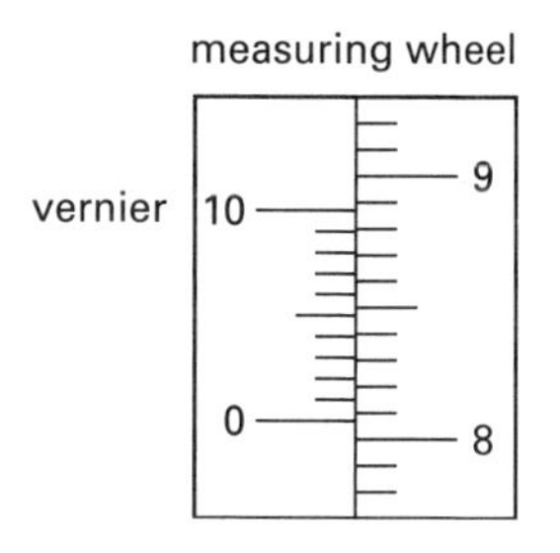

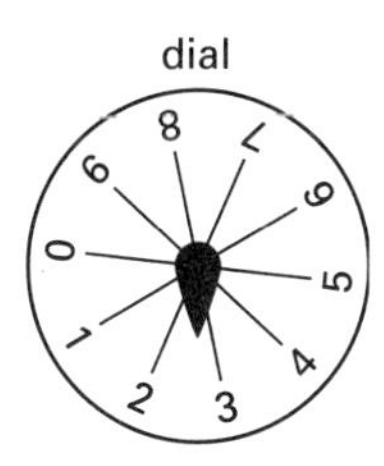

What number is represented?

(A) 2800

(B) 2804

(C) 2916

(D) 3809

26. A planimeter is used to trace the shoreline contour lines corresponding to difference depths of a reservoir. The map being traced has a scale of 1 in = 100 ft.

depth contour (ft)	planimeter area (in^2)
0	220
5	160
10	110
15	85
20	20
23 (maximum depth)	0

Most nearly, what is the volume of the reservoir?

(A) 12×10^6 ft^3

(B) 17×10^6 ft^3

(C) 24×10^6 ft^3

(D) 32×10^6 ft^3

27. By doubling the length of the tracer arm on an adjustable planimeter, the wheel roll vernier dial will

(A) decrease by a factor of 4

(B) decrease by a factor of 2

(C) increase by a factor of 2

(D) increase by a factor of 4

SOLUTIONS

1. All three dimensions appear to converge, so the drawing is in three-point perspective.

The answer is (C).

2. Both views project the cube's front face as a square, so the views are oblique. However, neither view is isometric, principal, or orthographic. The left view uses equal-length lines for the x- , y-, and z-projectors, so it is a cavalier projection. The right view shortens the z-projectors to 50%, so it is a cabinet projection.

The answer is (D).

3. Near the top of the illustration, the top railing, the floor, and the vertical poles intersect with approximately 120° angles, so the view is isometric.

The answer is (A).

4. The complete set of views is

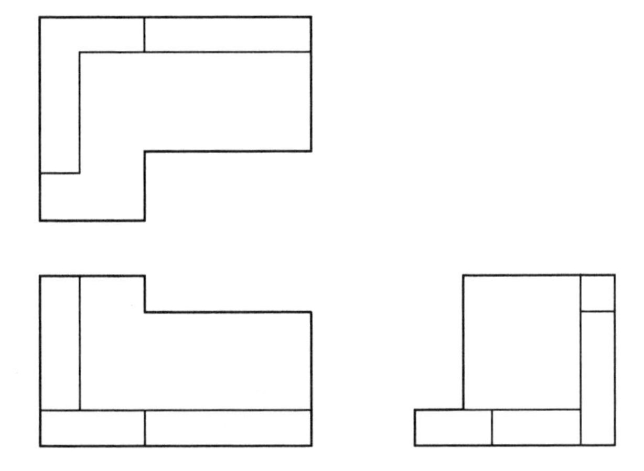

The answer is (D).

5. The complete set of views is

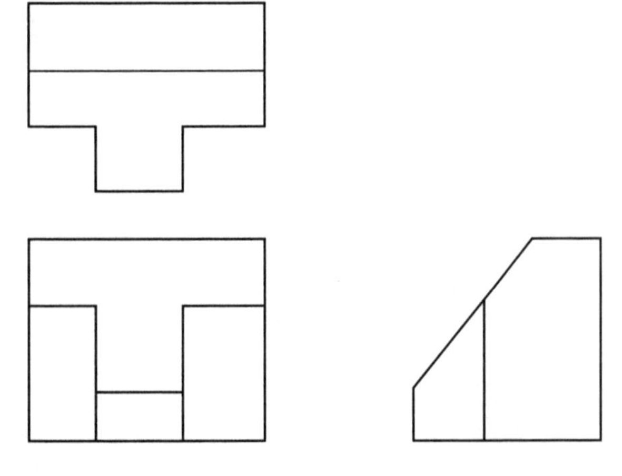

The answer is (B).

6. The complete set of views is

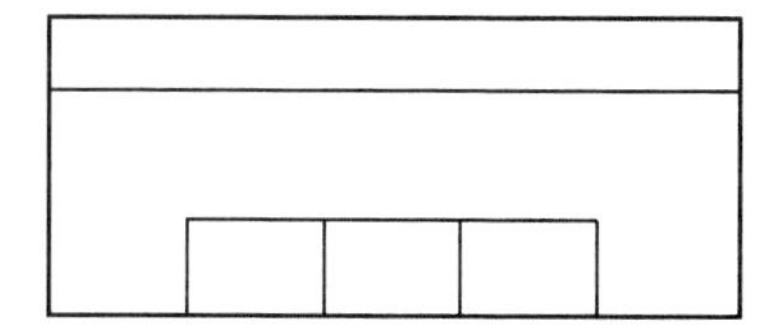

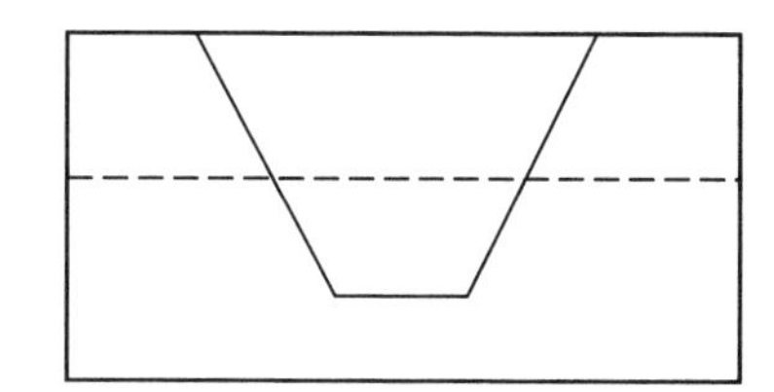

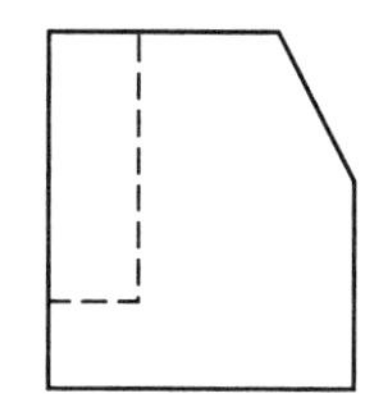

The answer is (C).

7. The complete set of views is

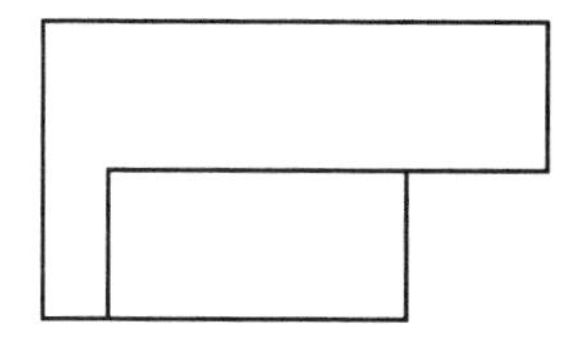

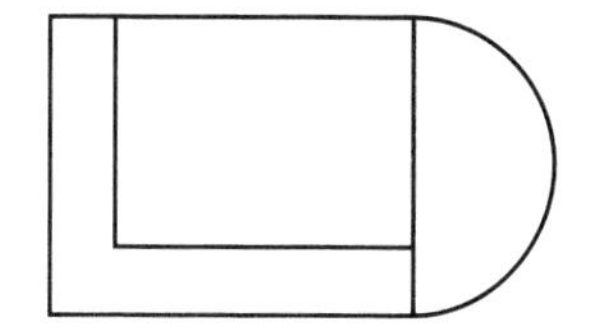

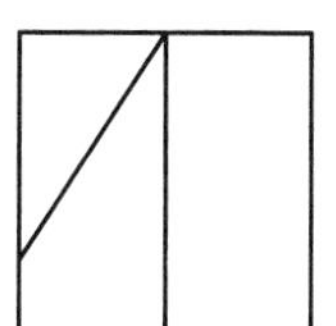

The answer is (C).

8. An isometric view of the object is shown.

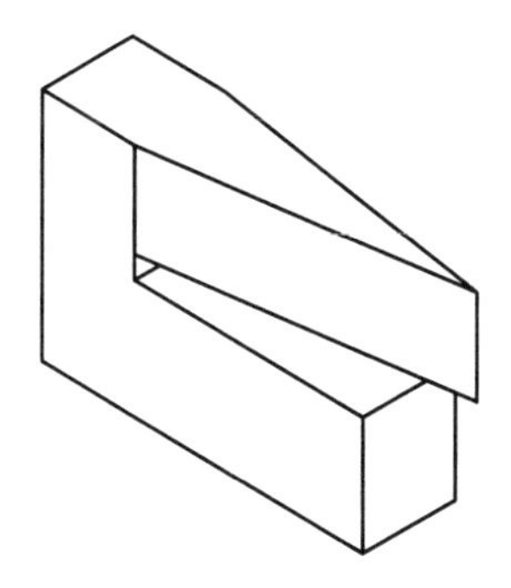

The answer is (A).

9. An isometric view of the object is shown.

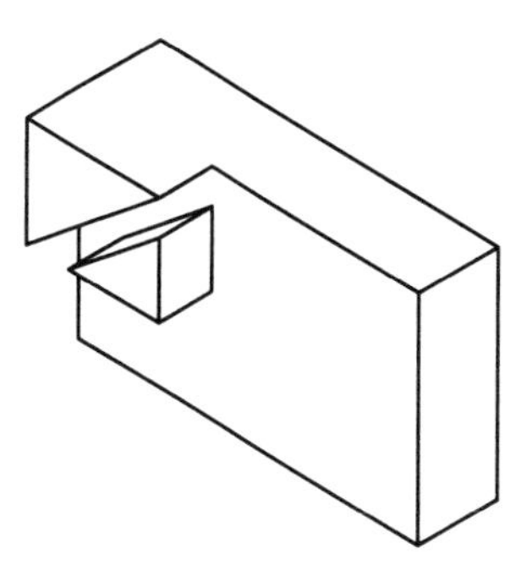

The answer is (D).

10. (a) The view shown is surface A's true shape.

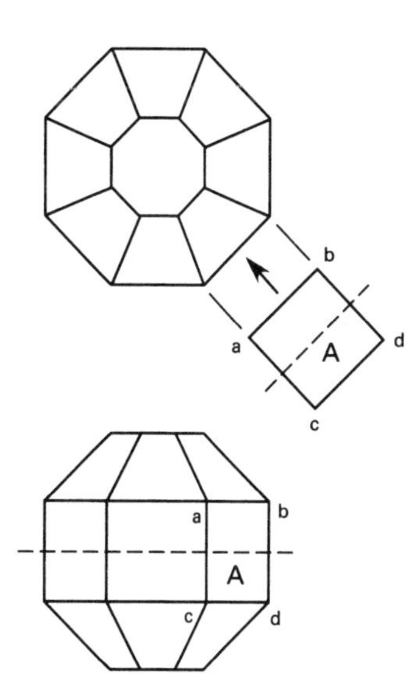

The answer is (A).

(b) An auxiliary elevation view shows surface A in its true shape.

The answer is (B).

(c) The view shown is surface B's true shape.

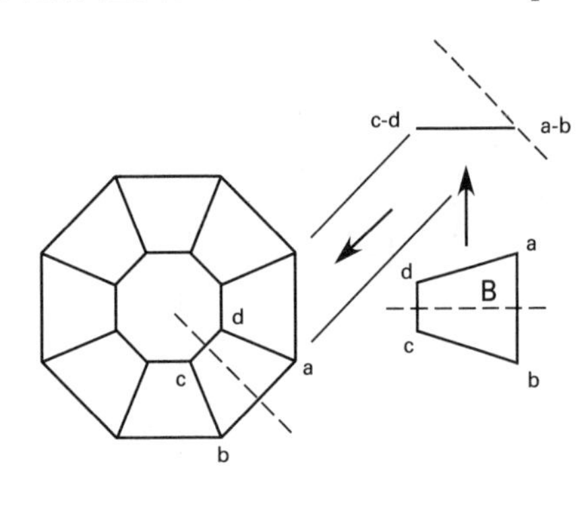

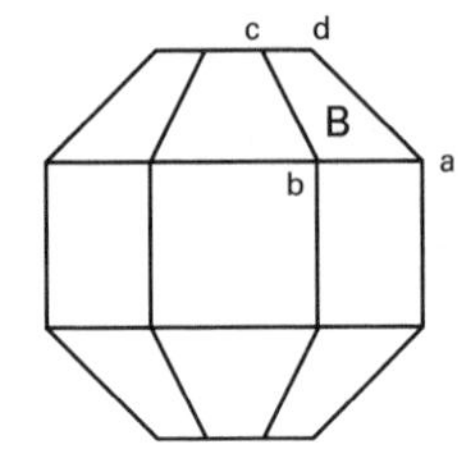

The answer is (A).

(d) An oblique view shows surface B in its true shape.

The answer is (B).

11. The isometric drawing of the object is as shown.

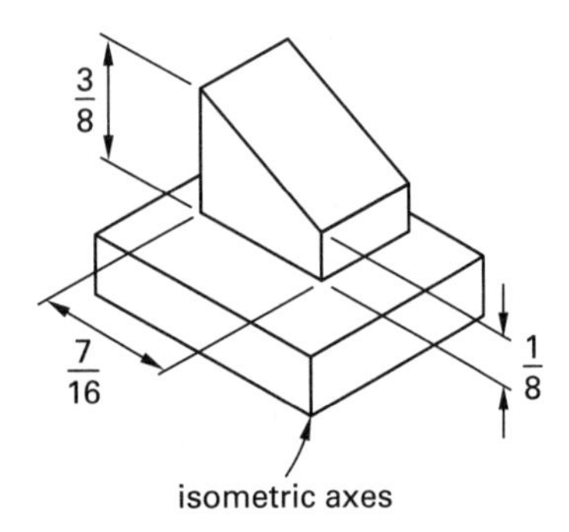

The answer is (B).

12. The right auxiliary normal view is as shown.

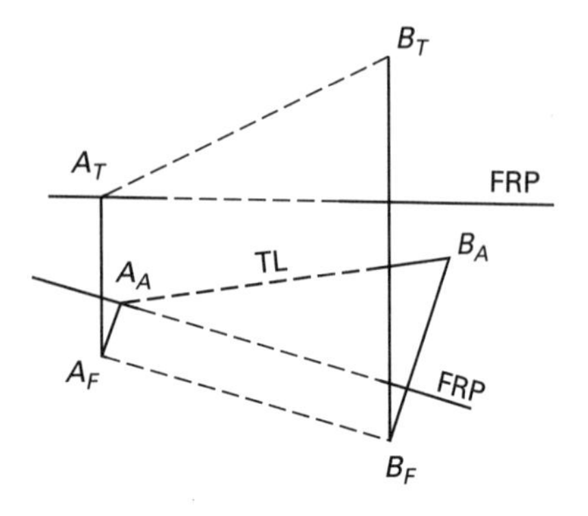

The answer is (A).

13. The auxiliary normal view is as shown.

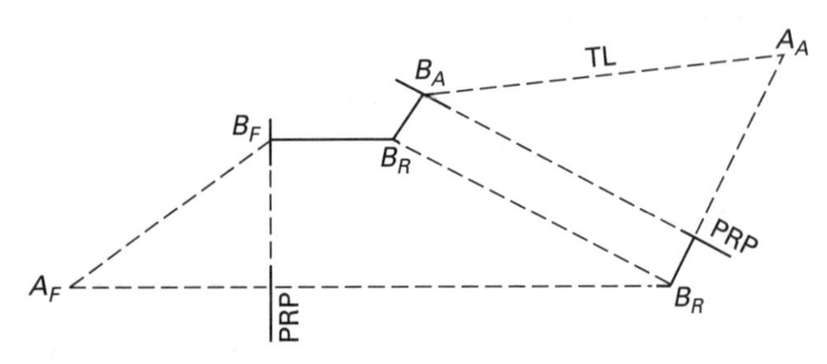

The answer is (C).

14. A straight line through the points $H = 10$ and $W = 40$ intersects the D-scale at approximately 22.

The answer is (B).

15. The values 0.0013 and 0.003 are too "fine" to be in micrometers. These values must be in microinches. The roughness height limit is 32 μin. The waviness height limit is 0.0013 μin. The roughness width limit is 0.003 μin.

The lay is perpendicular.

The answer is (B).

16. The circle in the crook of the surface finish arrow means that no resurfacing is permitted (i.e., it must be used exactly as it comes from the manufacturing process). The maximum height of the roughness is 0.6 μm. In addition, since no length of sampling is specified, the default limit of five sample lengths (i.e., between the five highest peaks and the five lowest valleys) applies. Also, since no other value is specified, a maximum of 16% (the default value) of the item may exceed the minimum surface roughness.

The answer is (C).

17. The triangle in the crook of the surface finish arrow means that the surface is to be machined. The maximum roughness height is 0.4 μm. The "max" means that the entire piece must satisfy the maximum specification (the 16% rule does not apply). No part of the item may exceed the specification.

The answer is (C).

18. ASME Standard Y14.5 defines the rules and symbols used with geometric dimensioning and tolerancing on engineering drawings. Per ASME Standard Y14.5, perpendicularity is applied using a datum reference. Straightness, flatness, and circularity are applied without using a datum reference.

The answer is (D).

19. The straightness zone is defined as the height, t, of the perpendicular plane encompassing all points.

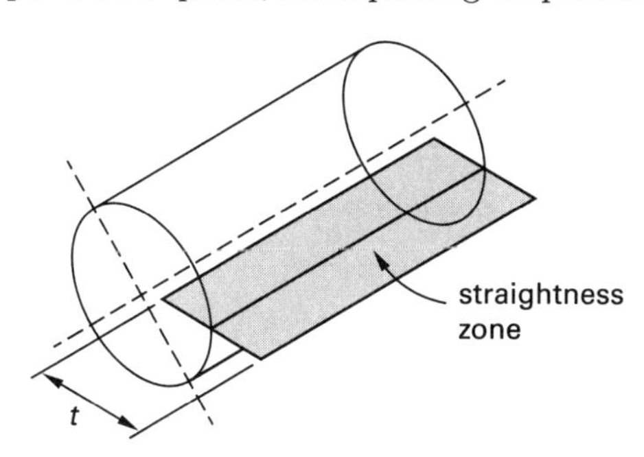

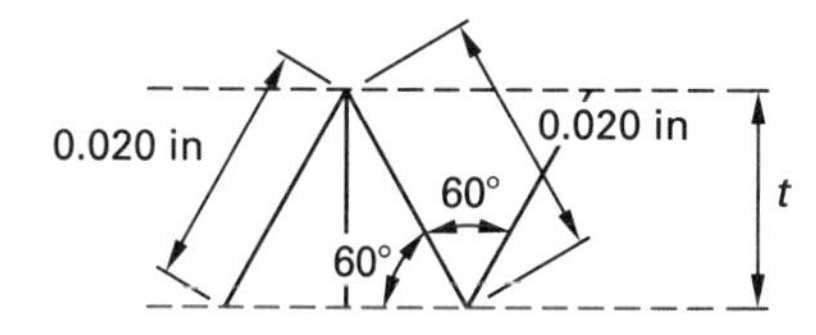

The height of the straightness zone is

$$t = (0.020 \text{ in})\sin 60° = \boxed{0.017 \text{ in}}$$

The answer is (B).

20. The manufacturing document uses *Dimensioning and Tolerancing* (ASME Standard Y14.5) nomenclature. The feature control frame indicates four 0.787 in diameter holes are to be created. The circled cross-hair symbol indicates dimensional tolerance. The hole's diametral tolerance is +0.01 in, −0.04 in. The hole center's locational tolerance is 0.050 in from the datum. The circle-M indicates that the tolerance is from a maximum material condition (MMC). The primary reference datum is datum A.

The pin diameter required to check the location tolerance of a hole is the *virtual size* of the hole. The virtual hole size is the diameter at maximum material condition minus the positional tolerance. The maximum material virtual size, MMVS, is

$$\begin{aligned} \text{MMVS} &= (0.787 \text{ in} - 0.040 \text{ in}) - 0.050 \text{ in} \\ &= \boxed{0.697 \text{ in} \quad (0.70 \text{ in})} \end{aligned}$$

The answer is (B).

21. The manufacturing document uses *Dimensioning and Tolerancing* (ASME Standard Y14.5) nomenclature. In the datum scheme, datum A is the primary datum (out of a total of three datums) in the ABC datum reference plane, because "A" is listed first in the feature control frame. Datum A represents a restrictive plane perpendicular to the page. The object is restrained from moving left-right, but it could slide vertically (up-down), and it could slide along the plane (toward and away from the viewer) without leaving datum plane A. The object could also rotate about a horizontal axis while maintaining contact with the plane. The object could not rotate about a vertical axis or about an axis perpendicular to the page. Datum A restricts movement in two rotational directions and in one translational direction, or three degrees of freedom.

Datum B is the secondary datum, because it is listed twice. It represents a restrictive axis, and since it is perpendicular to datum plane A, datum B is also a plane. While maintaining contact with datum plane B, the object should slide toward and away from the viewer. However, in addition to the restrictions from datum A, without losing contact with the datum plane B, the object is constrained from moving vertically up and down. The object is also constrained from rotating about a horizontal axis. Therefore, datum B restricts movement in one translational direction and in one rotational direction, or $\boxed{\text{two}}$ degrees of freedom.

Without a third datum, the object could still slide toward and away from the viewer without losing contact with datum planes A and B. This movement is prevented by datum C.

The answer is (B).

22. The design specification uses *Dimensioning and Tolerancing* (ASME Standard Y14.5) nomenclature. The feature control frame indicates that the required hole diameter is 0.800 in ± 0.010 in. Therefore, the range of acceptable diameters is [0.790 in, 0.810 in]. The locational tolerance of the hole is 0.020 in. However, if the hole is larger than 0.790 in (the minimum acceptable diameter), it might be possible to move the hole center more than 0.020 in and still have a part that meets the tolerance. The extra distance that the hole center can move is known as "bonus tolerance." Bonus tolerance is available whenever the maximum material condition (MMC) is specified. The maximum bonus tolerance is the difference between the MMC dimension and the least material condition (LMC) dimension.

Since the maximum material condition is specified in the feature control frame, the total amount of the maximum material positional tolerance, MMT, available for hole no. 1 is the positional tolerance of the hole at MMC plus the bonus tolerance.

$$\text{MMT} = 0.020 \text{ in} + (0.803 \text{ in} - 0.790 \text{ in}) = \boxed{0.033 \text{ in}}$$

The answer is (D).

23. The design specifications use *Dimensioning and Tolerancing* (ASME Standard Y14.5) nomenclature. The term "virtual condition" is used to describe the theoretical extreme boundary conditions that both mating objects possess in order to mate. For a shaft to fit into a hole, the hole virtual condition must be greater than the shaft virtual condition. If the hole virtual condition is less than the shaft virtual condition, then an interference fit (press fit) exists.

The virtual condition of the hole is the hole maximum material condition minus the positional tolerance at maximum material condition.

$$\text{MMVS}_h = (0.80 \text{ in} - 0.01 \text{ in}) - 0.01 \text{ in} = 0.78 \text{ in}$$

The virtual condition of the shaft is the shaft maximum material condition plus the straightness (i.e., the dimetral tolerance) at maximum material condition.

$$\text{MMVS}_s = (0.78 \text{ in} + 0.01 \text{ in}) + 0.01 \text{ in} = 0.80 \text{ in}$$

Since the hole virtual condition is less than the shaft virtual condition, an $\boxed{\text{interference (press) fit}}$ exists.

The answer is (B).

24. The area traced out by a planimeter is the length of the tracer arm times the circumference of the wheel times the number of revolutions that the wheel makes as the area is traced completely once. From Eq. 2.1,

$$\begin{aligned} A &= 2\pi rNL = \pi dNL = \pi(3 \text{ in})(7.65)(14 \text{ in}) \\ &= \boxed{1009 \text{ in}^2 \quad (1000 \text{ in}^2)} \end{aligned}$$

The answer is (D).

25.

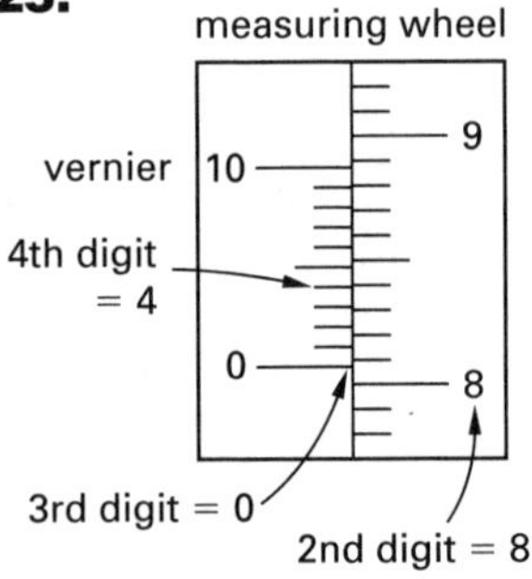

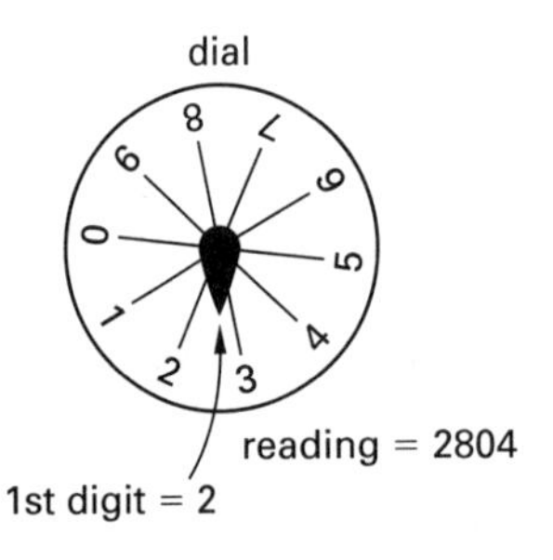

The dial has passed 2, but has not yet reached 3. Therefore, the most significant digit is 2. The "0" mark on the measuring wheel vernier has just passed 8, which is the next second most significant digit. The "0" mark has not reached the first graduation on the measuring wheel past "8," so the next most significant digit is 0. The two closest aligning lines on the vernier are 4, which is the least significant digit. The number is $\boxed{2804.}$

The answer is (B).

26. Use the average end-area method with the ends working vertically downward. Between a depth of 0 ft and 5 ft, the average lake area is

$$\begin{aligned} A_{0\text{-}5} &= \left(\frac{1}{2}\right)(220 \text{ in}^2 + 160 \text{ in}^2)\left(100 \ \frac{\text{ft}}{\text{in}}\right)^2 \\ &= 1{,}900{,}000 \text{ ft}^2 \end{aligned}$$

The volume of the water between depths 0 ft and 5 ft is

$$V_{0\text{-}5} = Ah = (1{,}900{,}000 \text{ ft}^2)(5 \text{ ft}) = 9{,}500{,}000 \text{ ft}^3$$

The following table is prepared similarly.

depth contour (in^2)	planimeter area (in^2)	area (ft^2)	volume (ft^3)
0	220		
		1,900,000	9,500,000
5	160		
		1,350,000	6,750,000
10	110		
		975,000	4,875,000
15	85		
		525,000	2,625,000
20	20		
		100,000	300,000
23 (maximum depth)	0		

The total volume is the sum of the layer volumes.

$$\begin{aligned} V_t &= 9{,}500{,}000 \text{ ft}^3 + 6{,}750{,}000 \text{ ft}^3 + 4{,}875{,}000 \text{ ft}^3 \\ &\quad + 2{,}625{,}000 \text{ ft}^3 + 300{,}000 \text{ ft}^3 \\ &= \boxed{24{,}050{,}000 \text{ ft}^3 \quad (24 \times 10^6 \text{ ft}^3)} \end{aligned}$$

The answer is (C).

27. From Eq. 2.1, the area traced out by a planimeter is

$$A = 2\pi rNL = \pi dNL$$

The area, A, of a measured section does not change, so if the tracer's arm length, L, is doubled, the number of wheel rolls, N, will $\boxed{\text{decrease by a factor of 2.}}$

The answer is (B).

3 Algebra

PRACTICE PROBLEMS

Roots of Quadratic Equations

1. What are the roots of the quadratic equation $x^2 - 7x - 44 = 0$?

(A) −11, 4

(B) −7.5, 3.5

(C) 7.5, −3.5

(D) 11, −4

Logarithm Identities

2. What is most nearly the value of x that satisfies the expression $17.3 = e^{1.1x}$?

(A) 0.17

(B) 2.6

(C) 5.8

(D) 15

Series

3. What is most nearly the following sum?

$$\sum_{j=1}^{5}\left((j+1)^2 - 1\right)$$

(A) 15

(B) 24

(C) 35

(D) 85

Logarithms

4. If a quantity increases by 0.1% of its current value every 0.1 sec, the doubling time is most nearly

(A) 14 sec

(B) 69 sec

(C) 690 sec

(D) 69,000 sec

SOLUTIONS

1. *method 1:* Use inspection.

Assuming that the roots are integers, there are only a few ways to arrive at 44 by multiplication: 1×44, 2×22, and 4×11. Of these, the difference of 4 and 11 is −7. Therefore, the quadratic equation can be factored into $(x - 11)(x + 4) = 0$. The roots are +11 and −4.

method 2: Use the quadratic formula, where $a = 1$, $b = -7$, and $c = -44$.

$$\begin{aligned} x_1, x_2 &= \frac{-b \pm \sqrt{b^2 - 4ac}}{2a} \\ &= \frac{-(-7) \pm \sqrt{(-7)^2 - (4)(1)(-44)}}{(2)(1)} \\ &= \frac{7 \pm 15}{2} \\ &= -4, 11 \end{aligned}$$

method 3: Use completing the square.

$$\begin{aligned} x^2 - 7x - 44 &= 0 \\ x^2 - 7x &= 44 \\ \left(x - \frac{7}{2}\right)^2 &= 44 + \left(\frac{7}{2}\right)^2 \\ \left(x - \frac{7}{2}\right)^2 &= 56.25 \\ x - 3.5 &= \sqrt{56.25} \\ x &= \pm 7.5 + 3.5 \\ &= \boxed{11, -4} \end{aligned}$$

The answer is (D).

2. Take the natural logarithm of both sides, using the identity $\log_b b^n = n$.

$$\begin{aligned} 17.3 &= e^{1.1x} \\ \ln 17.3 &= \ln e^{1.1x} \\ &= 1.1x \\ x &= \frac{\ln 17.3}{1.1} \\ &= \boxed{\ln e^{1.1x} \quad (2.6)} \end{aligned}$$

The answer is (B).

3. Let $S_n = (j+1)^2 - 1$.

For $j = 1$,

$$S_1 = (1+1)^2 - 1 = 3$$

For $j = 2$,

$$S_2 = (2+1)^2 - 1 = 8$$

For $j = 3$,

$$S_3 = (3+1)^2 - 1 = 15$$

For $j = 4$,

$$S_4 = (4+1)^2 - 1 = 24$$

For $j = 5$,

$$S_5 = (5+1)^2 - 1 = 35$$

Substituting the above expressions gives

$$\begin{aligned} \sum_{j=1}^{5}\left((j+1)^2 - 1\right) &= \sum_{j=1}^{5} S_j \\ &= S_1 + S_2 + S_3 + S_4 + S_5 \\ &= 3 + 8 + 15 + 24 + 35 \\ &= \boxed{85} \end{aligned}$$

The answer is (D).

4. Let n represent the number of elapsed periods of 0.1 sec, and let y_n represent the amount present after n periods.

y_0 represents the initial quantity.

$$\begin{aligned} y_1 &= 1.001 y_0 \\ y_2 &= 1.001 y_1 = (1.001)(1.001 y_0) = (1.001)^2 y_0 \end{aligned}$$

Therefore, by induction,

$$y_n = (1.001)^n y_0$$

The expression for a doubling of the original quantity is

$$2y_0 = y_n$$

Substitute for y_n.

$$\begin{aligned} 2y_0 &= (1.001)^n y_0 \\ 2 &= (1.001)^n \end{aligned}$$

Take the logarithm of both sides.

$$\log 2 = \log(1.001)^n = n \log 1.001$$

Solve for n.

$$n = \frac{\log 2}{\log 1.001} = 693.5$$

Since each period is 0.1 sec, the time is

$$\begin{aligned} t &= n(0.1 \text{ sec}) = (693.5)(0.1 \text{ sec}) \\ &= \boxed{69.35 \text{ sec} \quad (69 \text{ sec})} \end{aligned}$$

The answer is (B).

Linear Algebra

PRACTICE PROBLEMS

Determinants

1. What is most nearly the determinant of matrix **A**?

$$\mathbf{A} = \begin{bmatrix} 8 & 2 & 0 & 0 \\ 2 & 8 & 2 & 0 \\ 0 & 2 & 8 & 2 \\ 0 & 0 & 2 & 4 \end{bmatrix}$$

(A) 459

(B) 832

(C) 1552

(D) 1776

Simultaneous Linear Equations

2. Use Cramer's rule to solve for the values of x, y, and z that simultaneously satisfy the following equations.

$$x + y = -4$$
$$x + z - 1 = 0$$
$$2z - y + 3x = 4$$

(A) $(x, y, z) = (3, 2, 1)$

(B) $(x, y, z) = (-3, -1, 2)$

(C) $(x, y, z) = (3, -1, -3)$

(D) $(x, y, z) = (-1, -3, 2)$

SOLUTIONS

1. Expand by cofactors of the first column since there are two zeros in that column.

$$|\mathbf{A}| = 8\begin{vmatrix} 8 & 2 & 0 \\ 2 & 8 & 2 \\ 0 & 2 & 4 \end{vmatrix} - 2\begin{vmatrix} 2 & 0 & 0 \\ 2 & 8 & 2 \\ 0 & 2 & 4 \end{vmatrix} + 0 - 0$$

By first column:

$$\begin{aligned} \begin{vmatrix} 8 & 2 & 0 \\ 2 & 8 & 2 \\ 0 & 2 & 4 \end{vmatrix} &= (8)\big((8)(4) - (2)(2)\big) - (2)\big((2)(4) - (2)(0)\big) \\ &= (8)(28) - (2)(8) \\ &= 208 \end{aligned}$$

By first column:

$$\begin{aligned} \begin{vmatrix} 2 & 0 & 0 \\ 2 & 8 & 2 \\ 0 & 2 & 4 \end{vmatrix} &= (2)\big((8)(4) - (2)(2)\big) \\ &= 56 \end{aligned}$$

$$|\mathbf{A}| = (8)(208) - (2)(56) = \boxed{1552}$$

The answer is (C).

2. Rearrange the equations.

$$x + y = -4$$
$$x + z = 1$$
$$3x - y + 2z = 4$$

Write the set of equations in matrix form: $\mathbf{AX} = \mathbf{B}$.

$$\begin{bmatrix} 1 & 1 & 0 \\ 1 & 0 & 1 \\ 3 & -1 & 2 \end{bmatrix} \begin{bmatrix} x \\ y \\ z \end{bmatrix} = \begin{bmatrix} -4 \\ 1 \\ 4 \end{bmatrix}$$

Find the determinant of the matrix $\mathbf{A}$.

$$\begin{aligned} |\mathbf{A}| &= \begin{vmatrix} 1 & 1 & 0 \\ 1 & 0 & 1 \\ 3 & -1 & 2 \end{vmatrix} \\ &= 1\begin{vmatrix} 0 & 1 \\ -1 & 2 \end{vmatrix} - 1\begin{vmatrix} 1 & 0 \\ -1 & 2 \end{vmatrix} + 3\begin{vmatrix} 1 & 0 \\ 0 & 1 \end{vmatrix} \\ &= (1)\big((0)(2) - (1)(-1)\big) \\ &\quad - (1)\big((1)(2) - (-1)(0)\big) \\ &\quad + (3)\big((1)(1) - (0)(0)\big) \\ &= (1)(1) - (1)(2) + (3)(1) \\ &= 1 - 2 + 3 \\ &= 2 \end{aligned}$$

Find the determinant of the substitutional matrix $\mathbf{A}_1$.

$$\begin{aligned} |\mathbf{A}_1| &= \begin{vmatrix} -4 & 1 & 0 \\ 1 & 0 & 1 \\ 4 & -1 & 2 \end{vmatrix} \\ &= -4\begin{vmatrix} 0 & 1 \\ -1 & 2 \end{vmatrix} - 1\begin{vmatrix} 1 & 0 \\ -1 & 2 \end{vmatrix} + 4\begin{vmatrix} 1 & 0 \\ 0 & 1 \end{vmatrix} \\ &= (-4)\big((0)(2) - (1)(-1)\big) \\ &\quad - (1)\big((1)(2) - (-1)(0)\big) \\ &\quad + (4)\big((1)(1) - (0)(0)\big) \\ &= (-4)(1) - (1)(2) + (4)(1) \\ &= -4 - 2 + 4 \\ &= -2 \end{aligned}$$

Find the determinant of the substitutional matrix $\mathbf{A}_2$.

$$\begin{aligned} |\mathbf{A}_2| &= \begin{vmatrix} 1 & -4 & 0 \\ 1 & 1 & 1 \\ 3 & 4 & 2 \end{vmatrix} \\ &= 1\begin{vmatrix} 1 & 1 \\ 4 & 2 \end{vmatrix} - 1\begin{vmatrix} -4 & 0 \\ 4 & 2 \end{vmatrix} + 3\begin{vmatrix} -4 & 0 \\ 1 & 1 \end{vmatrix} \\ &= (1)\big((1)(2) - (4)(1)\big) \\ &\quad - (1)\big((-4)(2) - (4)(0)\big) \\ &\quad + (3)\big((-4)(1) - (1)(0)\big) \\ &= (1)(-2) - (1)(-8) + (3)(-4) \\ &= -2 + 8 - 12 \\ &= -6 \end{aligned}$$

Find the determinant of the substitutional matrix $\mathbf{A}_3$.

$$\begin{aligned} |\mathbf{A}_3| &= \begin{vmatrix} 1 & 1 & -4 \\ 1 & 0 & 1 \\ 3 & -1 & 4 \end{vmatrix} \\ &= 1\begin{vmatrix} 0 & 1 \\ -1 & 4 \end{vmatrix} - 1\begin{vmatrix} 1 & -4 \\ -1 & 4 \end{vmatrix} + 3\begin{vmatrix} 1 & -4 \\ 0 & 1 \end{vmatrix} \\ &= (1)\big((0)(4) - (-1)(1)\big) \\ &\quad - (1)\big((1)(4) - (-1)(-4)\big) \\ &\quad + (3)\big((1)(1) - (0)(-4)\big) \\ &= (1)(1) - (1)(0) + (3)(1) \\ &= 1 - 0 + 3 \\ &= 4 \end{aligned}$$

Use Cramer's rule.

$$x = \frac{|\mathbf{A}_1|}{|\mathbf{A}|} = \frac{-2}{2} = \boxed{-1}$$

$$y = \frac{|\mathbf{A}_2|}{|\mathbf{A}|} = \frac{-6}{2} = \boxed{-3}$$

$$z = \frac{|\mathbf{A}_3|}{|\mathbf{A}|} = \frac{4}{2} = \boxed{2}$$

The answer is (D).

5 Vectors

PRACTICE PROBLEMS

1. What are the dot products for the following vector pairs?

(a) $\mathbf{V}_1 = 2\mathbf{i} + 3\mathbf{j}$; $\mathbf{V}_2 = 5\mathbf{i} - 2\mathbf{j}$

(A) −4

(B) 4

(C) 8

(D) 11

(b) $\mathbf{V}_1 = 1\mathbf{i} + 4\mathbf{j}$; $\mathbf{V}_2 = 9\mathbf{i} - 3\mathbf{j}$

(A) −23

(B) −11

(C) −3

(D) 33

(c) $\mathbf{V}_1 = 7\mathbf{i} - 3\mathbf{j}$; $\mathbf{V}_2 = 3\mathbf{i} + 4\mathbf{j}$

(A) −21

(B) 9

(C) 11

(D) 21

(d) $\mathbf{V}_1 = 2\mathbf{i} - 3\mathbf{j} + 6\mathbf{k}$; $\mathbf{V}_2 = 8\mathbf{i} + 2\mathbf{j} - 3\mathbf{k}$

(A) −12

(B) −8

(C) 37

(D) 40

(e) $\mathbf{V}_1 = 6\mathbf{i} + 2\mathbf{j} + 3\mathbf{k}$; $\mathbf{V}_2 = \mathbf{i} + \mathbf{k}$

(A) −11

(B) 9

(C) 11

(D) 13

2. (a) The angle between the vectors in Prob. 1(a) is most nearly

(A) 60°

(B) 70°

(C) 80°

(D) 100°

(b) The angle between the vectors in Prob. 1(b) is most nearly

(A) 70°

(B) 85°

(C) 90°

(D) 95°

(c) The angle between the vectors in Prob. 1(c) is most nearly

(A) 73°

(B) 76°

(C) 100°

(D) 120°

3. (a) The cross product for the vector pair in Prob. 1(a) is

(A) $-19\mathbf{k}$

(B) $4\mathbf{k}$

(C) $11\mathbf{k}$

(D) $19\mathbf{k}$

(b) The cross product for the vector pair in Prob. 1(b) is

(A) $-39\mathbf{k}$

(B) $-23\mathbf{k}$

(C) $-3\mathbf{k}$

(D) $33\mathbf{k}$

(c) The cross product for the vector pair in Prob. 1(c) is most nearly

(A) $-9\mathbf{k}$

(B) $11\mathbf{k}$

(C) $21\mathbf{k}$

(D) $37\mathbf{k}$

(d) The cross product for the vector pair in Prob. 1(d) is most nearly

(A) $-24\mathbf{i} - 2\mathbf{j} + 10\mathbf{k}$

(B) $-3\mathbf{i} + 54\mathbf{j} + 28\mathbf{k}$

(C) $3\mathbf{i} - 42\mathbf{j} + 28\mathbf{k}$

(D) $21\mathbf{i} - 42\mathbf{j} - 20\mathbf{k}$

(e) The cross product for the vector pair in Prob. 1(e) is most nearly

(A) $-2\mathbf{i} - 3\mathbf{j} + \mathbf{k}$

(B) $-9\mathbf{j} + 2\mathbf{k}$

(C) $2\mathbf{i} - 3\mathbf{j} - 2\mathbf{k}$

(D) $-5\mathbf{i} - 9\mathbf{j} + 4\mathbf{k}$

SOLUTIONS

1. (a) The dot product is

$$\begin{aligned}\mathbf{V}_1 \cdot \mathbf{V}_2 &= \mathbf{V}_{1x}\mathbf{V}_{2x} + \mathbf{V}_{1y}\mathbf{V}_{2y} \\ &= (2)(5) + (3)(-2) \\ &= \boxed{4}\end{aligned}$$

The answer is (B).

(b) The dot product is

$$\begin{aligned}\mathbf{V}_1 \cdot \mathbf{V}_2 &= \mathbf{V}_{1x}\mathbf{V}_{2x} + \mathbf{V}_{1y}\mathbf{V}_{2y} \\ &= (1)(9) + (4)(-3) \\ &= \boxed{-3}\end{aligned}$$

The answer is (C).

(c) The dot product is

$$\begin{aligned}\mathbf{V}_1 \cdot \mathbf{V}_2 &= \mathbf{V}_{1x}\mathbf{V}_{2x} + \mathbf{V}_{1y}\mathbf{V}_{2y} \\ &= (7)(3) + (-3)(4) \\ &= \boxed{9}\end{aligned}$$

The answer is (B).

(d) The dot product is

$$\begin{aligned}\mathbf{V}_1 \cdot \mathbf{V}_2 &= \mathbf{V}_{1x}\mathbf{V}_{2x} + \mathbf{V}_{1y}\mathbf{V}_{2y} + \mathbf{V}_{1z}\mathbf{V}_{2z} \\ &= (2)(8) + (-3)(2) + (6)(-3) \\ &= \boxed{-8}\end{aligned}$$

The answer is (B).

(e) The dot product is

$$\begin{aligned}\mathbf{V}_1 \cdot \mathbf{V}_2 &= \mathbf{V}_{1x}\mathbf{V}_{2x} + \mathbf{V}_{1y}\mathbf{V}_{2y} + \mathbf{V}_{1z}\mathbf{V}_{2z} \\ &= (6)(1) + (2)(0) + (3)(1) \\ &= \boxed{9}\end{aligned}$$

The answer is (B).

2. (a) The angle between the vectors is

$$\cos\phi = \frac{\mathbf{V}_1 \cdot \mathbf{V}_2}{|\mathbf{V}_1||\mathbf{V}_2|} = \frac{4}{\sqrt{(2)^2+(3)^2}\sqrt{(5)^2+(-2)^2}}$$
$$= 0.206$$
$$\phi = \arccos 0.206 = \boxed{78.1^\circ \quad (80^\circ)}$$

The answer is (C).

(b) The angle between the vectors is

$$\cos\phi = \frac{\mathbf{V}_1 \cdot \mathbf{V}_2}{|\mathbf{V}_1||\mathbf{V}_2|} = \frac{-3}{\sqrt{(1)^2+(4)^2}\sqrt{(9)^2+(-3)^2}}$$
$$= -0.0767$$
$$\phi = \arccos(-0.0767) = \boxed{94.4^\circ \quad (95^\circ)}$$

The answer is (D).

(c) The angle between the vectors is

$$\cos\phi = \frac{\mathbf{V}_1 \cdot \mathbf{V}_2}{|\mathbf{V}_1||\mathbf{V}_2|} = \frac{9}{\sqrt{(7)^2+(-3)^2}\sqrt{(3)^2+(4)^2}}$$
$$= 0.236$$
$$\phi = \arccos 0.236 = \boxed{76.3^\circ \quad (76^\circ)}$$

The answer is (B).

3. (a) The cross product is

$$\mathbf{V}_1 \times \mathbf{V}_2 = \begin{vmatrix} \mathbf{i} & \mathbf{V}_{1x} & \mathbf{V}_{2x} \\ \mathbf{j} & \mathbf{V}_{1y} & \mathbf{V}_{2y} \\ \mathbf{k} & \mathbf{V}_{1z} & \mathbf{V}_{2z} \end{vmatrix} = \begin{vmatrix} \mathbf{i} & 2 & 5 \\ \mathbf{j} & 3 & -2 \\ \mathbf{k} & 0 & 0 \end{vmatrix}$$

Expand by the third row.

$$\mathbf{V}_1 \times \mathbf{V}_2 = \mathbf{k}\begin{vmatrix} 2 & 5 \\ 3 & -2 \end{vmatrix} = \big((2)(-2) - (3)(5)\big)\mathbf{k}$$
$$= \boxed{-19\mathbf{k}}$$

The answer is (A).

(b) The cross product is

$$\mathbf{V}_1 \times \mathbf{V}_2 = \begin{vmatrix} \mathbf{i} & 1 & 9 \\ \mathbf{j} & 4 & -3 \\ \mathbf{k} & 0 & 0 \end{vmatrix}$$

Expand by the third row.

$$\mathbf{V}_1 \times \mathbf{V}_2 = \mathbf{k}\begin{vmatrix} 1 & 9 \\ 4 & -3 \end{vmatrix} = \big((1)(-3) - (4)(9)\big)\mathbf{k}$$
$$= \boxed{-39\mathbf{k}}$$

The answer is (A).

(c) The cross product is

$$\mathbf{V}_1 \times \mathbf{V}_2 = \begin{vmatrix} \mathbf{i} & \mathbf{V}_{1x} & \mathbf{V}_{2x} \\ \mathbf{j} & \mathbf{V}_{1y} & \mathbf{V}_{2y} \\ \mathbf{k} & \mathbf{V}_{1z} & \mathbf{V}_{2z} \end{vmatrix} = \begin{vmatrix} \mathbf{i} & 7 & 3 \\ \mathbf{j} & -3 & 4 \\ \mathbf{k} & 0 & 0 \end{vmatrix}$$

Expand by the third row.

$$\mathbf{V}_1 \times \mathbf{V}_2 = \mathbf{k}\begin{vmatrix} 7 & 3 \\ -3 & 4 \end{vmatrix} = \big((7)(4) - (-3)(3)\big)\mathbf{k}$$
$$= \boxed{37\mathbf{k}}$$

The answer is (D).

(d) The cross product is

$$\mathbf{V}_1 \times \mathbf{V}_2 = \begin{vmatrix} \mathbf{i} & \mathbf{V}_{1x} & \mathbf{V}_{2x} \\ \mathbf{j} & \mathbf{V}_{1y} & \mathbf{V}_{2y} \\ \mathbf{k} & \mathbf{V}_{1z} & \mathbf{V}_{2z} \end{vmatrix} = \begin{vmatrix} \mathbf{i} & 2 & 8 \\ \mathbf{j} & -3 & 2 \\ \mathbf{k} & 6 & -3 \end{vmatrix}$$

Expand by the first column.

$$\mathbf{V}_1 \times \mathbf{V}_2 = \mathbf{i}\begin{vmatrix} -3 & 2 \\ 6 & -3 \end{vmatrix} - \mathbf{j}\begin{vmatrix} 2 & 8 \\ 6 & -3 \end{vmatrix} + \mathbf{k}\begin{vmatrix} 2 & 8 \\ -3 & 2 \end{vmatrix}$$
$$= \boxed{-3\mathbf{i} + 54\mathbf{j} + 28\mathbf{k}}$$

The answer is (B).

(e) The cross product is

$$\mathbf{V}_1 \times \mathbf{V}_2 = \begin{vmatrix} \mathbf{i} & \mathbf{V}_{1x} & \mathbf{V}_{2x} \\ \mathbf{j} & \mathbf{V}_{1y} & \mathbf{V}_{2y} \\ \mathbf{k} & \mathbf{V}_{1z} & \mathbf{V}_{2z} \end{vmatrix} = \begin{vmatrix} \mathbf{i} & 6 & 1 \\ \mathbf{j} & 2 & 0 \\ \mathbf{k} & 3 & 1 \end{vmatrix}$$

Expand by the second row.

$$\begin{aligned} \mathbf{V}_1 \times \mathbf{V}_2 &= -\mathbf{j}\begin{vmatrix} 6 & 1 \\ 3 & 1 \end{vmatrix} + (2)\begin{vmatrix} \mathbf{i} & 1 \\ \mathbf{k} & 1 \end{vmatrix} \\ &= \boxed{2\mathbf{i} - 3\mathbf{j} - 2\mathbf{k}} \end{aligned}$$

The answer is (C).

6 Trigonometry

PRACTICE PROBLEMS

1. A 5 lbm (5 kg) block sits on a 20° incline without slipping. (a) Draw the free-body diagram with respect to the axes parallel and perpendicular to the surface of the incline. (b) What is most nearly the magnitude of the frictional force (holding the block stationary) on the block?

(A) 1.7 lbf (17 N)

(B) 3.4 lbf (33 N)

(C) 4.7 lbf (46 N)

(D) 5.0 lbf (49 N)

2. Complete the following calculation related to a catenary cable.

$$S = c\left(\cosh\frac{a}{h} - 1\right) = (245\text{ ft})\left(\cosh\frac{50\text{ ft}}{245\text{ ft}} - 1\right)$$

(A) 3.9 ft

(B) 4.5 ft

(C) 5.1 ft

(D) 7.4 ft

3. Part of a turn-around area in a parking lot is shaped as a circular segment. The segment has a central angle of 120° and a radius of 75 ft. If the circular segment is to receive a special surface treatment, what area will be treated?

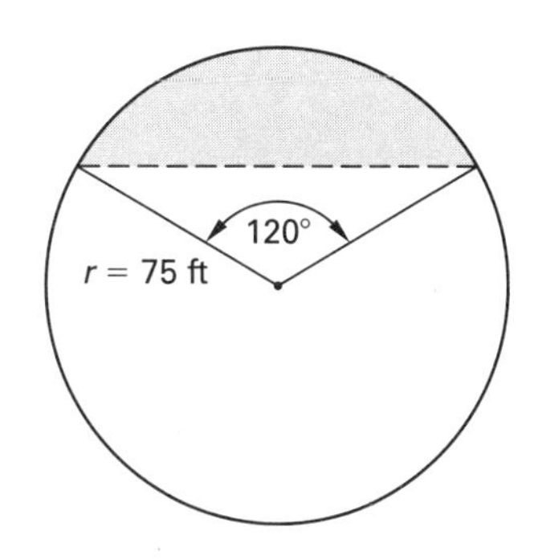

(A) 49 ft^2

(B) 510 ft^2

(C) 3500 ft^2

(D) 5900 ft^2

SOLUTIONS

1. (a) The free-body diagram is

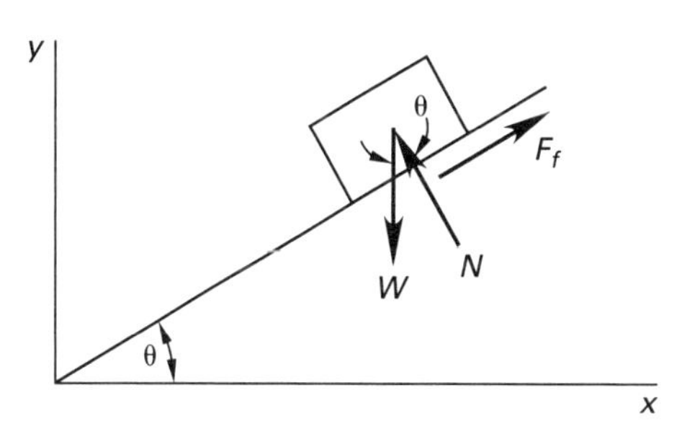

Customary U.S. Solution

(b) The mass of the block is $m = 5$ lbm. The angle of inclination is $\theta = 20°$. The weight is

$$W = \frac{mg}{g_c} = \frac{(5\text{ lbm})\left(32.2\ \frac{\text{ft}}{\text{sec}^2}\right)}{32.2\ \frac{\text{lbm-ft}}{\text{lbf-sec}^2}} = 5\text{ lbf}$$

The frictional force is

$$\begin{aligned} F_f &= W\sin\theta = (5\text{ lbf})\sin 20° \\ &= \boxed{1.71\text{ lbf}\quad(1.7\text{ lbf})} \end{aligned}$$

The answer is (A).

SI Solution

(b) The mass of the block is $m = 5$ kg. The angle of inclination is $\theta = 20°$. The gravitational force is

$$W = mg = (5\text{ kg})\left(9.81\ \frac{\text{m}}{\text{s}^2}\right) = 49.1\text{ N}$$

The frictional force is

$$\begin{aligned} F_f &= W\sin\theta = (49.1\text{ N})\sin 20° \\ &= \boxed{16.8\text{ N}\quad(17\text{ N})} \end{aligned}$$

The answer is (A).

2. This calculation contains a hyperbolic cosine function.

$$\begin{aligned} S &= c\left(\cosh\frac{a}{h} - 1\right) \\ &= (245 \text{ ft})\left(\cosh\frac{50 \text{ ft}}{245 \text{ ft}} - 1\right) \\ &= (245 \text{ ft})(1.0209 - 1) \\ &= \boxed{5.12 \text{ ft} \quad (5.1 \text{ ft})} \end{aligned}$$

The answer is (C).

3. From App. 7.A, the area of a circular segment is

$$A = \tfrac{1}{2}r^2(\phi - \sin\phi)$$

Since ϕ appears in the expression by itself, it must be expressed in radians.

$$\phi = \frac{(120°)(2\pi)}{360°} = 2.094 \text{ rad}$$

$$\begin{aligned} A &= \tfrac{1}{2}r^2(\phi - \sin\phi) = \left(\tfrac{1}{2}\right)(75 \text{ ft})^2(2.094 \text{ rad} - \sin 120°) \\ &= \boxed{3453.7 \text{ ft}^2 \quad (3500 \text{ ft}^2)} \end{aligned}$$

The answer is (C).

Analytic Geometry

PRACTICE PROBLEMS

1. The diameter of a sphere and the base of a cone are equal. What approximate percentage of that diameter must the cone's height be so that both volumes are equal?

(A) 133%

(B) 150%

(C) 166%

(D) 200%

2. The distance between the entrance and exit points on a horizontal circular roadway curve is 747 ft. The radius of the curve is 400 ft. The central angle between the entrance and exit points is most nearly

(A) 0.27°

(B) 54°

(C) 110°

(D) 340°

3. A vertical parabolic roadway crest curve starts deviating from a constant grade at station 103 (i.e., 10,300 ft from an initial benchmark). At sta 103+62, the curve is 2.11 ft lower than the tangent (i.e., from the straight line extension of the constant grade). Approximately how far will the curve be from the tangent at sta 103+87?

(A) 1.1 ft

(B) 1.5 ft

(C) 3.0 ft

(D) 4.2 ft

4. A pile driving hammer emits 143 W of sound power with each driving stroke. Assume isotropic emission and disregard reflected power. What is most nearly the maximum areal sound power density at the ground when 10.7 m of pile remains to be driven?

(A) 0.030 W/m^2

(B) 0.10 W/m^2

(C) 0.53 W/m^2

(D) 1.1 W/m^2

SOLUTIONS

1. Let d be the diameter of the sphere and the base of the cone. Use App. 7.B.

The volume of the sphere is

$$\begin{aligned} V_{\text{sphere}} &= \tfrac{4}{3}\pi r^3 = \tfrac{4}{3}\pi\left(\frac{d}{2}\right)^3 \\ &= \tfrac{1}{6}\pi d^3 \end{aligned}$$

The volume of the circular cone is

$$\begin{aligned} V_{\text{cone}} &= \tfrac{1}{3}\pi r^2 h = \tfrac{1}{3}\pi\left(\frac{d}{2}\right)^2 h \\ &= \tfrac{1}{12}\pi d^2 h \end{aligned}$$

Since the volume of the sphere and cone are equal,

$$\begin{aligned} V_{\text{cone}} &= V_{\text{sphere}} \\ \tfrac{1}{12}\pi d^2 h &= \tfrac{1}{6}\pi d^3 \\ h &= 2d \end{aligned}$$

The height of the cone must be $\boxed{200\%}$ of the diameter of the sphere.

The answer is (D).

2. Horizontal roadway curves are circular arcs. The circumference (perimeter) of an entire circle with a radius of 400 ft is

$$p = 2\pi r = (2\pi)(400 \text{ ft}) = 2513.3 \text{ ft}$$

From a ratio of curve length to angles,

$$\phi = \left(\frac{747 \text{ ft}}{2513.3 \text{ ft}}\right)(360°) = \boxed{107° \quad (110°)}$$

The answer is (C).

3. Vertical roadway curves are parabolic arcs. Parabolas are second-degree polynomials. Deviations, y, from a baseline are proportional to the square of the separation distance. That is, $y \propto x^2$.

$$\frac{y_1}{y_2} = \left(\frac{x_1}{x_2}\right)^2$$

$$y_2 = y_1\left(\frac{x_2}{x_1}\right)^2 = (2.11\text{ ft})\left(\frac{87\text{ ft}}{62\text{ ft}}\right)^2 = \boxed{4.15\text{ ft}\quad(4.2\text{ ft})}$$

The answer is (D).

4. The power is emitted isotropically, spherically, in all directions. The maximum sound power will occur at the surface, adjacent to the pile. The surface area of a sphere with a radius of 10.7 m is

$$A = 4\pi r^2 = (4\pi)(10.7\text{ m})^2 = 1438.7\text{ m}^2$$

The areal sound power density is

$$\rho_S = \frac{P}{A} = \frac{143\text{ W}}{1438.7\text{ m}^2} = \boxed{0.0994\text{ W/m}^2\quad(0.10\text{ W/m}^2)}$$

The answer is (B).

Differential Calculus

PRACTICE PROBLEMS

1. Find all minima, maxima, and inflection points for

$$y = x^3 - 9x^2 - 3$$

(A) inflection at $x = -3$
maximum at $x = 0$
minimum at $x = -6$

(B) inflection at $x = 3$
maximum at $x = 0$
minimum at $x = 6$

(C) inflection at $x = 3$
maximum at $x = 6$
minimum at $x = 0$

(D) inflection at $x = 0$
maximum at $x = -3$
minimum at $x = 3$

2. The equation for the elevation above mean sea level of a sag vertical roadway curve is

$$y(x) = 0.56x^2 - 3.2x + 708.28$$

y is measured in feet, and x is measured in 100 ft stations past the beginning of the curve. What is most nearly the elevation of the turning point (i.e., the lowest point on the curve)?

(A) 702 ft

(B) 704 ft

(C) 705 ft

(D) 706 ft

3. A car drives on a highway with a legal speed limit of 100 km/h. The fuel usage, Q (in liters per 100 kilometers driven), of a car driven at speed v (in km/h) is

$$Q(\mathrm{v}) = \frac{1750\mathrm{v}}{\mathrm{v}^2 + 6700}$$

At what approximate legal speed should the car travel in order to maximize the fuel efficiency?

(A) 82 km/h

(B) 87 km/h

(C) 93 km/h

(D) 100 km/h

4. A chemical feed storage tank is needed with a volume of 3000 ft^3 (gross of fittings). The tank will be formed as a circular cylinder with barrel length, L, capped by two hemispherical ends of radius, r. The manufacturing cost per unit area of hemispherical ends is double that of the cylinder. The dimensions that will minimize the manufacturing cost are most nearly

(A) radius = $4^1/_2$ ft; cylinder barrel length = 42 ft

(B) radius = 5 ft; cylinder barrel length = $31^1/_2$ ft

(C) radius = $5^1/_2$ ft; cylinder barrel length = $22^1/_2$ ft

(D) radius = 6 ft; cylinder barrel length = $18^1/_2$ ft

SOLUTIONS

1. Determine the critical points by taking the first derivative of the function and setting it equal to zero.

$$\frac{dy}{dx} = 3x^2 - 18x = 3x(x-6)$$
$$3x(x-6) = 0$$
$$x(x-6) = 0$$

The critical points are located at $x = 0$ and $x = 6$.

Determine the inflection points by setting the second derivative equal to zero. Take the second derivative.

$$\begin{aligned}\frac{d^2y}{dx^2} &= \left(\frac{d}{dx}\right)\left(\frac{dy}{dx}\right)\\ &= \frac{d}{dx}(3x^2 - 18x)\\ &= 6x - 18\end{aligned}$$

Set the second derivative equal to zero.

$$\frac{d^2y}{dx^2} = 0 = 6x - 18 = (6)(x-3)$$
$$(6)(x-3) = 0$$
$$x - 3 = 0$$
$$x = 3$$

The inflection point is at $\boxed{x = 3.}$

Determine the local maximum and minimum by substituting the critical points into the expression for the second derivative.

At the critical point $x = 0$,

$$\left.\frac{d^2y}{dx^2}\right|_{x=0} = (6)(x-3) = (6)(0-3) = -18$$

Since $-18 < 0$, $\boxed{x = 0}$ is a local maximum.

At the critical point $x = 6$,

$$\left.\frac{d^2y}{dx^2}\right|_{x=6} = (6)(x-3) = (6)(6-3) = 18$$

Since $18 > 0$, $\boxed{x = 6}$ is a local minimum.

The answer is (B).

2. Set the derivative of the curve's equation to zero.

$$\frac{dy(x)}{dx} = \frac{d}{dx}(0.56x^2 - 3.2x + 708.28) = 1.12x - 3.2$$

$$x_c = \frac{3.2}{1.12} = 2.857 \text{ sta}$$

Insert x_c into the elevation equation.

$$\begin{aligned}y_{\text{min}} &= y(x_c) = 0.56x^2 - 3.2x + 708.28\\ &= (0.56)(2.857)^2 - (3.2)(2.857) + 708.28\\ &= \boxed{703.71 \text{ ft} \quad (704 \text{ ft})}\end{aligned}$$

The answer is (B).

3. Use the quotient rule to calculate the derivative.

$$\mathbf{D}\left(\frac{f(x)}{g(x)}\right) = \frac{g(x)\mathbf{D}f(x) - f(x)\mathbf{D}g(x)}{\big(g(x)\big)^2}$$

$$\begin{aligned}\frac{dQ(\text{v})}{d\text{v}} &= \frac{d}{d\text{v}}\left(\frac{1750\text{v}}{\text{v}^2 + 6700}\right)\\ &= \frac{(\text{v}^2 + 6700)(1750) - (1750\text{v})(2\text{v})}{(\text{v}^2 + 6700)^2}\end{aligned}$$

Combining terms and simplifying,

$$\frac{dQ(\text{v})}{d\text{v}} = \frac{-1750\text{v}^2 + 11{,}725{,}000}{\text{v}^4 + 13{,}400\text{v}^2 + 44{,}890{,}000}$$

Set the derivative of the fuel consumption equation to zero. In order for the derivative to be zero, the numerator must be zero.

$$-1750\text{v}^2 + 11{,}725{,}000 = 0$$

$$\text{v} = \sqrt{\frac{11{,}725{,}000}{1750}} = 81.85 \quad (82 \text{ km/h})$$

Maximizing the fuel efficiency is the same as minimizing the fuel usage. It is not known if setting $dQ(\text{v})/dt = 0$ results in a minimum or maximum. While using $d^2Q(\text{v})/dt^2$ is possible, it is easier just to plot the points.

v	Q(v)
82 km/h	10.689 L
87 km/h	10.669 L
93 km/h	10.603 L
100 km/h	10.479 L

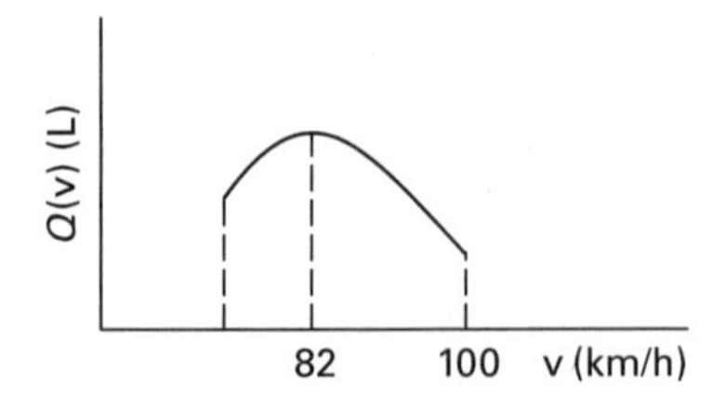

Clearly, Q(82 km/h) is a maximum, and the minimum fuel usage occurs at the endpoint of the range, at $\boxed{100 \text{ km/h.}}$

The answer is (D).

4. The volume of the tank will be the combined volume of a cylinder and a sphere. The cylinder and sphere have the same radius, r.

$$V = \pi r^2 L + \tfrac{4}{3}\pi r^3 = 3000 \text{ ft}^3$$

For any given cost per unit area (arbitrarily selected as $\$1/\text{ft}^2$), the cost function is

$$\begin{aligned} C(r, L) &= \left(1 \ \frac{\$}{\text{ft}^2}\right)(A_{\text{cylinder}} + 2A_{\text{sphere}}) = 2\pi r L + 2(4\pi r^2) \\ &= 2\pi r L + 8\pi r^2 \end{aligned}$$

Solve the volume equation for barrel length, L.

$$L = \frac{3000 - \frac{4}{3}\pi r^3}{\pi r^2}$$

Substitute L into the cost equation to get a cost function of a single variable.

$$\begin{aligned} C(r) &= 2\pi r L + 8\pi r^2 = 2\pi r\left(\frac{3000 - \frac{4}{3}\pi r^3}{\pi r^2}\right) + 8\pi r^2 \\ &= \frac{6000}{r} - \frac{8\pi r^2}{3} + 8\pi r^2 \\ &= \frac{6000}{r} + \frac{16\pi r^2}{3} \end{aligned}$$

Find the optimal value of r by setting the first derivative of the cost function equal to zero.

$$\frac{dC(r)}{dr} = \frac{d}{dr}\left(\frac{6000}{r} + \frac{16\pi r^2}{3}\right) = \frac{-6000}{r^2} + \frac{32\pi r}{3} = 0$$

$$\frac{6000}{r^2} = \frac{32\pi r}{3}$$

Cross multiply, and solve for the optimal value of radius, r.

$$\begin{aligned} 32\pi r^3 &= 18{,}000 \\ r &= \sqrt[3]{\frac{18{,}000}{32\pi}} \\ &= \boxed{5.636 \text{ ft} \quad (5^1/_2 \text{ ft})} \end{aligned}$$

Calculate the optimal value of the barrel length, L.

$$\begin{aligned} L &= \frac{3000 - \frac{4}{3}\pi r^3}{\pi r^2} \\ &= \frac{3000 - \frac{4}{3}\pi(5.636 \text{ ft})^3}{\pi(5.636 \text{ ft})^2} \\ &= \boxed{22.55 \text{ ft} \quad (22^1/_2 \text{ ft})} \end{aligned}$$

The answer is (C).

Integral Calculus

PRACTICE PROBLEMS

1. Find the indefinite integrals.

(a) $\int \sqrt{1-x}\,dx$

(A) $-\frac{2}{3}(1-x)^{3/2}+C$

(B) $-\frac{1}{2}(1-x)^{-1/2}+C$

(C) $\frac{3}{2}(1-x)^{3/2}+C$

(D) $2(1-x)^{3/2}+C$

(b) $\int \frac{x}{x^2+1}\,dx$

(A) $\frac{1}{x}\ln|(x^2+1)|+C$

(B) $\frac{1}{4}\ln|(x^2+1)|+C$

(C) $\frac{1}{2}\ln|(x^2+1)|+C$

(D) $\ln|(x^2+1)|+C$

(c) $\int \frac{x^2}{x^2+x-6}\,dx$

(A) $\ln|(x+3)|+\frac{4}{5}\ln|(x+2)|+C$

(B) $\ln|(x-3)|+\frac{4}{5}\ln|(x-2)|+C$

(C) $x-\frac{5}{9}\ln|(x-3)|+\frac{4}{5}\ln|(x-2)|+C$

(D) $x-\frac{9}{5}\ln|(x+3)|+\frac{4}{5}\ln|(x-2)|+C$

2. Calculate the definite integrals.

(a) $\int_1^3 (x^2+4x)\,dx$

(A) $16\frac{1}{2}$

(B) $24\frac{2}{3}$

(C) 27

(D) $42\frac{1}{3}$

(b) $\int_{-2}^{2} (x^3+1)\,dx$

(A) –4

(B) –2

(C) 0

(D) 4

(c) $\int_1^2 (4x^3-3x^2)\,dx$

(A) 8

(B) 16

(C) 24

(D) 32

3. Find the area bounded by $x=1$, $x=3$, $y+x+1=0$, and $y=6x-x^2$.

(A) $7\frac{1}{2}$

(B) $13\frac{1}{3}$

(C) $21\frac{1}{3}$

(D) $25\frac{1}{2}$

4. The velocity profile of a fluid experiencing laminar flow in a pipe of radius R is

$$\mathrm{v}(r)=\mathrm{v}_{\mathrm{max}}\left(1-\left(\frac{r}{R}\right)^2\right)$$

r is the distance from the centerline. $\mathrm{v}_{\mathrm{max}}$ is the centerline velocity.

(a) What is the volumetric flow rate?

(A) $\frac{\pi \mathrm{v}_{\mathrm{max}}R^2}{2}$

(B) $\pi \mathrm{v}_{\mathrm{max}}R^2$

(C) $\frac{3\pi \mathrm{v}_{\mathrm{max}}R^2}{2}$

(D) $2\pi \mathrm{v}_{\mathrm{max}}R^2$

(b) What is the average velocity?

(A) $\frac{\mathrm{v}_{\mathrm{max}}}{6}$

(B) $\frac{\mathrm{v}_{\mathrm{max}}}{4}$

(C) $\frac{\mathrm{v}_{\mathrm{max}}}{3}$

(D) $\frac{\mathrm{v}_{\mathrm{max}}}{2}$

5. Find a_0 (the first term of a Fourier series approximation, corresponding to the waveform's average value) for the two waveforms shown.

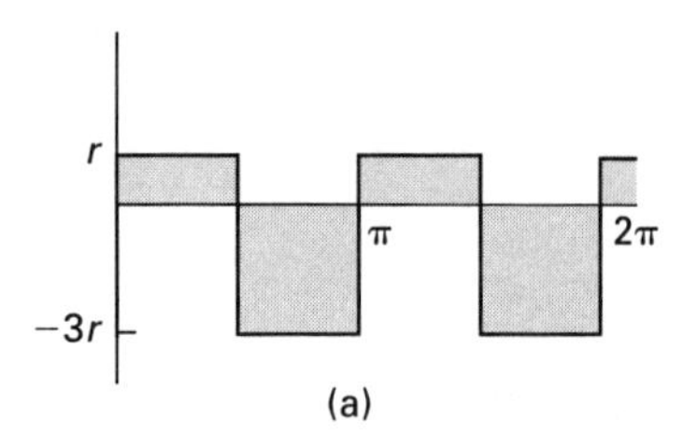

(a)

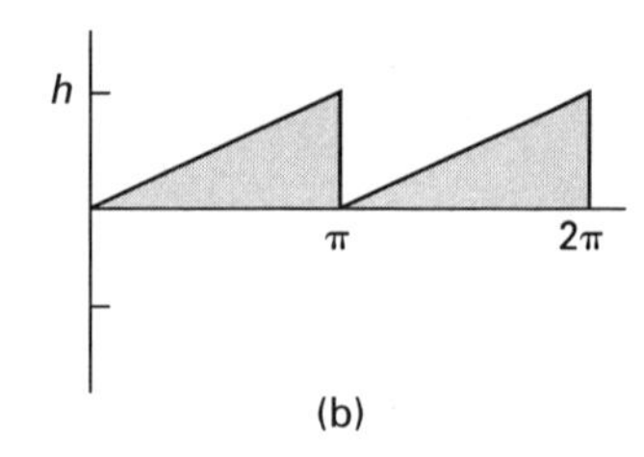

(b)

(a) For waveform (a),

(A) $-r$

(B) $-\frac{1}{2}r$

(C) r

(D) $2r$

(b) For waveform (b),

(A) $-\dfrac{h}{2}$

(B) $\dfrac{h}{2}$

(C) h

(D) $2h$

6. For each of the two waveforms shown, determine if its Fourier series is of type A, B, or C.

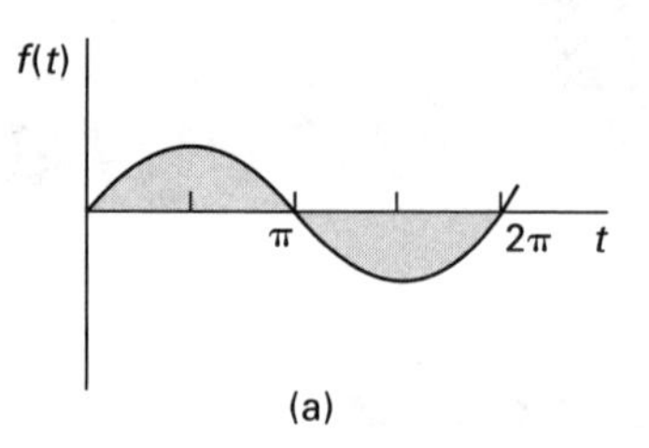

(a)

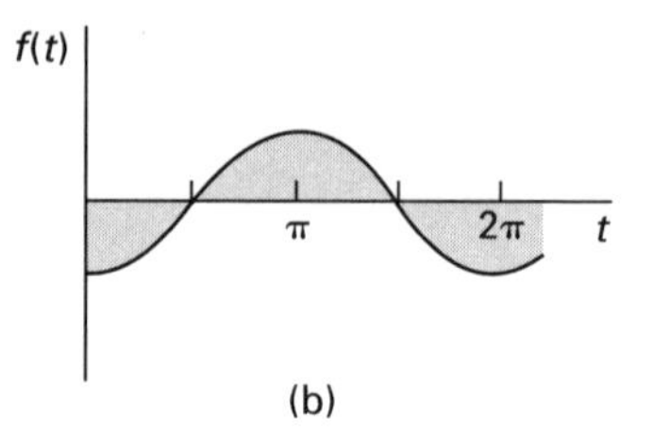

(b)

type A: $f(t) = a_0 + a_2 \cos 2t + b_2 \sin 2t + a_4 \cos 4t + b_4 \sin 4t + \cdots$

type B: $f(t) = a_0 + b_1 \sin t + b_2 \sin 2t + b_3 \sin 3t + \cdots$

type C: $f(t) = a_0 + a_1 \cos t + a_2 \cos 2t + a_3 \cos 3t + \cdots$

(a) For waveform (a),

(A) type A

(B) type B

(C) type C

(D) both type A and type C

(b) For waveform (b),

(A) type A

(B) type B

(C) type C

(D) both type A and type C

SOLUTIONS

1. (a) The indefinite integral is

$$\int \sqrt{1-x}\,dx = \int (1-x)^{1/2}dx = \boxed{-\tfrac{2}{3}(1-x)^{3/2}+C}$$

The answer is (A).

(b) The indefinite integral is

$$\int \frac{x}{x^2+1}\,dx = \tfrac{1}{2}\int \frac{2x}{x^2+1}\,dx = \boxed{\tfrac{1}{2}\ln|(x^2+1)|+C}$$

The answer is (C).

(c) The indefinite integral is

$$\begin{aligned}\frac{x^2}{x^2+x-6} &= \frac{x^2-x+x-6+6}{x^2+x-6}\\ &= 1-\frac{x-6}{x^2+x-6}\\ &= 1-\frac{x-6}{(x+3)(x-2)}\\ &= 1-\frac{\frac{9}{5}}{x+3}+\frac{\frac{4}{5}}{x-2}\end{aligned}$$

$$\begin{aligned}\int \frac{x^2}{x^2+x-6}\,dx &= \int\left(1-\frac{\frac{9}{5}}{x+3}+\frac{\frac{4}{5}}{x-2}\right)dx\\ &= \int dx - \int\frac{\frac{9}{5}}{x+3}\,dx+\int\frac{\frac{4}{5}}{x-2}\,dx\\ &= \boxed{x-\tfrac{9}{5}\ln|(x+3)|+\tfrac{4}{5}\ln|(x-2)|+C}\end{aligned}$$

The answer is (D).

2. (a) The definite integral is

$$\begin{aligned}\int_1^3 (x^2+4x)\,dx &= \left(\frac{x^3}{3}+2x^2\right)\Bigg|_1^3\\ &= \frac{(3)^3}{3}+(2)(3)^2-\left(\frac{(1)^3}{3}+(2)(1)^2\right)\\ &= \boxed{24\tfrac{2}{3}}\end{aligned}$$

The answer is (B).

(b) The definite integral is

$$\begin{aligned}\int_{-2}^2 (x^3+1)\,dx &= \left(\frac{x^4}{4}+x\right)\Bigg|_{-2}^2\\ &= \frac{(2)^4}{4}+2-\left(\frac{(-2)^4}{4}+(-2)\right)\\ &= \boxed{4}\end{aligned}$$

The answer is (D).

(c) The definite integral is

$$\begin{aligned}\int_1^2 (4x^3-3x^2)\,dx &= \left(x^4-x^3\right)\Big|_1^2\\ &= (2)^4-(2)^3-\left((1)^4-(1)^3\right)\\ &= \boxed{8}\end{aligned}$$

The answer is (A).

3. The bounded area is shown.

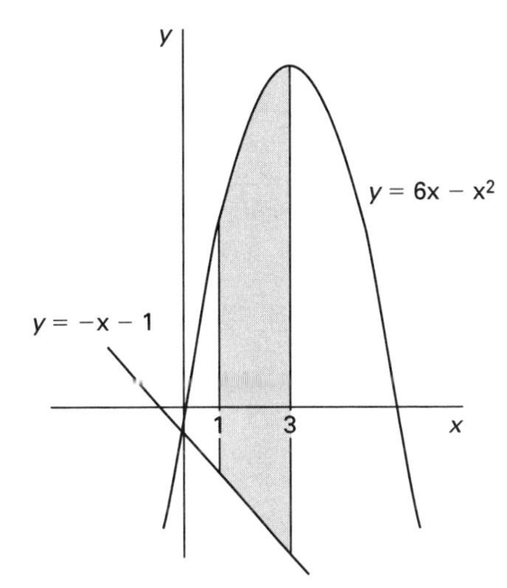

$$\begin{aligned}\text{area} &= \int_1^3 \left((6x-x^2)-(-x-1)\right)dx\\ &= \int_1^3 (-x^2+7x+1)\,dx\\ &= \left(-\frac{x^3}{3}+\tfrac{7}{2}x^2+x\right)\Bigg|_1^3\\ &= -\frac{(3)^3}{3}+\left(\tfrac{7}{2}\right)(3)^2+3-\left(-\frac{(1)^3}{3}+\left(\tfrac{7}{2}\right)(1)^2+1\right)\\ &= \boxed{21\tfrac{1}{3}}\end{aligned}$$

The answer is (C).

4. (a) Divide the circular internal area of the pipe into small annular rings. The radius at the ring is r; the differential thickness of the ring is dr; the differential area of the ring is $dA = 2\pi r\,dr$; and the velocity is $v(r)$. The volumetric flow rate through the annular ring is

$$dQ = v(r)dA = 2\pi r v_{max}\left(1 - \left(\frac{r}{R}\right)^2\right)dr$$

$$\begin{aligned} Q &= \int_{r=0}^{r=R} 2\pi r v_{max}\left(1 - \left(\frac{r}{R}\right)^2\right)dr \\ &= 2\pi v_{max}\int_{r=0}^{r=R}\left(r - \frac{r^3}{R^2}\right)dr \\ &= 2\pi v_{max}\left(\frac{r^2}{2} - \frac{r^4}{4R^2}\right)\Big|_0^R \\ &= 2\pi v_{max}\left(\frac{R^2}{2} - \frac{R^4}{4R^2} - 0 - 0\right) \\ &= \boxed{\frac{\pi v_{max}R^2}{2}} \end{aligned}$$

The answer is (A).

(b) The average velocity is

$$\bar{v} = \frac{Q}{A} = \frac{\frac{\pi v_{max}R^2}{2}}{\pi R^2} = \boxed{\frac{v_{max}}{2}}$$

The answer is (D).

5. (a) For waveform (a),

$$\begin{aligned} a_0 &= \frac{1}{2\pi}\int_0^{2\pi} f(t)dt = \frac{1}{\pi}\int_0^{\pi} f(t)dt \\ &= \frac{1}{\pi}\left(r\left(\frac{\pi}{2}\right) + (-3r)\left(\frac{\pi}{2}\right)\right) \\ &= \boxed{-r} \end{aligned}$$

The answer is (A).

(b) For waveform (b),

$$\begin{aligned} a_0 &= \frac{1}{2\pi}\int_0^{2\pi} f(t)dt = \frac{1}{\pi}\int_0^{\pi} f(t)dt \\ &= \left(\frac{1}{\pi}\right)\left(\tfrac{1}{2}\pi h\right) \\ &= \boxed{\frac{h}{2}} \end{aligned}$$

The answer is (B).

6. (a) For waveform (a),

Since $f(t) = -f(-t)$, it is $\boxed{\text{type B.}}$

The answer is (B).

(b) For waveform (b),

Since $f(t) = f(-t)$, it is $\boxed{\text{type C.}}$

The answer is (C).

10 Differential Equations

PRACTICE PROBLEMS

1. Solve the following differential equation for y.

$$y'' - 4y' - 12y = 0$$

(A) $A_1e^{6x} + A_2e^{-2x}$

(B) $A_1e^{-6x} + A_2e^{2x}$

(C) $A_1e^{6x} + A_2e^{2x}$

(D) $A_1e^{-6x} + A_2e^{-2x}$

2. Solve the following differential equation for y.

$$y' - y = 2xe^{2x} \quad y(0) = 1$$

(A) $y = 2e^{-2x}(x-1) + 3e^{-x}$

(B) $y = 2e^{2x}(x-1) + 3e^{x}$

(C) $y = -2e^{-2x}(x-1) + 3e^{-x}$

(D) $y = 2e^{2x}(x-1) + 3e^{-x}$

3. The oscillation exhibited by the top story of a certain building in free motion is given by the following differential equation.

$$x'' + 2x' + 2x = 0 \quad x(0) = 0 \quad x'(0) = 1$$

(a) What is x as a function of time?

(A) $e^{-2t}\sin t$

(B) $e^{t}\sin t$

(C) $e^{-t}\sin t$

(D) $e^{-t}\sin t + e^{-t}\cos t$

(b) The building's fundamental natural frequency of vibration is most nearly

(A) $1/2$

(B) 1

(C) $\sqrt{2}$

(D) 2

(c) The amplitude of oscillation is most nearly

(A) 0.32

(B) 0.54

(C) 1.7

(D) 6.6

(d) What is x as a function of time if a lateral wind load is applied with a form of $\sin t$?

(A) $\frac{6}{5}e^{-t}\sin t + \frac{2}{5}e^{-t}\cos t$

(B) $\frac{6}{5}e^{t}\sin t + \frac{2}{5}e^{t}\cos t$

(C) $\frac{2}{5}e^{t}\sin t + \frac{6}{5}e^{-t}\cos t + \frac{2}{5}\sin t - \frac{1}{5}\cos t$

(D) $\frac{2}{5}e^{-t}\cos t + \frac{6}{5}e^{-t}\sin t - \frac{2}{5}\cos t + \frac{1}{5}\sin t$

4. A 90 lbm (40 kg) bag of a chemical is accidentally dropped in an aerating lagoon. The chemical is water soluble and nonreacting. The lagoon is 120 ft (35 m) in diameter and filled to a depth of 10 ft (3 m). The aerators circulate and distribute the chemical evenly throughout the lagoon.

Water enters the lagoon at a rate of 30 gal/min (115 L/min). Fully mixed water is pumped into a reservoir at a rate of 30 gal/min (115 L/min). The established safe concentration of this chemical is 1 ppb (part per billion).

(a) The volume of the lagoon at time t is most nearly

(A) 28,000 ft^3 (720 m^3)

(B) 110,000 ft^3 (2900 m^3)

(C) 140,000 ft^3 (3700 m^3)

(D) 230,000 ft^3 (5600 m^3)

(b) The volumetric flow rate out of the lagoon is most nearly

(A) 2.3 ft^3/min (0.066 m^3/min)

(B) 3.6 ft^3/min (0.10 m^3/min)

(C) 4.0 ft^3/min (0.12 m^3/min)

(D) 4.6 ft^3/min (0.13 m^3/min)

(c) The initial mass of the water in the lagoon is most nearly

(A) 1.8×10^6 lbm (7.2×10^5 kg)

(B) 7.1×10^6 lbm (2.9×10^6 kg)

(C) 9.0×10^6 lbm (3.7×10^6 kg)

(D) 14×10^7 lbm (5.6×10^6 kg)

(d) The final mass of chemicals at a concentration of 1 ppb is most nearly

(A) 0.002 lbm (0.0007 kg)

(B) 0.007 lbm (0.003 kg)

(C) 0.009 lbm (0.004 kg)

(D) 0.014 lbm (0.006 kg)

(e) The number of days it will take for the concentration of the discharge water to reach this level is most nearly

(A) 25 days

(B) 50 days

(C) 100 days

(D) 200 days

5. A tank contains 100 gal (100 L) of brine made by dissolving 60 lbm (60 kg) of salt in pure water. Salt water with a concentration of 1 lbm/gal (1 kg/L) enters the tank at a rate of 2 gal/min (2 L/min). A well-stirred mixture is drawn from the tank at a rate of 3 gal/min (3 L/min). The mass of salt in the tank after one hour is most nearly

(A) 13 lbm (13 kg)

(B) 37 lbm (37 kg)

(C) 43 lbm (43 kg)

(D) 51 lbm (51 kg)

SOLUTIONS

1. Obtain the characteristic equation by replacing each derivative with a polynomial term of equal degree.

$$r^2 - 4r - 12 = 0$$

Factor the characteristic equation.

$$(r-6)(r+2) = 0$$

The roots are $r_1 = 6$ and $r_2 = -2$.

Since the roots are real and distinct, the solution is

$$y = A_1 e^{r_1 x} + A_2 e^{r_2 x} = \boxed{A_1 e^{6x} + A_2 e^{-2x}}$$

The answer is (A).

2. The equation is a first-order linear differential equation of the form

$$\begin{aligned} y' + p(x)y &= g(x) \\ p(x) &= -1 \\ g(x) &= 2xe^{2x} \end{aligned}$$

The integration factor $u(x)$ is

$$u(x) = \exp\left(\int p(x)\,dx\right) = \exp\left(\int(-1)\,dx\right) = e^{-x}$$

The closed form of the solution is

$$\begin{aligned} y &= \frac{1}{u(x)}\left(\int u(x)g(x)\,dx + C\right) \\ &= \frac{1}{e^{-x}}\left(\int (e^{-x})(2xe^{2x})\,dx + C\right) \\ &= e^x\big(2(xe^x - e^x) + C\big) \\ &= e^x\big(2e^x(x-1) + C\big) \\ &= 2e^{2x}(x-1) + Ce^x \end{aligned}$$

Apply the initial condition $y(0) = 1$ to obtain the integration constant, C.

$$\begin{aligned} y(0) = 2e^{(2)(0)}(0-1) + Ce^0 &= 1 \\ (2)(1)(-1) + C(1) &= 1 \\ -2 + C &= 1 \\ C &= 3 \end{aligned}$$

Substituting in the value for the integration constant, C, the solution is

$$\boxed{y = 2e^{2x}(x-1) + 3e^{x}}$$

The answer is (B).

3. (a) The differential equation is a homogeneous second-order linear differential equation with constant coefficients. Write the characteristic equation.

$$r^2 + 2r + 2 = 0$$

This is a quadratic equation of the form $ar^2 + br + c = 0$, where $a = 1$, $b = 2$, and $c = 2$.

Solve for r.

$$\begin{aligned} r &= \frac{-b \pm \sqrt{b^2 - 4ac}}{2a} = \frac{-2 \pm \sqrt{(2)^2 - (4)(1)(2)}}{(2)(1)} \\ &= \frac{-2 \pm \sqrt{4-8}}{2} \\ &= -1 \pm \sqrt{-1} \\ &= -1 \pm i \end{aligned}$$

$$r_1 = -1 + i, \text{ and } r_2 = -1 - i$$

Since the roots are imaginary and of the form $\alpha + i\omega$ and $\alpha - i\omega$, where $\alpha = -1$ and $\omega = 1$, the general form of the solution is

$$\begin{aligned} x(t) &= A_1 e^{\alpha t}\cos\omega t + A_2 e^{\alpha t}\sin\omega t \\ &= A_1 e^{-1t}\cos t + A_2 e^{-1t}\sin t \\ &= A_1 e^{-t}\cos t + A_2 e^{-t}\sin t \end{aligned}$$

Apply the initial conditions, $x(0) = 0$ and $x'(0) = 1$, to solve for A_1 and A_2.

First, apply the initial condition, $x(0) = 0$.

$$\begin{aligned} x(t) = A_1 e^0 \cos 0 + A_2 e^0 \sin 0 &= 0 \\ A_1(1)(1) + A_2(1)(0) &= 0 \\ A_1 &= 0 \end{aligned}$$

Substituting, the solution of the differential equation becomes

$$x(t) = A_2 e^{-t}\sin t$$

To apply the second initial condition, take the first derivative.

$$\begin{aligned} x'(t) &= \frac{d}{dt}(A_2 e^{-t}\sin t) = A_2 \frac{d}{dt}(e^{-t}\sin t) \\ &= A_2\left(\sin t \frac{d}{dt}(e^{-t}) + e^{-t}\frac{d}{dt}\sin t\right) \\ &= A_2\left(\sin t(-e^{-t}) + e^{-t}(\cos t)\right) \\ &= A_2(e^{-t})(-\sin t + \cos t) \end{aligned}$$

Apply the initial condition, $x'(0) = 1$.

$$\begin{aligned} x(0) = A_2 e^0(-\sin 0 + \cos 0) &= 1 \\ A_2(1)(0+1) &= 1 \\ A_2 &= 1 \end{aligned}$$

The solution is

$$\begin{aligned} x(t) &= A_2 e^{-t}\sin t = (1)e^{-t}\sin t \\ &= \boxed{e^{-t}\sin t} \end{aligned}$$

The answer is (C).

(b) To determine the natural frequency, set the damping term to zero. The equation has the form

$$x'' + 2x = 0$$

This equation has a general solution of the form

$$x(t) = x_0 \cos\omega t + \left(\frac{\mathrm{v}_0}{\omega}\right)\sin\omega t$$

ω is the natural frequency. Given the equation $x'' + 2x = 0$, the characteristic equation is

$$\begin{aligned} r^2 + 2 &= 0 \\ r &= \sqrt{-2} \\ &= \pm\sqrt{2}i \end{aligned}$$

Since the roots are imaginary and of the form $\alpha + i\omega$ and $\alpha - i\omega$, where $\alpha = 0$ and $\omega = \sqrt{2}$, the general form of the solution is

$$\begin{aligned} x(t) &= A_1 e^{\alpha t}\cos\omega t + A_2 e^{\alpha t}\sin\omega t \\ &= A_1 e^{0t}\cos\sqrt{2}t + A_2 e^{0t}\sin\sqrt{2}t \\ &= A_1(1)\cos\sqrt{2}t + A_2(1)\sin\sqrt{2}t \\ &= A_1\cos\sqrt{2}t + A_2\sin\sqrt{2}t \end{aligned}$$

Apply the initial conditions, $x(0) = 0$ and $x'(0) = 1$, to solve for A_1 and A_2. Applying the initial condition $x(0) = 0$ gives

$$\begin{aligned} x(0) = A_1 \cos\big((\sqrt{2})(0)\big) + A_2 \sin\big((\sqrt{2})(0)\big) &= 0 \\ A_1 \cos 0 + A_2 \sin 0 &= 0 \\ A_1(1) + A_2(0) &= 0 \\ A_1 &= 0 \end{aligned}$$

Substituting, the solution of the differential equation becomes

$$x(t) = A_2 \sin \sqrt{2}t$$

To apply the second initial condition, take the first derivative.

$$\begin{aligned} x'(t) &= \frac{d}{dt}(A_2 \sin \sqrt{2}t) \\ &= A_2\sqrt{2} \cos\sqrt{2}t \end{aligned}$$

Apply the second initial condition, $x'(0) = 1$.

$$\begin{aligned} x'(0) = A_2\sqrt{2} \cos\big((\sqrt{2})(0)\big) &= 1 \\ A_2\sqrt{2} \cos(0) &= 1 \\ A_2(\sqrt{2})(1) &= 1 \\ A_2\sqrt{2} &= 1 \\ A_2 = \frac{1}{\sqrt{2}} &= \frac{\sqrt{2}}{2} \end{aligned}$$

Substituting, the undamped solution becomes

$$x(t) = \frac{\sqrt{2}}{2} \sin\sqrt{2}t$$

This is of the form

$$x(t) = A \sin \omega t$$

Therefore, the undamped natural frequency is $\omega = \boxed{\sqrt{2}.}$

The answer is (C).

(c) The amplitude of the oscillation is the maximum displacement.

Take the derivative of the solution, $x(t) = e^{-t} \sin t$.

$$\begin{aligned} x'(t) &= \frac{d}{dt}(e^{-t} \sin t) = \sin t \frac{d}{dt}(e^{-t}) + e^{-t} \frac{d}{dt} \sin t \\ &= \sin t(-e^{-t}) + e^{-t} \cos t \\ &= e^{-t}(\cos t - \sin t) \end{aligned}$$

The maximum displacement occurs at $x'(t) = 0$.

Since $e^{-t} \neq 0$ except as t approaches infinity,

$$\begin{aligned} \cos t - \sin t &= 0 \\ \tan t &= 1 \\ t &= \tan^{-1}(1) \\ &= 0.785 \text{ rad} \end{aligned}$$

At $t = 0.785$ rad, the displacement is maximum. Substitute into the orginal solution to obtain a value for the maximum displacement.

$$x(0.785) = e^{-0.785} \sin 0.785 = 0.322$$

The amplitude is $\boxed{0.322\ (0.32).}$

The answer is (A).

(d) (An alternative solution using Laplace transforms follows this solution.) The application of a lateral wind load with the form $\sin t$ revises the differential equation to the form

$$x'' + 2x' + 2x = \sin t$$

Express the solution as the sum of the complementary x_c and particular x_p solutions.

$$x(t) = x_c(t) + x_p(t)$$

From part (a),

$$x_c(t) = A_1 e^{-t} \cos t + A_2 e^{-t} \sin t$$

The general form of the particular solution is given by

$$x_p(t) = x^s(A_3 \cos t + A_4 \sin t)$$

Determine the value of s; check to see if the terms of the particular solution solve the homogeneous equation.

Examine the term $A_3 \cos t$.

Take the first derivative.

$$\frac{d}{dx}(A_3 \cos t) = -A_3 \sin t$$

Take the second derivative.

$$\begin{aligned} \frac{d}{dx}\left(\frac{d}{dx}(A_3 \cos t)\right) &= \frac{d}{dx}(-A_3 \sin t) \\ &= -A_3 \cos t \end{aligned}$$

Substitute the terms into the homogeneous equation.

$$\begin{aligned} x'' + 2x' + 2x &= -A_3 \cos t + (2)(-A_3 \sin t) \\ &\quad + (2)(-A_3 \cos t) \\ &= A_3 \cos t - 2A_3 \sin t \\ &\neq 0 \end{aligned}$$

Except for the trival solution $A_3 = 0$, the term $A_3 \cos t$ does not solve the homogeneous equation.

Examine the second term, $A_4 \sin t$.

Take the first derivative.

$$\frac{d}{dx}(A_4 \sin t) = A_4 \cos t$$

Take the second derivative.

$$\frac{d}{dx}\left(\frac{d}{dx}(A_4 \sin t)\right) = \frac{d}{dx}(A_4 \cos t) = -A_4 \sin t$$

Substitute the terms into the homogeneous equation.

$$\begin{aligned} x'' + 2x' + 2x &= -A_4 \sin t + (2)(A_4 \cos t) \\ &\quad + (2)(A_4 \sin t) \\ &= A_4 \sin t + 2A_4 \cos t \\ &\neq 0 \end{aligned}$$

Except for the trivial solution $A_4 = 0$, the term $A_4 \sin t$ does not solve the homogeneous equation.

Neither of the terms satisfies the homogeneous equation, so $s = 0$. Therefore, the particular solution is of the form

$$x_p(t) = A_3 \cos t + A_4 \sin t$$

Use the method of undetermined coefficients to solve for A_3 and A_4. Take the first derivative.

$$\begin{aligned} x_p'(t) &= \frac{d}{dx}(A_3 \cos t + A_4 \sin t) \\ &= -A_3 \sin t + A_4 \cos t \end{aligned}$$

Take the second derivative.

$$\begin{aligned} x_p''(t) &= \frac{d}{dx}\left(\frac{d}{dx}(A_3 \cos t + A_4 \sin t)\right) \\ &= \frac{d}{dx}(-A_3 \sin t + A_4 \cos t) \\ &= -A_3 \cos t - A_4 \sin t \end{aligned}$$

Substitute the expressions for the derivatives into the differential equation.

$$\begin{aligned} x'' + 2x' + 2x &= (-A_3 \cos t - A_4 \sin t) \\ &\quad + (2)(-A_3 \sin t + A_4 \cos t) \\ &\quad + (2)(A_3 \cos t + A_4 \sin t) \\ &= \sin t \end{aligned}$$

Rearranging terms gives

$$\begin{aligned} (-A_3 + 2A_4 + 2A_3)\cos t & \\ + (-A_4 - 2A_3 + 2A_4)\sin t &= \sin t \\ (A_3 + 2A_4)\cos t + (-2A_3 + A_4)\sin t &= \sin t \end{aligned}$$

Equating coefficients gives

$$\begin{aligned} A_3 + 2A_4 &= 0 \\ -2A_3 + A_4 &= 1 \end{aligned}$$

Multiplying the first equation by 2 and adding equations gives

$$\begin{aligned} 2A_3 + 4A_4 &= 0 \\ +(-2A_3 + A_4) &= 1 \\ \hline 5A_4 = 1 \quad &\text{or} \quad A_4 = \tfrac{1}{5} \end{aligned}$$

From the first equation for $A_4 = 1/5$, $A_3 + (2)(1/5) = 0$, and $A_3 = -2/5$.

Substituting for the coefficients, the particular solution becomes

$$x_p(t) = -\tfrac{2}{5}\cos t + \tfrac{1}{5}\sin t$$

Combining the complementary and particular solutions gives

$$\begin{aligned} x(t) &= x_c(t) + x_p(t) \\ &= A_1 e^{-t} \cos t + A_2 e^{-t} \sin t - \tfrac{2}{5}\cos t + \tfrac{1}{5}\sin t \end{aligned}$$

Apply the initial conditions to solve for the coefficients A_1 and A_2, then apply the first initial condition, $x(0) = 0$.

$$\begin{aligned} x(t) = A_1 e^0 \cos 0 + A_2 e^0 \sin 0 & \\ \tfrac{2}{5}\cos 0 + \tfrac{1}{5}\sin 0 &= 0 \\ A_1(1)(1) + A_2(1)(0) + \left(-\tfrac{2}{5}\right)(1) + \left(\tfrac{1}{5}\right)(0) &= 0 \\ A_1 - \tfrac{2}{5} &= 0 \\ A_1 &= \tfrac{2}{5} \end{aligned}$$

Substituting for A_1, the solution becomes

$$x(t) = \tfrac{2}{5}e^{-t}\cos t + A_2 e^{-t}\sin t - \tfrac{2}{5}\cos t + \tfrac{1}{5}\sin t$$

Take the first derivative.

$$x'(t) = \frac{d}{dx}\left(\begin{array}{r}\left(\tfrac{2}{5}e^{-t}\cos t + A_2 e^{-t}\sin t\right)\\ +\left(-\tfrac{2}{5}\cos t + \tfrac{1}{5}\sin t\right)\end{array}\right)$$

$$\begin{aligned} &= \left(\tfrac{2}{5}\right)\left(-e^{-t}\cos t - e^{-t}\sin t\right)\\ &\quad + A_2\left(-e^{-t}\sin t + e^{-t}\cos t\right)\\ &\quad + \left(-\tfrac{2}{5}\right)(-\sin t) + \tfrac{1}{5}\cos t\end{aligned}$$

Apply the second initial condition, $x'(0) = 1$.

$$\begin{aligned} x'(0) &= \left(\tfrac{2}{5}\right)\left(-e^0\cos 0 - e^0\sin 0\right)\\ &\quad + A_2\left(-e^0\sin 0 + e^0\cos 0\right)\\ &\quad + \left(-\tfrac{2}{5}\right)(-\sin 0) + \tfrac{1}{5}\cos 0\\ &= 1\end{aligned}$$

$$\begin{aligned}\left(\tfrac{2}{5}\right)\big(-(1)(1) - (1)(0)\big) + A_2\big(-(1)(0) + (1)(1)\big)&\\ +\left(-\tfrac{2}{5}\right)(0) + \left(\tfrac{1}{5}\right)(1) &= 1\\ \left(\tfrac{2}{5}\right)(-1) + A_2(1) + \left(\tfrac{1}{5}\right) &= 1\\ A_2 &= \tfrac{6}{5}\end{aligned}$$

Substituting for A_2, the solution becomes

$$x(t) = \boxed{\tfrac{2}{5}e^{-t}\cos t + \tfrac{6}{5}e^{-t}\sin t - \tfrac{2}{5}\cos t + \tfrac{1}{5}\sin t}$$

The answer is (D).

(d) *Alternate solution:*

Use the Laplace transform method.

$$\begin{aligned} x'' + 2x' + 2x &= \sin t\\ \mathcal{L}(x'') + 2\mathcal{L}(x') + 2\mathcal{L}(x) &= \mathcal{L}(\sin t)\\ s^2\mathcal{L}(x) - 1 + 2s\mathcal{L}(x) + 2\mathcal{L}(x) &= \frac{1}{s^2+1}\\ \mathcal{L}(x)(s^2+2s+2) - 1 &= \frac{1}{s^2+1}\end{aligned}$$

$$\begin{aligned}\mathcal{L}(x) &= \frac{1}{s^2+2s+2} + \frac{1}{(s^2+1)(s^2+2s+2)}\\ &= \frac{1}{(s+1)^2+1} + \frac{1}{(s^2+1)(s^2+2s+2)}\end{aligned}$$

Use partial fractions to expand the second term.

$$\frac{1}{(s^2+1)(s^2+2s+2)} = \frac{A_1 + B_1 s}{s^2+1} + \frac{A_2 + B_2 s}{s^2+2s+2}$$

Cross multiply.

$$\begin{aligned} &A_1 s^2 + 2A_1 s + 2A_1 + B_1 s^3 + 2B_1 s^2\\ &= \frac{+\,A_2 s^2 + A_2 + B_2 s^3 + B_2 s}{(s^2+1)(s^2+2s+2)}\\ &s^3(B_1+B_2) + s^2(A_1 + A_2 + 2B_1)\\ &= \frac{+\,s(2A_1 + 2B_1 + B_2) + 2A_1 + A_2}{(s^2+1)(s^2+2s+2)}\end{aligned}$$

Compare numerators to obtain the following four simultaneous equations.

$$\begin{array}{lllllllll} & & & & B_1 & + & B_2 & = & 0\\ A_1 & + & A_2 & + & 2B_1 & & & = & 0\\ 2A_1 & & & + & 2B_1 & + & B_2 & = & 0\\ 2A_1 & + & A_2 & & & & & = & 1\end{array}$$

Use Cramer's rule to find A_1.

$$A_1 = \frac{\begin{vmatrix}0&0&1&1\\0&1&2&0\\0&0&2&1\\1&1&0&0\end{vmatrix}}{\begin{vmatrix}0&0&1&1\\1&1&2&0\\2&0&2&1\\2&1&0&0\end{vmatrix}} = \frac{-1}{-5} = \frac{1}{5}$$

The rest of the coefficients are found similarly.

$$\begin{aligned} A_1 &= \tfrac{1}{5}\\ A_2 &= \tfrac{3}{5}\\ B_1 &= -\tfrac{2}{5}\\ B_2 &= \tfrac{2}{5}\end{aligned}$$

Then,

$$\mathcal{L}(x) = \frac{1}{(s+1)^2+1} + \frac{\frac{1}{5}}{s^2+1} + \frac{-\frac{2}{5}s}{s^2+1} + \frac{\frac{3}{5}}{s^2+2s+2} + \frac{\frac{2}{5}s}{s^2+2s+2}$$

Take the inverse transform.

$$\begin{aligned} x(t) &= \mathcal{L}^{-1}\{\mathcal{L}(x)\} \\ &= e^{-t}\sin t + \tfrac{1}{5}\sin t - \tfrac{2}{5}\cos t + \tfrac{3}{5}e^{-t}\sin t \\ &\quad + \tfrac{2}{5}(e^{-t}\cos t - e^{-t}\sin t) \\ &= \boxed{\tfrac{6}{5}e^{-t}\sin t + \tfrac{2}{5}e^{-t}\cos t + \tfrac{1}{5}\sin t - \tfrac{2}{5}\cos t} \end{aligned}$$

The answer is (D).

4. *Customary U.S. Solution*

(a) The differential equation is

$$m'(t) = a(t) - \frac{m(t)o(t)}{V(t)}$$

$a(t)$ = rate of addition of chemical
$m(t)$ = mass of chemical at time t
$o(t)$ = volumetric flow out of the lagoon, 30 gpm
$V(t)$ = volume in the lagoon at time t

Water flows into the lagoon at a rate of 30 gpm, and a water-chemical mix flows out of the lagoon at a rate of 30 gpm. Therefore, the volume of the lagoon at time t is equal to the initial volume.

$$\begin{aligned} V(t) &= \left(\frac{\pi}{4}\right)(\text{diameter of lagoon})^2(\text{depth of lagoon}) \\ &= \left(\frac{\pi}{4}\right)(120\ \text{ft})^2(10\ \text{ft}) \\ &= \boxed{113{,}097\ \text{ft}^3 \quad (110{,}000\ \text{ft}^3)} \end{aligned}$$

The answer is (B).

(b) Using a conversion factor of 7.48 gal/ft^3 gives

$$o(t) = \frac{30\ \frac{\text{gal}}{\text{min}}}{7.48\ \frac{\text{gal}}{\text{ft}^3}} = \boxed{4.01\ \text{ft}^3/\text{min} \quad (4.0\ \text{ft}^3/\text{min})}$$

The answer is (C).

(c) Substituting into the general form of the differential equation gives

$$\begin{aligned} m'(t) &= a(t) - \frac{m(t)o(t)}{V(t)} \\ &= 0 - m(t)\left(\frac{4.01\ \frac{\text{ft}^3}{\text{min}}}{113{,}097\ \text{ft}^3}\right) \\ &= -\left(\frac{3.55 \times 10^{-5}}{\text{min}}\right)m(t) \end{aligned}$$

$$m'(t) + \left(\frac{3.55 \times 10^{-5}}{\text{min}}\right)m(t) = 0$$

The differential equation of the problem has the following characteristic equation.

$$r + \frac{3.55 \times 10^{-5}}{\text{min}} = 0$$

$$r = -3.55 \times 10^{-5}/\text{min}$$

The general form of the solution is

$$m(t) = Ae^{rt}$$

Substituting the root, r, gives

$$m(t) = Ae^{(-3.55 \times 10^{-5}/\text{min})t}$$

Apply the initial condition $m(0) = 90$ lbm at time $t = 0$.

$$\begin{aligned} m(0) = Ae^{(-3.55 \times 10^{-5}/\text{min})(0)} &= 90\ \text{lbm} \\ Ae^0 &= 90\ \text{lbm} \\ A &= 90\ \text{lbm} \end{aligned}$$

Therefore,

$$m(t) = (90\ \text{lbm})e^{(-3.55 \times 10^{-5}/\text{min})t}$$

Solve for t.

$$\begin{aligned} \frac{m(t)}{90\ \text{lbm}} &= e^{(-3.55 \times 10^{-5}/\text{min})t} \\ \ln\left(\frac{m(t)}{90\ \text{lbm}}\right) &= \ln\left(e^{(-3.55 \times 10^{-5}/\text{min})t}\right) \\ &= \left(\frac{-3.55 \times 10^{-5}}{\text{min}}\right)t \end{aligned}$$

$$t = \frac{\ln\left(\frac{m(t)}{90\ \text{lbm}}\right)}{\frac{-3.55 \times 10^{-5}}{\text{min}}}$$

The initial mass of the water in the lagoon is

$$m_i = V\rho = (113{,}097 \text{ ft}^3)\left(62.4 \ \frac{\text{lbm}}{\text{ft}^3}\right) = \boxed{7.06 \times 10^6 \text{ lbm} \quad (7.1 \times 10^6 \text{ lbm})}$$

The answer is (B).

(d) The final mass of chemicals at a concentration of 1 ppb is

$$m_f = \frac{7.06 \times 10^6 \text{ lbm}}{1 \times 10^9} = \boxed{7.06 \times 10^{-3} \text{ lbm} \quad (0.007 \text{ lbm})}$$

The answer is (B).

(e) Find the time required to achieve a mass of 7.06×10^{-3} lbm.

$$t = \left(\frac{\ln\left(\frac{m(t)}{90 \text{ lbm}}\right)}{\frac{-3.55 \times 10^{-5}}{\text{min}}}\right)\left(\frac{1 \text{ hr}}{60 \text{ min}}\right)\left(\frac{1 \text{ day}}{24 \text{ hr}}\right)$$

$$= \left(\frac{\ln\left(\frac{7.06 \times 10^{-3} \text{ lbm}}{90 \text{ lbm}}\right)}{\frac{-3.55 \times 10^{-5}}{\text{min}}}\right)\left(\frac{1 \text{ hr}}{60 \text{ min}}\right)\left(\frac{1 \text{ day}}{24 \text{ hr}}\right)$$

$$= \boxed{185 \text{ days} \quad (200 \text{ days})}$$

The answer is (D).

SI Solution

(a) The differential equation is

$$m'(t) = a(t) - \frac{m(t)o(t)}{V(t)}$$

$a(t)$ = rate of addition of chemical
$m(t)$ = mass of chemical at time t
$o(t)$ = volumetric flow out of the lagoon, 115 L/min
$V(t)$ = volume in the lagoon at time t

Water flows into the lagoon at a rate of 115 L/min, and a water-chemical mix flows out of the lagoon at a rate of 115 L/min. Therefore, the volume of the lagoon at time t is equal to the initial volume.

$$V(t) = \left(\frac{\pi}{4}\right)(\text{diameter of lagoon})^2(\text{depth of lagoon}) = \left(\frac{\pi}{4}\right)(35 \text{ m})^2(3 \text{ m}) = \boxed{2886 \text{ m}^3 \quad (2900 \text{ m}^3)}$$

The answer is (B).

(b) Using a conversion factor of 1000 L/m³ gives

$$o(t) = \frac{115 \ \frac{\text{L}}{\text{min}}}{1000 \ \frac{\text{L}}{\text{m}^3}} = \boxed{0.115 \text{ m}^3/\text{min} \quad (0.12 \text{ m}^3/\text{min})}$$

The answer is (C).

(c) Substituting into the general form of the differential equation gives

$$m'(t) = a(t) - \frac{m(t)o(t)}{V(t)} = 0 - m(t)\left(\frac{0.115 \ \frac{\text{m}^3}{\text{min}}}{2886 \text{ m}^3}\right) = -\left(\frac{3.985 \times 10^{-5}}{\text{min}}\right)m(t)$$

$$m'(t) + \left(\frac{3.985 \times 10^{-5}}{\text{min}}\right)m(t) = 0$$

The differential equation of the problem has the following characteristic equation.

$$r + \frac{3.985 \times 10^{-5}}{\text{min}} = 0$$
$$r = -3.985 \times 10^{-5}/\text{min}$$

The general form of the solution is

$$m(t) = Ae^{rt}$$

Substituting in for the root, r, gives

$$m(t) = Ae^{(-3.985 \times 10^{-5}/\text{min})t}$$

Apply the initial condition $m(0) = 40$ kg at time $t = 0$.

$$m(0) = Ae^{(-3.985 \times 10^{-5}/\text{min})(0)} = 40 \text{ kg}$$
$$Ae^0 = 40 \text{ kg}$$
$$A = 40 \text{ kg}$$

Therefore,

$$m(t) = (40 \text{ kg})e^{(-3.985 \times 10^{-5}/\text{min})t}$$

Solve for t.

$$\frac{m(t)}{40\text{ kg}} = e^{(-3.985\times10^{-5}/\text{min})t}$$

$$\ln\left(\frac{m(t)}{40\text{ kg}}\right) = \ln\left(e^{(-3.985\times10^{-5}/\text{min})t}\right) = \left(\frac{-3.985\times10^{-5}}{\text{min}}\right)t$$

$$t = \frac{\ln\left(\frac{m(t)}{40\text{ kg}}\right)}{\frac{-3.985\times10^{-5}}{\text{min}}}$$

The initial mass of water in the lagoon is

$$m_i = V\rho = (2886\text{ m}^3)\left(1000\ \frac{\text{kg}}{\text{m}^3}\right) = \boxed{2.886\times10^6\text{ kg}\quad(2.9\times10^6\text{ kg})}$$

The answer is (B).

(d) The final mass of chemicals at a concentration of 1 ppb is

$$m_f = \frac{2.886\times10^6\text{ kg}}{1\times10^9} = \boxed{2.886\times10^{-3}\text{ kg}\quad(0.003\text{ kg})}$$

The answer is (B).

(e) Find the time required to achieve a mass of 2.886×10^{-3} kg.

$$t = \left(\frac{\ln\left(\frac{m(t)}{40\text{ kg}}\right)}{\frac{-3.985\times10^{-5}}{\text{min}}}\right)\left(\frac{1\text{ h}}{60\text{ min}}\right)\left(\frac{1\text{ d}}{24\text{ h}}\right)$$

$$= \left(\frac{\ln\left(\frac{2.886\times10^{-3}\text{ kg}}{40\text{ kg}}\right)}{\frac{-3.985\times10^{-5}}{\text{min}}}\right)\left(\frac{1\text{ h}}{60\text{ min}}\right)\left(\frac{1\text{ d}}{24\text{ h}}\right)$$

$$= \boxed{166\text{ days}\quad(200\text{ days})}$$

The answer is (D).

5. Let

$$m(t) = \text{mass of salt in tank at time } t$$

$$m_0 = 60\text{ mass units (lbm or kg)}$$

$$m'(t) = \text{rate at which salt content is changing}$$

Two mass units of salt enter each minute, and three volumes leave each minute. The amount of salt leaving each minute is

$$\left(3\ \frac{\text{vol}}{\text{min}}\right)\left(\text{concentration in }\frac{\text{mass}}{\text{vol}}\right) = \left(3\ \frac{\text{vol}}{\text{min}}\right)\left(\frac{\text{salt mass}}{\text{volume}}\right) = \left(3\ \frac{\text{vol}}{\text{min}}\right)\left(\frac{m(t)}{100-t}\right)$$

$$m'(t) = 2 - (3)\left(\frac{m(t)}{100-t}\right)\quad\text{or}\quad m'(t) + \frac{3m(t)}{100-t} = 2\text{ mass units/min}$$

This is a first-order linear differential equation. The integrating factor is

$$m = \exp\left(3\int\frac{dt}{100-t}\right) = \exp\Big((3)\big(-\ln(100-t)\big)\Big) = (100-t)^{-3}$$

$$m(t) = (100-t)^3\left(2\int\frac{dt}{(100-t)^3} + k\right) = 100 - t + k(100-t)^3$$

$m = 60$ mass units at $t = 0$, so $k = -0.00004$.

$$m(t) = 100 - t - (0.00004)(100-t)^3$$

At $t = 60$ min,

$$m = 100 - 60\text{ min} - (0.00004)(100 - 60\text{ min})^3 = \boxed{37.44\ (37)\text{ mass units}}$$

The answer is (B).

11 Probability and Statistical Analysis of Data

PRACTICE PROBLEMS

Probability

1. Four military recruits whose respective shoe sizes are 7, 8, 9, and 10 report to the supply clerk to be issued boots. The supply clerk selects one pair of boots in each of the four required sizes and hands them at random to the recruits.

(a) Use exhaustive enumeration to determine the probability that all recruits will receive boots of an incorrect size.

(A) 0.25

(B) 0.38

(C) 0.45

(D) 0.61

(b) The probability that exactly three recruits will receive boots of the correct size is most nearly

(A) 0

(B) 0.063

(C) 0.17

(D) 0.25

Probability Distributions

2. The time taken by a toll taker to collect the toll from vehicles crossing a bridge is an exponential distribution with a mean of 23 sec. The probability that a random vehicle will be processed in 25 sec or more (i.e., will take longer than 25 sec) is most nearly

(A) 0.17

(B) 0.25

(C) 0.34

(D) 0.52

3. The number of cars entering a toll plaza on a bridge during the hour after midnight follows a Poisson distribution with a mean of 20.

(a) The probability that exactly 17 cars will pass through the toll plaza during that hour on any given night is most nearly

(A) 0.08

(B) 0.12

(C) 0.16

(D) 0.23

(b) The percent probability that three or fewer cars will pass through the toll plaza at that hour on any given night is most nearly

(A) 0.00032%

(B) 0.0019%

(C) 0.079%

(D) 0.11%

4. A survey field crew measures one leg of a traverse four times. The following results are obtained.

repetition	measurement	direction
1	1249.529	forward
2	1249.494	backward
3	1249.384	forward
4	1249.348	backward

The average of the measurements will be taken as the traverse length, and confidence limits will be determined for that value. The survey crew chief wants to be 90% confident that the true traverse length lies within the confidence interval.

(a) Most nearly, what is the upper confidence limit?

(A) 1249.510

(B) 1249.541

(C) 1249.581

(D) 1249.642

(b) Most nearly, what is the lower confidence limit?

(A) 1249.236

(B) 1249.281

(C) 1249.337

(D) 1249.368

(c) Which statement regarding the confidence limits and interval is true?

(A) There is a 90% probability that subsequent measurements of the traverse leg length will be within the confidence interval.

(B) There is a 90% probability that the average of the next four measurements of the traverse leg length will be within the confidence interval.

(C) There is a 90% probability that the true traverse leg length is within the confidence interval.

(D) There is a 90% probability that any measurements of the traverse leg length that fall outside of the interval have been affected by a systematic measurement error.

(d) The most probable value of the distance is most nearly

(A) 1249.399

(B) 1249.410

(C) 1249.439

(D) 1249.452

(e) The error in the most probable value (at 90% confidence) is most nearly

(A) 0.08

(B) 0.10

(C) 0.14

(D) 0.19

(f) If the distance is one side of a square traverse whose sides are all equal, the most probable closure error is most nearly

(A) 0.14

(B) 0.20

(C) 0.28

(D) 0.35

(g) The most probable error of part (f) expressed as a fraction of the total of four legs is most nearly

(A) 1:24,500

(B) 1:17,600

(C) 1:12,500

(D) 1:10,900

(h) Define accuracy and distinguish it from precision.

(A) If an experiment can be repeated with identical results, the results are considered accurate.

(B) If an experiment has a small bias, the results are considered precise.

(C) If an experiment is precise, it cannot also be accurate.

(D) If an experiment is unaffected by experimental error, the results are accurate.

(i) Which of the following is an example of a systematic error?

(A) measuring river depth as a motorized ski boat passes by

(B) using a steel tape that is too short to measure consecutive distances

(C) locating magnetic north while near a large iron ore deposit along an overland route

(D) determining local wastewater BOD after a toxic spill

Statistical Analysis

5. Two resistances, the meter resistor and a shunt resistor, are connected in parallel in an ammeter. Most of the current passing through the meter goes through the shunt resistor. In order to determine the accuracy of the resistance of shunt resistors being manufactured for a line of ammeters, a manufacturer tests a sample of 100 shunt resistors. The numbers of shunt resistors with the resistance indicated (to the nearest hundredth of an ohm) are as follows.

0.200 Ω, 1; 0.210 Ω, 3; 0.220 Ω, 5; 0.230 Ω, 10; 0.240 Ω, 17; 0.250 Ω, 40; 0.260 Ω, 13; 0.270 Ω, 6; 0.280 Ω, 3; 0.290 Ω, 2.

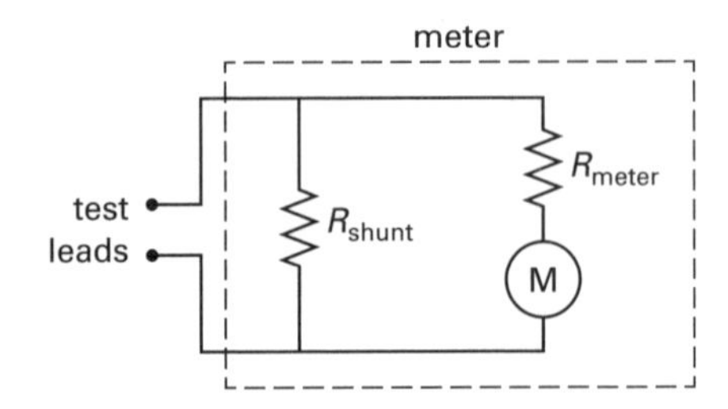

(a) The mean resistance is most nearly

(A) 0.235 Ω

(B) 0.247 Ω

(C) 0.251 Ω

(D) 0.259 Ω

(b) The sample standard deviation is most nearly

(A) 0.0003 Ω

(B) 0.010 Ω

(C) 0.016 Ω

(D) 0.24 Ω

(c) The median resistance is most nearly

(A) 0.22 Ω

(B) 0.24 Ω

(C) 0.25 Ω

(D) 0.26 Ω

(d) The sample variance is most nearly

(A) 0.00027 Ω^2

(B) 0.0083 Ω^2

(C) 0.0114 Ω^2

(D) 0.0163 Ω^2

6. California law requires a statistical analysis of the average speed driven by motorists on a road prior to the use of radar speed control. The following speeds (all in mph) were observed in a random sample of 40 cars.

44, 48, 26, 25, 20, 43, 40, 42, 29, 39, 23, 26, 24, 47, 45, 28, 29, 41, 38, 36, 27, 44, 42, 43, 29, 37, 34, 31, 33, 30, 42, 43, 28, 41, 29, 36, 35, 30, 32, 31

(a) Tabulate the frequency distribution and the cumulative frequency distribution of the data.

(b) Draw the frequency histogram.

(c) Draw the frequency polygon.

(d) Draw the cumulative frequency graph.

(e) The upper quartile speed is most nearly

(A) 30 mph

(B) 35 mph

(C) 40 mph

(D) 45 mph

(f) The mean speed is most nearly

(A) 31 mph

(B) 33 mph

(C) 35 mph

(D) 37 mph

(g) The standard deviation of the sample data is most nearly

(A) 2.1 mph

(B) 6.1 mph

(C) 6.8 mph

(D) 7.4 mph

(h) The sample standard deviation is most nearly

(A) 7.5 mph

(B) 18 mph

(C) 35 mph

(D) 56 mph

(i) The sample variance is most nearly

(A) 60 mi^2/hr^2

(B) 320 mi^2/hr^2

(C) 1200 mi^2/hr^2

(D) 3100 mi^2/hr^2

7. A spot speed study is conducted for a stretch of roadway. During a normal day, the speeds were found to be normally distributed with a mean of 46 and a standard deviation of 3.

(a) The 50th percentile speed is most nearly

(A) 39

(B) 43

(C) 46

(D) 49

(b) The 85th percentile speed is most nearly

(A) 47

(B) 48

(C) 49

(D) 52

(c) The upper two-standard deviation speed is most nearly

(A) 47

(B) 49

(C) 51

(D) 52

(d) The daily average speeds for the same stretch of roadway on consecutive normal days were determined by sampling 25 vehicles each day. The upper two-standard deviation average speed is most nearly

(A) 46

(B) 47

(C) 52

(D) 54

8. The diameters of bolt holes drilled in structural steel members are normally distributed with a mean of 0.502 in and a standard deviation of 0.005 in. Holes are out of specification if their diameters are less than 0.497 in or more than 0.507 in.

(a) The probability that a hole chosen at random will be out of specification is most nearly

(A) 0.16

(B) 0.22

(C) 0.32

(D) 0.68

(b) The probability that two holes out of a sample of 15 will be out of specification is most nearly

(A) 0.07

(B) 0.12

(C) 0.15

(D) 0.32

9. The length of a project's critical path is 43.83 days with a variance of 10.53 days^2. The length of the project is distributed normally.

(a) What is most nearly the probability of the project finishing in less than 42 days?

(A) 0.16

(B) 0.29

(C) 0.37

(D) 0.44

(b) Without using the standard normal table, what is most nearly the probability of the project finishing in less than 42 days?

(A) 0.16

(B) 0.29

(C) 0.37

(D) 0.44

10. The conductive transient temperature profile within an infinite solid block (uniform initial temperature of T_0, diffusivity of α) whose outer surface is maintained at a temperature T_s is

$$\frac{T(x,t) - T_s}{T_0 - T_s} = \operatorname{erf}\left(\frac{x}{2\sqrt{\alpha t}}\right)$$

A large, smooth block of aluminum (diffusivity of 4.6×10^{-5} m^2/s) initially at 25°C is placed on a 400°C flat, smooth surface. Approximately what will be the temperature in the block 8 cm from the plane of contact after 10 minutes?

(A) 210°C

(B) 270°C

(C) 300°C

(D) 340°C

11. A researcher wants to know if a sample of 20 normally distributed executive salaries has a standard deviation greater than \$150,000. The average of the sample is \$400,000. The sample standard deviation is \$195,000. At approximately what confidence level does a standard deviation greater than \$150,000 become likely?

(A) 50%

(B) 95%

(C) 97%

(D) 99%

Reliability

12. A mechanical component exhibits a negative exponential failure distribution with a mean time to failure of 1000 hr. The maximum operating time such that the reliability remains above 99% is most nearly

(A) 3.3 hr

(B) 5.6 hr

(C) 8.1 hr

(D) 10 hr

13. An electro-mechanical system is a fully redundant, 1-out-of-3 system. The mean service (repair) rate is 0.1 repairs per hour, and the mean failure rate is one failure per 300 hours. The operational availability of the system is most nearly

(A) 77%

(B) 89%

(C) 98%

(D) 99%

14. A system with three components in parallel has a mean time to failure of 10,000 hours. The reliabilities of the first and second components are 0.50 and 0.75, respectively. The system is required to operate for 1000 hours. The minimum reliability of the third component to satisfy the previously mentioned conditions is most nearly

(A) 0.2

(B) 0.3

(C) 0.4

(D) 0.6

15. A serial system consists of five identical components. The failure history of a number of these components has been recorded. What is most nearly the mean time to failure for the system?

elapsed time, t (years)	number of failures, f
0	0
1	0
2	0
3	0
4	1
5	2
6	2
7	2
8	3
9	3
10	3

(A) 1.7

(B) 2.5

(C) 3.2

(D) 18

16. The mean time between failures of a metal cutting lathe is 2770 hours. Assuming an exponential distribution, the reliability of the lathe after operating for 700 hours is most nearly

(A) 10%

(B) 23%

(C) 56%

(D) 78%

Hypothesis Testing

17. 100 bearings were tested to failure. The average life was 1520 hours, and the sample standard deviation was 120 hours. The manufacturer wants to claim an average 1600 hour life. Evaluate using confidence limits of 95% and 99%.

(A) The claim is accurate at both 95% and 99% confidence limits.

(B) The claim is inaccurate only at the 95% confidence limit.

(C) The claim is inaccurate only at the 99% confidence limit.

(D) The claim is inaccurate at both 95% and 99% confidence limits.

Curve Fitting

18. (a) What is most nearly the best equation for a straight line passing through the points given?

x	y
400	370
800	780
1250	1210
1600	1560
2000	1980
2500	2450
4000	3950

(A) $y = 0.276x + 259.6$

(B) $y = 0.768x + 62.8$

(C) $y = 0.994x - 25.0$

(D) $y = 1.210x - 114.0$

(b) The correlation coefficient is most nearly

(A) 0.284

(B) 0.501

(C) 0.537

(D) 1.000

19. What is most nearly the best equation for a line passing through the points given?

s	t
20	43
18	141
16	385
14	1099

(A) $\ln t = -22.80 + 1.324s$

(B) $\ln t = 5.57 - 0.00924s$

(C) $\ln t = 14.53 - 0.536s$

(D) $\ln t = 7.56 - 0.854s$

20. The number of vehicles lining up behind a flashing railroad crossing has been observed for five trains of different lengths, as given. What is most nearly the mathematical formula that relates the two variables?

no. of cars in train, x	no. of vehicles, y
2	14.8
5	18.0
8	20.4
12	23.0
27	29.9

(A) $y = -26.18 + 52.71 \log x$

(B) $y = 2.48 + 21.32 \log x$

(C) $y = 7.46 + 15.6 \log x$

(D) $y = 9.57 + 13.20 \log x$

21. The following yield data are obtained from five identical treatment plants.

treatment plant	average temperature, T	average yield, Y
1	207.1	92.30
2	210.3	92.58
3	200.4	91.56
4	201.1	91.63
5	203.4	91.83

(a) Develop a linear equation to correlate the yield and average temperature.

(A) $Y = -0.04T + 100$

(B) $Y = 0.019T + 88$

(C) $Y = 0.11T + 70$

(D) $Y = 0.44T + 1.2$

(b) The correlation coefficient is most nearly

(A) 0.80

(B) 0.87

(C) 0.90

(D) 1.00

22. The following data are obtained from a soil compaction test. What is most nearly the nonlinear formula that relates the two variables?

x	y
–1	0
0	1
1	1.4
2	1.7
3	2
4	2.2
5	2.4
6	2.6
7	2.8
8	3

(A) $y = \sqrt{x - 1}$

(B) $y = \sqrt{x}$

(C) $y = \sqrt{x + 0.5}$

(D) $y = \sqrt{x + 1}$

SOLUTIONS

1. (a) There are 4! = 24 different possible outcomes. By enumeration, there are 9 completely wrong combinations.

$$p\{\text{all wrong}\} = \frac{9}{24} = \boxed{0.375 \quad (0.38)}$$

sizes

correct → (sizes issued)	7	8	9	10	all wrong
	7	8	9	10	
	7	8	10	9	
	7	9	8	10	
	7	9	10	8	
	7	10	8	9	
	7	10	9	8	
	8	9	10	7	X
	8	9	7	10	
	8	10	9	7	
	8	10	7	9	X
	8	7	9	10	
	8	7	10	9	X
	9	10	7	8	X
	9	10	8	7	X
	9	7	10	8	X
	9	7	8	10	
	9	8	7	10	
	9	8	10	7	
	10	7	8	9	X
	10	7	9	8	
	10	8	7	9	
	10	8	9	7	
	10	9	8	7	X
	10	9	7	8	X

The answer is (B).

(b) If three recruits get the correct size, the fourth recruit will also since there will be only one pair remaining.

$$p\{\text{exactly 3}\} = \boxed{0}$$

The answer is (A).

2. For an exponential distribution function, the mean is

$$\mu = \frac{1}{\lambda}$$

Using Eq. 11.41, for a mean of 23,

$$\mu = 23 = \frac{1}{\lambda}$$

$$\lambda = \frac{1}{23} = 0.0435$$

Using Eq. 11.40, for an exponential distribution function,

$$p = F(x) = 1 - e^{-\lambda x}$$

$$p\{X < x\} = 1 - p$$

$$\begin{aligned} p\{X > x\} &= 1 - F(x) \\ &= 1 - (1 - e^{-\lambda x}) \\ &= e^{-\lambda x} \end{aligned}$$

The probability of a random vehicle being processed in 25 sec or more is

$$\begin{aligned} p\{x > 25\} &= e^{-(0.0435)(25)} \\ &= \boxed{0.337 \quad (0.34)} \end{aligned}$$

The answer is (C).

3. (a) The distribution is a Poisson distribution with an average of $\lambda = 20$.

The probability for a Poisson distribution is given by Eq. 11.63.

$$\begin{aligned} p\{x\} &= f(x) \\ &= \frac{e^{-\lambda}\lambda^x}{x!} \end{aligned}$$

Therefore, the probability of 17 cars is

$$\begin{aligned} p\{x = 17\} &= f(17) \\ &= \frac{e^{-20}20^{17}}{17!} \\ &= \boxed{0.076 \quad (0.08)} \end{aligned}$$

The answer is (A).

Background and Support

(b) The probability of three or fewer cars is

$$\begin{aligned} p\{x \le 3\} &= p\{x=0\} + p\{x=1\} + p\{x=2\} \\ &\quad + p\{x=3\} \\ &= f(0) + f(1) + f(2) + f(3) \\ &= \frac{e^{-20}20^0}{0!} + \frac{e^{-20}20^1}{1!} \\ &\quad + \frac{e^{-20}20^2}{2!} + \frac{e^{-20}20^3}{3!} \\ &= 2 \times 10^{-9} + 4.1 \times 10^{-8} \\ &\quad + 4.12 \times 10^{-7} + 2.75 \times 10^{-6} \\ &= 3.2 \times 10^{-6} \\ &= \boxed{0.0000032 \quad (0.00032\%)} \end{aligned}$$

The answer is (A).

4. (a) Find the average using Eq. 11.68.

$$\begin{aligned} \overline{x} &= \frac{\sum x_i}{n} \\ &= \frac{1249.529 + 1249.494 + 1249.384 + 1249.348}{4} \\ &= 1249.439 \end{aligned}$$

Since the sample population is small, use Eq. 11.75 to find the sample standard deviation.

$$\begin{aligned} s &= \sqrt{\frac{\sum (x_i - \overline{x})^2}{n-1}} \\ &= \sqrt{\frac{\begin{array}{l}(1249.529 - 1249.439)^2 \\ \quad + (1249.494 - 1249.439)^2 \\ \quad + (1249.384 - 1249.439)^2 \\ \quad + (1249.348 - 1249.439)^2\end{array}}{4-1}} \\ &= 0.08647 \end{aligned}$$

Although the measurements are obtained from a normal distribution, Student's t-distribution must be used because the number of measurements is less than 50 (some researchers say 30).

Since the true value could fall below or above the confidence interval, a two-tailed test is required. From App. 11.C, with degrees of freedom $\text{df} = n - 1 = 4 - 1 = 3$ and with $C = 90\%$, for a two-tailed confidence interval, the standard t-variable is 2.35. The t-distribution is symmetrical about its mean.

The upper confidence limit is

$$\begin{aligned} UCL &= \overline{x} + \frac{t_C s}{\sqrt{n}} = 1249.439 + \frac{(2.35)(0.08647)}{\sqrt{4}} \\ &= 1249.439 + 0.102 \\ &= 1249.541 \end{aligned}$$

The answer is (B).

(b) The lower confidence limit is

$$\begin{aligned} \text{LCL} &= \overline{x} - \frac{t_C s}{\sqrt{n}} = 1249.439 - \frac{(2.35)(0.08647)}{\sqrt{4}} \\ &= 1249.439 - 0.102 \\ &= 1249.337 \end{aligned}$$

The answer is (C).

(c) The confidence interval contains the true value of the traverse leg length with a 90% probability. The other three options are frequent incorrect interpretations of the confidence interval.

The answer is (C).

(d) The unbiased estimate of the most probable distance is $\boxed{1249.439.}$

The answer is (C).

(e) The error in the most probable value for the 90% confidence range is $\boxed{0.102\ (0.10).}$

The answer is (B).

(f) If the surveying crew places a marker, measures a distance x, places a second marker, and then measures the same distance x back to the original marker, the ending point should coincide with the original marker. If, due to measurement errors, the ending and starting points do not coincide, the difference is the closure error.

In this example, the survey crew moves around the four sides of a square, so there are two measurements in the x-direction and two measurements in the y-direction. If the errors E_1 and E_2 are known for two measurements, x_1 and x_2, the error associated with the sum or difference $x_1 \pm x_2$ is

$$E\{x_1 \pm x_2\} = \sqrt{E_1^2 + E_2^2}$$

In this case, the error in the x-direction is

$$E_x = \sqrt{(0.102)^2 + (0.102)^2} = 0.144$$

The error in the y-direction is calculated the same way and is also 0.144. E_x and E_y are combined by the Pythagorean theorem to yield

$$\begin{aligned} E_{\text{closure}} &= \sqrt{(0.144)^2 + (0.144)^2} \\ &= \boxed{0.204 \quad (0.20)} \end{aligned}$$

The answer is (B).

(g) In surveying, error may be expressed as a fraction of one or more legs of the traverse.

$$\frac{0.204}{(4)(1249)} = \boxed{\frac{1}{24,490} \quad (1{:}24{,}500)}$$

The answer is (A).

(h) An experiment is $\boxed{\text{accurate}}$ if it is unchanged by experimental error. Precision is concerned with the repeatability of the experimental results. If an experiment is repeated with identical results, the experiment is said to be precise. However, it is possible to have a highly precise experiment with a large bias.

The answer is (D).

(i) A systematic error is one that is always present and is unchanged from sample to sample. For example, a steel tape that is 0.02 ft $\boxed{\text{too short}}$ to measure consecutive distances introduces a systematic error.

The answer is (B).

5. (a) For convenience, tabulate the frequency-weighted values of R and R^2.

R	f	fR	fR^2
0.200	1	0.200	0.0400
0.210	3	0.360	0.1323
0.220	5	1.100	0.2420
0.230	10	2.300	0.5290
0.240	17	4.080	0.9792
0.250	40	10.000	2.5000
0.260	13	3.380	0.8788
0.270	6	1.620	0.4374
0.280	3	0.840	0.2352
0.290	2	0.580	0.1682
	100	24.730	6.1421

The mean resistance is

$$\overline{R} = \frac{\sum fR}{\sum f} = \frac{24.730\ \Omega}{100} = \boxed{0.2473\ \Omega \quad (0.247\ \Omega)}$$

The answer is (B).

(b) The sample standard deviation is given by Eq. 11.75.

$$s = \sqrt{\frac{\sum fR^2 - \frac{\left(\sum fR\right)^2}{n}}{n-1}} = \sqrt{\frac{6.1421\ \Omega^2 - \frac{(24.73\ \Omega)^2}{100}}{99}}$$

$$= \boxed{0.0163\ \Omega \quad (0.016\ \Omega)}$$

The answer is (C).

(c) The 50th and 51st values are both 0.25 Ω. The median is $\boxed{0.25\ \Omega.}$

The answer is (C).

(d) The sample variance is

$$s^2 = (0.0163\ \Omega)^2 = \boxed{0.0002656\ \Omega^2 \quad (0.00027\ \Omega^2)}$$

The answer is (A).

6. (a) Tabulate the frequency distribution and the cumulative frequency distribution of the data.

The lowest speed is 20 mph and the highest speed is 48 mph; therefore, the range is 28 mph. Choose 10 cells with a width of 3 mph.

midpoint	interval (mph)	frequency	cumulative frequency	cumulative percent
21	20–22	1	1	3
24	23–25	3	4	10
27	26–28	5	9	23
30	29–31	8	17	43
33	32–34	3	20	50
36	35–37	4	24	60
39	38–40	3	27	68
42	41–43	8	35	88
45	44–46	3	38	95
48	47–49	2	40	100

(b) Draw the frequency histogram.

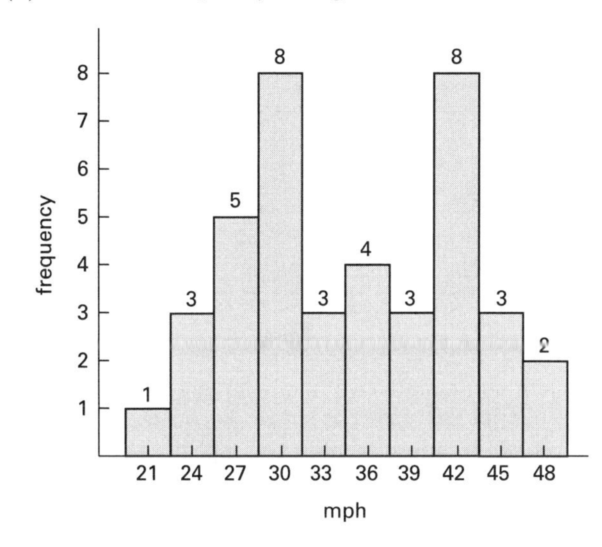

(c) Draw the frequency polygon.

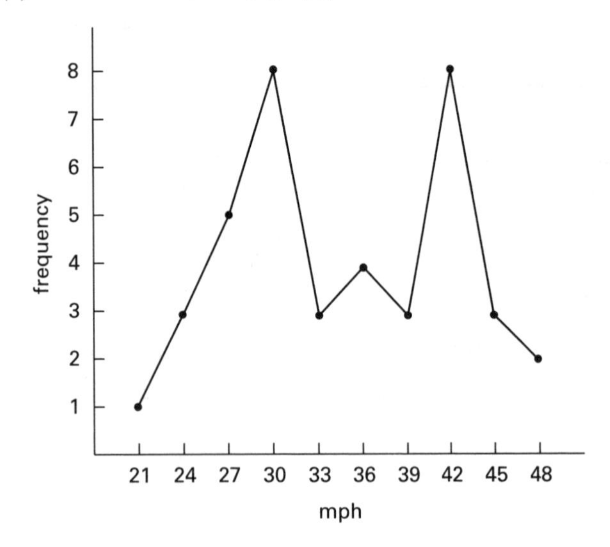

(d) Use the table in part (a).

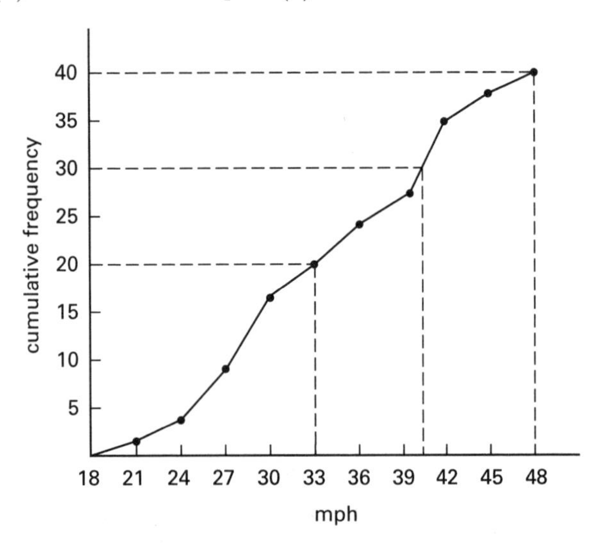

(e) From the cumulative frequency graph in part (d), the upper quartile speed occurs at 30 cars or 75%, which corresponds to approximately $\boxed{40 \text{ mph.}}$

The answer is (C).

(f) Calculate the following quantities.

$$\sum x_i = 1390 \text{ mi/hr}$$

$$n = 40$$

The mean is computed using Eq. 11.68.

$$\bar{x} = \frac{\sum x_i}{n} = \frac{1390 \ \frac{\text{mi}}{\text{hr}}}{40}$$

$$= \boxed{34.75 \text{ mi/hr} \quad (35 \text{ mph})}$$

The answer is (C).

(g) The standard deviation of the sample data is given by Eq. 11.74.

$$\sigma = \sqrt{\frac{\sum x^2}{n} - \mu^2}$$

$$\sum x^2 = 50{,}496 \text{ mi}^2/\text{hr}^2$$

Use the sample mean as an unbiased estimator of the population mean, μ.

$$\sigma = \sqrt{\frac{\sum x^2}{n} - \mu^2} = \sqrt{\frac{50{,}496 \ \frac{\text{mi}^2}{\text{hr}^2}}{40} - \left(34.75 \ \frac{\text{mi}}{\text{hr}}\right)^2}$$

$$= \boxed{7.405 \text{ mi/hr} \quad (7.4 \text{ mph})}$$

The answer is (D).

(h) The sample standard deviation is

$$s = \sqrt{\frac{\sum x^2 - \frac{\left(\sum x\right)^2}{n}}{n-1}}$$

$$= \sqrt{\frac{50{,}496 \ \frac{\text{mi}^2}{\text{hr}^2} - \frac{\left(1390 \ \frac{\text{mi}}{\text{hr}}\right)^2}{40}}{40-1}}$$

$$= \boxed{7.5 \text{ mi/hr} \quad (7.5 \text{ mph})}$$

The answer is (A).

(i) The sample variance is given by the square of the sample standard deviation.

$$s^2 = \left(7.5 \ \frac{\text{mi}}{\text{hr}}\right)^2$$

$$= \boxed{56.25 \text{ mi}^2/\text{hr}^2 \quad (60 \text{ mi}^2/\text{hr}^2)}$$

The answer is (A).

7. (a) The 50th percentile speed is the median speed, $\boxed{46,}$ which for a symmetrical normal distribution is the mean speed.

The answer is (C).

(b) The 85th percentile speed is the speed that is exceeded by only 15% of the measurements. Since this is a normal distribution, App. 11.A can be used. 15% in the upper tail corresponds to 35% between the mean and the 85th percentile. From App. 11.A, this occurs at approximately 1.04σ. The 85th percentile speed is

$$x_{85\%} = \mu + 1.04\sigma = 46 + (1.04)(3)$$

$$= \boxed{49.12 \quad (49)}$$

The answer is (C).

(c) The upper 2σ speed is

$$x_{2\sigma} = \mu + 2\sigma = 46 + (2)(3) = \boxed{52.0 \quad (52)}$$

The answer is (D).

(d) According to the central limit theorem, the mean of the average speeds is the same as the distribution mean, and the standard deviation of sample means (from Eq. 11.83) is

$$s_{\overline{x}} = \frac{\sigma_x}{\sqrt{n}} = \frac{3}{\sqrt{25}} = 0.6$$

The upper two-standard deviation average speed is

$$\overline{x}_{2\sigma} = \mu + 2\sigma_{\overline{x}} = 46 + (2)(0.6) = \boxed{47.2 \quad (47)}$$

The answer is (B).

8. (a) From Eq. 11.45,

$$z = \frac{x_0 - \mu}{\sigma}$$

$$z_{\text{upper}} = \frac{0.507 \text{ in} - 0.502 \text{ in}}{0.005 \text{ in}} = +1$$

From App. 11.A, the area outside $z = +1$ is

$$0.5 - 0.3413 = 0.1587$$

Since these are symmetrical limits, $z_{\text{lower}} = -1$.

$$\text{total fraction defective} = (2)(0.1587) = \boxed{0.3174 \quad (0.32)}$$

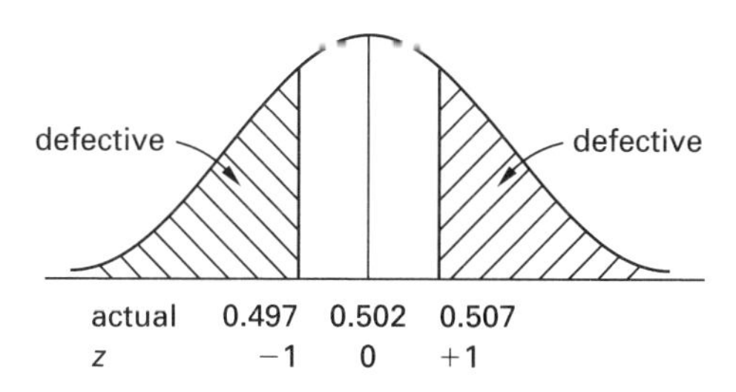

The answer is (C).

(b) This is a binomial problem.

$$p = p\{\text{defective}\} = 0.3174$$

$$q = 1 - p = 0.6826$$

From Eq. 11.28,

$$p\{x\} = f(x) = \binom{n}{x}\hat{p}^x\hat{q}^{n-x}$$

$$f(2) = \binom{15}{2}(0.3174)^2(0.6826)^{13} = \left(\frac{15!}{13!2!}\right)(0.3174)^2(0.6826)^{13} = \boxed{0.0739 \quad (0.07)}$$

The answer is (A).

9. (a) The standard deviation is

$$\sigma = \sqrt{\sigma^2} = \sqrt{10.53 \text{ days}^2} = 3.245 \text{ days}$$

Calculate the standard normal variable.

$$z = \left|\frac{\overline{x} - \mu}{\sigma}\right| = \left|\frac{42 \text{ days} - 43.83 \text{ days}}{3.245 \text{ days}}\right| = 0.564$$

From App. 11.A, the area under the normal curve between $z = 0.5$ and $z = 0.564$ is 0.2137 (interpolating between $z = 0.56$ and $z = 0.57$). The area under the curve between $z = 0.564$ and $z = +\infty$ (equivalent to the area under the curve between $t = -\infty$ and $t = 42$ days) is

$$p(t < 42 \text{ days}) - p(z < 0.564) = 0.5 - 0.2137 = \boxed{0.2863 \quad (0.29)}$$

The answer is (B).

(b) The standard normal variable was found to be 0.564 in part (a). From Eq. 11.57,

$$p\{0 < z < z_0\} = \tfrac{1}{2}\text{erf}\left(\frac{z_0}{\sqrt{2}}\right) = \tfrac{1}{2}\text{erf}\left(\frac{0.564}{\sqrt{2}}\right) = \tfrac{1}{2}\text{erf}(0.3988) \approx \tfrac{1}{2}(0.4284) = 0.2142$$

The area in the tail of the standard normal distribution is

$$p(t < 42 \text{ days}) = p(z < 0.564) = 0.5 - 0.2142 = \boxed{0.2858 \quad (0.29)}$$

The answer is (B).

10. The depth of the unknown temperature is

$$x = \frac{8 \text{ cm}}{100 \frac{\text{cm}}{\text{m}}} = 0.08 \text{ m}$$

$$\frac{T(x,t)-T_s}{T_0-T_s}=\text{erf}\left(\frac{x}{2\sqrt{\alpha t}}\right)$$

$$\frac{T-400^\circ\text{C}}{25^\circ\text{C}-400^\circ\text{C}}=\text{erf}\left(\frac{0.08\ \text{m}}{2\sqrt{\left(4.6\times10^{-5}\ \frac{\text{m}^2}{\text{s}}\right)(10\ \text{min})\times\left(60\ \frac{\text{s}}{\text{min}}\right)}}\right)$$

$$\frac{T-400^\circ\text{C}}{-375^\circ\text{C}}=\text{erf}(0.2408)$$

From App. 11.D, erf(0.24) = 0.2657.

$$\frac{T-400^\circ\text{C}}{-375^\circ\text{C}}=0.2657$$

$$T=\boxed{300.4^\circ\text{C}\quad(300^\circ\text{C})}$$

The answer is (C).

11. The chi-squared statistic for this hypothesis test is

$$\chi^2=\frac{(n-1)s^2}{\sigma^2}=\frac{(20-1)(\$195{,}000)^2}{(\$150{,}000)^2}=32.11$$

From App. 11.B, with $20-1=19$ degrees of freedom, $\chi^2=32.11$ at approximately $\alpha=0.03$ (3%). A standard deviation greater than \$150,000 can be supported with a $100\%-3\%=\boxed{97\%}$ confidence limit, but not a higher confidence limit.

The answer is (C).

12. Using Eq. 11.60,

$$\lambda=\frac{1}{\text{MTTF}}=\frac{1}{1000\ \text{hr}}=0.001\ \text{hr}^{-1}$$

Using Eq. 11.61, the reliability function is

$$R\{t\}=e^{-\lambda t}=e^{-0.001t}$$

Since the reliability is greater than 99%,

$$e^{-0.001t}>0.99$$

$$\ln e^{-0.001t}>\ln 0.99$$

$$-0.001t>\ln 0.99$$

$$t<-1000\ln 0.99$$

$$t<10.05\quad(10)$$

The maximum operating time such that the reliability remains above 99% is $\boxed{10\ \text{hr.}}$

The answer is (D).

13. From Table 11.1, the mean time to repair (MTTR) is the reciprocal of the mean service (repair) rate.

$$\text{MTTR}=\frac{1}{\mu}=\frac{1}{0.1\ \frac{\text{repairs}}{\text{hr}}}=10\ \text{hr/repair}$$

From Table 11.1, the mean time to failure (MTTF) of a 1-out-of-3 fully redundant system of identical components is 11/6 times the inverse of the mean failure rate.

$$\text{MTTF}=\left(\frac{11}{6}\right)\left(\frac{1}{\lambda}\right)=\frac{\frac{11}{6}}{\frac{1}{300\ \text{hr}}}$$

$$=550\ \text{hr}$$

The mean time between failures (MTBF) is the sum of the mean time to failure and the mean time to repair.

$$\text{MTBF}=\text{MTTR}+\text{MTTF}=10\ \text{hr}+550\ \text{hr}=560\ \text{hr}$$

The availability of the system is the ratio of the mean time between failures and the mean time to failure.

$$A_o=\frac{\text{MTTF}}{\text{MTBF}}=\frac{550\ \text{hr}}{560\ \text{hr}}\times100\%$$

$$=\boxed{98\%}$$

The answer is (C).

14. The total required reliability of the system is a function of the mean time to failure and the 1000 hour operational requirement.

$$R_t=e^{-(t/\text{MTTF})}$$

$$=e^{-(1000\ \text{hr}/10{,}000\ \text{hr})}$$

$$=0.90$$

The total reliability of the system can also be calculated as a function of the reliabilities of all three components.

$$R_t=1-(1-R_1)(1-R_2)(1-R_3)$$

$$0.90=1-(1-0.50)(1-0.75)(1-R_3)$$

$$R_3=\boxed{0.2}$$

The answer is (A).

15. The mean time to failure (MTTF) of the system is a function of the mean failure rate for the component and the number of components, n.

$$\text{MTTF}=\left(\frac{1}{\lambda}\right)\left(1+\frac{1}{2}+\frac{1}{3}+\cdots+\frac{1}{n}\right)$$

The mean failure rate of one of the identical components is computed based on the table provided in the problem.

$$\lambda = \frac{\text{total no. of failures}}{\text{total yrs operated}} = \frac{\sum_{i=1}^{\text{yr 10}} f_i}{\sum_{i=1}^{\text{yr 10}} t_i f_i}$$

$$= \frac{1+2+2+2+3+3+3}{\begin{array}{l}(0 \text{ yr})(0) + (1 \text{ yr})(0) + (2 \text{ yr})(0) + (3 \text{ yr})(0) \\ \quad + (4 \text{ yr})(1) + (5 \text{ yr})(2) + (6 \text{ yr})(2) + (7 \text{ yr})(2) \\ \quad + (8 \text{ yr})(3) + (9 \text{ yr})(3) + (10 \text{ yr})(3)\end{array}}$$

$$= 0.13 \text{ failures/yr}$$

The mean time to failure is calculated for the five-component system.

$$\begin{aligned}\text{MTTF} &= \left(\frac{1}{\lambda}\right)\left(1 + \frac{1}{2} + \frac{1}{3} + \frac{1}{4} + \frac{1}{5}\right) \\ &= \left(\frac{1}{0.13}\right)\left(1 + \frac{1}{2} + \frac{1}{3} + \frac{1}{4} + \frac{1}{5}\right) \\ &= \boxed{17.6 \quad (18)}\end{aligned}$$

The answer is (D).

16. The failure (hazard) rate is

$$\begin{aligned}\lambda &= \frac{1}{\text{mean time between failures}} = \frac{1}{2770 \text{ hr}} \\ &= 3.61 \times 10^{-4} \text{ hr}^{-1}\end{aligned}$$

The exponential reliability is

$$\begin{aligned}R(t) &= e^{-\lambda t} \\ R(700 \text{ hr}) &= e^{-(3.61 \times 10^{-4} \text{ hr}^{-1})(700 \text{ hr})} \\ &= \boxed{0.7766 \quad (78\%)}\end{aligned}$$

The answer is (D).

17. This is a typical hypothesis test of two sample population means. The two populations are the original population the manufacturer used to determine the 1600 hr average life value and the new population the sample was taken from. The mean ($\overline{x} = 1520$ hr) of the sample and its sample standard deviation ($s = 120$ hr) are known. The mean of the population is 1600 hr.

The standard deviation of the average lifetime population is

$$\sigma_{\overline{x}} = \frac{s}{\sqrt{n}} = \frac{120 \text{ hr}}{\sqrt{100}} = 12 \text{ hr}$$

The manufacturer can be reasonably sure that the claim of a 1600 hr average life is justified if 1600 hr is near the average test life of 1520 hr. "Reasonably sure" must be evaluated based on acceptable probability of being incorrect. If the manufacturer is willing to be wrong with a 5% probability, then a 95% confidence level is required.

Since the direction of bias is known, a one-tailed test is required. To determine if the mean has shifted downward, test the hypothesis that 1600 hr is within the 95% limit of a distribution with a mean of 1520 hr and a standard deviation of 12 hr. From a standard normal table, 5% of a standard normal distribution is outside of $z = 1.645$. Therefore, the 95% confidence limit is

$$1520 \text{ hr} + (1.645)(12 \text{ hr}) = 1540 \text{ hr}$$

The manufacturer can be 95% certain that the average lifetime of the bearings is less than 1600 hr.

If the manufacturer is willing to be wrong with a probability of only 1%, then a 99% confidence limit is required. From the normal table, $z = 2.33$, and the 99% confidence limit is

$$1520 \text{ hr} + (2.33)(12 \text{ hr}) = 1548 \text{ hr}$$

The manufacturer can be 99% certain that the average bearing life is less than 1600 hr.

Therefore, the claim is inaccurate at both 95% and 99% confidence limits.

The answer is (D).

18. (a) Plot the data points to determine if the relationship is linear.

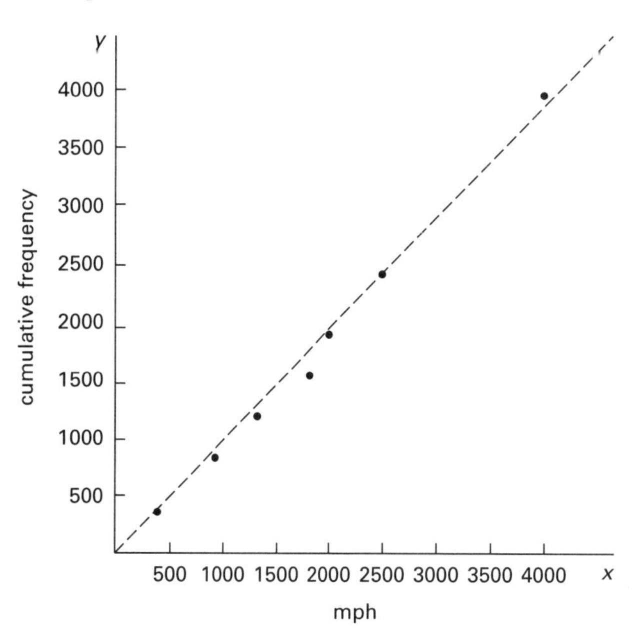

The data appear to be essentially linear. The slope, m, and the y-intercept, b, can be determined using linear regression.

The individual terms are

$$n = 7$$

$$\begin{aligned}\sum x_i &= 400 + 800 + 1250 + 1600 + 2000 \\ &\quad + 2500 + 4000 \\ &= 12{,}550\end{aligned}$$

$$\left(\sum x_i\right)^2 = (12{,}550)^2 = 1.575 \times 10^8$$

From Eq. 11.68,

$$\overline{x} = \frac{\sum x_i}{n} = \frac{12{,}550}{7} = 1792.9$$

$$\begin{aligned}\sum x_i^2 &= (400)^2 + (800)^2 + (1250)^2 + (1600)^2 \\ &\quad + (2000)^2 + (2500)^2 + (4000)^2 \\ &= 3.117 \times 10^7\end{aligned}$$

Similarly,

$$\begin{aligned}\sum y_i &= 370 + 780 + 1210 + 1560 + 1980 \\ &\quad + 2450 + 3950 \\ &= 12{,}300\end{aligned}$$

$$\left(\sum y_i\right)^2 = (12{,}300)^2 = 1.513 \times 10^8$$

$$\overline{y} = \frac{\sum y_i}{n} = \frac{12{,}300}{7} = 1757.1$$

$$\begin{aligned}\sum y_i^2 &= (370)^2 + (780)^2 + (1210)^2 + (1560)^2 \\ &\quad + (1980)^2 + (2450)^2 + (3950)^2 \\ &= 3.017 \times 10^7\end{aligned}$$

Also,

$$\begin{aligned}\sum x_i y_i &= (400)(370) + (800)(780) + (1250)(1210) \\ &\quad + (1600)(1560) + (2000)(1980) \\ &\quad + (2500)(2450) + (4000)(3950) \\ &= 3.067 \times 10^7\end{aligned}$$

Using Eq. 11.89, the slope is

$$\begin{aligned}m &= \frac{n\sum x_i y_i - \sum x_i \sum y_i}{n\sum x_i^2 - \left(\sum x_i\right)^2} \\ &= \frac{(7)(3.067 \times 10^7) - (12{,}550)(12{,}300)}{(7)(3.117 \times 10^7) - (12{,}550)^2} \\ &= 0.994\end{aligned}$$

Using Eq. 11.90, the y-intercept is

$$\begin{aligned}b &= \overline{y} - m\overline{x} = 1757.1 - (0.994)(1792.9) \\ &= -25.0\end{aligned}$$

The least squares equation of the line is

$$\begin{aligned}y &= mx + b \\ &= \boxed{0.994x - 25.0}\end{aligned}$$

The answer is (C).

(b) Using Eq. 11.91, the correlation coefficient is

$$\begin{aligned}r &= \frac{n\sum x_i y_i - \sum x_i \sum y_i}{\sqrt{\left(n\sum x_i^2 - \left(\sum x_i\right)^2\right)\left(n\sum y_i^2 - \left(\sum y_i\right)^2\right)}} \\ &= \frac{(7)(3.067 \times 10^7) - (12{,}500)(12{,}300)}{\sqrt{\begin{array}{l}\left((7)(3.117 \times 10^7) - (12{,}500)^2\right) \\ \quad \times \left((7)(3.017 \times 10^7) - (12{,}300)^2\right)\end{array}}} \\ &\approx \boxed{1.000}\end{aligned}$$

The answer is (D).

19. Plotting the data shows that the relationship is nonlinear.

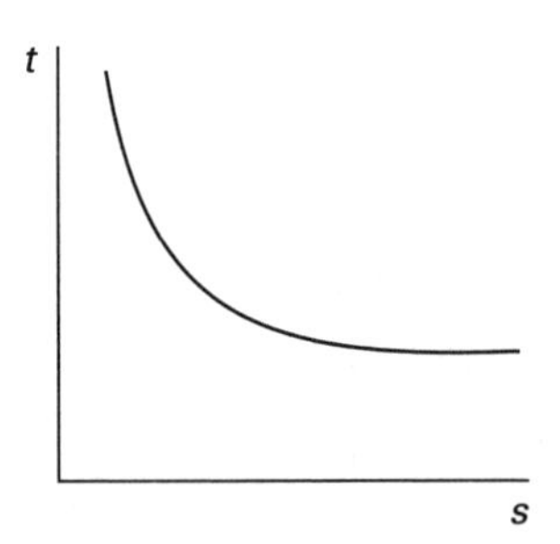

This appears to be an exponential with the form

$$t = ae^{bs}$$

Take the natural log of both sides.

$$\begin{aligned}\ln t &= \ln ae^{bs} = \ln a + \ln e^{bs} \\ &= \ln a + bs\end{aligned}$$

But, $\ln a$ is just a constant, c.

$$\ln t = c + bs$$

Make the transformation $R = \ln t$.

$$R = c + bs$$

s	R
20	3.76
18	4.95
16	5.95
14	7.00

This is linear.

$$n = 4$$

$$\sum s_i = 20 + 18 + 16 + 14 = 68$$

$$\bar{s} = \frac{\sum s}{n} = \frac{68}{4} = 17$$

$$\sum s_i^2 = (20)^2 + (18)^2 + (16)^2 + (14)^2 = 1176$$

$$\left(\sum s_i\right)^2 = (68)^2 = 4624$$

$$\sum R_i = 3.76 + 4.95 + 5.95 + 7.00 = 21.66$$

$$\bar{R} = \frac{\sum R_i}{n} = \frac{21.66}{4} = 5.415$$

$$\sum R_i^2 = (3.76)^2 + (4.95)^2 + (5.95)^2 + (7.00)^2 = 123.04$$

$$\left(\sum R_i\right)^2 = (21.66)^2 = 469.16$$

$$\sum s_i R_i = (20)(3.76) + (18)(4.95) + (16)(5.95) + (14)(7.00) = 357.5$$

The slope, b, of the transformed line is

$$b = \frac{n\sum s_i R_i - \sum s_i \sum R_i}{n\sum s_i^2 - \left(\sum s_i\right)^2} = \frac{(4)(357.5) - (68)(21.66)}{(4)(1176) - (68)^2} = -0.536$$

The intercept is

$$c = \bar{R} - b\bar{s} = 5.415 - (-0.536)(17) = 14.527$$

The transformed equation is

$$R = c + bs = 14.527 - 0.536s$$

$$\boxed{\ln t = 14.527 - 0.536s \quad (14.53 - 0.536s)}$$

The answer is (C).

20. The first step is to graph the data.

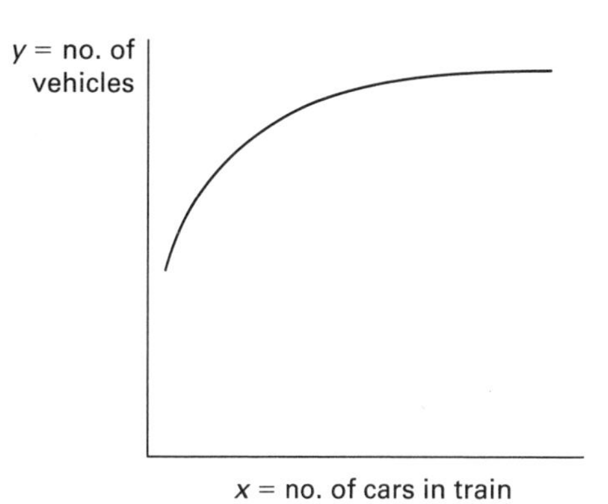

It is assumed that the relationship between the variables has the form $y = a + b\log x$. Therefore, the variable change $z = \log x$ is made, resulting in the following set of data.

z	y
0.301	14.8
0.699	18.0
0.903	20.4
1.079	23.0
1.431	29.9

$$\sum z_i = 4.413$$

$$\sum y_i = 106.1$$

$$\sum z_i^2 = 4.6082$$

$$\sum y_i^2 = 2382.2$$

$$\left(\sum z_i\right)^2 = 19.475$$

$$\left(\sum y_i\right)^2 = 11{,}257.2$$

$$\bar{z} = 0.8826$$

$$\bar{y} = 21.22$$

$$\sum z_i y_i = 103.06$$

$$n = 5$$

Using Eq. 11.89, the slope is

$$m = \frac{n\sum z_i y_i - \sum z_i \sum y_i}{n\sum z_i^2 - \left(\sum z_i\right)^2} = \frac{(5)(103.06) - (4.413)(106.1)}{(5)(4.6082) - 19.475} = 13.20$$

The y-intercept is

$$b = \bar{y} - m\bar{z} = 21.22 - (13.20)(0.8826) = 9.57$$

The resulting equation is

$$y = 9.57 + 13.20z$$

The relationship between x and y is approximately

$$\boxed{y = 9.57 + 13.20 \log x}$$

(This is not an optimal correlation, as better correlation coefficients can be obtained if other assumptions about the form of the equation are made. For example, $y = 9.1 + 4\sqrt{x}$ has a better correlation coefficient.)

The answer is (D).

21. (a) Plot the data to verify that they are linear.

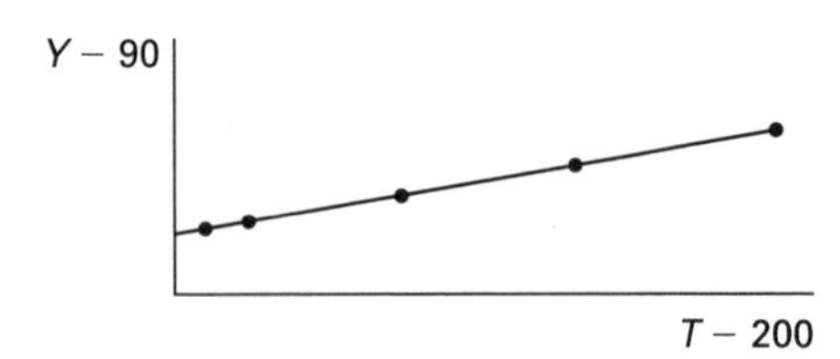

x $T-200$	y $Y-90$
7.1	2.30
10.3	2.58
0.4	1.56
1.1	1.63
3.4	1.83

step 1: Calculate the following quantities.

$$\sum x_i = 22.3 \qquad \sum y_i = 9.9$$

$$\sum x_i^2 = 169.43 \qquad \sum y_i^2 = 20.39$$

$$\left(\sum x_i\right)^2 = 497.29 \qquad \left(\sum y_i\right)^2 = 98.01$$

$$\overline{x} = \frac{22.3}{5} = 4.46 \qquad \overline{y} = 1.98$$

$$\sum x_i y_i = 51.54$$

step 2: From Eq. 11.89, the slope is

$$m = \frac{n\sum x_i y_i - \sum x_i \sum y_i}{n\sum x_i^2 - \left(\sum x_i\right)^2} = \frac{(5)(51.54) - (22.3)(9.9)}{(5)(169.43) - 497.29}$$
$$= 0.1055$$

step 3: From Eq. 11.90, the y-intercept is

$$b = \overline{y} - m\overline{x} = 1.98 - (0.1055)(4.46) = 1.509$$

The equation of the line is

$$y = mx + b = 0.1055x + 1.509$$
$$Y - 90 = (0.1055)(T - 200) + 1.509$$
$$\boxed{Y = 0.1055T + 70.409 \quad (0.11T + 70)}$$

The answer is (C).

(b) *step 4:* Use Eq. 11.91 to get the correlation coefficient.

$$r = \frac{n\sum x_i y_i - \sum x_i \sum y_i}{\sqrt{\left(n\sum x_i^2 - \left(\sum x_i\right)^2\right)\left(n\sum y_i^2 - \left(\sum y_i\right)^2\right)}}$$
$$= \frac{(5)(51.54) - (22.3)(9.9)}{\sqrt{\left((5)(169.43) - 497.29\right)\left((5)(20.39) - 98.01\right)}}$$
$$= \boxed{0.995 \quad (1.00)}$$

The answer is (D).

22. Plot the data to see if they are linear.

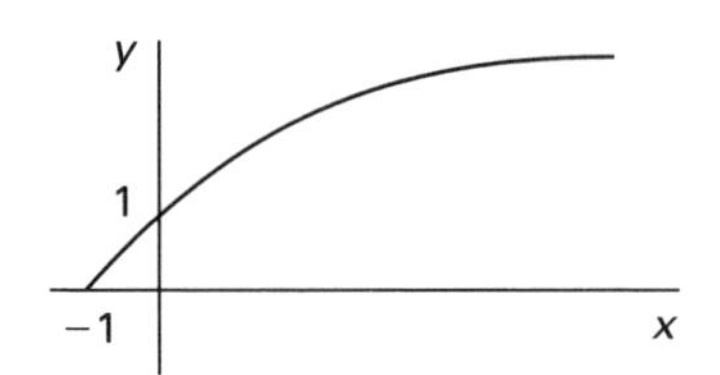

This looks like it could be of the form

$$y = a + b\sqrt{x}$$

However, when x is negative (as is the first point), the function is imaginary. Try shifting the curve to the right, replacing x with $x + 1$.

$$y = a + bz$$
$$z = \sqrt{x+1}$$

z	y
0	0
1	1
1.414	1.4
1.732	1.7
2	2
2.236	2.2
2.45	2.4
2.65	2.6
2.83	2.8
3	3

Since $y \approx z$, the relationship is

$$\boxed{y = \sqrt{x+1}}$$

In this problem, the answer was found accidentally. Usually, regression would be necessary.

The answer is (D).

12 Numbering Systems

PRACTICE PROBLEMS

Calculations in Other Numbering Systems

1. Perform the following binary (base-2) calculations.

(a) $101 + 011$

(A) 0110

(B) 1000

(C) 1001

(D) 1011

(b) $101 + 110$

(A) 0011

(B) 1010

(C) 1011

(D) 1100

(c) $101 + 100$

(A) 0100

(B) 0110

(C) 1000

(D) 1001

(d) $0100 - 1100$

(A) -1000

(B) -0100

(C) 0110

(D) 1000

(e) $1110 - 1000$

(A) 0010

(B) 0110

(C) 1100

(D) 1110

(f) $010 - 101$

(A) -100

(B) -011

(C) -010

(D) -001

(g) 111×11

(A) 10101

(B) 11011

(C) 11110

(D) 11111

(h) 100×11

(A) 0011

(B) 1000

(C) 1100

(D) 1110

(i) 1011×1101

(A) 10001111

(B) 11100111

(C) 11110000

(D) 11111111

2. Perform the following octal (base-8) calculations.

(a) $466 + 457$

(A) 923

(B) 1000

(C) 1105

(D) 1145

(b) $1007 + 6661$

(A) 7668

(B) 7670

(C) 7770

(D) 8670

(c) $321 + 465$

(A) 786

(B) 806

(C) 1006

(D) 1086

(d) 71 − 27

(A) 32

(B) 42

(C) 44

(D) 52

(e) 1143 − 367

(A) 554

(B) 754

(C) 776

(D) 876

(f) 646 − 677

(A) −131

(B) −31

(C) 31

(D) 131

(g) 77 × 66

(A) 5082

(B) 6512

(C) 6677

(D) 7766

(h) 325 × 36

(A) 11700

(B) 12346

(C) 14366

(D) 14936

(i) 3251 × 161

(A) 523411

(B) 550111

(C) 566011

(D) 570231

3. Perform the following calculations involving hexadecimal (base-16) numbers.

(a) BA + C

(A) B6

(B) BA

(C) C6

(D) CA

(b) BB + A

(A) B5

(B) B6

(C) C5

(D) C6

(c) BE + 10 + 1A

(A) B2

(B) BF

(C) E8

(D) EC

(d) FF − E

(A) F1

(B) F7

(C) FA

(D) FB

(e) 74 − 4A

(A) 2A

(B) 3A

(C) 3B

(D) 4B

(f) FB − BF

(A) 3B

(B) 3C

(C) 4A

(D) 4B

(g) 4A × 3E

(A) 7F

(B) 7AE

(C) 11EC

(D) 12EF

(h) FE × EF

(A) 22FE

(B) ED22

(C) EE11

(D) EF22

(i) 17 × 7A

(A) 8A

(B) 6AE

(C) AF6

(D) FE8

Conversions Between Bases

4. Convert the following numbers to decimal (base-10) numbers.

(a) $(674)_8$

(A) 136

(B) 444

(C) 1348

(D) 3552

(b) $(101101)_2$

(A) 45

(B) 46

(C) 48

(D) 64

(c) $(734.262)_8$

(A) 262.734

(B) 349.674

(C) 448.763

(D) 476.348

(d) $(1011.11)_2$

(A) 11.75

(B) 14.75

(C) 21.75

(D) 38.75

5. Convert the following decimal (base-10) numbers to octal (base-8) numbers.

(a) $(75)_{10}$

(A) $(113)_8$

(B) $(131)_8$

(C) $(311)_8$

(D) $(331)_8$

(b) $(0.375)_{10}$

(A) $(0.3)_8$

(B) $(3)_8$

(C) $(8)_8$

(D) $(24)_8$

(c) $(121.875)_{10}$

(A) $(151.7)_8$

(B) $(171.7)_8$

(C) $(177.1)_8$

(D) $(182.6)_8$

(d) $(1011100.01110)_2$

(A) $(134.34)_8$

(B) $(143.43)_8$

(C) $(341.41)_8$

(D) $(413.14)_8$

6. Convert the following numbers to binary (base-2) numbers.

(a) $(83)_{10}$

(A) $(1010001)_2$

(B) $(1010011)_2$

(C) $(1100100)_2$

(D) $(1100101)_2$

(b) $(100.3)_{10}$

(A) $(0001111.110111\cdots)_2$

(B) $(0011011.100011\cdots)_2$

(C) $(1000110.100010\cdots)_2$

(D) $(1100100.010011\cdots)_2$

(c) $(0.97)_{10}$

(A) $(0.101110\cdots)_2$

(B) $(0.111110\cdots)_2$

(C) $(110111.0)_2$

(D) $(111011.1)_2$

(d) $(321.422)_8$

(A) $(001100011.100001000)_2$

(B) $(010010100.001001100)_2$

(C) $(011000101.010010001)_2$

(D) $(011010001.100010010)_2$

SOLUTIONS

1. (a)

$$\begin{array}{r} 101 \\ +\ 011 \\ \hline \boxed{1000} \end{array}$$

The answer is (B).

(b)

$$\begin{array}{r} 101 \\ +\ 110 \\ \hline \boxed{1011} \end{array}$$

The answer is (C).

(c)

$$\begin{array}{r} 101 \\ +\ 100 \\ \hline \boxed{1001} \end{array}$$

The answer is (D).

(d)

$$-\left(\begin{array}{r} 1100 \\ -\ 0100 \\ \hline 1000 \end{array}\right) = \boxed{-1000}$$

The answer is (A).

(e)

$$\begin{array}{r} 1110 \\ -\ 1000 \\ \hline \boxed{0110} \end{array}$$

The answer is (B).

(f)

$$-\left(\begin{array}{r} 101 \\ -\ 010 \\ \hline 011 \end{array}\right) = \boxed{-011}$$

The answer is (B).

(g)

$$\begin{array}{r} 111 \\ \times\ \ 11 \\ \hline 111 \\ 111\ \ \\ \hline \boxed{10101} \end{array}$$

The answer is (A).

(h)

$$\begin{array}{r} 100 \\ \times\ \ 11 \\ \hline 100 \\ 100\ \ \\ \hline \boxed{1100} \end{array}$$

The answer is (C).

(i)

$$\begin{array}{r} 1011 \\ \times\ 1101 \\ \hline 1011 \\ 1011\ \ \\ 1011\ \ \ \ \\ \hline \boxed{10001111} \end{array}$$

The answer is (A).

2. (a)

$$\begin{array}{r} 466 \\ +\ 457 \\ \hline \boxed{1145} \end{array}$$

The answer is (D).

(b)

$$\begin{array}{r} 1007 \\ +\ 6661 \\ \hline \boxed{7670} \end{array}$$

The answer is (B).

(c)

$$\begin{array}{r} 321 \\ +\ 465 \\ \hline \boxed{1006} \end{array}$$

The answer is (C).

(d)

$$\begin{array}{r} 71 \\ -\ 27 \\ \hline \end{array} = \begin{array}{rrr} & 6 & 11 \\ - & 2 & 7 \\ \hline & \boxed{4} & \boxed{2} \end{array}$$

The answer is (B).

(e)

$$\begin{array}{r} 1143 \\ -\ \ 367 \\ \hline \end{array} = \begin{array}{rrr} & 113 & 13 \\ - & 36 & 7 \\ \hline \end{array} = \begin{array}{rrrr} & 10 & 13 & 13 \\ - & 3 & 6 & 7 \\ \hline & 5 & 5 & 4 \end{array}$$

The answer is (A).

(f)

$$-\left(\begin{array}{r} 677 \\ -\ 646 \\ \hline 31 \end{array}\right) = \boxed{-31}$$

The answer is (B).

(g)

$$\begin{array}{r} 77 \\ \times\ \ 66 \\ \hline 572 \\ 572\ \ \\ \hline \boxed{6512} \end{array}$$

The answer is (B).

(h)

$$\begin{array}{r} 325 \\ \times \quad 36 \\ \hline 2376 \\ 1177 \\ \hline \boxed{14366} \end{array}$$

The answer is (C).

(i)

$$\begin{array}{r} 3251 \\ \times \quad 161 \\ \hline 3251 \\ 23766 \\ 3251 \\ \hline \boxed{570231} \end{array}$$

The answer is (D).

3. (a)

$$\begin{array}{r} \text{BA} \\ + \quad \text{C} \\ \hline \boxed{\text{C6}} \end{array}$$

The answer is (C).

(b)

$$\begin{array}{r} \text{BB} \\ + \quad \text{A} \\ \hline \boxed{\text{C5}} \end{array}$$

The answer is (C).

(c)

$$\begin{array}{r} \text{BE} \\ 10 \\ + \quad \text{1A} \\ \hline \boxed{\text{E8}} \end{array}$$

The answer is (C).

(d)

$$\begin{array}{r} \text{FF} \\ - \quad \text{E} \\ \hline \boxed{\text{F1}} \end{array}$$

The answer is (A).

(e)

$$\begin{array}{r} 74 \\ - \text{ 4A} \\ \hline \end{array} = \begin{array}{rr} 6 & 14 \\ - \text{ } 4 & \text{A} \\ \hline \boxed{2 \quad \text{A}} \end{array}$$

The answer is (A).

(f)

$$\begin{array}{r} \text{FB} \\ - \text{ BF} \\ \hline \end{array} = \begin{array}{rr} \text{E} & \text{1B} \\ - \text{ B} & \text{F} \\ \hline \boxed{3 \quad \text{C}} \end{array}$$

The answer is (B).

(g)

$$\begin{array}{r} \text{4A} \\ \times \quad \text{3E} \\ \hline \text{40C} \\ \text{DE} \\ \hline \boxed{\text{11EC}} \end{array}$$

The answer is (C).

(h)

$$\begin{array}{r} \text{FE} \\ \times \quad \text{EF} \\ \hline \text{EE2} \\ \text{DE4} \\ \hline \boxed{\text{ED22}} \end{array}$$

The answer is (B).

(i)

$$\begin{array}{r} 17 \\ \times \quad \text{7A} \\ \hline \text{E6} \\ \text{A1} \\ \hline \boxed{\text{AF6}} \end{array}$$

The answer is (C).

4. (a)

$$\begin{aligned} (674)_8 &= (6)(8)^2 + (7)(8)^1 + (4)(8)^0 \\ &= \boxed{444} \end{aligned}$$

The answer is (B).

(b)

$$\begin{aligned} (101101)_2 &= (1)(2)^5 + (0)(2)^4 + (1)(2)^3 \\ &\quad + (1)(2)^2 + (0)(2)^1 + (1)(2)^0 \\ &= \boxed{45} \end{aligned}$$

The answer is (A).

(c)

$$\begin{aligned} (734.262)_8 &= (7)(8)^2 + (3)(8)^1 + (4)(8)^0 \\ &\quad + (2)(8)^{-1} + (6)(8)^{-2} + (2)(8)^{-3} \\ &= \boxed{476.348} \end{aligned}$$

The answer is (D).

(d)

$$\begin{aligned} (1011.11)_2 &= (1)(2)^3 + (0)(2)^2 + (1)(2)^1 \\ &\quad + (1)(2)^0 + (1)(2)^{-1} + (1)(2)^{-2} \\ &= \boxed{11.75} \end{aligned}$$

The answer is (A).

5. (a)

$$\begin{array}{rl} 75 \div 8 = 9 & \text{remainder } 3 \\ 9 \div 8 = 1 & \text{remainder } 1 \\ 1 \div 8 = 0 & \text{remainder } 1 \end{array}$$

$$(75)_{10} = \boxed{(113)_8}$$

The answer is (A).

(b)

$$0.375 \times 8 = 3 \quad \text{remainder } 0$$

$$(0.375)_{10} = \boxed{(0.3)_8}$$

The answer is (A).

(c)

$$121 \div 8 = 15 \quad \text{remainder } 1$$
$$15 \div 8 = 1 \quad \text{remainder } 7$$
$$1 \div 8 = 0 \quad \text{remainder } 1$$
$$0.875 \times 8 = 7 \quad \text{remainder } 0$$
$$(121.875)_{10} = \boxed{(171.7)_8}$$

The answer is (B).

(d) Since $(2)^3 = 8$, break the bits into groups of three starting at the decimal point and working outward in both directions.

$$001011100.011100 = 001\ 011\ 100.011\ 100$$

Convert each of the groups into its octal equivalent.

$$001\ 011\ 100.011\ 100 = 1\ 3\ 4.3\ 4$$
$$\boxed{(134.34)_8}$$

The answer is (A).

6. (a)

$$83 \div 2 = 41 \quad \text{remainder } 1$$
$$41 \div 2 = 20 \quad \text{remainder } 1$$
$$20 \div 2 = 10 \quad \text{remainder } 0$$
$$10 \div 2 = 5 \quad \text{remainder } 0$$
$$5 \div 2 = 2 \quad \text{remainder } 1$$
$$2 \div 2 = 1 \quad \text{remainder } 0$$
$$1 \div 2 = 0 \quad \text{remainder } 1$$
$$(83)_{10} = \boxed{(1010011)_2}$$

The answer is (B).

(b)

$$100 \div 2 = 50 \quad \text{remainder } 0$$
$$50 \div 2 = 25 \quad \text{remainder } 0$$
$$25 \div 2 = 12 \quad \text{remainder } 1$$
$$12 \div 2 = 6 \quad \text{remainder } 0$$
$$6 \div 2 = 3 \quad \text{remainder } 0$$
$$3 \div 2 = 1 \quad \text{remainder } 1$$
$$1 \div 2 = 0 \quad \text{remainder } 1$$
$$0.3 \times 2 = 0 \quad \text{remainder } 0.6$$
$$0.6 \times 2 = 1 \quad \text{remainder } 0.2$$
$$0.2 \times 2 = 0 \quad \text{remainder } 0.4$$
$$0.4 \times 2 = 0 \quad \text{remainder } 0.8$$
$$0.8 \times 2 = 1 \quad \text{remainder } 0.6$$
$$0.6 \times 2 = 1 \quad \text{remainder } 0.2$$
$$\vdots$$
$$(100.3)_{10} = \boxed{(1100100.010011\cdots)_2}$$

The answer is (D).

(c)

$$0.97 \times 2 = 1 \quad \text{remainder } 0.94$$
$$0.94 \times 2 = 1 \quad \text{remainder } 0.88$$
$$0.88 \times 2 = 1 \quad \text{remainder } 0.76$$
$$0.76 \times 2 = 1 \quad \text{remainder } 0.52$$
$$0.52 \times 2 = 1 \quad \text{remainder } 0.04$$
$$0.04 \times 2 = 0 \quad \text{remainder } 0.08$$
$$\vdots$$
$$(0.97)_{10} = \boxed{(0.111110\cdots)_2}$$

The answer is (B).

(d) Since $8 = (2)^3$, convert each octal digit into its binary equivalent.

$$321.422{:}\quad 3 = 011$$
$$2 = 010$$
$$1 = 001$$
$$4 = 100$$
$$2 = 010$$
$$2 = 010$$
$$\boxed{(321.422)_8 = (011010001.100010010)_2}$$

The answer is (D).

13 Numerical Analysis

PRACTICE PROBLEMS

1. A function is given as $y = 3x^{0.93} + 4.2$. What is most nearly the percent relative error if the value of y at $x = 2.7$ is found by using straight-line interpolation between $x = 2$ and $x = 3$?

(A) 0.06%

(B) 0.18%

(C) 2.5%

(D) 5.4%

2. Given the following data points, estimate y by straight-line interpolation for $x = 2.75$.

x	y
1	4
2	6
3	2
4	–14

(A) 2.1

(B) 2.4

(C) 2.7

(D) 3.0

3. Using the bisection method, find all of the roots of $f(x) = 0$ to the nearest 0.000005.

$$f(x) = x^3 + 2x^2 + 8x - 2$$

4. The increase in concentration of mixed-liquor suspended solids (MLSS) in an activated sludge aeration tank as a function of time is given in the table. Use a second-order Lagrangian interpolation to estimate the MLSS after 16 min of aeration.

t (min)	MLSS (mg/L)
0	0
10	227
15	362
20	517
22.5	602
30	901

(A) 350 mg/L

(B) 390 mg/L

(C) 540 mg/L

(D) 640 mg/L

SOLUTIONS

1. The actual value at $x = 2.7$ is

$$\begin{aligned} y(x) &= 3x^{0.93} + 4.2 \\ y(2.7) &= (3)(2.7)^{0.93} + 4.2 \\ &= 11.756 \end{aligned}$$

At $x = 3$,

$$y(3) = (3)(3)^{0.93} + 4.2 = 12.534$$

At $x = 2$,

$$y(2) = (3)(2)^{0.93} + 4.2 = 9.916$$

Use straight-line interpolation.

$$\begin{aligned} \frac{x_2 - x}{x_2 - x_1} &= \frac{y_2 - y}{y_2 - y_1} \\ \frac{3 - 2.7}{3 - 2} &= \frac{12.534 - y}{12.534 - 9.916} \\ y &= 11.749 \end{aligned}$$

The relative error is

$$\begin{aligned} \frac{\text{actual value} - \text{predicted value}}{\text{actual value}} &= \frac{11.756 - 11.749}{11.756} \\ &= \boxed{0.0006 \quad (0.06\%)} \end{aligned}$$

The answer is (A).

2. Let $x_1 = 2$; therefore, from the table of data points, $y_1 = 6$. Let $x_2 = 3$; therefore, from the table of data points, $y_2 = 2$.

Let $x = 2.75$. By straight-line interpolation,

$$\begin{aligned} \frac{x_2 - x}{x_2 - x_1} &= \frac{y_2 - y}{y_2 - y_1} \\ \frac{3 - 2.75}{3 - 2} &= \frac{2 - y}{2 - 6} \end{aligned}$$

$$\boxed{y = 3.0}$$

The answer is (D).

3. Use the equation $f(x) = x^3 + 2x^2 + 8x - 2$ to try to find an interval in which there is a root.

x	$f(x)$
0	–2
1	9

A root exists in the interval [0, 1].

Try $x = \frac{1}{2}(0+1) = 0.5$.

$$f(0.5) = (0.5)^3 + (2)(0.5)^2 + (8)(0.5) - 2 = 2.625$$

A root exists in [0, 0.5].

Try $x = 0.25$.

$$f(0.25) = (0.25)^3 + (2)(0.25)^2 + (8)(0.25) - 2 = 0.1406$$

A root exists in [0, 0.25].

Try $x = 0.125$.

$$\begin{aligned} f(0.125) &= (0.125)^3 + (2)(0.125)^2 + (8)(0.125) - 2 \\ &= -0.967 \end{aligned}$$

A root exists in [0.125, 0.25].

Try $x = \frac{1}{2}(0.125 + 0.25) = 0.1875$.

Continuing,

$$\begin{aligned} f(0.1875) &= -0.42 \quad [0.1875, 0.25] \\ f(0.21875) &= -0.144 \quad [0.21875, 0.25] \\ f(0.234375) &= -0.002 \quad \text{[This is close enough.]} \end{aligned}$$

One root is $x_1 \approx \boxed{0.234375\ (0.234).}$

Try to find the other two roots. Use long division to factor the polynomial.

$$\begin{array}{r} x^2 + 2.234375x + 8.52368 \\ x - 0.234375 \overline{\big)\, x^3 + \quad 2x^2 + \quad 8x - 2} \\ \underline{-(x^3 - 0.234375x^2)} \\ 2.234375x^2 + \quad 8x - 2 \\ \underline{-(2.234375x^2 - \; 0.52368x)} \\ 8.52368x - 2 \\ \underline{-\,(8.52368x - 1.9977)} \\ \approx 0 \end{array}$$

Use the quadratic equation to find the roots of $x^2 + 2.234375x + 8.52368$.

$$\begin{aligned} x_2, x_3 &= \frac{-2.234375 \pm \sqrt{(2.234375)^2 - (4)(1)(8.52368)}}{(2)(1)} \\ &= \boxed{-1.117189 \pm i2.697327} \quad \text{[both imaginary]} \end{aligned}$$

4. Choose the three data points that bracket $t = 16$ as closely as possible. These three points are $t_0 = 10$, $t_1 = 15$, and $t_2 = 20$.

	$i = 0$	$i = 1$	$i = 2$
$k = 0$: $S_0(16) = -0.08$	~~$\left(\frac{16-10}{10-10}\right)$~~	$\left(\frac{16-15}{10-15}\right)$	$\left(\frac{16-20}{10-20}\right)$
$k = 1$: $S_1(16) = 0.96$	$\left(\frac{16-10}{15-10}\right)$	~~$\left(\frac{16-15}{15-15}\right)$~~	$\left(\frac{16-20}{15-20}\right)$
$k = 2$: $S_2(16) = 0.12$	$\left(\frac{16-10}{20-10}\right)$	$\left(\frac{16-15}{20-15}\right)$	~~$\left(\frac{16-20}{20-20}\right)$~~

Use Eq. 13.3.

$$\begin{aligned} S &= S(10)S_0(16) + S(15)S_1(16) + S(20)S_2(16) \\ &= \left(227\ \frac{\text{mg}}{\text{L}}\right)(-0.08) + \left(362\ \frac{\text{mg}}{\text{L}}\right)(0.96) \\ &\quad + \left(517\ \frac{\text{mg}}{\text{L}}\right)(0.12) \\ &= \boxed{391\ \text{mg/L} \quad (390\ \text{mg/L})} \end{aligned}$$

The answer is (B).

14 Fluid Properties

PRACTICE PROBLEMS

(Use $g = 32.2$ ft/sec^2 or 9.81 m/s^2 unless told to do otherwise in the problem.)

Pressure

1. Atmospheric pressure is 14.7 lbf/in^2 (101.3 kPa). What is most nearly the absolute pressure in a tank if a gauge on the tank reads 8.7 lbf/in^2 (60 kPa) vacuum?

(A) 4 psia (27 kPa)

(B) 6 psia (41 kPa)

(C) 8 psia (55 kPa)

(D) 10 psia (68 kPa)

Viscosity

2. Air is considered to be an ideal gas with a specific gas constant of 53.3 ft-lbf/lbm-°R (287 J/kg·K). What is most nearly the kinematic viscosity of air at 80°F (27°C) and 70 psia (480 kPa)?

(A) 3.5×10^{-5} ft^2/sec (3.0×10^{-6} m^2/s)

(B) 4.0×10^{-5} ft^2/sec (4.0×10^{-6} m^2/s)

(C) 5.0×10^{-5} ft^2/sec (5.0×10^{-6} m^2/s)

(D) 6.0×10^{-5} ft^2/sec (6.0×10^{-6} m^2/s)

Solutions

3. Three solutions of nitric acid are combined: one with 8% nitric acid by volume, one with 10% nitric acid by volume, and one with 20% nitric acid by volume. The combined solutions produce 100 mL of a solution that is 12% nitric acid by volume. The 8% solution contributes half of the total volume of nitric acid contributed by the 10% and 20% solutions. The volume of 10% solution in the 12% solution is most nearly

(A) 20 mL

(B) 30 mL

(C) 50 mL

(D) 80 mL

Properties of Mixtures

4. A 25% (by volume) mixture of ethylene glycol and water is used in a solar heating application. The components are nonreacting. The mixture is intended to operate at standard atmospheric pressure and an average temperature of 140°F. What is most nearly the specific gravity of the mixture referred to 60°F water?

(A) 1.005

(B) 1.015

(C) 1.021

(D) 1.043

5. A blend contains equal masses (weights) of two nonreacting oils. The two oils have kinematic viscosities of 1 cSt and 1000 cSt, respectively. What is most nearly the kinematic viscosity of the mixture?

(A) 7 cSt

(B) 20 cSt

(C) 30 cSt

(D) 200 cSt

SOLUTIONS

1. *Customary U.S. Solution*

$$p_{\text{gage}} = -8.7 \text{ lbf/in}^2$$
$$p_{\text{atmospheric}} = 14.7 \text{ lbf/in}^2$$

The relationship between absolute, gage, and atmospheric pressure is

$$\begin{aligned} p_{\text{absolute}} &= p_{\text{gage}} + p_{\text{atmospheric}} = -8.7 \frac{\text{lbf}}{\text{in}^2} + 14.7 \frac{\text{lbf}}{\text{in}^2} \\ &= \boxed{6 \text{ lbf/in}^2 \quad (6 \text{ psia})} \end{aligned}$$

The answer is (B).

SI Solution

$$p_{\text{gage}} = -60 \text{ kPa}$$
$$p_{\text{atmospheric}} = 101.3 \text{ kPa}$$

The relationship between absolute, gage, and atmospheric pressure is

$$\begin{aligned} p_{\text{absolute}} &= p_{\text{gage}} + p_{\text{atmospheric}} = -60 \text{ kPa} + 101.3 \text{ kPa} \\ &= \boxed{41.3 \text{ kPa} \quad (41 \text{ kPa})} \end{aligned}$$

The answer is (B).

2. *Customary U.S. Solution*

From App. 14.D, for air at 14.7 psia and 80°F, the absolute viscosity (independent of pressure) is $\mu = 3.869 \times 10^{-7}$ lbf-sec/ft^2.

Determine the density of air at 70 psia and 80°F.

$$\begin{aligned} \rho &= \frac{p}{RT} = \frac{\left(70 \frac{\text{lbf}}{\text{in}^2}\right)\left(12 \frac{\text{in}}{\text{ft}}\right)^2}{\left(53.3 \frac{\text{ft-lbf}}{\text{lbm-°R}}\right)(80\text{°F} + 460°)} \\ &= 0.350 \text{ lbm/ft}^3 \end{aligned}$$

The kinematic viscosity, ν, is

$$\begin{aligned} \nu &= \frac{\mu g_c}{\rho} = \frac{\left(3.869 \times 10^{-7} \frac{\text{lbf-sec}}{\text{ft}^2}\right)\left(32.2 \frac{\text{lbm-ft}}{\text{lbf-sec}^2}\right)}{0.350 \frac{\text{lbm}}{\text{ft}^3}} \\ &= \boxed{3.56 \times 10^{-5} \text{ ft}^2\text{/sec} \quad (3.6 \times 10^{-5} \text{ ft}^2\text{/sec})} \end{aligned}$$

The answer is (A).

SI Solution

From App. 14.E, for air at 480 kPa and 27°C, the absolute viscosity (independent of pressure) is $\mu = 1.854 \times 10^{-5}$ Pa·s.

Determine the density of air at 480 kPa and 27°C.

$$\begin{aligned} \rho &= \frac{p}{RT} = \frac{(480 \text{ kPa})\left(1000 \frac{\text{Pa}}{\text{kPa}}\right)}{\left(287 \frac{\text{J}}{\text{kg·K}}\right)(27\text{°C} + 273°)} \\ &= 5.575 \text{ kg/m}^3 \end{aligned}$$

The kinematic viscosity, ν, is

$$\begin{aligned} \nu &= \frac{\mu}{\rho} = \frac{1.854 \times 10^{-5} \text{ Pa·s}}{5.575 \frac{\text{kg}}{\text{m}^3}} \\ &= \boxed{3.33 \times 10^{-6} \text{ m}^2\text{/s} \quad (3.0 \times 10^{-6} \text{ m}^2\text{/s})} \end{aligned}$$

The answer is (A).

3. Let

$$\begin{aligned} x &= \text{volume of 8\% solution} \\ 0.08x &= \text{volume of nitric acid contributed by 8\% solution} \\ y &= \text{volume of 10\% solution} \\ 0.10y &= \text{volume of nitric acid contributed by 10\% solution} \\ z &= \text{volume of 20\% solution} \\ 0.20z &= \text{volume of nitric acid contributed by 20\% solution} \end{aligned}$$

The three conditions that must be satisfied are

$$x + y + z = 100 \text{ mL}$$
$$0.08x + 0.10y + 0.20z = (0.12)(100 \text{ mL}) = 12 \text{ mL}$$
$$0.08x = \left(\tfrac{1}{2}\right)(0.10y + 0.20z)$$

Simplifying these equations,

$$x + y + z = 100$$
$$4x + 5y + 10z = 600$$
$$8x - 5y - 10z = 0$$

Adding the second and third equations gives

$$12x = 600$$
$$x = 50 \text{ mL}$$

Work with the first two equations to get

$$y + z = 100 - 50 = 50$$
$$5y + 10z = 600 - (4)(50) = 400$$

Multiplying the top equation by −5 and adding to the bottom equation,

$$5z = 150$$
$$z = 30 \text{ mL}$$

From the first equation,

$$y = \boxed{20 \text{ mL}}$$

The answer is (A).

4. From App. 14.A, the density of 140°F water is 61.38 lbm/ft^3. From App. 14.F, the specific gravity of ethylene glycol at 140°F is 1.107, referred to 60°F water, which has a density of 62.37 lbm/ft^3. Consider 100 ft^3 of mixture. Since the mixture percentages are volumetric, there will be 25 ft^3 of ethylene glycol and 75 ft^3 of water. The weight of 100 ft^3 will be

$$\begin{aligned} m &= \rho_{\text{water}} V_{\text{water}} + \rho_{\text{glycol}} V_{\text{glycol}} \\ &= \left(61.38 \ \frac{\text{lbm}}{\text{ft}^3}\right)(75 \text{ ft}^3) \\ &\quad + (1.107)\left(62.37 \ \frac{\text{lbm}}{\text{ft}^3}\right)(25 \text{ ft}^3) \\ &= 6329.59 \text{ lbm} \end{aligned}$$

The specific gravity is

$$\begin{aligned} \text{SG} &= \frac{\rho}{\rho_{\text{ref}}} = \frac{m}{V\rho_{\text{ref}}} = \frac{6329.59 \text{ lbm}}{(100 \text{ ft}^3)\left(62.37 \ \frac{\text{lbm}}{\text{ft}^3}\right)} \\ &= \boxed{1.0148 \quad (1.015)} \end{aligned}$$

The answer is (B).

5. The gravimetric fraction of each oil is $G = 0.50$.

From Eq. 14.43, the two viscosity blending indexes are

$$\begin{aligned} \text{VBI}_{1 \text{ cSt}} &= 10.975 + 14.534 \times \ln\big(\ln(\nu_{i,\text{cSt}} + 0.8)\big) \\ &= 10.975 + 14.534 \times \ln\big(\ln(1 \text{ cSt} + 0.8)\big) \\ &= 3.2518 \end{aligned}$$

$$\begin{aligned} \text{VBI}_{1000 \text{ cSt}} &= 10.975 + 14.534 \times \ln\big(\ln(\nu_{i,\text{cSt}} + 0.8)\big) \\ &= 10.975 + 14.534 \times \ln\big(\ln(1000 \text{ cSt} + 0.8)\big) \\ &= 39.0657 \end{aligned}$$

Using Eq. 14.44, the mixture VBI is

$$\begin{aligned} \text{VBI}_{\text{mixture}} &= \sum_i G_i \times \text{VBI}_i \\ &= (0.50)(3.2518) + (0.50)(39.0657) \\ &= 21.1588 \end{aligned}$$

From Eq. 14.45, the viscosity of the mixture is

$$\begin{aligned} \nu_{\text{mixture,cSt}} &= \exp\left(\exp\left(\frac{\text{VBI}_{\text{mixture}} - 10.975}{14.534}\right)\right) - 0.8 \\ &= \exp\left(\exp\left(\frac{21.1588 - 10.975}{14.534}\right)\right) - 0.8 \\ &= \boxed{6.70 \text{ cSt} \quad (7 \text{ cSt})} \end{aligned}$$

The answer is (A).

15 Fluid Statics

PRACTICE PROBLEMS

(Use $g = 32.2\ \text{ft/sec}^2$ or $9.81\ \text{m/s}^2$ unless told to do otherwise in the problem.)

1. A 4000 lbm (1800 kg) blimp contains 10,000 lbm (4500 kg) of hydrogen (specific gas constant = 766.5 ft-lbf/lbm-°R (4124 J/kg·K)) at 56°F (13°C) and 30.2 in Hg (770 mm Hg). If the hydrogen and air are in thermal and pressure equilibrium, what is most nearly the blimp's lift (lifting force)?

(A) 7.6×10^3 lbf (3.4×10^4 N)

(B) 1.2×10^4 lbf (5.3×10^4 N)

(C) 1.3×10^5 lbf (5.7×10^5 N)

(D) 1.7×10^5 lbf (7.7×10^5 N)

2. A hollow 6 ft (1.8 m) diameter sphere floats half-submerged in seawater. The mass of concrete that is required as an external anchor to just submerge the sphere completely is most nearly

(A) 2700 lbm (1200 kg)

(B) 4200 lbm (1900 kg)

(C) 5500 lbm (2500 kg)

(D) 6300 lbm (2700 kg)

3. Water removed from Lake Superior (elevation, 601 ft above mean sea level; water density, 62.4 lbm/ft^3) is transported by tanker ship to the Atlantic Ocean (elevation, 0 ft) through 16 Seaway locks. A tanker's displacement is 32,000 tonnes when loaded, and 5100 tonnes when empty. Each lock is 766 ft long and 80 ft wide. Water pumped from each lock flows to the Atlantic Ocean. Compared to a passage from Lake Superior to the Atlantic Ocean when empty, what is most nearly the change in water loss from Lake Superior when a ship passes through the locks fully loaded?

(A) 27,000 tonnes less loss

(B) no change in loss

(C) 27,000 tonnes additional loss

(D) 54,000 tonnes additional loss

SOLUTIONS

1. *Customary U.S. Solution*

The lift (lifting force) of the hydrogen-filled blimp, F_{lift}, is equal to the difference between the buoyant force, F_b, and the weight of the hydrogen contained in the blimp, W_{H}.

$$F_{\text{lift}} = F_b - W_{\text{H}} - W_{\text{blimp}}$$

The weight of the hydrogen is calculated from the mass of hydrogen.

$$W_{\text{H}} = \frac{m_{\text{H}}g}{g_c} = \frac{(10{,}000\ \text{lbm})\left(32.2\ \frac{\text{ft}}{\text{sec}^2}\right)}{32.2\ \frac{\text{lbm-ft}}{\text{lbf-sec}^2}} = 10{,}000\ \text{lbf}$$

The buoyant force is equal to the weight of the displaced air. The volume of the air displaced is equal to the volume of hydrogen enclosed in the blimp.

The absolute temperature of the hydrogen is

$$T = 56°\text{F} + 460° = 516°\text{R}$$

The pressure of the hydrogen is

$$p = \frac{(30.2\ \text{in Hg})\left(12\ \frac{\text{in}}{\text{ft}}\right)^2}{2.036\ \frac{\text{in Hg}}{\frac{\text{lbf}}{\text{in}^2}}} = 2136\ \text{lbf/ft}^2$$

Compute the volume of hydrogen from the ideal gas law.

$$V_{\text{H}} = \frac{m_{\text{H}}RT}{p} = \frac{(10{,}000\ \text{lbm})\left(766.5\ \frac{\text{ft-lbf}}{\text{lbm-°R}}\right)(516°\text{R})}{2136\ \frac{\text{lbf}}{\text{ft}^2}} = 1.85 \times 10^6\ \text{ft}^3$$

Since the volume of the hydrogen contained in the blimp is equal to the air displaced, the air displaced can be computed from the ideal gas equation. Since the air and hydrogen are in thermal and pressure equilibrium, the temperature and pressure are equal to the values given for the hydrogen.

For air, $R = 53.35$ ft-lbf/lbm-°R.

$$m_{air} = \frac{pV_H}{RT} = \frac{\left(2136\ \frac{\text{lbf}}{\text{ft}^2}\right)(1.85 \times 10^6\ \text{ft}^3)}{\left(53.35\ \frac{\text{ft-lbf}}{\text{lbm-°R}}\right)(516\text{°R})} = 1.435 \times 10^5\ \text{lbm}$$

The buoyant force is equal to the weight of the air.

$$F_b = W_{air} = \frac{m_{air}g}{g_c} = \frac{(1.435 \times 10^5\ \text{lbm})\left(32.2\ \frac{\text{ft}}{\text{sec}^2}\right)}{32.2\ \frac{\text{lbm-ft}}{\text{lbf-sec}^2}} = 1.435 \times 10^5\ \text{lbf}$$

The lift (lifting force) is

$$\begin{aligned} F_{lift} &= F_b - W_H - W_{blimp} \\ &= 1.435 \times 10^5\ \text{lbf} - 10{,}000\ \text{lbf} - 4000\ \text{lbf} \\ &= \boxed{1.295 \times 10^5\ \text{lbf}\quad (1.3 \times 10^5\ \text{lbf})} \end{aligned}$$

The answer is (C).

SI Solution

The lift (lifting force) of the hydrogen-filled blimp, F_{lift}, is equal to the difference between the buoyant force, F_b, and the weight of the hydrogen contained in the blimp, W_H.

$$F_{lift} = F_b - W_H - W_{blimp}$$

The weight of the hydrogen is calculated from the mass of hydrogen.

$$\begin{aligned} W_H &= m_H g \\ &= (4500\ \text{kg})\left(9.81\ \frac{\text{m}}{\text{s}^2}\right) \\ &= 44\,145\ \text{N} \end{aligned}$$

The buoyant force is equal to the weight of the displaced air. The volume of the air displaced is equal to the volume of hydrogen enclosed in the blimp.

The absolute temperature of the hydrogen is

$$T = 13\text{°C} + 273\text{°} = 286\text{K}$$

The absolute pressure of the hydrogen is

$$p = \frac{(770\ \text{mm Hg})\left(133.4\ \frac{\text{kPa}}{\text{m}}\right)}{1000\ \frac{\text{mm}}{\text{m}}} = 102.7\ \text{kPa}$$

Compute the volume of hydrogen from the ideal gas law.

$$V_H = \frac{m_H RT}{p} = \frac{(4500\ \text{kg})\left(4124\ \frac{\text{J}}{\text{kg·K}}\right)(286\text{K})}{(102.7\ \text{kPa})\left(1000\ \frac{\text{Pa}}{\text{kPa}}\right)} = 5.168 \times 10^4\ \text{m}^3$$

Since the volume of the hydrogen contained in the blimp is equal to the air displaced, the air displaced can be computed from the ideal gas equation. Since the air and hydrogen are assumed to be in thermal and pressure equilibrium, the temperature and pressure are equal to the values given for the hydrogen.

For air, $R = 287.03$ J/kg·K.

$$m_{air} = \frac{pV_H}{RT} = \frac{(102.7\ \text{kPa})\left(1000\ \frac{\text{Pa}}{\text{kPa}}\right)(5.168 \times 10^4\ \text{m}^3)}{\left(287.03\ \frac{\text{J}}{\text{kg·K}}\right)(286\text{K})} = 6.465 \times 10^4\ \text{kg}$$

The buoyant force is equal to the weight of the air.

$$F_b = W_{air} = m_{air}g = (6.465 \times 10^4\ \text{kg})\left(9.81\ \frac{\text{m}}{\text{s}^2}\right) = 6.34 \times 10^5\ \text{N}$$

The lift (lifting force) is

$$\begin{aligned} F_{lift} &= F_b - W_H - W_{blimp} \\ &= 6.34 \times 10^5\ \text{N} - 44\,145\ \text{N} - (1800\ \text{kg})\left(9.81\ \frac{\text{m}}{\text{s}^2}\right) \\ &= \boxed{5.7 \times 10^5\ \text{N}} \end{aligned}$$

The answer is (C).

2. *Customary U.S. Solution*

The weight of the sphere is equal to the weight of the displaced volume of water when floating.

The buoyant force is given by

$$F_b = \frac{\rho g V_{displaced}}{g_c}$$

Since the sphere is half submerged,

$$W_{\text{sphere}} = \tfrac{1}{2}\left(\frac{\rho g V_{\text{sphere}}}{g_c}\right)$$

For seawater, $\rho = 64.0\ \text{lbm/ft}^3$.

The volume of the sphere is

$$\begin{aligned} V_{\text{sphere}} &= \frac{\pi}{6}d^3 = \left(\frac{\pi}{6}\right)(6\ \text{ft})^3 \\ &= 113.1\ \text{ft}^3 \end{aligned}$$

The weight of the sphere is

$$\begin{aligned} W_{\text{sphere}} &= \tfrac{1}{2}\left(\frac{\rho g V_{\text{sphere}}}{g_c}\right) \\ &= \left(\tfrac{1}{2}\right)\left(\frac{\left(64.0\ \frac{\text{lbm}}{\text{ft}^3}\right)\left(32.2\ \frac{\text{ft}}{\text{sec}^2}\right)(113.1\ \text{ft}^3)}{32.2\ \frac{\text{lbm-ft}}{\text{lbf-sec}^2}}\right) \\ &= 3619\ \text{lbf} \end{aligned}$$

The equilibrium equation for a fully submerged sphere and anchor can be solved for the concrete volume.

$$\begin{aligned} W_{\text{sphere}} + W_{\text{concrete}} &= (V_{\text{sphere}} + V_{\text{concrete}})\rho_{\text{water}} \\ W_{\text{sphere}} + \rho_{\text{concrete}} V_{\text{concrete}}\left(\frac{g}{g_c}\right) \\ &= (V_{\text{sphere}} + V_{\text{concrete}})\rho_{\text{water}}\left(\frac{g}{g_c}\right) \end{aligned}$$

$$\begin{aligned} 3619\ \text{lbf} + \left(150\ \frac{\text{lbm}}{\text{ft}^3}\right)V_{\text{concrete}}\left(\frac{32.2\ \frac{\text{ft}}{\text{sec}^2}}{32.2\ \frac{\text{lbm-ft}}{\text{lbf-sec}^2}}\right) \\ = (113.1\ \text{ft}^3 + V_{\text{concrete}})\left(64.0\ \frac{\text{lbm}}{\text{ft}^3}\right)\left(\frac{32.2\ \frac{\text{ft}}{\text{sec}^2}}{32.2\ \frac{\text{lbm-ft}}{\text{lbf-sec}^2}}\right) \\ V_{\text{concrete}} &= 42.09\ \text{ft}^3 \\ m_{\text{concrete}} &= \rho_{\text{concrete}} V_{\text{concrete}} \\ &= \left(150\ \frac{\text{lbm}}{\text{ft}^3}\right)(42.09\ \text{ft}^3) \\ &= \boxed{6314\ \text{lbm}\quad(6300\ \text{lbm})} \end{aligned}$$

The answer is (D).

SI Solution

The weight of the sphere is equal to the weight of the displaced volume of water when floating.

The buoyant force is given by

$$F_b = \rho g V_{\text{displaced}}$$

Since the sphere is half submerged,

$$W_{\text{sphere}} = \tfrac{1}{2}\rho g V_{\text{sphere}}$$

For seawater, $\rho = 1025\ \text{kg/m}^3$.

The volume of the sphere is

$$V_{\text{sphere}} = \frac{\pi}{6}d^3 = \left(\frac{\pi}{6}\right)(1.8\ \text{m})^3 = 3.054\ \text{m}^3$$

The weight of the sphere required is

$$\begin{aligned} W_{\text{sphere}} &= \tfrac{1}{2}\rho g V_{\text{sphere}} \\ &= \left(\tfrac{1}{2}\right)\left(1025\ \frac{\text{kg}}{\text{m}^3}\right)\left(9.81\ \frac{\text{m}}{\text{s}^2}\right)(3.054\ \text{m}^3) \\ &= 15\,354\ \text{N} \end{aligned}$$

The equilibrium equation for a fully submerged sphere and anchor can be solved for the concrete volume.

$$\begin{aligned} W_{\text{sphere}} + W_{\text{concrete}} &= (V_{\text{sphere}} + V_{\text{concrete}})\rho_{\text{water}} \\ W_{\text{sphere}} + \rho_{\text{concrete}}\, g V_{\text{concrete}} &= g(V_{\text{sphere}} + V_{\text{concrete}})\rho_{\text{water}} \\ 15\,354\ \text{N} + \left(2400\ \frac{\text{kg}}{\text{m}^3}\right)\left(9.81\ \frac{\text{m}}{\text{s}^2}\right)V_{\text{concrete}} \\ &= (3.054\ \text{m}^3 + V_{\text{concrete}})\left(1025\ \frac{\text{kg}}{\text{m}^3}\right)\left(9.81\ \frac{\text{m}}{\text{s}^2}\right) \\ V_{\text{concrete}} &= 1.138\ \text{m}^3 \\ m_{\text{concrete}} &= \rho_{\text{concrete}} V_{\text{concrete}} \\ &= \left(2400\ \frac{\text{kg}}{\text{m}^3}\right)(1.138\ \text{m}^3) \\ &= \boxed{2731\ \text{kg}\quad(2700\ \text{kg})} \end{aligned}$$

The answer is (D).

3. From Archimedes' principle, each tonne of water carried in the tanker displaces a tonne of water in the lock. So, each tonne of water transported out of the lake results in a tonne less of lock loss. Compared to an empty tanker, the net result is zero.

The answer is (B).

Fluids

16 Fluid Flow Parameters

PRACTICE PROBLEMS

Use the following values unless told to do otherwise in the problem:

$$g = 32.2 \text{ ft/sec}^2 \ (9.81 \text{ m/s}^2)$$
$$\rho_{\text{water}} = 62.4 \text{ lbm/ft}^3 \ (1000 \text{ kg/m}^3)$$
$$p_{\text{atmospheric}} = 14.7 \text{ psia} \ (101.3 \text{ kPa})$$

Hydraulic Radius

1. A 10 in (25 cm) composition pipe is compressed by a tree root into an elliptical cross section until its inside height is only 7.2 in (18 cm). What is its approximate hydraulic radius when flowing half full?

(A) 2.2 in (5.5 cm)

(B) 2.7 in (6.9 cm)

(C) 3.2 in (8.1 cm)

(D) 4.5 in (11.4 cm)

2. A pipe with an inside diameter of 18.812 in contains water to a depth of 15.7 in. What is most nearly the hydraulic radius? (Work in customary U.S. units only.)

(A) 4.4 in

(B) 5.1 in

(C) 5.7 in

(D) 6.5 in

Pipe Ratings

3. A class IV, 60 in diameter C76 concrete pipe has a $D_{0.01}$ rating of 2000 lbf/ft^2 and a D_{ultimate} rating of 3000 lbf/ft^2. Approximately what vertical line force must be applied in three-point loading to a pipe section 8 ft long in order to induce a concrete crack at least 1 ft long?

(A) 80,000 lbf

(B) 120,000 lbf

(C) 240,000 lbf

(D) 680,000 lbf

SOLUTIONS

1. *Customary U.S. Solution*

The perimeter of the pipe is

$$p = \pi d = \pi(10 \text{ in}) = 31.42 \text{ in}$$

If the pipe is flowing half full, the wetted perimeter becomes

$$\text{wetted perimeter} = \tfrac{1}{2}p = \left(\tfrac{1}{2}\right)(31.42 \text{ in}) = 15.71 \text{ in}$$

The ellipse will have a minor axis, b, equal to one-half the height of the compressed pipe or

$$b = \frac{7.2 \text{ in}}{2} = 3.6 \text{ in}$$

When the pipe is compressed, the perimeter of the pipe will remain constant. The perimeter of an ellipse is given by

$$p \approx 2\pi\sqrt{\tfrac{1}{2}(a^2 + b^2)}$$

Solve for the major axis.

$$a = \sqrt{2\left(\frac{p}{2\pi}\right)^2 - b^2} = \sqrt{(2)\left(\frac{31.42 \text{ in}}{2\pi}\right)^2 - (3.6 \text{ in})^2}$$
$$= 6.09 \text{ in}$$

The flow area or area of the ellipse is given by

$$\text{flow area} = \tfrac{1}{2}\pi ab = \tfrac{1}{2}\pi(6.09 \text{ in})(3.6 \text{ in})$$
$$= 34.4 \text{ in}^2$$

The hydraulic radius is

$$r_h = \frac{\text{area in flow}}{\text{wetted perimeter}} = \frac{34.4 \text{ in}^2}{15.71 \text{ in}} = \boxed{2.19 \text{ in} \quad (2.2 \text{ in})}$$

The answer is (A).

SI Solution

The perimeter of the pipe is

$$p = \pi d = \pi(25 \text{ cm}) = 78.54 \text{ cm}$$

If the pipe is flowing half full, the wetted perimeter becomes

$$\text{wetted perimeter} = \tfrac{1}{2}p = \left(\tfrac{1}{2}\right)(78.54 \text{ cm}) = 39.27 \text{ cm}$$

Assume the compressed pipe is an elliptical cross section. The ellipse will have a minor axis, b, equal to one-half the height of the compressed pipe or

$$b = \frac{18 \text{ cm}}{2} = 9 \text{ cm}$$

When the pipe is compressed, the perimeter of the pipe will remain constant. The perimeter of an ellipse is given by

$$p \approx 2\pi\sqrt{\tfrac{1}{2}(a^2 + b^2)}$$

Solve for the major axis.

$$\begin{aligned} a &= \sqrt{2\left(\frac{p}{2\pi}\right)^2 - b^2} = \sqrt{(2)\left(\frac{78.54 \text{ cm}}{2\pi}\right)^2 - (9 \text{ cm})^2} \\ &= 15.2 \text{ cm} \end{aligned}$$

The flow area or area of the ellipse is given by

$$\begin{aligned} \text{flow area} &= \tfrac{1}{2}\pi ab = \tfrac{1}{2}\pi(15.2 \text{ cm})(9 \text{ cm}) \\ &= 214.9 \text{ cm}^2 \end{aligned}$$

The hydraulic radius is

$$\begin{aligned} r_h &= \frac{\text{area in flow}}{\text{wetted perimeter}} = \frac{214.9 \text{ cm}^2}{39.27 \text{ cm}} \\ &= \boxed{5.47 \text{ cm} \quad (5.5 \text{ cm})} \end{aligned}$$

The answer is (A).

2. *method 1:* Use App. 7.A for a circular segment.

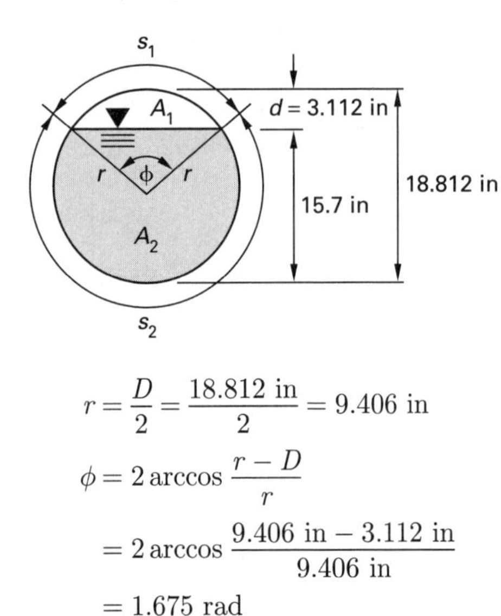

$$r = \frac{D}{2} = \frac{18.812 \text{ in}}{2} = 9.406 \text{ in}$$

$$\begin{aligned} \phi &= 2\arccos\frac{r - D}{r} \\ &= 2\arccos\frac{9.406 \text{ in} - 3.112 \text{ in}}{9.406 \text{ in}} \\ &= 1.675 \text{ rad} \end{aligned}$$

$$\begin{aligned} \sin\phi &= 0.9946 \\ A_1 &= \tfrac{1}{2}r^2(\phi - \sin\phi) \\ &= \left(\tfrac{1}{2}\right)(9.406 \text{ in})^2(1.675 - 0.9946) \\ &= 30.1 \text{ in}^2 \\ A_{\text{total}} &= A_1 + A_2 = \frac{\pi}{4}D^2 = \left(\frac{\pi}{4}\right)(18.812 \text{ in})^2 \\ &= 277.95 \text{ in}^2 \\ A_2 &= A_{\text{total}} - A_1 = 277.95 \text{ in}^2 - 30.1 \text{ in}^2 \\ &= 247.85 \text{ in}^2 \\ s_1 &= r\phi = (9.406 \text{ in})(1.675) = 15.76 \text{ in} \\ s_{\text{total}} &= s_1 + s_2 = \pi D = \pi(18.812 \text{ in}) \\ &= 59.1 \text{ in} \\ s_2 &= s_{\text{total}} - s_1 = 59.1 \text{ in} - 15.76 \text{ in} = 43.34 \text{ in} \\ r_h &= \frac{A_2}{s_2} = \frac{247.85 \text{ in}^2}{43.34 \text{ in}} = \boxed{5.719 \text{ in} \quad (5.7 \text{ in})} \end{aligned}$$

method 2: Use App. 16.A.

$$\frac{d}{D} = \frac{15.7 \text{ in}}{18.812 \text{ in}} = 0.83$$

From App. 16.A, $r_h/D = 0.3041$.

$$\begin{aligned} r_h &= (0.3041)(18.812 \text{ in}) \\ &= \boxed{5.72 \text{ in} \quad (5.7 \text{ in})} \end{aligned}$$

The answer is (C).

3. The definition of $D_{0.01}$ is the vertical force applied in three-point loading required to cause a crack at least 1 ft long. The dimensions used to calculate the D-load rating of 2000 lbf/ft^2 are pipe length and pipe diameter. Rearranging Eq. 16.35, the required line force is

$$\begin{aligned} F &= D_{0.01}D_{\text{ft}}L_{\text{ft}} = \frac{\left(2000 \dfrac{\text{lbf}}{\text{ft}^2}\right)(60 \text{ in})(8 \text{ ft})}{12 \dfrac{\text{in}}{\text{ft}}} \\ &= \boxed{80{,}000 \text{ lbf}} \end{aligned}$$

The answer is (A).

17 Fluid Dynamics

PRACTICE PROBLEMS

(Use $g = 32.2$ ft/sec^2 (9.81 m/s^2) and 60°F (16°C) water unless told to do otherwise in the problem.)

Basic Concepts

1. The capacity of a municipal, old, pressurized water supply pipe system must be doubled without increasing the average velocity. A replacement new pipe will have 40% less friction than the existing pipe. Approximately what increase in pipe diameter is required?

(A) 20%

(B) 30%

(C) 40%

(D) 100%

2. A pressurized water pipe system will be modified such that the pipe diameter will be doubled and the average flow velocity halved. Fluid properties are unchanged. When these changes are made, the Reynolds number will

(A) halve

(B) double

(C) quadruple

(D) remain the same

Conservation of Energy

3. 5 ft^3/sec (130 L/s) of water flows through a schedule-40 steel pipe that changes gradually in diameter from 6 in at point A to 18 in at point B. Point B is 15 ft (4.6 m) higher than point A. The respective pressures at points A and B are 10 psia (70 kPa) and 7 psia (48.3 kPa). All minor losses are insignificant. The velocity and direction of flow at point A are most nearly

(A) 3.2 ft/sec (1 m/s); from A to B

(B) 25 ft/sec (7 m/s); from A to B

(C) 3.2 ft/sec (1 m/s); from B to A

(D) 25 ft/sec (7 m/s); from B to A

4. Points A and B are separated by 3000 ft of new 6 in schedule-40 steel pipe. 750 gal/min of 60°F water flows from point A to point B. Point B is 60 ft above point A. Approximately what must be the pressure at point A if the pressure at B must be 50 psig?

(A) 90 psig

(B) 100 psig

(C) 120 psig

(D) 170 psig

5. A pipe network connects junctions A, B, C, and D as shown. All pipe sections have a Hazen-Williams C-value of 150. Water can be added and removed at any of the junctions to achieve the flows listed. Water flows from point A to point D. No flows are backward. All minor losses are insignificant. For simplicity, use the nominal pipe diameters.

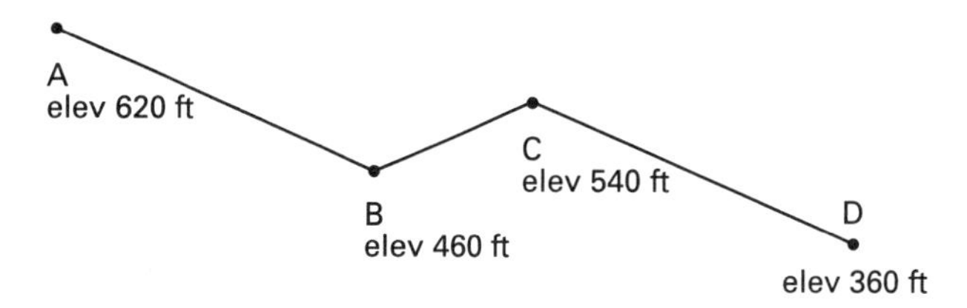

pipe section	length	diameter (in)	flow
A to B	20,000 ft	6	120 gal/min
B to C	10,000 ft	6	160 gal/min
C to D	30,000 ft	4	120 gal/min

(a) The friction loss from A to B is most nearly

(A) 0.1 ft

(B) 20 ft

(C) 40 ft

(D) 60 ft

(b) The velocity head in pipe AB is most nearly

(A) 0.007 ft

(B) 0.018 ft

(C) 0.020 ft

(D) 0.030 ft

(c) The friction loss from B to C is most nearly

(A) 0.3 ft

(B) 20 ft

(C) 30 ft

(D) 50 ft

(d) The friction loss from C to D is most nearly

(A) 240 ft

(B) 300 ft

(C) 310 ft

(D) 340 ft

(e) Assume the static pressure at point A is 20 psig. The pressure at point B is most nearly

(A) 30 psig

(B) 60 psig

(C) 80 psig

(D) 90 psig

(f) Assume the static pressure at point A is 20 psig. The pressure at point C is most nearly

(A) 20 psig

(B) 30 psig

(C) 40 psig

(D) 50 psig

(g) Assume the static pressure at point A is 20 psig. The pressure at point D is most nearly

(A) 10 psig

(B) 12 psig

(C) 18 psig

(D) 20 psig

(h) If the minimum static pressure anywhere in the system is 20 psig, the pressure at point A is most nearly

(A) 14 psig

(B) 17 psig

(C) 24 psig

(D) 30 psig

(i) If the minimum static pressure anywhere in the system is 20 psig, the elevation of the hydraulic grade line at point A referenced to point D is most nearly

(A) 280 ft

(B) 300 ft

(C) 480 ft

(D) 600 ft

Friction Loss

6. Based on the formulas commonly used by engineers, is friction head loss ever proportional to velocity, instead of velocity squared?

(A) yes, in laminar flow

(B) yes, for non-Newtonian fluids

(C) yes, in smooth pipes

(D) no, never

7. Water flows quietly through a pipe network consisting of 50 ft of new 3 in diameter, schedule-40 steel pipe, four 45° standard elbows, and a fully open gate valve. All fittings are flanged. The elevation of the network discharge is 20 ft lower than the elevation of the inlet. The element that contributes the most to specific energy loss is the

(A) four elbows

(B) gate valve

(C) pipe friction (excluding the fittings)

(D) elevation change

8. 300 ft of 18 in high-density polyethylene (HDPE) pipe (C=120) and 400 ft of 14 in HDPE pipe (C=120) are currently joined in series as part of a pipe network. The length of 16 in diameter HDPE pipe (C=120) that can replace the two pipes without increasing the pumping power required is most nearly

(A) 640 ft

(B) 790 ft

(C) 840 ft

(D) 940 ft

9. A 10 in diameter reinforced concrete pressure pipe (C=100) carries water flowing at 4 ft/sec. What is most nearly the theoretical friction loss per 100 ft of pipe length?

(A) 1.0 ft

(B) 4.2 ft

(C) 10 ft

(D) 14 ft

10. A pressurized supply line ($C = 140$) brings cold water to 600 residential connections. The line is 17,000 ft long and has a diameter of 10 in. The elevation is 1000 ft at the start of the line and 850 ft at the delivery end of the line. The average flow rate is 1.1 gal/min per residence, and the peaking factor is 2.5. The minimum required pressure at the delivery end is 60 psig. Approximately what must be the pressure at the start of the line during peak flow?

(A) 9.4 psig

(B) 40 psig

(C) 95 psig

(D) 110 psig

11. 1.5 ft^3/sec (40 L/s) of 70°F (20°C) water flows through 1200 ft (355 m) of 6 in (nominal) diameter new schedule-40 steel pipe. The friction loss is most nearly

(A) 4 ft (1.2 m)

(B) 18 ft (5.2 m)

(C) 36 ft (9.5 m)

(D) 70 ft (21 m)

12. 500 gal/min (30 L/s) of 100°F (40°C) water flows through 300 ft (90 m) of 6 in schedule-40 pipe. The pipe contains two 6 in flanged steel elbows, two full-open gate valves, a full-open 90° angle valve, and a swing check valve. The discharge is located 20 ft (6 m) higher than the entrance. The pressure difference between the two ends of the pipe is most nearly

(A) 12 psi (78 kPa)

(B) 21 psi (140 kPa)

(C) 45 psi (310 kPa)

(D) 87 psi (600 kPa)

13. 70°F (20°C) air is flowing at 60 ft/sec (18 m/s) through 300 ft (90 m) of 6 in schedule-40 pipe. The pipe contains two 6 in (0.15 m) flanged steel elbows, two full-open gate valves, a full-open 90° angle valve, and a swing check valve. The discharge is located 20 ft (6 m) higher than the entrance, and the average air density is 0.075 lbm/ft^3. The pressure difference between the two ends of the pipe is most nearly

(A) 0.26 psi (1.8 kPa)

(B) 0.49 psi (3.2 kPa)

(C) 1.5 psi (10 kPa)

(D) 13 psi (90 kPa)

Reservoirs

14. Three reservoirs (A, B, and C) are interconnected with a common junction (point D) at elevation 25 ft above an arbitrary reference point. The water levels for reservoirs A, B, and C are at elevations of 50 ft, 40 ft, and 22 ft, respectively. The pipe from reservoir A to the junction is 800 ft of 3 in (nominal) steel pipe. The pipe from reservoir B to the junction is 500 ft of 10 in (nominal) steel pipe. The pipe from reservoir C to the junction is 1000 ft of 4 in (nominal) steel pipe. All pipes are schedule-40 with a friction factor of 0.02. All minor losses and velocity heads can be neglected. The direction of flow and the pressure at point D are most nearly

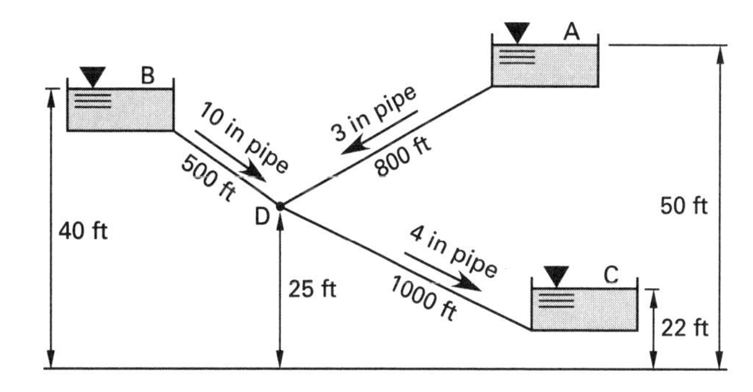

(A) out of reservoir B; 500 psf

(B) out of reservoir B; 930 psf

(C) into reservoir B; 1100 psf

(D) into reservoir B; 1260 psf

Water Hammer

15. A cast-iron pipe with expansion joints throughout has an inside diameter of 24 in (600 mm) and a wall thickness of 0.75 in (20 mm). The pipe's modulus of elasticity is 20×10^6 psi (140 GPa). The pipeline is 500 ft (150 m) long. 70°F (20°C) water is flowing at 6 ft/sec (2 m/s).

(a) The modulus of elasticity of the cast-iron pipe is most nearly

(A) 2.1×10^5 lbf/in^2 (1.5×10^9 Pa)

(B) 2.6×10^5 lbf/in^2 (1.8×10^9 Pa)

(C) 2.8×10^5 lbf/in^2 (2.0×10^9 Pa)

(D) 3.1×10^5 lbf/in^2 (2.2×10^9 Pa)

(b) The speed of sound in the pipe is most nearly

(A) 330 ft/sec (100 m/s)

(B) 1500 ft/sec (400 m/s)

(C) 4000 ft/sec (1200 m/s)

(D) 4500 ft/sec (1300 m/s)

Fluids

(c) If a valve is closed instantaneously, the pressure increase experienced in the pipe will be most nearly

(A) 48 psi (330 kPa)

(B) 140 psi (970 kPa)

(C) 320 psi (2.5 MPa)

(D) 470 psi (3.2 MPa)

(d) If the pipe is 500 ft (150 m) long, over what approximate length of time must the valve be closed to create a pressure equivalent to instantaneous closure?

(A) 0.25 sec

(B) 0.68 sec

(C) 1.6 sec

(D) 2.1 sec

Parallel Pipe Systems

16. 8 MGD (millions of gallons per day) (350 L/s) of 70°F (20°C) water flows into the new schedule-40 steel pipe network shown. Minor losses are insignificant.

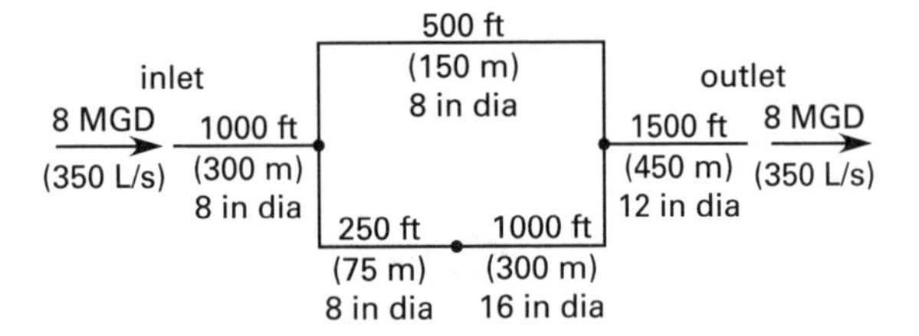

(a) The quantity of water flowing in the upper branch is most nearly

(A) 1.2 ft^3/sec (0.034 m^3/s)

(B) 2.9 ft^3/sec (0.081 m^3/s)

(C) 4.1 ft^3/sec (0.11 m^3/s)

(D) 5.3 ft^3/sec (0.15 m^3/s)

(b) The energy loss per unit mass between the inlet and the outlet is most nearly

(A) 120 ft-lbf/lbm (0.37 kJ/kg)

(B) 300 ft-lbf/lbm (0.90 kJ/kg)

(C) 480 ft-lbf/lbm (1.4 kJ/kg)

(D) 570 ft-lbf/lbm (1.7 kJ/kg)

Pipe Networks

17. A single-loop pipe network is shown. The distance between each junction is 1000 ft. All junctions are on the same elevation. All pipes have a Hazen-Williams C-value of 100. The volumetric flow rates are to be determined to within 2 gal/min. Start by assuming the following flows.

A to D: 300 gal/min

D to C: 100 gal/min

B to C: 200 gal/min

A to B: 400 gal/min

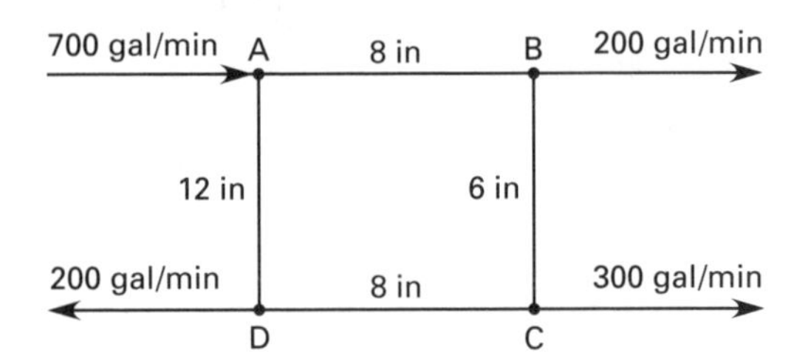

The flow rate between junctions B and C is most nearly

(A) 36 gal/min from B to C

(B) 58 gal/min from B to C

(C) 84 gal/min from C to B

(D) 110 gal/min from C to B

18. A double-loop pipe network is shown. The distance between each junction is 1000 ft. The water temperature is 60°F. Elevations and some pressure are known for the junctions. All pipes have a Hazen-Williams C-value of 100. Start by assuming the following flows.

A to F: 500 gal/min

A to B: 300 gal/min

B to E: 700 gal/min

C to B: 400 gal/min

C to D: 600 gal/min

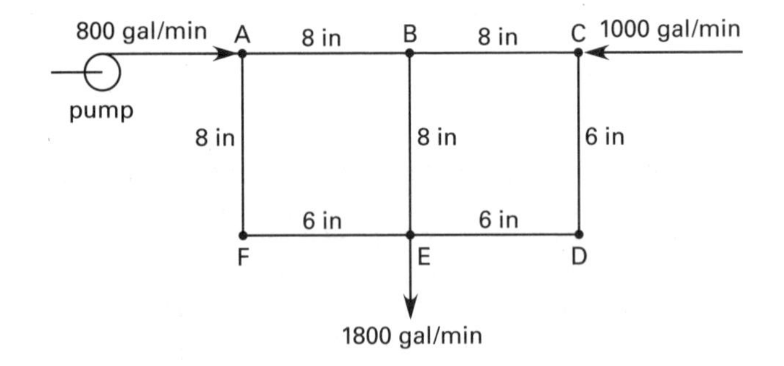

point	pressure	elevation
A		200 ft
B		150 ft
C	40 psig	300 ft
D		150 ft
E		200 ft
F		150 ft

(a) After one iteration, the corrections for loops 1 and 2, respectively, will be most nearly

(A) +100 gal/min, −116 gal/min

(B) +108 gal/min, −205 gal/min

(C) +112 gal/min, −196 gal/min

(D) +120 gal/min, −210 gal/min

(b) After two iterations, the corrections for loops 1 and 2, respectively, will most nearly be

(A) −70 gal/min, −9 gal/min

(B) −50 gal/min, −4 gal/min

(C) −10 gal/min, −6 gal/min

(D) 10 gal/min, 9 gal/min

(c) After three iterations, the corrections for loops 1 and 2, respectively, will most nearly be

(A) 0 gal/min, −8 gal/min

(B) −0.2 gal/min, −7 gal/min

(C) −0.3 gal/min, −8 gal/min

(D) −0.09 gal/min, −10 gal/min

(d) The flow rate between junctions B and E is most nearly

(A) 540 gal/min

(B) 620 gal/min

(C) 810 gal/min

(D) 980 gal/min

(e) When pressure at C is 40 psig, the friction head is most nearly

(A) 10 ft

(B) 40 ft

(C) 90 ft

(D) 110 ft

(f) The pressure at point D is most nearly

(A) 51 psig

(B) 73 psig

(C) 96 psig

(D) 120 psig

(g) If the pump receives water at 20 psig (140 kPa), the hydraulic power required is most nearly

(A) 14 hp

(B) 28 hp

(C) 35 hp

(D) 67 hp

19. The water distribution network shown consists of class A cast-iron pipe (specific roughness of 0.0008 ft) installed 1.5 years ago. 1.5 MGD of 50°F water enters at junction A and leaves at junction B. The minimum acceptable pressure at point D is 40 psig (280 kPa).

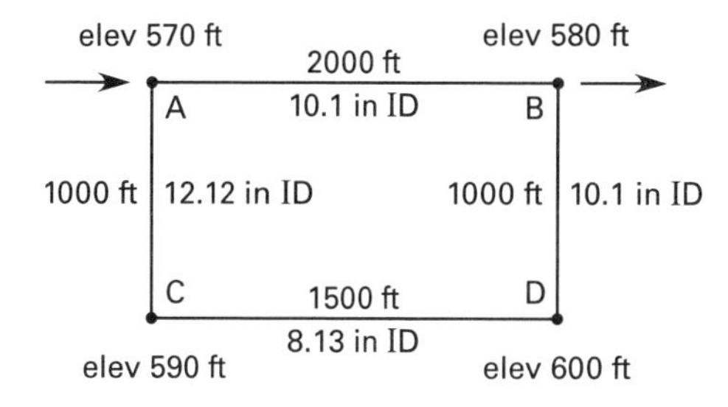

For parts (a) through (e), solve as a pipe network problem.

(a) The friction coefficient between junctions A and B is most nearly

(A) 6×10^{-6}

(B) 9×10^{-6}

(C) 12×10^{-6}

(D) 26×10^{-6}

(b) The friction coefficient between junctions B and D is most nearly

(A) 2×10^{-6}

(B) 6×10^{-6}

(C) 12×10^{-6}

(D) 26×10^{-6}

(c) The friction coefficient between junctions D and C is most nearly

(A) 6×10^{-6}

(B) 8×10^{-6}

(C) 12×10^{-6}

(D) 26×10^{-6}

(d) The friction coefficient between junctions C and A is most nearly

(A) 1×10^{-6}

(B) 2×10^{-6}

(C) 5×10^{-6}

(D) 6×10^{-6}

(e) The flow rate between junctions A and B is most nearly

(A) 320 gal/min

(B) 480 gal/min

(C) 590 gal/min

(D) 660 gal/min

For parts (f) through (h), solve as a parallel pipe problem. Assume $v_{max} = 5$ ft/sec.

(f) The flow rate between junctions A and B is most nearly

(A) 320 gal/min

(B) 480 gal/min

(C) 590 gal/min

(D) 660 gal/min

(g) Use a Darcy friction factor of 0.021. The pressure at point B is most nearly

(A) 32 psig

(B) 48 psig

(C) 57 psig

(D) 88 psig

(h) Use a Darcy friction factor of 0.021. The pressure at point A is most nearly

(A) 39 psig

(B) 55 psig

(C) 69 psig

(D) 110 psig

Discharge Through an Orifice

20. The velocity of discharge from a fire hose is 50 ft/sec (15 m/s). The hose is oriented 45° from the horizontal. Disregarding air friction, the maximum range of the discharge is most nearly

(A) 45 ft (14 m)

(B) 78 ft (23 m)

(C) 91 ft (27 m)

(D) 110 ft (33 m)

21. A full cylindrical tank that is 40 ft (12 m) high has a constant diameter of 20 ft. The tank has a 4 in (100 mm) diameter hole in its bottom. The coefficient of discharge for the hole is 0.98. Approximately how long will it take for the water level to drop from 40 ft to 20 ft (12 m to 6 m)?

(A) 950 sec

(B) 1200 sec

(C) 1450 sec

(D) 1700 sec

Siphons

22. A 24 in diameter siphon is used to transfer irrigation water from a water distribution canal to an irrigation ditch for a field of row crops below. The elevation of the water in the canal is 320 ft MSL (relative to mean sea level), and the elevation of the stilling basin for the row crops is 305 ft MSL. Counting fittings and minor losses, the siphon has a total equivalent length of 42 ft. The siphon is constructed of corrugated metal pipe with a standard galvanized surface. Counting entrance and exit losses, what is most nearly the total head loss experienced by the water?

(A) 7.0 ft

(B) 11 ft

(C) 15 ft

(D) 23 ft

Venturi Meters

23. A venturi meter with an 8 in diameter throat is installed in a 12 in diameter water line. The venturi is perfectly smooth, so that the discharge coefficient is 1.00. An attached mercury manometer registers a 4 in differential. The volumetric flow rate is most nearly

(A) 1.7 ft^3/sec

(B) 5.2 ft^3/sec

(C) 6.4 ft^3/sec

(D) 18 ft^3/sec

24. 60°F (15°C) benzene (specific gravity at 60°F (15°C) of 0.885) flows through an 8 in/3.5 in (200 mm/90 mm) venturi meter whose coefficient of discharge is 0.99. A mercury manometer indicates a 4 in difference in the heights of the mercury columns. The volumetric flow rate of the benzene is most nearly

(A) 1.2 ft^3/sec (34 L/s)

(B) 9.1 ft^3/sec (250 L/s)

(C) 13 ft^3/sec (360 L/s)

(D) 27 ft^3/sec (760 L/s)

Orifice Meters

25. A sharp-edged orifice meter with a 0.2 ft diameter opening is installed in a 1 ft diameter pipe. 70°F water approaches the orifice at 2 ft/sec. The indicated pressure drop across the orifice meter is most nearly

(A) 5.9 psi

(B) 13 psi

(C) 22 psi

(D) 47 psi

26. A mercury manometer is used to measure a pressure difference across an orifice meter in a water line. The difference in mercury levels is 7 in (17.8 cm). The pressure differential is most nearly

(A) 1.7 psi (12 kPa)

(B) 3.2 psi (22 kPa)

(C) 7.9 psi (55 kPa)

(D) 23 psi (160 kPa)

27. A sharp-edged ISA orifice is used in a schedule-40 steel 12 in (300 mm inside diameter) water line. (Figure 17.28 is applicable.) The water temperature is 70°F (20°C), and the flow rate is 10 ft^3/sec (250 L/s). The differential pressure change across the orifice (to the vena contracta) should be approximately 25 ft (7.5 m). The smallest orifice that can be used is most nearly

(A) 5.5 in (14 cm)

(B) 7.3 in (19 cm)

(C) 8.1 in (20 cm)

(D) 8.9 in (23 cm)

Impulse-Momentum

28. A pipe necks down from 24 in at point A to 12 in at point B. 8 ft^3/sec of 60°F water flows from point A to point B. The pressure head at point A is 20 ft. Friction is insignificant over the distance between points A and B. The magnitude and direction of the resultant force on the water are most nearly

(A) 2900 lbf, toward A

(B) 3500 lbf, toward A

(C) 2900 lbf, toward B

(D) 3500 lbf, toward B

29. A 2 in (50 mm) diameter horizontal water jet has an absolute velocity (with respect to a stationary point) of 40 ft/sec (12 m/s) as it strikes a curved blade. The blade is moving horizontally away with an absolute velocity of 15 ft/sec (4.5 m/s). Water is deflected 60° from the horizontal. The force on the blade is most nearly

(A) 18 lbf (80 N)

(B) 26 lbf (110 N)

(C) 35 lbf (160 N)

(D) 47 lbf (210 N)

Pumps

30. 2000 gal/min (125 L/s) of brine with a specific gravity of 1.2 passes through an 85% efficient pump. The centerlines of the pump's 12 in inlet and 8 in outlet are at the same elevation. The inlet suction gauge indicates 6 in (150 mm) of mercury below atmospheric. The discharge pressure gauge is located 4 ft (1.2 m) above the centerline of the pump's outlet and indicates 20 psig (138 kPa). All pipes are schedule-40. The input power to the pump is most nearly

(A) 12 hp (8.9 kW)

(B) 36 hp (26 kW)

(C) 52 hp (39 kW)

(D) 87 hp (65 kW)

Turbines

31. 100 ft^3/sec (2.6 m^3/s) of water passes through a horizontal turbine. The water's pressure is reduced from 30 psig (210 kPa) to 5 psig (35 kPa) vacuum. Disregarding friction, velocity, and other factors, the power generated is most nearly

(A) 110 hp (82 kW)

(B) 380 hp (280 kW)

(C) 730 hp (540 kW)

(D) 920 hp (640 kW)

Drag

32. A refrigeration truck is driven at 65 mph into a 15 mph headwind. The frontal area of the truck is 100 ft^2. The coefficient of drag is 0.5. When calculating the drag force on the truck, the velocity that should be used is most nearly

(A) 50 mph

(B) 65 mph

(C) 73 mph

(D) 80 mph

Fluids

33. A dish-shaped antenna faces directly into a 60 mph wind. The projected area of the antenna is 0.8 ft^2; the coefficient of drag, C_D, is 1.2; and the density of air is 0.076 lbm/ft^3. The total amount of drag force experienced by the antenna is most nearly

(A) 9.0 lbf

(B) 10 lbf

(C) 14 lbf

(D) 16 lbf

34. A 30 ft long smooth, straight, round log has been floated downstream in a 6 ft deep flume by upstream loggers. The outside diameter of the log is 12 in, and the saturated specific gravity of the wood is 0.72 relative to 50°F water. The 60°F flume water flows with an average velocity of 10 ft/sec. At the downstream collection point, one end of the log is chained to an excavator crane arm directly over the flume and lifted until the log is approximately vertical. The lower 2 ft of the log remain submerged in the flume. The upper end of the log is free to rotate, but the upper end cannot translate. Neglect buoyancy effects. What is most nearly the angle from the vertical that the log will deflect in the moving water?

(A) 5°

(B) 7°

(C) 11°

(D) 14°

35. A car traveling through 70°F (20°C) air has the following characteristics.

frontal area	28 ft^2 (2.6 m^2)
mass	3300 lbm (1500 kg)
drag coefficient	0.42
rolling resistance	1% of weight
engine thermal efficiency	28%
fuel heating value	115,000 Btu/gal (32 MJ/L)

For parts (a) through (e), assume the car is traveling at 55 mi/hr (90 km/h).

(a) The velocity of the car is most nearly

(A) 0.9 ft/sec (0.3 m/s)

(B) 14 ft/sec (4 m/s)

(C) 45 ft/sec (14 m/s)

(D) 80 ft/sec (25 m/s)

(b) The drag on the car is most nearly

(A) 1 lbf (5 N)

(B) 3 lbf (12 N)

(C) 30 lbf (130 N)

(D) 90 lbf (410 N)

(c) The total resisting force on the car is most nearly

(A) 60 lbf (280 N)

(B) 120 lbf (560 N)

(C) 3300 lbf (15 000 N)

(D) 3400 lbf (15 500 N)

(d) The power exhibited (manifested, delivered, etc.) by the car is most nearly

(A) 6 Btu/sec (7000 W)

(B) 13 Btu/sec (14 000 W)

(C) 22 Btu/sec (24 000 W)

(D) 350 Btu/sec (38 000 W)

(e) Considering only the drag and rolling resistance, the fuel consumption of the car is most nearly

(A) 0.026 gal/mi (0.062 L/km)

(B) 0.038 gal/mi (0.087 L/km)

(C) 0.051 gal/mi (0.12 L/km)

(D) 0.13 gal/mi (0.30 L/km)

For parts (f) through (j), assume the car is traveling at 65 mi/hr (105 km/h).

(f) The drag on the car is most nearly

(A) 30 lbf (130 N)

(B) 90 lbf (410 N)

(C) 120 lbf (560 N)

(D) 124 lbf (570 N)

(g) The total resisting force on the car is most nearly

(A) 120 lbf (560 N)

(B) 130 lbf (670 N)

(C) 150 lbf (710 N)

(D) 160 lbf (720 N)

(h) The power exhibited (manifested, delivered, etc.) by the car is most nearly

(A) 15 Btu/sec (16 000 W)

(B) 16 Btu/sec (18 000 W)

(C) 19 Btu/sec (21 000 W)

(D) 24 Btu/sec (26 000 W)

(i) The fuel consumption of the car is most nearly

(A) 0.03 gal/mi (0.08 L/km)

(B) 0.07 gal/mi (0.16 L/km)

(C) 0.10 gal/mi (0.23 L/km)

(D) 0.11 gal/mi (0.25 L/km)

(j) What is the approximate percentage increase in fuel consumption at 65 mi/hr (105 km/h) compared to 55 mi/hr (90 km/h)?

(A) 10%

(B) 20%

(C) 30%

(D) 40%

Similarity

36. A 1/20th airplane model is tested in a wind tunnel at full velocity and temperature. What is the approximate ratio of the wind tunnel pressure to normal ambient pressure?

(A) 5

(B) 10

(C) 20

(D) 40

37. 68°F (20°C) castor oil (kinematic viscosity at 68°F (20°C) of 1110×10^{-5} ft²/sec (103×10^{-5} m²/s)) flows through a pump whose impeller turns at 1000 rpm. A similar pump twice the first pump's size is tested with 68°F (20°C) air. Theoretically, what should be the approximate speed of the second pump's impeller to ensure similarity?

(A) 3.6 rpm

(B) 88 rpm

(C) 250 rpm

(D) 1600 rpm

SOLUTIONS

1. The decrease in friction will affect pumping power required, but it will not affect capacity. Capacity is a function of velocity and area only. The velocity is unchanged.

$$\frac{Q_2}{Q_1} = \frac{\text{v}A_2}{\text{v}A_1} = \frac{\frac{\pi \text{v} d_2^2}{4}}{\frac{\pi \text{v} d_1^2}{4}} = \frac{d_2^2}{d_1^2} = 2$$

$$d_2 = \sqrt{2}d_1 \approx \boxed{1.41d_1 \quad (40\% \text{ increase})}$$

The answer is (C).

2. The initial Reynolds number is

$$\text{Re}_1 = \frac{D\text{v}}{\nu}$$

The Reynolds number after modifications are made will be

$$\text{Re}_2 = \frac{(2D)\left(\frac{\text{v}}{2}\right)}{\nu} = \frac{D\text{v}}{\nu} = \text{Re}_1$$

The Reynolds number will $\boxed{\text{remain the same.}}$

The answer is (D).

3. *Customary U.S. Solution*

For schedule-40 pipe,

$$D_\text{A} = 0.5054 \text{ ft}$$
$$D_\text{B} = 1.4063 \text{ ft}$$

Let point A be at zero elevation.

The total energy at point A from Bernoulli's equation is

$$E_{t,\text{A}} = E_p + E_\text{v} + E_z = \frac{p_\text{A}}{\rho} + \frac{\text{v}_\text{A}^2}{2g_c} + \frac{z_\text{A} g}{g_c}$$

At point A, the diameter is 6 in. The velocity at point A is

$$\dot{V} = \text{v}_\text{A} A_\text{A} = \text{v}_\text{A}\left(\frac{\pi}{4}\right) D_\text{A}^2$$

$$\text{v}_\text{A} = \left(\frac{4}{\pi}\right)\left(\frac{\dot{V}}{D_\text{A}^2}\right) = \left(\frac{4}{\pi}\right)\left(\frac{5\ \frac{\text{ft}^3}{\text{sec}}}{(0.5054 \text{ ft})^2}\right) = \boxed{24.9 \text{ ft/sec} \quad (25 \text{ ft/sec})}$$

$$p_\text{A} = \left(10\ \frac{\text{lbf}}{\text{in}^2}\right)\left(12\ \frac{\text{in}}{\text{ft}}\right)^2 = 1440 \text{ lbf/ft}^2$$

$$z_\text{A} = 0$$

For water, $\rho \approx 62.4$ lbm/ft^3.

$$\begin{aligned}E_{t,\text{A}} &= \frac{p_\text{A}}{\rho} + \frac{\text{v}_\text{A}^2}{2g_c} + \frac{z_\text{A} g}{g_c} \\ &= \frac{1440 \ \frac{\text{lbf}}{\text{ft}^2}}{62.4 \ \frac{\text{lbm}}{\text{ft}^3}} + \frac{\left(24.9 \ \frac{\text{ft}}{\text{sec}}\right)^2}{(2)\left(32.2 \ \frac{\text{lbm-ft}}{\text{lbf-sec}^2}\right)} + 0 \\ &= 32.7 \text{ ft-lbf/lbm}\end{aligned}$$

Similarly, the total energy at point B is

$$\text{v}_\text{B} = \left(\frac{4}{\pi}\right)\left(\frac{\dot{V}}{D_\text{B}^2}\right) = \left(\frac{4}{\pi}\right)\left(\frac{5 \ \frac{\text{ft}^3}{\text{sec}}}{(1.4063 \text{ ft})^2}\right)$$

$$= 3.22 \text{ ft/sec}$$

$$p_\text{B} = \left(7 \ \frac{\text{lbf}}{\text{in}^2}\right)\left(12 \ \frac{\text{in}}{\text{ft}}\right)^2 = 1008 \text{ lbf/ft}^2$$

$$z_\text{B} = 15 \text{ ft}$$

$$\begin{aligned}E_{t,\text{B}} &= \frac{p_\text{B}}{\rho} + \frac{\text{v}_\text{B}^2}{2g_c} + \frac{z_\text{B} g}{g_c} \\ &= \frac{1008 \ \frac{\text{lbf}}{\text{ft}^2}}{62.4 \ \frac{\text{lbm}}{\text{ft}^3}} + \frac{\left(3.22 \ \frac{\text{ft}}{\text{sec}}\right)^2}{(2)\left(32.2 \ \frac{\text{lbm-ft}}{\text{lbf-sec}^2}\right)} \\ &\quad + \frac{(15 \text{ ft})\left(32.2 \ \frac{\text{ft}}{\text{sec}^2}\right)}{32.2 \ \frac{\text{lbm-ft}}{\text{lbf-sec}^2}} \\ &= 31.3 \text{ ft-lbf/lbm}\end{aligned}$$

Since $E_{t,\text{A}} > E_{t,\text{B}}$, $\boxed{\text{the flow is from point A to point B.}}$

The answer is (B).

SI Solution

Let point A be at zero elevation.

The total energy at point A from Bernoulli's equation is

$$E_{t,\text{A}} = E_p + E_\text{v} + E_z = \frac{p_\text{A}}{\rho} + \frac{\text{v}_\text{A}^2}{2} + z_\text{A} g$$

At point A, from App. 16.C, the diameter is 154 mm (0.154 m). The velocity at point A is

$$\dot{V} = \text{v}_\text{A} A_\text{A} = \text{v}_\text{A}\left(\frac{\pi}{4}\right) D_\text{A}^2$$

$$\text{v}_\text{A} = \left(\frac{4}{\pi}\right)\left(\frac{\dot{V}}{D_\text{A}^2}\right) = \left(\frac{4}{\pi}\right)\left(\frac{130 \ \frac{\text{L}}{\text{s}}}{(0.154 \text{ m})^2\left(1000 \ \frac{\text{L}}{\text{m}^3}\right)}\right)$$

$$= \boxed{6.98 \text{ m/s} \quad (7 \text{ m/s})}$$

$$p_\text{A} = 70 \text{ kPa} \quad (70\,000 \text{ Pa})$$

$$z_\text{A} = 0$$

For water, $\rho = 1000$ kg/m^3.

$$\begin{aligned}E_{t,\text{A}} &= \frac{p_\text{A}}{\rho} + \frac{\text{v}_\text{A}^2}{2} + z_\text{A} g \\ &= \frac{70\,000 \text{ Pa}}{1000 \ \frac{\text{kg}}{\text{m}^3}} + \frac{\left(6.98 \ \frac{\text{m}}{\text{s}}\right)^2}{2} + 0 \\ &= 94.36 \text{ J/kg}\end{aligned}$$

Similarly, at B, the diameter is 429 mm. The total energy at point B is

$$\begin{aligned}\text{v}_\text{B} &= \left(\frac{4}{\pi}\right)\left(\frac{\dot{V}}{D_\text{B}^2}\right) \\ &= \left(\frac{4}{\pi}\right)\left(\frac{130 \ \frac{\text{L}}{\text{s}}}{(0.429 \text{ m})^2\left(1000 \ \frac{\text{L}}{\text{m}^3}\right)}\right) \\ &= 0.90 \text{ m/s}\end{aligned}$$

$$p_\text{B} = 48.3 \text{ kPa} \quad (48\,300 \text{ Pa})$$

$$z_\text{B} = 4.6 \text{ m}$$

$$\begin{aligned}E_{t,\text{B}} &= \frac{p_\text{B}}{\rho} + \frac{\text{v}_\text{B}^2}{2} + z_\text{B} g \\ &= \frac{48\,300 \text{ Pa}}{1000 \ \frac{\text{kg}}{\text{m}^3}} + \frac{\left(0.90 \ \frac{\text{m}}{\text{s}}\right)^2}{2} \\ &\quad + (4.6 \text{ m})\left(9.81 \ \frac{\text{m}}{\text{s}^2}\right) \\ &= 93.8 \text{ J/kg}\end{aligned}$$

Since $E_{t,\text{A}} > E_{t,\text{B}}$, $\boxed{\text{the flow is from point A to point B.}}$

The answer is (B).

4.

$$\dot{V} = \frac{750 \ \frac{\text{gal}}{\text{min}}}{\left(7.4805 \ \frac{\text{gal}}{\text{ft}^3}\right)\left(60 \ \frac{\text{sec}}{\text{min}}\right)} = 1.671 \text{ ft}^3/\text{sec}$$

From App. 16.B, $D = 0.5054$ ft, and $A = 0.2006$ ft^2.

$$\text{v} = \frac{\dot{V}}{A} = \frac{1.671 \ \frac{\text{ft}^3}{\text{sec}}}{0.2006 \text{ ft}^2} = 8.33 \text{ ft/sec}$$

For 60°F water, from App. 14.A,

$$\rho = 62.37\ \text{lbm/ft}^3$$

$$\nu = 1.217 \times 10^{-5}\ \text{ft}^2/\text{sec}$$

$$\text{Re} = \frac{\text{v}D}{\nu} = \frac{\left(8.33\ \frac{\text{ft}}{\text{sec}}\right)(0.5054\ \text{ft})}{1.217 \times 10^{-5}\ \frac{\text{ft}^2}{\text{sec}}} = 3.46 \times 10^5$$

The specific weight is

$$\gamma = \frac{\rho g}{g_c} = \frac{\left(62.37\ \frac{\text{lbm}}{\text{ft}^3}\right)\left(32.2\ \frac{\text{ft}}{\text{sec}^2}\right)}{32.2\ \frac{\text{lbm-ft}}{\text{lbf-sec}^2}} = 62.37\ \text{lbf/ft}^3$$

For steel,

$$\epsilon = 0.0002$$

$$\frac{\epsilon}{D} = \frac{0.0002\ \text{ft}}{0.5054\ \text{ft}} \approx 0.0004$$

$$f = 0.0175$$

From Eq. 17.22,

$$h_f = \frac{fL\text{v}^2}{2Dg} = \frac{(0.0175)(3000\ \text{ft})\left(8.33\ \frac{\text{ft}}{\text{sec}}\right)^2}{(2)(0.5054\ \text{ft})\left(32.2\ \frac{\text{ft}}{\text{sec}^2}\right)}$$

$$= 111.9\ \text{ft}$$

Use the Bernoulli equation. Since velocity is the same at points A and B, it may be omitted.

$$\frac{p_1}{\gamma_1} = \frac{p_2}{\gamma_2} + (z_2 - z_1) + h_f$$

$$\frac{\left(12\ \frac{\text{in}}{\text{ft}}\right)^2 p_1}{62.37\ \frac{\text{lbf}}{\text{ft}^3}} = \frac{\left(50\ \frac{\text{lbf}}{\text{in}^2}\right)\left(12\ \frac{\text{in}}{\text{ft}}\right)^2}{62.37\ \frac{\text{lbf}}{\text{ft}^3}} + 60\ \text{ft} + 111.9\ \text{ft}$$

$$p_1 = \boxed{124.5\ \text{lbf/in}^2\quad(120\ \text{psig})}$$

The answer is (C).

5. (a) From Eq. 17.29, the friction loss from A to B is

$$h_{f,\text{ft,A-B}} = \frac{10.44 L_{\text{ft}} Q_{\text{gpm}}^{1.85}}{C^{1.85} d_{\text{in}}^{4.87}}$$

$$= \frac{(10.44)(20{,}000\ \text{ft})\left(120\ \frac{\text{gal}}{\text{min}}\right)^{1.85}}{(150)^{1.85}(6\ \text{in})^{4.87}}$$

$$= \boxed{22.4\ \text{ft}\quad(20\ \text{ft})}$$

The answer is (B).

(b) Calculate the velocity head.

$$\text{v} = \frac{\dot{V}}{A} = \frac{120\ \frac{\text{gal}}{\text{min}}}{\left(\frac{\pi}{4}\right)\left(\frac{6\ \text{in}}{12\ \frac{\text{in}}{\text{ft}}}\right)^2\left(7.4805\ \frac{\text{gal}}{\text{ft}^3}\right)\left(60\ \frac{\text{sec}}{\text{min}}\right)}$$

$$= 1.36\ \text{ft/sec}$$

$$h_\text{v} = \frac{\text{v}^2}{2g} = \frac{\left(1.36\ \frac{\text{ft}}{\text{sec}}\right)^2}{(2)\left(32.2\ \frac{\text{ft}}{\text{sec}^2}\right)}$$

$$= \boxed{0.029\ \text{ft}\quad(0.030\ \text{ft})}$$

The answer is (D).

(c) Velocity heads are low and can be disregarded. The friction loss from B to C is

$$h_{f,\text{B-C}} = \frac{(10.44)(10{,}000\ \text{ft})\left(160\ \frac{\text{gal}}{\text{min}}\right)^{1.85}}{(150)^{1.85}(6\ \text{in})^{4.87}}$$

$$= \boxed{19.10\ \text{ft}\quad(20\ \text{ft})}$$

The answer is (B).

(d) For C to D,

$$h_{f,\text{C-D}} = \frac{(10.44)(30{,}000\ \text{ft})\left(120\ \frac{\text{gal}}{\text{min}}\right)^{1.85}}{(150)^{1.85}(4\ \text{in})^{4.87}}$$

$$= \boxed{242.4\ \text{ft}\quad(240\ \text{ft})}$$

The answer is (A).

(e) Assume a pressure of 20 psig at point A.

$$h_{p,\text{A}} = \frac{\left(20\ \frac{\text{lbf}}{\text{in}^2}\right)\left(12\ \frac{\text{in}}{\text{ft}}\right)^2}{62.4\ \frac{\text{lbf}}{\text{ft}^3}}$$

$$= 46.2\ \text{ft}$$

Fluids

From the Bernoulli equation, ignoring velocity head,

$$h_{p,\mathrm{A}} + z_\mathrm{A} = h_{p,\mathrm{B}} + z_\mathrm{B} + h_{f,\text{A-B}}$$
$$46.2 \text{ ft} + 620 \text{ ft} = h_{p,\mathrm{B}} + 460 \text{ ft} + 22.4 \text{ ft}$$
$$h_{p,\mathrm{B}} = 183.8 \text{ ft}$$
$$p_\mathrm{B} = \gamma h_{p,\mathrm{B}} = \frac{\left(62.4 \frac{\text{lbf}}{\text{ft}^3}\right)(183.8 \text{ ft})}{\left(12 \frac{\text{in}}{\text{ft}}\right)^2} = \boxed{79.6 \text{ lbf/in}^2 \quad (80 \text{ psig})}$$

The answer is (C).

(f) For B to C,

$$h_{p,\mathrm{B}} + z_\mathrm{B} = h_{p,\mathrm{C}} + z_\mathrm{C} + h_{f,\text{B-C}}$$
$$183.8 \text{ ft} + 460 \text{ ft} = h_{p,\mathrm{C}} + 540 \text{ ft} + 19.10 \text{ ft}$$
$$h_{p,\mathrm{C}} = 84.7 \text{ ft}$$
$$p_\mathrm{C} = \gamma h_{p,\mathrm{C}} = \frac{\left(62.4 \frac{\text{lbf}}{\text{ft}^3}\right)(84.7 \text{ ft})}{\left(12 \frac{\text{in}}{\text{ft}}\right)^2} = \boxed{36.7 \text{ lbf/in}^2 \quad (40 \text{ psig})}$$

The answer is (C).

(g) For C to D,

$$h_{p,\mathrm{C}} + z_\mathrm{C} = h_{p,\mathrm{D}} + z_\mathrm{D} + h_{f,\text{C-D}}$$
$$84.7 \text{ ft} + 540 \text{ ft} = h_{p,\mathrm{D}} + 360 \text{ ft} + 242.4 \text{ ft}$$
$$h_{p,\mathrm{D}} = 22.3 \text{ ft}$$
$$p_\mathrm{D} = \gamma h_{p,\mathrm{D}} = \frac{\left(62.4 \frac{\text{lbf}}{\text{ft}^3}\right)(22.3 \text{ ft})}{\left(12 \frac{\text{in}}{\text{ft}}\right)^2} = \boxed{9.7 \text{ lbf/in}^2 \quad (10 \text{ psig})}$$

The answer is (A).

(h) $p_\mathrm{D} = 9.7 \text{ lbf/in}^2$ is too low; therefore, add $20 \text{ lbf/in}^2 - 9.7 \text{ lbf/in}^2 = 10.3 \text{ lbf/in}^2$ (psig) to each point.

$$p_\mathrm{A} = 20.0 \frac{\text{lbf}}{\text{in}^2} + 10.3 \frac{\text{lbf}}{\text{in}^2} = \boxed{30.3 \text{ lbf/in}^2 \quad (30 \text{ psig})}$$
$$p_\mathrm{B} = 79.6 \frac{\text{lbf}}{\text{in}^2} + 10.3 \frac{\text{lbf}}{\text{in}^2} = 89.9 \text{ lbf/in}^2 \quad (\text{psig})$$
$$p_\mathrm{C} = 36.7 \frac{\text{lbf}}{\text{in}^2} + 10.3 \frac{\text{lbf}}{\text{in}^2} = 47.0 \text{ lbf/in}^2 \quad (\text{psig})$$
$$p_\mathrm{D} = 9.7 \frac{\text{lbf}}{\text{in}^2} + 10.3 \frac{\text{lbf}}{\text{in}^2} = 20.0 \text{ lbf/in}^2 \quad (\text{psig})$$

The answer is (D).

(i) The elevation of the hydraulic grade line above point D is the sum of the potential and static heads.

$$\begin{aligned}\Delta h_{\text{A-D}} &= z_\mathrm{A} - z_\mathrm{D} + h_{p,\mathrm{A}} - h_{p,\mathrm{D}} \\ &= z_\mathrm{A} - z_\mathrm{D} + \frac{p_\mathrm{A} - p_\mathrm{D}}{\gamma} \\ &= 620 \text{ ft} - 360 \text{ ft} \\ &\quad + \frac{\left(30.3 \frac{\text{lbf}}{\text{in}^2} - 20 \frac{\text{lbf}}{\text{in}^2}\right)\left(12 \frac{\text{in}}{\text{ft}}\right)^2}{62.4 \frac{\text{lbf}}{\text{ft}^3}} \\ &= \boxed{283.8 \text{ ft} \quad (280 \text{ ft})}\end{aligned}$$

The answer is (A).

6. The Darcy equation is applicable to fluids in the laminar and turbulent regions.

$$h_f = \frac{fL\mathrm{v}^2}{2Dg}$$

In laminar flow in circular pipes, the friction factor is

$$f = \frac{64}{\mathrm{Re}} = \frac{64\nu}{D\mathrm{v}}$$

Combining these two equations,

$$h_f = \frac{fL\mathrm{v}^2}{2Dg} = \frac{\left(\frac{64\nu}{D\mathrm{v}}\right)L\mathrm{v}^2}{2Dg} = \frac{32\nu L\mathrm{v}}{D^2 g}$$

This is the Hagen-Poiseuille equation.

Friction head loss in laminar flow is proportional to velocity.

The answer is (A).

7. From App. 17.D, the equivalent length of four flanged 3 in elbows is

$$L_{e,\text{elbows}} = (4)(2.6 \text{ ft}) = 10.4 \text{ ft}$$

From App. 17.D, the equivalent length of the open gate valve is 2.8 ft.

The equivalent length of the pipe without fittings is 50 ft.

The elevation change is 20 ft, but this distance is not an equivalent length of pipe. Taking terms from the Bernoulli equation,

$$\Delta z = \frac{fL_e \mathrm{v}^2}{2Dg}$$

$$L_e = \frac{2Dg\Delta z}{f\mathrm{v}^2}$$

The flow rate is not given, so the velocity must be estimated. Since the water is said to flow quietly, the velocity is most likely less than 10 ft/sec. Since the velocity is not known, an exact determination of the friction factor, *f*, is not possible. Use a value of 0.02, which is appropriate for steel pipe and turbulent flow. Using nominal values for a quick estimate, the equivalent length of pipe equal to the elevation drop is

$$L_e = \frac{2Dg\Delta z}{f\mathrm{v}^2} = \frac{(2)(3\text{ in})\left(32.2\ \frac{\text{ft}}{\text{sec}^2}\right)(20\text{ ft})}{(0.02)\left(10\ \frac{\text{ft}}{\text{sec}}\right)^2\left(12\ \frac{\text{in}}{\text{ft}}\right)}$$

$$= 161\text{ ft}$$

If the velocity was 15 ft/sec, the equivalent length would be 71 ft. If the velocity was less than 10 ft/sec, the equivalent length would be even larger than 161 ft. Using the actual diameter of the pipe would also increase the equivalent length.

The answer is (D).

8. The pumping power will be the same if the head loss due to friction is the same. From Eq. 17.29, the friction loss in a section of pipe is

$$h_f = \frac{10.44LQ^{1.85}}{C^{1.85}d^{4.87}}$$

The flow rate is the same in all pipe sections, as is the Hazen-Williams roughness coefficient. These terms and the constant term cancel out.

$$\frac{300\text{ ft}}{(18\text{ in})^{4.87}} + \frac{400\text{ ft}}{(14\text{ in})^{4.87}} = \frac{L}{(16\text{ in})^{4.87}}$$

$$L = \boxed{935\text{ ft}\quad(940\text{ ft})}$$

The answer is (D).

9. Using App. 17.E, the friction loss is approximately 10 ft per 1000 ft of pipe, or approximately $\boxed{1.0\text{ ft per }100\text{ ft}}$ of pipe.

The answer is (A).

10. The total flow rate is

$$\dot{V} = (600\text{ res})(2.5)\left(1.1\ \frac{\frac{\text{gal}}{\text{min}}}{\text{res}}\right) = 1650\text{ gpm}$$

Write the energy equation, Eq. 17.64, in terms of specific weight, disregarding the velocity head (which is assumed to be small and constant in the line).

$$\frac{p_1}{\gamma} + z_1 = \frac{p_2}{\gamma} + z_2 + \frac{10.44LQ^{1.85}}{C^{1.85}d^{4.87}}$$

$$\frac{p_{1,\text{psig}}\left(12\ \frac{\text{in}}{\text{ft}}\right)^2}{62.4\ \frac{\text{lbf}}{\text{ft}^3}} + 1000\text{ ft} = \frac{\left(60\ \frac{\text{lbf}}{\text{in}^2}\right)\left(12\ \frac{\text{in}}{\text{ft}}\right)^2}{62.4\ \frac{\text{lbf}}{\text{ft}^3}} + 850\text{ ft} + \frac{(10.44)(17{,}000\text{ ft})\left(1650\ \frac{\text{gal}}{\text{min}}\right)^{1.85}}{(140)^{1.85}(10\text{ in})^{4.87}}$$

$$p_1 = \boxed{94.5\text{ psig}\quad(95\text{ psig})}$$

The answer is (C).

11. *Customary U.S. Solution*

For 6 in schedule-40 pipe, the internal diameter, D, is 0.5054 ft. The internal area is 0.2006 ft^2.

The velocity, v, is calculated from the volumetric flow, $\dot{V}$, and the flow area, A, by

$$\mathrm{v} = \frac{\dot{V}}{A} = \frac{1.5\ \frac{\text{ft}^3}{\text{sec}}}{0.2006\text{ ft}^2} = 7.48\text{ ft/sec}$$

Use App. 14.A. For water at 70°F, the kinematic viscosity, ν, is 1.059×10^{-5} ft^2/sec.

Calculate the Reynolds number.

$$\text{Re} = \frac{D\mathrm{v}}{\nu} = \frac{(0.5054\text{ ft})\left(7.48\ \frac{\text{ft}}{\text{sec}}\right)}{1.059\times 10^{-5}\ \frac{\text{ft}^2}{\text{sec}}} = 3.57\times 10^5$$

Since Re > 2100, the flow is turbulent. The friction loss coefficient can be determined from the Moody diagram.

For new steel pipe, the specific roughness, ϵ, is 0.0002 ft.

The relative roughness is

$$\frac{\epsilon}{D} = \frac{0.0002 \text{ ft}}{0.5054 \text{ ft}} = 0.0004$$

From the Moody diagram with Re = 3.57×10^5 and $\epsilon/D = 0.0004$, the friction factor, f, is 0.0174.

Use Darcy's equation to compute the frictional loss.

$$h_f = \frac{fLv^2}{2Dg} = \frac{(0.0174)(1200 \text{ ft})\left(7.48 \ \frac{\text{ft}}{\text{sec}}\right)^2}{(2)(0.5054 \text{ ft})\left(32.2 \ \frac{\text{ft}}{\text{sec}^2}\right)}$$
$$= \boxed{35.9 \text{ ft} \quad (36 \text{ ft})}$$

The answer is (C).

SI Solution

For 6 in pipe, the internal diameter is 154.1 mm, and the internal area is 186.5×10^{-4} m^2.

The velocity, v, is calculated from the volumetric flow, $\dot{V}$, and the flow area, A, by

$$v = \frac{\dot{V}}{A} = \frac{40 \ \frac{\text{L}}{\text{s}}}{(186.5 \times 10^{-4} \text{ m}^2)\left(1000 \frac{\text{L}}{\text{m}^3}\right)}$$
$$= 2.145 \text{ m/s}$$

From App. 14.B, for water at 20°C, the kinematic viscosity is

$$\nu = \frac{\mu}{\rho} = \frac{1.0050 \times 10^{-3} \text{ Pa·s}}{998.23 \ \frac{\text{kg}}{\text{m}^3}}$$
$$= 1.007 \times 10^{-6} \text{ m}^2/\text{s}$$

Calculate the Reynolds number.

$$\text{Re} = \frac{Dv}{\nu} = \frac{(154.1 \text{ mm})\left(2.145 \ \frac{\text{m}}{\text{s}}\right)}{\left(1.007 \times 10^{-6} \ \frac{\text{m}^2}{\text{s}}\right)\left(1000 \ \frac{\text{mm}}{\text{m}}\right)}$$
$$= 3.282 \times 10^5$$

Since Re > 2100, the flow is turbulent. The friction loss coefficient can be determined from the Moody diagram.

For new steel pipe, the specific roughness, ϵ, is 6.0×10^{-5} m.

The relative roughness is

$$\frac{\epsilon}{D} = \frac{6.0 \times 10^{-5} \text{ m}}{0.1541 \text{ m}} = 0.0004$$

From the Moody diagram with Re = 3.28×10^5 and $\epsilon/D = 0.0004$, the friction factor, f, is 0.0175.

Use Darcy's equation to compute the frictional loss.

$$h_f = \frac{fLv^2}{2Dg} = \frac{(0.0175)(355 \text{ m})\left(2.145 \ \frac{\text{m}}{\text{s}}\right)^2}{(2)(0.1541 \text{ m})\left(9.81 \ \frac{\text{m}}{\text{s}^2}\right)}$$
$$= \boxed{9.45 \text{ m} \quad (9.5 \text{ m})}$$

The answer is (C).

12. *Customary U.S. Solution*

For 6 in schedule-40 pipe, the internal diameter, D, is 0.5054 ft. The internal area is 0.2006 ft^2.

Convert the volumetric flow rate from gal/min to ft^3/sec.

$$\dot{V} = \frac{500 \ \frac{\text{gal}}{\text{min}}}{\left(60 \ \frac{\text{sec}}{\text{min}}\right)\left(7.4805 \ \frac{\text{gal}}{\text{ft}^3}\right)} = 1.114 \text{ ft}^3/\text{sec}$$

The velocity is

$$v = \frac{\dot{V}}{A} = \frac{1.114 \ \frac{\text{ft}^3}{\text{sec}}}{0.2006 \text{ ft}^2} = 5.55 \text{ ft/sec}$$

Use App. 14.A. For water at 100°F, the kinematic viscosity, ν, is 0.739×10^{-5} ft^2/sec, and the density is 62.00 lbm/ft^2.

Calculate the Reynolds number.

$$\text{Re} = \frac{Dv}{\nu} = \frac{(0.5054 \text{ ft})\left(5.55 \ \frac{\text{ft}}{\text{sec}}\right)}{0.739 \times 10^{-5} \ \frac{\text{ft}^2}{\text{sec}}}$$
$$= 3.80 \times 10^5$$

Since Re > 2100, the flow is turbulent. The friction loss coefficient can be determined from the Moody diagram.

For new steel pipe, the specific roughness, ϵ, is 0.0002 ft.

The relative roughness is

$$\frac{\epsilon}{D} = \frac{0.0002 \text{ ft}}{0.5054 \text{ ft}} = 0.0004$$

From the Moody diagram with Re = 3.80×10^5 and $\epsilon/D = 0.0004$, the friction factor, f, is 0.0173.

Use App. 17.D. The equivalent lengths of the valves and fittings are

standard radius elbow	2 × 8.9 ft =	17.8 ft
gate valve (fully open)	2 × 3.2 ft =	6.4 ft
90° angle valve (fully open)	1 × 63.0 ft =	63.0 ft
swing check valve	1 × 63.0 ft =	63.0 ft
		150.2 ft

The equivalent pipe length is the sum of the straight run of pipe and the equivalent length of pipe for the valves and fittings.

$$L_e = L + L_{\text{fittings}} = 300 \text{ ft} + 150.2 \text{ ft} = 450.2 \text{ ft}$$

Use Darcy's equation to compute the frictional loss.

$$h_f = \frac{fL_e\text{v}^2}{2Dg} = \frac{(0.0173)(450.2 \text{ ft})\left(5.55 \ \frac{\text{ft}}{\text{sec}}\right)^2}{(2)(0.5054 \text{ ft})\left(32.2 \ \frac{\text{ft}}{\text{sec}^2}\right)} = 7.37 \text{ ft}$$

The head loss is the sum of the head losses through the pipe, valves, and fittings and the change in elevation.

$$\Delta h = h_f + \Delta z = 7.37 \text{ ft} + 20 \text{ ft} = 27.37 \text{ ft}$$

The pressure difference between the entrance and discharge is

$$\Delta p = \gamma \Delta h = \rho \Delta h \times \frac{g}{g_c}$$

$$= \frac{\left(62.0 \ \frac{\text{lbm}}{\text{ft}^3}\right)(27.37 \text{ ft})}{\left(12 \ \frac{\text{in}}{\text{ft}}\right)^2} \times \frac{32.2 \ \frac{\text{ft}}{\text{sec}^2}}{32.2 \ \frac{\text{lbm-ft}}{\text{lbf-sec}^2}}$$

$$= \boxed{11.8 \text{ lbf/in}^2 \quad (12 \text{ psi})}$$

The answer is (A).

SI Solution

For 6 in pipe, the internal diameter is 154.1 mm (0.1541 m). The internal area is 186.5×10^{-4} m^2.

The velocity, v, is

$$\text{v} = \frac{\dot{V}}{A} = \frac{30 \ \frac{\text{L}}{\text{s}}}{(186.5 \times 10^{-4} \text{ m}^2)\left(1000 \ \frac{\text{L}}{\text{m}^3}\right)} = 1.61 \text{ m/s}$$

Use App. 14.B. For water at 40°C, the kinematic viscosity, ν, is 6.611×10^{-7} m^2/s, and the density is 992.25 kg/m^3.

Calculate the Reynolds number.

$$\text{Re} = \frac{D\text{v}}{\nu} = \frac{(0.1541 \text{ m})\left(1.61 \ \frac{\text{m}}{\text{s}}\right)}{6.611 \times 10^{-7} \ \frac{\text{m}^2}{\text{s}}} = 3.75 \times 10^5$$

Since Re > 2100, the flow is turbulent. The friction loss coefficient can be determined from the Moody diagram.

For new steel pipe, the specific roughness, ϵ, is 6.0×10^{-5} m.

The relative roughness is

$$\frac{\epsilon}{D} = \frac{6.0 \times 10^{-5} \text{ m}}{0.1541 \text{ m}} = 0.0004$$

From the Moody diagram with Re = 3.75×10^5 and $\epsilon/D = 0.0004$, the friction factor, f, is 0.0173.

Use App. 17.D. The equivalent lengths of the valves and fittings are

standard radius elbow	$2 \times (8.9 \text{ ft})\left(0.3048 \ \frac{\text{m}}{\text{ft}}\right) =$	5.4 m
gate valve (fully open)	$2 \times (3.2 \text{ ft})\left(0.3048 \ \frac{\text{m}}{\text{ft}}\right) =$	2.0 m
90° angle valve (fully open)	$1 \times (63.0 \text{ ft})\left(0.3048 \ \frac{\text{m}}{\text{ft}}\right) =$	19.2 m
swing check valve	$1 \times (63.0 \text{ ft})\left(0.3048 \ \frac{\text{m}}{\text{ft}}\right) =$	19.2 m
		45.8 m

The equivalent pipe length is the sum of the straight run of pipe and the equivalent length of pipe for the valves and fittings.

$$L_e = L + L_{\text{fittings}} = 90 \text{ m} + 45.8 \text{ m} = 135.8 \text{ m}$$

Use Darcy's equation to compute the frictional loss.

$$h_f = \frac{fL_e\text{v}^2}{2Dg} = \frac{(0.0173)(135.8\text{ m})\left(1.61\ \frac{\text{m}}{\text{s}}\right)^2}{(2)(0.1541\text{ m})\left(9.81\ \frac{\text{m}}{\text{s}^2}\right)}$$
$$= 2.01\text{ m}$$

The total head loss is the sum of the head losses through the pipe, valves, and fittings and the change in elevation.

$$\Delta h = h_f + \Delta z = 2.01\text{ m} + 6\text{ m}$$
$$= 8.01\text{ m}$$

The pressure difference between the entrance and discharge is

$$\Delta p = \rho\Delta hg = \left(992.25\ \frac{\text{kg}}{\text{m}^3}\right)(8.01\text{ m})\left(9.81\ \frac{\text{m}}{\text{s}^2}\right)$$
$$= \boxed{77\,969\text{ Pa}\quad(78\text{ kPa})}$$

The answer is (A).

13. *Customary U.S. Solution*

For 6 in schedule-40 pipe, the internal diameter, D, is 0.5054 ft. The internal area is 0.2006 ft^2.

Use App. 14.D. For air at 70°F and atmospheric pressure, the kinematic viscosity is $16.39 \times 10^{-5}\ \text{ft}^2/\text{sec}$.

Calculate the Reynolds number.

$$\text{Re} = \frac{D\text{v}}{\nu} = \frac{(0.5054\text{ ft})\left(60\ \frac{\text{ft}}{\text{sec}}\right)}{16.39 \times 10^{-5}\ \frac{\text{ft}^2}{\text{sec}}}$$
$$= 1.85 \times 10^5$$

Since Re > 2100, the flow is turbulent. The friction loss coefficient can be determined from the Moody diagram.

For new steel pipe, the specific roughness, ϵ, is 0.0002 ft.

The relative roughness is

$$\frac{\epsilon}{D} = \frac{0.0002\text{ ft}}{0.5054\text{ ft}} = 0.0004$$

From the Moody diagram with $\text{Re} = 1.85 \times 10^5$ and $\epsilon/D = 0.0004$, the friction factor, f, is 0.0184.

Use App. 17.D. The equivalent lengths of the valves and fittings are

standard radius elbow	2 × 8.9 ft =	17.8 ft
gate valve (fully open)	2 × 3.2 ft =	6.4 ft
90° angle valve (fully open)	1 × 63.0 ft =	63.0 ft
swing check valve	1 × 63.0 ft =	63.0 ft
		150.2 ft

The equivalent pipe length is the sum of the straight run of pipe and the equivalent lengths of pipe for the valves and fittings.

$$L_e = L + L_{\text{fittings}} = 300\text{ ft} + 150.2\text{ ft}$$
$$= 450.2\text{ ft}$$

Use Darcy's equation to compute the frictional loss.

$$h_f = \frac{fL_e\text{v}^2}{2Dg} = \frac{(0.0184)(450.2\text{ ft})\left(60\ \frac{\text{ft}}{\text{sec}}\right)^2}{(2)(0.5054\text{ ft})\left(32.2\ \frac{\text{ft}}{\text{sec}^2}\right)}$$
$$= 916.2\text{ ft}$$

The head loss is the sum of the head losses through the pipe, valves, and fittings and the change in elevation.

$$\Delta h = h_f + \Delta z = 916.2\text{ ft} + 20\text{ ft}$$
$$= 936.2\text{ ft}$$

The pressure difference between the entrance and discharge is

$$\Delta p = \gamma\Delta h = \rho\Delta h \times \frac{g}{g_c}$$
$$= \frac{\left(0.075\ \frac{\text{lbm}}{\text{ft}^3}\right)(936.2\text{ ft})}{\left(12\ \frac{\text{in}}{\text{ft}}\right)^2} \times \frac{32.2\ \frac{\text{ft}}{\text{sec}^2}}{32.2\ \frac{\text{lbm-ft}}{\text{lbf-sec}^2}}$$
$$= \boxed{0.49\text{ lbf/in}^2\quad(0.49\text{ psi})}$$

The answer is (B).

SI Solution

Use App. 16.C. For 6 in pipe, the internal diameter, D, is 154.1 mm (0.1541 m), and the internal area is $186.5 \times 10^{-4}\ \text{m}^2$.

Use App. 14.E. For air at 20°C, the kinematic viscosity, ν, is $1.512 \times 10^{-5}\ \text{m}^2/\text{s}$.

Calculate the Reynolds number.

$$\mathrm{Re} = \frac{D\mathrm{v}}{\nu} = \frac{(0.1541\ \mathrm{m})\left(18\ \frac{\mathrm{m}}{\mathrm{s}}\right)}{1.512 \times 10^{-5}\ \frac{\mathrm{m}^2}{\mathrm{s}}} = 1.83 \times 10^5$$

Since Re > 2100, the flow is turbulent. The friction loss coefficient can be determined from the Moody diagram.

For new steel pipe, the specific roughness, ϵ, is 6.0×10^{-5} m.

The relative roughness is

$$\frac{\epsilon}{D} = \frac{6.0 \times 10^{-5}}{0.1541\ \mathrm{m}} = 0.0004$$

From the Moody diagram with Re = 1.83×10^5 and $\epsilon/D = 0.0004$, the friction factor, *f*, is 0.0185.

Compute the equivalent lengths of the valves and fittings. (Convert from App. 17.D.)

standard radius elbow	$2 \times (8.9\ \mathrm{ft})\left(0.3048\ \frac{\mathrm{m}}{\mathrm{ft}}\right) =$	5.4 m
gate valve (fully open)	$2 \times (3.2\ \mathrm{ft})\left(0.3048\ \frac{\mathrm{m}}{\mathrm{ft}}\right) =$	2.0 m
90° angle valve (fully open)	$1 \times (63.0\ \mathrm{ft})\left(0.3048\ \frac{\mathrm{m}}{\mathrm{ft}}\right) =$	19.2 m
swing check valve	$1 \times (63.0\ \mathrm{ft})\left(0.3048\ \frac{\mathrm{m}}{\mathrm{ft}}\right) =$	19.2 m
		45.8 m

The equivalent pipe length is the sum of the straight run of pipe and the equivalent lengths of pipe for the valves and fittings.

$$L_e = L + L_{\mathrm{fittings}} = 90\ \mathrm{m} + 45.8\ \mathrm{m} = 135.8\ \mathrm{m}$$

Use Darcy's equation to compute the frictional loss.

$$h_f = \frac{fL_e\mathrm{v}^2}{2Dg} = \frac{(0.0185)(135.8\ \mathrm{m})\left(18\ \frac{\mathrm{m}}{\mathrm{s}}\right)^2}{(2)(0.1541\ \mathrm{m})\left(9.81\ \frac{\mathrm{m}}{\mathrm{s}^2}\right)} = 269.2\ \mathrm{m}$$

The head loss is the sum of the head losses through the pipe, valves, and fittings and the change in elevation.

$$\Delta h = h_f + \Delta z = 269.2\ \mathrm{m} + 6\ \mathrm{m} = 275.2\ \mathrm{m}$$

The density of the air, ρ, is approximately 1.20 kg/m^3.

The pressure difference between the entrance and discharge is

$$\Delta p = \rho\Delta hg = \left(1.20\ \frac{\mathrm{kg}}{\mathrm{m}^3}\right)(275.2\ \mathrm{m})\left(9.81\ \frac{\mathrm{m}}{\mathrm{s}^2}\right) = \boxed{3240\ \mathrm{Pa}\quad(3.2\ \mathrm{kPa})}$$

The answer is (B).

14. Assume that flows from reservoirs A and B are toward D and then toward C. From continuity,

$$\dot{V}_{\text{A-D}} + \dot{V}_{\text{B-D}} = \dot{V}_{\text{D-C}}$$

$$A_\mathrm{A}\mathrm{v}_{\text{A-D}} + A_\mathrm{B}\mathrm{v}_{\text{B-D}} - A_\mathrm{C}\mathrm{v}_{\text{D-C}} = 0$$

From App. 16.B, for schedule-40 pipe,

$$A_\mathrm{A} = 0.05134\ \mathrm{ft}^2 \qquad D_\mathrm{A} = 0.2557\ \mathrm{ft}$$
$$A_\mathrm{B} = 0.5476\ \mathrm{ft}^2 \qquad D_\mathrm{B} = 0.8350\ \mathrm{ft}$$
$$A_\mathrm{C} = 0.08841\ \mathrm{ft}^2 \qquad D_\mathrm{C} = 0.3355\ \mathrm{ft}$$

$$0.05134\mathrm{v}_{\text{A-D}} + 0.5476\mathrm{v}_{\text{B-D}} - 0.08841\mathrm{v}_{\text{D-C}} = 0 \quad \text{[Eq. I]}$$

Ignoring the velocity heads, the conservation of energy equation between A and D is

$$z_\mathrm{A} = \frac{p_\mathrm{D}}{\gamma} + z_\mathrm{D} + h_{f,\text{A-D}}$$

$$50\ \mathrm{ft} = \frac{p_\mathrm{D}}{62.4\ \frac{\mathrm{lbf}}{\mathrm{ft}^3}} + 25\ \mathrm{ft} + \frac{(0.02)(800\ \mathrm{ft})\mathrm{v}_{\text{A-D}}^2}{(2)(0.2557\ \mathrm{ft})\left(32.2\ \frac{\mathrm{ft}}{\mathrm{sec}^2}\right)}$$

$$\mathrm{v}_{\text{A-D}} = \mathrm{v}_{\text{A-D}} = \sqrt{25.73 - 0.0165p_\mathrm{D}} \quad \text{[Eq. II]}$$

Similarly, for B–D,

$$40\ \mathrm{ft} = \frac{p_\mathrm{D}}{62.4\ \frac{\mathrm{lbf}}{\mathrm{ft}^3}} + 25\ \mathrm{ft} + \frac{(0.02)(500\ \mathrm{ft})\mathrm{v}_{\text{B-D}}^2}{(2)(0.8350\ \mathrm{ft})\left(32.2\ \frac{\mathrm{ft}}{\mathrm{sec}^2}\right)}$$

$$\mathrm{v}_{\text{B-D}} = \sqrt{80.66 - 0.0862p_\mathrm{D}} \quad \text{[Eq. III]}$$

For D–C,

$$22\ \mathrm{ft} = \frac{p_\mathrm{D}}{62.4\ \frac{\mathrm{lbf}}{\mathrm{ft}^3}} + 25\ \mathrm{ft} - \frac{(0.02)(1000\ \mathrm{ft})\mathrm{v}_{\text{D-C}}^2}{(2)(0.3355\ \mathrm{ft})\left(32.2\ \frac{\mathrm{ft}}{\mathrm{sec}^2}\right)}$$

$$\mathrm{v}_{\text{D-C}} = \sqrt{3.24 + 0.0173p_\mathrm{D}} \quad \text{[Eq. IV]}$$

Equations I, II, III, and IV must be solved simultaneously. To do this, assume a value for p_D. This value then determines all three velocities in Eqs. II, III, and IV. These velocities are substituted into Eq. I. A trial and error solution yields

$$\text{v}_{\text{A-D}} = 3.21 \text{ ft/sec}$$

$$\text{v}_{\text{B-D}} = 0.408 \text{ ft/sec}$$

$$\text{v}_{\text{D-C}} = 4.40 \text{ ft/sec}$$

$$\boxed{p_D = 933.8 \text{ lbf/ft}^2 \quad (930 \text{ psf})}$$

$$\boxed{\text{Flow is from B to D.}}$$

The answer is (B).

15. *Customary U.S. Solution*

(a) For water at 70°F, $\rho = 62.3 \text{ lbm/ft}^3$, and $E_{\text{water}} = 320 \times 10^3 \text{ lbf/in}^2$.

For cast-iron pipe, $E_{\text{pipe}} = 20 \times 10^6 \text{ lbf/in}^2$. From Eq. 17.210, the composite modulus of elasticity of the pipe and water is

$$E = \frac{E_{\text{water}} t_{\text{pipe}} E_{\text{pipe}}}{t_{\text{pipe}} E_{\text{pipe}} + c_P D_{\text{pipe}} E_{\text{water}}}$$

$$= \frac{\left(320 \times 10^3 \ \frac{\text{lbf}}{\text{in}^2}\right)(0.75 \text{ in})\left(20 \times 10^6 \ \frac{\text{lbf}}{\text{in}^2}\right)}{(0.75 \text{ in})\left(20 \times 10^6 \ \frac{\text{lbf}}{\text{in}^2}\right) + (1)(24 \text{ in})\left(320 \times 10^3 \ \frac{\text{lbf}}{\text{in}^2}\right)}$$

$$= \boxed{2.12 \times 10^5 \text{ lbf/in}^2 \quad (2.1 \times 10^5 \text{ lbf/in}^2)}$$

The answer is (A).

(b) Using the value found in part (a) and from Eq. 17.210, the speed of sound in the pipe is

$$a = \sqrt{\frac{E g_c}{\rho}}$$

$$= \sqrt{\frac{\left(2.12 \times 10^5 \ \frac{\text{lbf}}{\text{in}^2}\right)\left(12 \ \frac{\text{in}}{\text{ft}}\right)^2\left(32.2 \ \frac{\text{lbm-ft}}{\text{lbf-sec}^2}\right)}{62.3 \ \frac{\text{lbm}}{\text{ft}^3}}}$$

$$= \boxed{3972 \text{ ft/sec} \quad (4000 \text{ ft/sec})}$$

The answer is (C).

(c) The maximum pressure is given by Eq. 17.208(b).

$$\Delta p = \frac{\rho a \Delta \text{v}}{g_c}$$

$$= \frac{\left(62.3 \ \frac{\text{lbm}}{\text{ft}^3}\right)\left(3972 \ \frac{\text{ft}}{\text{sec}}\right)\left(6 \ \frac{\text{ft}}{\text{sec}}\right)}{\left(32.2 \ \frac{\text{lbm-ft}}{\text{lbf-sec}^2}\right)\left(12 \ \frac{\text{in}}{\text{ft}}\right)^2}$$

$$= \boxed{320.2 \text{ lbf/in}^2 \quad (320 \text{ psi})}$$

The answer is (C).

(d) The length of time the pressure is constant at the valve is

$$t = \frac{2L}{a} = \frac{(2)(500 \text{ ft})}{3972 \ \frac{\text{ft}}{\text{sec}}} = \boxed{0.25 \text{ sec}}$$

The answer is (A).

SI Solution

(a) For water at 20°C, $\rho = 998.2 \text{ kg/m}^3$, and $E_{\text{water}} = 2.2 \times 10^9 \text{ Pa}$.

For cast-iron pipe, $E_{\text{pipe}} = 1.4 \times 10^{11} \text{ Pa}$. From Eq. 17.210, the composite modulus of elasticity of the pipe and water is

$$E = \frac{E_{\text{water}} t_{\text{pipe}} E_{\text{pipe}}}{t_{\text{pipe}} E_{\text{pipe}} + c_P D_{\text{pipe}} E_{\text{water}}}$$

$$= \frac{(2.2 \times 10^9 \text{ Pa})(0.02 \text{ m})(1.4 \times 10^{11} \text{ Pa})}{(0.02 \text{ m})(1.4 \times 10^{11} \text{ Pa}) + (1)(0.6 \text{ m})(2.2 \times 10^9 \text{ Pa})}$$

$$= \boxed{1.5 \times 10^9 \text{ Pa}}$$

The answer is (A).

(b) Using the value found in part (a) and from Eq. 17.210, the speed of sound in the pipe is

$$a = \sqrt{\frac{E}{\rho}} = \sqrt{\frac{1.50 \times 10^9 \text{ Pa}}{998.2 \ \frac{\text{kg}}{\text{m}^3}}}$$

$$= \boxed{1226 \text{ m/s} \quad (1200 \text{ m/s})}$$

The answer is (C).

(c) The maximum pressure is given by Eq. 17.208(a).

$$\Delta p = \rho a \Delta \text{v}$$

$$= \left(998.2 \ \frac{\text{kg}}{\text{m}^3}\right)\left(1226 \ \frac{\text{m}}{\text{s}}\right)\left(2 \ \frac{\text{m}}{\text{s}}\right)$$

$$= \boxed{2.45 \times 10^6 \text{ Pa} \quad (2.5 \text{ MPa})}$$

The answer is (C).

(d) The length of time the pressure is constant at the valve is

$$t = \frac{2L}{a} = \frac{(2)(150 \text{ m})}{1225 \ \frac{\text{m}}{\text{s}}} = \boxed{0.25 \text{ s}}$$

The answer is (A).

16. *Customary U.S. Solution*

(a) First, it is necessary to collect data on schedule-40 pipe and water. The fluid viscosity, pipe dimensions, and other parameters can be found in various appendices in Chap. 14 and Chap. 16. At 70°F water, $\nu = 1.059 \times 10^{-5} \text{ ft}^2/\text{sec}$.

From Table 17.2, $\epsilon = 0.0002$ ft. From App. 16.B,

$$\begin{aligned} &\text{8 in pipe} && D = 0.6651 \text{ ft} \\ & && A = 0.3474 \text{ ft}^2 \\ &\text{12 in pipe} && D = 0.9948 \text{ ft} \\ & && A = 0.7773 \text{ ft}^2 \\ &\text{16 in pipe} && D = 1.25 \text{ ft} \\ & && A = 1.2272 \text{ ft}^2 \end{aligned}$$

The flow quantity is converted from gallons per minute to cubic feet per second.

$$\dot{V} = \frac{(8 \text{ MGD})\left(10^6 \ \frac{\frac{\text{gal}}{\text{day}}}{\text{MGD}}\right)}{\left(24 \ \frac{\text{hr}}{\text{day}}\right)\left(60 \ \frac{\text{min}}{\text{hr}}\right)\left(7.4805 \ \frac{\text{gal}}{\text{ft}^3}\right)\left(60 \ \frac{\text{sec}}{\text{min}}\right)} = 12.378 \text{ ft}^3/\text{sec}$$

For the inlet pipe, the velocity is

$$\text{v} = \frac{\dot{V}}{A} = \frac{12.378 \ \frac{\text{ft}^3}{\text{sec}}}{0.3474 \text{ ft}^2} = 35.63 \text{ ft/sec}$$

The Reynolds number is

$$\text{Re} = \frac{D\text{v}}{\nu} = \frac{(0.6651 \text{ ft})\left(35.63 \ \frac{\text{ft}}{\text{sec}}\right)}{1.059 \times 10^{-5} \ \frac{\text{ft}^2}{\text{sec}}} = 2.24 \times 10^6$$

The relative roughness is

$$\frac{\epsilon}{D} = \frac{0.0002 \text{ ft}}{0.6651 \text{ ft}} = 0.0003$$

From the Moody diagram, $f = 0.015$.

Equation 17.23(b) is used to calculate the frictional energy loss.

$$\begin{aligned} E_{f,1} &= h_f \times \frac{g}{g_c} = \frac{fL\text{v}^2}{2Dg_c} \\ &= \frac{(0.015)(1000 \text{ ft})\left(35.63 \ \frac{\text{ft}}{\text{sec}}\right)^2}{(2)(0.6651 \text{ ft})\left(32.2 \ \frac{\text{lbm-ft}}{\text{lbf-sec}^2}\right)} \\ &= 444.6 \text{ ft-lbf/lbm} \end{aligned}$$

For the outlet pipe, the velocity is

$$\text{v} = \frac{\dot{V}}{A} = \frac{12.378 \ \frac{\text{ft}^3}{\text{sec}}}{0.7773 \text{ ft}^2} = 15.92 \text{ ft/sec}$$

The Reynolds number is

$$\text{Re} = \frac{D\text{v}}{\nu} = \frac{(0.9948 \text{ ft})\left(15.92 \ \frac{\text{ft}}{\text{sec}}\right)}{1.059 \times 10^{-5} \ \frac{\text{ft}^2}{\text{sec}}} = 1.5 \times 10^6$$

The relative roughness is

$$\frac{\epsilon}{D} = \frac{0.0002 \text{ ft}}{0.9948 \text{ ft}} = 0.0002$$

From the Moody diagram, $f = 0.014$.

Equation 17.23(b) is used to calculate the frictional energy loss.

$$\begin{aligned} E_{f,2} &= h_f \times \frac{g}{g_c} = \frac{fL\text{v}^2}{2Dg_c} \\ &= \frac{(0.014)(1500 \text{ ft})\left(15.92 \ \frac{\text{ft}}{\text{sec}}\right)^2}{(2)(0.9948 \text{ ft})\left(32.2 \ \frac{\text{lbm-ft}}{\text{lbf-sec}^2}\right)} \\ &= 83.1 \text{ ft-lbf/lbm} \end{aligned}$$

Assume a 50% split through the two branches. In the upper branch, the velocity is

$$\text{v} = \frac{\dot{V}}{A} = \frac{\left(\frac{1}{2}\right)\left(12.378 \ \frac{\text{ft}^3}{\text{sec}}\right)}{0.3474 \text{ ft}^2} = 17.82 \text{ ft/sec}$$

Fluids

The Reynolds number is

$$\mathrm{Re} = \frac{D\mathrm{v}}{\nu} = \frac{(0.6651\ \mathrm{ft})\left(17.82\ \frac{\mathrm{ft}}{\mathrm{sec}}\right)}{1.059\times10^{-5}\ \frac{\mathrm{ft}^2}{\mathrm{sec}}} = 1.1\times10^{6}$$

The relative roughness is

$$\frac{\epsilon}{D} = \frac{0.0002\ \mathrm{ft}}{0.6651\ \mathrm{ft}} = 0.0003$$

From the Moody diagram, $f = 0.015$.

For the 16 in pipe in the lower branch, the velocity is

$$\mathrm{v} = \frac{\dot{V}}{A} = \frac{\left(\frac{1}{2}\right)\left(12.378\ \frac{\mathrm{ft}^3}{\mathrm{sec}}\right)}{1.2272\ \mathrm{ft}^2} = 5.04\ \mathrm{ft/sec}$$

The Reynolds number is

$$\mathrm{Re} = \frac{D\mathrm{v}}{\nu} = \frac{(1.25\ \mathrm{ft})\left(5.04\ \frac{\mathrm{ft}}{\mathrm{sec}}\right)}{1.059\times10^{-5}\ \frac{\mathrm{ft}^2}{\mathrm{sec}}} = 5.95\times10^{5}$$

The relative roughness is

$$\frac{\epsilon}{D} = \frac{0.0002\ \mathrm{ft}}{1.25\ \mathrm{ft}} = 0.00016$$

From the Moody diagram, $f = 0.015$.

These values of f for the two branches are fairly insensitive to changes in $\dot{V}$, so they will be used for the rest of the problem in both branches.

Equation 17.23(b) is used to calculate the frictional energy loss in the upper branch.

$$\begin{aligned}E_{f,\mathrm{upper}} &= h_f\times\frac{g}{g_c} = \frac{fL\mathrm{v}^2}{2Dg_c}\\ &= \frac{(0.015)(500\ \mathrm{ft})\left(17.81\ \frac{\mathrm{ft}}{\mathrm{sec}}\right)^2}{(2)(0.6651\ \mathrm{ft})\left(32.2\ \frac{\mathrm{lbm\text{-}ft}}{\mathrm{lbf\text{-}sec}^2}\right)}\\ &= 55.5\ \mathrm{ft\text{-}lbf/lbm}\end{aligned}$$

To calculate a loss for any other flow in the upper branch,

$$\begin{aligned}E_{f,\mathrm{upper}\,2} &= E_{f,\mathrm{upper}}\left(\frac{\dot{V}}{\left(\frac{1}{2}\right)\left(12.378\ \frac{\mathrm{ft}^3}{\mathrm{sec}}\right)}\right)^2\\ &= \left(55.5\ \frac{\mathrm{ft\text{-}lbf}}{\mathrm{lbm}}\right)\left(\frac{\dot{V}}{6.189\ \frac{\mathrm{ft}^3}{\mathrm{sec}}}\right)^2\\ &= 1.45\,\dot{V}^2\end{aligned}$$

Similarly, for the lower branch, in the 8 in section,

$$\begin{aligned}E_{f,\mathrm{lower,8\,in}} &= h_f\times\frac{g}{g_c} = \frac{fL\mathrm{v}^2}{2Dg_c}\\ &= \frac{(0.015)(250\ \mathrm{ft})\left(17.81\ \frac{\mathrm{ft}}{\mathrm{sec}}\right)^2}{(2)(0.6651\ \mathrm{ft})\left(32.2\ \frac{\mathrm{lbm\text{-}ft}}{\mathrm{lbf\text{-}sec}^2}\right)}\\ &= 27.8\ \mathrm{ft\text{-}lbf/lbm}\end{aligned}$$

For the lower branch, in the 16 in section,

$$\begin{aligned}E_{f,\mathrm{lower,16\,in}} &= h_f\times\frac{g}{g_c} = \frac{fL\mathrm{v}^2}{2Dg_c}\\ &= \frac{(0.015)(1000\ \mathrm{ft})\left(5.04\ \frac{\mathrm{ft}}{\mathrm{sec}}\right)^2}{(2)(1.25\ \mathrm{ft})\left(32.2\ \frac{\mathrm{lbm\text{-}ft}}{\mathrm{lbf\text{-}sec}^2}\right)}\\ &= 4.7\ \mathrm{ft\text{-}lbf/lbm}\end{aligned}$$

The total loss in the lower branch is

$$\begin{aligned}E_{f,\mathrm{lower}} &= E_{f,\mathrm{lower,8\,in}} + E_{f,\mathrm{lower,16\,in}}\\ &= 27.8\ \frac{\mathrm{ft\text{-}lbf}}{\mathrm{lbm}} + 4.7\ \frac{\mathrm{ft\text{-}lbf}}{\mathrm{lbm}}\\ &= 32.5\ \mathrm{ft\text{-}lbf/lbm}\end{aligned}$$

To calculate a loss for any other flow in the lower branch,

$$\begin{aligned}E_{f,\mathrm{lower}\,2} &= E_{f,\mathrm{lower}}\left(\frac{\dot{V}}{\left(\frac{1}{2}\right)\left(12.378\ \frac{\mathrm{ft}^3}{\mathrm{sec}}\right)}\right)^2\\ &= \left(32.5\ \frac{\mathrm{ft\text{-}lbf}}{\mathrm{lbm}}\right)\left(\frac{\dot{V}}{6.189\ \frac{\mathrm{ft}^3}{\mathrm{sec}}}\right)^2\\ &= 0.85\,\dot{V}^2\end{aligned}$$

Let x be the fraction flowing in the upper branch. Then, because the friction losses are equal,

$$E_{f,\text{upper}\,2} = E_{f,\text{lower}\,2}$$
$$1.45x^2 = (0.85)(1-x)^2$$
$$x = 0.434$$
$$\dot{V}_{\text{upper}} = (0.434)\left(12.378\ \frac{\text{ft}^3}{\text{sec}}\right)$$
$$= \boxed{5.372\ \text{ft}^3/\text{sec}\quad(5.4\ \text{ft}^3/\text{sec})}$$

The answer is (D).

(b)
$$\dot{V}_{\text{lower}} = (1-0.432)\left(12.378\ \frac{\text{ft}^3}{\text{sec}}\right)$$
$$= 7.03\ \text{ft}^3/\text{sec}$$
$$E_{f,\text{total}} = E_{f,1} + E_{f,\text{lower}\,2} + E_{f,2}$$
$$E_{f,\text{lower}\,2} = 0.85\,\dot{V}^2_{\text{lower}}$$
$$= (0.85)\left(7.03\ \frac{\text{ft}^3}{\text{sec}}\right)^2$$
$$= 42.0\ \text{ft}$$
$$E_{f,\text{total}} = 444.6\ \frac{\text{ft-lbf}}{\text{lbm}} + 42.0\ \frac{\text{ft-lbf}}{\text{lbm}} + 83.1\ \frac{\text{ft-lbf}}{\text{lbm}}$$
$$= \boxed{569.7\ \text{ft-lbf/lbm}\quad(570\ \text{ft-lbf/lbm})}$$

The answer is (D).

SI Solution

(a) First, it is necessary to collect data on schedule-40 pipe and water. The fluid viscosity, pipe dimensions, and other parameters can be found in various appendices in Chap. 14 and Chap. 16. At 20°C water, $\nu = 1.007 \times 10^{-6}\ \text{m}^2/\text{s}$.

From Table 17.2, $\epsilon = 6 \times 10^{-5}$ m. From App. 16.C,

8 in pipe	$D = 202.7$ mm
	$A = 322.75\ \text{cm}^2$
12 in pipe	$D = 303.2$ mm
	$A = 721.9 \times 10^{-4}\ \text{m}^2$
16 in pipe	$D = 381$ mm
	$A = 1140 \times 10^{-4}\ \text{m}^2$

For the inlet pipe, the velocity is

$$\text{v} = \frac{\dot{V}}{A} = \frac{\left(350\ \frac{\text{L}}{\text{s}}\right)\left(100\ \frac{\text{cm}}{\text{m}}\right)^2}{(322.75\ \text{cm}^2)\left(1000\ \frac{\text{L}}{\text{m}^3}\right)} = 10.85\ \text{m/s}$$

The Reynolds number is

$$\text{Re} = \frac{D\text{v}}{\nu} = \frac{(0.2027\ \text{m})\left(10.85\ \frac{\text{m}}{\text{s}}\right)}{1.007 \times 10^{-6}\ \frac{\text{m}^2}{\text{s}}}$$
$$= 2.18 \times 10^6$$

The relative roughness is

$$\frac{\epsilon}{D} = \frac{6 \times 10^{-5}\ \text{m}}{0.2027\ \text{m}} = 0.0003$$

From the Moody diagram, $f = 0.015$.

Equation 17.23(a) is used to calculate the frictional energy loss.

$$E_{f,1} = h_f g = \frac{fL\text{v}^2}{2D}$$
$$= \frac{(0.015)(300\ \text{m})\left(10.85\ \frac{\text{m}}{\text{s}}\right)^2}{(2)(0.2027\ \text{m})}$$
$$= 1307\ \text{J/kg}$$

For the outlet pipe, the velocity is

$$\text{v} = \frac{\dot{V}}{A} = \frac{350\ \frac{\text{L}}{\text{s}}}{(721.9 \times 10^{-4}\ \text{m}^2)\left(1000\ \frac{\text{L}}{\text{m}^3}\right)} = 4.848\ \text{m/s}$$

The Reynolds number is

$$\text{Re} = \frac{D\text{v}}{\nu} = \frac{(0.3032\ \text{m})\left(4.848\ \frac{\text{m}}{\text{s}}\right)}{1.007 \times 10^{-6}\ \frac{\text{m}^2}{\text{s}}}$$
$$= 1.46 \times 10^6$$

The relative roughness is

$$\frac{\epsilon}{D} = \frac{6 \times 10^{-5}\ \text{m}}{0.3032\ \text{m}} = 0.0002$$

From the Moody diagram, $f = 0.014$.

Equation 17.23(a) is used to calculate the frictional energy loss.

$$E_{f,2} = h_f g = \frac{fL\text{v}^2}{2D}$$
$$= \frac{(0.014)(450\ \text{m})\left(4.848\ \frac{\text{m}}{\text{s}}\right)^2}{(2)(0.3032\ \text{m})}$$
$$= 244.2\ \text{J/kg}$$

Fluids

Assume a 50% split through the two branches. In the upper branch, the velocity is

$$\text{v} = \frac{\dot{V}}{A} = \frac{\left(\frac{1}{2}\right)\left(350\ \frac{\text{L}}{\text{s}}\right)}{(322.7 \times 10^{-4}\ \text{m}^2)\left(1000\ \frac{\text{L}}{\text{m}^3}\right)} = 5.423\ \text{m/s}$$

The Reynolds number is

$$\begin{aligned} \text{Re} &= \frac{D\text{v}}{\nu} = \frac{(0.2027\ \text{m})\left(5.423\ \frac{\text{m}}{\text{s}}\right)}{1.007 \times 10^{-6}\ \frac{\text{m}^2}{\text{s}}} \\ &= 1.1 \times 10^6 \end{aligned}$$

The relative roughness is

$$\frac{\epsilon}{D} = \frac{6 \times 10^{-5}\ \text{m}}{0.2027\ \text{m}} = 0.0003$$

From the Moody diagram, $f = 0.015$.

For the 16 in pipe in the lower branch, the velocity is

$$\text{v} = \frac{\dot{V}}{A} = \frac{\left(\frac{1}{2}\right)\left(350\ \frac{\text{L}}{\text{s}}\right)}{(1140 \times 10^{-4}\ \text{m}^2)\left(1000\ \frac{\text{L}}{\text{m}^3}\right)} = 1.535\ \text{m/s}$$

The Reynolds number is

$$\begin{aligned} \text{Re} &= \frac{D\text{v}}{\nu} = \frac{(0.381\ \text{m})\left(1.535\ \frac{\text{m}}{\text{s}}\right)}{1.007 \times 10^{-6}\ \frac{\text{m}^2}{\text{s}}} \\ &= 5.81 \times 10^5 \end{aligned}$$

The relative roughness is

$$\frac{\epsilon}{D} = \frac{6 \times 10^{-5}\ \text{m}}{0.381\ \text{m}} = 0.00016$$

From the Moody diagram, $f = 0.015$.

These values of f for the two branches are fairly insensitive to changes in $\dot{V}$, so they will be used for the rest of the problem in both branches.

Equation 17.23(a) is used to calculate the frictional energy loss in the upper branch.

$$\begin{aligned} E_{f,\text{upper}} &= h_f g = \frac{fL\text{v}^2}{2D} \\ &= \frac{(0.015)(150\ \text{m})\left(5.423\ \frac{\text{m}}{\text{s}}\right)^2}{(2)(0.2027\ \text{m})} \\ &= 163.2\ \text{J/kg} \end{aligned}$$

To calculate a loss for any other flow in the upper branch,

$$\begin{aligned} E_{f,\text{upper 2}} &= E_{f,\text{upper}}\left(\frac{\dot{V}}{\left(\frac{1}{2}\right)\left(0.350\ \frac{\text{m}^3}{\text{s}}\right)}\right)^2 \\ &= \left(163.2\ \frac{\text{J}}{\text{kg}}\right)\left(\frac{\dot{V}}{0.175\ \frac{\text{m}^3}{\text{s}}}\right)^2 \\ &= 5329\dot{V}^2 \end{aligned}$$

Similarly, for the lower branch, in the 8 in section,

$$\begin{aligned} E_{f,\text{lower,8 in}} &= h_f g = \frac{fL\text{v}^2}{2D} = \frac{(0.015)(75\ \text{m})\left(5.423\ \frac{\text{m}}{\text{s}}\right)^2}{(2)(0.2027\ \text{m})} \\ &= 81.61\ \text{J/kg} \end{aligned}$$

For the lower branch, in the 16 in section,

$$\begin{aligned} E_{f,\text{lower,16 in}} &= h_f g = \frac{fL\text{v}^2}{2D} = \frac{(0.015)(300\ \text{m})\left(1.585\ \frac{\text{m}}{\text{s}}\right)^2}{(2)(0.381\ \text{m})} \\ &= 14.84\ \text{J/kg} \end{aligned}$$

The total loss in the lower branch is

$$\begin{aligned} E_{f,\text{lower}} &= E_{f,\text{lower,8 in}} + E_{f,\text{lower,16 in}} \\ &= 81.61\ \frac{\text{J}}{\text{kg}} + 14.84\ \frac{\text{J}}{\text{kg}} \\ &= 96.45\ \text{J/kg} \end{aligned}$$

To calculate a loss for any other flow in the lower branch,

$$\begin{aligned} E_{f,\text{lower 2}} &= E_{f,\text{lower}}\left(\frac{\dot{V}}{\left(\frac{1}{2}\right)\left(0.350\ \frac{\text{m}^3}{\text{s}}\right)}\right)^2 \\ &= \left(96.45\ \frac{\text{J}}{\text{kg}}\right)\left(\frac{\dot{V}}{0.175\ \frac{\text{m}^3}{\text{s}}}\right)^2 \\ &= 3149\dot{V}^2 \end{aligned}$$

Let x be the fraction flowing in the upper branch. Then, because the friction losses are equal,

$$\begin{aligned} E_{f,\text{upper 2}} &= E_{f,\text{lower 2}} \\ 5329x^2 &= (3149)(1 - x)^2 \\ x &= 0.435 \end{aligned}$$

$$\dot{V}_{\text{upper}} = (0.435)\left(0.350\ \frac{\text{m}^3}{\text{s}}\right)$$
$$= \boxed{0.15\ \text{m}^3/\text{s}}$$

The answer is (D).

(b)

$$\dot{V}_{\text{lower}} = (1 - 0.435)\left(0.350\ \frac{\text{m}^3}{\text{s}}\right)$$
$$= 0.198\ \text{m}^3/\text{s}$$
$$E_{f,\text{total}} = E_{f,1} + E_{f,\text{lower}\,2} + E_{f,2}$$
$$E_{f,\text{lower}\,2} = 3149\dot{V}^2_{\text{lower}}$$
$$= (3149)\left(0.198\ \frac{\text{m}^3}{\text{s}}\right)^2$$
$$= 123.5\ \text{J/kg}$$
$$E_{f,\text{total}} = 1307\ \frac{\text{J}}{\text{kg}} + 123.5\ \frac{\text{J}}{\text{kg}} + 244.2\ \frac{\text{J}}{\text{kg}}$$
$$= \boxed{1675\ \text{J/kg}\quad (1.7\ \text{kJ/kg})}$$

The answer is (D).

17. *steps 1, 2, and 3:*

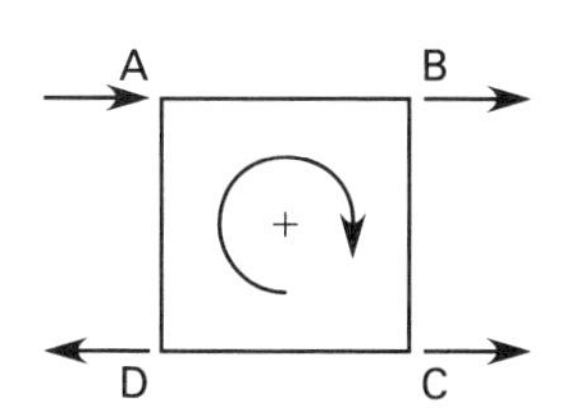

step 4: There is only one loop: ABCD.

step 5: Use Eq. 17.132.

$$K' = \frac{10.44L}{d^{4.87}C^{1.85}}$$

pipe AB:
$$K' = \frac{(10.44)(1000\ \text{ft})}{(8\ \text{in})^{4.87}(100)^{1.85}}$$
$$= 8.33 \times 10^{-5}$$

pipe BC:
$$K' = \frac{(10.44)(1000\ \text{ft})}{(6\ \text{in})^{4.87}(100)^{1.85}}$$
$$= 3.38 \times 10^{-4}$$

pipe CD:
$$K' = 8.33 \times 10^{-5}\quad [\text{same as AB}]$$

pipe DA:
$$K' = \frac{(10.44)(1000\ \text{ft})}{(12\ \text{in})^{4.87}(100)^{1.85}}$$
$$= 1.16 \times 10^{-5}$$

step 6: Assume the flows are as shown in the following illustration.

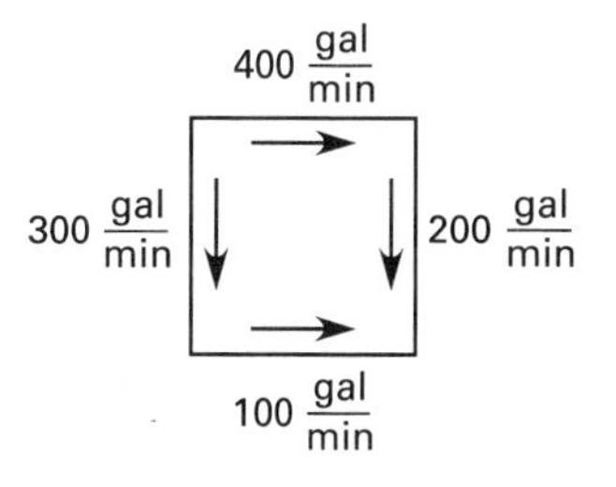

step 7: Use Eq. 17.140.

$$\delta = \frac{-\sum K'\dot{V}_a^n}{n\sum|K'\dot{V}_a^{n-1}|}$$

$$= \frac{-\left(\begin{array}{l}(8.33 \times 10^{-5})\left(400\ \frac{\text{gal}}{\text{min}}\right)^{1.85} \\ \quad + (3.38 \times 10^{-4})\left(200\ \frac{\text{gal}}{\text{min}}\right)^{1.85} \\ \quad - (8.33 \times 10^{-5})\left(100\ \frac{\text{gal}}{\text{min}}\right)^{1.85} \\ \quad - (1.16 \times 10^{-5})\left(300\ \frac{\text{gal}}{\text{min}}\right)^{1.85}\end{array}\right)}{(1.85)\left(\begin{array}{l}(8.33 \times 10^{-5})\left(400\ \frac{\text{gal}}{\text{min}}\right)^{0.85} \\ \quad + (3.38 \times 10^{-4})\left(200\ \frac{\text{gal}}{\text{min}}\right)^{0.85} \\ \quad + (8.33 \times 10^{-5})\left(100\ \frac{\text{gal}}{\text{min}}\right)^{0.85} \\ \quad + (1.16 \times 10^{-5})\left(300\ \frac{\text{gal}}{\text{min}}\right)^{0.85}\end{array}\right)}$$

$$= \frac{-10.67\ \frac{\text{gal}}{\text{min}}}{(1.85)\left(4.98 \times 10^{-2}\ \frac{\text{gal}}{\text{min}}\right)}$$
$$= -116\ \text{gal/min}$$

step 8: The adjusted flows are shown.

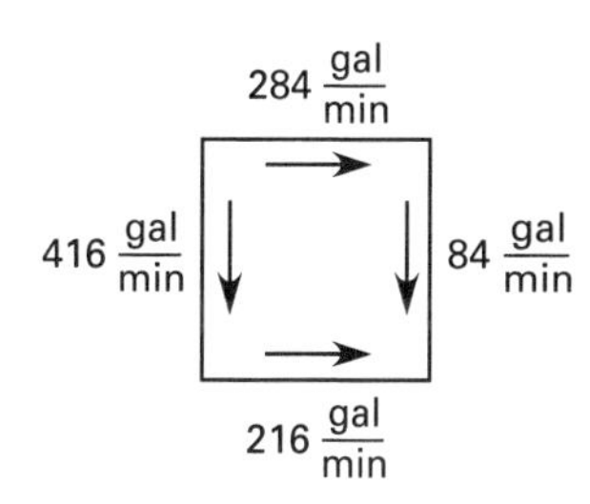

step 7: $\delta = -24$ gal/min

step 8: The adjusted flows are shown.

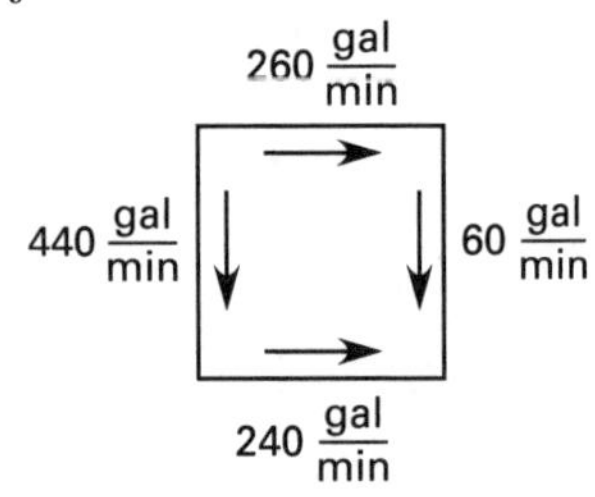

step 7: $\delta = -2$ gal/min [small enough]

step 8: The final adjusted flows are shown.

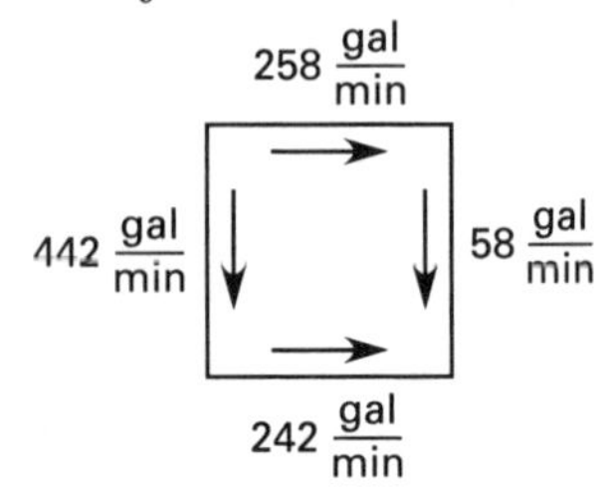

The answer is (B).

18. (a) This is a Hardy Cross problem. The pressure at point C does not change the solution procedure.

step 1: The Hazen-Williams roughness coefficient is given.

step 2: Choose clockwise as positive.

step 3: Nodes are already numbered.

step 4: Choose the loops as shown.

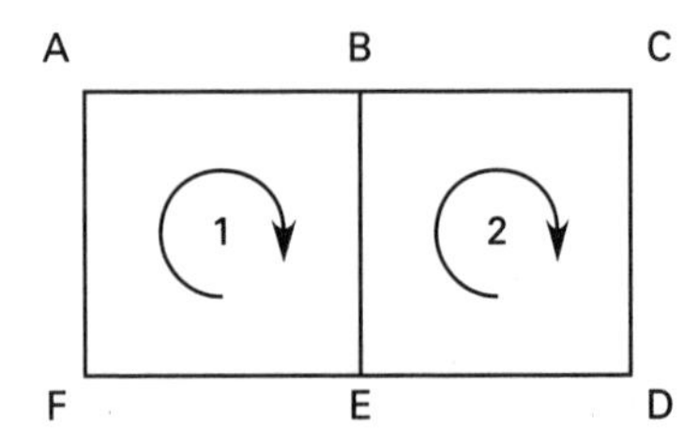

step 5: $\dot{V}$ is in gal/min, so use Eq. 17.132. d is in inches. L is in feet.

Each pipe has the same length.

$$K' = \frac{10.44L}{d^{4.87}C^{1.85}}$$

$$K'_{8\,\text{in}} = \frac{(10.44)(1000\text{ ft})}{(8\text{ in})^{4.87}(100)^{1.85}} = 8.33\times 10^{-5}$$

$$K'_{6\,\text{in}} = \frac{(10.44)(1000\text{ ft})}{(6\text{ in})^{4.87}(100)^{1.85}} = 3.38\times 10^{-4}$$

$$K'_{\text{CE}} = 2K'_{6\,\text{in}} = 6.76\times 10^{-4}$$

$$K'_{\text{EA}} = K'_{6\,\text{in}} + K'_{8\,\text{in}} = 4.21\times 10^{-4}$$

step 6: Assume the flows shown.

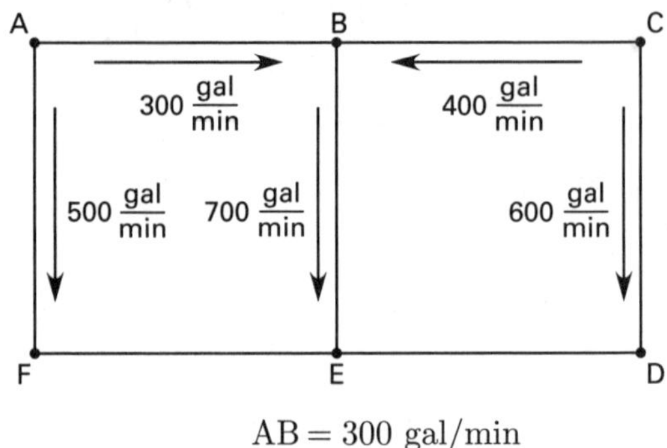

$$\text{AB} = 300\text{ gal/min}$$
$$\text{BE} = 700\text{ gal/min}$$
$$\text{AF} = \text{AE} = 500\text{ gal/min}$$
$$CD = \text{CE} = 600\text{ gal/min}$$
$$\text{CB} = 400\text{ gal/min}$$

step 7: If the elevations are included as part of the head loss,

$$\sum h = \sum K'\dot{V}_a^n + \delta\sum nK'\dot{V}_a^{n-1} + z_2 - z_1 = 0$$

However, since the loop closes on itself, $z_2 = z_1$, and the elevations can be omitted.

First iteration:

Loop 1:

$$\delta_1 = \frac{-\left(\begin{array}{l}(8.33\times 10^{-5})(300)^{1.85}\\ \quad + (8.33\times 10^{-5})(700)^{1.85}\\ \quad - (4.21\times 10^{-4})(500)^{1.85}\end{array}\right)}{(1.85)\left(\begin{array}{l}(8.33\times 10^{-5})(300)^{0.85}\\ \quad + (8.33\times 10^{-5})(700)^{0.85}\\ \quad + (4.21\times 10^{-4})(500)^{0.85}\end{array}\right)}$$

$$= \frac{-(-22.97)}{0.213}$$

$$= \boxed{+108\text{ gal/min}}$$

Loop 2:

$$\delta_2 = \frac{-\left(\begin{array}{l}(6.76\times 10^{-4})(600)^{1.85}\\ \quad - (8.33\times 10^{-5})(700)^{1.85}\\ \quad - (8.33\times 10^{-5})(400)^{1.85}\end{array}\right)}{(1.85)\left(\begin{array}{l}(6.76\times 10^{-4})(600)^{0.85}\\ \quad + (8.33\times 10^{-5})(700)^{0.85}\\ \quad + (8.33\times 10^{-5})(400)^{0.85}\end{array}\right)}$$

$$= \frac{-72.52}{0.353}$$

$$= \boxed{-205\text{ gal/min}}$$

The answer is (B).

(b) *Second iteration:*

$$\begin{aligned}
&\text{AB:}\ 300 + 108 = 408\ \text{gal/min}\\
&\text{BE:}\ 700 + 108 - (-205) = 1013\ \text{gal/min}\\
&\text{AE:}\ 500 - 108 = 392\ \text{gal/min}\\
&\text{CE:}\ 600 + (-205) = 395\ \text{gal/min}\\
&\text{CB:}\ 400 - (-205) = 605\ \text{gal/min}
\end{aligned}$$

Loop 1:

$$\delta_1 = \frac{-9.48}{0.205} = \boxed{-46\ \text{gal/min}\quad(-50\ \text{gal/min})}$$

Loop 2:

$$\delta_2 = \frac{-1.08}{0.292} = \boxed{-3.7\ \text{gal/min}\quad(-4\ \text{gal/min})}$$

The answer is (B).

(c) *Third iteration:*

$$\begin{aligned}
&\text{AB:}\ 408 + (-46) = 362\ \text{gal/min}\\
&\text{BE:}\ 1013 + (-46) - (-4) = 971\ \text{gal/min}\\
&\text{AE:}\ 392 - (-46) = 438\ \text{gal/min}\\
&\text{CE:}\ 395 + (-4) = 391\ \text{gal/min}\\
&\text{CB:}\ 605 - (-4) = 609\ \text{gal/min}
\end{aligned}$$

Loop 1:

$$\delta_1 = \frac{-0.066}{0.213} = \boxed{-0.31\ \text{gal/min}\quad(-0.3\ \text{gal/min})}$$

Loop 2:

$$\delta_2 = \frac{-2.4}{0.29} = \boxed{-8.3\ \text{gal/min}\quad(-8\ \text{gal/min})}$$

The answer is (C).

(d) Use the following flows.

$$\begin{aligned}
&\text{AB:}\ 362 + 0 = 362\ \text{gal/min}\\
&\text{BE:}\ 971 + 0 - (-8) = \boxed{\begin{array}{c}979\ \text{gal/min}\\(980\ \text{gal/min})\end{array}}\\
&\text{AE:}\ 438 - 0 = 438\ \text{gal/min}\\
&\text{CE:}\ 391 + (-8) = 383\ \text{gal/min}\\
&\text{CB:}\ 609 - (-8) = 617\ \text{gal/min}
\end{aligned}$$

The answer is (D).

(e) The friction loss in each section is

$$\begin{aligned}
h_{f,\text{AB}} &= (8.33\times10^{-5})(362)^{1.85} = 4.5\ \text{ft}\\
h_{f,\text{BE}} &= (8.33\times10^{-5})(979)^{1.85} = 28.4\ \text{ft}\\
h_{f,\text{AF}} &= (8.33\times10^{-5})(438)^{1.85} = 6.4\ \text{ft}\\
h_{f,\text{FE}} &= (3.38\times10^{-4})(438)^{1.85} = 26.0\ \text{ft}\\
h_{f,\text{CD}} = h_{f,\text{DE}} &= (3.38\times10^{-4})(383)^{1.85} = 20.3\ \text{ft}\\
h_{f,\text{CB}} &= (8.33\times10^{-5})(617)^{1.85} = 12.1\ \text{ft}
\end{aligned}$$

The pressure at C is 40 psig. Water density at 60°F is 62.37 lbf/ft^3.

$$h_\text{C} = \frac{p}{\gamma} = \frac{\left(40\ \frac{\text{lbf}}{\text{in}^2}\right)\left(12\ \frac{\text{in}}{\text{ft}}\right)^2}{62.37\ \frac{\text{lbf}}{\text{ft}^3}} = \boxed{92.4\ \text{ft}\quad(90\ \text{ft})}$$

The answer is (C).

(f) Use the energy continuity equation between adjacent points. The sign of the friction head depends on the loop and flow directions.

$$\begin{aligned}
h &= h_\text{C} + z_\text{C} - z - h_f\\
h_\text{D} &= 92.4\ \text{ft} + 300\ \text{ft} - 150\ \text{ft} - 20.3\ \text{ft} = 222.1\ \text{ft}\\
h_\text{E} &= 222.1\ \text{ft} + 150\ \text{ft} - 200\ \text{ft} - 20.3\ \text{ft} = 151.8\ \text{ft}\\
h_\text{F} &= 151.8\ \text{ft} + 200\ \text{ft} - 150\ \text{ft} + 26\ \text{ft} = 227.8\ \text{ft}\\
h_\text{B} &= 92.4\ \text{ft} + 300\ \text{ft} - 150\ \text{ft} - 12.1\ \text{ft} = 230.3\ \text{ft}\\
h_\text{A} &= 230.3\ \text{ft} + 150\ \text{ft} - 200\ \text{ft} + 4.5\ \text{ft} = 184.8\ \text{ft}
\end{aligned}$$

Using $p = \gamma h$,

$$p_\text{A} = \frac{\left(62.37\ \frac{\text{lbf}}{\text{ft}^3}\right)(184.8\ \text{ft})}{\left(12\ \frac{\text{in}}{\text{ft}}\right)^2} = 80.0\ \text{lbf/in}^2\quad(80.0\ \text{psig})$$

$$p_\text{B} = \frac{\left(62.37\ \frac{\text{lbf}}{\text{ft}^3}\right)(230.3\ \text{ft})}{\left(12\ \frac{\text{in}}{\text{ft}}\right)^2} = 99.7\ \text{psig}$$

$$p_\text{C} = 40\ \text{psig}\quad[\text{given}]$$

$$p_\text{D} = \frac{\left(62.37\ \frac{\text{lbf}}{\text{ft}^3}\right)(222.1\ \text{ft})}{\left(12\ \frac{\text{in}}{\text{ft}}\right)^2} = \boxed{96.2\ \text{psig}\quad(96\ \text{psig})}$$

$$p_\text{E} = \frac{\left(62.37\ \frac{\text{lbf}}{\text{ft}^3}\right)(151.8\ \text{ft})}{\left(12\ \frac{\text{in}}{\text{ft}}\right)^2} = 65.7\ \text{psig}$$

$$p_\text{F} = \frac{\left(62.37\ \frac{\text{lbf}}{\text{ft}^3}\right)(227.8\ \text{ft})}{\left(12\ \frac{\text{in}}{\text{ft}}\right)^2} = 98.7\ \text{psig}$$

The answer is (C).

(g) The pressure increase across the pump is

$$\Delta p = \left(80\ \frac{\text{lbf}}{\text{in}^2} - 20\ \frac{\text{lbf}}{\text{in}^2}\right)\left(12\ \frac{\text{in}}{\text{ft}}\right)^2 = 8640\ \text{lbf/ft}^2$$

Use Table 18.5 to find the hydraulic horsepower.

$$\begin{aligned} P &= \frac{\Delta p Q}{2.468\times 10^5} \\ &= \frac{\left(8640\ \frac{\text{lbf}}{\text{ft}^2}\right)\left(800\ \frac{\text{gal}}{\text{min}}\right)}{2.468\times 10^5\ \frac{\text{lbf-gal}}{\text{min-ft}^2\text{-hp}}} \\ &= \boxed{28\ \text{hp}} \end{aligned}$$

The answer is (B).

19. (a) *Pipe network solution*

step 1: Use the Darcy equation since Hazen-Williams coefficients are not given (or, assume C-values based on the age of the pipe). (See Eq. 17.133.)

step 2: Clockwise is positive.

step 3: All nodes are lettered.

step 4: There is only one loop.

step 5: $\epsilon = 0.0008$ ft. Assume full turbulence.

$$\frac{\epsilon}{D} = \frac{0.0008\ \text{ft}}{\frac{10.1\ \text{in}}{12\ \frac{\text{in}}{\text{ft}}}} = 0.00095 \quad \text{[use 0.001]}$$

$f \approx 0.020$ for full turbulence. (See Fig. 17.4.)

$$K' = \frac{1.251\times 10^{-7} fL}{D^5}$$

The friction coefficient between junctions A and B is

$$\begin{aligned} K'_{\text{AB}} &= (1.251\times 10^{-7})\left(\frac{(0.02)(2000\ \text{ft})}{\left(\frac{10.1\ \text{in}}{12\ \frac{\text{in}}{\text{ft}}}\right)^5}\right) \\ &= \boxed{11.8\times 10^{-6}\quad (12\times 10^{-6})} \end{aligned}$$

The answer is (C).

(b) The friction coefficient between junctions B and D is

$$\begin{aligned} K'_{\text{BD}} &= (1.251\times 10^{-7})\left(\frac{(0.02)(1000\ \text{ft})}{\left(\frac{10.1\ \text{in}}{12\ \frac{\text{in}}{\text{ft}}}\right)^5}\right) \\ &= \boxed{5.92\times 10^{-6}\quad (6\times 10^{-6})} \end{aligned}$$

The answer is (B).

(c) The friction coefficient between junctions D and C is

$$\begin{aligned} K'_{\text{DC}} &= (1.251\times 10^{-7})\left(\frac{(0.02)(1500\ \text{ft})}{\left(\frac{8.13\ \text{in}}{12\ \frac{\text{in}}{\text{ft}}}\right)^5}\right) \\ &= \boxed{26.3\times 10^{-6}\quad (26\times 10^{-6})} \end{aligned}$$

The answer is (D).

(d) The friction coefficient between junctions C and A is

$$\begin{aligned} K'_{\text{CA}} &= (1.251\times 10^{-7})\left(\frac{(0.02)(1000\ \text{ft})}{\left(\frac{12.12\ \text{in}}{12\ \frac{\text{in}}{\text{ft}}}\right)^5}\right) \\ &= \boxed{2.38\times 10^{-6}\quad (2\times 10^{-6})} \end{aligned}$$

The answer is (B).

(e) *step 6:* Assume $\dot{V}_{\text{AB}} = 1$ MGD.

$$\dot{V}_{\text{ACDB}} = 0.5\ \text{MGD}$$

$$\frac{\dot{V}_{\text{AB}}}{\dot{V}_{\text{ACDB}}} = \frac{1\ \text{MGD}}{0.5\ \text{MGD}} = 2$$

Convert $\dot{V}$ to gal/min.

$$\dot{V}_{\text{AB}} = \frac{(1\ \text{MGD})\left(1\times 10^6\ \frac{\text{gal}}{\text{MG}}\right)}{\left(24\ \frac{\text{hr}}{\text{day}}\right)\left(60\ \frac{\text{min}}{\text{hr}}\right)} = 694\ \text{gal/min}$$

$$\dot{V}_{\text{ACDB}} = \frac{(0.5\ \text{MGD})\left(1\times 10^6\ \frac{\text{gal}}{\text{MG}}\right)}{\left(24\ \frac{\text{hr}}{\text{day}}\right)\left(60\ \frac{\text{min}}{\text{hr}}\right)} = 347\ \text{gal/min}$$

step 7: There is only one loop.

$$694\ \frac{\text{gal}}{\text{min}} = (2)\left(347\ \frac{\text{gal}}{\text{min}}\right)$$

$$\left(694\ \frac{\text{gal}}{\text{min}}\right)^2 = (4)\left(347\ \frac{\text{gal}}{\text{min}}\right)^2$$

$$\delta = \frac{\begin{array}{l}(-1)(1\times 10^{-6})(347)^2 \\ \times\big((11.8)(4) - 5.92 - 26.3 - 2.38\big)\end{array}}{\begin{array}{l}(2)(1\times 10^{-6})(347) \\ \times\big((11.8)(2) + 5.92 + 26.3 + 2.38\big)\end{array}}$$

$$= -37.6\ \text{gal/min} \quad [\text{use } -38\ \text{gal/min}]$$

step 8:

$$\dot{V}_{\text{AB}} = 694\ \frac{\text{gal}}{\text{min}} + \left(-38\ \frac{\text{gal}}{\text{min}}\right) = 656\ \text{gal/min}$$

$$\dot{V}_{\text{ACDB}} = 347\ \frac{\text{gal}}{\text{min}} - \left(-38\ \frac{\text{gal}}{\text{min}}\right) = 385\ \text{gal/min}$$

Repeat step 7.

$$\frac{\dot{V}_{\text{AB}}}{\dot{V}_{\text{ACDB}}} = \frac{656\ \frac{\text{gal}}{\text{min}}}{385\ \frac{\text{gal}}{\text{min}}} = 1.7$$

$$(1.7)^2 = 2.90$$

$$656\ \frac{\text{gal}}{\text{min}} = (1.70)\left(385\ \frac{\text{gal}}{\text{min}}\right)$$

$$\left(656\ \frac{\text{gal}}{\text{min}}\right)^2 = (2.90)\left(385\ \frac{\text{gal}}{\text{min}}\right)^2$$

$$\delta = \frac{\begin{array}{l}(-1)(1\times 10^{-6})(385)^2 \\ \times\big((11.8)(2.90) - 5.92 - 26.3 - 2.38\big)\end{array}}{\begin{array}{l}(2)(1\times 10^{-6})(385) \\ \times\big((11.8)(1.70) + 5.92 + 26.3 + 2.38\big)\end{array}}$$

$$= 1.34 \quad (1.3)$$

$$\dot{V}_{\text{AB}} = \boxed{656\ \text{gal/min} \quad (660\ \text{gal/min})}$$

$$\dot{V}_{\text{ACDB}} = 385\ \text{gal/min}$$

Check the Reynolds number in leg AB to verify that $f = 0.02$ (from part (a)) was a good choice.

$$A_{10\,\text{in pipe}} = \left(\frac{\pi}{4}\right)\left(\frac{10.1\ \text{in}}{12\ \frac{\text{in}}{\text{ft}}}\right)^2 = 0.5564 \quad [\text{cast-iron pipe}]$$

The flow rate is

$$\frac{656\ \frac{\text{gal}}{\text{min}}}{\left(7.4805\ \frac{\text{gal}}{\text{ft}^3}\right)\left(60\ \frac{\text{sec}}{\text{min}}\right)} = 1.46\ \text{ft}^3/\text{sec}$$

$$\text{v} = \frac{\dot{V}}{A} = \frac{1.46\ \frac{\text{ft}^3}{\text{sec}}}{0.5564\ \text{ft}^2} = 2.62\ \text{ft/sec} \quad [\text{reasonable}]$$

For 50°F water,

$$\nu = 1.410\times 10^{-5}\ \text{ft}^2/\text{sec}$$

$$\text{Re} = \frac{\text{v}D}{\nu} = \frac{\left(2.62\ \frac{\text{ft}}{\text{sec}}\right)\left(\frac{10.1\ \text{in}}{12\ \frac{\text{in}}{\text{ft}}}\right)}{1.410\times 10^{-5}\ \frac{\text{ft}^2}{\text{sec}}}$$

$$= 1.56\times 10^5 \quad [\text{turbulent}]$$

The assumption of full turbulence made in step 5 is justified.

The answer is (D).

(f) *Alternative closed-form (parallel pipe) solution*

Use the Darcy equation. $\epsilon = 0.0008$ ft. The relative roughness is

$$\frac{\epsilon}{D} = \frac{0.0008\ \text{ft}}{\frac{10.1\ \text{in}}{12\ \frac{\text{in}}{\text{ft}}}} = 0.00095 \quad [\text{use } 0.001]$$

$\text{v}_{\text{max}} = 5$ ft/sec, and the temperature is 50°F. The Reynolds number is

$$\text{Re} = \frac{\text{v}D}{\nu} = \frac{\left(5\ \frac{\text{ft}}{\text{sec}}\right)\left(\frac{10.1\ \text{in}}{12\ \frac{\text{in}}{\text{ft}}}\right)}{1.410\times 10^{-5}\ \frac{\text{ft}^2}{\text{sec}}} = 2.98\times 10^5$$

From the Moody diagram, $f \approx 0.0205$.

$$h_{f,\text{AB}} = \frac{fL\text{v}^2}{2Dg} = \frac{fL\dot{V}^2}{2DA^2g}$$

$$= \frac{(0.0205)(2000\ \text{ft})\,\dot{V}_{\text{AB}}^2}{(2)\left(\frac{10.1\ \text{in}}{12\ \frac{\text{in}}{\text{ft}}}\right)(0.556\ \text{ft}^2)^2\left(32.2\ \frac{\text{ft}}{\text{sec}^2}\right)}$$

$$= 2.447\,\dot{V}_{\text{AB}}^2$$

$$h_{f,\mathrm{ACDB}} = \frac{fL\mathrm{v}^2}{2Dg} = \frac{fL\dot{V}^2}{2DA^2g}$$

$$= \frac{(0.0205)(1000 \text{ ft})\dot{V}^2_{\mathrm{ACDB}}}{(2)\left(\frac{12.12 \text{ in}}{12 \frac{\text{in}}{\text{ft}}}\right)(0.801 \text{ ft}^2)^2\left(32.2 \frac{\text{ft}}{\text{sec}^2}\right)}$$

$$+ \frac{(0.0205)(1500 \text{ ft})\dot{V}^2_{\mathrm{ACDB}}}{(2)\left(\frac{8.13 \text{ in}}{12 \frac{\text{in}}{\text{ft}}}\right)(0.360 \text{ ft}^2)^2\left(32.2 \frac{\text{ft}}{\text{sec}^2}\right)}$$

$$+ \frac{(0.0205)(1000 \text{ ft})\dot{V}^2_{\mathrm{ACDB}}}{(2)\left(\frac{10.1 \text{ in}}{12 \frac{\text{in}}{\text{ft}}}\right)(0.556 \text{ ft}^2)^2\left(32.2 \frac{\text{ft}}{\text{sec}^2}\right)}$$

$$= 0.4912\dot{V}^2_{\mathrm{ACDB}} + 5.438\dot{V}^2_{\mathrm{ACDB}} + 1.223\dot{V}^2_{\mathrm{ACDB}}$$

$$= 7.152\dot{V}^2_{\mathrm{ACDB}}$$

$$h_{f,\mathrm{AB}} = h_{f,\mathrm{ACDB}}$$

$$2.447\dot{V}^2_{\mathrm{AB}} = 7.152\dot{V}^2_{\mathrm{ACDB}}$$

$$\dot{V}_{\mathrm{AB}} = \sqrt{\frac{7.152}{2.447}}\dot{V}_{\mathrm{ACDB}} = 1.71\dot{V}_{\mathrm{ACDB}} \quad \text{[Eq. I]}$$

The total flow rate is

$$\frac{1.5 \text{ MGD}}{\left(24 \frac{\text{hr}}{\text{day}}\right)\left(60 \frac{\text{min}}{\text{hr}}\right)} = 1041.7 \text{ gal/min}$$

$$\dot{V}_{\mathrm{AB}} + \dot{V}_{\mathrm{ACDB}} = 1041.7 \text{ gal/min} \quad \text{[Eq. II]}$$

Solving Eqs. I and II simultaneously,

$$\dot{V}_{\mathrm{AB}} = \boxed{657.3 \text{ gal/min} \quad (660 \text{ gal/min})}$$

$$\dot{V}_{\mathrm{ACDB}} = 384.4 \text{ gal/min}$$

This answer is insensitive to the $\mathrm{v}_{\mathrm{max}} = 5$ ft/sec assumption. A second iteration using actual velocities from these flow rates does not change the answer.

The same technique can be used with the Hazen-Williams equation and an assumed value of C. If $C = 100$ is used, then

$$\dot{V}_{\mathrm{AB}} = 725.4 \text{ gal/min}$$

$$\dot{V}_{\mathrm{ACDB}} = 316.3 \text{ gal/min}$$

The answer is (D).

(g)

$$h_{f,\mathrm{AB}} = \frac{fL\mathrm{v}^2}{2Dg} = \frac{(0.021)(2000 \text{ ft})\left(2.62 \frac{\text{ft}}{\text{sec}}\right)^2}{(2)\left(\frac{10.1 \text{ in}}{12 \frac{\text{in}}{\text{ft}}}\right)\left(32.2 \frac{\text{ft}}{\text{sec}^2}\right)}$$

$$= 5.32 \text{ ft}$$

For leg BD,

$$\mathrm{v} = \frac{\dot{V}}{A} = \frac{\left(385 \frac{\text{gal}}{\text{min}}\right)\left(0.002228 \frac{\text{ft}^3\text{-min}}{\text{sec-gal}}\right)}{0.5564 \text{ ft}^2}$$

$$= 1.54 \text{ ft/sec}$$

$$h_{f,\mathrm{DB}} - \frac{fL\mathrm{v}^2}{2Dg}$$

$$= \frac{(0.021)(1000 \text{ ft})\left(1.54 \frac{\text{ft}}{\text{sec}}\right)^2}{(2)\left(\frac{10.1 \text{ in}}{12 \frac{\text{in}}{\text{ft}}}\right)\left(32.2 \frac{\text{ft}}{\text{sec}^2}\right)}$$

$$= 0.9 \text{ ft}$$

At 50°F, $\gamma = 62.4$ lbf/ft^3. From the Bernoulli equation (omitting the velocity term),

$$\frac{p_{\mathrm{B}}}{\gamma} + z_{\mathrm{B}} + h_{f,\mathrm{DB}} = \frac{p_{\mathrm{D}}}{\gamma} + z_{\mathrm{D}}$$

$$p_{\mathrm{B}} = \left(\frac{62.4 \frac{\text{lbf}}{\text{ft}^3}}{\left(12 \frac{\text{in}}{\text{ft}}\right)^2}\right) \times \left(\frac{\left(40 \frac{\text{lbf}}{\text{in}^2}\right)\left(12 \frac{\text{in}}{\text{ft}}\right)^2}{62.4 \frac{\text{lbf}}{\text{ft}^3}} + 600 \text{ ft} - 0.9 \text{ ft} - 580 \text{ ft}\right)$$

$$= \boxed{48.3 \text{ lbf/in}^2 \quad (48 \text{ psig})}$$

The answer is (B).

(h) The pressure at point A is

$$\frac{p_A}{\gamma} + z_A = \frac{p_B}{\gamma} + z_B + h_{f,AB}$$

$$p_A = \left(\frac{62.4\ \frac{\text{lbf}}{\text{ft}^3}}{\left(12\ \frac{\text{in}}{\text{ft}}\right)^2}\right) \times \left(\frac{\left(48.3\ \frac{\text{lbf}}{\text{in}^2}\right)\left(12\ \frac{\text{in}}{\text{ft}}\right)^2}{62.4\ \frac{\text{lbf}}{\text{ft}^3}} + 580\ \text{ft} + 5.32\ \text{ft} - 570\ \text{ft}\right)$$

$$= \boxed{54.9\ \text{lbf/in}^2 \quad (55\ \text{psig})}$$

The answer is (B).

20. *Customary U.S. Solution*

Use projectile equations.

From Table 57.2, the maximum range of the discharge is given by

$$R = \text{v}_o^2\left(\frac{\sin 2\phi}{g}\right) = \left(50\ \frac{\text{ft}}{\text{sec}}\right)^2\left(\frac{\sin(2)(45^\circ)}{32.2\ \frac{\text{ft}}{\text{sec}^2}}\right)$$

$$= \boxed{77.64\ \text{ft} \quad (78\ \text{ft})}$$

The answer is (B).

SI Solution

Use projectile equations.

The maximum range of the discharge is given by

$$R = \text{v}_o^2\left(\frac{\sin 2\phi}{g}\right) = \left(15\ \frac{\text{m}}{\text{s}}\right)^2\left(\frac{\sin(2)(45^\circ)}{9.81\ \frac{\text{m}}{\text{s}^2}}\right)$$

$$= \boxed{22.94\ \text{m} \quad (23\ \text{m})}$$

The answer is (B).

21.

$$A_o = \left(\frac{\pi}{4}\right)\left(\frac{4\ \text{in}}{12\ \frac{\text{in}}{\text{ft}}}\right)^2 = 0.08727\ \text{ft}^2$$

$$A_t = \left(\frac{\pi}{4}\right)(20\ \text{ft})^2 = 314.16\ \text{ft}^2$$

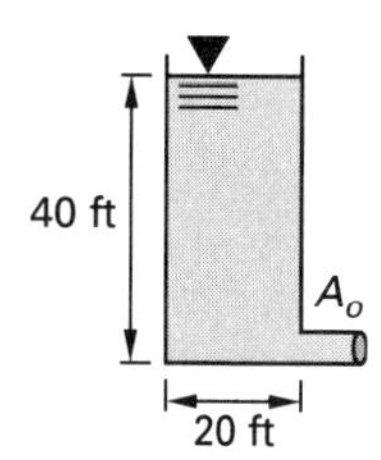

The time it takes to drop from 40 ft to 20 ft is given by Eq. 17.83.

$$t = \frac{2A_t(\sqrt{z_1} - \sqrt{z_2})}{C_d A_o \sqrt{2g}}$$

$$= \frac{(2)(314.16\ \text{ft}^2)(\sqrt{40\ \text{ft}} - \sqrt{20\ \text{ft}})}{(0.98)(0.08727\ \text{ft}^2)\sqrt{(2)\left(32.2\ \frac{\text{ft}}{\text{sec}^2}\right)}}$$

$$= \boxed{1696\ \text{sec} \quad (1700\ \text{sec})}$$

The answer is (D).

22. For low velocities, the total energy head of the water in the canal is 320 ft. Similarly, the total energy head of the water in the irrigation ditch is 305 ft. The head loss is 320 ft − 305 ft = $\boxed{15\ \text{ft.}}$

The answer is (C).

23.

$$C_d = 1.00 \quad [\text{given}]$$

$$F_{\text{va}} = \frac{1}{\sqrt{1 - \left(\frac{D_2}{D_1}\right)^4}} = \frac{1}{\sqrt{1 - \left(\frac{8\ \text{in}}{12\ \text{in}}\right)^4}}$$

$$= 1.116$$

$$A_2 = \left(\frac{\pi}{4}\right)\left(\frac{8\ \text{in}}{12\ \frac{\text{in}}{\text{ft}}}\right)^2 = 0.3491\ \text{ft}^2$$

The specific weight of mercury is 0.491 lbf/in^3; the specific weight of water is 0.0361 lbf/in^3.

$$p_1 - p_2 = \Delta(\gamma h)$$

$$= \left(\left(0.491\ \frac{\text{lbf}}{\text{in}^3}\right)(4\ \text{in}) - \left(0.0361\ \frac{\text{lbf}}{\text{in}^3}\right)(4\ \text{in})\right)\left(12\ \frac{\text{in}}{\text{ft}}\right)^2$$

$$= 262.0\ \text{lbf/ft}^2$$

From Eq. 17.154,

$$\dot{V} = F_{\text{va}} C_d A_2 \sqrt{\frac{2g(p_1 - p_2)}{\gamma}}$$

$$= (1.116)(1)(0.3491 \text{ ft}^2) \times \sqrt{\frac{(2)\left(32.2 \ \frac{\text{ft}}{\text{sec}^2}\right)\left(262 \ \frac{\text{lbf}}{\text{ft}^2}\right)}{62.4 \ \frac{\text{lbf}}{\text{ft}^3}}}$$

$$= \boxed{6.406 \text{ ft}^3/\text{sec} \quad (6.4 \text{ ft}^3/\text{sec})}$$

The answer is (C).

24. *Customary U.S. Solution*

The volumetric flow rate of benzene through the venturi meter is given by

$$\dot{V} = C_f A_2 \sqrt{\frac{2g(\rho_m - \rho)h}{\rho}}$$

The density of mercury, ρ_m, at 60°F is approximately 848 lbm/ft^3.

The density of the benzene at 60°F is

$$\rho = (\text{SG})\rho_{\text{water}} = (0.885)\left(62.4 \ \frac{\text{lbm}}{\text{ft}^3}\right) = 55.22 \text{ lbm/ft}^3$$

The throat area is

$$A_2 = \frac{\pi D_2^2}{4} = \frac{\pi\left(\frac{3.5 \text{ in}}{12 \ \frac{\text{in}}{\text{ft}}}\right)^2}{4} = 0.0668 \text{ ft}^2$$

$$\beta = \frac{3.5 \text{ in}}{8 \text{ in}} = 0.4375$$

$$C_f = \frac{C_d}{\sqrt{1 - \beta^4}} = \frac{0.99}{\sqrt{1 - (0.4375)^4}} = 1.00865$$

Find the volumetric flow of benzene.

$$\dot{V} = C_f A_2 \sqrt{\frac{2g(\rho_m - \rho)h}{\rho}}$$

$$= (1.00865)(0.0668 \text{ ft}^2) \times \sqrt{\frac{(2)\left(32.2 \ \frac{\text{ft}}{\text{sec}^2}\right)\left(848 \ \frac{\text{lbm}}{\text{ft}^3} - 55.22 \ \frac{\text{lbm}}{\text{ft}^3}\right) \times (4 \text{ in})}{\left(55.22 \ \frac{\text{lbm}}{\text{ft}^3}\right)\left(12 \ \frac{\text{in}}{\text{ft}}\right)}}$$

$$= \boxed{1.181 \text{ ft}^3/\text{sec} \quad (1.2 \text{ ft}^3/\text{sec})}$$

The answer is (A).

SI Solution

The volumetric flow rate of benzene through the venturi meter is given by

$$\dot{V} = C_f A_2 \sqrt{\frac{2g(\rho_m - \rho)h}{\rho}}$$

ρ_m is the density of mercury at 15°C; ρ_m is approximately 13 600 kg/m^3.

The density of the benzene at 15°C is

$$\rho = (\text{SG})\rho_{\text{water}} = (0.885)\left(1000 \ \frac{\text{kg}}{\text{m}^3}\right) = 885 \text{ kg/m}^3$$

The throat area is

$$A_2 = \frac{\pi D_2^2}{4} = \frac{\pi(0.09 \text{ m})^2}{4} = 0.0064 \text{ m}^2$$

$$\beta = \frac{9 \text{ cm}}{20 \text{ cm}} = 0.45$$

$$C_f = \frac{C_d}{\sqrt{1 - \beta^4}} = \frac{0.99}{\sqrt{1 - (0.45)^4}} = 1.01094$$

Find the volumetric flow of benzene.

$$\dot{V} = C_f A_2 \sqrt{\frac{2g(\rho_m - \rho)h}{\rho}}$$

$$= (1.01094)(0.0064 \text{ m}^2) \times \sqrt{\frac{(2)\left(9.81 \ \frac{\text{m}}{\text{s}^2}\right) \times \left(13\,600 \ \frac{\text{kg}}{\text{m}^3} - 885 \ \frac{\text{kg}}{\text{m}^3}\right)(0.1 \text{ m})}{885 \ \frac{\text{kg}}{\text{m}^3}}}$$

$$= \boxed{0.0344 \text{ m}^3/\text{s} \quad (34 \text{ L/s})}$$

The answer is (A).

25. From App. 14.A, for 70°F water,

$$\nu = 1.059 \times 10^{-5} \text{ ft}^2/\text{sec}$$

$$\gamma = 62.3 \text{ lbf/ft}^3$$

$$D_o = 0.2 \text{ ft}$$

$$\text{v}_o = \text{v}\left(\frac{D}{D_o}\right)^2 = \left(2 \ \frac{\text{ft}}{\text{sec}}\right)\left(\frac{1 \text{ ft}}{0.2 \text{ ft}}\right)^2 = 50 \text{ ft/sec}$$

$$\mathrm{Re} = \frac{D_o \mathrm{v}_o}{\nu} = \frac{(0.2\ \mathrm{ft})\left(50\ \frac{\mathrm{ft}}{\mathrm{sec}}\right)}{1.059 \times 10^{-5}\ \frac{\mathrm{ft}^2}{\mathrm{sec}}} = 9.44 \times 10^5$$

$$A_o = \left(\frac{\pi}{4}\right)(0.2\ \mathrm{ft})^2 = 0.0314\ \mathrm{ft}^2$$

$$A = \left(\frac{\pi}{4}\right)(1\ \mathrm{ft})^2 = 0.7854\ \mathrm{ft}^2$$

$$\frac{A_o}{A} = \frac{0.0314\ \mathrm{ft}^2}{0.7854\ \mathrm{ft}^2} = 0.040$$

From Fig. 17.28,

$$C_f \approx 0.60$$

$$\dot{V} = A\mathrm{v} = (0.7854\ \mathrm{ft}^2)\left(2\ \frac{\mathrm{ft}}{\mathrm{sec}}\right) = 1.571\ \mathrm{ft}^3/\mathrm{sec}$$

From Eq. 17.162(b), substituting $\gamma = \rho g/g_c$,

$$p_p - p_o = \left(\frac{\gamma}{2g}\right)\left(\frac{\dot{V}}{C_f A_o}\right)^2$$

$$= \frac{\left(\dfrac{62.3\ \frac{\mathrm{lbf}}{\mathrm{ft}^3}}{(2)\left(32.2\ \frac{\mathrm{ft}}{\mathrm{sec}^2}\right)}\right)\left(\dfrac{1.571\ \frac{\mathrm{ft}^3}{\mathrm{sec}}}{(0.60)(0.0314\ \mathrm{ft}^2)}\right)^2}{\left(12\ \frac{\mathrm{in}}{\mathrm{ft}}\right)^2}$$

$$= \boxed{46.7\ \mathrm{lbf/in}^2\quad (47\ \mathrm{psi})}$$

The answer is (D).

26. *Customary U.S. Solution*

The densities of mercury and water are

$$\rho_{\mathrm{mercury}} = 848\ \mathrm{lbm/ft}^3$$

$$\rho_{\mathrm{water}} = 62.4\ \mathrm{lbm/ft}^3$$

The manometer tube is filled with water above the mercury column. The pressure differential across the orifice meter is

$$\Delta p = p_1 - p_2 = (\rho_{\mathrm{mercury}} - \rho_{\mathrm{water}})h \times \frac{g}{g_c}$$

$$= \frac{\left(848\ \frac{\mathrm{lbm}}{\mathrm{ft}^3} - 62.4\ \frac{\mathrm{lbm}}{\mathrm{ft}^3}\right)(7\ \mathrm{in})}{12\ \frac{\mathrm{in}}{\mathrm{ft}}} \times \frac{32.2\ \frac{\mathrm{ft}}{\mathrm{sec}^2}}{32.2\ \frac{\mathrm{lbm\text{-}ft}}{\mathrm{lbf\text{-}sec}^2}}$$

$$= \boxed{458.3\ \mathrm{lbf/ft}^2\quad (3.2\ \mathrm{psi})}$$

The answer is (B).

SI Solution

The densities of mercury and water are

$$\rho_{\mathrm{mercury}} = 13\,600\ \mathrm{kg/m}^3$$

$$\rho_{\mathrm{water}} = 1000\ \mathrm{kg/m}^3$$

The manometer tube is filled with water above the mercury column. The pressure differential across the orifice meter is given by

$$\Delta p = p_1 - p_2 = (\rho_{\mathrm{mercury}} - \rho_{\mathrm{water}})hg$$

$$= \left(13\,600\ \frac{\mathrm{kg}}{\mathrm{m}^3} - 1000\ \frac{\mathrm{kg}}{\mathrm{m}^3}\right)(0.178\ \mathrm{m})\left(9.81\ \frac{\mathrm{m}}{\mathrm{s}^2}\right)$$

$$= \boxed{22\,002\ \mathrm{Pa}\quad (22\ \mathrm{kPa})}$$

The answer is (B).

27. *Customary U.S. Solution*

Use App. 16.B. For 12 in pipe,

$$D = 0.99483\ \mathrm{ft}$$

$$A = 0.7773\ \mathrm{ft}^2$$

The velocity is

$$\mathrm{v} = \frac{\dot{V}}{A} = \frac{10\ \frac{\mathrm{ft}^3}{\mathrm{sec}}}{0.7773\ \mathrm{ft}^2} = 12.87\ \mathrm{ft/sec}$$

For water at 70°F, $\nu = 1.059 \times 10^{-5}\ \mathrm{ft}^2/\mathrm{sec}$.

The Reynolds number in the pipe is

$$\mathrm{Re} = \frac{\mathrm{v}D}{\nu} = \frac{\left(12.87\ \frac{\mathrm{ft}}{\mathrm{sec}}\right)(0.99483\ \mathrm{ft})}{1.059 \times 10^{-5}\ \frac{\mathrm{ft}^2}{\mathrm{sec}}}$$

$$= 1.21 \times 10^6\quad \text{[fully turbulent]}$$

Flow through the orifice will have a higher Reynolds number and will also be turbulent.

The volumetric flow rate through a sharp-edged orifice is

$$\dot{V} = C_f A_o \sqrt{\frac{2g(\rho_m - \rho)h}{\rho}} = C_f A_o \sqrt{\frac{2g_c(p_1 - p_2)}{\rho}}$$

Rearranging,

$$C_f A_o = \frac{\dot{V}}{\sqrt{\dfrac{2g_c(p_1 - p_2)}{\rho}}}$$

The maximum head loss must not exceed 25 ft.

$$\frac{\frac{g_c}{g} \times (p_1 - p_2)}{\rho} = 25 \text{ ft}$$

$$\frac{g_c(p_1 - p_2)}{\rho} = (25 \text{ ft})g$$

Substituting,

$$C_f A_o = \frac{10 \ \frac{\text{ft}^3}{\text{sec}}}{\sqrt{(2)\left(32.2 \ \frac{\text{ft}}{\text{sec}^2}\right)(25 \text{ ft})}} = 0.249 \text{ ft}^2$$

Both C_f and A_o depend on the orifice diameter. For a 7 in diameter orifice,

$$A_o = \frac{\pi D_o^2}{4} = \frac{\pi\left(\frac{7 \text{ in}}{12 \ \frac{\text{in}}{\text{ft}}}\right)^2}{4} = 0.267 \text{ ft}^2$$

$$\frac{A_o}{A_1} = \frac{0.267 \text{ ft}^2}{0.7773 \text{ ft}^2} = 0.343$$

From Fig. 17.28, for $A_o/A_1 = 0.343$ and fully turbulent flow,

$$C_f = 0.645$$

$$C_f A_o = (0.645)(0.267 \text{ ft}^2) = 0.172 \text{ ft}^2 < 0.249 \text{ ft}^2$$

Therefore, a 7 in diameter orifice is too small.

Try a 9 in diameter orifice.

$$A_o = \frac{\pi D_o^2}{4} = \frac{\pi\left(\frac{9 \text{ in}}{12 \ \frac{\text{in}}{\text{ft}}}\right)^2}{4} = 0.442 \text{ ft}^2$$

$$\frac{A_o}{A_1} = \frac{0.442 \text{ ft}^2}{0.7773 \text{ ft}^2} = 0.569$$

From Fig. 17.28, for $A_o/A_1 = 0.569$ and fully turbulent flow,

$$C_f = 0.73$$

$$C_f A_o = (0.73)(0.442 \text{ ft}^2) = 0.323 \text{ ft}^2 > 0.249 \text{ ft}^2$$

Therefore, a 9 in orifice is too large.

Interpolating gives

$$D_o = 7 \text{ in} + \frac{(9 \text{ in} - 7 \text{ in})(0.249 \text{ ft}^2 - 0.172 \text{ ft}^2)}{0.323 \text{ ft}^2 - 0.172 \text{ ft}^2} = 8.0 \text{ in}$$

Further iterations yield

$$D_o \approx \boxed{8.1 \text{ in}}$$

$$C_f A_o = 0.243 \text{ ft}^2$$

The answer is (C).

SI Solution

For 300 mm inside diameter pipe, $D = 0.3$ m.

The velocity is

$$\text{v} = \frac{\dot{V}}{A} = \frac{\dot{V}}{\frac{\pi D^2}{4}} = \frac{250 \ \frac{\text{L}}{\text{s}}}{\left(\frac{\pi(0.3 \text{ m})^2}{4}\right)\left(1000 \ \frac{\text{L}}{\text{m}^3}\right)} = 3.54 \text{ m/s}$$

From App. 14.B, for water at 20°C, $\nu = 1.007 \times 10^{-6} \text{ m}^2/\text{s}$.

The Reynolds number in the pipe is

$$\text{Re} = \frac{\text{v}D}{\nu} = \frac{\left(3.54 \ \frac{\text{m}}{\text{s}}\right)(0.3 \text{ m})}{1.007 \times 10^{-6} \ \frac{\text{m}^2}{\text{s}}} = 1.05 \times 10^6 \quad \text{[fully turbulent]}$$

Flow through the orifice will have a higher Reynolds number and also be turbulent.

The volumetric flow rate through a sharp-edged orifice is

$$\dot{V} = C_f A_o \sqrt{\frac{2g(\rho_m - \rho)h}{\rho}} = C_f A_o \sqrt{\frac{2(p_1 - p_2)}{\rho}}$$

Rearranging,

$$C_f A_o = \frac{\dot{V}}{\sqrt{\frac{2(p_1 - p_2)}{\rho}}}$$

The maximum head loss must not exceed 7.5 m.

$$\frac{p_1 - p_2}{g\rho} = 7.5 \text{ m}$$

$$\frac{p_1 - p_2}{\rho} = (7.5 \text{ m})g$$

Substituting,

$$C_f A_o = \frac{0.25 \ \frac{\text{m}^3}{\text{s}}}{\sqrt{(2)\left(9.81 \ \frac{\text{m}}{\text{s}^2}\right)(7.5 \text{ m})}} = 0.021 \text{ m}^2$$

Both C_f and A_o depend on the orifice diameter. For an 18 cm diameter orifice,

$$A_o = \frac{\pi D_o^2}{4} = \frac{\pi(0.18\ \text{m})^2}{4} = 0.0254\ \text{m}^2$$

$$\frac{A_o}{A_1} = \frac{0.0254\ \text{m}^2}{0.0707\ \text{m}^2} = 0.359$$

From Fig. 17.28, for $A_o/A_1 = 0.359$ and fully turbulent flow,

$$C_f = 0.65$$

$$C_f A_o = (0.65)(0.0254\ \text{m}^2) = 0.0165\ \text{m}^2 < 0.021\ \text{m}^2$$

Therefore, an 18 cm diameter orifice is too small.

Try a 23 cm diameter orifice.

$$A_o = \frac{\pi D_o^2}{4} = \frac{\pi(0.23\ \text{m})^2}{4} = 0.0415\ \text{m}^2$$

$$\frac{A_o}{A_1} = \frac{0.0415\ \text{m}^2}{0.0707\ \text{m}^2} = 0.587$$

From Fig. 17.28, for $A_o/A_1 = 0.587$ and fully turbulent flow,

$$C_f = 0.73$$

$$C_f A_o = (0.73)(0.0415\ \text{m}^2) = 0.0303\ \text{m}^2 > 0.021\ \text{m}^2$$

Therefore, a 23 cm orifice is too large.

Interpolating gives

$$\begin{aligned} D_o &= 18\ \text{cm} \\ &\quad + (23\ \text{cm} - 18\ \text{cm})\left(\frac{0.021\ \text{m}^2 - 0.0165\ \text{m}^2}{0.0303\ \text{m}^2 - 0.0165\ \text{m}^2}\right) \\ &= 19.6\ \text{cm} \end{aligned}$$

Further iteration yields

$$D_o = \boxed{20\ \text{cm}}$$

$$C_f = 0.675$$

$$C_f A_o = 0.021\ \text{m}^2$$

The answer is (C).

28.

$$A_\text{A} = \left(\frac{\pi}{4}\right)\left(\frac{24\ \text{in}}{12\ \frac{\text{in}}{\text{ft}}}\right)^2 = 3.142\ \text{ft}^2$$

$$A_\text{B} = \left(\frac{\pi}{4}\right)\left(\frac{12\ \text{in}}{12\ \frac{\text{in}}{\text{ft}}}\right)^2 = 0.7854\ \text{ft}^2$$

$$\text{v}_\text{A} = \frac{\dot{V}}{A} = \frac{8\ \frac{\text{ft}^3}{\text{sec}}}{3.142\ \text{ft}^2} = 2.546\ \text{ft/sec}$$

$$p_\text{A} = \gamma h = \left(62.4\ \frac{\text{lbf}}{\text{ft}^3}\right)(20\ \text{ft}) = 1248\ \text{lbf/ft}^2$$

$$\text{v}_\text{B} = \frac{\dot{V}}{A} = \frac{8\ \frac{\text{ft}^3}{\text{sec}}}{0.7854\ \text{ft}^2} = 10.19\ \text{ft/sec}$$

Using the Bernoulli equation to solve for p_B,

$$\begin{aligned} p_\text{B} &= p_\text{A} + \left(\frac{\text{v}_\text{A}^2}{2g} - \frac{\text{v}_\text{B}^2}{2g}\right)\gamma \\ &= 1248\ \frac{\text{lbf}}{\text{ft}^2} + \left(\frac{\left(2.546\ \frac{\text{ft}}{\text{sec}}\right)^2 - \left(10.19\ \frac{\text{ft}}{\text{sec}}\right)^2}{(2)\left(32.2\ \frac{\text{ft}}{\text{sec}^2}\right)}\right) \\ &\quad \times \left(62.4\ \frac{\text{lbf}}{\text{ft}^3}\right) \\ &= 1153.7\ \text{lbf/ft}^2 \end{aligned}$$

With $\theta = 0°$, from Eq. 17.202(b),

$$\begin{aligned} F_x &= p_\text{B} A_\text{B} - p_\text{A} A_\text{A} + \frac{\dot{m}(\text{v}_\text{D} - \text{v}_\text{A})}{g_c} \\ &= \left(1153.7\ \frac{\text{lbf}}{\text{ft}^2}\right)(0.7854\ \text{ft}^2) - \left(1248\ \frac{\text{lbf}}{\text{ft}^2}\right)(3.142\ \text{ft}^2) \\ &\quad + \frac{\left(\left(8\ \frac{\text{ft}^3}{\text{sec}}\right)\left(62.4\ \frac{\text{lbm}}{\text{ft}^3}\right)\right)\left(10.19\ \frac{\text{ft}}{\text{sec}} - 2.546\ \frac{\text{ft}}{\text{sec}}\right)}{32.2\ \frac{\text{lbm-ft}}{\text{lbf-sec}^2}} \\ &= \boxed{-2897\ \text{lbf (2900 lbf) on the fluid (toward A)}} \end{aligned}$$

$$F_y = 0$$

The answer is (A).

29. *Customary U.S. Solution*

The mass flow rate of the water is

$$\dot{m} = \rho\dot{V} = \rho \mathrm{v} A = \frac{\rho \mathrm{v} \pi D^2}{4}$$
$$= \frac{\left(62.4\ \frac{\text{lbm}}{\text{ft}^3}\right)\left(40\ \frac{\text{ft}}{\text{sec}}\right)\pi\left(\frac{2\ \text{in}}{12\ \frac{\text{in}}{\text{ft}}}\right)^2}{4}$$
$$= 54.45\ \text{lbm/sec}$$

The effective mass flow rate of the water is

$$\dot{m}_{\text{eff}} = \left(\frac{\mathrm{v} - \mathrm{v}_b}{\mathrm{v}}\right)\dot{m}$$
$$= \left(\frac{40\ \frac{\text{ft}}{\text{sec}} - 15\ \frac{\text{ft}}{\text{sec}}}{40\ \frac{\text{ft}}{\text{sec}}}\right)\left(54.45\ \frac{\text{lbm}}{\text{sec}}\right)$$
$$= 34.0\ \text{lbm/sec}$$

The force in the (horizontal) x-direction is

$$F_x = \frac{\dot{m}_{\text{eff}}(\mathrm{v} - \mathrm{v}_b)(\cos\theta - 1)}{g_c}$$
$$= \frac{\left(34.0\ \frac{\text{lbm}}{\text{sec}}\right)\left(40\ \frac{\text{ft}}{\text{sec}} - 15\ \frac{\text{ft}}{\text{sec}}\right)(\cos 60^\circ - 1)}{32.2\ \frac{\text{lbm-ft}}{\text{lbf-sec}^2}}$$
$$= -13.2\ \text{lbf} \quad [\text{acting to the left}]$$

The force in the (vertical) y-direction is

$$F_y = \frac{\dot{m}_{\text{eff}}(\mathrm{v} - \mathrm{v}_b)\sin\theta}{g_c}$$
$$= \frac{\left(34.0\ \frac{\text{lbm}}{\text{sec}}\right)\left(40\ \frac{\text{ft}}{\text{sec}} - 15\ \frac{\text{ft}}{\text{sec}}\right)(\sin 60^\circ)}{32.2\ \frac{\text{lbm-ft}}{\text{lbf-sec}^2}}$$
$$= 22.9\ \text{lbf} \quad [\text{acting upward}]$$

The net resultant force is

$$F = \sqrt{F_x^2 + F_y^2} = \sqrt{(-13.2\ \text{lbf})^2 + (22.9\ \text{lbf})^2}$$
$$= \boxed{26.4\ \text{lbf} \quad (26\ \text{lbf})}$$

The answer is (B).

SI Solution

The mass flow rate of the water is

$$\dot{m} = \rho\dot{V} = \rho \mathrm{v} A = \frac{\rho \mathrm{v} \pi D^2}{4}$$
$$= \frac{\left(1000\ \frac{\text{kg}}{\text{m}^3}\right)\left(12\ \frac{\text{m}}{\text{s}}\right)\pi(0.05\ \text{m})^2}{4}$$
$$= 23.56\ \text{kg/s}$$

The effective mass flow rate of the water is

$$\dot{m}_{\text{eff}} = \left(\frac{\mathrm{v} - \mathrm{v}_b}{\mathrm{v}}\right)\dot{m}$$
$$= \left(\frac{12\ \frac{\text{m}}{\text{s}} - 4.5\ \frac{\text{m}}{\text{s}}}{12\ \frac{\text{m}}{\text{s}}}\right)\left(23.56\ \frac{\text{kg}}{\text{s}}\right)$$
$$= 14.73\ \text{kg/s}$$

The force in the (horizontal) x-direction is

$$F_x = \dot{m}_{\text{eff}}(\mathrm{v} - \mathrm{v}_b)(\cos\theta - 1)$$
$$= \left(14.73\ \frac{\text{kg}}{\text{s}}\right)\left(12\ \frac{\text{m}}{\text{s}} - 4.5\ \frac{\text{m}}{\text{s}}\right)(\cos 60^\circ - 1)$$
$$= -55.2\ \text{N} \quad [\text{acting to the left}]$$

The force in the (vertical) y-direction is

$$F_y = \dot{m}_{\text{eff}}(\mathrm{v} - \mathrm{v}_b)\sin\theta$$
$$= \left(14.73\ \frac{\text{kg}}{\text{s}}\right)\left(12\ \frac{\text{m}}{\text{s}} - 4.5\ \frac{\text{m}}{\text{s}}\right)(\sin 60^\circ)$$
$$= 95.7\ \text{N} \quad [\text{acting upward}]$$

The net resultant force is

$$F = \sqrt{F_x^2 + F_y^2} = \sqrt{(-55.2\ \text{N})^2 + (95.7\ \text{N})^2}$$
$$= \boxed{110.5\ \text{N} \quad (110\ \text{N})}$$

The answer is (B).

30. *Customary U.S. Solution*

For schedule-40 pipe,

$$D_i = 0.9948\ \text{ft}$$
$$A_i = 0.7773\ \text{ft}^2$$
$$\mathrm{v} = \frac{\dot{V}}{A_i} = \frac{2000\ \frac{\text{gal}}{\text{min}}}{(0.7773\ \text{ft}^2)\left(7.4805\ \frac{\text{gal}}{\text{ft}^3}\right)\left(60\ \frac{\text{sec}}{\text{min}}\right)}$$
$$= 5.73\ \text{ft/sec}$$

The pressures are in terms of gage pressure, and the density of mercury is 0.491 lbm/in³.

$$p_i = \left(14.7 \ \frac{\text{lbf}}{\text{in}^2} - \frac{(6 \text{ in})\left(0.491 \ \frac{\text{lbm}}{\text{in}^3}\right)\left(32.2 \ \frac{\text{ft}}{\text{sec}^2}\right)}{32.2 \ \frac{\text{lbm-ft}}{\text{lbf-sec}^2}} \right) \times \left(12 \ \frac{\text{in}}{\text{ft}}\right)^2$$

$$= 1692.6 \text{ lbf/ft}^2$$

$$E_{ti} = \frac{p_i}{\rho} + \frac{v_i^2}{2g_c} + \frac{z_i g}{g_c}$$

Since the pump inlet and outlet are at the same elevation, use $\Delta z = 0$. $\rho = (\text{SG})\rho_{\text{water}}$.

$$E_{ti} = \frac{p_i}{(\text{SG})\rho_{\text{water}}} + \frac{v_i^2}{2g_c} + 0$$

$$= \frac{1692.6 \ \frac{\text{lbf}}{\text{ft}^2}}{(1.2)\left(62.4 \ \frac{\text{lbm}}{\text{ft}^3}\right)} + \frac{\left(5.73 \ \frac{\text{ft}}{\text{sec}}\right)^2}{(2)\left(32.2 \ \frac{\text{lbm-ft}}{\text{lbf-sec}^2}\right)}$$

$$= 23.11 \text{ ft-lbf/lbm}$$

Calculate the total head at the inlet.

$$h_{ti} = E_{ti} \times \frac{g_c}{g} = 23.11 \ \frac{\text{ft-lbf}}{\text{lbm}} \times \frac{32.2 \ \frac{\text{lbm-ft}}{\text{lbf-sec}^2}}{32.2 \ \frac{\text{ft}}{\text{sec}^2}}$$

$$= 23.11 \text{ ft}$$

At the outlet side of the pump,

$$D_o = 0.6651 \text{ ft}$$

$$A_o = 0.3474 \text{ ft}^2$$

$$v_o = \frac{Q}{A_o} = \frac{2000 \ \frac{\text{gal}}{\text{min}}}{(0.3474 \text{ ft}^2)\left(7.4805 \ \frac{\text{gal}}{\text{ft}^3}\right)\left(60 \ \frac{\text{sec}}{\text{min}}\right)}$$

$$= 12.83 \text{ ft/sec}$$

The pressures are in terms of gage pressure. The gauge is located 4 ft above the pump outlet, which adds 4 ft of pressure head at the pump outlet.

$$p_o = \left(14.7 \ \frac{\text{lbf}}{\text{in}^2} + 20 \ \frac{\text{lbf}}{\text{in}^2}\right)\left(12 \ \frac{\text{in}}{\text{ft}}\right)^2 + 4 \text{ ft} \left(\frac{(1.2)\left(62.4 \ \frac{\text{lbm}}{\text{ft}^3}\right)\left(32.2 \ \frac{\text{ft}}{\text{sec}^2}\right)}{32.2 \ \frac{\text{lbm-ft}}{\text{lbf-sec}^2}} \right)$$

$$= 5296 \text{ lbf/ft}^2$$

$$E_{to} = \frac{p_o}{\rho} + \frac{v_o^2}{2g_c} + \frac{z_o g}{g_c}$$

Since the pump inlet and outlet are at the same elevation, $\Delta z = 0$. $\rho = (\text{SG})\rho_{\text{water}}$.

$$E_{to} = \frac{p_o}{(\text{SG})\rho_{\text{water}}} + \frac{v_o^2}{2g_c} + 0$$

$$= \frac{5296 \ \frac{\text{lbf}}{\text{ft}^2}}{(1.2)\left(62.4 \ \frac{\text{lbm}}{\text{ft}^3}\right)} + \frac{\left(12.83 \ \frac{\text{ft}}{\text{sec}}\right)^2}{(2)\left(32.2 \ \frac{\text{lbm-ft}}{\text{lbf-sec}^2}\right)}$$

$$= 73.28 \text{ ft-lbf/lbm}$$

Calculate the total head at the outlet.

$$h_{to} = E_{to} \times \frac{g_c}{g} = 73.28 \ \frac{\text{ft-lbf}}{\text{lbm}} \times \frac{32.2 \ \frac{\text{lbm-ft}}{\text{lbf-sec}^2}}{32.2 \ \frac{\text{ft}}{\text{sec}^2}}$$

$$= 73.28 \text{ ft}$$

Compute the total head across the pump.

$$\Delta h = h_{to} - h_{ti} = 73.28 \text{ ft} - 23.11 \text{ ft} = 50.17 \text{ ft}$$

The mass flow rate is

$$\dot{m} = \rho \dot{V} = (\text{SG})\rho_{\text{water}} \dot{V}$$

$$= \frac{(1.2)\left(62.4 \ \frac{\text{lbm}}{\text{ft}^3}\right)\left(2000 \ \frac{\text{gal}}{\text{min}}\right)}{\left(7.4805 \ \frac{\text{gal}}{\text{ft}^3}\right)\left(60 \ \frac{\text{sec}}{\text{min}}\right)}$$

$$= 333.7 \text{ lbm/sec}$$

The power input to the pump is

$$P = \frac{\Delta h \dot{m} \times \frac{g}{g_c}}{\eta}$$

$$= \frac{(50.17 \text{ ft})\left(333.7 \ \frac{\text{lbm}}{\text{sec}}\right) \times \frac{32.2 \ \frac{\text{ft}}{\text{sec}^2}}{32.2 \ \frac{\text{lbm-ft}}{\text{lbf-sec}^2}}}{(0.85)\left(550 \ \frac{\text{ft-lbf}}{\text{hp-sec}}\right)}$$

$$= \boxed{35.8 \text{ hp} \quad (36 \text{ hp})}$$

(It is not necessary to use absolute pressures as has been done in this solution.)

The answer is (B).

Fluids

SI Solution

For schedule-40 pipe,

$$D_i = 303.2 \text{ mm}$$

$$A_i = 0.0722 \text{ m}^2$$

$$\text{v} = \frac{\dot{V}}{A_i} = \frac{0.125 \ \frac{\text{m}^3}{\text{s}}}{0.0722 \text{ m}^2} = 1.73 \text{ m/s}$$

The pressures are in terms of gage pressure, and the density of mercury is 13 600 kg/m³.

$$p_i = 1.013 \times 10^5 \text{ Pa} - (0.15 \text{ m})\left(13\,600 \ \frac{\text{kg}}{\text{m}^3}\right)\left(9.81 \ \frac{\text{m}}{\text{s}^2}\right)$$

$$= 8.13 \times 10^4 \text{ Pa}$$

$$E_{ti} = \frac{p}{\rho} + \frac{\text{v}_i^2}{2} + z_i g$$

Since the pump inlet and outlet are at the same elevation, $\Delta z = 0$. $\rho = (\text{SG})\rho_{\text{water}}$.

$$E_{ti} = \frac{p}{(\text{SG})\rho_{\text{water}}} + \frac{\text{v}_i^2}{2} + 0$$

$$= \frac{8.13 \times 10^4 \text{ Pa}}{(1.2)\left(1000 \ \frac{\text{kg}}{\text{m}^3}\right)} + \frac{\left(1.73 \ \frac{\text{m}}{\text{s}}\right)^2}{2}$$

$$= 69.2 \text{ J/kg}$$

The total head at the inlet is

$$h_{ti} = \frac{E_{ti}}{g} = \frac{69.2 \ \frac{\text{J}}{\text{kg}}}{9.81 \ \frac{\text{m}}{\text{s}^2}} = 7.05 \text{ m}$$

Assume the pipe nominal diameter is equal to the internal diameter. On the outlet side of the pump,

$$D_i = 202.7 \text{ mm}$$

$$A_o = 0.0323 \text{ m}^2$$

$$\text{v}_o = \frac{\dot{V}}{A_o} = \frac{0.125 \ \frac{\text{m}^3}{\text{s}}}{0.0323 \text{ m}^2} = 3.87 \text{ m/s}$$

The pressures are in terms of gage pressure. The gauge is located 1.2 m above the pump outlet, which adds 1.2 m of pressure head at the pump outlet.

$$p_o = 1.013 \times 10^5 \text{ Pa} + 138 \times 10^3 \text{ Pa} + (1.2 \text{ m})\left((1.2)\left(1000 \ \frac{\text{kg}}{\text{m}^3}\right)\left(9.81 \ \frac{\text{m}}{\text{s}^2}\right)\right)$$

$$= 2.53 \times 10^5 \text{ Pa}$$

$$E_{to} = \frac{p_o}{\rho} + \frac{\text{v}_o^2}{2} + z_o g$$

Since the pump inlet and outlet are at the same elevation, $\Delta z = 0$. $\rho = (\text{SG})\rho_{\text{water}}$.

$$E_{to} = \frac{p_o}{(\text{SG})\rho_{\text{water}}} + \frac{\text{v}_o^2}{2} + 0$$

$$= \frac{2.53 \times 10^5 \text{ Pa}}{(1.2)\left(1000 \ \frac{\text{kg}}{\text{m}^3}\right)} + \frac{\left(3.87 \ \frac{\text{m}}{\text{s}}\right)^2}{2}$$

$$= 218.3 \text{ J/kg}$$

The total head at the outlet is

$$h_{to} = \frac{E_{to}}{g} = \frac{218.3 \ \frac{\text{J}}{\text{kg}}}{9.81 \ \frac{\text{m}}{\text{s}^2}} = 22.25 \text{ m}$$

The total head across the pump is

$$\Delta h = h_{to} - h_{ti} = 22.25 \text{ m} - 7.05 \text{ m} = 15.2 \text{ m}$$

The mass flow rate is

$$\dot{m} = \rho \dot{V} = (\text{SG})\rho_{\text{water}} Q$$

$$= (1.2)\left(1000 \ \frac{\text{kg}}{\text{m}^3}\right)\left(0.125 \ \frac{\text{m}^3}{\text{s}}\right)$$

$$= 150 \text{ kg/s}$$

The power input to the pump is

$$P = \frac{\Delta h \dot{m} g}{\eta} = \frac{(15.2 \text{ m})\left(150 \ \frac{\text{kg}}{\text{s}}\right)\left(9.81 \ \frac{\text{m}}{\text{s}^2}\right)}{0.85}$$

$$= \boxed{26\,314 \text{ W} \quad (26 \text{ kW})}$$

(It is not necessary to use absolute pressures as has been done in this solution.)

The answer is (B).

31. *Customary U.S. Solution*

The mass flow rate is

$$\dot{m} = \dot{V}\rho = \left(100\ \frac{\text{ft}^3}{\text{sec}}\right)\left(62.4\ \frac{\text{lbm}}{\text{ft}^3}\right) = 6240\ \text{lbm/sec}$$

The head loss across the horizontal turbine is

$$h_{\text{loss}} = \frac{\Delta p}{\rho} \times \frac{g_c}{g} = \frac{\left(30\ \frac{\text{lbf}}{\text{in}^2} - \left(-5\ \frac{\text{lbf}}{\text{in}^2}\right)\right)\left(12\ \frac{\text{in}}{\text{ft}}\right)^2}{62.4\ \frac{\text{lbm}}{\text{ft}^3}} \times \frac{32.2\ \frac{\text{lbm-ft}}{\text{lbf-sec}^2}}{32.2\ \frac{\text{ft}}{\text{sec}^2}} = 80.77\ \text{ft}$$

From Table 18.5, the power developed by the turbine is

$$P = \dot{m}h_{\text{loss}} \times \frac{g}{g_c} = \frac{\left(6240\ \frac{\text{lbm}}{\text{sec}}\right)(80.77\ \text{ft})}{550\ \frac{\text{ft-lbf}}{\text{hp-sec}}} \times \frac{32.2\ \frac{\text{ft}}{\text{sec}^2}}{32.2\ \frac{\text{lbm-ft}}{\text{lbf-sec}^2}} = \boxed{916\ \text{hp}\quad(920\ \text{hp})}$$

The answer is (D).

SI Solution

The mass flow rate is

$$\dot{m} = \dot{V}\rho = \left(2.6\ \frac{\text{m}^3}{\text{s}}\right)\left(1000\ \frac{\text{kg}}{\text{m}^3}\right) = 2600\ \text{kg/s}$$

The head loss across the horizontal turbine is

$$h_{\text{loss}} = \frac{\Delta p}{\rho g} = \frac{(210\ \text{kPa} - (-35\ \text{kPa}))\left(1000\ \frac{\text{Pa}}{\text{kPa}}\right)}{\left(1000\ \frac{\text{kg}}{\text{m}^3}\right)\left(9.81\ \frac{\text{m}}{\text{s}^2}\right)} = 25.0\ \text{m}$$

From Table 18.5, the power developed by the turbine is

$$P = \dot{m}h_{\text{loss}}g = \left(2600\ \frac{\text{kg}}{\text{s}}\right)(25.0\ \text{m})\left(9.81\ \frac{\text{m}}{\text{s}^2}\right) = \boxed{637\,650\ \text{W}\quad(640\ \text{kW})}$$

The answer is (D).

32. The speed of the truck relative to the ground is 65 mph. However, power is also required to overcome oncoming wind. The speed of the truck relative to the wind is 65 mph + 15 mph = $\boxed{80\ \text{mph.}}$ If the truck was stationary, the wind would exert a force on the truck that would push the truck backward, relative to the ground. The frictional forces between the tires and ground, as well as within the parking brake systems, perform work while preventing this motion.

The answer is (D).

33. From Eq. 17.218(b), the drag force on the antenna is

$$F_D = \frac{C_D A \rho \text{v}^2}{2g_c} = \frac{(1.2)(0.8\ \text{ft}^2)\left(0.076\ \frac{\text{lbm}}{\text{ft}^3}\right) \times \left(\frac{\left(60\ \frac{\text{mi}}{\text{hr}}\right)\left(5280\ \frac{\text{ft}}{\text{mi}}\right)}{3600\ \frac{\text{sec}}{\text{hr}}}\right)^2}{(2)\left(32.2\ \frac{\text{lbm-ft}}{\text{lbf-sec}^2}\right)} = \boxed{8.77\ \text{lbf}\quad(9.0\ \text{lbf})}$$

The answer is (A).

34. From App. 14.A, the density and kinematic viscosity of 60°F water are

$$\rho = 62.37\ \text{lbm/ft}^3$$
$$\nu = 1.217 \times 10^{-5}\ \text{ft}^2/\text{sec}$$

The Reynolds number is

$$\text{Re} = \frac{D\text{v}}{\nu} = \frac{(12\ \text{in})\left(10\ \frac{\text{ft}}{\text{sec}}\right)}{\left(1.217 \times 10^{-5}\ \frac{\text{ft}^2}{\text{sec}}\right)\left(12\ \frac{\text{in}}{\text{ft}}\right)} = 8.22 \times 10^5$$

Fluids

The projected area of the submerged portion of the log is

$$A = L_{\text{submerged}}D = \frac{(2\ \text{ft})(12\ \text{in})}{12\ \frac{\text{in}}{\text{ft}}} = 2\ \text{ft}^2$$

From Fig. 17.53, the drag coefficient, C_D, is approximately 0.35.

The total drag force on the log is

$$F_D = \frac{C_D A \rho \text{v}^2}{2g_c}$$

$$= \frac{(0.35)(2\ \text{ft}^2)\left(62.37\ \frac{\text{lbm}}{\text{ft}^3}\right)\left(10\ \frac{\text{ft}}{\text{sec}}\right)^2}{(2)\left(32.2\ \frac{\text{lbm-ft}}{\text{lbf-sec}^2}\right)}$$

$$= 67.8\ \text{lbf}$$

The volume of the log is

$$V = LA = L\pi r^2$$

$$= (30\ \text{ft})\left(12\ \frac{\text{in}}{\text{ft}}\right)\pi\left(\frac{12\ \text{in}}{2}\right)^2$$

$$= 40{,}715\ \text{in}^3$$

The density of 50°F water is 62.41 lbm/ft^3. The weight of the entire log is

$$W = \frac{\rho V g}{g_c} = \frac{(\text{SG})\rho_w V g}{g_c}$$

$$= \frac{(0.72)\left(62.41\ \frac{\text{lbm}}{\text{ft}^3}\right)(40{,}715\ \text{in}^3)\left(32.2\ \frac{\text{ft}}{\text{sec}^2}\right)}{\left(32.2\ \frac{\text{lbm-ft}}{\text{lbf-sec}^2}\right)\left(12\ \frac{\text{in}}{\text{ft}}\right)^3}$$

$$= 1058.8\ \text{lbf}$$

Choose the x- and y-directions as being perpendicular and parallel to the log, respectively. For small deflection angles, the drag force acts perpendicular to the log at half the submerged depth, 2 ft/2 = 1 ft from the lower, free end, and 30 ft − 1 ft = 29 ft from the suspended, upper end. For a small deflection angle of θ, the x-component of the drag force is

$$F_{D,x} = F_D \cos\theta$$

The log weight acts vertically downward at the midpoint of the log, 30 ft/2 = 15 ft from the suspended end. For a small deflection angle of θ, the x-component of the log weight is

$$W_x = W \sin\theta$$

Neglect buoyancy. Take moments about the suspended end.

$$\sum M = 0\text{:}\ F_{D,x}(29\ \text{ft}) - W_x(15\ \text{ft}) = 0$$

$$F_D \cos\theta(29\ \text{ft}) - W\sin\theta(15\ \text{ft}) = 0$$

$$(67.8\ \text{lbf})\cos\theta(29\ \text{ft}) - (1058.8\ \text{lbf})\sin\theta(15\ \text{ft}) = 0$$

$$\frac{\sin\theta}{\cos\theta} = \tan\theta = \frac{(67.8\ \text{lbf})(29\ \text{ft})}{(1058.8\ \text{lbf})(15\ \text{ft})} = 0.1238$$

$$\theta = \arctan 0.1238 = \boxed{7.06°\quad(7°)}$$

The answer is (B).

35. *Customary U.S. Solution*

(a) For air at 70°F,

$$\rho = \frac{p}{RT} = \frac{\left(14.7\ \frac{\text{lbf}}{\text{in}^2}\right)\left(12\ \frac{\text{in}}{\text{ft}}\right)^2}{\left(53.35\ \frac{\text{ft-lbf}}{\text{lbm-°R}}\right)(70°\text{F} + 460°)}$$

$$= 0.0749\ \text{lbm/ft}^3$$

$$\text{v} = \frac{\left(55\ \frac{\text{mi}}{\text{hr}}\right)\left(5280\ \frac{\text{ft}}{\text{mi}}\right)}{3600\ \frac{\text{sec}}{\text{hr}}}$$

$$= \boxed{80.67\ \text{ft/sec}\quad(80\ \text{ft/sec})}$$

The answer is (D).

(b) The drag on the car is

$$F_D = \frac{C_D A \rho \text{v}^2}{2g_c}$$

$$= \frac{(0.42)(28\ \text{ft}^2)\left(0.0749\ \frac{\text{lbm}}{\text{ft}^3}\right)\left(80.67\ \frac{\text{ft}}{\text{sec}}\right)^2}{(2)\left(32.2\ \frac{\text{lbm-ft}}{\text{lbf-sec}^2}\right)}$$

$$= \boxed{89.0\ \text{lbf}\quad(90\ \text{lbf})}$$

The answer is (D).

(c) The total resisting force is

$$F = F_D + \text{rolling resistance}$$

$$= 89.0\ \text{lbf} + (0.01)(3300\ \text{lbm}) \times \frac{g}{g_c}$$

$$= 89.0\ \text{lbf} + (0.01)(3300\ \text{lbm}) \times \frac{32.2\ \frac{\text{ft}}{\text{sec}^2}}{32.2\ \frac{\text{lbm-ft}}{\text{lbf-sec}^2}}$$

$$= \boxed{122.0\ \text{lbf}\quad(120\ \text{lbf})}$$

The answer is (B).

(d) The power manifested by virtue of the car's velocity is

$$P = F\text{v} = \frac{(122.0\ \text{lbf})\left(80.67\ \frac{\text{ft}}{\text{sec}}\right)}{778\ \frac{\text{ft-lbf}}{\text{Btu}}}$$

$$= \boxed{12.65\ \text{Btu/sec}\quad (13\ \text{Btu/sec})}$$

The answer is (B).

(e) The energy available from the fuel is

$$\begin{aligned} E_A &= (\text{engine thermal efficiency})(\text{fuel heating value}) \\ &= (0.28)\left(115{,}000\ \frac{\text{Btu}}{\text{gal}}\right) \\ &= 32{,}200\ \text{Btu/gal} \end{aligned}$$

The fuel consumption at 55 mi/hr is

$$\frac{P}{E_A\text{v}} = \frac{\left(12.65\ \frac{\text{Btu}}{\text{sec}}\right)\left(3600\ \frac{\text{sec}}{\text{hr}}\right)}{\left(32{,}200\ \frac{\text{Btu}}{\text{gal}}\right)\left(55\ \frac{\text{mi}}{\text{hr}}\right)}$$

$$= \boxed{0.0257\ \text{gal/mi}\quad (0.026\ \text{gal/mi})}$$

The answer is (A).

(f) At 65 mi/hr,

$$\text{v} = \frac{\left(65\ \frac{\text{mi}}{\text{hr}}\right)\left(5280\ \frac{\text{ft}}{\text{mi}}\right)}{3600\ \frac{\text{sec}}{\text{hr}}} = 95.33\ \text{ft/sec}$$

The drag on the car is

$$\begin{aligned} F_D &= \frac{C_D A \rho \text{v}^2}{2g_c} \\ &= \frac{(0.42)(28\ \text{ft}^2)\left(0.0749\ \frac{\text{lbm}}{\text{ft}^3}\right)\left(95.33\ \frac{\text{ft}}{\text{sec}}\right)^2}{(2)\left(32.2\ \frac{\text{lbm-ft}}{\text{lbf-sec}^2}\right)} \\ &= \boxed{124.3\ \text{lbf}\quad (124\ \text{lbf})} \end{aligned}$$

The answer is (D).

(g) The total resisting force is

$$\begin{aligned} F &= F_D + \text{rolling resistance} \\ &= 124.3\ \text{lbf} + (0.01)(3300\ \text{lbm}) \times \frac{g}{g_c} \\ &= 124.3\ \text{lbf} + (0.01)(3300\ \text{lbm}) \times \frac{32.2\ \frac{\text{ft}}{\text{sec}^2}}{32.2\ \frac{\text{lbm-ft}}{\text{lbf-sec}^2}} \\ &= \boxed{157.3\ \text{lbf}\quad (160\ \text{lbf})} \end{aligned}$$

The answer is (D).

(h) The power manifested by virtue of the car's velocity is

$$P = F\text{v} = \frac{(157.3\ \text{lbf})\left(95.33\ \frac{\text{ft}}{\text{sec}}\right)}{778\ \frac{\text{ft-lbf}}{\text{Btu}}}$$

$$= \boxed{19.27\ \text{Btu/sec}\quad (19\ \text{Btu/sec})}$$

The answer is (C).

(i) The fuel consumption at 65 mi/hr is

$$\frac{P}{E_A\text{v}} = \frac{\left(19.27\ \frac{\text{Btu}}{\text{sec}}\right)\left(3600\ \frac{\text{sec}}{\text{hr}}\right)}{\left(32{,}200\ \frac{\text{Btu}}{\text{gal}}\right)\left(65\ \frac{\text{mi}}{\text{hr}}\right)}$$

$$= \boxed{0.0331\ \text{gal/mi}\quad (0.03\ \text{gal/mi})}$$

The answer is (A).

(j) The relative difference between the fuel consumptions at 55 mi/hr and 65 mi/hr is

$$\frac{0.0331\ \frac{\text{gal}}{\text{mi}} - 0.0257\ \frac{\text{gal}}{\text{mi}}}{0.0257\ \frac{\text{gal}}{\text{mi}}} = \boxed{0.288\quad (30\%)}$$

The answer is (C).

SI Solution

(a) For air at 20°C,

$$\begin{aligned} \rho &= \frac{p}{RT} = \frac{1.013 \times 10^5\ \text{Pa}}{\left(287.03\ \frac{\text{J}}{\text{kg}\cdot\text{K}}\right)(20^\circ\text{C} + 273^\circ)} \\ &= 1.205\ \text{kg/m}^3 \end{aligned}$$

$$\begin{aligned} \text{v} &= \frac{\left(90\ \frac{\text{km}}{\text{h}}\right)\left(1000\ \frac{\text{m}}{\text{km}}\right)}{3600\ \frac{\text{s}}{\text{h}}} \\ &= \boxed{25.0\ \text{m/s}\quad (25\ \text{m/s})} \end{aligned}$$

The answer is (D).

(b) The drag on the car is

$$F_D = \frac{C_D A \rho \text{v}^2}{2} = \frac{(0.42)(2.6\ \text{m}^2)\left(1.205\ \frac{\text{kg}}{\text{m}^3}\right)\left(25.0\ \frac{\text{m}}{\text{s}}\right)^2}{2}$$

$$= \boxed{411.2\ \text{N}\quad (410\ \text{N})}$$

The answer is (D).

(c) The total resisting force is

$$\begin{aligned} F &= F_D + \text{rolling resistance} \times g \\ &= 411.2 \text{ N} + (0.01)(1500 \text{ kg})g \\ &= 411.2 \text{ N} + (0.01)(1500 \text{ kg})\left(9.81 \ \frac{\text{m}}{\text{s}^2}\right) \\ &= \boxed{558.4 \text{ N} \quad (560 \text{ N})} \end{aligned}$$

The answer is (B).

(d) The power required is

$$\begin{aligned} P = F\text{v} &= (558.4 \text{ N})\left(25 \ \frac{\text{m}}{\text{s}}\right) \\ &= \boxed{13\,960 \text{ W} \quad (14\,000 \text{ W})} \end{aligned}$$

The answer is (B).

(e) The energy available from the fuel is

$$\begin{aligned} E_A &= (\text{engine thermal efficiency})(\text{fuel heating value}) \\ &= (0.28)\left(32 \times 10^6 \ \frac{\text{J}}{\text{L}}\right) \\ &= 8.96 \times 10^6 \text{ J/L} \end{aligned}$$

The fuel consumption at 90 km/h is

$$\begin{aligned} \frac{P}{E_A\text{v}} &= \frac{(13\,960 \text{ W})\left(3600 \ \frac{\text{s}}{\text{h}}\right)}{\left(8.96 \times 10^6 \ \frac{\text{J}}{\text{L}}\right)\left(90 \ \frac{\text{km}}{\text{h}}\right)} \\ &= \boxed{0.0623 \text{ L/km} \quad (0.062 \text{ L/km})} \end{aligned}$$

The answer is (A).

(f) At 105 km/h,

$$\begin{aligned} \text{v} &= \frac{\left(105 \ \frac{\text{km}}{\text{h}}\right)\left(1000 \ \frac{\text{m}}{\text{km}}\right)}{3600 \ \frac{\text{s}}{\text{h}}} \\ &= 29.2 \text{ m/s} \end{aligned}$$

The drag on the car is

$$\begin{aligned} F_D &= \frac{C_D A \rho \text{v}^2}{2} = \frac{(0.42)(2.6 \text{ m}^2)\left(1.205 \ \frac{\text{kg}}{\text{m}^3}\right)\left(29.2 \ \frac{\text{m}}{\text{s}}\right)^2}{2} \\ &= \boxed{561.0 \text{ N} \quad (570 \text{ N})} \end{aligned}$$

The answer is (D).

(g) The total resisting force is

$$\begin{aligned} F &= F_D + \text{rolling resistance} \times g \\ &= 561.0 \text{ N} + (0.01)(1500 \text{ kg})g \\ &= 561.0 \text{ N} + (0.01)(1500 \text{ kg})\left(9.81 \ \frac{\text{m}}{\text{s}^2}\right) \\ &= \boxed{708.2 \text{ N} \quad (720 \text{ N})} \end{aligned}$$

The answer is (D).

(h) The power consumed by the car is

$$\begin{aligned} P = F\text{v} &= (708.2 \text{ N})\left(29.2 \ \frac{\text{m}}{\text{s}}\right) \\ &= \boxed{20\,679 \text{ W} \quad (21\,000 \text{ W})} \end{aligned}$$

The answer is (C).

(i) The fuel consumption at 105 km/h is

$$\begin{aligned} \frac{P}{E_A\text{v}} &= \frac{(20\,679 \text{ W})\left(3600 \ \frac{\text{s}}{\text{h}}\right)}{\left(8.96 \times 10^6 \ \frac{\text{J}}{\text{L}}\right)\left(105 \ \frac{\text{km}}{\text{h}}\right)} \\ &= \boxed{0.0791 \text{ L/km} \quad (0.08 \text{ L/km})} \end{aligned}$$

The answer is (A).

(j) The relative difference between the fuel consumptions at 90 km/h and 105 km/h is

$$\frac{0.0791 \ \frac{\text{L}}{\text{km}} - 0.0623 \ \frac{\text{L}}{\text{km}}}{0.0623 \ \frac{\text{L}}{\text{km}}} = \boxed{0.270 \quad (30\%)}$$

The answer is (C).

36. To ensure similarity between the model and the true conditions of the full-scale airplane, the Reynolds numbers must be equal.

$$\left(\frac{\text{v}L}{\nu}\right)_{\text{model}} = \left(\frac{\text{v}L}{\nu}\right)_{\text{true}}$$

Use the absolute viscosity.

$$\mu = \frac{\rho\nu}{g_c}$$

$$\nu = \frac{\mu g_c}{\rho}$$

$$\left(\frac{\text{v}L\rho}{\mu g_c}\right)_{\text{model}} = \left(\frac{\text{v}L\rho}{\mu g_c}\right)_{\text{true}}$$

Recall that the absolute viscosity is independent of pressure, so $\mu_{\text{model}} = \mu_{\text{true}}$.

Since g_c is a constant,

$$\left(\frac{\mathrm{v}L\rho}{\mu}\right)_{\text{model}} = \left(\frac{\mathrm{v}L\rho}{\mu}\right)_{\text{true}}$$

Assume the air behaves as an ideal gas.

$$\rho = \frac{p}{RT}$$

$$\left(\frac{\mathrm{v}Lp}{\mu RT}\right)_{\text{model}} = \left(\frac{\mathrm{v}Lp}{\mu RT}\right)_{\text{true}}$$

Since the tunnel operates with air at true velocity and temperature,

$$R_{\text{model}} = R_{\text{true}}$$

$$\mathrm{v}_{\text{model}} = \mathrm{v}_{\text{true}}$$

$$T_{\text{model}} = T_{\text{true}}$$

$$\left(\frac{Lp}{\mu}\right)_{\text{model}} = \left(\frac{Lp}{\mu}\right)_{\text{true}}$$

Therefore,

$$(Lp)_{\text{model}} = (Lp)_{\text{true}}$$

Since the scale of the model is 1/20,

$$L_{\text{model}} = \frac{L_{\text{true}}}{20}$$

Substituting gives

$$(Lp)_{\text{model}} = (Lp)_{\text{true}}$$

$$\left(\frac{L_{\text{true}}}{20}\right)p_{\text{model}} = L_{\text{true}}p_{\text{true}}$$

$$\boxed{p_{\text{model}} = 20p_{\text{true}}}$$

The answer is (C).

37. To ensure similarity between the two impellers, the Reynolds numbers, Re, must be equal.

$$\text{Re}_{\text{oil}} = \text{Re}_{\text{air}}$$

$$\frac{\mathrm{v}_{\text{oil}}D_{\text{oil}}}{\nu_{\text{oil}}} = \frac{\mathrm{v}_{\text{air}}D_{\text{air}}}{\nu_{\text{air}}}$$

$$\frac{\mathrm{v}_{\text{oil}}}{\mathrm{v}_{\text{air}}} = \left(\frac{\nu_{\text{oil}}}{\nu_{\text{air}}}\right)\left(\frac{D_{\text{air}}}{D_{\text{oil}}}\right)$$

v is the tangential velocity, and D is the impeller diameter.

$$\mathrm{v} \propto nD$$

$$\frac{\mathrm{v}_{\text{oil}}}{\mathrm{v}_{\text{air}}} = \left(\frac{n_{\text{oil}}}{n_{\text{air}}}\right)\left(\frac{D_{\text{oil}}}{D_{\text{air}}}\right)$$

Therefore,

$$\left(\frac{\nu_{\text{oil}}}{\nu_{\text{air}}}\right)\left(\frac{D_{\text{air}}}{D_{\text{oil}}}\right) = \left(\frac{n_{\text{oil}}}{n_{\text{air}}}\right)\left(\frac{D_{\text{oil}}}{D_{\text{air}}}\right)$$

$$n_{\text{air}} = n_{\text{oil}}\left(\frac{\nu_{\text{air}}}{\nu_{\text{oil}}}\right)\left(\frac{D_{\text{oil}}}{D_{\text{air}}}\right)^2$$

Since the air impeller is twice the size of the oil impeller, $D_{\text{air}} = 2D_{\text{oil}}$.

$$\begin{aligned} n_{\text{air}} &= n_{\text{oil}}\left(\frac{\nu_{\text{air}}}{\nu_{\text{oil}}}\right)\left(\frac{D_{\text{oil}}}{D_{\text{air}}}\right)^2 \\ &= n_{\text{oil}}\left(\frac{\nu_{\text{air}}}{\nu_{\text{oil}}}\right)\left(\frac{D_{\text{oil}}}{2D_{\text{oil}}}\right)^2 \\ &= \tfrac{1}{4}n_{\text{oil}}\left(\frac{\nu_{\text{air}}}{\nu_{\text{oil}}}\right) \end{aligned}$$

Customary U.S. Solution

From App. 14.D, for air at 68°F, $\nu = 15.72 \times 10^{-5}\ \text{ft}^2/\text{sec}$.

For castor oil at 68°F, $\nu = 1110 \times 10^{-5}\ \text{ft}^2/\text{sec}$ (given).

$$\begin{aligned} n_{\text{air}} &= \tfrac{1}{4}n_{\text{oil}}\left(\frac{\nu_{\text{air}}}{\nu_{\text{oil}}}\right) \\ &= \left(\tfrac{1}{4}\right)\left(1000\ \frac{\text{rev}}{\text{min}}\right)\left(\frac{15.72 \times 10^{-5}\ \frac{\text{ft}^2}{\text{sec}}}{1110 \times 10^{-5}\ \frac{\text{ft}^2}{\text{sec}}}\right) \\ &= \boxed{3.54\ \text{rpm}} \end{aligned}$$

The answer is (A).

SI Solution

From App. 14.E, for air at 20°C, $\nu = 1.512 \times 10^{-5}\ \text{m}^2/\text{s}$.

For castor oil at 20°C, $\nu = 103 \times 10^{-5}\ \text{m}^2/\text{s}$ (given).

$$\begin{aligned} n_{\text{air}} &= \tfrac{1}{4}n_{\text{oil}}\left(\frac{\nu_{\text{air}}}{\nu_{\text{oil}}}\right) \\ &= \left(\tfrac{1}{4}\right)\left(1000\ \frac{\text{rev}}{\text{min}}\right)\left(\frac{1.512 \times 10^{-5}\ \frac{\text{m}^2}{\text{s}}}{103 \times 10^{-5}\ \frac{\text{m}^2}{\text{s}}}\right) \\ &= \boxed{3.67\ \text{rpm}} \end{aligned}$$

The answer is (A).

Fluids

18 Hydraulic Machines and Fluid Distribution

PRACTICE PROBLEMS

1. Two centrifugal pumps used in a water pumping application have the characteristic curves shown. The pumps operate in parallel and discharge into a common header against a head of 40 ft. What is most nearly the discharge rate of the pumps operating in parallel?

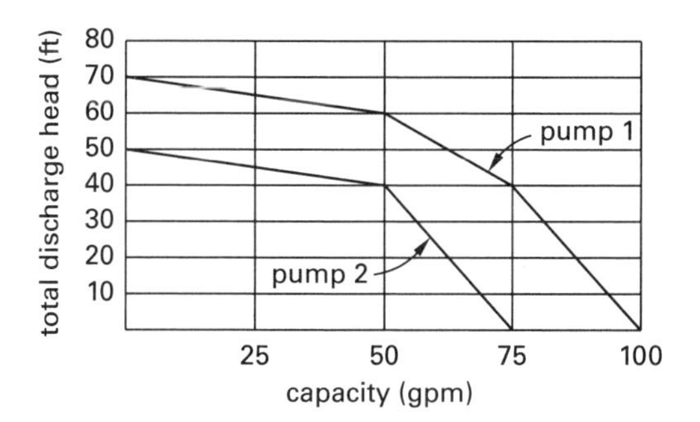

(A) 30 gpm

(B) 50 gpm

(C) 75 gpm

(D) 130 gpm

2. An electric motor drives a pump in a gasoline transfer network. The system and pump curves are defined by the points in the given table. The gasoline has a specific gravity of 0.7. What is most nearly the minimum horsepower for the motor?

	volume	
head (ft)	system curve (gpm)	pump curve (gpm)
10	0	1500
20	500	1200
30	1000	1000
40	1200	500
50	1500	0

(A) 4.0 hp

(B) 5.3 hp

(C) 5.5 hp

(D) 7.6 hp

3. A pump intended for occasional use in normal ambient conditions has an overall hydraulic efficiency of 0.85 and is required to develop a hydraulic horsepower of 4.5 hp. The pump is driven by an electric motor with a service factor of 1.80 and an electrical efficiency of 90%. The smallest NEMA standard motor size suitable for this application is

(A) 3 hp

(B) 5 hp

(C) 8 hp

(D) 10 hp

4. In a valve test bed, the fluid flow rate was measured as 800 gpm. The specific gravity of the fluid was 1.2. The pressure of the fluid one pipe diameter upstream of the valve was measured as 10 psig. The pressure ten pipe diameters downstream of the valve was measured as 0.1 psig. The pressure at the vena contracta was measured as −3 psig. The valve's pressure recovery factor for liquids is most nearly

(A) 0.3

(B) 0.6

(C) 0.8

(D) 0.9

5. Two centrifugal pumps used in a water pumping application have the characteristic curves shown. The pumps operate in series and have a combined discharge of 50 gpm. What is most nearly the total discharge head?

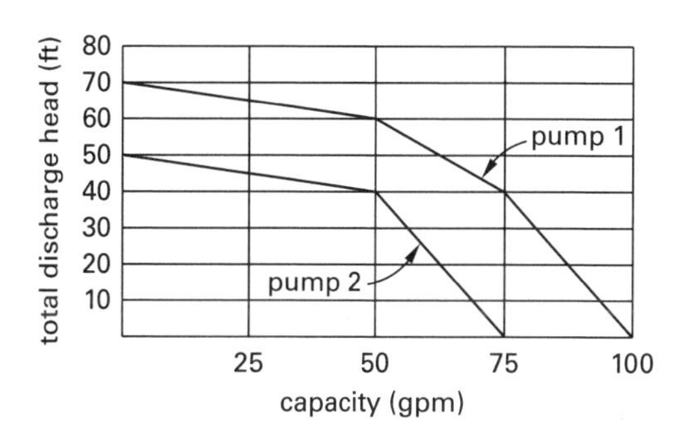

(A) 24 ft

(B) 50 ft

(C) 80 ft

(D) 100 ft

6. 2000 gal/min of 60°F thickened sludge with a specific gravity of 1.2 flows through a pump with an inlet diameter of 12 in and an outlet diameter of 8 in. The centerlines of the inlet and outlet are at the same elevation. The inlet pressure is 8 in of mercury (vacuum). A discharge pressure gauge located 4 ft above the pump discharge centerline reads 20 psig. The pump efficiency is 85%. All pipes are schedule-40. The input power of the pump is most nearly

(A) 26 hp

(B) 31 hp

(C) 37 hp

(D) 53 hp

7. 1.25 ft^3/sec (35 L/s) of 70°F (21°C) water is pumped from the bottom of a tank through 700 ft (230 m) of 4 in (102.3 mm) schedule-40 steel pipe. The line includes a 50 ft (15 m) rise in elevation, two right-angle elbows, a wide-open gate valve, and a swing check valve. All fittings and valves are regular screwed. The inlet pressure is 50 psig (345 kPa), and a working pressure of 20 psig (140 kPa) is needed at the end of the pipe. The hydraulic power for this pumping application is most nearly

(A) 16 hp (12 kW)

(B) 23 hp (17 kW)

(C) 49 hp (37 kW)

(D) 66 hp (50 kW)

8. 80 gal/min (5 L/s) of 80°F (27°C) water is lifted 12 ft (4 m) vertically by a pump through a total length of 50 ft (15 m) of a 2 in (5.1 cm) diameter smooth rubber hose. The discharge end of the hose is submerged in 8 ft (2.5 m) of water as shown.

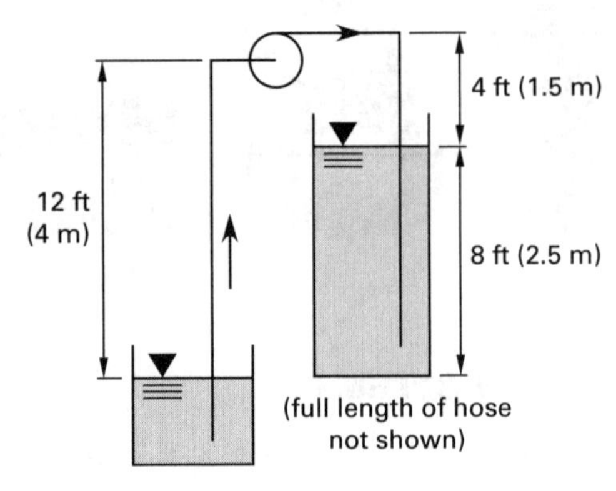

The head added by the pump is most nearly

(A) 10 ft (3.0 m)

(B) 13 ft (4.0 m)

(C) 22 ft (6.6 m)

(D) 31 ft (9.3 m)

9. A 20 hp motor drives a centrifugal pump. The pump discharges 60°F (16°C) water at a velocity of 12 ft/sec (4 m/s) into a 6 in (15.2 cm) steel schedule-40 line. The inlet is 8 in (20.3 cm) schedule-40 steel pipe. The pump suction is 5 psig (35 kPa) below standard atmospheric pressure. The friction and fitting head loss in the system is 10 ft (3.3 m). The pump efficiency is 70%. The suction and discharge lines are at the same elevation. The maximum height above the pump inlet that water is available with that velocity at standard atmospheric pressure is most nearly

(A) 28 ft (6.9 m)

(B) 37 ft (11 m)

(C) 49 ft (15 m)

(D) 81 ft (25 m)

10. An electrically driven pump is used to fill a tank on a hill from a lake below. The flow rate is 10,000 gal/hr (10.5 L/s) of 60°F (16°C) water. The atmospheric pressure is 14.7 psia (101 kPa). The pump is 12 ft (4 m) above the lake, and the tank surface level is 350 ft (115 m) above the pump. The suction and discharge lines are 4 in (10.2 cm) diameter schedule-40 steel pipe. The equivalent length of the inlet line between the lake and the pump is 300 ft (100 m). The total equivalent length between the lake and the tank is 7000 ft (2300 m), including all fittings, bends, screens, and valves. The cost of electricity is \$0.04 per kW·h. The overall efficiency of the pump and motor set is 70%.

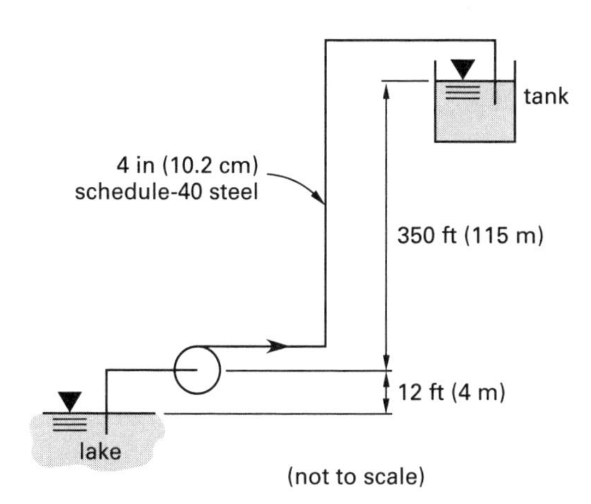

(a) The velocity in the pipe is most nearly

(A) 0.5 ft/sec (0.2 m/s)

(B) 4 ft/sec (1.0 m/s)

(C) 6 ft/sec (2.0 m/s)

(D) 10 ft/sec (5.0 m/s)

(b) The Reynolds number is most nearly

(A) 1.2×10^5

(B) 1.8×10^5

(C) 4.0×10^5

(D) 6.0×10^5

(c) The friction head is most nearly

(A) 28 ft (10 m)

(B) 120 ft (40 m)

(C) 290 ft (95 m)

(D) 950 ft (310 m)

(d) The hydraulic horsepower is most nearly

(A) 2.7 hp (2.2 kW)

(B) 5.4 hp (4.4 kW)

(C) 19 hp (15 kW)

(D) 20 hp (16 kW)

(e) The cost to operate the pump for one hour is most nearly

(A) $0.1

(B) $1

(C) $3

(D) $6

(f) The motor power required is approximately

(A) 10 hp (7.5 kW)

(B) 30 hp (25 kW)

(C) 50 hp (40 kW)

(D) 75 hp (60 kW)

(g) The NPSHA for this application is most nearly

(A) 4 ft (1.2 m)

(B) 8 ft (2.4 m)

(C) 12 ft (3.6 m)

(D) 16 ft (4.5 m)

11. A town with a stable, constant population of 10,000 produces sewage at an average rate of 100 gallons per capita day (gpcd), with peak flows of 250 gpcd. The pipe to the pumping station is 5000 ft in length and has a *C*-value of 130. The elevation drop along the length is 48 ft. Minor losses in infiltration are insignificant. The pump's maximum suction lift is 10 ft.

(a) If all whole-inch pipe diameters are available and the pipe flows 100% full under gravity flow, the minimum pipe diameter required is

(A) 8 in

(B) 12 in

(C) 14 in

(D) 18 in

(b) If constant-speed pumps are used, the minimum number of pumps that should be used, disregarding spares and backups, is

(A) 2

(B) 3

(C) 4

(D) 5

(c) If variable-speed pumps are used, the minimum number of pumps that should be used, disregarding spares and backups, is

(A) 1

(B) 2

(C) 3

(D) 4

(d) If three constant-speed pumps are used, with a fourth as backup, and the pump-motor set efficiency is 60%, the motor power required is most nearly

(A) 2 hp

(B) 3 hp

(C) 5 hp

(D) 8 hp

(e) If two variable-speed pumps are used, with a third as backup, and the pump-motor set efficiency is 80%, the motor power required is most nearly

(A) 3 hp

(B) 8 hp

(C) 12 hp

(D) 18 hp

(f) Which of the following are valid ways of controlling sump pump on-off cycles?

I. detecting sump levels

II. detecting pressure in the sump

III. detecting incoming flow rates

IV. using fixed run times

V. detecting outgoing flow rates

VI. operating manually

(A) I, II, and III

(B) I, II, IV, and V

(C) I, III, and V

(D) I, II, IV, V, and VI

12. A pump transfers 3.5 MGD of filtered water from the clear well of a 10 ft wide by 20 ft long rapid sand filter to a higher elevation. The pump efficiency is 85%, and the motor driving the pump has an efficiency of 90%. Minor losses are insignificant. Refer to the illustration shown for additional information.

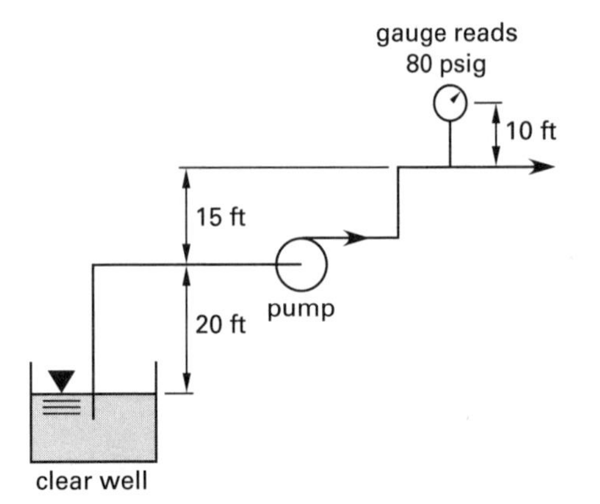

(a) The static suction lift is most nearly

(A) 15 ft

(B) 20 ft

(C) 35 ft

(D) 40 ft

(b) The static discharge head is most nearly

(A) 15 ft

(B) 20 ft

(C) 35 ft

(D) 40 ft

(c) Based on the information given, the approximate total dynamic head is most nearly

(A) 45 ft

(B) 185 ft

(C) 210 ft

(D) 230 ft

(d) The required motor power is most nearly

(A) 50 hp

(B) 100 hp

(C) 150 hp

(D) 200 hp

13. Gasoline with a specific gravity of 0.7 and kinematic viscosity of 6×10^{-6} ft^2/sec (5.6×10^{-10} m^2/s) is transferred from a tanker to a storage tank. The interior of the storage tank is maintained at atmospheric pressure by a vapor-recovery system. The free surface in the storage tank is 60 ft (20 m) above the tanker's free surface. The pipe consists of 500 ft (170 m) of 3 in (7.62 cm) schedule-40 steel pipe with six flanged elbows and two wide-open gate valves. The pump and motor both have individual efficiencies of 88%. Electricity costs \$0.045 per kW·h. The pump's performance data (based on cold, clear water) are known.

flow rate (gpm (L/s))	head (ft (m))
0 (0)	127 (42)
100 (6.3)	124 (41)
200 (12)	117 (39)
300 (18)	108 (36)
400 (24)	96 (32)
500 (30)	80 (27)
600 (36)	55 (18)

(a) The total equivalent length of pipe and fittings is most nearly

(A) 500 ft (169 m)

(B) 510 ft (170 m)

(C) 525 ft (178 m)

(D) 530 ft (180 m)

(b) If the flow rate is 100 gal/min (6.3 L/s), the velocity in the flow pipe is most nearly

(A) 2.8 ft/sec (0.86 m/s)

(B) 3.4 ft/sec (1.0 m/s)

(C) 4.3 ft/sec (1.3 m/s)

(D) 5.1 ft/sec (1.6 m/s)

(c) The friction head loss is most nearly

(A) 2.6 ft (1.0 m)

(B) 6.5 ft (2.2 m)

(C) 9.1 ft (3.1 m)

(D) 11 ft (3.8 m)

(d) The transfer rate is most nearly

(A) 150 gal/min (9.2 L/s)

(B) 180 gal/min (11 L/s)

(C) 200 gal/min (12 L/s)

(D) 230 gal/min (14 L/s)

(e) The total cost of operating the pump for one hour is most nearly

(A) \$0.20

(B) \$0.80

(C) \$1.30

(D) \$2.70

14. The pressure of 37 gal/min (65 L/s) of 80°F (27°C) SAE 40 oil is increased from 1 atm to 40 psig (275 kPa). The hydraulic power required is most nearly

(A) 0.45 hp (9 kW)

(B) 0.9 hp (18 kW)

(C) 1.8 hp (36 kW)

(D) 3.6 hp (72 kW)

15. A double-suction water pump moving 300 gal/sec (1.1 kL/s) turns at 900 rpm. The pump adds 20 ft (7 m) of head to the water. The specific speed is most nearly

(A) 3000 rpm (52 rpm)

(B) 6000 rpm (100 rpm)

(C) 9000 rpm (160 rpm)

(D) 12,000 rpm (210 rpm)

16. A two-stage centrifugal pump draws water from an inlet 10 ft below its eye. Each stage of the pump adds 150 ft of head. What is the approximate maximum suggested speed for this application?

(A) 900 rpm

(B) 1200 rpm

(C) 1700 rpm

(D) 2000 rpm

17. 100 gal/min (6.3 L/s) of pressurized hot water at 281°F and 80 psia (138°C and 550 kPa) is drawn through 30 ft (10 m) of 1.5 in (3.81 cm) schedule-40 steel pipe into a tank pressurized to a constant 2 psig (14 kPa). The inlet and outlet are both 20 ft (6 m) below the surface of the water when the tank is full. The inlet line contains a square mouth inlet, two wide-open gate valves, and two long-radius elbows. All components are regular screwed. The pump's NPSHR is 10 ft (3 m) for this application. The kinematic viscosity of 281°F (138°C) water is 0.239×10^{-5} ft^2/sec (0.222×10^{-6} m^2/s), and the vapor pressure is 50.02 psia (3.431 bar). Will the pump cavitate?

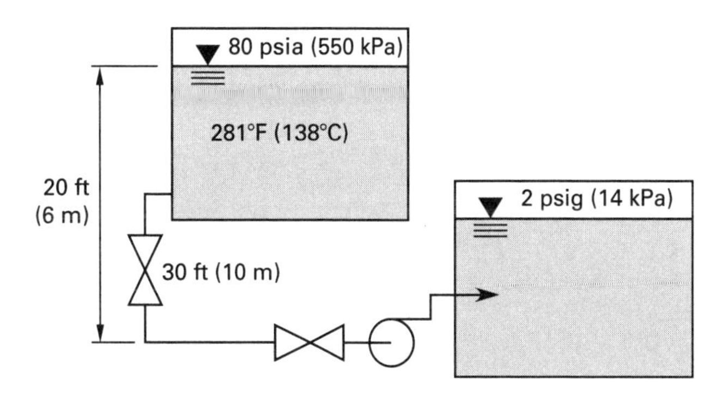

(A) yes; NPSHA = 4 ft (1.2 m)

(B) yes; NPSHA = 9 ft (2.7 m)

(C) no; NPSHA = 24 ft (7.2 m)

(D) no; NPSHA = 68 ft (21 m)

18. The velocity of the tip of a marine propeller is 4.2 times the velocity of the boat. The propeller is located 8 ft (3 m) below the surface. The temperature of the seawater is 68°F (20°C). The density of seawater is approximately 64.0 lbm/ft^3 (1024 kg/m^3), and the salt content is 2.5% by weight. The practical maximum boat velocity, as limited strictly by cavitation, is most nearly

(A) 9.1 ft/sec (2.7 m/s)

(B) 12 ft/sec (3.8 m/s)

(C) 15 ft/sec (4.5 m/s)

(D) 22 ft/sec (6.6 m/s)

19. The inlet of a centrifugal water pump is 7 ft (2.3 m) above the free surface from which it draws. The suction point is a submerged pipe. The suction line consists of 12 ft (4 m) of 2 in (5.08 cm) schedule-40 steel pipe and contains one long-radius elbow and one check valve. The discharge line is 2 in (5.08 cm) schedule-40 steel pipe and includes two long-radius elbows and an 80 ft (27 m) run. The discharge is 20 ft (6.3 m) above the free surface and is a jet to the open atmosphere. All components are regular screwed. The water temperature is 70°F (21°C). Use the following pump curve data.

flow rate (gpm (L/s))	head (ft (m))
0 (0)	110 (37)
10 (0.6)	108 (36)
20 (1.2)	105 (35)
30 (1.8)	102 (34)
40 (2.4)	98 (33)
50 (3.2)	93 (31)
60 (3.6)	87 (29)
70 (4.4)	79 (26)
80 (4.8)	66 (22)
90 (5.7)	50 (17)

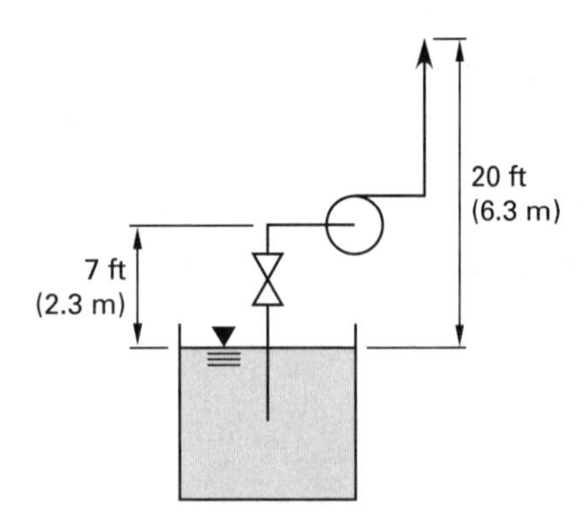

(a) The flow rate is most nearly

(A) 44 gal/min (2.9 L/s)

(B) 69 gal/min (4.5 L/s)

(C) 82 gal/min (5.5 L/s)

(D) 95 gal/min (6.2 L/s)

(b) What can be said about the use of this pump in this installation?

(A) A different pump should be used.

(B) The pump is operating near its most efficient point.

(C) Pressure fluctuations could result from surging.

(D) Overloading will not be a problem.

20. A pump was intended to run at 1750 rpm when driven by a 0.5 hp (0.37 kW) motor. The required power rating of a motor that will turn the pump at 2000 rpm is most nearly

(A) 0.25 hp (0.19 kW)

(B) 0.45 hp (0.34 kW)

(C) 0.65 hp (0.49 kW)

(D) 0.75 hp (0.55 kW)

21. A centrifugal pump running at 1400 rpm has the curve shown. The pump will be installed in an existing pipeline with known head requirements given by the formula $H = 30 + 2Q^2$. H is the system head in feet of water. Q is the flow rate in cubic feet per second.

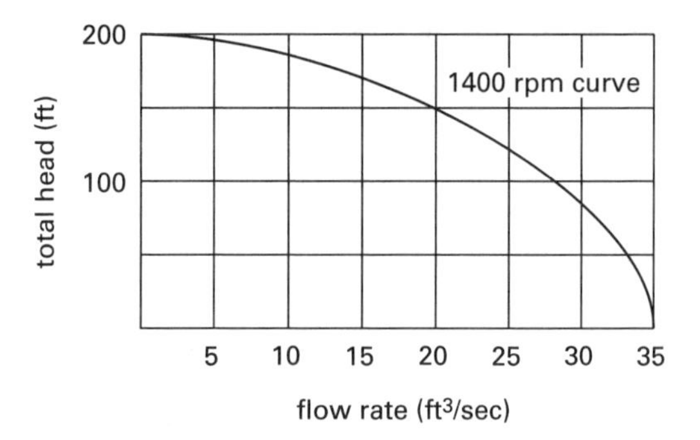

(a) If the pump is turned at 1400 rpm, the flow rate will most nearly be

(A) 2000 gal/min

(B) 3500 gal/min

(C) 4000 gal/min

(D) 4500 gal/min

(b) The power required to drive the pump is most nearly

(A) 190 hp

(B) 210 hp

(C) 230 hp

(D) 260 hp

(c) If the pump is turned at 1200 rpm, the flow rate will most nearly be

(A) 2000 gal/min

(B) 3500 gal/min

(C) 4000 gal/min

(D) 4500 gal/min

22. A horizontal turbine reduces 100 ft^3/sec of water from 30 psia to 5 psia. Friction is negligible. The power developed is most nearly

(A) 350 hp

(B) 500 hp

(C) 650 hp

(D) 800 hp

23. 1000 ft^3/sec of 60°F water flows from a high reservoir through a hydroelectric turbine installation, exiting 625 ft lower. The head loss due to friction is 58 ft. The turbine efficiency is 89%. The power developed in the turbines is most nearly

(A) 40 kW

(B) 18 MW

(C) 43 MW

(D) 71 MW

24. Water at 500 psig and 60°F (3.5 MPa and 16°C) drives a 250 hp (185 kW) turbine at 1750 rpm against a back pressure of 30 psig (210 kPa). The water discharges through a 4 in (100 mm) diameter nozzle at 35 ft/sec (10.5 m/s). The water is deflected 80° by a single blade moving directly away at 10 ft/sec (3 m/s).

(a) The specific speed is most nearly

(A) 4 (17)

(B) 25 (85)

(C) 75 (260)

(D) 230 (770)

(b) The total force acting on a single blade is most nearly

(A) 100 lbf (450 N)

(B) 140 lbf (570 N)

(C) 160 lbf (720 N)

(D) 280 lbf (1300 N)

25. A Francis-design hydraulic reaction turbine with 22 in (560 mm) diameter blades runs at 610 rpm. The turbine develops 250 hp (185 kW) when 25 ft^3/sec (700 L/s) of water flow through it. The pressure head at the turbine entrance is 92.5 ft (28.2 m). The elevation of the turbine above the tailwater level is 5.26 ft (1.75 m). The inlet and outlet velocities are both 12 ft/sec (3.6 m/s).

(a) The effective head is most nearly

(A) 90 ft (27 m)

(B) 95 ft (29 m)

(C) 100 ft (31 m)

(D) 105 ft (35 m)

(b) The overall turbine efficiency is most nearly

(A) 81%

(B) 88%

(C) 93%

(D) 96%

(c) If the effective head is 225 ft (70 m), the turbine speed will most nearly be

(A) 600 rpm

(B) 920 rpm

(C) 1100 rpm

(D) 1400 rpm

(d) If the effective head is 225 ft (70 m), the horsepower developed will most nearly be

(A) 560 hp (420 kW)

(B) 630 hp (470 kW)

(C) 750 hp (560 kW)

(D) 840 hp (640 kW)

(e) If the effective head is 225 ft (70 m), the flow rate will most nearly be

(A) 25 ft^3/sec (700 L/s)

(B) 38 ft^3/sec (1100 L/s)

(C) 56 ft^3/sec (1600 L/s)

(D) 64 ft^3/sec (1800 L/s)

Fluids

SOLUTIONS

1. Theoretically, when operating in parallel, each pump performs as if the other pump is not present. The capacities of each pump at a 40 ft discharge head are cumulative: 50 gpm for pump 2 and 75 gpm for pump 1.

$$Q_{\text{parallel}} = Q_2 + Q_1 = 50\ \frac{\text{gal}}{\text{min}} + 75\ \frac{\text{gal}}{\text{min}} = \boxed{125\ \text{gpm}\quad (130\ \text{gpm})}$$

The answer is (D).

2. The system and pump curves intersect at 1000 gpm and 30 ft. From Table 18.5, the hydraulic horsepower is

$$\text{WHP} = \frac{h_A Q(\text{SG})}{3956} = \frac{(30\ \text{ft})\left(1000\ \frac{\text{gal}}{\text{min}}\right)(0.7)}{3956\ \frac{\text{ft-gal}}{\text{hp-min}}} = \boxed{5.31\ \text{hp}\quad (5.3\ \text{hp})}$$

This is the minimum power that the electric motor can produce.

The answer is (B).

3. The motor efficiency is not used because NEMA motor power ratings are motor power output ratings. The motor is intended for occasional use, so the service factor should be included.

From Eq. 18.11, the smallest suitable motor size is

$$\text{BHP} = \frac{\text{WHP}}{\eta_p(\text{SF})} = \frac{4.5\ \text{hp}}{(0.85)(1.80)} = 2.94\ \text{hp}$$

From Table 18.7, the smallest NEMA standard motor size with a rating greater than 2.94 hp is $\boxed{3\ \text{hp.}}$

The answer is (A).

4. The lowest pressure will usually be found at the vena contracta. The liquid pressure recovery factor, F_L, of a valve is the square root of the ratio of the actual pressure loss to the maximum pressure loss.

$$F_L = \sqrt{\frac{p_1 - p_2}{(p_1 - p_{\text{vena contracta}})\,\text{SG}}} = \sqrt{\frac{10\ \frac{\text{lbf}}{\text{in}^2} - 0.1\ \frac{\text{lbf}}{\text{in}^2}}{\left(10\ \frac{\text{lbf}}{\text{in}^2} - \left(-3\ \frac{\text{lbf}}{\text{in}^2}\right)\right)(1.2)}} = \boxed{0.796\quad (0.8)}$$

The answer is (C).

5. When operated in series, the second pump receives water at the rate of the first pump's discharge, so both pumps experience the same flow rate. The second pump adds pressure head to the first pump's pressurization, so the discharge heads are cumulative. At 50 gpm, the discharge heads are 60 ft for pump 1 and 40 ft for pump 2, respectively.

$$h_{A,\text{series}} = h_{A,1} + h_{A,2} = 60\ \text{ft} + 40\ \text{ft} = \boxed{100\ \text{ft}}$$

The answer is (D).

6. The flow rate is

$$\dot{V} = \frac{2000\ \frac{\text{gal}}{\text{min}}}{\left(7.4805\ \frac{\text{gal}}{\text{ft}^3}\right)\left(60\ \frac{\text{sec}}{\text{min}}\right)} = 4.456\ \text{ft}^3/\text{sec}$$

From App. 16.B,

$$12\ \text{in:}\ D_1 = 0.99483\ \text{ft}\qquad A_1 = 0.7773\ \text{ft}^2$$

$$8\ \text{in:}\ D_2 = 0.6651\ \text{ft}\qquad A_2 = 0.3474\ \text{ft}^2$$

$$p_1 = \left(14.7\ \frac{\text{lbf}}{\text{in}^2} - (8\ \text{in})\left(0.491\ \frac{\text{lbf}}{\text{in}^3}\right)\right)\left(12\ \frac{\text{in}}{\text{ft}}\right)^2 = 1551.2\ \text{lbf/ft}^2$$

$$p_2 = \left(14.7\ \frac{\text{lbf}}{\text{in}^2} + 20\ \frac{\text{lbf}}{\text{in}^2}\right)\left(12\ \frac{\text{in}}{\text{ft}}\right)^2 + (4\ \text{ft})(1.2)\left(62.4\ \frac{\text{lbf}}{\text{ft}^3}\right) = 5296.3\ \text{lbf/ft}^2$$

$$\text{v}_1 = \frac{\dot{V}}{A_1} = \frac{4.456\ \frac{\text{ft}^3}{\text{sec}}}{0.7773\ \text{ft}^2} = 5.73\ \text{ft/sec}$$

$$\text{v}_2 = \frac{\dot{V}}{A_2} = \frac{4.456\ \frac{\text{ft}^3}{\text{sec}}}{0.3474\ \text{ft}^2} = 12.83\ \text{ft/sec}$$

From Eq. 18.8, the total heads (in feet of sludge) at points 1 and 2 are

$$h_{t,1} = h_{t,s} = \frac{p_1}{\gamma} + \frac{\text{v}_1^2}{2g} = \frac{1551.2\ \frac{\text{lbf}}{\text{ft}^2}}{\left(62.4\ \frac{\text{lbf}}{\text{ft}^3}\right)(1.2)} + \frac{\left(5.73\ \frac{\text{ft}}{\text{sec}}\right)^2}{(2)\left(32.2\ \frac{\text{ft}}{\text{sec}^2}\right)} = 21.23\ \text{ft}$$

$$h_{t,2} = h_{t,d} = \frac{p_2}{\gamma} + \frac{\text{v}_2^2}{2g} = \frac{5296.3\ \frac{\text{lbf}}{\text{ft}^2}}{\left(62.4\ \frac{\text{lbf}}{\text{ft}^3}\right)(1.2)} + \frac{\left(12.83\ \frac{\text{ft}}{\text{sec}}\right)^2}{(2)\left(32.2\ \frac{\text{ft}}{\text{sec}^2}\right)} = 73.29\ \text{ft}$$

The pump must add 73.29 ft − 21.23 ft = 52.06 ft of head (sludge head).

The power required is given in Table 18.5.

$$P_{\text{ideal}} = \frac{\Delta p\dot{V}}{550} = \frac{\Delta h\gamma\dot{V}}{550} = \frac{\Delta h(\text{SG})\gamma_w\dot{V}}{550}$$

$$= \frac{(52.06 \text{ ft})(1.2)\left(62.4 \frac{\text{lbf}}{\text{ft}^3}\right)\left(4.456 \frac{\text{ft}^3}{\text{sec}}\right)}{550 \frac{\text{ft-lbf}}{\text{hp-sec}}}$$

$$= 31.58 \text{ hp}$$

The input horsepower is

$$P_{\text{in}} = \frac{P_{\text{ideal}}}{\eta} = \frac{31.58 \text{ hp}}{0.85} = \boxed{37.15 \text{ hp} \quad (37 \text{ hp})}$$

The answer is (C).

7. *Customary U.S. Solution*

From App. 16.B, data for 4 in schedule-40 steel pipe are

$$D = 0.3355 \text{ ft}$$
$$A = 0.08841 \text{ ft}^2$$

The velocity in the pipe is

$$\text{v} = \frac{\dot{V}}{A} = \frac{1.25 \frac{\text{ft}^3}{\text{sec}}}{0.08841 \text{ ft}^2} = 14.139 \text{ ft/sec}$$

From App. 17.D, typical equivalent lengths for schedule-40, screwed steel fittings for 4 in pipes are

90° elbow: 13 ft

gate valve: 2.5 ft

check valve: 38 ft

The total equivalent length is

$$(2)(13 \text{ ft}) + (1)(2.5 \text{ ft}) + (1)(38 \text{ ft}) = 66.5 \text{ ft}$$

From App. 14.A, the density of water is 62.3 lbm/ft^3, and the kinematic viscosity of water, ν, at 70°F is 1.059×10^{-5} ft^2/sec. The Reynolds number is

$$\text{Re} = \frac{D\text{v}}{\nu} = \frac{(0.3355 \text{ ft})\left(14.139 \frac{\text{ft}}{\text{sec}}\right)}{1.059 \times 10^{-5} \frac{\text{ft}^2}{\text{sec}}}$$

$$= 4.479 \times 10^5$$

From App. 17.A, for steel, $\epsilon = 0.0002$ ft.

So,

$$\frac{\epsilon}{D} = \frac{0.0002 \text{ ft}}{0.3355 \text{ ft}} \approx 0.0006$$

Interpolating from App. 17.B, the friction factor is $f = 0.01835$.

The friction head is given by Eq. 18.5.

$$h_f = \frac{fL\text{v}^2}{2Dg}$$

$$= \frac{(0.01835)(700 \text{ ft} + 66.5 \text{ ft})\left(14.139 \frac{\text{ft}}{\text{sec}}\right)^2}{(2)(0.3355 \text{ ft})\left(32.2 \frac{\text{ft}}{\text{sec}^2}\right)}$$

$$= 130.1 \text{ ft}$$

The total dynamic head is given by Eq. 18.8. Point 1 is taken as the bottom of the supply tank. Point 2 is taken as the end of the discharge pipe.

$$h = \frac{(p_2 - p_1)g_c}{\rho g} + \frac{\text{v}_2^2 - \text{v}_1^2}{2g} + z_2 - z_1$$

$$\text{v}_1 \approx 0$$

$$z_2 - z_1 = 50 \text{ ft} \quad [\text{given as rise in elevation}]$$

The outlet and inlet pressures are

$$p_2 = 20 \text{ psig}$$
$$p_1 = 50 \text{ psig}$$

The pressure head added by the pump is

$$h = \frac{(p_2 - p_1)g_c}{\rho g} + \frac{\text{v}_2^2 - \text{v}_1^2}{2g} + z_2 - z_1$$

$$= \frac{\left(20 \frac{\text{lbf}}{\text{in}^2} - 50 \frac{\text{lbf}}{\text{in}^2}\right)\left(12 \frac{\text{in}}{\text{ft}}\right)^2\left(32.2 \frac{\text{lbm-ft}}{\text{lbf-sec}^2}\right)}{\left(62.3 \frac{\text{lbm}}{\text{ft}^3}\right)\left(32.2 \frac{\text{ft}}{\text{sec}^2}\right)}$$

$$+ \frac{\left(14.139 \frac{\text{ft}}{\text{sec}}\right)^2}{(2)\left(32.2 \frac{\text{ft}}{\text{sec}^2}\right)} + 50 \text{ ft}$$

$$= -16.2 \text{ ft}$$

The head added is

$$h_A = h + h_f$$
$$= -16.2 \text{ ft} + 130.1 \text{ ft}$$
$$= 113.9 \text{ ft}$$

The mass flow rate is

$$\dot{m} = \rho\dot{V}$$

$$= \left(62.3 \frac{\text{lbm}}{\text{ft}^3}\right)\left(1.25 \frac{\text{ft}^3}{\text{sec}}\right)$$

$$= 77.875 \text{ lbm/sec}$$

From Table 18.5, the hydraulic horsepower is

$$\begin{aligned} \text{WHP} &= \frac{h_A \dot{m}}{550} \times \frac{g}{g_c} \\ &= \frac{(113.9 \text{ ft})\left(77.875 \ \frac{\text{lbm}}{\text{sec}}\right)}{550 \ \frac{\text{ft-lbf}}{\text{hp-sec}}} \times \frac{32.2 \ \frac{\text{ft}}{\text{sec}^2}}{32.2 \ \frac{\text{lbm-ft}}{\text{lbf-sec}^2}} \\ &= \boxed{16.13 \text{ hp} \quad (16 \text{ hp})} \end{aligned}$$

The answer is (A).

SI Solution

From App. 16.C, data for 4 in schedule-40 steel pipe are

$$D = 102.26 \text{ mm}$$
$$A = 82.30 \text{ cm}^2$$

The velocity in the pipe is

$$\text{v} = \frac{\dot{V}}{A} = \frac{\left(35 \ \frac{\text{L}}{\text{s}}\right)\left(100 \ \frac{\text{cm}}{\text{m}}\right)^2}{(82.30 \text{ cm}^2)\left(1000 \ \frac{\text{L}}{\text{m}^3}\right)} = 4.25 \text{ m/s}$$

From App. 17.D, typical equivalent lengths for schedule-40, screwed steel fittings for 4 in pipes are

90° elbow: 13.0 ft

gate valve: 2.5 ft

check valve: 38.0 ft

The total equivalent length is

$$(2)(13.0 \text{ ft}) + (1)(2.5 \text{ ft}) + (1)(38.0 \text{ ft}) = 66.5 \text{ ft}$$
$$(66.5 \text{ ft})\left(0.3048 \ \frac{\text{m}}{\text{ft}}\right) = 20.27 \text{ m}$$

At 21°C, from App. 14.B, the water properties are

$$\rho = 998 \text{ kg/m}^3$$
$$\mu = 0.9827 \times 10^{-3} \text{ Pa·s}$$
$$\begin{aligned} \nu &= \frac{\mu}{\rho} = \frac{0.9827 \times 10^{-3} \text{ Pa·s}}{998 \ \frac{\text{kg}}{\text{m}^3}} \\ &= 9.85 \times 10^{-7} \text{ m}^2/\text{s} \end{aligned}$$

The Reynolds number is

$$\begin{aligned} \text{Re} &= \frac{D\text{v}}{\nu} = \frac{(102.26 \text{ mm})\left(4.25 \ \frac{\text{m}}{\text{s}}\right)}{\left(9.85 \times 10^{-7} \ \frac{\text{m}^2}{\text{s}}\right)\left(1000 \ \frac{\text{mm}}{\text{m}}\right)} \\ &= 4.417 \times 10^5 \end{aligned}$$

From Table 17.2, for steel, $\epsilon = 6.0 \times 10^{-5}$ m.

$$\begin{aligned} \frac{\epsilon}{D} &= \frac{(6.0 \times 10^{-5} \text{ m})\left(1000 \ \frac{\text{mm}}{\text{m}}\right)}{102.26 \text{ mm}} \\ &= 0.0006 \end{aligned}$$

Interpolating from App. 17.B, the friction factor is $f = 0.01836$.

From Eq. 18.5, the friction head is

$$\begin{aligned} h_f &= \frac{fL\text{v}^2}{2Dg} \\ &= \frac{(0.01836)(230 \text{ m} + 20.27 \text{ m})\left(4.25 \ \frac{\text{m}}{\text{s}}\right)^2\left(1000 \ \frac{\text{mm}}{\text{m}}\right)}{(2)(102.26 \text{ mm})\left(9.81 \ \frac{\text{m}}{\text{s}^2}\right)} \\ &= 41.4 \text{ m} \end{aligned}$$

The total dynamic head is given by Eq. 18.8. Point 1 is taken as the bottom of the supply tank. Point 2 is taken as the end of the discharge pipe.

$$h = \frac{p_2 - p_1}{\rho g} + \frac{\text{v}_2^2 - \text{v}_1^2}{2g} + z_2 - z_1$$
$$\text{v}_1 \approx 0$$
$$z_2 - z_1 = 15 \text{ m} \quad \text{[given as rise in elevation]}$$

The difference between outlet and inlet pressure is

$$p_2 - p_1 = 140 \text{ kPa} - 345 \text{ kPa} = -205 \text{ kPa}$$
$$\begin{aligned} h &= \frac{(-205 \text{ kPa})\left(1000 \ \frac{\text{Pa}}{\text{kPa}}\right)}{\left(998 \ \frac{\text{kg}}{\text{m}^3}\right)\left(9.81 \ \frac{\text{m}}{\text{s}^2}\right)} \\ &\quad + \frac{\left(4.25 \ \frac{\text{m}}{\text{s}}\right)^2}{(2)\left(9.81 \ \frac{\text{m}}{\text{s}^2}\right)} + 15 \text{ m} \\ &= -5.0 \text{ m} \end{aligned}$$

The head added by the pump is

$$h_A = h + h_f = -5.0 \text{ m} + 41.4 \text{ m} = 36.4 \text{ m}$$

The mass flow rate is

$$\begin{aligned} \dot{m} &= \rho \dot{V} = \frac{\left(998 \ \frac{\text{kg}}{\text{m}^3}\right)\left(35 \ \frac{\text{L}}{\text{s}}\right)}{1000 \ \frac{\text{L}}{\text{m}^3}} \\ &= 34.93 \text{ kg/s} \end{aligned}$$

From Table 18.6, the hydraulic power is

$$\mathrm{WkW} = \frac{9.81 h_A \dot{m}}{1000} = \frac{\left(9.81 \ \frac{\mathrm{m}}{\mathrm{s}^2}\right)(36.4 \ \mathrm{m})\left(34.93 \ \frac{\mathrm{kg}}{\mathrm{s}}\right)}{1000 \ \frac{\mathrm{W}}{\mathrm{kW}}} = \boxed{12.46 \ \mathrm{kW} \quad (12 \ \mathrm{kW})}$$

The answer is (A).

8. *Customary U.S. Solution*

The area of the rubber hose is

$$A = \frac{\pi D^2}{4} = \frac{\pi\left(\frac{2 \ \mathrm{in}}{12 \ \frac{\mathrm{in}}{\mathrm{ft}}}\right)^2}{4} = 0.0218 \ \mathrm{ft}^2$$

The velocity of water in the hose is

$$\mathrm{v} = \frac{\dot{V}}{A} = \frac{80 \ \frac{\mathrm{gal}}{\mathrm{min}}}{(0.0218 \ \mathrm{ft}^2)\left(7.4805 \ \frac{\mathrm{gal}}{\mathrm{ft}^3}\right)\left(60 \ \frac{\mathrm{sec}}{\mathrm{min}}\right)} = 8.176 \ \mathrm{ft/sec}$$

At 80°F from App. 14.A, the kinematic viscosity of water is $\nu = 0.930 \times 10^{-5} \ \mathrm{ft}^2/\mathrm{sec}$.

The Reynolds number is

$$\mathrm{Re} = \frac{\mathrm{v}D}{\nu} = \frac{\left(8.176 \ \frac{\mathrm{ft}}{\mathrm{sec}}\right)(2 \ \mathrm{in})}{\left(0.93 \times 10^{-5} \ \frac{\mathrm{ft}^2}{\mathrm{sec}}\right)\left(12 \ \frac{\mathrm{in}}{\mathrm{ft}}\right)} = 1.47 \times 10^5$$

Since the rubber hose is smooth, from App. 17.B, the friction factor is $f = 0.0166$.

From Eq. 18.5, the friction head is

$$h_f = \frac{f L \mathrm{v}^2}{2Dg} = \frac{(0.0166)(50 \ \mathrm{ft})\left(8.176 \ \frac{\mathrm{ft}}{\mathrm{sec}}\right)^2\left(12 \ \frac{\mathrm{in}}{\mathrm{ft}}\right)}{(2)(2 \ \mathrm{in})\left(32.2 \ \frac{\mathrm{ft}}{\mathrm{sec}^2}\right)} = 5.17 \ \mathrm{ft}$$

Neglecting entrance and exit losses, the head added by the pump is

$$h_A = h_f + h_z = 5.17 \ \mathrm{ft} + (12 \ \mathrm{ft} - 4 \ \mathrm{ft}) = \boxed{13.17 \ \mathrm{ft} \quad (13 \ \mathrm{ft})}$$

The answer is (B).

SI Solution

The area of the rubber hose is

$$A = \frac{\pi D^2}{4} = \frac{\pi(5.1 \ \mathrm{cm})^2}{(4)\left(100 \ \frac{\mathrm{cm}}{\mathrm{m}}\right)^2} = 0.00204 \ \mathrm{m}^2$$

The velocity of water in the hose is

$$\mathrm{v} = \frac{\dot{V}}{A} = \frac{5 \ \frac{\mathrm{L}}{\mathrm{s}}}{(0.00204 \ \mathrm{m}^2)\left(1000 \ \frac{\mathrm{L}}{\mathrm{m}^3}\right)} = 2.45 \ \mathrm{m/s}$$

At 27°C from App. 14.B, the kinematic viscosity of water is $\nu = 0.854 \times 10^{-6} \ \mathrm{m}^2/\mathrm{s}$.

The Reynolds number is

$$\mathrm{Re} = \frac{\mathrm{v}D}{\nu} = \frac{\left(2.45 \ \frac{\mathrm{m}}{\mathrm{s}}\right)(5.1 \ \mathrm{cm})}{\left(0.854 \times 10^{-6} \ \frac{\mathrm{m}^2}{\mathrm{s}}\right)\left(100 \ \frac{\mathrm{cm}}{\mathrm{m}}\right)} = 1.46 \times 10^5$$

Since the rubber hose is smooth, from App. 17.B, the friction factor is $f \approx 0.0166$.

From Eq. 18.5, the friction head is

$$h_f = \frac{f L \mathrm{v}^2}{2Dg} = \frac{(0.0166)(15 \ \mathrm{m})\left(2.45 \ \frac{\mathrm{m}}{\mathrm{s}}\right)^2\left(100 \ \frac{\mathrm{cm}}{\mathrm{m}}\right)}{(2)(5.1 \ \mathrm{cm})\left(9.81 \ \frac{\mathrm{m}}{\mathrm{s}^2}\right)} = 1.49 \ \mathrm{m}$$

Neglecting entrance and exit losses, the head added by the pump is

$$\begin{aligned} h_A &= h_f + h_z \\ &= 1.49 \ \mathrm{m} + 4 \ \mathrm{m} - 1.5 \ \mathrm{m} \\ &= \boxed{3.99 \ \mathrm{m} \quad (4.0 \ \mathrm{m})} \end{aligned}$$

The answer is (B).

9. *Customary U.S. Solution*

From App. 16.B, the diameters (inside) for 8 in and 6 in schedule-40 steel pipe are

$$\begin{aligned} D_1 &= 7.981 \ \mathrm{in} \\ D_2 &= 6.065 \ \mathrm{in} \end{aligned}$$

At 60°F from App. 14.A, the density of water is $62.37 \ \mathrm{lbm/ft}^3$.

The mass flow rate through 6 in pipe is

$$\dot{m} = A_2 \mathrm{v}_2 \rho = \frac{\pi\left(\frac{(6.065 \text{ in})^2}{4}\right)\left(12 \ \frac{\text{ft}}{\text{sec}}\right)\left(62.37 \ \frac{\text{lbm}}{\text{ft}^3}\right)}{\left(12 \ \frac{\text{in}}{\text{ft}}\right)^2}$$

$$= 150.2 \text{ lbm/sec}$$

The inlet (suction) pressure is

$$(14.7 \text{ psia} - 5 \text{ psig})\left(12 \ \frac{\text{in}}{\text{ft}}\right)^2 = 1397 \text{ lbf/ft}^2$$

From Table 18.5, the head added by the pump is

$$h_A = \frac{550(\text{BHP})\eta}{\dot{m}} \times \frac{g_c}{g}$$

$$= \frac{\left(550 \ \frac{\text{ft-lbf}}{\text{hp-sec}}\right)(20 \text{ hp})(0.70)}{150.2 \ \frac{\text{lbm}}{\text{sec}}} \times \frac{32.2 \ \frac{\text{lbm-ft}}{\text{lbf-sec}^2}}{32.2 \ \frac{\text{ft}}{\text{sec}^2}}$$

$$= 51.26 \text{ ft}$$

At 1 (pump inlet),

$$p_1 = 1397 \text{ lbf/ft}^2 \quad \text{[absolute]}$$

$$z_1 = 0$$

$$\mathrm{v}_1 = \frac{\mathrm{v}_2 A_2}{A_1} = \mathrm{v}_2\left(\frac{D_2}{D_1}\right)^2 = \left(12 \ \frac{\text{ft}}{\text{sec}}\right)\left(\frac{6.065 \text{ in}}{7.981 \text{ in}}\right)^2$$

$$= 6.93 \text{ ft/sec}$$

At 2 (pump outlet),

$$p_2 \quad \text{[unknown]}$$

$$\mathrm{v}_2 = 12 \text{ ft/sec} \quad \text{[given]}$$

$$z_2 = z_1 = 0$$

$$h_{f,1-2} = 0$$

Let z_3 be the additional head above atmospheric. From Eq. 18.8(b), the head added by the pump is

$$h_A = \frac{(p_2 - p_1)g_c}{\rho g} + \frac{\mathrm{v}_2^2 - \mathrm{v}_1^2}{2g} + z_2 - z_1 + h_{f,1-2}$$

$$51.26 \text{ ft} = \frac{\left(p_2 - 1397 \ \frac{\text{lbf}}{\text{ft}^2}\right)\left(32.2 \ \frac{\text{lbm-ft}}{\text{lbf-sec}^2}\right)}{\left(62.37 \ \frac{\text{lbm}}{\text{ft}^3}\right)\left(32.2 \ \frac{\text{ft}}{\text{sec}^2}\right)} + \frac{\left(12 \ \frac{\text{ft}}{\text{sec}}\right)^2 - \left(6.93 \ \frac{\text{ft}}{\text{sec}}\right)^2}{(2)\left(32.2 \ \frac{\text{ft}}{\text{sec}^2}\right)} + 0 \text{ ft} - 0 \text{ ft} + 0 \text{ ft}$$

$$p_2 = 4501.2 \text{ lbf/ft}^2$$

At 3 (discharge),

$$p_3 = \left(14.7 \ \frac{\text{lbf}}{\text{in}^2}\right)\left(12 \ \frac{\text{in}}{\text{ft}}\right)^2 = 2117 \text{ lbf/ft}^2$$

$$\mathrm{v}_3 = 12 \text{ ft/sec} \quad \text{[given]}$$

$$z_3 \quad \text{[unknown]}$$

$$h_{f,2-3} = 10 \text{ ft}$$

$$h_A = 0 \quad \text{[no pump between points 2 and 3]}$$

$$h_A = \frac{(p_3 - p_2)g_c}{\rho g} + \frac{\mathrm{v}_3^2 - \mathrm{v}_2^2}{2g} + z_3 - z_2 + h_{f,2-3}$$

$$0 = \frac{\left(2117 \ \frac{\text{lbf}}{\text{ft}^2} - 4501.2 \ \frac{\text{lbf}}{\text{ft}^2}\right)\left(32.2 \ \frac{\text{lbm-ft}}{\text{lbf-sec}^2}\right)}{\left(62.37 \ \frac{\text{lbm}}{\text{ft}^3}\right)\left(32.2 \ \frac{\text{ft}}{\text{sec}^2}\right)} + \frac{\left(12 \ \frac{\text{ft}}{\text{sec}}\right)^2 - \left(12 \ \frac{\text{ft}}{\text{sec}}\right)^2}{(2)\left(32.2 \ \frac{\text{ft}}{\text{sec}^2}\right)} + z_3 - 0 \text{ ft} + 10 \text{ ft}$$

$$z_3 = \boxed{28.2 \text{ ft} \quad (28 \text{ ft})}$$

z_3 could have been found directly without determining the intermediate pressure, p_2. This method is illustrated in the SI solution.

The answer is (A).

SI Solution

From App. 16.C, the inside diameters for 8 in and 6 in steel schedule-40 pipe are

$$D_1 = 202.717 \text{ mm}$$

$$D_2 = 154.051 \text{ mm}$$

At 16°C from App. 14.B, the density of water is 998.95 kg/m^3.

The mass flow rate through the 6 in pipe is

$$\dot{m} = A_2 \mathrm{v}_2 \rho = \frac{\pi\left(\frac{(154.051 \text{ mm})^2}{4}\right)\left(4 \ \frac{\text{m}}{\text{s}}\right)\left(998.95 \ \frac{\text{kg}}{\text{m}^3}\right)}{\left(1000 \ \frac{\text{mm}}{\text{m}}\right)^2}$$

$$= 74.5 \text{ kg/s}$$

The inlet (suction) pressure is

$$101.3 \text{ kPa} - 35 \text{ kPa} = 66.3 \text{ kPa}$$

From Table 18.6, the head added by the pump is

$$\begin{aligned} h_A &= \frac{1000(\text{BkW})\eta}{9.81\dot{m}} \\ &= \frac{\left(1000\ \frac{\text{W}}{\text{kW}}\right)(20\ \text{hp})\left(0.7457\ \frac{\text{kW}}{\text{hp}}\right)(0.70)}{\left(9.81\ \frac{\text{m}}{\text{s}^2}\right)\left(74.5\ \frac{\text{kg}}{\text{s}}\right)} \\ &= 14.29\ \text{m} \end{aligned}$$

At 1 (pump inlet),

$$\begin{aligned} p_1 &= 66.3\ \text{kPa} \\ z_1 &= 0 \\ \text{v}_1 &= \text{v}_2\left(\frac{A_2}{A_1}\right) = \text{v}_2\left(\frac{D_2}{D_1}\right)^2 = \left(4\ \frac{\text{m}}{\text{s}}\right)\left(\frac{154.051\ \text{mm}}{202.717\ \text{mm}}\right)^2 \\ &= 2.31\ \text{m/s} \end{aligned}$$

At 2 (pump outlet),

$$\begin{aligned} p_2 &= 101.3\ \text{kPa} \\ \text{v}_2 &= 4\ \text{m/s}\quad [\text{given}] \end{aligned}$$

From Eq. 18.8(a), the head added by the pump is

$$\begin{aligned} h_A &= \frac{p_2 - p_1}{\rho g} + \frac{\text{v}_2^2 - \text{v}_1^2}{2g} + z_2 - z_1 + h_f + z_3 \\ 14.29\ \text{m} &= \frac{(101.3\ \text{kPa} - 66.3\ \text{kPa})\left(1000\ \frac{\text{Pa}}{\text{kPa}}\right)}{\left(998.95\ \frac{\text{kg}}{\text{m}^3}\right)\left(9.81\ \frac{\text{m}}{\text{s}^2}\right)} \\ &\quad + \frac{\left(4\ \frac{\text{m}}{\text{s}}\right)^2 - \left(2.31\ \frac{\text{m}}{\text{s}}\right)^2}{(2)\left(9.81\ \frac{\text{m}}{\text{s}^2}\right)} \\ &\quad + 0\ \text{m} - 0\ \text{m} + 3.3\ \text{m} + z_3 \\ z_3 &= \boxed{6.87\ \text{m}\quad (6.9\ \text{m})} \end{aligned}$$

The answer is (A).

10. *Customary U.S. Solution*

(a) The flow rate is

$$\dot{V} = \left(10{,}000\ \frac{\text{gal}}{\text{hr}}\right)\left(0.1337\ \frac{\text{ft}^3}{\text{gal}}\right) = 1337\ \text{ft}^3/\text{hr}$$

From App. 16.B, for 4 in schedule-40 steel pipe,

$$\begin{aligned} D &= 0.3355\ \text{ft} \\ A &= 0.08841\ \text{ft}^2 \end{aligned}$$

The velocity in the pipe is

$$\begin{aligned} \text{v} &= \frac{\dot{V}}{A} = \frac{1337\ \frac{\text{ft}^3}{\text{hr}}}{(0.08841\ \text{ft}^2)\left(3600\ \frac{\text{sec}}{\text{hr}}\right)} \\ &= \boxed{4.20\ \text{ft/sec}\quad (4\ \text{ft/sec})} \end{aligned}$$

The answer is (B).

(b) From App. 14.A, the kinematic viscosity of water at 60°F is

$$\begin{aligned} \nu &= 1.217 \times 10^{-5}\ \text{ft}^2/\text{sec} \\ \rho &= 62.37\ \text{lbm/ft}^3 \end{aligned}$$

The Reynolds number is

$$\begin{aligned} \text{Re} &= \frac{D\text{v}}{\nu} = \frac{(0.3355\ \text{ft})\left(4.20\ \frac{\text{ft}}{\text{sec}}\right)}{1.217 \times 10^{-5}\ \frac{\text{ft}^2}{\text{sec}}} \\ &= \boxed{1.16 \times 10^5\quad (1.2 \times 10^5)} \end{aligned}$$

The answer is (A).

(c) From App. 17.A, for welded and seamless steel, $\epsilon = 0.0002$ ft.

$$\frac{\epsilon}{D} = \frac{0.0002\ \text{ft}}{0.3355\ \text{ft}} \approx 0.0006$$

From App. 17.B, the friction factor, f, is 0.0205. The 7000 ft of equivalent length includes the pipe between the lake and the pump. The friction head is

$$\begin{aligned} h_f &= \frac{fL\text{v}^2}{2Dg} = \frac{(0.0205)(7000\ \text{ft})\left(4.20\ \frac{\text{ft}}{\text{sec}}\right)^2}{(2)(0.3355\ \text{ft})\left(32.2\ \frac{\text{ft}}{\text{sec}^2}\right)} \\ &= \boxed{117.2\ \text{ft}\quad (120\ \text{ft})} \end{aligned}$$

The answer is (B).

(d) The head added by the pump is

$$\begin{aligned} h_A &= h_f + h_z = 117.2\ \text{ft} + (12\ \text{ft} + 350\ \text{ft}) \\ &= 479.2\ \text{ft} \end{aligned}$$

From Table 18.5, the hydraulic horsepower is

$$\begin{aligned} \text{WHP} &= \frac{h_A Q(\text{SG})}{3956} = \frac{(479.2\ \text{ft})\left(10{,}000\ \frac{\text{gal}}{\text{hr}}\right)(1)}{\left(3956\ \frac{\text{ft-gal}}{\text{hp-min}}\right)\left(60\ \frac{\text{min}}{\text{hr}}\right)} \\ &= \boxed{20.2\ \text{hp}\quad (20\ \text{hp})} \end{aligned}$$

The answer is (D).

Fluids

(e) From Eq. 18.15, the electrical horsepower is

$$\text{EHP} = \frac{\text{WHP}}{\eta} = \frac{20.2\ \text{hp}}{0.7} = 28.9\ \text{hp}$$

At \$0.04/kW-hr, power costs for 1 hr are

$$(28.9\ \text{hp})\left(0.7457\ \frac{\text{kW}}{\text{hp}}\right)(1\ \text{hr})\left(0.04\ \frac{\$}{\text{kW-hr}}\right) = \boxed{\$0.86\ \text{per hour}\quad(\$1\ \text{per hour})}$$

The answer is (B).

(f) The motor horsepower, EHP, is 28.9 hp. Select the next higher standard motor size. Use a $\boxed{30\ \text{hp motor.}}$

The answer is (B).

(g) From Eq. 18.4(b),

$$h_{\text{atm}} = \frac{p_{\text{atm}}}{\rho} \times \frac{g_c}{g} = \frac{\left(14.7\ \frac{\text{lbf}}{\text{in}^2}\right)\left(12\ \frac{\text{in}}{\text{ft}}\right)^2}{62.37\ \frac{\text{lbm}}{\text{ft}^3}} \times \frac{32.2\ \frac{\text{lbm-ft}}{\text{lbf-sec}^2}}{32.2\ \frac{\text{ft}}{\text{sec}^2}} = 33.94\ \text{ft}$$

The friction loss through 300 ft is provided.

$$h_{f(s)} = \left(\frac{300\ \text{ft}}{7000\ \text{ft}}\right)h_f = \left(\frac{300\ \text{ft}}{7000\ \text{ft}}\right)(117.2\ \text{ft}) = 5.0\ \text{ft}$$

From App. 14.A, the vapor pressure head at 60°F is 0.59 ft.

From Eq. 18.30(a), the NPSHA is

$$\begin{aligned}\text{NPSHA} &= h_{\text{atm}} + h_{z(s)} - h_{f(s)} - h_{\text{vp}} \\ &= 33.94\ \text{ft} - 12\ \text{ft} - 5.0\ \text{ft} - 0.59\ \text{ft} \\ &= \boxed{16.35\ \text{ft}\quad(16\ \text{ft})}\end{aligned}$$

The answer is (D).

SI Solution

(a) From App. 16.C, for 4 in schedule-40 steel pipe,

$$D = 102.26\ \text{mm}$$
$$A = 82.30 \times 10^{-4}\ \text{m}^2$$

The velocity in the pipe is

$$\text{v} = \frac{\dot{V}}{A} = \frac{10.5\ \frac{\text{L}}{\text{s}}}{(82.30 \times 10^{-4}\ \text{m}^2)\left(1000\ \frac{\text{L}}{\text{m}^3}\right)} = \boxed{1.28\ \text{m/s}\quad(1.0\ \text{m/s})}$$

The answer is (B).

(b) From App. 14.B, at 16°C the water data are

$$\rho = 998.95\ \text{kg/m}^3$$
$$\mu = 1.1081 \times 10^{-3}\ \text{Pa·s}$$

The Reynolds number is

$$\text{Re} = \frac{\rho \text{v} D}{\mu} = \frac{\left(998.95\ \frac{\text{kg}}{\text{m}^3}\right)\left(1.28\ \frac{\text{m}}{\text{s}}\right)(102.26\ \text{mm})}{(1.1081 \times 10^{-3}\ \text{Pa·s})\left(1000\ \frac{\text{mm}}{\text{m}}\right)} = \boxed{1.18 \times 10^5\quad(1.2 \times 10^5)}$$

The answer is (A).

(c) From Table 17.2, for welded and seamless steel, $\epsilon = 6.0 \times 10^{-5}$ m.

$$\frac{\epsilon}{D} = \frac{(6.0 \times 10^{-5}\ \text{m})\left(1000\ \frac{\text{mm}}{\text{m}}\right)}{102.26\ \text{mm}} \approx 0.0006$$

From App. 17.B, the friction factor is $f = 0.0205$.

From Eq. 18.5, the friction head is

$$h_f = \frac{fL\text{v}^2}{2Dg} = \frac{(0.0205)(2300\ \text{m})\left(1.28\ \frac{\text{m}}{\text{s}}\right)^2\left(1000\ \frac{\text{mm}}{\text{m}}\right)}{(2)(102.26\ \text{mm})\left(9.81\ \frac{\text{m}}{\text{s}^2}\right)} = \boxed{38.5\ \text{m}\quad(40\ \text{m})}$$

The answer is (B).

(d) The head added by the pump is

$$h_A = h_f + h_z = 38.5\ \text{m} + 4\ \text{m} + 115\ \text{m} = 157.5\ \text{m}$$

From Table 18.6, the hydraulic power is

$$\begin{aligned}\text{WkW} &= \frac{9.81 h_A Q(\text{SG})}{1000} \\ &= \frac{\left(9.81\ \frac{\text{m}}{\text{s}^2}\right)(157.5\ \text{m})\left(10.5\ \frac{\text{L}}{\text{s}}\right)(1)}{1000\ \frac{\text{W·L}}{\text{kW·kg}}} \\ &= \boxed{16.22\ \text{kW}\quad(16\ \text{kW})}\end{aligned}$$

The answer is (D).

(e) From Eq. 18.15, the electrical power is

$$\text{EHP} = \frac{\text{WkW}}{\eta} = \frac{16.22\ \text{kW}}{0.7} = 23.2\ \text{kW}$$

At \$0.04/kW·h, power costs for 1 h are

$$(23.2\ \text{kW})(1\ \text{h})\left(0.04\ \frac{\$}{\text{kW·h}}\right) = \boxed{\$0.93\quad(\$1)\ \text{per hour}}$$

The answer is (B).

(f) The required motor power is 23.2 kW. Select the next higher standard motor size. Use a 25 kW motor.

The answer is (B).

(g) From Eq. 18.4(a),

$$h_{\text{atm}} = \frac{p}{\rho g} = \frac{(101\text{ kPa})\left(1000\ \frac{\text{Pa}}{\text{kPa}}\right)}{\left(998.95\ \frac{\text{kg}}{\text{m}^3}\right)\left(9.81\ \frac{\text{m}}{\text{s}^2}\right)} = 10.31\text{ m}$$

The prorated friction loss through 100 m is

$$h_{f(s)} = \left(\frac{100\text{ m}}{2300\text{ m}}\right)h_f = \left(\frac{100\text{ m}}{2300\text{ m}}\right)(38.3\text{ m}) = 1.67\text{ m}$$

From App. 23.N, the vapor pressure at 16°C is 0.01819 bar.

From Eq. 18.4,

$$h_{\text{vp}} = \frac{p_{\text{vp}}}{g\rho} = \frac{(0.01819\text{ bar})\left(1\times 10^5\ \frac{\text{Pa}}{\text{bar}}\right)}{\left(9.81\ \frac{\text{m}}{\text{s}^2}\right)\left(998.95\ \frac{\text{kg}}{\text{m}^3}\right)} = 0.19\text{ m}$$

From Eq. 18.30(a), the NPSHA is

$$\begin{aligned}\text{NPSHA} &= h_{\text{atm}} + h_{z(s)} - h_{f(s)} - h_{\text{vp}} \\ &= 10.31\text{ m} - 4\text{ m} - 1.67\text{ m} - 0.19\text{ m} \\ &= \boxed{4.45\text{ m}\quad(4.5\text{ m})}\end{aligned}$$

The answer is (D).

11. (a) Sewers are usually gravity-flow (open channel) systems, $\Delta p = 0$ and $\Delta v = 0$, so $h_f = \Delta z = 48$ ft.

$$Q = \frac{\left(250\ \frac{\text{gal}}{\text{person-day}}\right)(10{,}000\text{ people})}{\left(24\ \frac{\text{hr}}{\text{day}}\right)\left(60\ \frac{\text{min}}{\text{hr}}\right)} = 1736\text{ gal/min}$$

Solving for d from Eq. 17.29,

$$\begin{aligned}d_{\text{in}}^{4.87} &= \frac{10.44 L_{\text{ft}} Q_{\text{gpm}}^{1.85}}{C^{1.85} h_f} \\ &= \frac{(10.44)(5000\text{ ft})\left(1736\ \frac{\text{gal}}{\text{min}}\right)^{1.85}}{(130)^{1.85}(48\text{ ft})} \\ &= 131{,}462\end{aligned}$$

$$d = \boxed{11.24\text{ in}\quad[\text{round to 12 in minimum}]}$$

The answer is (B).

(b) Without having a specific pump curve, the number of pumps can only be specified based on general rules. Use the *Ten States' Standards* (TSS), which states:

- No station will have less than two identical pumps.
- Capacity must be met with one pump out of service.
- Provision must be made in order to alternate pumps automatically.

Two pumps are required, plus spares.

The answer is (A).

(c) With a variable speed pump, it will be possible to adjust to the wide variations in flow (100–250 gpcd). It may be possible to operate with one pump. However, TSS still requires two.

The answer is (B).

(d) With three constant speed pumps,

$$Q = \frac{1736\ \frac{\text{gal}}{\text{min}}}{3} = 579\text{ gal/min at maximum capacity}$$

To get to the pump, the sewage must descend 48 ft under the influence of gravity. The pump only has to lift the sewage 10 ft. From Table 18.5, assuming specific gravity ≈ 1.00,

$$\begin{aligned}\text{rated motor power} &= \frac{h_A Q(\text{SG})}{3956\eta} = \frac{(10\text{ ft})\left(579\ \frac{\text{gal}}{\text{min}}\right)(1)}{\left(3956\ \frac{\text{gal-ft}}{\text{min-hp}}\right)(0.60)} \\ &= \boxed{2.44\text{ hp}\quad(3\text{ hp})}\end{aligned}$$

The answer is (B).

(e) With two variable-speed pumps,

$$Q = \frac{1736\ \frac{\text{gal}}{\text{min}}}{2} = 868\text{ gal/min}$$

$$\begin{aligned}\text{rated motor power} &= \frac{h_A Q(\text{SG})}{3956\eta} = \frac{(10\text{ ft})\left(868\ \frac{\text{gal}}{\text{min}}\right)(1)}{\left(3956\ \frac{\text{gal-ft}}{\text{min-hp}}\right)(0.80)} \\ &= \boxed{2.74\text{ hp}\quad(3\text{ hp})}\end{aligned}$$

The answer is (A).

(f) Incoming flow rate (choke III) cannot be controlled. It is independent of sump level. All other options are valid.

The answer is (D).

12. (a) The static suction lift, $h_{p(s)}$, is 20 ft.

The answer is (B).

(b) The static discharge head, $h_{p(d)}$, is $\boxed{15 \text{ ft.}}$

The answer is (A).

(c) There is no pipe size specified, so h_v cannot be calculated. Even so, v is typically in the 5–10 ft/sec range, and $h_v \approx 0$.

Since pipe lengths are not given, assume $h_f \approx 0$.

$$20 \text{ ft} + 15 \text{ ft} + \frac{\left(80 \ \frac{\text{lbf}}{\text{in}^2}\right)\left(12 \ \frac{\text{in}}{\text{ft}}\right)^2}{62.4 \ \frac{\text{lbf}}{\text{ft}^3}} + 10 \text{ ft}$$

$$= \boxed{229.6 \text{ ft (230 ft) of water}}$$

The answer is (D).

(d) The mass flow rate is

$$\frac{(3.5 \text{ MGD})\left(62.4 \ \frac{\text{lbm}}{\text{ft}^3}\right)\left(10^6 \ \frac{\text{gal}}{\text{MG}}\right)}{\left(7.4805 \ \frac{\text{gal}}{\text{ft}^3}\right)\left(24 \ \frac{\text{hr}}{\text{day}}\right) \times \left(60 \ \frac{\text{min}}{\text{hr}}\right)\left(60 \ \frac{\text{sec}}{\text{min}}\right)} = 337.9 \text{ lbm/sec}$$

The rated motor output power does not depend on the motor efficiency. The motor produces what it is rated to produce. From Table 18.5,

$$P = \frac{h_A \dot{m}}{550\eta_{\text{pump}}} \times \frac{g}{g_c}$$

$$= \frac{(229.6 \text{ ft})\left(337.9 \ \frac{\text{lbm}}{\text{sec}}\right)}{\left(550 \ \frac{\text{ft-lbf}}{\text{hp-sec}}\right)(0.85)} \times \frac{32.2 \ \frac{\text{ft}}{\text{sec}^2}}{32.2 \ \frac{\text{lbm-ft}}{\text{lbf-sec}^2}}$$

$$= 166.0 \text{ hp}$$

$$\boxed{\text{Use a 200 hp motor.}}$$

The answer is (D).

13. *Customary U.S. Solution*

(a) From App. 16.B, for 3 in schedule-40 steel pipe,

$$D = 0.2557 \text{ ft}$$

$$A = 0.05134 \text{ ft}^2$$

From App. 17.D, the equivalent lengths for various fittings are

$$\text{flanged elbow}, L_e = 4.4 \text{ ft}$$

$$\text{wide-open gate valve}, L_e = 2.8 \text{ ft}$$

The total equivalent length of pipe and fittings is

$$L_e = 500 \text{ ft} + (6)(4.4 \text{ ft}) + (2)(2.8 \text{ ft})$$

$$= \boxed{532 \text{ ft} \quad (530 \text{ ft})}$$

The answer is (D).

(b) The flow rate is 100 gal/min.

The velocity in the pipe is

$$\text{v} = \frac{\dot{V}}{A} = \frac{100 \ \frac{\text{gal}}{\text{min}}}{(0.05134 \text{ ft}^2)\left(7.4805 \ \frac{\text{gal}}{\text{ft}^3}\right)\left(60 \ \frac{\text{sec}}{\text{min}}\right)}$$

$$= \boxed{4.34 \text{ ft/sec} \quad (4.3 \text{ ft/sec})}$$

The answer is (C).

(c) The Reynolds number is

$$\text{Re} = \frac{\text{v}D}{\nu} = \frac{\left(4.34 \ \frac{\text{ft}}{\text{sec}}\right)(0.2557 \text{ ft})}{6 \times 10^{-6} \ \frac{\text{ft}^2}{\text{sec}}} = 1.85 \times 10^5$$

From App. 17.A, $\epsilon = 0.0002$ ft.

$$\frac{\epsilon}{D} = \frac{0.0002 \text{ ft}}{0.2557 \text{ ft}} \approx 0.0008$$

From the friction factor table, $f \approx 0.0204$.

For higher flow rates, f approaches 0.0186. Since the chosen flow rate was almost the lowest, $f = 0.0186$ should be used.

From Eq. 18.5, the friction head loss is

$$h_f = \frac{fL\text{v}^2}{2Dg} = \frac{(0.0186)(532 \text{ ft})\left(4.34 \ \frac{\text{ft}}{\text{sec}}\right)^2}{(2)(0.2557 \text{ ft})\left(32.2 \ \frac{\text{ft}}{\text{sec}^2}\right)}$$

$$= \boxed{11.3 \text{ ft (11 ft) of gasoline}}$$

The answer is (D).

(d) The friction head loss neglects the small velocity head. The other system points can be found using Eq. 18.43.

$$\frac{h_{f_1}}{h_{f_2}} = \left(\frac{Q_1}{Q_2}\right)^2$$

$$h_{f_2} = h_{f_1}\left(\frac{Q_2}{100 \ \frac{\text{gal}}{\text{min}}}\right)^2 = (11.3 \text{ ft})\left(\frac{Q_2}{100 \ \frac{\text{gal}}{\text{min}}}\right)^2$$

$$= 0.00113Q_2^2$$

Q (gal/min)	h_f (ft)	$h_f + 60$ (ft)
100	11.3	71.3
200	45.2	105.2
300	101.7	161.7
400	180.8	240.8
500	282.5	342.5
600	406.8	466.8

Fluids

Plot the system and pump curves. The pump's characteristic curve is independent of the liquid's specific gravity.

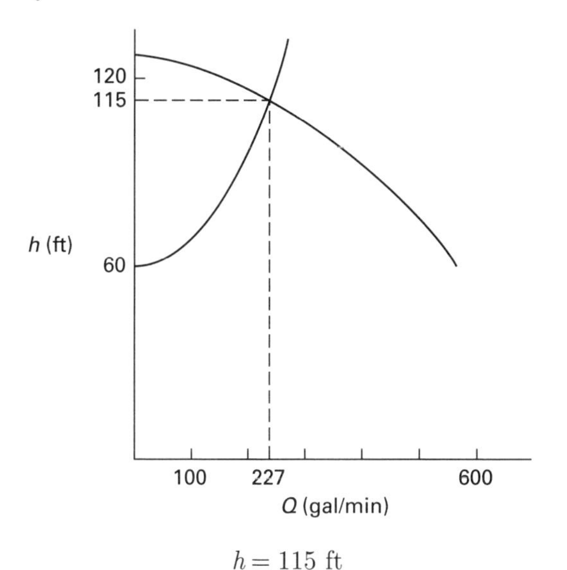

$$h = 115 \text{ ft}$$

The transfer rate is

$$\boxed{Q = 227 \text{ gal/min} \quad (230 \text{ gal/min})}$$

(This value could be used to determine a new friction factor.)

The answer is (D).

(e) From Table 18.5, the electrical power supplied to the motor is

$$\text{EHP} = \frac{h_A Q(\text{SG})}{3956\eta_{\text{pump}}\eta_{\text{motor}}} = \frac{(115 \text{ ft})\left(227 \ \frac{\text{gal}}{\text{min}}\right)(0.7)}{\left(3956 \ \frac{\text{ft-gal}}{\text{hp-min}}\right)(0.88)(0.88)}$$

$$= 5.96 \text{ hp}$$

The cost per hour is

$$(5.96 \text{ hp})\left(0.7457 \ \frac{\text{kW}}{\text{hp}}\right)(1 \text{ hr})\left(0.045 \ \frac{\$}{\text{kW·h}}\right) = \boxed{\$0.20}$$

The answer is (A).

SI Solution

(a) From App. 16.C, for 3 in schedule-40 pipe,

$$D = 77.92 \text{ mm}$$

$$A = 47.69 \times 10^{-4} \text{ m}^2$$

From App. 17.D, the equivalent lengths for various fittings are

$$\text{flanged elbow, } L_e = 4.4 \text{ ft}$$

$$\text{wide-open gate valve, } L_e = 2.8 \text{ ft}$$

The total equivalent length of pipe and fittings is

$$L_e = 170 \text{ m} + \big((6)(4.4 \text{ ft}) + (2)(2.8 \text{ ft})\big)\left(0.3048 \ \frac{\text{m}}{\text{ft}}\right)$$

$$= \boxed{179.8 \text{ m} \quad (180 \text{ m})}$$

The answer is (D).

(b) The flow rate is 6.3 L/s. The velocity in the pipe is

$$\text{v} = \frac{\dot{V}}{A} = \frac{6.3 \ \frac{\text{L}}{\text{s}}}{(47.69 \times 10^{-4} \text{ m}^2)\left(1000 \ \frac{\text{L}}{\text{m}^3}\right)}$$

$$= \boxed{1.32 \text{ m/s} \quad (1.3 \text{ m/s})}$$

The answer is (C).

(c) The Reynolds number is

$$\text{Re} = \frac{\text{v}D}{\nu} = \frac{\left(1.32 \ \frac{\text{m}}{\text{s}}\right)(77.92 \text{ mm})}{\left(5.6 \times 10^{-7} \ \frac{\text{m}^2}{\text{s}}\right)\left(1000 \ \frac{\text{mm}}{\text{m}}\right)} = 1.84 \times 10^5$$

From Table 17.2, $\epsilon = 6.0 \times 10^{-5}$ m.

$$\frac{\epsilon}{D} = \frac{(6.0 \times 10^{-5} \text{ m})\left(1000 \ \frac{\text{mm}}{\text{m}}\right)}{77.92 \text{ mm}} \approx 0.0008$$

From App. 17.B, $f = 0.0204$.

For higher flow rates, f approaches 0.0186. Since the chosen flow rate was almost the lowest, $f = 0.0186$ should be used.

From Eq. 18.5, the friction head loss is

$$h_f = \frac{fL\text{v}^2}{2Dg} = \frac{(0.0186)(179.8 \text{ m})\left(1.32 \ \frac{\text{m}}{\text{s}}\right)^2\left(1000 \ \frac{\text{mm}}{\text{m}}\right)}{(2)(77.92 \text{ mm})\left(9.81 \ \frac{\text{m}}{\text{s}^2}\right)}$$

$$= \boxed{3.8 \text{ m of gasoline}}$$

The answer is (D).

(d) The friction head loss neglects the small velocity head. The other system points can be found using Eq. 18.43.

$$\frac{h_{f1}}{h_{f2}} = \left(\frac{Q_1}{Q_2}\right)^2$$

$$h_{f2} = h_{f1}\left(\frac{Q_2}{Q_1}\right)^2 = (3.80 \text{ m})\left(\frac{Q_2}{6.3 \ \frac{\text{L}}{\text{s}}}\right)^2$$

$$= 0.0957Q_2^2$$

Q (L/s)	h_f (m)	$h_f + 20$ (m)
6.3	3.80	23.80
12	13.78	33.78
18	31.0	51.0
24	55.1	75.1
30	86.1	106.1
36	124.0	144.0

Plot the system and pump curves. The pump's characteristic curve is independent of the liquid's specific gravity.

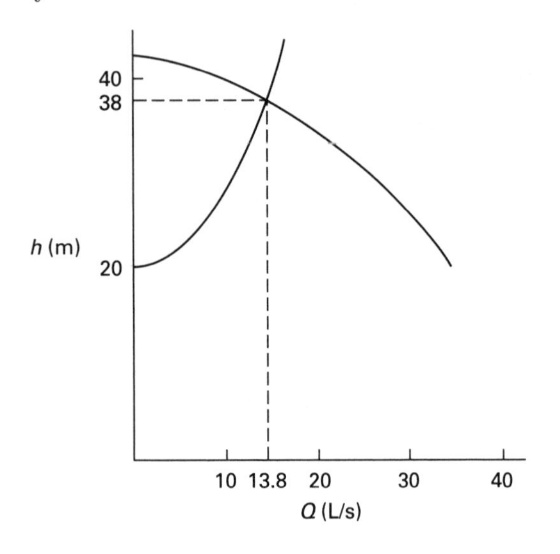

$$h = 38.0 \text{ m}$$

The transfer rate is

$$\boxed{Q = 13.6 \text{ L/s} \quad (14 \text{ L/s})}$$

(This value could be used to determine a new friction factor.)

The answer is (D).

(e) From Table 18.6, the electric power delivered to the motor is

$$\begin{aligned}\text{EkW} &= \frac{9.81 h_A Q(\text{SG})}{1000\eta_{\text{pump}}\eta_{\text{motor}}} \\ &= \frac{\left(9.81 \ \frac{\text{m}}{\text{s}^2}\right)(38.0 \text{ m})\left(13.8 \ \frac{\text{L}}{\text{s}}\right)(0.7)}{\left(1000 \ \frac{\text{W}}{\text{kW}}\right)(0.88)(0.88)} \\ &= 4.65 \text{ kW}\end{aligned}$$

The cost per hour is

$$(4.65 \text{ kW})(1 \text{ h})\left(0.045 \ \frac{\$}{\text{kW·h}}\right) = \boxed{\$0.21 \quad (\$0.20)}$$

The answer is (A).

14. *Customary U.S. Solution*

From Table 18.5, the hydraulic horsepower is

$$\text{WHP} = \frac{\Delta p Q}{1714}$$
$$\Delta p = p_d - p_s$$

The absolute pressures are

$$\begin{aligned}p_d &= 40 \text{ psig} + 14.7 \text{ psia} = 54.7 \text{ psia} \\ p_s &= 1 \text{ atm} = 14.7 \text{ psia} \\ \Delta p &= p_d - p_s = 54.7 \text{ psia} - 14.7 \text{ psia} = 40 \text{ psia}\end{aligned}$$

$$\text{WHP} = \frac{\left(40 \ \frac{\text{lbf}}{\text{in}^2}\right)\left(37 \ \frac{\text{gal}}{\text{min}}\right)}{1714 \ \frac{\text{lbf-gal}}{\text{in}^2\text{-min-hp}}} = \boxed{0.863 \text{ hp} \quad (0.9 \text{ hp})}$$

The answer is (B).

SI Solution

From Table 18.6, the hydraulic kilowatts are

$$\text{WkW} = \frac{\Delta p Q}{1000}$$
$$\Delta p = p_d - p_s$$

The absolute pressures are

$$\begin{aligned}p_d &= 275 \text{ kPa} + 101.3 \text{ kPa} = 376.3 \text{ kPa} \\ p_s &= 1 \text{ atm} = 101.3 \text{ kPa} \\ \Delta p &= p_d - p_s = 376.3 \text{ kPa} - 101.3 \text{ kPa} = 275 \text{ kPa}\end{aligned}$$

$$\text{WkW} = \frac{(275 \text{ kPa})\left(65 \ \frac{\text{L}}{\text{s}}\right)}{1000 \ \frac{\text{W}}{\text{kW}}} = \boxed{17.88 \text{ kW} \quad (18 \text{ kW})}$$

The answer is (B).

15. *Customary U.S. Solution*

From Eq. 18.28(b), the specific speed is

$$n_s = \frac{n\sqrt{Q}}{h_A^{0.75}}$$

For a double-suction pump, Q in the preceding equation is half of the full flow rate.

$$\begin{aligned}n_s &= \frac{\left(900 \ \frac{\text{rev}}{\text{min}}\right)\sqrt{\left(\frac{1}{2}\right)\left(300 \ \frac{\text{gal}}{\text{sec}}\right)\left(60 \ \frac{\text{sec}}{\text{min}}\right)}}{(20 \text{ ft})^{0.75}} \\ &= \boxed{9028 \text{ rpm} \quad (9000 \text{ rpm})}\end{aligned}$$

The answer is (C).

SI Solution

From Eq. 18.28(a), the specific speed is

$$n_s = \frac{n\sqrt{\dot{V}}}{h_A^{0.75}}$$

For a double-suction pump, $\dot{V}$ in the preceding equation is half of the full flow rate.

$$n_s = \frac{\left(900\ \frac{\text{rev}}{\text{min}}\right)\sqrt{\left(\frac{1}{2}\right)\left(1.1\ \frac{\text{kL}}{\text{s}}\right)\left(1\ \frac{\text{m}^3}{\text{kL}}\right)}}{(7\ \text{m})^{0.75}} = \boxed{155.1\ \text{rpm}\quad(160\ \text{rpm})}$$

The answer is (C).

16. *Customary U.S. Solution*

This problem is solved graphically using the charts of maximum suction lift from *Standards of the Hydraulic Institute.*

Each stage adds 150 ft of head, and the suction lift is 10 ft. Therefore, for a single-suction pump, $n \approx$ $\boxed{2050\ \text{rpm}\ (2000\ \text{rpm}).}$

The answer is (D).

17. *Customary U.S. Solution*

From App. 16.B, for 1.5 in schedule-40 steel pipe,

$$D = 0.1342\ \text{ft}$$
$$A = 0.01414\ \text{ft}^2$$

The velocity in the pipe is

$$\text{v} = \frac{\dot{V}}{A} = \frac{100\ \frac{\text{gal}}{\text{min}}}{(0.01414\ \text{ft}^2)\left(7.4805\ \frac{\text{gal}}{\text{ft}^3}\right)\left(60\ \frac{\text{sec}}{\text{min}}\right)} = 15.76\ \text{ft/sec}$$

From App. 17.D, the equivalent lengths for screwed steel fittings are

inlet (square mouth): $L_e = 3.1$ ft

long radius 90° elbow: $L_e = 3.4$ ft

wide-open gate valves: $L_e = 1.2$ ft

The total equivalent length is

$$30\ \text{ft} + 3.1\ \text{ft} + (2)(3.4\ \text{ft}) + (2)(1.2\ \text{ft}) = 42.3\ \text{ft}$$

From App. 17.A, for steel, $\epsilon = 0.0002$ ft.

$$\frac{\epsilon}{D} = \frac{0.0002\ \text{ft}}{0.1342\ \text{ft}} = 0.0015$$

At 281°F, $\nu = 0.239 \times 10^{-5}\ \text{ft}^2\text{/sec}$. The Reynolds number is

$$\text{Re} = \frac{D\text{v}}{\nu} = \frac{(0.1342\ \text{ft})\left(15.76\ \frac{\text{ft}}{\text{sec}}\right)}{0.239 \times 10^{-5}\ \frac{\text{ft}^2}{\text{sec}}} = 8.85 \times 10^5$$

From App. 17.B, the friction factor is $f = 0.022$.

From Eq. 18.5, the friction head is

$$h_f = \frac{fL\text{v}^2}{2Dg} = \frac{(0.022)(42.3\ \text{ft})\left(15.76\ \frac{\text{ft}}{\text{sec}}\right)^2}{(2)(0.1342\ \text{ft})\left(32.2\ \frac{\text{ft}}{\text{sec}^2}\right)} = 26.74\ \text{ft}$$

The density of the liquid is the reciprocal of the specific volume, taken from App. 23.A at 281°F.

$$\rho = \frac{1}{\upsilon_f} = \frac{1}{0.01727\ \frac{\text{ft}^3}{\text{lbm}}} = 57.9\ \text{lbm/ft}^3$$

From Eq. 18.4(b), the vapor pressure head is

$$h_{\text{vp}} = \frac{p_{\text{vapor}}}{\rho} \times \frac{g_c}{g} = \frac{\left(50.06\ \frac{\text{lbf}}{\text{in}^2}\right)\left(12\ \frac{\text{in}}{\text{ft}}\right)^2}{57.9\ \frac{\text{lbm}}{\text{ft}^3}} \times \frac{32.2\ \frac{\text{lbm-ft}}{\text{lbf-sec}^2}}{32.2\ \frac{\text{ft}}{\text{sec}^2}} = 124.5\ \text{ft}$$

From Eq. 18.4(b), the pressure head is

$$h_p = \frac{p}{\rho} \times \frac{g_c}{g} = \frac{\left(80\ \frac{\text{lbf}}{\text{in}^2}\right)\left(12\ \frac{\text{in}}{\text{ft}}\right)^2}{57.9\ \frac{\text{lbm}}{\text{ft}^3}} \times \frac{32.2\ \frac{\text{lbm-ft}}{\text{lbf-sec}^2}}{32.2\ \frac{\text{ft}}{\text{sec}^2}} = 199.0\ \text{ft}$$

From Eq. 18.30, the NPSHA is

$$\begin{aligned}\text{NPSHA} &= h_p + h_{z(s)} - h_{f(s)} - h_{\text{vp}} \\ &= 199.0\ \text{ft} + 20\ \text{ft} - 26.74\ \text{ft} - 124.5\ \text{ft} \\ &= \boxed{67.8\ \text{ft}\quad(68\ \text{ft})}\end{aligned}$$

Since NPSHR = 10 ft, $\boxed{\text{the pump will not cavitate.}}$

(A pump may not be needed in this configuration.)

The answer is (D).

Fluids

SI Solution

From App. 16.C, for 1.5 in schedule-40 steel pipe,

$$D = 40.89 \text{ mm}$$
$$A = 13.13 \times 10^{-4} \text{ m}^2$$

The velocity in the pipe is

$$\text{v} = \frac{\dot{V}}{A} = \frac{6.3 \frac{\text{L}}{\text{s}}}{(13.13 \times 10^{-4} \text{ m}^2)\left(1000 \frac{\text{L}}{\text{m}^3}\right)} = 4.80 \text{ m/s}$$

From App. 17.D, the equivalent lengths for screwed steel fittings are

inlet (square mouth): $L_e = 3.1$ ft

long radius 90° elbow: $L_e = 3.4$ ft

wide-open gate valves: $L_e = 1.2$ ft

The total equivalent length is

$$10 \text{ m} + \begin{pmatrix} 3.1 \text{ ft} + (2)(3.4 \text{ ft}) \\ + (2)(1.2 \text{ ft}) \end{pmatrix}\left(0.3048 \frac{\text{m}}{\text{ft}}\right) = 13.75 \text{ m}$$

From Table 17.2, for steel, $\epsilon = 6.0 \times 10^{-5}$ m.

$$\frac{\epsilon}{D} = \frac{(6.0 \times 10^{-5} \text{ m})\left(1000 \frac{\text{mm}}{\text{m}}\right)}{40.89 \text{ mm}} \approx 0.0015$$

At 138°C, $\nu = 0.222 \times 10^{-6}$ m²/s. The Reynolds number is

$$\text{Re} = \frac{D\text{v}}{\nu} = \frac{(40.89 \text{ mm})\left(4.80 \frac{\text{m}}{\text{s}}\right)}{\left(0.222 \times 10^{-6} \frac{\text{m}^2}{\text{s}}\right)\left(1000 \frac{\text{mm}}{\text{m}}\right)} = 8.84 \times 10^5$$

From App. 17.B, the friction factor is $f = 0.022$.

From Eq. 18.5, the friction head is

$$h_f = \frac{fL\text{v}^2}{2Dg} = \frac{(0.022)(13.75 \text{ m})\left(4.80 \frac{\text{m}}{\text{s}}\right)^2\left(1000 \frac{\text{mm}}{\text{m}}\right)}{(2)(40.89 \text{ mm})\left(9.81 \frac{\text{m}}{\text{s}^2}\right)} = 8.69 \text{ m}$$

The density of the liquid is the reciprocal of the specific volume. From App. 23.A at 281°F (138°C),

$$\rho = \frac{1}{\upsilon_f} = \frac{(1)\left(3.281 \frac{\text{ft}}{\text{m}}\right)^3}{\left(0.01727 \frac{\text{ft}^3}{\text{lbm}}\right)\left(2.205 \frac{\text{lbm}}{\text{kg}}\right)} = 927.5 \text{ kg/m}^3$$

From Eq. 18.4(a), the vapor pressure head is

$$h_{\text{vp}} = \frac{p_{\text{vapor}}}{\rho g} = \frac{(3.431 \text{ bar})\left(1 \times 10^5 \frac{\text{Pa}}{\text{bar}}\right)}{\left(927.5 \frac{\text{kg}}{\text{m}^3}\right)\left(9.81 \frac{\text{m}}{\text{s}^2}\right)} = 37.71 \text{ m}$$

From Eq. 18.4(a), the pressure head is

$$h_p = \frac{p}{\rho g} = \frac{(550 \text{ kPa})\left(1000 \frac{\text{Pa}}{\text{kPa}}\right)}{\left(927.5 \frac{\text{kg}}{\text{m}^3}\right)\left(9.81 \frac{\text{m}}{\text{s}^2}\right)} = 60.45 \text{ m}$$

From Eq. 18.30, the NPSHA is

$$\begin{aligned} \text{NPSHA} &= h_p + h_{z(s)} - h_{f(s)} - h_{\text{vp}} \\ &= 60.45 \text{ m} + 6 \text{ m} - 8.69 \text{ m} - 37.71 \text{ m} \\ &= \boxed{20.05 \text{ m} \quad (21 \text{ m})} \end{aligned}$$

Since NPSHR is 3 m, the pump will not cavitate.

(A pump may not be needed in this configuration.)

The answer is (D).

18. The solvent is the freshwater, and the solution is the seawater. Since seawater contains approximately 2½% salt (NaCl) by weight, 100 lbm of seawater will yield 2.5 lbm salt and 97.5 lbm water. The molecular weight of salt is 23.0 + 35.5 = 58.5 lbm/lbmol. The number of moles of salt in 100 lbm of seawater is

$$n_{\text{salt}} = \frac{m}{\text{MW}} = \frac{2.5 \text{ lbm}}{58.5 \frac{\text{lbm}}{\text{lbmol}}} = 0.043 \text{ lbmol}$$

Similarly, water's molecular weight is 18.016 lbm/lbmol. The number of moles of water is

$$n_{\text{water}} = \frac{97.5 \text{ lbm}}{18.016 \frac{\text{lbm}}{\text{lbmol}}} = 5.412 \text{ lbmol}$$

The mole fraction of water is

$$\frac{5.412 \text{ lbmol}}{5.412 \text{ lbmol} + 0.043 \text{ lbmol}} = 0.992$$

Customary U.S. Solution

Cavitation will occur when

$$h_{atm} - h_v < h_{vp}$$

The density of seawater is 64.0 lbm/ft^3.

From Eq. 18.4(b), the atmospheric head is

$$h_{atm} = \frac{p}{\rho} \times \frac{g_c}{g} = \frac{\left(14.7\ \frac{lbf}{in^2}\right)\left(12\ \frac{in}{ft}\right)^2}{64.0\ \frac{lbm}{ft^3}} \times \frac{32.2\ \frac{lbm\text{-}ft}{lbf\text{-}sec^2}}{32.2\ \frac{ft}{sec^2}}$$

$$= 33.075\ ft \quad \text{[ft of seawater]}$$

$$h_{depth} = 8\ ft \quad \text{[given]}$$

From Eq. 18.6, the velocity head is

$$h_v = \frac{v_{propeller}^2}{2g} = \frac{(4.2v_{boat})^2}{(2)\left(32.2\ \frac{ft}{sec^2}\right)} = 0.2739v_{boat}^2$$

From App. 23.A, the vapor pressure of freshwater at 68°F is $p_{vp} = 0.3393$ psia.

From App. 14.A, the density of water at 68°F is 62.32 lbm/ft^3. Raoult's law predicts the actual vapor pressure of the solution.

$$p_{vapor,solution} = \begin{pmatrix} \text{mole fraction} \\ \text{of the solvent} \end{pmatrix} p_{vapor,solvent}$$

$$p_{vapor,seawater} = (0.992)\left(0.3393\ \frac{lbf}{in^2}\right) = 0.3366\ lbf/in^2$$

From Eq. 18.4(b), the vapor pressure head is

$$h_{vapor,seawater} = \frac{p}{\rho} \times \frac{g_c}{g}$$

$$= \frac{\left(0.3366\ \frac{lbf}{in^2}\right)\left(12\ \frac{in}{ft}\right)^2}{64.0\ \frac{lbm}{ft^3}} \times \frac{32.2\ \frac{lbm\text{-}ft}{lbf\text{-}sec^2}}{32.2\ \frac{ft}{sec^2}}$$

$$= 0.7574\ ft$$

Solve for the boat velocity.

$$8\ ft + 33.075\ ft - 0.2739v_{boat}^2 = 0.7574\ ft$$

$$v_{boat} = \boxed{12.13\ ft/sec \quad (12\ ft/sec)}$$

The answer is (B).

SI Solution

Cavitation will occur when

$$h_{atm} - h_v < h_{vp}$$

The density of seawater is 1024 kg/m^3.

From Eq. 18.4(a), the atmospheric head is

$$h_{atm} = \frac{p}{\rho g} = \frac{(101.3\ kPa)\left(1000\ \frac{Pa}{kPa}\right)}{\left(1024\ \frac{kg}{m^3}\right)\left(9.81\ \frac{m}{s^2}\right)} = 10.08\ m$$

$$h_{depth} = 3\ m \quad \text{[given]}$$

From Eq. 18.6, the velocity head is

$$h_v = \frac{v_{propeller}^2}{2g} = \frac{(4.2v_{boat})^2}{(2)\left(9.81\ \frac{m}{s^2}\right)} = 0.899v_{boat}^2$$

The vapor pressure of 20°C freshwater is

$$p_{vp} = (0.02339\ bar)\left(100\ \frac{kPa}{bar}\right) = 2.339\ kPa$$

From App. 14.B, the density of water at 20°C is 998.23 kg/m^3. Raoult's law predicts the actual vapor pressure of the solution.

$$p_{vapor,solution} = p_{vapor,solvent} \begin{pmatrix} \text{mole fraction} \\ \text{of the solvent} \end{pmatrix}$$

The solvent is the freshwater and the solution is the seawater.

The mole fraction of water is 0.992.

$$p_{vapor,seawater} = (2.339\ kPa)(0.992) = 2.320\ kPa$$

From Eq. 18.4(a), the vapor pressure head is

$$h_{vapor,seawater} = \frac{p}{\rho g} = \frac{(2.320\ kPa)\left(1000\ \frac{Pa}{kPa}\right)}{\left(1024\ \frac{kg}{m^3}\right)\left(9.81\ \frac{m}{s^2}\right)} = 0.231\ m$$

Solve for the boat velocity.

$$3\ m + 10.08\ m - 0.899v_{boat}^2 = 0.231\ m$$

$$v_{boat} = \boxed{3.78\ m/s \quad (3.8\ m/s)}$$

The answer is (B).

19. *Customary U.S. Solution*

(a) From App. 17.D, the equivalent lengths of various screwed steel fittings are

inlet: $L_e = 8.5$ ft [essentially a reentrant inlet]

check valve: $L_e = 19$ ft

long radius elbow: $L_e = 3.6$ ft

Fluids

The total equivalent length of the 2 in line is

$$\begin{aligned} L_e &= 12 \text{ ft} + 8.5 \text{ ft} + 19 \text{ ft} + (3)(3.6 \text{ ft}) + 80 \text{ ft} \\ &= 130.3 \text{ ft} \end{aligned}$$

From App. 16.B, for 2 in schedule-40 pipe,

$$D = 0.1723 \text{ ft}$$
$$A = 0.02330 \text{ ft}^2$$

Since the flow rate is unknown, it must be assumed in order to find velocity. Assume 90 gal/min.

$$\dot{V} = \frac{90 \dfrac{\text{gal}}{\text{min}}}{\left(7.4805 \dfrac{\text{gal}}{\text{ft}^3}\right)\left(60 \dfrac{\text{sec}}{\text{min}}\right)} = 0.2005 \text{ ft}^3/\text{sec}$$

The velocity is

$$\text{v} = \frac{\dot{V}}{A} = \frac{0.2005 \dfrac{\text{ft}^3}{\text{sec}}}{0.02330 \text{ ft}^2} = 8.605 \text{ ft/sec}$$

From App. 14.A, the kinematic viscosity of water at 70°F is $\nu = 1.059 \times 10^{-5}$ ft^2/sec.

The Reynolds number is

$$\text{Re} = \frac{D\text{v}}{\nu} = \frac{(0.1723 \text{ ft})\left(8.605 \dfrac{\text{ft}}{\text{sec}}\right)}{1.059 \times 10^{-5} \dfrac{\text{ft}^2}{\text{sec}}} = 1.4 \times 10^5$$

From Table 17.2, the specific roughness of steel pipe is $\epsilon = 0.0002$ ft.

$$\frac{\epsilon}{D} = \frac{0.0002 \text{ ft}}{0.1723 \text{ ft}} \approx 0.0012$$

From App. 17.B, $f = 0.022$. At 90 gal/min, the friction loss in the line from Eq. 18.5 is

$$h_f = \frac{fL\text{v}^2}{2Dg} = \frac{(0.022)(130.3 \text{ ft})\left(8.605 \dfrac{\text{ft}}{\text{sec}}\right)^2}{(2)(0.1723 \text{ ft})\left(32.2 \dfrac{\text{ft}}{\text{sec}^2}\right)} = 19.1 \text{ ft}$$

From Eq. 18.6, the velocity head at 90 gal/min is

$$h_\text{v} = \frac{\text{v}^2}{2g} = \frac{\left(8.605 \dfrac{\text{ft}}{\text{sec}}\right)^2}{(2)\left(32.2 \dfrac{\text{ft}}{\text{sec}^2}\right)} = 1.1 \text{ ft}$$

In general, the friction head and velocity head are proportional to v^2 and Q^2.

$$h_f = (19.1 \text{ ft})\left(\frac{Q_2}{90 \dfrac{\text{gal}}{\text{min}}}\right)^2$$

$$h_\text{v} = (1.1 \text{ ft})\left(\frac{Q_2}{90 \dfrac{\text{gal}}{\text{min}}}\right)^2$$

(The 7 ft suction lift is included in the 20 ft static discharge head.)

The equation for the total system head is

$$h = h_z + h_\text{v} + h_f = 20 \text{ ft} + (1.1 \text{ ft} + 19.1 \text{ ft})\left(\frac{Q_2}{90 \dfrac{\text{gal}}{\text{min}}}\right)^2$$

Q_2 (gal/min)	system head, h (ft)
0	20.0
10	20.2
20	21.0
30	22.2
40	24.0
50	26.2
60	29.0
70	32.2
80	36.0
90	40.2
100	44.9
110	50.2

The intersection point of the system curve and the pump curve defines the operating flow rate.

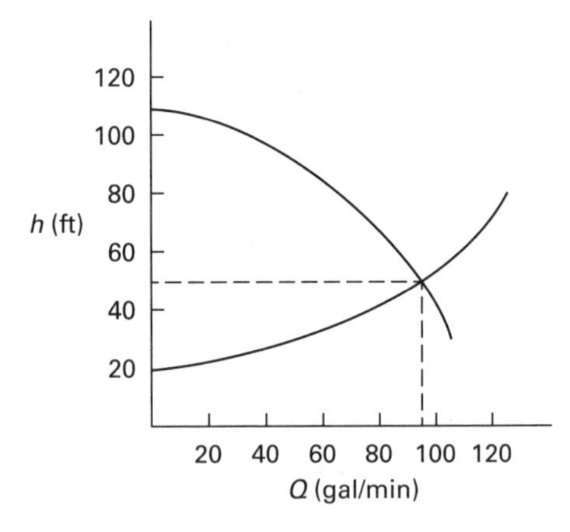

The flow rate is $\boxed{95 \text{ gal/min.}}$

The answer is (D).

(b) The intersection point is not in an efficient range for the pump because it is so far down on the system curve that the pumping efficiency will be low.

A different pump should be used.

The answer is (A).

SI Solution

(a) Use the equivalent lengths of various screwed steel fittings from the customary U.S. solution. The total equivalent length of 5.08 cm schedule-40 pipe is

$$L_e = 4\ \text{m} + \big(8.5\ \text{ft} + 19\ \text{ft} + (3)(3.6\ \text{ft})\big)\left(0.3048\ \frac{\text{m}}{\text{ft}}\right) + 27\ \text{m} = 42.67\ \text{m}$$

From App. 16.C, for 2 in schedule-40 pipe,

$$D = 52.501\ \text{mm}$$

$$A = 21.648 \times 10^{-4}\ \text{m}^2$$

Since the flow rate is unknown, it must be assumed in order to find velocity. Assume 6 L/s.

$$\dot{V} = \frac{6\ \frac{\text{L}}{\text{s}}}{1000\ \frac{\text{L}}{\text{m}^3}} = 6 \times 10^{-3}\ \text{m}^3/\text{s}$$

The velocity is

$$\text{v} = \frac{\dot{V}}{A} = \frac{6 \times 10^{-3}\ \frac{\text{m}^3}{\text{s}}}{21.648 \times 10^{-4}\ \text{m}^2} = 2.77\ \text{m/s}$$

From App. 14.B, the kinematic viscosity of water at 21°C is approximately $\nu = 9.849 \times 10^{-7}\ \text{m}^2/\text{s}$.

The Reynolds number is

$$\text{Re} = \frac{\text{v}D}{\nu} = \frac{\left(2.77\ \frac{\text{m}}{\text{s}}\right)(52.501\ \text{mm})}{\left(9.849 \times 10^{-7}\ \frac{\text{m}^2}{\text{s}}\right)\left(1000\ \frac{\text{mm}}{\text{m}}\right)} = 1.48 \times 10^5$$

From Table 17.2, the specific roughness of steel pipe is $\epsilon = 6.0 \times 10^{-5}$ m.

$$\frac{\epsilon}{D} = \frac{(6.0 \times 10^{-5}\ \text{m})\left(1000\ \frac{\text{mm}}{\text{m}}\right)}{52.501\ \text{mm}} \approx 0.0011$$

From App. 17.B, $f = 0.022$. At 6 L/s, the friction loss in the line from Eq. 18.5 is

$$h_f = \frac{fL\text{v}^2}{2Dg} = \frac{(0.022)(42.67\ \text{m})\left(2.77\ \frac{\text{m}}{\text{s}}\right)^2\left(1000\ \frac{\text{mm}}{\text{m}}\right)}{(2)(52.501\ \text{mm})\left(9.81\ \frac{\text{m}}{\text{s}^2}\right)} = 7.00\ \text{m}$$

At 6 L/s, the velocity head from Eq. 18.6 is

$$h_\text{v} = \frac{\text{v}^2}{2g} = \frac{\left(2.77\ \frac{\text{m}}{\text{s}}\right)^2}{(2)\left(9.81\ \frac{\text{m}}{\text{s}^2}\right)} = 0.39\ \text{m}$$

In general, the friction head and velocity head are proportional to v^2 and Q^2.

$$h_f = (7.00\ \text{m})\left(\frac{Q_2}{6\ \frac{\text{L}}{\text{s}}}\right)^2$$

$$h_\text{v} = (0.39\ \text{m})\left(\frac{Q_2}{6\ \frac{\text{L}}{\text{s}}}\right)^2$$

(The 2.3 m suction lift is included in the 6.3 m static discharge head.)

The equation for the total system head is

$$h = h_z + h_\text{v} + h_f = 6.3\ \text{m} + (0.39\ \text{m} + 7.00\ \text{m})\left(\frac{Q_2}{6\ \frac{\text{L}}{\text{s}}}\right)^2$$

Q_2 (L/s)	h (m)
0	6.3
0.6	6.37
1.2	6.60
1.8	6.96
2.4	7.48
3.2	8.40
3.6	8.96
4.4	10.27
4.8	11.03
5.7	12.97
6.0	13.69
6.5	14.98
7.0	16.36
7.5	17.85

The intersection point of the system curve and the pump curve defines the operating flow rate.

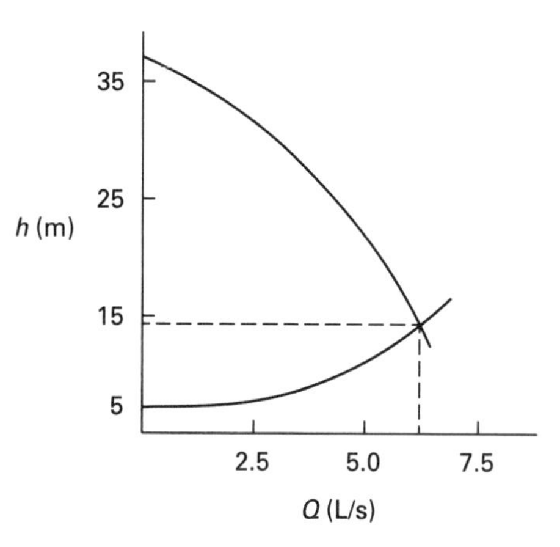

The flow rate is $\boxed{6.2\ \text{L/s.}}$

The answer is (D).

(b) The intersection point is not in an efficient range of the pump because it is so far down on the system curve that the pumping efficiency will be low.

A different pump should be used.

The answer is (A).

20. From Eq. 18.53,

$$P_2 = P_1\left(\frac{\rho_2 n_2^3 D_2^5}{\rho_1 n_1^3 D_1^5}\right) = P_1\left(\frac{n_2}{n_1}\right)^3 \quad [\rho_2 = \rho_1 \text{ and } D_2 = D_1]$$

Customary U.S. Solution

$$P_2 = (0.5 \text{ hp})\left(\frac{2000 \ \frac{\text{rev}}{\text{min}}}{1750 \ \frac{\text{rev}}{\text{min}}}\right)^3 = \boxed{0.75 \text{ hp}}$$

The answer is (D).

SI Solution

$$P_2 = (0.37 \text{ kW})\left(\frac{2000 \ \frac{\text{rev}}{\text{min}}}{1750 \ \frac{\text{rev}}{\text{min}}}\right)^3 = \boxed{0.55 \text{ kW}}$$

The answer is (D).

21. (a) Random values of Q are chosen, and the corresponding values of H are determined by the formula $H = 30 + 2Q^2$.

Q (ft^3/sec)	H (ft)
0	30
2.5	42.5
5	80
7.5	142.5
10	230
15	480
20	830
25	1280
30	1830

The intersection of the system curve and the 1400 rpm pump curve defines the operating point at that rpm.

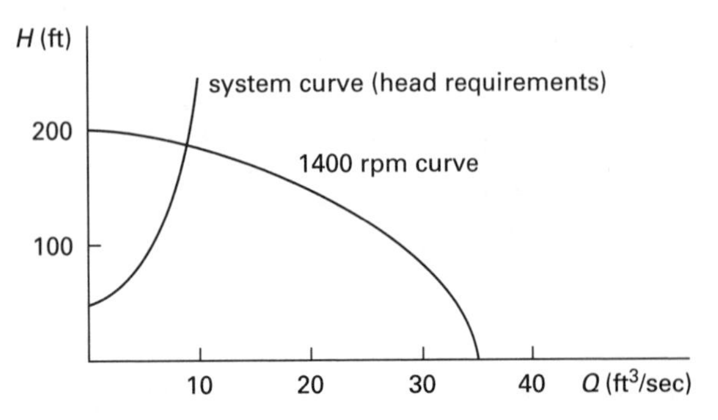

From the intersection of the graphs, at 1400 rpm the flow rate is approximately 9 ft^3/sec, and the corresponding head is $30 + (2)(9)^2 \approx 192$ ft.

$$Q = \left(9 \ \frac{\text{ft}^3}{\text{sec}}\right)\left(7.4805 \ \frac{\text{gal}}{\text{ft}^3}\right)\left(60 \ \frac{\text{sec}}{\text{min}}\right) = \boxed{4039 \text{ gal/min} \quad (4000 \text{ gal/min})}$$

The answer is (C).

(b) From Table 18.5, the hydraulic horsepower is

$$\text{WHP} = \frac{h_A \dot{V}(\text{SG})}{8.814} = \frac{(192 \text{ ft})\left(9 \ \frac{\text{ft}^3}{\text{sec}}\right)(1)}{8.814 \ \frac{\text{ft}^4}{\text{hp-sec}}} = 196 \text{ hp}$$

From Eq. 18.28(b), the specific speed is

$$n_s = \frac{n\sqrt{Q}}{h_A^{0.75}} = \frac{\left(1400 \ \frac{\text{rev}}{\text{min}}\right)\sqrt{4039 \ \frac{\text{gal}}{\text{min}}}}{(192 \text{ ft})^{0.75}} = 1725$$

From Fig. 18.8 with curve E, $\eta \approx 86\%$.

The minimum motor power should be

$$\frac{196 \text{ hp}}{0.86} = \boxed{228 \text{ hp} \quad (230 \text{ hp})}$$

The answer is (C).

(c) From Eq. 18.42,

$$Q_2 = Q_1\left(\frac{n_2}{n_1}\right) = \left(4039 \ \frac{\text{gal}}{\text{min}}\right)\left(\frac{1200 \ \frac{\text{rev}}{\text{min}}}{1400 \ \frac{\text{rev}}{\text{min}}}\right) = \boxed{3462 \text{ gal/min} \quad (3500 \text{ gal/min})}$$

The answer is (B).

22. Since turbines are essentially pumps running backward, use Table 18.5.

$$\Delta p = \left(30 \ \frac{\text{lbf}}{\text{in}^2} - 5 \ \frac{\text{lbf}}{\text{in}^2}\right)\left(12 \ \frac{\text{in}}{\text{ft}}\right)^2 = 3600 \text{ lbf/ft}^2$$

$$P = \frac{\Delta p \dot{V}}{550} = \frac{\left(3600 \ \frac{\text{lbf}}{\text{ft}^2}\right)\left(100 \ \frac{\text{ft}^3}{\text{sec}}\right)}{550 \ \frac{\text{ft-lbf}}{\text{hp-sec}}} = \boxed{654.5 \text{ hp} \quad (650 \text{ hp})}$$

The answer is (C).

23. The flow rate is

$$\dot{m} = \rho\dot{V} = \left(62.4\ \frac{\text{lbm}}{\text{ft}^3}\right)\left(1000\ \frac{\text{ft}^3}{\text{sec}}\right) = 6.24 \times 10^4\ \text{lbm/sec}$$

The head available for work is

$$\Delta h = 625\ \text{ft} - 58\ \text{ft} = 567\ \text{ft}$$

Use Table 18.5. The turbine efficiency is 89%. So, the power is

$$P = \frac{h_A\dot{m}}{550} \times \frac{g}{g_c} = \frac{(0.89)\left(6.24 \times 10^4\ \frac{\text{lbm}}{\text{sec}}\right)(567\ \text{ft})}{550\ \frac{\text{ft-lbf}}{\text{hp-sec}}} \times \frac{32.2\ \frac{\text{ft}}{\text{sec}^2}}{32.2\ \frac{\text{lbm-ft}}{\text{lbf-sec}^2}} = 57{,}253\ \text{hp}$$

Convert from hp to kW.

$$P = (57{,}253\ \text{hp})\left(0.7457\ \frac{\text{kW}}{\text{hp}}\right) = \boxed{4.27 \times 10^4\ \text{kW} \quad (43\ \text{MW})}$$

The answer is (C).

24. *Customary U.S. Solution*

(a) From App. 14.A, the density of water at 60°F is 62.37 lbm/ft^3. From Eq. 18.4(b), the head dropped is

$$h = \frac{\Delta p}{\rho} \times \frac{g_c}{g} = \frac{\left(500\ \frac{\text{lbf}}{\text{in}^2} - 30\ \frac{\text{lbf}}{\text{in}^2}\right)\left(12\ \frac{\text{in}}{\text{ft}}\right)^2}{62.37\ \frac{\text{lbm}}{\text{ft}^3}} \times \frac{32.2\ \frac{\text{lbm-ft}}{\text{lbf-sec}^2}}{32.2\ \frac{\text{ft}}{\text{sec}^2}} = 1085\ \text{ft}$$

From Eq. 18.56(b), the specific speed of a turbine is

$$n_s = \frac{n\sqrt{P_{\text{hp}}}}{h_t^{1.25}} = \frac{\left(1750\ \frac{\text{rev}}{\text{min}}\right)\sqrt{250\ \text{hp}}}{(1085\ \text{ft})^{1.25}} = \boxed{4.443 \quad (4)}$$

The answer is (A).

(b) The flow rate, $\dot{V}$, is

$$\dot{V} = A\text{v} = \left(\frac{\pi}{4}\right)\left(\frac{4\ \text{in}}{12\ \frac{\text{in}}{\text{ft}}}\right)^2\left(35\ \frac{\text{ft}}{\text{sec}}\right) = 3.054\ \text{ft}^3/\text{sec}$$

Since the analysis is for a single blade, not the entire turbine, only a portion of the water will catch up with the blade. From Eq. 17.195, the flow rate, considering the blade's movement away at 10 ft/sec, is

$$\dot{V}' = \frac{\left(35\ \frac{\text{ft}}{\text{sec}} - 10\ \frac{\text{ft}}{\text{sec}}\right)\left(3.054\ \frac{\text{ft}^3}{\text{sec}}\right)}{35\ \frac{\text{ft}}{\text{sec}}} = 2.181\ \text{ft}^3/\text{sec}$$

From Eq. 17.196(b) and Eq. 17.197(b), the forces in the x-direction and the y-direction are

$$F_x = \left(\frac{\dot{V}'\rho}{g_c}\right)(\text{v}_j - \text{v}_b)(\cos\theta - 1) = \left(\frac{\left(2.181\ \frac{\text{ft}^3}{\text{sec}}\right)\left(62.37\ \frac{\text{lbm}}{\text{ft}^3}\right)}{32.2\ \frac{\text{lbm-ft}}{\text{lbf-sec}^2}}\right) \times \left(35\ \frac{\text{ft}}{\text{sec}} - 10\ \frac{\text{ft}}{\text{sec}}\right)(\cos 80° - 1) = -87.27\ \text{lbf}$$

$$F_y = \left(\frac{\dot{V}'\rho}{g_c}\right)(\text{v}_j - \text{v}_b)\sin\theta = \left(\frac{\left(2.181\ \frac{\text{ft}^3}{\text{sec}}\right)\left(62.37\ \frac{\text{lbm}}{\text{ft}^3}\right)}{32.2\ \frac{\text{lbm-ft}}{\text{lbf-sec}^2}}\right) \times \left(35\ \frac{\text{ft}}{\text{sec}} - 10\ \frac{\text{ft}}{\text{sec}}\right)\sin 80° = 104.0\ \text{lbf}$$

The total force acting on the blade is

$$R = \sqrt{F_x^2 + F_y^2} = \sqrt{(-87.27\ \text{lbf})^2 + (104.0\ \text{lbf})^2} = \boxed{135.8\ \text{lbf} \quad (140\ \text{lbf})}$$

The answer is (B).

SI Solution

(a) From App. 14.B, the density of water at 16°C is 998.83 kg/m^3. From Eq. 18.4(a), the head dropped is

$$h = \frac{\Delta p}{\rho g} = \frac{(3.5\ \text{MPa})\left(1 \times 10^6\ \frac{\text{Pa}}{\text{MPa}}\right) - (210\ \text{kPa})\left(1000\ \frac{\text{Pa}}{\text{kPa}}\right)}{\left(998.83\ \frac{\text{kg}}{\text{m}^3}\right)\left(9.81\ \frac{\text{m}}{\text{s}^2}\right)} = 335.8\ \text{m}$$

From Eq. 18.56(a), the specific speed of a turbine is

$$n_s = \frac{n\sqrt{P_{\text{hp}}}}{h_t^{1.25}} = \frac{\left(1750\ \frac{\text{rev}}{\text{min}}\right)\sqrt{185\ \text{kW}}}{(335.8\ \text{m})^{1.25}}$$

$$= \boxed{16.56\quad(17)}$$

The answer is (A).

(b) The flow rate, $\dot{V}$, is

$$\dot{V} = A\text{v} = \left(\frac{\pi\left(\frac{100\ \text{mm}}{1000\ \frac{\text{mm}}{\text{m}}}\right)^2}{4}\right)\left(10.5\ \frac{\text{m}}{\text{s}}\right)$$

$$= 0.08247\ \text{m}^3/\text{s}$$

Since the analysis is for a single blade, not the entire turbine, only a portion of the water will catch up with the blade. From Eq. 17.195, the flow rate, considering the blade's movement away at 3 m/s, is

$$\dot{V}' = \frac{\left(10.5\ \frac{\text{m}}{\text{s}} - 3\ \frac{\text{m}}{\text{s}}\right)\left(0.08247\ \frac{\text{m}^3}{\text{s}}\right)}{10.5\ \frac{\text{m}}{\text{s}}}$$

$$= 0.05891\ \text{m}^3/\text{s}$$

From Eq. 17.196(a) and Eq. 17.197(a), the forces in the x-direction and the y-direction are

$$\begin{aligned}F_x &= \dot{V}'\rho(\text{v}_j - \text{v}_b)(\cos\theta - 1)\\ &= \left(0.05891\ \frac{\text{m}^3}{\text{s}}\right)\left(998.83\ \frac{\text{kg}}{\text{m}^3}\right)\\ &\quad\times\left(10.5\ \frac{\text{m}}{\text{s}} - 3\ \frac{\text{m}}{\text{s}}\right)(\cos 80^\circ - 1)\\ &= -364.7\ \text{N}\end{aligned}$$

$$\begin{aligned}F_y &= \dot{V}'\rho(\text{v}_j - \text{v}_b)\sin\theta\\ &= \left(0.05891\ \frac{\text{m}^3}{\text{s}}\right)\left(998.83\ \frac{\text{kg}}{\text{m}^3}\right)\\ &\quad\times\left(10.5\ \frac{\text{m}}{\text{s}} - 3\ \frac{\text{m}}{\text{s}}\right)\sin 80^\circ\\ &= 434.6\ \text{N}\end{aligned}$$

The total force acting on the blade is

$$R = \sqrt{F_x^2 + F_y^2} = \sqrt{(-364.7\ \text{N})^2 + (434.6\ \text{N})^2}$$

$$= \boxed{567.3\ \text{N}\quad(570\ \text{N})}$$

The answer is (B).

25. *Customary U.S. Solution*

(a) The total effective head is due to the pressure head, velocity head, and tailwater head.

$$h_{\text{eff}} = h_p + h_\text{v} - h_{z,\text{tailwater}} = h_p + \frac{\text{v}^2}{2g} - h_{z,\text{tailwater}}$$

$$= 92.5\ \text{ft} + \frac{\left(12\ \frac{\text{ft}}{\text{sec}}\right)^2}{(2)\left(32.2\ \frac{\text{ft}}{\text{sec}^2}\right)} - (-5.26\ \text{ft})$$

$$= \boxed{100\ \text{ft}}$$

The answer is (C).

(b) From Table 18.5, the theoretical hydraulic horsepower is

$$P_{\text{th}} = \frac{h_A\dot{V}(\text{SG})}{8.814} = \frac{(100\ \text{ft})\left(25\ \frac{\text{ft}^3}{\text{sec}}\right)(1)}{8.814\ \frac{\text{ft}^4}{\text{hp-sec}}}$$

$$= 283.6\ \text{hp}$$

The overall turbine efficiency is

$$\eta = \frac{P_{\text{brake}}}{P_{\text{th}}} = \frac{250\ \text{hp}}{283.6\ \text{hp}} = \boxed{0.882\quad(88\%)}$$

The answer is (B).

(c) From Eq. 18.43,

$$n_2 = n_1\sqrt{\frac{h_2}{h_1}} = \left(610\ \frac{\text{rev}}{\text{min}}\right)\sqrt{\frac{225\ \text{ft}}{100\ \text{ft}}}$$

$$= \boxed{915\ \text{rpm}\quad(920\ \text{rpm})}$$

The answer is (B).

(d) Combine Eq. 18.43 and Eq. 18.44.

$$P_2 = P_1\left(\frac{n_2}{n_1}\right)^3 = P_1\left(\left(\frac{h_2}{h_1}\right)^{1/2}\right)^3$$

$$= P_1\left(\frac{h_2}{h_1}\right)^{3/2}$$

$$= (250\ \text{hp})\left(\frac{225\ \text{ft}}{100\ \text{ft}}\right)^{3/2}$$

$$= \boxed{843.8\ \text{hp}\quad(840\ \text{hp})}$$

The answer is (D).

(e) From Eq. 18.43,

$$Q_2 = Q_1\sqrt{\frac{h_2}{h_1}} = \left(25\ \frac{\text{ft}^3}{\text{sec}}\right)\sqrt{\frac{225\ \text{ft}}{100\ \text{ft}}}$$

$$= \boxed{37.5\ \text{ft}^3/\text{sec}\quad(38\ \text{ft}^3/\text{sec})}$$

The answer is (B).

SI Solution

(a) The total effective head is due to the pressure head, velocity head, and tailwater head.

$$h_{\text{eff}} = h_p + h_{\text{v}} - h_{z,\text{tailwater}} = h_p + \frac{\text{v}^2}{2g} - h_{z,\text{tailwater}}$$

$$= 28.2 \text{ m} + \frac{\left(3.6 \ \frac{\text{m}}{\text{s}}\right)^2}{(2)\left(9.81 \ \frac{\text{m}}{\text{s}^2}\right)} - (-1.75 \text{ m})$$

$$= \boxed{30.61 \text{ m} \quad (31 \text{ m})}$$

The answer is (C).

(b) From Table 18.6, the theoretical hydraulic kilowatts are

$$P_{\text{th}} = \frac{9.81 h_A Q(\text{SG})}{1000}$$

$$= \frac{\left(9.81 \ \frac{\text{m}}{\text{s}^2}\right)(30.61 \text{ m})\left(700 \ \frac{\text{L}}{\text{s}}\right)(1)}{1000 \ \frac{\text{W}}{\text{kW}}}$$

$$= 210.2 \text{ kW}$$

The overall turbine efficiency is

$$\eta = \frac{P_{\text{brake}}}{P_{\text{th}}} = \frac{185 \text{ kW}}{210.2 \text{ kW}} = \boxed{0.88 \quad (88\%)}$$

The answer is (B).

(c) From Eq. 18.43,

$$n_2 = n_1\sqrt{\frac{h_2}{h_1}} = \left(610 \ \frac{\text{rev}}{\text{min}}\right)\sqrt{\frac{70 \text{ m}}{30.61 \text{ m}}}$$

$$= \boxed{922 \text{ rpm} \quad (920 \text{ rpm})}$$

The answer is (B).

(d) Combine Eq. 18.43 and Eq. 18.44.

$$P_2 = P_1\left(\frac{n_2}{n_1}\right)^3 = P_1\left(\left(\frac{h_2}{h_1}\right)^{1/2}\right)^3 = P_1\left(\frac{h_2}{h_1}\right)^{3/2}$$

$$= (185 \text{ kW})\left(\frac{70 \text{ m}}{30.61 \text{ m}}\right)^{3/2}$$

$$= \boxed{639.8 \text{ kW} \quad (640 \text{ kW})}$$

The answer is (D).

(e) From Eq. 18.43,

$$Q_2 = Q_1\sqrt{\frac{h_2}{h_1}} = \left(700 \ \frac{\text{L}}{\text{s}}\right)\sqrt{\frac{70 \text{ m}}{30.61 \text{ m}}}$$

$$= \boxed{1059 \text{ L/s} \quad (1100 \text{ L/s})}$$

The answer is (B).

19 Hydraulic and Pneumatic Systems

PRACTICE PROBLEMS

1. A valve is controlled by a pump and actuator connected by a tube with an internal diameter of 3.5 mm. The nominal working pressure in the tube is 28 MPa. The density of the hydraulic fluid is 860 kg/m^3. The effective bulk modulus is 1.5×10^9 Pa. When closing, the speed of a valve actuator is 10 m/s. The actuator piston diameter is 40 mm. The line length between the pump and the actuator is 11 m. The characteristic impedance of the actuator is most nearly

(A) 4×10^8 kg/s·m^4

(B) 9×10^8 kg/s·m^4

(C) 17×10^8 kg/s·m^4

(D) 31×10^8 kg/s·m^4

2. Given the situation described in Prob. 1, if the valve is suddenly closed, the maximum pressure in the transmission line will most nearly be

(A) 11 MPa

(B) 32 MPa

(C) 40 MPa

(D) 52 MPa

3. Given the situation described in Prob. 1, and assuming a lossless line, the fundamental frequency of the hydraulic line is most nearly

(A) 380 rad/s

(B) 620 rad/s

(C) 1700 rad/s

(D) 2500 rad/s

4. A steel hydraulic line is found to be leaking at the location of a threaded fitting. The first step an engineer should take to stop the leak should be to

(A) determine if the operating pressure in the line is within specifications

(B) shut down all production equipment drawing hydraulic fluid from the transmission line

(C) determine the type and torque rating of the connector

(D) de-energize the line and bring the hydraulic system to a zero-energy state

5. Which method should be used to locate a leak in a hydraulic system?

(A) Feel the area around the suspected leak area with a gloved hand.

(B) Inspect the area visually in normal light.

(C) Inject fluorescent dye into the hydraulic fluid and observe under black light.

(D) Determine that the operating pressure is within rated limits and tighten all connectors in the area to the correct torque rating while the system is running.

6. Maintenance of a hydraulic cylinder requires its disassembly. The hydraulic cylinder consists of a circular tube, piston, rod, and gland assembly that is retained by a metal retaining ring that fits into a groove machined around the inside diameter of cylinder tube. After removing the retaining ring, the gland assembly does not easily slide out. What method should be used to remove the stubborn gland assembly?

(A) Pressurize the cylinder with inert gas from a tank and allow the piston to push the gland assembly out.

(B) Pressurize the cylinder with compressed air from an air compressor and allow the piston to push the gland assembly out.

(C) Pressurize the cylinder with hydraulic fluid and allow the piston to push the gland assembly out.

(D) Use a rosebud end with an acetylene torch to thermally expand the gland-end of the tube.

7. Using a black iron pipe fitting to connect a hydraulic ram to a flexible rubber hose from a hydraulic pump will

(A) increase the temperature range over which the system can operate

(B) increase the natural frequency of vibration of the fluid transmission line

(C) decrease the resistance to fluid flow (i.e., decrease fluid friction)

(D) decrease the operating pressure range of the system

8. With regard to the testing of two-port fluid hydraulic components such as motors and cylinders, the phrase "testing to atmosphere" means

(A) life-testing a component in the humidity and temperature atmospheric conditions for which it is expected to be exposed

(B) life-testing a component in the humidity and temperature atmospheric conditions far in excess of those for which it is expected to be exposed

(C) applying hydraulic pressure to one port while leaving the other port open to the atmosphere

(D) exposing the interior of a heated component to pure oxygen for a standard test period in order to evaluate corrosion

9. After turning off the prime mover (motor and pump) and locking out the system, the safest method of confirming that a pressurized line is at a zero-energy state is by

(A) observing a pressure gage

(B) placing all moveable components in their lowest (least extended) positions

(C) placing all moveable components in their highest (most extended) positions

(D) cracking (i.e., loosening) a fitting in the line to bleed off residual pressure

10. Compressed air generally may be used to help drain liquid hydraulic fluid from a component or system

(A) when the hydraulic fluid is not being reused

(B) when the compressed air pressure is less than the pressure rating of the component or system

(C) when the compressed air first passes through a filter and moisture trap

(D) never

11. What type of component is represented by the illustration shown?

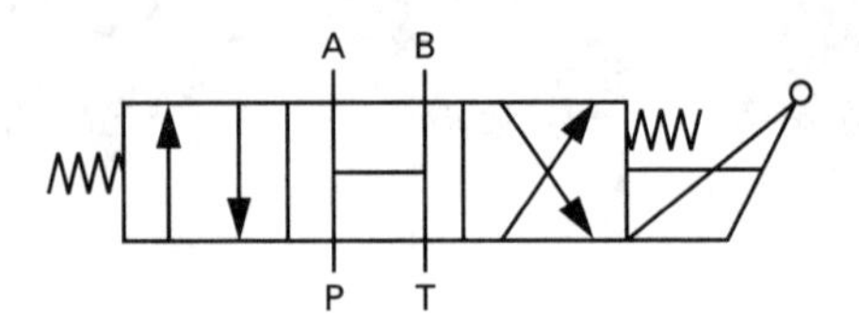

(A) open-center spool valve

(B) closed-center spool valve

(C) open-center spool valve with float

(D) closed-center spool valve with float

12. An uninsulated 4 ft^3 tank contains 60°F air. Air transfers into and out of the tank quickly. The tank's pneumatic compliance is most nearly

(A) 1.9×10^{-5} lbm-ft^2/lbf

(B) 1.0×10^{-4} lbm-ft^2/lbf

(C) 1.4×10^{-4} lbm-ft^2/lbf

(D) 8.7×10^{-4} lbm-ft^2/lbf

13. An electric motor drives a fixed displacement pump that operates a single rod cylinder to perform a weight-lifting operation. The hydraulic system is protected against overload by a pressure relief valve. A solenoid-operated 2/2 directional control valve controls the operation. Draw the system diagram using standard fluid power symbols.

14. 10 gpm of oil with kinematic viscosity 45 cSt flows through a pipe with an internal diameter of 1.0 in. Calculate the approximate Reynolds number, and determine if the flow is laminar or turbulent.

(A) 700; laminar

(B) 1700; laminar

(C) 5400; turbulent

(D) 19,200; turbulent

15. 1500 psig hydraulic fluid flows at the rate of 4 gpm into an actuating cylinder. The available power is most nearly

(A) 0.9 hp

(B) 2.6 hp

(C) 3.5 hp

(D) 7.2 hp

16. The volume of hydraulic fluid in a cylinder increases 2.0 in^3 when the hydraulic pressure is increased from 8000 psig to 8600 psig. The compliance of the cylinder is most nearly

(A) 1.3×10^{-8} in^5/lbf

(B) 2.3×10^{-5} in^5/lbf

(C) 0.0033 in^5/lbf

(D) 0.041 in^5/lbf

17. 30 inches of pipe with an inside diameter of 2 in is filled with SAE 30 oil. The oil's bulk modulus is 2.1×10^5 psi. The compliance of the hose is most nearly

(A) 1.8×10^{-9} ft^5/lbf

(B) 4.5×10^{-4} ft^5/lbf

(C) 5.2×10^{-2} ft^5/lbf

(D) 110 ft^5/lbf

18. A hydraulic motor displaces 10 cm^3 of hydraulic fluid per radian of rotation. Fluid pressure is increased by 8 MPa. The torque developed is most nearly

(A) 28 N·m

(B) 80 N·m

(C) 640 N·m

(D) 5120 N·m

19. The speed of sound in 25°C hydraulic fluid with a bulk modulus of 1.72 GPa and a density of 870 kg/m^3 is most nearly

(A) 790 m/s

(B) 950 m/s

(C) 1100 m/s

(D) 1400 m/s

20. A rod in an actuating cylinder travels at a maximum rate of 1 ft/sec when 30 gpm of hydraulic fluid flow through a hydraulic system at 2100 lbf/in^2. The cylinder is 85% efficient. A 1 in diameter hardened steel clevis pin is used to retain each end of the cylinder. Each pin is in double shear. Using the maximum shear stress theory with a factor of safety of 2.5, what should be the clevis pin's required minimum tensile strength?

(A) 55 kips/in^2

(B) 64 kips/in^2

(C) 110 kips/in^2

(D) 127 kips/in^2

21. A hydraulic wood splitter operating at a pressure of 2500 lbf/in^2 and flow rate of 8.8 gpm has a forward stroke velocity of 0.4 ft/sec. The hydraulic oil at normal operating temperature has a specific gravity of 1.0, a density of 62.4 lbm/ft^3, and a viscosity of 33.1 cS. The wood splitter has a 3 in bore. The cylinder efficiency is 85%. Fittings in the hydraulic path include a pressure relief valve ($C_v = 3.11$), a 4-way valve ($C_v = 2.90$), two flow control valves ($C_v = 2.5$), and a check valve ($C_v = 5.11$). There is also an equivalent of 40 ft of nominal $1/2$ in diameter tubing. The hydraulic system is driven by an 85% efficient pump. The system horsepower loss experienced in extending the cylinder is most nearly

(A) 0.6 hp

(B) 1.9 hp

(C) 2.5 hp

(D) 3.9 hp

22. A water tower feeds an irrigation system. To prevent soil erosion, a gate valve is used to limit the water discharge to 60 gpm. When fully open, the maximum pressure drop across the valve is limited to 5 lbf/in^2. The minimum size gate valve that can operate fully open is most nearly

(A) $1/2$ in

(B) $3/4$ in

(C) 1 in

(D) $1^1/4$ in

SOLUTIONS

1. The internal area of the actuator is

$$A = \frac{\pi D^2}{4} = \frac{\pi\left(\frac{40\ \text{mm}}{1000\ \frac{\text{mm}}{\text{m}}}\right)^2}{4} = 1.25664 \times 10^{-3}\ \text{m}^2$$

From Eq. 19.51, the characteristic impedance is

$$Z_o = \frac{1}{A}\sqrt{\rho B} = \frac{1}{1.25664 \times 10^{-3}\ \text{m}^2}\sqrt{\left(860\ \frac{\text{kg}}{\text{m}^3}\right)(1.5 \times 10^9\ \text{Pa})} = \boxed{9.03824 \times 10^8\ \text{kg/s}\cdot\text{m}^4 \quad (9 \times 10^8\ \text{kg/s}\cdot\text{m}^4)}$$

The answer is (B).

2. Using the internal area of the actuator as found in Prob. 1, the flow rate of control fluid in the valve is

$$\Delta Q = \Delta \text{v}_p A = \left(10\ \frac{\text{m}}{\text{s}}\right)(0.00125664\ \text{m}^2) = 0.0125664\ \text{m}^3/\text{s}$$

Use Joukowsky's equation, Eq. 19.58.

$$\Delta p = \Delta Q Z_o = \left(0.0125664\ \frac{\text{m}^3}{\text{s}}\right)\left(9.03824 \times 10^8\ \frac{\text{kg}}{\text{s}\cdot\text{m}^4}\right) = 1.13578 \times 10^7\ \text{Pa}$$

The maximum pressure includes the nominal pressure.

$$p_{\max} = p_o + \Delta p = 28 \times 10^6\ \text{Pa} + 1.13578 \times 10^7\ \text{Pa} = \boxed{3.94 \times 10^7\ \text{Pa} \quad (40\ \text{MPa})}$$

The answer is (C).

3. From Eq. 19.53, the effective speed of sound in the line is

$$a = \sqrt{\frac{B}{\rho}} = \sqrt{\frac{1.5 \times 10^9\ \text{Pa}}{860\ \frac{\text{kg}}{\text{m}^3}}} = 1320.7\ \text{m/s}$$

The time (i.e., the period) required for a pressure wave to travel from the pump to the actuator and return is

$$T = \frac{2L}{a} = \frac{(2)(11\ \text{m})}{1320.7\ \frac{\text{m}}{\text{s}}} = 1.6658 \times 10^{-2}\ \text{s}$$

The fundamental frequency is

$$\omega = 2\pi f = \frac{2\pi}{T} = \frac{2\pi}{1.6658 \times 10^{-2}\ \text{s}} = \boxed{377.2\ \text{rad/s} \quad (380\ \text{rad/s})}$$

The answer is (A).

4. Working on a pressurized system is highly dangerous. Both pressurized fluid and the subsequent sudden whipping movement of components spewing pressurized fluid can cause injury. The only safe procedure is to de-energize the system, lock it out (so that it cannot be turned on), complete the repair, and then check for leaks by re-energizing the system.

The answer is (D).

5. Physical contact with operating hydraulic systems is dangerous. To prevent injury from hot and pressurized hydraulic fluid, sharp edges, and moving components, tactile methods should not be used. Hands should never be inserted into spaces that cannot be observed. Fluorescent dye is not a typical evaluation methodology. Leaks should be observed visually using ambient light, flashlights, and/or bore scopes.

The answer is (B).

6. Filling the cylinder with hydraulic fluid and using a small hand pump to gradually increase the pressure (up to the maximum pressure rating of the cylinder, if necessary) is a common practice. Hydraulic fluid is incompressible, so a small movement of the gland assembly will eliminate the pressure in the cylinder. Use of compressed gas can result in explosive projectile motion of the gland, piston, and rod. Combining flame with flammable fluid and materials creates a fire and/ or explosion hazard.

The answer is (C).

7. Flexible rubber hoses and other components used in fluid power transmission lines operate at high pressures (1000 psi to 10,000 psi). The pressure ratings of black iron pipe fittings may be as low as 250 psi. Using a standard pipe fitting in a high pressure line will reduce the operating range severely and is extremely dangerous.

The answer is (D).

8. *Testing to atmosphere* is the dangerous practice of applying hydraulic pressure to only one port while leaving the other port open to the atmosphere. In addition to uncharacteristic performance of seals and glands, unexpected performance can include excessive hydraulic fluid flow (spewing) and uncontrolled movement of components (including projectile motion and whipping).

The answer is (C).

9. Due to fluid compliance in accumulators and flexible lines, pressurized fluid can remain when components are in any position. Cracking a high-pressure fitting can result in high-speed jets and sprays of fluid that can penetrate the body or create fire hazards, as well as allow pressurized components to move and whip about suddenly. All pressure systems should have a pressure gage to indicate their internal pressure.

The answer is (A).

10. Air pressure should never be applied to fluid power systems. Air pressure may remain trapped in components with integral pilot-controlled check valves, accumulators, and cylinders, and leave the components in an energized condition. Unlike with liquids, as a result of the compressibility of the gas, small movements of hydraulic component elements (e.g., pistons) pressurized with gas do not relieve the pressure.

The answer is (D).

11. An open-center (float) valve allows flow between all four ports when the valve is in the center (nonactuated) position. The actuator (downstream from the valve) is not held in position, but is free to float. An open-center valve also allows free flow from the inlet port to the return (or tank) port, but it blocks the actuator ports. The actuator cannot move (neglecting leakage) when the open-center valve is in the center position. A closed-center valve has all four ports blocked when it is in the center position. There is no pathway through the valve between any of the four ports.

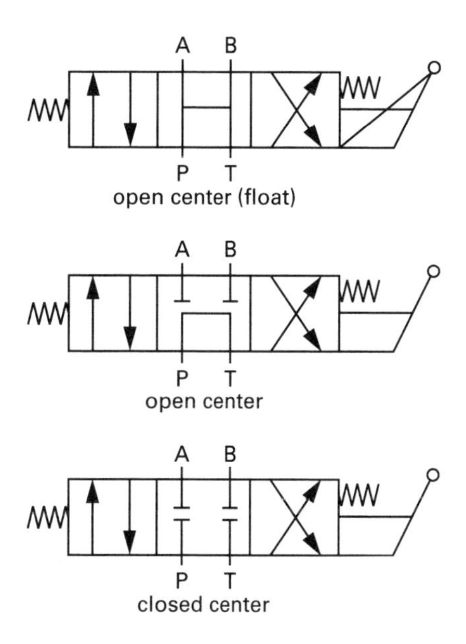

The answer is (C).

12. Since the air moves quickly, transfers are essentially adiabatic. The polytropic exponent is $n = k = 1.4$.

$$T = 60°\text{F} + 460° = 520°\text{R}$$

$$C_g = \frac{V}{nTR} = \frac{4\ \text{ft}^3}{(1.4)(520°\text{R})\left(53.35\ \frac{\text{ft-lbf}}{\text{lbm-°R}}\right)}$$

$$= \boxed{\begin{array}{l}1.03 \times 10^{-4}\ \text{lbm-ft}^2/\text{lbf} \\ \quad (1.0 \times 10^{-4}\ \text{lbm-ft}^2/\text{lbf})\end{array}}$$

The answer is (B).

13. The system diagram is as shown.

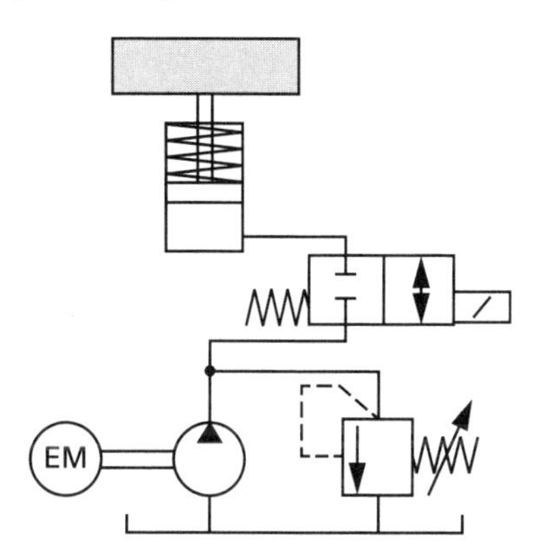

14. The flow rate is

$$Q = \frac{10\ \frac{\text{gal}}{\text{min}}}{\left(7.48\ \frac{\text{gal}}{\text{ft}^3}\right)\left(60\ \frac{\text{sec}}{\text{min}}\right)} = 0.0223\ \text{ft}^3/\text{sec}$$

The bulk velocity is

$$\text{v} = \frac{Q}{A} = \frac{0.0223\ \frac{\text{ft}^3}{\text{sec}}}{\left(\frac{\pi}{4}\right)\left(\frac{1\ \text{in}}{12\ \frac{\text{in}}{\text{ft}}}\right)^2} = 4.085\ \text{ft/sec}$$

Use Table 14.5 to convert the units of viscosity.

$$\nu = (45\ \text{cSt})\left(1.0764\times10^{-5}\ \frac{\text{ft}^2}{\text{sec-cSt}}\right) = 4.844\times10^{-4}\ \text{ft}^2/\text{sec}$$

The Reynolds number is

$$\text{Re} = \frac{D\text{v}}{\nu} = \frac{(1\ \text{in})\left(4.085\ \frac{\text{ft}}{\text{sec}}\right)}{\left(4.844\times10^{-4}\ \frac{\text{ft}^2}{\text{sec}}\right)\left(12\ \frac{\text{in}}{\text{ft}}\right)} = \boxed{702.8\quad(700)}$$

Since the Reynolds number is less than 2000, the flow is $\boxed{\text{laminar.}}$

The answer is (A).

15. The available power is

$$P = p\dot{V} = \frac{\left(1500\ \frac{\text{lbf}}{\text{in}^2}\right)\left(12\ \frac{\text{in}}{\text{ft}}\right)^2\left(4\ \frac{\text{gal}}{\text{min}}\right)}{\left(7.48\ \frac{\text{gal}}{\text{ft}^3}\right)\left(60\ \frac{\text{sec}}{\text{min}}\right)\left(550\ \frac{\text{ft-lbf}}{\text{hp-sec}}\right)} = \boxed{3.5\ \text{hp}}$$

The answer is (C).

16. The compliance of the cylinder is

$$C_f = \frac{dQ}{dp} = \frac{2\ \text{in}^3}{8600\ \frac{\text{lbf}}{\text{in}^2} - 8000\ \frac{\text{lbf}}{\text{in}^2}} = \boxed{0.00333\ \text{in}^5/\text{lbf}\quad(0.0033\ \text{in}^5/\text{lbf})}$$

The answer is (C).

17. The compliance of the hose is

$$C_{f,\text{compressibility}} = \frac{V}{B} = \frac{AL}{B} = \frac{\left(\frac{\pi}{4}\right)(2\ \text{in})^2(30\ \text{in})}{\left(2.1\times10^5\ \frac{\text{lbf}}{\text{in}^2}\right)\left(12\ \frac{\text{in}}{\text{ft}}\right)^5} = \boxed{1.80\times10^{-9}\ \text{ft}^5/\text{lbf}\quad(1.8\times10^{-9}\ \text{ft}^5/\text{lbf})}$$

The answer is (A).

18. The torque is

$$T = \Delta p\left(\frac{\text{displacement}}{2\pi}\right) = (8\ \text{MPa})\left(\frac{\left(10^6\ \frac{\text{Pa}}{\text{MPa}}\right)\left(10\ \frac{\text{cm}^3}{\text{rad}}\right)}{\left(100\ \frac{\text{cm}}{\text{m}}\right)^3}\right) = \boxed{80\ \text{N·m}}$$

The answer is (B).

19. The bulk modulus, B, is equivalent in concept to the modulus of elasticity, E, for solids. The speed of sound is

$$a = \sqrt{\frac{B}{\rho}} = \sqrt{\frac{(1.72\ \text{GPa})\left(10^9\ \frac{\text{Pa}}{\text{GPa}}\right)}{870\ \frac{\text{kg}}{\text{m}^3}}} = \boxed{1406\ \text{m/s}\quad(1400\ \text{m/s})}$$

The answer is (D).

20. Solve Eq. 19.22 for the area of the cylinder.

$$A_{\text{in}^2} = \frac{0.3208\,Q_{\text{gpm}}}{\text{v}_{\text{ft/sec}}}$$

The force exerted by the cylinder on the clevis pin is

$$F = \eta p A = \eta p \left(\frac{0.3208 Q_{\text{gpm}}}{\text{v}_{\text{ft/sec}}} \right)$$

$$= (0.85)\left(2100 \ \frac{\text{lbf}}{\text{in}^2}\right) \left(\frac{\left(0.3208 \ \frac{\text{min-in}^2\text{-ft}}{\text{sec-gal}}\right)\left(30 \ \frac{\text{gal}}{\text{min}}\right)}{1 \ \frac{\text{ft}}{\text{sec}}} \right)$$

$$= 17{,}179 \text{ lbf}$$

The pin is in double shear. The shear stress on each loaded pin face is

$$\tau = \frac{\frac{F}{2}}{\frac{\pi d^2}{4}} = \frac{\frac{17{,}179 \text{ lbf}}{2}}{\frac{\pi (1 \text{ in})^2}{4}} = 10{,}936 \text{ lbf/in}^2$$

From the maximum shear stress theory, the yield strength in shear is twice the yield strength in tension.

$$S_{yt} = 2S_{ys} \times \text{FS} = 2\tau \times \text{FS} = \frac{(2)\left(10{,}936 \ \frac{\text{lbf}}{\text{in}^2}\right)(2.5)}{1000 \ \frac{\text{lbf}}{\text{kip}}}$$

$$= \boxed{54.68 \text{ kips/in}^2 \quad (55 \text{ kips/in}^2)}$$

The answer is (A).

21. Determine the power losses associated with pipe friction, fittings, and operation of the cylinder.

From Eq. 19.22, the linear velocity of the cylinder is

$$\text{v}_{\text{ft/sec}} = 0.3208 \left(\frac{Q_{\text{gpm}}}{A_{\text{in}^2}} \right)$$

$$= \left(3.208 \ \frac{\text{ft-min-in}^2}{\text{gal-sec}}\right) \left(\frac{8.8 \ \frac{\text{gal}}{\text{min}}}{\frac{\pi (0.5 \text{ in})^2}{4}} \right)$$

$$= 14.38 \text{ ft/sec}$$

From Eq. 19.7, the Reynolds number is

$$\text{Re} = \frac{3162 Q_{\text{gpm}}}{d_{\text{in}} \nu_{\text{cS}}} = \frac{(3162)\left(8.8 \ \frac{\text{gal}}{\text{min}}\right)}{(0.5 \text{ in})(33.1 \text{ cS})}$$

$$= 1680.77$$

Since Re < 2000, the flow is laminar. The friction factor is

$$f = \frac{64}{\text{Re}} = \frac{64}{1680.77} = 0.03807$$

From Eq. 19.9, the pressure drop due to the pipe friction is

$$\Delta p_{f,\text{psi}} = \frac{2.15 \times 10^{-4} f \rho_{\text{lbm/ft}^3} L_{\text{ft}} Q_{\text{gpm}}^2}{d_{\text{in}}^5}$$

$$= \frac{\left(2.15 \times 10^{-4} \ \frac{\text{min}^2\text{-ft}^2\text{-in}^3\text{-lbf}}{\text{gal}^2\text{-lbm}}\right)(0.03807) \times \left(62.4 \ \frac{\text{lbm}}{\text{ft}^3}\right)(40 \text{ ft})\left(8.8 \ \frac{\text{gal}}{\text{min}}\right)^2}{(0.5 \text{ in})^5}$$

$$= 50.60 \text{ lbf/in}^2$$

The total pressure drop across all of the values is calculated from Eq. 19.10.

$$\Delta p_{v,\text{psi,total}} = \sum_{C_{v,1}}^{C_{v,5}} \left(\frac{Q_{\text{gpm}}^2 (\text{SG})}{C_{v,i}^2} \right)$$

$$= \left(8.8 \ \frac{\text{gal}}{\text{min}}\right)^2 (1.0) \left(\frac{1}{(3.11)^2} + \frac{1}{(2.90)^2} + \frac{1}{(2.50)^2} + \frac{1}{(2.50)^2} + \frac{1}{(5.11)^2} \right)$$

$$= 50.60 \text{ lbf/in}^2$$

The power required to overcome the pipe and fitting frictional losses is

$$P_{\text{friction,hp}} = \sum \frac{\Delta p_{\text{psi}} Q_{\text{gpm}}}{1714 \eta_{\text{pump}}}$$

$$= \left(\frac{Q_{\text{gpm}}}{1714 \eta_{\text{pump}}} \right) (\Delta p_{v,\text{total}} + \Delta p_f)$$

$$= \left(\frac{8.8 \ \frac{\text{gal}}{\text{min}}}{\left(1714 \ \frac{\text{lbf-gal}}{\text{hp-min-in}^2}\right)(0.85)} \right) \times \left(44.96 \ \frac{\text{lbf}}{\text{in}^2} + 50.60 \ \frac{\text{lbf}}{\text{in}^2}\right)$$

$$= 0.577 \text{ hp}$$

The frictional force that the cylinder must overcome is

$$F = \eta_{\text{cylinder}} pA = (1 - 0.85)\left(2500\ \frac{\text{lbf}}{\text{in}^2}\right)\left(\frac{\pi(3\ \text{in})^2}{4}\right)$$
$$= 2650.7\ \text{lbf}$$

The power required to overcome the friction force in the cylinder is

$$P_{\text{cylinder}} = Fv = \frac{(2650.7\ \text{lbf})\left(0.4\ \frac{\text{ft}}{\text{sec}}\right)}{550\ \frac{\text{ft-lbf}}{\text{hp-sec}}}$$
$$= 1.928\ \text{hp}$$

The total power required is

$$P_{\text{total}} = P_{\text{friction}} + P_{\text{cylinder}} = 0.577\ \text{hp} + 1.928\ \text{hp}$$
$$= \boxed{2.505\ \text{hp}\quad (2.5\ \text{hp})}$$

The answer is (C).

22. There are two unknowns: the valve coefficient, C_v, and the valve diameter, d.

From Eq. 19.10, a relationship that includes the valve coefficient and diameter is

$$Q_{\text{gpm}} = C_v\sqrt{\frac{\Delta p}{\text{SG}}}$$

From Eq. 17.50, a relationship that includes the valve coefficient and diameter is

$$C_v = \frac{29.9 d_{\text{in}}^2}{\sqrt{K}}$$

Substituting the valve coefficient into the flow rate equation,

$$Q_{\text{gpm}} = \left(\frac{29.9 d_{\text{in}}^2}{\sqrt{K}}\right)\sqrt{\frac{\Delta p}{\text{SG}}}$$

From Table 17.4, $K = 0.19$ for a fully open gate valve. Rearranging yields,

$$d_{\text{in}} = \sqrt{\frac{Q_{\text{gpm}}\sqrt{K}}{29.9\sqrt{\frac{\Delta p}{\text{SG}}}}} = \sqrt{\frac{\left(60\ \frac{\text{gal}}{\text{min}}\right)\sqrt{0.19}}{(29.9)\sqrt{\frac{5\ \frac{\text{lbf}}{\text{in}^2}}{1}}}}$$
$$= 0.625\ \text{in}$$

The valve size must be greater than 0.625 in. Of the options, $\boxed{3/4\ \text{in}}$ is the minimum size gate valve that can operate fully open.

The answer is (B).

20 Inorganic Chemistry

PRACTICE PROBLEMS

Empirical Formula Development

1. The gravimetric analysis of a compound is 40% carbon, 6.7% hydrogen, and 53.3% oxygen. The simplest formula for the compound is most nearly

(A) HCO

(B) HCO_2

(C) CH_2O

(D) CHO_2

Corrosion

2. When used together, which of the following metal pairs is LEAST likely to experience galvanic corrosion in sea water?

(A) zinc and platinum

(B) zinc and steel

(C) steel and lead

(D) brass and copper

3. When use of dissimilar metals is unavoidable, which of the following CANNOT be used to prevent or reduce the likelihood of galvanic corrosion?

(A) inert spacers between the dissimilar metals

(B) sacrificial anodes

(C) paints, coatings, and natural oxide buildups

(D) layers of cadmium, nickel, chromium, or zinc plating

4. A sacrificial galvanic protection system is proposed for a water storage tank made of low-carbon steel. Which of the following materials is most suitable as the sacrificial anode?

(A) brass

(B) nickel

(C) stainless steel

(D) zinc

5. A linear actuator mechanism incorporates a naval brass threaded rod that passes through a threaded metallic collar. The mechanism is exposed to both fresh and saltwater. To limit galvanic corrosion, which material would be best for the mechanism's collar?

(A) zinc-plated carbon steel

(B) nickel-plated carbon steel

(C) passivated 316 stainless steel

(D) non-passivated 316 stainless steel

Precipitation Softening

6. A municipal water supply has the following ionic concentrations.

Al^{+++}	0.5 mg/L
Ca^{++}	80.2 mg/L
Cl^-	85.9 mg/L
CO_2	19 mg/L
CO_3^{--}	(none)
Fe^{++}	1.0 mg/L
Fl^-	(none)
HCO_3^-	185 mg/L
Mg^{++}	24.3 mg/L
Na^+	46.0 mg/L
NO_3^-	(none)
SO_4	125 mg/L

(a) The total hardness is most nearly

(A) 160 mg/L as $CaCO_3$

(B) 200 mg/L as $CaCO_3$

(C) 260 mg/L as $CaCO_3$

(D) 300 mg/L as $CaCO_3$

(b) Approximately how much slaked lime is required to combine with the carbonate hardness?

(A) 45 mg/L as substance

(B) 90 mg/L as substance

(C) 130 mg/L as substance

(D) 150 mg/L as substance

(c) Approximately how much soda ash is required to react with the carbonate hardness?

(A) none

(B) 15 mg/L as substance

(C) 35 mg/L as substance

(D) 60 mg/L as substance

7. A municipal water supply contains the following ionic concentrations.

$Ca(HCO_3)_2$	137 mg/L as $CaCO_3$
$MgSO_4$	72 mg/L as $CaCO_3$
CO_2	(none)

(a) Approximately how much slaked lime is required to soften 1,000,000 gal of this water to a hardness of 100 mg/L if 30 mg/L of excess lime is used?

(A) 930 lbm

(B) 1200 lbm

(C) 1300 lbm

(D) 1700 lbm

(b) Approximately how much soda ash is required to soften 1,000,000 gal of this water to a hardness of 100 mg/L if 30 mg/L of excess lime is used?

(A) none

(B) 300 lbm

(C) 500 lbm

(D) 700 lbm

SOLUTIONS

1. Calculate the relative mole ratios of the atoms by assuming there are 100 g of sample.

For 100 g of sample,

substance	mass	$\frac{m}{\text{AW}}$ = no. moles	relative mole ratio
C	40 g	$\frac{40}{12} = 3.33$	1
H	6.7 g	$\frac{6.7}{1} = 6.7$	2
O	53.3 g	$\frac{53.3}{16} = 3.33$	1

The empirical formula is $\boxed{CH_2O.}$

The answer is (C).

2. Galvanic corrosion (galvanic action) results from a difference in oxidation potentials of metallic ions. The greater the difference in oxidation potentials, the greater the likelihood of galvanic action. To determine which of the options are least likely to experience galvanic action, use a galvanic series chart, such as Table 20.6. The closer the pair of metals are, the less likely they are to experience galvanic corrosion. Therefore, of the options, [brass and copper] are least likely to experience galvanic corrosion.

The answer is (D).

3. While [cadmium, nickel, chromium, and zinc] are often used as protective deposits on steel, porosities in the surfaces can act as small galvanic cells, resulting in invisible subsurface corrosion.

The answer is (D).

4. From Table 20.6, brass, nickel, and stainless steel have lower potentials than low-carbon steel; therefore, they will be cathodic, not anodic. [Zinc] has a higher potential than low-carbon steel, so it can be used as the sacrificial anode.

The answer is (D).

5. From Table 20.6, nickel is the closest to brass, so [nickel-plated carbon steel] should be used to make the actuator's threaded collar.

The answer is (B).

6. (a) Hardness is the sum of the concentrations of all doubly and triply charged positive ions, expressed as $CaCO_3$. Find the total hardness by multiplying the mg/L as substance by the factor from App. 20.C.

	mg/L as substance		factor from App. 20.C		
Ca^{++}:	80.2	×	2.50	=	200.5 mg/L
Mg^{++}:	24.3	×	4.10	=	99.63 mg/L
Fe^{++}:	1	×	1.79	=	1.79 mg/L
Al^{+++}:	0.5	×	5.56	=	2.78 mg/L

$$\text{hardness} = \boxed{\begin{array}{c}304.7 \text{ mg/L} \\ (300 \text{ mg/L})\end{array}}$$

The answer is (D).

(b) Add lime to remove the carbonate hardness. It does not matter whether the HCO_3^- comes from Mg^{++}, Ca^{++}, or Fe^{++}; adding lime will remove it.

There may be Mg^{++}, Ca^{++}, or Fe^{++} ions left over in the form of noncarbonate hardness, but the problem asked for carbonate hardness. Converting from mg/L of substance to mg/L as $CaCO_3$,

$$CO_2\text{:}\quad \left(19\ \frac{\text{mg}}{\text{L}}\right)(2.27) = 43.13 \text{ mg/L as } CaCO_3$$

$$HCO_3^-\text{:}\quad \left(185\ \frac{\text{mg}}{\text{L}}\right)(0.82) = 151.7 \text{ mg/L as } CaCO_3$$

The total equivalents to be neutralized are

$$43.13\ \frac{\text{mg}}{\text{L}} + 151.7\ \frac{\text{mg}}{\text{L}} = 194.83 \text{ mg/L as } CaCO_3$$

Convert $Ca(OH)_2$ to substance using App. 20.C.

$$\frac{\text{mg}}{\text{L}} \text{ of } Ca(OH)_2 = \frac{194.83\ \frac{\text{mg}}{\text{L}}}{1.35} = \boxed{\begin{array}{c}144.3 \text{ mg/L as substance} \\ (150 \text{ mg/L as substance})\end{array}}$$

The answer is (D).

(c) No soda ash is required since it is used to remove noncarbonate hardness.

The answer is (A).

7. (a) $Ca(HCO_3)_2$ and $MgSO_4$ both contribute to hardness. Since 100 mg/L of hardness is the goal, leave all $MgSO_4$ in the water. Take out 137 mg/L + 72 mg/L − 100 mg/L = 109 mg/L of $Ca(HCO_3)_2$. From App. 20.C (including the excess even though the reaction is not complete),

$$\text{pure } Ca(OH)_2 = 30\ \frac{\text{mg}}{\text{L}} + \frac{109\ \frac{\text{mg}}{\text{L}}}{1.35} = 110.74 \text{ mg/L}$$

$$\left(110.74\ \frac{\text{mg}}{\text{L}}\right)\left(8.345\ \frac{\text{lbm-L}}{\text{mg-MG}}\right) = \boxed{\begin{array}{c}924 \text{ lbm/MG} \\ (930 \text{ lbm})\end{array}}$$

The answer is (A).

(b) No soda ash is required.

The answer is (A).

21 Fuels and Combustion

PRACTICE PROBLEMS

1. Methane (MW = 16.043) with a heating value of 24,000 Btu/lbm (55.8 MJ/kg) is burned with a 50% efficiency. If the heat of vaporization of any water vapor formed is recovered, approximately how much water (specific heat of 1 Btu/lbm-°F (4.1868 kJ/kg·°C)) can be heated from 60°F to 200°F (15°C to 95°C) when 7 ft^3 (200 L) of methane at 60°F and 14.73 psia (288.9K and 101.51 kPa) are burned?

(A) 25 lbm (11 kg)

(B) 35 lbm (16 kg)

(C) 50 lbm (23 kg)

(D) 95 lbm (43 kg)

2. 15 lbm/hr (6.8 kg/h) of propane (C_3H_8, MW = 44.097) is burned stoichiometrically in air. Approximately what volume of dry carbon dioxide (CO_2, MW = 44.011) is formed after cooling to 70°F (21°C) and 14.7 psia (101 kPa)?

(A) 180 ft^3/hr (5.0 m^3/h)

(B) 270 ft^3/hr (7.6 m^3/h)

(C) 390 ft^3/hr (11 m^3/h)

(D) 450 ft^3/hr (13 m^3/h)

3. In a particular installation, 30% excess air at 15 psia (103 kPa) and 100°F (40°C) is needed for the combustion of methane. Approximately how much nitrogen (MW = 28.016) passes through the furnace if methane is burned at the rate of 4000 ft^3/hr (31 L/s)?

(A) 270 lbm/hr (0.033 kg/s)

(B) 930 lbm/hr (0.11 kg/s)

(C) 1800 lbm/hr (0.22 kg/s)

(D) 2700 lbm/hr (0.34 kg/s)

4. Approximately how much air is required to completely burn one unit mass of a fuel that is 84% carbon, 15.3% hydrogen, 0.3% sulfur, and 0.4% nitrogen by weight?

(A) 9 lbm air/lbm fuel (9 kg air/kg fuel)

(B) 12 lbm air/lbm fuel (12 kg air/kg fuel)

(C) 15 lbm air/lbm fuel (15 kg air/kg fuel)

(D) 18 lbm air/lbm fuel (18 kg air/kg fuel)

5. Propane (C_3H_8) is burned with 20% excess air. The gravimetric percentage of carbon dioxide in the flue gas is most nearly

(A) 8%

(B) 12%

(C) 15%

(D) 22%

6. The ultimate analysis of a coal is 80% carbon, 4% hydrogen, 2% oxygen, and the rest ash. The flue gases are 60°F and 14.7 psia (15.6°C and 101.3 kPa) when sampled, and are 80% nitrogen, 12% carbon dioxide, 7% oxygen, and 1% carbon monoxide by volume. The air required to burn 1 lbm (1 kg) of coal under these conditions is most nearly

(A) 11 lbm (11 kg)

(B) 15 lbm (15 kg)

(C) 19 lbm (19 kg)

(D) 23 lbm (23 kg)

7. What is the approximate heating value of an oil with a specific gravity of 40° API?

(A) 20,000 Btu/lbm (46 MJ/kg)

(B) 25,000 Btu/lbm (58 MJ/kg)

(C) 30,000 Btu/lbm (69 MJ/kg)

(D) 35,000 Btu/lbm (81 MJ/kg)

8. The ultimate analysis of a coal is 75% carbon, 5% hydrogen, 3% oxygen, 2% nitrogen, and the rest ash. Atmospheric air is 60°F (16°C) and at standard pressure.

(a) The theoretical temperature of the combustion products is most nearly

(A) 3500°F (1900°C)

(B) 4000°F (2200°C)

(C) 4500°F (2500°C)

(D) 5000°F (2800°C)

(b) Estimate the actual temperature of the combustion products at the boiler outlet. Neglect dissociation. Assume 40% excess air is used and 75% of the heat is transferred to the boiler.

(A) 650°F (340°C)

(B) 880°F (470°C)

(C) 970°F (520°C)

(D) 1300°F (700°C)

9. A fuel oil has the following ultimate analysis: 85.43% carbon, 11.31% hydrogen, 2.7% oxygen, 0.34% sulfur, and 0.22% nitrogen. The oil is burned with 60% excess air. Evaluate flue gas volumes at 600°F (320°C).

(a) The volume of wet flue gases that will be produced is most nearly

(A) 450 ft^3 (28 m^3)

(B) 500 ft^3 (32 m^3)

(C) 550 ft^3 (35 m^3)

(D) 600 ft^3 (38 m^3)

(b) The volume of dry flue gases that will be produced is most nearly

(A) 480 ft^3 (30 m^3)

(B) 560 ft^3 (35 m^3)

(C) 630 ft^3 (40 m^3)

(D) 850 ft^3 (79 m^3)

(c) The volumetric fraction of carbon dioxide is most nearly

(A) 6%

(B) 8%

(C) 10%

(D) 13%

10. The ultimate analysis of a coal is 51.45% carbon, 16.69% ash, 15.71% moisture, 7.28% oxygen, 4.02% hydrogen, 3.92% sulfur, and 0.93% nitrogen. 15,395 lbm (6923 kg) of the coal are burned, and 2816 lbm (1267 kg) of ash containing 20.9% carbon (by weight) are recovered. 13.3 lbm (kg) of dry gases are produced per pound (kilogram) of fuel burned. The air used per unit mass of fuel is most nearly

(A) 13 lbm air/lbm fuel (13 kg air/kg fuel)

(B) 14 lbm air/lbm fuel (14 kg air/kg fuel)

(C) 15 lbm air/lbm fuel (15 kg air/kg fuel)

(D) 17 lbm air/lbm fuel (17 kg air/kg fuel)

11. A coal is 65% carbon by weight. During combustion, 3% of the coal is lost in the ash pit. Combustion uses 9.87 lbm (kg) of air per pound (kg) of fuel. The flue gas analysis is 81.5% nitrogen, 9.5% carbon dioxide, and 9% oxygen. The percentage of excess air by mass is most nearly

(A) 10%

(B) 30%

(C) 70%

(D) 140%

12. A coal has an ultimate analysis of 67.34% carbon, 4.91% oxygen, 4.43% hydrogen, 4.28% sulfur, 1.08% nitrogen, and the rest ash. 3% of the carbon is lost during combustion. The flue gases are 81.9% nitrogen, 15.5% carbon dioxide, 1.6% carbon monoxide, and 1% oxygen by volume. The heat loss due to the formation of carbon monoxide is most nearly

(A) 600 Btu/lbm (1.4 MJ/kg)

(B) 800 Btu/lbm (1.9 MJ/kg)

(C) 1000 Btu/lbm (2.3 MJ/kg)

(D) 1200 Btu/lbm (2.8 MJ/kg)

13. (*Time limit: one hour*) A natural gas is 93% methane, 3.73% nitrogen, 1.82% hydrogen, 0.45% carbon monoxide, 0.35% oxygen, 0.25% ethylene, 0.22% carbon dioxide, and 0.18% hydrogen sulfide by volume. The gas is burned with 40% excess air. Atmospheric air is 60°F and at standard atmospheric pressure.

(a) The gas density is most nearly

(A) 0.017 lbm/ft^3

(B) 0.043 lbm/ft^3

(C) 0.069 lbm/ft^3

(D) 0.110 lbm/ft^3

(b) The theoretical air requirements are most nearly

(A) 5 ft^3 air/ft^3 fuel

(B) 7 ft^3 air/ft^3 fuel

(C) 9 ft^3 air/ft^3 fuel

(D) 13 ft^3 air/ft^3 fuel

(c) The percentage of CO_2 in the flue gas (wet basis) is most nearly

(A) 6.9%

(B) 7.7%

(C) 8.1%

(D) 11%

(d) The percentage of CO_2 in the flue gas (dry basis) is most nearly

(A) 6.9%

(B) 7.7%

(C) 8.1%

(D) 11%

14. A coal enters a steam generator at 73°F (23°C). The coal has an ultimate analysis of 78.42% carbon, 8.25% oxygen, 5.68% ash, 5.56% hydrogen, 1.09% nitrogen, and 1.0% sulfur. The coal's heating value is 14,000 Btu/lbm (32.6 MJ/kg). 7.03% of the coal's weight is lost in the ash pit. The ash contains 31.5% carbon. Air at 67°F (19°C) wet bulb and 73°F (23°C) dry bulb is supplied. The flue gases consist of 80.08% nitrogen, 14.0% carbon dioxide, 5.5% oxygen, and 0.42% carbon monoxide. The flue gases are at a temperature of 575°F (300°C). Saturated water at 212°F (100°C) and 1.0 atm enters the steam generator. 11.12 lbm (kg) of water are evaporated in the boiler per pound (kilogram) of dry coal consumed. From a complete heat balance, combustion losses in the furnace are most nearly

(A) 400 Btu/lbm (1020 kJ/kg)

(B) 650 Btu/lbm (1690 kJ/kg)

(C) 960 Btu/lbm (2500 kJ/kg)

(D) 1500 Btu/lbm (3900 kJ/kg)

15. A coal has a gravimetric analysis of 83% carbon, 5% hydrogen, 5% oxygen, and 7% noncombustible matter. 10% of the fired coal mass, including all of the noncombustible matter, is recovered in the ash pit. 26 lbm (26 kg) of air at 70°F (21°C) and 1 atmosphere are used per lbm (kg) of coal burned. When loaded into the furnace, the coal temperature is 60°F (15.6°C). The stack gas from coal combustion has a temperature of 550°F (290°C). The percentage of the combustion heat carried away by the stack gases, based on the coal's as-delivered properties, is most nearly

(A) 18%

(B) 24%

(C) 37%

(D) 49%

16. (*Time limit: one hour*) A utility boiler burns coal with an ultimate analysis of 76.56% carbon, 7.7% oxygen, 6.1% silicon, 5.5% hydrogen, 2.44% sulfur, and 1.7% nitrogen. 410 lbm/hr of refuse are removed with a composition of 30% carbon and 0% sulfur. All the sulfur and the remaining carbon are burned. The power plant has the following characteristics.

- coal feed rate: 15,300 lbm/hr
- electric power rating: 17 MW
- generator efficiency: 95%
- steam generator efficiency: 86%
- cooling water rate: 225 ft^3/sec

(a) The emission rate of solid particulates in lbm/hr is most nearly

(A) 23 lbm/hr

(B) 150 lbm/hr

(C) 810 lbm/hr

(D) 1700 lbm/hr

(b) The sulfur dioxide produced per hour is most nearly

(A) 220 lbm/hr

(B) 340 lbm/hr

(C) 750 lbm/hr

(D) 1100 lbm/hr

(c) The temperature rise of the cooling water is most nearly

(A) 2.4°F

(B) 6.5°F

(C) 9.8°F

(D) 13°F

(d) The efficiency the flue gas particulate collectors must have in order to meet a limit of 0.1 lbm of particulates per million Btus per hour (0.155 kg/MW) is most nearly

(A) 93.1%

(B) 97.4%

(C) 98.8%

(D) 99.1%

17. 250 SCFM (118 L/s) of propane are mixed with an oxidizer consisting of 60% oxygen and 40% nitrogen by volume in a proportion resulting in 40% excess oxygen by weight. The maximum velocity for the two reactants when combined is 400 ft/min (2 m/s) at 14.7 psia (101 kPa) and 80°F (27°C). Maximum velocity for the products is 800 ft/min (4 m/s) at 8 psia (55 kPa) and 460°F (240°C).

(a) The actual flow of oxygen is most nearly

(A) 150 lbm/min (1.1 kg/s)

(B) 180 lbm/min (1.3 kg/s)

(C) 220 lbm/min (1.6 kg/s)

(D) 270 lbm/min (2.0 kg/s)

(b) The minimum size of the inlet pipe is most nearly

(A) 8 ft^2 (0.8 m^2)

(B) 11 ft^2 (1.1 m^2)

(C) 14 ft^2 (1.3 m^2)

(D) 17 ft^2 (1.6 m^2)

(c) The volume of flue gases is most nearly

(A) 4000 ft^3/min (1.9 m^3/s)

(B) 7000 ft^3/min (3.3 m^3/s)

(C) 9000 ft^3/min (4.3 m^3/s)

(D) 11,000 ft^3/min (5.3 m^3/s)

(d) The minimum area of the stack is most nearly

(A) 8 ft^2 (0.8 m^2)

(B) 11 ft^2 (1.1 m^2)

(C) 14 ft^2 (1.3 m^2)

(D) 17 ft^2 (1.6 m^2)

(e) The dew point of the flue gases is most nearly

(A) 40°F (4°C)

(B) 100°F (38°C)

(C) 130°F (55°C)

(D) 180°F (80°C)

18. An industrial process uses hot gas at 3600°R (1980°C) and 14.7 psia (101 kPa). It is proposed that propane be burned stoichiometrically in a mixture of nitrogen and oxygen. After passing through the process, gas will be exhausted through a duct, being cooled slowly to 100°F (38°C) and 14.7 psia (101 kPa) before discharge. The following data are available.

- The enthalpies of formation (at the standard reference temperature) are

$C_3H_8(g)$:	$\Delta H_f = +28{,}800$ Btu/lbmol	(+67.0 GJ/kmol)
$CO_2(g)$:	$\Delta H_f = -169{,}300$ Btu/lbmol	(–393.8 GJ/kmol)
$H_2O(g)$:	$\Delta H_f = -104{,}040$ Btu/lbmol	(–242 GJ/kmol)

- The enthalpy increases from the standard reference temperature to 3600°R (1980°C) are

CO_2:	39,791 Btu/lbmol	(92.6 GJ/kmol)
H_2O:	31,658 Btu/lbmol	(73.6 GJ/kmol)
N_2:	24,471 Btu/lbmol	(56.9 GJ/kmol)

(a) What are most nearly the combining weights of the oxygen and nitrogen, respectively, per pound-mole of propane?

(A) 760 lbm/lbmol, 128 lbm/lbmol (760 kg/kmol, 128 kg/kmol)

(B) 810 lbm/lbmol, 160 lbm/lbmol (810 kg/kmol, 160 kg/kmol)

(C) 845 lbm/lbmol, 160 lbm/lbmol (845 kg/kmol, 160 kg/kmol)

(D) 850 lbm/lbmol, 160 lbm/lbmol (850 kg/kmol, 160 kg/kmol)

(b) The amount of water vapor, if any, present in the stack gas immediately after combustion is most nearly

(A) 0 lbmol/lbmol C_3H_8 (0 kmol H_2O/kmol C_3H_8)

(B) 1.2 lbmol/lbmol C_3H_8 (1.2 kmol H_2O/kmol C_3H_8)

(C) 2.3 lbmol/lbmol C_3H_8 (2.3 kmol H_2O/kmol C_3H_8)

(D) 3.5 lbmol/lbmol C_3H_8 (3.5 kmol H_2O/kmol C_3H_8)

(c) The amount of water removed from the stack gas is most nearly

(A) 0 lbm H_2O/lbmol C_3H_8 (0 kg H_2O/kmol C_3H_8)

(B) 9 lbm H_2O/lbmol C_3H_8 (9 kg H_2O/kmol C_3H_8)

(C) 30 lbm H_2O/lbmol C_3H_8 (30 kg H_2O/kmol C_3H_8)

(D) 50 lbm H_2O/lbmol C_3H_8 (50 kg H_2O/kmol C_3H_8)

19. An electrical power-generating plant burns refuse-derived fuel (RDF). After sorting, incoming refuse is shredded and compressed before being fed into the combustor. The raw refuse averages 7% by weight incombustible solids. 5000 lbm/hr of processed RDF produces 20,070 lbm/hr of saturated steam at 200 lbf/in^2. The combustion products are used to heat incoming feedwater to a saturated temperature of 160°F before entering the combustor. 2000 lbm/hr of water vapor condense in the feedwater heater at a partial pressure of 4 lbf/in^2 and are removed. All thermal losses are to be disregarded. What is most nearly the higher heating value of the RDF?

(A) 4300 Btu/lbm

(B) 4700 Btu/lbm

(C) 4900 Btu/lbm

(D) 5100 Btu/lbm

SOLUTIONS

1. *Customary U.S. Solution*

$$T = 60°\text{F} + 460° = 520°\text{R}$$
$$p = 14.73 \text{ psia}$$

The specific gas constant, R, is calculated from the universal gas constant, R^*, and the molecular weight.

$$R = \frac{R^*}{\text{MW}} = \frac{1545.35 \ \frac{\text{ft-lbf}}{\text{lbmol-°R}}}{16.043 \ \frac{\text{lbm}}{\text{lbmol}}} = 96.33 \text{ ft-lbf/lbm-°R}$$

$$m = \frac{pV}{RT} = \frac{\left(14.73 \ \frac{\text{lbf}}{\text{in}^2}\right)\left(12 \ \frac{\text{in}}{\text{ft}}\right)^2 (7 \text{ ft}^3)}{\left(96.33 \ \frac{\text{ft-lbf}}{\text{lbm-°R}}\right)(520°\text{R})} = 0.296 \text{ lbm}$$

The combustion energy available from methane is

$$Q = \eta m(\text{HHV}) = (0.5)(0.296 \text{ lbm})\left(24{,}000 \ \frac{\text{Btu}}{\text{lbm}}\right) = 3552 \text{ Btu}$$

This energy is used to heat water from 60°F to 200°F.

$$Q = m_{\text{water}} c_p (T_2 - T_1)$$

$$m_{\text{water}} = \frac{3552 \text{ Btu}}{\left(1 \ \frac{\text{Btu}}{\text{lbm-°F}}\right)(200°\text{F} - 60°\text{F})} = \boxed{25.37 \text{ lbm} \quad (25 \text{ lbm})}$$

The answer is (A).

SI Solution

The specific gas constant, R, is calculated from the universal gas constant, R^*, and the molecular weight.

$$R = \frac{R^*}{\text{MW}} = \frac{8314.5 \ \frac{\text{J}}{\text{kmol·K}}}{16.043 \ \frac{\text{kg}}{\text{kmol}}} = 518.26 \text{ J/kg·K}$$

$$m = \frac{pV}{RT} = \frac{(101.51 \text{ kPa})\left(1000 \ \frac{\text{Pa}}{\text{kPa}}\right)(200 \text{ L})}{\left(518.26 \ \frac{\text{J}}{\text{kg·K}}\right)(288.9\text{K})\left(1000 \ \frac{\text{L}}{\text{m}^3}\right)} = 0.136 \text{ kg}$$

Thermodynamics

The combustion energy available from methane is

$$Q = \eta m(\text{HHV})$$
$$= (0.5)(0.136\ \text{kg})\left(55.8\ \frac{\text{MJ}}{\text{kg}}\right)\left(1000\ \frac{\text{kJ}}{\text{MJ}}\right)$$
$$= 3794\ \text{kJ}$$

This energy is used to heat water from 15°C to 95°C.

$$Q = m_{\text{water}}\, c_p(T_2 - T_1)$$
$$m_{\text{water}} = \frac{3794\ \text{kJ}}{\left(4.1868\ \frac{\text{kJ}}{\text{kg}\cdot\text{C}}\right)(95°\text{C} - 15°\text{C})}$$
$$= \boxed{11.33\ \text{kg}\quad(11\ \text{kg})}$$

The answer is (A).

2. *Customary U.S. Solution*

From Table 21.7,

	C_3H_8	+	$5O_2$	$\rightarrow$	$3CO_2$	+	$4H_2O$
MW	44.097		(5)(32)		(3)(44.011)		
	44.097		160		132.033		

The amount of carbon dioxide produced is 132.033 lbm/44.097 lbm propane. For 15 lbm/hr of propane, the amount of carbon dioxide produced is

$$\left(\frac{132.033\ \text{lbm}}{44.097\ \text{lbm}}\right)\left(15\ \frac{\text{lbm}}{\text{hr}}\right) = 44.91\ \text{lbm/hr}$$

$$R = \frac{R^*}{\text{MW}} = \frac{1545.35\ \frac{\text{ft-lbf}}{\text{lbmol-°R}}}{44.011\ \frac{\text{lbm}}{\text{lbmol}}}$$
$$= 35.11\ \text{ft-lbf/lbm-°R}$$
$$T = 70°\text{F} + 460° = 530°\text{R}$$
$$\dot{V} = \frac{\dot{m}RT}{p}$$
$$= \frac{\left(44.91\ \frac{\text{lbm}}{\text{hr}}\right)\left(35.11\ \frac{\text{ft-lbf}}{\text{lbm-°R}}\right)(530°\text{R})}{\left(14.7\ \frac{\text{lbf}}{\text{in}^2}\right)\left(12\ \frac{\text{in}}{\text{ft}}\right)^2}$$
$$= \boxed{394.8\ \text{ft}^3/\text{hr}\quad(390\ \text{ft}^3/\text{hr})}$$

The answer is (C).

SI Solution

From Table 21.7,

	C_3H_8	+	$5O_2$	$\rightarrow$	$3CO_2$	+	$4H_2O$
MW	44.097		(5)(32)		(3)(44.011)		
	44.097		160		132.033		

The amount of carbon dioxide produced is 132.033 kg/44.097 kg propane. For 6.8 kg/h of propane, the amount of carbon dioxide produced is

$$\left(\frac{132.033\ \text{kg}}{44.097\ \text{kg}}\right)\left(6.8\ \frac{\text{kg}}{\text{h}}\right) = 20.36\ \text{kg/h}$$

$$R = \frac{R^*}{\text{MW}} = \frac{8314.5\ \frac{\text{J}}{\text{kmol}\cdot\text{K}}}{44.01\ \frac{\text{kg}}{\text{kmol}}} = 188.92\ \text{J/kg}\cdot\text{K}$$
$$T = 21°\text{C} + 273° = 294\text{K}$$
$$V = \frac{mRT}{p}$$
$$= \frac{\left(20.36\ \frac{\text{kg}}{\text{h}}\right)\left(188.92\ \frac{\text{J}}{\text{kg}\cdot\text{K}}\right)(294\text{K})}{(101\ \text{kPa})\left(1000\ \frac{\text{Pa}}{\text{kPa}}\right)}$$
$$= \boxed{11.20\ \text{m}^3/\text{h}\quad(11\ \text{m}^3/\text{h})}$$

The answer is (C).

3. Use the balanced chemical reaction equation from Table 21.7.

$$CH_4 + 2O_2 \rightarrow CO_2 + 2H_2O$$

Use Table 21.6 and Table 21.7. With 30% excess air and considering that there are 3.773 volumes of nitrogen for every volume of oxygen, the reaction equation is

$$CH_4 + (1.3)(2)O_2 + (1.3)(2)(3.773)N_2$$
$$\rightarrow CO_2 + 2H_2O + (1.3)(2)(3.773)N_2 + 0.6O_2$$
$$CH_4 + 2.6O_2 + 9.81N_2$$
$$\rightarrow CO_2 + 2H_2O + 9.81N_2 + 0.6O_2$$

Customary U.S. Solution

The volume of nitrogen that accompanies 4000 ft^3/hr of entering methane is

$$V_{N_2} = \left(9.81\ \frac{\text{ft}^3\ N_2}{\text{ft}^3\ CH_4}\right)\left(4000\ \frac{\text{ft}^3\ CH_4}{\text{hr}}\right)$$
$$= 39{,}240\ \text{ft}^3\ N_2/\text{hr}$$

This is the "partial volume" of nitrogen in the input stream.

$$R = \frac{R^*}{\text{MW}} = \frac{1545.35\ \frac{\text{ft-lbf}}{\text{lbmol-°R}}}{28.016\ \frac{\text{lbm}}{\text{lbmol}}} = 55.16\ \text{ft-lbf/lbm-°R}$$

The absolute temperature is

$$T = 100°\text{F} + 460° = 560°\text{R}$$

$$\dot{m}_{N_2} = \frac{p_{N_2}V_{N_2}}{RT} = \frac{\left(15\ \frac{\text{lbf}}{\text{in}^2}\right)\left(12\ \frac{\text{in}}{\text{ft}}\right)^2\left(39{,}240\ \frac{\text{ft}^3}{\text{hr}}\right)}{\left(55.16\ \frac{\text{ft-lbf}}{\text{lbm-°R}}\right)(560°\text{R})} = \boxed{2744\ \text{lbm/hr}\quad(2700\ \text{lbm/hr})}$$

The answer is (D).

SI Solution

The volume of nitrogen that accompanies 31 L/s of entering methane is

$$\frac{\left(9.81\ \frac{\text{m}^3\ \text{N}_2}{\text{m}^3\ \text{CH}_4}\right)\left(31\ \frac{\text{L CH}_4}{\text{s}}\right)}{1000\ \frac{\text{L}}{\text{m}^3}} = 0.3041\ \text{m}^3/\text{s}$$

This is the "partial volume" of nitrogen in the input stream.

$$R = \frac{R^*}{\text{MW}} = \frac{8314.5\ \frac{\text{J}}{\text{kmol·K}}}{28.016\ \frac{\text{kg}}{\text{kmol}}} = 296.8\ \text{J/kg·K}$$

The absolute temperature is

$$T = 40°\text{C} + 273° = 313\text{K}$$

$$\dot{m}_{N_2} = \frac{p_{N_2}V_{N_2}}{RT} = \frac{(103\ \text{kPa})\left(1000\ \frac{\text{Pa}}{\text{kPa}}\right)\left(0.3041\ \frac{\text{m}^3}{\text{s}}\right)}{\left(296.8\ \frac{\text{J}}{\text{kg·K}}\right)(313\text{K})} = \boxed{0.337\ \text{kg/s}\quad(0.34\ \text{kg/s})}$$

The answer is (D).

4. From Table 21.7, combustion reactions are

$$\begin{array}{lccccc} & \text{C} & + & \text{O}_2 & \rightarrow & \text{CO}_2 \\ \text{MW} & 12 & & 32 & & \end{array}$$

The mass of oxygen required per unit mass of carbon is

$$\frac{32}{12} = 2.67$$

$$\begin{array}{lccccc} & 2\text{H}_2 & + & \text{O}_2 & \rightarrow & 2\text{H}_2\text{O} \\ \text{MW} & (2)(2) & & 32 & & \end{array}$$

The mass of oxygen required per unit mass of hydro-gen is

$$\frac{32}{(2)(2)} = 8.0$$

$$\begin{array}{lccccc} & \text{S} & + & \text{O}_2 & \rightarrow & \text{SO}_2 \\ \text{MW} & 32.1 & & 32 & & \end{array}$$

The mass of oxygen required per unit mass of sulfur is

$$\frac{32}{32.1} = 1.0$$

Nitrogen does not burn.

The mass of oxygen required per unit mass of fuel is

$$(0.84)(2.67) + (0.153)(8) + (0.003)(1) = 3.47\ \text{units of mass of O}_2\text{/unit mass fuel}$$

From Table 21.6, air is 0.2315 O_2/unit mass.

Customary U.S. Solution

The air required is

$$\frac{3.47}{0.2315} = \boxed{14.99\quad(15\ \text{lbm air/lbm fuel})}$$

SI Solution

The air required is

$$\frac{3.47}{0.2315} = \boxed{14.99\quad(15\ \text{kg air/kg fuel})}$$

(Equation 21.6 could also have been used to solve this problem.)

The answer is (C).

5. From Table 21.7, the balanced chemical reaction equation is

$$C_3H_8 + 5O_2 \rightarrow 3CO_2 + 4H_2O$$

With 20% excess air, the oxygen volume is $(1.2)(5) = 6$.

$$C_3H_8 + 6O_2 \rightarrow 3CO_2 + 4H_2O + O_2$$

Thermodynamics

From Table 21.6, there are 3.773 volumes of nitrogen for every volume of oxygen.

$$(6)(3.773) = 22.6$$

$$C_3H_8 + 6O_2 + 22.6N_2 \rightarrow 3CO_2 + 4H_2O + O_2 + 22.6N_2$$

The percentage of carbon dioxide by weight in flue gas is

$$\begin{aligned}G_{CO_2} &= \frac{m_{CO_2}}{m_{total}} = \frac{B_{CO_2}(MW_{CO_2})}{\sum B_i(MW_i)} \\ &= \frac{(3)(44.011)}{(3)(44.011) + (4)(18.016) + 32 + (22.6)(28.016)} \\ &= \boxed{0.152 \quad (15\%)}\end{aligned}$$

The answer is (C).

6. The actual air/fuel ratio can be estimated from the flue gas analysis and the fraction of carbon in fuel.

From Eq. 21.9,

$$\begin{aligned}\frac{m_{air}}{m_{fuel}} &= \frac{3.04B_{N_2}G_C}{B_{CO_2} + B_{CO}} = \frac{(3.04)(0.80)(0.80)}{0.12 + 0.01} \\ &= 14.97\end{aligned}$$

Customary U.S. Solution

The air required to burn 1 lbm of coal is $\boxed{\text{14.97 lbm (15 lbm).}}$

SI Solution

The air required to burn 1 kg of coal is $\boxed{\text{14.97 kg (15 kg).}}$

The answer is (B).

Alternative Solution

The use of Eq. 21.9 obscures the process of finding the air/fuel ratio. (The SI solution is similar but is not presented here.)

step 1: Find the mass of oxygen in the stack gases.

$$R_{CO_2} = 35.11 \text{ ft-lbf/lbm-°R}$$
$$R_{CO} = 55.17 \text{ ft-lbf/lbm-°R}$$
$$R_{O_2} = 48.29 \text{ ft-lbf/lbm-°R}$$

The partial densities are

$$\begin{aligned}\rho_{CO_2} &= \frac{p}{RT} = \frac{(0.12)\left(14.7 \frac{\text{lbf}}{\text{in}^2}\right)\left(12 \frac{\text{in}}{\text{ft}}\right)^2}{\left(35.11 \frac{\text{ft-lbf}}{\text{lbm-°R}}\right)(60°\text{F} + 460°)} \\ &= 1.391 \times 10^{-2} \text{ lbm/ft}^3\end{aligned}$$

$$\begin{aligned}\rho_{CO} &= \frac{(0.01)\left(14.7 \frac{\text{lbf}}{\text{in}^2}\right)\left(12 \frac{\text{in}}{\text{ft}}\right)^2}{\left(55.11 \frac{\text{ft-lbf}}{\text{lbm-°R}}\right)(60°\text{F} + 460°)} \\ &= 7.387 \times 10^{-4} \text{ lbm/ft}^3\end{aligned}$$

$$\begin{aligned}\rho_{CO_2} &= \frac{(0.07)\left(14.7 \frac{\text{lbf}}{\text{in}^2}\right)\left(12 \frac{\text{in}}{\text{ft}}\right)^2}{\left(48.29 \frac{\text{ft-lbf}}{\text{lbm-°R}}\right)(60°\text{F} + 460°)} \\ &= 5.901 \times 10^{-3} \text{ lbm/ft}^3\end{aligned}$$

The fraction of oxygen in the three components is

$$CO_2\text{: } \frac{32.0}{44} = 0.7273$$
$$CO\text{: } \frac{16}{28} = 0.5714$$
$$O_2\text{: } 1.00$$

In 100 ft^3 of stack gases, the total oxygen mass will be

$$\begin{aligned}&(100 \text{ ft}^3)\left(\begin{array}{l}(0.7273)\left(1.391 \times 10^{-2} \frac{\text{lbm}}{\text{ft}^3}\right) \\ \quad + (0.5714)\left(7.387 \times 10^{-4} \frac{\text{lbm}}{\text{ft}^3}\right) \\ \quad + (1.00)\left(5.901 \times 10^{-3} \frac{\text{lbm}}{\text{ft}^3}\right)\end{array}\right) \\ &= 1.644 \text{ lbm}\end{aligned}$$

step 2: Since air is 23.15% oxygen by weight, the mass of air per 100 ft^3 of stack gases is

$$\frac{1.644 \text{ lbm}}{0.2315} = 7.102 \text{ lbm/100 ft}^3$$

step 3: Find the mass of carbon in the stack gases by a similar process.

$$CO_2\text{: } \frac{12}{44} = 0.2727$$
$$CO\text{: } \frac{12}{28} = 0.4286$$

$$\begin{aligned}&(100 \text{ ft}^3)\left(\begin{array}{l}(0.2727)\left(1.391 \times 10^{-2} \frac{\text{lbm}}{\text{ft}^3}\right) \\ \quad + (0.4286)\left(7.387 \times 10^{-4} \frac{\text{lbm}}{\text{ft}^3}\right)\end{array}\right) \\ &= 0.411 \text{ lbm}\end{aligned}$$

step 4: The coal is 80% carbon, so the air per lbm of coal for combustion of the carbon is

$$\left(\frac{0.80\ \dfrac{\text{lbm carbon}}{\text{lbm coal}}}{0.411\ \dfrac{\text{lbm carbon}}{100\ \text{ft}^3}}\right)\left(7.102\ \frac{\text{lbm}}{100\ \text{ft}^3}\right)$$
$$= 13.824\ \text{lbm air/lbm coal}$$

step 5: This does not include air to burn hydrogen, since Orsat is a dry analysis.

The theoretical air for the hydrogen is given by Eq. 21.6.

$$R_{a/f,\text{H}} = \left(34.5\ \frac{\text{lbm}}{\text{lbm}}\right)\left(G_\text{H} - \frac{G_\text{O}}{8}\right)$$
$$= \left(34.5\ \frac{\text{lbm}}{\text{lbm}}\right)\left(0.04 - \frac{0.02}{8}\right)$$
$$= 1.294\ \text{lbm air/lbm fuel}$$

Ignoring any excess air for the hydrogen, the total air per pound of coal is

$$13.824\ \frac{\text{lbm air}}{\text{lbm coal}} + 1.294\ \frac{\text{lbm air}}{\text{lbm coal}}$$
$$= \boxed{15.1\ \text{lbm air/lbm coal}\quad (15\ \text{lbm})}$$

The answer is (B).

7. From Eq. 14.10,

$$\text{SG} = \frac{141.5}{°\text{API} + 131.5} = \frac{141.5}{40 + 131.5} = 0.825$$

Customary U.S. Solution

From Eq. 21.18(b),

$$\text{HHV} = 22{,}320 - 3780(\text{SG})^2$$
$$= 22{,}320 - (3780)(0.825)^2$$
$$= \boxed{19{,}747\ \text{Btu/lbm}\quad (20{,}000\ \text{Btu/lbm})}$$

The answer is (A).

SI Solution

From Eq. 21.18(a),

$$\text{HHV} = 51.92 - 8.792(\text{SG})^2$$
$$= 51.92 - (8.792)(0.825)^2$$
$$= \boxed{45.94\ \text{MJ/kg}\quad (46\ \text{MJ/kg})}$$

The answer is (A).

8. *Customary U.S. Solution*

(a) *step 1:* From Eq. 21.16(b), substituting the lower heating value of hydrogen from App. 21.A, the lower heating value of coal is

$$\text{LHV} = 14{,}093G_\text{C} + (51{,}623)\left(G_\text{H} - \frac{G_\text{O}}{8}\right) + 3983G_\text{S}$$
$$= (14{,}093)(0.75) + (51{,}623)\left(0.05 - \frac{0.03}{8}\right) + (3983)(0)$$
$$= 12{,}957\ \text{Btu/lbm}$$

step 2: The gravimetric analysis of 1 lbm of coal is

carbon: 0.75 lbm

free hydrogen: $G_{\text{H,total}} - \dfrac{G_\text{O}}{8} = 0.05 - \dfrac{0.03}{8} = 0.0463$

The ratio of the molecular weight of water (18) to the molecular weight of hydrogen (2) is 9.

water: (9)(0.05 – 0.0463) = 0.0333

nitrogen: 0.02

step 3: From Table 21.8, the theoretical stack gases per lbm coal for 0.75 lbm of carbon are

$$CO_2 = (0.75)(3.667\ \text{lbm}) = 2.750\ \text{lbm}$$
$$N_2 = (0.75)(8.883\ \text{lbm}) = 6.662\ \text{lbm}$$

All products are calculated similarly (as the following table summarizes). All values are per pound of fuel.

	CO_2	N_2	H_2O
from C:	2.750 lbm	6.662 lbm	
from H_2:		1.217 lbm	0.414 lbm
from H_2O:			0.0333 lbm
from O_2:	shows up in	CO_2 and H_2	
from N_2:		0.02 lbm	
total:	2.750 lbm	7.899 lbm	0.4473 lbm

step 4: Assume the combustion gases leave at 1000°F.

$$T_\text{ave} = \left(\tfrac{1}{2}\right)(60°\text{F} + 1000°\text{F}) = 530°\text{F}$$
$$T_\text{ave} = 530°\text{F} + 460° = 990°\text{R}$$

The specific heat values are given in Table 21.1.

$$c_{p,CO_2} = 0.251\ \text{Btu/lbm-°F}$$
$$c_{p,N_2} = 0.255\ \text{Btu/lbm-°F}$$
$$c_{p,H_2O} = 0.475\ \text{Btu/lbm-°F}$$

Thermodynamics

The energy required to raise the combustion products (from 1 lbm of coal) 1°F is

$$\begin{aligned} m_{CO_2}c_{p,CO_2} + m_{N_2}c_{p,N_2} + m_{H_2O}c_{p,H_2O} \\ &= (2.750\text{ lbm})\left(0.251\ \frac{\text{Btu}}{\text{lbm-°F}}\right) \\ &\quad + (7.899\text{ lbm})\left(0.255\ \frac{\text{Btu}}{\text{lbm-°F}}\right) \\ &\quad + (0.4473\text{ lbm})\left(0.475\ \frac{\text{Btu}}{\text{lbm-°F}}\right) \\ &= 2.92\text{ Btu/°F} \end{aligned}$$

step 5: Assuming all combustion heat goes into the stack gases, the temperature is given by Eq. 21.19.

$$\begin{aligned} T_{\max} &= T_i + \frac{\text{lower heat of combustion}}{\text{energy required}} \\ &= 60°\text{F} + \frac{12{,}957\ \frac{\text{Btu}}{\text{lbm}}}{2.92\ \frac{\text{Btu}}{\text{lbm-°F}}} \\ &= \boxed{4497°\text{F}\quad(4500°\text{F})}\quad\text{[unreasonable]} \end{aligned}$$

The answer is (C).

(b) *step 6:* Nitrogen in the coal does not contribute to excess air. With 40% excess air and 75% of heat absorbed by the boiler, the excess air (based on 76.85% N_2 by weight) per pound of fuel is

$$\begin{aligned} &(0.40)\left(\frac{7.899\ \frac{\text{lbm}}{\text{lbm}} - 0.02\ \frac{\text{lbm}}{\text{lbm}}}{0.7685}\right) \\ &\quad = 4.101\text{ lbm/lbm} \end{aligned}$$

From Table 21.1, $c_{p,\text{air}} = 0.249$ Btu/lbm-°F.

Therefore,

$$\begin{aligned} T_{\max} &= 60°\text{F} + \frac{\left(12{,}957\ \frac{\text{Btu}}{\text{lbm}}\right)(1-0.75)}{\left(\begin{array}{c} 2.92\ \frac{\text{Btu}}{\text{lbm-°F}} + \left(4.101\ \frac{\text{lbm}}{\text{lbm}}\right) \\ \times\left(0.249\ \frac{\text{Btu}}{\text{lbm-°F}}\right)\end{array}\right)} \\ &= \boxed{881.9°\text{F}\quad(880°\text{F})} \end{aligned}$$

The answer is (B).

SI Solution

(a) *step 1:* From Eq. 21.16(a), and substituting the lower heating value of hydrogen from App. 21.A, the lower heating value of coal is

$$\begin{aligned} \text{LHV} &= 32.78G_C + \frac{\left(51{,}623\ \frac{\text{Btu}}{\text{lbm}}\right)\left(2.326\ \frac{\text{kJ-lbm}}{\text{kg-Btu}}\right)}{1000\ \frac{\text{kJ}}{\text{MJ}}} \\ &\quad\times\left(G_H - \frac{G_O}{8}\right) + 9.264G_S \\ &= (32.78)(0.75) + (120.1)\left(0.05 - \frac{0.03}{8}\right) \\ &\quad + (9.264)(0) \\ &= 30.14\text{ MJ/kg} \end{aligned}$$

Steps 2 and 3 are the same as for the customary U.S. solution except that all masses are in kg.

step 4: Assume the combustion gases leave at 550°C.

$$T_{\text{ave}} = \left(\tfrac{1}{2}\right)(16°\text{C} + 550°\text{C}) + 273° = 556\text{K}$$

Specific heat values are given in Table 21.1. Using the footnote for SI units,

$$\begin{aligned} c_{p,CO_2} &= 1.051\text{ kJ/kg·K} \\ c_{p,N_2} &= 1.068\text{ kJ/kg·K} \\ c_{p,H_2O} &= 1.989\text{ kJ/kg·K} \end{aligned}$$

The energy required to raise the combustion products (from 1 kg of coal) 1°C is

$$\begin{aligned} m_{CO_2}c_{p,CO_2} + m_{N_2}c_{p,N_2} + m_{H_2O}c_{p,H_2O} \\ &= (2.750\text{ kg})\left(1.051\ \frac{\text{kJ}}{\text{kg·K}}\right) \\ &\quad + (7.899\text{ kg})\left(1.068\ \frac{\text{kJ}}{\text{kg·K}}\right) \\ &\quad + (0.4473\text{ kg})\left(1.989\ \frac{\text{kJ}}{\text{kg·K}}\right) \\ &= 12.22\text{ kJ/K} \end{aligned}$$

step 5: Assuming all combustion heat goes into the stack gases, the temperature is given by Eq. 21.19.

$$T_{\max} = T_i + \frac{\text{lower heat of combustion}}{\text{energy required}}$$
$$= 16°\text{C} + \frac{\left(30.14\ \frac{\text{MJ}}{\text{kg}}\right)\left(1000\ \frac{\text{kJ}}{\text{MJ}}\right)}{12.22\ \frac{\text{kJ}}{\text{K}}}$$
$$= \boxed{2482°\text{C}\quad (2500°\text{C})}\quad [\text{unreasonable}]$$

The answer is (C).

(b) *step 6:* Nitrogen in the coal does not contribute to excess air. With 40% excess air and 75% of heat absorbed by the boiler, the excess air (based on 76.85% N_2 by weight) per kilogram of fuel is

$$(0.40)\left(\frac{7.899\ \frac{\text{kg}}{\text{kg}} - 0.02\ \frac{\text{kg}}{\text{kg}}}{0.7685}\right) = 4.101\ \text{kg/kg}$$

From Table 21.1, using the table footnote, $c_{p,\text{air}} =$ 1.043 kJ/kg·K.

Therefore,

$$T_{\max} = 16°\text{C} + \frac{\left(30.14\ \frac{\text{MJ}}{\text{kg}}\right)\left(1000\ \frac{\text{kJ}}{\text{MJ}}\right)(1-0.75)}{12.22\ \frac{\text{kJ}}{\text{K}} + (4.101\ \text{kg})\left(1.043\ \frac{\text{kJ}}{\text{kg·K}}\right)}$$
$$= \boxed{472.7°\text{C}\quad (470°\text{C})}$$

The answer is (B).

9. Assume the oxygen is in the form of moisture in the fuel.

The available hydrogen is

$$G_{\text{H,free}} = G_{\text{H}} - \frac{G_{\text{O}}}{8} = 0.1131 - \frac{0.027}{8} = 0.1097$$

Customary U.S. Solution

step 1: From Table 21.8, find the stoichiometric oxygen required per pound of fuel oil.

$$\text{C} \rightarrow \text{CO}_2\text{: O}_2 \text{ required} = (0.8543)(2.667\ \text{lbm}) = 2.2784\ \text{lbm}$$

$$\text{H}_2 \rightarrow \text{H}_2\text{O: O}_2 \text{ required} = (0.1097)(7.936\ \text{lbm}) = 0.8706\ \text{lbm}$$

$$\text{S} \rightarrow \text{SO}_2\text{: O}_2 \text{ required} = (0.0034)(0.998\ \text{lbm}) = 0.0034\ \text{lbm}$$

The total amount of oxygen required per pound of fuel oil is

$$2.2784\ \text{lbm} + 0.8706\ \text{lbm} + 0.0034\ \text{lbm} = 3.1524\ \text{lbm}$$

step 2: With 60% excess air, the excess oxygen per pound of fuel oil is

$$(0.6)(3.1524\ \text{lbm}) = 1.8914\ \text{lbm O}_2$$

From Eq. 23.60, this oxygen occupies a volume of

$$V = \frac{mRT}{p}$$

At standard conditions,

$$p = 14.7\ \text{psia}$$
$$T = 60°\text{F} + 460° = 520°\text{R}$$

From Table 23.7, for oxygen, $R = 48.29$ ft-lbf/lbm-°R.

$$V = \frac{(1.8914\ \text{lbm})\left(48.29\ \frac{\text{ft-lbf}}{\text{lbm-°R}}\right)(520°\text{R})}{\left(14.7\ \frac{\text{lbf}}{\text{in}^2}\right)\left(12\ \frac{\text{in}}{\text{ft}}\right)^2} = 22.44\ \text{ft}^3$$

step 3: The theoretical nitrogen per pound of fuel based on Table 21.6 is

$$\left(\frac{3.1524\ \text{lbm}}{0.2315}\right)(0.7685) = 10.465\ \text{lbm N}_2$$

The actual nitrogen per pound of fuel with 60% excess air and nitrogen in the fuel is

$$(10.465\ \text{lbm})(1.6) + 0.0022\ \text{lbm} = 16.746\ \text{lbm N}_2$$

From Eq. 23.60, this nitrogen occupies a volume of

$$V = \frac{mRT}{p}$$

From Table 23.7, R for nitrogen = 55.16 ft-lbf/lbm-°R.

$$V = \frac{(16.746\ \text{lbm})\left(55.16\ \frac{\text{ft-lbf}}{\text{lbm-°R}}\right)(520°\text{R})}{\left(14.7\ \frac{\text{lbf}}{\text{in}^2}\right)\left(12\ \frac{\text{in}}{\text{ft}}\right)^2}$$
$$= 226.91\ \text{ft}^3$$

step 4: From Table 21.8, the 60°F combustion product volumes per pound of fuel will be

CO_2:	(0.8543)(31.63 ft^3)	=	27.02 ft^3
H_2O:	(0.1131)(188.25 ft^3)	=	21.29 ft^3
SO_2:	(0.0034)(11.84 ft^3)	=	0.04 ft^3
N_2:	from step 3	=	226.91 ft^3
O_2:	from step 2	=	22.44 ft^3
	total	=	297.7 ft^3

(a) At 60°F, the wet volume per pound of fuel will be 297.7 ft^3.

At 600°F, the wet volume per pound of fuel is

$$V_{wet,600^\circ F} = (297.7\ ft^3)\left(\frac{600^\circ F + 460^\circ}{60^\circ F + 460^\circ}\right) = \boxed{606.9\ ft^3 \quad (600\ ft^3)}$$

The answer is (D).

(b) At 60°F, the dry volume per pound of fuel will be

$$297.7\ ft^3 - 21.29\ ft^3 = 276.4\ ft^3$$

At 600°F, the dry volume per pound of fuel is

$$V_{dry,600^\circ F} = (276.4\ ft^3)\left(\frac{600^\circ F + 460^\circ}{60^\circ F + 460^\circ}\right) = \boxed{563.4\ ft^3 \quad (560\ ft^3)}$$

The answer is (B).

(c) The volumetric fraction of dry carbon dioxide is

$$\frac{27.02\ ft^3}{276.41\ ft^3} = \boxed{0.098 \quad (10\%)}$$

The answer is (C).

SI Solution

step 1: From Table 21.8, find the stoichiometric oxygen required per kilogram of fuel oil.

$$C \rightarrow CO_2\text{: } O_2 \text{ required} = (0.8543)(2.667\ kg) = 2.2784\ kg$$

$$H_2 \rightarrow H_2O\text{: } O_2 \text{ required} = (0.1097)(7.936\ kg) = 0.8706\ kg$$

$$S \rightarrow SO_2\text{: } O_2 \text{ required} = (0.0034)(0.998\ kg) = 0.0034\ kg$$

The total amount of oxygen required per kilogram of fuel oil is

$$2.2784\ kg + 0.8706\ kg + 0.0034\ kg = 3.1524\ kg$$

step 2: With 60% excess air, the excess oxygen per kilogram of fuel oil is

$$(0.6)(3.1524\ kg) = 1.8914\ kg\ O_2$$

From Eq. 23.60, this oxygen occupies a volume of

$$V = \frac{mRT}{p}$$

At standard conditions,

$$p = 101.3\ kPa$$

$$T = 16^\circ C + 273^\circ = 289K$$

From Table 23.7, for oxygen, $R = 259.82$ J/kg·K.

$$V = \frac{(1.8914\ kg)\left(259.82\ \frac{J}{kg\cdot K}\right)(289K)}{(101.3\ kPa)\left(1000\ \frac{Pa}{kPa}\right)} = 1.402\ m^3$$

step 3: The theoretical nitrogen per kilogram of fuel based on Table 21.6 is

$$\left(\frac{3.1524\ kg}{0.2315}\right)(0.7685) = 10.465\ kg\ N_2$$

The actual nitrogen per kilogram of fuel with 60% excess air and nitrogen in the fuel is

$$(10.465\ kg)(1.6) + 0.0022\ kg = 16.746\ kg\ N_2$$

From Eq. 23.60, this nitrogen occupies a volume of

$$V = \frac{mRT}{p}$$

From Table 23.7, R for nitrogen = 296.77 J/kg·K.

$$V = \frac{(16.746\ kg)\left(296.77\ \frac{J}{kg\cdot K}\right)(289K)}{(101.3\ kPa)\left(1000\ \frac{Pa}{kPa}\right)} = 14.178\ m^3$$

step 4: From Table 21.8, the 16°C combustion product volumes per kilogram of fuel will be

CO_2:	$(0.8543)(31.63\ ft^3)(0.06243)$	=	$1.687\ m^3$
H_2O:	$(0.1131)(188.25\ ft^3)(0.06243)$	=	$1.329\ m^3$
SO_2:	$(0.0034)(11.84\ ft^3)(0.06243)$	=	$0.003\ m^3$
N_2:	from step 3	=	$14.178\ m^3$
O_2:	from step 2	=	$1.402\ m^3$
	total	=	$18.599\ m^3$

(a) At 16°C, the wet volume per kilogram of fuel will be 18.599 m^3.

At 320°C, the wet volume per kilogram of fuel is

$$V_{wet,320^\circ C} = (18.599\ m^3)\left(\frac{320^\circ C + 273^\circ}{16^\circ C + 273^\circ}\right) = \boxed{38.16\ m^3 \quad (38\ m^3)}$$

The answer is (D).

(b) At 16°C, the dry volume per kilogram of fuel will be

$$18.599 \text{ m}^3 - 1.329 \text{ m}^3 = 17.27 \text{ m}^3$$

At 320°C, the dry volume per kilogram of fuel is

$$V_{\text{dry},320°\text{C}} = (17.27 \text{ m}^3)\left(\frac{320°\text{C} + 273°}{16°\text{C} + 273°}\right) = \boxed{35.44 \text{ m}^3 \quad (35 \text{ m}^3)}$$

The answer is (B).

(c) The volumetric fraction of dry carbon dioxide is

$$\frac{1.687 \text{ m}^3}{17.27 \text{ m}^3} = \boxed{0.098 \quad (10\%)}$$

The answer is (C).

10. *Customary U.S. Solution*

step 1: Based on 15,395 lbm of coal burned producing 2816 lbm of ash containing 20.9% carbon by weight, the usable percentage of carbon per pound of fuel is

$$0.5145 - \frac{(2816 \text{ lbm})(0.209)}{15{,}395 \text{ lbm}} = 0.4763$$

step 2: Since moisture is reported separately, assume all of the oxygen and hydrogen are free. (This is not ordinarily the case.) From Table 21.8, find the stoichiometric oxygen required per pound of fuel.

$$\begin{aligned} C \rightarrow CO_2\text{: } O_2 \text{ required} &= (0.4763)(2.667 \text{ lbm}) \\ &= 1.2703 \text{ lbm} \\ H_2 \rightarrow H_2O\text{: } O_2 \text{ required} &= (0.0402)(7.936 \text{ lbm}) \\ &= 0.3190 \text{ lbm} \\ S \rightarrow SO_2\text{: } O_2 \text{ required} &= (0.0392)(0.998 \text{ lbm}) \\ &= 0.0391 \text{ lbm} \end{aligned}$$

The total amount of O_2 required per pound of fuel is

$$\begin{aligned} &1.2703 \text{ lbm} + 0.3190 \text{ lbm} \\ &\quad + 0.0391 \text{ lbm} - 0.0728 \text{ lbm} = 1.5556 \text{ lbm} \end{aligned}$$

step 3: The theoretical air per pound of fuel based on Table 21.6 is

$$\frac{1.5556 \text{ lbm}}{0.2315} = 6.720 \text{ lbm air}$$

step 4: Ignoring fly ash, the theoretical dry products per pound of fuel are given from Table 21.8.

CO_2:	(0.4763)(3.667 lbm)	=	1.7466 lbm
SO_2:	(0.0392)(1.998 lbm)	=	0.0783 lbm
N_2:	0.0093 lbm + (6.720 lbm)(0.7685)	=	5.1736 lbm
	total	=	6.999 lbm

step 5: The excess air per pound of fuel is

$$13.3 \text{ lbm} - 6.999 \text{ lbm} = 6.301 \text{ lbm}$$

step 6: The total air supplied per pound of fuel is

$$6.301 \text{ lbm} + 6.720 \text{ lbm} = \boxed{13.02 \text{ lbm} \quad (13 \text{ lbm})}$$

The answer is (A).

SI Solution

step 1: Based on 6923 kg of coal burned producing 1267 kg of ash containing 20.9% carbon by weight, the usable percentage of carbon per kilogram of fuel is

$$0.5145 - \frac{(1267 \text{ kg})(0.209)}{6923 \text{ kg}} = 0.4763$$

step 2: From Table 21.8, find the stoichiometric oxygen required per kilogram of fuel.

$$\begin{aligned} C \rightarrow CO_2\text{: } O_2 \text{ required} &= (0.4763)(2.667 \text{ kg}) \\ &= 1.2703 \text{ kg} \\ H_2 \rightarrow H_2O\text{: } O_2 \text{ required} &= (0.0402)(7.936 \text{ kg}) \\ &= 0.3190 \text{ kg} \\ S \rightarrow SO_2\text{: } O_2 \text{ required} &= (0.0392)(0.998 \text{ kg}) \\ &= 0.0391 \text{ kg} \end{aligned}$$

The total amount of O_2 required per kilogram of fuel is

$$\begin{aligned} &1.2703 \text{ kg} + 0.3190 \text{ kg} \\ &\quad + 0.0391 \text{ kg} - 0.0728 \text{ kg} = 1.5556 \text{ kg} \end{aligned}$$

step 3: The theoretical air per kilogram of fuel based on Table 21.6 is

$$\frac{1.5556 \text{ kg}}{0.2315} = 6.720 \text{ kg air}$$

step 4: Ignoring fly ash, the theoretical dry products per kilogram of fuel are given from Table 21.8.

CO_2:	(0.4763)(3.667 kg)	=	1.7466 kg
SO_2:	(0.0392)(1.998 kg)	=	0.0783 kg
N_2:	0.0093 kg + (6.720 kg)(0.7685)	=	5.1736 kg
	total	=	6.999 kg

step 5: The excess air per kilogram of fuel is

$$13.3 \text{ kg} - 6.999 \text{ kg} = 6.301 \text{ kg}$$

step 6: The total air supplied per kilogram of fuel is

$$6.301 \text{ kg} + 6.720 \text{ kg} = \boxed{13.02 \text{ kg} \quad (13 \text{ kg})}$$

The answer is (A).

11. (The customary U.S. and SI solutions are essentially identical.)

From Eq. 21.9, the actual air-fuel ratio can be estimated as

$$R_{a/f,\text{actual}} = \frac{3.04 B_{N_2} G_C}{B_{CO_2} + B_{CO}}$$

A fraction of carbon is reduced due to the percentage of coal lost in the ash pit.

$$G_C = (1 - 0.03)(0.65) = 0.6305$$

$$\begin{aligned} R_{a/f,\text{actual}} &= \frac{(3.04)(0.815)(0.6305)}{0.095 + 0} \\ &= 16.44 \text{ lbm air/lbm fuel (kg air/kg fuel)} \end{aligned}$$

Combustion uses 9.87 lbm of air/lbm of fuel.

$$(9.87 \text{ lbm})(1 - 0.03) = 9.57 \text{ lbm of air/lbm of fuel}$$

$$\begin{aligned} \%\text{ of excess air} &= \left(\frac{16.44 \ \frac{\text{lbm}}{\text{lbm}} - 9.57 \ \frac{\text{lbm}}{\text{lbm}}}{9.57 \ \frac{\text{lbm}}{\text{lbm}}} \right) \\ &\quad \times 100\% \\ &= \boxed{71.8\% \quad (70\%)} \end{aligned}$$

The answer is (C).

12. From Eq. 21.23, the heat loss due to the formation of carbon monoxide is

$$q = \frac{2.334 \text{HHV}_{CO} G_C B_{CO}}{B_{CO_2} + B_{CO}}$$

From App. 21.A, the heating value of carbon monoxide is

$$\text{HHV}_{CO} = 4347 \text{ Btu/lbm} \quad (10.11 \text{ MJ/kg})$$

$$G_C = (1 - 0.03)(0.6734) = 0.6532$$

Customary U.S. Solution

$$\begin{aligned} q &= \frac{(2.334)\left(4347 \ \frac{\text{Btu}}{\text{lbm}}\right)(0.6532)(0.016)}{0.155 + 0.016} \\ &= \boxed{620.1 \text{ Btu/lbm} \quad (600 \text{ Btu/lbm})} \end{aligned}$$

The answer is (A).

SI Solution

$$\begin{aligned} q &= \frac{(2.334)\left(10.11 \ \frac{\text{MJ}}{\text{kg}}\right)(0.6532)(0.016)}{0.155 + 0.016} \\ &= \boxed{1.44 \text{ MJ/kg} \quad (1.4 \text{ MJ/kg})} \end{aligned}$$

The answer is (A).

13. (a) For methane, $B = 0.93$.

From ideal gas laws (R for methane $= 96.32$ ft-lbf/lbm-°R),

$$\begin{aligned} \rho &= \frac{p}{RT} \\ &= \frac{\left(14.7 \ \frac{\text{lbf}}{\text{in}^2}\right)\left(12 \ \frac{\text{in}}{\text{ft}}\right)^2}{\left(96.32 \ \frac{\text{ft-lbf}}{\text{lbm-°R}}\right)(60°\text{F} + 460°)} \\ &= 0.0423 \text{ lbm/ft}^3 \end{aligned}$$

From Table 21.9, $K = 9.55$ ft^3 air/ft^3 fuel.

From Table 21.8,

$$\text{products:} \quad 1 \text{ ft}^3 \ CO_2, 2 \text{ ft}^3 \ H_2O$$

From App. 21.A, HHV = 1013 Btu/lbm.

Similar results for all the other fuel components are tabulated in the table.

Density is volumetrically weighted. The composite density is

$$\begin{aligned} \rho &= \sum B_i \rho_i \\ &= (0.93)\left(0.0423 \ \frac{\text{lbm}}{\text{ft}^3}\right) + (0.0373)\left(0.0738 \ \frac{\text{lbm}}{\text{ft}^3}\right) \\ &\quad + (0.0045)\left(0.0738 \ \frac{\text{lbm}}{\text{ft}^3}\right) + (0.0182)\left(0.0053 \ \frac{\text{lbm}}{\text{ft}^3}\right) \\ &\quad + (0.0025)\left(0.0739 \ \frac{\text{lbm}}{\text{ft}^3}\right) + (0.0018)\left(0.0900 \ \frac{\text{lbm}}{\text{ft}^3}\right) \\ &\quad + (0.0035)\left(0.0843 \ \frac{\text{lbm}}{\text{ft}^3}\right) + (0.0022)\left(0.1160 \ \frac{\text{lbm}}{\text{ft}^3}\right) \\ &= \boxed{0.0434 \text{ lbm/ft}^3 \quad (0.043 \text{ lbm/ft}^3)} \end{aligned}$$

The answer is (B).

Reaction Products for Prob. 13

gas	B	$\frac{\text{lbm}}{\text{ft}^3}$	ft^3 air	$\frac{\text{Btu}}{\text{lbm}}$	volumes of products		
					CO_2	H_2O	other
CH_4	0.93	0.0422	9.556	1013	1	2	–
N_2	0.0373	0.0738	–	–	–	–	1 N_2
CO	0.0045	0.0738	2.389	322	1	–	–
H_2	0.0182	0.0053	2.389	325	–	1	–
C_2H_4	0.0025	0.0739	14.33	1614	2	2	–
H_2S	0.0018	0.0900	7.167	647	–	1	1 SO_2
O_2	0.0035	0.0843	–	–	–	–	–
CO_2	0.0022	0.1160	–	–	1	–	–

(b) The air is 20.9% oxygen by volume. The theoretical air requirements are

$$\begin{aligned}\sum B_i V_{\text{air},i} - \frac{O_2 \text{ in fuel}}{0.209} &= (0.93)(9.556 \text{ ft}^3) + (0.0373)(0 \text{ ft}^3) \\ &\quad + (0.0045)(2.389 \text{ ft}^3) + (0.0182)(2.389 \text{ ft}^3) \\ &\quad + (0.0025)(14.33 \text{ ft}^3) + (0.0018)(7.167 \text{ ft}^3) \\ &\quad + (0.0035)(0 \text{ ft}^3) + (0.0022)(0 \text{ ft}^3) - \frac{0.0035}{0.209} \\ &= \boxed{8.9733 \text{ ft}^3 \text{ air/ft}^3 \text{ fuel} \quad (9 \text{ ft}^3 \text{ air/ft}^3 \text{ fuel})}\end{aligned}$$

The answer is (C).

((c) and (d)) The theoretical oxygen will be

$$\left(8.9733 \ \frac{\text{ft}^3}{\text{ft}^3}\right)(0.209) = 1.875 \text{ ft}^3/\text{ft}^3$$

The excess oxygen will be

$$\left(1.875 \ \frac{\text{ft}^3}{\text{ft}^3}\right)(0.4) = 0.75 \text{ ft}^3/\text{ft}^3$$

Similarly, with 40% excess air, the total nitrogen in the stack gases is

$$\begin{aligned}(1.4)(0.791)\left(8.9733 \ \frac{\text{ft}^3}{\text{ft}^3}\right) + 0.0373 \ \frac{\text{ft}^3}{\text{ft}^3} \\ = 9.974 \text{ ft}^3/\text{ft}^3 \text{ fuel}\end{aligned}$$

The wet stack gas volumes per ft^3 of fuel are

excess O_2:	=	0.7500 ft^3
excess N_2:	=	9.974 ft^3
excess SO_2:	=	0.0018 ft^3
excess CO_2: (0.93)(1) + (0.0045)(1)+ (0.0025)(2) + (0.0022)(1)	=	0.942 ft^3
excess H_2O: (0.93)(2) + (0.0182)(1)+ (0.0025)(2) + (0.0018)(1)	=	1.885 ft^3
total	=	13.55 ft^3

The total wet volume is 13.55 ft^3/ft^3 fuel.

The total dry volume is 11.67 ft^3/ft^3 fuel.

The volumetric analyses are

	O_2	N_2	SO_2	CO_2	H_2O
wet:	$\frac{0.7500 \text{ ft}^3}{13.55 \text{ ft}^3}$	$\frac{9.974 \text{ ft}^3}{13.55 \text{ ft}^3}$	$\frac{0.0018 \text{ ft}^3}{13.55 \text{ ft}^3}$	$\frac{0.942 \text{ ft}^3}{13.55 \text{ ft}^3}$	$\frac{1.885 \text{ ft}^3}{13.55 \text{ ft}^3}$
	= 0.0553	0.736	–	0.069	0.139
dry:	$\frac{0.7500 \text{ ft}^3}{11.67 \text{ ft}^3}$	$\frac{9.974 \text{ ft}^3}{11.67 \text{ ft}^3}$	$\frac{0.0018 \text{ ft}^3}{11.67 \text{ ft}^3}$	$\frac{0.942 \text{ ft}^3}{11.67 \text{ ft}^3}$	–
	= 0.0643	0.855	–	0.081	–

The percentage of CO_2 in the flue gas (wet basis) is

$$B_{CO_2,\text{wet}} = \boxed{6.9\%}$$

The answer is (A).

The percentage of CO_2 in the flue gas (dry basis) is

$$B_{CO_2,\text{dry}} = \boxed{8.1\%}$$

The answer is (C).

14. *Customary U.S. Solution*

step 1: From App. 23.B, the heat of vaporization at 212°F is $h_{fg} = 970.1$ Btu/lbm.

The heat absorbed in the boiler is

$$\begin{aligned}m_{H_2O} h_{fg} &= (11.12 \text{ lbm } H_2O)\left(970.1 \ \frac{\text{Btu}}{\text{lbm}}\right) \\ &= 10{,}787.5 \text{ Btu/lbm fuel}\end{aligned}$$

Thermodynamics

step 2: The losses for heating stack gases can be found as follows.

The burned carbon per lbm of fuel is

$$0.7842 - (0.315)(0.0703) = 0.7621 \text{ lbm/lbm fuel}$$

The mass ratio of dry flue gases to solid fuel is given by Eq. 21.12.

$$\begin{aligned}\frac{\text{mass of flue gas}}{\text{mass of solid fuel}} &= \frac{\left(\begin{array}{c}11B_{CO_2}+8B_{O_2}\\ +(7)(B_{CO}+B_{N_2})\end{array}\right)\left(G_C+\dfrac{G_S}{1.833}\right)}{(3)(B_{CO_2}+B_{CO})}\\ &= \frac{\left(\begin{array}{c}(11)(14.0)+(8)(5.5)\\ +(7)(0.42+80.08)\end{array}\right)\left(0.7621+\dfrac{0.01}{1.833}\right)}{(3)(14.0+0.42)}\\ &= 13.51 \text{ lbm stack gases/lbm fuel}\end{aligned}$$

Properties of nitrogen are commonly assumed for dry flue gas. From Table 21.1, for nitrogen at an average temperature of (575°F + 73°F)/2 = 324°F (784°R), $c_p \approx$ 0.252 Btu/lbm-°F.

The losses for heating stack gases are given by Eq. 21.20.

$$\begin{aligned}q_1 &= m_{\text{flue gas}} c_p (T_{\text{flue gas}} - T_{\text{incoming air}})\\ &= (13.51 \text{ lbm})\left(0.252\ \frac{\text{Btu}}{\text{lbm-°F}}\right)(575°\text{F} - 73°\text{F})\\ &= 1709.1 \text{ Btu/lbm fuel}\end{aligned}$$

The heat loss in the vapor formed during the combustion of hydrogen is given by Eq. 21.21.

$$q_2 = 8.94 G_H (h_g - h_f)$$

Assume that the partial pressure of the water vapor is below 1 psia (the lowest App. 23.C goes).

h_g at 575°F and 1 psia can be found from the superheat tables, App. 23.C.

$$h_g \approx 1324.3 \text{ Btu/lbm}$$

h_f at 73°F can be found from App. 23.A.

$$h_f = 41.07 \text{ Btu/lbm}$$

$$\begin{aligned}G_{H,\text{available}} &= G_{H,\text{total}} - \frac{G_O}{8} = 0.0556 - \frac{0.0825}{8}\\ &= 0.0453\\ q_2 &= (8.94)(0.0453)\left(1324.3\ \frac{\text{Btu}}{\text{lbm}} - 41.07\ \frac{\text{Btu}}{\text{lbm}}\right)\\ &= 519.7 \text{ Btu/lbm fuel}\end{aligned}$$

Heat is also lost when it is absorbed by the moisture that was originally in the combustion air.

$$q_3 = \omega m_{\text{combustion air}}(h_g - h'_g)$$

Assume that the partial pressure of the water vapor is below 1 psia (the lowest App. 23.C goes).

From App. 23.B, at 73°F and 0.4 psia, $h'_g \approx$ 1093 Btu/lbm. From the psychrometric chart,

$$\omega = 90 \text{ grains/lbm air} = 0.0129 \text{ lbm water/lbm air}$$

Considering the sulfur content, find the air/fuel ratio.

$$\begin{aligned}\frac{\text{lbm air}}{\text{lbm fuel}} &= \frac{3.04 B_{N_2}\left(G_C + \dfrac{G_S}{1.833}\right)}{B_{CO_2} + B_{CO}}\\ &= \frac{(3.04)(0.8008)\left(0.7621 + \dfrac{0.01}{1.833}\right)}{0.14 + 0.0042}\\ &= 12.96 \text{ lbm air/lbm fuel}\end{aligned}$$

$$\begin{aligned}q_3 &= \left(0.0129\ \frac{\text{lbm water}}{\text{lbm air}}\right)\left(12.96\ \frac{\text{lbm air}}{\text{lbm fuel}}\right)\\ &\quad\times\left(1324.3\ \frac{\text{Btu}}{\text{lbm}} - 1093\ \frac{\text{Btu}}{\text{lbm}}\right)\\ &= 38.7 \text{ Btu/lbm fuel}\end{aligned}$$

The energy lost in incomplete combustion is given by Eq. 21.23.

$$\begin{aligned}q_4 &= \frac{2.334 \text{HHV}_{CO} G_C B_{CO}}{B_{CO_2} + B_{CO}}\\ &= \frac{(2.334)\left(4347\ \dfrac{\text{Btu}}{\text{lbm}}\right)(0.7621)(0.0042)}{0.14 + 0.0042}\\ &= 225.2 \text{ Btu/lbm fuel}\end{aligned}$$

The energy lost in unburned carbon is given by Eq. 21.24.

$$\begin{aligned}q_5 &= \left(14{,}093\ \frac{\text{Btu}}{\text{lbm}}\right) m_{\text{ash}} G_{C,\text{ash}}\\ &= \left(14{,}093\ \frac{\text{Btu}}{\text{lbm}}\right)(0.0703)(0.315)\\ &= 312.1 \text{ Btu/lbm fuel}\end{aligned}$$

The energy lost in radiation and unaccounted losses per pound of fuel is

$$\begin{aligned}&14{,}000\ \frac{\text{Btu}}{\text{lbm}} - 10{,}787.5\ \frac{\text{Btu}}{\text{lbm}} - 1709.1\ \frac{\text{Btu}}{\text{lbm}} - 519.7\ \frac{\text{Btu}}{\text{lbm}}\\ &\quad - 38.7\ \frac{\text{Btu}}{\text{lbm}} - 225.2\ \frac{\text{Btu}}{\text{lbm}} - 312.1\ \frac{\text{Btu}}{\text{lbm}}\\ &= \boxed{407.7 \text{ Btu/lbm} \quad (400 \text{ Btu/lbm})}\end{aligned}$$

The answer is (A).

Thermodynamics

SI Solution

step 1: From App. 23.N, the heat of vaporization at 100°C is $h_{fg} = 2256.4$ kJ/kg.

The heat absorbed in the boiler is

$$m_{H_2O}h_{fg} = (11.12\ \text{kg})\left(2256.4\ \frac{\text{kJ}}{\text{kg}}\right) = 25\,091\ \text{kJ/kg fuel}$$

step 2: From step 2 of the U.S. solution,

$$\frac{\text{mass of flue gas}}{\text{mass of solid fuel}} = 13.51\ \text{kg stack gases/kg fuel}$$

Properties of nitrogen are commonly assumed for dry flue gas. From Table 21.1, for nitrogen at an average temperature of $(1/2)(300°\text{C} + 23°\text{C}) + 273° = 434.5\text{K}$ (782°R), $c_p \approx 0.252$ Btu/lbm-°R.

c_p from Table 21.1 can be found (using the table footnote) as

$$c_p = \left(0.252\ \frac{\text{Btu}}{\text{lbm-°R}}\right)\left(4.187\ \frac{\frac{\text{kJ}}{\text{kg·K}}}{\frac{\text{Btu}}{\text{lbm-°R}}}\right) = 1.055\ \text{kJ/kg·K}$$

The losses for heating stack gases are given by Eq. 21.20.

$$\begin{aligned} q_1 &= m_{\text{flue gas}}c_p(T_{\text{flue gas}} - T_{\text{incoming air}}) \\ &= (13.51\ \text{kg})\left(1.055\ \frac{\text{kJ}}{\text{kg·K}}\right)(300°\text{C} - 23°\text{C}) \\ &= 3948.1\ \text{kJ/kg fuel} \end{aligned}$$

The heat loss in the vapor formed during the combustion of hydrogen is given by Eq. 21.21.

$$q_2 = 8.94G_H(h_g - h_f)$$

Assume that the partial pressure of the water vapor is below 10 kPa (the lowest App. 23.P goes).

h_g at 300°C and 10 kPa can be found from the superheat tables, App. 23.P.

$$h_g = 3076.7\ \text{kJ/kg}$$

h_f at 23°C can be found from App. 23.N.

$$h_f = 96.47\ \text{kJ/kg}$$

$$G_{\text{H,available}} = G_{\text{H,total}} - \frac{G_O}{8} = 0.0556 - \frac{0.0825}{8} = 0.0453$$

$$\begin{aligned} q_2 &= (8.94)(0.0453)\left(3076.7\ \frac{\text{kJ}}{\text{kg}} - 96.47\ \frac{\text{kJ}}{\text{kg}}\right) \\ &= 1206.9\ \text{kJ/kg fuel} \end{aligned}$$

Heat is also lost when it is absorbed by the moisture originally in the combustion air.

$$q_3 = \omega m_{\text{combustion air}}(h_g - h'_g)$$

Assume that the partial pressure of the water vapor is low. At 23°C, from App. 23.N, $h'_g \approx 2542.9$ kJ/kg.

From the psychrometric chart for 19°C wet bulb and 23°C dry bulb, $\omega = 12.2$ g/kg dry air.

The air/fuel ratio from the customary U.S. solution is

$$\frac{\text{kg air}}{\text{kg fuel}} = 12.96\ \text{kg air/kg fuel}$$

$$\begin{aligned} q_3 &= \frac{\left(12.2\ \frac{\text{g}}{\text{kg}}\right)\left(12.96\ \frac{\text{kg}}{\text{kg}}\right) \times \left(3076.7\ \frac{\text{kJ}}{\text{kg}} - 2542.9\ \frac{\text{kJ}}{\text{kg}}\right)}{1000\ \frac{\text{g}}{\text{kg}}} \\ &= 84.40\ \text{kJ/kg} \end{aligned}$$

Energy lost in incomplete combustion is given by Eq. 21.23.

$$\begin{aligned} q_4 &= \frac{2.334\text{HHV}_{CO}G_C B_{CO}}{B_{CO_2} + B_{CO}} \\ &= \frac{(2.334)\left(10.11\ \frac{\text{MJ}}{\text{kg}}\right)(0.7621)(0.0042)\left(1000\ \frac{\text{kJ}}{\text{MJ}}\right)}{0.14 + 0.0042} \\ &= 523.8\ \text{kJ/kg fuel} \end{aligned}$$

The energy lost in unburned carbon is given by Eq. 21.24.

$$\begin{aligned} q_5 &= \left(32.8\ \frac{\text{MJ}}{\text{kg}}\right)m_{\text{ash}}G_{\text{C,ash}} \\ &= \left(32.8\ \frac{\text{MJ}}{\text{kg}}\right)\left(1000\ \frac{\text{kJ}}{\text{MJ}}\right)(0.0703)(0.315) \\ &= 726.3\ \text{kJ/kg fuel} \end{aligned}$$

The energy lost in radiation and unaccounted losses per kilogram of fuel is

$$\begin{aligned} &\left(32.6\ \frac{\text{MJ}}{\text{kg}}\right)\left(1000\ \frac{\text{kJ}}{\text{MJ}}\right) - 25\,091\ \frac{\text{kJ}}{\text{kg}} \\ &\quad - 3948.1\ \frac{\text{kJ}}{\text{kg}} - 1206.9\ \frac{\text{kJ}}{\text{kg}} - 84.40\ \frac{\text{kJ}}{\text{kg}} \\ &\quad - 523.8\ \frac{\text{kJ}}{\text{kg}} - 726.3\ \frac{\text{kJ}}{\text{kg}} \\ &\quad = \boxed{1019.5\ \text{kJ/kg}\quad (1020\ \text{kJ/kg})} \end{aligned}$$

The answer is (A).

Thermodynamics

15. *Customary U.S. Solution*

step 1: The incoming reactants on a per-pound basis are

0.07 lbm ash

0.05 lbm hydrogen

0.05 lbm oxygen

0.83 lbm carbon

This is an ultimate analysis. Assume that only the hydrogen that is not locked up with oxygen in the form of water is combustible. From Eq. 21.15, the available hydrogen is

$$G_{\text{H,available}} = G_{\text{H,total}} - \frac{G_{\text{O}}}{8} = 0.05 \text{ lbm} - \frac{0.05 \text{ lbm}}{8}$$
$$= 0.04375 \text{ lbm}$$

The mass of water produced is the hydrogen mass plus eight times as much oxygen. The locked hydrogen is

$$0.05 \text{ lbm} - 0.04375 \text{ lbm} = 0.00625 \text{ lbm}$$
$$\text{lbm of moisture} = G_{\text{H}} + G_{\text{O}} = G_{\text{H}} + 8G_{\text{H}}$$
$$= 0.00625 \text{ lbm} + (8)(0.00625 \text{ lbm})$$
$$= 0.05625 \text{ lbm}$$

The air is 23.15% oxygen by weight (see Table 21.6), so other reactants for 26 lbm of air are

$$(0.2315)(26 \text{ lbm}) = 6.019 \text{ lbm } O_2$$
$$(0.7685)(26 \text{ lbm}) = 19.981 \text{ lbm } N_2$$

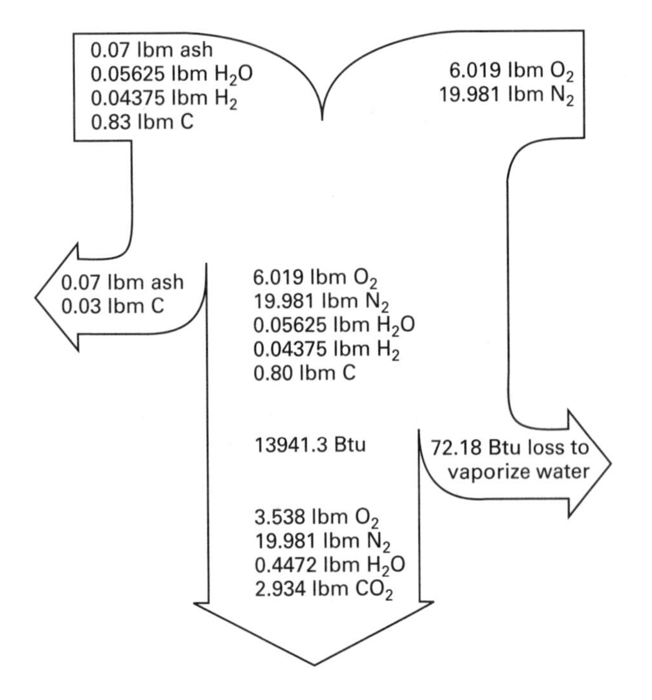

step 2: Ash pit material losses are 10%, or 0.1 lbm, which includes all of the ash.

0.07 lbm ash (noncombustible matter)

0.03 lbm unburned carbon

step 3: Determine what remains.

6.019 lbm oxygen

19.981 lbm nitrogen

0.05625 lbm water

0.04375 lbm hydrogen

0.80 lbm carbon

step 4: Determine the energy loss in vaporizing the moisture.

$$q = (\text{moisture})(h_g - h_f)$$

From App. 23.C, h_g at 550°F and 14.7 psia is 1311.4 Btu/lbm.

From App. 23.A, h_f at 60°F is 28.08 Btu/lbm.

$$q = (0.05625 \text{ lbm})\left(1311.4 \frac{\text{Btu}}{\text{lbm}} - 28.08 \frac{\text{Btu}}{\text{lbm}}\right)$$
$$= 72.19 \text{ Btu}$$

step 5: Calculate the heating value of the remaining fuel components using App. 21.A.

$$\text{HV}_{\text{C}} = (0.80 \text{ lbm})\left(14{,}093 \frac{\text{Btu}}{\text{lbm}}\right) = 11{,}274.4 \text{ Btu}$$
$$\text{HV}_{\text{H}} = (0.04375 \text{ lbm})\left(60{,}958 \frac{\text{Btu}}{\text{lbm}}\right) = 2666.9 \text{ Btu}$$

The heating value after the coal moisture is evaporated (i.e., the as-delivered heating value) is

$$11{,}274.4 \text{ Btu} + 2666.9 \text{ Btu} - 72.19 \text{ Btu}$$
$$= 13{,}869 \text{ Btu}$$

step 6: Using Table 21.8, determine the combustion products.

$$\text{oxygen required by carbon} = (0.80)(2.667 \text{ lbm})$$
$$= 2.134 \text{ lbm}$$
$$\text{oxygen required by hydrogen} = (0.04375)(7.936 \text{ lbm})$$
$$= 0.3472 \text{ lbm}$$
$$\begin{array}{r}\text{carbon dioxide} \\ \text{produced by carbon}\end{array} = (0.8)(3.667 \text{ lbm})$$
$$= 2.934 \text{ lbm}$$
$$\text{water produced by hydrogen} = (0.04375)(8.936 \text{ lbm})$$
$$= 0.3910 \text{ lbm}$$

The remaining oxygen is

$$6.019 \text{ lbm} - 2.134 \text{ lbm} - 0.3472 \text{ lbm} = 3.538 \text{ lbm}$$

step 7: The gaseous products must be heated from 70°F to 550°F. The average temperature is

$$\left(\tfrac{1}{2}\right)(70°\text{F} + 550°\text{F}) + 460° = 770°\text{R}$$

From Table 21.1, the specific heat of gaseous products is

gas	$\frac{\text{Btu}}{\text{lbm-°R}}$
oxygen	0.228
nitrogen	0.252
water	0.460
carbon dioxide	0.225

$$Q_{\text{heating}} = \begin{pmatrix} (3.538 \text{ lbm})\left(0.228 \ \frac{\text{Btu}}{\text{lbm-°R}}\right) \\ + (19.981 \text{ lbm})\left(0.252 \ \frac{\text{Btu}}{\text{lbm-°R}}\right) \\ + (0.3910 \text{ lbm})\left(0.460 \ \frac{\text{Btu}}{\text{lbm-°R}}\right) \\ + (2.934 \text{ lbm})\left(0.225 \ \frac{\text{Btu}}{\text{lbm-°R}}\right) \end{pmatrix} \times (550°\text{F} - 70°\text{F})$$
$$= 3207.3 \text{ Btu}$$

step 8: The percentage loss based on the coal's as-delivered heating value is

$$\frac{3207.3 \text{ Btu} + 72.19 \text{ Btu}}{13{,}869 \text{ Btu}} = \boxed{0.236 \quad (24\%)}$$

The answer is (B).

SI Solution

Steps 1 through 3 are the same as for the customary U.S. solution except that everything is based on kg.

step 4: Determine the energy loss in the vaporizing moisture.

$$q = (\text{moisture})(h_g - h_f)$$

From App. 23.P, h_g at 290°C and 101.3 kPa is 3054.5 kJ/kg.

h_f at 15.6°C from App. 23.N is 65.49 kJ/kg.

$$q = (0.05625 \text{ kg})\left(3054.5 \ \frac{\text{kJ}}{\text{kg}} - 65.49 \ \frac{\text{kJ}}{\text{kg}}\right)$$
$$= 168.1 \text{ kJ}$$

step 5: Calculate the heating value of the remaining fuel components using App. 21.A and the table footnote.

$$\text{HV}_\text{C} = (0.80 \text{ kg})\left(14\,093 \ \frac{\text{Btu}}{\text{lbm}}\right)\left(2.326 \ \frac{\frac{\text{kJ}}{\text{kg}}}{\frac{\text{Btu}}{\text{lbm}}}\right)$$
$$= 26\,224 \text{ kJ}$$

$$\text{HV}_\text{H} = (0.04375 \text{ kg})\left(60\,958 \ \frac{\text{Btu}}{\text{lbm}}\right)\left(2.326 \ \frac{\frac{\text{kJ}}{\text{kg}}}{\frac{\text{Btu}}{\text{lbm}}}\right)$$
$$= 6203 \text{ kJ}$$

The heating value after the coal moisture is evaporated is

$$26\,224 \text{ kJ} + 6203 \text{ kJ} - 168.1 \text{ kJ} = 32\,259 \text{ kJ}$$

step 6: This step is the same as for the customary U.S. solution except that all quantities are in kg.

step 7: The gaseous products must be heated from 21°C to 290°C. The average temperature is

$$\left(\tfrac{1}{2}\right)(21°\text{C} + 290°\text{C}) + 273° = 428.5\text{K}$$
$$(428.5\text{K})\left(1.8 \ \frac{°\text{R}}{\text{K}}\right) = 771°\text{R}$$

From Table 21.1, the specific heat of gaseous products is calculated using the table footnote.

gas	$\frac{\text{kJ}}{\text{kg·K}}$
oxygen	0.955
nitrogen	1.055
water	1.926
carbon dioxide	0.942

$$Q_{\text{heating}} = \begin{pmatrix} (3.538 \text{ kg})\left(0.955 \ \frac{\text{kJ}}{\text{kg·K}}\right) \\ + (19.981 \text{ kg})\left(1.055 \ \frac{\text{kJ}}{\text{kg·K}}\right) \\ + (0.3910 \text{ kg})\left(1.926 \ \frac{\text{kJ}}{\text{kg·K}}\right) \\ + (2.934 \text{ kg})\left(0.942 \ \frac{\text{kJ}}{\text{kg·K}}\right) \end{pmatrix} \times (290°\text{C} - 21°\text{C})$$
$$= 7525.4 \text{ kJ}$$

step 8: The percentage loss is

$$\frac{7525.4 \text{ kJ} + 168.1 \text{ kJ}}{32\,259 \text{ kJ}} = \boxed{0.238 \quad (24\%)}$$

The answer is (B).

16. (a) Silicon in ash is SiO_2 with a molecular weight of

$$28.09\ \frac{\text{lbm}}{\text{lbmol}} + (2)\left(16\ \frac{\text{lbm}}{\text{lbmol}}\right) = 60.09\ \text{lbm/lbmol}$$

The oxygen used with 6.1% by mass silicon is

$$\left(\frac{(2)(16\ \text{lbm})}{28.09\ \text{lbm}}\right)\left(0.061\ \frac{\text{lbm}}{\text{lbm coal}}\right) = 0.0695\ \text{lbm/lbm coal}$$

Silicon ash produced per lbm of coal is

$$0.061\ \frac{\text{lbm}}{\text{lbm coal}} + 0.0695\ \frac{\text{lbm}}{\text{lbm coal}} = 0.1305\ \text{lbm/lbm coal}$$

Silicon ash produced per hour is

$$\left(0.1305\ \frac{\text{lbm}}{\text{lbm coal}}\right)\left(15{,}300\ \frac{\text{lbm coal}}{\text{hr}}\right) = 1996.7\ \text{lbm/hr}$$

The silicon in 410 lbm/hr refuse is

$$\left(410\ \frac{\text{lbm}}{\text{hr}}\right)(1 - 0.3) = 287\ \text{lbm/hr}$$

The emission rate is

$$1996.7\ \frac{\text{lbm}}{\text{hr}} - 287\ \frac{\text{lbm}}{\text{hr}} = \boxed{1709.7\ \text{lbm/hr}\quad (1700\ \text{lbm/hr})}$$

The answer is (D).

(b) From Table 21.7, the stoichiometric reaction for sulfur is

$$\begin{array}{lccc} & S & + O_2 \rightarrow & SO_2 \\ \text{MW} & 32 & 32 & 64 \end{array}$$

Sulfur dioxide produced for 15,300 lbm/hr of coal feed is

$$\left(15{,}300\ \frac{\text{lbm fuel}}{\text{hr}}\right)\left(\frac{0.0244\ \text{lbm S}}{\text{lbm fuel}}\right)\left(\frac{64\ \frac{\text{lbm SO}_2}{\text{lbmol}}}{32\ \frac{\text{lbm S}}{\text{lbmol}}}\right) = \boxed{746.6\ \text{lbm SO}_2\text{/hr}\quad (750\ \text{lbm/hr})}$$

The answer is (C).

(c) From Eq. 21.16(b), the heating value of the fuel is

$$\begin{aligned} \text{HHV} &= 14{,}093 G_C + (60{,}958)\left(G_H - \frac{G_O}{8}\right) + 3983 G_S \\ &= \left(14{,}093\ \frac{\text{Btu}}{\text{lbm}}\right)(0.7656) \\ &\quad + \left(60{,}958\ \frac{\text{Btu}}{\text{lbm}}\right)\left(0.055 - \frac{0.077}{8}\right) \\ &\quad + \left(3983\ \frac{\text{Btu}}{\text{lbm}}\right)(0.0244) \\ &= 13{,}653\ \text{Btu/lbm} \end{aligned}$$

The gross available combustion power is

$$\dot{m}_f(\text{HV}) = \left(15{,}300\ \frac{\text{lbm}}{\text{hr}}\right)\left(13{,}653\ \frac{\text{Btu}}{\text{lbm}}\right) = 2.089 \times 10^8\ \text{Btu/hr}$$

The carbon in 410 lbm/hr refuse is

$$\left(410\ \frac{\text{lbm}}{\text{hr}}\right)(0.3) = 123\ \text{lbm/hr}$$

Power lost in unburned carbon in refuse is $\dot{m}_c$(HV).

From App. 21.A, the gross heat of combustion for carbon is 14,093 Btu/lbm.

$$\left(123\ \frac{\text{lbm}}{\text{hr}}\right)\left(14{,}093\ \frac{\text{Btu}}{\text{lbm}}\right) = 1.733 \times 10^6\ \text{Btu/hr}$$

The remaining combustion power is

$$2.089 \times 10^8\ \frac{\text{Btu}}{\text{hr}} - 1.733 \times 10^6\ \frac{\text{Btu}}{\text{hr}} = 2.072 \times 10^8\ \text{Btu/hr}$$

Losses in the steam generator and electrical generator will further reduce this to

$$(0.86)\left(2.072 \times 10^8\ \frac{\text{Btu}}{\text{hr}}\right) = 1.782 \times 10^8\ \text{Btu/hr}$$

With an electrical output of 17 MW, thermal energy removed by cooling water is

$$\begin{aligned} Q &= 1.782 \times 10^8\ \frac{\text{Btu}}{\text{hr}} \\ &\quad - (17\ \text{MW})\left(1000\ \frac{\text{kW}}{\text{MW}}\right)\left(3413\ \frac{\frac{\text{Btu}}{\text{hr}}}{\text{kW}}\right) \\ &= 1.202 \times 10^8\ \text{Btu/hr} \end{aligned}$$

At 60°F, the specific heat of water is $c_p = 1$ Btu/lbm-°F. The temperature rise of the cooling water is

$$\Delta T = \frac{Q}{\dot{m}c_p}$$
$$= \frac{1.202\times 10^8 \ \frac{\text{Btu}}{\text{hr}}}{\left(225 \ \frac{\text{ft}^3}{\text{sec}}\right)\left(62.4 \ \frac{\text{lbm}}{\text{ft}^3}\right)\left(3600 \ \frac{\text{sec}}{\text{hr}}\right)\left(1 \ \frac{\text{Btu}}{\text{lbm-°F}}\right)}$$
$$= \boxed{2.38\text{°F} \quad (2.4\text{°F})}$$

The electrical generation is not cooled by the cooling water. Therefore, it is not correct to include the generation efficiency in the calculation of losses.

The answer is (A).

(d) Limiting 0.1 lbm of particulates per million Btu per hour, the allowable emission rate is

$$\frac{\left(0.1 \ \frac{\text{lbm}}{\text{MBtu}}\right)\left(15{,}300 \ \frac{\text{lbm}}{\text{hr}}\right)\left(13{,}653 \ \frac{\text{Btu}}{\text{lbm}}\right)}{10^6 \ \frac{\text{MBtu}}{\text{Btu}}} = 20.89 \text{ lbm/hr}$$

The efficiency of the flue gas particulate collectors is

$$\eta = \frac{\text{actual emission rate} - \text{allowable emission rate}}{\text{actual emission rate}}$$
$$= \frac{1709.7 \ \frac{\text{lbm}}{\text{hr}} - 20.89 \ \frac{\text{lbm}}{\text{hr}}}{1709.7 \ \frac{\text{lbm}}{\text{hr}}}$$
$$= \boxed{0.988 \quad (98.8\%)}$$

The answer is (C).

17. *Customary U.S. Solution*

(a) The stoichiometric reaction for propane is given in Table 21.7.

	C_3H_8 +	$5O_2$	→ $3CO_2$	+ $4H_2O$
MW	44.097	(5)(32.000)	(3)(44.011)	(4)(18.016)

With 40% excess O_2 by weight,

$$\begin{array}{lll} & C_3H_8 & (1.4)(5)O_2 \\ \text{MW} & 44.097 + & (7)(32.000) \\ & 44.097 & 224 \end{array}$$

$$\begin{array}{lll} 3CO_2 & 4H_2O & 2O_2 \\ \rightarrow (3)(44.011) + & (4)(18.016) + & (2)(32.000) \\ 132.033 & 72.064 & 64 \end{array}$$

The excess oxygen is

$$(2)(32) = 64 \text{ lbm/lbmol } C_3H_8$$

The mass ratio of nitrogen to oxygen is

$$\frac{G_N}{G_O} = \left(\frac{B_N}{R_N}\right)\left(\frac{R_{O_2}}{B_{O_2}}\right)$$
$$= \left(\frac{0.40}{55.16 \ \frac{\text{ft-lbf}}{\text{lbm-°R}}}\right)\left(\frac{48.29 \ \frac{\text{ft-lbf}}{\text{lbm-°R}}}{0.60}\right)$$
$$= 0.584$$

(Values of R_N and R_{O_2} are taken from Table 23.7.)

The nitrogen accompanying the oxygen is

$$(7)(32)(0.584) = 130.8 \text{ lbm}$$

The mass balance per mole of propane is

$$\begin{array}{ll} & C_3H_8 + \ O_2 \ + \ N_2 \rightarrow CO_2 + H_2O + O_2 + N_2 \\ \frac{\text{lbm}}{\text{min}} & 29.05 + 147.57 + 86.17 \rightarrow 86.98 + 47.48 + 42.16 + 86.17 \end{array}$$

Standard conditions are 60°F and 1 atm. The propane density is given by Eq. 23.63.

$$\rho = \frac{p}{RT}$$

The absolute temperature, T, is

$$60\text{°F} + 460\text{°} = 520\text{°R}$$

R for propane from Table 23.7 is 35.04 ft-lbf/lbm-°R.

$$\rho = \frac{\left(14.7 \ \frac{\text{lbf}}{\text{in}^2}\right)\left(12 \ \frac{\text{in}}{\text{ft}}\right)^2}{\left(35.04 \ \frac{\text{ft-lbf}}{\text{lbm-°R}}\right)(520\text{°R})} = 0.1162 \text{ lbm/ft}^3$$

The mass flow rate of propane is

$$\left(250 \ \frac{\text{ft}^3}{\text{min}}\right)\left(0.1162 \ \frac{\text{lbm}}{\text{ft}^3}\right) = 29.05 \text{ lbm/min}$$

Scaling the other mass balance factors down by

$$\frac{29.05 \ \frac{\text{lbm}}{\text{min}}}{44.097 \ \frac{\text{lbm}}{\text{lbmol}}} = 0.6588 \text{ lbmol/min}$$

$$\begin{array}{ll} & C_3H_8 + \ O_2 \ + \ N_2 \rightarrow CO_2 + H_2O + O_2 + N_2 \\ \frac{\text{lbm}}{\text{min}} & 29.05 + 147.57 + 86.17 \rightarrow 86.98 + 47.48 + 42.16 + 86.17 \end{array}$$

The oxygen flow rate is $\boxed{147.57 \text{ lbm/min } (150 \text{ lbm/min}).}$

The answer is (A).

(b) Using R from Table 23.7, the specific volumes of the reactants are given by Eq. 23.63.

$$\upsilon_{C_3H_8} = \frac{RT}{p} = \frac{\left(35.04\ \frac{\text{ft-lbf}}{\text{lbm-°R}}\right)(80°\text{F} + 460°)}{\left(14.7\ \frac{\text{lbf}}{\text{in}^2}\right)\left(12\ \frac{\text{in}}{\text{ft}}\right)^2}$$
$$= 8.939\ \text{ft}^3/\text{lbm}$$

$$\upsilon_{O_2} = \frac{\left(48.29\ \frac{\text{ft-lbf}}{\text{lbm-°R}}\right)(80°\text{F} + 460°)}{\left(14.7\ \frac{\text{lbf}}{\text{in}^2}\right)\left(12\ \frac{\text{in}}{\text{ft}}\right)^2}$$
$$= 12.319\ \text{ft}^3/\text{lbm}$$

$$\upsilon_{N_2} = \frac{\left(55.16\ \frac{\text{ft-lbf}}{\text{lbm-°R}}\right)(80°\text{F} + 460°)}{\left(14.7\ \frac{\text{lbf}}{\text{in}^2}\right)\left(12\ \frac{\text{in}}{\text{ft}}\right)^2}$$
$$= 14.071\ \text{ft}^3/\text{lbm}$$

The total incoming volume is

$$\dot{V} = \left(29.05\ \frac{\text{lbm}}{\text{min}}\right)\left(8.939\ \frac{\text{ft}^3}{\text{lbm}}\right) + \left(147.57\ \frac{\text{lbm}}{\text{min}}\right)\left(12.319\ \frac{\text{ft}^3}{\text{lbm}}\right) + \left(86.17\ \frac{\text{lbm}}{\text{min}}\right)\left(14.071\ \frac{\text{ft}^3}{\text{lbm}}\right)$$
$$= 3290\ \text{ft}^3/\text{min}$$

Since the velocity must be kept below 400 ft/min, the minimum area of inlet pipe is

$$A_{\text{inlet}} = \frac{\dot{V}}{\text{v}} = \frac{3290\ \frac{\text{ft}^3}{\text{min}}}{400\ \frac{\text{ft}}{\text{min}}} = \boxed{8.23\ \text{ft}^2 \quad (8\ \text{ft}^2)}$$

The answer is (A).

(c) Similarly, the specific volumes of the products are

$$\upsilon = \frac{RT}{p}$$

$$\upsilon_{CO_2} = \frac{\left(35.11\ \frac{\text{ft-lbf}}{\text{lbm-°R}}\right)(460°\text{F} + 460°)}{\left(8\ \frac{\text{lbf}}{\text{in}^2}\right)\left(12\ \frac{\text{in}}{\text{ft}}\right)^2}$$
$$= 28.04\ \text{ft}^3/\text{lbm}$$

$$\upsilon_{H_2O} = \frac{\left(85.78\ \frac{\text{ft-lbf}}{\text{lbm-°R}}\right)(460°\text{F} + 460°)}{\left(8\ \frac{\text{lbf}}{\text{in}^2}\right)\left(12\ \frac{\text{in}}{\text{ft}}\right)^2}$$
$$= 68.50\ \text{ft}^3/\text{lbm}$$

$$\upsilon_{O_2} = \frac{\left(48.29\ \frac{\text{ft-lbf}}{\text{lbm-°R}}\right)(460°\text{F} + 460°)}{\left(8\ \frac{\text{lbf}}{\text{in}^2}\right)\left(12\ \frac{\text{in}}{\text{ft}}\right)^2}$$
$$= 38.56\ \text{ft}^3/\text{lbm}$$

$$\upsilon_{N_2} = \frac{\left(55.16\ \frac{\text{ft-lbf}}{\text{lbm-°R}}\right)(460°\text{F} + 460°)}{\left(8\ \frac{\text{lbf}}{\text{in}^2}\right)\left(12\ \frac{\text{in}}{\text{ft}}\right)^2}$$
$$= 44.05\ \text{ft}^3/\text{lbm}$$

The total exhaust volume is

$$\dot{V} = \left(86.98\ \frac{\text{lbm}}{\text{min}}\right)\left(28.04\ \frac{\text{ft}^3}{\text{lbm}}\right) + \left(47.48\ \frac{\text{lbm}}{\text{min}}\right)\left(68.50\ \frac{\text{ft}^3}{\text{lbm}}\right) + \left(42.16\ \frac{\text{lbm}}{\text{min}}\right)\left(38.56\ \frac{\text{ft}^3}{\text{lbm}}\right) + \left(86.17\ \frac{\text{lbm}}{\text{min}}\right)\left(44.05\ \frac{\text{ft}^3}{\text{lbm}}\right)$$
$$= \boxed{11{,}112.8\ \text{ft}^3/\text{min} \quad (11{,}000\ \text{ft}^3/\text{min})}$$

The answer is (D).

(d) Since the velocity of the products must be kept below 800 ft/min, the minimum area of the stack is

$$A_{\text{stack}} = \frac{Q}{\text{v}} = \frac{11{,}112.8\ \frac{\text{ft}^3}{\text{min}}}{800\ \frac{\text{ft}}{\text{min}}} = \boxed{13.89\ \text{ft}^2 \quad (14\ \text{ft}^2)}$$

The answer is (C).

(e) For ideal gases, the partial pressure is volumetrically weighted. The water vapor partial pressure in the stack is

$$(8\ \text{psi})\left(\frac{\left(47.48\ \frac{\text{lbm}}{\text{min}}\right)\left(68.50\ \frac{\text{ft}^3}{\text{lbm}}\right)}{11{,}112.8\ \frac{\text{ft}^3}{\text{min}}}\right) = 2.34\ \text{psia}$$

The saturation temperature corresponding to 2.34 psia is $T_{\text{dp}} = \boxed{131°\text{F}\ (130°\text{F}).}$

The answer is (C).

SI Solution

(a) Following the procedure for the customary U.S. solution, the mass balance per mole of propane is

	C_3H_8 +	O_2 +	N_2 →	CO_2 +	H_2O +	O_2 +	N_2
kg per mole	44.097 +	224 +	130.8 →	132.033 +	72.064 +	64 +	130.8

Thermodynamics

Standard conditions are 16°C and 101.3 kPa. The propane density is given by Eq. 23.63.

$$\rho = \frac{p}{RT}$$

The absolute temperature is

$$T = 16°\text{C} + 273° = 289\text{K}$$

From Table 23.7, R for propane is 188.55 J/kg·K.

$$\rho = \frac{(101.3\text{ kPa})\left(1000\ \frac{\text{Pa}}{\text{kPa}}\right)}{\left(188.55\ \frac{\text{J}}{\text{kg·K}}\right)(289\text{K})} = 1.86\text{ kg/m}^3$$

The mass flow rate of propane is

$$\frac{\left(118\ \frac{\text{L}}{\text{s}}\right)\left(1.86\ \frac{\text{kg}}{\text{m}^3}\right)}{1000\ \frac{\text{L}}{\text{m}^3}} = 0.2195\text{ kg/s}$$

Scale the other mass balance factors down.

$$\frac{0.2195\ \frac{\text{kg}}{\text{s}}}{44.097\ \frac{\text{kg}}{\text{kmol}}} = 0.004978\text{ kmol/s}$$

	C_3H_8 + O_2 + N_2	→	CO_2 + H_2O + O_2 + N_2
$\frac{\text{kg}}{\text{s}}$	0.2195 + 1.115 + 0.6511	→	0.6572 + 0.3587 + 0.3186 + 0.6511

The oxygen flow rate is $\boxed{1.115\text{ kg/s (1.1 kg/s).}}$

The answer is (A).

(b) Using R from Table 23.7, the specific volumes of the reactants are given by Eq. 23.63.

$$\upsilon_{C_3H_8} = \frac{RT}{p} = \frac{\left(188.55\ \frac{\text{J}}{\text{kg·K}}\right)(27°\text{C} + 273°)}{(101\text{ kPa})\left(1000\ \frac{\text{Pa}}{\text{kPa}}\right)} = 0.5600\text{ m}^3\text{/kg}$$

$$\upsilon_{O_2} = \frac{\left(259.82\ \frac{\text{J}}{\text{kg·K}}\right)(27°\text{C} + 273°)}{(101\text{ kPa})\left(1000\ \frac{\text{Pa}}{\text{kPa}}\right)} = 0.7717\text{ m}^3\text{/kg}$$

$$\upsilon_{N_2} = \frac{\left(296.77\ \frac{\text{J}}{\text{kg·K}}\right)(27°\text{C} + 273°)}{(101\text{ kPa})\left(1000\ \frac{\text{Pa}}{\text{kPa}}\right)} = 0.8815\text{ m}^3\text{/kg}$$

The total incoming volume, $\dot{V}$, is

$$\begin{aligned}\dot{V} &= \left(0.2195\ \frac{\text{kg}}{\text{s}}\right)\left(0.5600\ \frac{\text{m}^3}{\text{kg}}\right) \\ &\quad + \left(1.115\ \frac{\text{kg}}{\text{s}}\right)\left(0.7717\ \frac{\text{m}^3}{\text{kg}}\right) \\ &\quad + \left(0.6511\ \frac{\text{kg}}{\text{s}}\right)\left(0.8815\ \frac{\text{m}^3}{\text{kg}}\right) \\ &= 1.557\text{ m}^3\text{/s}\end{aligned}$$

Since the velocity for the reactants must be kept below 2 m/s, the minimum area of inlet pipe is

$$A_{\text{inlet}} = \frac{\dot{V}}{\text{v}} = \frac{1.557\ \frac{\text{m}^3}{\text{s}}}{2\ \frac{\text{m}}{\text{s}}} = \boxed{0.779\text{ m}^2\quad(0.8\text{ m}^2)}$$

The answer is (A).

(c) Similarly, the specific volumes of the products are

$$\upsilon = \frac{RT}{p}$$

$$\upsilon_{CO_2} = \frac{\left(188.92\ \frac{\text{J}}{\text{kg·K}}\right)(240°\text{C} + 273°)}{(55\text{ kPa})\left(1000\ \frac{\text{Pa}}{\text{kPa}}\right)} = 1.762\text{ m}^3\text{/kg}$$

$$\upsilon_{H_2O} = \frac{\left(461.5\ \frac{\text{J}}{\text{kg·K}}\right)(240°\text{C} + 273°)}{(55\text{ kPa})\left(1000\ \frac{\text{Pa}}{\text{kPa}}\right)} = 4.305\text{ m}^3\text{/kg}$$

$$\upsilon_{O_2} = \frac{\left(259.82\ \frac{\text{J}}{\text{kg·K}}\right)(240°\text{C} + 273°)}{(55\text{ kPa})\left(1000\ \frac{\text{Pa}}{\text{kPa}}\right)} = 2.423\text{ m}^3\text{/kg}$$

$$\upsilon_{N_2} = \frac{\left(296.77\ \frac{\text{J}}{\text{kg·K}}\right)(240°\text{C} + 273°)}{(55\text{ kPa})\left(1000\ \frac{\text{Pa}}{\text{kPa}}\right)} = 2.768\text{ m}^3\text{/kg}$$

The total exhaust volume is

$$\begin{aligned}\dot{V} &= \left(0.6572\ \frac{\text{kg}}{\text{s}}\right)\left(1.762\ \frac{\text{m}^3}{\text{kg}}\right) \\ &\quad + \left(0.3587\ \frac{\text{kg}}{\text{s}}\right)\left(4.305\ \frac{\text{m}^3}{\text{kg}}\right) \\ &\quad + \left(0.3186\ \frac{\text{kg}}{\text{s}}\right)\left(2.423\ \frac{\text{m}^3}{\text{kg}}\right) \\ &\quad + \left(0.6511\ \frac{\text{kg}}{\text{s}}\right)\left(2.768\ \frac{\text{m}^3}{\text{kg}}\right) \\ &= \boxed{5.276\text{ m}^3\text{/s}\quad(5.3\text{ m}^3\text{/s})}\end{aligned}$$

The answer is (D).

(d) Since the velocity of products must be kept below 4 m/s, the minimum area of the stack is

$$A_{\text{stack}} = \frac{\dot{V}}{\text{v}} = \frac{5.276\ \frac{\text{m}^3}{\text{s}}}{4\ \frac{\text{m}}{\text{s}}} = \boxed{1.319\ \text{m}^2 \quad (1.3\ \text{m}^2)}$$

The answer is (C).

(e) For ideal gases, the partial pressure is volumetrically weighted. The water vapor partial pressure in the stack is

$$(55\ \text{kPa})\left(\frac{\left(0.3587\ \frac{\text{kg}}{\text{s}}\right)\left(4.305\ \frac{\text{m}^3}{\text{kg}}\right)}{5.276\ \frac{\text{m}^3}{\text{s}}}\right) = 16.098\ \text{kPa}$$

The saturation temperature corresponding to 16.098 kPa is found from App. 23.O to be $T_{\text{dp}} = \boxed{54.5°\text{C}\ (55°\text{C}).}$

The answer is (C).

18. *Customary U.S. Solution*

(a) Since atmospheric air is not used, the nitrogen and oxygen can be varied independently. Furthermore, since enthalpy increase information is not given for oxygen, a 0% excess of oxygen can be assumed.

From Table 21.7,

	C_3H_8	+	$5O_2$	→	$3CO_2$	+	$4H_2O$
moles	(1)		(5)		(3)		(4)

Subtract the reactant enthalpies from the product enthalpies to calculate the heat of reaction. The enthalpy of formation of oxygen is zero since it is an element in its natural state. The heat of reaction is

$$\begin{aligned} & n_{CO_2}(\Delta H_f)_{CO_2} + n_{H_2O}(\Delta H_f)_{H_2O} \\ & \quad - n_{C_3H_8}(\Delta H_f)_{C_3H_8} - n_{O_2}(\Delta H_f)_{O_2} \\ & = \left(3\ \frac{\text{lbmol}}{\text{lbmol}}\right)\left(-169{,}300\ \frac{\text{Btu}}{\text{lbmol}}\right) \\ & \quad + \left(4\ \frac{\text{lbmol}}{\text{lbmol}}\right)\left(-104{,}040\ \frac{\text{Btu}}{\text{lbmol}}\right) \\ & \quad - \left(1\ \frac{\text{lbmol}}{\text{lbmol}}\right)\left(28{,}800\ \frac{\text{Btu}}{\text{lbmol}}\right) \\ & \quad - \left(5\ \frac{\text{lbmol}}{\text{lbmol}}\right)\left(0\ \frac{\text{Btu}}{\text{lbmol}}\right) \\ & = -952{,}860\ \text{Btu/lbmol of propane} \end{aligned}$$

The negative sign indicates an exothermal reaction.

Let x be the number of moles of nitrogen per mole of propane. Use the nitrogen to cool the combustion. The heat of reaction will increase the enthalpy of products from the standard reference temperature to 3600°R.

$$\begin{aligned} 952{,}860\ \frac{\text{Btu}}{\text{lbmol}} &= \left(3\ \frac{\text{lbmol}}{\text{lbmol}}\right)\left(39{,}791\ \frac{\text{Btu}}{\text{lbmol}}\right) \\ & \quad + \left(4\ \frac{\text{lbmol}}{\text{lbmol}}\right)\left(31{,}658\ \frac{\text{Btu}}{\text{lbmol}}\right) \\ & \quad + x\left(24{,}471\ \frac{\text{Btu}}{\text{lbmol}}\right) \\ x &= 28.89\ \text{lbmol/lbmol propane} \end{aligned}$$

The mass of nitrogen per pound-mole of propane is

$$m_{N_2} = \left(28.89\ \frac{\text{lbmol}}{\text{lbmol propane}}\right)\left(28.016\ \frac{\text{lbm}}{\text{lbmol}}\right) = \boxed{\begin{array}{c}809.4\ \text{lbm/lbmol propane} \\ (810\ \text{lbm/lbmol propane})\end{array}}$$

The mass of oxygen per pound-mole of propane is

$$m_{O_2} = \left(5\ \frac{\text{lbmol}}{\text{lbmol}}\right)\left(32\ \frac{\text{lbm}}{\text{lbmol}}\right) = \boxed{160\ \text{lbm/lbmol propane}}$$

The answer is (B).

(b) The partial pressure is volumetrically weighted. This is the same as molar (mole fraction) weighting.

product	lbmol	mole fraction
CO_2	3	3/35.89 = 0.0836
H_2O	4	4/35.89 = 0.1115
N_2	28.89	28.89/35.89 = 0.8049
O_2	0	0/35.89 = 0
	35.89 lbmol	1.000

The partial pressure of water vapor is

$$p_{H_2O} = \left(\frac{n_{H_2O}}{n}\right)p = (0.1115)(14.7\ \text{psia}) = 1.64\ \text{psia}$$

From App. 23.B, this pressure corresponds to approximately 118°F. Since the stack temperature is 100°F, some of the water will condense. From App. 23.A, the maximum vapor pressure of water at the stack temperature is 0.9505 psia. Let n be the number of moles of water per mole of propane in the stack gas.

$$n_{H_2O} = \left(\frac{p_{H_2O}}{p}\right)n = \left(\frac{0.9505\ \text{psia}}{14.7\ \text{psia}}\right)\left(35.89\ \frac{\text{lbmol}}{\text{lbmol}}\right) = \boxed{\begin{array}{c}2.321\ \text{lbmol/lbmol}\ C_3H_8 \\ (2.3\ \text{lbmol/lbmol}\ C_3H_8)\end{array}}$$

The answer is (C).

(c) The water removed is

$$4 - n_{H_2O} = 4\ \frac{\text{lbmol}}{\text{lbmol}} - 2.321\ \frac{\text{lbmol}}{\text{lbmol}}$$
$$= 1.679 \text{ lbmol of } H_2O/\text{lbmol of } C_3H_8$$
$$m = \left(1.679\ \frac{\text{lbmol}}{\text{lbmol}}\right)\left(18.016\ \frac{\text{lbm}}{\text{lbmol}}\right)$$
$$= \boxed{\begin{array}{l} 30.25 \text{ lbm } H_2O/\text{lbmol } C_3H_8 \\ \quad (30 \text{ lbm } H_2O/\text{lbmol } C_3H_8) \end{array}}$$

The answer is (C).

SI Solution

(a) From the customary U.S. solution, the heat of reaction is

$$\begin{aligned} & n_{CO_2}(\Delta H_f)_{CO_2} + n_{H_2O}(\Delta H_f)_{H_2O} \\ & \quad - n_{C_3H_8}(\Delta H_f)_{C_3H_8} - n_{O_2}(\Delta H_f)_{O_2} \\ & = \left(3\ \frac{\text{kmol}}{\text{kmol}}\right)\left(-393.8\ \frac{\text{GJ}}{\text{kmol}}\right) \\ & \quad + \left(4\ \frac{\text{kmol}}{\text{kmol}}\right)\left(-242\ \frac{\text{GJ}}{\text{kmol}}\right) \\ & \quad + \left(1\ \frac{\text{kmol}}{\text{kmol}}\right)\left(-67.0\ \frac{\text{GJ}}{\text{kmol}}\right) \\ & \quad - \left(5\ \frac{\text{kmol}}{\text{kmol}}\right)\left(0\ \frac{\text{GJ}}{\text{kmol}}\right) \\ & = -2216.4 \text{ GJ/kmol propane} \end{aligned}$$

The negative sign indicates an exothermal reaction.

Let x be the number of moles of nitrogen per mole of propane. Use the nitrogen to cool the combustion. The heat of reaction will increase the enthalpy of products from the standard reference temperature to 1980°C.

$$\begin{aligned} 2216.4\ \frac{\text{GJ}}{\text{mol}} &= \left(3\ \frac{\text{kmol}}{\text{kmol}}\right)\left(92.6\ \frac{\text{GJ}}{\text{kmol}}\right) \\ & \quad + \left(4\ \frac{\text{kmol}}{\text{kmol}}\right)\left(73.6\ \frac{\text{GJ}}{\text{kmol}}\right) \\ & \quad + x\left(56.9\ \frac{\text{GJ}}{\text{kmol}}\right) \\ x &= 28.90 \text{ kmol/kmol propane} \end{aligned}$$

The mass of nitrogen per mole of propane is

$$m_{N_2} = \left(28.90\ \frac{\text{kmol}}{\text{kmol}}\right)\left(28.016\ \frac{\text{kg}}{\text{kmol}}\right)$$
$$= \boxed{809.7 \text{ kg/kmol} \quad (810 \text{ kg/kmol})}$$

The mass of oxygen per kmol of propane is

$$m_{O_2} = \left(5\ \frac{\text{kmol}}{\text{kmol}}\right)\left(32\ \frac{\text{kg}}{\text{kmol}}\right) = \boxed{160 \text{ kg/kmol}}$$

The answer is (B).

(b) The partial pressure is volumetrically weighted. This is the same as molar (mole fraction) weighting.

product	kmol	mole fraction
CO_2	3	3/35.90 = 0.0836
H_2O	4	4/35.90 = 0.1114
N_2	28.90	28.89/35.90 = 0.8050
O_2	0	0/35.90 = 0
	35.90 lbmol	1.000

The partial pressure of water vapor is

$$p_{H_2O} = \left(\frac{n_{H_2O}}{n}\right)p = (0.1114)(101 \text{ kPa})$$
$$= 11.25 \text{ kPa}$$

From App. 23.O, this pressure corresponds to approximately 47.6°C. Since the stack temperature is 38°C, some of the water will condense. From App. 23.N, the maximum vapor pressure of water at the stack temperature is 6.633 kPa. Let n be the number of moles of water per mole of propane in the stack gas.

$$n_{H_2O} = \left(\frac{p_{H_2O}}{p}\right)n = \left(\frac{6.633 \text{ kPa}}{101 \text{ kPa}}\right)\left(35.90\ \frac{\text{kmol}}{\text{kmol}}\right)$$
$$= \boxed{\begin{array}{l} 2.358 \text{ kmol } H_2O/\text{kmol } C_3H_8 \\ \quad (2.3 \text{ kmol } H_2O/\text{kmol } C_3H_8) \end{array}}$$

The answer is (C).

(c) The liquid water removed is

$$4 - n_{H_2O} = 4\ \frac{\text{kmol}}{\text{kmol}} - 2.358\ \frac{\text{kmol}}{\text{kmol}}$$
$$= 1.642 \text{ kmol of } H_2O/\text{kmol propane}$$
$$m = \left(1.642\ \frac{\text{kmol}}{\text{kmol}}\right)\left(18.016\ \frac{\text{kg}}{\text{kmol}}\right)$$
$$= \boxed{\begin{array}{l} 29.58 \text{ kg } H_2O/\text{kmol } C_3H_8 \\ \quad (30 \text{ kg } H_2O/\text{kmol } C_3H_8) \end{array}}$$

The answer is (C).

19. The heat rate, Q, required to generate the steam is found from the generation rate of steam, $\dot{m}$, the enthalpy of the saturated vapor, h_g, and the enthalpy of the incoming saturated feedwater, h_f. From App. 23.B, for an absolute pressure of 200 lbf/in^2, h_g is 1198.8 Btu/lbm. From App. 23.A, for a temperature of 160°F, h_f is 128.0 Btu/lbm.

$$Q = \dot{m}(h_g - h_f)$$
$$= \left(20{,}070\ \frac{\text{lbm}}{\text{hr}}\right)\left(1198.8\ \frac{\text{Btu}}{\text{lbm}} - 128.0\ \frac{\text{Btu}}{\text{lbm}}\right)$$
$$= 21.5 \times 10^6 \text{ Btu/hr}$$

Thermodynamics

The lower heating value of the fuel, LHV, is

$$\text{LHV} = \frac{Q}{\dot{m}_{\text{fuel}}} = \frac{21.5 \times 10^6 \ \frac{\text{Btu}}{\text{hr}}}{5000 \ \frac{\text{lbm}}{\text{hr}}} = 4300 \ \text{Btu/lbm}$$

The higher heating value, HHV, includes the heat of vaporization (same as the heat of condensation) of the water in the fuel and generated by the combustion of the fuel. The mass of water in the combustion products produced per point of fuel is

$$m_{\text{water}} = \frac{2000 \ \frac{\text{lbm water}}{\text{hr}}}{5000 \ \frac{\text{lbm fuel}}{\text{hr}}} = 0.4 \ \text{lbm water/lbm fuel}$$

From App. 23.B, for a partial pressure of 4 lbf/in^2, the enthalpy of vaporization, h_{fg}, is 1006.0 Btu/lbm. From Eq. 21.14, the higher heating value is

$$\begin{aligned}\text{HHV} &= \text{LHV} + m_{\text{water}} h_{fg} \\ &= 4300 \ \frac{\text{Btu}}{\text{lbm}} + \left(0.4 \ \frac{\text{lbm}}{\text{lbm}}\right)\left(1006.0 \ \frac{\text{Btu}}{\text{lbm}}\right) \\ &= \boxed{4702.4 \ \text{Btu/lbm} \quad (4700 \ \text{Btu/lbm})}\end{aligned}$$

The answer is (B).

Thermodynamics

22 Energy, Work, and Power

PRACTICE PROBLEMS

Energy

1. A solid, cast-iron sphere (density of 0.256 lbm/in^3 (7090 kg/m^3)) of 10 in (25 cm) diameter travels without friction at 30 ft/sec (9 m/s) horizontally. Its kinetic energy is most nearly

(A) 900 ft-lbf (1.2 kJ)

(B) 1200 ft-lbf (1.6 kJ)

(C) 1600 ft-lbf (2.0 kJ)

(D) 1900 ft-lbf (2.3 kJ)

Work

2. The work done when a balloon carries a 12 lbm (5.2 kg) load to 40,000 ft (12 000 m) height is most nearly

(A) 2.4×10^5 ft-lbf (300 kJ)

(B) 4.8×10^5 ft-lbf (610 kJ)

(C) 7.7×10^5 ft-lbf (980 kJ)

(D) 9.9×10^5 ft-lbf (1.3 MJ)

3. What is most nearly the compression (deflection) if a 100 lbm (50 kg) weight is dropped from 8 ft (2 m) onto a spring with a stiffness of 33.33 lbf/in (5.837×10^3 N/m)?

(A) 27 in (0.67 m)

(B) 34 in (0.85 m)

(C) 39 in (0.90 m)

(D) 45 in (1.1 m)

4. A punch press flywheel with a moment of inertia of 483 lbm-ft^2 (20 kg·m^2) operates at 300 rpm. What is most nearly the speed in rpm to which the wheel will be reduced after a sudden punching requiring 4500 ft-lbf (6100 J) of work?

(A) 160 rpm

(B) 190 rpm

(C) 220 rpm

(D) 280 rpm

5. A force of 550 lbf (2500 N) making a 40° angle (upward) from the horizontal pushes a box 20 ft (6 m) across the floor. The work done is most nearly

(A) 2200 ft-lbf (3.0 kJ)

(B) 3700 ft-lbf (5.2 kJ)

(C) 4200 ft-lbf (6.0 kJ)

(D) 8400 ft-lbf (12 kJ)

6. A 1000 ft long (300 m long) cable has a mass of 2 lbm/ft (3 kg/m) and is fully suspended from a winding drum down into a vertical shaft. The work done to rewind the cable is most nearly

(A) 0.50×10^6 ft-lbf (0.6 MJ)

(B) 0.75×10^6 ft-lbf (0.9 MJ)

(C) 1×10^6 ft-lbf (1.3 MJ)

(D) 2×10^6 ft-lbf (2.6 MJ)

Power

7. Approximately what volume in ft^3 (m^3) of water can be pumped to a 130 ft (40 m) height in 1 hr by a 7 hp (5 kW) pump? Assume 85% efficiency.

(A) 1500 ft^3 (40 m^3)

(B) 1800 ft^3 (49 m^3)

(C) 2000 ft^3 (54 m^3)

(D) 2400 ft^3 (65 m^3)

8. The power in horsepower (kW) that is required to lift a 3300 lbm (1500 kg) mass 250 ft (80 m) vertically in 14 sec is most nearly

(A) 40 hp (30 kW)

(B) 70 hp (53 kW)

(C) 90 hp (68 kW)

(D) 110 hp (84 kW)

SOLUTIONS

1. *Customary U.S. Solution*

Since there is no friction, there is no rotation. The sphere slides.

$$\begin{aligned} E_{\text{kinetic}} &= \tfrac{1}{2}\left(\frac{m}{g_c}\right)v^2 = \tfrac{1}{2}\left(\frac{V\rho}{g_c}\right)v^2 \\ &= \left(\tfrac{1}{2}\right)\left(\tfrac{4}{3}\pi r^3\right)\left(\frac{\rho}{g_c}\right)v^2 \\ &= \tfrac{2}{3}\pi r^3\left(\frac{\rho}{g_c}\right)v^2 \\ &= \tfrac{2}{3}\pi\left(\frac{10\text{ in}}{2}\right)^3\left(\frac{0.256\ \frac{\text{lbm}}{\text{in}^3}}{32.2\ \frac{\text{lbm-ft}}{\text{lbf-sec}^2}}\right)\left(30\ \frac{\text{ft}}{\text{sec}}\right)^2 \\ &= \boxed{1873\text{ ft-lbf}\quad(1900\text{ ft-lbf})} \end{aligned}$$

The answer is (D).

SI Solution

Since there is no friction, there is no rotation. The sphere slides.

$$\begin{aligned} E_{\text{kinetic}} &= \tfrac{1}{2}mv^2 = \tfrac{1}{2}(\rho V)v^2 \\ &= \tfrac{1}{2}\rho\left(\tfrac{4}{3}\pi r^3\right)v^2 \\ &= \tfrac{2}{3}\pi r^3\rho v^2 \\ &= \tfrac{2}{3}\pi\left(\frac{0.25\text{ m}}{2}\right)^3\left(7090\ \frac{\text{kg}}{\text{m}^3}\right)\left(9\ \frac{\text{m}}{\text{s}}\right)^2 \\ &= \boxed{2349\text{ J}\quad(2.3\text{ kJ})} \end{aligned}$$

The answer is (D).

2. *Customary U.S. Solution*

From Eq. 22.11(b) and Eq. 22.12, the work done by the balloon is

$$\begin{aligned} W &= \Delta E_{\text{potential}} = \frac{mg\Delta h}{g_c} \\ &= \frac{(12\text{ lbm})\left(32.2\ \frac{\text{ft}}{\text{sec}^2}\right)(40{,}000\text{ ft})}{32.2\ \frac{\text{lbm-ft}}{\text{lbf-sec}^2}} \\ &= \boxed{4.8\times 10^5\text{ ft-lbf}} \end{aligned}$$

The answer is (B).

SI Solution

From Eq. 22.11(a) and Eq. 22.12, the work done by the balloon is

$$\begin{aligned} W &= \Delta E_{\text{potential}} = mg\Delta h \\ &= \frac{(5.2\text{ kg})\left(9.81\ \frac{\text{m}}{\text{s}^2}\right)(12\,000\text{ m})}{1000\ \frac{\text{J}}{\text{kJ}}} \\ &= \boxed{612.1\text{ kJ}\quad(610\text{ kJ})} \end{aligned}$$

The answer is (B).

3. *Customary U.S. Solution*

Equate the potential energy to the energy of the spring.

$$\begin{aligned} \Delta E_{\text{potential}} &= \Delta E_{\text{spring}} \\ W(\Delta h + \Delta x) &= \tfrac{1}{2}k(\Delta x)^2 \end{aligned}$$

Rearranging and using $W = m(g/g_c)$,

$$\tfrac{1}{2}k(\Delta x)^2 - W\Delta x - W\Delta h = 0$$

$$\begin{aligned} &\left(\tfrac{1}{2}\right)\left(33.33\ \frac{\text{lbf}}{\text{in}}\right)(\Delta x)^2 \\ &\quad - (100\text{ lbm})\left(\frac{32.2\ \frac{\text{ft}}{\text{sec}^2}}{32.2\ \frac{\text{lbm-ft}}{\text{lbf-sec}^2}}\right)\Delta x \\ &\quad - (100\text{ lbm})\left(\frac{32.2\ \frac{\text{ft}}{\text{sec}^2}}{32.2\ \frac{\text{lbm-ft}}{\text{lbf-sec}^2}}\right)(8\text{ ft})\left(12\ \frac{\text{in}}{\text{ft}}\right) = 0 \\ &16.665(\Delta x)^2 - 100\Delta x = 9600 \end{aligned}$$

Complete the square.

$$\begin{aligned} (\Delta x)^2 - 6\Delta x &= 576 \\ (\Delta x - 3)^2 &= 576 + 9 \\ \Delta x - 3 &= \pm\sqrt{585} = \pm 24.2 \\ \Delta x &= \boxed{27.2\text{ in}\quad(27\text{ in})} \end{aligned}$$

The answer is (A).

SI Solution

Equate the potential energy to the energy of the spring.

$$\begin{aligned} \Delta E_{\text{potential}} &= \Delta E_{\text{spring}} \\ mg(\Delta h + \Delta x) &= \tfrac{1}{2}k(\Delta x)^2 \end{aligned}$$

Rearrange.

$$\tfrac{1}{2}k(\Delta x)^2 - mg\Delta x - mg\Delta h = 0$$

$$\left(\tfrac{1}{2}\right)\left(5.837\times 10^3\ \frac{\text{N}}{\text{m}}\right)(\Delta x)^2 - (50\ \text{kg})\left(9.81\ \frac{\text{m}}{\text{s}^2}\right)\Delta x - (50\ \text{kg})\left(9.81\ \frac{\text{m}}{\text{s}^2}\right)(2\ \text{m}) = 0$$

$$2918.5(\Delta x)^2 - 490.5\Delta x - 981.0 = 0$$

$$(\Delta x)^2 - 0.1681\Delta x = 0.3361$$

Complete the square.

$$(\Delta x - 0.08403)^2 = 0.3361 + (0.08403)^2 = 0.3432$$

$$\Delta x - 0.08403 = \pm\sqrt{0.3432} = \pm 0.5858$$

$$\Delta x = \boxed{0.6699\ \text{m}\quad(0.67\ \text{m})}$$

The answer is (A).

4. *Customary U.S. Solution*

The work done by the wheel is equal to the change in the rotational energy.

$$W_{\text{done by wheel}} = \Delta E_{\text{rotational}} = \tfrac{1}{2}\left(\frac{I}{g_c}\right)\omega^2_{\text{initial}} - \tfrac{1}{2}\left(\frac{I}{g_c}\right)\omega^2_{\text{final}}$$

The final angular velocity is found from

$$\omega_{\text{final}} = \sqrt{\omega^2_{\text{initial}} - \frac{2g_c W}{I}} = 2\pi f$$

The final speed is

$$n_{\text{final,rpm}} = f_{\text{final,Hz}}\left(60\ \frac{\text{sec}}{\text{min}}\right) = \frac{\omega_{\text{final}}\left(60\ \frac{\text{sec}}{\text{min}}\right)}{2\pi}$$

$$= \left(\frac{1}{2\pi}\right)\left(60\ \frac{\text{sec}}{\text{min}}\right) \times \sqrt{\left(\left(2\pi\ \frac{\text{rad}}{\text{rev}}\right)\left(\frac{300\ \frac{\text{rev}}{\text{min}}}{60\ \frac{\text{sec}}{\text{min}}}\right)\right)^2 - \frac{(2)\left(32.2\ \frac{\text{lbm-ft}}{\text{lbf-sec}^2}\right)\times(4500\ \text{ft-lbf})}{483\ \text{lbm-ft}^2}}$$

$$= \boxed{187.8\ \text{rpm}\quad(190\ \text{rpm})}$$

The answer is (B).

SI Solution

The work done by the wheel is equal to the change in the rotational energy.

$$W_{\text{done by wheel}} = \Delta E_{\text{rotational}} = \tfrac{1}{2}I\omega^2_{\text{initial}} - \tfrac{1}{2}I\omega^2_{\text{final}}$$

The final angular velocity is found from

$$\omega_{\text{final}} = \sqrt{\omega^2_{\text{initial}} - \frac{2W}{I}} = 2\pi f$$

The final speed is

$$n_{\text{final,rpm}} = f_{\text{final,Hz}}\left(60\ \frac{\text{s}}{\text{min}}\right) = \frac{\omega_{\text{final}}\left(60\ \frac{\text{s}}{\text{min}}\right)}{2\pi}$$

$$= \left(\frac{1}{2\pi}\right)\left(60\ \frac{\text{s}}{\text{min}}\right)\sqrt{\left(\left(2\pi\ \frac{\text{rad}}{\text{rev}}\right)\left(\frac{300\ \frac{\text{rev}}{\text{min}}}{60\ \frac{\text{s}}{\text{min}}}\right)\right)^2 - \frac{(2)(6100\ \text{J})}{20\ \text{kg·m}^2}}$$

$$= \boxed{185.4\ \text{rpm}\quad(190\ \text{rpm})}$$

The answer is (B).

5. *Customary U.S. Solution*

The work done is

$$W_{\text{done on box}} = F_x\Delta x = F(\cos\theta)\Delta x = (550\ \text{lbf})(\cos 40°)(20\ \text{ft}) = \boxed{8426\ \text{ft-lbf}\quad(8400\ \text{ft-lbf})}$$

The answer is (D).

SI Solution

The work done is

$$W_{\text{done on box}} = F_x\Delta x = F(\cos\theta)\Delta x = \frac{(2500\ \text{N})(\cos 40°)(6\ \text{m})}{1000\ \frac{\text{J}}{\text{kJ}}} = \boxed{11.49\ \text{kJ}\quad(12\ \text{kJ})}$$

The answer is (D).

Thermodynamics

6. *Customary U.S. Solution*

The work done to rewind the cable is

$$\begin{aligned} W_{\text{to rewind cable}} &= \int_0^l F\,dh = \int_0^l ((l-h)w)\,dh \\ &= \tfrac{1}{2}wl^2 \\ &= \left(\tfrac{1}{2}\right)\left(2\ \frac{\text{lbf}}{\text{ft}}\right)(1000\ \text{ft})^2 \\ &= \boxed{1 \times 10^6\ \text{ft-lbf}} \end{aligned}$$

The answer is (C).

SI Solution

The work done to rewind the cable is

$$\begin{aligned} W_{\text{to rewind cable}} &= \int_0^l F\,dh = \int_0^l ((l-h)m_l g)\,dh \\ &= \tfrac{1}{2}m_l g l^2 \\ &= \left(\tfrac{1}{2}\right)\left(3\ \frac{\text{kg}}{\text{m}}\right)\left(9.81\ \frac{\text{m}}{\text{s}^2}\right)(300\ \text{m})^2 \\ &= \boxed{1.32 \times 10^6\ \text{J}\quad (1.3\ \text{MJ})} \end{aligned}$$

The answer is (C).

7. *Customary U.S. Solution*

The volume of water is found from the work performed.

$$\begin{aligned} P_{\text{actual}}\Delta t &= W_{\text{done by pump}} \\ \eta P_{\text{ideal}}\Delta t &= \Delta E_{\text{potential}} \\ &= \frac{mg\Delta h}{g_c} \\ &= \frac{(\rho V)g\Delta h}{g_c} \\ V &= \frac{\eta P_{\text{ideal}}\,\Delta t}{\frac{\rho g \Delta h}{g_c}} \\ &= \frac{(0.85)(7\ \text{hp})\left(550\ \frac{\text{ft-lbf}}{\text{hp-sec}}\right)(1\ \text{hr})\left(3600\ \frac{\text{sec}}{\text{hr}}\right)}{\frac{\left(62.4\ \frac{\text{lbm}}{\text{ft}^3}\right)\left(32.2\ \frac{\text{ft}}{\text{sec}^2}\right)(130\ \text{ft})}{32.2\ \frac{\text{lbm-ft}}{\text{lbf-sec}^2}}} \\ &= \boxed{1452\ \text{ft}^3\quad (1500\ \text{ft}^3)} \end{aligned}$$

The answer is (A).

SI Solution

The volume of water is found from the work performed.

$$\begin{aligned} P_{\text{actual}}\Delta t &= W_{\text{done by pump}} \\ \eta P_{\text{ideal}}\Delta t &= \Delta E_{\text{potential}} \\ &= mg\Delta h \\ &= (\rho V)g\Delta h \\ V &= \frac{\eta P_{\text{ideal}}\,\Delta t}{\rho g \Delta h} \\ &= \frac{(0.85)(5\ \text{kW})\left(1000\ \frac{\text{W}}{\text{kW}}\right)(1\ \text{h})\left(3600\ \frac{\text{s}}{\text{h}}\right)}{\left(1000\ \frac{\text{kg}}{\text{m}^3}\right)\left(9.81\ \frac{\text{m}}{\text{s}^2}\right)(40\ \text{m})} \\ &= \boxed{39.0\ \text{m}^3\quad (40\ \text{m}^3)} \end{aligned}$$

The answer is (A).

8. *Customary U.S. Solution*

The work required to lift the mass is

$$\begin{aligned} W &= P\Delta t = \frac{mg\Delta h}{g_c} \\ P &= \frac{mg\Delta h}{g_c \Delta t} \\ &= \frac{(3300\ \text{lbm})\left(32.2\ \frac{\text{ft}}{\text{sec}^2}\right)(250\ \text{ft})}{\left(32.2\ \frac{\text{lbm-ft}}{\text{sec}^2\text{-lbf}}\right)(14\ \text{sec})\left(550\ \frac{\text{ft-lbf}}{\text{hp-sec}}\right)} \\ &= \boxed{107\ \text{hp}\quad (110\ \text{hp})} \end{aligned}$$

The answer is (D).

SI Solution

The work required to lift the mass is

$$\begin{aligned} W &= P\Delta t = mg\Delta h \\ P &= \frac{mg\Delta h}{\Delta t} = \frac{(1500\ \text{kg})\left(9.81\ \frac{\text{m}}{\text{s}^2}\right)(80\ \text{m})}{(14\ \text{s})\left(1000\ \frac{\text{W}}{\text{kW}}\right)} \\ &= \boxed{84.1\ \text{kW}\quad (84\ \text{kW})} \end{aligned}$$

The answer is (D).

23 Thermodynamic Properties of Substances

PRACTICE PROBLEMS

1. The molar enthalpy of 250°F (120°C) steam with a quality of 92% is most nearly

(A) 16,000 Btu/lbmol (37 MJ/kmol)

(B) 18,000 Btu/lbmol (41 MJ/kmol)

(C) 20,000 Btu/lbmol (46 MJ/kmol)

(D) 22,000 Btu/lbmol (51 MJ/kmol)

2. The ratio of specific heats for air at 600°F (300°C) is most nearly

(A) 1.33

(B) 1.38

(C) 1.41

(D) 1.67

3. The density of helium at 600°F (300°C) and one standard atmosphere is most nearly

(A) 0.0052 lbm/ft^3 (0.085 kg/m^3)

(B) 0.0061 lbm/ft^3 (0.098 kg/m^3)

(C) 0.0076 lbm/ft^3 (0.12 kg/m^3)

(D) 0.0095 lbm/ft^3 (0.15 kg/m^3)

4. What is the thermodynamic state of water at 600°F and 300 psia?

(A) subcooled liquid

(B) saturated vapor

(C) superheated vapor

(D) real or ideal gas

5. 106°F water flows in a closed feedwater heater. At the pressure in the feedwater heater, water has a saturation temperature of 293°F and a saturation enthalpy of 262 Btu/lbm. What is most nearly the enthalpy of the water?

(A) 75 Btu/lbm

(B) 110 Btu/lbm

(C) 120 Btu/lbm

(D) 130 Btu/lbm

6. What is most nearly the enthalpy of saturated HFC-134a vapor at 180°F?

(A) 77 Btu/lbm

(B) 83 Btu/lbm

(C) 115 Btu/lbm

(D) 120 Btu/lbm

7. A 25% mixture (by volume) of ethylene glycol and water is used in a solar heating application. The mixture is intended to operate at standard atmospheric pressure and an average temperature of 150°F. At 150°F, ethylene glycol has a specific gravity of 1.11 (based on 60°F water) and a specific heat of 0.63 Btu/lbm-°F. What is most nearly the specific heat of the mixture at 150°F?

(A) 0.892 Btu/lbm-°F

(B) 0.899 Btu/lbm-°F

(C) 0.908 Btu/lbm-°F

(D) 0.913 Btu/lbm-°F

SOLUTIONS

1. *Customary U.S. Solution*

From App. 23.A, for 250°F steam, the enthalpy of saturated liquid, h_f, is 218.6 Btu/lbm. The heat of vaporization, h_{fg}, is 945.4 Btu/lbm. The enthalpy is given by Eq. 23.53.

$$h = h_f + xh_{fg} = 218.6\ \frac{\text{Btu}}{\text{lbm}} + (0.92)\left(945.4\ \frac{\text{Btu}}{\text{lbm}}\right)$$
$$= 1088.4\ \text{Btu/lbm}$$

The molecular weight of water is 18 lbm/lbmol. The molar enthalpy is given by Eq. 23.14.

$$H = \text{MW} \times h = \left(18\ \frac{\text{lbm}}{\text{lbmol}}\right)\left(1088.4\ \frac{\text{Btu}}{\text{lbm}}\right)$$
$$= \boxed{19{,}591\ \text{Btu/lbmol}\quad (20{,}000\ \text{Btu/lbmol})}$$

The answer is (C).

SI Solution

From App. 23.N, for 120°C steam, the enthalpy of saturated liquid, h_f, is 503.81 kJ/kg. The heat of vaporization, h_{fg}, is 2202.1 kJ/kg. The enthalpy is given by Eq. 23.53.

$$h = h_f + xh_{fg} = 503.81\ \frac{\text{kJ}}{\text{kg}} + (0.92)\left(2202.1\ \frac{\text{kJ}}{\text{kg}}\right)$$
$$= 2529.7\ \text{kJ/kg}$$

The molecular weight of water is 18 kg/kmol. Molar enthalpy is given by Eq. 23.14.

$$H = \text{MW} \times h = \left(18\ \frac{\text{kg}}{\text{kmol}}\right)\left(2529.7\ \frac{\text{kJ}}{\text{kg}}\right)$$
$$= \boxed{45\,535\ \text{kJ/kmol}\quad (46\ \text{MJ/kmol})}$$

The answer is (C).

2. *Customary U.S. Solution*

The absolute temperature is

$$600°\text{F} + 460° = 1060°\text{R}$$

From Table 21.1, the specific heat at constant pressure for air at 1060°R is $c_p = 0.250$ Btu/lbm-°R.

From Eq. 23.61, the specific gas constant is

$$R = \frac{R^*}{\text{MW}} = \frac{1545.33\ \frac{\text{ft-lbf}}{\text{lbmol-°R}}}{28.967\ \frac{\text{lbm}}{\text{lbmol}}}$$
$$= 53.35\ \text{ft-lbf/lbm-°R}$$

From Eq. 23.108(b),

$$c_v = c_p - \frac{R}{J} = 0.250\ \frac{\text{Btu}}{\text{lbm-°R}} - \frac{53.35\ \frac{\text{ft-lbf}}{\text{lbm-°R}}}{778\ \frac{\text{ft-lbf}}{\text{Btu}}}$$
$$= 0.1814\ \text{Btu/lbm-°R}$$

The ratio of specific heats is given by Eq. 23.28.

$$k = \frac{c_p}{c_v} = \frac{0.250\ \frac{\text{Btu}}{\text{lbm-°R}}}{0.1814\ \frac{\text{Btu}}{\text{lbm-°R}}}$$
$$= \boxed{1.378\quad (1.38)}$$

The answer is (B).

SI Solution

From Table 21.1, the specific heat at constant pressure for air is 0.250 Btu/lbm-°R. From the table footnote, the SI specific heat at constant pressure for air is

$$c_p = \left(0.250\ \frac{\text{Btu}}{\text{lbm-°R}}\right)\left(4.187\ \frac{\frac{\text{kJ}}{\text{kg·K}}}{\frac{\text{Btu}}{\text{lbm-°R}}}\right)$$
$$= 1.047\ \text{kJ/kg·K}$$

From Eq. 23.61, the specific gas constant is

$$R = \frac{R^*}{\text{MW}} = \frac{8314.3\ \frac{\text{J}}{\text{kmol·K}}}{28.967\ \frac{\text{kg}}{\text{kmol}}}$$
$$= 287.0\ \text{J/kg·K}$$

From Eq. 23.108(a),

$$c_v = c_p - R$$
$$= \left(1.047\ \frac{\text{kJ}}{\text{kg·K}}\right)\left(1000\ \frac{\text{J}}{\text{kJ}}\right) - 287.0\ \frac{\text{J}}{\text{kg·K}}$$
$$= 760\ \text{J/kg·K}$$

The ratio of specific heats is given by Eq. 23.28.

$$k = \frac{c_p}{c_v} = \frac{\left(1.047\ \frac{\text{kJ}}{\text{kg·K}}\right)\left(1000\ \frac{\text{J}}{\text{kJ}}\right)}{760\ \frac{\text{J}}{\text{kg·K}}}$$
$$= \boxed{1.377\quad (1.38)}$$

The answer is (B).

3. *Customary U.S. Solution*

From Eq. 23.61, the specific gas constant is

$$R = \frac{R^*}{\text{MW}} = \frac{1545.33\ \frac{\text{ft-lbf}}{\text{lbmol-°R}}}{4\ \frac{\text{lbm}}{\text{lbmol}}}$$
$$= 386.3\ \text{ft-lbf/lbm-°R}$$

The absolute temperature is

$$600°\text{F} + 460° = 1060°\text{R}$$

From Eq. 23.63, the density of helium is

$$\rho = \frac{p}{RT} = \frac{\left(14.7\ \frac{\text{lbf}}{\text{in}^2}\right)\left(12\ \frac{\text{in}}{\text{ft}}\right)^2}{\left(386.3\ \frac{\text{ft-lbf}}{\text{lbm-°R}}\right)(1060°\text{R})}$$
$$= \boxed{0.00517\ \text{lbm/ft}^3\quad (0.0052\ \text{lbm/ft}^3)}$$

The answer is (A).

SI Solution

From Eq. 23.61, the specific gas constant is

$$R = \frac{R^*}{\text{MW}} = \frac{8314.3\ \frac{\text{J}}{\text{kmol·K}}}{4\ \frac{\text{kg}}{\text{kmol}}}$$
$$= 2079\ \text{J/kg·K}$$

The absolute temperature is

$$300°\text{C} + 273° = 573\text{K}$$

From Eq. 23.63, the density of helium is

$$\rho = \frac{p}{RT} = \frac{1.013 \times 10^5\ \text{Pa}}{\left(2079\ \frac{\text{J}}{\text{kg·K}}\right)(573\text{K})}$$
$$= \boxed{0.0850\ \text{kg/m}^3\quad (0.085\ \text{kg/m}^3)}$$

The answer is (A).

4. The saturation temperature for 300 psia steam is 417°F, so since the water's temperature is higher than this, the water is either a superheated vapor or a gas. The critical temperature for water is 705°F, so the water can't be considered a gas. Therefore, it is a $\boxed{\text{superheated vapor.}}$

The answer is (C).

5. Since the pressure of the water in the feedwater heater is not given, a subcooled liquid table can't be used directly. (The saturation temperature could be used to find the pressure, however, if this approach was taken.) The specific heat of liquid water is approximately 1 Btu/lbm-°F, which is the reason that saturated liquid enthalpy and saturation temperature have essentially the same numerical values. Calculate the subcooled enthalpy by subtracting the sensible heat from 106°F to 293°F.

$$h_{106°\text{F}} = h_{\text{saturation}} - c_p(T_{\text{saturation}} - T)$$
$$= 262\ \frac{\text{Btu}}{\text{lbm}} - \left(1\ \frac{\text{Btu}}{\text{lbm-°F}}\right)(293°\text{F} - 106°\text{F})$$
$$= \boxed{75\ \text{Btu/lbm}}$$

The answer is (A).

6. Use App. 23.M. From the right side of the vapor dome on the 180°F isotherm, and dropping straight down to the enthalpy scale, the enthalpy is approximately $\boxed{120\ \text{Btu/lbm.}}$

The answer is (D).

7. Use a saturated steam table, such as App. 23.A, to get the properties of 150°F water. The specific volume is the reciprocal of the density. Since $\Delta h = c_p \Delta T$, the specific heat can be found from the change in enthalpy over a known temperature range. Use the saturation enthalpies at 140°F and 160°F.

$$\rho = \frac{1}{v_f}$$
$$= \frac{1}{0.01634\ \frac{\text{ft}^3}{\text{lbm}}}$$
$$= 61.20\ \text{lbm/ft}^3$$

$$c_{p,\text{water}} = \frac{h_{\text{sat},T_2} - h_{\text{sat},T_1}}{T_2 - T_1}$$
$$= \frac{128.00\ \frac{\text{Btu}}{\text{lbm}} - 107.99\ \frac{\text{Btu}}{\text{lbm}}}{160°\text{F} - 140°\text{F}}$$
$$= 1.00\ \text{Btu/lbm-°F}$$

Specific heats of liquid mixtures are gravimetrically weighted.

Thermodynamics

Consider 1 ft^3 of mixture, containing 0.25 ft^3 of ethylene glycol and $1 - 0.25\ ft^3 = 0.75\ ft^3$ water. The gravimetric fraction of ethylene glycol, G_{glycol}, in the mixture is

$$\begin{aligned} G_{glycol} &= \frac{m_{glycol}}{m_{glycol} + m_{water}} \\ &= \frac{SG_{glycol}\rho_{water,60^\circ F} V_{glycol}}{SG_{glycol}\rho_{water,60^\circ F} V_{glycol} + \rho_{water,150^\circ F} V_{water}} \\ &= \frac{(1.11)\left(62.4\ \frac{lbm}{ft^3}\right)(0.25\ ft^3)}{(1.11)\left(62.4\ \frac{lbm}{ft^3}\right)(0.25\ ft^3) + \left(61.2\ \frac{lbm}{ft^3}\right)(0.75\ ft^3)} \\ &= 0.274 \end{aligned}$$

The gravimetric fraction of water in the mixture is

$$G_{water} = 1 - G_{glycol} = 1 - 0.274 = 0.726$$

The specific heat of the mixture is

$$\begin{aligned} c_{p,mixture} &= G_{glycol}c_{p,glycol} + G_{water}c_{p,water} \\ &= (0.274)\left(0.63\ \frac{Btu}{lbm\text{-}^\circ F}\right) + (0.726)\left(1.00\ \frac{Btu}{lbm\text{-}^\circ F}\right) \\ &= \boxed{0.899\ Btu/lbm\text{-}^\circ F} \end{aligned}$$

The answer is (B).

Thermodynamics

24 Changes in Thermodynamic Properties

PRACTICE PROBLEMS

1. Which statement is FALSE?

(A) The availability of a system depends on the location of the system.

(B) In the absence of friction and other irreversibilities, a heat engine cycle can have a thermal efficiency of 100%.

(C) The gas temperature always decreases when the gas expands isentropically.

(D) Water in a liquid state cannot exist at any temperature if the pressure is less than the triple point pressure.

2. Which statement is true?

(A) Entropy does not change in an adiabatic process.

(B) The entropy of a closed system cannot decrease.

(C) Entropy increases when a refrigerant passes through a throttling valve.

(D) The entropy of air inside a closed room with a running, electrically driven fan will always increase over time.

3. Which process CANNOT be modeled as an isenthalpic process?

(A) viscous drag on an object moving through air

(B) an ideal gas accelerating to supersonic speed through a converging-diverging nozzle

(C) refrigerant passing through a pressure-reducing throttling valve

(D) high-pressure steam escaping through a spring-loaded pressure-relief (safety) valve

4. Air expands isentropically in a steady-flow process from 700°F and 400 psia to 50 psia.

(a) From an air table, what is most nearly the change in enthalpy?

(A) −680 Btu/lbm

(B) −130 Btu/lbm

(C) −90 Btu/lbm

(D) −14 Btu/lbm

(b) From the ideal gas laws for isentropic expansion, and without using any values for specific heat capacity, what is most nearly the change in enthalpy?

(A) −680 Btu/lbm

(B) −120 Btu/lbm

(C) −90 Btu/lbm

(D) −14 Btu/lbm

(c) Based on the specific heat capacity at room temperature, what is most nearly the change in enthalpy?

(A) −680 Btu/lbm

(B) −120 Btu/lbm

(C) −90 Btu/lbm

(D) −14 Btu/lbm

5. 0.60 kg of air ($c_p = 1.005$ kJ/kg·K; $c_v = 0.718$ kJ/kg·K) are contained in a perfectly insulated, rigid enclosure. The ambient conditions outside the enclosure are 95 kPa and 20°C. The air inside the enclosure is initially at 200 kPa and 20°C. Subsequently, an internal impeller within the enclosure raises the air's pressure to 230 kPa through a shaft from an external motor with a motor efficiency of 65%.

(a) What is most nearly the temperature inside the enclosure after the pressure is increased?

(A) 23°C

(B) 45°C

(C) 64°C

(D) 340°C

(b) What is most nearly the initial specific volume of the air within the enclosure?

(A) 0.42 m^3/kg

(B) 0.74 m^3/kg

(C) 1.8 m^3/kg

(D) cannot be determined

(c) What is most nearly the increase in density within the enclosure after the pressure increase?

(A) 0.00 kg/m^3

(B) 0.14 kg/m^3

(C) 0.21 kg/m^3

(D) 0.36 kg/m^3

(d) Approximately how much shaft work is required to raise the pressure from 200 kPa to 230 kPa?

(A) 19 kJ

(B) 25 kJ

(C) 32 kJ

(D) 140 kJ

(e) What is most nearly the increase in the available work potential of the air due to the pressure increase?

(A) 1.3 kJ

(B) 2.2 kJ

(C) 8.9 kJ

(D) 15 kJ

6. Cast iron is heated from 80°F to 780°F (27°C to 416°C). The heat required per unit mass is most nearly

(A) 70 Btu/lbm (160 kJ/kg)

(B) 120 Btu/lbm (280 kJ/kg)

(C) 170 Btu/lbm (390 kJ/kg)

(D) 320 Btu/lbm (740 kJ/kg)

7. The ventilation rate in a building is 3×10^5 ft^3/hr (2.4 m^3/s). The air is heated from 35°F to 75°F (2°C to 24°C) by water whose temperature decreases from 180°F to 150°F (82°C to 66°C). The water flow rate in gal/min (L/s) is most nearly

(A) 9 gal/min (0.58 L/s)

(B) 13 gal/min (0.83 L/s)

(C) 15 gal/min (0.96 L/s)

(D) 22 gal/min (1.4 L/s)

8. 8.0 ft^3 (0.25 m^3) of 180°F, 14.7 psia (82°C, 101.3 kPa) air are cooled to 100°F (38°C) in a constant-pressure process. The amount of work done is most nearly

(A) −2100 ft-lbf (−3.1 kJ)

(B) −1500 ft-lbf (−2.3 kJ)

(C) −1100 ft-lbf (−1.5 kJ)

(D) −900 ft-lbf (−1.3 kJ)

9. Most nearly, the availability of an isentropic process using steam with an initial quality of 95% and operating between 300 psia and 50 psia (2 MPa and 0.35 MPa) is

(A) 100 Btu/lbm (230 kJ/kg)

(B) 130 Btu/lbm (300 kJ/kg)

(C) 210 Btu/lbm (480 kJ/kg)

(D) 340 Btu/lbm (780 kJ/kg)

10. A closed air heater receives 540°F, 100 psia (280°C, 700 kPa) air and heats it to 1540°F (840°C). The outside temperature is 100°F (40°C). The pressure of the air drops 20 psi (150 kPa) as it passes through the heater.

(a) The absolute temperatures at the inlet, T_1, and outlet, T_2, of the air heater are most nearly

(A) $T_1 = 460$°R, $T_2 = 1500$°R
($T_1 = 260$K, $T_2 = 860$K)

(B) $T_1 = 540$°R, $T_2 = 1000$°R
($T_1 = 300$K, $T_2 = 550$K)

(C) $T_1 = 1000$°R, $T_2 = 2000$°R
($T_1 = 550$K, $T_2 = 1100$K)

(D) $T_1 = 1500$°R, $T_2 = 1000$°R
($T_1 = 860$K, $T_2 = 550$K)

(b) The maximum work is most nearly

(A) −240 Btu/lbm (−560 kJ/kg)

(B) −200 Btu/lbm (−460 kJ/kg)

(C) −170 Btu/lbm (−390 kJ/kg)

(D) −150 Btu/lbm (−360 kJ/kg)

(c) The percentage loss in available energy due to the pressure drop is most nearly

(A) 5.0%

(B) 12%

(C) 18%

(D) 34%

11. Xenon gas at 20 psia and 70°F (150 kPa and 21°C) is compressed to 3800 psia and 70°F (25 MPa and 21°C) by a compressor/heat exchanger combination. The compressed gas is stored at 70°F (21°C) in a 100 ft^3 (3 m^3) rigid tank initially charged with xenon gas at 20 psia (150 kPa).

(a) The mass of the xenon gas initially in the tank is most nearly

(A) 35 lbm (18 kg)

(B) 42 lbm (22 kg)

(C) 46 lbm (24 kg)

(D) 51 lbm (27 kg)

(b) The average mass flow rate of xenon gas into the tank if the compressor fills the tank in exactly one hour is most nearly

(A) 6300 lbm/hr (0.88 kg/s)

(B) 9700 lbm/hr (1.3 kg/s)

(C) 12,000 lbm/hr (1.6 kg/s)

(D) 14,000 lbm/hr (1.9 kg/s)

(c) If filling takes exactly one hour and electricity costs \$0.045 per kW-hr, the cost of filling the tank is most nearly

(A) \$8.00

(B) \$14

(C) \$27

(D) \$35

12. The mass of an insulated 20 ft^3 (0.6 m^3) steel tank is 40 lbm (20 kg). The steel has a specific heat of 0.11 Btu/lbm-°R (0.46 kJ/kg·K). The tank is placed in a room where the surrounding air is 70°F and 14.7 psia (21°C and 101.3 kPa). After the tank is evacuated to 1 psia and 70°F (7 kPa and 21°C), a valve is suddenly opened, allowing the tank to fill with room air. The air enters the tank in a well-mixed, turbulent condition.

(a) The volume of room air entering the tank is most nearly

(A) 1.4 ft^3 (0.04 m^3)

(B) 13 ft^3 (0.40 m^3)

(C) 14 ft^3 (0.44 m^3)

(D) 20 ft^3 (0.64 m^3)

(b) The work performed by the room air entering the tank is most nearly

(A) −42 Btu (−49 000 J)

(B) −34 Btu (−40 000 J)

(C) −8 Btu (−9500 J)

(D) −4 Btu (−4700 J)

(c) The air temperature of the air inside the tank after filling, after the gas and tank have reached thermal equilibrium, but before any heat loss from the tank to the room occurs is most nearly

(A) 70°F (20°C)

(B) 80°F (30°C)

(C) 100°F (40°C)

(D) 190°F (90°C)

SOLUTIONS

1. Availability depends on the temperature of the environment, T_L. Option A is true.

A heat engine's maximum efficiency is that of a Carnot engine cycle, which is always less than 100%. Option B is false.

An expansion includes a drop in enthalpy, and since $h = u + pv$, both u (manifested as temperature) and p decrease. Option C is true.

Liquid water cannot exist below the triple point pressure. Option D is true.

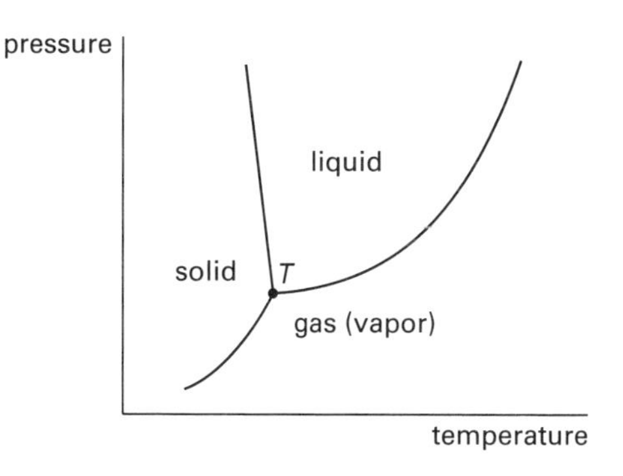

The answer is (B).

2. Entropy does not change in an isentropic (reversible adiabatic) process. However, not all adiabatic processes are reversible. Option A is false.

The entropy of a closed system can be decreased by decreasing the temperature. Option B is false.

When a refrigerant is throttled, the enthalpy remains constant, the pressure drops, and the entropy increases. Option C is true.

The fan and motor will certainly increase the entropy of the air inside the room. However, if the air temperature is decreased by heat loss to the outside, the entropy will decrease. Option D is false.

The answer is (C).

3. A throttling process (including flow through a control valve, safety relief valve, throttling valve, nozzle, or orifice) is modeled as an isenthalpic process. Consider steam expanding through a safety relief valve. To squeeze through the narrow restriction between the disk and the valve seat, steam has to accelerate to a high speed. It does so by converting enthalpy into kinetic energy. The process is not frictionless, but passage past the valve seat occurs so quickly as to be essentially so, and the process is considered to be isentropic. Once past the narrow restriction, the steam expands into the lower pressure region in the valve outlet. The steam decelerates as the flow area increases from the valve passageway to the downstream pipe. This decrease in velocity (kinetic energy) is manifested as an increase in temperature and enthalpy. The enthalpy drop associated with the initial increase in kinetic energy is reclaimed (except for a small portion lost due to the effects of friction). Therefore, the final process is essentially isenthalpic.

Viscous drag of an object in any fluid is an adiabatic process without work being performed on or by the fluid. The duration of the contact between fluid and object is short. Therefore, viscous drag is isenthalpic.

The increase in velocity in a supersonic nozzle comes entirely at the expense of enthalpy.

The answer is (B).

4. (a) From App. 23.F, for $T_1 = 700°\text{F} + 460° = 1160°\text{R}$ and 400 psia, $h_1 = 281.14$ Btu/lbm, and $p_{r,1} = 21.18$. Since the expansion is isentropic,

$$p_{r,2} = p_{r,1}\frac{p_2}{p_1} = (21.18)\left(\frac{50\ \frac{\text{lbf}}{\text{in}^2}}{400\ \frac{\text{lbf}}{\text{in}^2}}\right) = 2.6475$$

Interpolating from App. 23.F, the value of $p_{r,2}$ corresponds to $T_2 = 649.3°\text{R}$. For this temperature, $h_2 = 155.3$ Btu/lbm. The change in enthalpy is

$$\begin{aligned}\Delta h = h_2 - h_1 &= 155.3\ \frac{\text{Btu}}{\text{lbm}} - 281.14\ \frac{\text{Btu}}{\text{lbm}}\\ &= \boxed{-125.8\ \text{Btu/lbm}\quad(-130\ \text{Btu/lbm})}\end{aligned}$$

The answer is (B).

(b) From Eq. 24.107 (applicable to both closed and open systems),

$$T_2 = T_1\left(\frac{p_2}{p_1}\right)^{\frac{k-1}{k}} = (1160°\text{R})\left(\frac{50\ \frac{\text{lbf}}{\text{in}^2}}{400\ \frac{\text{lbf}}{\text{in}^2}}\right)^{\frac{1.4-1}{1.4}} = 640.4°\text{R}$$

The specific volumes are

$$\begin{aligned}\upsilon_1 = \frac{RT_1}{p_1} &= \frac{\left(53.35\ \frac{\text{ft-lbf}}{\text{lbm-°R}}\right)(1160°\text{R})}{\left(400\ \frac{\text{lbf}}{\text{in}^2}\right)\left(12\ \frac{\text{in}}{\text{ft}}\right)^2}\\ &= 1.074\ \text{ft}^3/\text{lbm}\end{aligned}$$

$$\begin{aligned}\upsilon_2 = \frac{RT_2}{p_2} &= \frac{\left(53.35\ \frac{\text{ft-lbf}}{\text{lbm-°R}}\right)(640.4°\text{R})}{\left(50\ \frac{\text{lbf}}{\text{in}^2}\right)\left(12\ \frac{\text{in}}{\text{ft}}\right)^2}\\ &= 4.745\ \text{ft}^3/\text{lbm}\end{aligned}$$

From Eq. 24.121,

$$\begin{aligned}h_2 - h_1 &= \frac{k(p_2\upsilon_2 - p_1\upsilon_1)}{(k-1)J}\\ &= \frac{(1.4)\begin{pmatrix}\left(50\ \frac{\text{lbf}}{\text{in}^2}\right)\left(4.745\ \frac{\text{ft}^3}{\text{lbm}}\right)\\ -\left(400\ \frac{\text{lbf}}{\text{in}^2}\right)\left(1.074\ \frac{\text{ft}^3}{\text{lbm}}\right)\end{pmatrix}\times\left(12\ \frac{\text{in}}{\text{ft}}\right)^2}{(1.4-1)\left(778.17\ \frac{\text{ft-lbf}}{\text{Btu}}\right)}\\ &= \boxed{-124.6\ \text{Btu/lbm}\quad(-120\ \text{Btu/lbm})}\end{aligned}$$

The answer is (B).

(c) From part (b), $T_2 = 640.4°\text{R}$. From Table 23.7, $c_p = 0.240$ Btu/lbm-°R.

From Eq. 24.120,

$$\begin{aligned}\Delta h = c_p\Delta T &= \left(0.240\ \frac{\text{Btu}}{\text{lbm-°R}}\right)(640.4°\text{R} - 1160°\text{R})\\ &= \boxed{-124.7\ \text{Btu/lbm}\quad(-120\ \text{Btu/lbm})}\end{aligned}$$

The answer is (B).

5. (a) Use Gay-Lussac's law.

$$T_2 = \frac{T_1p_2}{p_1} = \frac{(20°\text{C} + 273°)(230\ \text{kPa})}{200\ \text{kPa}} = 337\text{K}$$

$$T_{2,°\text{C}} = 337\text{K} - 273° = \boxed{64°\text{C}}$$

The answer is (C).

(b) The volume of the enclosure is not given, so specific volume cannot be calculated as $\upsilon = V/m$. Use the ideal gas law.

$$\begin{aligned}\upsilon_1 = \frac{RT_1}{p_1} &= \frac{\left(287.03\ \frac{\text{J}}{\text{kg·K}}\right)(20°\text{C} + 273°)}{(200\ \text{Pa})\left(1000\ \frac{\text{Pa}}{\text{kPa}}\right)}\\ &= \boxed{0.42\ \text{m}^3/\text{kg}}\end{aligned}$$

The answer is (A).

(c) Although the pressure and temperature of the air inside the enclosure change, the total air mass and enclosure volume do not change. Therefore, the specific volume does not change, and $\mu_2 = \mu_1$.

The answer is (A).

(d) This is a closed system. The first law of thermodynamics is applicable.

$$Q = \Delta U + W$$

Since the system is insulated, it is adiabatic, and $Q=0$.

$$W = -\Delta U$$

Work on a unit mass (not molar) basis. Use Eq. 24.51, valid for any process.

$$\begin{aligned} W &= -\Delta U = -c_v(T_2 - T_1) \\ &= (0.6 \text{ kg})\left(0.718 \ \frac{\text{kJ}}{\text{kg·K}}\right)(64°\text{C} - 20°\text{C}) \\ &= \boxed{19 \text{ kJ}} \end{aligned}$$

The answer is (A).

(e) From Eq. 24.81, with $\mu_2 = \mu_1$, the entropy change is

$$s_2 - s_1 = c_v \ln\frac{T_2}{T_1}$$

Use Eq. 24.179 to calculate the change in availability. Incorporate Eq. 24.50, valid for any process, and Eq. 24.81.

$$\begin{aligned} \Phi_2 - \Phi_1 &= h_2 - h_1 - T_L(s_2 - s_1) \\ &= c_p(T_2 - T_1) - T_L\left(c_v \ln\frac{T_2}{T_1}\right) \\ &= \left(1.005 \ \frac{\text{kJ}}{\text{kg·K}}\right)(64°\text{C} - 20°\text{C}) \\ &\quad - (20°\text{C} + 273°) \\ &\quad \times \left(\left(0.718 \ \frac{\text{kJ}}{\text{kg·K}}\right)\ln\left(\frac{337\text{K}}{20°\text{C} + 273°}\right)\right) \\ &= 14.79 \text{ kJ/kg} \end{aligned}$$

For the air mass within the enclosure, the increase in availability is

$$\begin{aligned} m(\Phi_2 - \Phi_1) &= (0.6 \text{ kg})\left(14.79 \ \frac{\text{kJ}}{\text{kg}}\right) \\ &= \boxed{8.87 \text{ kJ} \quad (8.9 \text{ kJ})} \end{aligned}$$

The answer is (C).

6. *Customary U.S. Solution*

From Table 23.2, the approximate value of specific heat for cast iron is $c_p = 0.10$ Btu/lbm-°F.

The heat required per unit mass is

$$\begin{aligned} q &= c_p(T_2 - T_1) = \left(0.10 \ \frac{\text{Btu}}{\text{lbm-°F}}\right)(780°\text{F} - 80°\text{F}) \\ &= \boxed{70 \text{ Btu/lbm}} \end{aligned}$$

The answer is (A).

SI Solution

From Table 23.2, the approximate value of specific heat of cast iron is $c_p = 0.42$ kJ/kg·K.

The heat required per unit mass is

$$\begin{aligned} q &= c_p(T_2 - T_1) \\ &= \left(0.42 \ \frac{\text{kJ}}{\text{kg·K}}\right)(416°\text{C} - 27°\text{C}) \\ &= \boxed{163.4 \text{ kJ/kg} \quad (160 \text{ kJ/kg})} \end{aligned}$$

The answer is (A).

7. *Customary U.S. Solution*

First calculate the mass flow rate of air to be heated by using the ideal gas law. (Usually the air mass would be evaluated at the entering conditions. This problem is ambiguous. The building conditions are used because a building ventilation rate was specified.)

$$\dot{m}_{\text{air}} = \frac{p\dot{V}}{RT}$$

The absolute temperature is

$$\begin{aligned} T &= 75°\text{F} + 460° = 535°\text{R} \\ \dot{m}_{\text{air}} &= \frac{\left(14.7 \ \frac{\text{lbf}}{\text{in}^2}\right)\left(12 \ \frac{\text{in}}{\text{ft}}\right)^2\left(3 \times 10^5 \ \frac{\text{ft}^3}{\text{hr}}\right)}{\left(53.3 \ \frac{\text{ft-lbf}}{\text{lbm-°R}}\right)(535°\text{R})} \\ &= 2.227 \times 10^4 \text{ lbm/hr} \end{aligned}$$

The heat lost by the water is equal to the heat gained by the air.

$$\begin{aligned} \dot{m}_w c_{p,w}(T_{1,w} - T_{2,w}) &= \dot{m}_{\text{air}} c_{p,\text{air}}(T_{2,\text{air}} - T_{1,\text{air}}) \\ \dot{m}_w\left(1 \ \frac{\text{Btu}}{\text{lbm-°F}}\right)(180°\text{F} - 150°\text{F}) &= \left(2.227 \times 10^4 \ \frac{\text{lbm}}{\text{hr}}\right)\left(0.241 \ \frac{\text{Btu}}{\text{lbm-°F}}\right)(75°\text{F} - 35°\text{F}) \\ \dot{m}_w &= 7156.1 \text{ lbm/hr} \end{aligned}$$

From App. 35.A, the density of water at 165°F is approximately 61 lbm/ft^3. The water volume flow rate is

$$\begin{aligned} \dot{V}_w &= \frac{\dot{m}}{\rho} = \frac{\left(7156.1 \ \frac{\text{lbm}}{\text{hr}}\right)\left(7.481 \ \frac{\text{gal}}{\text{ft}^3}\right)}{\left(61 \ \frac{\text{lbm}}{\text{ft}^3}\right)\left(60 \ \frac{\text{min}}{\text{hr}}\right)} \\ &= \boxed{14.63 \text{ gal/min} \quad (15 \text{ gal/min})} \end{aligned}$$

The answer is (C).

Thermodynamics

SI Solution

First calculate the mass flow rate of air to be heated by using the ideal gas law. (Usually the air mass would be evaluated at the entering conditions. This problem is ambiguous. The building conditions are used because a building ventilation rate was specified.)

$$\dot{m}_{\text{air}} = \frac{p\dot{V}}{RT}$$

The absolute temperature is

$$T = 24°\text{C} + 273° = 297\text{K}$$

$$\dot{m}_{\text{air}} = \frac{(1.013 \times 10^5 \text{ Pa})\left(2.4 \ \frac{\text{m}^3}{\text{s}}\right)}{\left(287 \ \frac{\text{J}}{\text{kg·K}}\right)(297\text{K})} = 2.85 \text{ kg/s}$$

The heat lost by the water is equal to the heat gained by the air.

$$\dot{m}_w c_{p,w}(T_{1,w} - T_{2,w}) = \dot{m}_{\text{air}} c_{p,\text{air}}(T_{2,\text{air}} - T_{1,\text{air}})$$

$$\dot{m}_w\left(4.190 \ \frac{\text{kJ}}{\text{kg·K}}\right)(82°\text{C} - 66°\text{C}) = \left(2.85 \ \frac{\text{kg}}{\text{s}}\right)\left(1.005 \ \frac{\text{kJ}}{\text{kg·K}}\right)(24°\text{C} - 2°\text{C})$$

$$\dot{m}_w = 0.940 \text{ kg/s}$$

From App. 35.B, the density of water at 74°C is approximately 976 kg/m^3. The water volume flow rate is

$$\dot{V}_w = \frac{\dot{m}}{\rho} = \frac{\left(0.940 \ \frac{\text{kg}}{\text{s}}\right)\left(1000 \ \frac{\text{L}}{\text{m}^3}\right)}{976 \ \frac{\text{kg}}{\text{m}^3}} = \boxed{0.963 \text{ L/s} \quad (0.96 \text{ L/s})}$$

The answer is (C).

8. *Customary U.S. Solution*

The mass of air is

$$m = \frac{p_1 V_1}{RT_1} = \frac{\left(14.7 \ \frac{\text{lbf}}{\text{in}^2}\right)\left(12 \ \frac{\text{in}}{\text{ft}}\right)^2(8.0 \text{ ft}^3)}{\left(53.3 \ \frac{\text{ft-lbf}}{\text{lbm-°R}}\right)(180°\text{F} + 460°)} = 0.4964 \text{ lbm}$$

For a constant pressure process from Eq. 24.65, on a per unit mass basis,

$$W = R(T_2 - T_1)$$

Since $\Delta T_{°\text{R}} = \Delta T_{°\text{F}}$, the total work for m in lbm is

$$\begin{aligned} W &= mR(T_2 - T_1) \\ &= (0.4964 \text{ lbm})\left(53.3 \ \frac{\text{ft-lbf}}{\text{lbm-°R}}\right)(100°\text{F} - 180°\text{F}) \\ &= \boxed{-2116.6 \text{ ft-lbf} \quad (-2100 \text{ ft-lbf})} \end{aligned}$$

This is negative because work is done on the system.

The answer is (A).

SI Solution

The mass of air is

$$m = \frac{p_1 V_1}{RT_1} = \frac{(101.3 \text{ kPa})\left(1000 \ \frac{\text{Pa}}{\text{kPa}}\right)(0.25 \text{ m}^3)}{\left(287 \ \frac{\text{J}}{\text{kg·K}}\right)(82°\text{C} + 273°)} = 0.2486 \text{ kg}$$

For a constant pressure process from Eq. 24.65, on a per unit mass basis,

$$W = R(T_2 - T_1)$$

The total work for m in kg is

$$\begin{aligned} W &= mR(T_2 - T_1) \\ &= (0.2486 \text{ kg})\left(287 \ \frac{\text{J}}{\text{kg·K}}\right)(38°\text{C} - 82°\text{C}) \\ &= \boxed{-3139.3 \text{ J} \quad (-3.1 \text{ kJ})} \end{aligned}$$

The answer is (A).

9. *Customary U.S. Solution*

From App. 23.B, for 300 psia, the enthalpy of saturated liquid, h_f, is 394.0 Btu/lbm. The heat of vaporization, h_{fg}, is 809.4 Btu/lbm. The enthalpy is given by Eq. 23.53.

$$h_1 = h_f + xh_{fg} = 394.0 \ \frac{\text{Btu}}{\text{lbm}} + (0.95)\left(809.4 \ \frac{\text{Btu}}{\text{lbm}}\right) = 1162.9 \text{ Btu/lbm}$$

From the Mollier diagram, for an isentropic process from 300 psia to 50 psia, $h_2 = 1031$ Btu/lbm.

The availability is calculated from Eq. 24.178 using an isentropic process ($s_1 = s_2$).

$$\text{availability} = h_1 - h_2 = 1162.9 \ \frac{\text{Btu}}{\text{lbm}} - 1031 \ \frac{\text{Btu}}{\text{lbm}} = \boxed{131.9 \text{ Btu/lbm} \quad (130 \text{ Btu/lbm})}$$

The answer is (B).

SI Solution

From App. 23.O, for 2 MPa, the enthalpy of saturated liquid, h_f, is 908.50 kJ/kg. The heat of vaporization, h_{fg}, is 1889.8 kJ/kg. The enthalpy is given by Eq. 23.53.

$$\begin{aligned} h_1 &= h_f + xh_{fg} = 908.50\ \frac{\text{kJ}}{\text{kg}} + (0.95)\left(1889.8\ \frac{\text{kJ}}{\text{kg}}\right) \\ &= 2703.8\ \text{kJ/kg} \end{aligned}$$

From the Mollier diagram, for an isentropic process from 2 MPa to 0.35 MPa, $h_2 = 2405$ kJ/kg.

The availability is calculated from Eq. 24.178 using an isentropic process ($s_1 = s_2$).

$$\begin{aligned} \text{availability} &= h_1 - h_2 = 2703.8\ \frac{\text{kJ}}{\text{kg}} - 2405\ \frac{\text{kJ}}{\text{kg}} \\ &= \boxed{298.8\ \text{kJ/kg} \quad (300\ \text{kJ/kg})} \end{aligned}$$

The answer is (B).

10. *Customary U.S. Solution*

(a) The absolute temperature at the inlet of the air heater is

$$T_1 = 540°\text{F} + 460° = \boxed{1000°\text{R}}$$

The absolute temperature at the outlet of the air heater is

$$T_2 = 1540°\text{F} + 460° = \boxed{2000°\text{R}}$$

The answer is (C).

(b) Since pressures are low and temperatures are high, use an air table.

From App. 23.F at 1000°R,

$$h_1 = 240.98\ \text{Btu/lbm}$$
$$\phi_1 = 0.75042\ \text{Btu/lbm-°R}$$

From App. 23.F at 2000°R,

$$h_2 = 504.71\ \text{Btu/lbm}$$
$$\phi_2 = 0.93205\ \text{Btu/lbm-°R}$$

The availability per unit mass is calculated from Eq. 24.178 using $T_L = 100°\text{F} + 460° = 560°\text{R}$.

$$W_{\text{max}} = h_1 - h_2 + T_L(s_2 - s_1)$$

For no pressure drop,

$$s_2 - s_1 = \phi_2 - \phi_1$$

$$\begin{aligned} W_{\text{max}} &= h_1 - h_2 + T_L(\phi_2 - \phi_1) \\ &= 240.98\ \frac{\text{Btu}}{\text{lbm}} - 504.71\ \frac{\text{Btu}}{\text{lbm}} \\ &\quad + (560°\text{R})\left(0.93205\ \frac{\text{Btu}}{\text{lbm-°R}} - 0.75042\ \frac{\text{Btu}}{\text{lbm-°R}}\right) \\ &= -162.02\ \text{Btu/lbm} \end{aligned}$$

With a pressure drop from 100 psia to 80 psia, from Eq. 23.52,

$$s_2 - s_1 = \phi_2 - \phi_1 - \frac{R}{J}\ln\left(\frac{p_2}{p_1}\right)$$

Therefore, the maximum total work due to the pressure drop is

$$\begin{aligned} W_{\text{max},p\ \text{loss}} &= h_1 - h_2 + T_L\left(\phi_2 - \phi_1 - \frac{R}{J}\ln\left(\frac{p_2}{p_1}\right)\right) \\ &= 240.98\ \frac{\text{Btu}}{\text{lbm}} - 504.71\ \frac{\text{Btu}}{\text{lbm}} \\ &\quad + (560°\text{R}) \\ &\quad \times \left(\begin{array}{l} 0.93205\ \frac{\text{Btu}}{\text{lbm-°R}} - 0.75042\ \frac{\text{Btu}}{\text{lbm-°R}} \\ \quad - \left(\dfrac{53.3\ \frac{\text{ft-lbf}}{\text{lbm-°R}}}{778\ \frac{\text{ft-lbf}}{\text{Btu}}}\right)\ln\left(\dfrac{80\ \text{psia}}{100\ \text{psia}}\right) \end{array}\right) \\ &= \boxed{-153.46\ \text{Btu/lbm} \quad (-150\ \text{Btu/lbm})} \end{aligned}$$

The answer is (D).

(c) The percentage loss in available energy is

$$\begin{aligned} &\frac{W_{\text{max}} - W_{\text{max},p\ \text{loss}}}{W_{\text{max}}} \times 100\% \\ &\quad = \frac{-162.02\ \frac{\text{Btu}}{\text{lbm}} - \left(-153.46\ \frac{\text{Btu}}{\text{lbm}}\right)}{-162.02\ \frac{\text{Btu}}{\text{lbm}}} \times 100\% \\ &\quad = \boxed{5.28\% \quad (5.0\%)} \end{aligned}$$

The answer is (A).

SI Solution

(a) The absolute temperature at the inlet of the air heater is

$$T_1 = 280°\text{C} + 273° = \boxed{553\text{K} \quad (550\text{K})}$$

The absolute temperature at the outlet of the air heater is

$$T_2 = 840°\text{C} + 273° = \boxed{1113\text{K} \quad (1100\text{K})}$$

The answer is (C).

(b) Since pressures are low and temperatures are high, use an air table.

From App. 23.S at 553K,

$$h_1 = 557.9 \text{ kJ/kg}$$
$$\phi_1 = 2.32372 \text{ kJ/kg·K}$$

From App. 23.S at 1113K,

$$h_2 = 1176.2 \text{ kJ/kg}$$
$$\phi_2 = 3.09092 \text{ kJ/kg·K}$$

The availability per unit mass is calculated from Eq. 24.178 using $T_L = 40°\text{C} + 273° = 313\text{K}$.

$$W_{\text{max}} = h_1 - h_2 + T_L(s_2 - s_1)$$

For no pressure drop,

$$\begin{aligned} s_2 - s_1 &= \phi_2 - \phi_1 \\ W_{\text{max}} &= h_1 - h_2 + T_L(\phi_2 - \phi_1) \\ &= 557.9 \ \frac{\text{kJ}}{\text{kg}} - 1176.2 \ \frac{\text{kJ}}{\text{kg}} \\ &\quad + (313\text{K})\left(3.09092 \ \frac{\text{kJ}}{\text{kg·K}} - 2.32372 \ \frac{\text{kJ}}{\text{kg·K}}\right) \\ &= -378.17 \text{ kJ/kg} \end{aligned}$$

With a pressure drop from 700 kPa to 550 kPa, from Eq. 23.52,

$$s_2 - s_1 = \phi_2 - \phi_1 - R \ln\left(\frac{p_2}{p_1}\right)$$

Therefore, the maximum total work due to the pressure drop is

$$\begin{aligned} W_{\text{max},p \text{ loss}} &= h_1 - h_2 + T_L\left(\phi_2 - \phi_1 - R \ln\left(\frac{p_2}{p_1}\right)\right) \\ &= 557.9 \ \frac{\text{kJ}}{\text{kg}} - 1176.2 \ \frac{\text{kJ}}{\text{kg}} \\ &\quad + (313\text{K}) \\ &\quad \times \left(\begin{array}{l} 3.09092 \ \frac{\text{kJ}}{\text{kg·K}} - 2.32372 \ \frac{\text{kJ}}{\text{kg·K}} \\ - \left(\frac{287 \ \frac{\text{J}}{\text{kg·K}}}{1000 \ \frac{\text{J}}{\text{kJ}}}\right) \ln\left(\frac{550 \text{ kPa}}{700 \text{ kPa}}\right) \end{array}\right) \\ &= \boxed{-356.50 \text{ kJ/kg} \quad (-360 \text{ kJ/kg})} \end{aligned}$$

The answer is (D).

(c) The percentage loss in available energy is

$$\begin{aligned} &\frac{W_{\text{max}} - W_{\text{max},p \text{ loss}}}{W_{\text{max}}} \times 100\% \\ &= \frac{-378.17 \ \frac{\text{kJ}}{\text{kg}} - \left(-356.50 \ \frac{\text{kJ}}{\text{kg}}\right)}{-378.17 \ \frac{\text{kJ}}{\text{kg}}} \times 100\% \\ &= \boxed{5.73\% \quad (5.0\%)} \end{aligned}$$

The answer is (A).

11. *Customary U.S. Solution*

(a) The absolute temperature is

$$T = 70°\text{F} + 460° = 530°\text{R}$$

From Table 23.7, $R = 11.77$ ft-lbf/lbm-°R. From Eq. 23.60,

$$\begin{aligned} m = \frac{pV}{RT} &= \frac{\left(20 \ \frac{\text{lbf}}{\text{in}^2}\right)\left(12 \ \frac{\text{in}}{\text{ft}}\right)^2 (100 \text{ ft}^3)}{\left(11.77 \ \frac{\text{ft-lbf}}{\text{lbm-°R}}\right)(530°\text{R})} \\ &= \boxed{46.17 \text{ lbm} \quad (46 \text{ lbm})} \end{aligned}$$

The answer is (C).

(b) From Table 23.4, the critical temperature and pressure of xenon are 521.9°R and 58.2 atm, respectively. The reduced variables are

$$T_r = \frac{T}{T_c} = \frac{530°\text{R}}{521.9°\text{R}} = 1.02$$

$$p_r = \frac{p}{p_c} = \frac{3800 \text{ psia}}{(58.2 \text{ atm})\left(14.7 \ \frac{\text{psia}}{\text{atm}}\right)} = 4.44$$

From App. 23.Z, Z is read as 0.61. Using Eq. 23.106,

$$\begin{aligned} m = \frac{pV}{ZRT} &= \frac{\left(3800 \ \frac{\text{lbf}}{\text{in}^2}\right)\left(12 \ \frac{\text{in}}{\text{ft}}\right)^2 (100 \text{ ft}^3)}{(0.61)\left(11.77 \ \frac{\text{ft-lbf}}{\text{lbm-°R}}\right)(530°\text{R})} \\ &= 14{,}380 \text{ lbm} \end{aligned}$$

The average mass flow rate of xenon is

$$\begin{aligned} \dot{m} &= \frac{14{,}380 \text{ lbm} - 46.17 \text{ lbm}}{1 \text{ hr}} \\ &= \boxed{14{,}334 \text{ lbm/hr} \quad (14{,}000 \text{ lbm/hr})} \end{aligned}$$

The answer is (D).

(c) For isothermal compression, the work is calculated from Eq. 24.93.

$$W = mRT\ln\left(\frac{p_1}{p_2}\right)$$

$$= \frac{(14{,}334\ \text{lbm})\left(11.77\ \frac{\text{ft-lbf}}{\text{lbm-°R}}\right)\times(530°\text{R})\ln\left(\frac{20\ \text{psia}}{3800\ \text{psia}}\right)}{\left(778\ \frac{\text{ft-lbf}}{\text{Btu}}\right)\left(3413\ \frac{\text{Btu}}{\text{kW-hr}}\right)}$$

$$= -176.7\ \text{kW-hr}\quad(\text{for 1 hr})$$

The cost of electricity is

$$\left(\frac{\$0.045}{\text{kW-hr}}\right)(176.7\ \text{kW-hr}) = \boxed{\$7.95\quad(\$8.00)}$$

The answer is (A).

SI Solution

(a) The absolute temperature is

$$T = 21°\text{C} + 273° = 294\text{K}$$

From Table 23.7, $R = 63.32$ J/kg·K. From Eq. 24.60,

$$m = \frac{pV}{RT} = \frac{(150\ \text{kPa})\left(1000\ \frac{\text{Pa}}{\text{kPa}}\right)(3\ \text{m}^3)}{\left(63.32\ \frac{\text{J}}{\text{kg·K}}\right)(294\text{K})}$$

$$= \boxed{24.17\ \text{kg}\quad(24\ \text{kg})}$$

The answer is (C).

(b) From Table 23.4, the critical temperature and pressure of xenon are 289.9K and 58.2 atm, respectively. The reduced variables are

$$T_r = \frac{T}{T_c} = \frac{294\text{K}}{289.9\text{K}} = 1.01$$

$$p_r = \frac{p}{p_c} = \frac{25\ \text{MPa}}{(58.2\ \text{atm})\left(0.1013\ \frac{\text{MPa}}{\text{atm}}\right)} = 4.24$$

From App. 23.Z, Z is read as 0.59. Using Eq. 23.106,

$$m = \frac{pV}{ZRT} = \frac{(25\ \text{MPa})\left(10^6\ \frac{\text{Pa}}{\text{MPa}}\right)(3\ \text{m}^3)}{(0.59)\left(63.32\ \frac{\text{J}}{\text{kg·K}}\right)(294\text{K})}$$

$$= 6828\ \text{kg}$$

The average mass flow rate of xenon is

$$\dot{m} = \frac{6828\ \text{kg} - 24.17\ \text{kg}}{(1\ \text{h})\left(3600\ \frac{\text{s}}{\text{h}}\right)} = \boxed{1.89\ \text{kg/s}\quad(1.9\ \text{kg/s})}$$

The answer is (D).

(c) For isothermal compression, the work is calculated from Eq. 24.93.

$$W = mRT\ln\left(\frac{p_1}{p_2}\right)$$

$$= \frac{(6828\ \text{kg} - 24.17\ \text{kg})\left(63.32\ \frac{\text{J}}{\text{kg·K}}\right)(294\text{K})\times\ln\left(\frac{150\ \text{kPa}}{(25\ \text{MPa})\left(1000\ \frac{\text{kPa}}{\text{MPa}}\right)}\right)}{\left(3600\ \frac{\text{s}}{\text{h}}\right)\left(1000\ \frac{\text{J}}{\text{kJ}}\right)}$$

$$= -178.0\ \text{kW·h}$$

The cost of electricity is

$$\left(\frac{\$0.045}{\text{kW·h}}\right)(178.0\ \text{kW·h}) = \boxed{\$8.01\quad(\$8.00)}$$

The answer is (A).

12. Choose the control volume to include the air outside the tank that is pushed into the tank (subscript "e" for "entering"), as well as the tank volume.

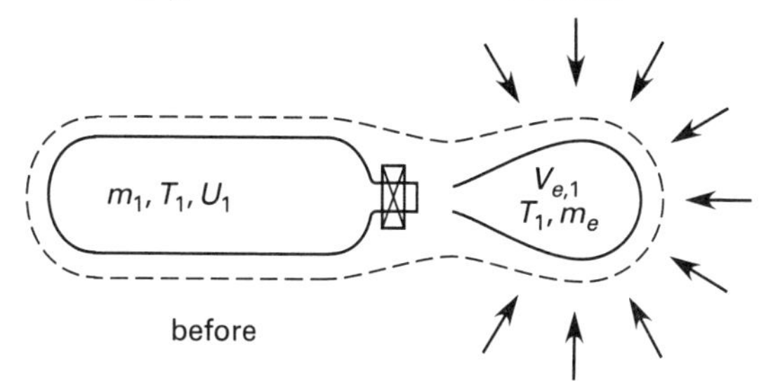

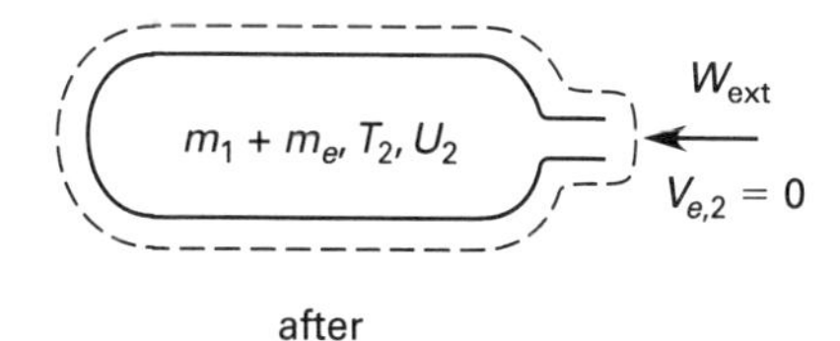

Customary U.S. Solution

(a) The absolute temperature of the air in the tank when evacuated is

$$T_1 = 70°\text{F} + 460° = 530°\text{R}$$

From Table 23.7, $R = 53.3$ ft-lbf/lbm-°R. From Eq. 24.60,

$$m = \frac{p_1V_1}{RT_1} = \frac{\left(1\ \frac{\text{lbf}}{\text{in}^2}\right)\left(12\ \frac{\text{in}}{\text{ft}}\right)^2(20\ \text{ft}^3)}{\left(53.3\ \frac{\text{ft-lbf}}{\text{lbm-°R}}\right)(530°\text{R})}$$

$$= 0.102\ \text{lbm}$$

Assume $T_2 = 300°F$. The absolute temperature is

$$T_2 = 300°F + 460° = 760°R$$

From Eq. 24.60,

$$m_2 = m_1 + m_e = \frac{p_2 V_2}{RT_2}$$

$$= \frac{\left(14.7\ \frac{\text{lbf}}{\text{in}^2}\right)\left(12\ \frac{\text{in}}{\text{ft}}\right)^2 (20\ \text{ft}^3)}{\left(53.3\ \frac{\text{ft-lbf}}{\text{lbm-°R}}\right)(760°R)}$$

$$m_1 + m_e \approx 1.045\ \text{lbm}$$

$$m_e \approx 1.045\ \text{lbm} - m_1$$

$$= 1.045\ \text{lbm} - 0.102\ \text{lbm}$$

$$= 0.943\ \text{lbm}$$

From Eq. 24.60, the initial volume of the external air is

$$V_{e,1} = \frac{mRT}{p}$$

$$= \frac{(0.943\ \text{lbm})\left(53.3\ \frac{\text{ft-lbf}}{\text{lbm-°R}}\right)(530°R)}{\left(14.7\ \frac{\text{lbf}}{\text{in}^2}\right)\left(12\ \frac{\text{in}}{\text{ft}}\right)^2}$$

$$= \boxed{12.58\ \text{ft}^3 \quad (13\ \text{ft}^3)}$$

The answer is (B).

(b) For a constant pressure, closed system, from Eq. 24.64, the total work is

$$W_{\text{ext}} = p(V_{e,2} - V_{e,1})$$

$$= \frac{\left(14.7\ \frac{\text{lbf}}{\text{in}^2}\right)\left(12\ \frac{\text{in}}{\text{ft}}\right)^2 (0\ \text{ft} - 12.58\ \text{ft}^3)}{778\ \frac{\text{ft-lbf}}{\text{Btu}}}$$

$$= \boxed{-34.23\ \text{Btu} \quad (-34\ \text{Btu})} \quad \left[\begin{array}{c}\text{surroundings do}\\ \text{work on the system}\end{array}\right]$$

The answer is (B).

(c) The energy in part (b) is used to raise the temperature of the air and tank. Consider air as an ideal gas. The process inside the tank is not constant-pressure, as was the compression of the external air mass. However, the process is adiabatic, so $\Delta U = -W$. From Eq. 23.61, $\Delta U = c_v \Delta T$.

$$W_{\text{ext}} = \left((m_1 + m_e)c_v\right)(T_2 - T_1)$$

$$34.23\ \text{Btu} = \left((1.045\ \text{lbm})\left(0.171\ \frac{\text{Btu}}{\text{lbm-°F}}\right)\right)(T_2 - 70°F)$$

$$T_2 = 261.6°F$$

(Since $T_2 = 300°F$ was assumed in part (a), a second iteration may be required.)

Now, allow the 261.6°F air and the 70°F tank to reach thermal equilibrium.

$$q_{\text{air}} = q_{\text{tank}} \quad \text{[adiabatic]}$$

$$c_{p,\text{air}} m_{\text{air}} (T_{\text{air}} - T_{\text{equilibrium}}) = c_{p,\text{tank}} m_{\text{tank}} (T_{\text{equilibrium}} - T_1)$$

$$\left(0.24\ \frac{\text{Btu}}{\text{lbm-°F}}\right)(1.045\ \text{lbm})(261.6°F - T_{\text{equilibrium}}) = \left(0.11\ \frac{\text{Btu}}{\text{lbm-°F}}\right)(40\ \text{lbm})(T_{\text{equilibrium}} - 70°F)$$

$$T_{\text{equilibrium}} = \boxed{80.3°F \quad (80°F)}$$

If a second iteration is performed starting with $T_2 = 261.6°F$, the following values are obtained:

$$T_2 = 722°R$$

$$m_e = 0.998\ \text{lbm}$$

$$V_{e,1} = 13.32\ \text{ft}^3$$

$$W_{\text{ext}} = -36.24\ \text{Btu}$$

$$T_2 = 262.6°F$$

$$T_{\text{equilibrium}} = 80.9°F$$

The answer is (B).

SI Solution

(a) The absolute temperature of the air in the tank when evacuated is

$$T_1 = 21°C + 273° = 294K$$

From Table 23.7, $R = 287$ J/kg·K. From Eq. 24.60,

$$m = \frac{p_1 V_1}{RT_1} = \frac{(7\ \text{kPa})\left(1000\ \frac{\text{Pa}}{\text{kPa}}\right)(0.6\ \text{m}^3)}{\left(287\ \frac{\text{J}}{\text{kg·K}}\right)(294K)} = 0.0498\ \text{kg}$$

Assume $T_2 = 127°C$. The absolute temperature is

$$T_2 = 127°C + 273° = 400K$$

From Eq. 24.60,

$$m_2 = m_1 + m_e = \frac{p_2 V_2}{RT_2}$$

$$= \frac{(101.3\ \text{kPa})\left(1000\ \frac{\text{Pa}}{\text{kPa}}\right)(0.6\ \text{m}^3)}{\left(287\ \frac{\text{J}}{\text{kg·K}}\right)(400K)}$$

$$= 0.5294\ \text{kg}$$

$$m_e \approx 0.5294\ \text{kg} - m_1$$

$$= 0.5294\ \text{kg} - 0.0498\ \text{kg}$$

$$= 0.4796\ \text{kg}$$

From Eq. 24.60, the initial volume of the external air is

$$V_{e,1} = \frac{mRT}{p} = \frac{(0.4796\ \text{kg})\left(287\ \frac{\text{J}}{\text{kg·K}}\right)(294\text{K})}{(101.3\ \text{kPa})\left(1000\ \frac{\text{Pa}}{\text{kPa}}\right)}$$

$$= \boxed{0.3995\ \text{m}^3 \quad (0.40\ \text{m}^3)}$$

The answer is (B).

(b) For a constant pressure, closed system, from Eq. 24.64, the total work is

$$W_{\text{ext}} = p(V_{e,2} - V_{e,1})$$

$$= (101.3\ \text{kPa})\left(1000\ \frac{\text{Pa}}{\text{kPa}}\right)(0\ \text{m}^3 - 0.3995\ \text{m}^3)$$

$$= \boxed{-40\,469\ \text{J} \quad (-40\,000\ \text{J})} \quad \left[\begin{array}{c}\text{surroundings do}\\ \text{work on the system}\end{array}\right]$$

The answer is (B).

(c) The energy in part (b) is used to raise the temperature of the air and tank. Consider air as an ideal gas. The process inside the tank is not constant-pressure, as was the compression of the external air mass. However, the process is adiabatic, so $\Delta U = -W$. From Eq. 24.61, $\Delta U = c_v \Delta T$.

$$W_{\text{ext}} = \big((m_1 + m_e)c_v\big)(T_2 - T_1)$$

$$40\,469\ \text{J} = \left((0.5294\ \text{kg})\left(718\ \frac{\text{J}}{\text{kg·K}}\right)\right)(T_2 - 21°\text{C})$$

$$T_2 = 127.5°\text{C}$$

This is close enough to the assumed value of T_2 that a second iteration is not necessary.

Now, allow the 127.5°C air and the 21°C tank to reach thermal equilibrium.

$$q_{\text{air}} = q_{\text{tank}} \quad [\text{adiabatic}]$$

$$c_{p,\text{air}} m_{\text{air}}(T_{\text{air}} - T_{\text{equilibrium}}) = c_{p,\text{tank}} m_{\text{tank}}(T_{\text{equilibrium}} - T_1)$$

$$\left(1005\ \frac{\text{J}}{\text{kg·K}}\right)(0.5294\ \text{kg})(127.5°\text{C} - T_{\text{equilibrium}}) = \left(460\ \frac{\text{J}}{\text{kg·K}}\right)(20\ \text{kg})(T_{\text{equilibrium}} - 21°\text{C})$$

$$T_{\text{equilibrium}} = \boxed{26.8°\text{C} \quad (30°\text{C})}$$

The answer is (B).

25 Compressible Fluid Dynamics

PRACTICE PROBLEMS

1. 150°F, 10 psia (65°C, 70 kPa) air flows at 750 ft/sec (225 m/s) through a converging section into a chamber whose back pressure is 5.5 psia (40 kPa).

(a) The entrance Mach number is most nearly

(A) 0.5

(B) 0.6

(C) 0.7

(D) 0.8

(b) The throat static pressure is most nearly

(A) 6.9 psia (48 kPa)

(B) 10 psia (69 kPa)

(C) 14 psia (98 kPa)

(D) 19 psia (130 kPa)

(c) The throat static temperature is most nearly

(A) 450°R (250K)

(B) 500°R (280K)

(C) 550°R (300K)

(D) 600°R (330K)

2. A round-nosed bullet travels at 2000 ft/sec (600 m/s) through 32°F, 14.7 psia (0°C, 101.3 kPa) air.

(a) Is the bullet speed subsonic or supersonic?

(A) subsonic (M = 0.9)

(B) supersonic (M = 1.1)

(C) supersonic (M = 1.4)

(D) supersonic (M = 1.8)

(b) The total pressure at the tip of the bullet is most nearly

(A) 50 psia (350 kPa)

(B) 70 psia (480 kPa)

(C) 90 psia (620 kPa)

(D) 110 psia (750 kPa)

(c) The total temperature at the tip of the bullet is most nearly

(A) 770°R (430K)

(B) 800°R (440K)

(C) 830°R (450K)

(D) 860°R (480K)

(d) The total enthalpy at the bullet face is most nearly

(A) 200 Btu/lbm (450 kJ/kg)

(B) 250 Btu/lbm (560 kJ/kg)

(C) 300 Btu/lbm (680 kJ/kg)

(D) 350 Btu/lbm (790 kJ/kg)

3. 4.5 lbm/sec (2 kg/s) of air with total properties of 160 psia and 240°F (1.1 MPa and 120°C) expands through a converging-diverging nozzle to 20 psia (140 kPa).

(a) The throat Mach number is most nearly

(A) 0.7

(B) 0.8

(C) 0.9

(D) 1.0

(b) The throat area is most nearly

(A) 0.005 ft^2 (0.0005 m^2)

(B) 0.01 ft^2 (0.0009 m^2)

(C) 0.02 ft^2 (0.002 m^2)

(D) 0.04 ft^2 (0.004 m^2)

(c) The exit Mach number is most nearly

(A) 1.4

(B) 1.7

(C) 2.0

(D) 2.6

(d) The exit area is most nearly

(A) 0.017 ft^2 (0.0015 m^2)

(B) 0.022 ft^2 (0.0019 m^2)

(C) 0.038 ft^2 (0.0033 m^2)

(D) 0.062 ft^2 (0.0054 m^2)

4. A wedge-shaped leading edge with a semivertex angle of 20° travels at 2700 ft/sec (800 m/s) through 60°F, 14.7 psia (16°C, 101.3 kPa) air. The semivertex shock angles are most nearly

(A) 26° (weak); 50° (strong)

(B) 33° (weak); 60° (strong)

(C) 39° (weak); 70° (strong)

(D) 46° (weak); 80° (strong)

5. A large tank contains 100 psia, 80°F (0.7 MPa, 27°C) air. The tank feeds a converging-diverging nozzle with a throat area of 1 in^2 (6.45×10^{-4} m^2). At a particular point in the nozzle, the Mach number is 2.

(a) The temperature at that point is most nearly

(A) 250°R (140K)

(B) 300°R (170K)

(C) 350°R (190K)

(D) 400°R (220K)

(b) The area at that point is most nearly

(A) 1.2 in^2 (0.0008 m^2)

(B) 1.7 in^2 (0.0011 m^2)

(C) 2.3 in^2 (0.0015 m^2)

(D) 2.8 in^2 (0.0018 m^2)

(c) The mass flow rate at that point is most nearly

(A) 2.3 lbm/sec (1.1 kg/s)

(B) 2.7 lbm/sec (1.2 kg/s)

(C) 3.1 lbm/sec (1.4 kg/s)

(D) 3.6 lbm/sec (1.6 kg/s)

6. The Mach number before a shock wave is 2. The temperature before the shock wave is 500°R (280K). The air velocity behind the shock wave is most nearly

(A) 730 ft/sec (220 m/s)

(B) 820 ft/sec (250 m/s)

(C) 900 ft/sec (270 m/s)

(D) 940 ft/sec (280 m/s)

7. Air with a total pressure of 100 psia (0.7 MPa) and a total temperature of 70°F (21°C) flows through a converging-diverging nozzle. Supersonic flow is achieved in the diverging section. At a particular point in the diverging section, the ratio of flow area to the critical throat area is 1.555.

(a) The Mach number at that point is most nearly

(A) 1.1

(B) 1.3

(C) 1.9

(D) 2.4

(b) The temperature at that point is most nearly

(A) 290°R (160K)

(B) 310°R (170K)

(C) 340°R (190K)

(D) 360°R (200K)

(c) The pressure at that point is most nearly

(A) 15 psia (0.10 MPa)

(B) 20 psia (0.14 MPa)

(C) 25 psia (0.18 MPa)

(D) 30 psia (0.21 MPa)

8. 20 lbm/sec (9 kg/s) of air with a static pressure of 10 psia (70 kPa) and a total temperature of 40°F (4°C) flow through a 1 ft^2 (0.09 m^2) round duct. The smallest area to which the duct can be reduced is most nearly

(A) 0.26 ft^2 (0.023 m^2)

(B) 0.54 ft^2 (0.049 m^2)

(C) 1.0 ft^2 (0.091 m^2)

(D) 1.3 ft^2 (0.17 m^2)

9. The static pressure in a supersonic wind tunnel is 1.38 psia (9.51 kPa). A pitot tube records a total pressure of 20 psia (140 kPa) behind the shock wave that forms at its entrance. The Mach number in the tunnel is most nearly

(A) 1.9

(B) 2.2

(C) 2.7

(D) 3.3

10. A boiler produces 100 psia, 800°F (0.7 MPa, 425°C) steam. An isentropic nozzle expands the steam to 60 psia (400 kPa). The steam velocity is most nearly

(A) 1100 ft/sec (330 m/s)

(B) 1500 ft/sec (450 m/s)

(C) 1800 ft/sec (580 m/s)

(D) 2400 ft/sec (720 m/s)

11. Air flows through a converging section. At a particular point where the flow area is 0.1 ft^2 (0.0009 m^2), the static pressure is 50 psia (350 kPa), the static temperature is 1000°R (560K), and the velocity is 600 ft/sec (180 m/s).

(a) The Mach number is most nearly

(A) 0.39

(B) 0.46

(C) 0.66

(D) 0.78

(b) The total temperature is most nearly

(A) 700°R (380K)

(B) 800°R (470K)

(C) 900°R (500K)

(D) 1000°R (580K)

(c) The total pressure is most nearly

(A) 53 psia (370 kPa)

(B) 56 psia (390 kPa)

(C) 65 psia (450 kPa)

(D) 82 psia (570 kPa)

(d) The critical exit area is most nearly

(A) 0.062 ft^2 (0.00054 m^2)

(B) 0.080 ft^2 (0.00072 m^2)

(C) 0.093 ft^2 (0.00085 m^2)

(D) 0.110 ft^2 (0.0010 m^2)

(e) The critical pressure is most nearly

(A) 23 psia (160 kPa)

(B) 29 psia (200 kPa)

(C) 38 psia (260 kPa)

(D) 63 psia (440 kPa)

(f) The critical temperature is most nearly

(A) 790°R (440K)

(B) 860°R (480K)

(C) 910°R (510K)

(D) 980°R (540K)

12. 3 lbm/sec (1.4 kg/s) of steam ($k = 1.31$) enters an 85% efficient transonic nozzle at 300 ft/sec (90 m/s). The steam expands from 200 psia and 600°F (1.5 MPa and 300°C) to 80 psia (500 kPa). The throat area is most nearly

(A) 0.0067 ft^2 (0.00060 m^2)

(B) 0.0081 ft^2 (0.00072 m^2)

(C) 0.010 ft^2 (0.00090 m^2)

(D) 0.013 ft^2 (0.0012 m^2)

13. 35,200 lbm/hr (4.4 kg/s) of steam ($k = 1.29$) flows through 95 ft (30 m) of 3 in (7.6 cm) inside diameter pipe. The pipe is insulated to prevent heat loss. The Fanning friction factor is 0.012. The average steam pressure along the pipe length is 200 psia (1.5 MPa). At the exit, the steam is at 115 psia and 540°F (0.7 MPa and 300°C). 30 ft (9 m) of pipe is subsequently added to the end of the pipe.

(a) The velocity 95 ft (30 m) from the entrance is most nearly

(A) 240 ft/sec (82 m/s)

(B) 1000 ft/sec (360 m/s)

(C) 2800 ft/sec (1000 m/s)

(D) 8000 ft/sec (3000 m/s)

(b) The sonic velocity 95 ft (30 m) from the entrance is most nearly

(A) 1300 ft/sec (400 m/s)

(B) 1400 ft/sec (420 m/s)

(C) 1900 ft/sec (580 m/s)

(D) 2000 ft/sec (600 m/s)

(c) The Mach number 95 ft (30 m) from the entrance is most nearly

(A) 0.55 (0.62)

(B) 0.77 (0.85)

(C) 0.80 (0.90)

(D) 1.50 (1.70)

(d) If 30 ft (9 m) of pipe is added to the original 95 ft (30 m), at what approximate distance from the original end of the pipe will the flow become choked?

(A) 3.0 ft (0.55 m)

(B) 3.3 ft (0.60 m)

(C) 3.8 ft (0.70 m)

(D) 4.1 ft (0.75 m)

(e) Will the steam flow be maintained after the additional 30 ft (9 m) of pipe is added?

(A) The flow will be choked in less than 30 ft (9 m); steam flow will not be maintained.

(B) The flow will be clear in less than 30 ft (9 m); steam flow will be maintained.

(C) The flow will be choked in more than 30 ft (9 m); steam flow will not be maintained.

(D) The flow will be clear in more than 30 ft (9 m); steam flow will be maintained.

14. 3600 lbm/hr (0.45 kg/s) of air flow through an adiabatic converging-diverging nozzle. The air enters with total properties of 160 psia and 660°R (1.1 MPa and 370K). The exit pressure is 14.7 psia (101.3 kPa). The ratio of specific heats for the air is 1.4 throughout the nozzle. The nozzle has a coefficient of discharge of 0.90. The total diverging angle is 6°. The length of the converging section is 5% of the diverging section.

(a) The sonic velocity at the throat is most nearly

(A) 1200 ft/sec (350 m/s)

(B) 1900 ft/sec (600 m/s)

(C) 3400 ft/sec (1000 m/s)

(D) 4000 ft/sec (1200 m/s)

(b) The sonic density at the throat is most nearly

(A) 0.40 lbm/ft^3 (6.3 kg/m^3)

(B) 0.42 lbm/ft^3 (6.6 kg/m^3)

(C) 0.43 lbm/ft^3 (6.8 kg/m^3)

(D) 0.44 lbm/ft^3 (7.0 kg/m^3)

(c) The throat area is most nearly

(A) 0.0014 ft^2 (0.00013 m^2)

(B) 0.0022 ft^2 (0.00020 m^2)

(C) 0.0023 ft^2 (0.00022 m^2)

(D) 0.0024 ft^2 (0.00023 m^2)

(d) The longitudinal distance from the nozzle's entrance to the throat is most nearly

(A) 0.0108 ft (0.00323 m)

(B) 0.0112 ft (0.00334 m)

(C) 0.0114 ft (0.00342 m)

(D) 0.0182 ft (0.00543 m)

15. The initial temperature and pressure of air inside a high-pressure reservoir are 68°F and 470 lbf/in^2, respectively. When a valve is suddenly opened, the air discharges through a simple square-edged orifice with an initial effective velocity of 875 ft/sec. The valve is closed when the reservoir pressure is reduced to 100 lbf/in^2. The orifice's area is 0.05 in^2. The back-pressure is 14.7 lbf/in^2 and is constant. The reservoir has a total volume of 100 ft^3, and the discharged air is not replenished. The discharge is adiabatic. The air's ratio of specific heats is 1.4 and remains constant.

(a) What is most nearly the initial static temperature of the air flowing through the orifice?

(A) 390°R

(B) 440°R

(C) 480°R

(D) 530°R

(b) What is most nearly the initial sonic velocity in the throat?

(A) 950 ft/sec

(B) 990 ft/sec

(C) 1000 ft/sec

(D) 1100 ft/sec

(c) What is most nearly the orifice coefficient?

(A) 0.53

(B) 0.62

(C) 0.69

(D) 0.85

(d) What is most nearly the initial discharge rate?

(A) 0.46 lbm/sec

(B) 0.69 lbm/sec

(C) 0.85 lbm/sec

(D) 1.2 lbm/sec

(e) The total mass of air discharged from the tank is most nearly

(A) 50 lbm

(B) 120 lbm

(C) 190 lbm

(D) 240 lbm

(f) Approximately how much time will elapse before the reservoir's internal pressure is reduced to 100 lbf/in^2 (690 kPa)?

(A) 8.3 min

(B) 10 min

(C) 12 min

(D) 16 min

SOLUTIONS

1. *Customary U.S. Solution*

(a) Draw the air flow converging into the chamber.

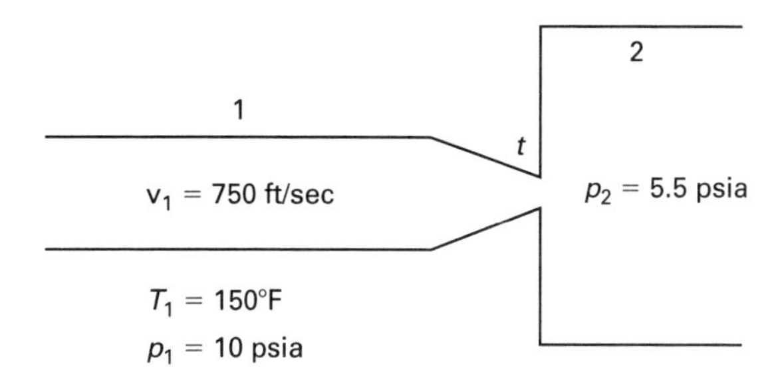

$$T_1 = 150°\text{F} + 460° = 610°\text{R}$$

The entrance Mach number is given by Eq. 25.4(b) as

$$\begin{aligned} M_1 &= \frac{v_1}{\sqrt{\frac{k g_c R^* T_1}{MW}}} \\ &= \frac{750\ \frac{\text{ft}}{\text{sec}}}{\sqrt{\frac{(1.4)\left(32.2\ \frac{\text{lbm-ft}}{\text{lbf-sec}^2}\right)\left(1545\ \frac{\text{ft-lbf}}{\text{lbmol-°R}}\right)(610°\text{R})}{29.0\ \frac{\text{lbm}}{\text{lbmol}}}}} \\ &= \boxed{0.62 \quad (0.6)} \end{aligned}$$

The answer is (B).

(b) From the M = 0.62 line in App. 25.A, the following factors can be read.

$$\frac{T_1}{T_0} = 0.9286$$

$$\frac{p_1}{p_0} = 0.7716$$

$$T_0 = \frac{T_1}{0.9286} = \frac{610°\text{R}}{0.9286} = 656.9°\text{R}$$

$$p_0 = \frac{p_1}{0.7716} = \frac{10\ \text{psia}}{0.7716} = 12.96\ \text{psia}$$

$$\frac{p_2}{p_0} = \frac{5.5\ \text{psia}}{12.96\ \text{psia}} = 0.424$$

Thermodynamics

The critical pressure ratio for air is 0.5283. Since $p_2/p_0 < 0.5283$, the flow is choked and the Mach number at the throat is $M_t = 1$. From App. 25.A, for M = 1,

$$\frac{p}{p_0} = 0.5283$$

$$\frac{T}{T_0} = 0.8333$$

$$p = (0.5283)p_0 = (0.5283)(12.96 \text{ psia}) = \boxed{6.85 \text{ psia} \quad (6.9 \text{ psia})}$$

The answer is (A).

(c) The throat static pressure is

$$T = (0.8333)T_0 = (0.8333)(656.9°\text{R}) = \boxed{547.4°\text{R} \quad (550°\text{R})}$$

The answer is (C).

SI Solution

(a) Draw the air flow converging into the chamber.

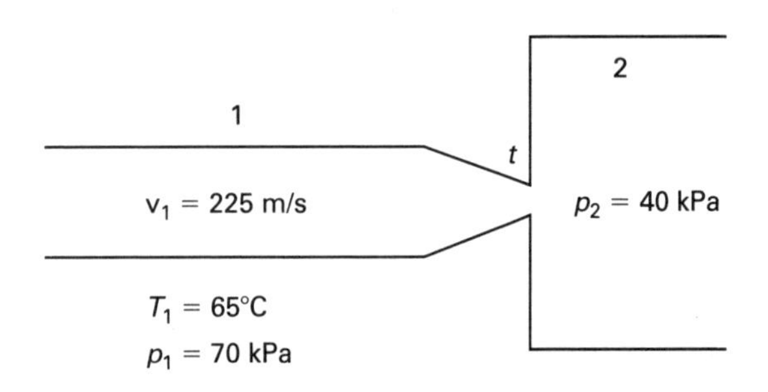

$$T_1 = 65°\text{C} + 273° = 338\text{K}$$

The entrance Mach number is given by Eq. 25.4(a) as

$$M_1 = \frac{v_1}{\sqrt{\frac{kR^* T_1}{MW}}} = \frac{225 \ \frac{\text{m}}{\text{s}}}{\sqrt{\frac{(1.4)\left(8314 \ \frac{\text{J}}{\text{kmol·K}}\right)(338\text{K})}{29.0 \ \frac{\text{kg}}{\text{kmol}}}}} = \boxed{0.61 \quad (0.6)}$$

The answer is (B).

(b) From the M = 0.61 line in App. 25.A, the following factors can be read.

$$\frac{T_1}{T_0} = 0.9307$$

$$\frac{p_1}{p_0} = 0.7778$$

$$T_0 = \frac{T_1}{0.9307} = \frac{338\text{K}}{0.9307} = 363.2\text{K}$$

$$p_0 = \frac{p_1}{0.7778} = \frac{70 \text{ kPa}}{0.7778} = 90.0 \text{ kPa}$$

$$\frac{p_2}{p_0} = \frac{40 \text{ kPa}}{90.0 \text{ kPa}} = 0.444$$

The critical pressure ratio for air is 0.5283. Since $p_2/p_0 < 0.5283$, the flow is choked and the Mach number at the throat is $M_t = 1$. From App. 25.A, for M = 1,

$$\frac{p}{p_0} = 0.5283$$

$$\frac{T}{T_0} = 0.8333$$

$$p = (0.5283)p_0 = (0.5283)(90.0 \text{ kPa}) = \boxed{47.5 \text{ kPa} \quad (48 \text{ kPa})}$$

The answer is (A).

(c) The throat static temperature is

$$T = (0.8333)T_0 = (0.8333)(363.2\text{K}) = \boxed{302.8\text{K} \quad (300\text{K})}$$

The answer is (C).

2. *Customary U.S. Solution*

The shock wave is normal to the direction of flight, so normal shock tables can be used.

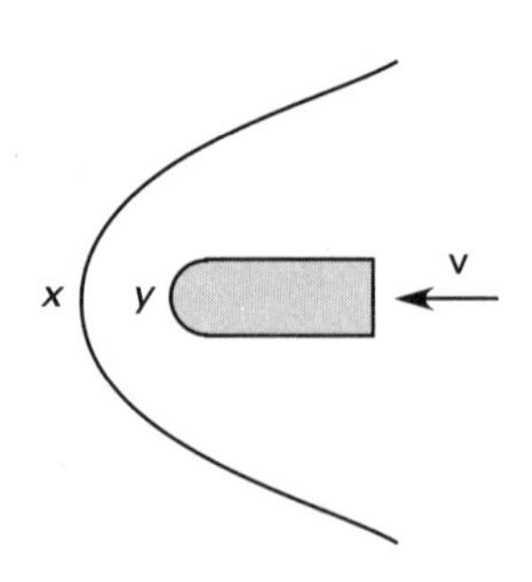

$$T_x = 32°\text{F} + 460° = 492°\text{R}$$

$$p_x = 14.7 \text{ psia}$$

(a) Using Eq. 25.4(b), the Mach number at y is

$$M_y = \frac{v}{\sqrt{\frac{k g_c R^* T}{MW}}} = \frac{2000\ \frac{ft}{sec}}{\sqrt{\frac{(1.4)\left(32.2\ \frac{lbm\text{-}ft}{lbf\text{-}sec^2}\right)\left(1545\ \frac{ft\text{-}lbf}{lbmol\text{-}°R}\right)(492°R)}{29.0\ \frac{lbm}{lbmol}}}} = \boxed{1.84\quad(1.8)}$$

The bullet speed is $\boxed{\text{supersonic.}}$

The answer is (D).

(b) Assume that the bullet is stationary and air is moving at M = 1.84.

The ratio of static pressure before the shock to the total pressure after the shock is read from the normal shock table (see App. 25.B) for $M_x = 1.84$ (interpolation).

$$\frac{p_x}{p_{0,y}} = 0.2060$$

Therefore, the total pressure is

$$p_{0,y} = \frac{p_x}{0.2060} = \frac{14.7\ \text{psia}}{0.2060} = \boxed{71.36\ \text{psia}\quad(70\ \text{psia})}$$

The answer is (B).

(c) The ratio of static temperature before the shock to the total temperature after the shock is read from the normal shock table (see App. 25.B) for $M_x = 1.84$ (interpolation).

$$\frac{T_x}{T_0} = 0.5963$$

Therefore, the total temperature is

$$T_0 = \frac{T_x}{0.5963} = \frac{492°R}{0.5963} = \boxed{825.1°R\quad(830°R)}$$

Since the shock wave is adiabatic, T_0 remains constant.

The answer is (C).

(d) For 825.1°R air, the enthalpy at the bullet face from the air tables (App. 23.F) is $h \approx \boxed{197.9\ \text{Btu/lbm}\ (200\ \text{Btu/lbm}).}$

The answer is (A).

SI Solution

The shock wave is normal to the direction of flight, so normal shock tables can be used.

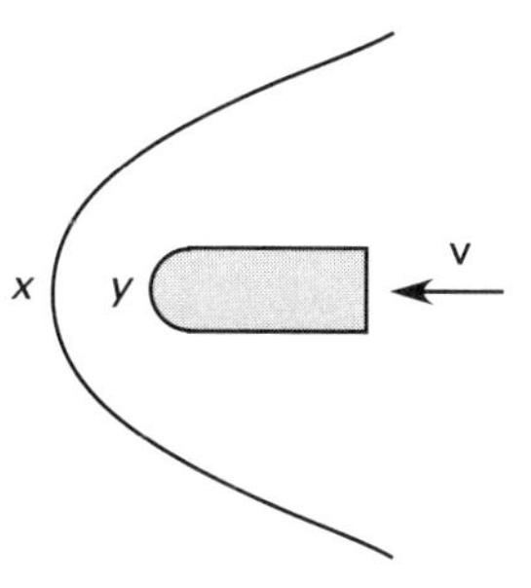

$$T_x = 0°C + 273° = 273K$$

$$p_x = 101.3\ \text{kPa}$$

(a) Using Eq. 25.4(a), the Mach number at y is

$$M_y = \frac{v}{\sqrt{\frac{kR^* T}{MW}}} = \frac{600\ \frac{m}{s}}{\sqrt{\frac{(1.4)\left(8314\ \frac{J}{kmol\cdot K}\right)(273K)}{29.0\ \frac{kg}{kmol}}}} = \boxed{1.81\quad(1.8)}$$

The bullet speed is $\boxed{\text{supersonic.}}$

The answer is (D).

(b) Assume that the bullet is stationary and air is moving at M = 1.81.

The ratio of static pressure before the shock to the total pressure after the shock is read from the normal shock table (see App. 25.B) for $M_x = 1.81$ (interpolation).

$$\frac{p_x}{p_{0,y}} = 0.2122$$

Therefore, the total pressure is

$$p_{0,y} = \frac{p_x}{0.2122} = \frac{101.3\ \text{kPa}}{0.2122} = \boxed{477.4\ \text{kPa}\quad(480\ \text{kPa})}$$

The answer is (B).

(c) The ratio of static temperature before the shock to the total temperature after the shock is read from the normal shock table (see App. 25.B) for $M_x = 1.81$ (interpolation).

$$\frac{T_x}{T_0} = 0.6042$$

Therefore, the total temperature is

$$T_0 = \frac{T_x}{0.6042} = \frac{273\text{K}}{0.6042} = \boxed{451.8\text{K} \quad (450\text{K})}$$

Since the shock wave is adiabatic, T_0 remains constant.

The answer is (C).

(d) For 451.8K air, the enthalpy at the bullet face from the air tables (App. 23.S) is $h \approx \boxed{454.0 \text{ kJ/kg } (450 \text{ kJ/kg}).}$

The answer is (A).

3. *Customary U.S. Solution*

(a) The absolute total temperature is

$$T_0 = 240°\text{F} + 460° = 700°\text{R}$$

The total density of the air is

$$\rho_0 = \frac{p_0}{RT_0} = \frac{\left(160 \ \frac{\text{lbf}}{\text{in}^2}\right)\left(12 \ \frac{\text{in}}{\text{ft}}\right)^2}{\left(53.3 \ \frac{\text{ft-lbf}}{\text{lbm-°R}}\right)(700°\text{R})} = 0.6175 \text{ lbm/ft}^3$$

$$\frac{p_{\text{back}}}{p_0} = \frac{20 \text{ psia}}{160 \text{ psia}} = 0.125$$

Since $p_{\text{back}}/p_0 < 0.5283$, the nozzle is supersonic and the throat flow is sonic. So, $\boxed{\text{M} = 1.0}$ at the throat.

The answer is (D).

(b) Read the property ratios for Mach 1 from the isentropic flow table: $[T/T_0] = 0.8333$ and $[\rho/\rho_0] = 0.6339$. The sonic properties at the throat are

$$T^* = \left[\frac{T}{T_0}\right] T_0 = (0.8333)(700°\text{R}) = 583.3°\text{R}$$

$$\rho^* = \left[\frac{\rho}{\rho_0}\right] \rho_0 = (0.6339)\left(0.6175 \ \frac{\text{lbm}}{\text{ft}^3}\right) = 0.3914 \text{ lbm/ft}^3$$

The sonic velocity at the throat is

$$a^* = \sqrt{kg_cRT^*} = \sqrt{(1.4)\left(32.2 \ \frac{\text{lbm-ft}}{\text{lbf-sec}^2}\right) \times \left(53.3 \ \frac{\text{ft-lbf}}{\text{lbm-°R}}\right)(583.3°\text{R})} = 1183.9 \text{ ft/sec}$$

The throat area is

$$A^* = \frac{\dot{m}}{\rho^* a^*} = \frac{4.5 \ \frac{\text{lbm}}{\text{sec}}}{\left(0.3914 \ \frac{\text{lbm}}{\text{ft}^3}\right)\left(1183.9 \ \frac{\text{ft}}{\text{sec}}\right)} = \boxed{0.00971 \text{ ft}^2 \quad (0.01 \text{ ft}^2)}$$

The answer is (B).

(c) At the exit,

$$\frac{p_e}{p_0} = \frac{20 \text{ psia}}{160 \text{ psia}} = 0.125$$

Searching the $[p/p_0]$ column of the isentropic flow tables for this value gives $\boxed{\text{M} \approx 2.01 \ (2.0).}$

The answer is (C).

(d) The corresponding ratio is $[A/A^*] = 1.7024$.

The exit area is

$$A_e = \left[\frac{A_e}{A^*}\right] A^* = (1.7024)(0.00971 \text{ ft}^2) = \boxed{0.01653 \text{ ft}^2 \quad (0.017 \text{ ft}^2)}$$

The answer is (A).

SI Solution

(a) The absolute total temperature is

$$T_0 = 120°\text{C} + 273° = 393\text{K}$$

The total density of the air is

$$\rho_0 = \frac{p_0}{RT_0} = \frac{(1.1 \text{ MPa})\left(10^6 \ \frac{\text{Pa}}{\text{MPa}}\right)}{\left(287 \ \frac{\text{J}}{\text{kg·K}}\right)(393\text{K})} = 9.753 \text{ kg/m}^3$$

$$\frac{p_{\text{back}}}{p_0} = \frac{140 \text{ kPa}}{(1.1 \text{ MPa})\left(10^3 \ \frac{\text{kPa}}{\text{MPa}}\right)} = 0.127$$

Since $p_{\text{back}}/p_0 < 0.5283$, the nozzle is supersonic and the throat flow is sonic. So, $\boxed{\text{M} = 1.0}$ at the throat.

The answer is (D).

(b) Read the property ratios for Mach 1 from the isentropic flow table: $[T/T_0] = 0.8333$ and $[\rho/\rho_0] = 0.6339$. The sonic properties at the throat are

$$T^* = \left[\frac{T}{T_0}\right] T_0 = (0.8333)(393\text{K})$$
$$= 327.5\text{K}$$
$$\rho^* = \left[\frac{\rho}{\rho_0}\right] \rho_0 = (0.6339)\left(9.753 \ \frac{\text{kg}}{\text{m}^3}\right)$$
$$= 6.182 \ \text{kg/m}^3$$

The sonic velocity at the throat is

$$a^* = \sqrt{kRT^*} = \sqrt{(1.4)\left(287 \ \frac{\text{J}}{\text{kg·K}}\right)(327.5\text{K})}$$
$$= 362.8 \ \text{m/s}$$

The throat area is

$$A^* = \frac{\dot{m}}{\rho^* a^*} = \frac{2 \ \frac{\text{kg}}{\text{s}}}{\left(6.182 \ \frac{\text{kg}}{\text{m}^3}\right)\left(362.8 \ \frac{\text{m}}{\text{s}}\right)}$$
$$= \boxed{8.92 \times 10^{-4} \ \text{m}^2 \quad (0.0009 \ \text{m}^2)}$$

The answer is (B).

(c) At the exit,

$$\frac{p_e}{p_0} = \frac{140 \ \text{kPa}}{(1.1 \ \text{MPa})\left(10^3 \ \frac{\text{kPa}}{\text{MPa}}\right)} = 0.127$$

Searching the $[p/p_0]$ column of the isentropic flow tables for this value gives $\boxed{\text{M} \approx 2.0.}$

The answer is (C).

(d) The corresponding ratio is $[A/A^*] = 1.6875$.

$$A_e = \left[\frac{A_e}{A^*}\right] A^* = (1.6875)(8.92 \times 10^{-4} \ \text{m}^2)$$
$$= \boxed{1.505 \times 10^{-3} \ \text{m}^2 \quad (0.0015 \ \text{m}^2)}$$

The answer is (A).

4. *Customary U.S. Solution*

The absolute total temperature is

$$T_0 = 60°\text{F} + 460° = 520°\text{R}$$

The sonic velocity is

$$a = \sqrt{kg_cRT_0} = \sqrt{\begin{array}{l}(1.4)\left(32.2 \ \frac{\text{lbm-ft}}{\text{lbf-sec}^2}\right) \\ \times \left(53.3 \ \frac{\text{ft-lbf}}{\text{lbm-°R}}\right)(520°\text{R})\end{array}}$$
$$= 1117.8 \ \text{ft/sec}$$

The Mach number is

$$\text{M} = \frac{\text{v}}{a} = \frac{2700 \ \frac{\text{ft}}{\text{sec}}}{1117.8 \ \frac{\text{ft}}{\text{sec}}} = 2.42$$

From Fig. 25.3, $\boxed{\theta \approx 46° \text{ (weak) and } 80° \text{ (strong).}}$

The answer is (D).

SI Solution

The absolute total temperature is

$$T_0 = 16°\text{C} + 273° = 289\text{K}$$

The sonic velocity is

$$a = \sqrt{kRT_0} = \sqrt{(1.4)\left(287 \ \frac{\text{J}}{\text{kg·K}}\right)(289\text{K})}$$
$$= 340.8 \ \text{m/s}$$

The Mach number is

$$\text{M} = \frac{\text{v}}{a} = \frac{800 \ \frac{\text{m}}{\text{s}}}{340.8 \ \frac{\text{m}}{\text{s}}} = 2.35$$

From Fig. 25.3, $\boxed{\theta \approx 46° \text{ (weak) and } 80° \text{ (strong).}}$

The answer is (D).

5. *Customary U.S. Solution*

(a) The absolute total temperature is

$$T_0 = 80°\text{F} + 460° = 540°\text{R}$$

The total pressure is $p_0 = 100$ psia.

From the isentropic flow tables at M = 2, $[p/p_0] = 0.1278$, $[T/T_0] = 0.5556$, and $[A/A^*] = 1.6875$.

The properties at Mach 2 are

$$T = \left[\frac{T}{T_0}\right] T_0 = (0.5556)(540°\text{R}) = \boxed{300°\text{R}}$$
$$p = \left[\frac{p}{p_0}\right] p_0 = (0.1278)(100 \ \text{psia}) = 12.78 \ \text{psia}$$

The answer is (B).

(b) The area at the point is

$$A = \left[\frac{A}{A^*}\right] A^* = (1.6875)(1 \ \text{in}^2)$$
$$= \boxed{1.6875 \ \text{in}^2 \quad (1.7 \ \text{in}^2)}$$

The answer is (B).

(c) Calculate the speed of sound, velocity, and density at the point.

$$a = \sqrt{kg_cRT} = \sqrt{\begin{array}{l}(1.4)\left(32.2\ \frac{\text{lbm-ft}}{\text{lbf-sec}^2}\right)\\ \quad\times\left(53.3\ \frac{\text{ft-lbf}}{\text{lbm-°R}}\right)(300\text{°R})\end{array}}$$
$$= 849\ \text{ft/sec}$$

$$\text{v} = \text{M}a = (2)\left(849\ \frac{\text{ft}}{\text{sec}}\right) = 1698\ \text{ft/sec}$$

$$\rho = \frac{p}{RT} = \frac{\left(12.78\ \frac{\text{lbf}}{\text{in}^2}\right)\left(12\ \frac{\text{in}}{\text{ft}}\right)^2}{\left(53.3\ \frac{\text{ft-lbf}}{\text{lbm-°R}}\right)(300\text{°R})}$$
$$= 0.1151\ \text{lbm/ft}^3$$

The mass flow rate is

$$\dot{m} = \rho A\text{v} = \frac{\left(0.1151\ \frac{\text{lbm}}{\text{ft}^3}\right)(1.6875\ \text{in}^2)\left(1698\ \frac{\text{ft}}{\text{sec}}\right)}{\left(12\ \frac{\text{in}}{\text{ft}}\right)^2}$$
$$= \boxed{2.29\ \text{lbm/sec}\quad(2.3\ \text{lbm/sec})}$$

The answer is (A).

SI Solution

(a) The absolute total temperature is

$$T_0 = 27\text{°C} + 273\text{°} = 300\text{K}$$

The total pressure is

$$p_0 = (0.7\ \text{MPa})\left(10^6\ \frac{\text{Pa}}{\text{MPa}}\right) = 7\times10^5\ \text{Pa}$$

From the isentropic flow tables at M = 2, $[p/p_0] = 0.1278$, $[T/T_0] = 0.5556$, and $[A/A^*] = 1.6875$.

The properties at Mach 2 are

$$T = \left[\frac{T}{T_0}\right]T_0 = (0.5556)(300\text{K})$$
$$= \boxed{166.7\text{K}\quad(170\text{K})}$$

$$p = \left[\frac{p}{p_0}\right]p_0 = (0.1278)(7\times10^5\ \text{Pa})$$
$$= 89\,460\ \text{Pa}$$

The answer is (B).

(b) The area at the point is

$$A = \left[\frac{A}{A^*}\right]A^* = (1.6875)(6.45\times10^{-4}\ \text{m}^2)$$
$$= \boxed{0.00109\ \text{m}^2\quad(0.0011\ \text{m}^2)}$$

The answer is (B).

(c) Calculate the speed of sound, velocity, and density at the point.

$$a = \sqrt{kRT} = \sqrt{(1.4)\left(287\ \frac{\text{J}}{\text{kg·K}}\right)(166.7\text{K})}$$
$$= 258.8\ \text{m/s}$$

$$\text{v} = \text{M}a = (2)\left(258.8\ \frac{\text{m}}{\text{s}}\right) = 517.6\ \text{m/s}$$

$$\rho = \frac{p}{RT} = \frac{89\,460\ \text{Pa}}{\left(287\ \frac{\text{J}}{\text{kg·K}}\right)(166.7\text{K})}$$
$$= 1.870\ \text{kg/m}^3$$

The mass flow rate is

$$\dot{m} = \rho A\text{v} = \left(1.870\ \frac{\text{kg}}{\text{m}^3}\right)(0.00109\ \text{m}^2)\left(517.6\ \frac{\text{m}}{\text{s}}\right)$$
$$= \boxed{1.055\ \text{kg/s}\quad(1.1\ \text{kg/s})}$$

The answer is (A).

6. *Customary U.S. Solution*

From the normal shock table at $\text{M}_x = 2$,

$$\text{M}_y = 0.5774$$
$$\frac{T_y}{T_x} = 1.687$$

The properties behind the shock wave are

$$T_y = \left[\frac{T_y}{T_x}\right]T_x = (1.687)(500\text{°R})$$
$$= 843.5\text{°R}$$

$$a_y = \sqrt{kg_cRT_y} = \sqrt{\begin{array}{l}(1.4)\left(32.2\ \frac{\text{lbm-ft}}{\text{lbf-sec}^2}\right)\\ \quad\times\left(53.3\ \frac{\text{ft-lbf}}{\text{lbm-°R}}\right)(843.5\text{°R})\end{array}}$$
$$= 1423.6\ \text{ft/sec}$$

The air velocity behind the shock wave is

$$\text{v}_y = \text{M}_y a_y = (0.5774)\left(1423.6\ \frac{\text{ft}}{\text{sec}}\right) = \boxed{822.0\ \text{ft/sec} \quad (820\ \text{ft/sec})}$$

The answer is (B).

SI Solution

From the normal shock table at $\text{M}_x = 2$,

$$\text{M}_y = 0.5774$$

$$\frac{T_y}{T_x} = 1.687$$

The properties behind the shock wave are

$$T_y = \left(\frac{T_y}{T_x}\right) T_x = (1.687)(280\text{K}) = 472.4\text{K}$$

$$a_y = \sqrt{kRT_y} = \sqrt{(1.4)\left(287\ \frac{\text{J}}{\text{kg·K}}\right)(472.4\text{K})} = 435.7\ \text{m/s}$$

The air velocity behind the shock wave is

$$\text{v}_y = \text{M}_y a_y = (0.5774)\left(435.7\ \frac{\text{m}}{\text{s}}\right) = \boxed{251.6\ \text{m/s} \quad (250\ \text{m/s})}$$

The answer is (B).

7. *Customary U.S. Solution*

(a) The absolute total temperature is

$$T_0 = 70°\text{F} + 460° = 530°\text{R}$$

The total pressure is $p_0 = 100$ psia.

The ratio of flow area to the critical throat area is

$$\frac{A}{A^*} = 1.555$$

From the isentropic flow tables at $A/A^* = 1.555$, $[p/p_0] = 0.1492$, $[T/T_0] = 0.5807$, and $\boxed{\text{M} = 1.9.}$

The answer is (C).

(b) The properties at $A/A^* = 1.555$ are

$$T = \left[\frac{T}{T_0}\right] T_0 = (0.5807)(530°\text{R}) = \boxed{307.8°\text{R} \quad (310°\text{R})}$$

The answer is (B).

(c) The pressure is

$$p = \left[\frac{p}{p_0}\right] p_0 = (0.1492)(100\ \text{psia}) = \boxed{14.92\ \text{psia} \quad (15\ \text{psia})}$$

The answer is (A).

SI Solution

(a) The absolute total temperature is

$$T_0 = 21°\text{C} + 273° = 294\text{K}$$

The total pressure is $p_0 = 0.7$ MPa.

The ratio of flow area to the critical throat area is

$$\frac{A}{A^*} = 1.555$$

From the isentropic flow tables at $A/A^* = 1.555$, $[p/p_0] = 0.1492$, $[T/T_0] = 0.5807$, and $\boxed{\text{M} = 1.9.}$

The answer is (C).

(b) The properties at $A/A^* = 1.555$ are

$$T = \left[\frac{T}{T_0}\right] T_0 = (0.5807)(294\text{K}) = \boxed{170.7\text{K} \quad (170\text{K})}$$

The answer is (B).

(c) The pressure is

$$p = \left[\frac{p}{p_0}\right] p_0 = (0.1492)(0.7\ \text{MPa}) = \boxed{0.104\ \text{MPa} \quad (0.10\ \text{MPa})}$$

The answer is (A).

8. *Customary U.S. Solution*

The absolute total temperature is

$$T_0 = 40°\text{F} + 460° = 500°\text{R}$$

Since v is unknown, assume that static temperature is 500°R. An iterative process may be required for this problem.

The density of air is

$$\rho = \frac{p}{RT} = \frac{\left(10\ \frac{\text{lbf}}{\text{in}^2}\right)\left(12\ \frac{\text{in}}{\text{ft}}\right)^2}{\left(53.3\ \frac{\text{ft-lbf}}{\text{lbm-°R}}\right)(500\text{°R})} = 0.054\ \text{lbm/ft}^3$$

The velocity of air is

$$\text{v} = \frac{\dot{m}}{A\rho} = \frac{20\ \frac{\text{lbm}}{\text{sec}}}{(1\ \text{ft}^2)\left(0.054\ \frac{\text{lbm}}{\text{ft}^3}\right)} = 370.4\ \text{ft/sec}$$

The sonic velocity is

$$a = \sqrt{kg_cRT} = \sqrt{(1.4)\left(32.2\ \frac{\text{lbm-ft}}{\text{lbf-sec}^2}\right) \times \left(53.3\ \frac{\text{ft-lbf}}{\text{lbm-°R}}\right)(500\text{°R})} = 1096.1\ \text{ft/sec}$$

The Mach number is

$$\text{M} = \frac{\text{v}}{a^*} = \frac{370.4\ \frac{\text{ft}}{\text{sec}}}{1096.1\ \frac{\text{ft}}{\text{sec}}} = 0.338$$

From the isentropic flow tables at M = 0.338, $[T/T_0] = 0.9777$. A closer approximation to static temperature is

$$T = \left[\frac{T}{T_0}\right] T_0 = (0.9777)(500\text{°R}) = 488.9\text{°R}$$

Recalculate all the properties at $T = 488.9$°R.

$$\rho = \frac{p}{RT} = \frac{\left(10\ \frac{\text{lbf}}{\text{in}^2}\right)\left(12\ \frac{\text{in}}{\text{ft}}\right)^2}{\left(53.3\ \frac{\text{ft-lbf}}{\text{lbm-°R}}\right)(488.9\text{°R})} = 0.0553\ \text{lbm/ft}^3$$

$$\text{v} = \frac{\dot{m}}{A\rho} = \frac{20\ \frac{\text{lbm}}{\text{sec}}}{(1\ \text{ft}^2)\left(0.0553\ \frac{\text{lbm}}{\text{ft}^3}\right)} = 361.7\ \text{ft/sec}$$

$$a = \sqrt{kg_cRT} = \sqrt{(1.4)\left(32.2\ \frac{\text{lbm-ft}}{\text{lbf-sec}^2}\right) \times \left(53.3\ \frac{\text{ft-lbf}}{\text{lbm-°R}}\right)(488.9\text{°R})} = 1083.8\ \text{ft/sec}$$

$$\text{M} = \frac{\text{v}}{a} = \frac{361.7\ \frac{\text{ft}}{\text{sec}}}{1083.8\ \frac{\text{ft}}{\text{sec}}} = 0.334$$

From the isentropic flow tables at M = 0.334, $[A/A^*] = 1.8516$.

$$A_{\text{smallest}} = \left[\frac{A^*}{A}\right] A = \left(\frac{1}{1.8516}\right)(1\ \text{ft}^2) = \boxed{0.54\ \text{ft}^2}$$

The answer is (B).

SI Solution

The absolute total temperature is

$$T_0 = 4\text{°C} + 273° = 277\text{K}$$

Since v is unknown, assume that static temperature is 277K. An iterative process may be required for this problem.

The density of air is

$$\rho = \frac{p}{RT} = \frac{(70\ \text{kPa})\left(1000\ \frac{\text{Pa}}{\text{kPa}}\right)}{\left(287\ \frac{\text{J}}{\text{kg·K}}\right)(277\text{K})} = 0.8805\ \text{kg/m}^3$$

The velocity of air is

$$\text{v} = \frac{\dot{m}}{A\rho} = \frac{9\ \frac{\text{kg}}{\text{s}}}{(0.09\ \text{m}^2)\left(0.8805\ \frac{\text{kg}}{\text{m}^3}\right)} = 113.6\ \text{m/s}$$

The sonic velocity is

$$a = \sqrt{kRT} = \sqrt{(1.4)\left(287\ \frac{\text{J}}{\text{kg·K}}\right)(277\text{K})} = 333.6\ \text{m/s}$$

The Mach number is

$$M = \frac{v}{a} = \frac{113.6\ \frac{m}{s}}{333.6\ \frac{m}{s}} = 0.34$$

From the isentropic flow tables at $M = 0.34$, $[T/T_0] = 0.9774$. A closer approximation to the static temperature is

$$T = \left[\frac{T}{T_0}\right] T_0 = (0.9774)(277K) = 270.7K$$

Recalculate all the properties at $T = 270.7K$.

$$\rho = \frac{p}{RT} = \frac{(70\ \text{kPa})\left(1000\ \frac{\text{Pa}}{\text{kPa}}\right)}{\left(287\ \frac{\text{J}}{\text{kg·K}}\right)(270.7K)} = 0.901\ \text{kg/m}^3$$

$$v = \frac{\dot{m}}{A\rho} = \frac{9\ \frac{\text{kg}}{\text{s}}}{(0.09\ \text{m}^2)\left(0.901\ \frac{\text{kg}}{\text{m}^3}\right)} = 111.0\ \text{m/s}$$

$$a = \sqrt{kRT} = \sqrt{(1.4)\left(287\ \frac{\text{J}}{\text{kg·K}}\right)(270.7K)} = 329.8\ \text{m/s}$$

$$M = \frac{v}{a} = \frac{111.0\ \frac{m}{s}}{329.8\ \frac{m}{s}} = 0.337$$

From the isentropic flow tables at $M = 0.337$, $[A/A^*] = 1.8372$.

$$A_{\text{smallest}} = \left[\frac{A^*}{A}\right] A = \left(\frac{1}{1.8372}\right)(0.09\ \text{m}^2) = \boxed{0.049\ \text{m}^2}$$

The answer is (B).

9. *Customary U.S. Solution*

The ratio of static pressure before the shock to total pressure after the shock is

$$\frac{p_x}{p_{0,y}} = \frac{1.38\ \text{psia}}{20\ \text{psia}} = 0.069$$

From the normal shock table (see App. 25.B) for $p_x/p_{0,y} = 0.069$, the Mach number in the tunnel is read directly as $\boxed{M_x = 3.3.}$

The answer is (D).

SI Solution

The ratio of static pressure before the shock to total pressure after the shock is

$$\frac{p_x}{p_{0,y}} = \frac{9.51\ \text{kPa}}{140\ \text{kPa}} = 0.068$$

From the normal shock table (see App. 25.B) for $p_x/p_{0,y} = 0.068$, the Mach number in the tunnel is interpolated as $\boxed{M_x = 3.32\ (3.3).}$

The answer is (D).

10. *Customary U.S. Solution*

The initial enthalpy is found from the superheat tables at 100 psia and 800°F.

$$h_1 = 1429.8\ \text{Btu/lbm}$$

Using the Mollier diagram and assuming isentropic expansion, the final enthalpy at 60 psia is $h_2 = 1362\ \text{Btu/lbm}$.

From Eq. 25.2(b), the steam exit velocity is

$$v_2 = \sqrt{2g_cJ(h_1 - h_2)} = \sqrt{(2)\left(32.2\ \frac{\text{lbm-ft}}{\text{lbf-sec}^2}\right)\left(778\ \frac{\text{ft-lbf}}{\text{Btu}}\right) \times \left(1429.8\ \frac{\text{Btu}}{\text{lbm}} - 1362\ \frac{\text{Btu}}{\text{lbm}}\right)} = \boxed{1843.1\ \text{ft/sec}\quad(1800\ \text{ft/sec})}$$

The answer is (C).

SI Solution

The initial enthalpy is found from the superheat tables at 0.7 MPa and 425°C.

$$h_1 = 3322.2\ \text{kJ/kg}$$

Using the Mollier diagram and assuming isentropic expansion, the final enthalpy at 400 kPa is $h_2 = 3154\ \text{kJ/kg}$.

From Eq. 25.2(a), the steam exit velocity is

$$\begin{aligned} v_2 &= \sqrt{2(h_1 - h_2)} \\ &= \sqrt{(2)\left(3322.2\ \frac{\text{kJ}}{\text{kg}} - 3154\ \frac{\text{kJ}}{\text{kg}}\right)\left(1000\ \frac{\text{J}}{\text{kJ}}\right)} \\ &= \boxed{580\ \text{m/s}} \end{aligned}$$

The answer is (C).

11. *Customary U.S. Solution*

(a) At the particular point, the sonic velocity is

$$\begin{aligned} a = \sqrt{kg_cRT} &= \sqrt{\begin{array}{l}(1.4)\left(32.2\ \frac{\text{lbm-ft}}{\text{lbf-sec}^2}\right) \\ \times\left(53.3\ \frac{\text{ft-lbf}}{\text{lbm-°R}}\right)(1000\text{°R})\end{array}} \\ &= 1550.1\ \text{ft/sec} \end{aligned}$$

The Mach number is

$$\text{M} = \frac{\text{v}}{a} = \frac{600\ \frac{\text{ft}}{\text{sec}}}{1550.1\ \frac{\text{ft}}{\text{sec}}} = \boxed{0.387\quad(0.39)}$$

The answer is (A).

(b) From the isentropic flow tables at $\text{M}=0.39$, $[p/p_0]=0.9004$, $[T/T_0]=0.9705$, and $[A/A^*]=1.6243$.

The total properties at $\text{M}=0.39$ are

$$\begin{aligned} T_0 &= \left[\frac{T_0}{T}\right]T = \left(\frac{1}{0.9705}\right)(1000\text{°R}) \\ &= \boxed{1030.4\text{°R}\quad(1000\text{°R})} \end{aligned}$$

The answer is (D).

(c) The total pressure is

$$\begin{aligned} p_0 &= \left[\frac{p_0}{p}\right]p = \left(\frac{1}{0.9004}\right)(50\ \text{psia}) \\ &= \boxed{55.5\ \text{psia}\quad(56\ \text{psia})} \end{aligned}$$

The answer is (B).

(d) The critical exit area is

$$\begin{aligned} A^* &= \left[\frac{A^*}{A}\right]A = \left(\frac{1}{1.6243}\right)(0.1\ \text{ft}^2) \\ &= \boxed{0.0616\ \text{ft}^2\quad(0.062\ \text{ft}^2)} \end{aligned}$$

The answer is (A).

(e) From the isentropic flow tables at $\text{M}=1$ (critical conditions), $[p^*/p_0] = 0.5283$. The critical pressure is

$$\begin{aligned} p^* &= \left[\frac{p^*}{p_0}\right]p_0 = (0.5283)(55.5\ \text{psia}) \\ &= \boxed{29.32\ \text{psia}\quad(29\ \text{psia})} \end{aligned}$$

The answer is (B).

(f) From the isentropic flow tables at $\text{M} = 1$ (critical conditions), $[T^*/T_0] = 0.8333$. The critical temperature is

$$\begin{aligned} T^* &= \left[\frac{T^*}{T_0}\right]T_0 = (0.8333)(1030.4\text{°R}) \\ &= \boxed{858.6\text{°R}\quad(860\text{°R})} \end{aligned}$$

The answer is (B).

SI Solution

(a) At the particular point, the sonic velocity is

$$\begin{aligned} a = \sqrt{kRT} &= \sqrt{(1.4)\left(287\ \frac{\text{J}}{\text{kg·K}}\right)(560\text{K})} \\ &= 474.4\ \text{m/s} \end{aligned}$$

The Mach number is

$$\text{M} = \frac{\text{v}}{a} = \frac{180\ \frac{\text{m}}{\text{s}}}{474.4\ \frac{\text{m}}{\text{s}}} = \boxed{0.38\quad(0.39)}$$

The answer is (A).

(b) From the isentropic flow tables at $\text{M}=0.38$, $[p/p_0]=0.9052$, $[T/T_0]=0.9719$, and $[A/A^*]=1.6587$.

The total properties at $\text{M}=0.38$ are

$$\begin{aligned} T_0 &= \left[\frac{T_0}{T}\right]T = \left(\frac{1}{0.9719}\right)(560\text{K}) \\ &= \boxed{576.2\text{K}\quad(580\text{K})} \end{aligned}$$

The answer is (D).

(c) The total pressure is

$$\begin{aligned} p_0 &= \left[\frac{p_0}{p}\right]p = \left(\frac{1}{0.9052}\right)(350\ \text{kPa}) \\ &= \boxed{386.7\ \text{kPa}\quad(390\ \text{kPa})} \end{aligned}$$

The answer is (B).

(d) The critical exit area is

$$A^* = \left[\frac{A^*}{A}\right]A = \left(\frac{1}{1.6587}\right)(0.0009 \text{ m}^2)$$
$$= \boxed{5.43 \times 10^{-4} \text{ m}^2 \quad (0.00054 \text{ m}^2)}$$

The answer Is (A).

(e) From the isentropic flow tables at M = 1 (critical conditions), $[p^*/p_0] = 0.5283$. The critical density is

$$p^* = \left[\frac{p^*}{p_0}\right]p_0 = (0.5283)(386.7 \text{ kPa})$$
$$= \boxed{204.3 \text{ kPa} \quad (200 \text{ kPa})}$$

The answer is (B).

(f) From the isentropic flow tables at M = 1 (critical conditions), $[T^*/T_0] = 0.8333$. The critical temperature is

$$T^* = \left[\frac{T^*}{T_0}\right]T_0 = (0.8333)(576.2\text{K})$$
$$= \boxed{480.1\text{K} \quad (480\text{K})}$$

The answer is (B).

12. *Customary U.S. Solution*

The initial enthalpy is found from the superheat tables at 200 psia and 600°F.

$$h_1 = 1322.3 \text{ Btu/lbm}$$

If the expansion through the nozzle had been isentropic, the exit enthalpy would have been approximately 1228 Btu/lbm. The nozzle efficiency is defined by Eq. 25.18.

$$\eta_{\text{nozzle}} = \frac{h_1 - h_2'}{h_1 - h_2}$$
$$h_2' = h_1 - \eta_{\text{nozzle}}(h_1 - h_2)$$
$$= 1322.3 \ \frac{\text{Btu}}{\text{lbm}} - (0.85)\left(1322.3 \ \frac{\text{Btu}}{\text{lbm}} - 1228 \ \frac{\text{Btu}}{\text{lbm}}\right)$$
$$= 1242.1 \text{ Btu/lbm}$$

Knowing $h = 1242.1$ Btu/lbm and $p = 80$ psia establishes (from the superheat tables) that

$$T_2' = 422.6°\text{F}$$
$$\upsilon_2' = 6.837 \text{ ft}^3/\text{lbm}$$

(By double interpolation from App. 23.C. Actual value is approximately 6.386 ft^3/lbm, quite a bit different from the value obtained from a complete superheat table.)

$$\rho_2' = \frac{1}{\upsilon_2'} = \frac{1}{6.837 \ \frac{\text{ft}^3}{\text{lbm}}} = 0.1463 \text{ lbm/ft}^3$$

The nozzle exit velocity can be calculated as

$$\text{v}_2' = \sqrt{2g_cJ(h_1 - h_2') + \text{v}_1^2}$$
$$= \sqrt{\begin{array}{l}(2)\left(32.2 \ \frac{\text{lbm-ft}}{\text{lbf-sec}^2}\right)\left(778 \ \frac{\text{ft-lbf}}{\text{Btu}}\right) \\ \times \left(1322.3 \ \frac{\text{Btu}}{\text{lbm}} - 1242.1 \ \frac{\text{Btu}}{\text{lbm}}\right) + \left(300 \ \frac{\text{ft}}{\text{sec}}\right)^2\end{array}}$$
$$= 2026.9 \text{ ft/sec}$$

The exit area of the nozzle is

$$A_e = \frac{\dot{m}}{\rho_2'\text{v}_2'} = \frac{3 \ \frac{\text{lbm}}{\text{sec}}}{\left(0.1463 \ \frac{\text{lbm}}{\text{ft}^3}\right)\left(2026.9 \ \frac{\text{ft}}{\text{sec}}\right)}$$
$$= 0.01012 \text{ ft}^2$$

The absolute temperature at the nozzle exit is

$$T_1 = 422.6°\text{F} + 460° = 882.6°\text{R}$$
$$k_{\text{steam}} = 1.31$$
$$R_{\text{steam}} = 85.8 \text{ ft-lbf/lbm-°R}$$

The sonic velocity is

$$a = \sqrt{kg_cRT}$$
$$= \sqrt{\begin{array}{l}(1.31)\left(32.2 \ \frac{\text{lbm-ft}}{\text{lbf-sec}^2}\right) \\ \times \left(85.8 \ \frac{\text{ft-lbf}}{\text{lbm-°R}}\right)(882.6°\text{R})\end{array}}$$
$$= 1787.3 \text{ ft/sec}$$

The Mach number is

$$\text{M} = \frac{\text{v}_2'}{a} = \frac{2026.9 \ \frac{\text{ft}}{\text{sec}}}{1787.3 \ \frac{\text{ft}}{\text{sec}}} = 1.13$$

Thermodynamics

Calculate the throat area. Since tables for $k = 1.31$ are not available,

$$\frac{A}{A^*} = \left(\frac{1}{\mathrm{M}}\right)\left(\frac{\left(\frac{1}{2}\right)(k-1)\mathrm{M}^2+1}{\left(\frac{1}{2}\right)(k-1)+1}\right)^{(k+1)/(2)(k-1)}$$

$$\frac{A_e}{A^*} = \left(\frac{1}{1.13}\right)\left(\frac{\left(\frac{1}{2}\right)(1.31-1)(1.13)^2+1}{\left(\frac{1}{2}\right)(1.31-1)+1}\right)^{(1.31+1)/(2)(1.31-1)} = 1.0138$$

$$A^* = \left[\frac{A^*}{A_e}\right]A_e = \left(\frac{1}{1.0138}\right)(0.01012\ \mathrm{ft}^2) = \boxed{0.00998\ \mathrm{ft}^2 \quad (0.010\ \mathrm{ft}^2)}$$

The answer is (C).

SI Solution

The initial enthalpy is found from the superheat tables at 1.5 MPa and 300°C.

$$h_1 = 3038.2\ \mathrm{kJ/kg}$$

If the expansion through the nozzle had been isentropic, the exit enthalpy would have been approximately 2790 kJ/kg. The nozzle efficiency is defined by Eq. 25.18.

$$\eta_{\mathrm{nozzle}} = \frac{h_1 - h_2'}{h_1 - h_2}$$

$$h_2' = h_1 - \eta_{\mathrm{nozzle}}(h_1 - h_2) = 3038.2\ \frac{\mathrm{kJ}}{\mathrm{kg}} - (0.85)\left(3038.2\ \frac{\mathrm{kJ}}{\mathrm{kg}} - 2790\ \frac{\mathrm{kJ}}{\mathrm{kg}}\right) = 2827\ \mathrm{kJ/kg}$$

Knowing $h = 2827$ kJ/kg and $p = 500$ kPa establishes (from the superheat tables) that

$$T_2' = 187.1°\mathrm{C}$$

$$\upsilon_2' = 0.4420\ \mathrm{m^3/kg}$$

$$\rho_2' = \frac{1}{\upsilon_2'} = \frac{1}{0.4420\ \frac{\mathrm{m}^3}{\mathrm{kg}}} = 2.262\ \mathrm{kg/m^3}$$

The nozzle exit velocity can be calculated as

$$\mathrm{v}_2' = \sqrt{2(h_1 - h_2') + \mathrm{v}_1^2} = \sqrt{\begin{array}{l}(2)\left(3038.2\ \frac{\mathrm{kJ}}{\mathrm{kg}} - 2827\ \frac{\mathrm{kJ}}{\mathrm{kg}}\right)\left(1000\ \frac{\mathrm{J}}{\mathrm{kJ}}\right) \\ \quad + \left(90\ \frac{\mathrm{m}}{\mathrm{s}}\right)^2\end{array}} = 656.1\ \mathrm{m/s}$$

The exit area of the nozzle is

$$A_e = \frac{\dot{m}}{\rho_2'\mathrm{v}_2'} = \frac{1.4\ \frac{\mathrm{kg}}{\mathrm{s}}}{\left(2.262\ \frac{\mathrm{kg}}{\mathrm{m}^3}\right)\left(656.1\ \frac{\mathrm{m}}{\mathrm{s}}\right)} = 9.433\times10^{-4}\ \mathrm{m}^2$$

The absolute temperature at the nozzle exit is

$$T_1 = 187.1°\mathrm{C} + 273° = 460.1\mathrm{K}$$

$$k_{\mathrm{steam}} = 1.31$$

$$R_{\mathrm{steam}} = 461.50\ \mathrm{J/kg{\cdot}K}$$

The sonic velocity is

$$a = \sqrt{kRT} = \sqrt{(1.31)\left(461.50\ \frac{\mathrm{J}}{\mathrm{kg{\cdot}K}}\right)(460.1\mathrm{K})} = 527.4\ \mathrm{m/s}$$

The Mach number is

$$\mathrm{M} = \frac{\mathrm{v}_2'}{a} = \frac{656.1\ \frac{\mathrm{m}}{\mathrm{s}}}{527.4\ \frac{\mathrm{m}}{\mathrm{s}}} = 1.24$$

Calculate the throat area. Since tables for $k = 1.31$ are not available,

$$\frac{A}{A^*} = \left(\frac{1}{\mathrm{M}}\right)\left(\frac{\left(\frac{1}{2}\right)(k-1)\mathrm{M}^2+1}{\left(\frac{1}{2}\right)(k-1)+1}\right)^{(k+1)/(2)(k-1)}$$

$$\frac{A_e}{A^*} = \left(\frac{1}{1.24}\right)\left(\frac{\left(\frac{1}{2}\right)(1.31-1)(1.24)^2+1}{\left(\frac{1}{2}\right)(1.31-1)+1}\right)^{(1.31+1)/(2)(1.31-1)} = 1.0454$$

The throat area is

$$A^* = \left[\frac{A^*}{A_e}\right]A_e = \left(\frac{1}{1.0454}\right)(9.433\times10^{-4}\ \mathrm{m}^2) = \boxed{9.023\times10^{-4}\ \mathrm{m}^2 \quad (0.00090\ \mathrm{m}^2)}$$

The answer is (C).

13. *Customary U.S. Solution*

(a) This is a Fanno flow problem. Check for choked flow.

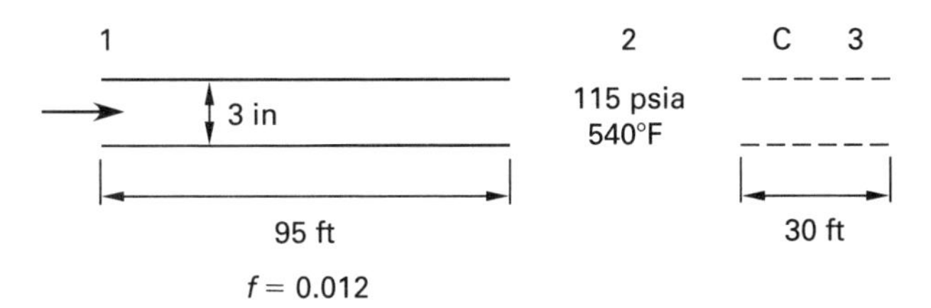

95 ft from the entrance (point 2), from the superheat tables at 115 psia and 540°F, $\upsilon_2 = 5.244$ ft³/lbm.

The velocity 95 ft from the entrance (point 2) is

$$\text{v}_2 = \frac{\dot{m}}{A\rho_2} = \frac{\dot{m}\upsilon_2}{A}$$

$$= \frac{\left(35{,}200\ \frac{\text{lbm}}{\text{hr}}\right)\left(5.244\ \frac{\text{ft}^3}{\text{lbm}}\right)\left(12\ \frac{\text{in}}{\text{ft}}\right)^2}{\left(\frac{\pi}{4}\right)(3\ \text{in})^2\left(3600\ \frac{\text{sec}}{\text{hr}}\right)}$$

$$= \boxed{1045\ \text{ft/sec}\quad(1000\ \text{ft/sec})}$$

The answer is (B).

(b) The absolute temperature 95 ft from the entrance (point 2) is

$$T_2 = 540°\text{F} + 460° = 1000°\text{R}$$

For 1000°R steam, $k = 1.29$ and $R = 85.8$ ft-lbf/lbm-°R.

The sonic velocity 95 ft from the entrance (point 2) is

$$a_2 = \sqrt{kg_cRT} = \sqrt{\begin{array}{l}(1.29)\left(32.2\ \frac{\text{lbm ft}}{\text{lbf-sec}^2}\right)\\ \quad\times\left(85.8\ \frac{\text{ft-lbf}}{\text{lbm-°R}}\right)(1000°\text{R})\end{array}}$$

$$= \boxed{1888\ \text{ft/sec}\quad(1900\ \text{ft/sec})}$$

The answer is (C).

(c) The Mach number 95 ft from the entrance (point 2) is

$$\text{M}_2 = \frac{\text{v}_2}{a_2} = \frac{1045\ \frac{\text{ft}}{\text{sec}}}{1888\ \frac{\text{ft}}{\text{sec}}} = \boxed{0.553\quad(0.55)}$$

The answer is (A).

(d) The distance from the original end of the pipe (point 2) to where the flow becomes choked is calculated from Eq. 25.36.

$$\frac{4f_{\text{Fanning}}L_{\text{max}}}{D} = \frac{1-\text{M}^2}{k\text{M}^2} + \left(\frac{1+k}{2k}\right)\ln\left(\frac{(1+k)\text{M}^2}{2\left(1+\frac{1}{2}(k-1)\text{M}^2\right)}\right)$$

$$L_{\text{max}} = \left(\frac{3\ \text{in}}{(4)(0.012)\left(12\ \frac{\text{in}}{\text{ft}}\right)}\right) \times\left(\begin{array}{l}\frac{1-(0.553)^2}{(1.29)(0.553)^2} + \left(\frac{1+1.29}{(2)(1.29)}\right)\\ \quad\times\ln\left(\frac{(1+1.29)(0.553)^2}{(2)\left(1+\left(\frac{1}{2}\right)(1.29-1)(0.553)^2\right)}\right)\end{array}\right)$$

$$= \boxed{4.11\ \text{ft}\quad(4.1\ \text{ft})}$$

The answer is (D).

(e) $\boxed{\text{The flow will be choked in less than 30 ft; steam flow is not maintained.}}$

The answer is (A).

SI Solution

(a) This is a Fanno flow problem. Check for choked flow.

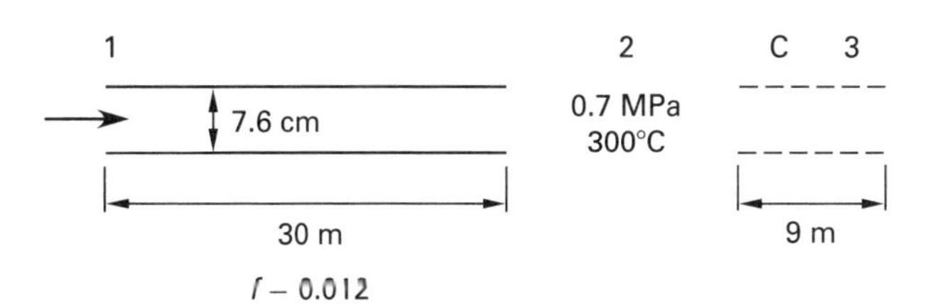

30 m from the entrance (point 2), from the superheat tables at 0.7 MPa and 300°C, $\upsilon_2 = 0.3714$ m³/kg.

The velocity 30 m from the entrance (point 2) is

$$\text{v}_2 = \frac{\dot{m}}{\rho_2 A} = \frac{\dot{m}\upsilon_2}{A} = \frac{\left(4.4\ \frac{\text{kg}}{\text{s}}\right)\left(0.3714\ \frac{\text{m}^3}{\text{kg}}\right)\left(100\ \frac{\text{cm}}{\text{m}}\right)^2}{\left(\frac{\pi}{4}\right)(7.6\ \text{cm})^2}$$

$$= \boxed{360.2\ \text{m/s}\quad(360\ \text{m/s})}$$

The answer is (B).

(b) The absolute temperature 30 m from the entrance (point 2) is

$$T_2 = 300°\text{C} + 273° = 573\text{K}$$

Thermodynamics

For 573K steam, $k = 1.29$ and $R = 461.5$ J/kg·K. The sonic velocity 30 m from the entrance (point 2) is

$$a_2 = \sqrt{kRT} = \sqrt{(1.29)\left(461.5\ \frac{\text{J}}{\text{kg·K}}\right)(573\text{K})} = \boxed{584.1\ \text{m/s}\quad (580\ \text{m/s})}$$

The answer is (C).

(c) The Mach number 30 m from the entrance (point 2) is

$$\text{M}_2 = \frac{\text{v}_2}{a_2} = \frac{360.2\ \frac{\text{m}}{\text{s}}}{584.1\ \frac{\text{m}}{\text{s}}} = \boxed{0.617\quad (0.62)}$$

The answer is (A).

(d) The distance from the original end of the pipe (point 2) to where the flow becomes choked is calculated from Eq. 25.36.

$$\frac{4f_{\text{Fanning}}L_{\text{max}}}{D} = \frac{1-\text{M}^2}{k\text{M}^2} + \left(\frac{1+k}{2k}\right)\ln\left(\frac{(1+k)\text{M}^2}{2\left(1+\frac{1}{2}(k-1)\text{M}^2\right)}\right)$$

$$L_{\text{max}} = \left(\frac{7.6\ \text{cm}}{(4)(0.012)\left(100\ \frac{\text{cm}}{\text{m}}\right)}\right) \times \left(\begin{array}{l}\dfrac{1-(0.617)^2}{(1.29)(0.617)^2} + \left(\dfrac{1+1.29}{(2)(1.29)}\right) \\ \times \ln\left(\dfrac{(1+1.29)(0.617)^2}{(2)\left(1+\left(\frac{1}{2}\right)(1.29-1)(0.617)^2\right)}\right)\end{array}\right) = \boxed{0.75\ \text{m}}$$

The answer is (D).

(e) $\boxed{\text{The flow will be choked in less than 9 m; steam flow is not maintained.}}$

The answer is (A).

14. *Customary U.S. Solution*

(a) Since $p_e/p_0 < 0.5283$, the flow must be supersonic. The throat flow is sonic. Read the property ratios for Mach 1 from the isentropic flow table: $[T/T_0] = 0.8333$ and $[p/p_0] = 0.5283$.

The sonic properties at the throat are

$$T^* = \left[\frac{T}{T_0}\right]T_0 = (0.8333)(660°\text{R}) = 550°\text{R}$$

$$p^* = \left[\frac{p}{p_0}\right]p_0 = (0.5283)(160\ \text{psia}) = 84.53\ \text{psia}$$

The sonic velocity is

$$a^* = \sqrt{kg_cRT^*} = \sqrt{(1.4)\left(32.2\ \frac{\text{lbm-ft}}{\text{lbf-sec}^2}\right) \times \left(53.3\ \frac{\text{ft-lbf}}{\text{lbm-°R}}\right)(550°\text{R})} = \boxed{1150\ \text{ft/sec}\quad (1200\ \text{ft/sec})}$$

The answer is (A).

(b) The sonic density is

$$\rho^* = \frac{p^*}{RT^*} = \frac{\left(84.53\ \frac{\text{lbf}}{\text{in}^2}\right)\left(12\ \frac{\text{in}}{\text{ft}}\right)^2}{\left(53.3\ \frac{\text{ft-lbf}}{\text{lbm-°R}}\right)(550°\text{R})} = \boxed{0.415\ \text{lbm/ft}^3\quad (0.42\ \text{lbm/ft}^3)}$$

The answer is (B).

(c) The overall nozzle efficiency is $C_d = 0.90$. From $\dot{m} = C_dA\text{v}\rho$, the throat area is

$$A^* = \frac{\dot{m}}{C_d\text{v}\rho} = \frac{3600\ \frac{\text{lbm}}{\text{hr}}}{(0.90)\left(1150\ \frac{\text{ft}}{\text{sec}}\right)\left(0.415\ \frac{\text{lbm}}{\text{ft}^3}\right) \times \left(3600\ \frac{\text{sec}}{\text{hr}}\right)} = \boxed{0.002328\ \text{ft}^2\quad (0.0023\ \text{ft}^2)}$$

The answer is (C).

(d) Find the diameter.

$$D^* = \sqrt{\frac{4A^*}{\pi}} = \sqrt{\frac{(4)(0.002328\ \text{ft}^2)}{\pi}} = 0.0544\ \text{ft}$$

At the exit,

$$p_3 = 14.7\ \text{psia}$$

$$\frac{p_e}{p_0} = \frac{14.7\ \text{psia}}{160\ \text{psia}} = 0.0919$$

From the isentropic flow tables at $p/p_0 = 0.0919$, M = 2.22 and $[A/A^*] = 2.041$.

$$A_e = \left[\frac{A}{A^*}\right]A^* = (2.041)(0.002328\ \text{ft}^2) = 0.004751\ \text{ft}^2$$

$$D_e = \sqrt{\frac{4A}{\pi}} = \sqrt{\frac{(4)(0.004751\ \text{ft}^2)}{\pi}} = 0.0778\ \text{ft}$$

The longitudinal distance from the throat to the exit is

$$x = \frac{0.0778\ \text{ft} - 0.0544\ \text{ft}}{(2)(\tan 3^\circ)} = 0.223\ \text{ft}$$

The entrance velocity is not known, so the entrance area cannot be found. However, the longitudinal distance from the entrance to the throat is

$$(0.05)(0.223\ \text{ft}) = \boxed{0.0112\ \text{ft}}$$

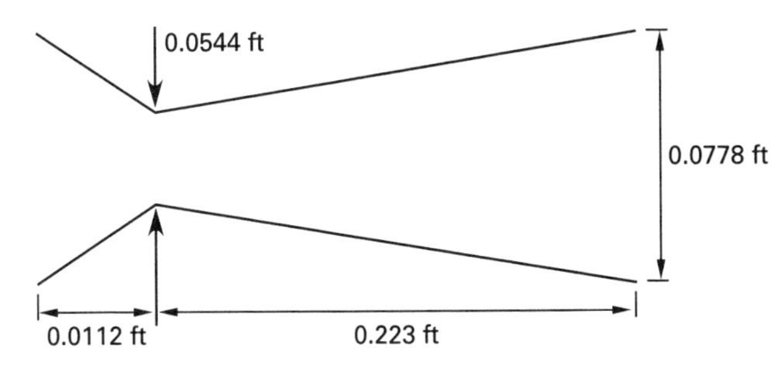

The answer is (B).

SI Solution

(a) Since $p_e/p_0 < 0.5283$, the flow must be supersonic. The throat flow is sonic. Read the property ratio for Mach 1 from the isentropic flow table: $[T/T_0] = 0.8333$ and $[p/p_0] = 0.5283$.

The sonic properties at the throat are

$$T^* = \left[\frac{T}{T_0}\right]T_0 = (0.8333)(370\text{K}) = 308.3\text{K}$$

$$p^* = \left[\frac{p}{p_0}\right]p_0 = (0.5283)(1.1\ \text{MPa}) = 0.581\ \text{MPa}$$

The sonic velocity is

$$\begin{aligned} a^* &= \sqrt{kRT^*} \\ &= \sqrt{(1.4)\left(287\ \frac{\text{J}}{\text{kg}\cdot\text{K}}\right)(308.3\text{K})} \\ &= \boxed{352\ \text{m/s}\quad (350\ \text{m/s})} \end{aligned}$$

The answer is (A).

(b) The sonic density is

$$\begin{aligned} \rho^* &= \frac{p^*}{RT^*} = \frac{(0.581\ \text{MPa})\left(10^6\ \frac{\text{Pa}}{\text{MPa}}\right)}{\left(287\ \frac{\text{J}}{\text{kg}\cdot\text{K}}\right)(308.3\text{K})} \\ &= \boxed{6.57\ \text{kg/m}^3\quad (6.6\ \text{kg/m}^3)} \end{aligned}$$

The answer is (B).

(c) The overall nozzle efficiency is $C_d = 0.90$. From $\dot{m} = C_d A v\rho$, the throat area is

$$\begin{aligned} A^* &= \frac{\dot{m}}{C_d v\rho} = \frac{0.45\ \frac{\text{kg}}{\text{s}}}{(0.90)\left(352\ \frac{\text{m}}{\text{s}}\right)\left(6.57\ \frac{\text{kg}}{\text{m}^3}\right)} \\ &= \boxed{0.000216\ \text{m}^2\quad (0.00022\ \text{m}^2)} \end{aligned}$$

The answer is (C).

(d) Find the diameter.

$$D^* = \sqrt{\frac{4A^*}{\pi}} = \sqrt{\frac{(4)(0.000216\ \text{m}^2)}{\pi}} = 0.0166\ \text{m}$$

At the exit,

$$p_e = 101.3\ \text{kPa}$$

$$\frac{p_e}{p_0} = \frac{(101.3\ \text{kPa})\left(10^3\ \frac{\text{Pa}}{\text{kPa}}\right)}{(1.1\ \text{MPa})\left(10^6\ \frac{\text{Pa}}{\text{MPa}}\right)} = 0.0921$$

From the isentropic flow tables at $p/p_0 = 0.0921$, M = 2.21 and $[A/A^*] = 2.024$.

$$\begin{aligned} A_e &= \left[\frac{A}{A^*}\right]A^* = (2.024)(0.000216\ \text{m}^2) \\ &= 0.000437\ \text{m}^2 \end{aligned}$$

$$D_e = \sqrt{\frac{4A}{\pi}} = \sqrt{\frac{(4)(0.000437\ \text{m}^2)}{\pi}} = 0.0236\ \text{m}$$

The longitudinal distance from the throat to the exit is

$$x = \frac{0.0236\ \text{m} - 0.0166\ \text{m}}{(2)(\tan 3^\circ)} = 0.0668\ \text{m}$$

The entrance velocity is not known, so the entrance area cannot be found. However, the longitudinal distance from the entrance to the throat is

$$(0.05)(0.0668\ \text{m}) = \boxed{0.00334\ \text{m}}$$

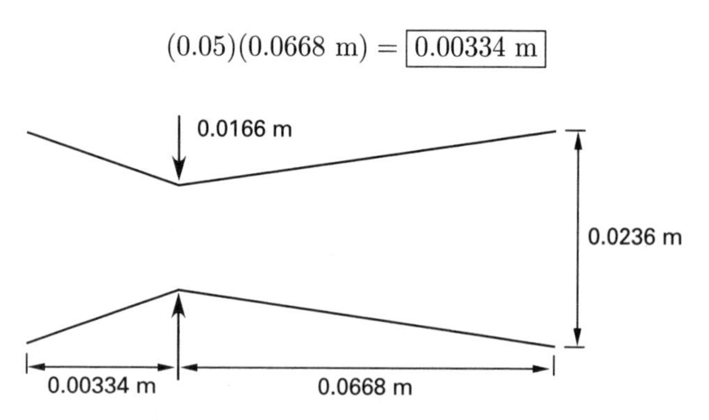

The answer is (B).

15. (a) The back-pressure ratio is

$$\frac{p_{\text{back}}}{p_{\text{reservoir}}} = \frac{14.7\ \frac{\text{lbf}}{\text{in}^2}}{470\ \frac{\text{lbf}}{\text{in}^2}} = 0.031$$

Since the back-pressure ratio is less than 0.5283 (the critical back-pressure ratio for gases with $k = 1.4$), flow in the orifice will be choked until the reservoir pressure reaches $14.7\ \text{lbf/in}^2/0.5283 = 27.8\ \text{lbf/in}^2$. Therefore, the flow will be choked for the entire duration of discharge from the initial pressure of $470\ \text{lbf/in}^2$ down to $100\ \text{lbf/in}^2$. The discharge will be sonic throughout the discharge.

The static temperature where sonic flow is achieved is

$$\begin{aligned} T_{\text{throat}} &= \left[\frac{T}{T_0}\right] T_{\text{reservoir}} = (0.8333)(68°\text{F} + 460°) \\ &= \boxed{440°\text{R}} \end{aligned}$$

The answer is (B).

(b) The discharge velocity will be limited by the speed of sound.

$$\begin{aligned} a &= \sqrt{kg_cRT_{\text{throat}}} \\ &= \sqrt{(1.4)\left(32.2\ \frac{\text{lbm-ft}}{\text{lbf-sec}^2}\right)\left(53.3\ \frac{\text{ft-lbf}}{\text{lbm-°R}}\right)(440°\text{R})} \\ &= \boxed{1028\ \text{ft/sec}\quad (1000\ \text{ft/sec})} \end{aligned}$$

The answer is (C).

(c) The term "effective discharge velocity" is nonstandard. The information is taken as a means to calculate the discharge coefficient (orifice coefficient).

$$\begin{aligned} C_d &= C_cC_v \approx C_v = \frac{\text{v}_{\text{actual}}}{\text{v}_{\text{ideal}}} \\ &= \frac{875\ \frac{\text{ft}}{\text{sec}}}{1028\ \frac{\text{ft}}{\text{sec}}} \\ &= \boxed{0.851\quad (0.85)} \end{aligned}$$

The answer is (D).

(d) The static pressure at the throat is

$$\begin{aligned} p_{\text{throat}} &= \left[\frac{p}{p_0}\right]_{\text{M}=1} p_{\text{reservoir}} = (0.5283)\left(470\ \frac{\text{lbf}}{\text{in}^2}\right) \\ &= 248.3\ \text{lbf/in}^2 \end{aligned}$$

The initial mass flow rate of the air discharging through the orifice, $\dot{m}_{\text{orifice}}$, is

$$\begin{aligned} \dot{m}_{\text{orifice}} &= C_d\dot{m}_{\text{ideal}} = C_d\rho_{\text{throat}}A\text{v} = \frac{C_dp_{\text{throat}}A\text{v}_{\text{ideal}}}{RT_{\text{throat}}} \\ &= \frac{(0.851)\left(248.3\ \frac{\text{lbf}}{\text{in}^2}\right)(0.05\ \text{in}^2)\left(1028\ \frac{\text{ft}}{\text{sec}}\right)}{\left(53.3\ \frac{\text{ft-lbf}}{\text{lbm-°R}}\right)(440°\text{R})} \\ &= \boxed{0.463\ \text{lbm/sec}\quad (0.46\ \text{lbm/sec})} \end{aligned}$$

The answer is (A).

(e) From the ideal gas law, the mass of air discharged from the reservoir is

$$\begin{aligned} m_{\text{discharged}} &= \frac{(p_1 - p_2)V}{RT_1} \\ &= \frac{\left(470\ \frac{\text{lbf}}{\text{in}^2} - 100\ \frac{\text{lbf}}{\text{in}^2}\right)\left(12\ \frac{\text{in}}{\text{ft}}\right)^2(100\ \text{ft}^3)}{\left(53.3\ \frac{\text{ft-lbf}}{\text{lbm-°R}}\right)(68°\text{F} + 460°)} \\ &= \boxed{189.3\ \text{lbm}\quad (190\ \text{lbm})} \end{aligned}$$

The answer is (C).

(f) Use Eq. 25.34. Since $m = pV/RT$, the mass of air in the tank is proportional to the pressure in the tank. $\rho = p/RT$ is substituted for the density term. The time to reduce the reservoir pressure to 470 lbf/in^2 is

$$t_2 - t_1 = \frac{\left(\left(\frac{m_2}{m_1}\right)^{(1-k)/2} - 1\right)\left(\frac{2V_{\text{tank}}}{C_d A_o (k-1)}\right)}{\sqrt{\frac{g_c k p_1}{\rho_1}\left(\frac{2}{k+1}\right)^{(k+1)/(k-1)}}}$$

$$= \frac{\left(\left(\frac{p_2}{p_1}\right)^{(1-k)/2} - 1\right)\left(\frac{2V_{\text{tank}}}{C_d A_o (k-1)}\right)}{\sqrt{g_c k R T_1\left(\frac{2}{k+1}\right)^{(k+1)/(k-1)}}}$$

$$= \frac{\left(\left(\frac{100\ \frac{\text{lbf}}{\text{in}^2}}{470\ \frac{\text{lbf}}{\text{in}^2}}\right)^{(1-1.4)/2} - 1\right) \times \left(\frac{(2)(100\ \text{ft}^3)\left(12\ \frac{\text{in}}{\text{ft}}\right)^2}{(0.851)(0.05\ \text{in}^2)(1.4-1)}\right)}{\sqrt{\begin{array}{l}\left(32.2\ \frac{\text{lbm-ft}}{\text{lbf-sec}^2}\right)(1.4)\left(53.3\ \frac{\text{ft-lbf}}{\text{lbm-°R}}\right) \\ \quad \times (68°\text{F} + 460°)\left(\frac{2}{1.4+1}\right)^{(1.4+1)/(1.4-1)}\end{array}} \times \left(60\ \frac{\text{sec}}{\text{min}}\right)}$$

$$= \boxed{15.7\ \text{min}\quad (16\ \text{min})}$$

The answer is (D).

Thermodynamics

26 Vapor Power Equipment

PRACTICE PROBLEMS

1. "BLEVE" is an acronym that is used to describe

(A) a boiler demand leveling valve

(B) a catastrophic boiler explosion

(C) the vapor from a bleed valve

(D) a burnout caused by insufficient feedwater

2. A small power generating plant has two furnace-boilers and two turbine-generators. During long periods of low electrical demand, capacity is achieved with a single boiler. To achieve a functional cross-coupled configuration, which valves should be opened and which should be closed?

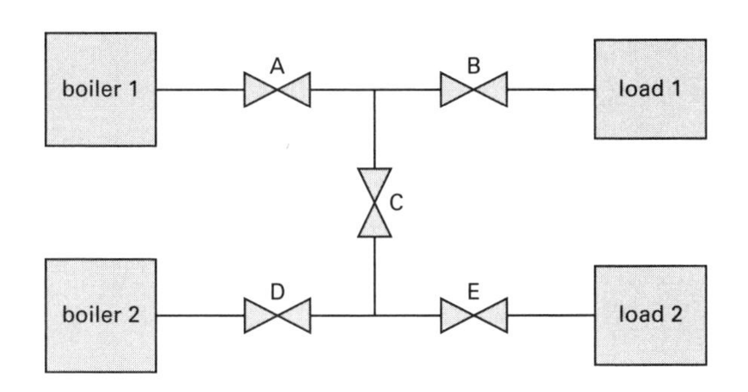

(A) open: A, B, D, and E
closed: C

(B) open: A, B, C, and D
closed: E

(C) open: A, B, C, and E
closed: D

(D) open: A, C, and E
closed: B, D

3. A steam generator receives liquid water at 80°F and produces saturated steam at 300 psia. Most nearly, what is the ratio of boiler sensible to boiler latent heat supplied to the water and steam?

(A) 0.06

(B) 0.12

(C) 0.27

(D) 0.43

4. What is most nearly the isentropic efficiency of a process that expands dry steam from 100 psia to 3 psia (700 kPa to 20 kPa) and 90% quality?

(A) 40%

(B) 50%

(C) 60%

(D) 70%

5. A 5000 kW steam turbine uses 200 psia (1.5 MPa) steam with 100°F (50°C) of superheat. The condenser is at 1 in Hg (3.4 kPa) absolute.

(a) The water rate at full load is most nearly

(A) 4.7 lbm/kW-hr (0.0006 kg/kW·s)

(B) 8.8 lbm/kW-hr (0.0011 kg/kW·s)

(C) 25 lbm/kW-hr (0.0031 kg/kW·s)

(D) 37 lbm/kW-hr (0.0046 kg/kW·s)

(b) If the actual load is only 2500 kW and the steam undergoes a pressure-reducing throttling process to reduce the availability, what is most nearly the loss in available energy per unit mass of steam?

(A) 80 Btu/lbm (190 kJ/kg)

(B) 130 Btu/lbm (310 kJ/kg)

(C) 190 Btu/lbm (460 kJ/kg)

(D) 370 Btu/lbm (890 kJ/kg)

6. A 10,000 kW steam turbine operates on 400 psia (3.0 MPa), 750°F (420°C) dry steam, expanding to 2 in Hg (6.8 kPa) absolute. The maximum adiabatic heat drop available for power production is most nearly

(A) 450 Btu/lbm (1100 kJ/kg)

(B) 600 Btu/lbm (1400 kJ/kg)

(C) 750 Btu/lbm (1800 kJ/kg)

(D) 900 Btu/lbm (2200 kJ/kg)

7. A 750 kW steam turbine has a water rate of 20 lbm/kW-hr (2.5×10^{-3} kg/kW·s). Steam with 50°F (30°C) of superheat is expanded from 165 psia (1.0 MPa absolute) to 26 in Hg (90 kPa) absolute. 65°F (18°C) cooling water is available. The terminal temperature difference is zero. The quantity of cooling water required is most nearly

(A) 45,000 lbm/hr (5.9 kg/s)

(B) 60,000 lbm/hr (7.8 kg/s)

(C) 80,000 lbm/hr (10 kg/s)

(D) 95,000 lbm/hr (12 kg/s)

8. 332,000 lbm/hr (41.8 kg/s) of 81°F (27°C) water enters a two-pass, counterflow heat exchanger. The heat exchanger is constructed of 1850 ft^2 (172 m^2) of $^5/_8$ in (15.9 mm) copper tubing. Saturated steam is bled from a turbine at 4.45 psia (30.6 kPa) and condenses to saturated liquid. The heated water leaves at 150°F (65.6°C) with an enthalpy of 1100 Btu/lbm (2.56 MJ/kg).

(a) The overall heat transfer coefficient is most nearly

(A) 1900 Btu/hr-ft^2-°F (11 kW/m^2·°C)

(B) 4400 Btu/hr-ft^2-°F (25 kW/m^2·°C)

(C) 6500 Btu/hr-ft^2-°F (37 kW/m^2·°C)

(D) 7700 Btu/hr-ft^2-°F (44 kW/m^2·°C)

(b) The steam extraction rate is most nearly

(A) 1.7×10^8 Btu/hr (48 MW)

(B) 3.9×10^8 Btu/hr (120 MW)

(C) 9.9×10^8 Btu/hr (270 MW)

(D) 2.5×10^9 Btu/hr (700 MW)

9. A two-pass surface condenser constructed of 1 in (25.4 mm) BWG tubing receives 82,000 lbm/hr (10.3 kg/s) of steam from a turbine. Steam enters the condenser with an enthalpy of 980 Btu/lbm (2.280 MJ/kg). The condenser operates at a pressure of 1 in Hg (3.4 kPa) absolute. Water is circulated at 8 ft/sec (2.4 m/s) through an equivalent length of 120 ft (36 m) of extra strong 30 in (76.2 cm) steel pipe. An additional head loss of 6 in wg (1.5 kPa) is incurred in the intake screens.

(a) The head added by the circulating water pump is most nearly

(A) 2.1 ft (0.62 m)

(B) 3.1 ft (0.93 m)

(C) 6.2 ft (1.9 m)

(D) 8.5 ft (2.6 m)

(b) If the water temperature increases 10°F (5.6°C) across the condenser, what is most nearly the circulation rate of cooling water?

(A) 9000 gal/min (34 kL/min)

(B) 11,000 gal/min (42 kL/min)

(C) 13,000 gal/min (49 kL/min)

(D) 15,000 gal/min (57 kL/min)

10. 100 lbm/hr (0.013 kg/s) of 60°F (16°C) water is turned into 14.7 psia (101.3 kPa) saturated steam in an electric boiler. Radiation losses are 35% of the supplied energy. If electricity is $0.04 per kW-hr, the cost is most nearly

(A) $2/hr

(B) $3/hr

(C) $4/hr

(D) $5/hr

11. A gas burner produces 250 lbm/hr (0.032 kg/s) of 98% dry steam at 40 psia (300 kPa) from 60°F (16°C) feedwater. The fuel gas enters at 80°F (26°C) and 4 in Hg (13.6 kPa) and has a heating value of 550 Btu/ft^3 (20.5 MJ/m^3) at standard industrial conditions. The barometric pressure is 30.2 in Hg (102.4 kPa). 13.5 ft^3/min (6.4 L/s) of fuel gas is consumed. The efficiency of the boiler is most nearly

(A) 37%

(B) 43%

(C) 57%

(D) 66%

12. A boiler evaporates 8.23 lbm (8.23 kg) of 120°F (50°C) water per pound (per kilogram) of coal fired, producing 100 psia (700 kPa) saturated steam. The coal is 2% moisture by weight as fired, and dry coal is 5% ash. 1% of the coal is removed from the ash pit. (The ash pit loss has the same composition as unfired, dry coal.) Coal is initially at 60°F (16°C), and combustion occurs at 14.7 psia (101.3 kPa). The combustion products leave at 600°F (315°C). The heating value of the dry coal is 12,800 Btu/lbm (29.80 MJ/kg). The efficiency of the boiler is most nearly

(A) 53%

(B) 68%

(C) 73%

(D) 82%

13. 500 psia (3.5 MPa) steam is superheated to 1000°F (500°C) before expanding through a 75% efficient turbine to 5 psia (30 kPa). No subcooling occurs. The pump work is negligible compared to the 200 MW generated.

(a) The quantity of steam required is most nearly

(A) 8.6×10^5 lbm/hr (120 kg/s)

(B) 1.4×10^6 lbm/hr (190 kg/s)

(C) 2.0×10^6 lbm/hr (270 kg/s)

(D) 4.2×10^6 lbm/hr (310 kg/s)

(b) The heat removed by the condenser is most nearly

(A) 3.2×10^8 Btu/hr (99 MW)

(B) 8.7×10^8 Btu/hr (270 MW)

(C) 2.1×10^9 Btu/hr (650 MW)

(D) 8.7×10^9 Btu/hr (2.7 GW)

14. 191,000 lbm/hr (24 kg/s) of 635°F (335°C) combustion gases flow through a 20 ft (6 m) wide boiler stack whose front and back plates are 5 ft 10 in (1.78 m) apart. An integral crossflow economizer is being designed to heat water from 212°F to 285°F (100°C to 140°C) by dropping the stack gas temperature to 470°F (240°C). Layers of 24 tubes with dimensions 0.957 in (24.3 mm) ID, 1.315 in (33.4 mm) OD, and 20 ft (6 m) length will be placed on a 2.315 in (58.8 mm) pitch in horizontal banks. The overall coefficient of heat transfer for the tubes is 10 Btu/hr-ft^2-°F (57 W/m^2·°C). How many 24-tube layers are required?

(A) 5

(B) 9

(C) 12

(D) 17

15. Water is used in an adiabatic steam desuperheater. 1000 lbm/hr (0.13 kg/s) of 200 psia (1.5 MPa), 600°F (300°C) steam enters with negligible velocity. 50 lbm/hr (0.0063 kg/s) of 82°F (28°C) water enters with negligible velocity. 100 psia (700 kPa) steam leaves the desuperheater at 2000 ft/sec (600 m/s).

(a) The temperature of the leaving steam is most nearly

(A) 330°F (165°C)

(B) 360°F (180°C)

(C) 400°F (200°C)

(D) 470°F (240°C)

(b) The quality of the leaving steam is most nearly

(A) 85%

(B) 88%

(C) 93%

(D) 98%

16. Waste steam at 400°F (200°C) and 100 psia (700 kPa) was originally used only for heating cold water. Cold water entered at 70°F (21°C) and 60 psia (400 kPa). 2000 lbm/hr (0.25 kg/s) of hot water at 180°F (80°C) and 20 psia (150 kPa) were produced. Now, the same quantity of steam is to be expanded through a low-pressure turbine. The low-pressure turbine has an isentropic efficiency of 60% and a mechanical efficiency of 96%. The steam will then flow through a mixing heater (see diagram). A pressure drop of 5 psi (30 kPa) occurs through the heater. The heater output must remain at 180°F (80°C) and 20 psia (150 kPa), but the output at point F may decrease.

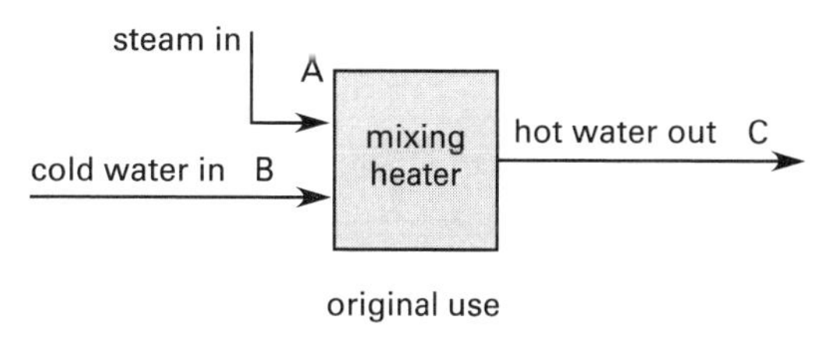

original use

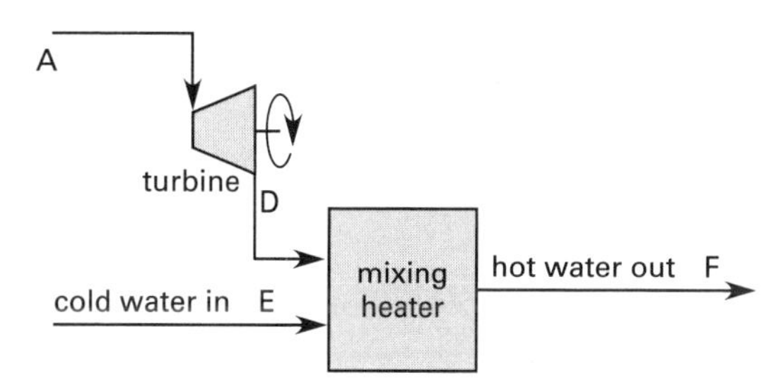

proposed use of waste steam

Power Cycles

(a) The steam flow is most nearly

(A) 160 lbm/hr (0.019 kg/s)

(B) 180 lbm/hr (0.022 kg/s)

(C) 210 lbm/hr (0.024 kg/s)

(D) 250 lbm/hr (0.029 kg/s)

(b) The pressure at point D is most nearly

(A) 15 psia (100 kPa)

(B) 20 psia (145 kPa)

(C) 25 psia (180 kPa)

(D) 30 psia (220 kPa)

(c) The power developed in the turbine is most nearly

(A) 4.7 hp (3.4 kW)

(B) 8.2 hp (6.3 kW)

(C) 14 hp (11 kW)

(D) 27 hp (21 kW)

(d) The flow rate at point F is most nearly

(A) 870 lbm/hr (0.10 kg/s)

(B) 1200 lbm/hr (0.14 kg/s)

(C) 1900 lbm/hr (0.24 kg/s)

(D) 3400 lbm/hr (0.41 kg/s)

SOLUTIONS

1. "BLEVE" is the acronym for "boiling liquid-expanding vapor explosion," essentially the catastrophic failure of a container whose walls cannot withstand the pressure of vaporization. BLEVEs are caused when a leak allows the pressurized liquid contents to flash into a vapor. The term is most frequently used to describe explosions of tanks containing flammable liquids, but it is equally applicable to boilers. Common elements of BLEVE events are weakening of tank walls (usually by fire) and inadequate or nonfunctional pressure-relief valves.

The answer is (B).

2. Cross-coupling provides flexibility integrating redundant, back-up, excess-capacity, and auxiliary units. Cross-coupling permits any available source to be connected to any load, facilitating operation when units are idle or down for maintenance. In addition, when both all loads are online, cross-coupling provides smoother transitions and faster response to fluctuating demand. With a two-two system (i.e., two sources and two loads) in cross-coupled mode, both sources are connected to both loads during periods of high demand; but, only one source is connected to either one load or the other during periods of low demand.

It is unlikely (unnecessary) that both sources would be connected to a single load. Also, it is unlikely that a single source would be sufficient for two loads. Operation with all cross-control valves closed is not a cross-controlled configuration.

The answer is (D).

3. Sensible heat changes temperature, while latent heat changes phase. The pressure throughout the steam generator is 300 psia. Water enters subcooled at 80°F and 300 psia. The water remains in liquid form as it is heated (sensible heat) to the saturation temperature corresponding to 300 psia, which (from App. 23.B) is 417.35°F. Additional energy (latent heat) converts the saturated water to saturated steam at 300 psia.

The enthalpy of 80°F liquid water is essentially a function of temperature only. From App. 23.A, $h_{80^\circ\text{F}} = 48.07$ Btu/lbm. From App. 23.B, $h_{f,300\text{ psia}} = 394.0$ Btu/lbm, and $h_{fg,300\text{ psia}} = 809.4$ Btu/lbm.

The ratio of sensible heat to latent heat is

$$\frac{q_{\text{sensible}}}{q_{\text{latent}}} = \frac{h_{f,300\text{ psia}} - h_{80^\circ\text{F}}}{h_{fg,300\text{ psia}}} = \frac{394.0\ \frac{\text{Btu}}{\text{lbm}} - 48.07\ \frac{\text{Btu}}{\text{lbm}}}{809.4\ \frac{\text{Btu}}{\text{lbm}}} = \boxed{0.427\quad(0.43)}$$

The answer is (D).

4. *Customary U.S. Solution*

From App. 23.B, for 100 psia, the enthalpy of dry steam is $h_1 = 1187.5$ Btu/lbm.

From the Mollier diagram for an isentropic process from 100 psia to 3 psia, $h_2 = 950$ Btu/lbm.

From App. 23.B, for 3 psia,

$$\begin{aligned} h_f &= 109.4 \text{ Btu/lbm} \\ h_{fg} &= 1012.8 \text{ Btu/lbm} \\ h_2' &= h_f + xh_{fg} \\ &= 109.4 \ \frac{\text{Btu}}{\text{lbm}} + (0.9)\left(1012.8 \ \frac{\text{Btu}}{\text{lbm}}\right) \\ &= 1020.9 \text{ Btu/lbm} \end{aligned}$$

From Eq. 26.17, the isentropic efficiency is

$$\eta_s = \frac{h_1 - h_2'}{h_1 - h_2} = \frac{1187.5 \ \frac{\text{Btu}}{\text{lbm}} - 1020.9 \ \frac{\text{Btu}}{\text{lbm}}}{1187.5 \ \frac{\text{Btu}}{\text{lbm}} - 950 \ \frac{\text{Btu}}{\text{lbm}}} = \boxed{0.701 \quad (70\%)}$$

The answer is (D).

SI Solution

From App. 23.O, for 700 kPa, the enthalpy of dry steam is $h_1 = 2762.8$ kJ/kg.

From the Mollier diagram for an isentropic process from 700 kPa to 20 kPa, $h_2 = 2245$ kJ/kg.

From App. 23.O, for 20 kPa,

$$\begin{aligned} h_f &= 251.42 \text{ kJ/kg} \\ h_{fg} &= 2357.5 \text{ kJ/kg} \\ h_2' &= h_f + xh_{fg} = 251.42 \ \frac{\text{kJ}}{\text{kg}} + (0.9)\left(2357.5 \ \frac{\text{kJ}}{\text{kg}}\right) \\ &= 2373.2 \text{ kJ/kg} \end{aligned}$$

From Eq. 26.17, the isentropic efficiency is

$$\eta_s = \frac{h_1 - h_2'}{h_1 - h_2} = \frac{2762.8 \ \frac{\text{kJ}}{\text{kg}} - 2373.2 \ \frac{\text{kJ}}{\text{kg}}}{2762.8 \ \frac{\text{kJ}}{\text{kg}} - 2245 \ \frac{\text{kJ}}{\text{kg}}} = \boxed{0.752 \quad (70\%)}$$

The answer is (D).

5. *Customary U.S. Solution*

(a) From App. 23.B, T_{sat} for 200 psia is 381.80°F.

The steam temperature is

$$381.80°\text{F} + 100°\text{F} = 481.80°\text{F}$$

From App. 23.C, $h_1 = 1258.4$ Btu/lbm.

1 in Hg is approximately 0.5 psia. From the Mollier diagram, assuming isentropic expansion and dropping straight down to the 0.5 psia line, $h_2 \approx 870$ Btu/lbm.

For isentropic expansion, $\eta_{turbine} = 1$, and the steam mass flow rate through the turbine is given by Eq. 26.22.

$$\begin{aligned} \dot{m} &= \frac{P_{turbine}}{h_1 - h_2} = \frac{(5000 \text{ kW})\left(3412 \ \frac{\text{Btu}}{\text{hr-kW}}\right)}{1258.4 \ \frac{\text{Btu}}{\text{lbm}} - 870 \ \frac{\text{Btu}}{\text{lbm}}} \\ &= 4.392 \times 10^4 \text{ lbm/hr} \end{aligned}$$

The water rate is

$$\begin{aligned} \text{WR} &= \frac{\dot{m}}{P_{turbine}} = \frac{4.392 \times 10^4 \ \frac{\text{lbm}}{\text{hr}}}{5000 \text{ kW}} \\ &= \boxed{8.784 \text{ lbm/kW-hr} \quad (8.8 \text{ lbm/kW-hr})} \end{aligned}$$

The answer is (B).

(b) The loss in available energy per unit mass is

$$\begin{aligned} \text{loss} &= \tfrac{1}{2}(h_1 - h_2) \\ &= \left(\tfrac{1}{2}\right)\left(1258.4 \ \frac{\text{Btu}}{\text{lbm}} - 870 \ \frac{\text{Btu}}{\text{lbm}}\right) \\ &= \boxed{194.2 \text{ Btu/lbm} \quad (190 \text{ Btu/lbm})} \end{aligned}$$

The answer is (C).

SI Solution

(a) From App. 23.A, T_{sat} for 1.5 MPa is 198.3°C.

The steam temperature is

$$198.3°\text{C} + 50°\text{C} = 248.3°\text{C}$$

From App. 23.P, $h_1 = 2920$ kJ/kg, and $s_1 = 6.7023$ kJ/kg·K.

Power Cycles

From App. 23.N, for 3.4 kPa, the entropy of saturated liquid, s_f, the entropy of saturated vapor, s_g, the enthalpy of saturated liquid, h_f, and the enthalpy of vaporization, h_{fg}, are

$$s_f = 0.3837 \text{ kJ/kg·K}$$
$$s_g = 8.5316 \text{ kJ/kg·K}$$
$$h_f = 109.84 \text{ kJ/kg}$$
$$h_{fg} = 2438.9 \text{ kJ/kg}$$

For isentropic expansion, $s_1 = s_2 = 6.7023$ kJ/kg·K.

Since $s_2 < s_g$, the expanded steam is in the liquid-vapor region. The quality of the mixture is given by Eq. 23.54.

$$\begin{aligned} x &= \frac{s - s_f}{s_{fg}} = \frac{s - s_f}{s_g - s_f} \\ &= \frac{6.7023 \ \frac{\text{kJ}}{\text{kg·K}} - 0.3837 \ \frac{\text{kJ}}{\text{kg·K}}}{8.5316 \ \frac{\text{kJ}}{\text{kg·K}} - 0.3837 \ \frac{\text{kJ}}{\text{kg·K}}} \\ &= 0.7755 \end{aligned}$$

The final enthalpy is given by Eq. 23.53.

$$\begin{aligned} h &= h_f + xh_{fg} = 109.84 \ \frac{\text{kJ}}{\text{kg}} + (0.7755)\left(2438.9 \ \frac{\text{kJ}}{\text{kg}}\right) \\ &= 2001.2 \text{ kJ/kg} \end{aligned}$$

For isentropic expansion, $\eta_{\text{turbine}} = 1$, and the steam mass flow rate through a turbine is given by Eq. 26.22.

$$\begin{aligned} \dot{m} &= \frac{P_{\text{turbine}}}{h_1 - h_2} = \frac{5000 \text{ kW}}{2920 \ \frac{\text{kJ}}{\text{kg}} - 2001.2 \ \frac{\text{kJ}}{\text{kg}}} \\ &= 5.442 \text{ kg/s} \end{aligned}$$

The water rate is

$$\begin{aligned} \text{WR} &= \frac{\dot{m}}{P_{\text{turbine}}} = \frac{5.442 \ \frac{\text{kg}}{\text{s}}}{5000 \text{ kW}} \\ &= \boxed{1.09 \times 10^{-3} \text{ kg/kW·s} \quad (0.0011 \text{ kg/kW·s})} \end{aligned}$$

The answer is (B).

(b) Loss in availability per unit mass is

$$\begin{aligned} \text{loss} &= \tfrac{1}{2}(h_1 - h_2) \\ &= \left(\tfrac{1}{2}\right)\left(2920 \ \frac{\text{kJ}}{\text{kg}} - 2001.2 \ \frac{\text{kJ}}{\text{kg}}\right) \\ &= \boxed{459.4 \text{ kJ/kg} \quad (460 \text{ kJ/kg})} \end{aligned}$$

The answer is (C).

6. *Customary U.S. Solution*

From App. 23.C, the enthalpy, h_1, of dry steam at 400 psia and 750°F is $h_1 = 1389.9$ Btu/lbm.

From the Mollier diagram, assuming isentropic expansion to 2 in Hg (about 1 psia), $h_2 = 935$ Btu/lbm.

The maximum adiabatic heat drop is

$$\begin{aligned} h_1 - h_2 &= 1389.9 \ \frac{\text{Btu}}{\text{lbm}} - 935 \ \frac{\text{Btu}}{\text{lbm}} \\ &= \boxed{454.9 \text{ Btu/lbm} \quad (450 \text{ Btu/lbm})} \end{aligned}$$

The answer is (A).

SI Solution

From App. 23.P, the enthalpy, h_1, of dry steam at 3.0 MPa and 420°C is $h_1 = 3277.1$ kJ/kg.

From the Mollier diagram, assuming isentropic expansion, $h_2 \approx 2210$ kJ/kg.

The maximum adiabatic heat drop is

$$\begin{aligned} h_1 - h_2 &= 3277.1 \ \frac{\text{kJ}}{\text{kg}} - 2210 \ \frac{\text{kJ}}{\text{kg}} \\ &= \boxed{1067.1 \text{ kJ/kg} \quad (1100 \text{ kJ/kg})} \end{aligned}$$

The answer is (A).

7. *Customary U.S. Solution*

The water rate is

$$\text{WR} = \frac{\dot{m}}{P_{\text{turbine}}}$$

The steam flow rate is

$$\begin{aligned} \dot{m}_{\text{steam}} &= (\text{WR})P_{\text{turbine}} \\ &= \left(20 \ \frac{\text{lbm}}{\text{kW-hr}}\right)(750 \text{ kW}) \\ &= 15{,}000 \text{ lbm/hr} \end{aligned}$$

From App. 23.B, for 165 psia, $T_{\text{sat}} = 366$°F.

The steam temperature is

$$366°\text{F} + 50°\text{F} = 416°\text{F}$$

From App. 23.C, $h_1 = 1226.0$ Btu/lbm.

Power Cycles

From Eq. 26.22,

$$P_{\text{turbine}} = \dot{m}(h_1 - h_2)$$

$$h_1 - h_2 = \frac{P_{\text{turbine}}}{\dot{m}} = \frac{(750 \text{ kW})\left(3412 \ \frac{\text{Btu}}{\text{kW-hr}}\right)}{15{,}000 \ \frac{\text{lbm}}{\text{hr}}}$$

$$= 170.6 \text{ Btu/lbm}$$

$$h_2 = h_1 - 170.6 \ \frac{\text{Btu}}{\text{lbm}} = 1226.0 \ \frac{\text{Btu}}{\text{lbm}} - 170.6 \ \frac{\text{Btu}}{\text{lbm}}$$

$$= 1055.4 \text{ Btu/lbm}$$

$$p_2 = (26 \text{ in Hg})\left(0.491 \ \frac{\text{lbf}}{\text{in}^3}\right)$$

$$= 12.77 \text{ psia}$$

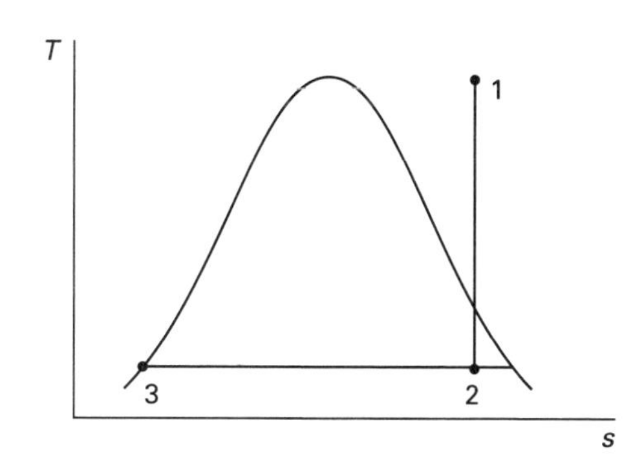

From App. 23.B, for $p_2 = 12.77$ psia,

$$T_{\text{sat}} = T_2 = T_3 \approx 204°\text{F}$$

$$h_{f,3} \approx 172.3 \text{ Btu/lbm}$$

From App. 35.A, the specific heat of water at 65°F is $c_{p,\text{water}} = 0.999$ Btu.

Assuming the water and steam leave in thermal equilibrium, the heat lost by steam is equal to the heat gained by water.

$$\dot{m}_{\text{water}} c_{p,\text{water}} (T_{\text{water,out}} - T_{\text{water,in}}) = \dot{m}_{\text{steam}} (h_2 - h_{f,3})$$

Since the terminal temperature difference is zero, $T_{\text{water,out}} = T_3$.

$$\dot{m}_{\text{water}} \left(0.999 \ \frac{\text{Btu}}{\text{lbm-°F}}\right)(204°\text{F} - 65°\text{F})$$

$$= \left(15{,}000 \ \frac{\text{lbm}}{\text{hr}}\right)\left(1055.3 \ \frac{\text{Btu}}{\text{lbm}} - 172.3 \ \frac{\text{Btu}}{\text{lbm}}\right)$$

$$\dot{m}_{\text{water}} = \boxed{9.54 \times 10^4 \text{ lbm/hr} \quad (95{,}000 \text{ lbm/hr})}$$

The answer is (D).

SI Solution

The steam flow rate is

$$\dot{m}_{\text{water}} = (\text{WR})P_{\text{turbine}}$$

$$= \left(2.5 \times 10^{-3} \ \frac{\text{kg}}{\text{kW·s}}\right)(750 \text{ kW})$$

$$= 1.875 \text{ kg/s}$$

From App. 23.O, for 1 MPa, $T_{\text{sat}} = 179.9°\text{C}$.

The steam temperature is

$$179.9°\text{C} + 30°\text{C} = 209.9°\text{C}$$

From App. 23.P, $h_1 = 2851.0$ kJ/kg.

From Eq. 26.22,

$$P_{\text{turbine}} = \dot{m}(h_1 - h_2)$$

$$h_1 - h_2 = \frac{P_{\text{turbine}}}{\dot{m}} = \frac{750 \text{ kW}}{1.875 \ \frac{\text{kg}}{\text{s}}}$$

$$= 400 \text{ kJ/kg}$$

$$h_2 = h_1 - 400 \ \frac{\text{kJ}}{\text{kg}} = 2851.0 \ \frac{\text{kJ}}{\text{kg}} - 400 \ \frac{\text{kJ}}{\text{kg}}$$

$$= 2451.0 \text{ kJ/kg}$$

From App. 23.O, for $p_2 = 90$ kPa,

$$T_{\text{sat}} = T_2 = T_3 = 96.69°\text{C}$$

$$h_{f,3} = 405.20 \text{ kJ/kg}$$

From App. 35.B, the specific heat of water at 18°C is $c_p = 4.186$ kJ/kg·K.

Assuming water and steam leave in thermal equilibrium, the heat lost by steam is equal to the heat gained by water.

$$\dot{m}_{\text{water}} c_{p,\text{water}} (T_{\text{water,out}} - T_{\text{water,in}}) = \dot{m}_{\text{steam}} (h_2 - h_{f,3})$$

Since the terminal difference is zero, $T_{\text{water,out}} = T_3$.

$$\dot{m}_{\text{water}} \left(4.186 \ \frac{\text{kJ}}{\text{kg·K}}\right)(96.71°\text{C} - 18°\text{C})$$

$$= \left(1.875 \ \frac{\text{kg}}{\text{s}}\right)\left(2451.0 \ \frac{\text{kJ}}{\text{kg}} - 405.20 \ \frac{\text{kJ}}{\text{kg}}\right)$$

$$\dot{m}_{\text{water}} = \boxed{11.64 \text{ kg/s} \quad (12 \text{ kg/s})}$$

The answer is (D).

8. *Customary U.S. Solution*

81°F — water vaporizing → 150°F, 1100 Btu/lbm
A → B
steam condensing ←
157.1°F, saturated liquid — 157.1°F, saturated vapor

(a) From App. 23.A, at 81°F, $h_{\text{water},1} \approx 49.06$ Btu/lbm.

From App. 23.B, for 4.45 psia, $T_{\text{sat}} = 157.1°\text{F}$.

The heat transferred to the water is

$$\begin{aligned} Q &= \dot{m}_{\text{water}}(h_{\text{water},2} - h_{\text{water},1}) \\ &= \left(332{,}000\ \frac{\text{lbm}}{\text{hr}}\right)\left(1100\ \frac{\text{Btu}}{\text{lbm}} - 49.06\ \frac{\text{Btu}}{\text{lbm}}\right) \\ &= 3.489 \times 10^8\ \text{Btu/hr} \end{aligned}$$

The two end temperature differences are

$$\Delta T_A = 157.1°\text{F} - 81°\text{F} = 76.1°\text{F}$$
$$\Delta T_B = 157.1°\text{F} - 150°\text{F} = 7.1°\text{F}$$

The logarithmic mean temperature difference, T_{lm}, from Eq. 36.68 is

$$\begin{aligned} \Delta T_{\text{lm}} &= \frac{\Delta T_A - \Delta T_B}{\ln \frac{\Delta T_A}{\Delta T_B}} = \frac{76.1°\text{F} - 7.1°\text{F}}{\ln \frac{76.1°\text{F}}{7.1°\text{F}}} \\ &= 29.09°\text{F} \end{aligned}$$

Since $T_{\text{steam},A} = T_{\text{steam},B}$, the correction factor, F_c, for ΔT_{lm} is 1.

The overall heat transfer coefficient is calculated from Eq. 36.69.

$$Q = U_o A_o F_c \Delta T_{\text{lm}}$$

$$\begin{aligned} U_o &= \frac{Q}{A_o F_c \Delta T_{\text{lm}}} = \frac{3.489 \times 10^8\ \frac{\text{Btu}}{\text{hr}}}{(1850\ \text{ft}^2)(1)(29.09°\text{F})} \\ &= \boxed{6483\ \text{Btu/hr-ft}^2\text{-°F} \quad (6500\ \text{Btu/hr-ft}^2\text{-°F})} \end{aligned}$$

The answer is (C).

(b) From App. 23.B, the enthalpy of saturated steam at 4.45 psia is $h_1 = 1128.6$ Btu/lbm.

From App. 23.B, the enthalpy of saturated water at 4.45 psia is $h_2 = 125.1$ Btu/lbm.

The enthalpy change for steam during condensation is

$$\begin{aligned} \Delta h &= h_1 - h_2 = 1128.6\ \frac{\text{Btu}}{\text{lbm}} - 125.1\ \frac{\text{Btu}}{\text{lbm}} \\ &= 1003.5\ \text{Btu/lbm} \end{aligned}$$

The heat transferred, Q, from the steam is equal to the heat gained by water.

$$Q = \dot{m}_{\text{steam}} \Delta h$$

$$\begin{aligned} \dot{m}_{\text{steam}} &= \frac{Q}{\Delta h} = \frac{3.489 \times 10^8\ \frac{\text{Btu}}{\text{hr}}}{1003.5\ \frac{\text{Btu}}{\text{lbm}}} \\ &= 3.477 \times 10^5\ \text{lbm/hr} \end{aligned}$$

The energy extraction rate is

$$\begin{aligned} \dot{m}_{\text{steam}} h_1 &= \left(3.477 \times 10^5\ \frac{\text{lbm}}{\text{hr}}\right)\left(1128.6\ \frac{\text{Btu}}{\text{lbm}}\right) \\ &= \boxed{3.92 \times 10^8\ \text{Btu/hr} \quad (3.9 \times 10^8\ \text{Btu/hr})} \end{aligned}$$

The answer is (B).

SI Solution

27°C — water vaporizing → 65.6°C, 2.56 MJ/kg
A → B
69.51°C ← steam condensing — 69.51°C

(a) From App. 23.N, at 27°C, $h_{\text{water},1} = 113.19$ kJ/kg.

From App. 23.O, for 30.6 kPa (0.306 bar), $T_{\text{sat}} = 69.51°\text{C}$.

The heat transferred to the water is

$$\begin{aligned} Q &= \dot{m}_{\text{water}}(h_{\text{water},2} - h_{\text{water},1}) \\ &= \left(41.8\ \frac{\text{kg}}{\text{s}}\right)\left(\left(2.56\ \frac{\text{MJ}}{\text{kg}}\right)\left(1000\ \frac{\text{kJ}}{\text{MJ}}\right) - 113.19\ \frac{\text{kJ}}{\text{kg}}\right) \\ &= 102\,277\ \text{kJ/s} \end{aligned}$$

The two end temperature differences are

$$\Delta T_A = 69.51°\text{C} - 27°\text{C} = 42.5°\text{C}$$
$$\Delta T_B = 69.51°\text{C} - 65.6°\text{C} = 3.9°\text{C}$$

The logarithmic mean temperature difference, T_{lm}, is calculated from Eq. 36.68.

$$\Delta T_{\text{lm}} = \frac{\Delta T_A - \Delta T_B}{\ln \frac{\Delta T_A}{\Delta T_B}} = \frac{42.5°\text{C} - 3.9°\text{C}}{\ln \frac{42.5°\text{C}}{3.9°\text{C}}} = 16.16°\text{C}$$

Since $T_{\text{steam},A} = T_{\text{steam},B}$, the correction factor, F_c, for ΔT_{lm} is 1.

The overall heat transfer coefficient is calculated from Eq. 36.69.

$$Q = U_o A_o F_c \Delta T_{\text{lm}}$$

$$U_o = \frac{Q}{A_o F_c \Delta T_{\text{lm}}} = \frac{\left(102\,277\ \frac{\text{kJ}}{\text{s}}\right)\left(1000\ \frac{\text{W}}{\text{kW}}\right)}{(172\ \text{m}^2)(1)(16.16°\text{C})} = \boxed{36\,797\ \text{W/m}^2\cdot°\text{C} \quad (37\ \text{kW/m}^2\cdot°\text{C})}$$

The answer is (C).

(b) From App. 23.O, the enthalpy of saturated steam at 30.6 kPa is $h_1 = 2625.2$ kJ/kg.

From App. 23.O, the enthalpy of saturated water at 30.6 kPa is $h_2 = 290.97$ kJ/kg.

The enthalpy change for steam during condensation is

$$\Delta h = h_1 - h_2 = 2625.2\ \frac{\text{kJ}}{\text{kg}} - 290.97\ \frac{\text{kJ}}{\text{kg}} = 2334.2\ \text{kJ/kg}$$

The heat transfer, Q, from the steam is equal to the heat gained by water.

$$Q = \dot{m}_{\text{steam}} \Delta h$$

$$\dot{m}_{\text{steam}} = \frac{Q}{\Delta h} = \frac{102\,277\ \frac{\text{kJ}}{\text{s}}}{2334.2\ \frac{\text{kJ}}{\text{kg}}} = 43.82\ \text{kg/s}$$

The energy extraction rate is

$$\dot{m}_{\text{steam}} h_1 = \left(43.82\ \frac{\text{kg}}{\text{s}}\right)\left(2625.2\ \frac{\text{kJ}}{\text{kg}}\right) = \boxed{115\,036\ \text{kW} \quad (120\ \text{MW})}$$

The answer is (B).

9.

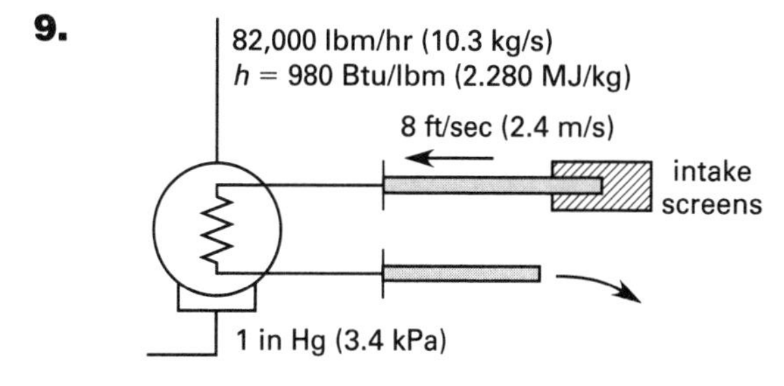

Customary U.S. Solution

(a) From App. 16.B, the dimensions of extra strong 30 in steel pipe are

$$D_i = 2.4167\ \text{ft}$$
$$A_i = 4.5869\ \text{ft}^2$$

From Table 17.2, the specific roughness for steel is $\epsilon = 0.0002$ ft.

$$\frac{\epsilon}{D} = \frac{0.0002\ \text{ft}}{2.4167\ \text{ft}} = 0.000083$$

Assume fully turbulent flow. (The Reynolds number can also be calculated.)

From the Moody friction factor chart for fully turbulent flow, $f \approx 0.012$.

The head loss from Eq. 17.22 is

$$h_f = \frac{fLv^2}{2Dg} = \frac{(0.012)(120\ \text{ft})\left(8\ \frac{\text{ft}}{\text{sec}}\right)^2}{(2)(2.4167\ \text{ft})\left(32.2\ \frac{\text{ft}}{\text{sec}^2}\right)} = 0.59\ \text{ft}$$

The screen loss is

$$6\ \text{in wg} = \left(\frac{6\ \text{in}}{12\ \frac{\text{in}}{\text{ft}}}\right) = 0.5\ \text{ft}$$

$$\text{velocity head} = \frac{v^2}{2g} = \frac{\left(8\ \frac{\text{ft}}{\text{sec}}\right)^2}{(2)\left(32.2\ \frac{\text{ft}}{\text{sec}^2}\right)} = 0.99\ \text{ft}$$

There is inadequate information to evaluate pressure and elevation heads. The total head added by a coolant pump (not shown) not including losses inside the condenser is

$$0.59\ \text{ft} + 0.5\ \text{ft} + 0.99\ \text{ft} = \boxed{2.08\ \text{ft} \quad (2.1\ \text{ft})}$$

The answer is (A).

Power Cycles

(b) The condenser pressure is

$$(1\ \text{in Hg})\left(0.491\ \frac{\frac{\text{lbf}}{\text{in}^2}}{\text{in Hg}}\right) = 0.5\ \text{psia}$$

From App. 23.B, the enthalpy of the saturated liquid is $h_f = 47.08$ Btu/lbm.

The specific heat of water is $c_{p,\text{water}} = 1$ Btu/lbm-°F.

The heat lost by the steam is equal to the heat gained by the water.

$$\dot{m}_{\text{water}} c_{p,\text{water}} \Delta T = \dot{m}_{\text{steam}}(h_2 - h_f)$$

$$\dot{m}_{\text{water}}\left(1\ \frac{\text{Btu}}{\text{lbm-°F}}\right)(10°\text{F}) = \left(82{,}000\ \frac{\text{lbm}}{\text{hr}}\right) \times \left(980\ \frac{\text{Btu}}{\text{lbm}} - 47.08\ \frac{\text{Btu}}{\text{lbm}}\right)$$

$$\dot{m}_{\text{water}} = 7.6499 \times 10^6\ \text{lbm/hr}$$

The density of water, ρ, is 62.4 lbm/ft^3.

$$Q = \frac{\dot{m}_{\text{water}}}{\rho} = \frac{\left(7.6499 \times 10^6\ \frac{\text{lbm}}{\text{hr}}\right)\left(7.48\ \frac{\text{gal}}{\text{ft}^3}\right)}{\left(62.4\ \frac{\text{lbm}}{\text{ft}^3}\right)\left(60\ \frac{\text{min}}{\text{hr}}\right)} = \boxed{1.528 \times 10^4\ \text{gal/min}\quad (15{,}000\ \text{gal/min})}$$

(The flow rate can also be determined from the velocity and pipe area. However, this does not use the 10° data or perform an energy balance.)

The answer is (D).

SI Solution

(a) From App. 16.B, the dimensions of extra strong 30 in steel pipe, using the footnote from the table, are

$$D_i = \frac{(29.00\ \text{in})\left(25.4\ \frac{\text{mm}}{\text{in}}\right)}{1000\ \frac{\text{mm}}{\text{m}}} = 0.7366\ \text{m}$$

$$A_i = \frac{(660.52\ \text{in}^2)\left(25.4\ \frac{\text{mm}}{\text{in}}\right)^2}{\left(1000\ \frac{\text{mm}}{\text{m}}\right)^2} = 0.4261\ \text{m}^2$$

From Table 17.2, the specific roughness for steel is $\epsilon = 6.0 \times 10^{-5}$ m.

$$\frac{\epsilon}{D} = \frac{6.0 \times 10^{-5}\ \text{m}}{0.7366\ \text{m}} = 0.0000815$$

Assume fully turbulent flow. (The Reynolds number can also be calculated.)

From the Moody friction factor chart for fully turbulent flow, $f \approx 0.012$.

The head loss from Eq. 17.22 is

$$h_f = \frac{fL\text{v}^2}{2Dg} = \frac{(0.012)(36\ \text{m})\left(2.4\ \frac{\text{m}}{\text{s}}\right)^2}{(2)(0.7366\ \text{m})\left(9.81\ \frac{\text{m}}{\text{s}^2}\right)} = 0.1722\ \text{m}$$

The screen loss is 1.5 kPa.

$$h_{\text{screen}} = \frac{\Delta p}{\rho g} = \frac{(1.5\ \text{kPa})\left(1000\ \frac{\text{Pa}}{\text{kPa}}\right)}{\left(1000\ \frac{\text{kg}}{\text{m}^3}\right)\left(9.81\ \frac{\text{m}}{\text{s}^2}\right)} = 0.1529\ \text{m}$$

The velocity head is

$$\frac{\text{v}^2}{2g} = \frac{\left(2.4\ \frac{\text{m}}{\text{s}}\right)^2}{(2)\left(9.81\ \frac{\text{m}}{\text{s}^2}\right)} = 0.2936\ \text{m}$$

There is inadequate information to evaluate pressure and elevation heads. The total head added by a coolant pump (not shown) not including losses inside the condenser is

$$0.1722\ \text{m} + 0.1529\ \text{m} + 0.2936\ \text{m} = \boxed{0.6187\ \text{m}\ (0.62\ \text{m})}$$

The answer is (A).

(b) The condenser pressure is 3.4 kPa.

From App. 23.N, the enthalpy of saturated liquid is $h_f = 109.8$ kJ/kg.

The specific heat of water is $c_{p,\text{water}} = 4.187$ kJ/kg·K.

The heat lost by the steam is equal to the heat gained by the water.

$$\dot{m}_{\text{water}} c_{p,\text{water}} \Delta T = \dot{m}_{\text{steam}}(h_2 - h_f)$$

$$\dot{m}_{\text{water}}\left(4.187\ \frac{\text{kJ}}{\text{kg·K}}\right)(5.6°\text{C}) = \left(10.3\ \frac{\text{kg}}{\text{s}}\right) \times \left(\left(2.280\ \frac{\text{MJ}}{\text{kg}}\right)\left(1000\ \frac{\text{kJ}}{\text{MJ}}\right) - 109.8\ \frac{\text{kJ}}{\text{kg}}\right)$$

$$\dot{m}_{\text{water}} = 953.3\ \text{kg/s}$$

The density of water, ρ, is 1000 kg/m^3.

$$Q = \frac{\dot{m}_{\text{water}}}{\rho} = \left(\frac{953.3\ \frac{\text{kg}}{\text{s}}}{1000\ \frac{\text{kg}}{\text{m}^3}}\right)\left(1000\ \frac{\text{L}}{\text{m}^3}\right)\left(60\ \frac{\text{s}}{\text{min}}\right) = \boxed{57\,200\ \text{L/min}\quad (57\ \text{kL/min})}$$

The answer is (D).

10. *Customary U.S. Solution*

From App. 23.A, the enthalpy of saturated liquid at 60°F is $h_1 = 28.08$ Btu/lbm.

From App. 23.B, the enthalpy of saturated steam at 14.7 psia is $h_2 = 1150.3$ Btu/lbm.

The heat transfer rate to the water is

$$\begin{aligned} Q &= \dot{m}(h_2 - h_1) \\ &= \left(100\ \frac{\text{lbm}}{\text{hr}}\right)\left(1150.3\ \frac{\text{Btu}}{\text{lbm}} - 28.08\ \frac{\text{Btu}}{\text{lbm}}\right) \\ &= 1.122 \times 10^5\ \text{Btu/hr} \\ &= \frac{1.122 \times 10^5\ \frac{\text{Btu}}{\text{hr}}}{\left(1000\ \frac{\text{W}}{\text{kW}}\right)\left(3.412\ \frac{\text{Btu}}{\text{W-hr}}\right)} \\ &= 32.89\ \text{kW} \end{aligned}$$

$$\begin{aligned} \text{cost} &= \frac{(32.89\ \text{kW})\left(\frac{\$0.04}{\text{kW-hr}}\right)}{1 - 0.35} \\ &= \boxed{\$2.02/\text{hr}\quad(\$2/\text{hr})} \end{aligned}$$

The answer is (A).

SI Solution

From App. 23.N, for saturated liquid at 16°C, $h_1 = 67.17$ kJ/kg.

From App. 23.N, the enthalpy of saturated steam at 101.3 kPa is $h_2 = 2675.4$ kJ/kg.

The heat transfer rate to the water is

$$\begin{aligned} Q &= \dot{m}(h_2 - h_1) \\ &= \left(0.013\ \frac{\text{kg}}{\text{s}}\right)\left(2675.4\ \frac{\text{kJ}}{\text{kg}} - 67.17\ \frac{\text{kJ}}{\text{kg}}\right) \\ &= 33.907\ \text{kW} \end{aligned}$$

$$\begin{aligned} \text{cost} &= \frac{(33.907\ \text{kW})\left(\frac{\$0.04}{\text{kW·h}}\right)}{1 - 0.35} \\ &= \boxed{\$2.09/\text{h}\quad(\$2/\text{h})} \end{aligned}$$

The answer is (A).

11. *Customary U.S. Solution*

From App. 23.A, the enthalpy of saturated liquid at 60°F is $h_1 = 28.08$ Btu/lbm.

From App. 23.B, for 40 psia steam, the enthalpy of saturated liquid, h_f, is 236.1 Btu/lbm. The heat of vaporization, h_{fg}, is 933.7 Btu/lbm. The enthalpy is given by Eq. 23.53.

$$\begin{aligned} h_2 &= h_f + xh_{fg} \\ &= 236.1\ \frac{\text{Btu}}{\text{lbm}} + (0.98)\left(933.7\ \frac{\text{Btu}}{\text{lbm}}\right) \\ &= 1151.13\ \text{Btu/lbm} \end{aligned}$$

The heat transfer rate is

$$\begin{aligned} Q &= \dot{m}(h_2 - h_1) \\ &= \frac{\left(250\ \frac{\text{lbm}}{\text{hr}}\right)\left(1151.13\ \frac{\text{Btu}}{\text{lbm}} - 28.08\ \frac{\text{Btu}}{\text{lbm}}\right)}{60\ \frac{\text{min}}{\text{hr}}} \\ &= 4679.4\ \text{Btu/min} \end{aligned}$$

Find the volume of gas used at standard conditions for a heating gas (60°F).

$$\begin{aligned} \dot{V}_{\text{std}} &= \dot{V}\left(\frac{T_0}{T}\right)\left(\frac{p}{p_0}\right) \\ &= \left(13.5\ \frac{\text{ft}^3}{\text{min}}\right)\left(\frac{60°\text{F} + 460°}{80°\text{F} + 460°}\right) \\ &\quad \times \left(\frac{(4\ \text{in Hg} + 30.2\ \text{in Hg})\left(0.491\ \frac{\text{lbf}}{\text{in}^3}\right)}{14.7\ \frac{\text{lbf}}{\text{in}^2}}\right) \\ &= 14.85\ \text{SCFM} \end{aligned}$$

The efficiency of the boiler is

$$\begin{aligned} \eta &= \frac{Q}{\text{heat input}} = \frac{4679.4\ \frac{\text{Btu}}{\text{min}}}{\left(14.85\ \frac{\text{ft}^3}{\text{min}}\right)\left(550\ \frac{\text{Btu}}{\text{ft}^3}\right)} \\ &= \boxed{0.573\quad(57\%)} \end{aligned}$$

The answer is (C).

SI Solution

From App. 23.N, the enthalpy of saturated liquid at 16°C is $h_1 = 67.17$ kJ/kg.

From App. 23.O, for 300 kPa steam, the enthalpy of saturated liquid, h_f, is 561.43 kJ/kg. The enthalpy of vaporization, h_{fg}, is 2163.5 kJ/kg. The enthalpy is given by Eq. 23.53.

$$\begin{aligned} h_2 &= h_f + xh_{fg} = 561.43\ \frac{\text{kJ}}{\text{kg}} + (0.98)\left(2163.5\ \frac{\text{kJ}}{\text{kg}}\right) \\ &= 2681.7\ \text{kJ/kg} \end{aligned}$$

The heat transfer rate is

$$\begin{aligned} Q &= \dot{m}(h_2 - h_1) \\ &= \left(0.032\ \frac{\text{kg}}{\text{s}}\right)\left(2681.7\ \frac{\text{kJ}}{\text{kg}} - 67.17\ \frac{\text{kJ}}{\text{kg}}\right) \\ &= 83.66\ \text{kJ/s} \end{aligned}$$

Find the volume of the gas used at standard conditions for a heating gas (16°C).

$$\begin{aligned} \dot{V}_{\text{std}} &= \dot{V}\left(\frac{T_0}{T}\right)\left(\frac{p}{p_0}\right) \\ &= \frac{\left(6.4\ \frac{\text{L}}{\text{s}}\right)\left(\frac{16^\circ\text{C} + 273^\circ}{26^\circ\text{C} + 273^\circ}\right)\left(\frac{13.6\ \text{kPa} + 102.4\ \text{kPa}}{101.3\ \text{kPa}}\right)}{1000\ \frac{\text{L}}{\text{m}^3}} \\ &= 0.00708\ \text{m}^3/\text{s} \end{aligned}$$

The efficiency of the boiler is

$$\begin{aligned} \eta &= \frac{Q}{\text{heat input}} \\ &= \frac{83.66\ \frac{\text{kJ}}{\text{s}}}{\left(0.00708\ \frac{\text{m}^3}{\text{s}}\right)\left(20.5\ \frac{\text{MJ}}{\text{m}^3}\right)\left(1000\ \frac{\text{kJ}}{\text{MJ}}\right)} \\ &= \boxed{0.576\quad(57\%)} \end{aligned}$$

The answer is (C).

12. *Customary U.S. Solution*

step 1: Determine the actual gravimetric analysis of the coal as fired. 1 lbm of coal contains 0.02 lbm moisture, leaving 0.98 lbm dry coal. 1% is lost to the ash pit. The remainder is 0.98 lbm − 0.01 lbm = 0.97 lbm.

step 2: Calculate the heating value of the remaining coal.

$$\text{HV} = (0.97\ \text{lbm})\left(12{,}800\ \frac{\text{Btu}}{\text{lbm}}\right) = 12{,}416\ \text{Btu}$$

step 3: Find the energy, Q, required to produce steam. From App. 23.A, the enthalpy of water at 120°F is $h_1 = 88.00$ Btu/lbm.

From App. 23.B, the enthalpy of saturated steam at 100 psia is $h_2 = 1187.5$ Btu/lbm.

$$\begin{aligned} Q &= m(h_2 - h_1) \\ &= (8.23\ \text{lbm})\left(1187.5\ \frac{\text{Btu}}{\text{lbm}} - 88.00\ \frac{\text{Btu}}{\text{lbm}}\right) \\ &= 9048.9\ \text{Btu} \end{aligned}$$

step 4: The combustion efficiency is

$$\eta = \frac{Q}{\text{HV}} = \frac{9048.9\ \text{Btu}}{12{,}416\ \text{Btu}} = \boxed{0.729\quad(73\%)}$$

The answer is (C).

SI Solution

Since boiler data are based on 1 unit mass of coal fired, step 1 will be the same as for the customary U.S. solution. Repeat the rest of the steps as follows.

step 2: Calculate the heating value of the remaining coal.

$$\text{HV} = (0.97\ \text{kg})\left(29.80\ \frac{\text{MJ}}{\text{kg}}\right)\left(1000\ \frac{\text{kJ}}{\text{MJ}}\right) = 28\,906\ \text{kJ}$$

step 3: Find the energy, Q, required to produce steam. From App. 23.N, the enthalpy of water at 50°C is $h_1 = 209.34$ kJ/kg.

From App. 23.O, the enthalpy of saturated steam at 700 kPa is $h_2 = 2762.8$ kJ/kg.

$$\begin{aligned} Q &= m(h_2 - h_1) \\ &= (8.23\ \text{kg})\left(2762.8\ \frac{\text{kJ}}{\text{kg}} - 209.34\ \frac{\text{kJ}}{\text{kg}}\right) \\ &= 21\,015\ \text{kJ} \end{aligned}$$

step 4: The combustion efficiency is

$$\eta = \frac{Q}{\text{HV}} = \frac{21\,015\ \text{kJ}}{28\,906\ \text{kJ}} = \boxed{0.727\quad(73\%)}$$

The answer is (C).

13. *Customary U.S. Solution*

(a) Refer to Fig. 27.1.

At point D (leaving the superheater and entering the turbine), the enthalpy, h_D, and entropy, s_D, can be obtained from App. 23.C.

$$\begin{aligned} h_D &= 1521.0\ \text{Btu/lbm} \\ s_D &= 1.7376\ \text{Btu/lbm-}^\circ\text{F} \end{aligned}$$

For isentropic expansion through the turbine, $s_E = s_D$. From App. 23.B, the enthalpy of saturated liquid, h_f, the enthalpy of evaporation, h_{fg}, the entropy of saturated liquid, s_f, and the entropy of evaporation, s_{fg}, are

$$h_f = 130.2 \text{ Btu/lbm}$$
$$h_{fg} = 1000.5 \text{ Btu/lbm}$$
$$s_f = 0.2349 \text{ Btu/lbm-°F}$$
$$s_{fg} = 1.6089 \text{ Btu/lbm-°F}$$

The quality of the mixture for an isentropic process is given by Eq. 23.54 (at point E).

$$x_E = \frac{s_E - s_f}{s_{fg}} = \frac{1.7376 \frac{\text{Btu}}{\text{lbm-°F}} - 0.2349 \frac{\text{Btu}}{\text{lbm-°F}}}{1.6089 \frac{\text{Btu}}{\text{lbm-°F}}} = 0.9340$$

The isentropic enthalpy is given by Eq. 23.53 (at point E).

$$\begin{aligned} h_E &= h_f + x_E h_{fg} \\ &= 130.2 \frac{\text{Btu}}{\text{lbm}} + (0.9340)\left(1000.5 \frac{\text{Btu}}{\text{lbm}}\right) \\ &= 1064.7 \text{ Btu/lbm} \end{aligned}$$

From Eq. 26.19, the actual enthalpy of steam at point E is

$$\begin{aligned} h'_E &= h_D - \eta_s(h_D - h_E) \\ &= 1521 \frac{\text{Btu}}{\text{lbm}} - (0.75)\left(1521 \frac{\text{Btu}}{\text{lbm}} - 1064.7 \frac{\text{Btu}}{\text{lbm}}\right) \\ &= 1178.8 \text{ Btu/lbm} \end{aligned}$$

Since the pump work is negligible, the mass flow rate of steam is

$$\begin{aligned} \dot{m} &= \frac{P}{W_{\text{turbine}}} = \frac{P}{h_D - h'_E} \\ &= \frac{(200 \text{ MW})\left(10^6 \frac{\text{W}}{\text{MW}}\right)\left(3.412 \frac{\text{Btu}}{\text{hr-W}}\right)}{1521 \frac{\text{Btu}}{\text{lbm}} - 1178.8 \frac{\text{Btu}}{\text{lbm}}} \\ &= \boxed{1.994 \times 10^6 \text{ lbm/hr} \quad (2.0 \times 10^6 \text{ lbm/hr})} \end{aligned}$$

The answer is (C).

(b) The heat removed by the condenser is given by Eq. 26.30.

$$\dot{Q} = \dot{m}(h'_E - h_F)$$

From App. 23.B, at 5 psia, $h_F = 130.2$ Btu/lbm.

$$\begin{aligned} \dot{Q} &= \left(1.994 \times 10^6 \frac{\text{lbm}}{\text{hr}}\right)\left(1178.8 \frac{\text{Btu}}{\text{lbm}} - 130.2 \frac{\text{Btu}}{\text{lbm}}\right) \\ &= \boxed{2.090 \times 10^9 \text{ Btu/hr} \quad (2.1 \times 10^9 \text{ Btu/hr})} \end{aligned}$$

The answer is (C).

SI Solution

(a) Refer to Fig. 27.1.

At point D (leaving the superheater and entering the turbine), the enthalpy, h_D, and entropy, s_D, can be obtained from the Mollier diagram.

$$h_D \approx 3430 \text{ kJ/kg}$$
$$s_D \approx 7.25 \text{ kJ/kg·K}$$

For isentropic expansion through the turbine, $s_E = s_D$. From App. 23.O, the enthalpy of saturated liquid, h_f, the enthalpy of evaporation, h_{fg}, the entropy of saturated liquid, s_f, and the entropy of saturated vapor, s_g, are

$$h_f = 289.27 \text{ kJ/kg}$$
$$h_{fg} = 2335.3 \text{ kJ/kg}$$
$$s_f = 0.9441 \text{ kJ/kg·K}$$
$$s_g = 7.7675 \text{ kJ/kg·K}$$

The quality of the mixture for an isentropic process is given by Eq. 23.54 (at point E) as

$$\begin{aligned} x_E &= \frac{s_E - s_f}{s_{fg}} = \frac{s_E - s_f}{s_g - s_f} \\ &= \frac{7.25 \frac{\text{kJ}}{\text{kg·K}} - 0.9441 \frac{\text{kJ}}{\text{kg·K}}}{7.7675 \frac{\text{kJ}}{\text{kg·K}} - 0.9441 \frac{\text{kJ}}{\text{kg·K}}} \\ &= 0.9242 \end{aligned}$$

The isentropic enthalpy is given by Eq. 23.53 (at point E) as

$$\begin{aligned} h_E &= h_f + x_E h_{fg} \\ &= 289.27 \frac{\text{kJ}}{\text{kg}} + (0.9242)\left(2335.3 \frac{\text{kJ}}{\text{kg}}\right) \\ &= 2447.6 \text{ kJ/kg} \end{aligned}$$

From Eq. 26.19, the actual enthalpy of steam at point E is

$$\begin{aligned} h'_E &= h_D - \eta_s(h_D - h_E) \\ &= 3430\ \frac{\text{kJ}}{\text{kg}} - (0.75)\left(3430\ \frac{\text{kJ}}{\text{kg}} - 2447.6\ \frac{\text{kJ}}{\text{kg}}\right) \\ &= 2693.2\ \text{kJ/kg} \end{aligned}$$

Since the pump work is negligible, the mass flow rate of steam is

$$\begin{aligned} \dot{m} &= \frac{P}{W_{\text{turbine}}} = \frac{P}{h_D - h'_E} = \frac{(200\ \text{MW})\left(10^3\ \frac{\text{kW}}{\text{MW}}\right)}{3430\ \frac{\text{kJ}}{\text{kg}} - 2693.2\ \frac{\text{kJ}}{\text{kg}}} \\ &= \boxed{271.4\ \text{kg/s}\quad (270\ \text{kg/s})} \end{aligned}$$

The answer is (C).

(b) The heat removed by the condenser is given by Eq. 26.30.

$$\dot{Q} = \dot{m}(h'_E - h_F)$$

From App. 23.O at 30 kPa, $h_F = 289.27$ kJ/kg.

$$\begin{aligned} \dot{Q} &= \left(271.4\ \frac{\text{kg}}{\text{s}}\right)\left(2693.2\ \frac{\text{kJ}}{\text{kg}} - 289.27\ \frac{\text{kJ}}{\text{kg}}\right) \\ &= \boxed{6.524 \times 10^5\ \text{kW}\quad (650\ \text{MW})} \end{aligned}$$

The answer is (C).

14. *Customary U.S. Solution*

The illustration shows one of N layers. Each layer consists of $n = 24$ tubes, only 3 of which are shown.

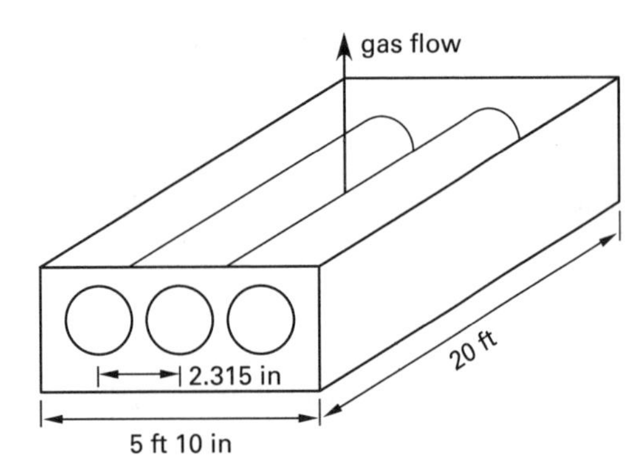

Assume stack gases consist primarily of nitrogen. The average gas temperature is

$$T_{\text{ave}} = \left(\tfrac{1}{2}\right)(635°\text{F} + 470°\text{F}) + 460° = 1012.5°\text{R}$$

The specific heat of nitrogen is calculated from Eq. 23.107 using constants given in Table 23.9.

$$\begin{aligned} c_p &= A + BT + CT^2 + \frac{D}{\sqrt{T}} \\ &= 0.2510\ \frac{\text{Btu}}{\text{lbm-°R}} + \left(-1.63 \times 10^{-5}\ \frac{\text{Btu}}{\text{lbm-°R}}\right)(1012.5°\text{R}) \\ &\quad + \left(20.4 \times 10^{-9}\ \frac{\text{Btu}}{\text{lbm-°R}}\right)(1012.5°\text{R})^2 + 0\ \frac{\text{Btu}}{\text{lbm-°R}} \\ &= 0.2554\ \text{Btu/lbm-°R} \end{aligned}$$

For the purpose of calculating the logarithmic temperature difference, assume counterflow operation. The two end temperature differences are

$$\Delta T_A = 635°\text{F} - 285°\text{F} = 350°\text{F}$$

$$\Delta T_B = 470°\text{F} - 212°\text{F} = 258°\text{F}$$

	gases →	
635°F		470°F
$\Delta T = 350°$F		$\Delta T = 258°$F
285°F	← water	212°F

The logarithmic temperature difference (see Eq. 36.68) is

$$\begin{aligned} \Delta T_{\text{lm}} &= \frac{\Delta T_A - \Delta T_B}{\ln \frac{\Delta T_A}{\Delta T_B}} = \frac{350°\text{F} - 258°\text{F}}{\ln \frac{350°\text{F}}{258°\text{F}}} \\ &= 301.7°\text{F} \end{aligned}$$

Determine F_c for crossflow.

$$R = \frac{635°\text{F} - 470°\text{F}}{285°\text{F} - 212°\text{F}} = 2.26$$

$$x = \frac{285°\text{F} - 212°\text{F}}{635°\text{F} - 212°\text{F}} = 0.17$$

From App. 36.D, F_c for all heat exchanger configurations is close to 1.0 for this set of parameters.

The heat transfer from the temperature gain of water is equal to the heat transfer based on the logarithmic mean temperature difference. From Eq. 36.69,

$$Q = \dot{m}c_p\Delta T = U_oA_oF_c\Delta T_{\text{lm}}$$

$$\begin{aligned} \left(191{,}000\ \frac{\text{lbm}}{\text{hr}}\right)\left(0.2554\ \frac{\text{Btu}}{\text{lbm-°R}}\right)(635°\text{F} - 470°\text{F}) \\ = \left(10\ \frac{\text{Btu}}{\text{hr-ft}^2\text{-°F}}\right)A_o(1.0)(301.7°\text{F}) \end{aligned}$$

$$A_o = 2667.9\ \text{ft}^2$$

The tube area per bank is

$$A_{\text{bank}} = n\pi D_o L = \frac{24\pi(1.315\text{ in})(20\text{ ft})}{12\ \frac{\text{in}}{\text{ft}}} = 165.2\text{ ft}^2$$

$$N = \frac{A_o}{A_{\text{bank}}} = \frac{2667.9\text{ ft}^2}{165.2\text{ ft}^2} = \boxed{16.1\quad(17)}$$

The answer is (D).

SI Solution

The illustration shows one of N layers. Each layer consists of $n = 24$ tubes, only 3 of which are shown.

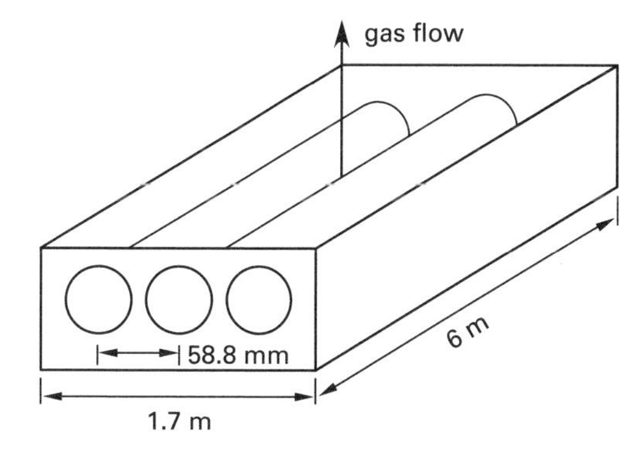

Assume stack gas consists primarily of nitrogen. The average gas temperature is

$$\begin{aligned} T_{\text{ave}} &= \left(\tfrac{1}{2}\right)(335^\circ\text{C} + 240^\circ\text{C}) + 273^\circ \\ &= 560.5\text{K} \quad (1009^\circ\text{R}) \end{aligned}$$

Use the value of the specific heat of nitrogen that was calculated for the customary U.S. solution, since the two temperatures are almost the same.

From the footnote in Table 23.9,

$$\begin{aligned} c_p &= \left(0.2554\ \frac{\text{Btu}}{\text{lbm-}^\circ\text{R}}\right)\left(4.1868\ \frac{\frac{\text{kJ}}{\text{kg}\cdot^\circ\text{C}}}{\frac{\text{Btu}}{\text{lbm-}^\circ\text{R}}}\right)\left(1000\ \frac{\text{J}}{\text{kJ}}\right) \\ &= 1069\ \text{J/kg}\cdot^\circ\text{C} \end{aligned}$$

For the purpose of calculating the logarithmic temperature difference, assume counterflow operation. The two end temperature differences are

$$\Delta T_A = 335^\circ\text{C} - 140^\circ\text{C} = 195^\circ\text{C}$$

$$\Delta T_B = 240^\circ\text{C} - 100^\circ\text{C} = 140^\circ\text{C}$$

	gases	
335°C	→	240°C
ΔT = 195°C		ΔT = 140°C
140°C	← water	100°C

The logarithmic mean temperature difference (see Eq. 36.68) is

$$\begin{aligned} \Delta T_{\text{lm}} &= \frac{\Delta T_A - \Delta T_B}{\ln\frac{\Delta T_A}{\Delta T_B}} = \frac{195^\circ\text{C} - 140^\circ\text{C}}{\ln\frac{195^\circ\text{C}}{140^\circ\text{C}}} \\ &= 166.0^\circ\text{C} \end{aligned}$$

$$R = \frac{335^\circ\text{C} - 240^\circ\text{C}}{140^\circ\text{C} - 100^\circ\text{C}} = 2.38$$

$$x = \frac{140^\circ\text{C} - 100^\circ\text{C}}{335^\circ\text{C} - 100^\circ\text{C}} = 0.17$$

From App. 36.D, F_c for all heat exchanger configurations is close to 1.0 for this set of conditions.

The heat transfer from the temperature gain of water is equal to the heat transfer based on the logarithmic mean temperature difference. From Eq. 36.69,

$$Q = \dot{m}c_p\Delta T = U_oA_oF_c\Delta T_{lm}$$

$$\begin{aligned} &\left(24\ \frac{\text{kg}}{\text{s}}\right)\left(1069\ \frac{\text{J}}{\text{kg}\cdot^\circ\text{C}}\right)(335^\circ\text{C} - 240^\circ\text{C}) \\ &\quad = \left(57\ \frac{\text{W}}{\text{m}^2\cdot^\circ\text{C}}\right)A_o(1.0)(166.0^\circ\text{C}) \\ &A_o = 257.6\text{ m}^2 \end{aligned}$$

The tube area per bank is

$$\begin{aligned} A_{\text{bank}} &= n\pi D_o L = \frac{24\pi(33.4\text{ mm})(6\text{ m})}{1000\ \frac{\text{mm}}{\text{m}}} \\ &= 15.1\text{ m}^2 \end{aligned}$$

$$N = \frac{A_o}{A_{\text{bank}}} = \frac{257.6\text{ m}^2}{15.1\text{ m}^2} = \boxed{17.1\quad(17)}$$

The answer is (D).

15.

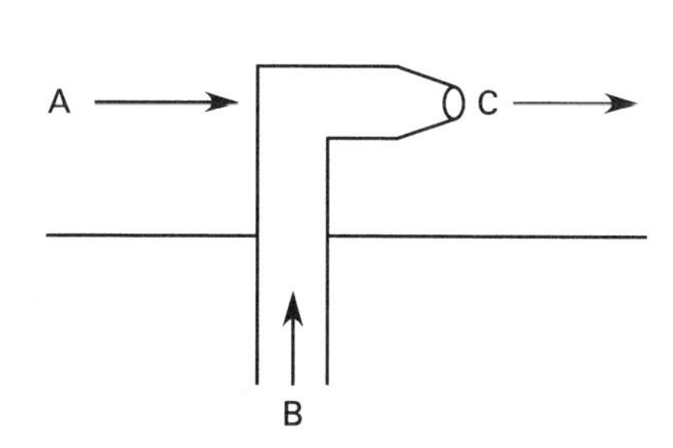

Customary U.S. Solution

(a) The enthalpy, h_A, and entropy, s_A, for steam at 600°F and 200 psia can be obtained from App. 23.C.

$$h_\text{A} = 1322.3\ \text{Btu/lbm}$$

$$s_\text{A} = 1.6771\ \text{Btu/lbm-}^\circ\text{R}$$

Power Cycles

For water at 82°F, enthalpy can be obtained for saturated water from App. 23.A and corrected for compression to 200 psia using App. 23.D. (Correction for compression is optional.)

$$\begin{aligned} h_B &= 50.06 \frac{\text{Btu}}{\text{lbm}} + 0.55 \frac{\text{Btu}}{\text{lbm}} \\ &= 50.61 \text{ Btu/lbm} \end{aligned}$$

From App. 23.A, $s_B = 0.09697$ Btu/lbm-°F. From an energy balance equation (with $Q=0$, $v_1=0$, $\Delta z=0$, and $W=0$), Eq. 26.34 can be written as

$$\begin{aligned} 0 &= h_C(m_A + m_B) - (h_A m_A + h_B m_B) \\ &\quad + \left(\frac{v_C^2}{2g_c J}\right)(m_A + m_B) \\ &= h_C\left(1000 \frac{\text{lbm}}{\text{hr}} + 50 \frac{\text{lbm}}{\text{hr}}\right) \\ &\quad - \left(\begin{array}{l} \left(1322.3 \frac{\text{Btu}}{\text{lbm}}\right)\left(1000 \frac{\text{lbm}}{\text{hr}}\right) \\ \quad + \left(50.61 \frac{\text{Btu}}{\text{lbm}}\right)\left(50 \frac{\text{lbm}}{\text{hr}}\right) \end{array}\right) \\ &\quad + \frac{\left(2000 \frac{\text{ft}}{\text{sec}}\right)^2 \left(1000 \frac{\text{lbm}}{\text{hr}} + 50 \frac{\text{lbm}}{\text{hr}}\right)}{(2)\left(32.2 \frac{\text{lbm-ft}}{\text{lbf-sec}^2}\right)\left(778 \frac{\text{ft-lbf}}{\text{Btu}}\right)} \end{aligned}$$

$$h_C = 1181.9 \text{ Btu/lbm}$$

$$p_C = 100 \text{ psia}$$

Since the enthalpy of saturated steam at 100 psia, h_g, from App. 23.B, is greater than h_C, steam leaving the desuperheater is not 100% saturated. From App. 23.B, the temperature of the saturated steam mix is **327.81°F (330°F).**

The answer is (A).

(b) The quality of the steam leaving can be determined using Eq. 23.53.

$$h_C = h_f + x h_{fg}$$

From App. 23.B, for 100 psia, $h_f = 298.5$ Btu/lbm and $h_{fg} = 889.0$ Btu/lbm.

$$\begin{aligned} x &= \frac{1181.9 \frac{\text{Btu}}{\text{lbm}} - 298.5 \frac{\text{Btu}}{\text{lbm}}}{889.0 \frac{\text{Btu}}{\text{lbm}}} \\ &= \boxed{0.994 \quad (99\%)} \end{aligned}$$

The answer is (D).

SI Solution

(a) The enthalpy, h_A, and entropy, s_A, for steam at 300°F and 1.5 MPa can be obtained from App. 23.P.

$$h_A = 3038.2 \text{ kJ/kg}$$

$$s_A = 6.9198 \text{ kJ/kg·K}$$

Appendix 23.Q does not go down to 1.5 MPa, so disregard the effect of compression. For water at 28°C from App. 23.N, $h_B = 117.37$ kJ/kg.

Similarly, from App. 23.N, $s_B = 0.4091$ kJ/kg·K.

From an energy balance equation (with $Q=0$, $v_1=0$, $\Delta z=0$, and $W=0$), Eq. 26.34 can be written as

$$\begin{aligned} 0 &= h_C(m_A + m_B) - (m_A h_A + m_B h_B) \\ &\quad + \left(\frac{v_C^2}{2}\right)(m_A + m_B) \\ &= h_C\left(0.13 \frac{\text{kg}}{\text{s}} + 0.0063 \frac{\text{kg}}{\text{s}}\right) \\ &\quad - \left(\begin{array}{l} \left(0.13 \frac{\text{kg}}{\text{s}}\right)\left(3038.2 \frac{\text{kJ}}{\text{kg}}\right) \\ \quad + \left(0.0063 \frac{\text{kg}}{\text{s}}\right)\left(117.37 \frac{\text{kJ}}{\text{kg}}\right) \end{array}\right) \\ &\quad + \left(\frac{\left(600 \frac{\text{m}}{\text{s}}\right)^2}{2}\right)\left(0.13 \frac{\text{kg}}{\text{s}} + 0.0063 \frac{\text{kg}}{\text{s}}\right) \\ &\quad \times \left(\frac{1 \text{ kJ}}{1000 \text{ J}}\right) \end{aligned}$$

$$h_C = 2723.2 \text{ kJ/kg}$$

$$p_C = 700 \text{ kPa}$$

Since the enthalpy of saturated steam at 700 kPa, from App. 23.O, is greater than h_C, steam leaving the desuperheater is not 100% saturated. From App. 23.O, the temperature of the steam mix is **164.9°C (165°C).**

The answer is (A).

(b) The quality of steam leaving can be determined using Eq. 23.53.

$$h_C = h_f + x h_{fg}$$

From App. 23.O, for 700 kPa, $h_f = 697.00$ kJ/kg and $h_{fg} = 2065.8$ kJ/kg.

$$x = \frac{2723.2\ \frac{\text{kJ}}{\text{kg}} - 697.00\ \frac{\text{kJ}}{\text{kg}}}{2065.8\ \frac{\text{kJ}}{\text{kg}}} = \boxed{0.9808\quad(99\%)}$$

The answer is (D).

16. *Customary U.S. Solution*

(a) Work with the original system to find the steam flow.

From App. 23.C, at 100 psia and 400°F, $h_A = 1227.7$ Btu/lbm.

From App. 23.A at 70°F, $h_B = 38.08$ Btu/lbm.

From App. 23.A at 180°F, $h_C = 148.04$ Btu/lbm.

Let x = fraction of steam in mixture.

From the energy balance equation,

$$m_A h_A + m_B h_B = m_C h_C$$

$$x\left(1227.7\ \frac{\text{Btu}}{\text{lbm}}\right) + (1-x)\left(38.08\ \frac{\text{Btu}}{\text{lbm}}\right) = (1)\left(148.04\ \frac{\text{Btu}}{\text{lbm}}\right)$$

$$x = 0.0924$$

The steam flow is

$$\dot{m} = x\left(2000\ \frac{\text{lbm}}{\text{hr}}\right) = (0.0924)\left(2000\ \frac{\text{lbm}}{\text{hr}}\right) = \boxed{184.8\ \text{lbm/hr}\quad(180\ \text{lbm/hr})}$$

The answer is (B).

(b) Since the pressure drop across the heater is 5 psi,

$$p_D = p_F + 5\ \text{psi} = 20\ \text{psia} + 5\ \text{psi} = \boxed{25\ \text{psia}}$$

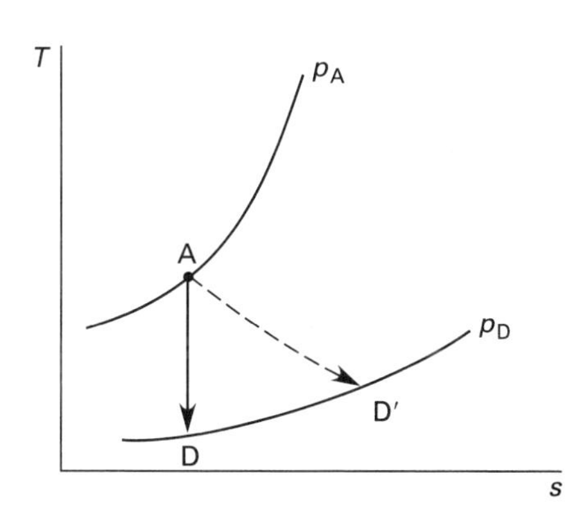

The answer is (C).

(c) With isentropic expansion through the turbine, using the Mollier diagram, $h_D \approx 1115$ Btu/lbm (liquid-vapor mixture).

From Eq. 26.19,

$$\begin{aligned} h'_D &= h_A - \eta_s(h_A - h_D) \\ &= 1227.7\ \frac{\text{Btu}}{\text{lbm}} - (0.60)\left(1227.7\ \frac{\text{Btu}}{\text{lbm}} - 1115\ \frac{\text{Btu}}{\text{lbm}}\right) \\ &= 1160\ \text{Btu/lbm} \end{aligned}$$

The power output is

$$P = \eta\dot{m}(h_A - h'_D) = \frac{(0.96)\left(184.8\ \frac{\text{lbm}}{\text{hr}}\right) \times \left(1227.7\ \frac{\text{Btu}}{\text{lbm}} - 1160\ \frac{\text{Btu}}{\text{lbm}}\right)\left(778\ \frac{\text{ft-lbf}}{\text{Btu}}\right)}{\left(3600\ \frac{\text{sec}}{\text{hr}}\right)\left(550\ \frac{\text{ft-lbf}}{\text{hp-sec}}\right)} = \boxed{4.72\ \text{hp}\quad(4.7\ \text{hp})}$$

The answer is (A).

(d) Let x equal the fraction of steam entering the heater.

From the energy balance equation,

$$m_A h'_D + m_B h_B = m_C h_C$$

$$x\left(1160\ \frac{\text{Btu}}{\text{lbm}}\right) + (1-x)\left(38.08\ \frac{\text{Btu}}{\text{lbm}}\right) = (1)\left(148.04\ \frac{\text{Btu}}{\text{lbm}}\right)$$

$$x = 0.098$$

$$\dot{m}_F = \frac{184.8\ \frac{\text{lbm}}{\text{hr}}}{0.098} = \boxed{1886\ \text{lbm/hr}\quad(1900\ \text{lbm/hr})}$$

The answer is (C).

SI Solution

(a) Work with the original system to find the steam flow.

From App. 23.P, at 700 kPa and 200°C, $h_A = 2845.3$ kJ/kg.

From App. 23.N, at 21°C, $h_B = 88.10$ kJ/kg.

From App. 23.N, at 80°C, $h_C = 335.01$ kJ/kg.

Let x = fraction of steam in mixture.

From the energy balance equation,

$$m_A h_A + m_B h_B = m_C h_C$$

$$x\left(2845.3\ \frac{\text{kJ}}{\text{kg}}\right) + (1-x)\left(88.10\ \frac{\text{kJ}}{\text{kg}}\right) = (1)\left(335.01\ \frac{\text{kJ}}{\text{kg}}\right)$$

$$x = 0.0896$$

The steam flow is

$$\dot{m} = x\left(0.25\ \frac{\text{kg}}{\text{s}}\right) = (0.0896)\left(0.25\ \frac{\text{kg}}{\text{s}}\right) = \boxed{0.0224\ \text{kg/s}\quad (0.022\ \text{kg/s})}$$

The answer is (B).

(b) Since the pressure drop across the heater is 30 kPa,

$$p_D = p_F + 30\ \text{kPa} = 150\ \text{kPa} + 30\ \text{kPa} = \boxed{180\ \text{kPa}}$$

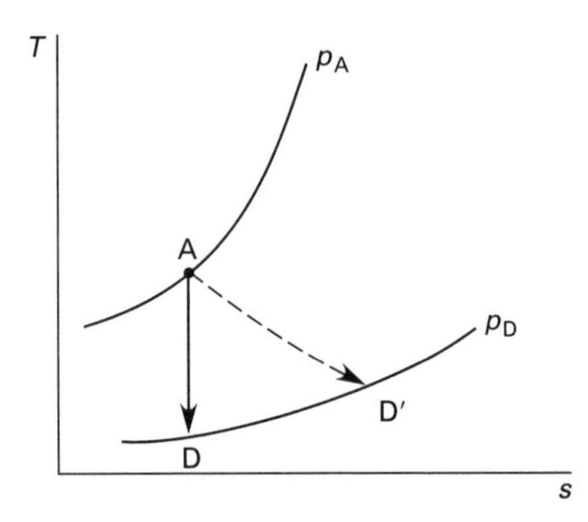

The answer is (C).

(c) For isentropic expansion through the turbine from the Mollier diagram, $h_D \approx 2583$ kJ/kg (liquid-vapor mixture).

From Eq. 26.19,

$$h'_D = h_A - \eta_s(h_A - h_D) = 2845.3\ \frac{\text{kJ}}{\text{kg}} - (0.60)\left(2845.3\ \frac{\text{kJ}}{\text{kg}} - 2583\ \frac{\text{kJ}}{\text{kg}}\right) = 2687.9\ \text{kJ/kg}$$

The power output is

$$P = \eta\dot{m}(h_A - h'_D) = (0.96)\left(0.0224\ \frac{\text{kg}}{\text{s}}\right)\left(2845.3\ \frac{\text{kJ}}{\text{kg}} - 2687.9\ \frac{\text{kJ}}{\text{kg}}\right) = \boxed{3.38\ \text{kW}\quad (3.4\ \text{kW})}$$

The answer is (A).

(d) Let x equal the fraction of steam entering the heater.

From the energy balance equation,

$$m_A h'_D + m_B h_B = m_C h_C$$

$$x\left(2687.9\ \frac{\text{kJ}}{\text{kg}}\right) + (1-x)\left(88.10\ \frac{\text{kJ}}{\text{kg}}\right) = (1)\left(335.01\ \frac{\text{kJ}}{\text{kg}}\right)$$

$$x = 0.0950$$

$$\dot{m}_F = \frac{0.0224\ \frac{\text{kg}}{\text{s}}}{0.0950} = \boxed{0.236\ \text{kg/s}\quad (0.24\ \text{kg/s})}$$

The answer is (C).

27 Vapor Power Cycles

PRACTICE PROBLEMS

1. A steam cycle operates between 650°F and 100°F (340°C and 38°C). The maximum possible thermal efficiency is most nearly

(A) 42%

(B) 49%

(C) 54%

(D) 58%

2. A steam Carnot cycle operates between 650°F and 100°F (340°C and 38°C). The turbine and compressor (pump) isentropic efficiencies are 90% and 80%, respectively. The thermal efficiency is most nearly

(A) 32%

(B) 37%

(C) 42%

(D) 48%

3. A steam turbine cycle produces 600 MWe (megawatts of electrical power). The condenser load is 3.07×10^9 Btu/hr (900 MW). The thermal efficiency is most nearly

(A) 32%

(B) 37%

(C) 40%

(D) 44%

4. A steam Rankine cycle operates with 100 psia (700 kPa) saturated steam that is reduced to 1 atm through expansion in a turbine with an isentropic efficiency of 80%. There is significant subcooling, and water is at 80°F (27°C) and 1 atm when it enters the boilerfeed pump. The pump's isentropic efficiency is 60%. The cycle's thermal efficiency is most nearly

(A) 10%

(B) 14%

(C) 21%

(D) 26%

5. A turbine and the boilerfeed pumps in a reheat cycle have isentropic efficiencies of 88% and 96%, respectively. The cycle starts with water at 60°F (16°C) at the entrance to the boilerfeed pump and produces 600°F (300°C), 600 psia (4 MPa) steam. The steam is reheated when its pressure drops during the first expansion to 20 psia (150 kPa). The thermal efficiency of the cycle is most nearly

(A) 28%

(B) 34%

(C) 39%

(D) 42%

6. The cycle in Prob. 5 is modified to include a bleed of 4.5 psia (37.76 kPa) steam (within the second expansion) for feedwater heating in a closed feedwater heater. The condenser pressure is unchanged. Steam leaves the feedwater heater as saturated liquid. The terminal temperature difference in the heater is 6°F (3°C). Condensate and drip pump work are both negligible. The thermal efficiency of the cycle is most nearly

(A) 23%

(B) 28%

(C) 35%

(D) 44%

7. A reheat steam cycle operates as shown. The pump work between points F and G is 0.15 Btu/lbm (0.3 kJ/kg).

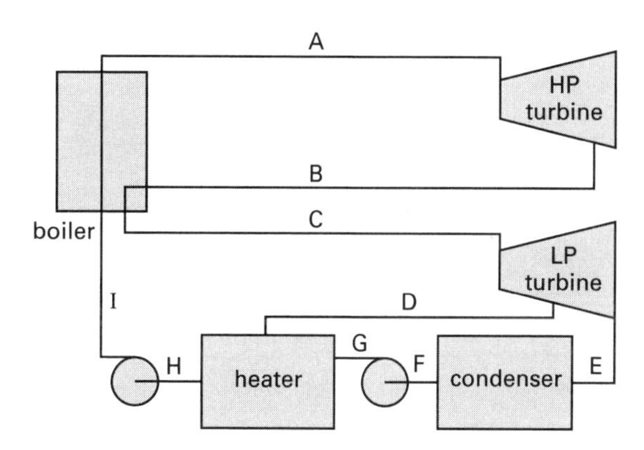

At A: 900 psia (6.2 MPa)
800°F (420°C)

At B: 200 psia (1.5 MPa)
1270 Btu/lbm (2960 kJ/kg)

At C: 190 psia (1.4 MPa)
800°F (420°C)

At D: 50 psia (350 kPa)
1280 Btu/lbm (2980 kJ/kg)

At E: 2 in Hg absolute (6.8 kPa)
1075 Btu/lbm (2500 kJ/kg)

At F: 69.73 Btu/lbm (162.5 kJ/kg)

At H: 250.2 Btu/lbm (583.0 kJ/kg)
0.0173 ft^3/lbm (0.0011 m^3/kg)

At I: 253.1 Btu/lbm (589.7 kJ/kg)

(a) The isentropic efficiency of the high-pressure turbine is most nearly

(A) 54%

(B) 61%

(C) 73%

(D) 76%

(b) The enthalpy at point G (after the pump) is most nearly

(A) 70 Btu/lbm (160 kJ/kg)

(B) 250 Btu/lbm (580 kJ/kg)

(C) 1100 Btu/lbm (2500 kJ/kg)

(D) 1300 Btu/lbm (3000 kJ/kg)

(c) The thermal efficiency of the cycle is most nearly

(A) 22%

(B) 29%

(C) 34%

(D) 41%

Power Cycles

SOLUTIONS

1. *Customary U.S. Solution*

The absolute temperatures are

$$T_{\text{high}} = 650°\text{F} + 460° = 1110°\text{R}$$
$$T_{\text{low}} = 100°\text{F} + 460° = 560°\text{R}$$

The maximum possible thermal efficiency is given by the Carnot cycle. From Eq. 27.8,

$$\eta_{\text{th}} = \frac{T_{\text{high}} - T_{\text{low}}}{T_{\text{high}}} = \frac{1110°\text{R} - 560°\text{R}}{1110°\text{R}}$$
$$= \boxed{0.495 \quad (49\%)}$$

The answer is (B).

SI Solution

The absolute temperatures are

$$T_{\text{high}} = 340°\text{C} + 273° = 613\text{K}$$
$$T_{\text{low}} = 38°\text{C} + 273° = 311\text{K}$$

The maximum possible thermal efficiency is given by the Carnot cycle. From Eq. 27.8,

$$\eta_{\text{th}} = \frac{T_{\text{high}} - T_{\text{low}}}{T_{\text{high}}} = \frac{613\text{K} - 311\text{K}}{613\text{K}}$$
$$= \boxed{0.493 \quad (49\%)}$$

The answer is (B).

2. *Customary U.S. Solution*

Refer to the Carnot cycle (see Fig. 27.2).

At point A, from App. 23.A, for saturated liquid at $T_A = 650°\text{F}$,

$$h_A = 696.0 \text{ Btu/lbm}$$
$$s_A = 0.8833 \text{ Btu/lbm-°F}$$

At point B, from App. 23.A, for saturated vapor at $T_B = 650°\text{F}$,

$$h_B = 1119.7 \text{ Btu/lbm}$$
$$s_B = 1.2651 \text{ Btu/lbm-°F}$$

At point C,

$$T_C = 100°F$$
$$s_C = s_B = 1.2651 \text{ Btu/lbm-°F}$$

From App. 23.A,

$$s_f = 0.1296 \text{ Btu/lbm-°F}$$
$$s_g = 1.9819 \text{ Btu/lbm-°F}$$
$$h_f = 68.03 \text{ Btu/lbm}$$
$$h_{fg} = 1036.7 \text{ Btu/lbm}$$

$$x_C = \frac{s_C - s_f}{s_g - s_f} = \frac{1.2651\ \frac{\text{Btu}}{\text{lbm-°F}} - 0.1296\ \frac{\text{Btu}}{\text{lbm-°F}}}{1.9819\ \frac{\text{Btu}}{\text{lbm-°F}} - 0.1296\ \frac{\text{Btu}}{\text{lbm-°F}}} = 0.613$$

$$h_C = h_f + x_C h_{fg} = 68.03\ \frac{\text{Btu}}{\text{lbm}} + (0.613)\left(1036.7\ \frac{\text{Btu}}{\text{lbm}}\right) = 703.5 \text{ Btu/lbm}$$

At point D,

$$T_D = 100°F$$
$$s_D = s_A = 0.8833 \text{ Btu/lbm-°F}$$

$$x_D = \frac{s_D - s_f}{s_g - s_f} = \frac{0.8833\ \frac{\text{Btu}}{\text{lbm-°F}} - 0.1296\ \frac{\text{Btu}}{\text{lbm-°F}}}{1.9819\ \frac{\text{Btu}}{\text{lbm-°F}} - 0.1296\ \frac{\text{Btu}}{\text{lbm-°F}}} = 0.407$$

$$h_D = h_f + x_D h_{fg} = 68.03\ \frac{\text{Btu}}{\text{lbm}} + (0.407)\left(1036.7\ \frac{\text{Btu}}{\text{lbm}}\right) = 489.8 \text{ Btu/lbm}$$

From Eq. 27.9, due to the inefficiency of the turbine,

$$h'_C = h_B - \eta_{s,\text{turbine}}(h_B - h_C) = 1119.7\ \frac{\text{Btu}}{\text{lbm}} - (0.9)\left(1119.7\ \frac{\text{Btu}}{\text{lbm}} - 703.5\ \frac{\text{Btu}}{\text{lbm}}\right) = 745.1 \text{ Btu/lbm}$$

From Eq. 27.10, due to the inefficiency of the pump,

$$h'_A = h_D + \frac{h_A - h_D}{\eta_{s,\text{pump}}} = 489.8\ \frac{\text{Btu}}{\text{lbm}} + \frac{696.0\ \frac{\text{Btu}}{\text{lbm}} - 489.8\ \frac{\text{Btu}}{\text{lbm}}}{0.8} = 747.6 \text{ Btu/lbm}$$

From Eq. 27.8, the thermal efficiency of the entire cycle is

$$\eta_{\text{th}} = \frac{(h_B - h'_C) - (h'_A - h_D)}{h_B - h'_A} = \frac{\left(1119.7\ \frac{\text{Btu}}{\text{lbm}} - 745.1\ \frac{\text{Btu}}{\text{lbm}}\right) - \left(747.6\ \frac{\text{Btu}}{\text{lbm}} - 489.8\ \frac{\text{Btu}}{\text{lbm}}\right)}{1119.7\ \frac{\text{Btu}}{\text{lbm}} - 747.6\ \frac{\text{Btu}}{\text{lbm}}} = \boxed{0.314\quad(32\%)}$$

The answer is (A).

SI Solution

At point A, $T_A = 340°C$.

From App. 23.N, for saturated liquid,

$$h_A = 1594.5 \text{ kJ/kg}$$
$$s_A = 3.6601 \text{ kJ/kg·K}$$

At point B, $T_B = 340°C$.

From App. 23.N, for saturated vapor,

$$h_B = 2621.9 \text{ kJ/kg}$$
$$s_B = 5.3356 \text{ kJ/kg·K}$$

At point C,

$$T_C = 38°C$$
$$s_C = s_B = 5.3356 \text{ kJ/kg·K}$$

Power Cycles

From App. 23.N,

$$s_f = 0.5456 \text{ kJ/kg·K}$$
$$s_g = 8.2935 \text{ kJ/kg·K}$$
$$h_f = 159.17 \text{ kJ/kg}$$
$$h_{fg} = 2410.7 \text{ kJ/kg}$$

$$x_C = \frac{s_C - s_f}{s_g - s_f} = \frac{5.3356 \ \frac{\text{kJ}}{\text{kg·K}} - 0.5456 \ \frac{\text{kJ}}{\text{kg·K}}}{8.2935 \ \frac{\text{kJ}}{\text{kg·K}} - 0.5456 \ \frac{\text{kJ}}{\text{kg·K}}} = 0.618$$

$$h_C = h_f + x_C h_{fg} = 159.17 \ \frac{\text{kJ}}{\text{kg}} + (0.618)\left(2410.7 \ \frac{\text{kJ}}{\text{kg}}\right) = 1649.0 \text{ kJ/kg}$$

At point D,

$$T_D = 38°\text{C}$$
$$s_D = s_A = 3.6601 \text{ kJ/kg·K}$$

$$x_D = \frac{s_D - s_f}{s_g - s_f} = \frac{3.6601 \ \frac{\text{kJ}}{\text{kg·K}} - 0.5456 \ \frac{\text{kJ}}{\text{kg·K}}}{8.2935 \ \frac{\text{kJ}}{\text{kg·K}} - 0.5456 \ \frac{\text{kJ}}{\text{kg·K}}} = 0.402$$

$$h_D = h_f + x_D h_{fg} = 159.17 \ \frac{\text{kJ}}{\text{kg}} + (0.402)\left(2410.7 \ \frac{\text{kJ}}{\text{kg}}\right) = 1128.3 \text{ kJ/kg}$$

From Eq. 27.9, due to the inefficiency of the turbine,

$$h'_C = h_B - \eta_{s,\text{turbine}}(h_B - h_C) = 2621.9 \ \frac{\text{kJ}}{\text{kg}} - (0.9)\left(2621.9 \ \frac{\text{kJ}}{\text{kg}} - 1649.0 \ \frac{\text{kJ}}{\text{kg}}\right) = 1746.3 \text{ kJ/kg}$$

From Eq. 27.10, due to the inefficiency of the pump,

$$h'_A = h_D + \frac{h_A - h_D}{\eta_{s,\text{pump}}} = 1128.3 \ \frac{\text{kJ}}{\text{kg}} + \frac{1594.5 \ \frac{\text{kJ}}{\text{kg}} - 1128.3 \ \frac{\text{kJ}}{\text{kg}}}{0.8} = 1711.1 \text{ kJ/kg}$$

From Eq. 27.8, the thermal efficiency of the entire cycle is

$$\eta_{th} = \frac{(h_B - h'_C) - (h'_A - h_D)}{h_B - h'_A} = \frac{\left(2621.9 \ \frac{\text{kJ}}{\text{kg}} - 1746.3 \ \frac{\text{kJ}}{\text{kg}}\right) - \left(1711.1 \ \frac{\text{kJ}}{\text{kg}} - 1128.3 \ \frac{\text{kJ}}{\text{kg}}\right)}{2621.9 \ \frac{\text{kJ}}{\text{kg}} - 1711.1 \ \frac{\text{kJ}}{\text{kg}}} = \boxed{0.321 \quad (32\%)}$$

The answer is (A).

3. *Customary U.S. Solution*

The condenser load is $Q_{out} = 3.07 \times 10^9$ Btu/hr.

The net work is

$$W_{net} = Q_{in} - Q_{out}$$

The boiler load is

$$Q_{in} = W_{net} + Q_{out} = (600 \text{ MW})\left(1000 \ \frac{\text{kW}}{\text{MW}}\right)\left(3412 \ \frac{\text{Btu}}{\text{kW-hr}}\right) + 3.07 \times 10^9 \ \frac{\text{Btu}}{\text{hr}} = 5.12 \times 10^9 \text{ Btu/hr}$$

From Eq. 27.1, the thermal efficiency is

$$\eta_{th} = \frac{Q_{in} - Q_{out}}{Q_{in}} = \frac{5.12 \times 10^9 \ \frac{\text{Btu}}{\text{hr}} - 3.07 \times 10^9 \ \frac{\text{Btu}}{\text{hr}}}{5.12 \times 10^9 \ \frac{\text{Btu}}{\text{hr}}} = \boxed{0.400 \quad (40\%)}$$

The answer is (C).

SI Solution

The condenser load is $Q_{out} = 900$ MW.

The boiler load is

$$Q_{in} = W_{net} + Q_{out} = 600 \text{ MW} + 900 \text{ MW} = 1500 \text{ MW}$$

From Eq. 27.1, the thermal efficiency is

$$\eta_{th} = \frac{Q_{in} - Q_{out}}{Q_{in}} = \frac{1500\ \text{MW} - 900\ \text{MW}}{1500\ \text{MW}} = \boxed{0.400\quad(40\%)}$$

The answer is (C).

4. *Customary U.S. Solution*

At point A, $p_A = 100$ psia.

From App. 23.B, the enthalpy of saturated liquid is $h_A = 298.5$ Btu/lbm.

At point B, $p_B = 100$ psia.

From App. 23.B, the enthalpy and entropy of saturated vapor are

$$h_B = 1187.5\ \text{Btu/lbm}$$
$$s_B = 1.6032\ \text{Btu/lbm-°R}$$

At point C,

$$p_C = 1\ \text{atm}$$
$$s_C = s_B = 1.6032\ \text{Btu/lbm-°R}$$

From App. 23.B, the entropy and enthalpy of saturated liquid and entropy and enthalpy of vaporization are

$$s_f = 0.3122\ \text{Btu/lbm-°R}$$
$$h_f = 180.2\ \text{Btu/lbm}$$
$$s_{fg} = 1.4445\ \text{Btu/lbm-°R}$$
$$h_{fg} = 970.1\ \text{Btu/lbm}$$

$$x_C = \frac{s_C - s_f}{s_{fg}} = \frac{1.6032\ \frac{\text{Btu}}{\text{lbm-°R}} - 0.3122\ \frac{\text{Btu}}{\text{lbm-°R}}}{1.4445\ \frac{\text{Btu}}{\text{lbm-°R}}} = 0.894$$

$$h_C = h_f + x_C h_{fg} = 180.2\ \frac{\text{Btu}}{\text{lbm}} + (0.894)\left(970.1\ \frac{\text{Btu}}{\text{lbm}}\right) = 1047.5\ \text{Btu/lbm}$$

At point D, $T = 80°\text{F}$ and $p_D = 1$ atm (subcooled). h and v are essentially independent of pressure.

From App. 23.A, the enthalpy and specific volume of saturated liquid are

$$h_D = 48.07\ \text{Btu/lbm}$$
$$v_D = 0.01607\ \text{ft}^3/\text{lbm}$$

At point E, $p_E = p_A = 100$ psia.

From Eq. 27.14,

$$\begin{aligned} h_E &\approx h_D + v_D(p_E - p_D) \\ &= 48.07\ \frac{\text{Btu}}{\text{lbm}} + \left(0.01607\ \frac{\text{ft}^3}{\text{lbm}}\right) \times \frac{\left(100\ \frac{\text{lbf}}{\text{in}^2} - 14.7\ \frac{\text{lbf}}{\text{in}^2}\right)\left(12\ \frac{\text{in}}{\text{ft}}\right)^2}{778\ \frac{\text{ft-lbf}}{\text{Btu}}} \\ &= 48.32\ \text{Btu/lbm} \end{aligned}$$

From Eq. 27.18, due to the inefficiency of the turbine,

$$\begin{aligned} h'_C &= h_B - \eta_{s,\text{turbine}}(h_B - h_C) \\ &= 1187.5\ \frac{\text{Btu}}{\text{lbm}} - (0.80)\left(1187.5\ \frac{\text{Btu}}{\text{lbm}} - 1047.5\ \frac{\text{Btu}}{\text{lbm}}\right) \\ &= 1075.5\ \text{Btu/lbm} \end{aligned}$$

From Eq. 27.19, due to the inefficiency of the pump,

$$\begin{aligned} h'_E &= h_D + \frac{h_E - h_D}{\eta_{s,\text{pump}}} \\ &= 48.07\ \frac{\text{Btu}}{\text{lbm}} + \frac{48.32\ \frac{\text{Btu}}{\text{lbm}} - 48.07\ \frac{\text{Btu}}{\text{lbm}}}{0.6} \\ &= 48.49\ \text{Btu/lbm} \end{aligned}$$

From Eq. 27.17, the thermal efficiency of the cycle is

$$\begin{aligned} \eta_{th} &= \frac{(h_B - h'_C) - (h'_E - h_D)}{h_B - h'_E} \\ &= \frac{\left(1187.5\ \frac{\text{Btu}}{\text{lbm}} - 1075.5\ \frac{\text{Btu}}{\text{lbm}}\right) - \left(48.49\ \frac{\text{Btu}}{\text{lbm}} - 48.07\ \frac{\text{Btu}}{\text{lbm}}\right)}{1187.5\ \frac{\text{Btu}}{\text{lbm}} - 48.49\ \frac{\text{Btu}}{\text{lbm}}} \\ &= \boxed{0.098\quad(10\%)} \end{aligned}$$

The answer is (A).

SI Solution

At point A, $p_A = 700$ kPa (7 bars).

From App. 23.O, the enthalpy of saturated liquid is $h_A = 697.00$ kJ/kg.

At point B, $p_B = 700$ kPa.

From App. 23.O, the enthalpy and entropy of saturated vapor are

$$h_B = 2762.8 \text{ kJ/kg}$$
$$s_B = 6.7071 \text{ kJ/kg·K}$$

At point C,

$$p_C = 1 \text{ atm}$$
$$s_C = s_B = 6.7071 \text{ kJ/kg·K}$$

From App. 23.O, the entropy and enthalpy of saturated liquid, the entropy of saturated vapor, and the enthalpy of vaporization are

$$s_f = 1.3062 \text{ kJ/kg·K}$$
$$h_f = 418.79 \text{ kJ/kg}$$
$$s_g = 7.3553 \text{ kJ/kg·K}$$
$$h_{fg} = 2256.6 \text{ kJ/kg}$$

$$x_C = \frac{s_C - s_f}{s_g - s_f} = \frac{6.7071 \frac{\text{kJ}}{\text{kg·K}} - 1.3062 \frac{\text{kJ}}{\text{kg·K}}}{7.3553 \frac{\text{kJ}}{\text{kg·K}} - 1.3062 \frac{\text{kJ}}{\text{kg·K}}}$$
$$= 0.893$$

$$h_C = h_f + x_C h_{fg} = 418.79 \frac{\text{kJ}}{\text{kg}} + (0.893)\left(2256.6 \frac{\text{kJ}}{\text{kg}}\right)$$
$$= 2433.9 \text{ kJ/kg}$$

At point D, $T = 27°\text{C}$ and $p_D = 1$ atm (subcooled). h and v are essentially independent of pressure.

From App. 23.N, the enthalpy and specific volume of saturated liquid are

$$h_D = 113.19 \text{ kJ/kg}$$
$$v_D = 1.0035 \text{ cm}^3\text{/g}$$

At point E, $p_E = p_A = 700$ kPa.

From Eq. 27.14,

$$h_E \approx h_D + v_D(p_E - p_D)$$
$$= 113.19 \frac{\text{kJ}}{\text{kg}} + \frac{\left(1.0035 \frac{\text{cm}^3}{\text{g}}\right)\left(1000 \frac{\text{g}}{\text{kg}}\right) \times (700 \text{ kPa} - 101 \text{ kPa})}{10^6 \frac{\text{cm}^3}{\text{m}^3}}$$
$$= 113.79 \text{ kJ/kg}$$

From Eq. 27.18, due to the inefficiency of the turbine,

$$h'_C = h_B - \eta_{s,\text{turbine}}(h_B - h_C)$$
$$= 2762.8 \frac{\text{kJ}}{\text{kg}} - (0.80)\left(2762.8 \frac{\text{kJ}}{\text{kg}} - 2433.9 \frac{\text{kJ}}{\text{kg}}\right)$$
$$= 2499.7 \text{ kJ/kg}$$

From Eq. 27.19, due to the inefficiency of the pump,

$$h'_E = h_D + \frac{h_E - h_D}{\eta_{s,\text{pump}}}$$
$$= 113.19 \frac{\text{kJ}}{\text{kg}} + \frac{113.79 \frac{\text{kJ}}{\text{kg}} - 113.19 \frac{\text{kJ}}{\text{kg}}}{0.6}$$
$$= 114.19 \text{ kJ/kg}$$

From Eq. 27.17, the thermal efficiency of the cycle is

$$\eta_{th} = \frac{(h_B - h'_C) - (h'_E - h_D)}{h_B - h'_E}$$
$$= \frac{\left(2762.8 \frac{\text{kJ}}{\text{kg}} - 2499.7 \frac{\text{kJ}}{\text{kg}}\right) - \left(114.19 \frac{\text{kJ}}{\text{kg}} - 113.19 \frac{\text{kJ}}{\text{kg}}\right)}{2762.8 \frac{\text{kJ}}{\text{kg}} - 114.19 \frac{\text{kJ}}{\text{kg}}}$$
$$= \boxed{0.10 \quad (10\%)}$$

The answer is (A).

5. *Customary U.S. Solution*

Refer to the reheat cycle (see Fig. 27.8).

At point B, $p_B = 600$ psia.

From App. 23.B, the enthalpy of the saturated liquid is $h_B = 471.7$ Btu/lbm.

At point C, $p_C = 600$ psia.

From App. 23.B, the enthalpy of the saturated vapor is $h_C = 1203.8$ Btu/lbm.

At point D,

$$T_D = 600°\text{F}$$
$$p_D = 600 \text{ psia}$$

From App. 23.C, the enthalpy and entropy of superheated vapor are

$$h_D = 1289.9 \text{ Btu/lbm}$$
$$s_D = 1.5326 \text{ Btu/lbm-°R}$$

At point E,

$$p_E = 20 \text{ psia}$$
$$s_E = s_D = 1.5326 \text{ Btu/lbm-°R}$$

From App. 23.B, the various saturation properties are

$$s_f = 0.3358 \text{ Btu/lbm-°R}$$
$$s_{fg} = 1.3961 \text{ Btu/lbm-°R}$$
$$h_f = 196.3 \text{ Btu/lbm}$$
$$h_{fg} = 959.9 \text{ Btu/lbm}$$

$$x_E = \frac{s_E - s_f}{s_g - s_f} = \frac{1.5326 \frac{\text{Btu}}{\text{lbm-°R}} - 0.3358 \frac{\text{Btu}}{\text{lbm-°R}}}{1.3961 \frac{\text{Btu}}{\text{lbm-°R}}}$$
$$= 0.857$$
$$h_E = h_f + x_E h_{fg}$$
$$= 196.3 \frac{\text{Btu}}{\text{lbm}} + (0.857)\left(959.9 \frac{\text{Btu}}{\text{lbm}}\right)$$
$$= 1018.9 \text{ Btu/lbm}$$

From Eq. 27.38, due to the inefficiency of the turbine,

$$h'_E = h_D - \eta_{s,\text{turbine}}(h_D - h_E)$$
$$= 1289.9 \frac{\text{Btu}}{\text{lbm}} - (0.88)\left(1289.9 \frac{\text{Btu}}{\text{lbm}} - 1018.9 \frac{\text{Btu}}{\text{lbm}}\right)$$
$$= 1051.4 \text{ Btu/lbm}$$

At point F, the temperature has been returned to 600°F, but the pressure stays at the expansion pressure, p_E.

$$p_F = 20 \text{ psia}$$
$$T_F = 600\text{°F}$$

From App. 23.C, the enthalpy and entropy of the superheated vapor are

$$h_F = 1334.9 \text{ Btu/lbm}$$
$$s_F = 1.9399 \text{ Btu/lbm-°R}$$

At point G,

$$T_G = 60\text{°F}$$
$$s_G = s_F$$
$$= 1.9399 \text{ Btu/lbm-°R}$$

From App. 23.A, the various saturation properties are

$$s_f = 0.05554 \text{ Btu/lbm-°R}$$
$$s_g = 2.0940 \text{ Btu/lbm-°R}$$
$$h_f = 28.08 \text{ Btu/lbm}$$
$$h_{fg} = 1059.3 \text{ Btu/lbm}$$
$$x_G = \frac{s_G - s_f}{s_g - s_f}$$
$$= \frac{1.9399 \frac{\text{Btu}}{\text{lbm-°R}} - 0.05554 \frac{\text{Btu}}{\text{lbm-°R}}}{2.0940 \frac{\text{Btu}}{\text{lbm-°R}} - 0.05554 \frac{\text{Btu}}{\text{lbm-°R}}}$$
$$= 0.924$$
$$h_G = h_f + x_G h_{fg}$$
$$= 28.08 \frac{\text{Btu}}{\text{lbm}} + (0.924)\left(1059.3 \frac{\text{Btu}}{\text{lbm}}\right)$$
$$= 1006.9 \text{ Btu/lbm}$$

From Eq. 27.39, due to the inefficiency of the turbine,

$$h'_G = h_F - \eta_{s,\text{turbine}}(h_F - h_G)$$
$$= 1334.9 \frac{\text{Btu}}{\text{lbm}} - (0.88)\left(1334.9 \frac{\text{Btu}}{\text{lbm}} - 1006.9 \frac{\text{Btu}}{\text{lbm}}\right)$$
$$= 1046.3 \text{ Btu/lbm}$$

At point H, $T_H = 60\text{°F}$.

From App. 23.A, the saturation pressure, enthalpy, and specific volume of the saturated liquid are

$$p_H = 0.2564 \text{ psia}$$
$$h_H = 28.08 \text{ Btu/lbm}$$
$$\upsilon_H = 0.01604 \text{ ft}^3\text{/lbm}$$

At point A, $p_A = 600$ psia.

Power Cycles

From Eq. 27.14,

$$\begin{aligned} h_A &\approx h_H + \upsilon_H(p_A - p_H) \\ &= 28.08\ \frac{\text{Btu}}{\text{lbm}} + \frac{\left(0.01604\ \frac{\text{ft}^3}{\text{lbm}}\right) \times \left(600\ \frac{\text{lbf}}{\text{in}^2} - 0.2564\ \frac{\text{lbf}}{\text{in}^2}\right)\left(12\ \frac{\text{in}}{\text{ft}}\right)^2}{778\ \frac{\text{ft-lbf}}{\text{Btu}}} \\ &= 29.86\ \text{Btu/lbm} \end{aligned}$$

From Eq. 27.40, due to the inefficiency of the pump,

$$\begin{aligned} h'_A &= h_H + \frac{h_A - h_H}{\eta_{s,\text{pump}}} \\ &= 28.08\ \frac{\text{Btu}}{\text{lbm}} + \frac{29.86\ \frac{\text{Btu}}{\text{lbm}} - 28.08\ \frac{\text{Btu}}{\text{lbm}}}{0.96} \\ &= 29.93\ \text{Btu/lbm} \end{aligned}$$

From Eq. 27.37, the thermal efficiency of the cycle for a non-isentropic process for the turbine and the pump is

$$\begin{aligned} \eta_{th} &= \frac{(h_D - h'_A) + (h_F - h'_E) - (h'_G - h_H)}{(h_D - h'_A) + (h_F - h'_E)} \\ &= \frac{\left(1289.9\ \frac{\text{Btu}}{\text{lbm}} - 29.93\ \frac{\text{Btu}}{\text{lbm}}\right) + \left(1334.9\ \frac{\text{Btu}}{\text{lbm}} - 1051.4\ \frac{\text{Btu}}{\text{lbm}}\right) - \left(1046.3\ \frac{\text{Btu}}{\text{lbm}} - 28.08\ \frac{\text{Btu}}{\text{lbm}}\right)}{\left(1289.9\ \frac{\text{Btu}}{\text{lbm}} - 29.93\ \frac{\text{Btu}}{\text{lbm}}\right) + \left(1334.9\ \frac{\text{Btu}}{\text{lbm}} - 1051.4\ \frac{\text{Btu}}{\text{lbm}}\right)} \\ &= \boxed{0.340 \quad (34\%)} \end{aligned}$$

The answer is (B).

SI Solution

Refer to the reheat cycle (see Fig. 27.8).

At point B, $p_B = 4$ MPa.

From App. 23.O, the enthalpy of saturated liquid is $h_B = 1082.5$ kJ/kg.

At point C, $p_C = 4$ MPa.

From App. 23.O, the enthalpy of saturated vapor is $h_C = 2800.8$ kJ/kg.

At point D, $p_D = 4$ MPa and $T_D = 300°C$.

From the Mollier diagram, the enthalpy of superheated vapor is $h_D = 2980$ kJ/kg.

At point E, from the Mollier diagram, assuming isentropic expansion, $h_E = 2395$ kJ/kg.

From Eq. 27.38, due to the inefficiency of the turbine,

$$\begin{aligned} h'_E &= h_D - \eta_{s,\text{turbine}}(h_D - h_E) \\ &= 2980\ \frac{\text{kJ}}{\text{kg}} - (0.88)\left(2980\ \frac{\text{kJ}}{\text{kg}} - 2395\ \frac{\text{kJ}}{\text{kg}}\right) \\ &= 2465.2\ \text{kJ/kg} \end{aligned}$$

At point F,

$$\begin{aligned} p_F &= 150\ \text{kPa} \\ T_F &= 300°\text{C} \end{aligned}$$

From App. 23.P, the enthalpy and entropy of the superheated vapor are

$$\begin{aligned} h_F &= 3073.3\ \text{kJ/kg} \\ s_F &= 8.0284\ \text{kJ/kg·K} \end{aligned}$$

At point G,

$$\begin{aligned} T_G &= 16°\text{C} \\ s_G &= s_F = 8.0284\ \text{kJ/kg·K} \end{aligned}$$

From App. 23.N, the various saturation properties are

$$\begin{aligned} s_f &= 0.2390\ \text{kJ/kg·K} \\ s_g &= 8.7570\ \text{kJ/kg·K} \\ h_f &= 67.17\ \text{kJ/kg} \\ h_{fg} &= 2463.0\ \text{kJ/kg} \end{aligned}$$

$$\begin{aligned} x_G &= \frac{s_G - s_f}{s_g - s_f} = \frac{8.0284\ \frac{\text{kJ}}{\text{kg·K}} - 0.2390\ \frac{\text{kJ}}{\text{kg·K}}}{8.7570\ \frac{\text{kJ}}{\text{kg·K}} - 0.2390\ \frac{\text{kJ}}{\text{kg·K}}} \\ &= 0.914 \end{aligned}$$

$$\begin{aligned} h_G &= h_f + x_G h_{fg} = 67.17\ \frac{\text{kJ}}{\text{kg}} + (0.914)\left(2463.0\ \frac{\text{kJ}}{\text{kg}}\right) \\ &= 2318.4\ \text{kJ/kg} \end{aligned}$$

From Eq. 27.39, due to the inefficiency of the turbine,

$$\begin{aligned} h'_G &= h_F - \eta_{s,\text{turbine}}(h_F - h_G) \\ &= 3073.3\ \frac{\text{kJ}}{\text{kg}} - (0.88)\left(3073.3\ \frac{\text{kJ}}{\text{kg}} - 2318.4\ \frac{\text{kJ}}{\text{kg}}\right) \\ &= 2409.0\ \text{kJ/kg} \end{aligned}$$

At point H, $T_H = 16°C$.

From App. 23.N, the saturation pressure, enthalpy, and specific volume of the saturated liquid are

$$p_H = (0.01819 \text{ bar})\left(100 \ \frac{\text{kPa}}{\text{bar}}\right) = 1.819 \text{ kPa}$$

$$h_H = 67.17 \text{ kJ/kg}$$

$$\upsilon_H = \frac{\left(1.0011 \ \frac{\text{cm}^3}{\text{g}}\right)\left(1000 \ \frac{\text{g}}{\text{kg}}\right)}{\left(100 \ \frac{\text{cm}}{\text{m}}\right)^3} = 1.0011 \times 10^{-3} \text{ m}^3/\text{kg}$$

At point A,

$$p_A = (4 \text{ MPa})\left(1000 \ \frac{\text{kPa}}{\text{MPa}}\right) = 4000 \text{ kPa}$$

From Eq. 27.14,

$$\begin{aligned} h_A &\approx h_H + \upsilon_H(p_A - p_H) \\ &= 67.17 \ \frac{\text{kJ}}{\text{kg}} + \left(1.0011 \times 10^{-3} \ \frac{\text{m}^3}{\text{kg}}\right) \\ &\quad \times (4000 \text{ kPa} - 1.819 \text{ kPa}) \\ &= 71.17 \text{ kJ/kg} \end{aligned}$$

From Eq. 27.40, due to the inefficiency of the pump,

$$\begin{aligned} h'_A &= h_H + \frac{h_A - h_H}{\eta_{s,\text{pump}}} = 67.17 \ \frac{\text{kJ}}{\text{kg}} + \frac{71.17 \ \frac{\text{kJ}}{\text{kg}} - 67.17 \ \frac{\text{kJ}}{\text{kg}}}{0.96} \\ &= 71.34 \text{ kJ/kg} \end{aligned}$$

From Eq. 27.37, the thermal efficiency of the cycle for a non-isentropic process for the turbine and the pump is

$$\begin{aligned} \eta_{th} &= \frac{(h_D - h'_A) + (h_F - h'_E) - (h'_G - h_H)}{(h_D - h'_A) + (h_F - h'_E)} \\ &= \frac{\begin{array}{l}\left(2980 \ \frac{\text{kJ}}{\text{kg}} - 71.34 \ \frac{\text{kJ}}{\text{kg}}\right) \\ \quad + \left(3073.3 \ \frac{\text{kJ}}{\text{kg}} - 2465.2 \ \frac{\text{kJ}}{\text{kg}}\right) \\ \quad - \left(2409.0 \ \frac{\text{kJ}}{\text{kg}} - 67.17 \ \frac{\text{kJ}}{\text{kg}}\right)\end{array}}{\begin{array}{l}\left(2980 \ \frac{\text{kJ}}{\text{kg}} - 71.34 \ \frac{\text{kJ}}{\text{kg}}\right) \\ \quad + \left(3073.3 \ \frac{\text{kJ}}{\text{kg}} - 2465.2 \ \frac{\text{kJ}}{\text{kg}}\right)\end{array}} \\ &= \boxed{0.334 \quad (34\%)} \end{aligned}$$

The answer is (B).

6. *Customary U.S. Solution*

Refer to the following diagram.

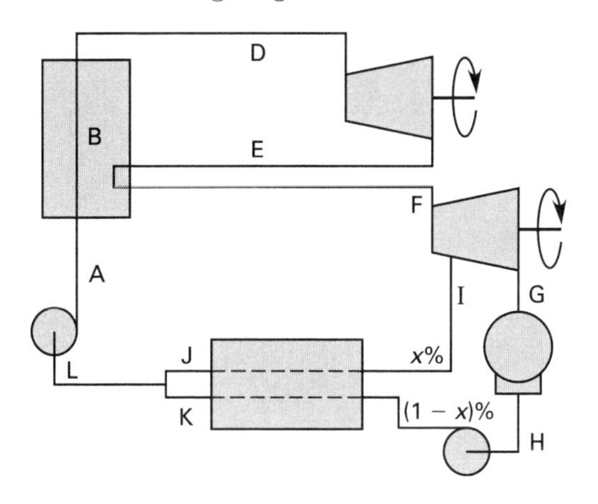

From Prob. 5,

$$\begin{aligned} h_B &= 486.24 \text{ Btu/lbm} \\ h_D &= 1289.9 \text{ Btu/lbm} \\ h'_E &= 1051.4 \text{ Btu/lbm} \\ h_F &= 1334.9 \text{ Btu/lbm} \\ s_F &= 1.9399 \text{ Btu/lbm-°R} \\ h'_G &= 1046.3 \text{ Btu/lbm} \\ h_H &= 28.08 \text{ Btu/lbm} \\ p_H &= 0.2564 \text{ psia} \\ T_H &= 60°\text{F} \end{aligned}$$

At point I,

$$\begin{aligned} p_I &= 4.5 \text{ psia} \\ s_I &= s_F = 1.9399 \text{ Btu/lbm-°R} \end{aligned}$$

From App. 23.C, using double interpolation,

$$h_I = 1181.2 \text{ Btu/lbm}$$

From Eq. 27.39, due to the inefficiency of the turbine,

$$\begin{aligned} h'_I &= h_F - \eta_{s,\text{turbine}}(h_F - h_I) \\ &= 1334.9 \ \frac{\text{Btu}}{\text{lbm}} - (0.88)\left(1334.9 \ \frac{\text{Btu}}{\text{lbm}} - 1181.2 \ \frac{\text{Btu}}{\text{lbm}}\right) \\ &= 1199.6 \text{ Btu/lbm} \end{aligned}$$

At point J, from App. 23.B, the saturated liquid enthalpy and temperature at 4.5 psia are approximately $h_J = 125.6$ Btu/lbm and 157.5°F, respectively.

Power Cycles

At point K, the temperature is

$$157.5°F - 6°F \approx 152°F$$

Since the water is subcooled, enthalpy is a function of temperature only.

$$h_K = 119.99 \text{ Btu/lbm}$$

The condensate pump will match the pressure at J: $p_K = 4.5$ psia.

From the energy balance in the heater,

$$(1-x)(h_K - h_H) = x(h'_I - h_J)$$

$$(1-x)\left(119.99 \frac{\text{Btu}}{\text{lbm}} - 28.08 \frac{\text{Btu}}{\text{lbm}}\right) = x\left(1199.6 \frac{\text{Btu}}{\text{lbm}} - 125.6 \frac{\text{Btu}}{\text{lbm}}\right)$$

$$1166.2x = 91.91$$

$$x = 0.0788$$

At point L, $p_L = 4.5$ psia.

$$\begin{aligned} h_L &= xh_J + (1-x)h_K \\ &= (0.0788)\left(125.6 \frac{\text{Btu}}{\text{lbm}}\right) + (1-0.0788)\left(119.99 \frac{\text{Btu}}{\text{lbm}}\right) \\ &= 120.4 \text{ Btu/lbm} \end{aligned}$$

Since this is a subcooled liquid, enthalpy is a function of temperature only, and from App. 23.A,

$$T_L = 152.4°F$$
$$\upsilon_L = 0.01635 \text{ ft}^3\text{/lbm}$$

At point A, $p_A = 600$ psia.

From Eq. 27.14,

$$\begin{aligned} h_A &\approx h_L + \upsilon_L(p_A - p_L) \\ &= 120.4 \frac{\text{Btu}}{\text{lbm}} + \frac{\left(0.01635 \frac{\text{ft}^3}{\text{lbm}}\right)\left(600 \frac{\text{lbf}}{\text{in}^2} - 4.5 \frac{\text{lbf}}{\text{in}^2}\right)\left(12 \frac{\text{in}}{\text{ft}}\right)^2}{778 \frac{\text{ft-lbf}}{\text{lbm}}} \\ &= 122.2 \text{ Btu/lbm} \end{aligned}$$

From Eq. 27.40, due to the inefficiency of the pump,

$$\begin{aligned} h'_A &= h_L + \frac{h_A - h_L}{\eta_{s,\text{pump}}} \\ &= 120.4 \frac{\text{Btu}}{\text{lbm}} + \frac{122.2 \frac{\text{Btu}}{\text{lbm}} - 120.4 \frac{\text{Btu}}{\text{lbm}}}{0.96} \\ &= 122.3 \text{ Btu/lbm} \end{aligned}$$

From Eq. 27.37, the thermal efficiency of the entire cycle neglecting condensation and drip pump, is

$$\begin{aligned} \eta_{th} &= \frac{W_{\text{turbines}} - W_{\text{pump}}}{Q_{\text{in}}} \\ &= \frac{(h_D - h'_E) + (h_F - h'_I) + (1-x)(h'_I - h'_G) - (h'_A - h_L)}{(h_D - h'_A) + (h_F - h'_E)} \\ &= \frac{\begin{array}{l}\left(1289.9 \frac{\text{Btu}}{\text{lbm}} - 1051.4 \frac{\text{Btu}}{\text{lbm}}\right) \\ + \left(1334.9 \frac{\text{Btu}}{\text{lbm}} - 1199.6 \frac{\text{Btu}}{\text{lbm}}\right) \\ + (1-0.0788)\left(1199.6 \frac{\text{Btu}}{\text{lbm}} - 1046.3 \frac{\text{Btu}}{\text{lbm}}\right) \\ - \left(122.3 \frac{\text{Btu}}{\text{lbm}} - 120.4 \frac{\text{Btu}}{\text{lbm}}\right)\end{array}}{\begin{array}{l}\left(1289.9 \frac{\text{Btu}}{\text{lbm}} - 122.3 \frac{\text{Btu}}{\text{lbm}}\right) \\ + \left(1334.9 \frac{\text{Btu}}{\text{lbm}} - 1051.4 \frac{\text{Btu}}{\text{lbm}}\right)\end{array}} \\ &= \boxed{0.353 \quad (35\%)} \end{aligned}$$

The answer is (C).

SI Solution

From Prob. 5,

$$h_B = 1082.5 \text{ kJ/kg}$$
$$h_D = 2980 \text{ kJ/kg}$$
$$h'_E = 2465.2 \text{ kJ/kg}$$
$$h_F = 3073.3 \text{ kJ/kg}$$
$$s_F = 8.0284 \text{ kJ/kg·K}$$
$$h'_G = 2318.4 \text{ kJ/kg}$$
$$h_H = 67.17 \text{ kJ/kg}$$

At point I,

$$p_I = 37.76 \text{ kPa} \quad [0.3776 \text{ bars}]$$
$$s_I = s_F = 8.0284 \text{ kJ/kg·K}$$

From App. 23.P, using double interpolation,

$$h_{\text{I}} = 2724.7 \text{ kJ/kg}$$

From Eq. 27.39, due to inefficiency of the turbine,

$$\begin{aligned} h'_{\text{I}} &= h_{\text{F}} - \eta_{s,\text{turbine}}(h_{\text{F}} - h_{\text{I}}) \\ &= 3073.3 \ \frac{\text{kJ}}{\text{kg}} - (0.88)\left(3073.3 \ \frac{\text{kJ}}{\text{kg}} - 2724.7 \ \frac{\text{kJ}}{\text{kg}}\right) \\ &= 2766.5 \text{ kJ/kg} \end{aligned}$$

At point J, from App. 23.O, the saturated liquid enthalpy and temperature at 37.76 kPa are $h_{\text{J}} =$ 311.2 kJ/kg and 74.3°C, respectively.

At point K, the temperature is

$$74.3°\text{C} - 3°\text{C} = 71.3°\text{C}$$

Since the water is subcooled, enthalpy is a function of temperature only.

$$h_{\text{K}} = 298.5 \text{ kJ/kg}$$

The condensate pump will match the pressure at J: $p_{\text{K}} = 37.76$ kPa.

From an energy balance in the heater,

$$\begin{aligned} (1-x)(h_{\text{K}} - h_{\text{H}}) &= x(h'_{\text{I}} - h_{\text{J}}) \\ (1-x)\left(298.5 \ \frac{\text{kJ}}{\text{kg}} - 67.17 \ \frac{\text{kJ}}{\text{kg}}\right) &= x\left(2766.5 \ \frac{\text{kJ}}{\text{kg}} - 325.8 \ \frac{\text{kJ}}{\text{kg}}\right) \\ 2686.7x &= 246.02 \\ x &= 0.0866 \end{aligned}$$

At point L, $p_{\text{L}} = 37.76$ kPa.

$$\begin{aligned} h_{\text{L}} &= xh_{\text{J}} + (1-x)h_{\text{K}} \\ &= (0.0866)\left(311.2 \ \frac{\text{kJ}}{\text{kg}}\right) + (1 - 0.0866)\left(298.5 \ \frac{\text{kJ}}{\text{kg}}\right) \\ &= 299.6 \text{ kJ/kg} \end{aligned}$$

Since this is a subcooled liquid, enthalpy is a function of temperature only, and from App. 23.N,

$$\begin{aligned} T_{\text{L}} &= 71.6°\text{C} \\ \upsilon_{\text{L}} &= 1.0237 \text{ cm}^3/\text{g} \end{aligned}$$

At point A, $p_{\text{A}} = 4$ MPa.

From Eq. 27.14,

$$\begin{aligned} h_{\text{A}} &\approx h_{\text{L}} + \upsilon_{\text{L}}(p_{\text{A}} - p_{\text{L}}) \\ &= 299.6 \ \frac{\text{kJ}}{\text{kg}} + \left(\frac{1.0237 \ \frac{\text{cm}^3}{\text{g}}}{\left(100 \ \frac{\text{cm}}{\text{m}}\right)^3}\right)\left(1000 \ \frac{\text{g}}{\text{kg}}\right) \\ &\quad \times \left((4 \text{ MPa})\left(1000 \ \frac{\text{kPa}}{\text{MPa}}\right) - (0.3776 \text{ bar})\left(1000 \ \frac{\text{kPa}}{\text{bar}}\right)\right) \\ &= 303.3 \text{ kJ/kg} \end{aligned}$$

From Eq. 27.40, due to the inefficiency in the pump,

$$\begin{aligned} h'_{\text{A}} &= h_{\text{L}} + \frac{h_{\text{A}} - h_{\text{L}}}{\eta_{s,\text{pump}}} \\ &= 299.6 \ \frac{\text{kJ}}{\text{kg}} + \frac{303.3 \ \frac{\text{kJ}}{\text{kg}} - 299.6 \ \frac{\text{kJ}}{\text{kg}}}{0.96} \\ &= 303.5 \text{ kJ/kg} \end{aligned}$$

From Eq. 27.37, the thermal efficiency of the entire cycle, neglecting condensation and drip pump, is

$$\begin{aligned} \eta_{\text{th}} &= \frac{W_{\text{turbine}} - W_{\text{pump}}}{Q_{\text{in}}} \\ &= \frac{\left((h_{\text{D}} - h'_{\text{E}}) + (h_{\text{F}} - h'_{\text{I}}) + (1-x)(h'_{\text{I}} - h'_{\text{G}})\right) - (h'_{\text{A}} - h_{\text{L}})}{(h_{\text{D}} - h'_{\text{A}}) + (h_{\text{F}} - h'_{\text{E}})} \\ &= \frac{\left(\begin{array}{l}\left(2980 \ \frac{\text{kJ}}{\text{kg}} - 2465.2 \ \frac{\text{kJ}}{\text{kg}}\right) \\ \quad + \left(3073.3 \ \frac{\text{kJ}}{\text{kg}} - 2766.5 \ \frac{\text{kJ}}{\text{kg}}\right) \\ \quad + (1 - 0.0866)\left(2766.5 \ \frac{\text{kJ}}{\text{kg}} - 2318.4 \ \frac{\text{kJ}}{\text{kg}}\right)\end{array}\right) - \left(303.5 \ \frac{\text{kJ}}{\text{kg}} - 299.6 \ \frac{\text{kJ}}{\text{kg}}\right)}{\left(2980 \ \frac{\text{kJ}}{\text{kg}} - 303.5 \ \frac{\text{kJ}}{\text{kg}}\right) + \left(3073.3 \ \frac{\text{kJ}}{\text{kg}} - 2465.2 \ \frac{\text{kJ}}{\text{kg}}\right)} \\ &= \boxed{0.374 \quad (35\%)} \end{aligned}$$

The answer is (C).

Power Cycles

7. *Customary U.S. Solution*

(a) Refer to the given illustration for Prob. 7 and to the following diagram.

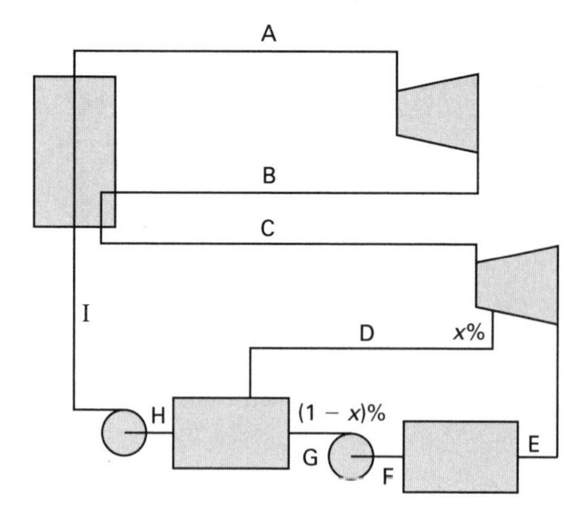

At point A,

$$p_A = 900 \text{ psia}$$
$$T_A = 800°\text{F}$$

Using the superheated steam table, App. 23.C,

$$h_A = 1393.9 \text{ Btu/lbm}$$
$$s_A = 1.5816 \text{ Btu/lbm-°R}$$

From the Mollier diagram, assuming isentropic expansion to 200 psia, $h_B = 1230$ Btu/lbm.

Isentropic efficiency of the high pressure turbine is

$$\eta_{s,\text{turbine}} = \frac{h_A - h'_B}{h_A - h_B} = \left(\frac{1393.9 \,\frac{\text{Btu}}{\text{lbm}} - 1270 \,\frac{\text{Btu}}{\text{lbm}}}{1393.9 \,\frac{\text{Btu}}{\text{lbm}} - 1230 \,\frac{\text{Btu}}{\text{lbm}}}\right) \times 100\% = \boxed{75.59\% \quad (76\%)}$$

The answer is (D).

(b) At point C,

$$p_C = 190 \text{ psia}$$
$$T_C = 800°\text{F}$$

Using the superheated steam table from App. 23.C, $h_C = 1426.0$ Btu/lbm.

At D: $h'_D = 1280$ Btu/lbm

At E: $h'_E = 1075$ Btu/lbm

At F: $h_F = 69.73$ Btu/lbm

At G: $W_{\text{pump}} = 0.15$ Btu/lbm

$$W_{\text{pump}} = h'_G - h_F$$
$$h'_G = W_{\text{pump}} + h_F = 0.15 \,\frac{\text{Btu}}{\text{lbm}} + 69.73 \,\frac{\text{Btu}}{\text{lbm}} = \boxed{69.88 \text{ Btu/lbm} \quad (70 \text{ Btu/lbm})}$$

At H: $h_H = 250.2$ Btu/lbm

At I: $h'_I = 253.1$ Btu/lbm

The answer is (A).

(c) From an energy balance in the heater,

$$xh'_D + (1-x)h'_G = h_H$$
$$x(h'_D - h'_G) = h_H - h'_G$$
$$x = \frac{h_H - h'_G}{h'_D - h'_G} = \frac{250.2 \,\frac{\text{Btu}}{\text{lbm}} - 69.88 \,\frac{\text{Btu}}{\text{lbm}}}{1280 \,\frac{\text{Btu}}{\text{lbm}} - 69.88 \,\frac{\text{Btu}}{\text{lbm}}} = 0.149$$

The thermal efficiency of the cycle is

$$\eta_{\text{th}} = \frac{W_{\text{out}} - W_{\text{in}}}{Q_{\text{in}}}$$
$$= \frac{(h_A - h'_B) + (h_C - h'_D) + (1-x)(h'_D - h'_E) - (h'_I - h_H) - (1-x)(h'_G - h_F)}{(h_A - h'_I) + (h_C - h'_B)}$$
$$= \frac{\begin{array}{l}\left(1393.9 \,\frac{\text{Btu}}{\text{lbm}} - 1270 \,\frac{\text{Btu}}{\text{lbm}}\right) + \left(1426.0 \,\frac{\text{Btu}}{\text{lbm}} - 1280 \,\frac{\text{Btu}}{\text{lbm}}\right) \\ + (1-0.149)\left(1280 \,\frac{\text{Btu}}{\text{lbm}} - 1075 \,\frac{\text{Btu}}{\text{lbm}}\right) - \left(253.1 \,\frac{\text{Btu}}{\text{lbm}} - 250.2 \,\frac{\text{Btu}}{\text{lbm}}\right) \\ - (1-0.149)\left(69.88 \,\frac{\text{Btu}}{\text{lbm}} - 69.73 \,\frac{\text{Btu}}{\text{lbm}}\right)\end{array}}{\left(1393.9 \,\frac{\text{Btu}}{\text{lbm}} - 253.1 \,\frac{\text{Btu}}{\text{lbm}}\right) + \left(1426.0 \,\frac{\text{Btu}}{\text{lbm}} - 1270 \,\frac{\text{Btu}}{\text{lbm}}\right)}$$
$$= \boxed{0.340 \quad (34\%)}$$

The answer is (C).

SI Solution

(a) Refer to the illustration for Prob. 7 and to the diagram in the customary U.S. solution.

At point A,

$$p_{\text{A}} = 6.2\ \text{MPa}$$
$$T_{\text{A}} = 420^\circ\text{C}$$

Using the Mollier diagram for steam, App. 23.R,

$$h_{\text{A}} = 3235.0\ \text{kJ/kg}$$
$$s_{\text{A}} = 6.65\ \text{kJ/kg·K}$$

From the Mollier diagram, assuming isentropic expansion to 1.5 MPa, $h_{\text{B}} = 2860$ kJ/kg.

Isentropic efficiency of the high pressure turbine is

$$\eta_{s,\text{turbine}} = \frac{h_{\text{A}} - h'_{\text{B}}}{h_{\text{A}} - h_{\text{B}}} = \frac{3235.0\ \frac{\text{kJ}}{\text{kg}} - 2960\ \frac{\text{kJ}}{\text{kg}}}{3235.0\ \frac{\text{kJ}}{\text{kg}} - 2860\ \frac{\text{kJ}}{\text{kg}}}$$
$$= \boxed{0.733 \quad (73\%)}$$

The answer is (C).

(b) At point C,

$$p_{\text{C}} = 1.4\ \text{MPa}$$
$$T_{\text{C}} = 420^\circ\text{C}$$

Using the superheated steam table from App. 23.P, $h_{\text{C}} = 3301.3$ kJ/kg.

At point D, $h'_{\text{D}} = 2980$ kJ/kg.

At point E, $h'_{\text{E}} = 2500$ kJ/kg.

At point F, $h'_{\text{F}} = 162.5$ kJ/kg.

At point G,

$$W_{\text{pump}} = 0.3\ \text{kJ/kg}$$
$$= h'_{\text{G}} - h_{\text{F}}$$
$$h'_{\text{G}} = W_{\text{pump}} + h_{\text{F}}$$
$$= 0.3\ \frac{\text{kJ}}{\text{kg}} + 162.5\ \frac{\text{kJ}}{\text{kg}}$$
$$= \boxed{162.8\ \text{kJ/kg} \quad (160\ \text{kJ/kg})}$$

At point H, $h_{\text{H}} = 583.0$ kJ/kg.

At point I, $h'_{\text{I}} = 589.7$ kJ/kg.

The answer is (A).

(c) From an energy balance in the heater,

$$xh'_{\text{D}} + (1-x)h'_{\text{G}} = h_{\text{H}}$$
$$x(h'_{\text{D}} - h'_{\text{G}}) = h_{\text{H}} - h'_{\text{G}}$$
$$x = \frac{h_{\text{H}} - h_{\text{G}}}{h'_{\text{D}} - h'_{\text{G}}}$$
$$= \frac{583.0\ \frac{\text{kJ}}{\text{kg}} - 162.8\ \frac{\text{kJ}}{\text{kg}}}{2980\ \frac{\text{kJ}}{\text{kg}} - 162.8\ \frac{\text{kJ}}{\text{kg}}}$$
$$= 0.149$$

The thermal efficiency of the cycle is

$$\eta_{\text{th}} = \frac{W_{\text{out}} - W_{\text{in}}}{Q_{\text{in}}}$$
$$= \frac{\begin{array}{c}(h_{\text{A}} - h'_{\text{B}}) + (h_{\text{C}} - h'_{\text{D}}) + (1-x)(h'_{\text{D}} - h'_{\text{E}}) \\ - (h'_{\text{I}} - h_{\text{H}}) - (1-x)(h'_{\text{G}} - h_{\text{F}})\end{array}}{(h_{\text{A}} - h'_{\text{I}}) + (h_{\text{C}} - h'_{\text{B}})}$$
$$= \frac{\begin{array}{c}\left(3235.0\ \frac{\text{kJ}}{\text{kg}} - 2960\ \frac{\text{kJ}}{\text{kg}}\right) \\ + \left(3301.3\ \frac{\text{kJ}}{\text{kg}} - 2980\ \frac{\text{kJ}}{\text{kg}}\right) \\ + (1-0.149)\left(2980\ \frac{\text{kJ}}{\text{kg}} - 2500\ \frac{\text{kJ}}{\text{kg}}\right) \\ - \left(598.7\ \frac{\text{kJ}}{\text{kg}} - 583.0\ \frac{\text{kJ}}{\text{kg}}\right) \\ - (1-0.149)\left(162.8\ \frac{\text{kJ}}{\text{kg}} - 162.5\ \frac{\text{kJ}}{\text{kg}}\right)\end{array}}{\begin{array}{c}\left(3235.0\ \frac{\text{kJ}}{\text{kg}} - 598.7\ \frac{\text{kJ}}{\text{kg}}\right) \\ + \left(3301.3\ \frac{\text{kJ}}{\text{kg}} - 2960\ \frac{\text{kJ}}{\text{kg}}\right)\end{array}}$$
$$= \boxed{0.332 \quad (34\%)}$$

The answer is (C).

Power Cycles

28 Reciprocating Combustion Engine Cycles

PRACTICE PROBLEMS

1. A 10 in (250 mm) bore, 18 in (460 mm) stroke, two-cylinder, four-stroke internal combustion engine operates with an indicated mean effective pressure of 95 psig (660 kPa) at 200 rpm. The actual torque developed is 600 ft-lbf (820 N·m). The friction horsepower is most nearly

(A) 17 hp (13 kW)

(B) 22 hp (17 kW)

(C) 45 hp (33 kW)

(D) 78 hp (60 kW)

2. A four-cycle internal combustion engine has a bore of 3.1 in (80 mm) and a stroke of 3.8 in (97 mm). When the engine is running at 4000 rpm, the air-fuel mixture enters the cylinders during the intake stroke with an average velocity of 100 ft/sec (30 m/s). The intake valve opens at top dead center (TDC) and closes at 40° past bottom dead center (BDC). The volumetric efficiency is 65%. The effective area of the intake valve is most nearly

(A) 0.012 ft^2 (0.0012 m^2)

(B) 0.023 ft^2 (0.0023 m^2)

(C) 0.031 ft^2 (0.0031 m^2)

(D) 0.058 ft^2 (0.0058 m^2)

3. An engine runs on the Otto cycle. The compression ratio is 10:1. The total intake volume of the cylinders is 11 ft^3 (0.3 m^3). Air enters at 14.2 psia and 80°F (98 kPa and 27°C). In two revolutions of the crank, after the compression strokes, 160 Btu (179 kJ) of energy is added.

(a) The temperature after the heat addition is most nearly

(A) 1290°R (720K)

(B) 1550°R (860K)

(C) 1750°R (970K)

(D) 2320°R (1340K)

(b) The thermal efficiency of the cycle is most nearly

(A) 45%

(B) 52%

(C) 57%

(D) 64%

4. A 4.25 in × 6 in (110 mm × 150 mm), six-cylinder four-stroke diesel engine runs at 1200 rpm while consuming 28 lbm of fuel per hour (0.0035 kg/s). The actual volumetric fraction of carbon dioxide in the exhaust is 9% (dry basis). At another throttle setting, when the air-fuel ratio is 15:1, there is 13.7% (dry basis) carbon dioxide in the exhaust. The atmospheric conditions are 14.7 psia and 70°F (101.3 kPa and 21°C). The volumetric efficiency at 1200 rpm is most nearly

(A) 59%

(B) 65%

(C) 72%

(D) 80%

5. At standard atmospheric conditions, a fully loaded diesel engine runs at 2000 rpm and develops 1000 bhp (750 kW). Metered fuel injection is used. The brake specific fuel consumption is 0.45 lbm/bhp-hr (76 kg/GJ). The mechanical efficiency is 80%, independent of the altitude. International Standard Atmosphere (ISA) conditions are applicable.

(a) At an altitude of 5000 ft (1500 m), the brake horsepower will be most nearly

(A) 810 hp (630 kW)

(B) 860 hp (650 kW)

(C) 890 hp (690 kW)

(D) 950 hp (740 kW)

(b) At an altitude of 5000 ft (1500 m), the brake specific fuel consumption will be most nearly

(A) 0.26 lbm/hp-hr (44 kg/GJ)

(B) 0.52 lbm/hp-hr (88 kg/GJ)

(C) 1.4 lbm/hp-hr (240 kg/GJ)

(D) 2.2 lbm/hp-hr (370 kg/GJ)

6. A gasoline-fueled internal combustion engine runs at 4600 rpm. The engine is four-stroke and V-8 in configuration with a displacement of 265 in^3 (4.3 L). The indicated work required to compress the air-fuel mixture is 1200 ft-lbf (1.6 kJ) per cycle. The indicated work done by the exhaust gases in expansion is 1500 ft-lbf (2.0 kJ) per cycle. The input energy from fuel combustion is 1.27 Btu (1.33 kJ) per cycle. Atmospheric air is at 14.7 psia and 70°F (101.3 kPa and 21°C). The air-fuel ratio is 20:1. The heating value of gasoline is 18,900 Btu/lbm (44 MJ/kg). Neglect the effects of friction.

(a) The indicated horsepower is most nearly

(A) 140 hp (110 kW)

(B) 170 hp (120 kW)

(C) 200 hp (160 kW)

(D) 240 hp (190 kW)

(b) The thermal efficiency is most nearly

(A) 30%

(B) 35%

(C) 39%

(D) 46%

(c) The mass per hour of gasoline consumed is most nearly

(A) 37 lbm/hr (0.0046 kg/s)

(B) 52 lbm/hr (0.0065 kg/s)

(C) 74 lbm/hr (0.0093 kg/s)

(D) 99 lbm/hr (0.012 kg/s)

(d) The specific fuel consumption is most nearly

(A) 0.055 lbm/hp-hr (10 kg/GJ)

(B) 0.11 lbm/hp-hr (19 kg/GJ)

(C) 0.22 lbm/hp-hr (38 kg/GJ)

(D) 0.44 lbm/hp-hr (76 kg/GJ)

7. A mixture of carbon dioxide and helium in an engine undergoes the following processes in a cycle.

A to B: compression and heat removal
B to C: constant volume heating
C to A: isentropic expansion

point	temperature	pressure
A	520°R (290K)	14.7 psia (101.3 kPa)
B	1240°R (690K)	unknown
C	1600°R (890K)	568.6 psia (3.920 MPa)

Both gases are ideal gases.

(a) The molar specific heat of the mixture is most nearly

(A) 3.6 Btu/lbmol-°R (15 kJ/kmol·K)

(B) 4.1 Btu/lbmol-°R (17 kJ/kmol·K)

(C) 5.3 Btu/lbmol-°R (22 kJ/kmol·K)

(D) 6.5 Btu/lbmol-°R (27 kJ/kmol·K)

(b) What is most nearly the molecular weight of the mixture?

(A) 13 lbm/lbmol (13 kg/kmol)

(B) 19 lbm/lbmol (19 kg/kmol)

(C) 22 lbm/lbmol (22 kg/kmol)

(D) 26 lbm/lbmol (26 kg/kmol)

(c) The gravimetric fractions of the helium and carbon dioxide in the mixture are most nearly

(A) He, 0.14; CO_2, 0.86

(B) He, 0.23; CO_2, 0.77

(C) He, 0.35; CO_2, 0.65

(D) He, 0.46; CO_2, 0.54

(d) The work done during the isentropic expansion process is most nearly

(A) 190 Btu/lbm (440 kJ/kg)

(B) 260 Btu/lbm (610 kJ/kg)

(C) 330 Btu/lbm (760 kJ/kg)

(D) 450 Btu/lbm (1000 kJ/kg)

(e) Draw the temperature-entropy and pressure-volume diagrams for the cycle.

8. When the atmospheric conditions are 14.7 psia and 80°F (101.3 kPa and 27°C), a diesel engine with metered fuel injection has the following operating characteristics.

brake horsepower:	200 bhp (150 kW)
brake specific fuel consumption:	0.48 lbm/hp-hr (81 kg/GJ)
air-fuel ratio:	22:1
mechanical efficiency:	86%

The engine is moved to an altitude where the atmospheric conditions are 12.2 psia and 60°F (84 kPa and 16°C). The running speed is unchanged.

(a) The frictionless power is most nearly

(A) 200 hp (150 kW)

(B) 230 hp (170 kW)

(C) 340 hp (260 kW)

(D) 420 hp (310 kW)

(b) The friction power is most nearly

(A) 0 hp (0 kW)

(B) 33 hp (24 kW)

(C) 140 hp (110 kW)

(D) 220 hp (160 kW)

(c) Using the ideal gas law, what are most nearly the air densities, ρ_{a1} and ρ_{a2}, respectively?

(A) 0.0005 lbm/ft^3, 0.0044 lbm/ft^3 (0.008 kg/m^3, 0.070 kg/m^3)

(B) 0.0032 lbm/ft^3, 0.0038 lbm/ft^3 (0.050 kg/m^3, 0.061 kg/m^3)

(C) 0.074 lbm/ft^3, 0.063 lbm/ft^3 (1.18 kg/m^3, 1.01 kg/m^3)

(D) 0.50 lbm/ft^3, 0.55 lbm/ft^3 (8.0 kg/m^3, 8.8 kg/m^3)

(d) The new net power is most nearly

(A) 140 hp (100 kW)

(B) 170 hp (130 kW)

(C) 260 hp (200 kW)

(D) 330 hp (240 kW)

(e) The new efficiency is most nearly

(A) 0.72

(B) 0.84

(C) 0.86

(D) 1.00

(f) The new air mass flow rate is most nearly

(A) 1680 lbm/hr (0.213 kg/s)

(B) 1770 lbm/hr (0.224 kg/s)

(C) 1820 lbm/hr (0.230 kg/s)

(D) 1960 lbm/hr (0.247 kg/s)

(g) The new fuel consumption is most nearly

(A) 0.57 lbm/hp-hr (97 kg/GJ)

(B) 0.67 lbm/hp-hr (110 kg/GJ)

(C) 0.89 lbm/hp-hr (150 kg/GJ)

(D) 1.1 lbm/hp-hr (190 kg/GJ)

(h) The new air/fuel ratio is most nearly

(A) 18.5

(B) 18.6

(C) 18.7

(D) 19.0

9. An internal combustion engine is normally fuel-injected with gasoline (C_8H_{18}; lower heating value of 23,200 Btu/lbm; 54 MJ/kg). It is desired to switch to alcohol (C_2H_5OH; lower heating value of 11,930 Btu/lbm; 27.7 MJ/kg). Minimal changes to the engine are to be made. The indicated and mechanical efficiencies are unchanged. Sufficient oxygen exists for complete combustion in all configurations.

(a) If the power output is to be unchanged, the percentage change in specific fuel consumption is most nearly

(A) 45%

(B) 65%

(C) 85%

(D) 95%

(b) If the power output and fuel-injection velocity are to be unchanged, the percentage change in injection port area is most nearly

(A) 35%

(B) 55%

(C) 81%

(D) 95%

(c) If no changes are made to the engine, the percentage decrease in power output is most nearly

(A) −75%

(B) −60%

(C) −45%

(D) −25%

Power Cycles

SOLUTIONS

1. *Customary U.S. Solution*

The actual brake horsepower from Eq. 28.13(b) is

$$\text{BHP} = \frac{nT}{5252} = \frac{(200\ \text{rpm})(600\ \text{ft-lbf})}{5252\ \frac{\text{ft-lbf}}{\text{hp-min}}} = 22.85\ \text{hp}$$

From Eq. 28.44, the number of power strokes per minute is

$$N = \frac{(2n)(\text{no. cylinders})}{\text{no. strokes per cycle}} = \frac{\left(2\ \frac{\text{strokes}}{\text{rev}}\right)\left(200\ \frac{\text{rev}}{\text{min}}\right)(2)}{4\ \frac{\text{strokes}}{\text{power strokes}}} = 200\ \text{power strokes/min}$$

The stroke is

$$L = \frac{18\ \text{in}}{12\ \frac{\text{in}}{\text{ft}}} = 1.5\ \text{ft}$$

The bore area is

$$A = \left(\frac{\pi}{4}\right)(10\ \text{in})^2 = 78.54\ \text{in}^2$$

From Eq. 28.43(b), the ideal (indicated) horsepower is

$$P_{\text{hp}} = \frac{pLAN}{33{,}000} = \frac{\left(95\ \frac{\text{lbf}}{\text{in}^2}\right)(1.5\ \text{ft})(78.54\ \text{in}^2)\left(200\ \frac{\text{power strokes}}{\text{min}}\right)}{33{,}000\ \frac{\text{ft-lbf}}{\text{hp-min}}} = 67.83\ \text{hp}$$

The friction horsepower is

$$\begin{aligned}\text{FHP} &= \text{IHP} - \text{BHP} \\ &= \text{ideal hp} - \text{actual BHP} \\ &= 67.83\ \text{hp} - 22.85\ \text{hp} \\ &= \boxed{44.98\ \text{hp}\quad (45\ \text{hp})}\end{aligned}$$

The answer is (C).

SI Solution

The actual brake power from Eq. 28.13(a) is

$$\text{BkW} = \frac{nT}{9549} = \frac{\left(200\ \frac{\text{rev}}{\text{min}}\right)(820\ \text{N·m})}{9549\ \frac{\text{N·m}}{\text{kW·min}}} = 17.17\ \text{kW}$$

From Eq. 28.44, the number of power strokes per minute is

$$N = \frac{(2n)(\text{no. cylinders})}{\text{no. of strokes per cycle}} = \frac{\left(2\ \frac{\text{strokes}}{\text{rev}}\right)\left(200\ \frac{\text{rev}}{\text{min}}\right)(2)}{4\ \frac{\text{strokes}}{\text{power strokes}}} = 200\ \text{power strokes/min}$$

The stroke is

$$L = \frac{460\ \text{mm}}{1000\ \frac{\text{mm}}{\text{m}}} = 0.46\ \text{m}$$

The bore area is

$$A = \left(\frac{\pi}{4}\right)\left(\frac{250\ \text{mm}}{1000\ \frac{\text{mm}}{\text{m}}}\right)^2 = 4.909 \times 10^{-2}\ \text{m}^2$$

From Eq. 28.43(a), the ideal (indicated) power is

$$P_{\text{kW}} = \frac{pLAN}{60} = \frac{(660\ \text{kPa})(0.46\ \text{m}) \times (4.909 \times 10^{-2}\ \text{m}^2)\left(200\ \frac{\text{power strokes}}{\text{min}}\right)}{60\ \frac{\text{s}}{\text{min}}} = 49.68\ \text{kW}$$

The friction power is

$$\text{ideal kW} - \text{actual kW} = 49.68\ \text{kW} - 17.17\ \text{kW} = \boxed{32.51\ \text{kW}\quad (33\ \text{kW})}$$

The answer is (C).

2. *Customary U.S. Solution*

The number of degrees that the valve is open is

$$180^\circ + 40^\circ = 220^\circ$$

The time that the valve is open is

$$\left(\frac{220^\circ}{360^\circ}\right)\left(\frac{\text{time}}{\text{rev}}\right) = \left(\frac{220^\circ}{360^\circ}\right)\left(\frac{60\ \frac{\text{sec}}{\text{min}}}{4000\ \text{rpm}}\right)$$
$$= 9.167 \times 10^{-3}\ \text{sec/rev}$$

(If the duration of an entire four-stroke cycle (2 revolutions) was used, the ratio would be 220°/720°.)

The displacement is

$$\left(\frac{\pi}{4}\right)(\text{bore})^2(\text{stroke}) = \left(\frac{\left(\frac{\pi}{4}\right)(3.1\ \text{in})^2}{\left(12\ \frac{\text{in}}{\text{ft}}\right)^2}\right)\left(\frac{3.8\ \text{in}}{12\ \frac{\text{in}}{\text{ft}}}\right)$$
$$= 0.0166\ \text{ft}^3$$

The actual incoming volume per intake stroke is

$$\begin{aligned} V &= (\text{volumetric efficiency})(\text{displacement}) \\ &= (0.65)(0.0166\ \text{ft}^3) \\ &= 0.01079\ \text{ft}^3 \end{aligned}$$

The area is

$$A = \frac{V}{\text{v}t} = \frac{0.01079\ \text{ft}^3}{\left(100\ \frac{\text{ft}}{\text{sec}}\right)(9.167 \times 10^{-3}\ \text{sec})}$$
$$= \boxed{0.0118\ \text{ft}^2 \quad (0.012\ \text{ft}^2)}$$

The answer is (A).

SI Solution

The number of degrees that the valve is open is

$$180^\circ + 40^\circ = 220^\circ$$

The time that the valve is open is

$$\left(\frac{220^\circ}{360^\circ}\right)\left(\frac{\text{time}}{\text{rev}}\right) = \left(\frac{220^\circ}{360^\circ}\right)\left(\frac{60\ \frac{\text{s}}{\text{min}}}{4000\ \frac{\text{rev}}{\text{min}}}\right)$$
$$= 9.167 \times 10^{-3}\ \text{s/rev}$$

(If the duration of an entire four-stroke cycle (2 revolutions) was used, the ratio would be 220°/720°.)

The displacement is

$$\left(\frac{\pi}{4}\right)(\text{bore})^2(\text{stroke}) = \left(\frac{\pi}{4}\right)(0.08\ \text{m})^2(0.097\ \text{m})$$
$$= 4.876 \times 10^{-4}\ \text{m}^3$$

The actual incoming volume per intake stroke is

$$\begin{aligned} V &= (\text{volumetric efficiency})(\text{displacement}) \\ &= (0.65)(4.876 \times 10^{-4}\ \text{m}^3) \\ &= 3.169 \times 10^{-4}\ \text{m}^3 \end{aligned}$$

The area is

$$A = \frac{V}{\text{v}t} = \frac{3.169 \times 10^{-4}\ \text{m}^3}{\left(30\ \frac{\text{m}}{\text{s}}\right)(9.167 \times 10^{-3}\ \text{s})}$$
$$= \boxed{1.152 \times 10^{-3}\ \text{m}^2 \quad (0.0012\ \text{m}^2)}$$

The answer is (A).

3. *Customary U.S. Solution*

Refer to the air-standard Otto cycle diagram (see Fig. 28.3).

(a) At A:

$$V_\text{A} = 11\ \text{ft}^3$$

The absolute temperature is

$$T_\text{A} = 80^\circ\text{F} + 460^\circ = 540^\circ\text{R}$$

For an ideal gas, the mass of the air in the intake volume is

$$m = \frac{pV}{RT} = \frac{\left(14.2\ \frac{\text{lbf}}{\text{in}^2}\right)\left(12\ \frac{\text{in}}{\text{ft}}\right)^2(11\ \text{ft}^3)}{\left(53.3\ \frac{\text{ft-lbf}}{\text{lbm-}^\circ\text{R}}\right)(540^\circ\text{R})}$$
$$= 0.781\ \text{lbm}$$

From air tables (see App. 23.F) at 540°R,

$$v_{r,\text{A}} = 144.32$$
$$u_\text{A} = 92.04\ \text{Btu/lbm}$$

At B:

The compression ratio is a ratio of volumes.

$$V_\text{B} = \left(\frac{1}{10}\right)V_\text{A} = \left(\frac{1}{10}\right)(11\ \text{ft}^3)$$
$$= 1.1\ \text{ft}^3$$

Power Cycles

Since the compression from A to B is isentropic,

$$\begin{aligned}\upsilon_{r,\text{B}} &= \frac{\upsilon_{r,\text{A}}}{10} = \frac{144.32}{10} \\ &= 14.432\end{aligned}$$

From the air tables (see App. 23.F) for $\upsilon_r = 14.432$,

$$T_\text{B} \approx 1314°\text{R}$$
$$u_\text{B} \approx 230.5 \text{ Btu/lbm}$$

At C:

$$\begin{aligned}u_\text{C} &= u_\text{B} + \frac{Q_{\text{in,B-C}}}{m} \\ &= 230.5\ \frac{\text{Btu}}{\text{lbm}} + \frac{160 \text{ Btu}}{0.781 \text{ lbm}} \\ &= 435.4 \text{ Btu/lbm}\end{aligned}$$

From air tables (see App. 23.F) at $u = 435.4$ Btu/lbm,

$$T_\text{C} = \boxed{2319°\text{R} \quad (2320°\text{R})}$$
$$\upsilon_{r,\text{C}} = 2.694$$

At D:

Since expansion is isentropic and the ratio of volumes is the same,

$$\begin{aligned}\upsilon_{r,\text{D}} &= 10\upsilon_{r,\text{C}} = (10)(2.694) \\ &= 26.94\end{aligned}$$

From air tables (see App. 23.F), at $\upsilon_r = 26.94$,

$$T_\text{D} = 1044°\text{R}$$
$$u_\text{D} = 180.38 \text{ Btu/lbm}$$

The answer is (D).

(b) The heat input is

$$\begin{aligned}q_\text{in} &= \frac{Q}{m} = \frac{160 \text{ Btu}}{0.781 \text{ lbm}} \\ &= 204.9 \text{ Btu/lbm}\end{aligned}$$

The heat rejected during a constant volume process is $q_\text{out} = \Delta u$.

Heat is rejected between D and A. Therefore,

$$\begin{aligned}q_\text{out} &= u_\text{D} - u_\text{A} = 180.38\ \frac{\text{Btu}}{\text{lbm}} - 92.04\ \frac{\text{Btu}}{\text{lbm}} \\ &= 88.34 \text{ Btu/lbm}\end{aligned}$$

From Eq. 28.41, the thermal efficiency is

$$\begin{aligned}\eta_\text{th} &= \frac{q_\text{in} - q_\text{out}}{q_\text{in}} = \frac{204.9\ \frac{\text{Btu}}{\text{lbm}} - 88.34\ \frac{\text{Btu}}{\text{lbm}}}{204.9\ \frac{\text{Btu}}{\text{lbm}}} \\ &= \boxed{0.569 \quad (57\%)}\end{aligned}$$

The answer is (C).

SI Solution

(a) At A:

$$V_\text{A} = 0.3 \text{ m}^3$$

The absolute temperature is

$$T_\text{A} = 27°\text{C} + 273° = 300\text{K}$$

From the ideal gas law, the mass of air in the intake volume is

$$\begin{aligned}m &= \frac{pV}{RT} = \frac{(98 \text{ kPa})\left(1000\ \frac{\text{Pa}}{\text{kPa}}\right)(0.3 \text{ m}^3)}{\left(287\ \frac{\text{J}}{\text{kg·K}}\right)(300\text{K})} \\ &= 0.3415 \text{ kg}\end{aligned}$$

From air tables (see App. 23.S) at 300K,

$$\upsilon_{r,\text{A}} = 621.2$$
$$u_\text{A} = 214.07 \text{ kJ/kg}$$

At B:

The compression ratio is a ratio of volumes.

$$\begin{aligned}V_\text{B} &= \left(\frac{1}{10}\right)V_\text{A} = \frac{0.3 \text{ m}^3}{10} \\ &= 0.03 \text{ m}^3\end{aligned}$$

Since the compression from A to B is isentropic,

$$\begin{aligned}\upsilon_{r,\text{B}} &= \frac{\upsilon_{r,\text{A}}}{10} = \frac{621.2}{10} \\ &= 62.12\end{aligned}$$

From air tables (see App. 23.S) for this value of $\upsilon_{r,\text{B}}$,

$$T_\text{B} = 730\text{K}$$
$$u_\text{B} = 536.0 \text{ kJ/kg}$$

At C:

$$\begin{aligned} u_\text{C} &= u_\text{B} + \frac{Q_{\text{in,B-C}}}{m} \\ &= 536.0\ \frac{\text{kJ}}{\text{kg}} + \frac{179\ \text{kJ}}{0.3415\ \text{kg}} \\ &= 1060.2\ \text{kJ/kg} \end{aligned}$$

From air tables (see App. 23.S) at $u = 1060.2$ kJ/kg,

$$\begin{aligned} T_\text{C} &= \boxed{1341.4\text{K} \quad (1340\text{K})} \\ v_{r,\text{C}} &= 10.215 \end{aligned}$$

At D:

Since expansion is isentropic and the ratio of volumes is the same,

$$\begin{aligned} v_{r,\text{D}} &= 10 v_{r,\text{C}} = (10)(10.215) \\ &= 102.15 \end{aligned}$$

From air tables (see App. 23.S), at $v_r = 102.15$,

$$\begin{aligned} T_\text{D} &= 607.9\text{K} \\ u_\text{D} &= 440.8\ \text{kJ/kg} \end{aligned}$$

The answer is (D).

(b) The heat input is

$$\begin{aligned} q_\text{in} &= \frac{Q}{m} = \frac{179\ \text{kJ}}{0.3415\ \text{kg}} \\ &= 524.2\ \text{kJ/kg} \end{aligned}$$

The heat rejected during a constant volume process is

$$q_\text{out} = \Delta u$$

Heat is rejected between D and A. Therefore,

$$\begin{aligned} q_\text{out} &= u_\text{D} - u_\text{A} = 440.8\ \frac{\text{kJ}}{\text{kg}} - 214.07\ \frac{\text{kJ}}{\text{kg}} \\ &= 226.7\ \text{kJ/kg} \end{aligned}$$

From Eq. 28.41, the thermal efficiency is

$$\begin{aligned} \eta_\text{th} &= \frac{q_\text{in} - q_\text{out}}{q_\text{in}} = \frac{524.2\ \frac{\text{kJ}}{\text{kg}} - 226.7\ \frac{\text{kJ}}{\text{kg}}}{524.2\ \frac{\text{kJ}}{\text{kg}}} \\ &= \boxed{0.568 \quad (57\%)} \end{aligned}$$

The answer is (C).

4. *Customary U.S. Solution*

step 1: Find the ideal mass of air ingested.

From Eq. 28.44, the number of power strokes per second is

$$\begin{aligned} N &= \frac{(2n)(\text{no. cylinders})}{\text{no. strokes per cycle}} \\ &= \frac{\left(2\ \frac{\text{strokes}}{\text{rev}}\right)\left(1200\ \frac{\text{rev}}{\text{min}}\right)(6\ \text{cylinders})}{\left(4\ \frac{\text{strokes}}{\text{power stroke}}\right)\left(60\ \frac{\text{sec}}{\text{min}}\right)} \\ &= 60\ \text{power strokes/sec} \end{aligned}$$

The swept volume is

$$\begin{aligned} V_s &= \left(\frac{\pi}{4}\right)(\text{bore})^2(\text{stroke}) = \frac{\left(\frac{\pi}{4}\right)(4.25\ \text{in})^2(6\ \text{in})}{\left(12\ \frac{\text{in}}{\text{ft}}\right)^3} \\ &= 0.04926\ \text{ft}^3 \end{aligned}$$

The ideal volume of air taken in per second is

$$\begin{aligned} \dot{V}_i &= (\text{swept volume})\left(\frac{\text{intake strokes}}{\text{sec}}\right) \\ &= V_s N \\ &= (0.04926\ \text{ft}^3)\left(60\ \frac{1}{\text{sec}}\right) \\ &= 2.956\ \text{ft}^3/\text{sec} \end{aligned}$$

The absolute temperature is

$$70°\text{F} + 460° = 530°\text{R}$$

From the ideal gas law, the ideal mass of air in the swept volume is

$$\begin{aligned} \dot{m} &= \frac{p\dot{V}}{RT} = \frac{\left(14.7\ \frac{\text{lbf}}{\text{in}^2}\right)\left(12\ \frac{\text{in}}{\text{ft}}\right)^2\left(2.956\ \frac{\text{ft}^3}{\text{sec}}\right)}{\left(53.35\ \frac{\text{ft-lbf}}{\text{lbm-°R}}\right)(530°\text{R})} \\ &= 0.2213\ \text{lbm/sec} \end{aligned}$$

step 2: Find the carbon dioxide volume in the exhaust assuming complete combustion when the air/fuel ratio is 15 ($\%CO_2 = 13.7\%$, dry basis).

Air is 76.85% (by weight) nitrogen, so the nitrogen/fuel ratio is

$$(0.7685)(15) = 11.528\ \text{lbm}\ N_2/\text{lbm fuel}$$

Power Cycles

From the ideal gas law, the nitrogen per pound of fuel burned is

$$V_{N_2} = \frac{mRT}{p} = \frac{\left(11.528\ \frac{\text{lbm N}_2}{\text{lbm fuel}}\right)\left(55.16\ \frac{\text{ft-lbf}}{\text{lbm-°R}}\right)(530°\text{R})}{\left(14.7\ \frac{\text{lbf}}{\text{in}^2}\right)\left(12\ \frac{\text{in}}{\text{ft}}\right)^2} = 159.2\ \text{ft}^3/\text{lbm fuel}$$

Similarly, air is 23.15% oxygen by weight, so the oxygen/fuel ratio is

$$(0.2315)(15) = 3.473\ \text{lbm O}_2/\text{lbm fuel}$$

From the ideal gas law, the oxygen per pound of fuel burned is

$$V_{O_2} = \frac{mRT}{p} = \frac{\left(3.473\ \frac{\text{lbm O}_2}{\text{lbm fuel}}\right)\left(48.29\ \frac{\text{ft-lbf}}{\text{lbm-°R}}\right)(530°\text{R})}{\left(14.7\ \frac{\text{lbf}}{\text{in}^2}\right)\left(12\ \frac{\text{in}}{\text{ft}}\right)^2} = 41.99\ \text{ft}^3/\text{lbm fuel}$$

When oxygen forms carbon dioxide, the chemical equation is $C + O_2 \rightarrow CO_2$.

It takes one volume of oxygen to form one volume of carbon dioxide. Considering nitrogen and excess oxygen in the exhaust, the percentage (fraction) of carbon dioxide in the exhaust is found from

$$\%CO_2 = \frac{\text{vol CO}_2}{\text{vol CO}_2 + \text{vol O}_2 + \text{vol N}_2}$$

$$\text{vol CO}_2 = x\ [\text{unknown}],\ \text{in ft}^3$$

$$\text{vol O}_2 = 41.99\ \text{ft}^3 - \text{oxygen used to make CO}_2 = 41.99\ \text{ft}^3 - x$$

$$\%CO_2 = 0.137\quad [\text{given}]$$

$$0.137 = \frac{x}{x + \left(41.99\ \frac{\text{ft}^3}{\text{lbm}} - x\right) + 159.2\ \frac{\text{ft}^3}{\text{lbm}}}$$

$$x = 27.56\ \text{ft}^3/\text{lbm fuel}$$

Assuming complete combustion, the volume of CO_2 will be constant regardless of the amount of air used.

step 3: Calculate the excess air if the percentage of carbon dioxide in the exhaust is 9%.

$$\%CO_2 = \frac{\text{vol CO}_2}{\text{vol CO}_2 + \text{vol O}_2 + \text{vol N}_2 + \text{vol excess air}}$$

$$0.09 = \frac{27.56\ \frac{\text{ft}^3}{\text{lbm}}}{27.56\ \frac{\text{ft}^3}{\text{lbm}} + \left(41.99\ \frac{\text{ft}^3}{\text{lbm}} - 27.56\ \frac{\text{ft}^3}{\text{lbm}}\right) + 159.2\ \frac{\text{ft}^3}{\text{lbm}} + \text{vol excess air}}$$

$$\text{vol excess air} = 105.0\ \text{ft}^3/\text{lbm fuel}$$

From the ideal gas law, the mass of excess air is

$$m_{\text{excess}} = \frac{\left(14.7\ \frac{\text{lbf}}{\text{in}^2}\right)\left(12\ \frac{\text{in}}{\text{ft}}\right)^2(105.0\ \text{ft}^3)}{\left(53.35\ \frac{\text{lbf-ft}}{\text{lbm-°R}}\right)(530°\text{R})} = 7.861\ \text{lbm/lbm fuel}$$

step 4: The actual air/fuel ratio is

$$15\ \frac{\text{lbm air}}{\text{lbm fuel}} + 7.861\ \frac{\text{lbm air}}{\text{lbm fuel}} = 22.861\ \text{lbm air/lbm fuel}$$

The actual air mass per second is

$$\frac{\left(22.861\ \frac{\text{lbm air}}{\text{lbm fuel}}\right)\left(28\ \frac{\text{lbm fuel}}{\text{hr}}\right)}{3600\ \frac{\text{sec}}{\text{hr}}} = 0.178\ \text{lbm/sec}$$

step 5: The volumetric efficiency is

$$\eta_v = \frac{0.178\ \frac{\text{lbm}}{\text{sec}}}{0.2213\ \frac{\text{lbm}}{\text{sec}}} = \boxed{0.803\quad (80\%)}$$

The answer is (D).

SI Solution

step 1: Find the ideal mass of air ingested.

From Eq. 28.44, the number of power strokes per second is

$$N = \frac{(2n)(\text{no. cylinders})}{\text{no. strokes per cycle}} = \frac{\left(2\ \frac{\text{strokes}}{\text{rev}}\right)\left(1200\ \frac{\text{rev}}{\text{min}}\right)(6\ \text{cylinders})}{\left(4\ \frac{\text{strokes}}{\text{power stroke}}\right)\left(60\ \frac{\text{s}}{\text{min}}\right)} = 60\ \text{power strokes/s}$$

The swept volume is

$$V_s = \left(\frac{\pi}{4}\right)(\text{bore})^2(\text{stroke})$$
$$= \left(\frac{\pi}{4}\right)(0.110\ \text{m})^2(0.150\ \text{m})$$
$$= 1.425 \times 10^{-3}\ \text{m}^3$$

The ideal volume of air taken in per second is

$$\dot{V}_i = (\text{swept volume})\left(\frac{\text{intake strokes}}{\text{s}}\right)$$
$$= V_s N$$
$$= (1.425 \times 10^{-3}\ \text{m}^3)\left(60\ \frac{1}{\text{s}}\right) = 0.0855\ \text{m}^3/\text{s}$$

The absolute temperature is

$$21°\text{C} + 273° = 294\text{K}$$

From the ideal gas law, the ideal mass of air in the swept volume is

$$\dot{m} = \frac{p\dot{V}}{RT} = \frac{(101.3\ \text{kPa})\left(1000\ \frac{\text{Pa}}{\text{kPa}}\right)\left(0.0855\ \frac{\text{m}^3}{\text{s}}\right)}{\left(287.03\ \frac{\text{J}}{\text{kg·K}}\right)(294\text{K})}$$
$$= 0.1026\ \text{kg/s}$$

step 2: Find the carbon dioxide volume in the exhaust assuming complete combustion when the air/fuel ratio is 15 ($\%CO_2 = 13.7\%$ dry basis).

Air is 76.85% nitrogen by weight, so the nitrogen/fuel ratio is

$$(0.7685)(15) = 11.528\ \text{kg N}_2/\text{kg fuel}$$

From the ideal gas law, the nitrogen per kg of fuel burned is

$$V_{N_2} = \frac{mRT}{p}$$
$$= \frac{\left(11.528\ \frac{\text{kg N}_2}{\text{kg fuel}}\right)\left(296.77\ \frac{\text{J}}{\text{kg·K}}\right)(294\text{K})}{(101.3\ \text{kPa})\left(1000\ \frac{\text{Pa}}{\text{kPa}}\right)}$$
$$= 9.929\ \text{m}^3/\text{kg fuel}$$

Similarly, air is 23.15% oxygen by weight, so the oxygen/fuel ratio is

$$(0.2315)(15) = 3.473\ \text{kg O}_2/\text{kg fuel}$$

From the ideal gas law, the oxygen per kg of fuel burned is

$$V_{O_2} = \frac{mRT}{p}$$
$$= \frac{\left(3.473\ \frac{\text{kg O}_2}{\text{kg fuel}}\right)\left(259.82\ \frac{\text{J}}{\text{kg·K}}\right)(294\text{K})}{(101.3\ \text{kPa})\left(1000\ \frac{\text{Pa}}{\text{kPa}}\right)}$$
$$= 2.619\ \text{m}^3/\text{kg fuel}$$

From the customary U.S. solution,

$$\%CO_2 = \frac{\text{vol CO}_2}{\text{vol CO}_2 + \text{vol O}_2 + \text{vol N}_2}$$
$$\text{vol CO}_2 = x\ [\text{unknown}],\ \text{in m}^3$$
$$\text{vol O}_2 = 2.619\ \text{m}^3 - \text{O}_2\ \text{used to make CO}_2$$
$$= 2.619\ \text{m}^3 - x$$
$$\%CO_2 = 0.137\quad [\text{given}]$$
$$0.137 = \frac{x}{x + \left(2.619\ \frac{\text{m}^3}{\text{kg}} - x\right) + 9.929\ \frac{\text{m}^3}{\text{kg}}}$$
$$x = 1.719\ \text{m}^3/\text{kg fuel}$$

Assuming complete combustion, the volume of carbon dioxide will be constant regardless of the amount of air used.

step 3: Calculate the excess air if the percentage of carbon dioxide in the exhaust is 9%. Therefore,

$$\%CO_2 = \frac{\text{vol CO}_2}{\text{vol CO}_2 + \text{vol O}_2 + \text{vol N}_2 + \text{vol excess air}}$$

$$0.09 = \frac{1.719\ \frac{\text{m}^3}{\text{kg}}}{1.719\ \frac{\text{m}^3}{\text{kg}} + \left(2.619\ \frac{\text{m}^3}{\text{kg}} - 1.719\ \frac{\text{m}^3}{\text{kg}}\right) + 9.929\ \frac{\text{m}^3}{\text{kg}} + \text{vol excess air}}$$

$$\text{vol excess air} = 6.552\ \text{m}^3/\text{kg fuel}$$

From the ideal gas law, the mass of the excess air is

$$m_{\text{excess}} = \frac{(101.3\ \text{kPa})\left(1000\ \frac{\text{Pa}}{\text{kPa}}\right)(6.552\ \text{m}^3)}{\left(287.03\ \frac{\text{J}}{\text{kg·K}}\right)(294\text{K})}$$
$$= 7.865\ \text{kg/kg fuel}$$

step 4: The actual air/fuel ratio is

$$15\ \frac{\text{kg air}}{\text{kg fuel}} + 7.865\ \frac{\text{kg air}}{\text{kg fuel}} = 22.865\ \text{kg air/kg fuel}$$

Power Cycles

The actual air mass per second is

$$\left(22.865\ \frac{\text{kg air}}{\text{kg fuel}}\right)\left(0.0035\ \frac{\text{kg}}{\text{s}}\right) = 0.0800\ \text{kg/s}$$

step 5: The volumetric efficiency is

$$\eta_v = \frac{0.0800\ \frac{\text{kg}}{\text{s}}}{0.1026\ \frac{\text{kg}}{\text{s}}} = \boxed{0.780\quad(80\%)}$$

The answer is (D).

5. *Customary U.S. Solution*

In this problem, the efficiency is known as the new altitude. This can be used to calculate the brake horsepower from the new indicated horsepower. The friction horsepower is not needed in this problem, since it is only used to calculate the indicated horsepower. It is not possible for the friction horsepower and the efficiency to both be unchanged between the original and new altitudes.

(a) *step 1:* From App. 25.E,

Altitude 1: standard atmospheric condition, 14.696 psia, 518.7°R.

Altitude 2: $z = 5000$ ft, 12.225 psia, 500.9°R.

step 2: Calculate the old frictionless power (see Eq. 28.71).

$$\text{IHP}_1 = \frac{\text{BHP}_1}{\eta_{m1}} = \frac{1000\ \text{hp}}{0.80} = 1250\ \text{hp}$$

step 3: Not needed since η_m is known to be constant with altitude.

step 4: From the ideal gas law,

$$\rho_{a1} = \frac{p}{RT} = \frac{\left(14.696\ \frac{\text{lbf}}{\text{in}^2}\right)\left(12\ \frac{\text{in}}{\text{ft}}\right)^2}{\left(53.35\ \frac{\text{lbf-ft}}{\text{lbm-°R}}\right)(518.7\text{°R})}$$
$$= 0.0765\ \text{lbm/ft}^3$$

Similarly,

$$\rho_{a2} = \frac{p}{RT} = \frac{\left(12.225\ \frac{\text{lbf}}{\text{in}^2}\right)\left(12\ \frac{\text{in}}{\text{ft}}\right)^2}{\left(53.35\ \frac{\text{ft-lbf}}{\text{lbm-°R}}\right)(500.9\text{°R})}$$
$$= 0.0659\ \text{lbm/ft}^3$$

step 5: Calculate the new frictionless power (see Eq. 28.73).

$$\text{IHP}_2 = \text{IHP}_1\left(\frac{\rho_{a2}}{\rho_{a1}}\right) = (1250\ \text{hp})\left(\frac{0.0659\ \frac{\text{lbm}}{\text{ft}^3}}{0.0765\ \frac{\text{lbm}}{\text{ft}^3}}\right)$$
$$= 1076.8\ \text{hp}$$

steps 6 and 7: Calculate the new net power using Eq. 28.75.

$$\text{BHP}_2 = \eta_{m2}(\text{IHP}_2) = (0.80)(1076.8\ \text{hp})$$
$$= \boxed{861.4\ \text{hp}\quad(860\ \text{hp})}$$

The answer is (B).

(b) *step 8:* Not needed.

step 9: The original fuel rate (see Eq. 28.77) is

$$\dot{m}_{f1} = (\text{BSFC}_1)(\text{BHP}_1)$$
$$= \left(0.45\ \frac{\text{lbm}}{\text{bhp-hr}}\right)(1000\ \text{bhp})$$
$$= 450\ \text{lbm/hr}$$

step 10: Not needed.

step 11: $\dot{m}_{f2} = \dot{m}_{f1} = 450$ lbm/hr

step 12: The new fuel consumption (see Eq. 28.82) is

$$\text{BSFC}_2 = \frac{\dot{m}_{f2}}{\text{BHP}_2} = \frac{450\ \frac{\text{lbm}}{\text{hr}}}{861.4\ \text{hp}}$$
$$= \boxed{0.522\ \text{lbm/hp-hr}\quad(0.52\ \text{lbm/hp-hr})}$$

The answer is (B).

SI Solution

In this problem, the efficiency is known as the new altitude. This can be used to calculate the brake horsepower from the new indicated horsepower. The friction horsepower is not needed in this problem, since it is only used to calculate the indicated horsepower. It is not possible for the friction horsepower and the efficiency to both be unchanged between the original and new altitudes.

(a) *step 1:* From App. 25.E,

Altitude 1: standard atmospheric condition, 1.01325 bar, 288.15K.

Altitude 2: 0.8456 bar, 278.4K.

$$z = 1500\ \text{m}$$

step 2: Calculate the friction power (see Eq. 28.71).

$$\text{IHP}_1 = \frac{\text{BHP}_1}{\eta_{m1}} = \frac{750 \text{ kW}}{0.80} = 937.5 \text{ kW}$$

step 3: Not needed since η_m is known to be constant with altitude.

step 4: From the ideal gas law, the air densities are

$$\rho_{a1} = \frac{p}{RT} = \frac{(1.01325 \text{ bar})\left(10^5 \ \frac{\text{Pa}}{\text{bar}}\right)}{\left(287.03 \ \frac{\text{J}}{\text{kg·K}}\right)(288.15\text{K})} = 1.225 \text{ kg/m}^3$$

Similarly,

$$\rho_{a2} = \frac{p}{RT} = \frac{(0.8456 \text{ bar})\left(10^5 \ \frac{\text{Pa}}{\text{bar}}\right)}{\left(287.03 \ \frac{\text{J}}{\text{kg·K}}\right)(278.4\text{K})} = 1.058 \text{ kg/m}^3$$

step 5: Calculate the new frictionless power (see Eq. 28.73).

$$\text{IHP}_2 = \text{IHP}_1\left(\frac{\rho_{a2}}{\rho_{a1}}\right) = (937.5 \text{ kW})\left(\frac{1.058 \ \frac{\text{kg}}{\text{m}^3}}{1.225 \ \frac{\text{kg}}{\text{m}^3}}\right) = 809.7 \text{ kW}$$

steps 6 and 7: Calculate the new net power (see Eq. 28.75).

$$\text{BHP}_2 = \eta_{m2}(\text{IHP}_2) = (0.80)(809.7 \text{ kW}) = \boxed{647.8 \text{ kW} \quad (650 \text{ kW})}$$

The answer is (B).

(b) *step 8:* Not needed.

step 9: The original fuel rate (see Eq. 28.77) is

$$\dot{m}_{f1} = (\text{BSFC}_1)(\text{BHP}_1) = \frac{\left(76 \ \frac{\text{kg}}{\text{GJ}}\right)(750 \text{ kW})}{10^6 \ \frac{\text{kJ}}{\text{GJ}}} = 0.057 \text{ kg/s}$$

step 10: Not needed.

step 11: $\dot{m}_{f2} = \dot{m}_{f1} = 0.057 \text{ kg/s}$

step 12: The new fuel consumption (see Eq. 28.82) is

$$\text{BSFC}_2 = \frac{\dot{m}_{f2}}{\text{BHP}_2} = \frac{\left(0.057 \ \frac{\text{kg}}{\text{s}}\right)\left(10^6 \ \frac{\text{kW}}{\text{GW}}\right)}{647.8 \text{ kW}} = \boxed{88 \text{ kg/GJ}}$$

The answer is (B).

6. *Customary U.S. Solution*

(a) From Eq. 28.44, the number of power strokes per minute is

$$N = \frac{(2n)(\text{no. cylinders})}{\text{no. strokes per cycle}} = \frac{\left(2 \ \frac{\text{strokes}}{\text{rev}}\right)\left(4600 \ \frac{\text{rev}}{\text{min}}\right)(8 \text{ cylinders})}{4 \ \frac{\text{strokes}}{\text{power stroke}}} = 18{,}400 \text{ power strokes/min}$$

The net work per cycle is

$$W_\text{net} = W_\text{out} - W_\text{in} = 1500 \text{ ft-lbf} - 1200 \text{ ft-lbf} = 300 \text{ ft-lbf/cycle}$$

The indicated horsepower is

$$\text{IHP} = \frac{\left(18{,}400 \ \frac{\text{power strokes}}{\text{min}}\right)(300 \text{ ft lbf})}{33{,}000 \ \frac{\text{ft-lbf}}{\text{hp-min}}} = \boxed{167.27 \text{ hp} \quad (170 \text{ hp})}$$

The answer is (B).

(b) From Eq. 28.6, the thermal efficiency is

$$\eta_\text{th} = \frac{W_\text{out} - W_\text{in}}{Q_\text{in}} = \frac{W_\text{net}}{Q_\text{in}} = \frac{300 \ \frac{\text{ft-lbf}}{\text{cycle}}}{\left(1.27 \ \frac{\text{Btu}}{\text{cycle}}\right)\left(778 \ \frac{\text{ft-lbf}}{\text{Btu}}\right)} = \boxed{0.304 \quad (30\%)}$$

The answer is (A).

(c) The lower heating value of gasoline is LHV = 18,900 Btu/lbm.

The fuel consumption is

$$\dot{m}_f = \frac{\left(1.27\ \frac{\text{Btu}}{\text{power stroke}}\right)\left(18{,}400\ \frac{\text{power strokes}}{\text{min}}\right)\left(60\ \frac{\text{min}}{\text{hr}}\right)}{18{,}900\ \frac{\text{Btu}}{\text{lbm}}}$$
$$= \boxed{74.18\ \text{lbm/hr}\quad (74\ \text{lbm/hr})}$$

The answer is (C).

(d) The specific fuel consumption is

$$\text{SFC} = \frac{\text{fuel usage rate}}{\text{power generated}} = \frac{74.18\ \frac{\text{lbm}}{\text{hr}}}{167.27\ \text{hp}}$$
$$= \boxed{0.443\ \text{lbm/hp-hr}\quad (0.44\ \text{lbm/hp-hr})}$$

The answer is (D).

SI Solution

(a) From Eq. 28.44, the number of power strokes per minute is

$$N = \frac{(2n)(\text{no. cylinders})}{\text{no. strokes per cycle}}$$
$$= \frac{\left(2\ \frac{\text{strokes}}{\text{rev}}\right)\left(4600\ \frac{\text{rev}}{\text{min}}\right)(8\ \text{cylinders})}{4\ \frac{\text{strokes}}{\text{power stroke}}}$$
$$= 18\,400\ \text{power strokes/min}$$

The net work per cycle is

$$W_{\text{net}} = W_{\text{out}} - W_{\text{in}} = 2.0\ \text{kJ} - 1.6\ \text{kJ}$$
$$= 0.4\ \text{kJ/cycle}$$

The indicated power is

$$\text{IkW} = \frac{\left(18\,400\ \frac{\text{power strokes}}{\text{min}}\right)(0.4\ \text{kJ})}{60\ \frac{\text{s}}{\text{min}}}$$
$$= \boxed{122.7\ \text{kW}\quad (120\ \text{kW})}$$

The answer is (B).

(b) From Eq. 28.6, the thermal efficiency is

$$\eta_{\text{th}} = \frac{W_{\text{out}} - W_{\text{in}}}{Q_{\text{in}}} = \frac{W_{\text{net}}}{Q_{\text{in}}}$$
$$= \frac{0.4\ \frac{\text{kJ}}{\text{cycle}}}{1.33\ \frac{\text{kJ}}{\text{cycle}}}$$
$$= \boxed{0.301\quad (30\%)}$$

The answer is (A).

(c) The heating value of gasoline is LHV = 44 MJ/kg.

The fuel consumption is

$$\dot{m}_f = \frac{\left(1.33\ \frac{\text{kJ}}{\text{power stroke}}\right)\left(18\,400\ \frac{\text{power strokes}}{\text{min}}\right)}{\left(44\ \frac{\text{MJ}}{\text{kg}}\right)\left(1000\ \frac{\text{kJ}}{\text{MJ}}\right)\left(60\ \frac{\text{s}}{\text{min}}\right)}$$
$$= \boxed{0.00927\ \text{kg/s}\quad (0.0093\ \text{kg/s})}$$

The answer is (C).

(d) The specific fuel consumption is

$$\text{SFC} = \frac{\text{fuel usage rate}}{\text{power generated}} = \frac{\left(0.00927\ \frac{\text{kg}}{\text{s}}\right)\left(10^6\ \frac{\text{kW}}{\text{kg}}\right)}{122.7\ \text{kW}}$$
$$= \boxed{75.55\ \text{kg/GJ}\quad (76\ \text{kg/GJ})}$$

The answer is (D).

7. *Customary U.S. Solution*

(a) At A:

$$T_{\text{A}} = 520°\text{R}$$
$$p_{\text{A}} = 14.7\ \text{psia}$$

At C:

$$T_{\text{C}} = 1600°\text{R}$$
$$p_{\text{C}} = 568.6\ \text{psia}$$

For isentropic process C-A, from Eq. 24.105,

$$p_{\text{A}} = p_{\text{C}}\left(\frac{T_{\text{A}}}{T_{\text{C}}}\right)^{\frac{k}{k-1}}$$

Therefore,

$$14.7\text{ psia} = (568.6\text{ psia})\left(\frac{520°\text{R}}{1600°\text{R}}\right)^{\frac{k}{k-1}}$$
$$0.02585 = (0.325)^{\frac{k}{k-1}}$$
$$\log 0.02585 = \left(\frac{k}{k-1}\right)\log 0.325$$
$$\frac{k}{k-1} = 3.252$$
$$k = 1.444$$

From Eq. 23.67, the molar specific heat of the mixture is

$$\begin{aligned} C_{p,\text{mixture}} &= \frac{R^*k}{k-1} \\ &= \frac{\left(1545\ \frac{\text{ft-lbf}}{\text{lbmol-°R}}\right)(1.444)}{\left(778\ \frac{\text{ft-lbf}}{\text{Btu}}\right)(1.444-1)} \\ &= \boxed{6.459\text{ Btu/lbmol-°R}\quad (6.5\text{ Btu/lbmol-°R})} \end{aligned}$$

The answer is (D).

(b) From Table 23.7,

$$\begin{aligned} (c_p)_{\text{He}} &= 1.240\text{ Btu/lbm-°R} \\ (\text{MW})_{\text{He}} &= 4.003\text{ lbm/lbmol} \\ (c_p)_{\text{CO}_2} &= 0.207\text{ Btu/lbm-°R} \\ (\text{MW})_{\text{CO}_2} &= 44.011\text{ lbm/lbmol} \\ C_{p,\text{He}} &= (\text{MW})_{\text{He}}(c_p)_{\text{He}} \\ &= \left(4.003\ \frac{\text{lbm}}{\text{lbmol}}\right)\left(1.240\ \frac{\text{Btu}}{\text{lbm-°R}}\right) \\ &= 4.96\text{ Btu/lbmol-°R} \\ C_{p,\text{CO}_2} &= (\text{MW})_{\text{CO}_2}(c_p)_{\text{CO}_2} \\ &= \left(44.011\ \frac{\text{lbm}}{\text{lbmol}}\right)\left(0.207\ \frac{\text{Btu}}{\text{lbm-°R}}\right) \\ &= 9.11\text{ Btu/lbmol-°R} \end{aligned}$$

Let x be the mole fraction of helium in the mixture.

$$x = \frac{n_{\text{He}}}{n_{\text{He}} + n_{\text{CO}_2}}$$

From Eq. 23.95, on a mole basis,

$$\begin{aligned} c_{p,\text{mixture}} &= x(c_p)_{\text{He}} + (1-x)(c_p)_{\text{CO}_2} \\ 6.459\ \frac{\text{Btu}}{\text{lbmol-°R}} &= x\left(4.96\ \frac{\text{Btu}}{\text{lbmol-°R}}\right) + (1-x)\left(9.11\ \frac{\text{Btu}}{\text{lbmol-°R}}\right) \\ x &= 0.639\text{ lbmol} \end{aligned}$$

On a per mole basis, the mass of helium would be

$$\begin{aligned} m_{\text{He}} &= x(\text{MW})_{\text{He}} = (0.639\text{ lbmol})\left(4.003\ \frac{\text{lbm}}{\text{lbmol}}\right) \\ &= 2.558\text{ lbm} \end{aligned}$$

Similarly, the mass of carbon dioxide on a per mole basis would be

$$\begin{aligned} m_{\text{CO}_2} &= (1-x)(\text{MW})_{\text{CO}_2} \\ &= (1\text{ lbmol} - 0.639\text{ lbmol})\left(44.011\ \frac{\text{lbm}}{\text{lbmol}}\right) \\ &= 15.888\text{ lbm} \end{aligned}$$

The molecular weight of the mixture is

$$\begin{aligned} (\text{MW})_{\text{mixture}} &= 2.558\text{ lbm} + 15.888\text{ lbm} \\ &= \boxed{18.446\text{ lbm/lbmol}\quad (19\text{ lbm/lbmol})} \end{aligned}$$

The answer is (B).

(c) The gravimetric (mass) fraction of the gases is

$$\begin{aligned} G_{\text{He}} &= \frac{m_{\text{He}}}{m_{\text{He}} + m_{\text{CO}_2}} \\ &= \frac{2.558\text{ lbm}}{2.558\text{ lbm} + 15.888\text{ lbm}} \\ &= \boxed{0.139\quad (0.14)} \\ G_{\text{CO}_2} &= 1 - G_{\text{He}} = 1 - 0.139 \\ &= \boxed{0.861\quad (0.86)} \end{aligned}$$

The answer is (A).

(d) From Eq. 23.65,

$$\begin{aligned} C_{v,\text{mixture}} &= C_{p,\text{mixture}} - R^* \\ &= 6.459\ \frac{\text{Btu}}{\text{lbmol-°R}} - \frac{1545\ \frac{\text{ft-lbf}}{\text{lbmol-°R}}}{778\ \frac{\text{ft-lbf}}{\text{Btu}}} \\ &= 4.469\text{ Btu/lbmol-°R} \end{aligned}$$

From Eq. 24.116, the work done during the isentropic expansion process is

$$\begin{aligned} W &= c_v(T_1 - T_2) = \left(\frac{C_v}{\text{MW}}\right)_{\text{mixture}} \times (T_{\text{C}} - T_{\text{A}}) \\ &= \left(\frac{4.469\ \frac{\text{Btu}}{\text{lbmol-°R}}}{18.446\ \frac{\text{lbm}}{\text{lbmol}}}\right)(1600°\text{R} - 520°\text{R}) \\ &= \boxed{261.7\text{ Btu/lbm}\quad (260\text{ Btu/lbm})} \end{aligned}$$

The answer is (B).

Power Cycles

(e) The temperature-entropy and pressure-volume diagrams for the cycle are

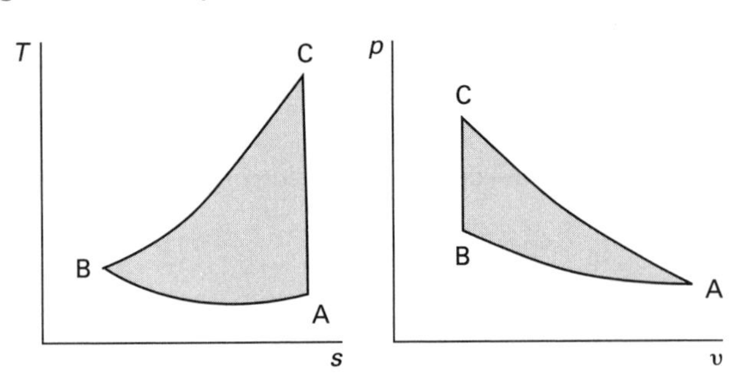

SI Solution

(a) At A:

$$T_A = 290\text{K}$$
$$p_A = 101.3 \text{ kPa}$$

At C:

$$T_C = 890\text{K}$$
$$p_C = 3.920 \text{ MPa}$$

For isentropic process C-A from Eq. 24.105,

$$p_A = p_C\left(\frac{T_A}{T_C}\right)^{\frac{k}{k-1}}$$

Therefore,

$$101.3 \text{ kPa} = (3.920 \text{ MPa})\left(1000 \ \frac{\text{kPa}}{\text{MPa}}\right)\left(\frac{290\text{K}}{890\text{K}}\right)^{\frac{k}{k-1}}$$
$$0.02584 = (0.3258)^{\frac{k}{k-1}}$$
$$\log 0.02584 = \left(\frac{k}{k-1}\right)\log 0.3258$$
$$\frac{k}{k-1} = 3.2599$$
$$k = 1.442$$

From Eq. 23.67, the molar specific heat of the mixture is

$$C_{p,\text{mixture}} = \frac{R^* k}{k-1} = \frac{\left(8.3143 \ \frac{\text{kJ}}{\text{kmol·K}}\right)(1.442)}{1.442 - 1} = \boxed{27.125 \text{ kJ/kmol·K} \quad (27 \text{ kJ/kmol·K})}$$

The answer is (D).

(b) From Table 23.7,

$$(c_p)_{\text{He}} = \frac{5192 \ \frac{\text{J}}{\text{kg·K}}}{1000 \ \frac{\text{J}}{\text{kJ}}} = 5.192 \text{ kJ/kg·K}$$
$$(\text{MW})_{\text{He}} = 4.003 \text{ kg/kmol}$$
$$(c_p)_{CO_2} = \frac{867 \ \frac{\text{J}}{\text{kg·K}}}{1000 \ \frac{\text{J}}{\text{kJ}}} = 0.867 \text{ kJ/kg·K}$$
$$(\text{MW})_{CO_2} = 44.011 \text{ kg/kmol}$$

$$C_{p,\text{He}} = (\text{MW})_{\text{He}}(c_p)_{\text{He}} = \left(4.003 \ \frac{\text{kg}}{\text{kmol}}\right)\left(5.192 \ \frac{\text{kJ}}{\text{kg·K}}\right) = 20.784 \text{ kJ/kmol·K}$$
$$C_{p,CO_2} = (\text{MW})_{CO_2}(c_p)_{CO_2} = \left(44.011 \ \frac{\text{kg}}{\text{kmol}}\right)\left(0.867 \ \frac{\text{kJ}}{\text{kg·K}}\right) = 38.158 \text{ kJ/kmol·K}$$

Let x be the mole fraction of helium in the mixture.

$$x = \frac{n_{\text{He}}}{n_{\text{He}} + n_{CO_2}}$$

From Eq. 23.95 on a mole basis,

$$C_{p,\text{mixture}} = x(C_p)_{\text{He}} + (1-x)(C_p)_{CO_2}$$
$$27.082 \ \frac{\text{kJ}}{\text{kmol·K}} = x\left(20.784 \ \frac{\text{kJ}}{\text{kmol·K}}\right) + (1-x)\left(38.158 \ \frac{\text{kJ}}{\text{kmol·K}}\right)$$
$$x = 0.638 \text{ kmol}$$

On a per mole basis, the mass of helium would be

$$m_{\text{He}} = x(\text{MW})_{\text{He}} = (0.638 \text{ kmol})\left(4.003 \ \frac{\text{kg}}{\text{kmol}}\right) = 2.554 \text{ kg}$$

Similarly, the mass of CO_2 on a per mole basis would be

$$m_{CO_2} = (1-x)(\text{MW})_{CO_2} = (1 \text{ kmol} - 0.638 \text{ kmol})\left(44.011 \ \frac{\text{kg}}{\text{kmol}}\right) = 15.932 \text{ kg}$$

Power Cycles

The molecular weight of the mixture is

$$(\text{MW})_{\text{mixture}} = 2.554 \text{ kg} + 15.932 \text{ kg} = \boxed{18.486 \text{ kg/kmol} \quad (19 \text{ kg/kmol})}$$

The answer is (B).

(c) The gravimetric (mass) fraction of the gases is

$$G_{\text{He}} = \frac{m_{\text{He}}}{m_{\text{He}} + m_{\text{CO}_2}} = \frac{2.554 \text{ kg}}{2.554 \text{ kg} + 15.932 \text{ kg}} = \boxed{0.138 \quad (0.14)}$$

$$G_{\text{CO}_2} = 1 - G_{\text{He}} = 1 - 0.138 = \boxed{0.862 \quad (0.86)}$$

The answer is (A).

(d) From Eq. 23.65,

$$C_{v,\text{mixture}} = C_{p,\text{mixture}} - R^* = 27.125 \ \frac{\text{kJ}}{\text{kmol·K}} - 8.3143 \ \frac{\text{kJ}}{\text{kmol·K}} = 18.811 \text{ kJ/kmol·K}$$

From Eq. 24.116, the work done during the isentropic expansion process is

$$W = c_v(T_1 - T_2) = \left(\frac{C_v}{\text{MW}}\right)_{\text{mixture}} \times (T_{\text{C}} - T_{\text{A}}) = \left(\frac{18.811 \ \frac{\text{kJ}}{\text{kmol·K}}}{18.486 \ \frac{\text{kg}}{\text{kmol}}}\right)(890\text{K} - 290\text{K}) = \boxed{610.5 \text{ kJ/kg} \quad (610 \text{ kJ/kg})}$$

The answer is (B).

(e) The temperature-entropy and pressure-volume diagrams for the cycle are

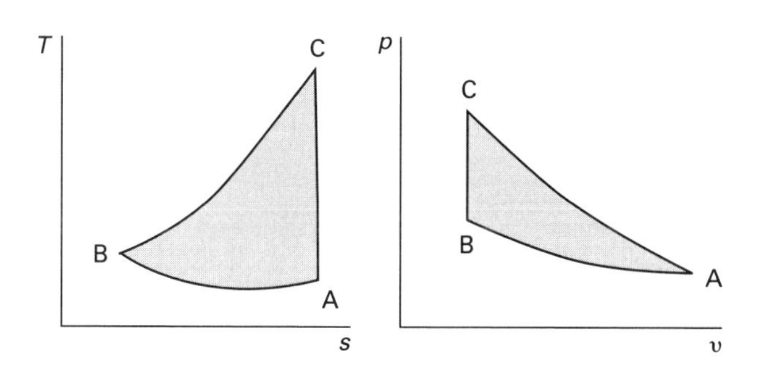

8. *Customary U.S. Solution*

(a) *step 1:* Not needed.

step 2: From Eq. 28.71, calculate the frictionless power.

$$\text{IHP}_1 = \frac{\text{BHP}_1}{\eta_{m1}} = \frac{200 \text{ hp}}{0.86} = \boxed{232.6 \text{ hp} \quad (230 \text{ hp})}$$

The answer is (B).

(b) *step 3:* From Eq. 28.72, calculate the friction power, which is assumed to be constant at constant speed.

$$\text{FHP} = \text{IHP}_1 - \text{BHP}_1 = 232.6 \text{ hp} - 200 \text{ hp} = \boxed{32.6 \text{ hp} \quad (33 \text{ hp})}$$

The answer is (B).

(c) *step 4:* Calculate the air densities ρ_{a1} and ρ_{a2} from the ideal gas law. The absolute temperatures are

$$T_1 = 80°\text{F} + 460° = 540°\text{R}$$

$$T_2 = 60°\text{F} + 460° = 520°\text{R}$$

$$\rho_{a1} = \frac{p_1}{RT_1} = \frac{\left(14.7 \ \frac{\text{lbf}}{\text{in}^2}\right)\left(12 \ \frac{\text{in}}{\text{ft}}\right)^2}{\left(53.3 \ \frac{\text{lbf-ft}}{\text{lbm-°R}}\right)(540°\text{R})} = \boxed{0.0735 \text{ lbm/ft}^3 \quad (0.074 \text{ lbm/ft}^3)}$$

$$\rho_{a2} = \frac{p_2}{RT_1} = \frac{\left(12.2 \ \frac{\text{lbf}}{\text{in}^2}\right)\left(12 \ \frac{\text{in}}{\text{ft}}\right)^2}{\left(53.3 \ \frac{\text{lbf-ft}}{\text{lbm-°R}}\right)(520°\text{R})} = \boxed{0.0634 \text{ lbm/ft}^3 \quad (0.063 \text{ lbm/ft}^3)}$$

The answer is (C).

(d) *step 5:* From Eq. 28.73, calculate the new frictionless power.

$$\text{IHP}_2 = \text{IHP}_1\left(\frac{\rho_{a2}}{\rho_{a1}}\right) = (232.6 \text{ hp})\left(\frac{0.0634 \ \frac{\text{lbm}}{\text{ft}^3}}{0.0735 \ \frac{\text{lbm}}{\text{ft}^3}}\right) = 200.6 \text{ hp}$$

step 6: From Eq. 28.74, calculate the new net power.

$$\text{BHP}_2 = \text{IHP}_2 - \text{FHP} = 200.6 \text{ hp} - 32.6 \text{ hp} = \boxed{168.0 \text{ hp} \quad (170 \text{ hp})}$$

The answer is (B).

Power Cycles

(e) *step 7:* From Eq. 28.75, calculate the new efficiency.

$$\eta_{m2} = \frac{\text{BHP}_2}{\text{IHP}_2} = \frac{168.0 \text{ hp}}{200.6 \text{ hp}} = \boxed{0.837 \quad (0.84)}$$

The answer is (B).

(f) *step 8:* From Eq. 28.76, the volumetric air flow rates are the same since the engine speed is constant.

$$\dot{V}_{a2} = \dot{V}_{a1}$$

step 9: The original air and fuel rates from Eq. 28.77 through Eq. 28.79 are

$$\begin{aligned}\dot{m}_{f1} &= (\text{BSFC}_1)(\text{BHP}_1) = \left(0.48 \ \frac{\text{lbm}}{\text{hp-hr}}\right)(200 \text{ hp}) \\ &= 96 \text{ lbm/hr} \\ \dot{m}_{a1} &= (\text{AFR})(\dot{m}_{f1}) = (22)\left(96 \ \frac{\text{lbm}}{\text{hr}}\right) \\ &= 2112 \text{ lbm/hr}\end{aligned}$$

$$\begin{aligned}\dot{V}_{a1} &= \frac{\dot{m}_{a1}}{\rho_{a1}} = \frac{2112 \ \frac{\text{lbm}}{\text{hr}}}{0.0735 \ \frac{\text{lbm}}{\text{ft}^3}} \\ &= 28{,}735 \text{ ft}^3/\text{hr} \\ \dot{V}_{a2} &= \dot{V}_{a1} = 28{,}735 \text{ ft}^3/\text{hr}\end{aligned}$$

step 10: From Eq. 28.80, the new air mass flow rate is

$$\begin{aligned}\dot{m}_{a2} &= \dot{V}_{a2}\rho_{a2} \\ &= \left(28{,}735 \ \frac{\text{ft}^3}{\text{hr}}\right)\left(0.0634 \ \frac{\text{lbm}}{\text{ft}^3}\right) \\ &= \boxed{1821.8 \text{ lbm/hr} \quad (1820 \text{ lbm/hr})}\end{aligned}$$

The answer is (C).

(g) *step 11:* For engines with metered injection,

$$\dot{m}_{f2} = \dot{m}_{f1} = 96 \text{ lbm/hr}$$

step 12: From Eq. 28.82, the new fuel consumption is

$$\begin{aligned}\text{BSFC}_2 &= \frac{\dot{m}_{f2}}{\text{BHP}_2} = \frac{96 \ \frac{\text{lbm}}{\text{hr}}}{168.0 \text{ hp}} \\ &= \boxed{0.571 \text{ lbm/hp-hr} \quad (0.57 \text{ lbm/hp-hr})}\end{aligned}$$

The answer is (A).

(h) From Eq. 28.78, the new air/fuel ratio is

$$\text{AFR}_2 = \frac{\dot{m}_{a2}}{\dot{m}_{f2}} = \frac{1821.8 \ \frac{\text{lbm}}{\text{hr}}}{96 \ \frac{\text{lbm}}{\text{hr}}} = \boxed{18.98 \quad (19.0)}$$

The answer is (D).

SI Solution

(a) *step 1:* Not needed.

step 2: From Eq. 28.71, calculate the frictionless power.

$$\begin{aligned}\text{IkW}_1 &= \frac{\text{BkW}_1}{\eta_{m1}} = \frac{150 \text{ kW}}{0.86} \\ &= \boxed{174.4 \text{ kW} \quad (170 \text{ kW})}\end{aligned}$$

The answer is (B).

(b) *step 3:* From Eq. 28.72, calculate the friction power, which is assumed to be constant at constant speed.

$$\begin{aligned}\text{FkW} &= \text{IkW}_1 - \text{BkW}_1 = 174.4 \text{ kW} - 150 \text{ kW} \\ &= \boxed{24.4 \text{ kW} \quad (24 \text{ kW})}\end{aligned}$$

The answer is (B).

(c) *step 4:* Calculate the air densities ρ_{a1} and ρ_{a2} from the ideal gas law. The absolute temperatures are

$$\begin{aligned}T_1 &= 27°\text{C} + 273° = 300\text{K} \\ T_2 &= 16°\text{C} + 273° = 289\text{K}\end{aligned}$$

$$\begin{aligned}\rho_{a1} &= \frac{p_1}{RT_1} = \frac{(101.3 \text{ kPa})\left(1000 \ \frac{\text{Pa}}{\text{kPa}}\right)}{\left(287.03 \ \frac{\text{kJ}}{\text{kg·K}}\right)(300\text{K})} \\ &= \boxed{1.176 \text{ kg/m}^3 \quad (1.18 \text{ kg/m}^3)}\end{aligned}$$

$$\begin{aligned}\rho_{a2} &= \frac{p_2}{RT_2} = \frac{(84 \text{ kPa})\left(1000 \ \frac{\text{Pa}}{\text{kPa}}\right)}{\left(287.03 \ \frac{\text{kJ}}{\text{kg·K}}\right)(289\text{K})} \\ &= \boxed{1.013 \text{ kg/m}^3 \quad (1.01 \text{ kg/m}^3)}\end{aligned}$$

The answer is (C).

(d) *step 5:* From Eq. 28.73, calculate the new frictionless power.

$$\begin{aligned}\text{IkW}_2 &= \text{IkW}_1\left(\frac{\rho_{a2}}{\rho_{a1}}\right) = (174.4 \text{ kW})\left(\frac{1.013 \ \frac{\text{kg}}{\text{m}^3}}{1.176 \ \frac{\text{kg}}{\text{m}^3}}\right) \\ &= 150.2 \text{ kW}\end{aligned}$$

Power Cycles

step 6: From Eq. 28.74, calculate the new net power.

$$\begin{aligned}\text{BkW}_2 &= \text{IkW}_2 - \text{FkW} = 150.2\ \text{kW} - 24.4\ \text{kW}\\ &= \boxed{125.8\ \text{kW}\quad (130\ \text{kW})}\end{aligned}$$

The answer is (B).

(e) *step 7:* From Eq. 28.75, calculate the new efficiency.

$$\eta_{m2} = \frac{\text{BkW}_2}{\text{IkW}_2} = \frac{125.8\ \text{kW}}{150.2\ \text{kW}} = \boxed{0.838\quad (0.84)}$$

The answer is (B).

(f) *step 8:* From Eq. 28.76, the volumetric air flow rates are the same since the engine speed is constant.

$$\dot{V}_{a2} = \dot{V}_{a1}$$

step 9: The original air and fuel rates from Eq. 28.77 through Eq. 28.79 are

$$\begin{aligned}\dot{m}_{f1} &= (\text{BSFC}_1)(\text{BkW}_1) = \frac{\left(81\ \frac{\text{kg}}{\text{GJ}}\right)(150\ \text{kW})}{10^6\ \frac{\text{kJ}}{\text{GJ}}}\\ &= 0.01215\ \text{kg/s}\\ \dot{m}_{a1} &= (\text{AFR})\dot{m}_{f1} = (22)\left(0.01215\ \frac{\text{kg}}{\text{s}}\right)\\ &= 0.2673\ \text{kg/s}\\ \dot{V}_{a1} &= \frac{\dot{m}_{a1}}{\rho_{a1}} = \frac{0.2673\ \frac{\text{kg}}{\text{s}}}{1.176\ \frac{\text{kg}}{\text{m}^3}} = 0.2273\ \text{m}^3/\text{s}\\ \dot{V}_{a2} &= \dot{V}_{a1} = 0.2273\ \text{m}^3/\text{s}\end{aligned}$$

step 10: From Eq. 28.80, the new air mass flow rate is

$$\begin{aligned}\dot{m}_{a2} &= \dot{V}_{a2}\rho_{a2} = \left(0.2273\ \frac{\text{m}^3}{\text{s}}\right)\left(1.013\ \frac{\text{kg}}{\text{m}^3}\right)\\ &= \boxed{0.2303\ \text{kg/s}\quad (0.230\ \text{kg/s})}\end{aligned}$$

The answer is (C).

(g) *step 11:* For engines with metered injection,

$$\dot{m}_{f2} = \dot{m}_{f1} = 0.01215\ \text{kg/s}$$

step 12: From Eq. 28.82, the new fuel consumption is

$$\begin{aligned}\text{BSFC}_2 &= \frac{\dot{m}_{f2}}{\text{BkW}_2} = \frac{\left(0.01215\ \frac{\text{kg}}{\text{s}}\right)\left(10^6\ \frac{\text{kJ}}{\text{GJ}}\right)}{125.8\ \text{kW}}\\ &= \boxed{96.58\ \text{kg/GJ}\quad (97\ \text{kg/GJ})}\end{aligned}$$

The answer is (A).

(h) From Eq. 28.78, the new air/fuel ratio is

$$\text{AFR}_2 = \frac{\dot{m}_{a2}}{\dot{m}_{f2}} = \frac{0.2303\ \frac{\text{kg}}{\text{s}}}{0.01215\ \frac{\text{kg}}{\text{s}}} = \boxed{18.95\quad (19.0)}$$

The answer is (D).

9. *Customary U.S. Solution*

(a) If the power output is to be unchanged,

$$\begin{aligned}\text{BHP}_1 &= \text{BHP}_2\\ \dot{m}_{f1}(\text{LHV}_1) &= \dot{m}_{f2}(\text{LHV}_2)\end{aligned}$$

From Eq. 28.11,

$$\begin{aligned}\dot{m}_f &= (\text{BSFC})(\text{BHP})\\ (\text{BSFC}_1)(\text{LHV}_1) &= (\text{BSFC}_2)(\text{LHV}_2)\\ \frac{\text{BSFC}_2}{\text{BSFC}_1} &= \frac{\text{LHV}_1}{\text{LHV}_2} = \frac{23{,}200\ \frac{\text{Btu}}{\text{lbm}}}{11{,}930\ \frac{\text{Btu}}{\text{lbm}}} = 1.945\\ \frac{\text{BSFC}_2 - \text{BSFC}_1}{\text{BSFC}_1} &= \frac{\text{BSFC}_2}{\text{BSFC}_1} - 1 = 1.945 - 1\\ &= \boxed{0.944\quad (95\%)}\end{aligned}$$

The answer is (D).

(b) If the fuel injection velocity is to be unchanged ($\text{v}_2 = \text{v}_1$), then the injection point size must be increased to allow more fuel to be injected. For a fixed injection duration,

$$\begin{aligned}\dot{m} &\propto \rho A\text{v}\\ A &\propto \frac{\dot{m}}{\rho\text{v}}\\ \dot{m}_2 &= 1.945\dot{m}_1\quad [\text{part (a)}]\\ \frac{A_2 - A_1}{A_1} &= \frac{\frac{\dot{m}_2}{\rho_2\text{v}_2} - \frac{\dot{m}_1}{\rho_1\text{v}_1}}{\frac{\dot{m}_1}{\rho_1\text{v}_1}} = \frac{\frac{\dot{m}_2}{\rho_2} - \frac{\dot{m}_1}{\rho_1}}{\frac{\dot{m}_1}{\rho_1}}\\ &= \frac{\frac{1.945\dot{m}_1}{\rho_2} - \frac{\dot{m}_1}{\rho_1}}{\frac{\dot{m}_1}{\rho_1}}\\ &= \frac{\frac{1.945}{\rho_2} - \frac{1}{\rho_1}}{\frac{1}{\rho_1}}\\ \frac{A_2 - A_1}{A_1} &= (1.945)\left(\frac{\rho_1}{\rho_2}\right) - 1\end{aligned}$$

Power Cycles

From Table 21.3, for gasoline, $SG_1 = 0.74$. For ethanol, $SG_2 = 0.794$.

$$\frac{\rho_1}{\rho_2} = \frac{SG_1}{SG_2} = \frac{0.74}{0.794}$$

$$\frac{A_2 - A_1}{A_1} = (1.945)\left(\frac{0.74}{0.794}\right) - 1 = \boxed{0.812 \quad (81\%)}$$

The answer is (C).

(c) If no changes are made to the engine, the volume of alcohol injected will be the same as the volume of gasoline. The power output is proportional to the fuel mass flow rate and heating value.

$$\frac{P_2 - P_1}{P_1} = \frac{\dot{m}_{f2}(\text{LHV}_2) - \dot{m}_{f1}(\text{LHV}_1)}{\dot{m}_{f1}(\text{LHV}_1)}$$

$$\dot{m}_f \propto \rho A \text{v}$$

$$\frac{P_2 - P_1}{P_1} = \frac{(\rho_2 A_2 \text{v}_2)(\text{LHV}_2) - (\rho_1 A_1 \text{v}_1)(\text{LHV}_1)}{(\rho_1 A_1 \text{v}_1)(\text{LHV}_1)}$$

Since no changes are made to the engine, $A_2 = A_1$, and $\text{v}_2 = \text{v}_1$.

$$\frac{P_2 - P_1}{P_1} = \frac{\rho_2(\text{LHV}_2) - \rho_1(\text{LHV}_1)}{\rho_1(\text{LHV}_1)} = \frac{\text{LHV}_2 - \left(\frac{\rho_1}{\rho_2}\right)(\text{LHV}_1)}{\left(\frac{\rho_1}{\rho_2}\right)(\text{LHV}_1)}$$

From part (b),

$$\frac{\rho_1}{\rho_2} = \frac{0.74}{0.794}$$

$$\frac{P_2 - P_1}{P_1} = \frac{11{,}930\ \frac{\text{Btu}}{\text{lbm}} - \left(\frac{0.74}{0.794}\right)\left(23{,}200\ \frac{\text{Btu}}{\text{lbm}}\right)}{\left(\frac{0.74}{0.794}\right)\left(23{,}200\ \frac{\text{Btu}}{\text{lbm}}\right)} = \boxed{-0.45 \quad (-45\%)}$$

The answer is (C).

SI Solution

(a) From part (a) of the customary U.S. solution,

$$\frac{\text{BSFC}_2}{\text{BSFC}_1} = \frac{\text{LHV}_1}{\text{LHV}_2} = \frac{54\ \frac{\text{MJ}}{\text{kg}}}{27.7\ \frac{\text{MJ}}{\text{kg}}} = 1.949$$

$$\frac{\text{BSFC}_2 - \text{BSFC}_1}{\text{BSFC}_1} = \frac{\text{BSFC}_2}{\text{BSFC}_1} - 1 = 1.949 - 1 = \boxed{0.949 \quad (95\%)}$$

The answer is (D).

(b) From part (b) of the customary U.S. solution,

$$\frac{A_2 - A_1}{A_1} = \frac{\frac{\dot{m}_2}{\rho_2} - \frac{\dot{m}_1}{\rho_1}}{\frac{\dot{m}_1}{\rho_1}}$$

$$\dot{m}_2 = 1.949\dot{m}_1 \quad \text{[part (a)]}$$

$$\frac{A_2 - A_1}{A_1} = \frac{\frac{1.949\dot{m}_1}{\rho_2} - \frac{\dot{m}_1}{\rho_1}}{\frac{\dot{m}_1}{\rho_1}} = (1.949)\left(\frac{\rho_1}{\rho_2}\right) - 1 = (1.949)\left(\frac{0.74}{0.794}\right) - 1 = \boxed{0.816 \quad (81\%)}$$

The answer is (C).

(c) From part (c) of the customary U.S. solution,

$$\frac{P_2 - P_1}{P_1} = \frac{\text{LHV}_2 - \left(\frac{\rho_1}{\rho_2}\right)(\text{LHV}_1)}{\left(\frac{\rho_1}{\rho_2}\right)(\text{LHV}_1)} = \frac{27.7\ \frac{\text{MJ}}{\text{kg}} - \left(\frac{0.74}{0.794}\right)\left(54\ \frac{\text{MJ}}{\text{kg}}\right)}{\left(\frac{0.74}{0.794}\right)\left(54\ \frac{\text{MJ}}{\text{kg}}\right)} = \boxed{-0.45 \quad (-45\%)}$$

The answer is (C).

29 Combustion Turbine Cycles

PRACTICE PROBLEMS

1. Air expands isentropically at the rate of 10 ft^3/sec (280 L/s) from 200 psia and 1500°F (1.4 MPa and 820°C) to 50 psia (350 kPa).

(a) Using air tables, the air's final temperature is most nearly

(A) 1275°R (710K)

(B) 1325°R (740K)

(C) 1375°R (770K)

(D) 1425°R (790K)

(b) Using air tables, the air's final volumetric flow rate is most nearly

(A) 28 ft^3/sec (780 L/s)

(B) 31 ft^3/sec (860 L/s)

(C) 39 ft^3/sec (1100 L/s)

(D) 45 ft^3/sec (1300 L/s)

(c) Using air tables, the air's enthalpy change is most nearly

(A) −350 Btu/lbm (−770 kJ/kg)

(B) −230 Btu/lbm (−530 kJ/kg)

(C) −160 Btu/lbm (−370 kJ/kg)

(D) −110 Btu/lbm (−250 kJ/kg)

2. In an air-standard gas turbine, air at 14.7 psia and 60°F (101.3 kPa and 16°C) enters a compressor and is compressed through a volume ratio of 5:1. The compressor efficiency is 83%. Air enters the turbine at 1500°F (820°C) and expands to 14.7 psia (101.3 kPa). The turbine efficiency is 92%. The thermal efficiency of the cycle is most nearly

(A) 33%

(B) 39%

(C) 44%

(D) 51%

3. A 65% efficient regenerator is added to the gas turbine described in Prob. 2. Assume the specific heat remains constant. The new thermal efficiency is most nearly

(A) 24%

(B) 28%

(C) 34%

(D) 41%

4. A gas turbine operating on the Brayton cycle with an 8:1 pressure ratio is located at 7000 ft (2100 m) altitude. The conditions at that altitude are 12 psia and 35°F (82 kPa and 2°C). While consuming 0.609 lbm/hp-hr (100 kg/GJ) of fuel and 50,000 cfm (23 500 L/s) of air, the turbine develops 6000 bhp (4.5 MW). The turbine efficiency is 80%, and the compressor efficiency is 85%. The fuel has a lower heating value of 19,000 Btu/lbm (44 MJ/kg). The fuel mass is small compared to the air mass. There is no pressure loss in the combustor. The turbine receives combustor gases at 1800°F (980°C). The air inlet filter area is 254 ft^2 (22.9 m^2). The turbine is moved to sea level where the conditions are 14.7 psia and 70°F (101.3 kPa and 21°C). The combustion efficiency and combustor temperature remain the same.

(a) At 7000 ft, the air density is most nearly

(A) 0.065 lbm/ft^3 (1.0 kg/m^3)

(B) 0.071 lbm/ft^3 (1.1 kg/m^3)

(C) 0.083 lbm/ft^3 (1.3 kg/m^3)

(D) 0.100 lbm/ft^3 (1.6 kg/m^3)

(b) At 7000 ft, the air mass flow rate is most nearly

(A) 3000 lbm/min (22 kg/s)

(B) 3200 lbm/min (23 kg/s)

(C) 3300 lbm/min (24 kg/s)

(D) 3400 lbm/min (25 kg/s)

(c) At 7000 ft, the ideal fuel rate is most nearly

(A) 3000 lbm/hr (0.37 kg/s)

(B) 3200 lbm/hr (0.40 kg/s)

(C) 3400 lbm/hr (0.42 kg/s)

(D) 3600 lbm/hr (0.44 kg/s)

(d) At 7000 ft, the combustor efficiency is most nearly

(A) 82%

(B) 87%

(C) 93%

(D) 98%

(e) At 7000 ft, the friction horsepower is most nearly

(A) 730 hp (0.43 MW)

(B) 780 hp (0.46 MW)

(C) 840 hp (0.50 MW)

(D) 910 hp (0.54 MW)

(f) At sea level, the new brake horsepower is most nearly

(A) 2800 hp (2.1 MW)

(B) 4400 hp (3.4 MW)

(C) 5100 hp (4.0 MW)

(D) 6300 hp (4.8 MW)

(g) At sea level, the new brake specific fuel consumption is most nearly

(A) 0.42 lbm/hp-hr (67 kg/GJ)

(B) 0.51 lbm/hp-hr (82 kg/GJ)

(C) 0.63 lbm/hp-hr (100 kg/GJ)

(D) 0.87 lbm/hp-hr (140 kg/GJ)

5. A precision air turbine is used to drive a small dentist's drill. 140°F (60°C) air enters the turbine at the rate of 15 lbm/hr (1.9 g/s). The output of the turbine is 0.25 hp (0.19 kW). The turbine exhausts to 15 psia (103.5 kPa). The flow is steady. The isentropic efficiency of the expansion process is 60%.

(a) If the expansion was isentropic, the enthalpy of the air exiting the turbine would be most nearly

(A) 70 Btu/lbm (170 kJ/kg)

(B) 100 Btu/lbm (230 kJ/kg)

(C) 140 Btu/lbm (320 kJ/kg)

(D) 200 Btu/lbm (460 kJ/kg)

(b) The actual exhaust temperature is most nearly

(A) 370°R (210K)

(B) 420°R (230K)

(C) 450°R (240K)

(D) 610°R (340K)

(c) The inlet air pressure is most nearly

(A) 160 psia (1.2 MPa)

(B) 210 psia (1.5 MPa)

(C) 240 psia (1.7 MPa)

(D) 290 psia (2.0 MPa)

(d) The change in entropy through the turbine is most nearly

(A) 0.078 Btu/lbm-°R (0.35 kJ/kg·K)

(B) 0.10 Btu/lbm-°R (0.46 kJ/kg·K)

(C) 0.18 Btu/lbm-°R (0.82 kJ/kg·K)

(D) 0.32 Btu/lbm-°R (1.5 kJ/kg·K)

SOLUTIONS

1. *Customary U.S. Solution*

(a) The absolute temperature is

$$T_1 = 1500°\text{F} + 460° = 1960°\text{R}$$

From air tables (see App. 23.F) at 1960°R,

$$h_1 = 493.64 \text{ Btu/lbm}$$
$$p_{r,1} = 160.48$$
$$v_{r,1} = 4.53$$

After expansion,

$$p_{r,2} = p_{r,1}\left(\frac{p_2}{p_1}\right) = (160.48)\left(\frac{50 \text{ psia}}{200 \text{ psia}}\right) = 40.12$$

From air tables (see App. 23.F) at $p_{r,2} = 40.12$,

$$T_2 = \boxed{1375°\text{R}}$$
$$h_2 = 336.39 \text{ Btu/lbm}$$
$$v_{r,2} = 12.721$$

(Air tables are based on correlations from actual measured air properties. Ideal gas laws assume constant specific heats. If an ideal gas relationship is used, the final temperature will be 1319°R.)

The answer is (C).

(b) The air's final volumetric flow rate is

$$\dot{V}_2 = \dot{V}_1\left(\frac{v_{r,2}}{v_{r,1}}\right) = \left(10 \ \frac{\text{ft}^3}{\text{sec}}\right)\left(\frac{12.721}{4.53}\right) = \boxed{28.1 \text{ ft}^3/\text{sec} \quad (28 \text{ ft}^3/\text{sec})}$$

The answer is (A).

(c) The air's enthalpy change is

$$\Delta h = h_2 - h_1 = 336.39 \ \frac{\text{Btu}}{\text{lbm}} - 493.64 \ \frac{\text{Btu}}{\text{lbm}} = \boxed{-157.25 \text{ Btu/lbm} \quad (-160 \text{ Btu/lbm}) \quad [\text{decrease}]}$$

The answer is (C).

SI Solution

(a) The absolute temperature is

$$T_1 = 820°\text{C} + 273° = 1093\text{K}$$

From air tables (see App. 23.S) at 1093K,

$$h_1 = 1153 \text{ kJ/kg}$$
$$p_{r,1} = 162.94$$
$$v_{r,1} = 19.275$$

After expansion,

$$p_{r,2} = p_{r,1}\left(\frac{p_2}{p_1}\right) = (162.94)\left(\frac{350 \text{ kPa}}{(1.4 \text{ MPa})\left(1000 \ \frac{\text{kPa}}{\text{MPa}}\right)}\right) = 40.74$$

From air tables (see App. 23.S) at $p_{r,2} = 40.74$,

$$T_2 = \boxed{767.2\text{K} \quad (770\text{K})}$$
$$h_2 = 786.05 \text{ kJ/kg}$$
$$v_{r,2} = 53.99$$

(Air tables are based on correlations from actual measured air properties. Ideal gas laws assume constant specific heats. If an ideal gas relationship is used, the final temperature will be 736K.)

The answer is (C).

(b) The air's final volumetric flow rate is

$$\dot{V}_2 = \dot{V}_1\left(\frac{v_{r,2}}{v_{r,1}}\right) = \left(280 \ \frac{\text{L}}{\text{s}}\right)\left(\frac{53.99}{19.275}\right) = \boxed{784.3 \text{ L/s} \quad (780 \text{ L/s})}$$

The answer is (A).

(c) The air's enthalpy change is

$$\Delta h = h_2 - h_1 = 786.05 \ \frac{\text{kJ}}{\text{kg}} - 1153 \ \frac{\text{kJ}}{\text{kg}} = \boxed{-366.95 \text{ kJ/kg} \quad (-370 \text{ kJ/kg}) \quad [\text{decrease}]}$$

The answer is (C).

2. *Customary U.S. Solution*

(Use an air table. The SI solution assumes an ideal gas.)

Refer to Fig. 29.3.

At A:

$$T_\text{A} = 60°\text{F} + 460° = 520°\text{R} \quad [\text{given}]$$
$$p_\text{A} = 14.7 \text{ psia} \quad [\text{given}]$$

Power Cycles

From the air table (see App. 23.F),

$$v_{r,A} = 158.58$$
$$p_{r,A} = 1.2147$$
$$h_A = 124.27 \text{ Btu/lbm}$$

At B:

The process from A to B is isentropic.

$$v_{r,B} = v_{r,A}\left(\frac{V_B}{V_A}\right) = (158.58)\left(\frac{1}{5}\right) = 31.716$$

Locate this volume ratio in the air table (see App. 23.F).

$$T_B \approx 980°\text{R}$$
$$h_B \approx 236.02 \text{ Btu/lbm}$$
$$p_{r,B} = 11.430$$

Since process A-B is isentropic,

$$p_B = \left(\frac{p_{r,B}}{p_{r,A}}\right)p_A = \left(\frac{11.430}{1.2147}\right)(14.7 \text{ psia}) = 138.3 \text{ psia}$$

At C:

$$T_C = 1500°\text{F} + 460° = 1960°\text{R} \quad \text{[given]}$$
$$p_C = p_B = 138.3 \text{ psia}$$

Locate the temperature in the air table.

$$h_C = 493.64 \text{ Btu/lbm}$$
$$p_{r,C} = 160.48$$

At D:

$$p_D = 14.7 \text{ psia}$$

Since process C-D is isentropic,

$$p_{r,D} = p_{r,C}\left(\frac{p_D}{p_C}\right) = (160.48)\left(\frac{14.7 \text{ psia}}{138.3 \text{ psia}}\right) = 17.058$$

Locate this pressure ratio in the air table.

$$T_D = 1094°\text{R}$$
$$h_D = 264.49 \text{ Btu/lbm}$$

Since the efficiency of compression is 83%, from Eq. 29.20,

$$h'_B = h_A + \frac{h_B - h_A}{\eta_{s,\text{compressor}}} = 124.27 \ \frac{\text{Btu}}{\text{lbm}} + \frac{236.02 \ \frac{\text{Btu}}{\text{lbm}} - 124.27 \ \frac{\text{Btu}}{\text{lbm}}}{0.83} = 258.9 \text{ Btu/lbm}$$

Since the efficiency of the expansion process is 92%, from Eq. 29.22,

$$h'_D = h_C - \eta_{s,\text{turbine}}(h_C - h_D) = 493.64 \ \frac{\text{Btu}}{\text{lbm}} - (0.92)\left(493.64 \ \frac{\text{Btu}}{\text{lbm}} - 264.49 \ \frac{\text{Btu}}{\text{lbm}}\right) = 282.8 \text{ Btu/lbm}$$

From Eq. 29.18, the thermal efficiency is

$$\eta_{th} = \frac{(h_C - h'_B) - (h'_D - h_A)}{h_C - h'_B} = \frac{\left(493.64 \ \frac{\text{Btu}}{\text{lbm}} - 258.9 \ \frac{\text{Btu}}{\text{lbm}}\right) - \left(282.8 \ \frac{\text{Btu}}{\text{lbm}} - 124.27 \ \frac{\text{Btu}}{\text{lbm}}\right)}{493.64 \ \frac{\text{Btu}}{\text{lbm}} - 258.9 \ \frac{\text{Btu}}{\text{lbm}}} = \boxed{0.325 \quad (33\%)}$$

The answer is (A).

SI Solution

Refer to Fig. 29.3.

At A:

$$T_A = 16°\text{C} + 273° = 289\text{K} \quad \text{[given]}$$
$$p_A = 101.3 \text{ kPa} \quad \text{[given]}$$

At B:

$$T_B = T_A\left(\frac{v_A}{v_B}\right)^{k-1} = (289\text{K})(5)^{1.4-1} = 550.2\text{K}$$
$$p_B = p_A\left(\frac{v_A}{v_B}\right)^{k} = (101.3 \text{ kPa})(5)^{1.4} = 964.2 \text{ kPa}$$

At C:

$$T_C = 820°\text{C} + 273° = 1093\text{K} \quad \text{[given]}$$
$$p_C = p_B = 964.2 \text{ kPa}$$

Power Cycles

At D:

$$p_D = 101.3 \text{ kPa} \quad \text{[given]}$$

$$T_D = T_C\left(\frac{p_D}{p_C}\right)^{(k-1)/k} = (1093\text{K})\left(\frac{101.3 \text{ kPa}}{964.2 \text{ kPa}}\right)^{(1.4-1)/1.4}$$
$$= 574.2\text{K}$$

For ideal gases, the specific heats are constant. Therefore, the change in internal energy (and enthalpy, approximately) is proportional to the change in temperature. The actual temperature (see Eq. 29.21) is

$$T'_B = T_A + \frac{T_B - T_A}{\eta_{s,\text{compressor}}}$$
$$= 289\text{K} + \frac{550.2\text{K} - 289\text{K}}{0.83}$$
$$= 603.7\text{K}$$

From Eq. 29.23,

$$T'_D = T_C - \eta_{s,\text{turbine}}(T_C - T_D)$$
$$= 1093\text{K} - (0.92)(1093\text{K} - 574.2\text{K})$$
$$= 615.7\text{K}$$

From Eq. 29.19, the thermal efficiency is

$$\eta_{th} = \frac{(T_C - T'_B) - (T'_D - T_A)}{T_C - T'_B}$$
$$= \frac{(1093\text{K} - 603.7\text{K}) - (615.7\text{K} - 289\text{K})}{1093\text{K} - 603.7\text{K}}$$
$$= \boxed{0.332 \quad (33\%)}$$

The answer is (A).

3. *Customary U.S. Solution*

Since specific heats are constant, use ideal gas equations rather than air tables.

Refer to Fig. 29.4.

At A:

$$T_A = 60°\text{F} + 460° = 520°\text{R} \quad \text{[given]}$$
$$p_A = 14.7 \text{ psia} \quad \text{[given]}$$

At B:

$$T_B = T_A\left(\frac{v_A}{v_B}\right)^{k-1} = (520°\text{R})(5)^{1.4-1} = 989.9°\text{R}$$

$$p_B = p_A\left(\frac{v_A}{v_B}\right)^{k} = (14.7 \text{ psia})(5)^{1.4}$$
$$= 139.9 \text{ psia}$$

At D:

$$T_D = 1500°\text{F} + 460° = 1960°\text{R} \quad \text{[given]}$$
$$p_D = p_B = 139.9 \text{ psia}$$

At E:

$$p_E = 14.7 \text{ psia} \quad \text{[given]}$$

$$T_E = T_D\left(\frac{p_E}{p_D}\right)^{(k-1)/k} = (1960°\text{R})\left(\frac{14.7 \text{ psia}}{139.9 \text{ psia}}\right)^{(1.4-1)/1.4}$$
$$= 1029.6°\text{R}$$

From Eq. 29.21,

$$T'_B = T_A + \frac{T_B - T_A}{\eta_{s,\text{compressor}}} = 520°\text{R} + \frac{989.9°\text{R} - 520°\text{R}}{0.83}$$
$$= 1086.1°\text{R}$$

From Eq. 29.23,

$$T'_E = T_D - \eta_{s,\text{turbine}}(T_D - T_E)$$
$$= 1960°\text{R} - (0.92)(1960°\text{R} - 1029.6°\text{R})$$
$$= 1104.0°\text{R}$$

From Eq. 29.24,

$$\eta_{\text{regenerator}} = \frac{h_C - h'_B}{h'_E - h'_B}$$

For constant c_p,

$$\eta_{\text{regenerator}} = \frac{T_C - T'_B}{T'_E - T'_B}$$
$$0.65 = \frac{T_C - 1086.1°\text{R}}{1104.0°\text{R} - 1086.1°\text{R}}$$
$$T_C = 1097.7°\text{R}$$

From Eq. 29.25 with constant specific heats,

$$\eta_{th} = \frac{(T_D - T'_E) - (T'_B - T_A)}{T_D - T_C}$$
$$= \frac{(1960°\text{R} - 1104.0°\text{R}) - (1086.1°\text{R} - 520°\text{R})}{1960°\text{R} - 1097.7°\text{R}}$$
$$= \boxed{0.336 \quad (34\%)}$$

The answer is (C).

Power Cycles

SI Solution

Refer to Fig. 29.4. From Prob. 2,

$$T_A = 289\text{K}$$
$$T'_B = 603.7\text{K}$$
$$T_D = T_C = 1093\text{K}$$
$$T'_E = T'_D = 615.7\text{K}$$

For constant specific heats and from Eq. 29.24,

$$\eta_{\text{regenerator}} = \frac{T_C - T'_B}{T'_E - T'_B}$$
$$0.65 = \frac{T_C - 603.7\text{K}}{615.7\text{K} - 603.7\text{K}}$$
$$T_C = 611.5\text{K}$$

From Eq. 29.25 with constant specific heats,

$$\eta_{\text{th}} = \frac{(T_D - T'_E) - (T'_B - T_A)}{T_D - T_C}$$
$$= \frac{(1093\text{K} - 615.7\text{K}) - (603.7\text{K} - 289\text{K})}{1093\text{K} - 611.5\text{K}}$$
$$= \boxed{0.338 \quad (34\%)}$$

The answer is (C).

4. Use the illustration shown for both the customary U.S. and SI solutions.

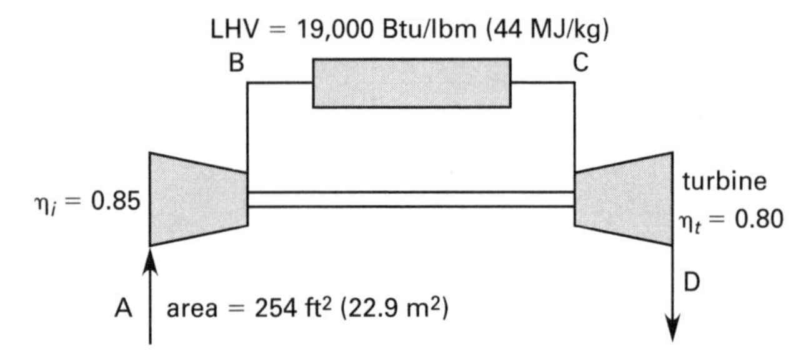

Customary U.S. Solution

(a) *At 7000 ft altitude:*

$$\dot{V}_{a1} = 50{,}000 \text{ ft}^3/\text{min}$$
$$\text{BHP}_1 = 6000 \text{ hp}$$
$$\text{BSCF}_1 = 0.609 \text{ lbm/hp-hr}$$

At A:

$$p_A = 12 \text{ psia}$$
$$T_A = 35°\text{F}$$

The absolute temperature at A is

$$T_A = 35°\text{F} + 460° = 495°\text{R}$$

Interpolating from App. 23.F (air table),

$$h_A = 118.28 \text{ Btu/lbm}$$
$$p_{r,A} = 1.0238$$

From the ideal gas law, the air density is

$$\rho_{a1} = \frac{p_A}{RT_A} = \frac{\left(12 \ \frac{\text{lbf}}{\text{in}^2}\right)\left(12 \ \frac{\text{in}}{\text{ft}}\right)^2}{\left(53.35 \ \frac{\text{ft-lbf}}{\text{lbm-°R}}\right)(495°\text{R})}$$
$$= \boxed{0.0654 \text{ lbm/ft}^3 \quad (0.065 \text{ lbm/ft}^3)}$$

The answer is (A).

(b) From Eq. 28.79, the air mass flow rate is

$$\dot{m}_{a1} = \dot{V}_{a1}\rho_{a1}$$
$$= \left(50{,}000 \ \frac{\text{ft}^3}{\text{min}}\right)\left(0.0654 \ \frac{\text{lbm}}{\text{ft}^3}\right)$$
$$= \boxed{3270 \text{ lbm/min} \quad (3300 \text{ lbm/min})}$$

The answer is (C).

(c) At B:

$$p_B = 8p_A = (8)(12 \text{ psia}) = 96 \text{ psia}$$

Assuming isentropic compression,

$$p_{r,B} = 8p_{r,A} = (8)(1.0238)$$
$$= 8.1904$$

From App. 23.F (air table), this $p_{r,B}$ corresponds to

$$T_B = 893.3°\text{R}$$
$$h_B = 214.62 \text{ Btu/lbm}$$

Due to the inefficiency of the compressor, from Eq. 29.20,

$$h'_B = h_A + \frac{h_B - h_A}{\eta_{s,\text{compression}}}$$
$$= 118.28 \ \frac{\text{Btu}}{\text{lbm}} + \frac{214.62 \ \frac{\text{Btu}}{\text{lbm}} - 118.28 \ \frac{\text{Btu}}{\text{lbm}}}{0.85}$$
$$= 231.62 \text{ Btu/lbm}$$

From App. 23.F, this corresponds to $T'_B = 962.3°R$.

$$\begin{aligned} W_{\text{compression}} &= h'_B - h_A \\ &= 231.62\ \frac{\text{Btu}}{\text{lbm}} - 118.28\ \frac{\text{Btu}}{\text{lbm}} \\ &= 113.34\ \text{Btu/lbm} \end{aligned}$$

At C:

The absolute temperature is

$$T_C = 1800°F + 460° = 2260°R \quad \text{[no change if moved]}$$

Since there is no pressure drop across the combustor, $p_C = 96$ psia.

From App. 23.F (air table),

$$h_C = 577.52\ \text{Btu/lbm}$$
$$p_{r,C} = 286.7$$

The energy requirement from the fuel is

$$\begin{aligned} \dot{m}\Delta h &= \dot{m}_{a1}(h_C - h'_B) \\ &= \left(3270\ \frac{\text{lbm}}{\text{min}}\right)\left(577.52\ \frac{\text{Btu}}{\text{lbm}} - 231.62\ \frac{\text{Btu}}{\text{lbm}}\right) \\ &= 1.131 \times 10^6\ \text{Btu/min} \end{aligned}$$

The ideal fuel rate is

$$\begin{aligned} \frac{\dot{m}\Delta h}{\text{LHV}} &= \frac{\left(1.131 \times 10^6\ \frac{\text{Btu}}{\text{min}}\right)\left(60\ \frac{\text{min}}{\text{hr}}\right)}{19{,}000\ \frac{\text{Btu}}{\text{lbm}}} \\ &= \boxed{3571.6\ \text{lbm/hr} \quad (3600\ \text{lbm/hr})} \end{aligned}$$

The answer is (D).

(d) From Eq. 28.77, the ideal BSFC is

$$(\text{BSFC})_{\text{ideal}} = \frac{3571.6\ \frac{\text{lbm}}{\text{hr}}}{6000\ \text{hp}} = 0.595\ \text{lbm/hp-hr}$$

The combustor efficiency is

$$\begin{aligned} \eta_{\text{combustor}} &= \frac{(\text{BSFC})_{\text{ideal}}}{(\text{BSFC})_{\text{actual}}} = \frac{0.595\ \frac{\text{lbm}}{\text{hp-hr}}}{0.609\ \frac{\text{lbm}}{\text{hp-hr}}} \\ &= \boxed{0.977 \quad (98\%)} \end{aligned}$$

The answer is (D).

(e) At D:

As determined by atmospheric conditions, $p_D = 12$ psia.

If expansion is isentropic,

$$p_{r,D} = \frac{p_{r,C}}{8} = \frac{286.7}{8} = 35.8375$$

From App. 23.F (air table), this $p_{r,D}$ corresponds to

$$T_D = 1334.8°R$$
$$h_D = 325.95\ \text{Btu/lbm}$$

Due to the inefficiency of the turbine, from Eq. 29.22,

$$\begin{aligned} h'_D &= h_C - \eta_{s,\text{turbine}}(h_C - h_D) \\ &= 577.52\ \frac{\text{Btu}}{\text{lbm}} - (0.80)\left(\begin{array}{r} 577.52\ \frac{\text{Btu}}{\text{lbm}} \\ -\ 325.95\ \frac{\text{Btu}}{\text{lbm}} \end{array}\right) \\ &= 376.26\ \text{Btu/lbm} \end{aligned}$$

$$\begin{aligned} W_{\text{turbine}} &= h_C - h'_D = 577.52\ \frac{\text{Btu}}{\text{lbm}} - 376.26\ \frac{\text{Btu}}{\text{lbm}} \\ &= 201.26\ \text{Btu/lbm} \end{aligned}$$

The theoretical net horsepower is

$$\begin{aligned} \text{IHP} &= \dot{m}_{a1}(W_{\text{turbine}} - W_{\text{compression}}) \\ &= \left(3270\ \frac{\text{lbm}}{\text{min}}\right)\left(\begin{array}{c} \left(201.26\ \frac{\text{Btu}}{\text{lbm}} - 113.34\ \frac{\text{Btu}}{\text{lbm}}\right) \\ \times \left(\frac{778\ \frac{\text{ft-lbf}}{\text{Btu}}}{33{,}000\ \frac{\text{ft-lbf}}{\text{hp-min}}}\right) \end{array}\right) \\ &= 6778\ \text{hp} \end{aligned}$$

From Eq. 28.72, the friction horsepower is

$$\begin{aligned} \text{FHP} &= \text{IHP} - \text{BHP} = 6778\ \text{hp} - 6000\ \text{hp} \\ &= \boxed{778\ \text{hp} \quad (780\ \text{hp})} \end{aligned}$$

The answer is (B).

(f) *At sea level (zero altitude):*

At A:

The absolute temperature is

$$T_A = 70°F + 460° = 530°R$$
$$p_A = 14.7\ \text{psia}$$

From App. 23.F (air table),

$$h_A = 126.86\ \text{Btu/lbm}$$
$$p_{r,A} = 1.2998$$

Power Cycles

From the ideal gas law, the air density is

$$\rho_{a2} = \frac{p_A}{RT_A} = \frac{\left(14.7\ \frac{\text{lbf}}{\text{in}^2}\right)\left(12\ \frac{\text{in}}{\text{ft}}\right)^2}{\left(53.35\ \frac{\text{lbf-ft}}{\text{lbm-°R}}\right)(530°\text{R})} = 0.0749\ \text{lbm/ft}^3$$

From Eq. 28.80, the air mass flow rate is

$$\dot{m}_{a2} = \dot{V}_{a2}\rho_{a2} = \left(50{,}000\ \frac{\text{ft}^3}{\text{min}}\right)\left(0.0749\ \frac{\text{lbm}}{\text{ft}^3}\right) = 3745\ \text{lbm/min}$$

At B:

$$p_B = 8p_A = (8)(14.7\ \text{psia}) = 117.6\ \text{psia}$$

Assuming isentropic compression,

$$p_{r,B} = 8p_{r,A} = (8)(1.2998) = 10.3984$$

From App. 23.F (air table), this $p_{r,B}$ corresponds to

$$T_B = 954.5°\text{R}$$
$$h_B = 229.70\ \text{Btu/lbm}$$

Due to the inefficiency of the compressor, from Eq. 29.20,

$$h'_B = h_A + \frac{h_B - h_A}{\eta_{s,\text{compression}}} = 126.86\ \frac{\text{Btu}}{\text{lbm}} + \frac{229.70\ \frac{\text{Btu}}{\text{lbm}} - 126.86\ \frac{\text{Btu}}{\text{lbm}}}{0.85} = 247.85\ \text{Btu/lbm}$$

From App. 23.F (air table), $h'_B = 247.85$ Btu/lbm corresponds to $T'_B = 1027.6°$R.

$$W_{\text{compression}} = h'_B - h_A = 247.85\ \frac{\text{Btu}}{\text{lbm}} - 126.86\ \frac{\text{Btu}}{\text{lbm}} = 120.99\ \text{Btu/lbm}$$

At C:

The absolute temperature is

$$T_C = 2260°\text{R} \quad [\text{no change}]$$

Since there is no pressure drop across the combustor, $p_C = 117.6$ psia.

From App. 23.F (air table),

$$h_C = 577.52\ \text{Btu/lbm}$$
$$p_{r,C} = 286.7$$

The power provided by the fuel is $\dot{m}\Delta h$.

$$\dot{m}_{a2}(h_C - h'_B) = \left(3745\ \frac{\text{lbm}}{\text{min}}\right)\left(577.52\ \frac{\text{Btu}}{\text{lbm}} - 247.85\ \frac{\text{Btu}}{\text{lbm}}\right) = 1.235 \times 10^6\ \text{Btu/min}$$

Assuming a constant combustion efficiency of 97.9%, the actual fuel rate is

$$\frac{\dot{m}\Delta h}{\text{LHV}} = \frac{\left(1.235 \times 10^6\ \frac{\text{Btu}}{\text{min}}\right)\left(60\ \frac{\text{min}}{\text{hr}}\right)}{\left(19{,}000\ \frac{\text{Btu}}{\text{lbm}}\right)(0.979)} = 3983.7\ \text{lbm/hr}$$

At D:

$$p_D = 14.7\ \text{psia}$$

If expansion is isentropic,

$$p_{r,D} = \frac{p_{r,C}}{8} = \frac{286.7}{8} = 35.8375 \quad [\text{no change}]$$

From App. 23.F (air table), this corresponds to

$$T_D = 1334.8°\text{R}$$
$$h_D = 325.95\ \text{Btu/lbm}$$
$$h'_D = 376.26\ \text{Btu/lbm} \quad [\text{no change}]$$
$$W_{\text{turbine}} = 201.26\ \text{Btu/lbm} \quad [\text{no change}]$$

The theoretical net horsepower is

$$\text{IHP} = \dot{m}_{a2}(W_{\text{turbine}} - W_{\text{compression}}) = \left(3745\ \frac{\text{lbm}}{\text{min}}\right)\left(201.26\ \frac{\text{Btu}}{\text{lbm}} - 120.99\ \frac{\text{Btu}}{\text{lbm}}\right) \times \left(\frac{778\ \frac{\text{ft-lbf}}{\text{Btu}}}{33{,}000\ \frac{\text{ft-lbf}}{\text{hp-min}}}\right) = 7087\ \text{hp}$$

Assuming the frictional horsepower is constant,

$$\text{BHP} = \text{IHP} - \text{FHP} = 7087\ \text{hp} - 778\ \text{hp} = \boxed{6309\ \text{hp}\quad(6300\ \text{hp})}$$

The answer is (D).

(g) From Eq. 28.77, BSFC is

$$\text{BSFC} = \frac{3983.7\ \frac{\text{lbm}}{\text{hr}}}{6309\ \text{hp}} = \boxed{\begin{array}{r}0.631\ \text{lbm/hp-hr} \\ (0.63\ \text{lbm/hp-hr})\end{array}}$$

The answer is (C).

SI Solution

(a) *At 2100 m altitude:*

At A:

$$p_{\text{A}} = 82\ \text{kPa} \quad \text{[given]}$$

The absolute temperature is

$$T_{\text{A}} = 2°\text{C} + 273° = 275\text{K} \quad \text{[given]}$$

From App. 23.S (air table),

$$h_{\text{A}} = 275.12\ \text{kJ/kg}$$
$$p_{r,\text{A}} = 1.0240$$

From the ideal gas law, the air density is

$$\rho_{a1} = \frac{p_{\text{A}}}{RT_{\text{A}}} = \frac{(82\ \text{kPa})\left(1000\ \frac{\text{Pa}}{\text{kPa}}\right)}{\left(287.03\ \frac{\text{J}}{\text{kg·K}}\right)(275\text{K})} = \boxed{1.0389\ \text{kg/m}^3 \quad (1.0\ \text{kg/m}^3)}$$

The answer is (A).

(b) From Eq. 28.79, the air mass flow rate is

$$\dot{m}_{a1} = \dot{V}_{a1}\rho_{a1} = \frac{\left(23{,}500\ \frac{\text{L}}{\text{s}}\right)\left(1.0389\ \frac{\text{kg}}{\text{m}^3}\right)}{1000\ \frac{\text{L}}{\text{m}^3}} = \boxed{24.41\ \text{kg/s} \quad (24\ \text{kg/s})}$$

The answer is (C).

(c) At B:

$$p_{\text{C}} = 8p_{\text{A}} = (8)(82\ \text{kPa}) = 656\ \text{kPa}$$

Assuming isentropic compression,

$$p_{r,\text{B}} = 8p_{r,\text{A}} = (8)(1.0240) = 8.192$$

From App. 23.S (air table), this $p_{r,\text{B}}$ corresponds to

$$T_{\text{B}} = 496.3\text{K}$$
$$h_{\text{B}} = 499.33\ \text{kJ/kg}$$

Due to the inefficiency of the compressor, from Eq. 29.21,

$$h'_{\text{B}} = h_{\text{A}} + \frac{h_{\text{B}} - h_{\text{A}}}{\eta_{s,\text{compression}}} = 275.12\ \frac{\text{kJ}}{\text{kg}} + \frac{499.33\ \frac{\text{kJ}}{\text{kg}} - 275.12\ \frac{\text{kJ}}{\text{kg}}}{0.85} = 538.90\ \text{kJ/kg}$$

From App. 23.S (air table), this value of $h = 538.90$ kJ/kg corresponds to $T'_{\text{B}} = 534.7\text{K}$.

$$W_{\text{compression}} = h'_{\text{B}} - h_{\text{A}} = 538.90\ \frac{\text{kJ}}{\text{kg}} - 275.12\ \frac{\text{kJ}}{\text{kg}} = 263.78\ \text{kJ/kg}$$

At C:

The absolute temperature is

$$T_{\text{C}} = 980°\text{C} + 273° = 1253\text{K}$$

Since there is no pressure drop across the combustor, $p_{\text{C}} = 656$ kPa.

From App. 23.S (air table),

$$h_{\text{C}} = 1340.28\ \text{kJ/kg}$$
$$p_{r,\text{C}} = 284.3$$

The power provided by the fuel is

$$\dot{m}\Delta h = \dot{m}_{a1}(h_{\text{C}} - h'_{\text{B}}) = \left(24.41\ \frac{\text{kg}}{\text{s}}\right)\left(1340.28\ \frac{\text{kJ}}{\text{kg}} - 538.90\ \frac{\text{kJ}}{\text{kg}}\right) = 19\,562\ \text{kJ/s}$$

The ideal fuel rate is

$$\frac{\dot{m}\Delta h}{\text{LHV}} = \frac{19\,562\ \frac{\text{kJ}}{\text{s}}}{\left(44\ \frac{\text{MJ}}{\text{kg}}\right)\left(1000\ \frac{\text{kJ}}{\text{MJ}}\right)} = \boxed{0.4446\ \text{kg/s} \quad (0.44\ \text{kg/s})}$$

The answer is (D).

(d) From Eq. 28.77, the ideal BSFC is

$$(\text{BSFC})_{\text{ideal}} = \left(\frac{0.4446\ \frac{\text{kg}}{\text{s}}}{4.5\ \text{MW}}\right)\left(1000\ \frac{\text{MW}}{\text{GW}}\right) = 98.8\ \text{kg/GJ}$$

Power Cycles

The combustor efficiency is

$$\eta_{\text{combustor}} = \frac{(\text{BSFC})_{\text{ideal}}}{(\text{BSFC})_{\text{actual}}} = \frac{98.8 \ \frac{\text{kg}}{\text{GJ}}}{100 \ \frac{\text{kg}}{\text{GJ}}} = \boxed{0.988 \quad (98\%)}$$

The answer is (D).

(e) At D:

Determined by atmospheric conditions, $p_\text{D} = 82$ kPa.

If expansion is isentropic,

$$p_{r,\text{D}} = \frac{p_{r,\text{C}}}{8} = \frac{284.3}{8} = 35.54$$

From App. 23.S (air table), this $p_{r,\text{D}}$ corresponds to

$$T_\text{D} = 740\text{K}$$
$$h_\text{D} = 756.44 \ \text{kJ/kg}$$

From Eq. 29.22, due to the inefficiency of the turbine,

$$\begin{aligned} h'_\text{D} &= h_\text{C} - \eta_{s,\text{turbine}}(h_\text{C} - h_\text{D}) \\ &= 1340.28 \ \frac{\text{kJ}}{\text{kg}} - (0.80) \\ &\quad \times \left(1340.28 \ \frac{\text{kJ}}{\text{kg}} - 756.44 \ \frac{\text{kJ}}{\text{kg}}\right) \\ &= 873.21 \ \text{kJ/kg} \end{aligned}$$

$$\begin{aligned} W_{\text{turbine}} &= h_\text{C} - h'_\text{D} = 1340.28 \ \frac{\text{kJ}}{\text{kg}} - 873.21 \ \frac{\text{kJ}}{\text{kg}} \\ &= 467.07 \ \text{kJ/kg} \end{aligned}$$

The theoretical net power is

$$\begin{aligned} \text{IkW} &= \dot{m}_{a1}(W_{\text{turbine}} - W_{\text{compression}}) \\ &= \left(24.41 \ \frac{\text{kg}}{\text{s}}\right)\left(\frac{467.07 \ \frac{\text{kJ}}{\text{kg}} - 263.78 \ \frac{\text{kJ}}{\text{kg}}}{1000 \ \frac{\text{kW}}{\text{MW}}}\right) \\ &= 4.962 \ \text{MW} \end{aligned}$$

From Eq. 28.74, the friction horsepower is

$$\begin{aligned} \text{FkW} &= \text{IkW} - \text{BkW} = 4.962 \ \text{MW} - 4.5 \ \text{MW} \\ &= \boxed{0.462 \ \text{MW} \quad (0.46 \ \text{MW})} \end{aligned}$$

The answer is (B).

(f) *At sea level (zero altitude):*

At A:

The absolute temperature is

$$T_\text{A} = 21^\circ\text{C} + 273^\circ = 294\text{K}$$
$$p_\text{A} = 101.3 \ \text{kPa}$$

From App. 23.S,

$$h_\text{A} = 294.17 \ \text{kJ/kg}$$
$$p_{r,\text{A}} = 1.2917$$

From the ideal gas law, the air density is

$$\begin{aligned} \rho_{a2} &= \frac{p_\text{A}}{RT_\text{A}} = \frac{(101.3 \ \text{kPa})\left(1000 \ \frac{\text{Pa}}{\text{kPa}}\right)}{\left(287.03 \ \frac{\text{J}}{\text{kg}\cdot\text{K}}\right)(294\text{K})} \\ &= 1.2004 \ \text{kg/m}^3 \end{aligned}$$

From Eq. 28.80, the air mass flow rate is

$$\begin{aligned} \dot{m}_{a2} &= \dot{V}_{a2}\rho_{a2} \\ &= \frac{\left(23\,500 \ \frac{\text{L}}{\text{s}}\right)\left(1.2004 \ \frac{\text{kg}}{\text{m}^3}\right)}{1000 \ \frac{\text{L}}{\text{m}^3}} \\ &= 28.21 \ \text{kg/s} \end{aligned}$$

At B:

$$p_\text{B} = 8p_\text{A} = (8)(101.3 \ \text{kPa}) = 810.4 \ \text{kPa}$$

Assuming isentropic expansion,

$$p_{r,\text{B}} = 8p_{r,\text{A}} = (8)(1.2917) = 10.3336$$

From App. 23.S (air table), this $p_{r,\text{B}}$ corresponds to

$$T_\text{B} = 529.5\text{K}$$
$$h_\text{B} = 533.46 \ \text{kJ/kg}$$

Due to the inefficiency of the compressor, from Eq. 29.20,

$$\begin{aligned} h'_\text{B} &= h_\text{A} + \frac{h_\text{B} - h_\text{A}}{\eta_{s,\text{compression}}} \\ &= 294.17 \ \frac{\text{kJ}}{\text{kg}} + \frac{533.46 \ \frac{\text{kJ}}{\text{kg}} - 294.17 \ \frac{\text{kJ}}{\text{kg}}}{0.85} \\ &= 575.69 \ \text{kJ/kg} \end{aligned}$$

Power Cycles

From App. 23.S (air table), this value of h_B corresponds to

$$T'_B = 570\text{K}$$

$$W_{\text{compression}} = h'_B - h_A = 575.69\ \frac{\text{kJ}}{\text{kg}} - 294.17\ \frac{\text{kJ}}{\text{kg}} = 281.52\ \text{kJ/kg}$$

At C:

The absolute temperature is

$$T_C = 1253\text{K} \quad [\text{no change}]$$

Assuming there is no pressure drop across the combustor, $p_C = 810.4$ kPa.

From App. 23.S (air table),

$$h_C = 1340.28\ \text{kJ/kg}$$

$$p_{r,C} = 284.3$$

The energy requirement from the fuel is

$$\dot{m}\Delta h = \dot{m}_{a2}(h_C - h'_B) = \left(28.21\ \frac{\text{kg}}{\text{s}}\right)\left(1340.28\ \frac{\text{kJ}}{\text{kg}} - 575.69\ \frac{\text{kJ}}{\text{kg}}\right) = 21\,569\ \text{kJ/s}$$

Assuming a constant combustion efficiency of 98.8%, the actual fuel rate is

$$\frac{\dot{m}\Delta h}{\text{LHV}} = \frac{21\,569\ \frac{\text{kJ}}{\text{s}}}{\left(44\ \frac{\text{MJ}}{\text{kg}}\right)\left(1000\ \frac{\text{kJ}}{\text{MJ}}\right)(0.988)} = 0.496\ \text{kg/s}$$

At D:

$$p_D = 101.3\ \text{kPa}$$

If expansion is isentropic,

$$p_{r,D} = \frac{p_{r,C}}{8} = \frac{284.3}{8} = 35.538 \quad [\text{no change}]$$

From App. 23.S (air table), this $p_{r,D}$ corresponds to

$$T_D = 740\text{K}$$

$$h_D = 756.44\ \text{kJ/kg}$$

$$h'_D = 873.21\ \text{kJ/kg} \quad [\text{no change}]$$

$$W_{\text{turbine}} = 467.07\ \text{kJ/kg}$$

The theoretical net power is

$$\text{IkW} = \dot{m}_{a2}(W_{\text{turbine}} - W_{\text{compression}}) = \frac{\left(28.21\ \frac{\text{kg}}{\text{s}}\right)\left(467.07\ \frac{\text{kJ}}{\text{kg}} - 281.52\ \frac{\text{kJ}}{\text{kg}}\right)}{1000\ \frac{\text{kW}}{\text{MW}}} = 5.234\ \text{MW}$$

Assuming the frictional horsepower is constant,

$$\text{BkW} = \text{IkW} - \text{FkW} = 5.234\ \text{MW} - 0.462\ \text{MW} = \boxed{4.772\ \text{MW} \quad (4.8\ \text{MW})}$$

The answer is (D).

(g) From Eq. 28.77,

$$\text{BSFC} = \frac{\left(0.496\ \frac{\text{kg}}{\text{s}}\right)\left(1000\ \frac{\text{MW}}{\text{GW}}\right)}{4.772\ \text{MW}} = \boxed{103.9\ \text{kg/GJ} \quad (100\ \text{kg/GJ})}$$

The answer is (C).

5. Use the illustration shown for both the customary U.S. and SI solutions.

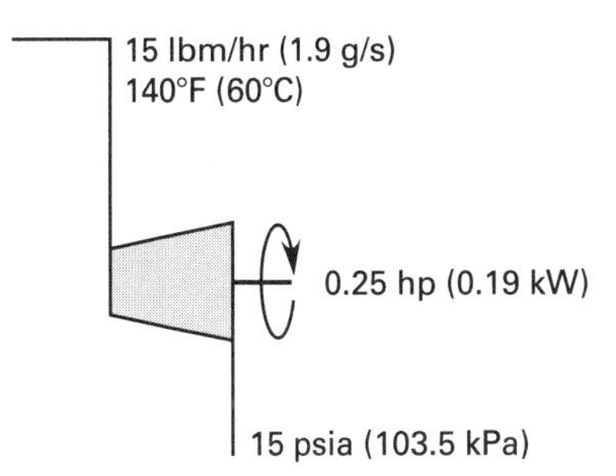

Customary U.S. Solution

(a) The drill power is

$$P = 0.25\ \text{hp} \quad [\text{given}] = (0.25\ \text{hp})\left(2545\ \frac{\text{Btu}}{\text{hp-hr}}\right) = 636.25\ \text{Btu/hr}$$

The absolute inlet temperature is

$$T_1 = 140°\text{F} + 460° = 600°\text{R}$$

Power Cycles

From App. 23.F, the properties of air entering the turbine are

$$h_1 = 143.47 \text{ Btu/lbm}$$
$$p_{r,1} = 2.005$$
$$\phi_1 = 0.62607 \text{ Btu/lbm-°R}$$

From Eq. 26.18,

$$P = \dot{m}(h_1 - h'_2)$$
$$\eta_{s,\text{turbine}} = \frac{h_1 - h'_2}{h_1 - h_2}$$

So,

$$P = \dot{m}\eta_{s,\text{turbine}}(h_1 - h_2)$$
$$h_2 = h_1 - \frac{P}{\dot{m}\eta_{s,\text{turbine}}}$$
$$= 143.47 \ \frac{\text{Btu}}{\text{lbm}} - \frac{636.25 \ \frac{\text{Btu}}{\text{hr}}}{\left(15 \ \frac{\text{lbm}}{\text{hr}}\right)(0.60)}$$
$$= \boxed{72.776 \text{ Btu/lbm} \quad (70 \text{ Btu/lbm})}$$

The answer is (A).

(b) Appendix 23.F doesn't go low enough. From the Keenan and Kayes *Gas Tables*, for $h = 72.776$ Btu/lbm,

$$T_2 = 305\text{°R}$$
$$p_{r,2} = 0.18851$$

Due to the irreversibility of the expansion from Eq. 26.19,

$$h'_2 = h_1 - \eta_s(h_1 - h_2)$$
$$= 143.47 \ \frac{\text{Btu}}{\text{lbm}} - (0.60)\left(143.47 \ \frac{\text{Btu}}{\text{lbm}} - 72.776 \ \frac{\text{Btu}}{\text{lbm}}\right)$$
$$= 101.05 \text{ Btu/lbm}$$

From App. 23.F for $h = 101.05$ Btu/lbm,

$$T'_2 = \boxed{423\text{°R} \quad (420\text{°R})}$$
$$\phi_2 = 0.54225 \text{ Btu/lbm-°R}$$

The answer is (B).

(c) The isentropic efficiency does not change the entrance and exit pressures, so the air tables can be used assuming $s_1 = s_2$. Since $p_1/p_2 = p_{r,1}/p_{r,2}$,

$$p_1 = p_2\left(\frac{p_{r,1}}{p_{r,2}}\right) = (15 \text{ psia})\left(\frac{2.005}{0.18851}\right)$$
$$= \boxed{159.5 \text{ psia} \quad (160 \text{ psia})}$$

The answer is (A).

(d) From Eq. 23.52, the entropy change is

$$s_2 - s_1 = \phi_2 - \phi_1 - R \ln\left(\frac{p_2}{p_1}\right)$$

From Table 23.7, $R = 53.35$ ft-lbf/lbm-°R.

$$s_2 - s_1 = 0.54225 \ \frac{\text{Btu}}{\text{lbm-°R}} - 0.62607 \ \frac{\text{Btu}}{\text{lbm-°R}} - \left(\frac{53.35 \ \frac{\text{ft-lbf}}{\text{lbm-°R}}}{778 \ \frac{\text{ft-lbf}}{\text{Btu}}}\right) \ln\left(\frac{15 \text{ psia}}{159.5 \text{ psia}}\right)$$
$$= \boxed{0.07829 \text{ Btu/lbm-°R} \quad (0.078 \text{ Btu/lbm-°R})}$$

The answer is (A).

SI Solution

(a) The absolute inlet temperature is

$$T_1 = 60\text{°C} + 273\text{°} = 333\text{K}$$

From App. 23.S, the properties of air entering the turbine are

$$h_1 = 333.40 \text{ kJ/kg}$$
$$p_{r,1} = 2.0064$$
$$\phi_1 = 1.80784 \text{ kJ/kg·K}$$

From Eq. 26.18,

$$P = \dot{m}(h_1 - h'_2)$$
$$\eta_{s,\text{turbine}} = \frac{h_1 - h'_2}{h_1 - h_2}$$
$$P = \dot{m}\eta_{s,\text{turbine}}(h_1 - h_2)$$
$$h_2 = h_1 - \frac{P}{\dot{m}\eta_{s,\text{turbine}}}$$
$$= 333.40 \ \frac{\text{kJ}}{\text{kg}} - \frac{(0.19 \text{ kW})\left(1000 \ \frac{\text{g}}{\text{kg}}\right)}{\left(1.9 \ \frac{\text{g}}{\text{s}}\right)(0.60)}$$
$$= \boxed{166.73 \text{ kJ/kg} \quad (170 \text{ kJ/kg})}$$

The answer is (A).

(b) Appendix 23.S doesn't go low enough. From the gas tables, for $h = 167.03$ kJ/kg,

$$T_2 = 164.4\text{K}$$
$$p_{r,2} = 0.16980$$

Due to the irreversibility of the expansion from Eq. 26.19,

$$\begin{aligned} h_2' &= h_1 - \eta_s(h_1 - h_2) \\ &= 333.40\ \frac{\text{kJ}}{\text{kg}} - (0.60)\left(333.40\ \frac{\text{kJ}}{\text{kg}} - 167.03\ \frac{\text{kJ}}{\text{kg}}\right) \\ &= 233.58\ \text{kJ/kg} \end{aligned}$$

From App. 23.S, for $h = 233.58$ kJ/kg,

$$\begin{aligned} T_2' &= \boxed{233.56\text{K} \quad (230\text{K})} \\ \phi_2 &= 1.45076\ \text{kJ/kg·K} \end{aligned}$$

The answer is (B).

(c) Since $p_1/p_2 = p_{r,1}/p_{r,2}$,

$$\begin{aligned} p_1 &= \frac{(103.5\ \text{kPa})\left(\dfrac{2.0064}{0.16980}\right)}{1000\ \dfrac{\text{kPa}}{\text{MPa}}} \\ &= \boxed{1.223\ \text{MPa} \quad (1.2\ \text{MPa})} \end{aligned}$$

The answer is (A).

(d) From Table 23.7, $R = 287.03$ J/kg·K. From Eq. 23.52, the entropy change is

$$\begin{aligned} s_2 - s_1 &= \phi_2 - \phi_1 - R \ln \frac{p_2}{p_1} \\ &= 1.45076\ \frac{\text{kJ}}{\text{kg·K}} - 1.80784\ \frac{\text{kJ}}{\text{kg·K}} \\ &\quad - \left(\frac{287.03\ \dfrac{\text{J}}{\text{kg·K}}}{1000\ \dfrac{\text{J}}{\text{kJ}}}\right) \ln\left(\frac{103.5\ \text{kPa}}{1223\ \text{kPa}}\right) \\ &= \boxed{0.3517\ \text{kJ/kg·K} \quad (0.35\ \text{kJ/kg·K})} \end{aligned}$$

The answer is (A).

30 Nuclear Power Cycles

PRACTICE PROBLEMS

1. A nuclear reactor develops 4000 MW thermal (1200 MW electric) when operating at full power. The fuel rods are submerged in pressurized water within the pressure vessel. Each fuel rod's average linear power density is 19 kW/m. The pressure vessel is modeled as a cylinder with an internal diameter of 5 m and an average internal length of 12 m. Fuel rods and control assemblies occupy 20% of the internal volume of the pressure vessel. Fuel rods are 3.6 m in length. The tops of the fuel rods are 6.1 m below the pool surface. At full capacity, conditions within the pressure vessel are 17.5 MPa and 350°C. After a loss of pressurization, gravity-fed emergency flooding will maintain the water temperature at 95°C for 12 hours. If the reactor stays in the full capacity configuration, and if electrical power to the pumps and other emergency back-up cooling and control systems is lost, assuming no heat loss to the surroundings, and neglecting the heat capacity of the fuel rods and control assemblies, approximately how long after a loss of pressurization event will the tops of the fuel rods become uncovered?

(A) 0.02 h

(B) 12 h

(C) 37 h

(D) 72 h

2. Radioactive sodium emits 2.75 MeV gamma rays to initiate the photodisintegration of deuterium.

$$\lambda + {}_1D^2 \rightarrow {}_1H^1 + {}_0n^1$$

The kinetic energy of the neutron is most nearly

(A) 0.15 MeV

(B) 0.26 MeV

(C) 0.31 MeV

(D) 0.55 MeV

3. The half-life of Cs-132 is approximately 6.47 days. Approximately how long will it take to reduce the activity of a Cs-132 sample to 5% of the original value?

(A) 28 days

(B) 39 days

(C) 64 days

(D) 81 days

4. A source with an activity of 2 curies emits 2 MeV gamma rays. For this intensity, the linear attenuation coefficient and build-up factor for a lead shield are 0.5182 cm^{-1} and 2.78, respectively. Approximately how thick must a lead shield be to reduce the activity to 1%?

(A) 5 cm

(B) 8 cm

(C) 11 cm

(D) 25 cm

5. A 10 cm thick lead plate shields a 2 MeV gamma source. The source flux density is 1,000,000 $\lambda/cm^2{\cdot}s$. For this intensity, the linear attenuation coefficient and build-up factor for a lead shield are approximately 0.5182 cm^{-1} and 2.78, respectively. Use 0.0238 cm^2/g as the energy absorption coefficient in air.

(a) The approximate total exit flux is most nearly

(A) 5.6×10^3 $1/cm^2{\cdot}s$

(B) 9.9×10^3 $1/cm^2{\cdot}s$

(C) 1.1×10^4 $1/cm^2{\cdot}s$

(D) 1.6×10^4 $1/cm^2{\cdot}s$

(b) The approximate dose rate for exposure in air is most nearly

(A) 20 mR/hr

(B) 30 mR/hr

(C) 40 mR/hr

(D) 50 mR/hr

6. A neutron flux of 1×10^8 neutrons/$cm^2{\cdot}s$ irradiates a 50°C gold-foil target for 24 hours. At 20°C, gold's absorption cross section is 98 b and its density is 19.32 g/cm^3. The half-life of activated gold, Au-198, is 2.7 days. If the target is a thin disk with a diameter of 25 mm and a volume of 0.4909 cm^3, the removal activity is most nearly

(A) 1.1×10^{-3} curie

(B) 1.5×10^{-3} curie

(C) 3.2×10^{-3} curie

(D) 8.4×10^{-3} curie

7. A neutron point source surrounded on all sides by 20°C water emits 1×10^7 neutrons/s. The average diffusion coefficient for water is approximately 0.16 cm, and its diffusion length is approximately 2.85 cm. The neutron flux 20 cm from the point source is most nearly

(A) 92 n/cm^2·s

(B) 150 n/cm^2·s

(C) 220 n/cm^2·s

(D) 350 n/cm^2·s

8. The thermal fission and absorption cross sections for natural uranium are 4.18 b and 7.68 b, respectively. The probability that a 20°C neutron that has been absorbed will cause fission in natural uranium is most nearly

(A) 8%

(B) 35%

(C) 45%

(D) 54%

9. A source of 1 MeV gamma rays has an activity of 1×10^8 λ/s. For this intensity, the linear attenuation coefficient and build-up factor for an iron shield are approximately 0.4683 cm^{-1} and 14.93, respectively. 0.0280 cm^2/g is the energy absorption coefficient in air. The thickness of spherical iron shield that will reduce the exposure rate to 1 mR/hr outside the shield is most nearly

(A) 8 cm

(B) 15 cm

(C) 37 cm

(D) 64 cm

10. A 1 GW breeder reactor is being designed to process 2000 kg of a 20% Pu-239 and 80% U-238 mixture. The specific fuel usage is 1.23 g/MW·day. For fast neutrons, Pu-239 has an absorption cross section of 1.95 b, has a fission cross section of 1.8 b, and releases 2.95 neutrons per fission. For fast neutrons, U-238 has an absorption cross section of 0.59 b, has a fission cross section of 0.5 b, and releases 2.45 neutrons per fission. The fast fission factor is 1.05. The exponential doubling time is most nearly

(A) 1400 days

(B) 2100 days

(C) 3600 days

(D) 4500 days

11. The maximum neutron flux in a bare spherical reactor is 4.5×10^{15} n/cm^2·s. The radius of the reactor is 40 cm, and the macroscopic fission cross section of the fuel is 0.005 cm^{-1}. It takes 3.1×10^{10} fissions to generate one watt of power. The power generated in the reactor is most nearly

(A) 18 MW

(B) 31 MW

(C) 59 MW

(D) 140 MW

12. A nuclear reactor produces 500,000 kW thermal. The cylindrical core is 10 ft in diameter and 10 ft high. The core contains 100,000 lbm of uranium enriched to 2% U-235. The U-235 fission cross section is 547 barns. The average thermal neutron flux is most nearly

(A) 1.2×10^{13} n/cm^2·s

(B) 7.5×10^{13} n/cm^2·s

(C) 3.4×10^{14} n/cm^2·s

(D) 6.9×10^{14} n/cm^2·s

SOLUTIONS

1. The volume of water within the pressure vessel is

$$\begin{aligned} V_{\text{water}} &= (1-0.2)AL = (1-0.2)\left(\frac{\pi}{4}\right)D^2L \\ &= (0.8)\left(\frac{\pi}{4}\right)(5\text{ m})^2(12\text{ m}) \\ &= 188.5\text{ m}^3 \end{aligned}$$

The volume of water above the fuel rods is

$$\begin{aligned} V_{\text{above}} &= AL = (1-0.2)\left(\frac{\pi}{4}\right)D^2L \\ &= \left(\frac{\pi}{4}\right)(5\text{ m})^2(6.1\text{ m}) \\ &= 119.8\text{ m}^3 \end{aligned}$$

At 95°C, water has a density of

$$\begin{aligned} \rho_{\text{sat},95°\text{C}} &= \frac{1}{\upsilon_{f,95°\text{C}}} = \frac{\left(100\ \frac{\text{cm}}{\text{m}}\right)^3}{\left(1.0396\ \frac{\text{cm}^3}{\text{g}}\right)\left(1000\ \frac{\text{g}}{\text{kg}}\right)} \\ &= 961.9\text{ kg/m}^3 \end{aligned}$$

The mass of all of the water in the tank is

$$\begin{aligned} m_{\text{water}} &= \rho_{\text{water}}V_{\text{water}} = \left(961.9\ \frac{\text{kg}}{\text{m}^3}\right)(188.5\text{ m}^3) \\ &= 1.813\times 10^5\text{ kg} \end{aligned}$$

The mass of the water above the fuel rods is

$$\begin{aligned} m_{\text{above}} &= \rho_{\text{water}}V_{\text{above}} = \left(961.9\ \frac{\text{kg}}{\text{m}^3}\right)(119.8\text{ m}^3) \\ &= 1.152\times 10^5\text{ kg} \end{aligned}$$

Near saturation, the specific heat of water is

$$\begin{aligned} c_p &\approx \frac{h_2-h_1}{T_2-T_1} = \frac{419.17\ \frac{\text{kJ}}{\text{kg}} - 398.09\ \frac{\text{kJ}}{\text{kg}}}{100°\text{C}-95°\text{C}} \\ &= 4.216\text{ kJ/kg}\cdot°\text{C} \end{aligned}$$

The time required to raise the temperature of all of the water from 95°C to 100°C is

$$\begin{aligned} t_1 &= \frac{m_{\text{water}}c_p\Delta T}{\dot{q}} \\ &= \frac{(1.813\times 10^5\text{ kg})\left(4.216\ \frac{\text{kJ}}{\text{kg}\cdot°\text{C}}\right)\times\left(1000\ \frac{\text{J}}{\text{kJ}}\right)(100°\text{C}-95°\text{C})}{(4000\text{ MW})\left(10^6\ \frac{\text{W}}{\text{MW}}\right)} \\ &= 0.96\text{ s} \end{aligned}$$

From App. 23.O, the latent heat of vaporization, h_{fg}, at one atmosphere is 2256.4 kJ/kg. The energy required to vaporize a mass of water equal to the water above the fuel rods is

$$\begin{aligned} q &= m_{\text{above}}h_{fg} = (1.152\times 10^5\text{ kg})\left(2256.4\ \frac{\text{kJ}}{\text{kg}}\right) \\ &= 2.599\times 10^8\text{ kJ} \end{aligned}$$

The time (starting from saturation) to vaporize the water above the fuel rods is

$$t_2 = \frac{q}{\dot{q}} = \frac{2.599\times 10^8\text{ kJ}}{(4000\text{ MW})\left(1000\ \frac{\text{kW}}{\text{MW}}\right)} = 64.98\text{ s}$$

The total time since loss of pressurization is

$$\begin{aligned} t_{\text{total}} &= 12\text{ h} + t_1 + t_2 = 12\text{ h} + \frac{0.96\text{ s} + 64.98\text{ s}}{\left(60\ \frac{\text{min}}{\text{h}}\right)\left(60\ \frac{\text{s}}{\text{min}}\right)} \\ &= \boxed{12.02\text{ h}\quad(12\text{ h})} \end{aligned}$$

The answer is (B).

2. Use carbon-based atomic masses. The mass increase is

$$\begin{aligned} \Delta m &= m_{\text{H}} + m_{\text{n}} - m_{\text{D}} \\ &= 1.007825\text{ amu} + 1.008665\text{ amu} - 2.01410\text{ amu} \\ &= 0.00239\text{ amu} \end{aligned}$$

Express the mass increase as energy. Convert to electron volts.

$$\begin{aligned} \Delta E &= (0.00239\text{ amu})\left(931.46\ \frac{\text{MeV}}{\text{amu}}\right) \\ &= 2.226\text{ MeV} \end{aligned}$$

Thus, 2.75 MeV − 2.226 MeV = 0.524 MeV are shared by the neutron and the hydrogen atom. Assuming the energy is initially manifested as kinetic energy, and since the neutron and hydrogen atom have roughly the same mass, the neutron receives half.

$$\begin{aligned} \Delta E_n &= \frac{0.524\text{ MeV}}{2} \\ &= \boxed{0.262\text{ MeV}\quad(0.26\text{ MeV})} \end{aligned}$$

The answer is (B).

Power Cycles

3. The decay constant is

$$\lambda = \frac{0.693}{6.47 \text{ days}} = 0.1071 \text{ day}^{-1}$$

The activity can be predicted by

$$\frac{A}{A_0} = e^{-\lambda t}$$

$$0.05 = e^{(-0.1071 \text{ 1/day})t}$$

Take the natural logarithm of both sides to determine t.

$$\boxed{t = 27.97 \text{ days} \quad (28 \text{ days})}$$

The answer is (A).

4. The linear attenuation coefficient and build-up factor for 2 MeV gamma radiation are, respectively, approximately

$$\mu_l = 0.5182 \text{ cm}^{-1};\ B = 2.78$$

(If necessary, an initial estimate, say 10 cm, of shield thickness can be used with a table of mass attenuation coefficients. In that case, the solution will be iterative.)

$$\frac{\phi}{\phi_0} = Be^{-\mu_l x}$$

$$0.01 = 2.78e^{-(0.5182 \text{ 1/cm})x}$$

Solve for x by taking the natural logarithm of both sides.

$$\boxed{x = 10.86 \text{ cm} \quad (11 \text{ cm})}$$

The answer is (C).

5. (a) The uncollided exit flux is

$$\begin{aligned}\phi = \phi_0 e^{-\mu x} &= \left(10^6 \ \frac{1}{\text{cm}^2\cdot\text{s}}\right) e^{-(0.5182 \text{ 1/cm})(10 \text{ cm})} \\ &= 5.62 \times 10^3 \text{ 1/cm}^2\cdot\text{s}\end{aligned}$$

The build-up flux is

$$\begin{aligned}\phi_B = BI &= (2.78)\left(5.62 \times 10^3 \ \frac{1}{\text{cm}^2\cdot\text{s}}\right) \\ &= \boxed{1.56 \times 10^4 \text{ 1/cm}^2\cdot\text{s} \quad (1.6 \times 10^4 \text{ 1/cm}^2\cdot\text{s})}\end{aligned}$$

The answer is (D).

(b) There are many approximate correlations between dose rate and intensity in air. One of them is

$$\begin{aligned}D' &= \left(0.0659 \ \frac{\text{g}\cdot\text{s}\cdot\text{mR}}{\text{MeV}\cdot\text{hr}}\right) E_0 \phi_B \mu_{a,\text{air}} \\ &= (0.0659E)(2 \text{ MeV})\left(1.56 \times 10^4 \ \frac{1}{\text{cm}^2\cdot\text{s}}\right) \\ &\quad \times \left(0.0238 \ \frac{\text{cm}^2}{\text{g}}\right) \\ &= \boxed{48.9 \text{ mR/hr} \quad (50 \text{ mR/hr})}\end{aligned}$$

The answer is (D).

6. The absorption cross section at 50°C is

$$\begin{aligned}\overline{\sigma} &= \frac{\sigma_{a,\text{thermal}}}{1.128} \sqrt{\frac{20°\text{C} + 273°}{T_{°\text{C}} + 273°}} \\ &= \frac{98 \text{ b}}{1.128} \sqrt{\frac{20°\text{C} + 273°}{50°\text{C} + 273°}} \\ &= 82.75 \text{ b}\end{aligned}$$

The number of atoms in the target foil is

$$\begin{aligned}N &= \frac{mN_o}{A} = \frac{\rho V N_o}{A} \\ &= \frac{\left(19.32 \ \frac{\text{g}}{\text{cm}^3}\right)(0.4909 \text{ cm}^3) \times \left(6.022 \times 10^{23} \ \frac{\text{atoms}}{\text{mole}}\right)}{197 \ \frac{\text{g}}{\text{mole}}} \\ &= 2.9 \times 10^{22} \text{ atoms}\end{aligned}$$

The decay constant for the activated gold is

$$\lambda = \frac{0.693}{2.7 \text{ days}} = 0.2567 \text{ day}^{-1}$$

The removal activity is

$$\begin{aligned}A_R &= J\overline{\sigma}_a N(1 - e^{-\lambda t}) \\ &= \frac{\left(10^8 \ \frac{1}{\text{cm}^2\cdot\text{s}}\right)(82.75 \text{ b})\left(1 \times 10^{-24} \ \frac{\text{cm}^2}{\text{b}}\right) \times (2.9 \times 10^{22} \text{ atoms})\left(1 - e^{(-0.2567 \text{ 1/day})(1 \text{ day})}\right)}{3.7 \times 10^{10} \ \frac{1}{\text{curie}\cdot\text{s}}} \\ &= \boxed{1.47 \times 10^{-3} \text{ curie} \quad (1.5 \times 10^{-3} \text{ curie})}\end{aligned}$$

The answer is (B).

7. For an isotropic point source, the diffused flux is

$$\phi = \frac{A_{0,\text{isotropic}} e^{-r/L}}{4\pi \overline{D} r} = \frac{\left(1\times 10^7\ \frac{1}{\text{s}}\right) e^{-20\ \text{cm}/2.85\ \text{cm}}}{4\pi(0.16\ \text{cm})(20\ \text{cm})} = \boxed{223\ 1/\text{cm}^2\cdot\text{s}\quad (220\ \text{n/cm}^2\cdot\text{s})}$$

The answer is (C).

8. The probability is

$$p\{f\} = \frac{\sigma_f}{\sigma_a} = \frac{4.18\ \text{b}}{7.68\ \text{b}} = \boxed{0.544\quad (54\%)}$$

The answer is (D).

9. The build-up flux in a spherical shield from an isotropic source is

$$\phi_B = \frac{BA_{0,\text{isotropic}} e^{-\mu_l r}}{4\pi r^2} = \frac{(14.93)\left(1\times 10^8\ \frac{1}{\text{s}}\right) e^{-(0.4683\ 1/\text{cm})r}}{4\pi r^2} = \frac{\left(1.19\times 10^8\ \frac{1}{\text{s}}\right) e^{-(0.4683\ 1/\text{cm})r}}{r^2}$$

The approximate dose (exposure) rate equation used is

$$D' = \left(0.0659\ \frac{\text{g}\cdot\text{s}\cdot\text{mR}}{\text{MeV}\cdot\text{hr}}\right) E_0 \phi_B \mu_{a,\text{air}}$$

$$1\ \frac{\text{mR}}{\text{hr}} = \frac{\left(0.0659\ \frac{\text{g}\cdot\text{s}\cdot\text{mR}}{\text{MeV}\cdot\text{hr}}\right)(1\ \text{MeV}) \times \left(1.19\times 10^8\ \frac{1}{\text{cm}^2\cdot\text{s}}\right) e^{-(0.4683\ 1/\text{cm})r} \times \left(0.0280\ \frac{\text{cm}^2}{\text{g}}\right)}{r^2}$$

By trial and error, $r = \boxed{14.8\ \text{cm}\ (15\ \text{cm}).}$

The answer is (B).

10. Calculate the average number of neutrons generated per fast neutron released.

$$\eta = G_{\text{Pu-239}} n_{\text{Pu-239}} \left(\frac{\sigma_{f,\text{Pu-239}}}{\sigma_{a,\text{Pu-239}}}\right) + G_{\text{U-238}} n_{\text{U-238}} \left(\frac{\sigma_{f,\text{U-238}}}{\sigma_{a,\text{U-238}}}\right) = (0.2)(2.95)\left(\frac{1.8\ \text{b}}{1.95\ \text{b}}\right) + (0.8)(2.45)\left(\frac{0.5\ \text{b}}{0.59\ \text{b}}\right) = 2.206$$

The conversion ratio is

$$\text{CR} = \eta\epsilon - 1 = (2.206)(1.05) - 1 = 1.316$$

The linear doubling time is

$$t_{d,\text{linear}} = \frac{m}{(\text{CR}-1)m'P} = \frac{(2000\ \text{kg})\left(1000\ \frac{\text{g}}{\text{kg}}\right)}{(1.316-1)\left(1.23\ \frac{\text{g}}{\text{MW}\cdot\text{day}}\right)(1\ \text{GW}) \times \left(1000\ \frac{\text{MW}}{\text{GW}}\right)} = 5146\ \text{days}$$

The exponential doubling time is

$$t_{d,\text{exponential}} = 0.693 t_{d,\text{linear}} = (0.693)(5146\ \text{days}) = \boxed{3566\ \text{days}\quad (3600\ \text{days})}$$

The answer is (C).

11. In a spherical reactor, the ratio of maximum to average neutron flux is 3.29.

Reactor power is predicted by

$$P = \frac{\overline{\phi}\sum_f V}{3.1\times 10^{10}\ \frac{1}{\text{W}\cdot\text{s}}} = \frac{\left(\frac{4.5\times 10^{15}\ \frac{1}{\text{cm}^2\cdot\text{s}}}{3.29}\right)\left(0.005\ \frac{1}{\text{cm}}\right) \times \left(\frac{4\pi}{3}\right)(40\ \text{cm})^3}{3.1\times 10^{10}\ \frac{1}{\text{W}\cdot\text{s}}} = \boxed{5.9\times 10^7\ \text{W}\quad (59\ \text{MW})}$$

The answer is (C).

Power Cycles

12. The mass of uranium is

$$\begin{aligned} m &= (100{,}000 \text{ lbm})\left(0.4536 \ \frac{\text{kg}}{\text{lbm}}\right)\left(1000 \ \frac{\text{g}}{\text{kg}}\right) \\ &= 4.54 \times 10^7 \text{ g} \end{aligned}$$

The number of fuel molecules per unit volume (cm^3) is

$$\begin{aligned} N_f &= \frac{G_f m N_o}{V(\text{MW})} \\ &= \frac{(0.02)(4.54 \times 10^7 \text{ g})\left(6.022 \times 10^{23} \ \frac{\text{atoms}}{\text{mol}}\right)}{V\left(235 \ \frac{\text{g}}{\text{mol}}\right)} \\ &= 2.33 \times 10^{27}/V \end{aligned}$$

From the thermal power equation,

$$\begin{aligned} \overline{\phi} &= \frac{\left(3.1 \times 10^{10} \ \frac{1}{\text{W·s}}\right) P_{\text{thermal}}}{\Sigma_f V} \\ &= \frac{\left(3.1 \times 10^{10} \ \frac{1}{\text{W·s}}\right) P_{\text{thermal}}}{\sigma_f N_f V} \\ &= \frac{\left(3.1 \times 10^{10} \ \frac{\text{fissions}}{\text{W·s}}\right)(5 \times 10^8 E)}{(547 \text{ b})\left(1 \times 10^{-24} \ \frac{\text{cm}^2}{\text{b}}\right)(2.33 \times 10^{27})} \\ &= \boxed{1.22 \times 10^{13} \text{ fissions/cm}^2\text{·s} \quad (1.2 \times 10^{13} \text{ n/cm}^2\text{·s})} \end{aligned}$$

The answer is (A).

Power Cycles

31 Advanced and Alternative Power-Generating Systems

PRACTICE PROBLEMS

1. An ocean thermal energy conversion plant draws 40°F (4.4°C) water from a depth of 1200 ft (360 m). The water temperature at the surface is 82°F (27.8°C). The maximum achievable thermal efficiency is most nearly

(A) 7.8%

(B) 11%

(C) 15%

(D) 21%

2. A bank of solar cells supplies power to a 95% efficient electric water heater. The area of the solar cell bank is 500 ft^2. The solar cell bank has a maximum theoretical efficiency of 31% and receives an incident solar flux of 150 W/m^2. The heater boils water to make saturated steam at 30 psig to supply a turbine. Feedwater enters the boiler as saturated liquid at 140°F. What is most nearly the maximum steam flow rate to the turbine?

(A) 6.6 lbm/hr

(B) 9.5 lbm/hr

(C) 11 lbm/hr

(D) 21 lbm/hr

3. A self-contained solar light consists of a 4 in^2 solar cell, a nickel-cadmium battery that is 70% efficient in both charging and discharging cycles, a photocell that detects ambient light and that acts as a switch, a low-voltage cutoff relay to protect the battery, and a light emitting diode (LED) that provides illumination. While operating, the LED draws 80 mA of current at 1.2 V. During an average day, the sunlight density varies with time, as shown. After an average day, the LED operates for eight hours before it is shut off by the low-voltage cutoff. Each charging cycle begins at the low-voltage cutoff. What is most nearly the efficiency of the solar light?

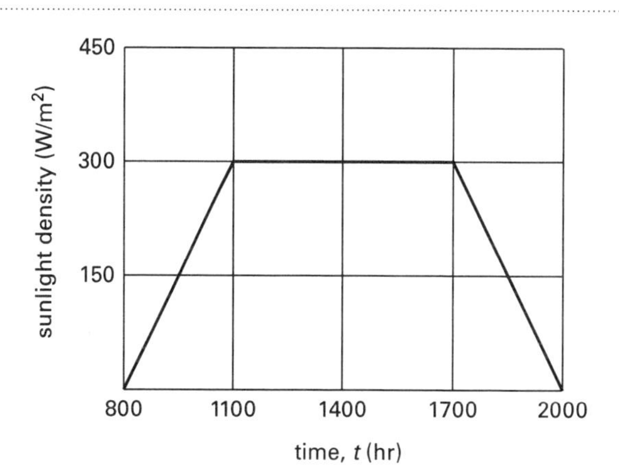

(A) 8.3%

(B) 9.2%

(C) 11%

(D) 16%

Wind Turbines

4. A 20 mi/hr wind at standard temperature and pressure passes through a horizontal wind turbine. The wind is perpendicular to the plane of the blades. The turbine has a blade diameter of 30 ft. Air temperature and pressure are 70°F and 1 atm. What is most nearly the maximum power that the turbine can generate?

(A) 12 hp

(B) 38 hp

(C) 150 hp

(D) 1200 hp

5. A wind turbine and generator powers an off-grid load. The turbine has a power coefficient of 0.45 and a rotor diameter of 35 ft. The load is three phase at 480 V and draws 20 A with a power factor of 0.90. The air is at standard conditions. The wind velocity is 20 mi/hr. Air temperature and pressure are 70°F and 1 atm. What is most nearly the efficiency of the turbine-generator system?

(A) 73%

(B) 81%

(C) 85%

(D) 87%

6. In a 20 mi/hr wind, a turbine produces 200 kW. The power coefficient, turbine, and generator efficiencies are constant for all wind speeds. The turbine is required to supply 50 kW to the electrical grid and 250 kW to a local load. What is most nearly the wind speed required for the turbine to supply the load and the electric grid?

(A) 23 mi/hr

(B) 25 mi/hr

(C) 29 mi/hr

(D) 32 mi/hr

7. A wind turbine located has 50 ft diameter rotor blades and generates alternating current. The turbine uses a rectifier to convert the alternating current to direct current, which is then converted to 480 V, three-phase alternating current in an inverter. In standard conditions and with a 30 mi/hr wind, the wind turbine supplies a load that draws 115 A at a power factor of 0.88. The generator efficiency is 90%, and the rectifier and inverter efficiencies are each 92%. Air temperature and pressure are 70°F and 1 atm. What is most nearly the wind turbine's power coefficient?

(A) 0.33

(B) 0.42

(C) 0.48

(D) 0.57

8. A wind turbine blade has a radius of 90 ft. The mass per unit length of the blade is 20 lbm/ft for the first 15 ft from the center of rotation. As the blade tapers out to the tip, the mass per unit length varies linearly from 20 lbm/ft to 2 lbm/ft. The blade turns at 250 rev/min. What is most nearly the tension in the blade root caused from the centrifugal force at the hub?

(A) 3.7×10^5 lbf

(B) 7.9×10^5 lbf

(C) 5.2×10^6 lbf

(D) 1.6×10^7 lbf

9. The teetering rotor of a wind turbine is 90 ft in diameter and turns at 200 rev/min. The average mass per unit length of each of the two blades is 20 lbm/ft. The center of gravity of each blade is located 19 ft from the rotational axis. At one moment in time, the plane of rotation is displaced 9° from the plane normal to the axis of rotation. What is most nearly the magnitude of the tensile centrifugal force at the axis of rotation due to both halves of the blade?

(A) 1.5×10^4 lbf

(B) 3.6×10^4 lbf

(C) 2.3×10^5 lbf

(D) 2.4×10^5 lbf

10. Wind turbine A is subject to 20 mi/hr winds and has a blade diameter of 60 ft. Wind turbine B is subject to 30 mi/hr winds and has a blade diameter of 40 ft. The ratio of the available power from turbine A to the available power from turbine B is most nearly

(A) 0.50

(B) 0.67

(C) 1.5

(D) 2.0

11. A wind turbine rotor, with two tapered 20 ft long blades, is locked in position for maintenance. Each blade is modeled as a flat projection 3 ft wide for the first 11 ft from the hub, followed by 9 ft that tapers uniformly from a flat projected width of 3 ft to a width of 1.5 ft. The blades are vertical in the locked position, with the lower blade in the lee of the tower. Tower interference reduces the effective wind speed on the lower blade by 25%. The average drag coefficient is 0.4. The tower extends 40 ft vertically above its foundation. Air temperature and pressure are 70°F and 1 atm. Neglecting drag on the tower, what is most nearly the moment at the tower base when there is a 20 mi/hr wind in standard conditions?

(A) 370 ft-lbf

(B) 760 ft-lbf

(C) 1300 ft-lbf

(D) 1400 ft-lbf

12. A wind turbine rotor with three equally spaced, 20 ft long blades is locked in position for maintenance. One blade is vertical, extending above the tower. Each rotor blade is modeled as a flat projecting triangle and is 20 ft long with a 3 ft base. The average drag coefficient for the blades is 0.3. The tower does not occlude any of the blades. The tower height is 120 ft. Air temperature and pressure are 70°F and 1 atm. Neglecting drag on the tower, what is most nearly the moment at the base of the tower in a 30 mi/hr wind and standard conditions?

(A) 5500 ft-lbf

(B) 7300 ft-lbf

(C) 9300 ft-lbf

(D) 11,000 ft-lbf

Batteries

13. A 1.2 V, 600 milliamp-hour battery is fully uncharged. A matched charger consists of a transformer-rectifier unit that has a transformation efficiency of 65%. The transformer primary uses 120 V, single-phase AC power. During the charge cycle, the primary draws 1 mA with a power factor of 0.8. The charger secondary delivers 1.5 V to the battery. The battery takes 18 hours to fully charge from the fully uncharged condition. The electrical transformation efficiency within the battery during the charge cycle is most nearly

(A) 42%

(B) 64%

(C) 88%

(D) 98%

14. Two identical 12 V batteries are connected in a series-parallel arrangement as shown to supply a resistive 250 W load. The current through the load is 10.9 A.

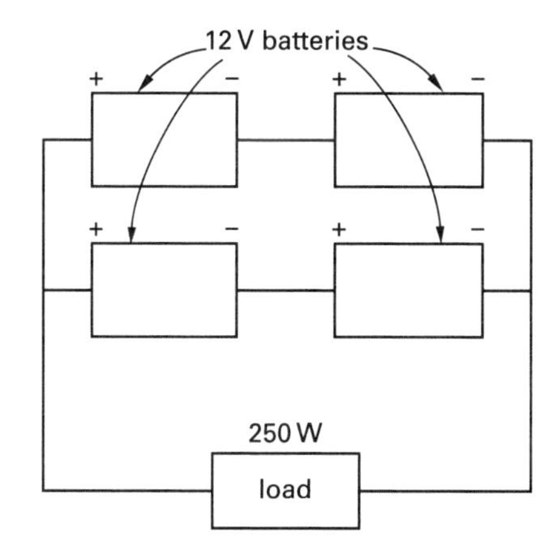

The internal resistance of each battery is most nearly

(A) 0.01 Ω

(B) 0.02 Ω

(C) 0.1 Ω

(D) 0.2 Ω

15. A 150 W computer is supplied with 120 V AC power through an uninterruptible power supply (UPS). The UPS serves as an emergency power supply to the computer, and in the event of power loss, the UPS will power the computer for 30 minutes. The UPS consists of a battery charger to charge the internal 12 V battery and an inverter/regulator to supply 120 V AC power to the load. The inverter/regulator is 92% efficient. What is most nearly the amp-hour rating of the UPS battery?

(A) 5700 mA-hr

(B) 6800 mA-hr

(C) 11,000 mA-hr

(D) 12,000 mA-hr

Geothermal Power

16. A geothermal well supplies steam at 200 psia and 95% quality to two power generating plants. One plant has a non-condensing, isentropic turbine that exhausts to the atmosphere. The backpressure on this turbine is 2.0 psig. The other plant has an isentropic turbine that exhausts to a condenser at 1.0 psia. The steam flow rate is the same to both turbines, and the turbines have the same efficiency. The ratio of the power output of the condensing turbine to the non-condensing turbine is most nearly

(A) 1.1

(B) 1.5

(C) 1.8

(D) 2.1

Fuel Cells

17. A proposed residential fuel cell system runs on bottled natural gas. The system consists of a fuel reformer, a power stack, and a power conditioner. The overall efficiency of the system will be 45%. The average load supplied by the fuel cell system is 1.7 kW, but the proposed system can supply a maximum of 2.3 kW. The cost of natural gas is \$0.40/therm, and by comparison, the cost of electricity from the electrical grid is \$0.08/kW-hr. The fuel cell system will cost \$3500, including installation. What is the pay-back period for the system?

(A) 1.6 yr

(B) 2.1 yr

(C) 3.5 yr

(D) 4.7 yr

18. A molten carbonate fuel cell system supplies electricity to a manufacturing plant. Waste heat from the system is used to generate steam for use in a manufacturing process. The fuel cell system consists of a fuel cell stack with internal reformer, an inverter to convert the DC power generated to AC power, and a heat exchanger to heat water to make steam. Methane gas fuels the system. The efficiency of the inverter is 90%, and the efficiency of the heat exchanger is 75%. The lower heating value (LHV) of methane is 913 Btu/ft^3.

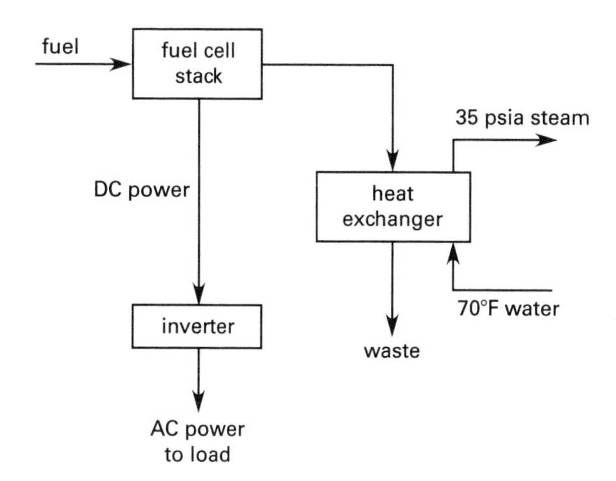

Power Cycles

At full power, the system consumes 1200 ft^3/hr of methane gas at 60°F and atmospheric pressure, has an output of 150 kW, and heats 100 lbm/hr of 70°F water to generate 35 psia saturated steam. Feedwater enters the steam generator at 70°F. What is most nearly the fuel cell stack's efficiency?

(A) 48%

(B) 55%

(C) 65%

(D) 87%

19. An experimental solid polymer fuel cell runs on hydrogen and powers an electric motor in a car. The 4000 lbm car has a maximum speed of 55 mi/hr. At that speed, the fuel cell consumes 0.2 ft^3/sec of hydrogen at 60°F and atmospheric pressure. The hydrogen has a lower heating value (LHV) of 275 Btu/ft^3. The rolling coefficient of friction is 0.01. The drag force on the car is 136 lbf. The motor and drive train are 80% efficient. The efficiency of the fuel cell is most nearly

(A) 23%

(B) 28%

(C) 33%

(D) 41%

SOLUTIONS

1. *Customary U.S. Solution*

The maximum achievable thermal efficiency is achieved with the Carnot cycle and is given by Eq. 27.8.

$$\eta_{th} = \frac{T_{high} - T_{low}}{T_{high}}$$

$$T_{high} = 82°F + 460° = 542°R$$

$$T_{low} = 40°F + 460° = 500°R$$

$$\eta_{th} = \frac{542°R - 500°R}{542°R} = \boxed{0.0775 \quad (7.8\%)}$$

The answer is (A).

SI Solution

The maximum achievable thermal efficiency is achieved with the Carnot cycle and is given by Eq. 27.8.

$$\eta_{th} = \frac{T_{high} - T_{low}}{T_{high}}$$

$$T_{high} = 27.8°C + 273° = 300.8K$$

$$T_{low} = 4.4°C + 273° = 277.4K$$

$$\eta_{th} = \frac{300.8K - 277.4K}{300.8K} = \boxed{0.0778 \quad (7.8\%)}$$

The answer is (A).

2. The total power from the bank is

$$\begin{aligned} P_t &= \eta_{cell} SA \\ &= (0.31)\left(150\ \frac{W}{m^2}\right)(500\ ft^2) \\ &\quad \times \left(0.3048\ \frac{m}{ft}\right)^2 \left(3.412\ \frac{Btu}{W\text{-}hr}\right) \\ &= 7370\ Btu/hr \end{aligned}$$

The power available to boil water, P_{water}, is

$$\begin{aligned} P_{water} &= P_t\eta_{heater} = \left(7370\ \frac{Btu}{hr}\right)(0.95) \\ &= 7004\ Btu/hr \end{aligned}$$

From App. 23.B, the saturation temperature for water at 30 psig (45 psia) is 274.41°F with an enthalpy of 243.5 Btu/lbm. The total energy, h_t, required to heat 1 lbm of water from 140°F to 274.41°F and vaporize it is

$$\begin{aligned} h_t &= h_{f,45\ psia} - h_{f,140°F} + h_{fg,45\ psia} \\ &= 243.5\ \frac{Btu}{lbm} - 108.0\ \frac{Btu}{lbm} + 928.8\ \frac{Btu}{lbm} \\ &= 1064.3\ Btu/lbm \end{aligned}$$

The mass flow rate of steam, $\dot{m}$, to the turbine is

$$\dot{m} = \frac{P_{\text{water}}}{h_t} = \frac{7004 \ \frac{\text{Btu}}{\text{hr}}}{1064.3 \ \frac{\text{Btu}}{\text{lbm}}} = \boxed{6.58 \text{ lbm/hr} \quad (6.6 \text{ lbm/hr})}$$

The answer is (A).

3. The energy supplied by the battery and used by the LED is

$$E_{\text{used}} = \frac{VIt}{\eta_{\text{battery}}} = \frac{(1.2 \text{ V})(0.08 \text{ A})(8 \text{ h})}{0.70} = 1.097 \text{ W·h}$$

The total energy captured is

$$E_{\text{captured}} = A\dot{E} = A_{\text{cell}}\left(\frac{b_1h_1}{2} + b_2h_2 + \frac{b_3h_3}{2}\right)$$

$$= \frac{(4 \text{ in}^2)\left(\frac{(3 \text{ h})\left(300 \ \frac{\text{W}}{\text{m}^2}\right)}{2} + (6 \text{ h})\left(300 \ \frac{\text{W}}{\text{m}^2}\right) + \frac{(3 \text{ h})\left(300 \ \frac{\text{W}}{\text{m}^2}\right)}{2}\right)}{\left(12 \ \frac{\text{in}}{\text{ft}}\right)^2\left(3.281 \ \frac{\text{ft}}{\text{m}}\right)^2}$$

$$= 6.97 \text{ W·h}$$

The efficiency of the solar cell is

$$\eta_{\text{cell}} = \frac{E_{\text{used}}}{E_{\text{captured}}} = \frac{1.097 \text{ W·h}}{6.97 \text{ W·h}} = \boxed{0.157 \quad (16\%)}$$

The energy stored by the battery is irrelevant.

$$E_{\text{stored}} = \eta_{\text{battery}}E_{\text{captured}} = (0.70)(1.097 \text{ W·h}) = 0.768 \text{ W·h}$$

The answer is (D).

4. Air at 70°F and 1 atm has a density, ρ, of approximately 0.075 lbm/ft^3. From Eq. 31.6(b), the maximum power density, P_{max}, of the airstream is

$$P_{\text{max}} = \frac{\pi r_{\text{rotor}}^2 \rho \text{v}^3}{2g_c}$$

$$= \frac{\pi\left(\frac{30 \text{ ft}}{2}\right)^2\left(0.075 \ \frac{\text{lbm}}{\text{ft}^3}\right)\left(20 \ \frac{\text{mi}}{\text{hr}}\right)^3\left(5280 \ \frac{\text{ft}}{\text{mi}}\right)^3}{(2)\left(32.2 \ \frac{\text{lbm-ft}}{\text{lbf-sec}^2}\right)\left(3600 \ \frac{\text{sec}}{\text{hr}}\right)^3\left(550 \ \frac{\text{ft-lbf}}{\text{hp-sec}}\right)}$$

$$= \boxed{37.8 \text{ hp} \quad (38 \text{ hp})}$$

The answer is (B).

5. Air at 70°F and 1 atm has a density of approximately 0.075 lbm/ft^3. From Eq. 31.6(b), the total ideal power produced by the turbine is

$$P_{\text{turbine}} = \frac{A\rho\text{v}^3}{2g_c} = \frac{\pi\left(\frac{d}{2}\right)^2\rho\text{v}^3}{2g_c}$$

$$= \frac{\pi\left(\frac{35 \text{ ft}}{2}\right)^2\left(0.075 \ \frac{\text{lbm}}{\text{ft}^3}\right)\left(20 \ \frac{\text{mi}}{\text{hr}}\right)^3 \times \left(5280 \ \frac{\text{ft}}{\text{mi}}\right)^3\left(746 \ \frac{\text{W}}{\text{hp}}\right)}{(2)\left(32.2 \ \frac{\text{lbm-ft}}{\text{lbf-sec}^2}\right)\left(3600 \ \frac{\text{sec}}{\text{hr}}\right)^3 \times \left(550 \ \frac{\text{ft-lbf}}{\text{hp-sec}}\right)}$$

$$= 38{,}358 \text{ W}$$

The actual power generated depends on the power coefficient, C_P. From Eq. 31.8,

$$P_{\text{actual}} = C_P P_{\text{ideal}} = (0.45)(38{,}358 \text{ W}) = 1.726 \times 10^4 \text{ W}$$

Equation 72.67 gives the total power, P_t, for a three-phase system.

$$P_t = \sqrt{3}VI\cos\phi = \sqrt{3}VI(\text{pf}) = \sqrt{3}(480 \text{ V})(20 \text{ A})(0.90) = 1.500 \times 10^4 \text{ W}$$

The turbine-generator system efficiency, η, is

$$\eta = \frac{P_t}{P_{\text{turbine}}} = \frac{1.500 \times 10^4 \text{ W}}{1.726 \times 10^4 \text{ W}} = \boxed{0.87 \quad (87\%)}$$

The answer is (D).

6. From Eq. 31.6(b), the power generated is

$$P = \frac{A\rho\text{v}^3}{2g_c}$$

Since the area, density, and all efficiencies are unchanged, the required wind velocity is

$$\text{v}_2 = \text{v}_1\sqrt[3]{\frac{P_2}{P_1}} = \left(20 \ \frac{\text{mi}}{\text{hr}}\right)\sqrt[3]{\frac{300 \text{ kW}}{200 \text{ kW}}} = \boxed{22.89 \text{ mi/hr} \quad (23 \text{ mi/hr})}$$

The answer is (A).

Power Cycles

7. From Eq. 72.57 and Eq. 72.68, the electrical power delivered to the load is

$$\begin{aligned} P_{\text{load}} &= \sqrt{3}VI(\text{pf}) = \sqrt{3}(480\ \text{V})(115\ \text{A})(0.88) \\ &= 8.41 \times 10^4\ \text{W} \end{aligned}$$

The power developed by the wind turbine is

$$\begin{aligned} P_{\text{developed}} &= \frac{P_{\text{load}}}{\eta_{\text{generator}}\,\eta_{\text{rectifier}}\,\eta_{\text{inverter}}} \\ &= \frac{8.41 \times 10^4\ \text{W}}{(0.90)(0.92)(0.92)} \\ &= 1.10 \times 10^5\ \text{W} \end{aligned}$$

From Eq. 31.6(b), the ideal wind turbine power, P_{ideal}, is

$$P_{\text{ideal}} = \frac{\pi r_{\text{rotor}}^2 \rho \text{v}^3}{2g_c} = \frac{\pi\left(\frac{D}{2}\right)^2 \rho \text{v}^3}{2g_c}$$

Rearrange Eq. 31.8 to find the turbine power coefficient, C_P. Air at 70°F and 1 atm has a density of approximately 0.075 lbm/ft^3.

$$\begin{aligned} C_P &= \frac{P_{\text{developed}}}{P_{\text{ideal}}} = \frac{8P_{\text{developed}}\,g_c}{\pi D^2 \rho \text{v}^3} \\ &= \frac{(8)(1.1 \times 10^5\ \text{W})\left(32.2\ \frac{\text{lbm-ft}}{\text{lbf-sec}^2}\right) \times \left(3600\ \frac{\text{sec}}{\text{hr}}\right)^3}{\pi(50\ \text{ft})^2\left(0.075\ \frac{\text{lbm}}{\text{ft}^3}\right)\left(30\ \frac{\text{mi}}{\text{hr}}\right)^3 \times \left(5280\ \frac{\text{ft}}{\text{mi}}\right)^3\left(1.356\ \frac{\text{W}}{\frac{\text{ft-lbf}}{\text{sec}}}\right)} \\ &= \boxed{0.416\quad(0.42)} \end{aligned}$$

The answer is (B).

8. The mass of the root of the blade is

$$m_{\text{root}} = (15\ \text{ft})\left(20\ \frac{\text{lbm}}{\text{ft}}\right) = 300\ \text{lbm}$$

Measured from the center of rotation, the centroid of the root is located at

$$\bar{x}_{\text{root}} = \frac{15\ \text{ft}}{2} = 7.5\ \text{ft}$$

The mass of the tapered part of the blade is

$$\begin{aligned} m_{\text{tapered}} &= (90\ \text{ft} - 15\ \text{ft})\left(\frac{20\ \frac{\text{lbm}}{\text{ft}} + 2\ \frac{\text{lbm}}{\text{ft}}}{2}\right) \\ &= 825\ \text{lbm} \end{aligned}$$

The mass per length of the tapered part of the blade is distributed as a trapezoid, which can be subdivided into triangular and rectangular shapes. Measured from the center of rotation, the centroid of the tapered part of the blade is located at

$$\begin{aligned} \bar{x}_{\text{tapered}} &= 15\ \text{ft} + \frac{\left(\left(\tfrac{1}{3}\right)(75\ \text{ft})\right) \times \left(\left(\tfrac{1}{2}\right)(75\ \text{ft}) \times \left(20\ \frac{\text{lbm}}{\text{ft}} - 2\ \frac{\text{lbm}}{\text{ft}}\right)\right) + \left(\left(\tfrac{1}{2}\right)(75\ \text{ft})\right)(75\ \text{ft})\left(2\ \frac{\text{lbm}}{\text{ft}}\right)}{825\ \text{lbm}} \\ &= 42.27\ \text{ft} \end{aligned}$$

The total mass of the blade is

$$m_t = m_{\text{root}} + m_{\text{tapered}} = 300\ \text{lbm} + 825\ \text{lbm} = 1125\ \text{lbm}$$

The center of gravity, referenced to the root, of the blade is located at

$$\begin{aligned} \bar{x} &= \frac{\sum \bar{x}_i m_i}{\sum m_i} \\ &= \frac{(7.5\ \text{ft})(300\ \text{lbm}) + (42.27\ \text{ft})(825\ \text{lbm})}{1125\ \text{lbm}} \\ &= 33.0\ \text{ft} \end{aligned}$$

The centrifugal force is

$$\begin{aligned} F_c &= \frac{mr\omega^2}{g_c} \\ &= \frac{(1125\ \text{lbm})(33.0\ \text{ft})\left(\frac{\left(250\ \frac{\text{rev}}{\text{min}}\right)\left(2\pi\ \frac{\text{rad}}{\text{rev}}\right)}{60\ \frac{\text{sec}}{\text{min}}}\right)^2}{32.2\ \frac{\text{lbm-ft}}{\text{lbf-sec}^2}} \\ &= \boxed{7.9 \times 10^5\ \text{lbf}} \end{aligned}$$

(Both blades experience this centrifugal force. However, the tensile force at the hub is not doubled.)

The answer is (B).

Power Cycles

9. The mass of each blade is

$$m_{\text{blade}} = \left(\frac{90\text{ ft}}{2}\right)\left(20\ \frac{\text{lbm}}{\text{ft}}\right) = 900\text{ lbm}$$

Since the blade is teetering, the radius is foreshortened.

$$\begin{aligned}F_c &= \frac{m(r\cos\theta)\omega^2}{g_c} \\ &= \frac{(900\text{ lbm})(19\text{ ft})(\cos 9^\circ)\left(\dfrac{\left(200\ \dfrac{\text{rev}}{\text{min}}\right)\left(2\pi\ \dfrac{\text{rad}}{\text{rev}}\right)}{60\ \dfrac{\text{sec}}{\text{min}}}\right)^2}{32.2\ \dfrac{\text{lbm-ft}}{\text{lbf-sec}^2}} \\ &= \boxed{2.30\times 10^5\text{ lbf}\quad (2.3\times 10^5\text{ lbf})}\end{aligned}$$

(Both blades experience this centrifugal force. However, the tensile force at the hub is not doubled.)

The answer is (C).

10. From Eq. 31.6(b), the available power is

$$P = \frac{\pi r_{\text{rotor}}^2 \rho \text{v}^3}{2g_c} = \frac{\pi\left(\frac{D}{2}\right)^2}{2g_c} = \frac{\pi D^2 \rho \text{v}^3}{8g_c}$$

With all of the other factors being equal, the ratio of the power available from turbine A and turbine B is

$$\begin{aligned}\frac{P_{\text{A}}}{P_{\text{B}}} &= \frac{D_{\text{A}}^2\text{v}_{\text{A}}^3}{D_{\text{B}}^2\text{v}_{\text{B}}^3} = \frac{(60\text{ ft})^2\left(20\ \frac{\text{mi}}{\text{hr}}\right)^3}{(40\text{ ft})^2\left(30\ \frac{\text{mi}}{\text{hr}}\right)^3} \\ &= \boxed{0.67}\end{aligned}$$

The answer is (B).

11. Use Eq. 17.218(b) to find the force on each blade. Air at 70°F and 1 atm has a density of approximately 0.075 lbm/ft^3.

The force on the untapered section of the top blade is

$$\begin{aligned}F_{\text{top,untapered}} &= \frac{C_D A \rho \text{v}^2}{2g_c} \\ &= \frac{(0.4)(11\text{ ft})(3\text{ ft})\left(0.075\ \frac{\text{lbm}}{\text{ft}^3}\right)\times\left(20\ \frac{\text{mi}}{\text{hr}}\right)^2\left(5280\ \frac{\text{ft}}{\text{mi}}\right)^2}{(2)\left(32.2\ \frac{\text{lbm-ft}}{\text{lbf-sec}^2}\right)\left(3600\ \frac{\text{sec}}{\text{hr}}\right)^2} \\ &= 13.2\text{ lbf}\end{aligned}$$

The resultant of this force is located 40 ft + (1/2)(11 ft) = 45.5 ft above the tower foundation.

The area of the tapered top blade section is

$$\begin{aligned}A_{\text{tapered}} &= \frac{(b_1+b_2)h}{2} = \frac{(3\text{ ft}+1.5\text{ ft})(9\text{ ft})}{2} \\ &= 20.25\text{ ft}^2\end{aligned}$$

The force on the tapered top blade section is

$$\begin{aligned}F_{\text{top,tapered}} &= \frac{C_D A \rho \text{v}^2}{2g_c} \\ &= \frac{(0.4)(20.25\text{ ft}^2)\left(0.075\ \frac{\text{lbm}}{\text{ft}^3}\right)\times\left(20\ \frac{\text{mi}}{\text{hr}}\right)^2\left(5280\ \frac{\text{ft}}{\text{mi}}\right)^2}{(2)\left(32.2\ \frac{\text{lbm-ft}}{\text{lbf-sec}^2}\right)\left(3600\ \frac{\text{sec}}{\text{hr}}\right)^2} \\ &= 8.1\text{ lbf}\end{aligned}$$

The centroid of the tapered section, measured from the 3 ft wide end, is located at

$$\frac{(9\text{ ft})(1.5\text{ ft})\left(\frac{9\text{ ft}}{2}\right)+\left(\frac{1}{2}\right)(1.5\text{ ft})(9\text{ ft})\left(\frac{9\text{ ft}}{3}\right)}{20.25\text{ ft}^2} = 4\text{ ft}$$

The resultant of this force is located 40 ft + 11 ft + 4 ft = 55 ft above the tower foundation.

The wind speed on the bottom blade sections is 100% − 25% = 75% of the top blade. Since velocity is squared in the drag force equation, the force on the untapered part of the bottom blade is

$$F_{\text{bottom,untapered}} = (0.75)^2(13.2\text{ lbf}) = 7.4\text{ lbf}$$

The resultant of this force is located 40 ft + (1/2)(11 ft) = 34.5 ft above the tower foundation.

The force on the tapered part of the bottom blade is

$$F_{\text{bottom,tapered}} = (0.75)^2(8.1\text{ lbf}) = 4.6\text{ lbf}$$

The resultant of this force is located 40 ft − 11 ft − 4 ft = 25 ft above the tower foundation.

The moment at the base of the tower is

$$\begin{aligned}M &= \sum Fr \\ &= (13.2\text{ lbf})(45.5\text{ ft}) + (8.1\text{ lbf})(55\text{ ft}) \\ &\quad + (7.4\text{ lbf})(34.5\text{ ft}) + (4.6\text{ lbf})(25\text{ ft}) \\ &= \boxed{1416\text{ ft-lbf}\quad (1400\text{ ft-lbf})}\end{aligned}$$

The answer is (D).

12. Use Eq. 17.218(b) to find the force on the vertical (top) blade. At 70°F and 1 atm, the density of air is aproximately 0.075 lbm/ft^3.

$$F_{\text{top}} = \frac{C_D A \rho v^2}{2g_c}$$

$$= \frac{(0.3)\left(\frac{(20 \text{ ft})(3 \text{ ft})}{2}\right)\left(0.075 \ \frac{\text{lbm}}{\text{ft}^3}\right) \times \left(30 \ \frac{\text{mi}}{\text{hr}}\right)^2\left(5280 \ \frac{\text{ft}}{\text{mi}}\right)^2}{(2)\left(32.2 \ \frac{\text{lbm-ft}}{\text{lbf-sec}^2}\right)\left(3600 \ \frac{\text{sec}}{\text{hr}}\right)^2}$$

$$= 20.29 \text{ lbf}$$

Use Eq. 45.7 to find the moment of the top blade. The centroid of the top blade section is located at one third its height.

$$M_{\text{top}} = r_{\text{top}} F_{\text{top}} = \left(120 \text{ ft} + \frac{20 \text{ ft}}{3}\right)(20.29 \text{ lbf}) = 2570 \text{ ft-lbf}$$

The force on the bottom blades is the same as the force on the top blade. The moment arm of the bottom blades, r_{bottom}, is

$$r_{\text{bottom}} = h_{\text{tower}} - \left(\frac{L_{\text{blade}}}{3}\right)\sin\theta = 120 \text{ ft} - \left(\frac{20 \text{ ft}}{3}\right)\sin 30° = 116.7 \text{ ft}$$

The moment due to each bottom blade is

$$M_{\text{bottom}} = r_{\text{bottom}} F_{\text{bottom}} = (116.7 \text{ ft})(20.29 \text{ lbf}) = 2368 \text{ ft-lbf}$$

The total overturning moment, M, is given by

$$M = M_{\text{top}} + 2M_{\text{bottom}} = 2570 \text{ ft-lbf} + (2)(2368 \text{ ft-lbf}) = \boxed{7306 \text{ ft-lbf} \quad (7300 \text{ ft-lbf})}$$

The answer is (B).

13. The energy used by the primary side of the transformer is

$$E = Pt = VI(\text{pf})t = \frac{(120 \text{ V})(1 \text{ mA})(0.8)(18 \text{ hr})}{1000 \ \frac{\text{mA}}{\text{A}}} = 1.728 \text{ W-hr}$$

The charger-rectifier efficiency is 65%, so the energy supplied to the battery is

$$E_{\text{supplied}} = \eta E = (0.65)(1.728 \text{ W-hr}) = 1.12 \text{ W-hr}$$

The energy storage of the battery is

$$E_{\text{battery}} = V_{\text{battery}} I_{\text{battery}} = (1.2 \text{ V})(0.600 \text{ A-hr}) = 0.72 \text{ W-hr}$$

The efficiency of the battery during the charge cycle is

$$\eta_{\text{charge}} = \frac{E_{\text{battery}}}{E_{\text{supplied}}} = \frac{0.72 \text{ W-hr}}{1.12 \text{ W-hr}} = \boxed{0.64 \quad (64\%)}$$

The answer is (B).

14. The effective voltage across the load is

$$V = \frac{P}{I} = \frac{250 \text{ W}}{10.9 \text{ A}} = 22.94 \text{ V}$$

22.94 V is less than (2)(12 V) = 24 V, so the batteries have internal resistance. Since there are two batteries in series in each leg, the voltage loss in each battery is half the total voltage loss.

$$V_{\text{loss}} = \frac{V_{\text{source}} - V}{2} = \frac{(2)(12 \text{ V}) - 22.94 \text{ V}}{2} = 0.53 \text{ V}$$

Since there are two battery legs in parallel, the current through each battery is one half the load current.

$$I_{\text{battery}} = \frac{I}{2} = \frac{10.9 \text{ A}}{2} = 5.45 \text{ A}$$

The internal resistance of each battery is

$$R_{\text{battery}} = \frac{V_{\text{loss}}}{I_{\text{battery}}} = \frac{0.53 \text{ V}}{5.45 \text{ A}} = \boxed{0.097 \ \Omega \quad (0.1 \ \Omega)}$$

The answer is (C).

15. The total energy used by the computer is

$$E = Pt = \frac{(150 \text{ W})(30 \text{ min})}{60 \ \frac{\text{min}}{\text{hr}}} = 75 \text{ W-hr}$$

The energy that must be delivered by the battery, given the efficiency of the rectifier/regulator, is

$$E_{\text{battery}} = \frac{E}{\eta} = \frac{75 \text{ W-hr}}{0.92} = 81.5 \text{ W-hr}$$

The amp-hour rating of the battery is

$$C_{\text{battery}} = \frac{E_{\text{battery}}}{V_{\text{battery}}} = \frac{(81.5\ \text{W-hr})\left(1000\ \frac{\text{mW}}{\text{W}}\right)}{12\ \text{V}}$$
$$= \boxed{6790\ \text{mA-hr}\quad(6800\ \text{mA-hr})}$$

The answer is (B).

16. From App. 23.B, for saturated steam at 200 psia,

$$h_f = 355.5\ \text{Btu/lbm}$$
$$h_{fg} = 843.3\ \text{Btu/lbm}$$
$$s_f = 0.5438\ \text{Btu/lbm-°R}$$
$$s_{fg} = 1.0022\ \text{Btu/lbm-°R}$$

From Eq. 23.53 and Eq. 23.54, the enthalpy and entropy of the steam entering the turbine are

$$\begin{aligned} h_{\text{entering}} &= h_f + xh_{fg} \\ &= 355.5\ \frac{\text{Btu}}{\text{lbm}} + (0.95)\left(843.3\ \frac{\text{Btu}}{\text{lbm}}\right) \\ &= 1156.6\ \text{Btu/lbm} \end{aligned}$$

$$\begin{aligned} s_{\text{entering}} &= s_f + xs_{fg} \\ &= 0.5438\ \frac{\text{Btu}}{\text{lbm-°R}} + (0.95)\left(1.0022\ \frac{\text{Btu}}{\text{lbm-°R}}\right) \\ &= 1.4959\ \text{Btu/lbm-°R} \end{aligned}$$

The pressure at the exhaust of the non-condensing turbine is 2 psig, or 16.7 psia. Interpolating from App. 23.B, for saturated steam at 16.7 psia,

$$h_f = 186.3\ \text{Btu/lbm}$$
$$h_{fg} = 966.2\ \text{Btu/lbm}$$
$$s_f = 0.3212\ \text{Btu/lbm-°R}$$
$$s_{fg} = 1.4259\ \text{Btu/lbm-°R}$$

With isentropic expansion, the quality of the steam at the exhaust of the non-condensing turbine is

$$\begin{aligned} x_{\text{non-cond}} &= \frac{s_{\text{entering}} - s_f}{s_{fg}} \\ &= \frac{1.4959\ \frac{\text{Btu}}{\text{lbm-°R}} - 0.3212\ \frac{\text{Btu}}{\text{lbm-°R}}}{1.4259\ \frac{\text{Btu}}{\text{lbm-°R}}} \\ &= 0.8238 \end{aligned}$$

The enthalpy at the exhaust is

$$\begin{aligned} h_{\text{non-cond}} &= h_f + x_{\text{non-cond}}h_{fg} \\ &= 186.3\ \frac{\text{Btu}}{\text{lbm}} + (0.8238)\left(966.2\ \frac{\text{Btu}}{\text{lbm}}\right) \\ &= 982.3\ \text{Btu/lbm} \end{aligned}$$

The specific work of the non-condensing turbine is

$$\begin{aligned} W_{\text{non-cond}} &= h_{\text{entering}} - h_{\text{non-cond}} \\ &= 1156.6\ \frac{\text{Btu}}{\text{lbm}} - 982.3\ \frac{\text{Btu}}{\text{lbm}} \\ &= 174.3\ \text{Btu/lbm} \end{aligned}$$

The pressure at the exhaust of the condensing turbine is 1 psia. From App. 23.B, for saturated steam at 1.0 psia,

$$h_f = 69.72\ \text{Btu/lbm}$$
$$h_{fg} = 1035.7\ \text{Btu/lbm}$$
$$s_f = 0.1326\ \text{Btu/lbm-°R}$$
$$s_{fg} = 1.8450\ \text{Btu/lbm-°R}$$

With isentropic expansion, the quality of the steam at the exhaust of the condensing turbine is

$$\begin{aligned} x_{\text{cond}} &= \frac{s_{\text{entering}} - s_f}{s_{fg}} \\ &= \frac{1.4959\ \frac{\text{Btu}}{\text{lbm-°R}} - 0.1326\ \frac{\text{Btu}}{\text{lbm-°R}}}{1.8450\ \frac{\text{Btu}}{\text{lbm-°R}}} \\ &= 0.7389 \end{aligned}$$

The enthalpy at the exhaust is

$$\begin{aligned} h_{\text{cond}} &= h_f + x_{\text{cond}}h_{fg} \\ &= 69.72\ \frac{\text{Btu}}{\text{lbm}} + (0.7389)\left(1035.7\ \frac{\text{Btu}}{\text{lbm}}\right) \\ &= 835.0\ \text{Btu/lbm} \end{aligned}$$

The specific work of the condensing turbine is

$$\begin{aligned} W_{\text{cond}} &= h_{\text{entering}} - h_{\text{cond}} = 1156.6\ \frac{\text{Btu}}{\text{lbm}} - 835.0\ \frac{\text{Btu}}{\text{lbm}} \\ &= 321.6\ \text{Btu/lbm} \end{aligned}$$

The ratio of the condensing to non-condensing work is

$$\frac{W_{\text{cond}}}{W_{\text{non-cond}}} = \frac{321.6\ \frac{\text{Btu}}{\text{lbm}}}{174.3\ \frac{\text{Btu}}{\text{lbm}}} = \boxed{1.845\quad(1.8)}$$

The answer is (C).

Power Cycles

17. The average total energy used in a year is

$$E = (1.7\text{ kW})\left(365\ \frac{\text{days}}{\text{yr}}\right)\left(24\ \frac{\text{hr}}{\text{day}}\right)$$
$$= 14{,}892\text{ kW-hr/yr}$$

The cost of supplying this energy from the electrical grid is

$$C_{\text{elec}} = \left(14{,}892\ \frac{\text{kW-hr}}{\text{yr}}\right)\left(\frac{\$0.08}{\text{kW-hr}}\right) = \$1191.40/\text{yr}$$

Convert the energy to Btu.

$$E = \left(14{,}892\ \frac{\text{kW-hr}}{\text{yr}}\right)\left(3412\ \frac{\text{Btu}}{\text{kW-hr}}\right)$$
$$= 5.081 \times 10^7\text{ Btu/yr}$$

The heating value of the natural gas needed to supply the average load for a year is

$$E_{\text{system}} = \frac{E}{\eta_{\text{system}}} = \frac{5.081 \times 10^7\ \frac{\text{Btu}}{\text{yr}}}{0.45}$$
$$= 1.13 \times 10^8\text{ Btu/yr}$$

The cost of supplying natural gas to the fuel cell system for a year is

$$C_{\text{system}} = \left(1.13 \times 10^8\ \frac{\text{Btu}}{\text{yr}}\right)\left(\frac{\frac{\$0.40}{\text{therm}}}{10^5\ \frac{\text{Btu}}{\text{therm}}}\right) = \$452/\text{yr}$$

The fuel cell system savings are

$$C_{\text{elec}} - C_{\text{system}} = \frac{\$1191.40}{\text{yr}} - \frac{\$452}{\text{yr}} = \$739.40/\text{yr}$$

The pay-back period is

$$\frac{C_{\text{initial}}}{\text{savings}} = \frac{\$3500}{\frac{\$739.40}{\text{yr}}} = \boxed{4.73\text{ yr}\quad(4.7\text{ yr})}$$

The answer is (D).

18. The total power generated by the methane is

$$P_{\text{methane}} = \dot{V}_{\text{methane}}(\text{LHV}) = \left(1200\ \frac{\text{ft}^3}{\text{hr}}\right)\left(913\ \frac{\text{Btu}}{\text{ft}^3}\right)$$
$$= 1.1 \times 10^6\text{ Btu/hr}$$

The AC electrical power output from the inverter is

$$P_{\text{inverter,out}} = (150\text{ kW})\left(3412\ \frac{\text{Btu}}{\text{kW-hr}}\right)$$
$$= 5.12 \times 10^5\text{ Btu/hr}$$

The DC electrical power input to the inverter is

$$P_{\text{inverter,in}} = \frac{P_{\text{inverter,out}}}{\eta_{\text{inverter}}} = \frac{5.12 \times 10^5\ \frac{\text{Btu}}{\text{hr}}}{0.9}$$
$$= 5.69 \times 10^5\text{ Btu/hr}$$

From App. 23.A, the enthalpy of the feedwater at 70°F is 38.08 Btu/lbm. From App. 23.B, the enthalpy of saturated steam at 35 psia is 1167.2 Btu/lbm. The power needed to generate steam is

$$P_{\text{exchanger,out}} = \dot{m}(h_g - h_f)$$
$$= \left(100\ \frac{\text{lbm}}{\text{hr}}\right)\left(1167.2\ \frac{\text{Btu}}{\text{lbm}} - 38.08\ \frac{\text{Btu}}{\text{lbm}}\right)$$
$$= 1.13 \times 10^5\text{ Btu/hr}$$

The power delivered to the heat exchanger is

$$P_{\text{exhanger,in}} = \frac{P_{\text{exchanger,out}}}{\eta_{\text{exchanger}}} = \frac{1.13 \times 10^5\ \frac{\text{Btu}}{\text{hr}}}{0.75}$$
$$= 1.51 \times 10^5\text{ Btu/hr}$$

The overall efficiency of the fuel cell stack is

$$\eta_{\text{system}} = \frac{P_{\text{inverter,in}} + P_{\text{exchanger,in}}}{P_{\text{methane}}}$$
$$= \frac{5.69 \times 10^5\ \frac{\text{Btu}}{\text{hr}} + 1.51 \times 10^5\ \frac{\text{Btu}}{\text{hr}}}{1.1 \times 10^6\ \frac{\text{Btu}}{\text{hr}}}$$
$$= \boxed{0.6545\quad(65\%)}$$

The answer is (C).

19. The force on the car due to rolling friction is

$$\begin{aligned} F_r &= f_r m \times \frac{g}{g_c} \\ &= (0.01)(4000\ \text{lbm})\left(\frac{32.2\ \frac{\text{ft}}{\text{sec}^2}}{32.2\ \frac{\text{lbm-ft}}{\text{lbf-sec}^2}}\right) \\ &= 40\ \text{lbf} \end{aligned}$$

The total resisting force on the car is the drag force plus the force due to rolling friction.

$$F = F_D + F_r = 136\ \text{lbf} + 40\ \text{lbf} = 176\ \text{lbf}$$

The power required to maintain the car's velocity is

$$\begin{aligned} P_{\text{car}} = F\text{v} &= \frac{(176\ \text{lbf})\left(55\ \frac{\text{mi}}{\text{hr}}\right)\left(5280\ \frac{\text{ft}}{\text{mi}}\right)}{\left(3600\ \frac{\text{sec}}{\text{hr}}\right)\left(778\ \frac{\text{ft-lbf}}{\text{Btu}}\right)} \\ &= 18.25\ \text{Btu/sec} \end{aligned}$$

The power delivered to the drive train is

$$P_{\text{drive train}} = \frac{P_{\text{car}}}{\eta_{\text{drive train}}} = \frac{18.25\ \frac{\text{Btu}}{\text{sec}}}{0.8} = 22.8\ \text{Btu/sec}$$

Energy is released by the hydrogen at a rate of

$$Q = \dot{V}(\text{LHV}) = \left(0.2\ \frac{\text{ft}^3}{\text{sec}}\right)\left(275\ \frac{\text{Btu}}{\text{ft}^3}\right) = 55\ \text{Btu/sec}$$

The efficiency of the fuel cell is

$$\eta = \frac{P_{\text{drive train}}}{Q} = \frac{22.8\ \frac{\text{Btu}}{\text{sec}}}{55\ \frac{\text{Btu}}{\text{sec}}} = \boxed{0.414\quad(41\%)}$$

The answer is (D).

Power Cycles

32 Gas Compression Cycles

PRACTICE PROBLEMS

1. The isentropic efficiency of a compressor that takes in air at 8 psia and −10°F (55 kPa and −20°C) and discharges it at 40 psia and 315°F (275 kPa and 160°C) is most nearly

(A) 61%

(B) 66%

(C) 74%

(D) 81%

2. A reciprocating air compressor has a 7% clearance and discharges 65 psia (450 kPa) air at the rate of 48 lbm/min (0.36 kg/s). Ambient air is at 14.7 psia and 65°F. The polytropic exponent is 1.33. The mass of air that is compressed each minute is most nearly

(A) 42 lbm/min (0.32 kg/s)

(B) 51 lbm/min (0.38 kg/s)

(C) 56 lbm/min (0.42 kg/s)

(D) 74 lbm/min (0.55 kg/s)

3. Air at 14.7 psia and 500°F (101.3 kPa and 260°C) is compressed in a centrifugal compressor to 6 atm. The isentropic efficiency of the compression process is 65%.

(a) The compression work is most nearly

(A) 140 Btu/lbm (320 kJ/kg)

(B) 190 Btu/lbm (440 kJ/kg)

(C) 230 Btu/lbm (540 kJ/kg)

(D) 350 Btu/lbm (810 kJ/kg)

(b) The final temperature is most nearly

(A) 1550°R (860K)

(B) 1650°R (920K)

(C) 1750°R (970K)

(D) 1850°R (1030K)

(c) The increase in entropy is most nearly

(A) 0.048 Btu/lbm-°R (0.21 kJ/kg·K)

(B) 0.099 Btu/lbm-°R (0.47 kJ/kg·K)

(C) 0.23 Btu/lbm-°R (1.0 kJ/kg·K)

(D) 0.44 Btu/lbm-°R (2.1 kJ/kg·K)

4. Compressors A and B both discharge into a common tank. The storage tank contains 100 psia, 90°F air (700 kPa, 32°C). Both compressors receive air at 14.7 psia and 80°F (101.3 kPa and 27°C). The flow rate through compressor A is 600 cfm (300 L/s). Process C uses 100 cfm of 80 psia, 85°F (50 L/s of 550 kPa, 29°C) air. Process D uses 120 cfm of 85 psia, 80°F (60 L/s of 590 kPa, 27°C) air. Process E uses 8 lbm/min (0.06 kg/s) of 85°F (29°C) air. Air is an ideal gas, and compressibility effects are to be disregarded.

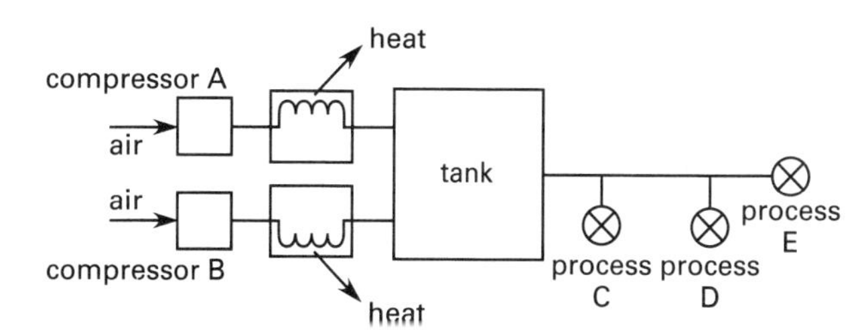

(a) The total mass flow rate for all three processes is most nearly

(A) 77 lbm/min (0.61 kg/s)

(B) 85 lbm/min (0.68 kg/s)

(C) 99 lbm/min (0.79 kg/s)

(D) 105 lbm/min (0.84 kg/s)

(b) The required input for compressor B is most nearly

(A) 33 lbm/min (0.26 kg/s)

(B) 44 lbm/min (0.35 kg/s)

(C) 55 lbm/min (0.44 kg/s)

(D) 77 lbm/min (0.61 kg/s)

(c) The volumetric flow rate through compressor B is most nearly

(A) 620 ft^3/min (320 L/s)

(B) 740 ft^3/min (370 L/s)

(C) 900 ft^3/min (450 L/s)

(D) 1100 ft^3/min (550 L/s)

5. 300 cfm (150 L/s) of air at 14.7 psia and 90°F (101.3 kPa and 32°C) enter a compressor. The compressor discharges air into a water-cooled heat exchanger. The compressed air is stored at 300 psig and 90°F (2.07 MPa and 32°C) in a 1000 ft^3 (27 m^3) tank. The tank feeds three air-driven tools with the flow rates and properties listed. The pressures to the air-driven tools are regulated and remain constant at the minimum required operating pressure as the tank pressure changes. Air is an ideal gas, and compressibility effects are to be disregarded.

	tool 1	tool 2	tool 3
flow rate			
(cfm)	40	15	unknown
(L/s)	19	7	unknown
flow rate			
(lbm/min)	unknown	unknown	6
(kg/s)	unknown	unknown	0.045
minimum pressure			
(psig)	90	50	80
(kPa)	620	350	550
temperature			
(°F)	90	85	80
(°C)	32	29	27

(a) The mass flow rate of air into the compressor is most nearly

(A) 18 lbm/min (0.14 kg/s)

(B) 22 lbm/min (0.17 kg/s)

(C) 25 lbm/min (0.20 kg/s)

(D) 32 lbm/min (0.25 kg/s)

(b) The mass of the stored compressed air in a 1000 ft^3 tank is most nearly

(A) 1360 lbm (560 kg)

(B) 1420 lbm (585 kg)

(C) 1490 lbm (615 kg)

(D) 1550 lbm (639 kg)

(c) The total mass flow leaving the system is most nearly

(A) 31 lbm/min (0.21 kg/s)

(B) 37 lbm/min (0.24 kg/s)

(C) 43 lbm/min (0.28 kg/s)

(D) 48 lbm/min (0.32 kg/s)

(d) The net flow rate of air to the tank is most nearly

(A) −23 lbm/min (−0.08 kg/s)

(B) −17 lbm/min (−0.06 kg/s)

(C) −10 lbm/min (−0.03 kg/s)

(D) −3.0 lbm/min (−0.01 kg/s)

(e) Approximately how long can the system run?

(A) 0.6 hr (1.2 h)

(B) 1.8 hr (3.6 h)

(C) 3.4 hr (6.8 h)

(D) 6.6 hr (13 h)

SOLUTIONS

1.

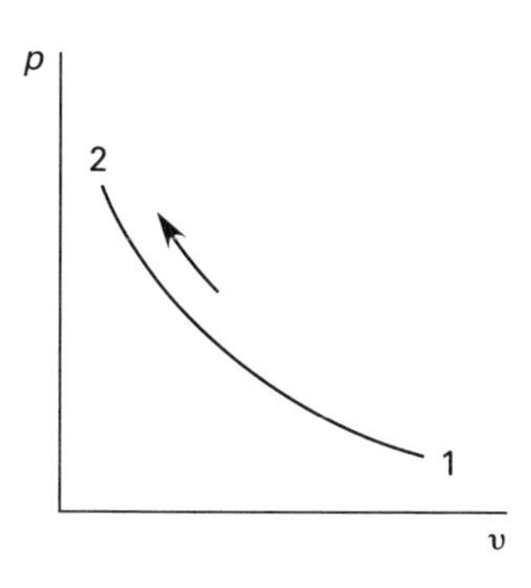

Customary U.S. Solution

Assume air is an ideal gas.

From Eq. 28.21, the isentropic temperature at point 2 is

$$T_2 = T_1\left(\frac{p_2}{p_1}\right)^{(k-1)/k}$$

The absolute temperature at point 1 is

$$\begin{aligned} T_1 &= -10°\text{F} + 460° = 450°\text{R} \\ T_2 &= (450°\text{R})\left(\frac{40 \text{ psia}}{8 \text{ psia}}\right)^{(1.4-1)/1.4} \\ &= 712.7°\text{R} \\ T_2 &= 712.7°\text{R} - 460° = 252.7°\text{F} \end{aligned}$$

The efficiency of the compressor is given by Eq. 29.21.

$$\begin{aligned} \eta_{s,\text{compressor}} &= \frac{T_2 - T_1}{T_2' - T_1} \\ &= \frac{252.7°\text{F} - (-10°\text{F})}{315°\text{F} - (-10°\text{F})} \\ &= \boxed{0.808 \quad (81\%)} \end{aligned}$$

The answer is (D).

SI Solution

Assume air is an ideal gas.

From Eq. 28.21, the isentropic temperature at point 2 is

$$T_2 = T_1\left(\frac{p_2}{p}\right)^{(k-1)/k}$$

The absolute temperature at point 1 is

$$\begin{aligned} T_1 &= -20°\text{C} + 273° = 253\text{K} \\ T_2 &= (253\text{K})\left(\frac{275 \text{ kPa}}{55 \text{ kPa}}\right)^{(1.4-1)/1.4} \\ &= 400.7\text{K} \\ T_2 &= 400.7\text{K} - 273° = 127.7°\text{C} \end{aligned}$$

The efficiency of the compressor is given by Eq. 29.21.

$$\begin{aligned} \eta_{s,\text{compressor}} &= \frac{T_2 - T_1}{T_2' - T_1} \\ &= \frac{127.7°\text{C} - (-20°\text{C})}{160°\text{C} - (-20°\text{C})} \\ &= \boxed{0.821 \quad (81\%)} \end{aligned}$$

The answer is (D).

2.

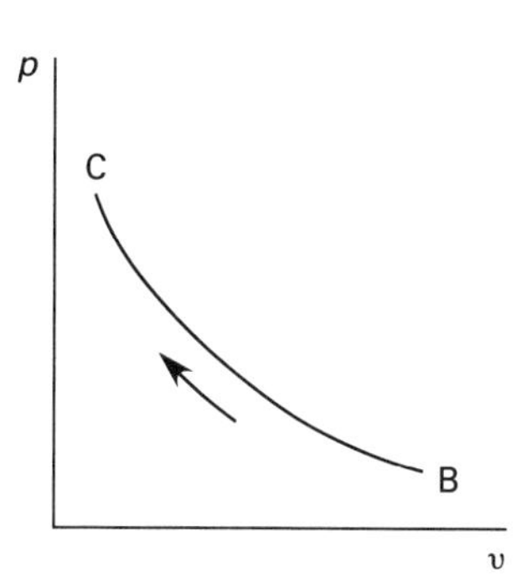

Customary U.S. Solution

The compression ratio for a reciprocating compressor is defined by Eq. 32.4.

$$r_p = \frac{p_\text{C}}{p_\text{B}} = \frac{65 \text{ psia}}{14.7 \text{ psia}} = 4.42$$

The volumetric efficiency is given by Eq. 32.6.

$$\begin{aligned} \eta_v &= 1 - \left(r_p^{1/n} - 1\right)\left(\frac{c}{100\%}\right) \\ &= 1 - \left(4.42^{1/1.33} - 1\right)\left(\frac{7\%}{100\%}\right) \\ &= 0.856 \quad (85.6\%) \end{aligned}$$

Power Cycles

The mass of air compressed per minute from Eq. 32.8 is

$$\dot{m} = \frac{\dot{m}_{\text{actual}}}{\eta_v} = \frac{48\ \frac{\text{lbm}}{\text{min}}}{0.856}$$

$$= \boxed{56.07\ \text{lbm/min} \quad (56\ \text{lbm/min})}$$

The answer is (C).

SI Solution

The compression ratio for the reciprocating compressor is defined by Eq. 32.4.

$$r_p = \frac{p_C}{p_B} = \frac{450\ \text{kPa}}{101\ \text{kPa}} = 4.46$$

The volumetric efficiency is given by Eq. 32.6.

$$\eta_v = 1 - \left(r_p^{1/n} - 1\right)\left(\frac{c}{100\%}\right)$$

$$= 1 - \left(4.46^{1/1.33} - 1\right)\left(\frac{7\%}{100\%}\right)$$

$$= 0.855 \quad (85.5\%)$$

The mass of air compressed per minute from Eq. 32.8 is

$$\dot{m} = \frac{\dot{m}_{\text{actual}}}{\eta_v} = \frac{0.36\ \frac{\text{kg}}{\text{s}}}{0.855}$$

$$= \boxed{0.4211\ \text{kg/s} \quad (0.42\ \text{kg/s})}$$

The answer is (C).

3. *Customary U.S. Solution*

((a) and (b)) Although the ideal gas laws can be used, it is more expedient to use air tables for the low pressures.

The absolute inlet temperature is

$$T_1 = 500°\text{F} + 460° = 960°\text{R}$$

From App. 23.F,

$$h_1 = 231.06\ \text{Btu/lbm}$$

$$p_{r,1} = 10.61$$

$$\phi_1 = 0.74030\ \text{Btu/lbm-°R}$$

Since $p_1/p_2 = p_{r,1}/p_{r,2}$ and the compression ratio is 6, for isentropic compression,

$$p_{r,2} = 6p_{r,1} = (6)(10.61) = 63.66$$

From App. 23.F, at $p_{r,2} = 63.66$,

$$T_2 = 1552°\text{R}$$

$$h_2 = 382.95\ \text{Btu/lbm}$$

The actual enthalpy from the definition of isentropic efficiency is

$$h_2' = h_1 + \frac{h_2 - h_1}{\eta_s}$$

$$= 231.06\ \frac{\text{Btu}}{\text{lbm}} + \frac{382.95\ \frac{\text{Btu}}{\text{lbm}} - 231.06\ \frac{\text{Btu}}{\text{lbm}}}{0.65}$$

$$= 464.74\ \text{Btu/lbm}$$

From App. 23.F, this corresponds to $T_2' = 1855°\text{R}$ and $\phi_2' = 0.91129\ \text{Btu/lbm-°R}$.

The compression work is

$$W = h_2' - h_1 = 464.74\ \frac{\text{Btu}}{\text{lbm}} - 231.06\ \frac{\text{Btu}}{\text{lbm}}$$

$$= \boxed{233.68\ \text{Btu/lbm} \quad (230\ \text{Btu/lbm})}$$

The answer is (C).

(b) The final temperature was found in part (a) to be $\boxed{1855°\text{R}\ (1850°\text{R}).}$

The answer is (D).

(c) From Eq. 23.52, the increase in entropy is

$$\Delta s = \phi_2' - \phi_1 - R\ln\left(\frac{p_2}{p_1}\right)$$

$$= 0.91129\ \frac{\text{Btu}}{\text{lbm-°R}} - 0.74030\ \frac{\text{Btu}}{\text{lbm-°R}} - \left(\frac{53.3\ \frac{\text{ft-lbf}}{\text{lbm-°R}}}{778\ \frac{\text{ft-lbf}}{\text{Btu}}}\right)\ln\left(\frac{6\ \text{atm}}{1\ \text{atm}}\right)$$

$$= \boxed{0.04824\ \text{Btu/lbm-°R} \quad (0.048\ \text{Btu/lbm-°R})}$$

The answer is (A).

SI Solution

((a) and (b)) Although the ideal gas laws can be used, it is more expedient to use air tables for the lower pressures.

The absolute inlet temperature is

$$T_1 = 260°\text{C} + 273° = 533\text{K}$$

From App. 23.S,

$$h_1 = 537.09 \text{ kJ/kg}$$
$$p_{r,1} = 10.59$$
$$\phi_1 = 2.28161 \text{ kJ/kg·K}$$

Since $p_1/p_2 = p_{r,1}/p_{r,2}$ and the compression ratio is 6, for isentropic compression,

$$p_{r,2} = 6p_{r,1} = (6)(10.59) = 63.54$$

From App. 23.S, at $p_{r,2} = 63.54$,

$$T_2 = 861.5\text{K}$$
$$h_2 = 889.9 \text{ kJ/kg}$$

The actual enthalpy from the definition of isentropic efficiency is

$$\begin{aligned} h_2' &= h_1 + \frac{h_2 - h_1}{\eta_s} \\ &= 537.09 \ \frac{\text{kJ}}{\text{kg}} + \frac{889.9 \ \frac{\text{kJ}}{\text{kg}} - 537.09 \ \frac{\text{kJ}}{\text{kg}}}{0.65} \\ &= 1079.87 \text{ kJ/kg} \end{aligned}$$

From App. 23.S, this corresponds to $T_2' = 1029.6\text{K}$ and $\phi_2' = 3.00099 \text{ kJ/kg·K}$.

The compression work is

$$\begin{aligned} W &= h_2' - h_1 \\ &= 1079.87 \ \frac{\text{kJ}}{\text{kg}} - 537.09 \ \frac{\text{kJ}}{\text{kg}} \\ &= \boxed{542.78 \text{ kJ/kg} \quad (540 \text{ kJ/kg})} \end{aligned}$$

The answer is (C).

(b) The final temperature was found in part (a) to be $\boxed{1029.6\text{K } (1030\text{K}).}$

The answer is (D).

(c) From Eq. 23.52, the increase in entropy is

$$\begin{aligned} \Delta s &= \phi_2' - \phi_1 - R \ln\left(\frac{p_2}{p_1}\right) \\ &= 3.00099 \ \frac{\text{kJ}}{\text{kg·K}} - 2.28161 \ \frac{\text{kJ}}{\text{kg·K}} \\ &\quad - \frac{\left(287.03 \ \frac{\text{J}}{\text{kg·K}}\right) \ln\left(\frac{6 \text{ atm}}{1 \text{ atm}}\right)}{1000 \ \frac{\text{J}}{\text{kJ}}} \\ &= \boxed{0.20509 \text{ kJ/kg·K} \quad (0.21 \text{ kJ/kg·K})} \end{aligned}$$

The answer is (A).

4. *Customary U.S. Solution*

(a) The absolute temperature for process C is 85°F + 460° = 545°R. Using ideal gas laws, the mass flow rate for process C is

$$\begin{aligned} \dot{m}_\text{C} &= \frac{p_\text{C} \dot{V}_\text{C}}{RT_\text{C}} = \frac{\left(80 \ \frac{\text{lbf}}{\text{in}^2}\right)\left(12 \ \frac{\text{in}}{\text{ft}}\right)^2 \left(100 \ \frac{\text{ft}^3}{\text{min}}\right)}{\left(53.3 \ \frac{\text{ft-lbf}}{\text{lbm-°R}}\right)(545\text{°R})} \\ &= 39.66 \text{ lbm/min} \end{aligned}$$

Similarly, the mass flow rate for process D with an absolute temperature of 80°F + 460° = 540°R is

$$\begin{aligned} \dot{m}_\text{D} &= \frac{p_\text{D} \dot{V}_\text{D}}{RT_\text{D}} = \frac{\left(85 \ \frac{\text{lbf}}{\text{in}^2}\right)\left(12 \ \frac{\text{in}}{\text{ft}}\right)^2 \left(120 \ \frac{\text{ft}^3}{\text{min}}\right)}{\left(53.3 \ \frac{\text{ft-lbf}}{\text{lbm-°R}}\right)(540\text{°R})} \\ &= 51.03 \text{ lbm/min} \end{aligned}$$

For process E, the mass flow rate is given as $\dot{m}_\text{E} = 8 \text{ lbm/min}$.

The total mass flow rate for all three processes is

$$\begin{aligned} \dot{m}_\text{total} &= \dot{m}_\text{C} + \dot{m}_\text{D} + \dot{m}_\text{E} \\ &= 39.66 \ \frac{\text{lbm}}{\text{min}} + 51.03 \ \frac{\text{lbm}}{\text{min}} + 8 \ \frac{\text{lbm}}{\text{min}} \\ &= \boxed{98.69 \text{ lbm/min} \quad (99 \text{ lbm/min})} \end{aligned}$$

The answer is (C).

(b) The mass flow rate into compressor A is with an absolute temperature of 80°F + 460° = 540°R.

$$\dot{m}_A = \frac{p_A \dot{V}_A}{RT_A} = \frac{\left(14.7\ \frac{\text{lbf}}{\text{in}^2}\right)\left(12\ \frac{\text{in}}{\text{ft}}\right)^2\left(600\ \frac{\text{ft}^3}{\text{min}}\right)}{\left(53.3\ \frac{\text{ft-lbf}}{\text{lbm-°R}}\right)(540\text{°R})} = 44.13\ \text{lbm/min}$$

The required input for compressor B is

$$\dot{m}_B = \dot{m}_{\text{total}} - \dot{m}_A = 98.69\ \frac{\text{lbm}}{\text{min}} - 44.13\ \frac{\text{lbm}}{\text{min}} = \boxed{54.56\ \text{lbm/min}\quad (55\ \text{lbm/min})}$$

The answer is (C).

(c) The volumetric flow rate for compressor B is

$$\dot{V}_B = \frac{\dot{m}_B R T_B}{p_B} = \frac{\left(54.56\ \frac{\text{lbm}}{\text{min}}\right)\left(53.3\ \frac{\text{ft-lbf}}{\text{lbm-°R}}\right)(540\text{°R})}{\left(14.7\ \frac{\text{lbf}}{\text{in}^2}\right)\left(12\ \frac{\text{in}}{\text{ft}}\right)^2} = \boxed{742\ \text{ft}^3/\text{min}\quad (740\ \text{ft}^3/\text{min})}$$

The answer is (B).

SI Solution

(a) The absolute temperature for process C is 29°C + 273° = 302K. Using ideal gas laws, the mass flow rate for process C is

$$\dot{m}_C = \frac{p_C \dot{V}_C}{RT_C} = \frac{(550\ \text{kPa})\left(1000\ \frac{\text{Pa}}{\text{kPa}}\right)\left(50\ \frac{\text{L}}{\text{s}}\right)}{\left(287\ \frac{\text{J}}{\text{kg·K}}\right)(302\text{K})\left(1000\ \frac{\text{L}}{\text{m}^3}\right)} = 0.3173\ \text{kg/s}$$

The absolute temperature for process D is 27°C + 273° = 300K. Using ideal gas laws, the mass flow rate for process D is

$$\dot{m}_D = \frac{p_D \dot{V}_D}{RT_D} = \frac{(590\ \text{kPa})\left(1000\ \frac{\text{Pa}}{\text{kPa}}\right)\left(60\ \frac{\text{L}}{\text{s}}\right)}{\left(287\ \frac{\text{J}}{\text{kg·K}}\right)(300\text{K})\left(1000\ \frac{\text{L}}{\text{m}^3}\right)} = 0.4111\ \text{kg/s}$$

For process E, the mass flow rate is given as $\dot{m}_E = 0.06$ kg/s.

The total mass flow rate for all three processes is

$$\dot{m}_{\text{total}} = \dot{m}_C + \dot{m}_D + \dot{m}_E = 0.3173\ \frac{\text{kg}}{\text{s}} + 0.4111\ \frac{\text{kg}}{\text{s}} + 0.06\ \frac{\text{kg}}{\text{s}} = \boxed{0.7884\ \text{kg/s}\quad (0.79\ \text{kg/s})}$$

The answer is (C).

(b) The absolute temperature for air into compressors A and B is 27°C + 273° = 300K.

The mass flow rate into compressor A is

$$\dot{m}_A = \frac{p_A \dot{V}_A}{RT_A} = \frac{(101.3\ \text{kPa})\left(1000\ \frac{\text{Pa}}{\text{kPa}}\right)\left(300\ \frac{\text{L}}{\text{s}}\right)}{\left(287\ \frac{\text{J}}{\text{kg·K}}\right)(300\text{K})\left(1000\ \frac{\text{L}}{\text{m}^3}\right)} = 0.3530\ \text{kg/s}$$

The required input for compressor B is

$$\dot{m}_B = \dot{m}_{\text{total}} - \dot{m}_A = 0.7884\ \frac{\text{kg}}{\text{s}} - 0.3530\ \frac{\text{kg}}{\text{s}} = \boxed{0.4354\ \text{kg/s}\quad (0.44\ \text{kg/s})}$$

The answer is (C).

(c) The volumetric flow rate for compressor B is

$$\begin{aligned}\dot{V}_\text{B} &= \frac{\dot{m}_\text{B} R T_\text{B}}{p_\text{B}} \\ &= \frac{\left(0.4354\ \frac{\text{kg}}{\text{s}}\right)\left(287\ \frac{\text{J}}{\text{kg·K}}\right)(300\text{K})\left(1000\ \frac{\text{L}}{\text{m}^3}\right)}{(101.3\ \text{kPa})\left(1000\ \frac{\text{Pa}}{\text{kPa}}\right)} \\ &= \boxed{370\ \text{L/s}\quad (370\ \text{L/s})}\end{aligned}$$

The answer is (B).

5.

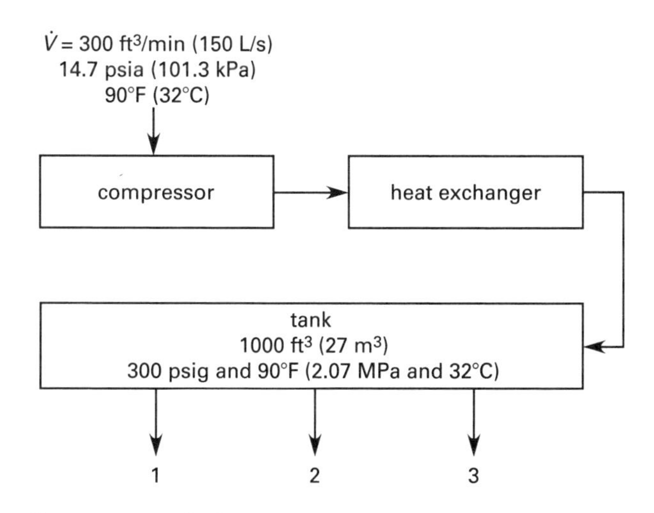

Customary U.S. Solution

(a) Assume steady flow and constant properties.

The absolute temperature for compressor air is 90°F + 460° = 550°R.

Using ideal gas laws, the mass flow rate of air into the compressor is

$$\begin{aligned}\dot{m} &= \frac{p\dot{V}}{RT} \\ &= \frac{\left(14.7\ \frac{\text{lbf}}{\text{in}^2}\right)\left(12\ \frac{\text{in}}{\text{ft}}\right)^2\left(300\ \frac{\text{ft}^3}{\text{min}}\right)}{\left(53.3\ \frac{\text{ft-lbf}}{\text{lbm-°R}}\right)(550\text{°R})} \\ &= \boxed{21.7\ \text{lbm/min}\quad (22\ \text{lbm/min})}\end{aligned}$$

The answer is (B).

(b) The absolute temperature and absolute pressure of stored compressed air are

$$\begin{aligned}T &= 90\text{°F} + 460\text{°} \\ &= 550\text{°R} \\ p &= 300\ \text{psig} + 14.7\ \text{psia} \\ &= 314.7\ \text{psia}\end{aligned}$$

The mass of stored compressed air in a 1000 ft^3 tank is

$$\begin{aligned}m_\text{tank} &= \frac{pV}{RT} \\ &= \frac{\left(314.7\ \frac{\text{lbf}}{\text{in}^2}\right)\left(12\ \frac{\text{in}}{\text{ft}}\right)^2(1000\ \text{ft}^3)}{\left(53.3\ \frac{\text{ft-lbf}}{\text{lbm-°R}}\right)(550\text{°R})} \\ &= \boxed{1545.9\ \text{lbm}\quad (1550\ \text{lbm})}\end{aligned}$$

The answer is (D).

(c) Assuming that each tool operates at its minimum pressure, the mass leaving the system can be calculated as follows.

Tool 1:

The absolute pressure is

$$90\ \text{psig} + 14.7\ \text{psia} = 104.7\ \text{psia}$$

The absolute temperature is

$$\begin{aligned}90\text{°F} + 460\text{°} &= 550\text{°R} \\ \dot{m}_\text{tool 1} &= \frac{p\dot{V}}{RT} \\ &= \frac{\left(104.7\ \frac{\text{lbf}}{\text{in}^2}\right)\left(12\ \frac{\text{in}}{\text{ft}}\right)^2\left(40\ \frac{\text{ft}^3}{\text{min}}\right)}{\left(53.3\ \frac{\text{ft-lbf}}{\text{lbm-°R}}\right)(550\text{°R})} \\ &= 20.57\ \text{lbm/min}\end{aligned}$$

Tool 2:

The absolute pressure is

$$50\ \text{psig} + 14.7\ \text{psia} = 64.7\ \text{psia}$$

Power Cycles

The absolute temperature is

$$85°\text{F} + 460° = 545°\text{R}$$

$$\begin{aligned}\dot{m}_{\text{tool 2}} &= \frac{p\dot{V}}{RT} \\ &= \frac{\left(64.7\ \frac{\text{lbf}}{\text{in}^2}\right)\left(12\ \frac{\text{in}}{\text{ft}}\right)^2\left(15\ \frac{\text{ft}^3}{\text{min}}\right)}{\left(53.3\ \frac{\text{ft-lbf}}{\text{lbm-°R}}\right)(545°\text{R})} \\ &= 4.81\ \text{lbm/min}\end{aligned}$$

Tool 3:

$$\dot{m}_{\text{tool 3}} = 6\ \text{lbm/min} \quad \text{[given]}$$

The total mass flow leaving the system is

$$\begin{aligned}\dot{m}_{\text{total}} &= \dot{m}_{\text{tool 1}} + \dot{m}_{\text{tool 2}} + \dot{m}_{\text{tool 3}} \\ &= 20.57\ \frac{\text{lbm}}{\text{min}} + 4.81\ \frac{\text{lbm}}{\text{min}} + 6\ \frac{\text{lbm}}{\text{min}} \\ &= \boxed{31.38\ \text{lbm/min} \quad (31\ \text{lbm/min})}\end{aligned}$$

The answer is (A).

(d) The critical pressure of 104.7 psia is required for tool 1 operation. The mass in the tank when critical pressure is achieved is

$$\begin{aligned}m_{\text{tank,critical}} &= \frac{pV}{RT} \\ &= \frac{\left(104.7\ \frac{\text{lbf}}{\text{in}^2}\right)\left(12\ \frac{\text{in}}{\text{ft}}\right)^2(1000\ \text{ft}^3)}{\left(53.3\ \frac{\text{ft-lbf}}{\text{lbm-°R}}\right)(550°\text{R})} \\ &= 514.3\ \text{lbm}\end{aligned}$$

The amount in the tank to be depleted is

$$\begin{aligned}m_{\text{depleted}} &= m_{\text{tank}} - m_{\text{tank,critical}} = 1545.9\ \text{lbm} - 514.3\ \text{lbm} \\ &= 1031.6\ \text{lbm}\end{aligned}$$

The net flow rate of air to the tank is

$$\begin{aligned}\dot{m}_{\text{net}} &= \dot{m}_{\text{in}} - \dot{m}_{\text{out}} = 21.7\ \frac{\text{lbm}}{\text{min}} - 31.38\ \frac{\text{lbm}}{\text{min}} \\ &= \boxed{-9.68\ \text{lbm/min} \quad (-10\ \text{lbm/min})}\end{aligned}$$

The answer is (C).

(e) The time the system can run is

$$\frac{m_{\text{depleted}}}{\dot{m}_{\text{net}}} = \frac{1031.6\ \text{lbm}}{9.68\ \frac{\text{lbm}}{\text{min}}} = \boxed{106.6\ \text{min} \quad (1.8\ \text{hr})}$$

The answer is (B).

SI Solution

(a) The absolute temperature of the compressor air is

$$32°\text{C} + 273° = 305\text{K}$$

Using ideal gas laws, the mass flow rate of air into the compressor is

$$\begin{aligned}\dot{m} &= \frac{p\dot{V}}{RT} \\ &= \frac{(101.3\ \text{kPa})\left(1000\ \frac{\text{Pa}}{\text{kPa}}\right)\left(150\ \frac{\text{L}}{\text{s}}\right)}{\left(287\ \frac{\text{J}}{\text{kg·K}}\right)(305\text{K})\left(1000\ \frac{\text{L}}{\text{m}^3}\right)} \\ &= \boxed{0.1736\ \text{kg/s} \quad (0.17\ \text{kg/s})}\end{aligned}$$

The answer is (B).

(b) The absolute temperature and pressure of stored compressed air are

$$\begin{aligned}T &= 32°\text{C} + 273° = 305\text{K} \\ p &= (2.07\ \text{MPa})\left(10^6\ \frac{\text{Pa}}{\text{MPa}}\right) \\ &= 2.07 \times 10^6\ \text{Pa}\end{aligned}$$

The mass of stored compressed air in a 1000 ft^3 tank is

$$\begin{aligned}m_{\text{tank}} &= \frac{pV}{RT} \\ &= \frac{(2.07 \times 10^6\ \text{Pa})(27\ \text{m}^3)}{\left(287\ \frac{\text{J}}{\text{kg·K}}\right)(305\text{K})} \\ &= \boxed{638.5\ \text{kg} \quad (639\ \text{kg})}\end{aligned}$$

The answer is (D).

(c) Assuming that each tool operates at its minimum pressure, the mass leaving the system can be calculated as follows.

Tool 1:

The absolute temperature is $32°C + 273° = 305K$.

$$\dot{m}_{\text{tool 1}} = \frac{p\dot{V}}{RT} = \frac{(620 \text{ kPa})\left(1000 \ \frac{\text{Pa}}{\text{kPa}}\right)\left(19 \ \frac{\text{L}}{\text{s}}\right)}{\left(287 \ \frac{\text{J}}{\text{kg·K}}\right)(305\text{K})\left(1000 \ \frac{\text{L}}{\text{m}^3}\right)} = 0.1346 \text{ kg/s}$$

Tool 2:

The absolute temperature is $29°C + 273° = 302K$.

$$\dot{m}_{\text{tool 2}} = \frac{p\dot{V}}{RT} = \frac{(350 \text{ kPa})\left(1000 \ \frac{\text{Pa}}{\text{kPa}}\right)\left(7 \ \frac{\text{L}}{\text{s}}\right)}{\left(287 \ \frac{\text{J}}{\text{kg·K}}\right)(302\text{K})\left(1000 \ \frac{\text{L}}{\text{m}^3}\right)} = 0.02827 \text{ kg/s}$$

Tool 3:

$$\dot{m}_{\text{tool 3}} = 0.045 \text{ kg/s} \quad \text{[given]}$$

The total mass flow rate leaving the system is

$$\begin{aligned} \dot{m}_{\text{total}} &= \dot{m}_{\text{tool 1}} + \dot{m}_{\text{tool 2}} + \dot{m}_{\text{tool 3}} \\ &= 0.1346 \ \frac{\text{kg}}{\text{s}} + 0.02827 \ \frac{\text{kg}}{\text{s}} + 0.045 \ \frac{\text{kg}}{\text{s}} \\ &= \boxed{0.2079 \text{ kg/s} \quad (0.21 \text{ kg/s})} \end{aligned}$$

The answer is (A).

(d) The critical pressure of 620 kPa is required for tool 1 operation. The mass in the tank when critical pressure is achieved is

$$m_{\text{tank,critical}} = \frac{pV}{RT} = \frac{(620 \text{ kPa})\left(1000 \ \frac{\text{Pa}}{\text{kPa}}\right)(27 \text{ m}^3)}{\left(287 \ \frac{\text{J}}{\text{kg·K}}\right)(305\text{K})} = 191.2 \text{ kg}$$

The amount in the tank to be depleted is

$$\begin{aligned} m_{\text{depleted}} &= m_{\text{tank}} - m_{\text{tank,critical}} = 638.5 \text{ kg} - 191.2 \text{ kg} \\ &= 447.3 \text{ kg} \end{aligned}$$

The net flow rate of air into the tank is

$$\begin{aligned} \dot{m}_{\text{net}} &= \dot{m}_{\text{in}} - \dot{m}_{\text{out}} = 0.1736 \ \frac{\text{kg}}{\text{s}} - 0.2079 \ \frac{\text{kg}}{\text{s}} \\ &= \boxed{-0.0343 \text{ kg/s} \quad (-0.03 \text{ kg/s})} \end{aligned}$$

The answer is (C).

(e) The time the system can run is

$$\frac{m_{\text{depleted}}}{\dot{m}_{\text{net}}} = \frac{447.3 \text{ kg}}{0.0343 \ \frac{\text{kg}}{\text{s}}} = \boxed{13\,041 \text{ s} \quad (3.6 \text{ h})}$$

The answer is (B).

33 Refrigeration Cycles

PRACTICE PROBLEMS

Carnot Refrigeration Cycle

1. A heat pump operates on the Carnot cycle between 40°F and 700°F (4°C and 370°C). The coefficient of performance is most nearly

(A) 1.5

(B) 1.8

(C) 2.2

(D) 2.7

2. A heat pump using refrigerant R-12 operates on the Carnot cycle. The refrigerant evaporates and is compressed at 35.7 psia and 172.4 psia (246 kPa and 1190 kPa), respectively. The coefficient of performance is most nearly

(A) 3.2

(B) 3.9

(C) 4.3

(D) 5.8

3. A refrigerator uses refrigerant R-12. The input power is 585 W. Heat absorbed from the cooled space is 450 Btu/hr (0.13 kW). The coefficient of performance is most nearly

(A) 0.2

(B) 0.4

(C) 0.7

(D) 0.9

4. Ammonia is used in a reversed Carnot cycle refrigerator with reservoirs at 110°F (45°C) and 10°F (−10°C). 1000 Btu/hr (1000 kJ/h) are to be removed.

(a) The coefficient of performance is most nearly

(A) 2.3

(B) 2.9

(C) 4.7

(D) 5.2

(b) The work input is most nearly

(A) 210 Btu/hr (210 kJ/h)

(B) 340 Btu/hr (340 kJ/h)

(C) 400 Btu/hr (400 kJ/h)

(D) 630 Btu/hr (630 kJ/h)

(c) The rejected heat is most nearly

(A) 1000 Btu/hr (1000 kJ/h)

(B) 1200 Btu/hr (1200 kJ/h)

(C) 1400 Btu/hr (1400 kJ/h)

(D) 1600 Btu/hr (1600 kJ/h)

5. A refrigerator cools a continuous aqueous solution ($c_p = 1$ Btu/lbm-°F; 4.19 kJ/kg·°C) flow of 100 gal/min (0.4 m^3/min) from 80°F (25°C) to 20°F (−5°C) in an 80°F (25°C) environment. The minimum power requirement is most nearly

(A) 82 hp (63 kW)

(B) 100 hp (74 kW)

(C) 130 hp (90 kW)

(D) 150 hp (94 kW)

Vapor Compression Cycle

6. An ammonia compressor is used in a heat pump cycle. Suction pressure is 30 psia (200 kPa). Discharge pressure is 160 psia (1.0 MPa). Saturated liquid ammonia enters the throttle valve. The refrigeration effect is 500 Btu/lbm (1200 kJ/kg) ammonia. The coefficient of performance as a heat pump is most nearly

(A) 2.1

(B) 4.5

(C) 5.3

(D) 5.8

7. A refrigeration cycle uses Freon-12 as a refrigerant between a 70°F (20°C) environment and a −30°F (−30°C) heat source. If the vapor leaving the evaporator and liquid leaving the condenser are both saturated, what is most nearly the volume flow of refrigerant leaving the evaporator per ton (kW) of refrigeration? Assume isentropic compression.

(A) 9.8 ft^3/min-ton (0.00081 m^3/s-kW)

(B) 12 ft^3/min-ton (0.0013 m^3/s-kW)

(C) 25 ft^3/min-ton (0.0021 m^3/s-kW)

(D) 44 ft^3/min-ton (0.0036 m^3/s-kW)

Air Refrigeration Cycle

8. An air refrigeration cycle compresses air from 70°F (20°C) and 14.7 psia (101 kPa) to 60 psia (400 kPa) in a 70% efficient compressor. The air is cooled to 25°F (−4.0°C) in a constant pressure process before entering a turbine with isentropic efficiency of 0.80. Assume air is an ideal gas.

(a) The temperature of the air leaving the compressor is most nearly

(A) 720°R (400K)

(B) 790°R (440K)

(C) 860°R (470K)

(D) 900°R (490K)

(b) The temperature of the air leaving the turbine is most nearly

(A) 360°R (200K)

(B) 430°R (240K)

(C) 490°R (270K)

(D) 560°R (310K)

(c) The coefficient of performance of the cycle is most nearly

(A) 0.7

(B) 0.8

(C) 0.9

(D) 1.1

Power Cycles

SOLUTIONS

1. *Customary U.S. Solution*

The coefficient of performance for a heat pump operating on the Carnot cycle is given by Eq. 33.9.

$$\text{COP}_{\text{heat pump}} = \frac{T_{\text{high}}}{T_{\text{high}} - T_{\text{low}}}$$

The absolute temperatures are

$$T_{\text{high}} = 700°\text{F} + 460° = 1160°\text{R}$$

$$T_{\text{low}} = 40°\text{F} + 460° = 500°\text{R}$$

$$\text{COP}_{\text{heat pump}} = \frac{1160°\text{R}}{1160°\text{R} - 500°\text{R}} = \boxed{1.76 \quad (1.8)}$$

The answer is (B).

SI Solution

The coefficient of performance for a heat pump operating on the Carnot cycle is given by Eq. 33.9.

$$\text{COP}_{\text{heat pump}} = \frac{T_{\text{high}}}{T_{\text{high}} - T_{\text{low}}}$$

The absolute temperatures are

$$T_{\text{high}} = 370°\text{C} + 273° = 643\text{K}$$

$$T_{\text{low}} = 4°\text{C} + 273° = 277\text{K}$$

$$\text{COP}_{\text{heat pump}} = \frac{643\text{K}}{643\text{K} - 277\text{K}} = \boxed{1.76 \quad (1.8)}$$

The answer is (B).

2. *Customary U.S. Solution*

The coefficient of performance for a heat pump operating on the Carnot cycle is given by Eq. 33.9.

$$\text{COP}_{\text{heat pump}} = \frac{T_{\text{high}}}{T_{\text{high}} - T_{\text{low}}}$$

The temperature T_{high} is the saturation temperature at 172.4 psia, and the temperature T_{low} is the saturation temperature at 35.7 psia. From App. 23.H,

$$T_{\text{high}} = 120.2°\text{F} + 460° = 580.2°\text{R}$$

$$T_{\text{low}} = 19.5°\text{F} + 460° = 479.5°\text{R}$$

$$\text{COP}_{\text{heat pump}} = \frac{580.2°\text{R}}{580.2°\text{R} - 479.5°\text{R}} = \boxed{5.76 \quad (5.8)}$$

The answer is (D).

SI Solution

The coefficient of performance for a heat pump operating on the Carnot cycle is given by Eq. 33.9.

$$\text{COP}_{\text{heat pump}} = \frac{T_{\text{high}}}{T_{\text{high}} - T_{\text{low}}}$$

The temperature T_{high} is the saturation temperature at 1190 kPa, and the temperature T_{low} is the saturation temperature at 246 kPa. From App. 23.T,

$$T_{\text{high}} = 49°\text{C} + 273° = 322\text{K}$$
$$T_{\text{low}} = -6.8°\text{C} + 273° = 266.2\text{K}$$
$$\text{COP}_{\text{heat pump}} = \frac{322\text{K}}{322\text{K} - 266.2\text{K}} = \boxed{5.77 \quad (5.8)}$$

The answer is (D).

3. *Customary U.S. Solution*

The coefficient of performance for a refrigerator is given by Eq. 33.1.

$$\text{COP}_{\text{refrigerator}} = \frac{Q_{\text{in}}}{W_{\text{in}}} = \frac{450 \ \frac{\text{Btu}}{\text{hr}}}{(585 \text{ W})\left(3.4121 \ \frac{\frac{\text{Btu}}{\text{hr}}}{\text{W}}\right)}$$
$$= \boxed{0.225 \quad (0.2)}$$

The answer is (A).

SI Solution

From Eq. 33.1, the coefficient of performance for a refrigerator is

$$\text{COP}_{\text{refrigerator}} = \frac{Q_{\text{in}}}{W_{\text{in}}} = \frac{(0.13 \text{ kW})\left(1000 \ \frac{\text{W}}{\text{kW}}\right)}{585 \text{ W}}$$
$$= \boxed{0.222 \quad (0.2)}$$

The answer is (A).

4. *Customary U.S. Solution*

(a) The coefficient of performance is

$$\text{COP} = \frac{T_{\text{low}}}{T_{\text{high}} - T_{\text{low}}} = \frac{460° + 10°\text{F}}{110°\text{F} - 10°\text{F}}$$
$$= \boxed{4.7}$$

The answer is (C).

(b) The work input is

$$W_{\text{in}} = \frac{Q_{\text{in}}}{\text{COP}} = \frac{1000 \ \frac{\text{Btu}}{\text{hr}}}{4.7}$$
$$= \boxed{212.77 \text{ Btu/hr} \quad (210 \text{ Btu/hr})}$$

The answer is (A).

(c) The rejected heat is

$$Q_{\text{out}} = Q_{\text{in}} + W_{\text{in}} = 1000 \ \frac{\text{Btu}}{\text{hr}} + 212.77 \ \frac{\text{Btu}}{\text{hr}}$$
$$= \boxed{1212.77 \text{ Btu/hr} \quad (1200 \text{ Btu/hr})}$$

The answer is (B).

SI Solution

(a) The coefficient of performance is

$$\text{COP} = \frac{T_{\text{low}}}{T_{\text{high}} - T_{\text{low}}} = \frac{-10°\text{C} + 273.15°}{45°\text{C} - (-10°\text{C})}$$
$$= \boxed{4.785 \quad (4.7)}$$

The answer is (C).

(b) The work input is

$$W_{\text{in}} = \frac{Q_{\text{in}}}{\text{COP}} = \frac{1000 \ \frac{\text{kJ}}{\text{h}}}{4.785}$$
$$= \boxed{208.99 \text{ kJ/h} \quad (210 \text{ kJ/h})}$$

The answer is (A).

(c) The work input is

$$Q_{\text{out}} = Q_{\text{in}} + W_{\text{in}} = 1000 \ \frac{\text{kJ}}{\text{h}} + 208.99 \ \frac{\text{kJ}}{\text{h}}$$
$$= \boxed{1208.99 \text{ kJ/h} \quad (1200 \text{ kJ/h})}$$

The answer is (B).

5. *Customary U.S. Solution*

The density of an aqueous solution is essentially the same as water.

$$\dot{m} = \dot{V}\rho = \left(100 \ \frac{\text{gal}}{\text{min}}\right)\left(0.1337 \ \frac{\text{ft}^3}{\text{gal}}\right)\left(62.4 \ \frac{\text{lbm}}{\text{ft}^3}\right)$$
$$= 834.3 \text{ lbm/min}$$

$$\dot{Q}_{in} = \dot{m}c_p\Delta T$$
$$= \left(834.3\ \frac{\text{lbm}}{\text{min}}\right)\left(1\ \frac{\text{Btu}}{\text{lbm-°F}}\right)(80°\text{F} - 20°\text{F})$$
$$= 50{,}058\ \text{Btu/min}$$

$$\text{COP} = \frac{T_{low}}{T_{high} - T_{low}} = \frac{20°\text{F} + 460°}{80°\text{F} - 20°\text{F}} = 8$$

$$W_{in,hp} = \frac{4.715 Q_{in,tons}}{\text{COP}}$$
$$= \frac{(4.715)\left(\dfrac{50{,}058\ \dfrac{\text{Btu}}{\text{min}}}{200\ \dfrac{\text{Btu}}{\text{min-ton}}}\right)}{8}$$
$$= \boxed{147.5\ \text{hp}\quad(150\ \text{hp})}$$

The answer is (D).

SI Solution

The density of an aqueous solution is essentially the same as water.

$$\dot{m} = \dot{V}\rho = \left(0.4\ \frac{\text{m}^3}{\text{min}}\right)\left(1000\ \frac{\text{kg}}{\text{m}^3}\right)$$
$$= 400\ \text{kg/min}$$
$$\dot{Q}_{in} = \dot{m}c_p\Delta T$$
$$= \left(400\ \frac{\text{kg}}{\text{min}}\right)\left(4.19\ \frac{\text{kJ}}{\text{kg·°C}}\right)(25°\text{C} - (-5°\text{C}))$$
$$= 50\,280\ \text{kJ/min}$$
$$\text{COP} = \frac{T_{low}}{T_{high} - T_{low}} = \frac{-5°\text{C} + 273.15°}{25°\text{C} - (-5°\text{C})}$$
$$= 8.938$$
$$\dot{W}_{in} = \frac{\dot{Q}_{in}}{\text{COP}} = \frac{50\,280\ \dfrac{\text{kJ}}{\text{min}}}{(8.938)\left(60\ \dfrac{\text{s}}{\text{min}}\right)}$$
$$= \boxed{93.76\ \text{kW}\quad(94\ \text{kW})}$$

The answer is (D).

6. *Customary U.S. Solution*

At A: $p_A = 160$ psia

Interpolating between 140 psia and 170 psia,

$$\frac{h_A - 126\ \dfrac{\text{Btu}}{\text{lbm}}}{139.3\ \dfrac{\text{Btu}}{\text{lbm}} - 126\ \dfrac{\text{Btu}}{\text{lbm}}} = \frac{160\ \text{psia} - 140\ \text{psia}}{170\ \text{psia} - 140\ \text{psia}}$$
$$h_A = 134.9\ \text{Btu/lbm}$$

At B: $h_B = h_A = 134.9$ Btu/lbm

At C:

$$h_C = h_B + q_{in} = 134.9\ \frac{\text{Btu}}{\text{lbm}} + 500\ \frac{\text{Btu}}{\text{lbm}}$$
$$= 634.9\ \text{Btu/lbm}$$

From a superheated ammonia table at 30 psia,

$$s_C = 1.3845\ \text{Btu/lbm-°F}$$

At D:

$$s_D = s_C = 1.3845\ \text{Btu/lbm-°F}$$
$$p_D = p_A = 160\ \text{psia}$$

Interpolating between 250°F and 300°F,

$$\frac{h_D - 737.6\ \dfrac{\text{Btu}}{\text{lbm}}}{1.3845\ \dfrac{\text{Btu}}{\text{lbm-°F}} - 1.3675\ \dfrac{\text{Btu}}{\text{lbm-°F}}} = \frac{767.1\ \dfrac{\text{Btu}}{\text{lbm}} - 737.6\ \dfrac{\text{Btu}}{\text{lbm}}}{1.4076\ \dfrac{\text{Btu}}{\text{lbm-°F}} - 1.3675\ \dfrac{\text{Btu}}{\text{lbm-°F}}}$$
$$h_D = 750.1\ \text{Btu/lbm}$$

$$\text{COP} = \frac{q_{out}}{W_{in}} = \frac{h_C - h_B}{h_D - h_C} + 1$$
$$= \frac{634.9\ \dfrac{\text{Btu}}{\text{lbm}} - 134.9\ \dfrac{\text{Btu}}{\text{lbm}}}{750.1\ \dfrac{\text{Btu}}{\text{lbm}} - 634.9\ \dfrac{\text{Btu}}{\text{lbm}}} + 1$$
$$= \boxed{5.34\quad(5.3)}$$

The answer is (C).

SI Solution

At A: $p_A = 1.0$ MPa

From a saturated ammonia table,

$$h_A = 443.65\ \text{kJ/kg}$$

At B: $h_B = h_A = 443.65$ kJ/kg

At C:

$$h_C = h_B + q_{in} = 443.65\ \frac{\text{kJ}}{\text{kg}} + 1200\ \frac{\text{kJ}}{\text{kg}}$$
$$= 1643.65\ \text{kJ/kg}\quad[\text{superheated}]$$
$$p_C = 200\ \text{kPa}$$

Power Cycles

Interpolating from a superheated ammonia table at 200 kPa (0.20 MPa),

$$s_C = 6.4960\ \frac{kJ}{kg{\cdot}K} + \left(\frac{1643.65\ \frac{kJ}{kg} - 1625.18\ \frac{kJ}{kg}}{1647.94\ \frac{kJ}{kg} - 1625.18\ \frac{kJ}{kg}}\right) \times \left(6.5759\ \frac{kJ}{kg{\cdot}K} - 6.4960\ \frac{kJ}{kg{\cdot}K}\right)$$
$$= 6.5608\ \text{kJ/kg·K}$$

At D:

$$s_D = s_C = 6.5608\ \text{kJ/kg·K}$$
$$p_D = p_A = 1.0\ \text{MPa}$$

Interpolating from a superheated ammonia table at 1 MPa,

$$h_D = 1900.02\ \frac{kJ}{kg} + \left(\frac{6.5608\ \frac{kJ}{kg{\cdot}K} - 6.5376\ \frac{kJ}{kg{\cdot}K}}{6.5965\ \frac{kJ}{kg{\cdot}K} - 6.5376\ \frac{kJ}{kg{\cdot}K}}\right) \times \left(1924.43\ \frac{kJ}{kg} - 1900.02\ \frac{kJ}{kg}\right)$$
$$= 1909.63\ \text{kJ/kg}$$

$$\text{COP} = \frac{q_{out}}{W_{in}} = \frac{h_C - h_B}{h_D - h_C} + 1 = \frac{1643.65\ \frac{kJ}{kg} - 443.65\ \frac{kJ}{kg}}{1909.63\ \frac{kJ}{kg} - 1643.65\ \frac{kJ}{kg}} + 1 = \boxed{5.512\quad (5.3)}$$

The answer is (C).

7. *Customary U.S. Solution*

At A:

$T_A = 70°\text{F}$ from saturated Freon-12 table

$$h_A = 23.9\ \text{Btu/lbm}$$

At B: $h_B = h_A = 23.9\ \text{Btu/lbm}$

At C:

$$T_C = T_B = -30°\text{F}$$
$$h_C = 74.7\ \text{Btu/lbm}$$
$$\upsilon_C = 3.088\ \text{ft}^3\text{/lbm}$$

$$q_{in} = h_C - h_B = 74.7\ \frac{Btu}{lbm} - 23.9\ \frac{Btu}{lbm} = 50.8\ \text{Btu/lbm}$$

$$\dot{m} = \frac{(1\ \text{ton})\left(200\ \frac{Btu}{min\text{-}ton}\right)}{50.8\ \frac{Btu}{lbm}} = 3.94\ \text{lbm/min}$$

$$\dot{V} = \dot{m}\upsilon_C = \left(3.94\ \frac{lbm}{min}\right)\left(3.088\ \frac{ft^3}{lbm}\right) = \boxed{12.16\ \text{ft}^3\text{/min}\quad (12\ \text{ft}^3\text{/min-ton})}$$

The answer is (B).

SI Solution

At A: $T_A = 20°\text{C}$

From the saturated Freon-12 table,

$$h_A = 83.57\ \text{kJ/kg}$$

At B: $h_B = h_A = 219.18\ \text{kJ/kg}$

At C:

$$T_C = T_B = -30°\text{C}$$
$$h_C \approx 338.76\ \text{kJ/kg}$$
$$\upsilon_C \approx 0.15993\ \text{m}^3\text{/kg}$$

$$q_{in} = h_C - h_B = 338.76\ \frac{kJ}{kg} - 219.18\ \frac{kJ}{kg} = 119.58\ \text{kJ/kg}$$

$$\dot{m} = \frac{Q}{q_{in}} = \frac{1\ \text{kW}}{119.58\ \frac{kJ}{kg}} = 0.008363\ \text{kg/s}$$

$$\dot{V} = \dot{m}\upsilon_C = \left(0.008363\ \frac{kg}{s}\right)\left(0.15993\ \frac{m^3}{kg}\right) = \boxed{0.001337\ \text{m}^3\text{/s}\quad (0.0013\ \text{m}^3\text{/s-kW})}$$

The answer is (B).

8. *Customary U.S. Solution*

(a) Assuming ideal gas,

$$\frac{T_D}{T_C} = \left(\frac{p_{high}}{p_{low}}\right)^{(k-1)/k}$$

Power Cycles

For air, $k=1.4$.

$$\begin{aligned} T_D &= T_C\left(\frac{60\text{ psia}}{14.7\text{ psia}}\right)^{(1.4-1)/1.4} \\ &= (70°\text{F}+460°)(1.495) \\ &= 792.1°\text{R} \end{aligned}$$

The temperature leaving the compressor if the process is not isentropic is

$$\begin{aligned} T'_D &= T_C + \frac{T_D - T_C}{\eta_{\text{compressor}}} \\ &= 530°\text{R} + \frac{792.1°\text{R} - 530°\text{R}}{0.7} \\ &= \boxed{904.4°\text{R}\quad(900°\text{R})} \end{aligned}$$

The answer is (D).

(b) Find the temperature at B.

$$\begin{aligned} \frac{T_A}{T_B} &= \left(\frac{p_{\text{high}}}{p_{\text{low}}}\right)^{(k-1)/k} \\ T_B &= \frac{T_A}{\left(\frac{p_{\text{high}}}{p_{\text{low}}}\right)^{(k-1)/k}} = \frac{25°\text{F}+460°}{\left(\frac{60\text{ psia}}{14.7\text{ psia}}\right)^{(1.4-1)/1.4}} \\ &= 324.5°\text{R} \end{aligned}$$

The temperature leaving the turbine if the process is not isentropic is

$$\begin{aligned} T'_B &= T_A - \eta_{\text{turbine}}(T_A - T_B) \\ &= 485°\text{R} - (0.80)(485°\text{R} - 324.5°\text{R}) \\ &= \boxed{356.6°\text{R}\quad(360°\text{R})} \end{aligned}$$

The answer is (A).

(c) The coefficient of performance of the cycle is

$$\begin{aligned} \text{COP} &= \frac{T_C - T'_B}{(T'_D - T_A) - (T_C - T'_B)} \\ &= \frac{530°\text{R} - 356.6°\text{R}}{(904.5°\text{R} - 485°\text{R}) - (530°\text{R} - 356.6°\text{R})} \\ &= \boxed{0.705\quad(0.7)} \end{aligned}$$

The answer is (A).

SI Solution

(a) Assuming air is an ideal gas with $k=1.4$,

$$\frac{T_D}{T_C} = \left(\frac{p_{\text{high}}}{p_{\text{low}}}\right)^{(k-1)/k}$$

$$\begin{aligned} T_C &= 20\text{K} + 273.15° = 293.15\text{K} \\ T_D &= T_C\left(\frac{p_{\text{high}}}{p_{\text{low}}}\right)^{(k-1)/k} = (293.15\text{K})\left(\frac{400\text{ kPa}}{101\text{ kPa}}\right)^{(1.4-1)/1.4} \\ &= 434.4\text{K} \end{aligned}$$

The temperature leaving the compressor if the process is not isentropic is

$$\begin{aligned} T'_D &= T_C + \frac{T_D - T_C}{\eta_{\text{compressor}}} \\ &= 293.15\text{K} + \frac{434.4\text{K} - 293.15\text{K}}{0.7} \\ &= \boxed{494.9\text{K}\quad(490\text{K})} \end{aligned}$$

The answer is (D).

(b) Find the temperature at B.

$$\begin{aligned} T_A &= -4°\text{C} + 273.15° = 269.15\text{K} \\ T_B &= \frac{T_A}{\left(\frac{p_{\text{high}}}{p_{\text{low}}}\right)^{(k-1)/k}} = \frac{269.15\text{K}}{\left(\frac{400\text{ kPa}}{101\text{ kPa}}\right)^{(1.4-1)/1.4}} \\ &= 181.6\text{K} \end{aligned}$$

The temperature leaving the turbine if the process is not isentropic is

$$\begin{aligned} T'_B &= T_A - \eta_{\text{turbine}}(T_A - T_B) \\ &= 269.15\text{K} - (0.80)(269.15\text{K} - 181.6\text{K}) \\ &= \boxed{199.1\text{K}\quad(200\text{K})} \end{aligned}$$

The answer is (A).

(c) The coefficient of performance of the cycle is

$$\begin{aligned} \text{COP} &= \frac{T_C - T'_B}{(T'_D - T_A) - (T_C - T'_B)} \\ &= \frac{293.15\text{K} - 199.1\text{K}}{(494.9\text{K} - 269.15\text{K}) - (293.15\text{K} - 199.1\text{K})} \\ &= \boxed{0.714\quad(0.7)} \end{aligned}$$

The answer is (A).

34 Fundamental Heat Transfer

PRACTICE PROBLEMS

Thermal Conductivity

1. Experiments have shown that the thermal conductivity, k, of a particular material varies with temperature, T, according to the following relationship.

$$k_T = (0.030)(1 + 0.0015T)$$

What is most nearly the value of k that should be used for a transfer of heat through the material if the hot-side temperature is 350° (°F or °C) and the cold-side temperature is 150° (°F or °C)?

(A) 0.04

(B) 0.06

(C) 0.10

(D) 0.20

Conduction

2. What is most nearly the heat flow through insulating brick 1.0 ft (30 cm) thick in an oven wall with a thermal conductivity of 0.038 Btu-ft/hr-ft^2-°F (0.066 W/m·K)? The thermal gradient is 350°F (195°C).

(A) 10 Btu/hr-ft^2 (30 W/m^2)

(B) 15 Btu/hr-ft^2 (40 W/m^2)

(C) 20 Btu/hr-ft^2 (60 W/m^2)

(D) 30 Btu/hr-ft^2 (100 W/m^2)

3. A composite wall is made up of 3.0 in (7.6 cm) of material A exposed to 1000°F (540°C), 5.0 in (13 cm) of material B, and 6.0 in (15 cm) of material C exposed to 200°F (90°C). The mean thermal conductivities for materials A, B, and C are 0.06, 0.5, and 0.8 Btu-ft/hr-ft^2-°F (0.1, 0.9, and 1.4 W/m·K), respectively.

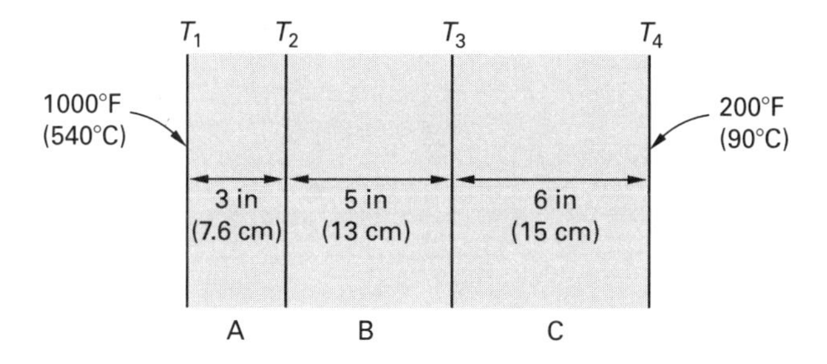

(a) The temperature at the A-B material interface is most nearly

(A) 300°F (100°C)

(B) 350°F (150°C)

(C) 400°F (200°C)

(D) 500°F (250°C)

(b) The temperature at the B-C material interface is most nearly

(A) 300°F (150°C)

(B) 350°F (175°C)

(C) 400°F (200°C)

(D) 500°F (250°C)

4. In an old factory, a hot liquid flows through a 30 ft long copper tube that has a 1 in inner diameter and a 1/3 in wall thickness. The copper tube is covered with a 1/2 in thick layer of glass wool insulation, and the glass wool is covered with a 1/2 in thick layer of asbestos. The outside temperatures of the copper tube and asbestos layer are 500°F and 150°F, respectively. The thermal conductivities of copper, glass wool, and asbestos are 220 Btu/hr-ft-°F, 0.0315 Btu/hr-ft-°F, and 0.115 Btu/hr-ft-°F, respectively. The temperature at the interface of the glass wool and asbestos is most nearly

(A) 190°F

(B) 200°F

(C) 260°F

(D) 350°F

5. In an old factory, a hot liquid flows through a 30 ft long copper tube that has a 1 in inner diameter and a 1/3 in wall thickness. The copper tube is covered with a 1/2 in thick layer of glass wool insulation, and the glass wool is covered with a layer of asbestos of unknown thickness. The outside temperatures of the copper tube and asbestos layer are 500°F and 80°F, respectively. The heat transfer rate per unit length is 110 Btu/hr-ft. The thermal conductivities of copper, glass wool, and asbestos are 220 Btu/hr-ft-°F, 0.0315 Btu/hr-ft-°F, and

Heat Transfer

0.115 Btu/hr-ft-°F, respectively. The thickness of the asbestos layer is most nearly

(A) 0.2 in

(B) 0.3 in

(C) 1.0 in

(D) 2.5 in

6. A chromium spherical tank with a 4 ft outer diameter contains hot oil. The tank is covered with a 5 in thick spherical layer of magnesia insulation. The magnesia insulation is covered with a 2 in thick spherical layer of styrofoam. The outside temperatures of the chromium sphere and the styrofoam layers are 250°F and 70°F, respectively. The thermal conductivities of magnesia and styrofoam are 0.04 Btu/hr-ft-°F and 0.02 Btu/hr-ft-°F, respectively. The temperature at the interface of the magnesia and styrofoam layers is most nearly

(A) 100°F

(B) 110°F

(C) 140°F

(D) 220°F

7. A metallic, thin-walled spherical tank is covered with a layer of insulation. The thickness of the insulation is equal to the radius of the spherical tank. To increase the volumetric capacity, the original tank will be replaced with a new, insulated, thin-walled spherical tank that has a new tank radius equal to twice the original tank radius. The insulation layer material and thickness are unchanged. Simultaneously, the process is modified so that the outside temperatures for the sphere and its insulated layer are also doubled. The ratio of the heat transfer of the new tank to the heat transfer of the original tank is most nearly

(A) 1:2

(B) 2:1

(C) 3:1

(D) 6:1

Unsteady Heat Flow

8. The heat supply of a large building is turned off at 5:00 p.m. when the interior temperature is 70°F (21°C). The outdoor temperature is constant at 40°F (4°C). The thermal capacity of the building and its contents is 100,000 Btu/°F (60 MJ/K), and the conductance is 6500 Btu/hr-°F (1.1 kW/K). The interior temperature at 1:00 a.m. is most nearly

(A) 45°F (7°C)

(B) 50°F (10°C)

(C) 60°F (15°C)

(D) 65°F (20°C)

9. Steel ball bearings varying in diameter, d, from $1/4$ in to $1\,1/2$ in (6.35 mm to 38.1 mm) are quenched from 1800°F (980°C) in an oil bath that remains at 110°F (43°C). The ball bearings are removed when their internal (center) temperature reaches 250°F (120°C). The average film coefficient is 56 Btu/hr-ft^2-°F (320 W/m^2·K).

(a) The time constant in hours for the cooling process as a function of diameter is most nearly

(A) $0.010d$ hr ($1400d$ h)

(B) $0.013d$ hr ($1900d$ h)

(C) $0.017d$ hr ($2300d$ h)

(D) $0.020d$ hr ($2800d$ h)

(b) A linear equation for the time the ball bearings remain in the oil bath as a function of diameter is most nearly

(A) $0.013d$ ($1800d$)

(B) $0.023d$ ($3000d$)

(C) $0.033d$ ($4700d$)

(D) $0.042d$ ($5900d$)

10. Trays of spherical oranges (8 cm diameter; thermal conductivity of 0.45 W/m·°C) initially at 15°C are cooled by 5°C air moving at 0.4 m/s.

(a) The initial convective surface film temperature is most nearly

(A) 2°C

(B) 4°C

(C) 7°C

(D) 10°C

(b) The Reynolds number is most nearly

(A) 1500

(B) 2200

(C) 4200

(D) 7600

(c) The initial cooling rate is most nearly

(A) 4.1 W

(B) 27 W

(C) 93 W

(D) 120 W

(d) The initial conductive temperature gradient at the surfaces of the oranges is most nearly

(A) 10°C/m

(B) 20°C/m

(C) 130°C/m

(D) 460°C/m

(e) The Nusselt number is most nearly

(A) 26

(B) 42

(C) 66

(D) 150

Internal Heat Generation

11. A 0.4 in (1.0 cm) diameter uranium dioxide fuel rod with a thermal conductivity of 1.1 Btu-ft/hr-ft^2-°F (1.9 W/m·K) is clad with 0.020 in (0.5 mm) of stainless steel. The fuel rod generates 4×10^7 Btu/hr-ft^3 (4.1×10^8 W/m^3) internally. A coolant at 500°F (260°C) circulates around the clad rod. The outside film coefficient is 10,000 Btu/hr-ft^2-°F (57 kW/m^2·K). The temperature at the longitudinal centerline of the rod is most nearly

(A) 2100°F (1200°C)

(B) 2400°F (1300°C)

(C) 2700°F (1500°C)

(D) 3100°F (1700°C)

Finned Radiators

12. Two long pieces of $^1/_{16}$ in (1.6 mm) copper wire are connected end-to-end with a hot soldering iron. The minimum melting temperature of the solder is 450°F (230°C). The surrounding air temperature is 80°F (27°C). The unit film coefficient is 3 Btu/hr-ft^2-°F (17 W/m^2·K). Disregard radiation losses. The minimum rate of heat application to keep the solder molten is most nearly

(A) 4.3 Btu/hr (1.3 W)

(B) 11 Btu/hr (3.3 W)

(C) 20 Btu/hr (6.0 W)

(D) 47 Btu/hr (14 W)

SOLUTIONS

1. Use the value of k at an average temperature of $^1/_2(T_1 + T_2)$.

$$\begin{aligned} T &= \left(\tfrac{1}{2}\right)(150° + 350°) = 250° \\ k &= (0.030)(1 + 0.0015T) \\ &= (0.030)\big(1 + (0.0015)(250°)\big) \\ &= \boxed{0.04125 \quad (0.04)} \end{aligned}$$

The answer is (A).

2. *Customary U.S. Solution*

Use Eq. 34.24. From Fourier's law of heat conduction,

$$\begin{aligned} \frac{Q_{1-2}}{A} &= \frac{k\Delta T}{L} = \frac{\left(0.038 \; \frac{\text{Btu-ft}}{\text{hr-ft}^2\text{-°F}}\right)(350°\text{F})}{1.0 \text{ ft}} \\ &= \boxed{13.3 \text{ Btu/hr-ft}^2 \quad (15 \text{ Btu/hr-ft}^2)} \end{aligned}$$

The answer is (B).

SI Solution

Use Eq. 34.24. From Fourier's law of heat conduction,

$$\begin{aligned} \frac{Q_{1-2}}{A} &= \frac{k\Delta T}{L} = \frac{\left(0.066 \; \frac{\text{W}}{\text{m·K}}\right)(195\text{K})\left(100 \; \frac{\text{cm}}{\text{m}}\right)}{30 \text{ cm}} \\ &= \boxed{42.9 \text{ W/m}^2 \quad (40 \text{ W/m}^2)} \end{aligned}$$

The answer is (B).

3. *Customary U.S. Solution*

Since the wall temperatures are given, it is not necessary to consider films.

From Eq. 34.28, the heat flow through the composite wall is

$$Q = \frac{A(T_1 - T_4)}{\sum\limits_{i=1}^{n} \frac{L_i}{k_i}}$$

On a per unit area basis,

$$\frac{Q}{A} = \frac{T_1 - T_4}{\sum_{i=1}^{n} \frac{L_i}{k_i}} = \frac{T_1 - T_2}{\frac{L_A}{k_A}} = \frac{T_2 - T_3}{\frac{L_B}{k_B}}$$

$$= \frac{1000^\circ\text{F} - 200^\circ\text{F}}{\frac{3 \text{ in}}{\left(0.06 \frac{\text{Btu-ft}}{\text{hr-ft}^2\text{-}^\circ\text{F}}\right)\left(12 \frac{\text{in}}{\text{ft}}\right)} + \frac{5 \text{ in}}{\left(0.5 \frac{\text{Btu-ft}}{\text{hr-ft}^2\text{-}^\circ\text{F}}\right)\left(12 \frac{\text{in}}{\text{ft}}\right)} + \frac{6 \text{ in}}{\left(0.8 \frac{\text{Btu-ft}}{\text{hr-ft}^2\text{-}^\circ\text{F}}\right)\left(12 \frac{\text{in}}{\text{ft}}\right)}}$$

$$= 142.2 \text{ Btu/hr-ft}^2$$

(a) To find the temperature at the A-B interface, T_2, use

$$\frac{Q}{A} = \frac{T_1 - T_2}{\frac{L_A}{k_A}}$$

$$T_2 = T_1 - \left(\frac{Q}{A}\right)\left(\frac{L_A}{k_A}\right)$$

$$= 1000^\circ\text{F} - \left(142.2 \frac{\text{Btu}}{\text{hr-ft}^2}\right) \times \left(\frac{3 \text{ in}}{\left(0.06 \frac{\text{Btu-ft}}{\text{hr-ft}^2\text{-}^\circ\text{F}}\right)\left(12 \frac{\text{in}}{\text{ft}}\right)}\right)$$

$$= \boxed{407.5^\circ\text{F} \quad (400^\circ\text{F})}$$

The answer is (C).

(b) To find the temperature at the B-C interface, T_3, use

$$\frac{Q}{A} = \frac{T_2 - T_3}{\frac{L_B}{k_B}}$$

$$T_3 = T_2 - \left(\frac{Q}{A}\right)\left(\frac{L_B}{k_B}\right)$$

$$= 407.5^\circ\text{F} - \left(142.2 \frac{\text{Btu}}{\text{hr-ft}^2}\right) \times \left(\frac{5 \text{ in}}{\left(0.5 \frac{\text{Btu-ft}}{\text{hr-ft}^2\text{-}^\circ\text{F}}\right)\left(12 \frac{\text{in}}{\text{ft}}\right)}\right)$$

$$= \boxed{289^\circ\text{F} \quad (300^\circ\text{F})}$$

The answer is (A).

SI Solution

Since the wall temperatures are given, it is not necessary to consider films.

From Eq. 34.28, the heat flow through the composite wall is

$$Q = \frac{A(T_1 - T_4)}{\sum_{i=1}^{n} \frac{L_i}{k_i}}$$

On a per unit area basis,

$$\frac{Q}{A} = \frac{T_1 - T_4}{\sum_{i=1}^{n} \frac{L_i}{k_i}} = \frac{T_1 - T_2}{\frac{L_A}{k_A}} = \frac{T_2 - T_3}{\frac{L_B}{k_B}}$$

$$= \frac{540^\circ\text{C} - 90^\circ\text{C}}{\frac{7.6 \text{ cm}}{\left(0.1 \frac{\text{W}}{\text{m}\cdot\text{K}}\right)\left(100 \frac{\text{cm}}{\text{m}}\right)} + \frac{13 \text{ cm}}{\left(0.9 \frac{\text{W}}{\text{m}\cdot\text{K}}\right)\left(100 \frac{\text{cm}}{\text{m}}\right)} + \frac{15 \text{ cm}}{\left(1.4 \frac{\text{W}}{\text{m}\cdot\text{K}}\right)\left(100 \frac{\text{cm}}{\text{m}}\right)}}$$

$$= 444.8 \text{ W/m}^2$$

(a) To find the temperature at the A-B interface, T_2, use

$$\frac{Q}{A} = \frac{T_1 - T_2}{\frac{L_A}{k_A}}$$

$$T_2 = T_1 - \left(\frac{Q}{A}\right)\left(\frac{L_A}{k_A}\right)$$

$$= 540^\circ\text{C} - \left(444.8 \frac{\text{W}}{\text{m}^2}\right)\left(\frac{7.6 \text{ cm}}{\left(0.1 \frac{\text{W}}{\text{m}\cdot\text{K}}\right)\left(100 \frac{\text{cm}}{\text{m}}\right)}\right)$$

$$= \boxed{202.0^\circ\text{C} \quad (200^\circ\text{C})}$$

The answer is (C).

(b) To find the temperature at the B-C interface, T_3, use

$$\frac{Q}{A} = \frac{T_2 - T_3}{\frac{L_B}{k_B}}$$

$$T_3 = T_2 - \left(\frac{Q}{A}\right)\left(\frac{L_B}{k_B}\right)$$

$$= 202.0^\circ\text{C} - \left(444.8 \frac{\text{W}}{\text{m}^2}\right)\left(\frac{13 \text{ cm}}{\left(0.9 \frac{\text{W}}{\text{m}\cdot\text{K}}\right)\left(100 \frac{\text{cm}}{\text{m}}\right)}\right)$$

$$= \boxed{137.8^\circ\text{C} \quad (150^\circ\text{C})}$$

The answer is (A).

4. Because the temperature outside the copper tube is known, the thermal conductivity of the copper and the pipe's inside radius can be disregarded. r_a is the radius at the interface between the copper and the first layer of insulation. r_b is the radius at the interface between the glass wool insulation and the asbestos layer. r_c is the outer radius of the glass wool layer plus the thickness of the asbestos layer.

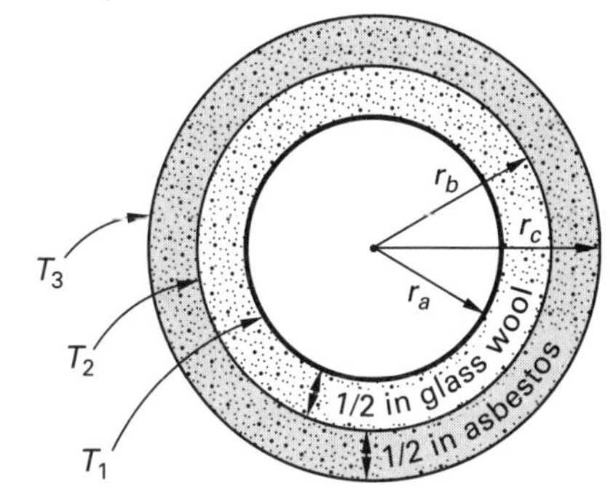

$$r_a = r_{\text{inside}} + t_{\text{pipe}} = \tfrac{1}{2}\text{ in} + \tfrac{1}{3}\text{ in} = 0.833\text{ in}$$

$$r_b = r_a + t_{\text{glass wool}} = 0.833\text{ in} + \tfrac{1}{2}\text{ in} = 1.333\text{ in}$$

$$r_c = r_b + t_{\text{asbestos}} = 1.333\text{ in} + \tfrac{1}{2}\text{ in} = 1.833\text{ in}$$

The copper tube's length is not necessary for the solution. Use Eq. 34.40 and the known temperatures to calculate the heat transfer per unit length from the outside of the copper tube through the asbestos.

$$\frac{Q}{L} = \frac{2\pi(T_1 - T_3)}{\dfrac{\ln \dfrac{r_b}{r_a}}{k_{\text{glass wool}}} + \dfrac{\ln \dfrac{r_c}{r_b}}{k_{\text{asbestos}}}}$$

$$= \frac{2\pi(500°\text{F} - 150°\text{F})}{\dfrac{\ln \dfrac{1.333\text{ in}}{0.833\text{ in}}}{0.0315\ \dfrac{\text{Btu}}{\text{hr-ft-°F}}} + \dfrac{\ln \dfrac{1.833\text{ in}}{1.333\text{ in}}}{0.115\ \dfrac{\text{Btu}}{\text{hr-ft-°F}}}}$$

$$= 124.28\text{ Btu/hr-ft}$$

Use Eq. 34.40 again to calculate the temperature at the interface of glass wool and the asbestos from the heat layer through a single layer.

$$\frac{Q}{L} = \frac{2\pi(T_2 - T_3)}{\dfrac{\ln \dfrac{r_c}{r_b}}{k_{\text{asbestos}}}}$$

$$T_2 = \frac{\left(\dfrac{Q}{L}\right)\left(\dfrac{\ln \dfrac{r_c}{r_b}}{k_{\text{asbestos}}}\right)}{2\pi} + T_3$$

$$= \frac{\left(124.28\ \dfrac{\text{Btu}}{\text{hr-ft}}\right)\left(\dfrac{\ln \dfrac{1.833\text{ in}}{1.333\text{ in}}}{0.115\ \dfrac{\text{Btu}}{\text{hr-ft-°F}}}\right)}{2\pi} + 150°\text{F}$$

$$= \boxed{204.79°\text{F}\quad(200°\text{F})}$$

The answer is (B).

5. r_a is the radius at the interface between the copper pipe and the first layer of insulation. r_b is the radius at the interface between the glass wool insulation and the asbestos layer.

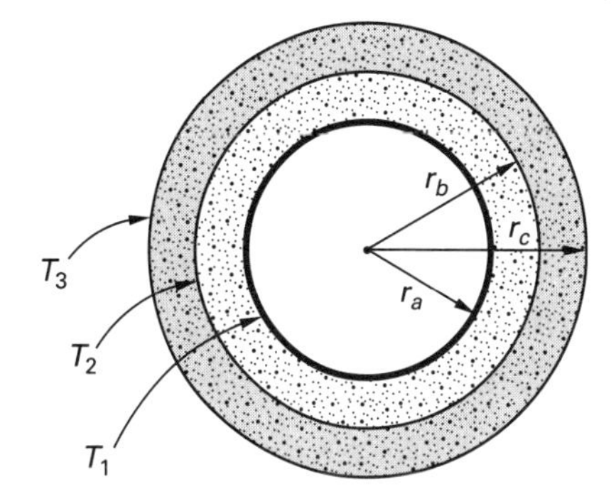

$$r_a = r_{\text{inside}} + t_{\text{pipe}} = \tfrac{1}{2}\text{ in} + \tfrac{1}{3}\text{ in} = 0.833\text{ in}$$

$$r_b = r_a + t_{\text{glass wool}} = 0.833\text{ in} + \tfrac{1}{2}\text{ in} = 1.333\text{ in}$$

The copper tube's length is not necessary for the solution. Use Eq. 34.40 to calculate the unknown radius, r_c, which is the outer radius of the glass wool layer plus the thickness of the asbestos layer.

$$\frac{Q}{L} = \frac{2\pi(T_1 - T_3)}{\dfrac{1}{k_{\text{glass wool}}}\ln \dfrac{r_b}{r_a} + \dfrac{1}{k_{\text{asbestos}}}\ln \dfrac{r_c}{r_b}}$$

$$110\ \frac{\text{Btu}}{\text{hr-ft}} = \frac{2\pi(500°\text{F} - 80°\text{F})}{\left(\dfrac{1}{0.0315\ \dfrac{\text{Btu}}{\text{hr-ft-°F}}}\right)\ln \dfrac{1.333\text{ in}}{0.833\text{ in}} + \left(\dfrac{1}{0.115\ \dfrac{\text{Btu}}{\text{hr ft °F}}}\right)\ln \dfrac{r_c}{1.333\text{ in}}}$$

$$\ln \frac{r_c}{1.333\text{ in}} = 1.0425$$

$$r_c = (1.333\text{ in})e^{1.0425}$$

$$= 3.781\text{ in}$$

The thickness of the asbestos is

$$t_{\text{asbestos}} = r_c - r_b = 3.781\text{ in} - 1.333\text{ in}$$

$$= \boxed{2.448\text{ in}\quad(2.5\text{ in})}$$

The answer is (D).

6. r_a is the outer radius of the tank. r_b is the radius at the interface between the magnesia insulation and the styrofoam layer. r_c is the outer radius of the magnesia insulation layer plus the thickness of the styrofoam layer.

Heat Transfer

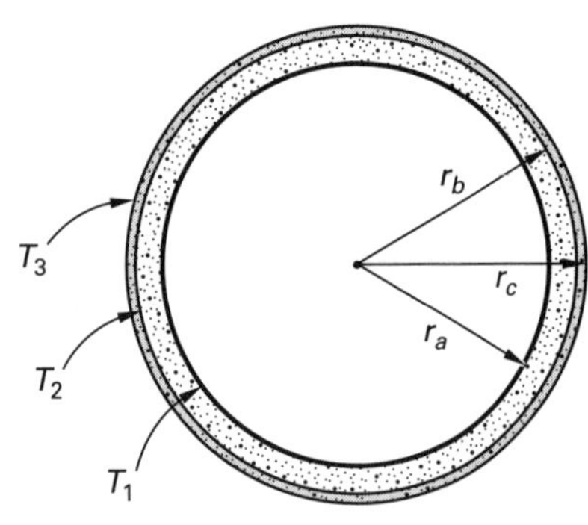

$$r_a = \frac{r_o}{2} = \frac{4 \text{ ft}}{2} = 2 \text{ ft}$$

$$r_b = r_a + t_{\text{magnesia}} = 2 \text{ ft} + \frac{5 \text{ in}}{12 \frac{\text{in}}{\text{ft}}} = 2.42 \text{ ft}$$

$$r_c = r_b + t_{\text{styrofoam}} = 2.42 \text{ ft} + \frac{2 \text{ in}}{12 \frac{\text{in}}{\text{ft}}} = 2.58 \text{ ft}$$

Use Eq. 34.49 to calculate the heat transfer from the chromium sphere through the styrofoam.

$$Q = \frac{4\pi(T_1 - T_3)}{\frac{\frac{1}{r_a} - \frac{1}{r_b}}{k_{\text{magnesia}}} + \frac{\frac{1}{r_b} - \frac{1}{r_c}}{k_{\text{styrofoam}}}}$$

$$= \frac{4\pi(250°\text{F} - 70°\text{F})}{\frac{\frac{1}{2 \text{ ft}} - \frac{1}{2.42 \text{ ft}}}{0.04 \frac{\text{Btu}}{\text{hr-ft-°F}}} + \frac{\frac{1}{2.42 \text{ ft}} - \frac{1}{2.58 \text{ ft}}}{0.02 \frac{\text{Btu}}{\text{hr-ft-°F}}}}$$

$$= 655.5 \text{ Btu/hr-ft}$$

Use Eq. 34.49 again to calculate the temperature at the interface of the magnesia and the styrofoam from the heat transfer through a single layer.

$$T_1 - T_2 = \frac{Q\left(\frac{\frac{1}{r_a} - \frac{1}{r_b}}{k_{\text{magnesia}}}\right)}{4\pi}$$

$$T_2 = T_1 - \frac{Q\left(\frac{\frac{1}{r_a} - \frac{1}{r_b}}{k_{\text{magnesia}}}\right)}{4\pi}$$

$$= 250°\text{F} - \frac{\left(655.5 \frac{\text{Btu}}{\text{hr-ft}}\right)\left(\frac{\frac{1}{2 \text{ ft}} - \frac{1}{2.42 \text{ ft}}}{0.04 \frac{\text{Btu}}{\text{hr-ft-°F}}}\right)}{4\pi}$$

$$= \boxed{136.8°\text{F} \quad (140°\text{F})}$$

The answer is (C).

7. $r_{a,1}$ is the outside radius of the original tank.

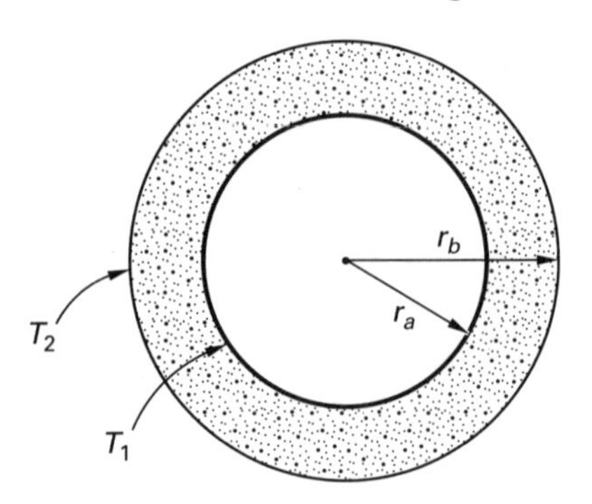

$$r_{a,1} = r$$

$$r_{b,1} = r + r = 2r$$

The heat transfer rate through an insulated, spherical tank is found from Eq. 34.49. For the original tank,

$$Q_1 = \frac{4\pi(T_{a,1} - T_{b,1})}{\frac{\frac{1}{r_{a,1}} - \frac{1}{r_{b,1}}}{k}} = \frac{4\pi(T_{a,1} - T_{b,1})}{\frac{\frac{1}{r} - \frac{1}{2r}}{k}}$$

$$= 8\pi k r(T_{a,1} - T_{b,1})$$

For the new tank,

$$r_{a,2} = 2r$$

$$r_{b,2} = 2r + r = 3r$$

The outside temperatures of the tank and insulation are doubled. From Eq. 34.48, the heat transfer rate for the new tank, Q_2, can be simplified to

$$Q_2 = \frac{4\pi(T_{a,2} - T_{b,2})}{\frac{\frac{1}{r_{a,2}} - \frac{1}{r_{b,2}}}{k}} = \frac{4\pi(2T_{a,1} - 2T_{b,1})}{\frac{\frac{1}{2r} - \frac{1}{3r}}{k}}$$

$$= 48\pi k r(T_{a,1} - T_{b,1})$$

The ratio of the heat transfer of the new tank to the heat transfer of the original tank is

$$\frac{Q_2}{Q_1} = \frac{48\pi k r(T_{a,1} - T_{b,1})}{8\pi k r(T_{a,1} - T_{b,1})} = \boxed{6 \quad (6{:}1)}$$

The answer is (D).

8. *Customary U.S. Solution*

This is a transient problem. The total time is from 5 p.m. to 1 a.m., which is eight hours.

The thermal capacitance (capacity), C_e, is given as 100,000 Btu/°F.

Resistance and conductance are reciprocals. The thermal resistance is

$$R_e = \frac{1}{\text{thermal conductance}} = \frac{1}{6500\ \frac{\text{Btu}}{\text{hr-°F}}}$$
$$= 0.0001538\ \text{hr-°F/Btu}$$

From Eq. 34.59,

$$T_t = T_\infty + (T_0 - T_\infty)e^{-t/R_eC_e}$$
$$T_{8\ \text{hr}} = 40°\text{F} + (70°\text{F} - 40°\text{F}) \times \exp\left(\frac{-8\ \text{hr}}{\left(0.0001538\ \frac{\text{hr-°F}}{\text{Btu}}\right)\left(100{,}000\ \frac{\text{Btu}}{\text{°F}}\right)}\right)$$
$$= \boxed{57.8°\text{F}\quad(60°\text{F})}$$

The answer is (C).

SI Solution

The thermal capacitance (capacity) is

$$C_e = \left(60\ \frac{\text{MJ}}{\text{K}}\right)\left(1000\ \frac{\text{kJ}}{\text{MJ}}\right) = 60\,000\ \text{kJ/K}\quad\text{[given]}$$

Resistance and conductance are reciprocals. The thermal resistance is

$$R_e = \frac{1}{\text{thermal conductance}} = \frac{1}{1.1\ \frac{\text{kW}}{\text{K}}} = 0.909\ \text{K/kW}$$

From Eq. 34.59,

$$T_t = T_\infty + (T_0 - T_\infty)e^{-t/R_eC_e}$$
$$= 4°\text{C} + (21°\text{C} - 4°\text{C}) \times \exp\left(\frac{(-8\ \text{h})\left(3600\ \frac{\text{s}}{\text{h}}\right)}{\left(0.909\ \frac{\text{K}}{\text{kW}}\right)\left(60\,000\ \frac{\text{kJ}}{\text{K}}\right)}\right)$$
$$= \boxed{14.0°\text{C}\quad(15°\text{C})}$$

The answer is (C).

9. *Customary U.S. Solution*

This is a transient problem. Check the Biot number to see if the lumped parameter method can be used.

The characteristic length from Eq. 34.14 is

$$L_c = \frac{V}{A_s} = \frac{\left(\frac{\pi}{6}\right)d^3}{\pi d^2} = \frac{d}{6}$$

For the largest ball, $d = 1.5$ in.

$$L_c = \frac{d}{6} = \frac{1.5\ \text{in}}{(6)\left(12\ \frac{\text{in}}{\text{ft}}\right)} = 0.0208\ \text{ft}$$

Evaluate the thermal conductivity, k, of steel at

$$\left(\tfrac{1}{2}\right)(1800°\text{F} + 250°\text{F}) = 1025°\text{F}$$

From App. 34.B, for steel, $k \approx 22.0$ Btu-ft/hr-ft^2-°F.

From Eq. 34.16, the Biot number is

$$\text{Bi} = \frac{hL_c}{k} = \frac{\left(56\ \frac{\text{Btu}}{\text{hr-ft}^2\text{-°F}}\right)(0.0208\ \text{ft})}{22.0\ \frac{\text{Btu-ft}}{\text{hr-ft}^2\text{-°F}}} = 0.053$$

(a) For small balls, Bi will be even smaller.

Since Bi < 0.10, the lumped parameter method can be used.

The assumptions are

- homogeneous body temperature
- minimal radiation losses
- oil bath temperature remains constant
- h remains constant

From Eq. 34.57 and Eq. 34.58, the time constant is

$$C_eR_e = c_p\rho V\left(\frac{1}{hA_s}\right) = \left(\frac{c_p\rho}{h}\right)\left(\frac{V}{A_s}\right) = \left(\frac{c_p\rho}{h}\right)L_c$$
$$= \left(\frac{c_p\rho}{h}\right)\left(\frac{d}{6}\right)$$

From App. 34.B, $\rho = 490$ lbm/ft^3 and $c_p = 0.11$ Btu/lbm-°F, even though those values are for 32°F.

The time constant in hours is (measuring d in inches)

$$C_eR_e = \frac{\left(\frac{\left(0.11\ \frac{\text{Btu}}{\text{lbm-°F}}\right)\left(490\ \frac{\text{lbm}}{\text{ft}^3}\right)}{56\ \frac{\text{Btu}}{\text{hr-ft}^2\text{-°F}}}\right)\left(\frac{d}{6}\right)}{12\ \frac{\text{in}}{\text{ft}}}$$
$$= \boxed{0.01337d\quad(0.013d)}$$

The answer is (B).

Heat Transfer

(b) Taking the natural log of the transient equation, Eq. 34.60,

$$T_t = 250°\text{F}$$
$$T_\infty = 110°\text{F}$$
$$\Delta T = 1800°\text{F} - 110°\text{F} = 1690°\text{F}$$
$$\begin{aligned}\ln(T_t - T_\infty) &= \ln\left(\Delta T e^{\frac{-t}{R_e C_e}}\right) \\ &= \ln \Delta T + \ln e^{\frac{-t}{R_e C_e}} \\ &= \ln \Delta T - \frac{t}{R_e C_e}\end{aligned}$$
$$\ln(250°\text{F} - 110°\text{F}) = \ln(1690°\text{F}) - \frac{t}{0.01337d}$$
$$4.942 = 7.432 - \frac{t}{0.01337d}$$
$$t = \boxed{0.0333d \quad (0.033d)}$$

The answer is (C).

SI Solution

For the largest ball, the characteristic length is

$$L_c = \frac{d}{6} = \frac{38.1\ \text{mm}}{(6)\left(1000\ \frac{\text{mm}}{\text{m}}\right)} = 6.35 \times 10^{-3}\ \text{m}$$

Evaluate the thermal conductivity, k, of steel at

$$\left(\tfrac{1}{2}\right)(980°\text{C} + 120°\text{C}) = 550°\text{C}$$

From App. 34.B and its footnote,

$$\begin{aligned}k &\approx \left(22.0\ \frac{\text{Btu-ft}}{\text{hr-ft}^2\text{-°F}}\right)\left(\frac{1.7307\ \text{W·hr·ft}^2\text{·°F}}{\text{m·K·Btu·ft}}\right) \\ &= 38.08\ \text{W/m·K}\end{aligned}$$

From Eq. 34.16, the Biot number is

$$\text{Bi} = \frac{hL_c}{k} = \frac{\left(320\ \frac{\text{W}}{\text{m}^2\text{·K}}\right)(6.35 \times 10^{-3}\ \text{m})}{38.08\ \frac{\text{W}}{\text{m·K}}} = 0.053$$

(a) For small balls, Bi will be even smaller.

Since Bi < 0.10, the lumped method can be used. The assumptions are given in the customary U.S. solution. From Eq. 34.57 and Eq. 34.58, the time constant in hours is

$$\begin{aligned}C_e R_e &= c_p \rho V\left(\frac{1}{hA_s}\right) = \left(\frac{c_p\rho}{h}\right)\left(\frac{V}{A_s}\right) = \left(\frac{c_p\rho}{h}\right)L_c \\ &= \left(\frac{c_p\rho}{h}\right)\left(\frac{d}{6}\right)\end{aligned}$$

From App. 34.B and its footnote,

$$\begin{aligned}\rho &= \left(490\ \frac{\text{lbm}}{\text{ft}^3}\right)\left(16.0185\ \frac{\text{kg·ft}^3}{\text{m}^3\text{·lbm}}\right) \\ &= 7849.1\ \text{kg/m}^3\end{aligned}$$

$$\begin{aligned}c_p &= \left(0.11\ \frac{\text{Btu}}{\text{lbm-°F}}\right)\left(4186.8\ \frac{\text{J·lbm·°F}}{\text{kg·K·Btu}}\right) \\ &= 460.5\ \text{J/kg·K}\end{aligned}$$

$$\begin{aligned}C_e R_e &= \left(\frac{\left(460.5\ \frac{\text{J}}{\text{kg·K}}\right)\left(7849.1\ \frac{\text{kg}}{\text{m}^3}\right)}{320\ \frac{\text{W}}{\text{m}^2\text{·K}}}\right)\left(\frac{d}{6}\right) \\ &= \boxed{1882.6d \quad (1900d)}\end{aligned}$$

The answer is (B).

(b) Taking the natural log of the transient equation, Eq. 34.60,

$$T_t = 120°\text{C}$$
$$T_\infty = 43°\text{C}$$
$$\Delta T = 980°\text{C} - 43°\text{C} = 937°\text{C}$$
$$\ln(T_t - T_\infty) = \ln \Delta T - \frac{t}{1882.6d}$$
$$\ln(120°\text{C} - 43°\text{C}) = \ln(937°\text{C}) - \frac{t}{1882.6d}$$
$$4.344 = 6.843 - \frac{t}{R_e C_e}$$
$$t = \boxed{4704.6d \quad (4700d)}$$

The answer is (C).

10. (a) The initial film temperature is

$$T_{\text{film}} = \frac{T_s + T_\infty}{2} = \frac{15°\text{C} + 5°\text{C}}{2} = \boxed{10°\text{C}}$$

The answer is (D).

(b) From App. 35.D, the kinematic viscosity of the cooling medium (air) at 10°C is

$$\nu = \frac{\mu}{\rho} = \frac{1.78 \times 10^{-5}\ \frac{\text{kg}}{\text{m·s}}}{1.246\ \frac{\text{kg}}{\text{m}^3}} = 1.43 \times 10^{-5}\ \text{m}^2/\text{s}$$

The Reynolds number is

$$\begin{aligned}\text{Re} &= \frac{\text{v}L}{\nu} = \frac{\left(0.4\ \frac{\text{m}}{\text{s}}\right)(8\ \text{cm})}{\left(1.43 \times 10^{-5}\ \frac{\text{m}^2}{\text{s}}\right)\left(100\ \frac{\text{cm}}{\text{m}}\right)} \\ &= \boxed{2238 \quad (2200)}\end{aligned}$$

The answer is (B).

(c) From App. 35.D, the thermal conductivity of the cooling medium (air) at 10°C is

$$k_{\text{air}} = 0.02492 \text{ W/m·K}$$

From App. 34.D, the Nusselt correlation constants are $C = 5.05$, $a = 0$, and $b = 0.333$.

$$\text{Nu} = C\text{Pr}^a\text{Re}^b = 5.05\text{Pr}^0\text{Re}^{0.333}$$
$$\text{Nu} = \frac{hL}{k_{\text{air}}}$$
$$\frac{hL}{k_{\text{air}}} = 5.05\text{Re}^{0.333}$$

The convective film coefficient is

$$h = \frac{5.05k_{\text{air}}\text{Re}^{0.333}}{L} = \frac{(5.05)\left(0.02492 \ \frac{\text{W}}{\text{m·K}}\right)(2238)^{0.333}}{\frac{8 \text{ cm}}{100 \ \frac{\text{cm}}{\text{m}}}}$$
$$= 20.52 \text{ W/m}^2\text{·°C}$$

The surface area of the sphere is

$$A_{\text{surface}} = 4\pi r^2 = \pi d^2 = \pi\left(\frac{8 \text{ cm}}{100 \ \frac{\text{cm}}{\text{m}}}\right)^2$$
$$= 0.0201 \text{ m}^2$$

The initial cooling rate is

$$Q = hA_{\text{surface}}(T_\infty - T_s)$$
$$= \left(20.52 \ \frac{\text{W}}{\text{m}^2\text{·°C}}\right)(0.0201 \text{ m}^2)(15\text{°C} - 5\text{°C})$$
$$= \boxed{4.12 \text{ W} \quad (4.1 \text{ W})}$$

The answer is (A).

(d) The conductive heat loss near the interior surface of the orange is equal to the convective heat loss near the exterior surface of the orange. On a per unit surface area basis,

$$\frac{Q_{\text{conductive}}}{A_{\text{surface}}} = \frac{Q_{\text{convective}}}{A_{\text{surface}}}$$
$$\frac{Q}{A_{\text{surface}}} = \frac{-k_{\text{orange}}\frac{\partial T}{\partial r}}{A_{\text{surface}}} = \frac{h(T_s - T_\infty)}{A_{\text{surface}}}$$

The thermal conductivity of the orange is given as 0.45 W/m·°C. The initial conductive temperature gradient at the surfaces of the oranges is

$$\frac{\partial T}{\partial r} = \frac{-h(T_s - T_\infty)}{k_{\text{orange}}} = \frac{-\left(20.52 \ \frac{\text{W}}{\text{m}^2\text{·°C}}\right)(5\text{°C} - 15\text{°C})}{0.45 \ \frac{\text{W}}{\text{m·°C}}}$$
$$= \boxed{456\text{°C/m} \quad (460\text{°C/m})}$$

The answer is (D).

(e) The Nusselt number is

$$\text{Nu} = \frac{hL}{k_{\text{air}}} = \frac{\left(20.52 \ \frac{\text{W}}{\text{m}^2\text{·°C}}\right)\left(\frac{8 \text{ cm}}{100 \ \frac{\text{cm}}{\text{m}}}\right)}{0.02492 \ \frac{\text{W}}{\text{m·°C}}}$$
$$= \boxed{65.9 \quad (66)}$$

The answer is (C).

11. *Customary U.S. Solution*

The volume of the rod is

$$V = AL = \left(\frac{\pi}{4}\right)d^2L$$

The volume per unit length is

$$\frac{V}{L} = \frac{\pi}{4}d^2 = \frac{\left(\frac{\pi}{4}\right)(0.4 \text{ in})^2}{\left(12 \ \frac{\text{in}}{\text{ft}}\right)^2}$$
$$= 8.727 \times 10^{-4} \text{ ft}^3\text{/ft}$$

The heat output per unit length of rod is

$$\frac{Q}{L} = \left(\frac{V}{L}\right)G = \left(8.727 \times 10^{-4} \ \frac{\text{ft}^3}{\text{ft}}\right)\left(4 \times 10^7 \ \frac{\text{Btu}}{\text{hr-ft}^3}\right)$$
$$= 3.491 \times 10^4 \text{ Btu/hr-ft}$$

The diameter of the cladding is

$$d_o = 0.4 \text{ in} + (2)(0.020 \text{ in}) = 0.44 \text{ in}$$

The surface area per unit length of cladding is

$$A = \pi d_o = \frac{\pi(0.44 \text{ in})}{12 \ \frac{\text{in}}{\text{ft}}} = 0.1152 \text{ ft}^2\text{/ft}$$

Heat Transfer

From Eq. 34.30, the surface temperature of the cladding is

$$T_s = \frac{Q}{hA} + T_\infty = \frac{3.491 \times 10^4 \ \frac{\text{Btu}}{\text{hr-ft}}}{\left(10{,}000 \ \frac{\text{Btu}}{\text{hr-ft}^2\text{-°F}}\right)\left(0.1152 \ \frac{\text{ft}^2}{\text{ft}}\right)} + 500\text{°F} = 530.3\text{°F}$$

For the cladding,

$$r_o = \frac{d_o}{2} = \frac{0.44 \text{ in}}{(2)\left(12 \ \frac{\text{in}}{\text{ft}}\right)} = 0.01833 \text{ ft}$$

$$r_i = \frac{d}{2} = \frac{0.4 \text{ in}}{(2)\left(12 \ \frac{\text{in}}{\text{ft}}\right)} = 0.01667 \text{ ft}$$

From App. 34.B, k for stainless steel (at 572°F) is 11 Btu-ft/hr-ft^2-°F. This is reasonable because the inside cladding temperature is greater than the surface temperature (530.3°F).

From Eq. 34.7 and Eq. 34.26, for a cylinder,

$$T_{\text{inside}} - T_{\text{outside}} = \frac{Q \ln \frac{r_o}{r_i}}{2\pi k L} = \frac{\frac{Q}{L} \ln \frac{r_o}{r_i}}{2\pi k} = \frac{\left(3.491 \times 10^4 \ \frac{\text{Btu}}{\text{hr-ft}}\right) \ln \frac{0.01833 \text{ ft}}{0.01667 \text{ ft}}}{2\pi\left(11 \ \frac{\text{Btu-ft}}{\text{hr-ft}^2\text{-°F}}\right)} = 47.9\text{°F}$$

$$T_{\text{inside cladding}} = T_{\text{outside fuel rod}} + 47.9\text{°F} = 530.3\text{°F} + 47.9\text{°F} = 578.2\text{°F}$$

From Eq. 34.65, using $r_{o,\text{fuel rod}} = r_{i,\text{cladding}}$, and using k for the uranium dioxide (not the stainless steel),

$$T_{\text{center}} = T_o + \frac{G r_o^2}{4k} = 578.2\text{°F} + \frac{\left(4 \times 10^7 \ \frac{\text{Btu}}{\text{hr-ft}^3}\right)(0.01667 \text{ ft})^2}{(4)\left(1.1 \ \frac{\text{Btu-ft}}{\text{hr-ft}^2\text{-°F}}\right)} = \boxed{3104\text{°F} \quad (3100\text{°F})}$$

The answer is (D).

SI Solution

The rod volume per unit length is

$$\frac{V}{L} = \left(\frac{\pi}{4}\right) d^2 = \frac{\left(\frac{\pi}{4}\right)(1.0 \text{ cm})^2}{\left(100 \ \frac{\text{cm}}{\text{m}}\right)^2} = 7.854 \times 10^{-5} \text{ m}^3/\text{m}$$

The heat output per unit length of rod is

$$\frac{Q}{L} = \left(\frac{V}{L}\right) G = \left(7.854 \times 10^{-5} \ \frac{\text{m}^3}{\text{m}}\right)\left(4.1 \times 10^8 \ \frac{\text{W}}{\text{m}^3}\right) = 32\,201.4 \text{ W/m}$$

The diameter of the cladding is

$$d_o = \frac{1.0 \text{ cm} + \frac{(2)(0.5 \text{ mm})}{10 \ \frac{\text{mm}}{\text{cm}}}}{100 \ \frac{\text{cm}}{\text{m}}} = 0.011 \text{ m}$$

The surface area per unit length of cladding is

$$A = \pi d_o = \pi(0.011 \text{ m}) = 0.0346 \text{ m}^2/\text{m}$$

From Eq. 34.30, the surface temperature of the cladding is

$$T_s = \frac{Q}{hA} + T_\infty = \frac{32\,201.4 \ \frac{\text{W}}{\text{m}}}{\left(57 \ \frac{\text{kW}}{\text{m}^2\cdot\text{K}}\right)\left(1000 \ \frac{\text{W}}{\text{kW}}\right)\left(0.0346 \ \frac{\text{m}^2}{\text{m}}\right)} + 260\text{°C} = 276.3\text{°C}$$

For the cladding,

$$r_o = \frac{d_o}{2} = \frac{0.011 \text{ m}}{2} = 0.0055 \text{ m}$$

$$r_i = \frac{d_i}{2} = \frac{0.01 \text{ m}}{2} = 0.0050 \text{ m}$$

From App. 34.B and the table's footnote, k for stainless steel at 300°C is

$$k \approx \left(11 \ \frac{\text{Btu-ft}}{\text{hr-ft}^2\text{-°F}}\right)\left(1.7307 \ \frac{\text{W}\cdot\text{hr}\cdot\text{ft}^2\cdot\text{°F}}{\text{m}\cdot\text{K}\cdot\text{Btu}\cdot\text{ft}}\right) = 19.038 \text{ W/m}\cdot\text{K}$$

This is reasonable because the inside cladding temperature is greater than the surface temperature (276.3°C).

Heat Transfer

For a cylinder, from Eq. 34.7 and Eq. 34.26,

$$T_{\text{inside}} - T_{\text{outside}} = \frac{Q \ln \frac{r_o}{r_i}}{2\pi k L} = \left(\frac{Q}{L}\right)\left(\frac{\ln \frac{r_o}{r_i}}{2\pi k}\right)$$

$$= \left(32\,201.4 \ \frac{\text{W}}{\text{m}}\right)\left(\frac{\ln \frac{0.0055 \text{ m}}{0.0050 \text{ m}}}{(2\pi)\left(19.038 \ \frac{\text{W}}{\text{m}\cdot\text{K}}\right)}\right)$$

$$= 25.7°\text{C}$$

$$T_{\text{inside cladding}} = T_{\text{outside fuel rod}} + 25.7°\text{C}$$

$$= 276.3°\text{C} + 25.7°\text{C}$$

$$= 302°\text{C}$$

From Eq. 34.65, using $r_{o,\text{fuel rod}} = r_{i,\text{cladding}}$, and using k for the uranium dioxide (not the stainless steel),

$$T_{\text{center}} = T_o + \frac{Gr_o^2}{4k}$$

$$= 302°\text{C} + \frac{\left(4.1 \times 10^8 \ \frac{\text{W}}{\text{m}^3}\right)(0.0050 \text{ m})^2}{(4)\left(1.9 \ \frac{\text{W}}{\text{m}\cdot\text{K}}\right)}$$

$$= \boxed{1651°\text{C} \quad (1700°\text{C})}$$

The answer is (D).

12. *Customary U.S. Solution*

Consider this to be two infinite cylindrical fins with

$$T_b = 450°\text{F}$$
$$T_\infty = 80°\text{F}$$
$$h = 3 \text{ Btu/hr-ft}^2\text{-°F}$$

From Eq. 34.75, the perimeter length is

$$P = \pi d = \frac{\pi\left(\frac{1}{16} \text{ in}\right)}{12 \ \frac{\text{in}}{\text{ft}}} = 0.01636 \text{ ft}$$

From App. 34.B, k at 450°F is approximately 215 Btu-ft/hr-ft^2-°F.

From Eq. 34.74, the cross-sectional area of the fin at its base is

$$A_b = \pi r^2 = \pi\left(\frac{d}{2}\right)^2 = \left(\frac{\pi}{4}\right)d^2$$

$$= \frac{\left(\frac{\pi}{4}\right)\left(\frac{1}{16} \text{ in}\right)^2}{\left(12 \ \frac{\text{in}}{\text{ft}}\right)^2}$$

$$= 2.131 \times 10^{-5} \text{ ft}^2$$

From Eq. 34.73, with two fins joined at the middle,

$$Q = 2\sqrt{hPkA_b}(T_b - T_\infty)$$

$$= (2)\sqrt{\begin{array}{l}\left(3 \ \frac{\text{Btu}}{\text{hr-ft}^2\text{-°F}}\right)(0.01636 \text{ ft}) \\ \times \left(215 \ \frac{\text{Btu-ft}}{\text{hr-ft}^2\text{-°F}}\right)(2.131 \times 10^{-5} \text{ ft}^2)\end{array}}$$
$$\times (450°\text{F} - 80°\text{F})$$

$$= \boxed{11.1 \text{ Btu/hr} \quad (11 \text{ Btu/hr})}$$

This disregards radiation.

The answer is (B).

SI Solution

Consider this an infinite cylindrical fin with

$$T_b = 230°\text{C}$$
$$T_\infty = 27°\text{C}$$
$$h = 17 \text{ W/m}^2\cdot\text{K}$$

From Eq. 34.75, the perimeter length is

$$P = \pi d = \frac{\pi(1.6 \text{ mm})}{1000 \ \frac{\text{mm}}{\text{m}}} = 5.027 \times 10^{-3} \text{ m}$$

From App. 34.B and the table footnote, k at 230°C is

$$k \approx \left(215 \ \frac{\text{Btu-ft}}{\text{hr-ft}^2\text{-°F}}\right)\left(1.7307 \ \frac{\text{W}\cdot\text{hr}\cdot\text{ft}^2\cdot\text{°F}}{\text{m}\cdot\text{K}\cdot\text{Btu}\cdot\text{ft}}\right)$$

$$= 372.1 \text{ W/m}\cdot\text{K}$$

From Eq. 34.74, the cross-sectional area of the fin at its base is

$$A_b = \pi r^2 = \pi\left(\frac{d}{2}\right)^2 = \left(\frac{\pi}{4}\right)d^2$$

$$= \frac{\left(\frac{\pi}{4}\right)(1.6 \text{ mm})^2}{\left(1000 \ \frac{\text{mm}}{\text{m}}\right)^2}$$

$$= 2.011 \times 10^{-6} \text{ m}^2$$

From Eq. 34.73, with two fins joined at the middle,

$$Q = 2\sqrt{hPkA_b}(T_b - T_\infty)$$

$$= (2)\sqrt{\begin{array}{l}\left(17 \ \frac{\text{W}}{\text{m}^2\cdot\text{K}}\right)(5.027 \times 10^{-3} \text{ m}) \\ \times \left(372.1 \ \frac{\text{W}}{\text{m}\cdot\text{K}}\right)(2.011 \times 10^{-6} \text{ m}^2)\end{array}}$$
$$\times (230°\text{C} - 27°\text{C})$$

$$= \boxed{3.25 \text{ W} \quad (3.3 \text{ W})}$$

This disregards radiation.

The answer is (B).

Heat Transfer

35 Natural Convection, Evaporation, and Condensation

PRACTICE PROBLEMS

1. The density of 87% wet steam at 50 psia (350 kPa) is most nearly

(A) 0.75 lbm/ft^3 (12 kg/m^3)

(B) 0.89 lbm/ft^3 (14 kg/m^3)

(C) 0.94 lbm/ft^3 (15 kg/m^3)

(D) 1.07 lbm/ft^3 (17 kg/m^3)

2. The viscosity of 100°F (38°C) water in units of lbm/hr-ft (kg/s·m) is most nearly

(A) 1.2 lbm/ft-hr (0.00052 kg/m·s)

(B) 1.4 lbm/ft-hr (0.00060 kg/m·s)

(C) 1.6 lbm/ft-hr (0.00068 kg/m·s)

(D) 1.8 lbm/ft-hr (0.00077 kg/m·s)

3. A fluid in a tank is maintained at 85°F (29°C) by a copper tube carrying hot water. The water decreases in temperature from 190°F (88°C) to 160°F (71°C) as it flows through the tube. The temperature at which the fluid's film coefficient should be evaluated is most nearly

(A) 130°F (54°C)

(B) 160°F (71°C)

(C) 175°F (79°C)

(D) 190°F (88°C)

4. A bare, horizontal conductor with circular cross section with an outside diameter of 0.6 in (1.5 cm) dissipates 8.0 W/ft (25 W/m). The conductor is cooled by free convection, and the surrounding air temperature is 60°F (15°C). The film temperature is 100°F (38°C). The conductor's surface temperature is most nearly

(A) 85°F (29°C)

(B) 110°F (43°C)

(C) 130°F (54°C)

(D) 160°F (67°C)

5. A pot of water sits on a 4 in diameter hot stove burner. The initial water temperature is 70°F. The temperature at the bottom of the pot is the same as the water, except for a 4 in diameter circle that has a temperature of 320°F. After the water temperature has risen to 80°F, what is most nearly the heat transfer rate?

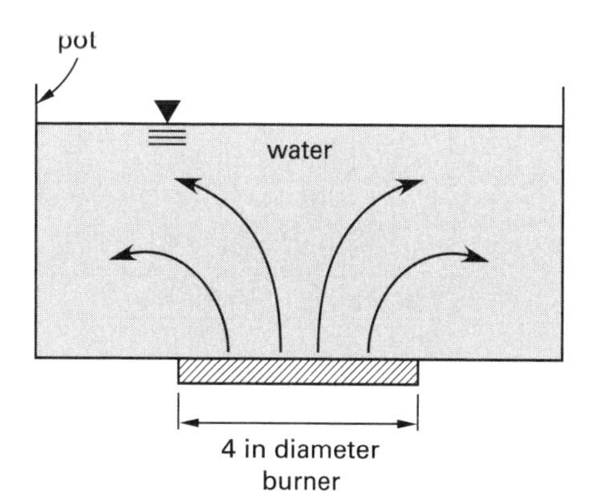

(A) 8500 Btu/hr

(B) 9200 Btu/hr

(C) 9400 Btu/hr

(D) 11,000 Btu/hr

Heat Transfer

SOLUTIONS

1. *Customary U.S. Solution*

The steam is 87% wet, so the quality is $x = 0.13$.

From App. 23.B at 50 psia,

$$\upsilon_f = 0.01727 \text{ ft}^3/\text{lbm}$$
$$\upsilon_g = 8.517 \text{ ft}^3/\text{lbm}$$

From Eq. 23.56, the specific volume of steam is

$$\begin{aligned}\upsilon &= \upsilon_f + x\upsilon_{fg}\\ &= 0.01727 \ \frac{\text{ft}^3}{\text{lbm}} + (0.13)\left(8.517 \ \frac{\text{ft}^3}{\text{lbm}} - 0.01727 \ \frac{\text{ft}^3}{\text{lbm}}\right)\\ &= 1.122 \text{ ft}^3/\text{lbm}\end{aligned}$$

The density is

$$\begin{aligned}\rho &= \frac{1}{\upsilon} = \frac{1}{1.122 \ \frac{\text{ft}^3}{\text{lbm}}}\\ &= \boxed{0.891 \text{ lbm/ft}^3 \quad (0.89 \text{ lbm/ft}^3)}\end{aligned}$$

The answer is (B).

SI Solution

The steam is 87% wet, so the quality is $x = 0.13$.

From App. 23.O at 350 kPa (350 bars),

$$\upsilon_f = 1.0786 \text{ cm}^3/\text{g}$$
$$\upsilon_g = 524.2 \text{ cm}^3/\text{g}$$

From Eq. 23.56, the specific volume of steam is

$$\begin{aligned}\upsilon &= \upsilon_f + x\upsilon_{fg}\\ &= 1.0786 \ \frac{\text{cm}^3}{\text{g}} + (0.13)\left(524.2 \ \frac{\text{cm}^3}{\text{g}}\right)\\ &= 69.22 \text{ cm}^3/\text{g}\end{aligned}$$

The density is

$$\begin{aligned}\rho &= \frac{1}{\upsilon} = \left(\frac{1}{69.22 \ \frac{\text{cm}^3}{\text{g}}}\right)\left(\frac{\left(100 \ \frac{\text{cm}}{\text{m}}\right)^3}{1000 \ \frac{\text{g}}{\text{kg}}}\right)\\ &= \boxed{14.45 \text{ kg/m}^3 \quad (14 \text{ kg/m}^3)}\end{aligned}$$

The answer is (B).

2. *Customary U.S. Solution*

From App. 35.A, the viscosity of water at 100°F is

$$\begin{aligned}\mu &= \left(0.458 \times 10^{-3} \ \frac{\text{lbm}}{\text{ft-sec}}\right)\left(3600 \ \frac{\text{sec}}{\text{hr}}\right)\\ &= \boxed{1.6488 \text{ lbm/ft-hr} \quad (1.6 \text{ lbm/ft-hr})}\end{aligned}$$

The answer is (C).

SI Solution

From App. 35.B, the viscosity of water at 38°C (use 37.8°C) is

$$\mu = \boxed{0.682 \times 10^{-3} \text{ kg/m·s} \quad (0.00068 \text{ kg/m·s})}$$

The answer is (C).

3. *Customary U.S. Solution*

The midpoint tube temperature is

$$T_s = \left(\tfrac{1}{2}\right)(190°\text{F} + 160°\text{F}) = 175°\text{F}$$
$$T_\infty = 85°\text{F} \quad \text{[given]}$$

Per Eq. 35.11, the film coefficient should be evaluated at a temperature of

$$\begin{aligned}T_h &= \tfrac{1}{2}(T_s + T_\infty)\\ &= \left(\tfrac{1}{2}\right)(175°\text{F} + 85°\text{F})\\ &= \boxed{130°\text{F}}\end{aligned}$$

The answer is (A).

SI Solution

The midpoint tube temperature is

$$T_s = \left(\tfrac{1}{2}\right)(88°\text{C} + 71°\text{C}) = 79.5°\text{C}$$
$$T_\infty = 29°\text{C} \quad \text{[given]}$$

Per Eq. 35.11, the film coefficient should be evaluated at a temperature of

$$\begin{aligned}T_h &= \tfrac{1}{2}(T_s + T_\infty)\\ &= \left(\tfrac{1}{2}\right)(79.5°\text{F} + 29°\text{C})\\ &= \boxed{54.3°\text{C} \quad (54°\text{C})}\end{aligned}$$

The answer is (A).

4. *Customary U.S. Solution*

The heat loss per unit length is

$$\frac{Q}{L} = \left(8.0\ \frac{\text{W}}{\text{ft}}\right)\left(3.413\ \frac{\text{Btu}}{\text{hr-W}}\right) = 27.3\ \text{Btu/hr-ft}$$

From App. 35.C, the air properties at 100°F are

$$\text{Pr} = 0.72$$
$$\frac{g\beta\rho^2}{\mu^2} = 1.76 \times 10^6\ \frac{1}{\text{ft}^3\text{-°F}}$$

From Eq. 35.4, the Grashof number is

$$\text{Gr} = L^3\left(\frac{g\beta\rho^2}{\mu^2}\right)\Delta T$$

The characteristic length is the wire diameter.

$$L = \frac{0.6\ \text{in}}{12\ \frac{\text{in}}{\text{ft}}} = 0.05\ \text{ft}$$

The temperature gradient is

$$\Delta T = T_s - T_\infty = T_{\text{wire}} - 60°\text{F}$$

T_{wire} is unknown. Start by assuming $T_{\text{wire}} = 150°\text{F}$.

$$\begin{aligned}\text{Gr} &= (0.05\ \text{ft})^3\left(1.76 \times 10^6\ \frac{1}{\text{ft}^3\text{-°F}}\right)(150°\text{F} - 60°\text{F}) \\ &= 19{,}800\end{aligned}$$
$$\text{PrGr} = (0.72)(19{,}800) = 14{,}256$$

Use the characteristic length for the diameter, d. From Table 35.3,

$$\begin{aligned}h &= 0.27\left(\frac{T_{\text{wire}} - T_\infty}{d}\right)^{1/4} \\ &= (0.27)\left(\frac{150°\text{F} - 60°\text{F}}{0.05\ \text{ft}}\right)^{1/4} \\ &= 1.76\ \text{Btu/hr-ft}^2\text{-°F}\end{aligned}$$

The heat transfer from the wire is found from Eq. 35.1.

$$Q = \pi dLh(T_{\text{wire}} - T_\infty)$$
$$\begin{aligned}T_{\text{wire}} &= \frac{\frac{Q}{L}}{\pi dh} + T_\infty \\ &= \frac{27.3\ \frac{\text{Btu}}{\text{hr-ft}}}{\pi(0.05\ \text{ft})\left(1.76\ \frac{\text{Btu}}{\text{hr-ft}^2\text{-°F}}\right)} + 60°\text{F} \\ &= 158.7°\text{F}\end{aligned}$$

Perform one more iteration with $T_{\text{wire}} = 158°\text{F}$.

$$\begin{aligned}\text{Gr} &= (0.05)^3\left(1.76 \times 10^6\ \frac{1}{\text{ft}^3\text{-°F}}\right)(158°\text{F} - 60°\text{F}) \\ &= 21{,}560\end{aligned}$$
$$\text{PrGr} = (0.72)(21{,}560) = 15{,}523$$

From Table 35.3,

$$\begin{aligned}h &= 0.27\left(\frac{T_{\text{wire}} - T_\infty}{d}\right)^{1/4} \\ &= (0.27)\left(\frac{158°\text{F} - 60°\text{F}}{0.05\ \text{ft}}\right)^{1/4} \\ &= 1.80\ \text{Btu/hr-ft}^2\text{-°F}\end{aligned}$$
$$\begin{aligned}T_{\text{wire}} &= \frac{\frac{Q}{L}}{\pi dh} + T_\infty \\ &= \frac{27.3\ \frac{\text{Btu}}{\text{hr-ft}}}{\pi(0.05\ \text{ft})\left(1.80\ \frac{\text{Btu}}{\text{hr-ft}^2\text{-°F}}\right)} + 60°\text{F} \\ &= \boxed{156.6°\text{F}\quad(160°\text{F})}\end{aligned}$$

There is no need to repeat iterations since the assumed temperature and the calculated temperature are about the same.

The answer is (D).

SI Solution

From App. 35.D, the air properties at 38°C are

$$\text{Pr} = 0.705$$
$$\frac{g\beta\rho^2}{\mu^2} = 1.12 \times 10^8\ \frac{1}{\text{K·m}^3}$$

From Eq. 35.4, the Grashof number is

$$\text{Gr} = L^3\left(\frac{g\beta\rho^2}{\mu^2}\right)\Delta T$$

The characteristic length is the wire diameter.

$$L = \frac{1.5\ \text{cm}}{100\ \frac{\text{cm}}{\text{m}}} = 0.015\ \text{m}$$

The temperature gradient is

$$\Delta T = T_s - T_\infty = T_{\text{wire}} - 15°\text{C}$$
$$\text{Gr} = (0.015\ \text{m})^3\left(1.12 \times 10^8\ \frac{1}{\text{K·m}^3}\right)(T_{\text{wire}} - 15°\text{C})$$

Heat Transfer

T_{wire} is unknown. Start by assuming $T_{\text{wire}} = 70°\text{C}$.

$$\begin{aligned}\text{Gr} &= (0.015\ \text{m})^3\left(1.12\times10^8\ \frac{1}{\text{K·m}^3}\right)(70°\text{C} - 15°\text{C})\\ &= 20\,790\\ \text{PrGr} &= (0.705)(20\,790)\\ &= 14\,657\end{aligned}$$

Use the characteristic length for the diameter, d. From Table 35.3,

$$\begin{aligned}h &= 1.32\left(\frac{T_{\text{wire}} - T_\infty}{d}\right)^{1/4}\\ &= (1.32)\left(\frac{70°\text{C} - 15°\text{C}}{0.015\ \text{m}}\right)^{1/4}\\ &= 10.27\ \text{W/m}^2\text{·K}\end{aligned}$$

The heat transfer from the wire is found from Eq. 35.1.

$$\begin{aligned}Q &= \pi dLh(T_{\text{wire}} - T_\infty)\\ T_{\text{wire}} &= \frac{\frac{Q}{L}}{\pi dh} + T_\infty\\ &= \frac{25\ \frac{\text{W}}{\text{m}}}{\pi(0.015\ \text{m})\left(10.27\ \frac{\text{W}}{\text{m}^2\text{·K}}\right)} + 15°\text{C}\\ &= \boxed{66.7°\text{C}\quad(67°\text{C})}\end{aligned}$$

There is no need to perform another iteration since the assumed temperature and the calculated temperature are about the same.

The answer is (D).

5. From Table 35.2, for a hot surface facing up, the characteristic length is $0.9d$, where d is the diameter of the hot surface.

$$\begin{aligned}L &= 0.9d = (0.9)\left(\frac{4\ \text{in}}{12\ \frac{\text{in}}{\text{ft}}}\right)\\ &= 0.3\ \text{ft}\end{aligned}$$

The temperature gradient is

$$\begin{aligned}T_s - T_\infty &= 320°\text{F} - 80°\text{F}\\ &= 240°\text{F}\end{aligned}$$

The film temperature is

$$\begin{aligned}T_{\text{film}} &= \tfrac{1}{2}(T_s + T_\infty) = \left(\tfrac{1}{2}\right)(320°\text{F} + 80°\text{F})\\ &= 200°\text{F}\end{aligned}$$

The film properties at 200°F are found from App. 35.A.

$$\begin{aligned}k &= 0.394\ \text{Btu-ft/hr-ft}^2\text{-°F}\\ \rho &= 60.1\ \text{lbm/ft}^3\\ \mu &= 0.205\times10^{-3}\ \text{lbm/ft-sec}\\ \beta &= 4.0\times10^{-4}\ 1/°\text{F}\\ \text{Pr} &= 1.88\end{aligned}$$

The Grashof number is

$$\begin{aligned}\text{Gr} &= \frac{L^3 g\beta\rho^2(T_s - T_\infty)}{\mu^2}\\ &= \frac{(0.3\ \text{ft})^3\left(32.2\ \frac{\text{ft}}{\text{sec}^2}\right)\left(4.0\times10^{-4}\ \frac{1}{°\text{F}}\right)\times\left(60.1\ \frac{\text{lbm}}{\text{ft}^3}\right)^2(240°\text{F})}{\left(0.205\times10^{-3}\ \frac{\text{lbm}}{\text{ft-sec}}\right)^2}\\ &= 7.17\times10^9\end{aligned}$$

From Eq. 35.12 and Table 35.2,

$$\begin{aligned}h &= \frac{kC(\text{GrPr})^n}{L}\\ &= \frac{\left(0.394\ \frac{\text{Btu-ft}}{\text{hr-ft}^2\text{-°F}}\right)(0.14)\left((7.17\times10^9)(1.88)\right)^{1/3}}{0.3\ \text{ft}}\\ &= 437.6\ \text{Btu/hr-ft}^2\text{-°F}\end{aligned}$$

The heat transfer rate is

$$\begin{aligned}Q &= hA(T_s - T_\infty)\\ &= \frac{\left(437.6\ \frac{\text{Btu}}{\text{hr-ft}^2\text{-°F}}\right)\left(\pi\left(\frac{4\ \text{in}}{2}\right)^2\right)(320°\text{F} - 80°\text{F})}{\left(12\ \frac{\text{in}}{\text{ft}}\right)^2}\\ &= \boxed{9164\ \text{Btu/hr}\quad(9200\ \text{Btu/hr})}\end{aligned}$$

The answer is (B).

36 Forced Convection and Heat Exchangers

PRACTICE PROBLEMS

Logarithmic Mean Temperature Difference

1. Water is heated from 55°F to 87°F (15°C to 30°C) by stack gases that are cooled from 350°F to 270°F (175°C to 130°C). The logarithmic mean temperature difference is most nearly

(A) 190°F (105°C)

(B) 210°F (117°C)

(C) 235°F (130°C)

(D) 270°F (150°C)

2. A fluid in a tank is maintained at 85°F (30°C) by an immersed hot water coil. The hot water enters at 190°F (90°C) and leaves at 160°F (70°C). The logarithmic mean temperature difference is most nearly

(A) 89°F (49°C)

(B) 96°F (53°C)

(C) 111°F (62°C)

(D) 127°F (71°C)

Film Temperature

3. Approximately what film temperature should be used for the heated fluid in Prob. 2?

(A) 110°F (43°C)

(B) 130°F (55°C)

(C) 140°F (60°C)

(D) 150°F (66°C)

4. If the fluid in Prob. 2 is gradually heated from 85°F (30°C) to 110°F (45°C), the initial film temperature is most nearly

(A) 110°F (43°C)

(B) 130°F (55°C)

(C) 140°F (60°C)

(D) 150°F (66°C)

Reynolds Number

5. Light no. 10 oil is heated from 95°F to 105°F (35°C to 41°C) in a 0.6 in (1.52 cm) inside diameter tube. The average velocity of the oil is 2.0 ft/sec (0.6 m/s). The viscosity of the oil at 100°F (38°C) is 45 centistokes (cS). The Reynolds number is most nearly

(A) 200

(B) 2500

(C) 4600

(D) 19,000

Convective Heat Transfer

6. A white, uninsulated, rectangular duct passes through a 50 ft (15 m) wide room. The duct is 18 in (45 cm) wide and 12 in (30 cm) high. The room and its contents are at 70°F (21°C). Air at 100°F (40°C) enters the duct flowing at 800 ft/min (4.0 m/s). The film coefficient for the outside of the duct is 2.0 Btu/hr-ft^2-°F (11 W/m^2·K). All theoretical hydraulic properties are applicable.

(a) What is most nearly the heat transfer to the room?

(A) 4300 Btu/hr (1.4 kW)

(B) 5800 Btu/hr (1.9 kW)

(C) 7400 Btu/hr (2.4 kW)

(D) 9100 Btu/hr (2.9 kW)

(b) What is most nearly the temperature of the air after it has traveled in the duct the full 50 ft (15 m)?

(A) 85°F (29°C)

(B) 88°F (31°C)

(C) 91°F (33°C)

(D) 94°F (36°C)

(c) What is most nearly the pressure drop in in (cm) of water due to friction in the duct?

(A) 0.020 in wg (4.6 Pa)

(B) 0.033 in wg (7.5 Pa)

(C) 0.041 in wg (9.4 Pa)

(D) 0.055 in wg (13 Pa)

Laminar Flow in Tubes

7. A steel pipe carrying 350°F (175°C) air is 100 ft (30 m) long. The outside and inside diameters are 4.00 in and 3.50 in (10 cm and 9.0 cm), respectively. The pipe is covered with 2.0 in (5.0 cm) of insulation with a thermal conductivity of 0.05 Btu-ft/hr-ft^2-°F (0.086 W/m·K). The pipe passes through a 50°F (10°C) basement. Flow is laminar and fully developed. The heat loss is most nearly

(A) 3100 Btu/hr (0.93 kW)

(B) 3500 Btu/hr (1.1 kW)

(C) 4100 Btu/hr (1.3 kW)

(D) 8700 Btu/hr (2.0 kW)

Turbulent Flow in Tubes

8. An uninsulated horizontal pipe with 4.00 in (10 cm) outside diameter carries saturated 300 psia (2.1 MPa) steam through a 70°F (21°C) room. The steam flow rate is 5000 lbm/hr (0.63 kg/s). The pipe emissivity is 0.80. Approximately what decrease in quality will occur in the first 50 ft (15 m) of length?

(A) 1.6%

(B) 2.2%

(C) 2.8%

(D) 4.3%

Heat Exchangers

9. A crossflow tubular feedwater heater is being designed to heat 2940 lbm/hr (0.368 kg/s) of water from 70°F (21°C) to 190°F (90°C). Saturated steam at 134 psia (923 kPa) is condensing on the outside of the tubes. The tubes are a copper alloy containing 70% Cu and 30% Ni. Each tube has a 1 in (2.54 cm) outside diameter and a 0.9 in (2.29 cm) inside diameter. The water velocity inside the tubes is 3 ft/sec (0.9 m/s). The outside tube surface area required is most nearly

(A) 3.0 ft^2 (0.30 m^2)

(B) 4.0 ft^2 (0.39 m^2)

(C) 10 ft^2 (0.97 m^2)

(D) 18 ft^2 (1.7 m^2)

10. A U-tube surface feedwater heater with one shell pass and two tube passes is being designed to heat 500,000 lbm/hr (60 kg/s) of water from 200°F to 390°F (100°C to 200°C). The water flows at 5 ft/sec (1.5 m/s) through the tubes. Dry, saturated steam at 400°F (205°C) is to be used as the heating medium. The heater is to operate straight condensing (i.e., the condensed steam will not be mixed with the heated water). Saturated water at 400°F (205°C) is removed from the heater. The tubes in the heater are 7/8 in (2.2 cm) outside diameter with 1/16 in (1.6 mm) walls. The overall heat transfer coefficient is estimated as 700 Btu/hr-ft^2-°F (4 kW/m^2·°C).

(a) Approximately how many tubes are required?

(A) 80

(B) 110

(C) 140

(D) 170

(b) The length of the tube bundle is most nearly

(A) 30 ft (9.0 m)

(B) 40 ft (12 m)

(C) 60 ft (17 m)

(D) 80 ft (24 m)

11. A single-pass heat exchanger is tested in a clean condition and is found to heat 100 gal/min (6.3 L/s) of 70°F (21°C) water to 140°F (60°C). The hot side uses 230°F (110°C) steam. The tube's inner surface area is 50 ft^2 (4.7 m^2). After being used in the field for several months, the exchanger heats 100 gal/min (6.3 L/s) of 70°F (21°C) water to 122°F (50°C). The fouling factor is most nearly

(A) 0.0004 hr-ft^2-°F/Btu (0.00008 m^2·K/W)

(B) 0.0008 hr-ft^2-°F/Btu (0.0001 m^2·K/W)

(C) 0.001 hr-ft^2-°F/Btu (0.0002 m^2·K/W)

(D) 0.002 hr-ft^2-°F/Btu (0.0004 m^2·K/W)

12. Cold water is used to cool hot water in a single-pass, tube-in-tube, counterflow heat exchanger. Compared to a plot of heat transfer rate-versus-distance traveled for the hot water, the plot of heat transfer rate-versus-time traveled for the hot water is

(A) identical

(B) reversed end-to-end

(C) shifted (delayed)

(D) inverted top-to-bottom

13. Cold water is used to cool hot water in a single-pass, tube-in-tube, counterflow heat exchanger. Compared to a plot of heat transfer rate-versus-distance traveled by the cold water, the heat transfer rate-versus-distance traveled plot for the hot water is

(A) identical

(B) reversed end-to-end

(C) shifted (delayed)

(D) inverted top-to-bottom

14. Cold water is used to cool hot water in a single-pass, tube-in-tube, parallel-flow (co-current) heat exchanger. Compared to a plot of temperature-versus-location by the hot water, the temperature-versus-location traveled plot of the cold water is

(A) identical

(B) reversed end-to-end

(C) shifted (delayed)

(D) inverted top-to-bottom

15. Cold water is used to cool hot water in a single-pass, tube-in-tube, counterflow heat exchanger. The temperature gradient experienced by the hot water along the length of the heat exchanger will

(A) essentially be constant

(B) decrease nonlinearly

(C) increase nonlinearly

(D) decrease linearly

16. Cold water is used to condense saturated steam in a single-pass, tube-in-tube, counterflow heat exchanger without subcooling. The temperature of the steam along the length of the heat exchanger from entrance to exit

(A) decreases linearly

(B) decreases parabolically

(C) decreases logarithmically

(D) is constant

17. Heat exchanger "duty" is best defined as the

(A) entering enthalpy difference of hot and cold fluids

(B) non-adiabatic heat loss

(C) heat transfer rate per unit mass of cold fluid

(D) heat transfer rate per unit time

18. A 2 ft long cylindrical fin with a diameter of 0.5 in is attached at one end to a heat source. The temperature of the heat source is 350°F, and the temperature of the ambient air is 75°F. The thermal conductivity of the fin is 128 Btu/hr-ft-°F, and the average film coefficient along the length of fin is 1.3 Btu/hr-ft^2-°F. No heat is transferred from the exposed fin tip.

(a) Heat transfer from the heat source into the base of the fin is most nearly

(A) 22 Btu/hr

(B) 29 Btu/hr

(C) 36 Btu/hr

(D) 46 Btu/hr

(b) The efficiency of the fin is

(A) 20%

(B) 35%

(C) 48%

(D) 65%

Tubes in Crossflow

19. A heat-conducting rod has an outside diameter of 0.35 in (8.9 mm). Its uniform temperature is 100°F (38°C). The rod is inserted perpendicularly into a 100 ft/sec (30 m/s) airflow. The air temperature is 150°F (66°C). The film coefficient on the outside of the rod is most nearly

(A) 36 Btu/hr-ft^2-°F (210 W/m^2·K)

(B) 45 Btu/hr-ft^2-°F (260 W/m^2·K)

(C) 66 Btu/hr-ft^2-°F (380 W/m^2·K)

(D) 91 Btu/hr-ft^2-°F (530 W/m^2·K)

SOLUTIONS

1. *Customary U.S. Solution*

The logarithmic mean temperature difference will be different for different types of flow.

Parallel flow:

55°F → 87°F

A B

350°F → 270°F

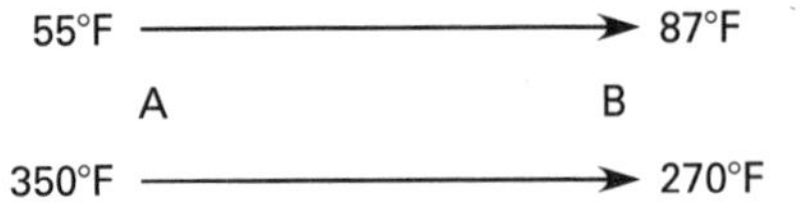

$$\Delta T_A = 350°\text{F} - 55°\text{F} = 295°\text{F}$$

$$\Delta T_B = 270°\text{F} - 87°\text{F} = 183°\text{F}$$

From Eq. 36.68, the logarithmic mean temperature difference is

$$\Delta T_{\text{lm}} = \frac{\Delta T_A - \Delta T_B}{\ln \dfrac{\Delta T_A}{\Delta T_B}} = \frac{295°\text{F} - 183°\text{F}}{\ln \dfrac{295°\text{F}}{183°\text{F}}} = \boxed{234.6°\text{F} \quad (235°\text{F})}$$

Counterflow:

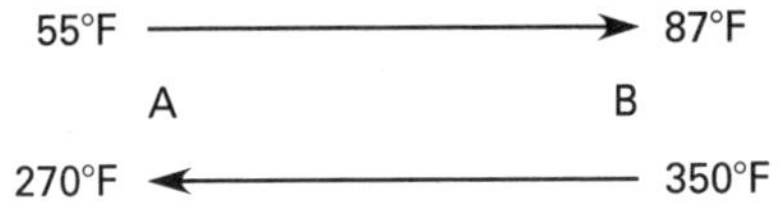

55°F → 87°F

A B

270°F ← 350°F

$$\Delta T_A = 270°\text{F} - 55°\text{F} = 215°\text{F}$$

$$\Delta T_B = 350°\text{F} - 87°\text{F} = 263°\text{F}$$

From Eq. 36.68, the logarithmic mean temperature difference is

$$\Delta T_{\text{lm}} = \frac{\Delta T_A - \Delta T_B}{\ln \dfrac{\Delta T_A}{\Delta T_B}} = \frac{215°\text{F} - 263°\text{F}}{\ln \dfrac{215°\text{F}}{263°\text{F}}} = \boxed{238.2°\text{F} \quad (235°\text{F})}$$

The answer is (C).

SI Solution

The logarithmic mean temperature difference will be different for different types of flow.

Parallel flow:

15°C → 30°C

A B

175°C → 130°C

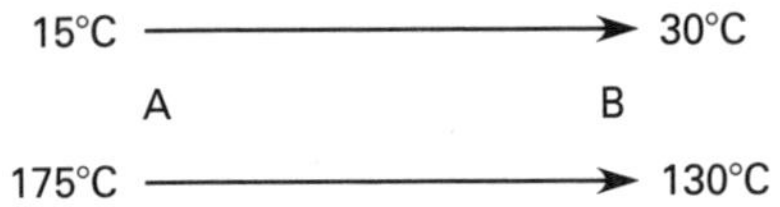

$$\Delta T_A = 175°\text{C} - 15°\text{C} = 160°\text{C}$$

$$\Delta T_B = 130°\text{C} - 30°\text{C} = 100°\text{C}$$

From Eq. 36.68, the logarithmic mean temperature difference is

$$\Delta T_{\text{lm}} = \frac{\Delta T_A - \Delta T_B}{\ln \dfrac{\Delta T_A}{\Delta T_B}} = \frac{160°\text{C} - 100°\text{C}}{\ln \dfrac{160°\text{C}}{100°\text{C}}} = \boxed{127.7°\text{C} \quad (130°\text{C})}$$

Counterflow:

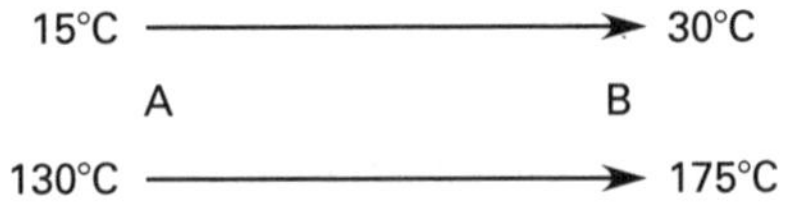

15°C → 30°C

A B

130°C → 175°C

$$\Delta T_A = 130°\text{C} - 15°\text{C} = 115°\text{C}$$

$$\Delta T_B = 175°\text{C} - 30°\text{C} = 145°\text{C}$$

From Eq. 36.68, the logarithmic mean temperature difference is

$$\Delta T_{\text{lm}} = \frac{\Delta T_A - \Delta T_B}{\ln \dfrac{\Delta T_A}{\Delta T_B}} = \frac{115°\text{C} - 145°\text{C}}{\ln \dfrac{115°\text{C}}{145°\text{C}}} = \boxed{129.4°\text{C} \quad (130°\text{C})}$$

The answer is (C).

2. *Customary U.S. Solution*

$$\Delta T_A = 190°\text{F} - 85°\text{F} = 105°\text{F}$$

$$\Delta T_B = 160°\text{F} - 85°\text{F} = 75°\text{F}$$

From Eq. 36.68, the logarithmic mean temperature difference is

$$\Delta T_{\text{lm}} = \frac{\Delta T_A - \Delta T_B}{\ln \dfrac{\Delta T_A}{\Delta T_B}} = \frac{105°\text{F} - 75°\text{F}}{\ln \dfrac{105°\text{F}}{75°\text{F}}} = \boxed{89.2°\text{F} \quad (89°\text{F})}$$

The answer is (A).

SI Solution

$$\Delta T_A = 90°\text{C} - 30°\text{C} = 60°\text{C}$$

$$\Delta T_B = 70°\text{C} - 30°\text{C} = 40°\text{C}$$

From Eq. 36.68, the logarithmic mean temperature difference is

$$\Delta T_{\text{lm}} = \frac{\Delta T_A - \Delta T_B}{\ln \dfrac{\Delta T_A}{\Delta T_B}} = \frac{60°\text{C} - 40°\text{C}}{\ln \dfrac{60°\text{C}}{40°\text{C}}} = \boxed{49.3°\text{C} \quad (49°\text{C})}$$

The answer is (A).

3. *Customary U.S. Solution*

$$T_{\text{tank}} = 85°\text{F}$$
$$T_{\text{coil}} = \left(\tfrac{1}{2}\right)(190°\text{F} + 160°\text{F}) = 175°\text{F}$$

The film temperature is

$$\begin{aligned} T_f &= \tfrac{1}{2}(T_{\text{tank}} + T_{\text{coil}}) = \left(\tfrac{1}{2}\right)(85°\text{F} + 175°\text{F}) \\ &= \boxed{130°\text{F}} \end{aligned}$$

The answer is (B).

SI Solution

$$T_{\text{tank}} = 30°\text{C}$$
$$T_{\text{coil}} = \left(\tfrac{1}{2}\right)(90°\text{C} + 70°\text{C}) = 80°\text{C}$$

The film temperature is

$$\begin{aligned} T_f &= \tfrac{1}{2}(T_{\text{tank}} + T_{\text{coil}}) = \left(\tfrac{1}{2}\right)(30°\text{C} + 80°\text{C}) \\ &= \boxed{55°\text{C}} \end{aligned}$$

The answer is (B).

4. *Customary U.S. Solution*

$$T_{\text{coil,initial}} = \left(\tfrac{1}{2}\right)(190°\text{F} + 160°\text{F}) = 175°\text{F}$$

The initial film temperature is

$$\begin{aligned} T_{f,\text{initial}} &= \tfrac{1}{2}(T_{\text{tank,initial}} + T_{\text{coil,initial}}) \\ &= \left(\tfrac{1}{2}\right)(85°\text{F} + 175°\text{F}) \\ &= \boxed{130°\text{F}} \end{aligned}$$

The answer is (B).

SI Solution

$$T_{\text{coil,initial}} = \left(\tfrac{1}{2}\right)(90°\text{C} + 70°\text{C}) = 80°\text{C}$$

The initial film temperature is

$$\begin{aligned} T_{f,\text{initial}} &= \tfrac{1}{2}(T_{\text{tank,initial}} + T_{\text{coil,initial}}) \\ &= \left(\tfrac{1}{2}\right)(30°\text{C} + 80°\text{C}) \\ &= \boxed{55°\text{C}} \end{aligned}$$

The answer is (B).

5. *Customary U.S. Solution*

The Reynolds number is

$$\begin{aligned} \text{Re}_d &= \frac{\text{v}D}{\nu} \\ D &= \frac{0.6\ \text{in}}{12\ \frac{\text{in}}{\text{ft}}} = 0.05\ \text{ft} \\ \text{v} &= 2\ \text{ft/sec} \\ \nu &= \frac{45\ \text{cS}}{\left(100\ \frac{\text{cS}}{\text{S}}\right)\left(929\ \frac{\text{sec-stoke}}{\text{ft}^2}\right)} \\ &= 4.84 \times 10^{-4}\ \text{ft}^2/\text{sec} \\ \text{Re} &= \frac{\left(2\ \frac{\text{ft}}{\text{sec}}\right)(0.05\ \text{ft})}{4.84 \times 10^{-4}\ \frac{\text{ft}^2}{\text{sec}}} = \boxed{206.6\quad(200)} \end{aligned}$$

The answer is (A).

SI Solution

The Reynolds number is

$$\begin{aligned} \text{Re}_d &= \frac{\text{v}D}{\nu} \\ D &= \frac{1.52\ \text{cm}}{100\ \frac{\text{cm}}{\text{m}}} = 0.0152\ \text{m} \\ \text{v} &= 0.6\ \text{m/s} \\ \nu &= \frac{(45\ \text{cS})\left(1\ \frac{\frac{\mu\text{m}^2}{\text{s}}}{\text{cS}\cdot\text{s}}\right)}{10^6\ \frac{\mu\text{m}^2}{\text{m}^2}} \\ &= 45 \times 10^{-6}\ \text{m}^2/\text{s} \\ \text{Re}_d &= \frac{\left(0.6\ \frac{\text{m}}{\text{s}}\right)(0.0152\ \text{m})}{45 \times 10^{-6}\ \frac{\text{m}^2}{\text{s}}} = \boxed{202.7\quad(200)} \end{aligned}$$

The answer is (A).

6. *Customary U.S. Solution*

(a) The exposed duct area is

$$\begin{aligned} A &= (2W + 2H)L \\ &= \frac{\big((2)(18\ \text{in}) + (2)(12\ \text{in})\big)(50\ \text{ft})}{12\ \frac{\text{in}}{\text{ft}}} \\ &= 250\ \text{ft}^2 \end{aligned}$$

Heat Transfer

The duct is noncircular; therefore, the hydraulic diameter of the duct will be used. From Eq. 36.48, the theoretical hydraulic diameter is

$$d_H = 4\left(\frac{\text{area in flow}}{\text{wetted perimeter}}\right) = 4\left(\frac{WH}{2(W+H)}\right)$$

$$= \frac{(4)\left(\frac{(18 \text{ in})(12 \text{ in})}{(2)(18 \text{ in} + 12 \text{ in})}\right)}{12 \ \frac{\text{in}}{\text{ft}}}$$

$$= 1.2 \text{ ft}$$

From App. 35.C, for air at 100°F,

$$\nu = 18.0 \times 10^{-5} \text{ ft}^2/\text{sec}$$

$$\rho = 0.0710 \text{ lbm/ft}^3$$

The Reynolds number is

$$\text{Re} = \frac{\text{v}D}{\nu} = \frac{\left(800 \ \frac{\text{ft}}{\text{min}}\right)(1.2 \text{ ft})}{\left(18.0 \times 10^{-5} \ \frac{\text{ft}^2}{\text{sec}}\right)\left(60 \ \frac{\text{sec}}{\text{min}}\right)}$$

$$= 8.90 \times 10^4$$

This is a turbulent flow. From Eq. 36.39,

$$h_i \approx (0.00351 + 0.000001583 T_{°\text{F}})\left(\frac{G^{0.8}_{\text{lbm/hr-ft}^2}}{d^{0.2}_{\text{ft}}}\right)$$

$$G = \rho\text{v} = \left(0.0710 \ \frac{\text{lbm}}{\text{ft}^3}\right)\left(800 \ \frac{\text{ft}}{\text{min}}\right)\left(60 \ \frac{\text{min}}{\text{hr}}\right)$$

$$= 3408.0 \text{ lbm/hr-ft}^2$$

$$h_i = \big(0.00351 + (0.000001583)(100°\text{F})\big) \times \left(\frac{\left(3408.0 \ \frac{\text{lbm}}{\text{hr-ft}^2}\right)^{0.8}}{(1.2 \text{ ft})^{0.2}}\right)$$

$$= 2.37 \text{ Btu/hr-ft}^2\text{-°F}$$

Disregarding the duct thermal resistance, the overall heat transfer coefficient from Eq. 36.74 is

$$\frac{1}{U} = \frac{1}{h_i} + \frac{1}{h_o}$$

$$= \frac{1}{2.37 \ \frac{\text{Btu}}{\text{hr-ft}^2\text{-°F}}} + \frac{1}{2.0 \ \frac{\text{Btu}}{\text{hr-ft}^2\text{-°F}}}$$

$$= 0.922 \text{ hr-ft}^2\text{-°F/Btu}$$

$$U = 1.08 \text{ Btu/hr-ft}^2\text{-°F}$$

The heat transfer to the room is

$$Q = UA(T_{\text{ave}} - T_\infty) = UA\Big(\left(\tfrac{1}{2}\right)(T_{\text{in}} + T_{\text{out}}) - T_\infty\Big)$$

Since T_{out} is unknown, an iteration procedure may be required. Assume $T_{\text{out}} \approx 95°\text{F}$.

$$Q = \left(1.08 \ \frac{\text{Btu}}{\text{hr-ft}^2\text{-°F}}\right)(250 \text{ ft}^2) \times \Big(\left(\tfrac{1}{2}\right)(100°\text{F} + 95°\text{F}) - 70°\text{F}\Big)$$

$$= \boxed{7425 \text{ Btu/hr} \quad (7400 \text{ Btu/hr})}$$

(Notice that ΔT (not ΔT_{lm}) is used in accordance with standard conventions in the HVAC industry.)

The answer is (C).

(b) Temperature T_{out} can be verified by using

$$Q = \dot{m}c_p(T_{\text{in}} - T_{\text{out}})$$

$$\dot{m} = GA_{\text{flow}} = GWH$$

$$= \frac{\left(3408.0 \ \frac{\text{lbm}}{\text{hr-ft}^2}\right)(18 \text{ in})(12 \text{ in})}{\left(12 \ \frac{\text{in}}{\text{ft}}\right)^2}$$

$$= 5112 \text{ lbm/hr}$$

$$c_p = 0.240 \text{ Btu/lbm-°F} \quad \text{[from App. 35.C at 100°F]}$$

$$7425 \ \frac{\text{Btu}}{\text{hr}} = \left(5112 \ \frac{\text{lbm}}{\text{hr}}\right)\left(0.240 \ \frac{\text{Btu}}{\text{lbm-°F}}\right) \times (100°\text{F} - T_{\text{out}})$$

$$T_{\text{out}} = \boxed{94°\text{F}} \quad \text{[close enough]}$$

The answer is (D).

(c) Assume clean galvanized ductwork with $\epsilon = 0.0005$ ft and about 25 joints per 100 ft. From Eq. 41.38, the equivalent diameter of a rectangular duct is

$$d_H = \frac{1.3(\text{short side} \times \text{long side})^{5/8}}{(\text{short side} + \text{long side})^{1/4}}$$

$$= \frac{(1.3)\big((12 \text{ in})(18 \text{ in})\big)^{5/8}}{(12 \text{ in} + 18 \text{ in})^{1/4}}$$

$$= 16 \text{ in}$$

The flow rate is

$$\dot{V} = \text{v}A = \frac{\left(800 \ \frac{\text{ft}}{\text{min}}\right)(12 \text{ in})(18 \text{ in})}{\left(12 \ \frac{\text{in}}{\text{ft}}\right)^2} = 1200 \text{ cfm}$$

From Fig. 41.5, the friction loss is 0.066 in wg/100 ft. Therefore,

$$\Delta p = (0.066 \text{ in wg})\left(\frac{50 \text{ ft}}{100 \text{ ft}}\right) = \boxed{0.033 \text{ in wg}}$$

(Notice that the flow rate and not the flow velocity must be used with D_e in Fig. 41.5.)

The answer is (B).

SI Solution

(a) The exposed duct area is

$$\begin{aligned} A &= (2W+2H)L \\ &= \frac{\big((2)(45\ \text{cm})+(2)(30\ \text{cm})\big)(15\ \text{m})}{100\ \frac{\text{cm}}{\text{m}}} \\ &= 22.5\ \text{m}^2 \end{aligned}$$

The duct is noncircular; therefore, the hydraulic diameter of the duct will be used. From Eq. 36.48, the hydraulic diameter is

$$\begin{aligned} d_H &= 4\left(\frac{\text{area in flow}}{\text{wetted perimeter}}\right) = 4\left(\frac{WH}{2(W+H)}\right) \\ &= \frac{(4)\left(\frac{(45\ \text{cm})(30\ \text{cm})}{(2)(45\ \text{cm}+30\ \text{cm})}\right)}{100\ \frac{\text{cm}}{\text{m}}} \\ &= 0.36\ \text{m} \end{aligned}$$

From App. 35.D, for air at 40°C,

$$\begin{aligned} \mu &= 1.91\times10^{-5}\ \text{kg/m·s} \\ \rho &= 1.130\ \text{kg/m}^3 \\ c_p &= 1.0051\ \text{kJ/kg·K} \\ k &= 0.02718\ \text{W/m·K} \end{aligned}$$

The Reynolds number is

$$\begin{aligned} \text{Re} &= \frac{\rho \text{v} D}{\mu} = \frac{\left(1.130\ \frac{\text{kg}}{\text{m}^3}\right)\left(4.0\ \frac{\text{m}}{\text{s}}\right)(0.36\ \text{m})}{1.91\times10^{-5}\ \frac{\text{kg}}{\text{m·s}}} \\ &= 8.52\times10^4 \end{aligned}$$

This is a turbulent flow. From Eq. 36.34, the Nusselt number is

$$\text{Nu} = 0.023\,\text{Re}^{0.8}$$

The film coefficient is

$$\begin{aligned} h &= 0.023\,\text{Re}^{0.8}\left(\frac{k}{d}\right) \\ &= (0.023)(8.52\times10^4)^{0.8}\left(\frac{0.02718\ \frac{\text{W}}{\text{m·K}}}{0.36\ \text{m}}\right) \\ &= 15.28\ \text{W/m}^2\text{·K} \end{aligned}$$

Disregarding the duct thermal resistance, the overall heat transfer coefficient from Eq. 36.73 is

$$\begin{aligned} \frac{1}{U} &= \frac{1}{h_i}+\frac{1}{h_o} \\ &= \frac{1}{11\ \frac{\text{W}}{\text{m}^2\text{·K}}}+\frac{1}{15.28\ \frac{\text{W}}{\text{m}^2\text{·K}}} \\ &= 0.1564\ \text{m}^2\text{·K/W} \\ U &= 6.39\ \text{W/m}^2\text{·K} \end{aligned}$$

The heat transfer to the room is

$$Q = UA(T_{\text{ave}}-T_\infty) = UA\Big(\left(\tfrac{1}{2}\right)(T_{\text{in}}+T_{\text{out}})-T_\infty\Big)$$

Since T_{out} is unknown, an iterative procedure may be required. Assume $T_{\text{out}} = 36°\text{C}$.

$$\begin{aligned} Q &= \left(6.39\ \frac{\text{W}}{\text{m}^2\text{·K}}\right)(22.5\ \text{m}^2) \\ &\quad\times\Big(\left(\tfrac{1}{2}\right)(40°\text{C}+36°\text{C})-21°\text{C}\Big) \\ &= \boxed{2444.2\ \text{W}\quad(2.4\ \text{kW})} \end{aligned}$$

(Notice that ΔT (not ΔT_{lm}) is used in accordance with standard conventions in the HVAC industry.)

The answer is (C).

(b) Temperature T_{out} can be verified by using

$$\begin{aligned} Q &= \dot{m}c_p(T_{\text{in}}-T_{\text{out}}) \\ \dot{m} &= \rho A_{\text{flow}}\text{v} \\ &= \frac{\left(1.130\ \frac{\text{kg}}{\text{m}^3}\right)(45\ \text{cm})(30\ \text{cm})\left(4\ \frac{\text{m}}{\text{s}}\right)}{\left(100\ \frac{\text{cm}}{\text{m}}\right)^2} \\ &= 0.6102\ \text{kg/s} \end{aligned}$$

$$\begin{aligned} 2444.2\ \text{W} &= \left(0.6102\ \frac{\text{kg}}{\text{s}}\right)\left(1.0051\ \frac{\text{kJ}}{\text{kg·K}}\right)\left(1000\ \frac{\text{J}}{\text{kJ}}\right) \\ &\quad\times(40°\text{C}-T_{\text{out}}) \\ T_{\text{out}} &= \boxed{36°\text{C}}\quad\text{[same as assumed]} \end{aligned}$$

The answer is (D).

(c) Assume clean galvanized ductwork with $\epsilon = 0.15$ mm and about 1 joint per meter. From Eq. 41.38, the hydraulic diameter of a rectangular duct is

$$\begin{aligned} D_e &= \frac{1.3(\text{short side}\times\text{long side})^{5/8}}{(\text{short side}+\text{long side})^{1/4}} \\ &= \left(\frac{(1.3)\big((30\ \text{cm})(45\ \text{cm})\big)^{5/8}}{(30\ \text{cm}+45\ \text{cm})^{1/4}}\right)\left(10\ \frac{\text{mm}}{\text{cm}}\right) \\ &= 400\ \text{mm} \end{aligned}$$

The flow rate is

$$\dot{V} = vA = \left(4\ \frac{\text{m}}{\text{s}}\right)(0.45\ \text{m})(0.30\ \text{m})\left(1000\ \frac{\text{L}}{\text{m}^3}\right) = 540\ \text{L/s}$$

From Fig. 41.6, the friction loss is 0.5 Pa/m. For 15 m,

$$\Delta p = \left(0.5\ \frac{\text{Pa}}{\text{m}}\right)(15\ \text{m}) = \boxed{7.5\ \text{Pa}}$$

(Notice that the flow rate and not the flow velocity must be used with D_e in Fig. 41.6.)

The answer is (B).

7. *Customary U.S. Solution*

Refer to Fig. 34.3. The corresponding radii are

$$r_a = \frac{d_i}{2} = \frac{3.5\ \text{in}}{(2)\left(12\ \frac{\text{in}}{\text{ft}}\right)} = 0.1458\ \text{ft}$$

$$r_b = \frac{d_o}{2} = \frac{4\ \text{in}}{(2)\left(12\ \frac{\text{in}}{\text{ft}}\right)} = 0.1667\ \text{ft}$$

$$r_c = r_b + t_{\text{insulation}} = 0.1667\ \text{ft} + \frac{2\ \text{in}}{12\ \frac{\text{in}}{\text{ft}}} = 0.3334\ \text{ft}$$

From App. 34.B, for steel at 350°F, $k_{\text{pipe}} \approx$ 25.6 Btu-ft/hr-ft^2-°F.

Initially assume a typical value of $h_c = 1.5$ Btu/hr-ft^2-°F.

For fully developed laminar flow from Eq. 36.28, $\text{Nu}_d = 3.658$.

From App. 35.C, for air at 350°F,

$$k_{\text{air}} \approx 0.0203\ \text{Btu/hr-ft-°F}$$

$$\text{Nu}_d = \frac{h_a d_i}{k_{\text{air}}} = 3.658$$

$$h_a = \frac{3.658 k_{\text{air}}}{d_i} = \frac{(3.658)\left(0.0203\ \frac{\text{Btu}}{\text{hr-ft-°F}}\right)\left(12\ \frac{\text{in}}{\text{ft}}\right)}{3.5\ \text{in}} = 0.255\ \text{Btu/hr-ft}^2\text{-°F}$$

Neglect thermal resistance between pipe and insulation. From Eq. 34.41, the heat transfer is

$$Q = \frac{2\pi L(T_i - T_\infty)}{\frac{1}{r_a h_a} + \frac{\ln\left(\frac{r_b}{r_a}\right)}{k_{\text{pipe}}} + \frac{\ln\left(\frac{r_c}{r_b}\right)}{k_{\text{insulation}}} + \frac{1}{r_c h_c}}$$

$$= \frac{2\pi(100\ \text{ft})(350°\text{F} - 50°\text{F})}{\frac{1}{(0.1458\ \text{ft})\left(0.255\ \frac{\text{Btu}}{\text{hr-ft}^2\text{-°F}}\right)} + \frac{\ln\left(\frac{0.1667\ \text{ft}}{0.1458\ \text{ft}}\right)}{25.6\ \frac{\text{Btu-ft}}{\text{hr-ft}^2\text{-°F}}} + \frac{\ln\left(\frac{0.3334\ \text{ft}}{0.1667\ \text{ft}}\right)}{0.05\ \frac{\text{Btu-ft}}{\text{hr-ft}^2\text{-°F}}} + \frac{1}{(0.3334\ \text{ft})\left(1.5\ \frac{\text{Btu}}{\text{hr-ft}^2\text{-°F}}\right)}}$$

$$= \frac{188{,}496\ \text{ft-°F}}{26.90\ \frac{\text{hr-ft-°F}}{\text{Btu}} + 0.00523\ \frac{\text{hr-ft-°F}}{\text{Btu}} + 13.863\ \frac{\text{hr-ft-°F}}{\text{Btu}} + 2.00\ \frac{\text{hr-ft-°F}}{\text{Btu}}}$$

$$= 4408\ \text{Btu/hr}$$

Use Eq. 34.41 to find T_2 by using all resistances except the outer $(T_2 - T_\infty)$ resistance.

$$T_i - T_2 = \left(\frac{4408\ \frac{\text{Btu}}{\text{hr}}}{2\pi(100\ \text{ft})}\right) \times \left(26.9\ \frac{\text{hr-ft-°F}}{\text{Btu}} + 0.00523\ \frac{\text{hr-ft-°F}}{\text{Btu}} + 13.863\ \frac{\text{hr-ft-°F}}{\text{Btu}}\right) = 286.0°\text{F}$$

$$T_2 = T_i - 286.0°\text{F} = 350°\text{F} - 286.0°\text{F} = 64.0°\text{F}$$

To evaluate h_c, use film temperature.

$$T_{\text{film}} = \tfrac{1}{2}(T_2 + T_\infty) = \left(\tfrac{1}{2}\right)(64.0°\text{F} + 50°\text{F}) = 57°\text{F}$$

From App. 35.C, for air at 57°F,

$$\text{Pr} = 0.72$$

$$\frac{g\beta\rho^2}{\mu^2} = 2.645 \times 10^6\ \frac{1}{\text{ft}^3\text{-°F}}$$

From Eq. 35.4, the Grashof number is

$$\mathrm{Gr} = \frac{L^3 g\beta\rho^2 (T_2 - T_\infty)}{\mu^2}$$

For pipe,

$$L = d_c = 2r_c = (2)(0.3334\ \text{ft}) - 0.6668\ \text{ft}$$

$$\begin{aligned}\mathrm{Gr} &= (0.6668\ \text{ft})^3 \left(2.645 \times 10^6\ \frac{1}{\text{ft}^3\text{-}^\circ\text{F}}\right) \\ &\quad \times (64.0^\circ\text{F} - 50^\circ\text{F}) \\ &= 1.1 \times 10^7\end{aligned}$$

$$\begin{aligned}\mathrm{Gr\,Pr} &= (1.1 \times 10^7)(0.72) \\ &= 7.9 \times 10^6\end{aligned}$$

From Table 35.3,

$$\begin{aligned}h_c &\approx 0.27 \left(\frac{T_2 - T_\infty}{d_c}\right)^{1/4} \\ &= (0.27)\left(\frac{64.0^\circ\text{F} - 50^\circ\text{F}}{0.6668\ \text{ft}}\right)^{1/4} \\ &= 0.578\ \text{Btu/hr-ft}^2\text{-}^\circ\text{F}\end{aligned}$$

At the second iteration, the heat transfer is

$$\begin{aligned}Q &= \frac{188{,}496\ \text{ft-}^\circ\text{F}}{\begin{array}{l}26.90\ \dfrac{\text{hr-ft-}^\circ\text{F}}{\text{Btu}} + 0.00523\ \dfrac{\text{hr-ft-}^\circ\text{F}}{\text{Btu}} \\ \quad + 13.863\ \dfrac{\text{hr-ft-}^\circ\text{F}}{\text{Btu}} \\ \quad + \dfrac{1}{(0.3334\ \text{ft})\left(0.578\ \dfrac{\text{Btu}}{\text{hr-ft-}^\circ\text{F}}\right)}\end{array}} \\ &= \boxed{4102\ \text{Btu/hr} \quad (4100\ \text{Btu/hr})}\end{aligned}$$

Additional iterations will improve the accuracy further.

The answer is (C).

SI Solution

Refer to Fig. 34.3. The corresponding radii are

$$r_a = \frac{d_i}{2} = \frac{9.0\ \text{cm}}{(2)\left(100\ \frac{\text{cm}}{\text{m}}\right)} = 0.045\ \text{m}$$

$$r_b = \frac{d_o}{2} = \frac{10\ \text{cm}}{(2)\left(100\ \frac{\text{cm}}{\text{m}}\right)} = 0.050\ \text{m}$$

$$\begin{aligned}r_c &= r_b + t_{\text{insulation}} = 0.050\ \text{m} + \frac{5.0\ \text{cm}}{100\ \frac{\text{cm}}{\text{m}}} \\ &= 0.100\ \text{m}\end{aligned}$$

From App. 34.B and the table footnote, for steel at 175°C (~347°F),

$$\begin{aligned}k_{\text{pipe}} &\approx \left(25.6\ \frac{\text{Btu}}{\text{hr-ft-}^\circ\text{F}}\right)\left(1.7307\ \frac{\text{W·hr·ft·}^\circ\text{F}}{\text{m·K·Btu}}\right) \\ &= 44.31\ \text{W/m·K}\end{aligned}$$

Initially assume a typical value of $h_c = 3.5\ \text{W/m}^2\text{·K}$.

For fully developed laminar flow from Eq. 36.28, $\mathrm{Nu}_d = 3.658$.

From App. 35.D, for air at 175°C,

$$k_{\text{air}} \approx 0.03709\ \text{W/m·K}$$

$$\mathrm{Nu}_d = \frac{h_a d_i}{k_{\text{air}}} = 3.658$$

$$\begin{aligned}h_a &= \frac{3.658 k_{\text{air}}}{d_i} = \frac{(3.658)\left(0.0379\ \frac{\text{W}}{\text{m·K}}\right)\left(100\ \frac{\text{cm}}{\text{m}}\right)}{9.0\ \text{cm}} \\ &= 1.540\ \text{W/m}^2\text{·K}\end{aligned}$$

Neglect thermal resistance between pipe and insulation. From Eq. 34.41, the heat transfer is

$$\begin{aligned}Q &= \frac{2\pi L (T_i - T_\infty)}{\dfrac{1}{r_a h_a} + \dfrac{\ln\left(\dfrac{r_b}{r_a}\right)}{k_{\text{pipe}}} + \dfrac{\ln\left(\dfrac{r_c}{r_b}\right)}{k_{\text{insulation}}} + \dfrac{1}{r_c h_c}} \\ &= \frac{2\pi(30\ \text{m})(175^\circ\text{C} - 10^\circ\text{C})}{\begin{array}{l}\dfrac{1}{(0.045\ \text{m})\left(1.540\ \dfrac{\text{W}}{\text{m}^2\text{·K}}\right)} + \dfrac{\ln\left(\dfrac{0.050\ \text{m}}{0.045\ \text{m}}\right)}{44.31\ \dfrac{\text{W}}{\text{m·K}}} \\ \quad + \dfrac{\ln\left(\dfrac{0.10\ \text{m}}{0.050\ \text{m}}\right)}{0.086\ \dfrac{\text{W}}{\text{m·K}}} + \dfrac{1}{(0.10\ \text{m})\left(3.5\ \dfrac{\text{W}}{\text{m}^2\text{·K}}\right)}\end{array}} \\ &= \frac{31\,101.8\ \text{m·}^\circ\text{C}}{\begin{array}{l}14.43\ \dfrac{\text{m·K}}{\text{W}} + 0.00238\ \dfrac{\text{m·K}}{\text{W}} \\ \quad + 8.06\ \dfrac{\text{m·K}}{\text{W}} + 2.86\ \dfrac{\text{m·K}}{\text{W}}\end{array}} \\ &= 1227\ \text{W}\end{aligned}$$

Use Eq. 34.41 to find T_2 by using all resistances except the outer $(T_2 - T_\infty)$ resistance.

$$\begin{aligned}T_i - T_2 &= \left(\frac{1227\ \text{W}}{2\pi(30\ \text{m})}\right) \\ &\quad \times \left(14.43\ \frac{\text{m·K}}{\text{W}} + 0.00238\ \frac{\text{m·K}}{\text{W}} + 8.06\ \frac{\text{m·K}}{\text{W}}\right) \\ &= 146.4^\circ\text{C}\end{aligned}$$

$$\begin{aligned}T_2 &= T_i - 146.4^\circ\text{C} = 175^\circ\text{C} - 146.4^\circ\text{C} \\ &= 28.6^\circ\text{C}\end{aligned}$$

Heat Transfer

To evaluate h_c, use film temperature.

$$T_{\text{film}} = \tfrac{1}{2}(T_2 + T_\infty) = \left(\tfrac{1}{2}\right)(28.6°\text{C} + 10°\text{C}) = 19.3°\text{C}$$

From App. 35.D, for air at 19.3°C,

$$\text{Pr} = 0.710$$
$$\frac{g\beta\rho^2}{\mu^2} = 1.52 \times 10^8\ \frac{1}{\text{K·m}^3}$$

From Eq. 35.4, the Grashof number is

$$\text{Gr} = \frac{L^3 g\beta\rho^2(T_2 - T_\infty)}{\mu^2}$$

For pipe,

$$L = d_c = 2r_c = (2)(0.10\ \text{m}) = 0.20\ \text{m}$$
$$\text{Gr} = (0.20\ \text{m})^3\left(1.52 \times 10^8\ \frac{1}{\text{K·m}^3}\right)(28.6°\text{C} - 10°\text{C}) = 2.26 \times 10^7$$
$$\text{Gr Pr} = (2.26 \times 10^7)(0.710) = 16.0 \times 10^7$$

From Table 35.3,

$$h_c \approx 1.37\left(\frac{T_2 - T_\infty}{d_c}\right)^{1/4} = (1.37)\left(\frac{28.6°\text{C} - 10°\text{C}}{0.20\ \text{m}}\right)^{1/4} = 4.25\ \text{W/m}^2\text{·K}\quad \left[\begin{array}{c}\text{versus assumed value of}\\ 3.5\ \text{W/m}^2\text{·K}\end{array}\right]$$

Further iteration will improve accuracy.

$$Q = \frac{31\,101.8\ \text{m·°C}}{14.43\ \frac{\text{m·K}}{\text{W}} + 0.00238\ \frac{\text{m·K}}{\text{W}} + 8.06\ \frac{\text{m·K}}{\text{W}} + \frac{1}{(0.10\ \text{m})\left(4.25\ \frac{\text{W}}{\text{m}^2\text{·K}}\right)}} = \boxed{1252\ \text{W}\quad (1.3\ \text{kW})}$$

The answer is (C).

8. *Customary U.S. Solution*

Neglecting pipe resistance (no information for pipe is given), $T_{\text{pipe}} = T_{\text{sat}}$.

From App. 23.B for 300 lbf/in^2 steam, $T_{\text{sat}} = 417.35°\text{F}$. When a vapor condenses, the vapor and condensed liquid are at the same temperature. Therefore, the entire pipe is assumed to be at 417.35°F. The outside film coefficient should be evaluated from Eq. 36.11.

$$T_{\text{film}} = \tfrac{1}{2}(T_s + T_\infty) = \left(\tfrac{1}{2}\right)(417.35°\text{F} + 70°\text{F}) = 243.7°\text{F}$$

From App. 35.C, for air at 243.7°F,

$$\text{Pr} = 0.715$$
$$\frac{g\beta\rho^2}{\mu^2} = 0.673 \times 10^6\ \frac{1}{\text{ft}^3\text{-°F}}$$

From Eq. 35.4, the Grashof number is

$$\text{Gr} = \frac{L^3 g\beta\rho^2(T_s - T_\infty)}{\mu^2}$$
$$L = d_{\text{outside}} = \frac{4\ \text{in}}{12\ \frac{\text{in}}{\text{ft}}} = 0.3333\ \text{ft}$$
$$\text{Gr} = (0.3333\ \text{ft})^3\left(0.673 \times 10^6\ \frac{1}{\text{ft}^3\text{-°F}}\right) \times (417.35°\text{F} - 70°\text{F}) = 8.66 \times 10^6$$
$$\text{Gr Pr} = (8.66 \times 10^6)(0.715) = 6.19 \times 10^6$$

From Table 35.3,

$$h_c \approx 0.27\left(\frac{T_s - T_\infty}{d_{\text{outside}}}\right)^{1/4} = (0.27)\left(\frac{417.35°\text{F} - 70°\text{F}}{0.3333\ \text{ft}}\right)^{1/4} = 1.53\ \text{Btu/hr-ft}^2\text{-°F}$$

From Eq. 35.1, the heat transfer for the first 50 ft due to convection is

$$Q = h_c A_{\text{outside}}(T_s - T_\infty) = h_c(\pi d_{\text{outside}} L)(T_s - T_\infty) = \left(1.53\ \frac{\text{Btu}}{\text{hr-ft}^2\text{-°F}}\right)\pi(0.3333\ \text{ft})(50\ \text{ft}) \times (417.35°\text{F} - 70°\text{F})$$
$$Q_{\text{convection}} = 27{,}824\ \text{Btu/hr}$$

To determine heat transfer due to radiation, assume oxidized steel pipe, completely enclosed.

$$F_a = 1$$

The absolute temperatures are

$$T_1 = 417.35°\text{F} + 460° = 877.35°\text{R}$$
$$T_2 = 70°\text{F} + 460° = 530°\text{R}$$
$$F_e = \epsilon_{\text{pipe}} = 0.80$$

The radiation heat transfer is

$$\begin{aligned}E_{\text{net}} &= \sigma F_a F_e \left(T_1^4 - T_2^4\right) \\ &= \left(0.1713 \times 10^{-8} \ \frac{\text{Btu}}{\text{hr-ft}^2\text{-}°\text{R}^4}\right) \\ &\quad \times (1)(0.80)\left((877.35°\text{R})^4 - (530°\text{R})^4\right) \\ &= 704 \ \text{Btu/hr-ft}^2\end{aligned}$$

$$\begin{aligned}Q_{\text{radiation}} &= E_{\text{net}}A = E_{\text{net}}(\pi d_{\text{outside}}L) \\ &= \left(704 \ \frac{\text{Btu}}{\text{hr-ft}^2}\right)\pi(0.3333 \ \text{ft})(50 \ \text{ft}) \\ &= 36{,}858 \ \text{Btu/hr}\end{aligned}$$

The total heat loss is

$$\begin{aligned}Q_{\text{total}} &= Q_{\text{convection}} + Q_{\text{radiation}} \\ &= 27{,}824 \ \frac{\text{Btu}}{\text{hr}} + 36{,}858 \ \frac{\text{Btu}}{\text{hr}} \\ &= 64{,}682 \ \text{Btu/hr}\end{aligned}$$

$$Q_{\text{total}} = \dot{m}\Delta h$$

The enthalpy decrease per pound is

$$\Delta h = \frac{Q_{\text{total}}}{\dot{m}_{\text{steam}}} = \frac{64{,}682 \ \frac{\text{Btu}}{\text{hr}}}{5000 \ \frac{\text{lbm}}{\text{hr}}} = 12.94 \ \text{Btu/lbm}$$

This is a quality loss of

$$\Delta x = \frac{\Delta h}{h_{fg}} = \frac{12.94 \ \frac{\text{Btu}}{\text{lbm}}}{809.4 \ \frac{\text{Btu}}{\text{lbm}}} = \boxed{0.0160 \quad (1.6\%)}$$

The answer is (A).

SI Solution

From the customary U.S. solution, T_{pipe} is the same for the entire length.

From App. 23.O for 2.1 MPa, $T_{\text{sat}} = 214.72°\text{C}$. The outside film coefficient should be evaluated from Eq. 36.11 as

$$\begin{aligned}T_{\text{film}} &= \tfrac{1}{2}(T_s + T_\infty) = \left(\tfrac{1}{2}\right)(214.72°\text{C} + 21°\text{C}) \\ &= 117.9°\text{C}\end{aligned}$$

From App. 35.D, for air at 117.9°C,

$$\text{Pr} = 0.692$$
$$\frac{g\beta\rho^2}{\mu^2} = 0.403 \times 10^8 \ \frac{1}{\text{K}\cdot\text{m}^3}$$

From Eq. 35.4, the Grashof number is

$$\text{Gr} = \frac{L^3 g\beta\rho^2(T_s - T_\infty)}{\mu^2}$$

$$L = d_{\text{outside}} = \frac{10 \ \text{cm}}{100 \ \frac{\text{cm}}{\text{m}}} = 0.10 \ \text{m}$$

$$\begin{aligned}\text{Gr} &= (0.10 \ \text{m})^3\left(0.403 \times 10^8 \ \frac{1}{\text{K}\cdot\text{m}^3}\right) \\ &\quad \times (214.72°\text{C} - 21°\text{C}) \\ &= 7.807 \times 10^6\end{aligned}$$

$$\begin{aligned}\text{Gr Pr} &= (7.807 \times 10^6)(0.692) \\ &= 5.40 \times 10^6\end{aligned}$$

From Table 35.3,

$$\begin{aligned}h_c &\approx 1.32\left(\frac{T_s - T_\infty}{d_{\text{outside}}}\right)^{1/4} \\ &= (1.32)\left(\frac{214.72°\text{C} - 21°\text{C}}{0.10 \ \text{m}}\right)^{1/4} \\ &= 8.76 \ \text{W/m}^2\cdot\text{K}\end{aligned}$$

From Eq. 35.1, the heat transfer for the first 15 m due to convection is

$$\begin{aligned}Q_{\text{convection}} &= h_c(\pi d_{\text{outside}}L)(T_s - T_\infty) \\ &= \left(8.76 \ \frac{\text{W}}{\text{m}^2\cdot\text{K}}\right)\pi(0.10 \ \text{m})(15 \ \text{m}) \\ &\quad \times (214.72°\text{C} - 21°\text{C}) \\ &= 7997 \ \text{W}\end{aligned}$$

To determine heat transfer due to radiation, assume oxidized steel pipe, completely enclosed.

$$F_a = 1$$

The absolute temperatures are

$$T_1 = 214.72°\text{C} + 273° = 487.72\text{K}$$
$$T_2 = 21°\text{C} + 273° = 294\text{K}$$
$$F_e = \epsilon_{\text{pipe}} = 0.80$$

The radiation heat transfer is

$$\begin{aligned} E_{net} &= \sigma F_a F_e \left(T_1^4 - T_2^4\right) \\ &= \left(5.67\times 10^{-8}\ \frac{W}{m^2{\cdot}K^4}\right) \\ &\quad \times (1)(0.80)\left((487.72K)^4 - (294K)^4\right) \\ &= 2228\ W/m^2 \end{aligned}$$

$$\begin{aligned} Q_{radiation} &= E_{net}(\pi d_{outside} L) \\ &= \left(2228\ \frac{W}{m^2}\right)\pi(0.10\ m)(15\ m) \\ &= 10\,499\ W \end{aligned}$$

The total heat loss is

$$\begin{aligned} Q_{total} &= Q_{convection} + Q_{radiation} = 7997\ W + 10\,499\ W \\ &= 18\,496\ W \end{aligned}$$

$$Q_{total} = \dot{m}\Delta h$$

The enthalpy decrease per kilogram is

$$\begin{aligned} \Delta h &= \frac{Q_{total}}{\dot{m}_{steam}} = \frac{18\,496\ W}{\left(0.63\ \frac{kg}{s}\right)\left(1000\ \frac{J}{kJ}\right)} \\ &= 29.36\ kJ/kg \end{aligned}$$

This is a quality loss of

$$\begin{aligned} \Delta x &= \frac{\Delta h}{h_{fg}} = \frac{29.36\ \frac{kJ}{kg}}{1879.8\ \frac{kJ}{kg}} \\ &= \boxed{0.0156\quad(1.6\%)} \end{aligned}$$

The answer is (A).

9. *Customary U.S. Solution*

The bulk temperature of the water is

$$\begin{aligned} T_b &= \tfrac{1}{2}(T_{in} + T_{out}) = \left(\tfrac{1}{2}\right)(70°F + 190°F) \\ &= 130°F \end{aligned}$$

From App. 35.A, the properties of water at 130°F are

$$\begin{aligned} c_p &= 0.999\ Btu/lbm\text{-}°F \\ \nu &= 0.582\times 10^{-5}\ ft^2/sec \\ Pr &= 3.45 \\ k &= 0.376\ Btu/hr\text{-}ft\text{-}°F \end{aligned}$$

The heat transfer is found from the temperature gain of the water.

$$\begin{aligned} Q &= \dot{m}c_p\Delta T \\ &= \left(2940\ \frac{lbm}{hr}\right)\left(0.999\ \frac{Btu}{lbm\text{-}°F}\right)(190°F - 70°F) \\ &= 352{,}447\ Btu/hr \end{aligned}$$

The Reynolds number is

$$\begin{aligned} Re &= \frac{vD}{\nu} = \frac{\left(3\ \frac{ft}{sec}\right)(0.9\ in)}{\left(0.582\times 10^{-5}\ \frac{ft^2}{sec}\right)\left(12\ \frac{in}{ft}\right)} \\ &= 3.87\times 10^4 \end{aligned}$$

From Eq. 36.33, the film coefficient is

$$\begin{aligned} h &= 0.023 Re^{0.8} Pr^n \left(\frac{k}{d}\right) \\ &= (0.023)(3.87\times 10^4)^{0.8}(3.45)^{0.4} \\ &\quad \times \left(\frac{\left(0.376\ \frac{Btu}{hr\text{-}ft\text{-}°F}\right)\left(12\ \frac{in}{ft}\right)}{0.9\ in}\right) \\ &= 885\ Btu/hr\text{-}ft^2\text{-}°F \end{aligned}$$

The saturation temperature for 134 psia steam is approximately 350°F. Assume the wall is 20°F lower (≈330°F). The film properties are evaluated at the average of the wall and saturation temperatures.

$$\begin{aligned} T_h &= \tfrac{1}{2}(T_{sat,v} + T_s) = \left(\tfrac{1}{2}\right)(350°F + 330°F) \\ &= 340°F \end{aligned}$$

Film properties are obtained from App. 35.A for liquid water and from App. 23.A and App. 23.B for vapor.

$$k_{340°F} = 0.392\ Btu/hr\text{-}ft\text{-}°F$$

$$\begin{aligned} \mu_{340°F} &= \left(0.109\times 10^{-3}\ \frac{lbm}{sec\text{-}ft}\right)\left(3600\ \frac{sec}{hr}\right) \\ &= 0.392\ lbm/ft\text{-}hr \end{aligned}$$

$$\begin{aligned} \rho_{l,340°F} &= \frac{1}{v_{f,340°F}} = \frac{1}{0.01787\ \frac{ft^3}{lbm}} \\ &= 55.96\ lbm/ft^3 \end{aligned}$$

$$\begin{aligned} \rho_{v,340°F} &= \frac{1}{v_{g,340°F}} = \frac{1}{3.788\ \frac{ft^3}{lbm}} \\ &= 0.2640\ lbm/ft^3 \end{aligned}$$

$$h_{fg,134\,\text{psia}} = 871.3 \text{ Btu/lbm}$$

$$d = \frac{1.0 \text{ in}}{12 \ \frac{\text{in}}{\text{ft}}} = 0.083 \text{ ft}$$

$$g = \left(32.2 \ \frac{\text{ft}}{\text{sec}^2}\right)\left(3600 \ \frac{\text{sec}}{\text{hr}}\right)^2 = 4.17 \times 10^8 \text{ ft/hr}^2$$

From Eq. 35.27, the film coefficient is

$$h_o = 0.725\left(\frac{\rho_l(\rho_l - \rho_v)gh_{fg}k_l^3}{d\mu_l(T_{\text{sat},v} - T_s)}\right)^{1/4}$$

$$= (0.725) \times \left(\frac{\left(55.96 \ \frac{\text{lbm}}{\text{ft}^3}\right)\left(55.96 \ \frac{\text{lbm}}{\text{ft}^3} - 0.2640 \ \frac{\text{lbm}}{\text{ft}^3}\right) \times \left(4.17 \times 10^8 \ \frac{\text{ft}}{\text{hr}^2}\right)\left(871.3 \ \frac{\text{Btu}}{\text{lbm}}\right) \times \left(0.392 \ \frac{\text{Btu}}{\text{hr-ft-°F}}\right)^3}{(0.083 \text{ ft})\left(0.392 \ \frac{\text{lbm}}{\text{ft-hr}}\right)(350°\text{F} - 330°\text{F})}\right)^{1/4}$$

$$= 2320 \text{ Btu/hr-ft}^2\text{-°F}$$

From Table 36.7, for copper alloy (70% Cu, 30% Ni), at 330°F, k = 19.6 Btu-ft/hr-ft²-°F.

From Eq. 36.70, the overall heat transfer coefficient based on the outside area is

$$\frac{1}{U_o} = \frac{1}{h_o} + \left(\frac{r_o}{k_{\text{tube}}}\right)\ln\frac{r_o}{r_i} + \frac{r_o}{r_i h_i}$$

$$r_o = \frac{d_o}{2} = \frac{1 \text{ in}}{(2)\left(12 \ \frac{\text{in}}{\text{ft}}\right)} = 0.0417 \text{ ft}$$

$$r_i = \frac{d_i}{2} = \frac{0.9 \text{ in}}{(2)\left(12 \ \frac{\text{in}}{\text{ft}}\right)} = 0.0375 \text{ ft}$$

$$U_o = \frac{1}{\frac{1}{2320 \ \frac{\text{Btu}}{\text{hr-ft}^2\text{-°F}}} + \left(\frac{0.0417 \text{ ft}}{19.6 \ \frac{\text{Btu-ft}}{\text{hr-ft}^2\text{-°F}}}\right)\ln\left(\frac{0.0417 \text{ ft}}{0.0375 \text{ ft}}\right) + \frac{0.0417 \text{ ft}}{(0.0375 \text{ ft})\left(885 \ \frac{\text{Btu}}{\text{hr-ft}^2\text{-°F}}\right)}}$$

$$= 523 \text{ Btu/hr-ft}^2\text{-°F}$$

For crossflow operation,

70°F ⟶ 190°F
A B
350°F ⟵ 350°F

$$\Delta T_A = 350°\text{F} - 70°\text{F} = 280°\text{F}$$

$$\Delta T_B = 350°\text{F} - 190°\text{F} = 160°\text{F}$$

From Eq. 36.68, the logarithmic mean temperature difference is

$$\Delta T_{\text{lm}} = \frac{\Delta T_A - \Delta T_B}{\ln \frac{\Delta T_A}{\Delta T_B}} = \frac{280°\text{F} - 160°\text{F}}{\ln \frac{280°\text{F}}{160°\text{F}}} = 214.4°\text{F}$$

The heat transfer is known; therefore, the outside area can be calculated from Eq. 36.69.

$$Q = U_o A_o F_c \Delta T_{\text{lm}}$$

For steam condensation, the temperature of steam remains constant. Therefore, $F_c = 1$.

$$A_o = \frac{352{,}447 \ \frac{\text{Btu}}{\text{hr}}}{\left(523 \ \frac{\text{Btu}}{\text{hr-ft}^2\text{-°F}}\right)(1)(214.4°\text{F})} = \boxed{3.14 \text{ ft}^2 \quad (3.0 \text{ ft}^2)}$$

At this point, the assumption that $T_{\text{sat},v} - T_s = 20°\text{F}$ could be checked using $Q = U_{\text{partial}} A_o \Delta T_{\text{partial}}$, working from the inside (at 130°F) to the outside (at T_s).

The answer is (A).

SI Solution

The bulk temperature of the water is

$$T_b = \tfrac{1}{2}(T_{\text{in}} + T_{\text{out}}) = \left(\tfrac{1}{2}\right)(21°\text{C} + 90°\text{C}) = 55.5°\text{C}$$

From App. 35.B, the properties of water at 55.5°C are

$$c_p = 4.186 \text{ kJ/kg·K}$$
$$\rho = 986.6 \text{ kg/m}^3$$
$$\mu = 0.523 \times 10^{-3} \text{ kg/m·s}$$
$$k = 0.6503 \text{ W/m·K}$$
$$\text{Pr} = 3.37$$

The heat transfer is found from the temperature gain of the water.

$$Q = \dot{m}c_p\Delta T = \left(0.368 \ \frac{\text{kg}}{\text{s}}\right)\left(4.186 \ \frac{\text{kJ}}{\text{kg·K}}\right)\left(1000 \ \frac{\text{J}}{\text{kJ}}\right) \times (90°\text{C} - 21°\text{C}) = 106\,291 \text{ W}$$

Heat Transfer

The Reynolds number is

$$\mathrm{Re} = \frac{\rho \mathrm{v} D}{\mu} = \frac{\left(986.6\ \frac{\mathrm{kg}}{\mathrm{m}^3}\right)\left(0.9\ \frac{\mathrm{m}}{\mathrm{s}}\right)(2.29\ \mathrm{cm})}{\left(0.523 \times 10^{-3}\ \frac{\mathrm{kg}}{\mathrm{m{\cdot}s}}\right)\left(100\ \frac{\mathrm{cm}}{\mathrm{m}}\right)} = 3.89 \times 10^4$$

From Eq. 36.33, the film coefficient is

$$\begin{aligned} h &= 0.023\,\mathrm{Re}^{0.8}\mathrm{Pr}^n\left(\frac{k}{d}\right) \\ &= (0.023)(3.89 \times 10^4)^{0.8}(3.37)^{0.4} \times \left(\frac{\left(0.6503\ \frac{\mathrm{W}}{\mathrm{m{\cdot}K}}\right)\left(100\ \frac{\mathrm{cm}}{\mathrm{m}}\right)}{2.29\ \mathrm{cm}}\right) \\ &= 4989\ \mathrm{W/m^2{\cdot}K} \end{aligned}$$

The saturation temperature for 923 kPa steam is 176.4°C. Assume the wall is 10°C lower (or 166.4°C). The film properties are evaluated at the average of the wall and saturation temperatures.

$$\begin{aligned} T_h &= \tfrac{1}{2}(T_{\mathrm{sat},v} + T_s) = \left(\tfrac{1}{2}\right)(176.4^\circ\mathrm{C} + 166.4^\circ\mathrm{C}) \\ &= 171.4^\circ\mathrm{C} \end{aligned}$$

Film properties are obtained from App. 35.B for liquid water and App. 23.N for vapor.

$$k_{171.4^\circ\mathrm{C}} = 0.6745\ \mathrm{W/m{\cdot}K}$$

$$\mu_{171.4^\circ\mathrm{C}} = 0.1712 \times 10^{-3}\ \mathrm{kg/m{\cdot}s}$$

$$\rho_{l,171.4^\circ\mathrm{C}} = \frac{1}{v_{f,171.4^\circ\mathrm{C}}} = \frac{(1)\left(100\ \frac{\mathrm{cm}}{\mathrm{m}}\right)^3}{\left(1.1161\ \frac{\mathrm{cm}^3}{\mathrm{g}}\right)\left(1000\ \frac{\mathrm{g}}{\mathrm{kg}}\right)} = 896.0\ \mathrm{kg/m^3}$$

$$\rho_{v,171.4^\circ\mathrm{C}} = \frac{1}{v_{g,171.4^\circ\mathrm{C}}} = \frac{(1)\left(100\ \frac{\mathrm{cm}}{\mathrm{m}}\right)^3}{\left(235.3\ \frac{\mathrm{cm}^3}{\mathrm{g}}\right)\left(1000\ \frac{\mathrm{g}}{\mathrm{kg}}\right)} = 4.25\ \mathrm{kg/m^3}$$

$$h_{fg,923\ \mathrm{kPa}} = \left(2026.8\ \frac{\mathrm{kJ}}{\mathrm{kg}}\right)\left(1000\ \frac{\mathrm{J}}{\mathrm{kg}}\right) = 2.0268 \times 10^6\ \mathrm{J/kg}$$

$$d = \frac{2.29\ \mathrm{cm}}{100\ \frac{\mathrm{cm}}{\mathrm{m}}} = 0.0229\ \mathrm{m}$$

From Eq. 35.27, the film coefficient is

$$\begin{aligned} h_o &= 0.725\left(\frac{\rho_l(\rho_l - \rho_v)g h_{fg} k_l^3}{d\mu_l(T_{\mathrm{sat},v} - T_s)}\right)^{1/4} \\ &= (0.725) \times \left(\frac{\left(896\ \frac{\mathrm{kg}}{\mathrm{m}^3}\right)\left(896\ \frac{\mathrm{kg}}{\mathrm{m}^3} - 4.25\ \frac{\mathrm{kg}}{\mathrm{m}^3}\right)\left(9.81\ \frac{\mathrm{m}}{\mathrm{s}^2}\right) \times \left(2.0268 \times 10^6\ \frac{\mathrm{J}}{\mathrm{kg}}\right)\left(0.6745\ \frac{\mathrm{W}}{\mathrm{m{\cdot}K}}\right)^3}{(0.0254\ \mathrm{m})\left(0.1712 \times 10^{-3}\ \frac{\mathrm{kg}}{\mathrm{m{\cdot}s}}\right) \times (176.4^\circ\mathrm{C} - 166.4^\circ\mathrm{C})}\right)^{1/4} \\ &= 13\,266\ \mathrm{W/m^2{\cdot}K} \end{aligned}$$

From Table 36.7 and the table footnote, for copper alloy (70% Cu, 30% Ni) at 166.4°C,

$$\begin{aligned} k &\approx \left(19.6\ \frac{\mathrm{Btu}}{\mathrm{hr{\cdot}ft{\cdot}^\circ F}}\right)\left(1.731\ \frac{\mathrm{W{\cdot}hr{\cdot}ft{\cdot}^\circ F}}{\mathrm{m{\cdot}K{\cdot}Btu}}\right) \\ &= 33.93\ \mathrm{W/m{\cdot}K} \end{aligned}$$

From Eq. 36.70, the overall heat transfer coefficient based on outside area is

$$\frac{1}{U_o} = \frac{1}{h_o} + \left(\frac{r_o}{k_{\mathrm{tube}}}\right)\ln\frac{r_o}{r_i} + \frac{r_o}{r_i h_i}$$

$$r_o = \frac{d_o}{2} = \frac{2.54\ \mathrm{cm}}{(2)\left(100\ \frac{\mathrm{cm}}{\mathrm{m}}\right)} = 0.0127\ \mathrm{m}$$

$$r_i = \frac{d_i}{2} = \frac{0.0229\ \mathrm{m}}{2} = 0.0115\ \mathrm{m}$$

$$\begin{aligned} U_o &= \frac{1}{\frac{1}{13\,266\ \frac{\mathrm{W}}{\mathrm{m^2{\cdot}K}}} + \left(\frac{0.0127\ \mathrm{m}}{33.93\ \frac{\mathrm{W}}{\mathrm{m{\cdot}K}}}\right)\ln\left(\frac{0.0127\ \mathrm{m}}{0.0115\ \mathrm{m}}\right) + \frac{0.0127\ \mathrm{m}}{(0.0115\ \mathrm{m})\left(4989\ \frac{\mathrm{W}}{\mathrm{m^2{\cdot}K}}\right)}} \\ &= 2995\ \mathrm{W/m^2{\cdot}K} \end{aligned}$$

Heat Transfer

For crossflow operation,

21°C → 90°C
A ... B
176.4°C ← 176.4°C

$$\Delta T_A = 176.4°\text{C} - 21°\text{C} = 155.4°\text{C}$$
$$\Delta T_B = 176.4°\text{C} - 90°\text{C} = 86.4°\text{C}$$

From Eq. 36.68, the logarithmic mean temperature difference is

$$\Delta T_{\text{lm}} = \frac{\Delta T_A - \Delta T_B}{\ln \frac{\Delta T_A}{\Delta T_B}} = \frac{155.4°\text{C} - 86.4°\text{C}}{\ln \frac{155.4°\text{C}}{86.4°\text{C}}} = 117.5°\text{C}$$

The heat transfer is known; therefore, the outside area can be calculated from Eq. 36.69.

$$Q = U_o A_o F_c \Delta T_{\text{lm}}$$

For steam condensation, the temperature of steam remains constant. Therefore, $F_c = 1$.

$$A_o = \frac{106\,291 \text{ W}}{\left(2995 \frac{\text{W}}{\text{m}^2\cdot\text{K}}\right)(1)(117.5°\text{C})} = \boxed{0.302 \text{ m}^2 \quad (0.30 \text{ m}^2)}$$

At this point, the assumption that $T_{\text{sat},v} - T_s = 10°\text{C}$ could be checked using $Q = U_{\text{partial}} A_o \Delta T_{\text{partial}}$, working from the inside (at 55°C) to the outside (at T_s).

The answer is (A).

10. Use the illustration shown for both the customary U.S. and SI solutions.

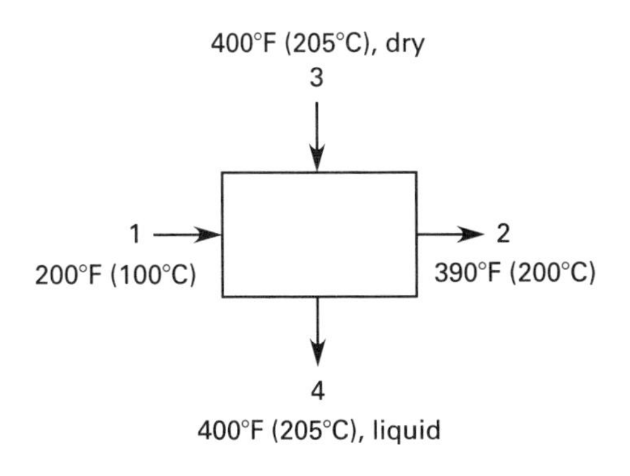

Customary U.S. Solution

(a) From App. 23.A, the enthalpy of each point is

$$h_1 = 168.13 \text{ Btu/lbm}$$
$$h_2 = 364.3 \text{ Btu/lbm}$$
$$h_3 = 1201.4 \text{ Btu/lbm}$$
$$h_4 = 375.1 \text{ Btu/lbm}$$

The heat transfer is due to the temperature gain of water.

$$Q = \dot{m}(h_2 - h_1) = \left(500{,}000 \frac{\text{lbm}}{\text{hr}}\right)\left(364.3 \frac{\text{Btu}}{\text{lbm}} - 168.13 \frac{\text{Btu}}{\text{lbm}}\right) = 9.809 \times 10^7 \text{ Btu/hr}$$

The mass flow rate per tube is

$$\dot{m}_{\text{tube}} = \rho A_{\text{tube}} \text{v}$$

Select ρ where the specific volume is greatest (at 390°F). From App. 23.A,

$$v_f = 0.01850 \text{ ft}^3/\text{lbm}$$
$$\rho = \frac{1}{v_f} = \frac{1}{0.01850 \frac{\text{ft}^3}{\text{lbm}}} = 54.05 \text{ lbm/ft}^3$$

The inside diameter of the tube is

$$d_i = d_o - 2(\text{wall}) = \frac{7}{8} \text{ in} - (2)\left(\frac{1}{16} \text{ in}\right) = 0.750 \text{ in}$$

The area per tube is

$$A_{\text{tube}} = \left(\frac{\pi}{4}\right) d_i^2 = \frac{\left(\frac{\pi}{4}\right)(0.750 \text{ in})^2}{\left(12 \frac{\text{in}}{\text{ft}}\right)^2} = 0.003068 \text{ ft}^2$$

$$\dot{m}_{\text{tube}} = \left(54.05 \frac{\text{lbm}}{\text{ft}^3}\right)(0.003068 \text{ ft}^2) \times \left(5 \frac{\text{ft}}{\text{sec}}\right)\left(3600 \frac{\text{sec}}{\text{hr}}\right) = 2985 \text{ lbm/hr}$$

Heat Transfer

The required number of tubes is

$$N = \frac{500{,}000\ \frac{\text{lbm}}{\text{hr}}}{2985\ \frac{\text{lbm}}{\text{hr}}} = \boxed{167.5\quad(170)}$$

The answer is (D).

(b) Consider counterflow or parallel flow. Since one fluid temperature remains constant, it will not make a difference. Also, from App. 36.A, $F_c = 1$.

$$\Delta T_A = 400°\text{F} - 200°\text{F} = 200°\text{F}$$
$$\Delta T_B = 400°\text{F} - 390°\text{F} = 10°\text{F}$$

From Eq. 36.68, the logarithmic mean temperature difference is

$$\Delta T_{\text{lm}} = \frac{\Delta T_A - \Delta T_B}{\ln\frac{\Delta T_A}{\Delta T_B}} = \frac{200°\text{F} - 10°\text{F}}{\ln\frac{200°\text{F}}{10°\text{F}}} = 63.4°\text{F}$$

The heat transfer is known. So,

$$Q = UA\Delta T_{\text{lm}}F_c$$

Let L be the length of the tube bundle, approximately one-half the length of a single tube. The surface area is

$$A = (\pi D_o L)N \times 2\ \text{passes}$$
$$Q = U(\pi D_o L)N\Delta T_{\text{lm}}F_c \times 2\ \text{passes}$$

The length of the bundle takes into consideration both passes (i.e., uses the total surface area).

$$\begin{aligned} L &= \frac{Q}{2U\pi D_o\Delta T_{\text{lm}}F_cN} \\ &= \frac{\left(9.809\times10^7\ \frac{\text{Btu}}{\text{hr}}\right)\left(12\ \frac{\text{in}}{\text{ft}}\right)}{(2\ \text{passes})\left(700\ \frac{\text{Btu}}{\text{hr-ft}^2\text{-°F}}\right)\pi\left(\frac{7}{8}\ \text{in}\right)\times(63.4°\text{F})(1)(168)} \\ &= \boxed{28.7\ \text{ft}\quad(30\ \text{ft})} \end{aligned}$$

This is the approximate length of the tube bundle and one half of the total straight length of the bent tubes.

The answer is (A).

SI Solution

(a) From App. 23.N, the enthalpy of each point is

$$h_1 = 419.17\ \text{kJ/kg}$$
$$h_2 = 852.27\ \text{kJ/kg}$$
$$h_3 = 2794.8\ \text{kJ/kg}$$
$$h_4 = 874.88\ \text{kJ/kg}$$

The heat transfer is due to the temperature gain of water.

$$\begin{aligned} Q &= \dot{m}(h_2 - h_1) \\ &= \left(60\ \frac{\text{kg}}{\text{s}}\right)\left(852.27\ \frac{\text{kJ}}{\text{kg}} - 419.17\ \frac{\text{kJ}}{\text{kg}}\right) \\ &= 25\,986\ \text{kW} \end{aligned}$$

The mass flow rate per tube is

$$\dot{m}_{\text{tube}} = \rho A_{\text{tube}}\text{v}$$

Select ρ where the specific volume is greatest (at 200°C). From App. 23.N,

$$\begin{aligned} \upsilon_f &= \frac{\left(1.1565\ \frac{\text{cm}^3}{\text{g}}\right)\left(1000\ \frac{\text{g}}{\text{kg}}\right)}{\left(100\ \frac{\text{cm}}{\text{m}}\right)^3} \\ &= 1.1565\times10^{-3}\ \text{m}^3/\text{kg} \end{aligned}$$

$$\begin{aligned} \rho &= \frac{1}{\upsilon_f} = \frac{1}{1.1565\times10^{-3}\ \frac{\text{m}^3}{\text{kg}}} \\ &= 864.7\ \text{kg/m}^3 \end{aligned}$$

The inside diameter of the tube is

$$\begin{aligned} d_i &= d_o - 2(\text{wall}) \\ &= \frac{2.2\ \text{cm} - \frac{(2)(1.6\ \text{mm})}{10\ \frac{\text{mm}}{\text{cm}}}}{100\ \frac{\text{cm}}{\text{m}}} \\ &= 0.0188\ \text{m} \end{aligned}$$

The area per tube is

$$\begin{aligned} A_{\text{tube}} &= \left(\frac{\pi}{4}\right)d_i^2 = \left(\frac{\pi}{4}\right)(0.0188\ \text{m})^2 \\ &= 2.776\times10^{-4}\ \text{m}^2 \end{aligned}$$

$$\begin{aligned} \dot{m}_{\text{tube}} &= \left(864.7\ \frac{\text{kg}}{\text{m}^3}\right)(2.776\times10^{-4}\ \text{m}^2)\left(1.5\ \frac{\text{m}}{\text{s}}\right) \\ &= 0.360\ \text{kg/s} \end{aligned}$$

Heat Transfer

The required number of tubes is

$$N = \frac{60 \ \frac{\text{kg}}{\text{s}}}{0.360 \ \frac{\text{kg}}{\text{s}}} = \boxed{166.7 \quad (170)}$$

The answer is (D).

(b) Consider counterflow or parallel flow. Since one fluid temperature remains constant, it will not make a difference. Also, from App. 36.A, $F_c = 1$.

$$\Delta T_A = 205°\text{C} - 100°\text{C} = 105°\text{C}$$

$$\Delta T_B = 205°\text{C} - 200°\text{C} = 5°\text{C}$$

From Eq. 36.68, the logarithmic mean temperature difference is

$$\Delta T_{\text{lm}} = \frac{\Delta T_A - \Delta T_B}{\ln \frac{\Delta T_A}{\Delta T_B}} = \frac{105°\text{C} - 5°\text{C}}{\ln \frac{105°\text{C}}{5°\text{C}}} = 32.8°\text{C}$$

The heat transfer is known. So,

$$Q = UA\Delta T_{\text{lm}} F_c$$

Let L be the length of the tube bundle, approximately one-half the length of a single tube. The surface area is

$$A = (\pi D_o L)N \times 2 \text{ passes}$$

$$Q = U(\pi D_o L)N\Delta T_{\text{lm}} F_c \times 2 \text{ passes}$$

The length of the bundle takes into consideration both passes (i.e., uses the total surface area).

$$\begin{aligned} L &= \frac{Q}{2U\pi D_o N \Delta T_{\text{lm}} F_c} \\ &= \frac{(25\,986 \text{ kW})\left(100 \ \frac{\text{cm}}{\text{m}}\right)}{(2)\left(4 \ \frac{\text{kW}}{\text{m}^2\text{-°C}}\right)\pi(2.2 \text{ cm})(167) \times (32.8°\text{C})(1)} \\ &= \boxed{8.63 \text{ m} \quad (9.0 \text{ m})} \end{aligned}$$

This is the approximate length of the tube bundle and one half of the total straight length of the bent tubes.

The answer is (A).

11. *Customary U.S. Solution*

The water's bulk temperature is

$$T_{b,\text{water}} = \left(\tfrac{1}{2}\right)(70°\text{F} + 140°\text{F}) = 105°\text{F}$$

The fluid properties at 105°F are obtained from App. 35.A.

$$\rho_{105°\text{F}} = 61.92 \text{ lbm/ft}^3$$

$$c_{p,105°\text{F}} = 0.998 \text{ Btu/lbm-°F}$$

The mass flow rate of water is

$$\begin{aligned} \dot{m}_{\text{water}} &= \dot{V}\rho \\ &= \frac{\left(100 \ \frac{\text{gal}}{\text{min}}\right)\left(60 \ \frac{\text{min}}{\text{hr}}\right)\left(61.92 \ \frac{\text{lbm}}{\text{ft}^3}\right)}{7.48 \ \frac{\text{gal}}{\text{ft}^3}} \\ &= 49{,}668 \text{ lbm/hr} \end{aligned}$$

The heat transfer is found from the temperature gain of the water.

$$\begin{aligned} Q_{\text{clean}} &= \dot{m}c_p\Delta T \\ &= \left(49{,}668 \ \frac{\text{lbm}}{\text{hr}}\right)\left(0.998 \ \frac{\text{Btu}}{\text{lbm-°F}}\right) \times (140°\text{F} - 70°\text{F}) \\ &= 3.470 \times 10^6 \text{ Btu/hr} \end{aligned}$$

$$\Delta T_A = 230°\text{F} - 70°\text{F} = 160°\text{F}$$

$$\Delta T_B = 230°\text{F} - 140°\text{F} = 90°\text{F}$$

From Eq. 36.68, the logarithmic mean temperature difference is

$$\Delta T_{\text{lm}} = \frac{\Delta T_A - \Delta T_B}{\ln \frac{\Delta T_A}{\Delta T_B}} = \frac{160°\text{F} - 90°\text{F}}{\ln \frac{160°\text{F}}{90°\text{F}}} = 121.66°\text{F}$$

The heat transfer is known. Therefore, the overall heat transfer coefficient can be calculated from Eq. 36.69.

$$Q_{\text{clean}} = U_{\text{clean}} A F_c \Delta T_{\text{lm}}$$

For steam condensation, the temperature of steam remains constant. Therefore, $F_c = 1$.

$$\begin{aligned} U_{\text{clean}} &= \frac{Q_{\text{clean}}}{AF_c\Delta T_{\text{lm}}} = \frac{3.470 \times 10^6 \ \frac{\text{Btu}}{\text{hr}}}{(50 \text{ ft}^2)(1)(121.66°\text{F})} \\ &= 570.4 \text{ Btu/hr-ft}^2\text{-°F} \end{aligned}$$

After fouling,

$$\begin{aligned} Q_{\text{fouled}} &= \left(49{,}668 \ \frac{\text{lbm}}{\text{hr}}\right)\left(0.998 \ \frac{\text{Btu}}{\text{lbm-°F}}\right) \times (122°\text{F} - 70°\text{F}) \\ &= 2.578 \times 10^6 \text{ Btu/hr} \end{aligned}$$

$$\Delta T_B = 230°\text{F} - 122°\text{F} = 108°\text{F}$$

$$\Delta T_A = 230°\text{F} - 70°\text{F} = 160°\text{F}$$

From Eq. 36.68, the logarithmic mean temperature difference is

$$\Delta T_{\text{lm}} = \frac{\Delta T_A - \Delta T_B}{\ln \frac{\Delta T_A}{\Delta T_B}} = \frac{160^\circ\text{F} - 108^\circ\text{F}}{\ln \frac{160^\circ\text{F}}{108^\circ\text{F}}} = 132.3^\circ\text{F}$$

$$U_{\text{fouled}} = \frac{2.578 \times 10^6 \ \frac{\text{Btu}}{\text{hr}}}{(50 \text{ ft}^2)(1)(132.3^\circ\text{F})} = 389.7 \text{ Btu/hr-ft}^2\text{-}^\circ\text{F}$$

From Eq. 36.75, the fouling factor is

$$\begin{aligned} R_f &= \frac{1}{U_{\text{fouled}}} - \frac{1}{U_{\text{clean}}} \\ &= \frac{1}{389.7 \ \frac{\text{Btu}}{\text{hr-ft}^2\text{-}^\circ\text{F}}} - \frac{1}{570.4 \ \frac{\text{Btu}}{\text{hr-ft}^2\text{-}^\circ\text{F}}} \\ &= \boxed{0.000813 \text{ hr-ft}^2\text{-}^\circ\text{F/Btu} \quad (0.0008 \text{ hr-ft}^2\text{-}^\circ\text{F/Btu})} \end{aligned}$$

The answer is (B).

SI Solution

The water's bulk temperature is

$$T_{b,\text{water}} = \left(\tfrac{1}{2}\right)(21^\circ\text{C} + 60^\circ\text{C}) = 40.5^\circ\text{C}$$

The fluid properties at 40.5°C are obtained from App. 35.B.

$$\rho_{40.5^\circ\text{C}} = 993.5 \text{ kg/m}^3$$

$$c_{p,40.5^\circ\text{C}} = 4.183 \text{ kJ/kg·K}$$

The mass flow rate of water is

$$\dot{m}_{\text{water}} = \dot{V}\rho = \frac{\left(6.3 \ \frac{\text{L}}{\text{s}}\right)\left(993.5 \ \frac{\text{kg}}{\text{m}^3}\right)}{1000 \ \frac{\text{L}}{\text{m}^3}} = 6.26 \text{ kg/s}$$

The heat transfer is found from the temperature gain of the water.

$$\begin{aligned} Q_{\text{clean}} &= \dot{m}c_p\Delta T \\ &= \left(6.26 \ \frac{\text{kg}}{\text{s}}\right)\left(4.183 \ \frac{\text{kJ}}{\text{kg·K}}\right)\left(1000 \ \frac{\text{J}}{\text{kg}}\right) \\ &\quad \times (60^\circ\text{C} - 21^\circ\text{C}) \\ &= 1.021 \times 10^6 \text{ W} \end{aligned}$$

$$\Delta T_A = 110^\circ\text{C} - 21^\circ\text{C} = 89^\circ\text{C}$$

$$\Delta T_B = 110^\circ\text{C} - 60^\circ\text{C} = 50^\circ\text{C}$$

From Eq. 36.68, the logarithmic mean temperature difference is

$$\Delta T_{\text{lm}} = \frac{\Delta T_A - \Delta T_B}{\ln \frac{\Delta T_A}{\Delta T_B}} = \frac{89^\circ\text{C} - 50^\circ\text{C}}{\ln \frac{89^\circ\text{C}}{50^\circ\text{C}}} = 67.64^\circ\text{C}$$

The heat transfer is known. Therefore, the overall heat transfer coefficient can be calculated from Eq. 36.69.

$$Q_{\text{clean}} = U_{\text{clean}} A F_c \Delta T_{\text{lm}}$$

For steam condensation, the temperature of steam remains constant. Therefore, $F_c = 1$.

$$\begin{aligned} U_{\text{clean}} &= \frac{Q_{\text{clean}}}{A F_c \Delta T_{\text{lm}}} \\ &= \frac{1.021 \times 10^6 \text{ W}}{(4.7 \text{ m}^2)(1)(67.64^\circ\text{C})} \\ &= 3211.6 \text{ W/m}^2\text{·K} \end{aligned}$$

After fouling,

$$\begin{aligned} Q_{\text{fouled}} &= \left(6.26 \ \frac{\text{kg}}{\text{s}}\right)\left(4.183 \ \frac{\text{kJ}}{\text{kg·K}}\right)\left(1000 \ \frac{\text{J}}{\text{kJ}}\right) \\ &\quad \times (50^\circ\text{C} - 21^\circ\text{C}) \\ &= 7.594 \times 10^5 \text{ W} \end{aligned}$$

$$\Delta T_A = 110^\circ\text{C} - 21^\circ\text{C} = 89^\circ\text{C}$$

$$\Delta T_B = 110^\circ\text{C} - 50^\circ\text{C} = 60^\circ\text{C}$$

From Eq. 36.68, the logarithmic mean temperature difference is

$$\Delta T_{\text{lm}} = \frac{\Delta T_A - \Delta T_B}{\ln \frac{\Delta T_A}{\Delta T_B}} = \frac{89^\circ\text{C} - 60^\circ\text{C}}{\ln \frac{89^\circ\text{C}}{60^\circ\text{C}}} = 73.55^\circ\text{C}$$

$$U_{\text{fouled}} = \frac{7.594 \times 10^5 \text{ W}}{(4.7 \text{ m}^2)(1)(73.55^\circ\text{C})} = 2196.8 \text{ W/m}^2\text{·K}$$

From Eq. 36.75, the fouling factor is

$$\begin{aligned} R_f &= \frac{1}{U_{\text{fouled}}} - \frac{1}{U_{\text{clean}}} \\ &= \frac{1}{2196.8 \ \frac{\text{W}}{\text{m}^2\text{·K}}} - \frac{1}{3211.6 \ \frac{\text{W}}{\text{m}^2\text{·K}}} \\ &= \boxed{0.000144 \text{ m}^2\text{·K/W} \quad (0.0001 \text{ m}^2\text{·K/W})} \end{aligned}$$

The answer is (B).

12. Heat exchanger tubing has a constant cross-sectional area, so the flow velocity within a heat exchanger is constant. Since distance traveled at constant velocity is proportional to time, and since both fluids have the same specific heat, the heat transfer rate-distance and heat transfer rate-time profiles are $\boxed{\text{identical.}}$

The answer is (A).

Heat Transfer

13. At any point in the heat exchanger, the instantaneous heat gain by the cold water is equal to the instantaneous heat loss by the hot water. The heat transfer rates are the same. At that point, the two fluids will have traveled different distances. For example, at one entrance, the hot water will have traveled very little distance, while the cold water will have traveled almost the entire length of the heat exchanger (e.g., reversed end-to-end). However, the heat transfer rates will be the same. (It could be argued that from the standpoint of the thermodynamic sign convention, hot water has a negative heat transfer rate while cold water has a positive heat transfer rate. In that case, the temperature profiles would be reversed end-to-end *and* inverted top-to-bottom. However, that is not one of the options.)

The answer is (B).

14. At the entrance, as a result of the large temperature difference, the hot water loses heat rapidly, and the cold water gains heat rapidly. At the exit, the hot water loses heat more slowly while the cold water gains heat more slowly. Thus, the plot is inverted top-to-bottom.

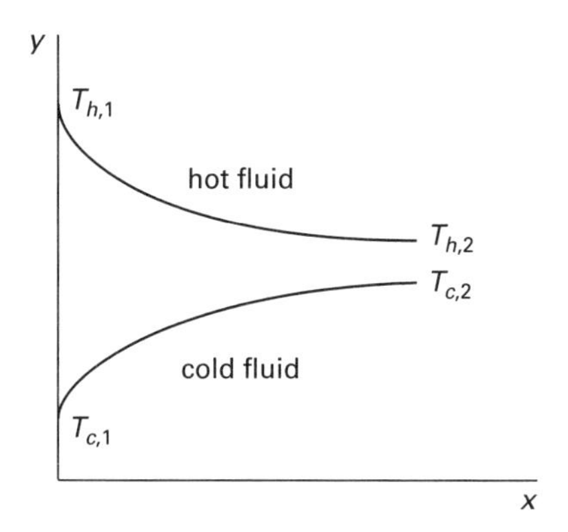

The answer is (D).

15. The temperature difference between the two water flows will essentially be constant.

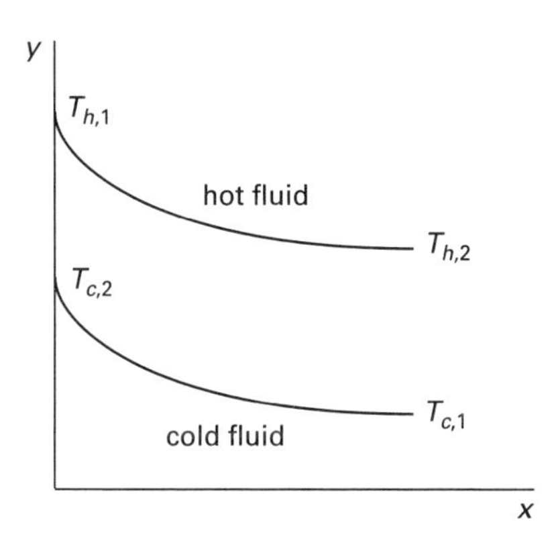

The answer is (A).

16. Since the steam enters the heat exchanger saturatedand is not subcooled, the steam temperature remains constant at the saturation temperature throughout.

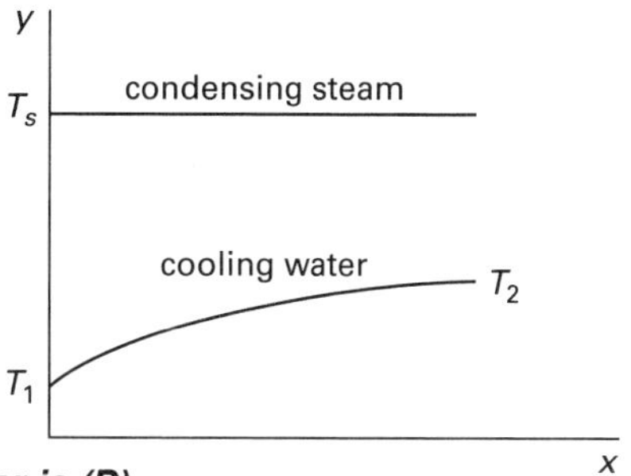

The answer is (D).

17. "Duty" is the amount of energy that the heat exchanger transfers per unit time.

The answer is (D).

18. (a) The perimeter of the cylindrical fin is

$$P = \pi d = \frac{\pi(0.5 \text{ in})}{12 \ \frac{\text{in}}{\text{ft}}} = 0.13 \text{ ft}$$

The cross-sectional area at the base of the cylindrical fin is

$$A_b = \frac{\pi d^2}{4} = \frac{\pi(0.5 \text{ in})^2}{(4)\left(12 \ \frac{\text{in}}{\text{ft}}\right)^2} = 0.0014 \text{ ft}^2$$

From Eq. 34.79, the value of m is

$$m = \sqrt{\frac{2h}{kr}} = \sqrt{\frac{(2)\left(1.3 \ \frac{\text{Btu}}{\text{hr-ft}^2\text{-°F}}\right)}{\left(128 \ \frac{\text{Btu}}{\text{hr-ft-°F}}\right)\left(\frac{0.5 \text{ in}}{2}\right)}{12 \ \frac{\text{in}}{\text{ft}}}}}$$

$$= 0.98 \text{ ft}^{-1} \quad [\text{say } 1.00 \text{ ft}^{-1}]$$

The heat transfer from the heat source into the base of the fin is the fin's heat dissipation rate. From Eq. 34.80, the heat transfer from the fin is

$$Q = \sqrt{hPkA_b}(T_b - T_\infty)\tanh mL$$

$$= \sqrt{\begin{array}{l}\left(1.3 \ \frac{\text{Btu}}{\text{hr-ft}^2\text{-°F}}\right)(0.13 \text{ ft}) \\ \times \left(128 \ \frac{\text{Btu}}{\text{hr-ft-°F}}\right)(0.0014 \text{ ft}^2)\end{array}} \times (350\text{°F} - 75\text{°F})\tanh\left(\left(1.0 \ \frac{1}{\text{ft}}\right)(2 \text{ ft})\right)$$

$$= \boxed{45.7 \text{ Btu/hr} \quad (46 \text{ Btu/hr})}$$

The answer is (D).

Heat Transfer

(b) From Eq. 34.87, the fin efficiency is

$$\eta_f = \frac{\tanh mL}{mL} = \frac{\tanh\left(\left(1.0\ \frac{1}{\text{ft}}\right)(2\ \text{ft})\right)}{\left(1.0\ \frac{1}{\text{ft}}\right)(2\ \text{ft})}$$
$$= \boxed{0.482\quad(48\%)}$$

The answer is (C).

19. *Customary U.S. Solution*

From Eq. 35.11, the film temperature of the air is

$$T_h = \tfrac{1}{2}(T_s + T_\infty) = \left(\tfrac{1}{2}\right)(100^\circ\text{F} + 150^\circ\text{F}) = 125^\circ\text{F}$$

From App. 35.C, the air properties at 125°F are

$$\nu = 0.195 \times 10^{-3}\ \text{ft}^2/\text{sec}$$
$$k = 0.0159\ \text{Btu/hr-ft-}^\circ\text{F}$$
$$\text{Pr} = 0.72$$

The Reynolds number based on the diameter, d, is

$$\text{Re}_d = \frac{\text{v}d}{\nu} = \frac{\left(100\ \frac{\text{ft}}{\text{sec}}\right)(0.35\ \text{in})}{\left(0.195 \times 10^{-3}\ \frac{\text{ft}^2}{\text{sec}}\right)\left(12\ \frac{\text{in}}{\text{ft}}\right)}$$
$$= 1.50 \times 10^4$$

From Eq. 36.47 and Eq. 36.51, the film coefficient is

$$h = C_1(\text{Re}_d)^n\text{Pr}^{1/3}\left(\frac{k}{d}\right)$$

From Table 36.3, $C_1 = 0.193$ and $n = 0.618$.

$$h = (0.193)(1.50 \times 10^4)^{0.618}(0.72)^{1/3} \times \left(\frac{\left(0.0159\ \frac{\text{Btu}}{\text{hr-ft-}^\circ\text{F}}\right)\left(12\ \frac{\text{in}}{\text{ft}}\right)}{0.35\ \text{in}}\right)$$
$$= \boxed{35.9\ \text{Btu/hr-ft}^2\text{-}^\circ\text{F}\quad(36\ \text{Btu/hr-ft}^2\text{-}^\circ\text{F})}$$

The answer is (A).

SI Solution

From Eq. 35.11, the film temperature of the air is

$$T_h = \tfrac{1}{2}(T_s + T_\infty) = \left(\tfrac{1}{2}\right)(38^\circ\text{C} + 66^\circ\text{C}) = 52^\circ\text{C}$$

From App. 35.D, the air properties at 52°C are

$$\rho = 1.09\ \text{kg/m}^3$$
$$\mu = 1.966 \times 10^{-5}\ \text{kg/m·s}$$
$$k = 0.02815\ \text{W/m·K}$$
$$\text{Pr} = 0.703$$

The Reynolds number based on the diameter, d, is

$$\text{Re}_d = \frac{\rho \text{v} d}{\mu}$$
$$= \frac{\left(1.09\ \frac{\text{kg}}{\text{m}^3}\right)\left(30\ \frac{\text{m}}{\text{s}}\right)(8.9\ \text{mm})}{\left(1.966 \times 10^{-5}\ \frac{\text{kg}}{\text{m·s}}\right)\left(1000\ \frac{\text{mm}}{\text{m}}\right)}$$
$$= 1.48 \times 10^4$$

From Eq. 36.47 and Eq. 36.51, the film coefficient is

$$h = C_1(\text{Re}_d)^h\text{Pr}^{1/3}\left(\frac{k}{d}\right)$$

From Table 36.3, $C_1 = 0.193$ and $n = 0.618$.

$$h = (0.193)(1.48 \times 10^4)^{0.618}(0.703)^{1/3} \times \left(\frac{\left(0.02815\ \frac{\text{W}}{\text{m·K}}\right)\left(1000\ \frac{\text{mm}}{\text{m}}\right)}{8.9\ \text{mm}}\right)$$
$$= \boxed{205.0\ \text{W/m}^2\text{·K}\quad(210\ \text{W/m}^2\text{·K})}$$

The answer is (A).

37 Radiation and Combined Heat Transfer

PRACTICE PROBLEMS

Arrangement Factors

1. A 6 in (15 cm) thick furnace wall has a 3 in (8 cm) square peephole. The interior of the furnace is at 2200°F (1200°C). The surrounding air temperature is 70°F (20°C). The heat loss due to radiation when the peephole is open is most nearly

(A) 450 Btu/hr (150 W)

(B) 1300 Btu/hr (440 W)

(C) 2000 Btu/hr (680 W)

(D) 7900 Btu/hr (2.7 kW)

Combined Heat Transfer

2. A 9 in (23 cm) diameter duct is painted with white enamel. The surface of the duct is at 200°F (95°C). The duct carries hot air through a room whose walls are 70°F (20°C). The air in the room is at 80°F (27°C). The unit heat transfer is most nearly

(A) 650 Btu/hr-ft length (650 W/m length)

(B) 900 Btu/hr-ft length (900 W/m length)

(C) 1100 Btu/hr-ft length (1100 W/m length)

(D) 1500 Btu/hr-ft length (1500 W/m length)

3. The walls of a cold storage unit have the cross section shown.

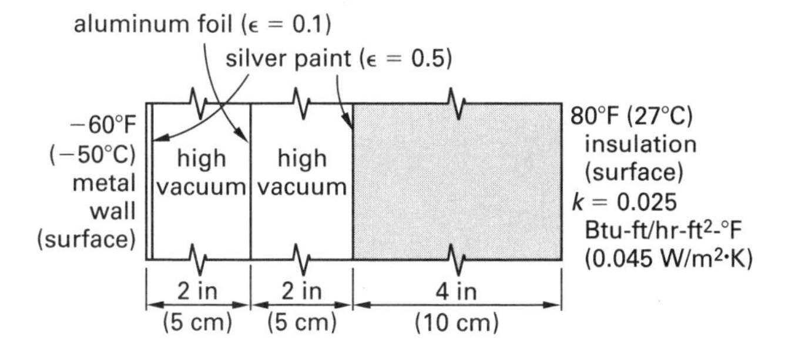

(a) The heat transfer per unit area of wall is most nearly

(A) 3 Btu/hr-ft^2 (10 W/m^2)

(B) 7 Btu/hr-ft^2 (23 W/m^2)

(C) 14 Btu/hr-ft^2 (46 W/m^2)

(D) 53 Btu/hr-ft^2 (180 W/m^2)

(b) The temperature of the aluminum foil is most nearly

(A) 420°R (230K)

(B) 460°R (260K)

(C) 475°R (264K)

(D) 490°R (270K)

4. Dry air at 1 atmospheric pressure flows at 500 ft^3/min (0.25 m^3/s) through 50 ft (15 m) of 12 in (30 cm) diameter uninsulated duct. The emissivity of the duct surface is 0.28. Air enters the duct at 45°F (7°C). The walls, air, and contents of the room through which the duct passes are at 80°F (27°C). An engineer states that the air leaving the duct will be at 50°F (10°C).

(a) The film coefficient for air flowing inside the duct is most nearly

(A) 2 Btu/hr-ft^2-°F (12 W/m^2·K)

(B) 3 Btu/hr-ft^2-°F (17 W/m^2·K)

(C) 5 Btu/hr-ft^2-°F (28 W/m^2·K)

(D) 9 Btu/hr-ft^2-°F (51 W/m^2·K)

(b) The heat transfer due to convection is most nearly

(A) 640 Btu/hr (200 W)

(B) 1600 Btu/hr (490 W)

(C) 2000 Btu/hr (610 W)

(D) 2200 Btu/hr (670 W)

(c) The total heat transfer to the air is most nearly

(A) 2200 Btu/hr (690 W)

(B) 2500 Btu/hr (780 W)

(C) 2700 Btu/hr (840 W)

(D) 3600 Btu/hr (1120 W)

(d) Considering both convection and radiation, is the engineer's statement that the air leaving the duct will be at 50°F (10°C) correct?

(A) The temperature is 45°F (7.2°C); this does not agree with the engineer's estimate.

(B) The temperature is 49°F (9.5°C); this does agree with the engineer's estimate.

(C) The temperature is 51°F (11°C); this does agree with the engineer's estimate.

(D) The temperature is 53°F (12°C); this does not agree with the engineer's estimate.

5. A steel pipe is painted on the outside with dull gray (oil-based) paint. The pipe is 35 ft (10 m) long. The pipe has a 4.00 in (10.2 cm) inside diameter and 4.25 in (10.8 cm) outside diameter. The pipe carries 200 ft^3/min (0.1 m^3/s) of 500°F (260°C), 25 psig (170 kPa) air through a 70°F (20°C) room. The conditions of the air at the end of the pipe are 350°F (180°C) and 15 psig (100 kPa).

(a) The overall coefficient of heat transfer is most nearly

(A) 1.0 Btu/hr-ft^2-°F (6.1 W/m^2·K)

(B) 1.8 Btu/hr-ft^2-°F (11 W/m^2·K)

(C) 3.6 Btu/hr-ft^2-°F (22 W/m^2·K)

(D) 49 Btu/hr-ft^2-°F (300 W/m^2·K)

(b) Using theoretical methods or empirical correlations, the calculated overall coefficient of heat transfer is most nearly

(A) 1.0 Btu/hr-ft^2-°F (6.1 W/m^2·K)

(B) 1.8 Btu/hr-ft^2-°F (11 W/m^2·K)

(C) 3.8 Btu/hr-ft^2-°F (22 W/m^2·K)

(D) 54 Btu/hr-ft^2-°F (330 W/m^2·K)

(c) Which of the following is NOT a possible reason for differences between the actual and calculated overall coefficients of heat transfer?

(A) The internal film coefficient was disregarded.

(B) h_r and h_o are additive.

(C) Pipe thermal resistance was disregarded.

(D) Dirt was on the outside of the duct.

6. A semiconductor device is modeled as a circular cylinder 0.75 in (19 mm) in diameter and 1.5 in (38 mm) high standing on one end. The device emits 5.0 W and is cooled by a combination of natural convection and radiation. The surface emissivity is 0.65. The base is insulated and transmits no heat. The air and surroundings are at 14.7 psia (101 kPa) and 75°F (24°C).

(a) The surface temperature of the device is most nearly

(A) 650°R (360K)

(B) 700°R (390K)

(C) 740°R (410K)

(D) 750°R (415K)

(b) The percentages of heat that are lost through convection and radiation, respectively, are most nearly

(A) 38%, 61%

(B) 41%, 59%

(C) 59%, 41%

(D) 61%, 39%

7. The temperature of a gas in a duct with 600°F (315°C) walls is evaluated with a 0.5 in (13 mm) diameter thermocouple probe. The emissivity of the probe is 0.8. The gas flow rate is 3480 lbm/hr-ft^2 (4.7 kg/s·m^2). The gas velocity is 400 ft/min (2 m/s). The film coefficient on the probe is given empirically as

$$h = \frac{0.024G^{0.8}}{D^{0.4}} \qquad h = \frac{17G^{0.8}}{D^{0.4}}$$

h in Btu/hr-ft^2-°F	h in W/m^2·K
G in lbm/hr-ft^2	G in kg/s·m^2
D in ft	D in m

(a) If the actual gas temperature is 300°F (150°C), the probe's reading is most nearly

(A) 700°R (390K)

(B) 740°R (410K)

(C) 760°R (420K)

(D) 780°R (430K)

(b) If the probe reading indicates that the gas temperature is 300°F (150°C), the actual gas temperature is most nearly

(A) 650°R (360K)

(B) 740°R (410K)

(C) 770°R (430K)

(D) 810°R (450K)

SOLUTIONS

1. *Customary U.S. Solution*

The absolute temperatures are

$$T_{\text{furnace}} = 2200°\text{F} + 460° = 2660°\text{R}$$
$$T_{\infty} = 70°\text{F} + 460° = 530°\text{R}$$

Assuming that the walls are reradiating, nonconducting, and varying in temperature from 2200°F at the inside to 70°F at the outside, Fig. 37.3, curve 6, can be used to find F_{12} using $x = 3\text{ in}/6\text{ in} = 0.5$; $F_{12} = 0.38$.

The radiation heat loss is

$$Q = AE_{\text{net}} = A\sigma F_{12}\left(T_{\text{furnace}}^4 - T_{\infty}^4\right)$$
$$= \frac{(3\text{ in})^2\left(0.1713\times 10^{-8}\ \frac{\text{Btu}}{\text{hr-ft}^2\text{-°R}^4}\right)\times(0.38)\left((2660°\text{R})^4 - (530°\text{R})^4\right)}{\left(12\ \frac{\text{in}}{\text{ft}}\right)^2}$$
$$= \boxed{2033.6\text{ Btu/hr}\quad(2000\text{ Btu/hr})}$$

The answer is (C).

SI Solution

The absolute temperatures are

$$T_{\text{furnace}} = 1200°\text{C} + 273° = 1473\text{K}$$
$$T_{\infty} = 20°\text{C} + 273° = 293\text{K}$$

Making the same assumptions as for the customary U.S. solution, Fig. 37.3, curve 6, can be used to find F_{12} using $x = 8\text{ cm}/15\text{ cm} = 0.533$; $F_{12} = 0.4$.

The radiation heat loss is

$$Q = AE_{\text{net}} = A\sigma F_{12}\left(T_{\text{furnace}}^4 - T_{\infty}^4\right)$$
$$= \frac{(8\text{ cm})^2\left(5.67\times 10^{-8}\ \frac{\text{W}}{\text{m}^2\cdot\text{K}^4}\right)\times(0.4)\left((1473\text{K})^4 - (293\text{K})^4\right)}{\left(100\ \frac{\text{cm}}{\text{m}}\right)^2}$$
$$= \boxed{682.3\text{ W}\quad(680\text{ W})}$$

The answer is (C).

2. *Customary U.S. Solution*

The absolute temperatures are

$$T_{\infty} = 80°\text{F} + 460° = 540°\text{R}$$
$$T_{\text{duct}} = 200°\text{F} + 460° = 660°\text{R}$$
$$T_{\text{wall}} = 70°\text{F} + 460° = 530°\text{R}$$

The characteristic length of a cylinder is its diameter. Assume laminar flow. From Table 35.3, the convective film coefficient on the outside of the duct is approximately

$$h_{\text{convective}} = 0.27\left(\frac{T_{\text{duct}} - T_{\infty}}{D}\right)^{1/4}$$
$$= (0.27)\left(\frac{(660°\text{R} - 540°\text{R})\left(12\ \frac{\text{in}}{\text{ft}}\right)}{9\text{ in}}\right)^{1/4}$$
$$= 0.96\text{ Btu/hr-ft}^2\text{-°F}$$

The duct area per unit length is

$$\frac{A}{L} = \pi D = \frac{\pi(9\text{ in})}{12\ \frac{\text{in}}{\text{ft}}} = 2.356\text{ ft}^2/\text{ft}$$

The convection losses (per unit length) are

$$\frac{Q_{\text{convection}}}{L} = h\left(\frac{A}{L}\right)\Delta T = h\left(\frac{A}{L}\right)\left(T_{\text{duct}} - T_{\infty}\right)$$
$$= \left(0.96\ \frac{\text{Btu}}{\text{hr-ft}^2\text{-°F}}\right)\left(2.356\ \frac{\text{ft}^2}{\text{ft}}\right)\times(660°\text{R} - 540°\text{R})$$
$$= 271.4\text{ Btu/hr-ft}$$

Assume $\epsilon_{\text{duct}} \approx 0.90$. Then $F_e = \epsilon_{\text{duct}} = 0.90$. $F_a = 1$ since the duct is enclosed. The radiation losses (per unit length) are

$$\frac{E_{\text{net}}}{L} = \left(\frac{A}{L}\right)\sigma F_a F_e\left(T_{\text{duct}}^4 - T_{\text{wall}}^4\right)$$
$$= (2.356\text{ ft}^2)\left(0.1713\times 10^{-8}\ \frac{\text{Btu}}{\text{hr-ft}^2\text{-°R}^4}\right)\times(1)(0.90)\left((660°\text{R})^4 - (530°\text{R})^4\right)$$
$$= 402.6\text{ Btu/hr-ft}$$

The total heat transfer per unit length is

$$\frac{Q_{\text{total}}}{L} = \frac{Q_{\text{convection}}}{L} + \frac{E_{\text{net}}}{L} = 271.4\ \frac{\text{Btu}}{\text{hr-ft}} + 402.6\ \frac{\text{Btu}}{\text{hr-ft}}$$
$$= \boxed{674\text{ Btu/hr-ft length}\quad(650\text{ Btu/hr-ft length})}$$

The answer is (A).

SI Solution

The absolute temperatures are

$$T_{\text{duct}} = 95°\text{C} + 273° = 368\text{K}$$
$$T_{\text{wall}} = 20°\text{C} + 273° = 293\text{K}$$
$$T_{\infty} = 27°\text{C} + 273° = 300\text{K}$$

The characteristic length of a cylinder is its diameter. Assume laminar flow. From Table 35.3, the convective film coefficient on the outside of the duct is approximately

$$h_{\text{convective}} = 1.32\left(\frac{T_{\text{duct}} - T_\infty}{D}\right)^{1/4}$$
$$= (1.32)\left(\frac{(368\text{K} - 300\text{K})\left(100\ \frac{\text{cm}}{\text{m}}\right)}{23\ \text{cm}}\right)^{1/4}$$
$$= 5.47\ \text{W/m}^2\text{·K}$$

The duct area per unit length is

$$\frac{A}{L} = \pi D = \frac{\pi(23\ \text{cm})}{100\ \frac{\text{cm}}{\text{m}}} = 0.723\ \text{m}^2/\text{m}$$

The convective losses per unit length are

$$\frac{Q_{\text{convective}}}{L} = h\left(\frac{A}{L}\right)(T_{\text{duct}} - T_\infty)$$
$$= \left(5.47\ \frac{\text{W}}{\text{m}^2\text{·K}}\right)\left(0.723\ \frac{\text{m}^2}{\text{m}}\right)(368\text{K} - 300\text{K})$$
$$= 268.9\ \text{W/m}$$

Assuming $\epsilon_{\text{duct}} \approx 0.90$, $F_e = \epsilon_{\text{duct}} = 0.90$. $F_a = 1$ since the duct is enclosed. The radiation losses per unit length are

$$\frac{E_{\text{net}}}{L} = \left(\frac{A}{L}\right)\sigma F_a F_e\left(T_{\text{duct}}^4 - T_{\text{wall}}^4\right)$$
$$= \left(0.723\ \frac{\text{m}^2}{\text{m}}\right)\left(5.67\times 10^{-8}\ \frac{\text{W}}{\text{m}^2\text{·K}^4}\right)$$
$$\times (1)(0.90)\left((368\text{K})^4 - (293\text{K})^4\right)$$
$$= 404.7\ \text{W/m}$$

The total heat transfer per unit length is

$$\frac{Q_{\text{total}}}{L} = \frac{Q_{\text{convection}}}{L} + \frac{E_{\text{net}}}{L} = 268.9\ \frac{\text{W}}{\text{m}} + 404.7\ \frac{\text{W}}{\text{m}}$$
$$= \boxed{673.6\ \text{W/m length}\quad (650\ \text{W/m length})}$$

The answer is (A).

3. Label the cross sections A to D.

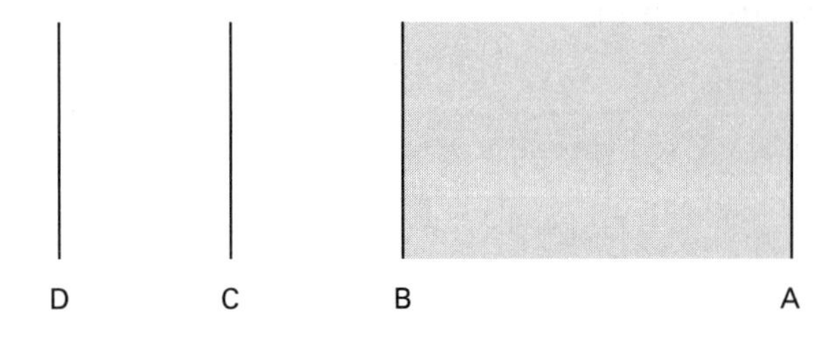

Customary U.S. Solution

(a) The absolute temperatures are

$$T_{\text{D}} = -60°\text{F} + 460° = 400°\text{R}$$
$$T_{\text{A}} = 80°\text{F} + 460° = 540°\text{R}$$

The conductive heat transfer per unit area from A to B is

$$\frac{Q_{\text{A-B}}}{A} = \frac{k(T_{\text{A}} - T_{\text{B}})}{L}$$
$$= \frac{\left(0.025\ \frac{\text{Btu-ft}}{\text{hr-ft}^2\text{-°F}}\right)(540°\text{R} - T_{\text{B}})\left(12\ \frac{\text{in}}{\text{ft}}\right)}{4\ \text{in}}$$
$$= 40.5 - 0.075T_{\text{B}} \quad [\text{Eq. 1}]$$

Since the spaces are evacuated, only radiation should be considered from B to C and from C to D.

The radiation heat transfer per unit area from B to C is

$$\frac{E_{\text{B-C}}}{A} = \sigma F_e F_a (T_{\text{B}}^4 - T_{\text{C}}^4)$$

Since the freezer is assumed to be large, $F_a = 1$.

From Table 37.1 for infinite parallel planes,

$$F_e = \frac{1}{\frac{1}{\epsilon_1} + \frac{1}{\epsilon_2} - 1} = \frac{1}{\frac{1}{0.5} + \frac{1}{0.1} - 1}$$
$$= 0.0909$$
$$\frac{E_{\text{B-C}}}{A} = \left(0.1713\times 10^{-8}\ \frac{\text{Btu}}{\text{hr-ft}^2\text{-°R}^4}\right)$$
$$\times (0.0909)(1)(T_{\text{B}}^4 - T_{\text{C}}^4)$$
$$= (1.56\times 10^{-10})T_{\text{B}}^4 - (1.56\times 10^{-10})T_{\text{C}}^4 \quad [\text{Eq. 2}]$$

Similarly, radiation heat transfer per unit area from C to D is

$$\frac{E_{\text{C-D}}}{A} = \left(0.1713\times 10^{-8}\ \frac{\text{Btu}}{\text{hr-ft}^2\text{-°R}^4}\right)$$
$$\times (0.0909)(1)\left(T_{\text{C}}^4 - (400°\text{R})^4\right)$$
$$= (1.56\times 10^{-10})T_{\text{C}}^4 - 3.99 \quad [\text{Eq. 3}]$$

The heat transfer from B to C is equal to the heat transfer from C to D.

$$\frac{E_{\text{B-C}}}{A} = \frac{E_{\text{C-D}}}{A}$$

$$(1.56\times10^{-10})T_B^4 - (1.56\times10^{-10})T_C^4 = (1.56\times10^{-10})T_C^4 - 3.99$$
$$T_B^4 - T_C^4 = T_C^4 - 2.56\times10^{10}$$
$$T_B^4 + 2.56\times10^{10} = 2T_C^4$$
$$T_C^4 = \tfrac{1}{2}T_B^4 + 1.28\times10^{10} \quad \text{[Eq. 4]}$$

The heat transfer from A to B is equal to the heat transfer from B to C.

$$\frac{Q_{A-B}}{A} = \frac{E_{B-C}}{A}$$
$$40.5 - 0.075T_B = (1.56\times10^{-10})T_B^4 - (1.56\times10^{-10})T_C^4$$
$$T_C^4 = T_B^4 + (4.81\times10^8)T_B - 2.60\times10^{11} \quad \text{[Eq. 5]}$$

Since Eq. 4 = Eq. 5,

$$\tfrac{1}{2}T_B^4 + 1.28\times10^{10} = T_B^4 + (4.81\times10^8)T_B - 2.60\times10^{11}$$
$$T_B^4 + (9.62\times10^8)T_B = 5.456\times10^{11}$$

By trial and error, $T_B = 501.4°R$.

From Eq. 1,

$$\frac{Q_{A-B}}{A} = 40.5 - 0.075T_B = 40.5 - (0.075)(501.4°\text{R}) = \boxed{2.895\ \text{Btu/hr-ft}^2\quad(3\ \text{Btu/hr-ft}^2)}$$

The answer is (A).

(b) From Eq. 4,

$$T_C = \left(\tfrac{1}{2}T_B^4 + 1.28\times10^{10}\right)^{1/4} = \left(\left(\tfrac{1}{2}\right)(501.4)^4 + 1.28\times10^{10}\right)^{1/4} = \boxed{459°\text{R}\quad(460°\text{R})}$$

The answer is (B).

SI Solution

(a) The absolute temperatures are

$$T_D = -50°\text{C} + 273° = 223\text{K}$$
$$T_A = 27°\text{C} + 273° = 300\text{K}$$

The conductive heat transfer per unit area from A to B is

$$\frac{Q_{A-B}}{A} = \frac{k(T_A - T_B)}{L} = \frac{\left(0.045\ \frac{\text{W}}{\text{m·K}}\right)(300\text{K} - T_B)\left(100\ \frac{\text{cm}}{\text{m}}\right)}{10\ \text{cm}} = 135 - 0.45T_B \quad \text{[Eq. 1]}$$

Since spaces are evacuated, only radiation should be considered from B to C and from C to D.

The radiation heat transfer per unit area from B to C is

$$\frac{E_{B-C}}{A} = \sigma F_e F_a (T_B^4 - T_C^4)$$

From the customary U.S. solution, $F_e = 0.0909$ and $F_a = 1$.

$$\frac{E_{B-C}}{A} = \left(5.67\times10^{-8}\ \frac{\text{W}}{\text{m}^2\text{·K}^4}\right)\times(0.0909)(1)(T_B^4 - T_C^4) = (5.15\times10^{-9})T_B^4 - (5.15\times10^{9})T_C^4 \quad \text{[Eq. 2]}$$

Similarly, the radiation heat transfer per unit area from C to D is

$$\frac{E_{C-D}}{A} = \left(5.67\times10^{-8}\ \frac{\text{W}}{\text{m}^2\text{·K}^4}\right)\times(0.0909)(1)\left(T_C^4 - (223\text{K})^4\right) = (5.15\times10^{-9})T_C^4 - 12.746 \quad \text{[Eq. 3]}$$

The heat transfer from B to C is equal to the heat transfer from C to D.

$$\frac{E_{D-C}}{A} = \frac{E_{C-D}}{A}$$
$$(5.15\times10^{-9})T_B^4 - (5.15\times10^{-9})T_C^4 = (5.15\times10^{-9})T_C^4 - 12.746$$
$$T_B^4 - T_C^4 = T_C^4 - 2.475\times10^9$$
$$T_B^4 + 2.745\times10^9 = 2T_C^4$$
$$T_C^4 = \tfrac{1}{2}T_B^4 + 1.237\times10^9 \quad \text{[Eq. 4]}$$

The heat transfer from A to B is equal to the heat transfer from B to C.

$$\frac{Q_{A-B}}{A} = \frac{E_{B-C}}{A}$$
$$135 - 0.45T_B = (5.15\times10^{-9})T_B^4 - (5.15\times10^{-9})T_C^4$$
$$T_C^4 = T_B^4 + (8.74\times10^7)T_B - 2.62\times10^{10} \quad \text{[Eq. 5]}$$

Heat Transfer

Since Eq. 4 = Eq. 5,

$$\tfrac{1}{2}T_B^4 + 1.237\times10^9 = T_B^4 + (8.74\times10^7)T_B - 2.62\times10^{10}$$

$$T_B^4 + (17.48\times10^7)T_B = 5.49\times10^{10}$$

By trial and error, $T_B \approx 278\text{K}$.

From Eq. 1,

$$\begin{aligned}\frac{Q_{\text{A-B}}}{A} &= 135 - 0.45T_B \\ &= 135 - (0.45)(278\text{K}) \\ &= \boxed{9.9\ \text{W/m}^2\quad(10\ \text{W/m}^2)}\end{aligned}$$

The answer is (A).

(b) From Eq. 4,

$$\begin{aligned}T_C &= \left(\tfrac{1}{2}T_B^4 + 1.237\times10^9\right)^{1/4} \\ &= \left(\left(\tfrac{1}{2}\right)(278\text{K})^4 + 1.237\times10^9\right)^{1/4} \\ &= \boxed{254.9\text{K}\quad(260\text{K})}\end{aligned}$$

The answer is (B).

4. *Customary U.S. Solution*

The absolute temperature of air entering the duct is

$$45°\text{F} + 460° = 505°\text{R}$$

From the ideal gas law, the density of air entering the duct is

$$\begin{aligned}\rho = \frac{p}{RT} &= \frac{\left(14.7\ \frac{\text{lbf}}{\text{in}^2}\right)\left(12\ \frac{\text{in}}{\text{ft}}\right)^2}{\left(53.35\ \frac{\text{ft-lbf}}{\text{lbm-°R}}\right)(505°\text{R})} \\ &= 0.07857\ \text{lbm/ft}^3\end{aligned}$$

The mass flow rate of entering air is

$$\begin{aligned}\dot{m} = \rho\dot{V} &= \left(0.07857\ \frac{\text{lbm}}{\text{ft}^3}\right)\left(500\ \frac{\text{ft}^3}{\text{min}}\right)\left(60\ \frac{\text{min}}{\text{hr}}\right) \\ &= 2357.1\ \text{lbm/hr}\end{aligned}$$

The mass velocity entering the duct is

$$\begin{aligned}G = \frac{\dot{m}}{A_{\text{flow}}} &= \frac{\left(2357.1\ \frac{\text{lbm}}{\text{hr}}\right)\left(12\ \frac{\text{in}}{\text{ft}}\right)^2}{\left(\frac{\pi}{4}\right)(12\ \text{in})^2} \\ &= 3001.1\ \text{lbm/hr-ft}^2\end{aligned}$$

(a) To calculate the initial film coefficients, estimate the temperature based on the claim. The film coefficients are not highly sensitive to small temperature differences.

$$\begin{aligned}T_{\text{bulk,air}} &= \tfrac{1}{2}(T_{\text{air,in}} + T_{\text{air,out}}) = \left(\tfrac{1}{2}\right)(45°\text{F} + 50°\text{F}) \\ &= 47.5°\text{F}\end{aligned}$$

$$T_{\text{surface}} = 70°\text{F}\quad[\text{estimate}]$$

The film coefficient for air flowing inside the duct is given by Eq. 36.39 as

$$\begin{aligned}h_i &\approx (0.00351 + 0.000001583T_{°\text{F}})\left(\frac{G^{0.8}_{\text{lbm/hr-ft}^2}}{d^{0.2}_{\text{ft}}}\right) \\ &= (0.00351 + (0.000001583)(47.5°\text{F})) \times \left(\frac{\left(3001.1\ \frac{\text{lbm}}{\text{hr-ft}^2}\right)^{0.8}}{\left(\frac{12\ \text{in}}{12\ \frac{\text{in}}{\text{ft}}}\right)^{0.2}}\right) \\ &= \boxed{2.17\ \text{Btu/hr-ft}^2\text{-°F}\quad(2\ \text{Btu/hr-ft}^2\text{-°F})}\end{aligned}$$

If Eq. 36.40(b) is used instead, the value of h_i is approximately 2.01 Btu/hr-ft^2-°F, indicating that the temperature is not important.

The answer is (A).

(b) For natural convection on the outside of the duct, estimate the film temperature.

$$\begin{aligned}T_{\text{film}} &= \tfrac{1}{2}(T_{\text{surface}} + T_\infty) = \left(\tfrac{1}{2}\right)(70°\text{F} + 80°\text{F}) \\ &= 75°\text{F}\end{aligned}$$

From App. 35.C, the properties of air at 75°F are

$$\text{Pr} \approx 0.72$$

$$\frac{g\beta\rho^2}{\mu^2} = 2.27\times10^6\ \frac{1}{\text{ft}^3\text{-°F}}$$

The characteristic length is the diameter of the duct.

$$L = \frac{12\ \text{in}}{12\ \frac{\text{in}}{\text{ft}}} = 1\ \text{ft}$$

The Grashof number is

$$\begin{aligned}\text{Gr} &= L^3\left(\frac{\rho^2\beta g}{\mu^2}\right)(T_\infty - T_s) \\ &= (1\ \text{ft})^3\left(2.27\times10^6\ \frac{1}{\text{ft}^3\text{-°F}}\right)(80°\text{F} - 70°\text{F}) \\ &= 2.27\times10^7\end{aligned}$$

$$\text{PrGr} = (0.72)(2.27\times10^7) = 1.63\times10^7$$

From Table 35.3, the film coefficient for a horizontal cylinder is

$$\begin{aligned} h_o &= 0.27\left(\frac{T_\infty - T_s}{d}\right)^{1/4} = (0.27)\left(\frac{80°F - 70°F}{1 \text{ ft}}\right)^{1/4} \\ &= 0.48 \text{ Btu/hr-ft}^2\text{-°F} \end{aligned}$$

Neglecting the wall resistance, the overall film coefficient from Eq. 36.72 is

$$\begin{aligned} \frac{1}{U} &= \frac{1}{h_o} + \frac{1}{h_i} = \frac{1}{2.17 \frac{\text{Btu}}{\text{hr-ft}^2\text{-°F}}} + \frac{1}{0.48 \frac{\text{Btu}}{\text{hr-ft}^2\text{-°F}}} \\ &= 2.544 \text{ hr-ft}^2\text{-°F/Btu} \\ U &= \frac{1}{2.544 \frac{\text{hr-ft}^2\text{-°F}}{\text{Btu}}} = 0.393 \text{ Btu/hr-ft}^2\text{-°F} \end{aligned}$$

The heat transfer due to convection is

$$\begin{aligned} Q_{\text{convection}} &= UA_{\text{surface}}(T_\infty - T_{\text{bulk,air}}) \\ &= U(\pi dL)(T_\infty - T_{\text{bulk,air}}) \\ &= \left(0.393 \frac{\text{Btu}}{\text{hr-ft}^2\text{-°F}}\right) \\ &\quad \times \pi(1 \text{ ft})(50 \text{ ft})(80°F - 47.5°F) \\ &= \boxed{2006.3 \text{ Btu/hr} \quad (2000 \text{ Btu/hr})} \end{aligned}$$

The answer is (C).

(c) The heat transfer due to radiation is

$$Q_{\text{radiation}} = \sigma F_e F_a A_{\text{surface}}(T_\infty^4 - T_{\text{surface}}^4)$$

Assume the room and duct have an unobstructed view of each other. Then $F_a = 1.0$ and $F_e = \epsilon = 0.28$. The absolute temperatures are

$$\begin{aligned} T_\infty &= 80°F + 460° = 540°R \\ T_{\text{surface}} &= 70°F + 460° = 530°R \\ Q_{\text{radiation}} &= \left(0.1713 \times 10^{-8} \frac{\text{Btu}}{\text{hr-ft}^2\text{-°R}^4}\right)(0.28)(1.0) \\ &\quad \times \pi(1 \text{ ft})(50 \text{ ft})\left((540°R)^4 - (530°R)^4\right) \\ &= 461.5 \text{ Btu/hr} \end{aligned}$$

The total heat transfer to the air is

$$\begin{aligned} Q_{\text{total}} &= Q_{\text{convection}} + Q_{\text{radiation}} \\ &= 2006.3 \frac{\text{Btu}}{\text{hr}} + 461.5 \frac{\text{Btu}}{\text{hr}} \\ &= \boxed{2467.8 \text{ Btu/hr} \quad (2500 \text{ Btu/hr})} \end{aligned}$$

(A second iteration could begin by using Eq. 36.2 to calculate a more accurate surface temperature for the duct.)

The answer is (B).

(d) At 47.5°F, the specific heat of air is approximately 0.240 Btu/lbm-°F. Since the heat transfer is known, the temperature of air leaving the duct can be calculated from

$$\begin{aligned} Q_{\text{total}} &= \dot{m}c_p(T_{\text{air,out}} - T_{\text{air,in}}) \\ T_{\text{air,out}} &= T_{\text{air,in}} + \frac{Q_{\text{total}}}{\dot{m}c_p} \\ &= 45°F + \frac{2467.8 \frac{\text{Btu}}{\text{hr}}}{\left(2357.1 \frac{\text{lbm}}{\text{hr}}\right)\left(0.24 \frac{\text{Btu}}{\text{lbm-°F}}\right)} \\ &= \boxed{49.4°F \quad (49°F)} \end{aligned}$$

This $\boxed{\text{agrees}}$ with the engineer's estimate.

The answer is (B).

SI Solution

The absolute temperature of air entering the duct is

$$7°C + 273° = 280K$$

From the ideal gas law, the density of air entering the duct is

$$\begin{aligned} \rho &= \frac{p}{RT} = \frac{(101.3 \text{ kPa})\left(1000 \frac{\text{Pa}}{\text{kPa}}\right)}{\left(287.03 \frac{\text{J}}{\text{kg·K}}\right)(280K)} \\ &= 1.2604 \text{ kg/m}^3 \end{aligned}$$

The mass flow rate of air entering the duct is

$$\dot{m} = \rho\dot{V} = \left(1.2604 \frac{\text{kg}}{\text{m}^3}\right)\left(0.25 \frac{\text{m}^3}{\text{s}}\right) = 0.3151 \text{ kg/s}$$

The diameter of the duct is

$$d = \frac{30 \text{ cm}}{100 \frac{\text{cm}}{\text{m}}} = 0.30 \text{ m}$$

The velocity of air entering the duct is

$$\begin{aligned} \text{v} &= \frac{\dot{V}}{A_{\text{flow}}} = \frac{0.25 \frac{\text{m}^3}{\text{s}}}{\left(\frac{\pi}{4}\right)(0.30 \text{ m})^2} \\ &= 3.537 \text{ m/s} \end{aligned}$$

(a) To calculate the initial film coefficients, estimate the temperatures based on the claim. The film coefficients are not highly sensitive to small temperature differences.

$$T_{\text{bulk,air}} = \tfrac{1}{2}(T_{\text{air,in}} + T_{\text{air,out}}) = \left(\tfrac{1}{2}\right)(7^\circ\text{C} + 10^\circ\text{C}) = 8.5^\circ\text{C}$$

$$T_{\text{surface}} = 20^\circ\text{C} \quad \text{[estimate]}$$

The film coefficient for air flowing inside the duct is given from Eq. 36.40(a) (independent of temperature).

$$h_i \approx \frac{3.52\text{v}_{\text{m/s}}^{0.8}}{d_{\text{m}}^{0.2}} = \frac{(3.52)\left(3.537\ \frac{\text{m}}{\text{s}}\right)^{0.8}}{(0.30\ \text{m})^{0.2}} = \boxed{12.3\ \text{W/m}^2\text{·K} \quad (12\ \text{W/m}^2\text{·K})}$$

The answer is (A).

(b) For natural convection on the outside of the duct, estimate the film coefficient.

$$T_{\text{film}} = \tfrac{1}{2}(T_{\text{surface}} + T_\infty) = \left(\tfrac{1}{2}\right)(20^\circ\text{C} + 27^\circ\text{C}) = 23.5^\circ\text{C}$$

From App. 35.D, the properties of air at 23.5°C are

$$\text{Pr} = 0.709$$

$$\frac{g\beta\rho^2}{\mu^2} = 1.43 \times 10^8\ \frac{1}{\text{K·m}^3}$$

The characteristic length, L, is the diameter of the duct, which is 0.30 m. The Grashof number is

$$\begin{aligned}\text{Gr} &= L^3\left(\frac{\rho^2\beta g}{\mu^2}\right)(T_\infty - T_{\text{surface}}) \\ &= (0.30\ \text{m})^3\left(1.43 \times 10^8\ \frac{1}{\text{K·m}^3}\right)(27^\circ\text{C} - 20^\circ\text{C}) \\ &= 2.70 \times 10^7\end{aligned}$$

$$\text{PrGr} = (0.709)(2.70 \times 10^7) = 1.9 \times 10^7$$

From Table 35.3, the film coefficient for a horizontal cylinder is

$$\begin{aligned}h_o &\approx 1.32\left(\frac{T_\infty - T_{\text{surface}}}{d}\right)^{1/4} \\ &= (1.32)\left(\frac{27^\circ\text{C} - 20^\circ\text{C}}{0.30\ \text{m}}\right)^{1/4} \\ &= 2.90\ \text{W/m}^2\text{·K}\end{aligned}$$

Neglecting the wall resistance, the overall film coefficient from Eq. 36.72 is

$$\begin{aligned}\frac{1}{U} &= \frac{1}{h_o} + \frac{1}{h_i} = \frac{1}{12.3\ \frac{\text{W}}{\text{m}^2\text{·K}}} + \frac{1}{2.90\ \frac{\text{W}}{\text{m}^2\text{·K}}} \\ &= 0.426\ \text{m}^2\text{·K/W}\end{aligned}$$

$$U = \frac{1}{0.426\ \frac{\text{m}^2\text{·K}}{\text{W}}} = 2.35\ \text{W/m}^2\text{·K}$$

The heat transfer due to convection is

$$\begin{aligned}Q_{\text{convection}} &= UA_{\text{surface}}(T_\infty - T_{\text{bulk,air}}) \\ &= U\pi dL(T_\infty - T_{\text{bulk,air}}) \\ &= \left(2.35\ \frac{\text{W}}{\text{m}^2\text{·K}}\right)\pi(0.30\ \text{m})(15\ \text{m}) \times (27^\circ\text{C} - 8.5^\circ\text{C}) \\ &= \boxed{614.6\ \text{W} \quad (610\ \text{W})}\end{aligned}$$

The answer is (C).

(c) The heat transfer due to radiation is

$$Q_{\text{radiation}} = \sigma F_e F_a A_{\text{surface}}(T_\infty^4 - T_{\text{surface}}^4)$$

Assume the room and duct have an unobstructed view of each other. Then $F_a = 1.0$ and $F_e = \epsilon = 0.28$. The absolute temperatures are

$$T_\infty = 27^\circ\text{C} + 273^\circ = 300\text{K}$$

$$T_{\text{surface}} = 20^\circ\text{C} + 273^\circ = 293\text{K}$$

$$\begin{aligned}Q_{\text{radiation}} &= \left(5.67 \times 10^{-8}\ \frac{\text{W}}{\text{m}^2\text{·K}^4}\right)(0.28)(1)\pi \times (0.30\ \text{m})(15\ \text{m})\left((300\text{K})^4 - (293\text{K})^4\right) \\ &= 163.8\ \text{W}\end{aligned}$$

The total heat transfer to the air is

$$\begin{aligned}Q_{\text{total}} &= Q_{\text{convection}} + Q_{\text{radiation}} = 614.6\ \text{W} + 163.8\ \text{W} \\ &= \boxed{778.4\ \text{W} \quad (780\ \text{W})}\end{aligned}$$

(A second iteration could begin by using Eq. 36.2 to calculate a more accurate surface temperature for the duct.)

The answer is (B).

(d) At 8.5°C, the specific heat of air is

$$\left(1.0048\ \frac{\text{kJ}}{\text{kg·K}}\right)\left(1000\ \frac{\text{J}}{\text{kJ}}\right) = 1004.8\ \text{J/kg·K}$$

Since the heat transfer is known, the temperature of air leaving the duct can be calculated from

$$Q_{\text{total}} = \dot{m}c_p(T_{\text{air,out}} - T_{\text{air,in}})$$

$$\begin{aligned} T_{\text{air,out}} &= T_{\text{air,in}} + \frac{Q_{\text{total}}}{\dot{m}c_p} \\ &= 7^\circ\text{C} + \frac{778.4 \text{ W}}{\left(0.3151 \ \frac{\text{kg}}{\text{s}}\right)\left(1004.8 \ \frac{\text{J}}{\text{kg·K}}\right)} \\ &= \boxed{9.5^\circ\text{C}} \end{aligned}$$

This $\boxed{\text{agrees}}$ with the engineer's estimate.

The answer is (B).

5. *Customary U.S. Solution*

(a) The absolute temperature of entering air is

$$500^\circ\text{F} + 460^\circ = 960^\circ\text{R}$$

The absolute pressure of entering air is

$$25 \text{ psig} + 14.7 \ \frac{\text{lbf}}{\text{in}^2} = 39.7 \text{ psia}$$

The density of air entering, from the ideal gas law, is

$$\begin{aligned} \rho &= \frac{p}{RT} \\ &= \frac{\left(39.7 \ \frac{\text{lbf}}{\text{in}^2}\right)\left(12 \ \frac{\text{in}}{\text{ft}}\right)^2}{\left(53.35 \ \frac{\text{ft-lbf}}{\text{lbm-}^\circ\text{R}}\right)(960^\circ\text{R})} \\ &= 0.1116 \text{ lbm/ft}^3 \end{aligned}$$

The mass flow rate of entering air is

$$\begin{aligned} \dot{m} = \rho\dot{V} &= \left(0.1116 \ \frac{\text{lbm}}{\text{ft}^3}\right)\left(200 \ \frac{\text{ft}^3}{\text{min}}\right)\left(60 \ \frac{\text{min}}{\text{hr}}\right) \\ &= 1339.2 \text{ lbm/hr} \end{aligned}$$

At low pressures, the air enthalpy is found from air tables. (See App. 23.F.)

The absolute temperature of leaving air is

$$350^\circ\text{F} + 460^\circ = 810^\circ\text{R}$$

From App. 23.F,

$$h_1 = 231.06 \text{ Btu/lbm at } 960^\circ\text{R}$$

$$h_2 = 194.25 \text{ Btu/lbm at } 810^\circ\text{R}$$

The heat loss is

$$\begin{aligned} Q &= \dot{m}(h_1 - h_2) \\ &= \left(1339.2 \ \frac{\text{lbm}}{\text{hr}}\right)\left(231.06 \ \frac{\text{Btu}}{\text{lbm}} - 194.25 \ \frac{\text{Btu}}{\text{lbm}}\right) \\ &= 49{,}296 \text{ Btu/hr} \end{aligned}$$

Assuming midpoint pipe surface temperature,

$$\tfrac{1}{2}(T_{\text{in}} + T_{\text{out}}) = \left(\tfrac{1}{2}\right)(500^\circ\text{F} + 350^\circ\text{F}) = 425^\circ\text{F}$$

Since the heat loss is known, the overall heat transfer coefficient can be determined from

$$Q = UA\Delta T = U(\pi dL)\Delta T$$

$$\begin{aligned} U &= \frac{Q}{(\pi dL)\Delta T} \\ &= \frac{\left(49{,}296 \ \frac{\text{Btu}}{\text{hr}}\right)\left(12 \ \frac{\text{in}}{\text{ft}}\right)}{\pi(4.25 \text{ in})(35 \text{ ft})(425^\circ\text{F} - 70^\circ\text{F})} \\ &= \boxed{3.57 \text{ Btu/hr-ft}^2\text{-}^\circ\text{F} \quad (3.6 \text{ Btu/hr-ft}^2\text{-}^\circ\text{F})} \end{aligned}$$

The answer is (C).

(b) To calculate the overall heat transfer coefficient, disregard the pipe thermal resistance and the inside film coefficient (small compared with outside film and radiation).

Work with the midpoint pipe temperature of 425°F.

The absolute temperatures are

$$T_1 = 425^\circ\text{F} + 460^\circ = 885^\circ\text{R}$$

$$T_2 = 70^\circ\text{F} + 460^\circ = 530^\circ\text{R}$$

For radiation heat loss, assume $F_a = 1$. For 500°F enamel paint of any color,

$$F_e = \epsilon \approx 0.9$$

$$\begin{aligned} \frac{Q_{\text{net}}}{A} = E_{\text{net}} &= \sigma F_e F_a (T_1^4 - T_2^4) \\ &= \left(0.1713 \times 10^{-8} \ \frac{\text{Btu}}{\text{hr-ft}^2\text{-}^\circ\text{R}^4}\right) \\ &\quad \times (0.9)(1.0)\left((885^\circ\text{R})^4 - (530^\circ\text{R})^4\right) \\ &= 824.1 \text{ Btu/hr-ft}^2 \end{aligned}$$

From Eq. 37.23, the radiant heat transfer coefficient is

$$\begin{aligned} h_{\text{radiation}} &= \frac{E_{\text{net}}}{T_1 - T_2} = \frac{824.1 \ \frac{\text{Btu}}{\text{hr}}}{885^\circ\text{R} - 530^\circ\text{R}} \\ &= 2.32 \text{ Btu/hr-ft}^2\text{-}^\circ\text{F} \end{aligned}$$

For the outside film coefficient, evaluate the film at the pipe midpoint. The film temperature is

$$T_f = \left(\tfrac{1}{2}\right)(425°\text{F} + 70°\text{F}) = 247.5°\text{F}$$

From App. 35.C,

$$\text{Pr} = 0.72$$

$$\frac{g\beta\rho^2}{\mu^2} = 0.657 \times 10^6 \ \frac{1}{\text{ft}^3\text{-°F}}$$

The characteristic length, L, is $d_o = 4.25$ in. The Grashof number is

$$\begin{aligned}\text{Gr} &= L^3\left(\frac{\rho^2\beta g}{\mu^2}\right)\Delta T \\ &= \frac{(4.25 \text{ in})^3\left(0.657 \times 10^6 \ \frac{1}{\text{ft}^3\text{-°F}}\right)(425°\text{F} - 70°\text{F})}{\left(12 \ \frac{\text{in}}{\text{ft}}\right)^3} \\ &= 1.04 \times 10^7\end{aligned}$$

$$\text{PrGr} = (0.72)(1.04 \times 10^7) = 7.5 \times 10^6$$

From Table 35.3, the film coefficient for a horizontal cylinder is

$$\begin{aligned}h_o &= (0.27)\left(\frac{\Delta T}{d}\right)^{1/4} = (0.27)\left(\frac{425°\text{F} - 70°\text{F}}{\frac{4.25 \text{ in}}{12 \ \frac{\text{in}}{\text{ft}}}}\right)^{1/4} \\ &= 1.52 \text{ Btu/hr-ft}^2\text{-°F}\end{aligned}$$

The overall film coefficient is

$$\begin{aligned}U = h_{\text{total}} &= h_{\text{radiation}} + h_o \\ &= 2.32 \ \frac{\text{Btu}}{\text{hr-ft}^2\text{-°F}} + 1.52 \ \frac{\text{Btu}}{\text{hr-ft}^2\text{-°F}} \\ &= \boxed{3.84 \text{ Btu/hr-ft}^2\text{-°F} \quad (3.8 \text{ Btu/hr-ft}^2\text{-°F})}\end{aligned}$$

This is not too far from the actual value.

The answer is (C).

(c) The following are possible reasons for differences between actual and calculated overall coefficients of heat transfer.

- The internal film coefficient was disregarded.
- The emissivity could be lower due to dirt on the outside of the duct.
- h_r and h_o are *not* really additive.
- Pipe thermal resistance was disregarded.
- The midpoint calculations should be replaced with integration along the length.

The answer is (B).

SI Solution

(a) The absolute temperature of entering air is

$$260°\text{C} + 273° = 533\text{K}$$

The absolute pressure of entering air is

$$170 \text{ kPa} + 101.3 \text{ kPa} = 271.3 \text{ kPa}$$

The density of air entering, from the ideal gas law, is

$$\begin{aligned}\rho = \frac{p}{RT} &= \frac{(271.3 \text{ kPa})\left(1000 \ \frac{\text{Pa}}{\text{kPa}}\right)}{\left(287.03 \ \frac{\text{J}}{\text{kg·K}}\right)(533\text{K})} \\ &= 1.773 \text{ kg/m}^3\end{aligned}$$

The mass flow rate of entering air is

$$\begin{aligned}\dot{m} = \rho\dot{V} &= \left(1.773 \ \frac{\text{kg}}{\text{m}^3}\right)\left(0.1 \ \frac{\text{m}^3}{\text{s}}\right) \\ &= 0.1773 \text{ kg/s}\end{aligned}$$

At low pressures, the air enthalpy is found from air tables. (See App. 23.F.) The absolute temperature of leaving air is

$$180°\text{C} + 273° = 453\text{K}$$

From App. 23.S,

$$h_1 = 537.09 \text{ kJ/kg at } 533\text{K}$$

$$h_2 = 454.87 \text{ kJ/kg at } 453\text{K}$$

The heat loss is

$$\begin{aligned}Q &= \dot{m}(h_1 - h_2) \\ &= \left(0.1773 \ \frac{\text{kg}}{\text{s}}\right)\left(537.09 \ \frac{\text{kJ}}{\text{kg}} - 454.87 \ \frac{\text{kJ}}{\text{kg}}\right) \\ &\quad \times \left(1000 \ \frac{\text{W}}{\text{kW}}\right) \\ &= 14578 \text{ W}\end{aligned}$$

Assuming midpoint pipe surface temperature,

$$\tfrac{1}{2}(T_{\text{in}} - T_{\text{out}}) = \left(\tfrac{1}{2}\right)(260°\text{C} + 180°\text{C}) = 220°\text{C}$$

Since the heat loss is known, the overall heat transfer coefficient can be determined from

$$Q = UA\Delta T = U(\pi dL)\Delta T$$

$$U = \frac{Q}{(\pi dL)\Delta T}$$
$$= \frac{(14578\ \text{W})\left(100\ \frac{\text{cm}}{\text{m}}\right)}{\pi(10.8\ \text{cm})(10\ \text{m})(220^\circ\text{C} - 20^\circ\text{C})}$$
$$= \boxed{21.5\ \text{W/m}^2\text{·K}\quad (22\ \text{W/m}^2\text{·K})}$$

The answer is (C).

(b) From the customary U.S. solution, work with the midpoint pipe temperature of 220°C.

The absolute temperatures are

$$T_1 = 220^\circ\text{C} + 273^\circ = 493\text{K}$$
$$T_2 = 20^\circ\text{C} + 273^\circ = 293\text{K}$$

For radiation heat loss, assume $F_a = 1$. For 260°C enamel paint of any color,

$$F_e = \epsilon \approx 0.9$$

From Eq. 37.23, the radiant heat transfer coefficient is

$$h_{\text{radiation}} = \frac{\sigma F_a F_e (T_1^4 - T_2^4)}{T_1 - T_2}$$
$$= \frac{\left(5.67 \times 10^{-8}\ \frac{\text{W}}{\text{m}^2\text{·K}^4}\right)(1)(0.9) \times \left((493\text{K})^4 - (293\text{K})^4\right)}{493\text{K} - 293\text{K}}$$
$$= 13.2\ \text{W/m}^2\text{·K}$$

For the outside film coefficient, evaluate the film at the pipe midpoint. The film temperature is

$$T_f = \left(\tfrac{1}{2}\right)(220^\circ\text{C} + 20^\circ\text{C}) = 120^\circ\text{C}$$

From App. 35.D, at 120°C,

$$\text{Pr} \approx 0.692$$
$$\frac{g\beta\rho^2}{\mu^2} = 0.528 \times 10^8\ \frac{1}{\text{K·m}^3}$$

The characteristic length is

$$L = \text{outside diameter} = \frac{10.8\ \text{cm}}{100\ \frac{\text{cm}}{\text{m}}} = 0.108\ \text{m}$$

The Grashof number is

$$\text{Gr} = L^3\left(\frac{\rho^2 g\beta}{\mu^2}\right)\Delta T$$
$$= (0.108\ \text{m})^3\left(0.528 \times 10^8\ \frac{1}{\text{K·m}^3}\right)(220^\circ\text{C} - 20^\circ\text{C})$$
$$= 1.33 \times 10^7$$

$$\text{PrGr} = (0.692)(1.33 \times 10^7) = 9.20 \times 10^6$$

From Table 35.3, the film coefficient for a horizontal cylinder is

$$h_o = 1.32\left(\frac{\Delta T}{d}\right)^{1/4} = (1.32)\left(\frac{220^\circ\text{C} - 20^\circ\text{C}}{0.108\ \text{m}}\right)^{1/4}$$
$$= 8.66\ \text{W/m}^2\text{·K}$$

The overall film coefficient is

$$U = h_{\text{total}} = h_{\text{radiation}} + h_o$$
$$= 13.2\ \frac{\text{W}}{\text{m}^2\text{·K}} + 8.66\ \frac{\text{W}}{\text{m}^2\text{·K}}$$
$$= \boxed{21.86\ \text{W/m}^2\text{·K}\quad (22\ \text{W/m}^2\text{·K})}$$

This is not too far from the actual value.

The answer is (C).

(c) See the customary U.S. solution.

The answer is (B).

6. *Customary U.S. Solution*

(a) Heat is lost from the top and sides by radiation and convection. The absolute temperature of the surroundings is

$$75^\circ\text{F} + 460^\circ = 535^\circ\text{R}$$

$$A_{\text{sides}} = \pi dL = \frac{\pi(0.75\ \text{in})(1.5\ \text{in})}{\left(12\ \frac{\text{in}}{\text{ft}}\right)^2}$$
$$= 0.0245\ \text{ft}^2$$

$$A_{\text{top}} = \left(\frac{\pi}{4}\right)d^2 = \frac{\left(\frac{\pi}{4}\right)(0.75\ \text{in})^2}{\left(12\ \frac{\text{in}}{\text{ft}}\right)^2}$$
$$= 0.003068\ \text{ft}^2$$

$$Q_{\text{total}} = Q_{\text{convection}} + Q_{\text{radiation}}$$
$$= h_{\text{sides}}A_{\text{sides}}(T_s - T_\infty) + h_{\text{top}}A_{\text{top}}(T_s - T_\infty) + \sigma F_e F_a (A_{\text{sides}} + A_{\text{top}})(T_s^4 - T_\infty^4) \quad \text{[Eq. 1]}$$

Heat Transfer

For the first approximation of T_s, assume $h_{sides} = h_{top} = 1.65$ Btu/hr-ft²-°F, $F_a = 1$, and $F_e = \epsilon = 0.65$.

$$(5.0\ \text{W})\left(\frac{3.142\ \frac{\text{Btu}}{\text{hr}}}{1\ \text{W}}\right) = \left(1.65\ \frac{\text{Btu}}{\text{hr-ft}^2\text{-°F}}\right) \times (0.0245\ \text{ft}^2)(T_s - 535°\text{R}) + \left(1.65\ \frac{\text{Btu}}{\text{hr-ft}^2\text{-°F}}\right) \times (0.003068\ \text{ft}^2)(T_s - 535°\text{R}) + \left(0.1713\times 10^{-8}\ \frac{\text{Btu}}{\text{hr-ft}^2\text{-°R}^4}\right) \times (0.65)(1)\left(\begin{array}{l}0.0245\ \text{ft}^2\\ \quad + 0.003068\ \text{ft}^2\end{array}\right) \times \left(T_s^4 - (535°\text{R})^4\right)$$

By trial and error, $T_s \approx 750°\text{R}$.

For natural convection on the outside, estimate the film temperature.

$$\begin{aligned} T_{film} &= \tfrac{1}{2}(T_s + T_\infty) \\ &= \left(\tfrac{1}{2}\right)(750°\text{R} + 535°\text{R}) - 460° \\ &= 182.5°\text{F} \end{aligned}$$

From App. 35.C, the properties of air at 182.5°F are

$$\text{Pr} \approx 0.72$$

$$\frac{g\beta\rho^2}{\mu^2} = 1.01\times 10^6\ \frac{1}{\text{ft}^3\text{-°F}}$$

For the sides, the characteristic length is 1.5 in. The Grashof number is

$$\begin{aligned} \text{Gr} &= L^3\left(\frac{\rho^2\beta g}{\mu^2}\right)(T_s - T_\infty) \\ &= \frac{(1.5\ \text{in})^3\left(1.01\times 10^6\ \frac{1}{\text{ft}^3\text{-°F}}\right)(750°\text{R} - 535°\text{R})}{\left(12\ \frac{\text{in}}{\text{ft}}\right)^3} \\ &= 4.24\times 10^5 \end{aligned}$$

$$\begin{aligned} \text{PrGr} &= (0.72)(4.24\times 10^5) \\ &= 3.05\times 10^5 \end{aligned}$$

From Table 35.3, the film coefficient for a vertical surface is

$$\begin{aligned} h_{sides} &= 0.29\left(\frac{T_s - T_\infty}{L}\right)^{1/4} \\ &= 0.29\left(\frac{(750°\text{R} - 535°\text{R})\left(12\ \frac{\text{in}}{\text{ft}}\right)}{1.5\ \text{in}}\right)^{1/4} \\ &= 1.87\ \text{Btu/hr-ft}^2\text{-°F} \end{aligned}$$

(This application is slightly outside the useful range of this correlation.)

For the top, the characteristic length is 0.75 in, so

$$\begin{aligned} \text{Gr} &= \frac{(0.75\ \text{in})^3\left(1.01\times 10^6\ \frac{1}{\text{ft}^3\text{-°F}}\right)(750°\text{R} - 535°\text{R})}{\left(12\ \frac{\text{in}}{\text{ft}}\right)^3} \\ &= 5.30\times 10^4 \end{aligned}$$

$$\begin{aligned} \text{PrGr} &= (0.72)(5.30\times 10^4) \\ &= 3.8\times 10^4 \end{aligned}$$

From Table 35.3, the film coefficient for a horizontal surface is

$$\begin{aligned} h_{top} &= 0.27\left(\frac{T_s - T_\infty}{L}\right)^{1/4} \\ &= (0.27)\left(\frac{(750°\text{R} - 535°\text{R})\left(12\ \frac{\text{in}}{\text{ft}}\right)}{(0.9)(0.75\ \text{in})}\right)^{1/4} \\ &= 2.12\ \text{Btu/hr-ft}^2\text{-°F} \end{aligned}$$

Substituting the calculated values of h_{top} and h_{sides} into Eq. 1 gives

$$\begin{aligned} Q_{total} &= (5.0\ \text{W})\left(\frac{3.412\ \frac{\text{Btu}}{\text{hr}}}{1\ \text{W}}\right) \\ &= \left(1.87\ \frac{\text{Btu}}{\text{hr-ft}^2\text{-°F}}\right)(0.0245\ \text{ft}^2)(T_s - 535°\text{R}) \\ &\quad + \left(2.12\ \frac{\text{Btu}}{\text{hr-ft}^2\text{-°F}}\right)(0.003068\ \text{ft}^2)(T_s - 535°\text{R}) \\ &\quad + \left(0.1713\times 10^{-8}\ \frac{\text{Btu}}{\text{hr-ft}^2\text{-°F}}\right)(0.65)(1) \\ &\quad \times (0.0245\ \text{ft}^2 + 0.003068\ \text{ft}^2)\left(T_s^4 - (535°\text{R})^4\right) \end{aligned}$$

By trial and error, $T_s \approx \boxed{736°\text{R}\ (740°\text{R}).}$

The answer is (C).

(b) Substituting $T_s = 736°R$ in the preceding equation,

$$Q_{\text{total}} = 9.209\ \frac{\text{Btu}}{\text{hr}} + 1.277\ \frac{\text{Btu}}{\text{hr}} + 6.429\ \frac{\text{Btu}}{\text{hr}}$$
$$= 16.92\ \text{Btu/hr}$$

$$\frac{Q_{\text{convection}}}{Q_{\text{total}}} = \frac{9.209\ \frac{\text{Btu}}{\text{hr}} + 1.277\ \frac{\text{Btu}}{\text{hr}}}{16.92\ \frac{\text{Btu}}{\text{hr}}}$$
$$= \boxed{0.620\quad(62\%)}$$

$$\frac{Q_{\text{radiation}}}{Q_{\text{total}}} = \frac{6.492\ \frac{\text{Btu}}{\text{hr}}}{16.92\ \frac{\text{Btu}}{\text{hr}}} = \boxed{0.380\quad(38\%)}$$

The answer is (D).

SI Solution

(a) Heat is lost from the top and sides by radiation and convection. The absolute temperature of the surrounding is

$$24°\text{C} + 273° = 297\text{K}$$

$$A_{\text{sides}} = \frac{\pi dL = \pi(19\ \text{mm})(38\ \text{mm})}{\left(1000\ \frac{\text{mm}}{\text{m}}\right)^2}$$
$$= 0.00227\ \text{m}^2$$

$$A_{\text{top}} = \left(\frac{\pi}{4}\right)d^2 = \frac{\left(\frac{\pi}{4}\right)(19\ \text{mm})^2}{\left(1000\ \frac{\text{mm}}{\text{m}}\right)^2}$$
$$= 0.000284\ \text{m}^2$$

$$Q_{\text{total}} = Q_{\text{convection}} + Q_{\text{radiation}}$$
$$= h_{\text{sides}}A_{\text{sides}}(T_s - T_\infty) + h_{\text{top}}A_{\text{top}}(T_s - T_\infty)$$
$$+ \sigma F_e F_a(A_{\text{sides}} + A_{\text{top}})(T_s^4 - T_\infty^4) \quad \text{[Eq. 1]}$$

For a first approximation of T_s, assume $h_{\text{sides}} = h_{\text{top}} = 9.4\ \text{W/m}^2{\cdot}\text{K}$, $F_a = 1$, and $F_e = \epsilon = 0.65$.

$$5.0\ \text{W} = \left(9.4\ \frac{\text{W}}{\text{m}^2{\cdot}\text{K}}\right)(0.00227\ \text{m}^2)(T_s - 297\text{K})$$
$$+ \left(9.4\ \frac{\text{W}}{\text{m}^2{\cdot}\text{K}}\right)(0.000284\ \text{m}^2)(T_s - 297\text{K})$$
$$+ \left(5.67\times 10^{-8}\ \frac{\text{W}}{\text{m}^2{\cdot}\text{K}^4}\right)(0.65)(1)$$
$$\times (0.00227\ \text{m}^2 + 0.000284\ \text{m}^2)$$
$$\times \left(T_s^4 - (297\text{K})^4\right)$$

By trial and error, $T_s = 416.75\text{K}$.

For natural convection on the outside, estimate the film temperature.

$$T_{\text{film}} = \tfrac{1}{2}(T_s + T_\infty)$$
$$= \tfrac{1}{2}(416.75\text{K} + 297\text{K}) - 273°$$
$$= 83.9°\text{C}$$

From App. 35.D, the properties of air at 83.9°C are

$$\text{Pr} \approx 0.697$$
$$\frac{g\beta\rho^2}{\mu^2} \approx 0.616\times 10^8\ \frac{1}{\text{K}{\cdot}\text{m}^3}$$

For the sides, the characteristic length is 38 mm.

$$\text{Gr} = \frac{(38\ \text{mm})^3\left(0.616\times 10^8\ \frac{1}{\text{K}{\cdot}\text{m}^3}\right)(416.75\text{K} - 297\text{K})}{\left(1000\ \frac{\text{mm}}{\text{m}}\right)^3}$$
$$= 4.048\times 10^5$$

$$\text{PrGr} = (0.697)(4.048\times 10^5) = 2.82\times 10^5$$

For the top, the characteristic length is 19 mm.

$$\text{Gr} = \frac{(19\ \text{mm})^3\left(0.616\times 10^8\ \frac{1}{\text{K}{\cdot}\text{m}^3}\right) \times (416.75\text{K} - 297\text{K})}{\left(1000\ \frac{\text{mm}}{\text{m}}\right)^3}$$
$$= 5.06\times 10^4$$
$$\text{PrGr} = (0.697)(5.06\times 10^4) = 3.53\times 10^4$$

From Table 35.3, the film coefficient for a horizontal surface is

$$h_{\text{top}} = 1.32\left(\frac{T_s - T_\infty}{d}\right)^{1/4}$$
$$= (1.32)\left(\frac{(416.75\text{K} - 297\text{K})\left(1000\ \frac{\text{mm}}{\text{m}}\right)}{(0.9)(19\ \text{mm})}\right)^{1/4}$$
$$= 12.08\ \text{W/m}^2{\cdot}\text{K}$$

From Table 35.3, the film coefficient for a vertical surface is

$$h_{\text{sides}} = 1.37\left(\frac{T_s - T_\infty}{L}\right)^{1/4}$$
$$= (1.37)\left(\frac{(416.75\text{K} - 297\text{K})\left(1000\ \frac{\text{mm}}{\text{m}}\right)}{38\ \text{mm}}\right)^{1/4}$$
$$= 10.26\ \text{W/m}^2{\cdot}\text{K}$$

(This application is slightly outside the useful range of this correlation.)

Using the calculated values of h_{top} and h_{sides} into Eq. 1,

$$\begin{aligned} Q_{total} &= 5\ \text{W} \\ &= \left(10.26\ \frac{\text{W}}{\text{m}^2\cdot\text{K}}\right)(0.00227\ \text{m}^2)(T_s - 297\text{K}) \\ &\quad + \left(12.08\ \frac{\text{W}}{\text{m}^2\cdot\text{K}}\right)(0.000284\ \text{m}^2)(T_s - 297\text{K}) \\ &\quad + \left(5.67\times 10^{-8}\ \frac{\text{W}}{\text{m}^2\cdot\text{K}^4}\right)(0.65)(1) \\ &\quad \times (0.00227\ \text{m}^2 + 0.000284\ \text{m}^2) \\ &\quad \times \left(T_s^4 - (297\text{K})^4\right) \end{aligned}$$

By trial and error, $T_s = \boxed{411\text{K}\ (410\text{K}).}$

The answer is (C).

(b) Substitute $T_s = 411\text{K}$ into the preceding equation.

$$\begin{aligned} Q_{total} &= 2.655\ \text{W} + 0.381\ \text{W} + 1.953\ \text{W} \\ &= 4.989\ \text{W} \end{aligned}$$

$$\frac{Q_{convection}}{Q_{total}} = \frac{2.655\ \text{W} + 0.381\ \text{W}}{4.989\ \text{W}} = \boxed{0.609\quad(61\%)}$$

$$\frac{Q_{radiation}}{Q_{total}} = \frac{1.953\ \text{W}}{4.989\ \text{W}} = \boxed{0.391\quad(39\%)}$$

The answer is (D).

7. *Customary U.S. Solution*

The velocity is relatively low, so incompressible flow can be assumed.

The film coefficient on the probe is

$$\begin{aligned} h &= \frac{0.024G^{0.8}}{D^{0.4}} = \frac{(0.024)\left(3480\ \frac{\text{lbm}}{\text{hr-ft}^2}\right)^{0.8}}{\left(\frac{0.5\ \text{in}}{12\ \frac{\text{in}}{\text{ft}}}\right)^{0.4}} \\ &= 58.3\ \text{Btu/hr-ft}^2\text{-°F} \end{aligned}$$

The absolute temperature of the walls is

$$T_{walls} = 600°\text{F} + 460° = 1060°\text{R}$$

Neglect conduction and the insignificant kinetic energy loss. The thermocouple gains heat through radiation from the walls and loses heat through convection to the gas.

$$\begin{aligned} Q_{convection} &= AE_{radiation} \\ hA(T_{probe} - T_{gas}) &= A\sigma\epsilon(T_{walls}^4 - T_{probe}^4) \\ h(T_{probe} - T_{gas}) &= \sigma\epsilon(T_{walls}^4 - T_{probe}^4) \end{aligned}$$

(a) If the actual gas temperature is $300°\text{F} + 460° = 760°\text{R}$,

$$\begin{aligned} &\left(58.3\ \frac{\text{Btu}}{\text{hr-ft}^2\text{-°F}}\right) \\ &\quad\times (T_{probe} - 760°\text{R}) = \left(0.1713\times 10^{-8}\ \frac{\text{Btu}}{\text{hr-ft}^2\text{-°R}^4}\right) \\ &\quad\quad\times (0.8)\left((1060°\text{R})^4 - T_{probe}^4\right) \end{aligned}$$

$$\begin{aligned} &(1.37\times 10^{-9})T_{probe}^4 \\ &\quad + (58.3)T_{probe} = 46{,}038 \end{aligned}$$

By trial and error, $T_{probe} = \boxed{781°\text{R}\ (780°\text{R}).}$

The answer is (D).

(b) If $T_{probe} = 300°\text{F} + 460° = 760°\text{R}$,

$$\begin{aligned} &\left(58.3\ \frac{\text{Btu}}{\text{hr-ft}^2\text{-°F}}\right) \\ &\quad\times (760°\text{R} - T_{gas}) = \left(0.1713\times 10^{-8}\ \frac{\text{Btu}}{\text{hr-ft}^2\text{-°R}^4}\right) \\ &\quad\quad\times (0.8)\left((1060°\text{R})^4 - (760°\text{R})^4\right) \\ &T_{gas} = \boxed{738.2°\text{R}\quad(740°\text{R})} \end{aligned}$$

The answer is (B).

SI Solution

The velocity is relatively low, so incompressible flow can be assumed.

The film coefficient on the probe is

$$\begin{aligned} h &= \frac{17G^{0.8}}{D^{0.4}} = \frac{(17)\left(4.7\ \frac{\text{kg}}{\text{s}\cdot\text{m}^2}\right)^{0.8}}{\left(\frac{13\ \text{mm}}{1000\ \frac{\text{mm}}{\text{m}}}\right)^{0.4}} \\ &= 333.1\ \text{W/m}^2\cdot\text{K} \end{aligned}$$

The absolute temperature of the walls is

$$T_{walls} = 315°\text{C} + 273° = 588\text{K}$$

Neglect conduction and the insignificant kinetic energy loss. The thermocouple gains heat through radiation from the walls and loses heat through convection to the gas.

$$\frac{Q_{\text{convection}}}{A} = E_{\text{radiation}}$$

$$h(T_{\text{probe}} - T_{\text{gas}}) = \sigma\epsilon(T_{\text{walls}}^4 - T_{\text{probe}}^4)$$

(a) If the actual gas temperature is 150°C + 273° = 423K,

$$\left(333.1 \ \frac{\text{W}}{\text{m}^2\cdot\text{K}}\right) \times (T_{\text{probe}} - 423\text{K}) = \left(5.67 \times 10^{-8} \ \frac{\text{W}}{\text{m}^2\cdot\text{K}^4}\right)(0.8) \times \left((588\text{K})^4 - T_{\text{probe}}^4\right)$$

$$(4.536 \times 10^{-8})T_{\text{probe}}^4 + (333.1)T_{\text{probe}} = 146\,323$$

By trial and error, $T_{\text{probe}} = \boxed{435\text{K}\ (430\text{K}).}$

The answer is (D).

(b) If $T_{\text{probe}} = 150°\text{C} + 273° = 423\text{K}$,

$$\left(333.1 \ \frac{\text{W}}{\text{m}^2\cdot\text{K}}\right) \times (423\text{K} - T_{\text{gas}}) = \left(5.67 \times 10^{-8} \ \frac{\text{W}}{\text{m}^2\cdot\text{K}^4}\right)(0.8) \times \left((588\text{K})^4 - (423\text{K})^4\right)$$

$$T_{\text{gas}} = \boxed{411.1\text{K} \quad (410\text{K})}$$

The answer is (B).

Heat Transfer

38 Psychrometrics

PRACTICE PROBLEMS

1. A room contains air at 80°F (27°C) dry-bulb and 67°F (19°C) wet-bulb. The total pressure is 1 atm.

(a) The humidity ratio is most nearly

(A) 0.009 lbm/lbm (0.009 kg/kg)

(B) 0.011 lbm/lbm (0.011 kg/kg)

(C) 0.014 lbm/lbm (0.014 kg/kg)

(D) 0.018 lbm/lbm (0.018 kg/kg)

(b) The enthalpy is most nearly

(A) 30.2 Btu/lbm (51.3 kJ/kg)

(B) 30.8 Btu/lbm (52.4 kJ/kg)

(C) 31.5 Btu/lbm (53.9 kJ/kg)

(D) 31.9 Btu/lbm (54.2 kJ/kg)

(c) The specific heat is most nearly

(A) 0.234 Btu/lbm-°F (0.979 kJ/kg·K)

(B) 0.237 Btu/lbm-°F (0.991 kJ/kg·K)

(C) 0.239 Btu/lbm-°F (0.999 kJ/kg·K)

(D) 0.242 Btu/lbm-°F (1.012 kJ/kg·K)

2. If one layer of cooling coils effectively bypasses one-third of the air passing through it, the theoretical bypass factor for four layers of identical cooling coils in series is most nearly

(A) 0.01

(B) 0.09

(C) 0.33

(D) 0.67

3. 1000 ft^3/min (0.5 m^3/s) of air at 50°F (10°C) dry-bulb and 95% relative humidity are mixed with 1500 ft^3/min (0.75 m^3/s) of air at 76°F (24°C) dry-bulb and 45% relative humidity.

(a) The dry-bulb temperature of the mixture is most nearly

(A) 55°F (13°C)

(B) 65°F (18°C)

(C) 68°F (20°C)

(D) 70°F (21°C)

(b) The specific humidity of the mixture is most nearly

(A) 0.008 lbm/lbm (0.008 kg/kg)

(B) 0.009 lbm/lbm (0.009 kg/kg)

(C) 0.010 lbm/lbm (0.010 kg/kg)

(D) 0.011 lbm/lbm (0.011 kg/kg)

(c) The dew point of the mixture is most nearly

(A) 45°F (7.2°C)

(B) 48°F (8.9°C)

(C) 51°F (11°C)

(D) 54°F (12°C)

4. Air at 60°F (16°C) dry-bulb and 45°F (7°C) wet-bulb passes through an air washer with a humidifying efficiency of 70%.

(a) The effective bypass factor of the system is most nearly

(A) 0.30

(B) 0.50

(C) 0.67

(D) 0.70

(b) The dry-bulb temperature of the air leaving the washer is most nearly

(A) 45°F (7.2°C)

(B) 50°F (9.7°C)

(C) 54°F (12°C)

(D) 57°F (14°C)

5. 95°F (35°C) dry-bulb, 75°F (24°C) wet-bulb air passes through a cooling tower and leaves at 85°F (29°C) dry-bulb and 90% relative humidity.

(a) The enthalpy change per cubic foot (meter) of air is most nearly

(A) 0.57 Btu/ft^3 (18 kJ/m^3)

(B) 1.3 Btu/ft^3 (40 kJ/m^3)

(C) 3.2 Btu/ft^3 (99 kJ/m^3)

(D) 7.6 Btu/ft^3 (240 kJ/m^3)

(b) The change in moisture content per cubic foot (meter) of air is most nearly

(A) 1.8×10^{-4} lbm/ft^3 (2.7×10^{-3} kg/m^3)

(B) 3.3×10^{-4} lbm/ft^3 (4.5×10^{-3} kg/m^3)

(C) 6.7×10^{-4} lbm/ft^3 (9.9×10^{-3} kg/m^3)

(D) 9.2×10^{-4} lbm/ft^3 (14×10^{-3} kg/m^3)

6. An air washer receives 1800 ft^3/min (0.85 m^3/s) of air at 70°F (21°C) and 40% relative humidity and discharges the air at 75% relative humidity. A recirculating water spray with a constant temperature of 50°F (10°C) is used.

(a) What are the conditions of the discharged air?

(b) What mass of makeup water is required per minute?

7. Repeat Prob. 6(a) and Prob. 6(b) using saturated steam at atmospheric pressure in place of the 50°F (10°C) water spray.

8. During performances, a theater experiences a sensible heat load of 500,000 Btu/hr (150 kW) and a moisture load of 175 lbm/hr (80 kg/h). Air enters the theater at 65°F (18°C) and 55% relative humidity and is removed when it reaches 75°F (24°C) or 60% relative humidity, whichever comes first.

(a) What is the ventilation rate in mass of air per hour?

(b) What are the conditions of the air leaving the theater?

9. 500 ft^3/min (0.25 m^3/s) of air at 80°F (27°C) dry-bulb and 70% relative humidity are removed from a room. 150 ft^3/min (0.075 m^3/s) pass through an air conditioner and leave saturated at 50°F (10°C). The remaining 350 ft^3/min (0.175 m^3/s) bypass the air conditioner and mix with the conditioned air at 1 atm.

(a) The mixture's temperature is most nearly

(A) 55°F (13°C)

(B) 66°F (19°C)

(C) 71°F (22°C)

(D) 74°F (23°C)

(b) The mixture's humidity ratio is most nearly

(A) 0.010 lbm/lbm (1.0 g/kg)

(B) 0.013 lbm/lbm (1.3 g/kg)

(C) 0.017 lbm/lbm (1.7 g/kg)

(D) 0.021 lbm/lbm (2.1 g/kg)

(c) The mixture's relative humidity is most nearly

(A) 45%

(B) 57%

(C) 73%

(D) 81%

(d) The heat load (in tons) of the air conditioner is most nearly

(A) 0.90 ton

(B) 1.3 ton

(C) 2.4 tons

(D) 2.9 tons

10. (*Time limit: one hour*) A dehumidifier takes 5000 ft^3/min (2.36 m^3/s) of air at 95°F (35°C) dry-bulb and 70% relative humidity and discharges it at 60°F (16°C) dry-bulb and 95% relative humidity. The dehumidifier uses a wet R-12 refrigeration cycle operating between 100°F (saturated) (38°C) and 50°F (10°C).

(a) Locate the air entering and leaving points on the psychrometric chart.

(b) Find the quantity of water removed from the air.

(c) Find the quantity of heat removed from the air.

(d) Draw the temperature-entropy and enthalpy-entropy diagrams for the refrigeration cycle.

(e) Find the temperature, pressure, enthalpy, entropy, and specific volume for each endpoint of the refrigeration cycle.

11. (*Time limit: one hour*) 1500 ft^3/min (0.71 m^3/s) of saturated 25 psia (170 kPa) air is heated from 200°F to 400°F (93°C to 204°C) in a constant pressure, constant moisture drying process.

(a) The final relative humidity is most nearly

(A) 4.7%

(B) 9.2%

(C) 13%

(D) 23%

(b) The final specific humidity is most nearly

(A) 0.41 lbm/lbm (0.41 kg/kg)

(B) 0.53 lbm/lbm (0.53 kg/kg)

(C) 0.66 lbm/lbm (0.66 kg/kg)

(D) 0.79 lbm/lbm (0.79 kg/kg)

(c) The heat required per unit mass of dry air is most nearly

(A) 31 Btu/lbm (71 kJ/kg)

(B) 57 Btu/lbm (130 kJ/kg)

(C) 99 Btu/lbm (230 kJ/kg)

(D) 120 Btu/lbm (280 kJ/kg)

(d) The final dew point is most nearly

(A) 180°F (82°C)

(B) 200°F (93°C)

(C) 220°F (100°C)

(D) 240°F (120°C)

12. (*Time limit: one hour*) 410 lbm/hr (0.052 kg/s) of dry 800°F (427°C) air pass through a scrubber to reduce particulate emissions. To protect the elastomeric seals in the scrubber, the air temperature is reduced to 350°F (177°C) by passing the air through a spray of 80°F (27°C) water. The pressure in the spray chamber is 20 psia (140 kPa).

(a) Approximately how much water is evaporated per hour?

(A) 18 lbm/hr (0.0023 kg/s)

(B) 27 lbm/hr (0.0035 kg/s)

(C) 31 lbm/hr (0.0040 kg/s)

(D) 39 lbm/hr (0.0050 kg/s)

(b) The relative humidity of the air leaving the spray chamber is most nearly

(A) 1.3%

(B) 2.0%

(C) 3.1%

(D) 4.4%

13. Combustion products leaving a gas turbine combustor are released at a temperature of 180°F into the atmosphere. The combustion products have a relative humidity of 20%. What is most nearly the dew point temperature of the combustion products?

(A) 110°F

(B) 120°F

(C) 130°F

(D) 140°F

14. Each hour, 100 lbm of methane are burned with excess air. The combustion products pass through a heat exchanger used to heat water. The combustion products enter the heat exchanger at 340°F, and they leave the heat exchanger at 110°F. The total pressure of the combustion products in the heat exchanger is 19 psia. The humidity ratio of the combustion products is 560 gr/lbm. The dew point temperature of the combustion products is most nearly

(A) 90°F

(B) 100°F

(C) 120°F

(D) 130°F

HVAC

SOLUTIONS

1. *Customary U.S. Solution*

((a) and (b)) Locate the intersection of 80°F dry-bulb and 67°F wet-bulb on the psychrometric chart (see App. 38.A). Read the value of humidity and enthalpy.

$$\boxed{\begin{array}{l} \omega = 0.0112 \text{ lbm moisture/lbm dry air} \\ \quad (0.011 \text{ lbm/lbm}) \\ h = 31.5 \text{ Btu/lbm dry air} \end{array}}$$

The answer is (B) for Prob. 1(a).

The answer is (C) for Prob. 1(b).

(c) c_p is gravimetrically weighted. c_p for air is 0.240 Btu/lbm-°F, and c_p for steam is approximately 0.40 Btu/lbm-°F.

$$G_{\text{air}} = \frac{1}{1+0.0112} = 0.989$$

$$G_{\text{steam}} = \frac{0.0112}{1+0.0112} = 0.011$$

$$\begin{aligned} c_{p,\text{mixture}} &= G_{\text{air}} c_{p,\text{air}} + G_{\text{steam}} c_{p,\text{steam}} \\ &= (0.989)\left(0.240 \ \frac{\text{Btu}}{\text{lbm-°F}}\right) \\ &\quad + (0.011)\left(0.40 \ \frac{\text{Btu}}{\text{lbm-°F}}\right) \\ &= \boxed{0.242 \text{ Btu/lbm-°F}} \end{aligned}$$

The answer is (D).

SI Solution

((a) and (b)) Locate the intersection of 27°C dry-bulb and 19°C wet-bulb on the psychrometric chart (see App. 38.B). Read the value of humidity and enthalpy.

$$\boxed{\begin{array}{l} \omega = \dfrac{10.5 \ \dfrac{\text{g}}{\text{kg dry air}}}{1000 \ \dfrac{\text{g}}{\text{kg}}} = 0.0105 \text{ kg/kg dry air} \\ \quad (0.011 \text{ kg/kg}) \\ h = 53.9 \text{ kJ/kg dry air} \end{array}}$$

The answer is (B) for Prob. 1(a).

The answer is (C) for Prob. 1(b).

(c) c_p is gravimetrically weighted. c_p for air is 1.0048 kJ/kg·K, and c_p for steam is approximately 1.675 kJ/kg·K.

$$G_{\text{air}} = \frac{1}{1+0.0105} = 0.990$$

$$G_{\text{steam}} = \frac{0.0105}{1+0.0105} = 0.010$$

$$\begin{aligned} c_{p,\text{mixture}} &= G_{\text{air}} c_{p,\text{air}} + G_{\text{air}} c_{p,\text{steam}} \\ &= (0.990)\left(1.0048 \ \frac{\text{kJ}}{\text{kg·K}}\right) \\ &\quad + (0.010)\left(1.675 \ \frac{\text{kJ}}{\text{kg·K}}\right) \\ &= \boxed{1.0115 \text{ kJ/kg·K} \quad (1.012 \text{ kJ/kg·K})} \end{aligned}$$

The answer is (D).

2. $\text{BF}_{n \text{ layers}} = (\text{BF}_{1 \text{ layer}})^n = \left(\frac{1}{3}\right)^4 = \boxed{0.0123 \quad (0.01)}$

Only 1% of the air will be untreated.

The answer is (A).

3. Use the illustration shown for both the customary U.S. and SI solutions.

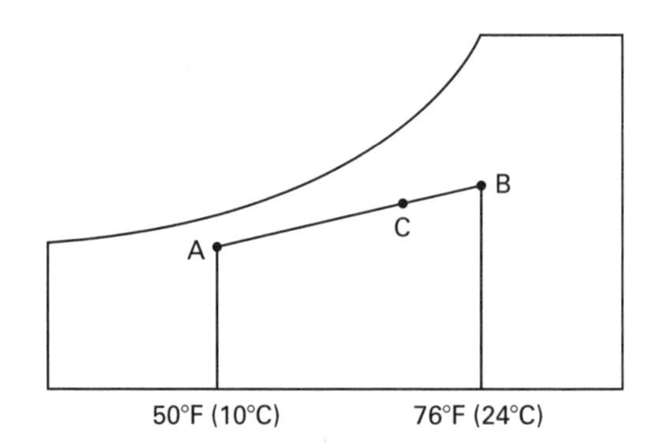

Customary U.S. Solution

(a) Locate the two points on the psychrometric chart and draw a line between them.

Reading from the chart (specific volumes),

$$\upsilon_{\text{A}} = 13.0 \text{ ft}^3\text{/lbm}$$

$$\upsilon_{\text{B}} = 13.7 \text{ ft}^3\text{/lbm}$$

The density at each point is

$$\rho_{\text{A}} = \frac{1}{\upsilon_{\text{A}}} = \frac{1}{13.0 \ \dfrac{\text{ft}^3}{\text{lbm}}} = 0.0769 \text{ lbm/ft}^3$$

$$\rho_{\text{B}} = \frac{1}{\upsilon_{\text{B}}} = \frac{1}{13.7 \ \dfrac{\text{ft}^3}{\text{lbm}}} = 0.0730 \text{ lbm/ft}^3$$

The mass flow at each point is

$$\dot{m}_A = \rho_A \dot{V}_A = \left(0.0769\ \frac{\text{lbm}}{\text{ft}^3}\right)\left(1000\ \frac{\text{ft}^3}{\text{min}}\right) = 76.9\ \text{lbm/min}$$

$$\dot{m}_B = \rho_B \dot{V}_B = \left(0.0730\ \frac{\text{lbm}}{\text{ft}^3}\right)\left(1500\ \frac{\text{ft}^3}{\text{min}}\right) = 109.5\ \text{lbm/min}$$

The gravimetric fraction of flow A is

$$\frac{76.9\ \frac{\text{lbm}}{\text{min}}}{76.9\ \frac{\text{lbm}}{\text{min}} + 109.5\ \frac{\text{lbm}}{\text{min}}} = 0.413$$

Since the scales are all linear,

$$0.413 = \frac{T_B - T_C}{T_B - T_A}$$

$$\begin{aligned} T_C &= T_B - (0.413)(T_B - T_A) \\ &= 76°\text{F} - (0.413)(76°\text{F} - 50°\text{F}) \\ &= \boxed{65.3°\text{F}\quad(65°\text{F})} \end{aligned}$$

The answer is (B).

(b) $\omega = \boxed{\begin{array}{c}0.0082\ \text{lbm moisture/lbm dry air} \\ (0.008\ \text{lbm/lbm})\end{array}}$

The answer is (A).

(c) $T_{dp} = \boxed{51°\text{F}}$

The answer is (C).

SI Solution

(a) Locate the two points on the psychrometric chart and draw a line between them.

Reading from the chart (specific volumes),

$$\upsilon_A = 0.813\ \text{m}^3/\text{kg dry air}$$

$$\upsilon_B = 0.856\ \text{m}^3/\text{kg dry air}$$

The density at each point is

$$\rho_A = \frac{1}{\upsilon_A} = \frac{1}{0.813\ \frac{\text{m}^3}{\text{kg}}} = 1.23\ \text{kg/m}^3$$

$$\rho_B = \frac{1}{\upsilon_B} = \frac{1}{0.856\ \frac{\text{m}^3}{\text{kg}}} = 1.17\ \text{kg/m}^3$$

The mass flow at each point is

$$\dot{m}_A = \rho_A \dot{V}_A = \left(1.23\ \frac{\text{kg}}{\text{m}^3}\right)\left(0.5\ \frac{\text{m}^3}{\text{s}}\right) = 0.615\ \text{kg/s}$$

$$\dot{m}_B = \rho_B \dot{V}_B = \left(1.17\ \frac{\text{kg}}{\text{m}^3}\right)\left(0.75\ \frac{\text{m}^3}{\text{s}}\right) = 0.878\ \text{kg/s}$$

The gravimetric fraction of flow A is

$$\frac{0.615\ \frac{\text{kg}}{\text{s}}}{0.615\ \frac{\text{kg}}{\text{s}} + 0.878\ \frac{\text{kg}}{\text{s}}} = 0.412$$

Since the scales are linear,

$$0.412 = \frac{T_B - T_C}{T_B - T_A}$$

$$\begin{aligned} T_C &= T_B - (0.412)(T_B - T_A) \\ &= 24°\text{C} - (0.412)(24°\text{C} - 10°\text{C}) \\ &= \boxed{18.2°\text{C}\quad(18°\text{C})} \end{aligned}$$

The answer is (B).

(b) $\omega = \dfrac{8.0\ \frac{\text{g}}{\text{kg dry air}}}{1000\ \frac{\text{g}}{\text{kg}}} = \boxed{\begin{array}{c}0.008\ \text{kg moisture/kg dry air} \\ (0.008\ \text{kg/kg})\end{array}}$

The answer is (A).

(c) $T_{dp} = \boxed{10.6°\text{C}\quad(11°\text{C})}$

The answer is (C).

4. *Customary U.S. Solution*

(a) From Eq. 38.24, the bypass factor is

$$\text{BF} = 1 - \eta_{sat} = 1 - 0.70 = \boxed{0.30}$$

The answer is (A).

(b) From Eq. 38.31, the dry-bulb temperature of air leaving the washer can be determined.

$$\eta_{sat} = \frac{T_{db,in} - T_{db,out}}{T_{db,in} - T_{wb,in}}$$

$$0.70 = \frac{60°F - T_{db,out}}{60°F - 45°F}$$

$$T_{db,out} = \boxed{49.5°F \quad (50°F)}$$

The answer is (B).

SI Solution

(a) From Eq. 38.24, the bypass factor is

$$BF = 1 - \eta_{sat} = 1 - 0.70 = \boxed{0.30}$$

The answer is (A).

(b) From Eq. 38.31, the dry-bulb temperature of air leaving the washer can be determined.

$$\eta_{sat} = \frac{T_{db,in} - T_{db,out}}{T_{db,in} - T_{wb,in}}$$

$$0.70 = \frac{16°C - T_{db,out}}{16°C - 7°C}$$

$$T_{db,out} = \boxed{9.7°C}$$

The answer is (B).

5. *Customary U.S. Solution*

(a) Refer to the psychrometric chart (see App. 38.A).

At point 1, properties of air at $T_{db} = 95°F$ and $T_{wb} = 75°F$ are

$$\omega_1 = 0.0141 \text{ lbm moisture/lbm air}$$

$$h_1 = 38.4 \text{ Btu/lbm air}$$

$$\upsilon_1 = 14.3 \text{ ft}^3\text{/lbm air}$$

At point 2, properties of air at $T_{db} = 85°F$ and 90% relative humidity are

$$\omega_2 = 0.0237 \text{ lbm moisture/lbm air}$$

$$h_2 = 46.6 \text{ Btu/lbm air}$$

The enthalpy change is

$$\frac{h_2 - h_1}{\upsilon_1} = \frac{46.6 \ \frac{\text{Btu}}{\text{lbm air}} - 38.4 \ \frac{\text{Btu}}{\text{lbm air}}}{14.3 \ \frac{\text{ft}^3}{\text{lbm air}}} = \boxed{0.573 \text{ Btu/ft}^3 \text{ air} \quad (0.57 \text{ Btu/ft}^3)}$$

The answer is (A).

(b) The moisture added is

$$\frac{\omega_2 - \omega_1}{\upsilon_1} = \frac{0.0237 \ \frac{\text{lbm moisture}}{\text{lbm air}} - 0.0141 \ \frac{\text{lbm moisture}}{\text{lbm air}}}{14.3 \ \frac{\text{ft}^3}{\text{lbm air}}} = \boxed{\begin{array}{r} 6.71 \times 10^{-4} \text{ lbm/ft}^3 \text{ air} \\ (6.7 \times 10^{-4} \text{ lbm/ft}^3) \end{array}}$$

The answer is (C).

SI Solution

(a) Refer to the psychrometric chart (see App. 38.B).

At point 1, properties of air at $T_{db} = 35°C$ and $T_{wb} = 24°C$ are

$$\omega_1 = 14.3 \text{ g/kg air}$$

$$h_1 = 71.8 \text{ kJ/kg air}$$

$$\upsilon_1 = 0.8893 \text{ m}^3\text{/kg air}$$

At point 2, properties of air at $T_{db} = 29°C$ and 90% relative humidity are

$$\omega_2 = 23.1 \text{ g/kg air}$$

$$h_2 = 88 \text{ kJ/kg air}$$

The enthalpy change is

$$\frac{h_2 - h_1}{\upsilon_1} = \frac{88 \ \frac{\text{kJ}}{\text{kg air}} - 71.8 \ \frac{\text{kJ}}{\text{kg air}}}{0.8893 \ \frac{\text{m}^3}{\text{kg air}}} = \boxed{18.2 \text{ kJ/m}^3 \text{ air} \quad (18 \text{ kJ/m}^3)}$$

The answer is (A).

HVAC

(b) The moisture added is

$$\frac{\omega_2 - \omega_1}{\upsilon_1} = \frac{23.1\ \frac{\text{g}}{\text{kg air}} - 14.3\ \frac{\text{kg}}{\text{kg air}}}{\left(0.8893\ \frac{\text{m}^3}{\text{kg air}}\right)\left(1000\ \frac{\text{g}}{\text{kg}}\right)}$$

$$= \boxed{\begin{array}{c} 9.90 \times 10^{-3}\ \text{kg/m}^3\ \text{air} \\ (9.9 \times 10^{-3}\ \text{kg/m}^3) \end{array}}$$

The answer is (C).

6. Use the illustration shown for both the customary U.S. and SI solutions.

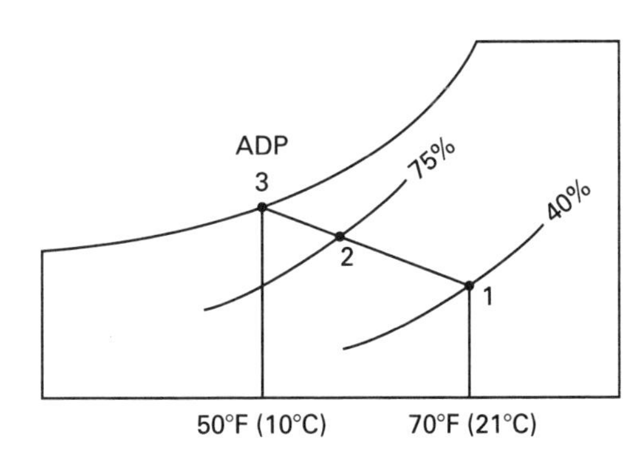

Customary U.S. Solution

(a) Refer to the psychrometric chart (see App. 38.A).

At point 1, properties of air at $T_{\text{db}} = 70°\text{F}$ and $\phi = 40\%$ are

$$h_1 = 23.6\ \text{Btu/lbm air}$$
$$\omega_1 = 0.00623\ \text{lbm moisture/lbm air}$$
$$\upsilon_1 = 13.48\ \text{ft}^3\text{/lbm air}$$

The mass flow rate of incoming air is

$$\dot{m}_{a,1} = \frac{\dot{V}_1}{\upsilon_1} = \frac{1800\ \frac{\text{ft}^3}{\text{min}}}{13.48\ \frac{\text{ft}^3}{\text{lbm air}}} = 133.53\ \text{lbm air/min}$$

Locate point 1 on the psychrometric chart.

Notice that the temperature of the recirculating water is constant but not equal to the air's entering wet-bulb temperature. Therefore, this is not an adiabatic process.

Locate point 3 as 50°F saturated condition (water being sprayed) on the psychrometric chart.

Draw a line from point 1 to point 3. The intersection of this line with 75% relative humidity defines point 2 as

$$\boxed{\begin{array}{l} h_2 = 21.4\ \text{Btu/lbm air} \\ \omega_2 = 0.0072\ \text{lbm moisture/lbm air} \\ T_{\text{db},2} = 56°\text{F} \\ T_{\text{wb},2} = 51.8°\text{F} \end{array}}$$

(b) The moisture (water) added is

$$\dot{m}_w = \dot{m}_{a,1}(\omega_2 - \omega_1)$$
$$= \left(133.53\ \frac{\text{lbm air}}{\text{min}}\right) \times \left(0.0072\ \frac{\text{lbm moisture}}{\text{lbm air}} - 0.00625\ \frac{\text{lbm moisture}}{\text{lbm air}}\right)$$
$$= \boxed{0.127\ \text{lbm/min}}$$

SI Solution

(a) Refer to the psychrometric chart (see App. 38.B).

At point 1, properties of air at $T_{\text{db}} = 21°\text{C}$ and $\phi = 40\%$ are

$$h_1 = 36.75\ \text{kJ/kg air}$$
$$\omega_1 = 6.2\ \text{g moisture/kg air}$$
$$\upsilon_1 = 0.842\ \text{m}^3\text{/kg air}$$

The mass flow rate of incoming air is

$$\dot{m}_{a,1} = \frac{\dot{V}_1}{\upsilon_1} = \frac{0.85\ \frac{\text{m}^3}{\text{s}}}{0.842\ \frac{\text{m}^3}{\text{kg air}}}$$
$$= 1.010\ \text{kg air/s}$$

Locate point 1 on the psychrometric chart.

Notice that the temperature of the recirculating water is constant but not equal to the air's entering wet-bulb temperature. Therefore, this is not an adiabatic process.

Locate point 3 as 10°C saturated condition (water being sprayed) on the psychrometric chart.

Draw a line from point 1 to point 3. The intersection of this line with 75% relative humidity defines point 2 as

$$\boxed{\begin{aligned} h_2 &= 34.1 \text{ kJ/kg air} \\ \omega_2 &= 7.6 \text{ g moisture/kg air} \\ T_{\text{db},2} &= 14.7^\circ\text{C} \\ T_{\text{wb},2} &= 12.0^\circ\text{C} \end{aligned}}$$

(b) The water added is

$$\dot{m}_w = \dot{m}_{a,1}(\omega_2 - \omega_1)$$

$$= \frac{\left(1.010 \ \frac{\text{kg air}}{\text{s}}\right)\left(7.6 \ \frac{\text{g moisture}}{\text{kg air}} - 6.2 \ \frac{\text{g moisture}}{\text{kg air}}\right)}{1000 \ \frac{\text{g}}{\text{kg}}}$$

$$= \boxed{0.00141 \text{ kg/s}}$$

7. *Customary U.S. Solution*

(See the SI solution for a trial-and-error solution procedure.)

((a) and (b)) From Prob. 6,

$$\begin{aligned} \omega_1 &= 0.00623 \text{ lbm moisture/lbm air} \\ h_1 &= 23.6 \text{ Btu/lbm air} \\ \dot{m}_{a,1} &= 133.53 \text{ lbm air/min} \end{aligned}$$

From the steam table (see App. 23.B) for 1 atm steam, $h_{\text{steam}} = 1150.3$ Btu/lbm.

From the conservation of energy equation (see Eq. 38.33),

$$\dot{m}_{a,1}h_1 + \dot{m}_{\text{steam}}h_{\text{steam}} = \dot{m}_{a,1}h_2$$

$$\left(133.53 \ \frac{\text{lbm air}}{\text{min}}\right)\left(23.6 \ \frac{\text{Btu}}{\text{lbm air}}\right) + \dot{m}_{\text{steam}}\left(1150.3 \ \frac{\text{Btu}}{\text{lbm}}\right) = \left(133.53 \ \frac{\text{lbm air}}{\text{min}}\right)h_2 \quad \text{[Eq. 1]}$$

Solve for $\dot{m}_{\text{steam}}$.

$$\dot{m}_{\text{steam}} = 0.1161h_2 - 2.740 \quad \text{[Eq. 3]}$$

From a conservation of mass for the water (see Eq. 38.34),

$$\dot{m}_{a,1}\omega_1 + \dot{m}_{\text{steam}} = \dot{m}_{a,2}\omega_2$$

$$\left(133.53 \ \frac{\text{lbm air}}{\text{min}}\right)\left(0.00623 \ \frac{\text{lbm moisture}}{\text{lbm air}}\right) + \dot{m}_{\text{steam}} = \left(133.53 \ \frac{\text{lbm air}}{\text{min}}\right)\omega_2 \quad \text{[Eq. 2]}$$

Solve for $\dot{m}_{\text{steam}}$.

$$\dot{m}_{\text{steam}} = 133.53\omega_2 - 0.8319 \quad \text{[Eq. 4]}$$

Equate Eq. 3 and Eq. 4.

$$\begin{aligned} 0.1161h_2 - 2.740 &= 133.53\omega_2 - 0.8319 \\ h_2 &= 1150.1\omega_2 + 16.43 \end{aligned}$$

Plot this line on the psychrometric chart, and determine properties where it crosses the 75% relative humidity line.

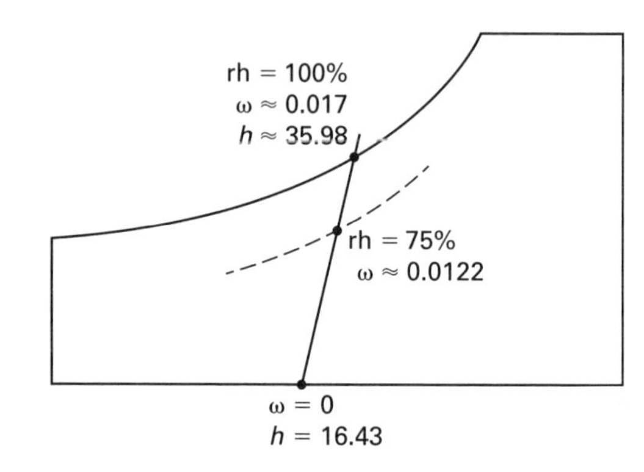

At 75% relative humidity, from the psychrometric chart,

$$\boxed{\begin{aligned} \omega_2 &= 0.0122 \text{ lbm moisture/lbm air} \\ h_2 &= 30.5 \text{ Btu/lbm} \\ T_{\text{db}} &= 72^\circ\text{F} \\ T_{\text{wb}} &= 66^\circ\text{F} \end{aligned}}$$

The mass flow rate is

$$\dot{m}_{\text{steam}} = (0.1161)\left(30.5 \ \frac{\text{Btu}}{\text{lbm}}\right) - 2.740 = \boxed{0.801 \text{ lbm/min}}$$

SI Solution

(See the customary U.S. solution for a graphical method.)

((a) and (b)) From Prob. 6,

$$\begin{aligned} \omega_1 &= 6.2 \text{ g moisture/kg air} \\ h_1 &= 36.75 \text{ kJ/kg air} \\ \dot{m}_{a,1} &= 1.010 \text{ kg air/s} \end{aligned}$$

From the steam table (see App. 23.O), for 1 atm steam, $h_{\text{steam}} = 2675.4$ kJ/kg.

From the conservation of energy equation (see Eq. 38.33),

$$\dot{m}_{a,1}h_1 + \dot{m}_{\text{steam}}h_{\text{steam}} = \dot{m}_{a,1}h_2$$

$$\left(1.010\ \frac{\text{kg air}}{\text{s}}\right)\left(36.75\ \frac{\text{kJ}}{\text{kg air}}\right) + \dot{m}_{\text{steam}}\left(2675.4\ \frac{\text{kJ}}{\text{kg}}\right) = \left(1.010\ \frac{\text{kg air}}{\text{s}}\right)h_2 \quad \text{[Eq. 1]}$$

From a conservation of mass for the water (see Eq. 38.34),

$$\dot{m}_{a,1}\omega_1 + \dot{m}_{\text{steam}} = \dot{m}_{a,2}\omega_2$$

$$\left(1.010\ \frac{\text{kg air}}{\text{s}}\right)\left(6.2\ \frac{\text{g moisture}}{\text{kg air}}\right) + \dot{m}_{\text{steam}} = \left(1.010\ \frac{\text{kg air}}{\text{s}}\right)\omega_2 \quad \text{[Eq. 2]}$$

Since no single relationship exists between ω_2, $\dot{m}_{\text{steam}}$, and h_2, a trial-and-error solution can be used. Once $\dot{m}_{\text{steam}}$ is selected, ω_2 and h_2 can be found from Eq. 1 and Eq. 2 as

$$h_2 = 36.75 + 2648.9\dot{m}_{\text{steam}}$$

$$\omega_2 = 0.0062 + 0.99\dot{m}_{\text{steam}}$$

Once h_2 and ω_2 are known, the relative humidity can be determined from the psychrometric chart. Continue the process until a relative humidity of 75% is achieved.

$\dot{m}_{\text{steam}}$ $\left(\frac{\text{kg}}{\text{s}}\right)$	ω_2 $\left(\frac{\text{kg moisture}}{\text{kg air}}\right)$	h_2 $\left(\frac{\text{kJ}}{\text{kg air}}\right)$	ϕ_2 (%)
0.005	0.0112	49.99	69.5
0.0055	0.0116	51.32	74.5
0.0056	0.0117	51.58	75.0

$$\dot{m}_{\text{steam}} = 0.0056\ \text{kg/s}$$

$$\omega_2 = \left(0.0117\ \frac{\text{kg moisture}}{\text{kg air}}\right)\left(1000\ \frac{\text{g}}{\text{kg}}\right) = 11.7\ \text{g moisture/kg air}$$

$$T_{\text{db}} = 21.3°\text{C}$$

$$T_{\text{wb}} = 18.2°\text{C}$$

8. *Customary U.S. Solution*

(a) From the psychrometric chart (see App. 38.A), for incoming air at 65°F and 55% relative humidity, $\omega_1 = 0.0072$ lbm moisture/lbm air.

With sensible heating as a limiting factor, calculate the mass flow rate of air entering the theater from Eq. 38.25 (ventilation rate).

$$\dot{q} = \dot{m}_a(c_{p,\text{air}} + \omega c_{p,\text{moisture}})(T_2 - T_1)$$

$$500{,}000\ \frac{\text{Btu}}{\text{hr}} = \dot{m}_a \times \left(\begin{array}{l} 0.240\ \frac{\text{Btu}}{\text{lbm-°F}} \\ \quad + 0.0072\ \frac{\text{lbm moisture}}{\text{lbm air}} \\ \quad \times \left(0.444\ \frac{\text{Btu}}{\text{lbm-°F}}\right) \end{array}\right) \times (75°\text{F} - 65°\text{F})$$

$$\dot{m}_a = \boxed{2.056 \times 10^5\ \text{lbm air/hr}}$$

(b) Assume that this air absorbs all the moisture. Then, the final humidity ratio is given by

$$\dot{m}_w = \dot{m}_a(\omega_2 - \omega_1)$$

$$\omega_2 = \left(\frac{\dot{m}_w}{\dot{m}_a}\right) + \omega_1 = \frac{175\ \frac{\text{lbm moisture}}{\text{hr}}}{2.056 \times 10^5\ \frac{\text{lbm air}}{\text{hr}}} + 0.0072\ \frac{\text{lbm moisture}}{\text{lbm air}}$$

$$= 0.00805\ \text{lbm moisture/lbm air}$$

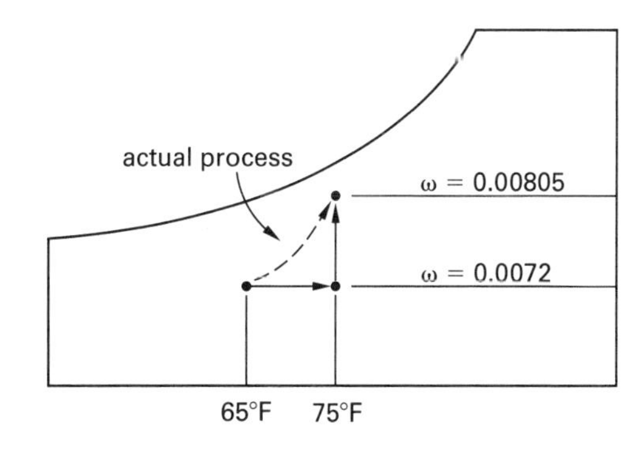

The final conditions are

$$T_{\text{db}} = 75°\text{F} \quad \text{[given]}$$

$$\omega_2 = 0.00805\ \text{lbm moisture/lbm air}$$

From the psychrometric chart (see App. 38.A), the relative humidity is 44%. This is below 60%.

SI Solution

(a) From the psychrometric chart (see App. 38.B), for incoming air at 18°C and 55% relative humidity, $\omega_1 = 7.1$ g moisture/kg air.

With sensible heating as a limiting factor, calculate the mass flow rate of air entering the theater from Eq. 38.25 (ventilation rate).

$$\dot{q} = \dot{m}_a(c_{p,\text{air}} + \omega c_{p,\text{moisture}})(T_2 - T_1)$$

$$\begin{aligned}(150\ \text{kW})\\ \times\left(1000\ \frac{\text{W}}{\text{kW}}\right)\end{aligned} = \dot{m}_a\left(\begin{array}{c}\left(1.005\ \frac{\text{kJ}}{\text{kg}\cdot{}^\circ\text{C}}\right)\left(1000\ \frac{\text{J}}{\text{kJ}}\right)\\ +\left(1.805\ \frac{\text{kJ}}{\text{kg}\cdot{}^\circ\text{C}}\right)\left(1000\ \frac{\text{J}}{\text{kJ}}\right)\\ \times\left(\dfrac{7.1\ \frac{\text{g moisture}}{\text{kg air}}}{1000\ \frac{\text{g}}{\text{kg}}}\right)\end{array}\right) \times (24^\circ\text{C} - 18^\circ\text{C})$$

$$\dot{m}_a = \boxed{24.56\ \text{kg/s}}$$

(b) Assume that this air absorbs all the moisture. Then, the final humidity ratio is given by

$$\dot{m}_w = \dot{m}_a(\omega_2 - \omega_1)$$

$$\omega_2 = \frac{\dot{m}_w}{\dot{m}_a} + \omega_1$$

$$= \frac{80\ \frac{\text{kg}}{\text{h}}}{\left(24.56\ \frac{\text{kg}}{\text{s}}\right)\left(3600\ \frac{\text{s}}{\text{h}}\right)} + \frac{7.0\ \frac{\text{g moisture}}{\text{kg air}}}{1000\ \frac{\text{g}}{\text{kg}}}$$

$$= 0.00790\ \text{kg moisture/kg air}$$

The final conditions are

$$\boxed{\begin{aligned} T_{\text{db}} &= 24^\circ\text{C}\\ \omega_2 &= \left(0.00790\ \frac{\text{kg moisture}}{\text{kg air}}\right)\left(1000\ \frac{\text{g}}{\text{kg}}\right)\\ &= 7.9\ \text{g moisture/kg air}\end{aligned}}$$

From the psychrometric chart (see App. 38.B), the relative humidity is 44%. This is below 60%.

9. Use the illustration shown for both customary U.S. and SI solutions.

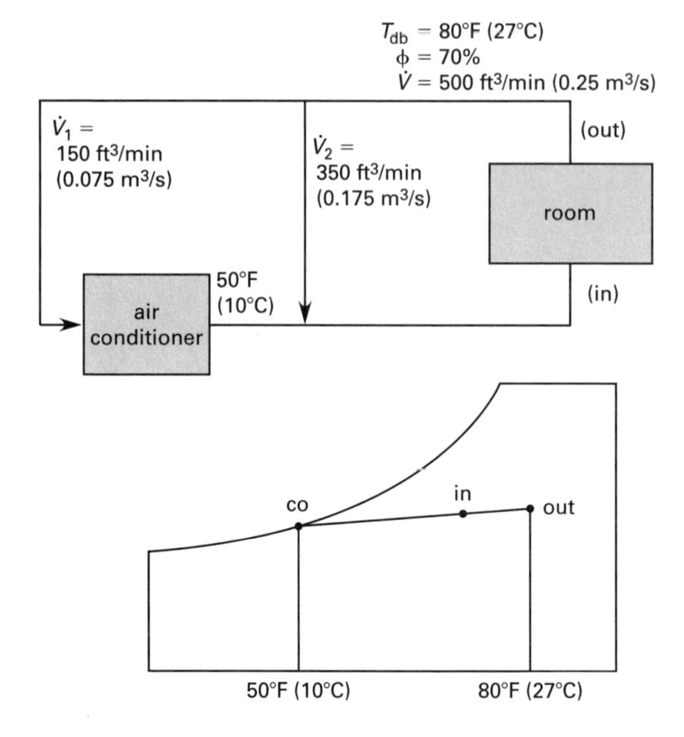

Customary U.S. Solution

Locate point "out" ($T_{\text{db}} = 80$°F and $\phi = 70\%$) and point "co" (saturated at 50°F) on the psychrometric chart. At point "out" from App. 38.A,

$$\upsilon_{\text{out}} = 13.95\ \text{ft}^3/\text{lbm air}$$

$$h_{\text{out}} = 36.2\ \text{Btu/lbm air}$$

At point "co," $h_{\text{co}} = 20.3$ Btu/lbm air.

The air mass flow rate through the air conditioner is

$$\dot{m}_1 = \frac{\dot{V}_1}{\upsilon_1} = \frac{\dot{V}_1}{\upsilon_{\text{out}}} = \frac{150\ \frac{\text{ft}^3}{\text{min}}}{13.95\ \frac{\text{ft}^3}{\text{lbm air}}} = 10.75\ \text{lbm air/min}$$

The mass flow rate of the bypass air is

$$\dot{m}_2 = \frac{\dot{V}_2}{\upsilon} = \frac{350\ \frac{\text{ft}^3}{\text{min}}}{13.95\ \frac{\text{ft}^3}{\text{lbm air}}} = 25.09\ \text{lbm air/min}$$

The percentage of bypass air is

$$x = \frac{25.09\ \frac{\text{lbm air}}{\text{min}}}{10.75\ \frac{\text{lbm air}}{\text{min}} + 25.09\ \frac{\text{lbm air}}{\text{min}}} = 0.70 \quad (70\%)$$

Using the lever rule and the fact that all of the temperature scales are linear,

$$\begin{aligned} T_{\text{db,in}} &= T_{\text{co}} + (0.70)(T_{\text{out}} - T_{\text{co}}) \\ &= 50°\text{F} + (0.70)(80°\text{F} - 50°\text{F}) \\ &= 71°\text{F} \end{aligned}$$

At that point,

(a) $\boxed{T_{\text{db,in}} = 71°\text{F}}$

The answer is (C).

(b) $\boxed{\begin{array}{c}\omega_{\text{in}} = 0.0132 \text{ lbm moisture/lbm air} \\ (0.013 \text{ lbm/lbm})\end{array}}$

The answer is (B).

(c) $\boxed{\phi_{\text{in}} = 81\%}$

The answer is (D).

(d) The air conditioner capacity is given by

$$\begin{aligned} \dot{Q} &= \dot{m}_{\text{air}}(h_{t,2} - h_{t,1}) = \dot{m}_1(h_{\text{out}} - h_{\text{co}}) \\ &= \left(10.75 \ \frac{\text{lbm air}}{\text{min}}\right)\left(36.2 \ \frac{\text{Btu}}{\text{lbm air}} - 20.3 \ \frac{\text{Btu}}{\text{lbm air}}\right) \\ &\quad \times \left(\frac{1 \text{ ton}}{200 \ \frac{\text{Btu}}{\text{min}}}\right) \\ &= \boxed{0.85 \text{ ton} \quad (0.90 \text{ ton})} \end{aligned}$$

The answer is (A).

SI Solution

Locate point "out" ($T_{\text{db}} = 27°\text{C}$, $\phi = 70\%$) and point "co" (saturated at 10°C) on the psychrometric chart. At point "out" from App. 38.B,

$$\begin{aligned} \upsilon_{\text{out}} &= 0.872 \text{ m}^3\text{/kg air} \\ h_{\text{out}} &= 67.3 \text{ kJ/kg air} \end{aligned}$$

At point "co" from App. 38.B, $h_{\text{co}} = 29.26$ kJ/kg air.

At mass flow rate through the air conditioner,

$$\dot{m}_1 = \frac{\dot{V}_1}{\upsilon} = \frac{0.075 \ \frac{\text{m}^3}{\text{s}}}{0.872 \ \frac{\text{m}^3}{\text{kg air}}} = 0.0860 \text{ kg air/s}$$

The flow rate of bypass air is

$$\dot{m}_2 = \frac{\dot{V}_2}{\upsilon} = \frac{0.175 \ \frac{\text{m}^3}{\text{s}}}{0.872 \ \frac{\text{m}^3}{\text{kg air}}} = 0.2007 \text{ kg air/s}$$

The percentage bypass air is

$$x = \frac{0.2007 \ \frac{\text{kg air}}{\text{s}}}{0.0860 \ \frac{\text{kg air}}{\text{s}} + 0.2007 \ \frac{\text{kg air}}{\text{s}}} = 0.70 \quad (70\%)$$

Using the lever rule and the fact that all of the temperature scales are linear,

$$\begin{aligned} T_{\text{db,in}} &= T_{\text{co}} + (0.70)(T_{\text{out}} - T_{\text{co}}) \\ &= 10°\text{C} + (0.70)(27°\text{C} - 10°\text{C}) \\ &= 21.9°\text{C} \end{aligned}$$

At that point,

(a) $\boxed{T_{\text{db,in}} = 21.9°\text{C} \quad (22°\text{C})}$

The answer is (C).

(b) $\boxed{\omega_{\text{in}} = 1.34 \text{ g moisture/kg air} \quad (1.3 \text{ g/kg})}$

The answer is (B).

(c) $\boxed{\phi_{\text{in}} = 81\%}$

The answer is (D).

(d) The air conditioner capacity is given by

$$\begin{aligned} \dot{Q} &= \dot{m}_{\text{air}}(h_{t,2} - h_{t,1}) = \dot{m}_1(h_{\text{out}} - h_{\text{co}}) \\ &= \left(0.0860 \ \frac{\text{kg air}}{\text{s}}\right)\left(67.3 \ \frac{\text{kJ}}{\text{kg air}} - 29.26 \ \frac{\text{kJ}}{\text{kg air}}\right) \\ &\quad \times \left(0.2843 \ \frac{\text{ton}}{\text{kW}}\right) \\ &= \boxed{0.93 \text{ ton} \quad (0.90 \text{ ton})} \end{aligned}$$

The answer is (A).

10. Use the illustration shown for both customary U.S. and SI solutions.

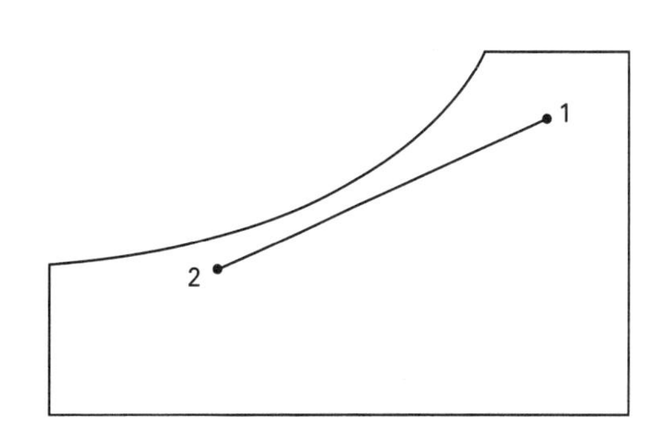

Customary U.S. Solution

(a) At point 1, from the psychrometric chart (see App. 38.A) at $T_{\text{db}} = 95°\text{F}$ and $\phi = 70\%$,

$$\boxed{\begin{aligned} h_1 &= 50.7 \text{ Btu/lbm air} \\ \upsilon_1 &= 14.56 \text{ ft}^3/\text{lbm air} \\ \omega_1 &= 0.0253 \text{ lbm water/lbm air} \end{aligned}}$$

At point 2, from the psychrometric chart (see App. 38.A) at $T_{\text{db}} = 60°\text{F}$ and $\phi = 95\%$,

$$\boxed{\begin{aligned} h_2 &= 25.8 \text{ Btu/lbm air} \\ \omega_2 &= 0.0105 \text{ lbm water/lbm air} \end{aligned}}$$

(b) The air mass flow rate is

$$\dot{m}_a = \frac{\dot{V}}{\upsilon_1} = \frac{5000 \ \frac{\text{ft}^3}{\text{min}}}{14.56 \ \frac{\text{ft}^3}{\text{lbm air}}} = 343.4 \text{ lbm air/min}$$

From Eq. 38.26, the water removed is

$$\begin{aligned} \dot{m}_w &= \dot{m}_a(\omega_1 - \omega_2) \\ &= \left(343.4 \ \frac{\text{lbm air}}{\text{min}}\right) \\ &\quad \times \left(0.0253 \ \frac{\text{lbm water}}{\text{lbm air}} - 0.0105 \ \frac{\text{lbm water}}{\text{lbm air}}\right) \\ &= \boxed{5.08 \text{ lbm water/min}} \end{aligned}$$

(c) From Eq. 38.27, the quantity of heat removed is

$$\begin{aligned} \dot{q} &= \dot{m}_a(h_1 - h_2) \\ &= \left(343.4 \ \frac{\text{lbm air}}{\text{min}}\right)\left(50.7 \ \frac{\text{Btu}}{\text{lbm air}} - 25.8 \ \frac{\text{Btu}}{\text{lbm air}}\right) \\ &= \boxed{8551 \text{ Btu/min}} \end{aligned}$$

(d) Considering an R-12 refrigeration cycle operating at saturated conditions at 100°F, the T-s and h-s diagrams are shown.

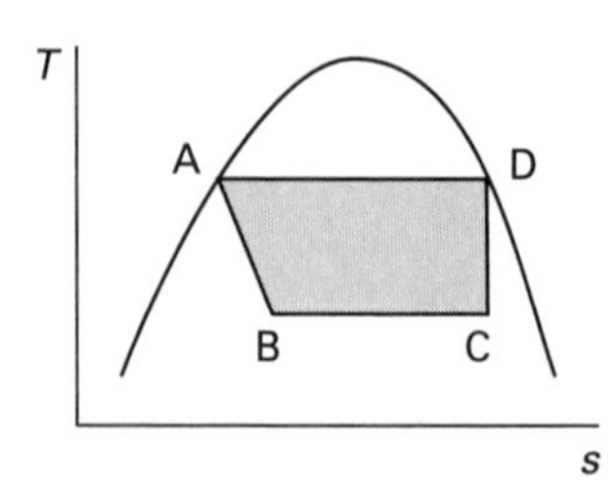

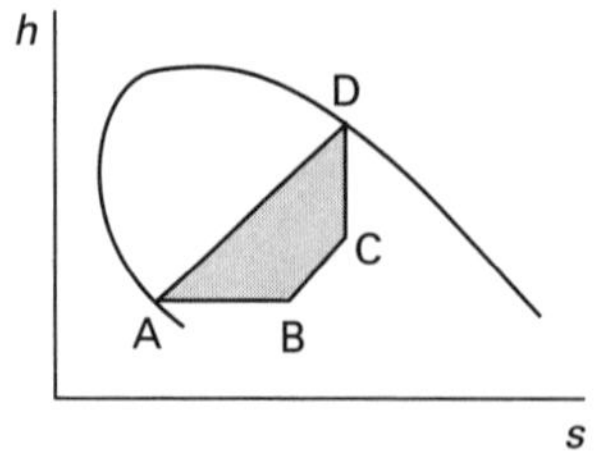

(e) Use App. 23.G for saturated conditions. At point A,

$$\begin{aligned} T &= \boxed{100°\text{F}} \quad [\text{given}] \\ p &= p_{\text{sat}} \text{ at } 100°\text{F} = \boxed{131.6 \text{ psia}} \\ h_{\text{A}} &= h_f \text{ at } 100°\text{F} = \boxed{31.16 \text{ Btu/lbm}} \\ s_{\text{A}} &= s_f \text{ at } 100°\text{F} = \boxed{0.06316 \text{ Btu/lbm-°R}} \\ \upsilon_{\text{A}} &= \upsilon_f \text{ at } 100°\text{F} = \boxed{0.0127 \text{ ft}^3/\text{lbm}} \end{aligned}$$

At point B,

$$\begin{aligned} T &= \boxed{50°\text{F}} \quad [\text{given}] \\ p &= p_{\text{sat}} \text{ at } 50°\text{F} = \boxed{61.39 \text{ psia}} \\ h_{\text{B}} &= h_{\text{A}} = \boxed{31.16 \text{ Btu/lbm}} \\ h_{f,\text{B}} &= 19.27 \text{ Btu/lbm} \\ h_{fg,\text{B}} &= 64.51 \text{ Btu/lbm} \end{aligned}$$

$$x = \frac{h_{\text{B}} - h_{f,\text{B}}}{h_{fg,\text{B}}} = \frac{31.16 \ \frac{\text{Btu}}{\text{lbm}} - 19.27 \ \frac{\text{Btu}}{\text{lbm}}}{64.51 \ \frac{\text{Btu}}{\text{lbm}}} = 0.184$$

$$\begin{aligned} s_{\text{B}} &= s_{f,\text{B}} + x s_{fg,\text{B}} \\ &= 0.04126 \ \frac{\text{Btu}}{\text{lbm-°R}} + (0.184)\left(0.12659 \ \frac{\text{Btu}}{\text{lbm-°R}}\right) \\ &= \boxed{0.06455 \text{ Btu/lbm-°R}} \\ \upsilon_{\text{B}} &= \upsilon_{f,\text{B}} + x(\upsilon_g - \upsilon_f) \\ &= 0.0118 \ \frac{\text{ft}^3}{\text{lbm}} + (0.184)\left(0.673 \ \frac{\text{ft}^3}{\text{lbm}} - 0.0118 \ \frac{\text{ft}^3}{\text{lbm}}\right) \\ &= \boxed{0.1335 \text{ ft}^3/\text{lbm}} \end{aligned}$$

At point D,

$$\begin{aligned} T &= \boxed{100°\text{F}} \quad [\text{given}] \\ p &= p_{\text{sat}} \text{ at } 100°\text{F} = \boxed{131.6 \text{ psia}} \\ h_{\text{D}} &= h_g \text{ at } 100°\text{F} = \boxed{88.62 \text{ Btu/lbm}} \\ s_{\text{D}} &= s_g \text{ at } 100°\text{F} = \boxed{0.16584 \text{ Btu/lbm-°R}} \\ \upsilon_{\text{D}} &= \upsilon_g \text{ at } 100°\text{F} = \boxed{0.319 \text{ ft}^3/\text{lbm}} \end{aligned}$$

At point C,

$$\begin{aligned} T &= \boxed{50°\text{F}} \quad [\text{given}] \\ p &= p_{\text{sat}} \text{ at } 50°\text{F} = \boxed{61.39 \text{ psia}} \\ s_{\text{C}} &= s_{\text{D}} = \boxed{0.16584 \text{ Btu/lbm-°R}} \\ s_{f,\text{C}} &= 0.04126 \text{ Btu/lbm-°R} \\ s_{fg,\text{C}} &= 0.12659 \text{ Btu/lbm-°R} \end{aligned}$$

$$x = \frac{s_C - s_{f,C}}{s_{fg,C}} = \frac{0.16584\ \frac{\text{Btu}}{\text{lbm-°R}} - 0.04126\ \frac{\text{Btu}}{\text{lbm-°R}}}{0.12659\ \frac{\text{Btu}}{\text{lbm-°R}}}$$

$$= 0.984$$

$$h_C = h_{f,C} + xh_{fg,C} = 19.27\ \frac{\text{Btu}}{\text{lbm}} + (0.984)\left(64.51\ \frac{\text{Btu}}{\text{lbm}}\right)$$

$$= \boxed{82.75\ \text{Btu/lbm}}$$

$$v_C = v_{f,C} + x(v_{g,C} - v_{f,C})$$

$$= 0.0118\ \frac{\text{ft}^3}{\text{lbm}} + (0.984)\left(0.673\ \frac{\text{ft}^3}{\text{lbm}} - 0.0118\ \frac{\text{ft}^3}{\text{lbm}}\right)$$

$$= \boxed{0.662\ \text{ft}^3/\text{lbm}}$$

SI Solution

(a) At point 1, from the psychrometric chart (see App. 38.B) at $T_{\text{db}} = 35°\text{C}$ and $\phi = 70\%$,

$$\boxed{\begin{aligned} h_1 &= 99.9\ \text{kJ/kg air} \\ v_1 &= 0.91\ \text{m}^3/\text{kg air} \\ \omega_1 &= \frac{25.3\ \frac{\text{g moisture}}{\text{kg air}}}{1000\ \frac{\text{g}}{\text{kg}}} = 0.0253\ \text{kg moisture/kg air} \end{aligned}}$$

At point 2, from the psychrometric chart (see App. 38.B) at $T_{\text{db}} = 16°\text{C}$ and $\phi = 95\%$,

$$\boxed{\begin{aligned} h_2 &= 43.4\ \text{kJ/kg air} \\ \omega_2 &= \frac{10.8\ \frac{\text{g moisture}}{\text{kg air}}}{1000\ \frac{\text{g}}{\text{kg}}} = 0.0108\ \text{kg moisture/kg air} \end{aligned}}$$

(b) The air mass flow rate is

$$\dot{m}_a = \frac{\dot{V}_1}{v_1} = \frac{\dot{V}_1}{v_{\text{out}}} = \frac{2.36\ \frac{\text{m}^3}{\text{s}}}{0.91\ \frac{\text{m}^3}{\text{kg air}}} = 2.593\ \text{kg air/s}$$

From Eq. 38.26, the water removed is

$$\dot{m}_w = \dot{m}_a(\omega_1 - \omega_2)$$

$$= \left(2.593\ \frac{\text{kg air}}{\text{s}}\right)\left(\begin{array}{l} 0.0253\ \frac{\text{kg moisture}}{\text{kg air}} \\ \quad - 0.0108\ \frac{\text{kg moisture}}{\text{kg air}} \end{array}\right)$$

$$= \boxed{0.0376\ \text{kg moisture/s}}$$

(c) From Eq. 38.27, the quantity of heat removed is

$$\dot{q} = \dot{m}_a(h_1 - h_2)$$

$$= \left(2.593\ \frac{\text{kg air}}{\text{s}}\right)\left(99.9\ \frac{\text{kJ}}{\text{kg air}} - 43.4\ \frac{\text{kJ}}{\text{kg air}}\right)$$

$$= \boxed{146.5\ \text{kW}}$$

(d) The T-s and h-s diagrams are shown in the customary U.S. solution.

(e) Use App. 23.T for saturated conditions. At point A,

$$T = \boxed{38°\text{C}} \quad \text{[given]}$$

$$p = p_{\text{sat}}\ \text{at}\ 38°\text{C} = \boxed{0.91324\ \text{MPa}}$$

$$h_A = h_f\ \text{at}\ 38°\text{C} = \boxed{237.23\ \text{kJ/kg}}$$

$$s_A = s_f\ \text{at}\ 38°\text{C} = \boxed{1.1259\ \text{kJ/kg·°C}}$$

$$v_A = v_f\ \text{at}\ 38°\text{C} = \frac{1}{\rho_f} = \frac{1}{1261.9\ \frac{\text{kg}}{\text{m}^3}}$$

$$= \boxed{0.0007925\ \text{m}^3/\text{kg}}$$

At point B,

$$T = \boxed{10°\text{C}} \quad \text{[given]}$$

$$p = p_{\text{sat}}\ \text{at}\ 10°\text{C} = \boxed{0.42356\ \text{MPa}}$$

$$h_B = h_A = \boxed{237.23\ \text{kJ/kg}}$$

$$h_{f,B} = 209.48\ \text{kJ/kg}$$

$$h_{g,B} = 356.79\ \text{kJ/kg}$$

$$x = \frac{h_B - h_{f,B}}{h_{g,B} - h_{f,B}} = \frac{237.23\ \frac{\text{kJ}}{\text{kg}} - 209.48\ \frac{\text{kJ}}{\text{kg}}}{356.79\ \frac{\text{kJ}}{\text{kg}} - 209.48\ \frac{\text{kJ}}{\text{kg}}}$$

$$= 0.188$$

$$s_B = s_{f,B} + x(s_{g,B} - s_{f,B})$$

$$= 1.0338\ \frac{\text{kJ}}{\text{kg·K}} + (0.188)\left(1.5541\ \frac{\text{kJ}}{\text{kg·K}} - 1.0338\ \frac{\text{kJ}}{\text{kg·K}}\right)$$

$$= \boxed{1.1316\ \text{kJ/kg·K}}$$

$$v_B = v_{f,B} + x(v_{g,B} - v_{f,B})$$

$$v_{f,B} = \frac{1}{\rho_{f,B}} = \frac{1}{1363.0\ \frac{\text{kg}}{\text{m}^3}} = 0.0007337\ \text{m}^3/\text{kg}$$

$$v_{g,B} = 0.04119\ \text{m}^3/\text{kg}$$

HVAC

$$\begin{aligned} v_B &= 0.0007337\ \frac{m^3}{kg} \\ &\quad + (0.188)\left(0.04119\ \frac{m^3}{kg} - 0.0007337\ \frac{m^3}{kg}\right) \\ &= \boxed{0.008339\ m^3/kg} \end{aligned}$$

At point D,

$$\begin{aligned} T &= \boxed{38°C} \quad \text{[given]} \\ p = p_{sat} \text{ at } 38°C &= \boxed{0.91324\ MPa} \\ h_D = h_g \text{ at } 38°C &= \boxed{367.95\ kJ/kg} \\ s_D = s_g \text{ at } 38°C &= \boxed{1.5461\ kJ/kg{\cdot}K} \\ v_D = v_g \text{ at } 38°C &= \boxed{0.01931\ m^3/kg} \end{aligned}$$

At point C,

$$\begin{aligned} T &= \boxed{10°C} \\ p = p_{sat} \text{ at } 10°C &= \boxed{0.42356\ MPa} \\ s_C = s_D &= \boxed{1.5461\ kJ/kg{\cdot}K} \\ s_{f,C} &= 1.0338\ kJ/kg{\cdot}K \\ s_{g,C} &= 1.5541\ kJ/kg{\cdot}K \end{aligned}$$

$$x = \frac{s_C - s_{f,C}}{s_{g,C} - s_{f,C}} = \frac{1.5461\ \frac{kJ}{kg{\cdot}K} - 1.0338\ \frac{kJ}{kg{\cdot}K}}{1.5541\ \frac{kJ}{kg{\cdot}K} - 1.0338\ \frac{kJ}{kg{\cdot}K}} = 0.985$$

$$\begin{aligned} h_C &= h_{f,C} + x(h_{g,C} - h_{f,C}) \\ &= 209.48\ \frac{kJ}{kg} + (0.985)\left(356.79\ \frac{kJ}{kg} - 209.48\ \frac{kJ}{kg}\right) \\ &= \boxed{354.58\ kJ/kg} \end{aligned}$$

$$\begin{aligned} v_C &= v_{f,C} + x(v_{g,C} - v_{f,C}) \\ v_{g,C} &= 0.04119\ m^3/kg \\ v_C &= 0.0007337\ \frac{m^3}{kg} \\ &\quad + (0.985)\left(0.04119\ \frac{m^3}{kg} - 0.0007337\ \frac{m^3}{kg}\right) \\ &= \boxed{0.04058\ m^3/kg} \end{aligned}$$

11. *Customary U.S. Solution*

(a) The saturation pressure at 200°F from App. 23.A is $p_{sat,1} = 11.538$ psia.

Since air is saturated (100% relative humidity), the water vapor pressure is equal to the saturation pressure.

$$p_{w,1} = p_{sat,1} = 11.538\ \text{psia}$$

The partial pressure of the air is

$$\begin{aligned} p_{a,1} &= p_1 - p_{w,1} = 25\ \text{psia} - 11.538\ \text{psia} \\ &= 13.462\ \text{psia} \end{aligned}$$

Use Eq. 38.7 to determine the humidity ratio.

$$\omega = 0.622\left(\frac{p_{w,1}}{p_{a,1}}\right) = (0.622)\left(\frac{11.538\ \text{psia}}{13.462\ \text{psia}}\right) = 0.533$$

Since it is a constant pressure, constant moisture drying process, mole fractions and partial pressures do not change.

$$p_{w,2} = p_{w,1} = 11.538\ \text{psia}$$

The saturation pressure at 400°F from App. 23.A is $p_{sat,2} = 247.3$ psia.

The relative humidity at state 2 is

$$\phi_2 = \frac{p_{w,2}}{p_{sat,2}} = \frac{11.538\ \text{psia}}{247.3\ \text{psia}} = \boxed{0.0467 \quad (4.7\%)}$$

The answer is (A).

(b) Although the volume may change, the mass does not. The specific humidity remains constant.

$$\omega_2 = \omega_1 = \boxed{0.533\ \text{lbm water/lbm air} \quad (0.53\ \text{lbm/lbm})}$$

The answer is (B).

(c) The heat required consists of two parts.

Obtain enthalpy for air from App. 23.F.

The absolute temperatures are

$$\begin{aligned} T_1 &= 200°F + 460° = 660°R \\ T_2 &= 400°F + 460° = 860°R \\ h_1 &= 157.92\ \text{Btu/lbm} \\ h_2 &= 206.46\ \text{Btu/lbm} \end{aligned}$$

The heat absorbed by the air is

$$\begin{aligned} q_1 = h_2 - h_1 &= 206.46\ \frac{Btu}{lbm} - 157.92\ \frac{Btu}{lbm} \\ &= 48.54\ \text{Btu/lbm dry air} \end{aligned}$$

(There will be a small error if constant specific heat is used instead.)

For water, use the Mollier diagram. From App. 23.E, h_1 at 200°F and 11.529 psia is 1146 Btu/lbm (almost saturated).

HVAC

Follow a constant 11.529 psia pressure curve up to 400°F.

$$h_2 = 1240 \text{ Btu/lbm}$$

(There will be a small error if Eq. 38.19(b) is used instead.)

The heat absorbed by the steam is

$$\begin{aligned} q_2 &= \omega(h_2 - h_1) \\ &= \left(0.532 \ \frac{\text{lbm water}}{\text{lbm air}}\right)\left(1240 \ \frac{\text{Btu}}{\text{lbm}} - 1146 \ \frac{\text{Btu}}{\text{lbm}}\right) \\ &= 50.01 \text{ Btu/lbm air} \end{aligned}$$

The total heat absorbed is

$$\begin{aligned} q_{\text{total}} &= q_1 + q_2 = 48.54 \ \frac{\text{Btu}}{\text{lbm air}} + 50.01 \ \frac{\text{Btu}}{\text{lbm air}} \\ &= \boxed{98.55 \text{ Btu/lbm air} \quad (99 \text{ Btu/lbm})} \end{aligned}$$

(d) The dew point is the temperature at which water starts to condense out in a constant pressure process. Following the constant 11.538 psia pressure line back to the saturation line, $\boxed{T_{\text{dp}} = 200°\text{F.}}$

The answer is (B).

SI Solution

(a) From App. 23.N, the saturation pressure at 93°C is

$$p_{\text{sat},1} = (0.7884 \text{ bar})\left(100 \ \frac{\text{kPa}}{\text{bar}}\right) = 78.84 \text{ kPa}$$

Since air is saturated (100% relative humidity), water vapor pressure is equal to saturation pressure.

$$p_{w,1} = p_{\text{sat},1} = 78.84 \text{ kPa}$$

The partial pressure of air is

$$\begin{aligned} p_{a,1} &= p_1 - p_{w,1} = 170 \text{ kPa} - 78.84 \text{ kPa} \\ &= 91.16 \text{ kPa} \end{aligned}$$

Use Eq. 38.7 to determine the humidity ratio.

$$\begin{aligned} \omega &= 0.622\left(\frac{p_{w,1}}{p_{a,1}}\right) = (0.622)\left(\frac{78.84 \text{ kPa}}{91.16 \text{ kPa}}\right) \\ &= 0.538 \text{ kg water/kg air} \end{aligned}$$

Since it is a constant pressure, constant moisture drying process, mole fractions and partial pressure do not change.

$$p_{w,2} = p_{w,1} = 78.84 \text{ kPa}$$

The saturation pressure at 204°C from App. 23.N is

$$p_{\text{sat},2} = (16.90 \text{ bar})\left(100 \ \frac{\text{kPa}}{\text{bar}}\right) = 1690 \text{ kPa}$$

The relative humidity at state 2 is

$$\phi_2 = \frac{p_{w,2}}{p_{\text{sat},2}} = \frac{78.84 \text{ kPa}}{1690 \text{ kPa}} = \boxed{0.0467 \quad (4.7\%)}$$

The answer is (A).

(b) Although the volume may change, the mass does not. The specific humidity remains constant.

$$\omega_2 = \omega_1 = \boxed{0.538 \text{ kg water/kg air} \quad (0.53 \text{ kg/kg})}$$

The answer is (B).

(c) The heat required consists of two parts.

Obtain enthalpy for air from App. 23.S.

The absolute temperatures are

$$\begin{aligned} T_1 &= 93°\text{C} + 273° = 366\text{K} \\ T_2 &= 204°\text{C} + 273° = 477\text{K} \\ h_1 &= 366.63 \text{ kJ/kg} \\ h_2 &= 479.42 \text{ kJ/kg} \end{aligned}$$

The heat absorbed by air is

$$\begin{aligned} q_1 &= h_2 - h_1 = 479.42 \ \frac{\text{kJ}}{\text{kg}} - 366.63 \ \frac{\text{kJ}}{\text{kg}} \\ &= 112.79 \text{ kJ/kg air} \end{aligned}$$

(There will be a small error if constant specific heat is used instead.)

For water, use the Mollier diagram. From App. 23.R, h_1 at 93°C and 78.79 kPa is 2670 kJ/kg (almost saturated).

Follow a constant 78.79 kPa pressure curve up to 204°C.

$$h_2 = 2890 \text{ kJ/kg}$$

(There will be a small error if Eq. 38.19(a) is used instead.)

The heat absorbed by steam is

$$\begin{aligned} q_2 &= \omega(h_2 - h_1) \\ &= \left(0.537 \ \frac{\text{kg water}}{\text{kg air}}\right)\left(2890 \ \frac{\text{kJ}}{\text{kg}} - 2670 \ \frac{\text{kJ}}{\text{kg}}\right) \\ &= 118.14 \text{ kJ/kg air} \end{aligned}$$

HVAC

The total heat absorbed is

$$q_{\text{total}} = q_1 + q_2 = 112.79\ \frac{\text{kJ}}{\text{kg air}} + 118.14\ \frac{\text{kJ}}{\text{kg air}}$$
$$= \boxed{230.93\ \text{kJ/kg air}\quad (230\ \text{kJ/kg})}$$

The answer is (C).

(d) The dew point is the temperature at which water starts to condense out in a constant pressure process. Following the constant 78.84 kPa pressure line back to the saturation line, $\boxed{T_{\text{dp}} \approx 93°\text{C}.}$

The answer is (B).

12. *Customary U.S. Solution*

(a) The absolute air temperatures are

$$T_1 = 800°\text{F} + 460° = 1260°\text{R}$$
$$T_2 = 350°\text{F} + 460° = 810°\text{R}$$

At low pressures, use air tables. From App. 23.F,

$$h_1 = 306.65\ \text{Btu/lbm}$$
$$h_2 = 194.25\ \text{Btu/lbm}$$

From App. 23.A, the enthalpy of water at 80°F is $h_{w,1} = 48.07$ Btu/lbm (48 Btu/lbm).

From Eq. 38.19(b), the enthalpy of steam at 350°F is

$$h_{w,2} \approx \left(0.444\ \frac{\text{Btu}}{\text{lbm-°F}}\right)(350°\text{F}) + 1061\ \frac{\text{Btu}}{\text{lbm}}$$
$$= 1216.4\ \text{Btu/lbm}$$

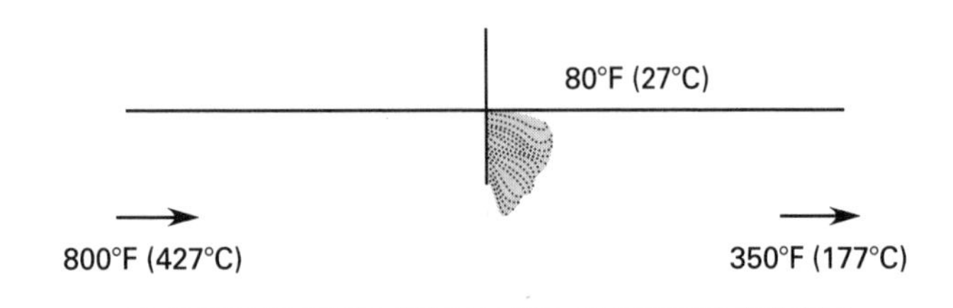

Air temperature is reduced from 800°F to 350°F, and this energy is used to change water at 80°F to steam at 350°F. From the energy balance equation,

$$\dot{m}_w(h_{w,2} - h_{w,1}) = \dot{m}_a(h_1 - h_2)$$
$$\dot{m}_w = \frac{\dot{m}_a(h_1 - h_2)}{h_{w,2} - h_{w,1}}$$
$$= \frac{\left(410\ \frac{\text{lbm}}{\text{hr}}\right)\left(306.65\ \frac{\text{Btu}}{\text{lbm}} - 194.25\ \frac{\text{Btu}}{\text{lbm}}\right)}{1216.4\ \frac{\text{Btu}}{\text{lbm}} - 48.07\ \frac{\text{Btu}}{\text{lbm}}}$$
$$= \boxed{39.4\ \text{lbm/hr water}\quad (39\ \text{lbm/hr})}$$

The answer is (D).

(b) The number of moles of water evaporated (in the air mixture) is

$$\dot{n}_w = \frac{39.4\ \frac{\text{lbm}}{\text{hr}}}{18.016\ \frac{\text{lbm}}{\text{lbmol}}} = 2.19\ \text{lbmol/hr}$$

The number of moles of air in the mixture at the exit is

$$\dot{n}_a = \frac{410\ \frac{\text{lbm}}{\text{hr}}}{28.967\ \frac{\text{lbm}}{\text{lbmol}}} = 14.15\ \text{lbmol/hr}$$

The mole fraction of water in the mixture is

$$x_w = \frac{\dot{n}_w}{\dot{n}_a + \dot{n}_w} = \frac{2.19\ \frac{\text{lbmol}}{\text{hr}}}{14.15\ \frac{\text{lbmol}}{\text{hr}} + 2.19\ \frac{\text{lbmol}}{\text{hr}}}$$
$$= 0.134$$

The partial pressure of water vapor is

$$p_w = xp_{\text{chamber}} = (0.134)(20\ \text{psia}) = 2.68\ \text{psia}$$

From App. 23.A, the saturation pressure at 350°F is $p_{\text{sat}} = 134.63$ psia.

The relative humidity is

$$\phi = \frac{p_w}{p_{\text{sat}}} = \frac{2.68\ \text{psia}}{134.63\ \text{psia}} = \boxed{0.020\quad (2.0\%)}$$

The answer is (B).

SI Solution

(a) The absolute temperatures are

$$T_1 = 427°\text{C} + 273° = 700\text{K}$$
$$T_2 = 177°\text{C} + 273° = 450\text{K}$$

Air tables can be used at low pressures. From App. 23.S,

$$h_1 = 713.27\ \text{kJ/kg}$$
$$h_2 = 451.80\ \text{kJ/kg}$$

From App. 23.N, the enthalpy of water at 27°C is $h_{w,1} = 113.19$ kJ/kg (113 kJ/kg).

From Eq. 38.19(a), the enthalpy of steam at 177°C is

$$h_{w,2} = \left(1.805\ \frac{\text{kJ}}{\text{kg·°C}}\right)(177°\text{C}) + 2501\ \frac{\text{kJ}}{\text{kg}}$$
$$= 2820.5\ \text{kJ/kg}$$

HVAC

Air temperature is reduced from 427°C to 177°C, and this energy is used to change water at 27°C to steam at 177°C. From the energy balance equation,

$$\begin{aligned}\dot{m}_w &= \frac{\dot{m}_a(h_1 - h_2)}{h_{w,2} - h_{w,1}} \\ &= \frac{\left(0.052\ \frac{\text{kg}}{\text{s}}\right)\left(713.27\ \frac{\text{kJ}}{\text{kg}} - 451.80\ \frac{\text{kJ}}{\text{kg}}\right)}{2820.5\ \frac{\text{kJ}}{\text{kg}} - 113.19\ \frac{\text{kJ}}{\text{kg}}} \\ &= \boxed{0.00502\ \text{kg/s}\quad(0.0050\ \text{kg/s})}\end{aligned}$$

The answer is (D).

(b) The number of moles of water evaporated (in the air mixture) is

$$\dot{n}_w = \frac{0.00502\ \frac{\text{kg}}{\text{s}}}{18.016\ \frac{\text{kg}}{\text{kmol}}} = 2.79 \times 10^{-4}\ \text{kmol/s}$$

The number of moles of air in the mixture at the exit is

$$\dot{n}_a = \frac{0.052\ \frac{\text{kg}}{\text{s}}}{28.967\ \frac{\text{kg}}{\text{kmol}}} = 1.80 \times 10^{-3}\ \text{kmol/s}$$

The mole fraction of water in the mixture is

$$\begin{aligned}x_w &= \frac{\dot{n}_w}{\dot{n}_a + \dot{n}_w} \\ &= \frac{2.79 \times 10^{-4}\ \frac{\text{kmol}}{\text{s}}}{1.80 \times 10^{-3}\ \frac{\text{kmol}}{\text{s}} + 2.79 \times 10^{-4}\ \frac{\text{kmol}}{\text{s}}} \\ &= 0.134\end{aligned}$$

The partial pressure of water vapor is

$$\begin{aligned}p_w &= xp_{\text{chamber}} = (0.134)(140\ \text{kPa}) \\ &= 18.76\ \text{kPa}\end{aligned}$$

From App. 23.N, the saturation pressure at 177°C is

$$p_{\text{sat}} = (9.368\ \text{bar})\left(100\ \frac{\text{kPa}}{\text{bar}}\right) = 936.8\ \text{kPa}$$

The relative humidity is

$$\phi = \frac{p_w}{p_{\text{sat}}} = \frac{18.76\ \text{kPa}}{936.8\ \text{kPa}} = \boxed{0.020\quad(2.0\%)}$$

The answer is (B).

13. *Method 1*: Use a high temperature psychrometric chart (such as App. 38.D). Locate the point where the 180°F line intersects the 20% relative humidity line. From that point, move horizontally to the left to intersect the saturation curve. Read a dew point temperature of $\boxed{116°\text{F}\ (120°\text{F}).}$

Method 2: Interpolating from App. 23.B, the saturation pressure, p_{sat}, of 180°F water vapor is 7.515 psia. Use Eq. 38.9 to find the partial pressure of the water vapor, p_w. The relative humidity, ϕ, is given as 20%.

$$\phi = \frac{p_w}{p_{\text{sat}}}$$

$$\begin{aligned}p_w &= \phi p_{\text{sat}} = (0.20)\left(7.515\ \frac{\text{lbf}}{\text{in}^2}\right) \\ &= 1.5\ \text{lbf/in}^2\end{aligned}$$

From App. 23.B, the dew point temperature corresponding to 1.5 lbf/in^2 is $\boxed{115.64°\text{F}\ (120°\text{F}).}$

The answer is (B).

14. With a pressure of 19 psia, standard conditions do not apply, so a high-temperature psychrometric chart cannot not be used.

There are 7000 grains per pound. The humidity ratio expressed in lbm/lbm is

$$\omega = \frac{560\ \frac{\text{gr}}{\text{lbm}}}{7000\ \frac{\text{gr}}{\text{lbm}}} = 0.08\ \frac{\text{lbm}}{\text{lbm}}$$

The total pressure, p_t, is 19 psia. From Eq. 38.16, the partial pressure of the water vapor in the combustion products is

$$\begin{aligned}p_w &= \frac{p_t\,\omega}{0.622 + \omega} = \frac{(19\ \text{psia})\left(0.08\ \frac{\text{lbm}}{\text{lbm}}\right)}{0.622 + 0.08\ \frac{\text{lbm}}{\text{lbm}}} \\ &= 2.17\ \text{psia}\end{aligned}$$

The dew point is found from steam tables (see App. 23.B) as the saturation temperature corresponding to the partial pressure of the water vapor. Use 2.0 psia for convenience. The saturation temperature of 2.0 psia water vapor is $\boxed{126.03°\text{F}\ (130°\text{F}).}$

The answer is (D).

HVAC

39 Cooling Towers and Fluid Coolers

PRACTICE PROBLEMS

1. (*Time limit: one hour*) An evaporative counterflow air cooling tower removes 1×10^6 Btu/hr (290 kW) from a water flow. The temperature of the water is reduced from 120°F to 110°F (49°C to 43°C). Air enters the cooling tower at 91°F (33°C) and 60% relative humidity, and air leaves at 100°F (38°C) and 82% relative humidity.

(a) Calculate the air flow rate.

(b) Calculate the quantity of makeup water.

SOLUTIONS

1. *Customary U.S. Solution*

(a) The cooled water flow rate is given by

$$Q = \dot{m}_w c_p \Delta T$$

$$1 \times 10^6 \text{ Btu} = \dot{m}_w \left(1.0 \; \frac{\text{Btu}}{\text{lbm-°F}}\right)(120\text{°F} - 110\text{°F})$$

$$\dot{m}_w = 1.0 \times 10^5 \text{ lbm/hr}$$

From the psychrometric chart (see App. 38.A), for air in at $T_{\text{db}} = 91\text{°F}$ and $\phi = 60\%$,

$$h_{\text{in}} \approx 42.7 \text{ Btu/lbm}$$

$$\omega_{\text{in}} = 0.0190 \text{ lbm moisture/lbm air}$$

For air out, the normal psychrometric chart (offscale) cannot be used. Use Eq. 38.11 to find the humidity ratio and Eq. 38.17, Eq. 38.18(b), and Eq. 38.19(b) to calculate the enthalpy of air. (Appendix 38.D could also be used as a simpler solution.)

From App. 23.A, the saturated steam pressure at 100°F is 0.9505 psia.

$$p_w = \phi p_{\text{sat}} = (0.82)(0.9505 \text{ psia}) = 0.7794 \text{ psia}$$

From Eq. 38.1,

$$p_a = p - p_w = 14.696 \text{ psia} - 0.7794 \text{ psia} = 13.9166 \text{ psia}$$

From Eq. 38.11, the humidity ratio for air out is

$$\phi = 1.608\omega_{\text{out}}\left(\frac{p_a}{p_{\text{sat}}}\right)$$

$$0.82 = 1.608\omega_{\text{out}}\left(\frac{13.9166 \text{ psia}}{0.9505 \text{ psia}}\right)$$

$$\omega_{\text{out}} = 0.0348 \text{ lbm moisture/lbm air}$$

HVAC

From Eq. 38.17, Eq. 38.18(b), and Eq. 38.19(b), the enthalpy of air out is

$$\begin{aligned} h_2 &= h_a + \omega_2 h_w \\ &= \left(0.240\ \frac{\text{Btu}}{\text{lbm-°F}}\right) T_{2,\text{°F}} + \omega_2 \\ &\quad \times \left(\left(0.444\ \frac{\text{Btu}}{\text{lbm-°F}}\right) T_{2,\text{°F}} + 1061\ \frac{\text{Btu}}{\text{lbm}}\right) \\ &= \left(0.240\ \frac{\text{Btu}}{\text{lbm-°F}}\right)(100\text{°F}) \\ &\quad + \left(0.0348\ \frac{\text{lbm moisture}}{\text{lbm air}}\right) \left(\begin{array}{c} \left(0.444\ \frac{\text{Btu}}{\text{lbm-°F}}\right) \\ \times (100\text{°F}) \\ + 1061\ \frac{\text{Btu}}{\text{lbm}} \end{array}\right) \\ &= 62.47\ \text{Btu/lbm air} \end{aligned}$$

The mass flow rate of air can be determined from Eq. 38.25.

$$q = \dot{m}_a (h_2 - h_1)$$

$$\dot{m}_{\text{air}} = \frac{q}{h_2 - h_1} = \frac{1 \times 10^6\ \frac{\text{Btu}}{\text{hr}}}{62.47\ \frac{\text{Btu}}{\text{lbm air}} - 42.7\ \frac{\text{Btu}}{\text{lbm air}}} = \boxed{5.058 \times 10^4\ \text{lbm air/hr}}$$

(b) From conservation of water vapor,

$$\begin{aligned} \omega_1 \dot{m}_{\text{air}} + \dot{m}_{\text{make-up}} &= \omega_2 \dot{m}_{\text{air}} \\ \dot{m}_{\text{make-up}} &= \dot{m}_{\text{air}} (\omega_{\text{out}} - \omega_{\text{in}}) \\ &= \left(5.058 \times 10^4\ \frac{\text{lbm air}}{\text{hr}}\right) \\ &\quad \times \left(\begin{array}{c} 0.0348\ \frac{\text{lbm moisture}}{\text{lbm air}} \\ - 0.0191\ \frac{\text{lbm moisture}}{\text{lbm air}} \end{array}\right) \\ &= \boxed{794\ \text{lbm water/hr}} \end{aligned}$$

SI Solution

(a) The cooled water flow rate is given by

$$\begin{aligned} Q &= \dot{m}_w c_p \Delta T \\ 290\ \text{kW} &= \dot{m}_w \left(4.187\ \frac{\text{kJ}}{\text{kg·°C}}\right)(49\text{°C} - 43\text{°C}) \\ \dot{m}_w &= 11.54\ \text{kg/s} \end{aligned}$$

From the psychrometric chart (see App. 38.B), for air in at $T_{\text{db}} = 33\text{°C}$ and $\phi = 60\%$,

$$\begin{aligned} h_{\text{in}} &= 82.3\ \text{kJ/kg air} \\ \omega_{\text{in}} &= \frac{19.2\ \frac{\text{g moisture}}{\text{kg air}}}{1000\ \frac{\text{g}}{\text{kg}}} \\ &= 0.0192\ \text{kg moisture/kg air} \end{aligned}$$

For air out, the psychrometric chart (off scale) cannot be used. Use Eq. 38.11 to find the humidity ratio and Eq. 38.17, Eq. 38.18(a), and Eq. 38.19(a) to calculate enthalpy of air.

From App. 23.N, the saturated steam pressure at 38°C is 0.06633 bars.

$$p_w = \phi p_{\text{sat}} = (0.82)(0.06633\ \text{bar}) = 0.05439\ \text{bar}$$

From Eq. 38.1,

$$p_a = p - p_w = 1\ \text{bar} - 0.05439\ \text{bar} = 0.94561\ \text{bar}$$

From Eq. 38.11, the humidity ratio for air out is

$$\begin{aligned} \phi &= 1.608 \omega_{\text{out}} \left(\frac{p_a}{p_{\text{sat}}}\right) \\ 0.82 &= 1.608 \omega_{\text{out}} \left(\frac{0.94561\ \text{bar}}{0.06633\ \text{bar}}\right) \\ \omega_{\text{out}} &= 0.0358\ \text{kg moisture/kg air} \end{aligned}$$

From Eq. 38.17, Eq. 38.18(a), and Eq. 38.19(a), the enthalpy of air out is

$$\begin{aligned} h_2 &= h_a + \omega_2 h_w \\ &= \left(1.005\ \frac{\text{kJ}}{\text{kg·°C}}\right) T_{\text{°C}} \\ &\quad + \omega_{\text{out}} \left(\left(1.805\ \frac{\text{kJ}}{\text{kg·°C}}\right) T_{\text{°C}} + 2501\ \frac{\text{kJ}}{\text{kg}}\right) \\ &= \left(1.005\ \frac{\text{kJ}}{\text{kg·°C}}\right)(38\text{°C}) + \left(0.0358\ \frac{\text{kg moisture}}{\text{kg air}}\right) \\ &\quad \times \left(\left(1.805\ \frac{\text{kJ}}{\text{kg·°C}}\right)(38\text{°C}) + 2501\ \frac{\text{kJ}}{\text{kg}}\right) \\ &= 130.2\ \text{kJ/kg air} \end{aligned}$$

The mass flow rate of air can be determined from Eq. 38.25.

$$q = \dot{m}_a(h_2 - h_1)$$

$$\dot{m}_{\text{air}} = \frac{q}{h_2 - h_1} = \frac{290 \text{ kW}}{130.2 \ \frac{\text{kJ}}{\text{kg air}} - 82.3 \ \frac{\text{kJ}}{\text{kg air}}} = \boxed{6.054 \text{ kg air/s}}$$

(b) From conservation of water vapor,

$$\omega_1 \dot{m}_{\text{air}} + \dot{m}_{\text{make-up}} = \omega_2 \dot{m}_{\text{air}}$$

$$\begin{aligned} \dot{m}_{\text{make-up}} &= \dot{m}_{\text{air}}(\omega_{\text{out}} - \omega_{\text{in}}) \\ &= \left(6.054 \ \frac{\text{kg air}}{\text{s}}\right) \times \left(\begin{array}{l} 0.0358 \ \frac{\text{kg moisture}}{\text{kg air}} \\ \quad - 0.0192 \ \frac{\text{kg moisture}}{\text{kg air}} \end{array}\right) \\ &= \boxed{0.100 \text{ kg water/s}} \end{aligned}$$

HVAC

40 Ventilation

PRACTICE PROBLEMS

1. An office room has floor dimensions of 60 ft by 95 ft (18 m by 29 m) and a ceiling height of 10 ft (3 m). Cool air is supplied from ceiling diffusers. 45 people occupy the office.

(a) What is most nearly the ventilation rate based on six air changes per hour?

(A) 1300 ft^3/min (35 m^3/min)

(B) 5700 ft^3/min (160 m^3/min)

(C) 13,000 ft^3/min (350 m^3/min)

(D) 34,000 ft^3/min (900 m^3/min)

(b) What is most nearly the ventilation rate at the breathing zone based on ASHRAE Standard 62.1?

(A) 520 ft^3/min (250 L/s)

(B) 570 ft^3/min (270 L/s)

(C) 700 ft^3/min (330 L/s)

(D) 850 ft^3/min (400 L/s)

2. 150 ppm of methanol (TLV = 200 ppm; MW = 32.04; SG = 0.792) and 285 ppm of methylene chloride (TLV = 500 ppm; MW = 84.94; SG = 1.336) are found in the air in a plating booth. Two pints (1 L) of each are evaporated per hour. Use a mixing safety factor (i.e., a *K*-value) of 6. The minimum ventilation rate required for dilution is most nearly

(A) 2500 ft^3/min (1200 L/s)

(B) 5200 ft^3/min (2500 L/s)

(C) 10,000 ft^3/min (4800 L/s)

(D) 12,000 ft^3/min (6200 L/s)

3. An auditorium is designed to seat 4500 people. The ventilation rate is 1.62×10^7 ft^3/hr (4.54×10^5 m^3/h) of outside air. The outside temperature is 0°F (−18°C) dry-bulb, and the outside pressure is 14.6 psia (100.6 kPa). Air leaves the auditorium at 70°F (21°C) dry-bulb. There is no recirculation. The furnace has a capacity of 1,250,000 Btu/hr (370 kW).

(a) The temperature at which the air should enter the auditorium is most nearly

(A) 52°F (11°C)

(B) 55°F (13°C)

(C) 62°F (17°C)

(D) 67°F (19°C)

(b) Approximately how much heat should be supplied to the ventilation air?

(A) 1.5×10^7 Btu/hr (4.5 MW)

(B) 2.2×10^7 Btu/hr (6.5 MW)

(C) 3.9×10^7 Btu/hr (12 MW)

(D) 5.1×10^7 Btu/hr (16 MW)

(c) Has the furnace been sized properly?

(A) The furnace is less than half the required capacity.

(B) The furnace is more than half the required capacity.

(C) The furnace is less than twice the required capacity.

(D) The furnace is more than twice the required capacity.

4. A room is maintained at design conditions of 75°F (23.9°C) dry-bulb and 50% relative humidity. The air outside is at 95°F (35°C) dry-bulb and 75°F (23.9°C) wet-bulb. The outside air is conditioned and mixed with some room exhaust air. The mixed, conditioned air enters the room and increases 20°F (11.1°C) in temperature before being removed from the room. The sensible and latent loads are 200,000 Btu/hr (60 kW) and 50,000 Btu/hr (15 kW), respectively. Air leaves the coil at 50.8°F (10°C). The volume of air flowing through the coil is most nearly

(A) 4200 ft^3/min (120 m^3/min)

(B) 5800 ft^3/min (160 m^3/min)

(C) 7600 ft^3/min (220 m^3/min)

(D) 9300 ft^3/min (260 m^3/min)

5. According to ASHRAE Standard 62.1, the occupant diversity as it applies to the calculation of outdoor air in multi-zone systems accounts for the

(A) variation in total system occupancy over time

(B) variation in zonal occupancy over time

(C) fractional mixture of women and children in the system population

(D) likelihood of simultaneous peak zone populations

6. Modern ventilation standards consider both population and floor area when calculating outdoor air requirements because

(A) exact occupancy numbers are difficult to predict

(B) contaminants originate from both the population and the floor

(C) the floor area component ensures a minimum ventilation rate

(D) modern buildings are more "tight," and infiltration cannot be counted on

7. How should the pressurization (relative to the surroundings) in clean rooms and laboratories be maintained?

(A) Clean rooms and laboratories should both be at positive pressures.

(B) Clean rooms and laboratories should both be at negative pressures.

(C) Clean rooms should be at positive pressures, while laboratories should be at negative pressures.

(D) Clean rooms should be at negative pressures, while laboratories should be at positive pressures.

8. Assuming atmospheric air is a mixture of oxygen and nitrogen only, above what approximate ambient volumetric nitrogen concentration should supplemental oxygen be provided by an employer to employees?

(A) 77% nitrogen

(B) 79% nitrogen

(C) 81% nitrogen

(D) 88% nitrogen

9. The volumetric fraction of carbon dioxide in a submarine is measured to be 0.06%. What is most nearly the concentration measured in parts per million (ppm)?

(A) 0.6 ppm

(B) 60 ppm

(C) 600 ppm

(D) 6000 ppm

SOLUTIONS

1. *Customary U.S. Solution*

(a) The office volume is

$$V = (60 \text{ ft})(95 \text{ ft})(10 \text{ ft}) = 57{,}000 \text{ ft}^3$$

Based on six air changes per hour, the flow rate is

$$\dot{V} = \frac{\left(57{,}000 \ \frac{\text{ft}^3}{\text{air change}}\right)\left(6 \ \frac{\text{air changes}}{\text{hr}}\right)}{60 \ \frac{\text{min}}{\text{hr}}} = \boxed{5700 \text{ ft}^3/\text{min}}$$

The answer is (B).

(b) The floor area is

$$A = (60 \text{ ft})(95 \text{ ft}) = 5700 \text{ ft}^2$$

From Table 40.1 (office building: offices), $R_p = 5 \text{ cfm/person}$, and $R_a = 0.06 \text{ cfm/ft}^2$. From Eq. 40.1,

$$\begin{aligned}\dot{V}_{\text{bz}} &= R_p P_z + R_a A_z \\ &= \left(5 \ \frac{\text{ft}^3}{\text{min-person}}\right)(45 \text{ people}) \\ &\quad + \left(0.06 \ \frac{\text{ft}^3}{\text{min-ft}^2}\right)(5700 \text{ ft}^2) \\ &= 567 \text{ ft}^3/\text{min}\end{aligned}$$

Since cool air is supplied from the ceiling, from Table 40.2, $E_z = 1.0$. Using Eq. 40.2, the total outdoor air to the zone is

$$\dot{V}_{\text{oz}} = \frac{\dot{V}_{\text{bz}}}{E_z} = \frac{567 \ \frac{\text{ft}^3}{\text{min}}}{1.0} = \boxed{567 \text{ ft}^3/\text{min} \quad (570 \text{ ft}^3/\text{min})}$$

The answer is (B).

SI Solution

(a) The office volume is

$$V = (18 \text{ m})(29 \text{ m})(3 \text{ m}) = 1566 \text{ m}^3$$

Based on six air changes per hour, the flow rate is

$$\dot{V} = \frac{\left(1566 \ \frac{\text{m}^3}{\text{air change}}\right)\left(6 \ \frac{\text{air changes}}{\text{h}}\right)}{60 \ \frac{\text{min}}{\text{h}}} = \boxed{156.6 \text{ m}^3/\text{min} \quad (160 \text{ m}^3/\text{min})}$$

The answer is (B).

(b) The floor area is

$$A = (18 \text{ m})(29 \text{ m}) = 522 \text{ m}^2$$

From Table 40.1 (office building: offices), $R_p = 2.5 \text{ L/s·person}$, and $R_a = 0.3 \text{ L/s·m}^2$. From Eq. 40.1,

$$\begin{aligned}\dot{V}_{\text{bz}} &= R_p P_z + R_a A_z \\ &= \left(2.5 \ \frac{\text{L}}{\text{s·person}}\right)(45 \text{ people}) \\ &\quad + \left(0.3 \ \frac{\text{L}}{\text{s·m}^2}\right)(522 \text{ m}^2) \\ &= 269.1 \text{ L/s}\end{aligned}$$

Since cool air is supplied from the ceiling, from Table 40.2, $E_z = 1.0$. Using Eq. 40.2, the total outdoor air to the zone is

$$\dot{V}_{\text{oz}} = \frac{\dot{V}_{\text{bz}}}{E_z} = \frac{269.1 \ \frac{\text{L}}{\text{s}}}{1.0} = \boxed{269.1 \text{ L/s} \quad (270 \text{ L/s})}$$

The answer is (B).

2. *Customary U.S. Solution*

Use Eq. 40.25(b) to find the required dilution ventilation rate. The maximum concentration, C, is equal to the threshold limit value, TLV.

$$\dot{V}_{\text{ft}^3/\text{min}} = \frac{(4.03 \times 10^8) K (\text{SG}) R_{\text{pints/min}}}{(\text{MW})\text{TLV}_{\text{ppm}}}$$

For the methanol,

$$\begin{aligned}\dot{V}_{\text{ft}^3/\text{min}} &= \frac{(4.03 \times 10^8)(6)(0.792)\left(2 \ \frac{\text{pints}}{\text{hr}}\right)}{(32.04)(200 \text{ ppm})\left(60 \ \frac{\text{min}}{\text{hr}}\right)} \\ &= 9962 \text{ ft}^3/\text{min}\end{aligned}$$

For the methylene chloride,

$$\begin{aligned}\dot{V}_{\text{ft}^3/\text{min}} &= \frac{(4.03 \times 10^8)(6)(1.336)\left(2 \ \frac{\text{pints}}{\text{hr}}\right)}{(84.94)(500 \text{ ppm})\left(60 \ \frac{\text{min}}{\text{hr}}\right)} \\ &= 2535 \text{ ft}^3/\text{min}\end{aligned}$$

The total dilution ventilation rate is

$$\begin{aligned}&9962 \ \frac{\text{ft}^3}{\text{min}} + 2535 \ \frac{\text{ft}^3}{\text{min}} \\ &= \boxed{12{,}497 \text{ ft}^3/\text{min} \quad (12{,}000 \text{ ft}^3/\text{min})}\end{aligned}$$

The answer is (D).

SI Solution

Use Eq. 40.25(a) to find the required dilution ventilation rate. The maximum concentration, C, is equal to the threshold limit value, TLV.

$$\dot{V}_{\text{L/s}} = \frac{(4.02 \times 10^8)K(\text{SG})R_{\text{L/min}}}{(\text{MW})C_{\text{max}}}$$

For the methanol,

$$\begin{aligned}\dot{V}_{\text{L/s}} &= \frac{(4.02 \times 10^8)(6)(0.792)\left(1\ \frac{\text{L}}{\text{h}}\right)}{(32.04)(200\ \text{ppm})\left(60\ \frac{\text{min}}{\text{h}}\right)} \\ &= 4969\ \text{L/s}\end{aligned}$$

For the methylene chloride,

$$\begin{aligned}\dot{V}_{\text{L/s}} &= \frac{(4.02 \times 10^8)(6)(1.336)\left(1\ \frac{\text{L}}{\text{h}}\right)}{(84.94)(500\ \text{ppm})\left(60\ \frac{\text{min}}{\text{h}}\right)} \\ &= 1265\ \text{L/s}\end{aligned}$$

The total dilution ventilation rate is

$$4969\ \frac{\text{L}}{\text{s}} + 1265\ \frac{\text{L}}{\text{s}} = \boxed{6234\ \text{L/s} \quad (6200\ \text{L/s})}$$

The answer is (D).

3. *Customary U.S. Solution*

(a) The absolute temperature of outside air is

$$T = 0°\text{F} + 460° = 460°\text{R}$$

From the ideal gas law, the density of outside air is

$$\begin{aligned}\rho = \frac{p}{RT} &= \frac{\left(14.6\ \frac{\text{lbf}}{\text{in}^2}\right)\left(12\ \frac{\text{in}}{\text{ft}}\right)^2}{\left(53.35\ \frac{\text{ft-lbf}}{\text{lbm-°R}}\right)(460°\text{R})} \\ &= 0.08567\ \text{lbm/ft}^3\end{aligned}$$

The mass flow rate is

$$\begin{aligned}\dot{m} = \dot{V}\rho &= \left(1.62 \times 10^7\ \frac{\text{ft}^3}{\text{hr}}\right)\left(0.08567\ \frac{\text{lbm}}{\text{ft}^3}\right) \\ &= 1.388 \times 10^6\ \text{lbm/hr}\end{aligned}$$

Assume there can be no latent heat (no moisture) at 0°F.

From Table 40.4, the sensible heat generated by each person seated in the theater is 225 Btu/hr. Therefore,

$$\begin{aligned}\dot{q}_{\text{in from people}} &= \left(225\ \frac{\frac{\text{Btu}}{\text{hr}}}{\text{person}}\right)(4500\ \text{persons}) \\ &= 1.01 \times 10^6\ \text{Btu/hr}\end{aligned}$$

The air leaves the auditorium at 70°F. From App. 35.C, the specific heat of dry air is 0.240 Btu/lbm-°F. (This remains fairly constant over normal temperature ranges.) Since $\dot{q}$ is known, the air temperature entering the auditorium is

$$\dot{q} = \dot{m}c_p(T_{\text{out,air}} - T_{\text{in,air}})$$

$$\begin{aligned}T_{\text{in,air}} &= T_{\text{out,air}} - \frac{\dot{q}}{\dot{m}c_p} \\ &= 70°\text{F} - \frac{1.01 \times 10^6\ \frac{\text{Btu}}{\text{hr}}}{\left(1.388 \times 10^6\ \frac{\text{lbm}}{\text{hr}}\right)\left(0.240\ \frac{\text{Btu}}{\text{lbm-°F}}\right)} \\ &= \boxed{67°\text{F}}\end{aligned}$$

The answer is (D).

(b) The heat needed to heat dry ventilation air from 0°F to 67°F is

$$\begin{aligned}\dot{q} &= \dot{m}c_p\Delta T \\ &= \left(1.388 \times 10^6\ \frac{\text{lbm}}{\text{hr}}\right)\left(0.240\ \frac{\text{Btu}}{\text{lbm-°F}}\right)(67°\text{F} - 0°\text{F}) \\ &= \boxed{2.23 \times 10^7\ \text{Btu/hr} \quad (2.2 \times 10^7\ \text{Btu/hr})}\end{aligned}$$

The answer is (B).

(c) Since 1.25×10^6 Btu/hr $< 2.23 \times 10^7$ Btu/hr, the furnace is $\boxed{\text{less than half}}$ the required capacity, so it is too small.

The answer is (A).

SI Solution

(a) The absolute temperature of outside air is

$$T = -18°\text{C} + 273° = 255\text{K}$$

From the ideal gas law, the density of outside air is

$$\rho = \frac{p}{RT} = \frac{(100.6\ \text{kPa})\left(1000\ \frac{\text{Pa}}{\text{kPa}}\right)}{\left(287.03\ \frac{\text{J}}{\text{kg·K}}\right)(255\text{K})} = 1.374\ \text{kg/m}^3$$

The mass flow rate is

$$\dot{m} = \rho\dot{V} = \left(1.374\ \frac{\text{kg}}{\text{m}^3}\right)\left(4.54 \times 10^5\ \frac{\text{m}^3}{\text{h}}\right)$$
$$= 6.238 \times 10^5\ \text{kg/h}$$

Assume there can be no latent heat (no moisture) at −18°C.

From Table 40.4 and the table footnote, the sensible heat generated by people seated in the theater is

$$\left(225\ \frac{\frac{\text{Btu}}{\text{h}}}{\text{person}}\right)\left(0.293\ \frac{\text{W}}{\frac{\text{Btu}}{\text{h}}}\right) = 65.93\ \text{W/person}$$

Therefore,

$$\dot{q}_{\text{in from people}} = \frac{\left(65.93\ \frac{\text{W}}{\text{person}}\right)(4500\ \text{persons})}{1000\ \frac{\text{W}}{\text{kW}}}$$
$$= 296.7\ \text{kW}$$

The air leaves the auditorium at 21°C. From App. 35.D, the specific heat of dry air is 1.0048 kJ/kg·K. (This remains fairly constant over normal temperature ranges.) Since $\dot{q}$ is known, the air temperature entering the auditorium is

$$\dot{q} = \dot{m}c_p(T_{\text{out,air}} - T_{\text{in,air}})$$
$$T_{\text{in,air}} = T_{\text{out,air}} - \frac{\dot{q}}{\dot{m}c_p}$$
$$= 21°\text{C} - \frac{(296.7\ \text{kW})\left(3600\ \frac{\text{s}}{\text{h}}\right)}{\left(6.238 \times 10^5\ \frac{\text{kg}}{\text{h}}\right)\left(1.0048\ \frac{\text{kJ}}{\text{kg·K}}\right)}$$
$$= \boxed{19.3°\text{C}\quad(19°\text{C})}$$

The answer is (D).

(b) The heat needed to heat dry ventilation air from −18°C to 19.3°C is

$$\dot{q} = \dot{m}c_p\Delta T$$
$$= \frac{\left(6.238 \times 10^5\ \frac{\text{kg}}{\text{h}}\right)\left(1.0048\ \frac{\text{kJ}}{\text{kg·°C}}\right) \times (19.3°\text{C} - (-18°\text{C}))}{3600\ \frac{\text{s}}{\text{h}}}$$
$$= \boxed{6494.28\ \text{kW}\quad(6.5\ \text{MW})}$$

The answer is (B).

(c) Since 370 kW < 6494 kW, the furnace is $\boxed{\text{less than half}}$ the required capacity, so it is too small.

The answer is (A).

4. *Customary U.S. Solution*

The mixed conditioned air enters the room at

$$T_{\text{db,in}} = 75°\text{F} - 20°\text{F} = 55°\text{F}$$

From Eq. 40.20(b), the volumetric flow rate of air entering the room is

$$\dot{V}_{\text{in,ft}^3/\text{min}} = \frac{\dot{q}_{s,\text{Btu/hr}}}{\left(1.08\ \frac{\text{Btu-min}}{\text{ft}^3\text{-hr-°F}}\right)(T_{\text{id,°F}} - T_{\text{in,°F}})}$$
$$= \frac{200{,}000\ \frac{\text{Btu}}{\text{hr}}}{\left(1.08\ \frac{\text{Btu-min}}{\text{ft}^3\text{-hr-°F}}\right)(75°\text{F} - 55°\text{F})}$$
$$= 9259\ \text{ft}^3/\text{min}$$

This is a mixing problem.

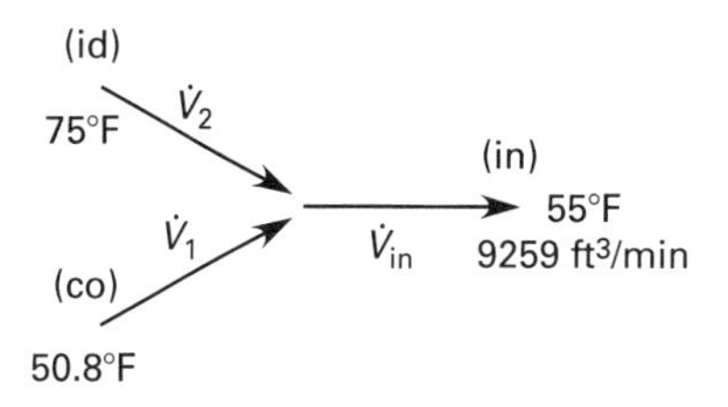

Using the lever rule, and since the temperature scales are all linear, the fraction of air passing through the coil is

$$\frac{T_{\text{id}} - T_{\text{in}}}{T_{\text{id}} - T_{\text{co}}} = \frac{75°\text{F} - 55°\text{F}}{75°\text{F} - 50.8°\text{F}} = 0.826$$
$$\dot{V}_1 = (0.826)\left(9259\ \frac{\text{ft}^3}{\text{min}}\right)$$
$$= \boxed{7648\ \text{ft}^3/\text{min}\quad(7600\ \text{ft}^3/\text{min})}$$

The answer is (C).

SI Solution

The mixed conditioned air enters the room at

$$T_{\text{db,in}} = 23.9°\text{C} - 11.1°\text{C} = 12.8°\text{C}$$

HVAC

From Eq. 40.20(a),

$$\dot{V}_{\text{in,m}^3/\text{min}} = \frac{\dot{q}_{s,\text{kW}}}{\left(0.02 \ \frac{\text{kJ}\cdot\text{min}}{\text{m}^3\cdot\text{s}\cdot{}^\circ\text{C}}\right)(T_{\text{id},^\circ\text{C}} - T_{\text{in},^\circ\text{C}})}$$
$$= \frac{60 \ \text{kW}}{\left(0.02 \ \frac{\text{kJ}\cdot\text{min}}{\text{m}^3\cdot\text{s}\cdot{}^\circ\text{C}}\right)(23.9^\circ\text{C} - 12.8^\circ\text{C})}$$
$$= 270.3 \ \text{m}^3/\text{min}$$

This is a mixing problem.

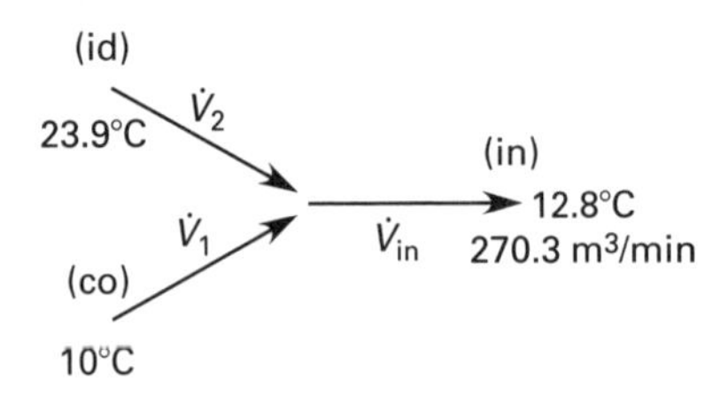

Using the lever rule, and since the temperature scales are all linear, the fraction of air passing through the coil is

$$\frac{T_{\text{id}} - T_{\text{in}}}{T_{\text{id}} - T_{\text{co}}} = \frac{23.9^\circ\text{C} - 12.8^\circ\text{C}}{23.9^\circ\text{C} - 10^\circ\text{C}} = 0.799$$
$$\dot{V}_1 = (0.799)\left(270.3 \ \frac{\text{m}^3}{\text{min}}\right)$$
$$= \boxed{216.0 \ \text{m}^3/\text{min} \quad (220 \ \text{m}^3/\text{min})}$$

The answer is (C).

5. The occupancy diversity, D, is the ratio of the total system population to the sum of the zonal peak populations. Diversity will have a value of 1.0 if all zones simultaneously operate at peak occupancy, in which case, the total system population will be the same as the sum of the zonal peak populations. In most cases, the zones will experience peak occupancies at different times, and the diversity will be less than 1.0.

The answer is (D).

6. The occupants (i.e., the population) and the building both contribute contaminants independently to the occupied space. Occupant contaminants include moisture, heat, odors, carbon dioxide, and in some cases, tobacco smoke. Buildings produce odors, dust, organic vapors, and mold and fungi spores.

The answer is (B).

7. Clean rooms should be maintained at positive pressures with filtered forced air in order to prevent the incursion of airborne particles from the atmosphere. The intake air is filtered and cleaned, and the interior of the clean room is protected. Most laboratories should be maintained at negative pressures in order to keep airborne hazardous materials in the air distribution system. The exhaust is filtered and cleaned, and the environment is protected.

The answer is (C).

8. OSHA defines an oxygen-deficient atmosphere as one having less than 19.5% oxygen. In this scenario, the maximum nitrogen content would be

$$100\% - 19.5\% = \boxed{80.5\% \quad (81\%)}$$

The answer is (C).

9. Air contaminants are measured by volume, hence "ppm" are parts per million by volume (ppmv). In this problem, there are 0.06 volumes of carbon dioxide for every 100 volumes of air (i.e., 0.06 parts per 100 parts). In one million volumes of air, there would be $10^6/10^2 = 10^4$ more volumes of carbon dioxide.

$$(0.06 \ \text{pphv})\left(10^4 \ \frac{\text{ppmv}}{\text{pphv}}\right) = \boxed{600 \ \text{ppmv} \quad (600 \ \text{ppm})}$$

The answer is (C).

HVAC

41 Fans, Ductwork, and Terminal Devices

PRACTICE PROBLEMS

(Note: Round all duct dimensions to the next larger whole inch or multiples of 25 mm.)

1. A round 18 in (457 mm) duct is to be replaced by a rectangular duct with an aspect ratio of 4:1. The dimensions of the rectangular duct are most nearly

(A) 8 in × 30 in (200 mm × 760 mm)

(B) 9 in × 35 in (220 mm × 880 mm)

(C) 12 in × 18 in (300 mm × 460 mm)

(D) 18 in × 72 in (460 mm × 1800 mm)

2. A fan moves 10,000 SCFM (5700 L/s) through ductwork that has a dynamic friction pressure of 4 in wg (1 kPa). If the fan speed is decreased so that the flow rate becomes 8000 SCFM (4700 L/s), the total pressure in the duct after the fan will most nearly be

(A) 2.2 in wg (0.53 kPa)

(B) 2.4 in wg (0.58 kPa)

(C) 2.6 in wg (0.62 kPa)

(D) 2.8 in wg (0.68 kPa)

3. A $^1/_8$ scale model fan is tested at 300 rpm. At standard air conditions and at that speed, the model fan moves 40 ft^3/min (19 L/s) against 0.5 in wg (125 Pa). If a full-sized fan operates at 5000 ft (1500 m) altitude at half the model's speed, the air power is most nearly

(A) 11 hp (8.6 kW)

(B) 14 hp (11 kW)

(C) 19 hp (15 kW)

(D) 24 hp (19 kW)

4. A duct system consists of 750 ft (230 m) of galvanized 20 in (508 mm) diameter duct with four round elbows (radius-to-diameter ratio of 1.5). A 2 in (51 mm) diameter pipe passes perpendicularly through the duct at two locations. The flow rate is 6000 SCFM (4200 L/s). The friction loss is most nearly

(A) 1.7 in wg (1.1 kPa)

(B) 2.9 in wg (2.0 kPa)

(C) 3.7 in wg (2.4 kPa)

(D) 4.8 in wg (3.2 kPa)

5. 1500 SCFM (700 L/s) of air flows in an 18 in (457 mm) diameter round duct. After a branch reduction of 300 SCFM (140 L/s), the fitting reduces to 14 in (356 mm) in the through direction. The static regain in the through direction is most nearly

(A) −0.037 in wg loss (−8.7 Pa loss)

(B) −0.073 in wg loss (−17 Pa loss)

(C) −0.11 in wg loss (−26 Pa loss)

(D) −0.19 in wg loss (−45 Pa loss)

6. A fan in a small theater delivers 1500 ft^3/min (700 L/s) through the system shown. The duct is rectangular with an aspect ratio of 1.5:1 and the shorter side vertical. All elbows have a radius-to-width ratio of 1.5. The terminal pressure at each outlet must be 0.25 in wg (63 Pa) or higher. Takeoff fitting losses are to be disregarded. Use the equal-friction method to size the system.

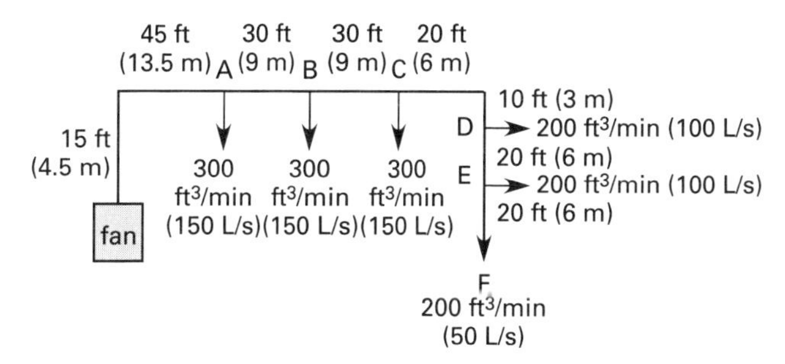

(a) The static pressure at the fan is most nearly

(A) 0.5 in wg (110 Pa)

(B) 0.7 in wg (160 Pa)

(C) 0.8 in wg (200 Pa)

(D) 1.0 in wg (260 Pa)

(b) The total pressure at the fan is most nearly

(A) 1.0 in wg (230 Pa)

(B) 1.2 in wg (280 Pa)

(C) 1.6 in wg (380 Pa)

(D) 2.0 in wg (500 Pa)

7. A fan in a theater delivers 300 ft^3/min (140 L/s) of air through round ducts to each of the twelve outlets shown. The minimum terminal pressure at the outlets is

0.15 in wg (38 Pa). All elbows have a radius-to-diameter ratio of 1.25. Disregard branch takeoff fitting losses. Use the equal-friction method to size the system.

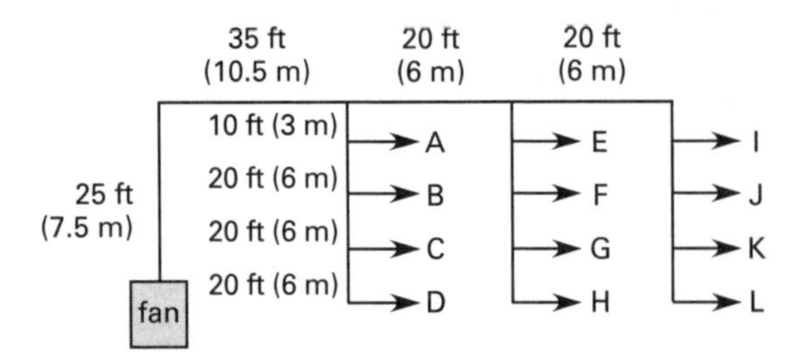

(a) The static pressure at the fan is most nearly

(A) 0.5 in wg (130 Pa)

(B) 0.7 in wg (160 Pa)

(C) 0.8 in wg (210 Pa)

(D) 1.0 in wg (260 Pa)

(b) The total pressure at the fan is most nearly

(A) 0.5 in wg (110 Pa)

(B) 0.7 in wg (170 Pa)

(C) 0.8 in wg (210 Pa)

(D) 1.0 in wg (260 Pa)

8. Size the longest run in Prob. 7 using the static regain method with a regain coefficient of 0.75. (Customary U.S. solution only required.)

(a) The diameter of the main duct is most nearly

(A) 10 in

(B) 20 in

(C) 25 in

(D) 35 in

(b) The equivalent length of the main duct from the fan to the first takeoff and bend is most nearly

(A) 75 ft

(B) 85 ft

(C) 95 ft

(D) 110 ft

(c) The friction loss in the main duct up to the first takeoff is most nearly

(A) 0.12 in wg

(B) 0.14 in wg

(C) 0.15 in wg

(D) 0.17 in wg

(d) The total pressure supplied by the fan is most nearly

(A) 0.14 in wg

(B) 0.27 in wg

(C) 0.29 in wg

(D) 0.32 in wg

9. 3000 ft^3/min (1400 L/s) of air with a density of 0.075 lbm/ft^3 (1.2 kg/m^3) enters a 12 in (305 mm) diameter duct at section A. The four-piece 90° elbows have a radius-to-diameter ratio of 1.5 and an equivalent length of 14 diameters. The Darcy friction factor is 0.02 everywhere in the system. Use the static regain method with a static regain coefficient of 0.65.

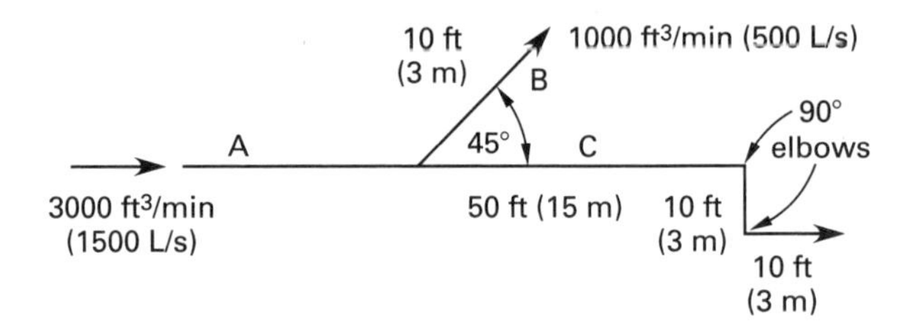

(a) Assuming a 10 in (250 mm) diameter duct, the total equivalent length of run C is most nearly

(A) 75 ft (21 m)

(B) 85 ft (24 m)

(C) 95 ft (29 m)

(D) 110 ft (31 m)

(b) The friction loss in section C is most nearly

(A) 8.0 ft (2.0 m)

(B) 26 ft (9.0 m)

(C) 52 ft (16 m)

(D) 88 ft (27 m)

(c) The diameter at section B is most nearly

(A) 11 in (280 mm)

(B) 14 in (340 mm)

(C) 18 in (460 mm)

(D) 21 in (530 mm)

(d) The regain (in feet of air) in sections B and C is most nearly

(A) 26 ft (8.0 m)

(B) 31 ft (9.0 m)

(C) 34 ft (10 m)

(D) 40 ft (12 m)

(e) Ideally, which sections will require dampers?

(A) no sections

(B) section B only

(C) sections B and C

(D) sections A, B, and C

10. A motor rated at 5 hp delivers 4.3 hp to a fan. The fan rotates at 920 rpm and moves 160°F air at standard pressure into a drying process. The static pressure across the fan is 2.3 in wg. The fan efficiency is 62%. The fan flow rate at standard conditions is most nearly

(A) 5100 ft^3/min

(B) 5800 ft^3/min

(C) 6300 ft^3/min

(D) 7400 ft^3/min

11. Which of the following would NOT normally be used to vary the flow rate of a fan-driven ventilation system?

(A) system dampers

(B) variable pitch blades

(C) inlet vanes

(D) variable speed motors

12. Which of the following parameters is normally sensed in order to control a variable frequency drive (VFD) in a variable air volume (VAV) system?

(A) duct velocity pressure

(B) supply air fan rate of flow

(C) fan motor current

(D) VAV box airflow

13. Fan A and fan B are identical variable-speed fans operating in parallel. They are of the same rating, and they have the same efficiency. Both fans draw from the same source. Fan A always turns at 2000 rev/min and has a volumetric flow rate of 4000 ft^3/min. Fan B's speed is varied according to demand. When fan B turns at 1800 rev/min, its volumetric flow rate is 5000 ft^3/min. At a particular moment, the total flow rate supplied by fans A and B is 10,000 ft^3/min. At that moment, when operating in parallel with fan A, the speed of fan B is most nearly

(A) 1500 rev/min

(B) 2100 rev/min

(C) 2200 rev/min

(D) 4200 rev/min

SOLUTIONS

1. *Customary U.S. Solution*

With an aspect ratio of $R = 4$, the short side is given by Eq. 41.36.

$$\text{short side} = \frac{D(1+R)^{1/4}}{1.3R^{5/8}} = \frac{(18 \text{ in})(1+4)^{1/4}}{(1.3)(4)^{5/8}} = 8.7 \text{ in} \quad (9 \text{ in})$$

From Eq. 41.37, the long side is

$$\text{long side} = R(\text{short side}) = (4)(8.7 \text{ in}) = 34.8 \text{ in} \quad (35 \text{ in})$$

The dimensions of the rectangular duct are $\boxed{9 \text{ in} \times 35 \text{ in.}}$

The answer is (B).

SI Solution

With an aspect ratio of $R = 4$, the short side is given by Eq. 41.36.

$$\text{short side} = \frac{D(1+R)^{1/4}}{1.3R^{5/8}} = \frac{(457 \text{ mm})(1+4)^{1/4}}{(1.3)(4)^{5/8}} = 221 \text{ mm} \quad (220 \text{ mm})$$

From Eq. 41.37, the long side is

$$\text{long side} = R(\text{short side}) = (4)(221 \text{ mm}) = 884 \text{ mm} \quad (880 \text{ mm})$$

The dimensions of the rectangular duct are $\boxed{220 \text{ mm} \times 880 \text{ mm.}}$

The answer is (B).

2. *Customary U.S. Solution*

Use Eq. 41.21 to calculate the friction pressure at the reduced flow rate.

$$\text{FP}_2 = \text{FP}_1\left(\frac{Q_2}{Q_1}\right)^2 = (4 \text{ in wg})\left(\frac{8000 \text{ SCFM}}{10{,}000 \text{ SCFM}}\right)^2 = 2.56 \text{ in wg}$$

The dynamic loss in the duct at the reduced flow rate is 2.56 in wg. However, the pressure in the duct after the fan includes the terminal pressure of 0.1–0.3 in wg. The duct pressure will be 2.7–2.9 in wg. Therefore, the total pressure in the duct after the fan will be $\boxed{2.8 \text{ in wg.}}$

The answer is (D).

HVAC

SI Solution

Use Eq. 41.21 to calculate the friction pressure at the reduced flow rate.

$$\text{FP}_2 = \text{FP}_1\left(\frac{Q_2}{Q_1}\right)^2 = (1\ \text{kPa})\left(\frac{3700\ \frac{\text{L}}{\text{s}}}{4700\ \frac{\text{L}}{\text{s}}}\right)^2 = 0.619\ \text{kPa}$$

The dynamic loss in the duct at the reduced flow rate is 0.619 kPa. However, the pressure in the duct after the fan includes the terminal pressure of 25–75 Pa. The duct pressure will be 0.64–0.69 kPa. Therefore, the total pressure in the duct after the fan will be $\boxed{0.68\ \text{kPa.}}$

The answer is (D).

3. *Customary U.S. Solution*

The air horsepower is given by Eq. 41.13(b).

$$\text{AHP}_1 = \frac{Q_{\text{ft}^3/\text{min}}(\text{TP}_{\text{in wg}})}{6356} = \frac{\left(40\ \frac{\text{ft}^3}{\text{min}}\right)(0.5\ \text{in wg})}{6356\ \frac{\text{in-ft}^3}{\text{hp-min}}}$$

$$= 3.147 \times 10^{-3}\ \text{hp}$$

In order to predict the performance of a dynamically similar fan, use Eq. 41.28.

$$\frac{\text{AHP}_1}{\text{AHP}_2} = \left(\frac{D_1}{D_2}\right)^5\left(\frac{n_1}{n_2}\right)^3\left(\frac{\gamma_1}{\gamma_2}\right)$$

$$\text{AHP}_2 = \text{AHP}_1\left(\frac{D_2}{D_1}\right)^5\left(\frac{n_2}{n_1}\right)^3\left(\frac{\gamma_2}{\gamma_1}\right)$$

$$\frac{D_2}{D_1} = \frac{1}{\frac{1}{8}} = 8$$

$$\frac{n_2}{n_1} = \frac{1}{2}$$

At standard air conditions, $\rho_1 = 0.075\ \text{lbm/ft}^3$. Use standard atmospheric data from App. 25.E for 5000 ft altitude.

$$\rho_{5000\ \text{ft}} = \rho_2 = \frac{p}{RT} = \frac{\left(12.225\ \frac{\text{lbf}}{\text{in}^2}\right)\left(12\ \frac{\text{in}}{\text{ft}}\right)^2}{\left(53.3\ \frac{\text{ft-lbf}}{\text{lbm-°R}}\right)(500.9°\text{R})}$$

$$= 0.06594\ \text{lbf/ft}^3$$

$$\frac{\gamma_2}{\gamma_1} = \frac{\frac{\rho_2 g}{g_c}}{\frac{\rho_1 g}{g_c}} = \frac{\rho_2}{\rho_1} = \frac{0.06594\ \frac{\text{lbm}}{\text{ft}^3}}{0.075\ \frac{\text{lbm}}{\text{ft}^3}} = 0.8792$$

$$\text{AHP}_2 = \left(3.147 \times 10^{-3}\ \text{hp}\right)(8)^5\left(\tfrac{1}{2}\right)^3(0.8792)$$

$$= \boxed{11.33\ \text{hp} \quad (11\ \text{hp})}$$

The answer is (A).

SI Solution

The air power is given by Eq. 41.13(a).

$$\text{AkW}_1 = \frac{Q_{\text{L/s}}(\text{TP}_{\text{Pa}})}{10^6} = \frac{\left(19\ \frac{\text{L}}{\text{s}}\right)(125\ \text{Pa})}{10^6\ \frac{\text{L·W}}{\text{m}^3\text{·kW}}}$$

$$= 2.375 \times 10^{-3}\ \text{kW}$$

In order to predict the performance of a dynamically similar fan, use Eq. 41.28.

$$\frac{\text{AkW}_1}{\text{AkW}_2} = \left(\frac{D_1}{D_2}\right)^5\left(\frac{n_1}{n_2}\right)^3\left(\frac{\gamma_1}{\gamma_2}\right)$$

$$\text{AkW}_2 = \text{AkW}_1\left(\frac{D_2}{D_1}\right)^5\left(\frac{n_2}{n_1}\right)^3\left(\frac{\gamma_2}{\gamma_1}\right)$$

$$\frac{D_2}{D_1} = \frac{1}{\frac{1}{8}} = 8$$

$$\frac{n_2}{n_1} = \frac{1}{2}$$

At standard air conditions, $\rho = 1.2\ \text{kg/m}^3$. Use standard atmospheric data from App. 25.E for 1500 m altitude.

$$\rho_{1500\ \text{m}} = \rho_2 = \frac{p}{RT} = \frac{(0.8456\ \text{bar})\left(10^5\ \frac{\text{Pa}}{\text{bar}}\right)}{\left(287\ \frac{\text{J}}{\text{kg·K}}\right)(278.4\text{K})}$$

$$= 1.058\ \text{kg/m}^3$$

$$\frac{\gamma_2}{\gamma_1} = \frac{\rho_2 g}{\rho_1 g} = \frac{\rho_2}{\rho_1} = \frac{1.058\ \frac{\text{kg}}{\text{m}^3}}{1.2\ \frac{\text{kg}}{\text{m}^3}}$$

$$= 0.882$$

$$\text{AkW}_2 = (2.375 \times 10^{-3}\ \text{kW})(8)^5\left(\tfrac{1}{2}\right)^3(0.882)$$

$$= \boxed{8.58\ \text{kW} \quad (8.6\ \text{kW})}$$

The answer is (A).

HVAC

4. *Customary U.S. Solution*

From the standard friction loss chart (see Fig. 41.5), the friction loss is 0.42 in of water per 100 ft of duct and v = 2700 ft/min. The friction loss in 750 ft of duct is

$$\text{FP}_{f,1} = \left(0.42\ \frac{\text{in wg}}{100\ \text{ft}}\right)\left(\frac{750\ \text{ft}}{100\ \text{ft}}\right) = 3.15\ \text{in wg}$$

The equivalent length of each round elbow (radius-to-diameter ratio of 1.5) can be found from Table 41.5.

$$L_e = 12D$$

For four round elbows,

$$L_e = (4)(12D) = \frac{(4)(12)(20\ \text{in})}{12\ \frac{\text{in}}{\text{ft}}} = 80\ \text{ft}$$

The friction loss is

$$\text{FP}_{f,2} = \left(0.42\ \frac{\text{in wg}}{100\ \text{ft}}\right)\left(\frac{80\ \text{ft}}{100\ \text{ft}}\right) = 0.336\ \text{in wg}$$

For a cross duct,

$$\frac{D_1}{D_2} = \frac{2\ \text{in}}{20\ \text{in}} = 0.1$$

From Table 41.4, the loss coefficient is $K_{\text{up}} = 0.2$. For two cross pipes, the friction loss is given by Eq. 41.39(b).

$$\text{FP}_{f,3} = 2K_{\text{up}}\left(\frac{\text{v}_{\text{ft/min}}}{4005}\right)^2 = (2)(0.2)\left(\frac{2700\ \frac{\text{ft}}{\text{min}}}{4005\ \frac{\text{ft}}{\text{min}}}\right)^2$$
$$= 0.182\ \text{in wg}$$

The total loss is

$$\text{FP}_{f,1} + \text{FP}_{f,2} + \text{FP}_{f,3} = 3.15\ \text{in wg} + 0.336\ \text{in wg} + 0.182\ \text{in wg}$$
$$= \boxed{3.668\ \text{in wg}\quad (3.7\ \text{in wg})}$$

The answer is (C).

SI Solution

From the standard friction loss chart (see Fig. 41.6), the friction loss is 9 Pa/m and v = 21.5 m/s. The friction loss in 230 m of duct is

$$\text{FP}_{f,1} = \left(9\ \frac{\text{Pa}}{\text{m}}\right)(230\ \text{m}) = 2070\ \text{Pa}$$

The equivalent length of each round elbow (radius-to-diameter ratio of 1.5) can be found in Table 41.5.

$$L_e = 12D$$

For four round elbows,

$$L_e = (4)(12D) = \frac{(4)(12)(508\ \text{mm})}{1000\ \frac{\text{mm}}{\text{m}}} = 24.38\ \text{m}$$

The friction loss is

$$\text{FP}_{f,2} = \left(9\ \frac{\text{Pa}}{\text{m}}\right)(24.38\ \text{m}) = 219.4\ \text{Pa}$$

For a cross duct,

$$\frac{D_1}{D_2} = \frac{51\ \text{mm}}{508\ \text{mm}} = 0.1$$

From Table 41.4, the loss coefficient K_{up} is 0.2. For two cross pipes, the friction loss is given by Eq. 41.39(a).

$$\text{FP}_{f,3} = (2)0.6K_{\text{up}}\text{v}_{\text{m/s}}^2$$
$$= (2)\left(0.6\ \frac{\text{Pa}\cdot\text{s}^2}{\text{m}^2}\right)(0.2)\left(21.5\ \frac{\text{m}}{\text{s}}\right)^2$$
$$= 110.9\ \text{Pa}$$

The total loss is

$$\text{FP}_{f,1} + \text{FP}_{f,2} + \text{FP}_{f,3} = 2070\ \text{Pa} + 219.4\ \text{Pa} + 110.9\ \text{Pa}$$
$$= \boxed{2400.3\ \text{Pa}\quad (2.4\ \text{kPa})}$$

The answer is (C).

5. *Customary U.S. Solution*

For the 18 in duct,

$$A = \frac{\pi}{4}D^2 = \left(\frac{\pi}{4}\right)\left(\frac{18\ \text{in}}{12\ \frac{\text{in}}{\text{ft}}}\right)^2 = 1.767\ \text{ft}^2$$

$$\text{v}_1 = \frac{Q}{A} = \frac{1500\ \frac{\text{ft}^3}{\text{min}}}{1.767\ \text{ft}^2} = 848.9\ \text{ft/min}$$

HVAC

For the 14 in duct,

$$A = \frac{\pi}{4}D^2 = \left(\frac{\pi}{4}\right)\left(\frac{14 \text{ in}}{12 \ \frac{\text{in}}{\text{ft}}}\right)^2 = 1.069 \text{ ft}^2$$

$$v_2 = \frac{Q}{A} = \frac{1500 \ \frac{\text{ft}^3}{\text{min}} - 300 \ \frac{\text{ft}^3}{\text{min}}}{1.069 \text{ ft}^2} = 1122.5 \text{ ft/min}$$

Since $v_2 > v_1$, there will not be an increase in static pressure. There will be a static pressure loss. Use $R = 1.1$. From Eq. 41.45(b), the static pressure loss is

$$\begin{aligned} SR_{\text{actual}} &= R\left(\frac{v_{\text{up}}^2 - v_{\text{down}}^2}{(4005)^2}\right) \\ &= (1.1)\left(\frac{\left(848.9 \ \frac{\text{ft}}{\text{min}}\right)^2 - \left(1122.5 \ \frac{\text{ft}}{\text{min}}\right)^2}{\left(4005 \ \frac{\text{ft}}{\text{min}}\right)^2}\right) \\ &= \boxed{-0.037 \text{ in wg loss}} \end{aligned}$$

The answer is (A).

SI Solution

For the 457 mm duct,

$$A = \frac{\pi}{4}D^2 = \left(\frac{\pi}{4}\right)\left(\frac{457 \text{ mm}}{1000 \ \frac{\text{mm}}{\text{m}}}\right)^2 = 0.164 \text{ m}^2$$

$$v_1 = \frac{Q}{A} = \frac{700 \ \frac{\text{L}}{\text{s}}}{(0.164 \text{ m}^2)\left(1000 \ \frac{\text{L}}{\text{m}^3}\right)} = 4.27 \text{ m/s}$$

For the 356 mm duct,

$$A = \frac{\pi}{4}D^2 = \left(\frac{\pi}{4}\right)\left(\frac{356 \text{ mm}}{1000 \ \frac{\text{mm}}{\text{m}}}\right)^2 = 0.10 \text{ m}^2$$

$$v_2 = \frac{Q}{A} = \frac{700 \ \frac{\text{L}}{\text{s}} - 140 \ \frac{\text{L}}{\text{s}}}{(0.10 \text{ m}^2)\left(1000 \ \frac{\text{L}}{\text{m}^3}\right)} = 5.60 \text{ m/s}$$

Since $v_2 > v_1$, there will not be an increase in static pressure. There will be a static pressure loss. Use $R = 1.1$. From Eq. 41.45(a), the static pressure loss is

$$\begin{aligned} SR &= R0.6(v_{\text{up}}^2 - v_{\text{down}}^2) \\ &= (1.1)\left(0.6 \ \frac{\text{Pa·s}^2}{\text{m}^2}\right)\left(\left(4.27 \ \frac{\text{m}}{\text{s}}\right)^2 - \left(5.60 \ \frac{\text{m}}{\text{s}}\right)^2\right) \\ &= \boxed{-8.66 \text{ Pa loss} \quad (-8.7 \text{ Pa loss})} \end{aligned}$$

The answer is (A).

6. *Customary U.S. Solution*

Use the equal-friction method to size the system.

step 1: From Table 41.8, choose the main duct velocity as approximately 1600 ft/min.

step 2: The total air flow from the fan is 1500 ft^3/min. From Fig. 41.5, the main duct diameter is 13 in. The friction loss is 0.27 in wg per 100 ft.

step 3: After the first takeoff at A, the flow rate in section A–B is

$$1500 \ \frac{\text{ft}^3}{\text{min}} - 300 \ \frac{\text{ft}^3}{\text{min}} = 1200 \text{ ft}^3/\text{min}$$

From Fig. 41.5, for 1200 ft^3/min and 0.27 in wg per 100 ft, the diameter is 11.8 in (say 12 in). The velocity from the chart, 1500 ft/min, cannot be used because the duct is not circular. Similarly, the diameters for the other sections are obtained and listed in the following table.

section	Q (ft^3/min)	D (in)
fan–A	1500	13.0
A–B	1200	12.0
B–C	900	11.0
C–D	600	9.5
D–E	400	8.0
E–F	200	6.3

These diameters are for round duct. The equivalent rectangular duct sides with an aspect ratio of 1.5 are found as follows.

$$\begin{aligned} a &= \frac{D(1.5+1)^{0.25}}{(1.3)(1.5)^{5/8}} \\ &= 0.75D \\ b &= Ra = (1.5)(0.75D) \\ &= 1.125D \end{aligned}$$

HVAC

Convert the diameters to sides a and b.

$$a_{\text{fan-A}} = (0.75)(13 \text{ in}) = 9.75 \text{ in}$$
$$b_{\text{fan-A}} = (1.125)(13 \text{ in}) = 14.63 \text{ in}$$

Use a 10 in × 15 in duct.

The following table is prepared similarly, rounding up as appropriate.

section	D (in)	a (in)	b (in)	v (ft/min)
fan–A	13.0	10	15	1440
A–B	12.0	9	14	1370
B–C	11.0	9	13	1110
C–D	9.5	8	11	980
D–E	8.0	6	9	1070
E–F	6.3	5	8	720

The velocity in section fan–A is

$$\text{v}_{\text{fan-A}} = \frac{\left(1500\ \frac{\text{ft}^3}{\text{min}}\right)\left(12\ \frac{\text{in}}{\text{ft}}\right)^2}{(10 \text{ in})(15 \text{ in})} = 1440 \text{ ft/min}$$

The velocities in the other sections are found similarly.

(a) *step 4:* By inspection, the longest run is fan–F.

Use Eq. 41.41 to estimate the equivalent lengths of the bends. Since both bends have the same radius-width and width-height ratios, the equivalent lengths can be combined.

$$L_e = (W_{\text{fan-A}} + W_{\text{C-D}})\left(0.33\frac{r}{W}\right)^{-2.13(H/W)^{0.126}}$$
$$= \left(\frac{15 \text{ in} + 11 \text{ in}}{12\ \frac{\text{in}}{\text{ft}}}\right)\left((0.33)(1.5)\right)^{-(2.13)(1/1.5)^{0.126}}$$
$$= 9.0 \text{ ft}$$

The equivalent length of the entire run is

	15 ft	(fan to first bend)
	45 ft	(first bend to point A)
	30 ft	(section A–B)
	30 ft	(section B–C)
	20 ft	(point A to second bend)
	9.0 ft	(equivalent length of two bends)
	10 ft	(second bend to point D)
	20 ft	(section D–E)
	20 ft	(section E–F)
total:	199 ft	

The straight-through friction loss in the longest run is

$$\left(\frac{199 \text{ ft}}{100 \text{ ft}}\right)\left(0.27\ \frac{\text{in wg}}{100 \text{ ft}}\right) = 0.54 \text{ in wg}$$

The fan must be able to supply a static pressure of

$$\text{SP}_{\text{fan}} = 0.54 \text{ in wg} + 0.25 \text{ in wg}$$
$$= \boxed{0.79 \text{ in wg} \quad (0.8 \text{ in wg})}$$

The answer is (C).

(b) The total pressure supplied by the fan is

$$\text{TP}_{\text{fan}} = 0.79 \text{ in wg} + \left(\frac{1440\ \frac{\text{ft}}{\text{min}}}{4005\ \frac{\text{ft}}{\text{min}}}\right)^2$$
$$= \boxed{0.92 \text{ in wg} \quad (1.0 \text{ in wg})}$$

The answer is (A).

SI Solution

Use the equal-friction method to size the system.

step 1: From Table 41.8, choose the main duct velocity as approximately 1600 ft/min. From the table footnote, the SI velocity is

$$\text{v}_{\text{main}} = \left(1600\ \frac{\text{ft}}{\text{min}}\right)\left(0.00508\ \frac{\text{m·min}}{\text{s·ft}}\right) = 8.1 \text{ m/s}$$

step 2: The total air flow from the fan is 700 L/s. From Fig. 41.6, the main duct diameter is approximately 340 mm. The friction loss is 2.3 Pa/m.

step 3. After the first takeoff at A, the flow rate in section A–B is

$$700\ \frac{\text{L}}{\text{s}} - 150\ \frac{\text{L}}{\text{s}} = 550 \text{ L/s}$$

From Fig. 41.6 for 550 L/s and 2.3 Pa/m, the diameter is 300 mm. The velocity from the chart, 7.7 m/s, cannot be used because the duct is not circular. Similarly, the diameters for the other sections are obtained and listed in the following table.

section	Q (L/s)	D (mm)
fan–A	700	340
A–B	550	300
B–C	400	275
C–D	250	225
D–E	150	180
E–F	50	125

These diameters are for round duct. The equivalent rectangular duct sides with an aspect ratio of 1.5 are found as follows.

$$a = \frac{D(1.5+1)^{0.25}}{(1.3)(1.5)^{5/8}} = 0.75D$$

$$b = Ra = (1.5)(0.75D) = 1.125D$$

Convert the diameters to sides a and b.

$$a_{\text{fan-A}} = (0.75)(340 \text{ mm}) = 255 \text{ mm}$$

$$b_{\text{fan-A}} = (1.125)(340 \text{ mm}) = 383 \text{ mm}$$

Use a 275 mm × 400 mm duct.

The following table is prepared similarly, rounding up as appropriate.

section	D (mm)	a (mm)	b (mm)	v (m/s)
fan–A	340	275	400	6.4
A–B	300	225	350	7.0
B–C	275	225	325	5.5
C–D	225	175	275	5.2
D–E	180	150	225	4.4
E–F	125	100	150	3.3

The velocity in section fan–A is

$$\text{v}_{\text{fan-A}} = \frac{700 \ \frac{\text{L}}{\text{s}}}{\left(\frac{(275 \text{ mm})(400 \text{ mm})}{\left(1000 \ \frac{\text{mm}}{\text{m}}\right)^2}\right)\left(1000 \ \frac{\text{L}}{\text{m}^3}\right)}$$

$$= 6.4 \text{ m/s}$$

The velocities in the other sections are found similarly.

(a) *step 4:* By inspection, the longest run is fan–F.

Use Eq. 41.41 to estimate the equivalent lengths of the bends. Since both bends have the same radius-width and width-height ratios, the equivalent lengths can be combined.

$$L_e = (W_{\text{fan-A}} + W_{\text{C-D}})\left(0.33\frac{r}{W}\right)^{-2.13(H/W)^{0.126}}$$

$$= \left(\frac{400 \text{ mm} + 275 \text{ mm}}{1000 \ \frac{\text{mm}}{\text{m}}}\right)((0.33)(1.5))^{-(2.13)(1/1.5)^{0.126}}$$

$$= 2.8 \text{ m}$$

The equivalent length of the entire run is

4.5 m	(fan to first bend)
13.5 m	(first bend to point A)
9 m	(section A–B)
9 m	(section B–C)
6 m	(point C to second bend)
2.8 m	(equivalent length of two bends)
3 m	(second bend to point D)
6 m	(section D–E)
6 m	(section E–F)
total: 59.8 m	

The straight-through friction loss in the longest run is

$$(59.8 \text{ m})\left(2.3 \ \frac{\text{Pa}}{\text{m}}\right) = 138 \text{ Pa}$$

The fan must be able to supply a static pressure of

$$\text{SP}_{\text{fan}} = 138 \text{ Pa} + 63 \text{ Pa} = \boxed{201 \text{ Pa} \quad (200 \text{ Pa})}$$

The answer is (C).

(b) The total pressure supplied by the fan is

$$\text{TP}_{\text{fan}} = 201 \text{ Pa} + \left(0.6 \ \frac{\text{Pa·s}^2}{\text{m}^2}\right)\left(6.4 \ \frac{\text{m}}{\text{s}}\right)^2$$

$$= \boxed{226 \text{ Pa} \quad (230 \text{ Pa})}$$

The answer is (A).

7. *Customary U.S. Solution*

Use the equal-friction method to size the system.

step 1: From Table 41.8, choose the main duct velocity as 1600 ft/min.

step 2: The total air flow from the fan is

$$(12)\left(300 \ \frac{\text{ft}^3}{\text{min}}\right) = 3600 \text{ ft}^3/\text{min}$$

From Fig. 41.5, the main duct diameter is 20 in. The friction loss is 0.16 in wg per 100 ft.

step 3: After the first takeoff, the flow rate in the main duct is

$$3600 \ \frac{\text{ft}^3}{\text{min}} - (4)\left(300 \ \frac{\text{ft}^3}{\text{min}}\right) = 2400 \text{ ft}^3/\text{min}$$

From Fig. 41.5, for 2400 ft^3/min and 0.16 in wg per 100 ft, the diameter is 17.2 in and the velocity is 1440 ft/min. Similarly, the diameters and the velocities for other sections are obtained and listed in the following table.

HVAC

section	Q (ft^3/min)	D (in)	v (ft/min)
fan–first takeoff	3600	20.0	1600
first–second takeoff	2400	17.2	1440
second–third takeoff	1200	13.2	1210
first–A	1200	13.2	1210
A–B	900	12.0	1150
B–C	600	10.2	1030
C–D	300	7.9	860
second–E	1200	13.2	1210
E–F	900	12.0	1150
F–G	600	10.2	1030
G–H	300	7.9	860
I–J	900	12.0	1150
J–K	600	10.2	1030
K–L	300	7.9	860

(a) *step 4:* By inspection, the longest run is fan–L. From Table 41.5, the equivalent length of each bend is $14.5D$ (interpolated). For two elbows,

$$\begin{aligned} L_{e,\text{bend}} &= 14.5D \\ &= (14.5)\left(\frac{20 \text{ in} + 13.2 \text{ in}}{12 \ \frac{\text{in}}{\text{ft}}}\right) \\ &= 40.1 \text{ ft} \end{aligned}$$

The equivalent length of the entire run is

	25 ft	(fan to first bend)
	35 ft	(first bend to first section)
	20 ft	(first section to second section)
	20 ft	(second section to third section)
	10 ft	(third section to point I)
	40.1 ft	(equivalent length of two bends)
	20 ft	(point I to point J)
	20 ft	(point J to point K)
	20 ft	(point K to point L)
total:	210.1 ft	

The straight-through friction loss in the longest run is

$$\left(\frac{210.1 \text{ ft}}{100 \text{ ft}}\right)\left(0.16 \ \frac{\text{in wg}}{100 \text{ ft}}\right) = 0.34 \text{ in wg}$$

The fan must be able to supply a static pressure of

$$\begin{aligned} \text{SP}_{\text{fan}} &= 0.34 \text{ in wg} + 0.15 \text{ in wg} \\ &= \boxed{0.49 \text{ in wg} \quad (0.5 \text{ in wg})} \end{aligned}$$

The answer is (A).

(b) The total pressure supplied by the fan is

$$\begin{aligned} \text{TP}_{\text{fan}} &= 0.49 \text{ in wg} + \left(\frac{1600 \ \frac{\text{ft}}{\text{min}}}{4005 \ \frac{\text{ft}}{\text{min}}}\right)^2 \\ &= \boxed{0.65 \text{ in wg} \quad (0.7 \text{ in wg})} \end{aligned}$$

The answer is (B).

SI Solution

Use the equal-friction method to size the system.

step 1: From Table 41.8, choose the main duct velocity as 1600 ft/min. From the table footnote, the SI velocity is

$$\begin{aligned} \text{v}_{\text{main}} &= \left(1600 \ \frac{\text{ft}}{\text{min}}\right)\left(0.00508 \ \frac{\text{m}\cdot\text{min}}{\text{s}\cdot\text{ft}}\right) \\ &= 8.1 \text{ m/s} \end{aligned}$$

step 2: The total air flow from the fan is

$$(12)\left(140 \ \frac{\text{L}}{\text{s}}\right) = 1680 \text{ L/s}$$

From Fig. 41.6, the main duct diameter is 500 mm. The friction loss is 1.5 Pa/m.

step 3: After the first takeoff, the flow rate in the main duct is

$$1680 \ \frac{\text{L}}{\text{s}} - (4)\left(140 \ \frac{\text{L}}{\text{s}}\right) = 1120 \text{ L/s}$$

From Fig. 41.6, for 1120 L/s and 1.5 Pa/m, the diameter is 440 mm and the velocity is 7.5 m/s. Similarly, the diameters and the velocities for other sections are obtained and listed in the following table.

section	Q (L/s)	D (mm)	v (m/s)
fan–first takeoff	1680	500	8.1
first–second takeoff	1120	440	7.5
second–third takeoff	560	335	6.4
first–A	560	335	6.4
A–B	420	305	5.9
B–C	280	260	5.4
C–D	140	195	4.5
second–E	560	335	6.4
E–F	420	305	5.9
F–G	280	260	5.4
G–H	140	195	4.5
I–J	420	305	5.9
J–K	280	260	5.4
K–L	140	195	4.5

(a) *step 4:* By inspection, the longest run is fan–L. From Table 41.5, the equivalent length of each bend is $14.5D$ (interpolated). For two elbows,

$$L_{e,\text{bend}} = 14.5D = (14.5)\left(\frac{500 \text{ mm} + 335 \text{ mm}}{1000 \frac{\text{mm}}{\text{m}}}\right)$$
$$= 12.1 \text{ m}$$

The equivalent length of the entire run is

	7.5 m	(fan to first bend)
	10.5 m	(first bend to first section)
	6 m	(first section to second section)
	6 m	(second section to third section)
	3 m	(third section to point I)
	12.1 m	(equivalent length of two bends)
	6 m	(point I to point J)
	6 m	(point J to point K)
	6 m	(point K to point L)
total:	63.1 m	

The straight-through friction loss in the longest run is

$$(63.1 \text{ m})\left(1.5 \frac{\text{Pa}}{\text{m}}\right) = 95 \text{ Pa}$$

The fan must be able to supply a static pressure of

$$\text{SP}_{\text{fan}} = 95 \text{ Pa} + 38 \text{ Pa} = \boxed{133 \text{ Pa} \quad (130 \text{ Pa})}$$

The answer is (A).

(b) The total pressure supplied by the fan is

$$\text{TP}_{\text{fan}} = 133 \text{ Pa} + (0.6)\left(8.1 \frac{\text{m}}{\text{s}}\right)^2 = \boxed{172 \text{ Pa} \quad (170 \text{ Pa})}$$

The answer is (B).

8. (a) *step 1:* From Table 41.8, choose the main duct velocity as 1600 ft/min.

step 2: The diameter of the main duct is

$$A = \frac{Q}{\text{v}} = \frac{3600 \frac{\text{ft}^3}{\text{min}}}{1600 \frac{\text{ft}}{\text{min}}} = 2.25 \text{ ft}^2$$

$$D = \sqrt{\frac{4A}{\pi}} = \sqrt{\frac{(4)(2.25 \text{ ft}^2)}{\pi}}\left(12 \frac{\text{in}}{\text{ft}}\right)$$
$$= \boxed{20.3 \text{ in} \quad (20 \text{ in})}$$

The answer is (B).

(b) *step 3:* From Table 41.5, the equivalent length of each bend is $14.5D$ (interpolated). For the first elbow,

$$L_{e,\text{bend}} = 14.5D = (14.5)\left(\frac{20.3 \text{ in}}{12 \frac{\text{in}}{\text{ft}}}\right) = 24.5 \text{ ft}$$

The equivalent length of the main duct from the fan to the first takeoff and bend is

$$L = 25 \text{ ft} + 24.5 \text{ ft} + 35 \text{ ft} = \boxed{84.5 \text{ ft} \quad (85 \text{ ft})}$$

The answer is (B).

(c) *step 4:* From Fig. 41.5, the friction loss in the main run up to the branch takeoff is approximately 0.16 in wg per 100 ft. The actual friction loss is

$$\text{FP}_{\text{main}} = \left(0.16 \frac{\text{in wg}}{100 \text{ ft}}\right)\left(\frac{84.5 \text{ ft}}{100 \text{ ft}}\right)$$
$$= \boxed{0.135 \text{ in wg} \quad (0.14 \text{ in wg})}$$

The answer is (B).

(d) *step 5:* After the first takeoff,

$$Q = 3600 \frac{\text{ft}^3}{\text{min}} - 1200 \frac{\text{ft}^3}{\text{min}} = 2400 \text{ ft}^3/\text{min}$$
$$L = 20 \text{ ft}$$
$$\frac{L}{Q^{0.61}} = \frac{20 \text{ ft}}{\left(2400 \frac{\text{ft}^3}{\text{min}}\right)^{0.61}} = 0.173$$

(This equation is dimensionally inconsistent.)

From Fig. 41.10, the velocity, v_2, is 1390 ft/min.

step 6: Solve for the duct size from $A = Q/\text{v}$.

$$D_2 = \sqrt{\frac{4A_2}{\pi}} = \sqrt{\frac{4Q_2}{\pi \text{v}_2}} = \sqrt{\frac{(4)\left(2400 \frac{\text{ft}^3}{\text{min}}\right)}{\pi\left(1390 \frac{\text{ft}}{\text{min}}\right)}}\left(12 \frac{\text{in}}{\text{ft}}\right)$$
$$= 17.8 \text{ in} \quad (18 \text{ in})$$

step 7: After the second takeoff,

$$Q = 2400 \frac{\text{ft}^3}{\text{min}} - 1200 \frac{\text{ft}^3}{\text{min}} = 1200 \text{ ft}^3/\text{min}$$
$$L = 20 \text{ ft} + 14.5D + 10 \text{ ft}$$
$$= 30 \text{ ft} + 14.5D$$

Since L contains the unknown diameter, D, this will require an iterative procedure. Using velocity $\text{v}_1 =$

1390 ft/min before the takeoff and using an iterative procedure, $D = 15$ in and $v = 980$ ft/min.

step 8: Proceeding similarly, the following table is developed for the remaining sizes.

section	L (ft)	Q (ft^3/min)	$\frac{L}{Q^{0.61}}$	v (ft/min)
I–J	20	900	0.32	830
J–K	20	600	0.40	660
K–L	20	300	0.62	500

step 9: Solve for the duct size from $A = Q/\mathrm{v}$.

$$D = \sqrt{\frac{4A}{\pi}} = \sqrt{\frac{4Q}{\pi \mathrm{v}}}$$

$$D_{\text{I–J}} = \sqrt{\frac{(4)\left(900 \ \frac{\text{ft}^3}{\text{min}}\right)}{\pi\left(830 \ \frac{\text{ft}}{\text{min}}\right)}}\left(12 \ \frac{\text{in}}{\text{ft}}\right)$$
$$= 14.1 \text{ in} \quad (14 \text{ in})$$

$$D_{\text{J–K}} = \sqrt{\frac{(4)\left(600 \ \frac{\text{ft}^3}{\text{min}}\right)}{\pi\left(660 \ \frac{\text{ft}}{\text{min}}\right)}}\left(12 \ \frac{\text{in}}{\text{ft}}\right)$$
$$= 12.9 \text{ in} \quad (13 \text{ in})$$

$$D_{\text{K–L}} = \sqrt{\frac{(4)\left(300 \ \frac{\text{ft}^3}{\text{min}}\right)}{\pi\left(500 \ \frac{\text{ft}}{\text{min}}\right)}}\left(12 \ \frac{\text{in}}{\text{ft}}\right)$$
$$= 10.5 \text{ in} \quad (11 \text{ in})$$

step 10: The total pressure supplied by the fan is

$$0.135 \text{ in wg} + 0.15 \text{ in wg} = \boxed{0.285 \text{ in wg} \quad (0.29 \text{ in wg})}$$

The answer is (C).

9. *Customary U.S. Solution*

(a) No dampers are needed in duct A. The area of section A is

$$A = \frac{\pi}{4}D^2 = \left(\frac{\pi}{4}\right)\left(\frac{12 \text{ in}}{12 \ \frac{\text{in}}{\text{ft}}}\right)^2 = 0.7854 \text{ ft}^2$$

The velocity in section A is

$$\mathrm{v}_{\text{A}} = \frac{Q}{A} = \frac{3000 \ \frac{\text{ft}^3}{\text{min}}}{(0.7854 \text{ ft}^2)\left(60 \ \frac{\text{sec}}{\text{min}}\right)} = 63.66 \text{ ft/sec}$$

For a four-piece elbow with a radius-to-diameter ratio of 1.5, $L_e = 15D$. The total equivalent length of run C is

$$L_e = 50 \text{ ft} + 10 \text{ ft} + 10 \text{ ft} + (2)\left((15)\left(\frac{10 \text{ in}}{12 \ \frac{\text{in}}{\text{ft}}}\right)\right)$$
$$= \boxed{95 \text{ ft}}$$

The answer is (C).

(b) For any diameter, D, in inches of section C, the velocity will be

$$\mathrm{v}_{\text{C}} = \frac{Q}{A} = \frac{2000 \ \frac{\text{ft}^3}{\text{min}}}{\left(\frac{\pi}{4}\right)\left(\frac{D}{12 \ \frac{\text{in}}{\text{ft}}}\right)^2\left(60 \ \frac{\text{sec}}{\text{min}}\right)}$$
$$= 6111.5/D^2$$

From Eq. 17.27, the friction loss in section C will be

$$h_{f,\text{C}} = \frac{fL\mathrm{v}^2}{2Dg} = \frac{(0.02)(93.3 \text{ ft})\left(\frac{6111.5}{D^2}\right)^2}{(2)\left(\frac{D}{12 \ \frac{\text{in}}{\text{ft}}}\right)\left(32.2 \ \frac{\text{ft}}{\text{sec}^2}\right)}$$
$$= \frac{1.3 \times 10^7}{D^5} \quad [\text{ft of air}]$$

The regain between A and C will be

$$h_{\text{regain}} = R\left(\frac{\mathrm{v}_{\text{A}}^{?} - \mathrm{v}_{\text{G}}^{?}}{2g}\right)$$
$$= (0.65)\left(\frac{\left(63.66 \ \frac{\text{ft}}{\text{sec}}\right)^2 - \left(\frac{6111.5}{D^2}\right)^2}{(2)\left(32.2 \ \frac{\text{ft}}{\text{sec}^2}\right)}\right)$$
$$= 40.9 - \frac{3.77 \times 10^5}{D^4} \quad [\text{ft of air}]$$

The principle of static regain is that

$$h_{f,\text{C}} = h_{\text{regain}}$$
$$\frac{1.3 \times 10^7}{D^5} = 40.9 - \frac{3.77 \times 10^5}{D^4}$$

HVAC

By trial and error, $D \approx 13.5$ in. Since the assumed value of D is different from the calculated value, this process should be repeated.

$$L_e = 50 \text{ ft} + 10 \text{ ft} + 10 \text{ ft} + (2)\left((14)\left(\frac{13.5 \text{ in}}{12 \frac{\text{in}}{\text{ft}}}\right)\right)$$
$$= 101.5 \text{ ft}$$

From Eq. 17.27, the friction loss in section C will be

$$h_{f,\text{C}} = \frac{fL\text{v}^2}{2Dg} = \frac{(0.02)(101.5 \text{ ft})\left(\frac{6111.5}{D^2}\right)^2}{(2)\left(\frac{D}{12 \frac{\text{in}}{\text{ft}}}\right)\left(32.2 \frac{\text{ft}}{\text{sec}^2}\right)}$$
$$= \frac{1.41 \times 10^7}{D^5} \quad [\text{ft of air}]$$

The principle of static regain is that

$$h_{f,\text{C}} = h_{\text{regain}}$$
$$\frac{1.41 \times 10^7}{D^5} = 40.9 - \frac{3.77 \times 10^5}{D^4}$$

By trial and error, $D_\text{C} = 13.63$ in (14 in). This results in a friction loss of

$$h_{f,\text{C}} = \frac{1.41 \times 10^7}{(14 \text{ in})^5} = \boxed{26.2 \text{ ft (26 ft) of air}}$$

Because the regain cancels friction loss, the pressure loss from A to C is zero. A damper is not needed in duct C.

The answer is (B).

(c) For any diameter, D, in inches in section B, the velocity will be

$$\text{v}_\text{B} = \frac{Q}{A} = \frac{1000 \frac{\text{ft}^3}{\text{min}}}{\left(\frac{\pi}{4}\right)\left(\frac{D}{12 \frac{\text{in}}{\text{ft}}}\right)^2\left(60 \frac{\text{sec}}{\text{min}}\right)}$$
$$= \frac{3055.8}{D^2} \quad [\text{ft/sec}]$$

From Eq. 17.27, the friction loss in section B will be

$$h_{f,\text{B}} = \frac{fL\text{v}^2}{2Dg} = \frac{(0.02)(10 \text{ ft})\left(\frac{3055.8}{D^2}\right)^2}{(2)\left(\frac{D}{12 \frac{\text{in}}{\text{ft}}}\right)\left(32.2 \frac{\text{ft}}{\text{sec}^2}\right)}$$
$$= \frac{3.48 \times 10^5}{D^5} \quad [\text{ft of air}]$$

From Eq. 41.48, the friction loss in the branch takeoff between sections A and B will be

$$\text{TP}_\text{A} - \text{TP}_\text{B} = K_{\text{br}}(\text{VP}_{\text{up}})$$

At this point, assume $\text{v}_\text{B}/\text{v}_\text{A} = 1.0$. From Table 41.7, for a 45° angle of takeoff, $K_{\text{br}} = 0.5$.

$$\text{TP}_\text{A} - \text{TP}_\text{B} = (0.5)\left(\frac{3820 \frac{\text{ft}}{\text{min}}}{4005 \frac{\text{ft}}{\text{min}}}\right)^2 = 0.455 \text{ in wg}$$

From Eq. 41.6,

$$p_{\text{psig}} = (0.455 \text{ in wg})\left(0.0361 \frac{\text{lbf}}{\text{in}^3}\right) = 0.01643 \text{ lbf/in}^2$$

From Eq. 41.5,

$$h_f = \frac{p_{\text{psig}}}{\gamma} = \frac{\left(0.01643 \frac{\text{lbf}}{\text{in}^2}\right)\left(12 \frac{\text{in}}{\text{ft}}\right)^2}{0.075 \frac{\text{lbf}}{\text{ft}^3}} = 31.5 \text{ ft of air}$$

The regain between A and B will be

$$h_{\text{regain}} = R\left(\frac{\text{v}_\text{A}^2 - \text{v}_\text{B}^2}{2g}\right)$$
$$= (0.65)\left(\frac{\left(63.66 \frac{\text{ft}}{\text{sec}}\right)^2 - \left(\frac{3055.8}{D^2}\right)^2}{(2)\left(32.2 \frac{\text{ft}}{\text{sec}^2}\right)}\right)$$
$$= 40.9 \text{ ft} - \frac{9.42 \times 10^4}{D^4} \quad [\text{ft of air}]$$

Set the regain equal to the loss.

$$h_{\text{regain}} = h_{f,\text{B}}$$
$$40.9 \text{ ft} - \frac{9.42 \times 10^4}{D^4} = \frac{3.48 \times 10^5}{D^5} + 31.5 \text{ ft}$$

By trial and error,

$$D_B = \boxed{10.77 \text{ in} \quad (11 \text{ in})}$$

The answer is (A).

(d) Calculate the regain in sections B and C.

$$v_B = \frac{3055.8}{D^2} = \frac{3055.8}{(11 \text{ in})^2} = 25.25 \text{ ft/sec}$$

$$\frac{v_B}{v_A} = \frac{25.25 \frac{\text{ft}}{\text{sec}}}{63.6 \frac{\text{ft}}{\text{sec}}} \approx 0.4$$

From Table 41.7, for a 45° angle of takeoff, $K_{br} = 0.5$. Since the value of K_{br} remains the same, there is no need to repeat the preceding procedure. For section B, the friction canceled by the regain is

$$h_{f,B} = 31.5 \text{ ft} + \frac{3.48 \times 10^5}{D^5} = 31.5 \text{ ft} + \frac{3.48 \times 10^5}{(11 \text{ in})^5}$$
$$= \boxed{33.7 \text{ ft (34 ft) of air}}$$

The answer is (C).

(e) From a practical standpoint, flow dampers would probably be placed at least in sections B and C to adjust for any deviations from ideal behavior, as well as to allow for future modifications to the airflow. A volume damper in section A will ensure that only 3000 cfm will pass through section A. From an ideal standpoint, the diameters have been selected to zero out the friction in both sections B and C, so the exit pressures at the ends of these two sections are the same as the pressure in section A just before section B branches off. Upon start-up, the transient flows in sections B and C will self-adjust to make the friction losses in those branches equal, which will occur when the design flows are achieved. Ideally, no dampers are required in any of the sections.

The answer is (A).

SI Solution

(a) No dampers are needed in duct A. The area of section A is

$$A = \frac{\pi}{4} D^2 = \left(\frac{\pi}{4}\right)\left(\frac{305 \text{ mm}}{1000 \frac{\text{mm}}{\text{m}}}\right)^2 = 0.0731 \text{ m}^2$$

The velocity in section A is

$$v_A = \frac{Q}{A} = \frac{1400 \frac{\text{L}}{\text{s}}}{(0.0731 \text{ m}^2)\left(1000 \frac{\text{L}}{\text{m}^3}\right)} = 19.15 \text{ m/s}$$

For a four-piece elbow with a radius-to-diameter ratio of 1.5, $L_e = 15D$. The total equivalent length of run C is

$$L_e = 15 \text{ m} + 3 \text{ m} + 3 \text{ m} + (2)\left((15)\left(\frac{250 \text{ mm}}{1000 \frac{\text{mm}}{\text{m}}}\right)\right)$$
$$= \boxed{28.5 \text{ m} \quad (29 \text{ m})}$$

The answer is (C).

(b) For any diameter, D, in mm of section C, the velocity will be

$$v_C = \frac{Q}{A} = \frac{900 \frac{\text{L}}{\text{s}}}{\left(\frac{\pi}{4}\right)\left(\frac{D}{1000 \frac{\text{mm}}{\text{m}}}\right)^2\left(1000 \frac{\text{L}}{\text{m}^3}\right)}$$
$$= \frac{1.146 \times 10^6}{D^2} \quad [\text{m/s}]$$

From Eq. 17.27, the friction loss in section C will be

$$h_{f,C} = \frac{fLv^2}{2Dg} = \frac{(0.02)(28 \text{ m})\left(\frac{1.146 \times 10^6}{D^2}\right)^2}{(2)\left(\frac{D}{1000 \frac{\text{mm}}{\text{m}}}\right)\left(9.81 \frac{\text{m}}{\text{s}^2}\right)}$$
$$= \frac{3.75 \times 10^{13}}{D^5} \quad [\text{m of air}]$$

The regain between A and C will be

$$h_{\text{regain}} = R\left(\frac{v_A^2 - v_C^2}{2g}\right)$$
$$= (0.65)\left(\frac{\left(19.15 \frac{\text{m}}{\text{s}}\right)^2 - \left(\frac{1.146 \times 10^6}{D^2}\right)^2}{(2)\left(9.81 \frac{\text{m}}{\text{s}^2}\right)}\right)$$
$$= 12.15 \text{ m} - \frac{4.35 \times 10^{10}}{D^4} \quad [\text{m of air}]$$

The principle of static regain is that

$$h_{f,C} = h_{\text{regain}}$$
$$\frac{3.75 \times 10^{13}}{D^5} = 12.15 \text{ m} - \frac{4.35 \times 10^{10}}{D^4}$$

HVAC

By trial and error, $D \approx 335$ mm. Since the assumed value of D is different from the calculated value, this process should be repeated.

$$L_e = 15 \text{ m} + 3 \text{ m} + 3 \text{ m} + (2)\left((14)\left(\frac{335 \text{ mm}}{1000 \ \frac{\text{mm}}{\text{m}}}\right)\right) = 30.4 \text{ m}$$

From Eq. 17.27, the friction loss in section C will be

$$h_{f,\text{C}} = \frac{fLv^2}{2Dg} = \frac{(0.02)(30.4 \text{ m})\left(\frac{1.146 \times 10^6}{D^2}\right)^2}{(2)\left(\frac{D}{1000 \ \frac{\text{mm}}{\text{m}}}\right)\left(9.81 \ \frac{\text{m}}{\text{s}^2}\right)} = \frac{4.07 \times 10^{13}}{D^5} \quad [\text{m of air}]$$

The principle of static regain is that

$$h_{f,\text{C}} = h_{\text{regain}}$$

$$\frac{4.07 \times 10^{13}}{D^5} = 12.15 - \frac{4.35 \times 10^{10}}{D^4}$$

By trial and error, $D = 340$ mm. This results in a friction loss of

$$h_{f,\text{C}} = \frac{4.07 \times 10^{13}}{(340 \text{ mm})^5} = \boxed{8.96 \text{ m } (9.0 \text{ m}) \text{ of air}}$$

Because the regain cancels this friction loss, the pressure loss from A to C is zero. A damper is not needed in duct C.

The answer is (B).

(c) For any diameter, D, in mm in section B, the velocity will be

$$v_\text{B} = \frac{Q}{A} = \frac{500 \ \frac{\text{L}}{\text{s}}}{\left(\frac{\pi}{4}\right)\left(\frac{D}{1000 \ \frac{\text{mm}}{\text{m}}}\right)^2\left(1000 \ \frac{\text{L}}{\text{m}^3}\right)} = \frac{6.366 \times 10^5}{D^2} \quad [\text{m/s}]$$

From Eq. 17.27, the friction loss in section B will be

$$h_{f,\text{B}} = \frac{fLv^2}{2Dg} = \frac{(0.02)(3 \text{ m})\left(\frac{6.366 \times 10^5}{D^2}\right)^2}{(2)\left(\frac{D}{1000 \ \frac{\text{mm}}{\text{m}}}\right)\left(9.81 \ \frac{\text{m}}{\text{s}^2}\right)} = \frac{1.239 \times 10^{12}}{D^5} \quad [\text{m of air}]$$

From Eq. 41.48, the friction loss in the branch takeoff between sections A and B will be

$$\text{TP}_\text{A} - \text{TP}_\text{B} = K_\text{br}(\text{VP}_\text{up})$$

At this point, assume $v_\text{B}/v_\text{A} = 1.0$. From Table 41.7, for a 45° angle of takeoff, $K_\text{br} = 0.5$.

$$\text{TP}_\text{A} - \text{TP}_\text{B} = (0.5)(0.6)\left(19.15 \ \frac{\text{m}}{\text{s}}\right)^2 = 110 \text{ Pa}$$

From Eq. 41.5,

$$h_f = \frac{p}{\rho g} = \frac{110 \text{ Pa}}{\left(1.2 \ \frac{\text{kg}}{\text{m}^3}\right)\left(9.81 \ \frac{\text{m}}{\text{s}^2}\right)} = 9.34 \text{ m of air}$$

The regain between A and B will be

$$h_\text{regain} = R\left(\frac{v_\text{A}^2 - v_\text{B}^2}{2g}\right) = (0.65)\left(\frac{\left(19.15 \ \frac{\text{m}}{\text{s}}\right)^2 - \left(\frac{6.366 \times 10^5}{D^2}\right)^2}{(2)\left(9.81 \ \frac{\text{m}}{\text{s}^2}\right)}\right) = 12.15 \text{ m} - \frac{1.343 \times 10^{10}}{D^4} \quad [\text{m of air}]$$

Set the regain equal to the loss.

$$12.15 \text{ m} - \frac{1.343 \times 10^{10}}{D^4} = \frac{1.239 \times 10^{12}}{D^5} + 9.34 \text{ m}$$

By trial and error,

$$D_\text{B} = \boxed{280 \text{ mm}}$$

The answer is (A).

HVAC

(d) Calculate the regain in sections B and C. By trial and error,

$$v_B = 8.12 \text{ m/s}$$

Then,

$$\frac{v_B}{v_A} = \frac{8.12 \ \frac{\text{m}}{\text{s}}}{19.15 \ \frac{\text{m}}{\text{s}}} \approx 0.4$$

From Table 41.7, for a 45° angle of takeoff, $K_{br} = 0.5$. Since the value of K_{br} remains the same, there is no need to repeat the preceding procedure. For section B, the friction canceled by the regain is

$$\begin{aligned} h_{f,B} &= 9.34 \text{ m} + \frac{1.239 \times 10^{12}}{D^5} \\ &= 9.34 \text{ m} + \frac{1.239 \times 10^{12}}{(280 \text{ mm})^5} \\ &= \boxed{10.06 \text{ m (10 m) of air}} \end{aligned}$$

The answer is (C).

(e) From a practical standpoint, flow dampers would probably be placed at least in sections B and C to adjust for any deviations from ideal behavior, as well as to allow for future modifications to the airflow. A volume damper in section A will ensure that only 3000 cfm will pass through section A. From an ideal standpoint, the diameters have been selected to zero out the friction in both sections B and C, so the exit pressures at the ends of these two sections are the same as the pressure in section A just before section B branches off. Upon start-up, the transient flows in sections B and C will self-adjust to make the friction losses in those branches equal, which will occur when the design flows are achieved. Ideally, no dampers are required in any of the sections.

The answer is (A).

10. Rearrange Eq. 41.13(b) to find the volumetric flow rate.

$$\begin{aligned} Q_{\text{ft}^3/\text{min}} &= \frac{6356(\text{AHP})}{\text{TP}_{\text{in wg}}} = \frac{\left(6356 \ \frac{\text{in-ft}^3}{\text{hp-min}}\right)(4.3 \text{ hp})}{2.3 \text{ in wg}} \\ &= 11{,}883 \text{ ft}^3/\text{min} \end{aligned}$$

At 62% efficiency, the actual cubic feet per minute, ACFM, is

$$\begin{aligned} \text{ACFM} &= \eta Q \\ &= (0.62)\left(11{,}883 \ \frac{\text{ft}^3}{\text{min}}\right) \\ &= 7367 \text{ ft}^3/\text{min} \end{aligned}$$

Use Eq. 41.1 and Eq. 41.2 to find the standard flow rate, SCFM. The pressure is standard.

$$\begin{aligned} \text{SCFM} &= \frac{\text{ACFM}}{K_d} = \frac{\text{ACFM}}{\frac{T_{\text{actual}}}{T_{\text{std}}}} = \frac{7367 \ \frac{\text{ft}^3}{\text{min}}}{\frac{160°\text{F} + 460°}{70°\text{F} + 460°}} \\ &= \boxed{6298 \text{ ft}^3/\text{min} \quad (6300 \text{ ft}^3/\text{min})} \end{aligned}$$

The answer is (C).

11. Variable flow rates are achieved with variable blade pitches, inlet vanes in the supply line, and variable speed motors. System dampers in the discharge lines should not be used because they increase friction loss, are noisy, and are nonlinear in their response.

The answer is (A).

12. Velocity pressure can be used to determine velocity and is an indication of total system flow. At the entrance to the VAV box, a pitot tube senses total pressure, and a static port senses static pressure. Velocity pressure is calculated and a signal sent to the controller by a differential pressure transmitter. The duct velocity pressure can be used to determine the fan speed needed.

The answer is (A).

13. The volumetric flow rate of fan A is 4000 ft^3/min. Therefore, the volumetric flow rate of fan B is

$$\begin{aligned} Q_B &= Q_{\text{total}} - Q_A = 10{,}000 \ \frac{\text{ft}^3}{\text{min}} - 4000 \ \frac{\text{ft}^3}{\text{min}} \\ &= 6000 \text{ ft}^3/\text{min} \end{aligned}$$

Since the fans are of the same size, have the same efficiency, and move air with the same density, Eq. 41.22 may be used.

$$\frac{Q_B}{Q_A} = \frac{n_B}{n_A}$$

$$\begin{aligned} n_B &= n_A\left(\frac{Q_B}{Q_A}\right) = \left(1800 \ \frac{\text{rev}}{\text{min}}\right)\left(\frac{6000 \ \frac{\text{ft}^3}{\text{min}}}{5000 \ \frac{\text{ft}^3}{\text{min}}}\right) \\ &= \boxed{2160 \text{ rev/min} \quad (2200 \text{ rev/min})} \end{aligned}$$

The answer is (C).

42 Heating Load

PRACTICE PROBLEMS

1. A conditioned room contains 12,000 W of fluorescent lights (with older coil ballasts) and twelve 90% efficient, 10 hp motors operating at 80% of their rated capacities. The lights are pendant-mounted on chains from the ceiling. The motors drive various pieces of machinery located in the conditioned space. The internal heat gain is most nearly

(A) 1.9×10^5 Btu/hr (55 kW)

(B) 3.2×10^5 Btu/hr (94 kW)

(C) 3.9×10^5 Btu/hr (110 kW)

(D) 4.6×10^5 Btu/hr (130 kW)

2. A building is located in the city of New York. There are 4772 (2651 in SI) degree-days (basis of 65°F (18°C)) during the October 15 to May 15 period for this area. The building is heated by fuel oil whose heating value is 153,600 Btu/gal (42 800 MJ/m^3) and which costs \$0.15/gal (\$0.04/L). The calculated design heat loss is 3.5×10^6 Btu/hr (1 MW) based on 70°F (21.1°C) inside and 0°F (−17.8°C) outside design temperatures. The furnace has an efficiency of 70%. The cost of heating this building during the winter (October 15 through May 15) is most nearly

(A) \$7000

(B) \$8000

(C) \$9000

(D) \$10,000

3. A flat roof consists of $1^1/_2$ in (38 mm) of insulation installed over 3 in (76 mm) of soft pine sheathing, and a $^3/_4$ in (19 mm) acoustical ceiling suspended 4 in (100 mm) below the pine. (The surface emissivity of soft pine is $\epsilon_{\text{wood}} = 0.90$. The emissivity of the acoustical ceiling is $\epsilon_{\text{paper}} = 0.95$.) The outside wind velocity is 15 mi/hr (24 km/h). The interior design temperature is 80°F (26.7°C). The exterior design temperature is 95°F (35°C). The overall coefficient of heat transfer is most nearly

(A) 0.08 Btu/hr-ft^2-°F (0.5 W/m^2·°C)

(B) 1.0 Btu/hr-ft^2-°F (6.0 W/m^2·°C)

(C) 4.1 Btu/hr-ft^2-°F (24 W/m^2·°C)

(D) 18 Btu/hr-ft^2-°F (110 W/m^2·°C)

4. A 12 ft × 12 ft (3.6 m × 3.6 m) floor is constructed as a concrete slab with two exposed edges. The slab edge heat loss coefficient for these two edges is 0.55 Btu/hr-ft-°F (0.95 W/m·°C). The other two edges form part of a basement wall exposed to 70°F air. The inside design temperature is 70°F (21.1°C). The outdoor design temperature is −10°F (−23.3°C). Using the slab edge method, the heat loss from the slab is most nearly

(A) 500 Btu/hr (140 W)

(B) 800 Btu/hr (240 W)

(C) 1100 Btu/hr (300 W)

(D) 4200 Btu/hr (1300 W)

5. A first-floor office in a remodeled historic building has floor dimensions of 100 ft × 40 ft (30 m × 12 m) and a ceiling height of 10 ft (3 m). One of the 40 ft (12 m) walls is shared with an adjacent heated space. The three remaining walls have one 4 ft (1.2 m) wide × 6 ft (1.8 m) high, double glass with $^1/_4$ in (6.4 mm) air space, weather-stripped, double-hung window per 10 ft (3 m). The crack coefficient for the windows is 32 ft^3/hr-ft (3.0 m^3/h·m). One wall of 10 windows is exfiltrating. The basement and second floor are heated to 70°F (21.1°C). The wall coefficient (exclusive of film resistance) is 0.2 Btu/hr-ft^2-°F (1.1 W/m^2·°C), and the outside wind velocity is 15 mi/hr (24 km/h). The inside design temperature is 70°F (21.1°C); the outside design temperature is −10°F (−23.3°C). Including infiltration but disregarding ventilation air, the heating load is most nearly

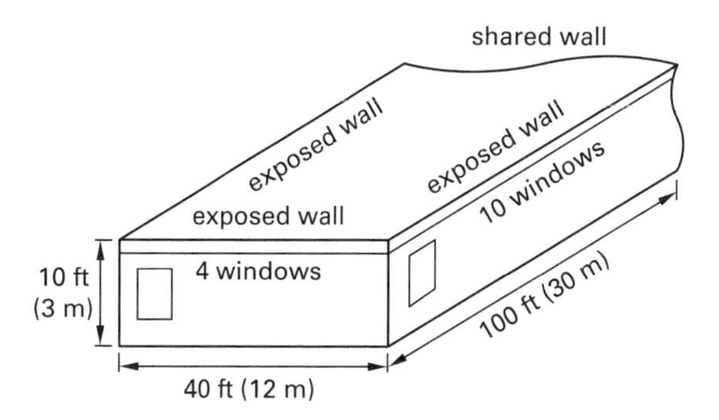

(A) 53,000 Btu/hr (15 kW)

(B) 66,000 Btu/hr (19 kW)

(C) 71,000 Btu/hr (20 kW)

(D) 79,000 Btu/hr (22 kW)

6. A building is located in Newark, New Jersey. The heating season lasts for 245 days. There are 5252 (2918 in SI) degree-days (basis of 65°F (18°C)), and the outside design temperature is 0°F (−17.8°C). The temperature in the building is maintained at 70°F (21.1°C) between the hours of 8:30 a.m. and 5:30 p.m. During the rest of the day and the night, the temperature is allowed to drop to 50°F (10°C). The building is heated with coal that has a heating value of 13,000 Btu/lbm (30.2 MJ/kg). A heat loss of 650,000 Btu/hr (0.19 MW) has been calculated based on 70°F (21.1°C) inside and 0°F (−17.8°C) outside temperatures. The furnace has an efficiency of 70%. The mass of coal required each year is most nearly

(A) 18,000 lbm/yr (4000 kg/yr)

(B) 43,000 lbm/yr (9500 kg/yr)

(C) 69,000 lbm/yr (15 000 kg/yr)

(D) 84,000 lbm/yr (40 000 kg/yr)

7. The design heat loss of a building is 200,000 Btu/hr (60 kW) originally based on 70°F (21.1°C) inside and 0°F (−17.8°C) outside design temperatures. At that location, there are 4200 (2333 in SI) degree-days (basis of 65°F (18°C)) over the 210-day heating season. The building is occupied 24 hr/day. If the thermostat is lowered from 70°F to 68°F (21.1°C to 20°C), the percentage reduction in heating fuel is most nearly

(A) 2.0%

(B) 4.0%

(C) 8.0%

(D) 14%

8. A building with the characteristics listed is located where the annual heating season lasts 21 weeks. The inside design temperature is 70°F (21.1°C). The average outside air temperature is 30°F (−1°C). The gas furnace has an efficiency of 75%. Fuel costs $0.25 per therm. The building is occupied only from 8 a.m. until 6 p.m., Monday through Friday. In the past, the building was maintained at 70°F (21.1°C) at all times and ventilated with one air change per hour (based on inside conditions). The thermostat is now being set back 12°F (6.7°C), and the ventilation is being reduced by 50% when the building is unoccupied. Infiltration through cracks and humidity changes are disregarded.

internal volume	801,000 ft^3 (22 700 m^3)
wall area	11,040 ft^2 (993 m^2)
wall overall heat transfer coefficient	0.15 Btu/hr-ft^2-°F (0.85 W/m^2·°C)
window area	2760 ft^2 (260 m^2)
window overall heat transfer coefficient	1.13 Btu/hr-ft^2-°F (6.42 W/m^2·°C)
roof area	26,700 ft^2 (2480 m^2)
roof overall heat transfer coefficient	0.05 Btu/hr-ft^2-°F (0.3 W/m^2·°C)
slab on grade	690 lineal ft (210 m) of exposed slab edge
slab edge coefficient	1.5 Btu/hr-ft-°F (2.6 W/m·°C)

(a) The unoccupied time for a 21-week period over a year is most nearly

(A) 1500 hr/yr

(B) 2100 hr/yr

(C) 2500 hr/yr

(D) 3600 hr/yr

(b) At 70°F (21.1°C), the energy savings due to the ventilation change are most nearly

(A) 5.6×10^8 Btu/yr (5.9×10^8 kJ/yr)

(B) 7.9×10^8 Btu/yr (8.3×10^8 kJ/yr)

(C) 9.3×10^8 Btu/yr (9.8×10^8 kJ/yr)

(D) 14×10^9 Btu/yr (15×10^9 kJ/yr)

(c) The savings due to reduced heat losses from the roof, floor, and walls is most nearly

(A) 4.1×10^7 Btu/yr (4.4×10^7 kJ/yr)

(B) 9.2×10^7 Btu/yr (9.8×10^7 kJ/yr)

(C) 1.8×10^8 Btu/yr (1.9×10^8 kJ/yr)

(D) 2.1×10^8 Btu/yr (2.3×10^8 kJ/yr)

(d) The energy saved per year is most nearly

(A) 11,000 therm/yr

(B) 13,000 therm/yr

(C) 15,000 therm/yr

(D) 18,000 therm/yr

(e) The annual savings is most nearly

(A) $3100/yr

(B) $3800/yr

(C) $4600/yr

(D) $6200/yr

SOLUTIONS

1. *Customary U.S. Solution*

Based on Sec. 42.10, for fluorescent lights, the rated wattage should be increased by 20–25% to account for ballast heating. Since the lights are pendant-mounted on chains from the ceiling, most of this heat enters the conditioned space. Assume rated wattage increases by 20%. η is given as 1. From Eq. 42.8(b), the internal heat gain due to lights (SF = 1.2) is

$$\begin{aligned}\dot{q}_{\text{lights}} &= \left(3412\ \frac{\text{Btu}}{\text{kW-hr}}\right)(\text{SF})\left(\frac{P_{\text{kW}}}{\eta}\right)\\ &= \left(3412\ \frac{\text{Btu}}{\text{kW-hr}}\right)(1.2)\left(\frac{12{,}000\ \text{W}}{(1)\left(1000\ \frac{\text{W}}{\text{kW}}\right)}\right)\\ &= 4.9133\times 10^4\ \text{Btu/hr}\end{aligned}$$

Use Eq. 42.8(b) to find the internal heat gain due to the twelve motors (SF = 0.8). η is given as 0.90.

$$\begin{aligned}\dot{q}_{\text{motors}} &= \left(2545\ \frac{\text{Btu}}{\text{hr-hp}}\right)(\text{SF})\left(\frac{P_{\text{hp}}}{\eta}\right)\\ &= (12\ \text{motors})\left(2545\ \frac{\text{Btu}}{\text{hp-hr}}\right)(0.8)\left(\frac{10\ \text{hp}}{0.90}\right)\\ &= 2.7147\times 10^5\ \text{Btu/hr}\end{aligned}$$

$$\begin{aligned}\dot{q}_{\text{total}} &= \dot{q}_{\text{lights}} + \dot{q}_{\text{motors}}\\ &= 4.9147\times 10^4\ \frac{\text{Btu}}{\text{hr}} + 2.7147\times 10^5\ \frac{\text{Btu}}{\text{hr}}\\ &= \boxed{3.206\times 10^5\ \text{Btu/hr}\quad (3.2\times 10^5\ \text{Btu/hr})}\end{aligned}$$

The answer is (B).

SI Solution

Based on the customary U.S. solution, SF = 1.2 for lights. η is given as 1. From Eq. 42.8(a), the internal heat gain due to lights is

$$\begin{aligned}\dot{q}_{\text{lights}} &= (\text{SF})\left(\frac{P_{\text{kW}}}{\eta}\right)\\ &= (1.2)\left(\frac{12\,000\ \text{W}}{(1)\left(1000\ \frac{\text{W}}{\text{kW}}\right)}\right)\\ &= 14.4\ \text{kW}\end{aligned}$$

Use Eq. 42.8(a), to find the internal heat gain due to the twelve motors (SF = 0.8). η is given as 0.90.

$$\begin{aligned}\dot{q}_{\text{motors}} &= \left(0.7457\ \frac{\text{kW}}{\text{hp}}\right)(\text{SF})\left(\frac{P_{\text{hp}}}{\eta}\right)\\ &= (12\ \text{motors})\left(0.7457\ \frac{\text{kW}}{\text{hp}}\right)(0.8)\left(\frac{10\ \text{hp}}{0.90}\right)\\ &= 79.5\ \text{kW}\end{aligned}$$

$$\begin{aligned}\dot{q}_{\text{total}} &= \dot{q}_{\text{lights}} + \dot{q}_{\text{motors}} = 14.4\ \text{kW} + 79.5\ \text{kW}\\ &= \boxed{93.9\ \text{kW}\quad (94\ \text{kW})}\end{aligned}$$

The answer is (B).

2. *Customary U.S. Solution*

From Eq. 42.13(b), the fuel consumption (in gal/heating season) is

$$\begin{aligned}&\frac{\left(24\ \frac{\text{hr}}{\text{day}}\right)\dot{q}_{\text{Btu/hr}}(\text{HDD})}{(T_i - T_o)(\text{HV}_{\text{Btu/gal}})\eta_{\text{furnace}}}\\ &\quad = \frac{\left(24\ \frac{\text{hr}}{\text{day}}\right)\left(3.5\times 10^6\ \frac{\text{Btu}}{\text{hr}}\right)(4772^\circ\text{F-days})}{(70^\circ\text{F} - 0^\circ\text{F})\left(153{,}600\ \frac{\text{Btu}}{\text{gal}}\right)(0.70)}\\ &\quad = 53{,}260\ \text{gal}\end{aligned}$$

The total cost of 53,260 gal fuel at \$0.15/gal is

$$(53{,}260\ \text{gal})\left(\frac{\$0.15}{\text{gal}}\right) = \boxed{\$7989\quad (\$8000)}$$

The answer is (D).

SI Solution

$$\text{HV}_{\text{kJ/L}} = \frac{\left(42\,800\ \frac{\text{MJ}}{\text{m}^3}\right)\left(1000\ \frac{\text{kJ}}{\text{MJ}}\right)}{1000\ \frac{\text{L}}{\text{m}^3}} = 42\,800\ \text{kJ/L}$$

From Eq. 42.13(a), fuel consumption (in L/heating season) is

$$\begin{aligned}&\frac{\left(86\,400\ \frac{\text{s}}{\text{d}}\right)\dot{q}_{\text{kW}}(\text{HDD})}{(T_i - T_o)(\text{HV}_{\text{kJ/L}})\eta_{\text{furnace}}}\\ &\quad = \frac{\left(86\,400\ \frac{\text{s}}{\text{d}}\right)(1\ \text{MW})\left(1000\ \frac{\text{kW}}{\text{MW}}\right)\times(2651^\circ\text{C·d})}{(21.1^\circ\text{C} - (-17.8^\circ\text{C}))\left(42\,800\ \frac{\text{kJ}}{\text{L}}\right)(0.70)}\\ &\quad = 196\,531\ \text{L}\end{aligned}$$

HVAC

The total cost at \$0.04/L is

$$(196\,531 \text{ L})\left(\frac{\$0.04}{\text{L}}\right) = \boxed{\$7861 \quad (\$8000)}$$

The answer is (B).

3. *Customary U.S. Solution*

From Table 42.2, the surface film coefficient for outside air (horizontal, heat flow down, 15 mph) is $h_o = 6.00 \text{ Btu/hr-ft}^2\text{-°F}$. The film thermal resistance is

$$R_1 = \frac{1}{h_o} = \frac{1}{6.00 \frac{\text{Btu}}{\text{hr-ft}^2\text{-°F}}} = 0.167 \text{ hr-ft}^2\text{-°F/Btu}$$

From App. 42.A, for roof insulation, the thermal resistance per inch is $2.78 \text{ hr-ft}^2\text{-°F/Btu-in}$.

$$R_2 = \left(2.78 \frac{\text{hr-ft}^2\text{-°F}}{\text{Btu-in}}\right)(1.5 \text{ in}) = 4.17 \text{ hr-ft}^2\text{-°F/Btu}$$

From App. 42.A, the thermal resistance per inch for softwood is $1.25 \text{ hr-ft}^2\text{-°F/Btu-in}$.

$$R_3 = \left(1.25 \frac{\text{hr-ft}^2\text{-°F}}{\text{Btu-in}}\right)(3 \text{ in}) = 3.75 \text{ hr-ft}^2\text{-°F/Btu}$$

Use Eq. 42.4 to find the effective surface emissivity for the space between soft wood and an acoustic ceiling. $\epsilon_1 = \epsilon_{\text{wood}} = 0.90$ (see Sec. 42.6). For acoustical tile, $\epsilon_2 = \epsilon_{\text{paper}} = 0.95$ (see App. 37.A).

$$E = \frac{1}{\frac{1}{\epsilon_1} + \frac{1}{\epsilon_2} - 1} = \frac{1}{\frac{1}{0.90} + \frac{1}{0.95} - 1} = 0.86$$

From Table 42.3, the thermal conductance of 4 in horizontal planar air space at $E = 0.86$ with heat flow down is approximately $0.81 \text{ Btu/hr-ft}^2\text{-°F}$. The thermal resistance is

$$R_4 = \frac{1}{0.81 \frac{\text{Btu}}{\text{hr-ft}^2\text{-°F}}} = 1.235 \text{ hr-ft}^2\text{-°F/Btu}$$

From App. 42.A, the thermal resistance for 3/4 in acoustical tile is $R_5 = 1.78 \text{ hr-ft}^2\text{-°F/Btu}$.

From Table 42.2, the surface film coefficient for inside air (v = 0) (horizontal, heat flow down) is $h_i = 1.08 \text{ Btu/hr-ft}^2\text{-°F}$. The film thermal resistance is

$$R_6 = \frac{1}{h_i} = \frac{1}{1.08 \frac{\text{Btu}}{\text{hr-ft}^2\text{-°F}}} = 0.926 \text{ hr-ft}^2\text{-°F/Btu}$$

The total resistance is

$$\begin{aligned} R_{\text{total}} &= R_1 + R_2 + R_3 + R_4 + R_5 + R_6 \\ &= 0.167 \frac{\text{hr-ft}^2\text{-°F}}{\text{Btu}} + 4.17 \frac{\text{hr-ft}^2\text{-°F}}{\text{Btu}} \\ &\quad + 3.75 \frac{\text{hr-ft}^2\text{-°F}}{\text{Btu}} + 1.235 \frac{\text{hr-ft}^2\text{-°F}}{\text{Btu}} \\ &\quad + 1.78 \frac{\text{hr-ft}^2\text{-°F}}{\text{Btu}} + 0.926 \frac{\text{hr-ft}^2\text{-°F}}{\text{Btu}} \\ &= 12.03 \text{ hr-ft}^2\text{-°F/Btu} \end{aligned}$$

The overall coefficient of heat transfer is

$$\begin{aligned} U &= \frac{1}{R_{\text{total}}} = \frac{1}{12.03 \frac{\text{hr-ft}^2\text{-°F}}{\text{Btu}}} \\ &= \boxed{0.0831 \text{ Btu/hr-ft}^2\text{-°F} \quad (0.08 \text{ Btu/hr-ft}^2\text{-°F})} \end{aligned}$$

The answer is (A).

SI Solution

From Table 42.2, the surface film coefficient for outside air (horizontal, heat flow down) at 15 mph is $h_o = 34.1 \text{ W/m}^2\text{·°C}$. The film thermal resistance is

$$R_1 = \frac{1}{h_o} = \frac{1}{34.1 \frac{\text{W}}{\text{m}^2\text{·°C}}} = 0.0293 \text{ m}^2\text{·°C/W}$$

From App. 42.A, for roof insulation, the thermal resistance per inch is $2.78 \text{ hr-ft}^2\text{-°F/Btu-in}$.

$$\begin{aligned} R_2 &= \frac{\left(2.78 \frac{\text{hr-ft}^2\text{-°F}}{\text{Btu-in}}\right)\left(6.93 \frac{\frac{\text{m·°C}}{\text{W}}}{\frac{\text{hr-ft}^2\text{-°F}}{\text{Btu-in}}}\right)(38 \text{ mm})}{1000 \frac{\text{mm}}{\text{m}}} \\ &= 0.732 \text{ m}^2\text{·°C/W} \end{aligned}$$

HVAC

From App. 42.A, the thermal resistance per inch for softwood is 1.25 hr-ft^2-°F/Btu-in.

$$R_3 = \frac{\left(1.25\ \frac{\text{hr-ft}^2\text{-°F}}{\text{Btu-in}}\right)\left(6.93\ \frac{\frac{\text{m·°C}}{\text{W}}}{\frac{\text{hr-ft}^2\text{-°F}}{\text{Btu-in}}}\right)(76\ \text{mm})}{1000\ \frac{\text{mm}}{\text{m}}}$$
$$= 0.658\ \text{m}^2\text{·°C/W}$$

From the customary U.S. solution, the effective surface emissivity between softwood and an acoustic ceiling is $E = 0.86$.

From Table 42.3, the thermal conductance of 100 mm horizontal planar air space at $E = 0.86$ with heat flow down is approximately 4.6 W/m^2·°C. The thermal resistance is

$$R_4 = \frac{1}{4.6\ \frac{\text{W}}{\text{m}^2\text{·°C}}} = 0.217\ \text{m}^2\text{·°C/W}$$

From App. 42.A, the thermal resistance for 19 mm (3/4 in) acoustical tile is

$$R_5 = \left(1.78\ \frac{\text{hr-ft}^2\text{-°F}}{\text{Btu}}\right)\left(0.176\ \frac{\frac{\text{m}^2\text{·°C}}{\text{W}}}{\frac{\text{hr-ft}^2\text{-°F}}{\text{Btu}}}\right)$$
$$= 0.313\ \text{m}^2\text{·°C/W}$$

From Table 42.2, the surface film coefficient for inside air ($v = 0$) (horizontal, heat flow down) is $h_i = 6.13$ W/m^2·°C. The film thermal resistance is

$$R_6 = \frac{1}{h_i} = \frac{1}{6.13\ \frac{\text{W}}{\text{m}^2\text{·°C}}} = 0.163\ \text{m}^2\text{·°C/W}$$

The total resistance is

$$\begin{aligned} R_{\text{total}} &= R_1 + R_2 + R_3 + R_4 + R_5 + R_6 \\ &= 0.0293\ \frac{\text{m}^2\text{·°C}}{\text{W}} + 0.732\ \frac{\text{m}^2\text{·°C}}{\text{W}} \\ &\quad + 0.658\ \frac{\text{m}^2\text{·°C}}{\text{W}} + 0.217\ \frac{\text{m}^2\text{·°C}}{\text{W}} \\ &\quad + 0.313\ \frac{\text{m}^2\text{·°C}}{\text{W}} + 0.163\ \frac{\text{m}^2\text{·°C}}{\text{W}} \\ &= 2.112\ \text{m}^2\text{·°C/W} \end{aligned}$$

The overall coefficient of heat transfer is

$$\begin{aligned} U &= \frac{1}{R_{\text{total}}} = \frac{1}{2.112\ \frac{\text{m}^2\text{·°C}}{\text{W}}} \\ &= \boxed{0.473\ \text{W/m}^2\text{·°C} \quad (0.5\ \text{W/m}^2\text{·°C})} \end{aligned}$$

The answer is (A).

4. *Customary U.S. Solution*

The slab edge coefficient is given as $F = 0.55$ Btu/hr-ft-°F.

The perimeter length is

$$\begin{aligned} p &= 12\ \text{ft} + 12\ \text{ft} \quad [\text{2 edges only}] \\ &= 24\ \text{ft} \end{aligned}$$

From Eq. 42.5, the heat loss from the slab is

$$\begin{aligned} \dot{q} &= pF(T_i - T_o) \\ &= (24\ \text{ft})\left(0.55\ \frac{\text{Btu}}{\text{hr-ft-°F}}\right)(70\text{°F} - (-10\text{°F})) \\ &= \boxed{1056\ \text{Btu/hr} \quad (1100\ \text{Btu/hr})} \end{aligned}$$

The answer is (C).

SI Solution

The slab edge coefficient is given as $F = 0.95$ W/m·°C.

The perimeter length is

$$\begin{aligned} p &= 3.6\ \text{m} + 3.6\ \text{m} \quad [\text{2 edges only}] \\ &= 7.2\ \text{m} \end{aligned}$$

From Eq. 42.5, the heat loss from the slab is

$$\begin{aligned} \dot{q} &= pF(T_i - T_o) \\ &= (7.2\ \text{m})\left(0.95\ \frac{\text{W}}{\text{m·°C}}\right)(21.1\text{°C} - (-23.3\text{°C})) \\ &= \boxed{303.7\ \text{W} \quad (300\ \text{W})} \end{aligned}$$

The answer is (C).

5. *Customary U.S. Solution*

For three unshared walls, the total exposed (walls + windows) area is

$$(10\ \text{ft})(40\ \text{ft} + 100\ \text{ft} + 100\ \text{ft}) = 2400\ \text{ft}^2$$

- There are two windows per 20 ft.
- The number of windows for a 40 ft wall is 4.
- The number of windows for each 100 ft wall is 10.
- The total number of windows is

$$4+10+10=24$$

- The total window area is

$$(24)(4\text{ ft})(6\text{ ft})=576\text{ ft}^2$$

- The total wall area is

$$2400\text{ ft}^2-576\text{ ft}^2=1824\text{ ft}^2$$

From Table 42.2, the outside film coefficient for a vertical wall in a 15 mph wind is $h_o=6.00$ Btu/hr-ft^2-°F.

From Table 42.2, the inside film coefficient for still air (vertical, heat flow horizontal) is $h_i=1.46$ Btu/hr-ft^2-°F.

Notice that L/k was given in the problem statement, not k. From Eq. 42.3, the overall coefficient of heat transfer for the wall is

$$\begin{aligned}U_{\text{wall}}&=\frac{1}{\frac{1}{h_i}+\frac{L_{\text{wall}}}{k_{\text{wall}}}+\frac{1}{h_o}}\\&=\frac{1}{\frac{1}{1.46\ \frac{\text{Btu}}{\text{hr-ft}^2\text{-°F}}}+\frac{1}{0.2\ \frac{\text{Btu}}{\text{hr-ft}^2\text{-°F}}}+\frac{1}{6.00\ \frac{\text{Btu}}{\text{hr-ft}^2\text{-°F}}}}\\&=0.1709\text{ Btu/hr-ft}^2\text{-°F}\end{aligned}$$

From Eq. 42.2, the heat loss from the wall is

$$\begin{aligned}\dot{q}_{\text{wall}}&=U_{\text{wall}}A_{\text{wall}}\Delta T\\&=\left(0.1709\ \frac{\text{Btu}}{\text{hr-ft}^2\text{-°F}}\right)(1824\text{ ft}^2)(70\text{°F}-(-10\text{°F}))\\&=24{,}938\text{ Btu/hr}\end{aligned}$$

From App. 42.A, for double, vertical (1/4 in air space) glass windows, the thermal resistance is 1.63 hr-ft^2-°F/Btu. From the footnote, this includes the inside and outside film coefficients.

$$U_{\text{windows}}=\frac{1}{1.63\ \frac{\text{hr-ft}^2\text{-°F}}{\text{Btu}}}=0.613\text{ Btu/hr-ft}^2\text{-°F}$$

The heat loss from the windows is

$$\begin{aligned}\dot{q}_{\text{windows}}&=U_{\text{windows}}A_{\text{windows}}\Delta T\\&=\left(0.613\ \frac{\text{Btu}}{\text{hr-ft}^2\text{-°F}}\right)(576\text{ ft}^2)(70\text{°F}-(-10\text{°F}))\\&=28{,}247\text{ Btu/hr}\end{aligned}$$

From Eq. 40.9, the infiltration flow rate is

$$\dot{V}=BL$$

The crack coefficient is given as $B=32$ ft^3/hr-ft.

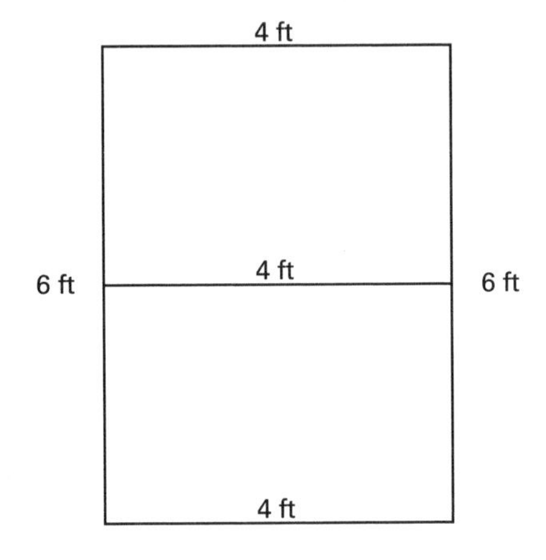

For double-hung windows, the crack length is

$$L=4\text{ ft}+4\text{ ft}+4\text{ ft}+6\text{ ft}+6\text{ ft}=24\text{ ft}$$

$$\dot{V}=BL=\left(32\ \frac{\text{ft}^3}{\text{hr-ft}}\right)(24\text{ ft})=768\text{ ft}^3\text{/hr per window}$$

Although there are 24 windows, one wall of 10 is exfiltrating.

$$\dot{V}=(24-10)\left(768\ \frac{\text{ft}^3}{\text{hr}}\right)=10{,}752\text{ ft}^3\text{/hr}$$

For atmospheric air, $p=14.7$ psia and $c_p=0.24$ Btu/lbm-°F.

The air density is

$$\begin{aligned}\rho=\frac{p}{RT}&=\frac{\left(14.7\ \frac{\text{lbf}}{\text{in}^2}\right)\left(12\ \frac{\text{in}}{\text{ft}}\right)^2}{\left(53.35\ \frac{\text{ft-lbf}}{\text{lbm-°R}}\right)(-10\text{°F}+460°)}\\&=0.088\text{ lbm/ft}^3\end{aligned}$$

HVAC

The mass flow rate of infiltrated air is

$$\dot{m} = \rho\dot{V} = \left(0.088\ \frac{\text{lbm}}{\text{ft}^3}\right)\left(10{,}752\ \frac{\text{ft}^3}{\text{hr}}\right)$$
$$= 946.2\ \text{lbm/hr}$$

The heat loss due to infiltration is

$$\dot{q}_{\text{infiltration}} = \dot{m}c_p\Delta T$$
$$= \left(946.2\ \frac{\text{lbm}}{\text{hr}}\right)\left(0.24\ \frac{\text{Btu}}{\text{lbm-°F}}\right) \times \left(70\text{°F} - (-10\text{°F})\right)$$
$$= 18{,}167\ \text{Btu/hr}$$

The total heat loss is

$$\dot{q}_{\text{total}} = \dot{q}_{\text{wall}} + \dot{q}_{\text{windows}} + \dot{q}_{\text{infiltration}}$$
$$= 24{,}938\ \frac{\text{Btu}}{\text{hr}} + 28{,}247\ \frac{\text{Btu}}{\text{hr}} + 18{,}167\ \frac{\text{Btu}}{\text{hr}}$$
$$= \boxed{71{,}352\ \text{Btu/hr}\quad (71{,}000\ \text{Btu/hr})}$$

The answer is (C).

SI Solution

For three unshared walls, the total exposed (walls + windows) area is

$$(3\ \text{m})(12\ \text{m} + 30\ \text{m} + 30\ \text{m}) = 216\ \text{m}^2$$

For 24 windows (from the customary U.S. solution), the total window area is

$$(24)(1.2\ \text{m})(1.8\ \text{m}) = 51.8\ \text{m}^2$$

The total wall area is

$$216\ \text{m}^2 - 51.8\ \text{m}^2 = 164.2\ \text{m}^2$$

From Table 42.2, the outside film coefficient for a vertical wall in a 15 mph wind is $h_o = 34.1\ \text{W/m}^2\cdot\text{°C}$.

From Table 42.2, the inside film coefficient for still air (vertical, heat flow horizontal) is $h_i = 8.29\ \text{W/m}^2\cdot\text{°C}$.

Notice that L/k was given in the problem statement, not k. From Eq. 42.3, the overall coefficient of heat transfer for the wall is

$$U_{\text{wall}} = \frac{1}{\frac{1}{h_i} + \frac{L_{\text{wall}}}{k_{\text{wall}}} + \frac{1}{h_o}}$$
$$= \frac{1}{\frac{1}{8.29\ \frac{\text{W}}{\text{m}^2\cdot\text{°C}}} + \frac{1}{1.1\ \frac{\text{W}}{\text{m}^2\cdot\text{°C}}} + \frac{1}{34.1\ \frac{\text{W}}{\text{m}^2\cdot\text{°C}}}}$$
$$= 0.944\ \text{W/m}^2\cdot\text{°C}$$

From Eq. 42.2, the heat loss from the wall is

$$\dot{q}_{\text{wall}} = U_{\text{wall}}A_{\text{wall}}\Delta T$$
$$= \left(0.944\ \frac{\text{W}}{\text{m}^2\cdot\text{°C}}\right)(164.2\ \text{m}^2)\left(21.1\text{°C} - (-23.3\text{°C})\right)$$
$$= 6882\ \text{W}$$

From the customary U.S. solution,

$$U_{\text{windows}} = \left(0.613\ \frac{\text{Btu}}{\text{hr-ft}^2\text{-°F}}\right)\left(5.68\ \frac{\frac{\text{W}}{\text{m}^2\cdot\text{°C}}}{\frac{\text{Btu}}{\text{hr-ft}^2\text{-°F}}}\right)$$
$$= 3.48\ \text{W/m}^2\cdot\text{°C}$$

The heat loss from the windows is

$$\dot{q}_{\text{windows}} = U_{\text{windows}}A_{\text{windows}}\Delta T$$
$$= \left(3.48\ \frac{\text{W}}{\text{m}^2\cdot\text{°C}}\right)(51.8\ \text{m}^2)\left(21.1\text{°C} - (-23.3\text{°C})\right)$$
$$= 8004\ \text{W}$$

From Eq. 40.9, the infiltration flow rate is

$$\dot{V} = BL$$

The crack coefficient is given as $B = 3.0\ \text{m}^3/\text{h}\cdot\text{m}$.

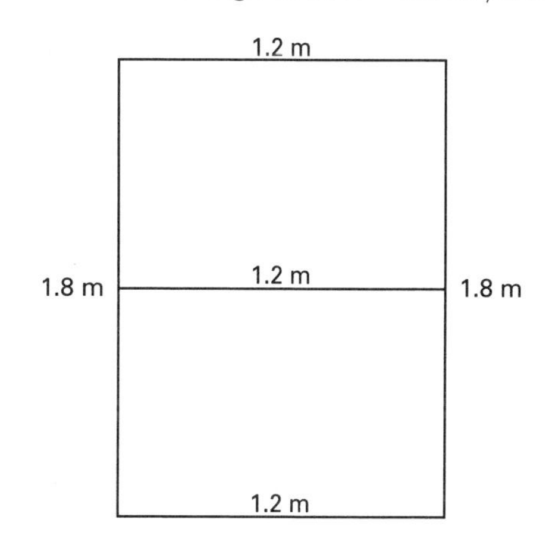

HVAC

For double-hung windows, the crack length is

$$L = 1.2 \text{ m} + 1.2 \text{ m} + 1.2 \text{ m} + 1.8 \text{ m} + 1.8 \text{ m} = 7.2 \text{ m}$$

$$\dot{V} = BL = \left(3.0 \ \frac{\text{m}^3}{\text{h·m}}\right)(7.2 \text{ m}) = 21.6 \text{ m}^3/\text{h per window}$$

Although there are 24 windows, one wall of 10 is exfiltrating.

$$\dot{V} = (24 - 10)\left(21.6 \ \frac{\text{m}^3}{\text{h}}\right) = 302.4 \text{ m}^3/\text{h}$$

For atmospheric air, $p = 100\,000$ Pa and $c_p = 1.005$ kJ/kg·°C.

The air density is

$$\rho = \frac{p}{RT} = \frac{100\,000 \text{ Pa}}{\left(287.03 \ \frac{\text{J}}{\text{kg·K}}\right)(-23.3°\text{C} + 273°)} = 1.4 \text{ kg/m}^3$$

The mass flow rate of infiltrated air is

$$\dot{m} = \rho\dot{V} = \left(1.4 \ \frac{\text{kg}}{\text{m}^3}\right)\left(302.4 \ \frac{\text{m}^3}{\text{h}}\right) = 423.4 \text{ kg/h}$$

The heat loss due to infiltration is

$$\dot{q}_{\text{infiltration}} = \dot{m}c_p\Delta T = \frac{\left(423.4 \ \frac{\text{kg}}{\text{h}}\right)\left(1.005 \ \frac{\text{kJ}}{\text{kg·°C}}\right)\left(1000 \ \frac{\text{J}}{\text{kJ}}\right) \times \left(21.1°\text{C} - (-23.3°\text{C})\right)}{3600 \ \frac{\text{s}}{\text{h}}} = 5248 \text{ W}$$

The total heat loss is

$$\dot{q}_{\text{total}} = \dot{q}_{\text{wall}} + \dot{q}_{\text{windows}} + \dot{q}_{\text{infiltration}} = 6882 \text{ W} + 8004 \text{ W} + 5248 \text{ W} = \boxed{20\,134 \text{ W} \quad (20 \text{ kW})}$$

The answer is (C).

6. *Customary U.S. Solution*

Use Hitchin's empirical formula to calculate the approximate average outside temperature from the degree-days. The standard base temperature for degree-days is 65°F. The empirical constant, k, for Fahrenheit temperatures is approximately 0.39.

$$\frac{\text{HDD}}{N} = \frac{T_b - \overline{T}}{1 - e^{-k(T_b - \overline{T})}}$$

$$\frac{5252 \text{ degree-days}}{245 \text{ days}} = \frac{65°\text{F} - \overline{T}}{1 - e^{-(0.39)(65°\text{F} - \overline{T})}}$$

By trial and error or any other method, $\overline{T} \approx 43.56°$F.

Since $\overline{T}$ is so much lower than T_b, the denominator is essentially 1.0, and the answer derived from Hitchin's formula is essentially what would have been derived directly from Eq. 42.12. However, Eq. 42.12 is less accurate when $\overline{T}$ and T_b are closer.

The heat loss per degree of temperature difference is

$$\frac{\dot{q}}{\Delta T} = \frac{650{,}000 \ \frac{\text{Btu}}{\text{hr}}}{70°\text{F} - 0°\text{F}} = 9286 \text{ Btu/hr-°F}$$

The period between 8:30 a.m. and 5:30 p.m. is 9 hr. For this period, the temperature in the building is 70°F. For the remaining 15 hr, the temperature in the building is 50°F. The total winter heat loss is

$$\dot{q}_{\text{total,winter}} = (245 \text{ days})\left(9286 \ \frac{\text{Btu}}{\text{hr-°F}}\right) \times \left(\begin{array}{l}\left(9 \ \frac{\text{hr}}{\text{day}}\right)(70°\text{F} - 43.56°\text{F}) \\ + \left(15 \ \frac{\text{hr}}{\text{day}}\right)(50°\text{F} - 43.56°\text{F})\end{array}\right) = 7.61 \times 10^8 \text{ Btu}$$

The fuel consumption (per year) is

$$\text{fuel consumption} = \frac{\dot{q}_{\text{total,winter}}}{(\text{HV})\eta_{\text{furnace}}} = \frac{7.61 \times 10^8 \text{ Btu}}{\left(13{,}000 \ \frac{\text{Btu}}{\text{lbm}}\right)(0.70)} = \boxed{83{,}626 \text{ lbm/yr} \quad (84{,}000 \text{ lbm/yr})}$$

The answer is (D).

HVAC

SI Solution

Use Hitchin's empirical formula to calculate the approximate average outside temperature from the degree-days. The standard base temperature for degree-days is 18°C. The empirical constant, k, for Celsius temperatures is approximately 0.71.

$$\frac{\text{HDD}}{N} = \frac{T_b - \overline{T}}{1 - e^{-k(T_b - \overline{T})}}$$

$$\frac{2918 \text{ degree-days}}{245 \text{ days}} = \frac{18°\text{C} - \overline{T}}{1 - e^{-(0.71)(18°\text{C} - \overline{T})}}$$

By trial and error or any other method, $\overline{T} \approx 6.09°\text{C}$.

Since $\overline{T}$ is so much lower than T_b, the denominator is essentially 1.0, and the answer derived from Hitchin's formula is essentially what would have been derived directly from Eq. 42.12. However, Eq. 42.12 is less accurate when $\overline{T}$ and T_b are closer.

The heat loss per degree of temperature difference is

$$\frac{\dot{q}}{\Delta T} = \frac{(0.19 \text{ MW})\left(10^6 \ \frac{\text{W}}{\text{MW}}\right)}{21.1°\text{C} - (-17.8°\text{C})} = 4884 \text{ W/°C}$$

The period between 8:30 a.m. and 5:30 p.m. is 9 h. For this period, the temperature in the building is 21.1°C. For the remaining 15 h, the temperature in the building is 10°C. The total winter heat loss is

$$\begin{aligned} \dot{q}_{\text{total,winter}} &= (245 \text{ d})\left(4884 \ \frac{\text{W}}{°\text{C}}\right) \\ &\quad \times \left(\left(\begin{array}{l} \left(9 \ \frac{\text{h}}{\text{d}}\right)\left(3600 \ \frac{\text{s}}{\text{h}}\right) \\ \quad \times (21.1°\text{C} - 6.09°\text{C}) \end{array} \right) + \left(\begin{array}{l} \left(15 \ \frac{\text{h}}{\text{d}}\right)\left(3600 \ \frac{\text{s}}{\text{h}}\right) \\ \quad \times (10°\text{C} - 6.09°\text{C}) \end{array} \right) \right) \\ &= 8.345 \times 10^5 \text{ MJ} \end{aligned}$$

The fuel consumption (per year) is

$$\begin{aligned} \text{fuel consumption} &= \frac{\dot{q}_{\text{total,winter}}}{(\text{HV})\eta_{\text{furnace}}} = \frac{8.345 \times 10^5 \text{ MJ}}{\left(30.2 \ \frac{\text{MJ}}{\text{kg}}\right)(0.70)} \\ &= \boxed{39\,475 \text{ kg/yr} \quad (40\,000 \text{ kg/yr})} \end{aligned}$$

The answer is (D).

7. *Customary U.S. Solution*

The heat loss per degree of temperature difference is

$$\frac{\dot{q}}{\Delta T} = \frac{200{,}000 \ \frac{\text{Btu}}{\text{hr}}}{70°\text{F} - 0°\text{F}} = 2857 \text{ Btu/hr-°F}$$

From Eq. 42.12, the average temperature, $\overline{T}$, over the entire heating season can be estimated.

$$\frac{\text{HDD}}{N} = T_b - \overline{T}$$

$$\text{HDD} = N(T_b - \overline{T}) = N(65°\text{F} - \overline{T})$$

The heating degree-days, HDD, are 4200.

The number of days in the heating season, N, is 210.

$$4200 \text{ days} = (210 \text{ days})(65°\text{F} - \overline{T})$$

$$\overline{T} = 45°\text{F}$$

The total original winter heat loss based on 70°F inside is

$$\begin{aligned} &(210 \text{ days})\left(24 \ \frac{\text{hr}}{\text{day}}\right)\left(2857 \ \frac{\text{Btu}}{\text{hr-°F}}\right) \\ &\quad \times (70°\text{F} - 45°\text{F}) = 3.60 \times 10^8 \text{ Btu} \end{aligned}$$

The reduced heat loss based on 68°F inside is

$$\begin{aligned} &(210 \text{ days})\left(24 \ \frac{\text{hr}}{\text{day}}\right)\left(2857 \ \frac{\text{Btu}}{\text{hr-°F}}\right) \\ &\quad \times (68°\text{F} - 45°\text{F}) = 3.31 \times 10^8 \text{ Btu} \end{aligned}$$

The reduction is

$$\frac{3.60 \times 10^8 \text{ Btu} - 3.31 \times 10^8 \text{ Btu}}{3.60 \times 10^8 \text{ Btu}} = \boxed{0.081 \quad (8.0\%)}$$

The answer is (C).

SI Solution

Since the problem only concerns the percentage reduction in heat loss, determine only the average temperature, $\overline{T}$, as all other parameters except inside temperature remain constant.

From Eq. 42.12, the average temperature, $\overline{T}$, over the entire heating season can be calculated.

$$\frac{\text{HDD}}{N} = T_b - \overline{T}$$
$$\text{HDD} = N(T_b - \overline{T}) = N(18°\text{C} - \overline{T})$$

The heating Kelvin degree-days, HDD, are 2333 K·d.

The number of days in the entire heating season, N, is 210.

$$2333 \text{ d} = (210 \text{ d})(18°\text{C} - \overline{T})$$
$$\overline{T} = 6.89°\text{C}$$

The reduction is

$$\frac{\Delta T_{\text{original}} - \Delta T_{\text{reduced}}}{\Delta T_{\text{original}}}$$
$$\Delta T_{\text{original}} = 21.1°\text{C} - 6.89°\text{C} = 14.21°\text{C}$$
$$\Delta T_{\text{reduced}} = 20°\text{C} - 6.89°\text{C} = 13.11°\text{C}$$

The reduction is

$$\frac{14.21°\text{C} - 13.11°\text{C}}{14.21°\text{C}} = \boxed{0.077 \quad (8.0\%)}$$

The answer is (C).

8. *Customary U.S. Solution*

(a) Prior to the thermostat change, the energy requirement for ventilation air is

$$q_{\text{air},1} = m_1 c_p (T_{i,1} - T_o) = N_1 V t \rho c_p (T_{i,1} - T_o)$$

After the change, the energy requirement is

$$q_{\text{air},2} = m_2 c_p (T_{i,2} - T_o) = N_2 V t \rho c_p (T_{i,2} - T_o)$$

The energy savings are

$$\Delta q_{\text{air}} = q_{\text{air},1} - q_{\text{air},2} = N_1 V t \rho c_p (T_{i,1} - T_o) - N_2 V t \rho c_p (T_{i,2} - T_o) = V t \rho c_p \big(N_1 (T_{i,1} - T_o) - N_2 (T_{i,2} - T_o)\big)$$

During an entire week, the building is occupied from 8:00 a.m. to 6:00 p.m. (for 10 hr) and is unoccupied 14 hr each day for 5 days, and it is unoccupied 24 hr each day for 2 days (over the weekend).

The unoccupied time for a 21-week period over a year is

$$t = \left(14 \ \frac{\text{hr}}{\text{day}}\right)\left(5 \ \frac{\text{days}}{\text{wk}}\right)\left(21 \ \frac{\text{wk}}{\text{yr}}\right) + \left(24 \ \frac{\text{hr}}{\text{day}}\right)\left(2 \ \frac{\text{days}}{\text{wk}}\right)\left(21 \ \frac{\text{wk}}{\text{yr}}\right) = \boxed{2478 \text{ hr/yr} \quad (2500 \text{ hr/yr})}$$

The answer is (C).

(b) With air at 70°F, $c_p \approx 0.24$ Btu/lbm-°F and $\rho \approx 0.075$ lbm/ft^3. From part (a), the energy savings are

$$\Delta q_{\text{air}} = V t \rho c_p \big(N_1 (T_{i,1} - T_o) - N_2 (T_{i,2} - T_o)\big)$$
$$= \left(801{,}000 \ \frac{\text{ft}^3}{\text{air change}}\right)\left(2478 \ \frac{\text{hr}}{\text{yr}}\right)\left(0.075 \ \frac{\text{lbm}}{\text{ft}^3}\right) \times \left(0.24 \ \frac{\text{Btu}}{\text{lbm-°F}}\right) \times \left(\begin{array}{l}\left(1 \ \frac{\text{air change}}{\text{hr}}\right) \\ \times (70°\text{F} - 30°\text{F}) - \left(0.5 \ \frac{\text{air change}}{\text{hr}}\right) \\ \times ((70°\text{F} - 12°\text{F}) - 30°\text{F})\end{array}\right)$$
$$= \boxed{9.289 \times 10^8 \text{ Btu/yr} \quad (9.3 \times 10^8 \text{ Btu/yr})}$$

The answer is (C).

(c) The savings due to reduced heat losses from the roof, floor, and walls (for a 12°F setback) are

$$\dot{q} = UA\Delta T$$
$$= \left(\begin{array}{l}\left(0.15 \ \frac{\text{Btu}}{\text{hr-ft}^2\text{-°F}}\right)(11{,}040 \text{ ft}^2) \\ + \left(1.13 \ \frac{\text{Btu}}{\text{hr-ft}^2\text{-°F}}\right)(2760 \text{ ft}^2) \\ + \left(0.05 \ \frac{\text{Btu}}{\text{hr-ft}^2\text{-°F}}\right)(26{,}700 \text{ ft}^2) \\ + \left(1.5 \ \frac{\text{Btu}}{\text{hr-ft-°F}}\right)(690 \text{ ft})\end{array}\right)(12°\text{F})$$
$$= 85{,}738 \text{ Btu/hr}$$

$$q = \dot{q}t = \left(85{,}738 \ \frac{\text{Btu}}{\text{hr}}\right)\left(2478 \ \frac{\text{hr}}{\text{yr}}\right) = \boxed{2.125 \times 10^8 \text{ Btu/yr} \quad (2.1 \times 10^8 \text{ Btu/yr})}$$

The answer is (D).

(d) The energy saved per year is

$$q_{\text{actual}} = \frac{q_{\text{ideal}}}{\eta_{\text{furnace}}} = \frac{\Delta q_{\text{air}} + q}{\eta_{\text{furnace}}} = \frac{9.289 \times 10^8 \ \frac{\text{Btu}}{\text{yr}} + 2.125 \times 10^8 \ \frac{\text{Btu}}{\text{yr}}}{(0.75)\left(100{,}000 \ \frac{\text{Btu}}{\text{therm}}\right)} = \boxed{15{,}218 \text{ therm/yr} \quad (15{,}000 \text{ therm/yr})}$$

The answer is (C).

(e) The annual cost savings are

$$\left(15{,}218 \ \frac{\text{therm}}{\text{yr}}\right)\left(\frac{\$0.25}{\text{therm}}\right) = \boxed{\$3805/\text{yr} \quad (\$3800/\text{yr})}$$

The answer is (B).

SI Solution

(a) Prior to the thermostat change, the energy requirement for ventilation air is

$$q_{\text{air},1} = m_1 c_p (T_{i,1} - T_o) = N_1 V t \rho c_p (T_{i,1} - T_o)$$

After the change, the energy requirement is

$$q_{\text{air},2} = m_2 c_p (T_{i,2} - T_o) = N_2 V t \rho c_p (T_{i,2} - T_o)$$

The energy savings are

$$\begin{aligned} \Delta q_{\text{air}} &= q_{\text{air},1} - q_{\text{air},2} \\ &= N_1 V t \rho c_p (T_{i,1} - T_o) - N_2 V t \rho c_p (T_{i,2} - T_o) \\ &= V t \rho c_p \big(N_1 (T_{i,1} - T_o) - N_2 (T_{i,2} - T_o)\big) \end{aligned}$$

During an entire week, the building is occupied from 8:00 a.m. to 6:00 p.m. (for 10 h) and is unoccupied 14 h each day for 5 d, and it is unoccupied 24 h each day for 2 d (over the weekend).

The unoccupied time for a 21-week period over a year is

$$\begin{aligned} t &= \left(14 \ \frac{\text{h}}{\text{d}}\right)\left(5 \ \frac{\text{d}}{\text{wk}}\right)\left(21 \ \frac{\text{wk}}{\text{yr}}\right) + \left(24 \ \frac{\text{h}}{\text{d}}\right)\left(2 \ \frac{\text{d}}{\text{wk}}\right)\left(21 \ \frac{\text{wk}}{\text{yr}}\right) \\ &= \boxed{2478 \text{ h/yr} \quad (2500 \text{ h/yr})} \end{aligned}$$

The answer is (C).

(b) With air at 21°C, $c_p \approx 1.005$ kJ/kg·°C and $\rho \approx 1.2$ kg/m^3.

From part (a), the energy savings are

$$\begin{aligned} \Delta q_{\text{air}} &= V t \rho c_p \big(N_1 (T_{i,1} - T_o) - N_2 (T_{i,2} - T_o)\big) \\ &= \left(22\,700 \ \frac{\text{m}^3}{\text{air change}}\right)\left(2478 \ \frac{\text{h}}{\text{yr}}\right)\left(1.2 \ \frac{\text{kg}}{\text{m}^3}\right) \\ &\quad \times \left(1.005 \ \frac{\text{kJ}}{\text{kg}\cdot{}^\circ\text{C}}\right) \\ &\quad \times \left(\begin{array}{l} \left(1 \ \frac{\text{air change}}{\text{h}}\right)\left(21.1^\circ\text{C} - (-1^\circ\text{C})\right) \\ - \left(0.5 \ \frac{\text{air change}}{\text{h}}\right)\left((21.1^\circ\text{C} - 6.7^\circ\text{C}) - (-1^\circ\text{C})\right) \end{array}\right) \\ &= \boxed{9.769 \times 10^8 \text{ kJ/yr} \quad (9.8 \times 10^8 \text{ kJ/yr})} \end{aligned}$$

The answer is (C).

(c) The savings due to reduced heat losses from the roof, floor, and walls for 6.7°C setback are

$$\begin{aligned} \dot{q} &= U A \Delta T \\ &= \left(\begin{array}{l} \left(0.85 \ \frac{\text{W}}{\text{m}^2\cdot{}^\circ\text{C}}\right)(993 \text{ m}^2) \\ + \left(6.42 \ \frac{\text{W}}{\text{m}^2\cdot{}^\circ\text{C}}\right)(260 \text{ m}^2) \\ + \left(0.3 \ \frac{\text{W}}{\text{m}^2\cdot{}^\circ\text{C}}\right)(2480 \text{ m}^2) \\ + \left(2.0 \ \frac{\text{W}}{\text{m}\cdot{}^\circ\text{C}}\right)(210 \text{ m}) \end{array}\right)(6.7^\circ\text{C}) \\ &= 25\,482 \text{ W} \\ q &= \dot{q} t \\ &= \frac{(25\,482 \text{ W})\left(2478 \ \frac{\text{h}}{\text{yr}}\right)\left(3600 \ \frac{\text{s}}{\text{h}}\right)}{1000 \ \frac{\text{J}}{\text{kJ}}} \\ &= \boxed{2.273 \times 10^8 \text{ kJ/yr} \quad (2.3 \times 10^8 \text{ kJ/yr})} \end{aligned}$$

The answer is (D).

HVAC

(d) The energy saved per year is

$$q_{\text{actual}} = \frac{q_{\text{ideal}}}{\eta_{\text{furnace}}} = \frac{\Delta q_{\text{air}} + q}{\eta_{\text{furnace}}}$$

$$= \frac{9.769 \times 10^8 \ \frac{\text{kJ}}{\text{yr}} + 2.273 \times 10^8 \ \frac{\text{kJ}}{\text{yr}}}{(0.75)\left(1000 \ \frac{\text{kJ}}{\text{MJ}}\right)\left(105.506 \ \frac{\text{MJ}}{\text{therm}}\right)}$$

$$= \boxed{15\,218 \text{ therm/yr} \quad (15\,000 \text{ therm/yr})}$$

The answer is (C).

(e) The annual cost savings are

$$\left(15\,218 \ \frac{\text{therm}}{\text{yr}}\right)\left(\frac{\$0.25}{\text{therm}}\right) = \boxed{\$3805/\text{yr} \quad (\$3800/\text{yr})}$$

The answer is (B).

HVAC

43 Cooling Load

PRACTICE PROBLEMS

1. An air conditioning unit has a SEER-13 rating and a cooling load of 8000 Btu/hr. The unit operates eight hours per day for 140 days each year. The average cost of electricity is $0.25/kW-hr.

(a) What is most nearly the air conditioning unit's average power usage?

(A) 450,000 W-hr/yr

(B) 690,000 W-hr/yr

(C) 9,000,000 W-hr/yr

(D) 120,000,000 W-hr/yr

(b) What is most nearly the hourly cost of operating the air conditioning unit?

(A) $0.15/hr

(B) $0.47/hr

(C) $1.10/hr

(D) $2.00/hr

2. A small, one-story building in Austin, TX, consists of a single room. The room is 40 ft wide × 50 ft long and has an 8 ft ceiling. A blower door has been used to determine the CFM50 as 1800 ft^3/min. The energy climate factor is 17.

(a) What is most nearly ACH50?

(A) 0.11

(B) 0.40

(C) 6.8

(D) 9.1

(b) What is most nearly ACHnat?

(A) 0.40 air changes/hr

(B) 0.52 air changes/hr

(C) 0.67 air changes/hr

(D) 0.79 air changes/hr

3. Multiple blower door leakage tests performed on a building are used to develop a CFM50 building leakage curve correlation with coefficient $C = 110.2$ and exponent $n = 0.702$. What airflow is needed to create a 5 Pa pressure difference in the building?

(A) 210 ft^3/min

(B) 340 ft^3/min

(C) 390 ft^3/min

(D) 430 ft^3/min

4. The bypass air conditioning system shown has the following operating characteristics.

total air pressure	14.7 psia (101.3 kPa)
supply temperature	58°F (14.4°C) dry-bulb
sensible load	200,000 Btu/hr (58.6 kW)
latent load	450,000 grains/hr (29 kg/h)
outside air	90°F (32.2°C) dry-bulb, 76°F (24.4°C) wet-bulb
make-up outside air	2000 ft^3/min (940 L/s)
condition of air leaving air handler	saturated

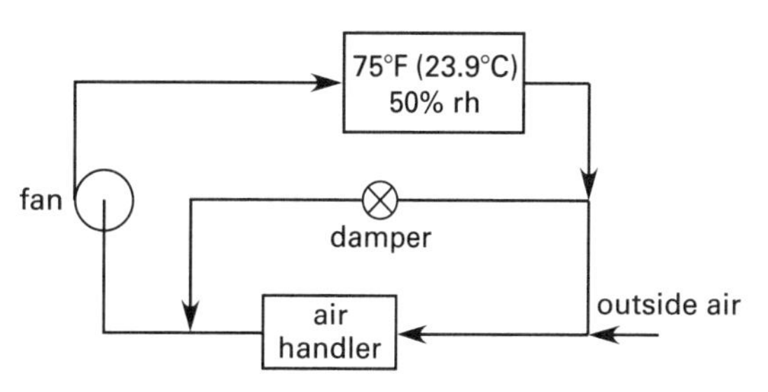

(a) The temperature of the air leaving the air handler is most nearly

(A) 34°F (1°C)

(B) 38°F (3°C)

(C) 49°F (9°C)

(D) 55°F (13°C)

(b) The rate of supply air is most nearly

(A) 7400 ft^3/min (3400 L/s)

(B) 11,000 ft^3/min (5100 L/s)

(C) 14,000 ft^3/min (6400 L/s)

(D) 19,000 ft^3/min (8700 L/s)

(c) The moisture content of the supply air is most nearly

(A) 0.0081 lbm/lbm (8.1 g/kg)

(B) 0.0097 lbm/lbm (9.7 g/kg)

(C) 0.0110 lbm/lbm (11 g/kg)

(D) 0.0130 lbm/lbm (13 g/kg)

(d) What is most nearly the efficiency of the process experienced by the air passing through the air handler's coils?

(A) 0%

(B) 35%

(C) 65%

(D) 100%

(e) The bypass factor of the system is most nearly

(A) 0

(B) 0.35

(C) 0.65

(D) 1.0

5. A building is located at 32°N latitude. The inside design temperature is 78°F (25.6°C). The outside design temperature is 95°F (35°C). The daily temperature range is 22°F (12.2°C). Energy gains through the floor, from lights, and from occupants are insignificant. The building's construction is as follows.

walls	1600 ft^2 (144 m^2) facing north 1400 ft^2 (126 m^2) facing south 1500 ft^2 (135 m^2) facing east 1400 ft^2 (126 m^2) facing west 4 in (100 mm) brick facing 3 in (75 mm) concrete block 1 in (25 mm) mineral wool (emissivity = 0.96) 2 in (50 mm) furring 3/8 in (9.5 mm) drywall gypsum (emissivity = 0.95) 1/2 in (12 mm) plaster
roof	6000 ft^2 (540 m^2) 4 in concrete (100 mm) 2 in insulation (50 mm) insulation conductivity 2.28 hr-ft^2-°F/Btu-in single felt layer 1 in (25 mm) air gap ($\epsilon = 0.90$) 1/2 in (12 mm) acoustical ceiling tile
windows	100 ft^2 (9 m^2) facing east 1/4 in (6.4 mm) thick, single glazing cream-colored Venetian shades (shading factor = 0.75) no exterior shading

The maximum temperature differences occur between 4 p.m. and 6 p.m. The equivalent temperature difference for the concrete roof is 74°F at 4 p.m. and 68°F at 6 p.m. The equivalent temperature differences for 4 in brick face walls are as follows.

time	facing direction	temperature difference
4 p.m.	N	17°F
	E	20°F
	S	28°F
	W	20°F
6 p.m.	N	20°F
	E	22°F
	S	28°F
	W	35°F

(a) What is the cooling load for mid-July at 4:00 p.m. sun time? (Solve using customary U.S. units.)

(A) 46,000 Btu/hr

(B) 55,000 Btu/hr

(C) 69,000 Btu/hr

(D) 79,000 Btu/hr

(b) Is this the peak cooling load? Explain. (Solve using customary U.S. units.)

(A) Yes. The instantaneous heat absorption by the building peaks around 12:00 p.m. (noon) local sun time; the cooling load peaks around 4:00 p.m. local sun time.

(B) Yes. The instantaneous heat absorption by the building peaks around 2:00 p.m. local sun time; the cooling load peaks around 4:00 p.m. local sun time.

(C) No. The instantaneous heat absorption by the building peaks around 12:00 p.m. (noon) local sun time; the cooling load peaks around 2:00 p.m. local sun time.

(D) No. The instantaneous heat absorption by the building peaks around 4:00 p.m. local sun time; the cooling load peaks around 6:00 p.m. local sun time.

SOLUTIONS

1. (a) The total annual cooling load is

$$\dot{q}_{c,\text{annual}} = \left(8000\ \frac{\text{Btu}}{\text{hr}}\right)\left(140\ \frac{\text{days}}{\text{yr}}\right)\left(8\ \frac{\text{hr}}{\text{day}}\right) = 8{,}960{,}000\ \text{Btu/yr}$$

With a SEER-13 rating, the annual electrical energy usage is

$$\begin{aligned}\text{annual energy} &= \frac{\dot{q}_{c,\text{annual}}}{\text{SEER}_{\text{Btu/W-hr}}} \\ &= \frac{8{,}960{,}000\ \frac{\text{Btu}}{\text{yr}}}{13\ \frac{\text{Btu}}{\text{W-hr}}} \\ &= \boxed{\begin{array}{l}689{,}231\ \text{W-hr/yr} \\ \quad (690{,}000\ \text{W-hr/yr})\end{array}}\end{aligned}$$

The answer is (B).

(b) The power usage while the air conditioning unit is operating is

$$\begin{aligned}P_{\text{kW}} &= \frac{\dot{q}_{c,\text{hourly}}}{\text{SEER}_{\text{Btu/kW-hr}}} \\ &= \frac{8000\ \frac{\text{Btu}}{\text{hr}}}{\left(13\ \frac{\text{Btu}}{\text{W-hr}}\right)\left(1000\ \frac{\text{W}}{\text{kW}}\right)} \\ &= 0.615\ \text{kW}\end{aligned}$$

The hourly cost is

$$\begin{aligned}\text{cost}_{\text{hourly}} &= P_{\text{kW}}(\text{cost}_{\text{avg}}) \\ &= (0.615\ \text{kW})\left(0.25\ \frac{\$}{\text{kW-hr}}\right) \\ &= \boxed{\$0.154/\text{hr} \quad (\$0.15/\text{hr})}\end{aligned}$$

The answer is (A).

2. (a) The building's volume is

$$\begin{aligned}V_{\text{structure,ft}^3} &= (40\ \text{ft})(50\ \text{ft})(8\ \text{ft}) \\ &= 16{,}000\ \text{ft}^3\end{aligned}$$

Rearranging Eq. 43.12, ACH50 is

$$\text{CFM50} = \frac{(\text{ACH50})\,V_{\text{structure,ft}^3}}{60\ \frac{\text{min}}{\text{hr}}}$$

$$\begin{aligned}\text{ACH50} &= \frac{(\text{CFM50})\left(60\ \frac{\text{min}}{\text{hr}}\right)}{V_{\text{structure,ft}^3}} \\ &= \frac{\left(1800\ \frac{\text{ft}^3}{\text{min}}\right)\left(60\ \frac{\text{min}}{\text{hr}}\right)}{16{,}000\ \text{ft}^3} \\ &= \boxed{6.75 \quad (6.8)}\end{aligned}$$

The answer is (C).

(b) Use Eq. 43.16. The energy climate factor is the same as the LBL factor or the N factor.

$$\begin{aligned}\text{ACHnat} &= \frac{\text{ACH50}}{\text{LBL}} = \frac{6.75}{17} \\ &= \boxed{\begin{array}{l}0.397\ \text{air changes/hr} \\ \quad (0.40\ \text{air changes/hr})\end{array}}\end{aligned}$$

The answer is (A).

3. Using Eq. 43.11,

$$\begin{aligned}Q_{\text{ft}^3/\text{min}} &= C\Delta p_{\text{Pa}}^n = (110.2)(5\ \text{Pa})^{0.702} \\ &= \boxed{341.1\ \text{ft}^3/\text{min} \quad (340\ \text{ft}^3/\text{min})}\end{aligned}$$

The answer is (B).

4. *Customary U.S. Solution*

(a) This is a standard bypass problem.

The indoor conditions are

$$\begin{aligned}T_i &= 75^\circ\text{F} \\ \phi_i &= 50\%\end{aligned}$$

The outdoor conditions are

$$\begin{aligned}T_{o,\text{db}} &= 90^\circ\text{F} \\ T_{o,\text{wb}} &= 76^\circ\text{F}\end{aligned}$$

The ventilation rate is 2000 ft^3/min.

HVAC

The loads are

$$q_s = 200{,}000 \text{ Btu/hr}$$
$$q_l = 450{,}000 \text{ gr/hr}$$

To find the sensible heat ratio, q_l must be expressed in Btu/hr.

From the psychrometric chart (see App. 38.A), for the room conditions, $\omega = 0.0095$ lbm moisture/lbm dry air.

From Eq. 38.7,

$$\begin{aligned}\omega &= 0.622\left(\frac{p_w}{p_a}\right)\\ &= 0.622\left(\frac{p_w}{p - p_w}\right)\\ 0.0095\ \frac{\text{lbm moisture}}{\text{lbm dry air}} &= (0.622)\left(\frac{p_{w,\text{psia}}}{14.7 \text{ psia} - p_{w,\text{psia}}}\right)\\ p_w &= 0.22 \text{ psia}\end{aligned}$$

From steam tables (see App. 23.A), for 0.22 psia,

$$\begin{aligned}h_{fg} &= 1061.7 \text{ Btu/lbm}\\ q_l &= \frac{\left(450{,}000\ \frac{\text{gr}}{\text{hr}}\right)\left(1061.7\ \frac{\text{Btu}}{\text{lbm}}\right)}{7000\ \frac{\text{gr}}{\text{lbm}}}\\ &= 68{,}252 \text{ Btu/hr}\end{aligned}$$

From Eq. 43.21, the room sensible heat ratio is

$$\begin{aligned}\text{RSHR} &= \frac{q_s}{q_s + q_l} = \frac{200{,}000\ \frac{\text{Btu}}{\text{hr}}}{200{,}000\ \frac{\text{Btu}}{\text{hr}} + 68{,}252\ \frac{\text{Btu}}{\text{hr}}}\\ &= 0.75\end{aligned}$$

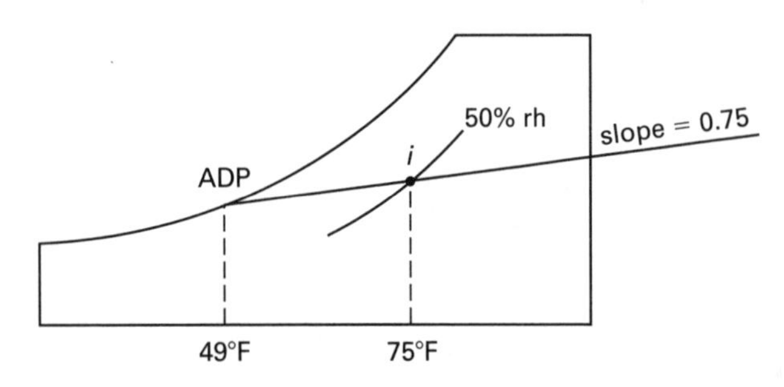

Locate 75°F and 50% relative humidity on the psychrometric chart (see App. 38.A). Draw a line with a slope of 0.75 through this point.

The left-hand intersection shows $T_{co} = 49°F$. Since the air leaves the conditioner saturated,

$$\text{BF}_{\text{coil}} = 0$$
$$T_{\text{co}} = \boxed{49°\text{F}}$$

The answer is (C).

(b) Calculate the air flow through the room from Eq. 43.23(b).

$$\begin{aligned}\dot{V}_{\text{in,ft}^3/\text{min}} &= \frac{\dot{q}_{s,\text{Btu/hr}}}{\left(1.08\ \frac{\text{Btu-min}}{\text{hr-ft}^3\text{-°F}}\right)(T_i - T_{\text{in}})}\\ &= \frac{200{,}000\ \frac{\text{Btu}}{\text{hr}}}{\left(1.08\ \frac{\text{Btu-min}}{\text{hr-ft}^3\text{-°F}}\right)(75°\text{F} - 58°\text{F})}\\ &= \boxed{10{,}893 \text{ ft}^3\text{/min} \quad (11{,}000 \text{ ft}^3\text{/min})}\end{aligned}$$

The answer is (B).

(c) Since $T_{\text{in}} = 58°F$, locate this dry-bulb temperature on the condition line. Reading from the chart (see App. 38.A),

$$\omega_{\text{in}} = \boxed{0.0081 \text{ lbm/lbm}}$$

The answer is (A).

(d) Since the air leaves the coils saturated, the efficiency of the air handler is $\boxed{100\%.}$ No air escapes being saturated.

The answer is (D).

(e) The bypass factor of the system can be calculated from the dry-bulb temperatures.

$$\begin{aligned}\text{BF} &= \frac{T_{\text{supply}} - T_{\text{air handler}}}{T_{\text{bypass}} - T_{\text{air hander}}} = \frac{58°\text{F} - 49°\text{F}}{75°\text{F} - 49°\text{F}}\\ &= \boxed{0.35}\end{aligned}$$

The answer is (B).

SI Solution

(a) The indoor conditions are

$$T_i = 23.9°\text{C}$$
$$\phi_i = 50\%$$

The outdoor conditions are

$$T_{o,\text{db}} = 32.2^\circ\text{C}$$
$$T_{o,\text{wb}} = 24.4^\circ\text{C}$$

The ventilation rate is 940 L/s.

The loads are

$$q_s = 58.6\ \text{kW}$$
$$q_l = 29\ \text{kg/h}$$

To find the sensible heat ratio, q_l must be expressed in kilowatts.

From the psychrometric chart (see App. 38.B), for the room conditions, $\omega \approx 9.3$ g moisture/kg dry air (9.3×10^{-3} kg moisture/kg dry air).

From Eq. 38.7,

$$\begin{aligned}\omega &= 0.622\left(\frac{p_w}{p_a}\right)\\ &= 0.622\left(\frac{p_w}{p - p_w}\right)\\ 9.3 \times 10^{-3}\ \frac{\text{kg moisture}}{\text{kg dry air}} &= (0.622)\left(\frac{p_w}{101.3\ \text{kPa} - p_w}\right)\\ p_w &= 1.49\ \text{kPa}\end{aligned}$$

From steam tables (see App. 23.N), for 1.49 kPa,

$$\begin{aligned}h_{fg} &= 2470.3\ \text{kJ/kg}\\ q_l &= \frac{\left(29\ \frac{\text{kg}}{\text{h}}\right)\left(2470.3\ \frac{\text{kJ}}{\text{kg}}\right)}{3600\ \frac{\text{s}}{\text{h}}}\\ &= 19.9\ \text{kW}\end{aligned}$$

From Eq. 38.22, the sensible heat ratio is

$$\begin{aligned}\text{SHR} &= \frac{q_s}{q_s + q_l}\\ &= \frac{58.6\ \text{kW}}{58.6\ \text{kW} + 19.9\ \text{kW}}\\ &= 0.75\end{aligned}$$

Locate 23.9°C and 50% relative humidity on the psychrometric chart (see App. 38.B). Draw a line with a slope of 0.75 through this point.

The left-hand intersection shows ADP = 9°C. Since the air leaves the conditioner saturated,

$$\text{BF}_{\text{coil}} = 0$$
$$T_{\text{co}} = \boxed{9^\circ\text{C}}$$

The answer is (C).

(b) Calculate the air flow through the room from Eq. 43.23(a).

$$\begin{aligned}\dot{V}_{\text{in,L/s}} &= \frac{\dot{q}_{s,\text{W}}}{\left(1.20\ \frac{\text{J}}{\text{L}\cdot^\circ\text{C}}\right)(T_i - T_{\text{in}})}\\ &= \frac{(58.6\ \text{kW})\left(1000\ \frac{\text{W}}{\text{kW}}\right)}{\left(1.20\ \frac{\text{J}}{\text{L}\cdot^\circ\text{C}}\right)(23.9^\circ\text{C} - 14.4^\circ\text{C})}\\ &= \boxed{5140\ \text{L/s}\quad(5100\ \text{L/s})}\end{aligned}$$

The answer is (B).

(c) Since $T_{\text{in}} = 14.4^\circ\text{C}$, locate this dry-bulb temperature on the condition line. Reading from the chart (see App. 38.B),

$$\omega_{\text{in}} = \boxed{8.1\ \text{g/kg dry air}}$$

The answer is (A).

(d) Since the air leaves the coils saturated, the efficiency of the air handler is $\boxed{100\%.}$ No air escapes being saturated.

The answer is (D).

(e) The bypass factor of the system can be calculated from the dry-bulb temperatures.

$$\begin{aligned}\text{BF} &= \frac{T_{\text{supply}} - T_{\text{air handler}}}{T_{\text{bypass}} - T_{\text{air handler}}} = \frac{58^\circ\text{C} - 49^\circ\text{C}}{75^\circ\text{C} - 49^\circ\text{C}}\\ &= \boxed{0.35}\end{aligned}$$

The answer is (B).

5. (a) *step 1:* Determine thermal resistances for the walls, roof, and windows.

For the walls:

Assume 2 in furring to be 2 in air space.

From Eq. 42.4, the effective space emissivity is

$$\frac{1}{E}=\frac{1}{\epsilon_1}+\frac{1}{\epsilon_2}-1=\frac{1}{0.96}+\frac{1}{0.95}-1$$
$$E=0.91$$

From Table 42.3, for vertical air space orientation, the thermal conductance (by extrapolating for $E=0.91$) is

$$\frac{1}{C}=\frac{1}{1.07\ \frac{\text{Btu}}{\text{hr-ft}^2\text{-}^\circ\text{F}}}=0.93\ \text{hr-ft}^2\text{-}^\circ\text{F/Btu}$$

Use App. 42.A.

walls

type of construction	$R, \frac{\text{hr-ft}^2\text{-}^\circ\text{F}}{\text{Btu}}$	notes
4 in brick spacing	0.44	
3 in concrete block	0.40	
1 in mineral wool	3.85	value given per in
2 in furring	0.93	see calculation above
3/8 in drywall	0.34	0.45 for 1/2 is given
gypsum		
1/2 in plaster (sand aggregate)	0.09	
surface outside (Table 42.3)	0.25	assume 7 1/2 mph air speed
surface inside (Table 42.3)	0.68	still air, $U=1.46\ \frac{\text{Btu}}{\text{hr-ft}^2\text{-}^\circ\text{F}}$

roof

type of construction	$R, \frac{\text{hr-ft}^2\text{-}^\circ\text{F}}{\text{Btu}}$	notes
4 in concrete	0.32	sand aggregate
2 in insulation	5.56	2.78/in
felt	0.06	
1 in air gap	0.87	$\epsilon=0.90$ for buildup material
1/2 in accoustical ceiling tile	1.19	
surface outside (Table 42.3)	0.25	assume 7 1/2 mph air speed and downward heat flow ~upward heat flow
surface inside (Table 42.3)	0.93	still air, $U=1.08\ \frac{\text{Btu}}{\text{hr-ft}^2\text{-}^\circ\text{F}}$

windows

type of construction	$R, \frac{\text{hr-ft}^2\text{-}^\circ\text{F}}{\text{Btu}}$	notes
1/4 in thick, single glazing	0.66	without shades, $R=0.88\ \frac{\text{hr-ft}^2\text{-}^\circ\text{F}}{\text{Btu}}$
cream colored venetian shades, no exterior shades		shading factor = 0.75

step 2: Determine the overall heat transfer coefficient for the walls, roof, and windows using Eq. 42.3.

$$U=\frac{1}{\sum R_i}$$

For the walls,

$$U=\frac{1}{\begin{array}{l}0.25\ \frac{\text{hr-ft}^2\text{-}^\circ\text{F}}{\text{Btu}}+0.44\ \frac{\text{hr-ft}^2\text{-}^\circ\text{F}}{\text{Btu}}\\ \quad+0.40\ \frac{\text{hr-ft}^2\text{-}^\circ\text{F}}{\text{Btu}}+3.85\ \frac{\text{hr-ft}^2\text{-}^\circ\text{F}}{\text{Btu}}\\ \quad+0.93\ \frac{\text{hr-ft}^2\text{-}^\circ\text{F}}{\text{Btu}}+0.34\ \frac{\text{hr-ft}^2\text{-}^\circ\text{F}}{\text{Btu}}\\ \quad+0.09\ \frac{\text{hr-ft}^2\text{-}^\circ\text{F}}{\text{Btu}}+0.68\ \frac{\text{hr-ft}^2\text{-}^\circ\text{F}}{\text{Btu}}\end{array}}$$
$$=0.143\ \text{Btu/hr-ft}^2\text{-}^\circ\text{F}$$

The cooling load is calculated from the equivalent temperature differences for the walls and roof. Tables of these values are not common; they are now often incorporated directly into HVAC computer application databases. Results from the remainder of this problem will vary with the temperature differences used.

For 4 in brick walls facing at 4 p.m.,

$$Q_{\text{walls}}=\left(0.143\ \frac{\text{Btu}}{\text{hr-ft}^2\text{-}^\circ\text{F}}\right)\times\left(\begin{array}{l}(1600\ \text{ft}^2)(17^\circ\text{F})+(1400\ \text{ft}^2)(28^\circ\text{F})\\ \quad+(1500\ \text{ft}^2)(20^\circ\text{F})+(1400\ \text{ft}^2)(20^\circ\text{F})\end{array}\right)$$
$$=17{,}789\ \text{Btu/hr}$$

For the roof,

$$U=\frac{1}{\begin{array}{l}0.25\ \frac{\text{hr-ft}^2\text{-}^\circ\text{F}}{\text{Btu}}+0.32\ \frac{\text{hr-ft}^2\text{-}^\circ\text{F}}{\text{Btu}}\\ \quad+5.56\ \frac{\text{hr-ft}^2\text{-}^\circ\text{F}}{\text{Btu}}+0.06\ \frac{\text{hr-ft}^2\text{-}^\circ\text{F}}{\text{Btu}}\\ \quad+0.87\ \frac{\text{hr-ft}^2\text{-}^\circ\text{F}}{\text{Btu}}+1.19\ \frac{\text{hr-ft}^2\text{-}^\circ\text{F}}{\text{Btu}}\\ \quad+0.93\ \frac{\text{hr-ft}^2\text{-}^\circ\text{F}}{\text{Btu}}\end{array}}$$
$$=0.109\ \text{Btu/hr-ft}^2\text{-}^\circ\text{F}$$

For a 4 in concrete roof at 4 p.m.,

$$\begin{aligned} Q_{\text{roof}} &= \left(0.109\ \frac{\text{Btu}}{\text{hr-ft}^2\text{-}^\circ\text{F}}\right)(6000\ \text{ft}^2)(74^\circ\text{F}) \\ &= 48{,}396\ \text{Btu/hr} \end{aligned}$$

For the windows:

Since all windows facing east will receive no direct sunlight at 4 p.m., consider $\Delta T = 95^\circ\text{F} - 78^\circ\text{F}$.

$$\begin{aligned} Q_{\text{windows}} &= \left(\frac{1}{0.66\ \frac{\text{hr-ft}^2\text{-}^\circ\text{F}}{\text{Btu}}}\right)(100\ \text{ft}^2)(95^\circ\text{F} - 78^\circ\text{F}) \\ &= 2576\ \text{Btu/hr} \end{aligned}$$

The total sensible transmission load (at 4 p.m.) is

$$\begin{aligned} Q_{\text{total}} &= 17{,}789\ \frac{\text{Btu}}{\text{hr}} + 48{,}396\ \frac{\text{Btu}}{\text{hr}} + 2576\ \frac{\text{Btu}}{\text{hr}} \\ &= \boxed{68{,}761\ \text{Btu/hr}\quad (69{,}000\ \text{Btu/hr})} \end{aligned}$$

The answer is (C).

(b) This $\boxed{\text{may not}}$ be the peak cooling load. Since Q_{walls} and Q_{roof} are the major contributors and ΔT equivalent temperature differences are maximum between 4 p.m. and 6 p.m., the peak cooling load will be maximum somewhere between $\boxed{\text{4 p.m. and 6 p.m.}}$ Consider Q_{walls} and Q_{roof} for 6 p.m.

For 4 in brick walls facing at 6 p.m.,

$$\begin{aligned} Q_{\text{walls}} &= \left(0.143\ \frac{\text{Btu}}{\text{hr-ft}^2\text{-}^\circ\text{F}}\right) \\ &\quad \times \left(\begin{array}{l}(1600\ \text{ft}^2)(20^\circ\text{F}) + (1400\ \text{ft}^2)(28^\circ\text{F}) \\ \quad + (1500\ \text{ft}^2)(22^\circ\text{F}) + (1400\ \text{ft}^2)(35^\circ\text{F})\end{array}\right) \\ &= 21{,}908\ \text{Btu/hr} \end{aligned}$$

For the roof,

$$\begin{aligned} Q_{\text{roof}} &= \left(0.109\ \frac{\text{Btu}}{\text{hr-ft}^2\text{-}^\circ\text{F}}\right)(6000\ \text{ft}^2)(68^\circ\text{F}) \\ &= 44{,}472\ \text{Btu/hr} \end{aligned}$$

The total sensible transmission load at 6 p.m. is

$$\begin{aligned} Q_{\text{total}} &= 21{,}908\ \frac{\text{Btu}}{\text{hr}} + 44{,}472\ \frac{\text{Btu}}{\text{hr}} + 2576\ \frac{\text{Btu}}{\text{hr}} \\ &= 68{,}956\ \text{Btu/hr} \end{aligned}$$

This is very close to the 4 p.m. load.

The answer is (D).

44 Air Conditioning Systems and Controls

PRACTICE PROBLEMS

1. What is the typical approximate supply air pressure in a pneumatic HVAC control system?

(A) 7 psig

(B) 18 psig

(C) 35 psig

(D) 140 psig

2. The control signal in a pneumatic control system is air pressure that falls within which approximate range?

(A) 0–4 psig

(B) 3–15 psig

(C) 12–45 psig

(D) 30–120 psig

3. What kind of device is a pneumatic thermostat?

(A) reverse acting

(B) normally closed

(C) normally open

(D) direct acting

4. What is the term for a device that converts a pneumatic signal of 15 psig to a control pressure of 20 psig?

(A) a pneumatic transformer

(B) a pneumatic amplifier

(C) a pneumatic relay

(D) a pneumatic booster

5. What kind of device is a pneumatic hot water valve?

(A) reverse acting

(B) normally closed

(C) normally open

(D) direct acting

6. What kind of device is a pneumatic steam valve?

(A) reverse acting

(B) normally closed

(C) normally open

(D) direct acting

Relay Logic Diagrams

7. The control relay diagram shown is missing a rung. What are the most likely elements of that rung?

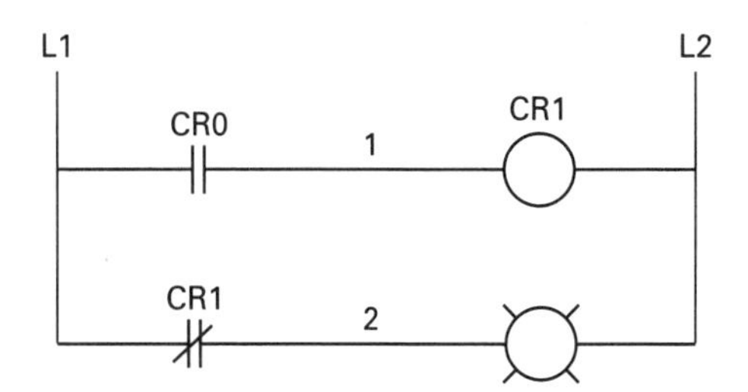

(A) a switch

(B) a relay coil

(C) a switch and a relay coil

(D) a switch and a relay contact

8. What is the most likely description of element M1?

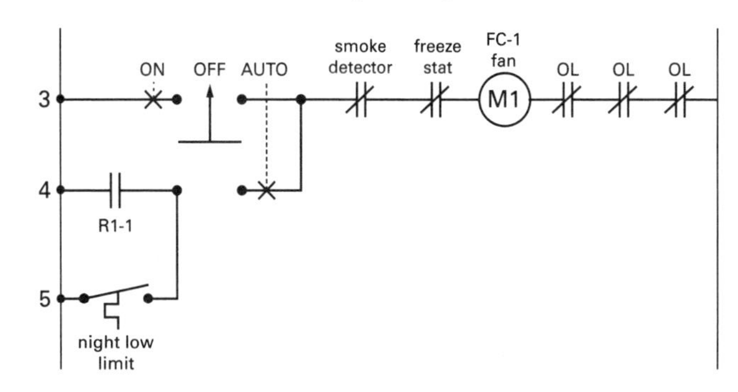

(A) a start/run enable solenoid or exciter circuit

(B) a fan motor

(C) a relay coil

(D) a fan interrupt circuit

HVAC

9. Which Boolean truth table describes the control of motor M in the relay logic diagram shown?

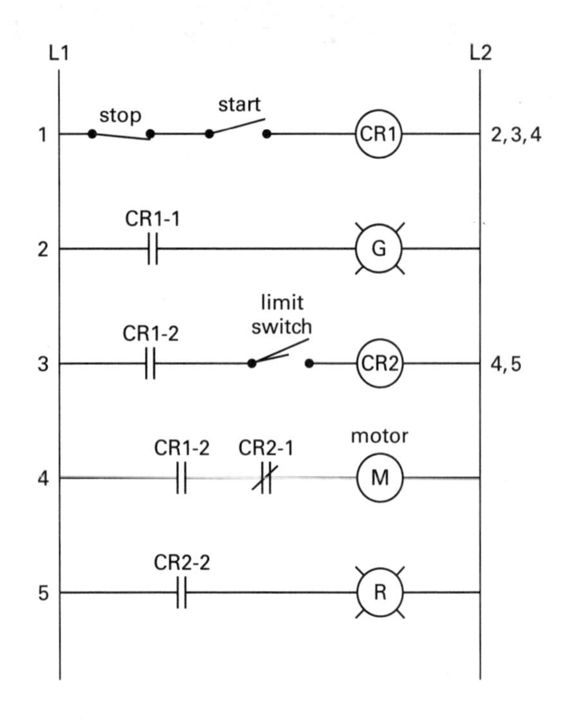

(A)

CR1	CR2	M
0	0	1
0	1	0
1	0	0
1	1	0

(B)

CR1	CR2	M
0	0	0
0	1	1
1	0	0
1	1	0

(C)

CR1	CR2	M
0	0	0
0	1	0
1	0	1
1	1	0

(D)

CR1	CR2	M
0	0	0
0	1	0
1	0	0
1	1	1

10. Given that the stop switch is closed as shown and motor M is not running, what are the functions of switches PB1 and PB2 in the relay logic diagram?

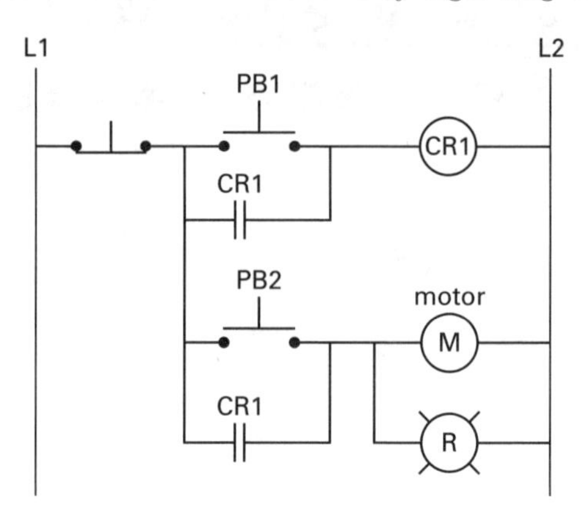

(A) PB1 starts the motor running continuously, and PB2 stops the motor.

(B) PB1 starts the motor for as long as PB1 is pushed, and PB2 starts the motor running continuously.

(C) PB1 stops the motor, and PB2 starts the motor running continuously.

(D) PB1 starts the motor running continuously, and PB2 starts the motor for as long as PB2 is pushed.

11. What are the functions of the (initially) NC M2 and M1 relay contacts in the two main rungs?

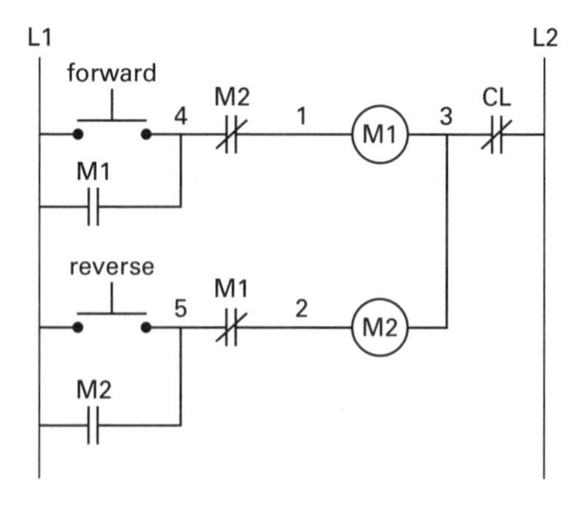

(A) When running forward, NC M2 latches and prevents the device from stopping.

(B) When running forward, NC M2 prevents the device from being placed in reverse.

(C) When running forward, NC M1 prevents the device from being placed in reverse.

(D) When running forward, NC M1 latches and prevents the device from stopping.

HVAC

12. What is most likely the purpose of the transformer in the control circuit shown?

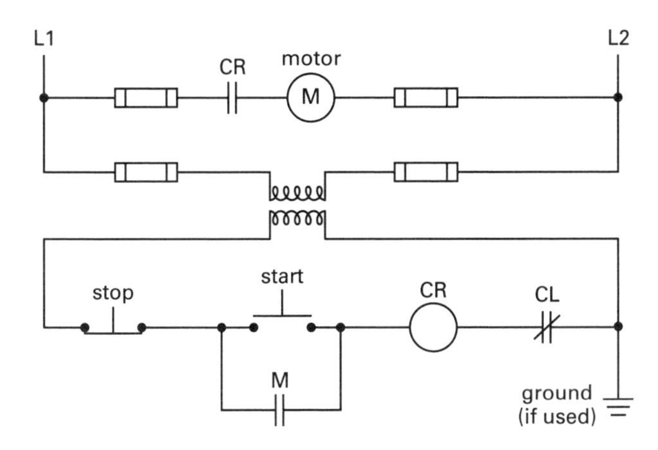

(A) isolate the motor M from start/stop transients

(B) step up the control voltage to motor M level

(C) step up the line voltage to control relay CR level

(D) step down the line voltage for safety

SOLUTIONS

1. Pneumatic air pressure is typically supplied at 18 psig to 20 psig, but may be as high as 25 psig.

The answer is (B).

2. A device's pneumatic signal cannot be less than the supply signal, and it must be sufficient for detection. 3–15 psig is a typical range of output pressures.

The answer is (B).

3. Thermostats are direct acting (DA) because, as the room cools, an increase in control pressure is needed.

The answer is (D).

4. A pneumatic relay has three ports: a signal input, a supply input, and a control output. When the full input signal is received, the control output valve is opened to the full supply input.

The answer is (C).

5. A hot water valve is normally open so that the valve will provide uncontrolled heat in the event of a power failure or air pressure outage.

The answer is (C).

6. A steam valve is normally closed so that uncontrolled steam will not be introduced into the HVAC system in the event of a power failure or air pressure outage.

The answer is (B).

7. Relay contact CR0 does not have any way to change state. The missing rung must have the CR0 relay coil, and the coil itself needs a control element, probably a switch of some type.

The answer is (C).

8. The smoke detector, freeze stat, and overload relay contacts are all NC, which means power is supplied through them to M1 during normal operation in order to keep fan FC-1 running.

M1 is probably a start/run enable solenoid or fan motor exciter circuit.

The answer is (A).

9. When the start switch in rung 1 is pushed, the coil in control relay CR1 is energized. The coil controls contacts in rungs 2, 3, and 4. The NO contacts CR1-2 in rung 3 close, making the rung live, but power is not supplied to coil CR2 until the limit switch is closed. In rung 4, the NO contacts CR1-2 also close, and power is supplied through the NC contacts CR2-1 to motor M. If the limit switch is closed by overtravel, relay coil CR2 will be energized, opening the NC CR2-1 contacts in rung 4, and stopping the motor, M. Therefore, in order for the motor to run, CR1 must be energized, and CR2 must not be energized.

The answer is (C).

10. If the motor is not running, when PB1 is pushed, (circled) relay coil CR1 is energized, and both NO CR1 contacts are closed. This supplies power through the lowest sub-rung to motor M, turning on the motor continuously (because contacts CR1 are latched on). If the motor is not running, pushing PB2 will supply power to the motor for as long as PB2 is pushed. PB2 does not control a relay coil and has no effect on the CR1 contacts.

The answer is (D).

11. When the forward switch is pressed, power is supplied through the switch and NC M2 to coil M1, opening the NC contacts M1 in rung 2.

This prevents the device from being placed in reverse.

The answer is (C).

12. The lower circuit does not contain a battery or connection to a power source. The voltage in the circuit is provided by the transformer. The transformer is probably a step-down transformer that allows the start/stop control circuit to operate at a lower (and, therefore, safer) voltage than the controlled fused upper circuit.

The answer is (D).

HVAC

45 Determinate Statics

PRACTICE PROBLEMS

1. Two towers are located on level ground 100 ft (30 m) apart. They support a transmission line with a mass of 2 lbm/ft (3 kg/m). The midpoint sag is 10 ft (3 m).

(a) The midpoint tension is most nearly

(A) 125 lbf (0.55 kN)

(B) 250 lbf (1.1 kN)

(C) 375 lbf (1.6 kN)

(D) 500 lbf (2.2 kN)

(b) The maximum tension in the transmission line is most nearly

(A) 170 lbf (0.70 kN)

(B) 210 lbf (0.86 kN)

(C) 270 lbf (1.2 kN)

(D) 330 lbf (1.4 kN)

(c) If the maximum tension is 500 lbf (2200 N), the sag in the cable is most nearly

(A) 1.3 ft (0.4 m)

(B) 3.2 ft (1.0 m)

(C) 4.0 ft (1.2 m)

(D) 5.1 ft (1.5 m)

2. Two legs of a tripod are mounted on a vertical wall. Both legs are horizontal. The apex is 12 distance units from the wall. The right leg is 13.4 units long. The wall mounting points are 10 units apart. A third leg is mounted on the wall 6 units to the left of the right upper leg and 9 units below the two top legs. A vertical downward load of 200 is supported at the apex. What is most nearly the reaction at the lowest mounting point?

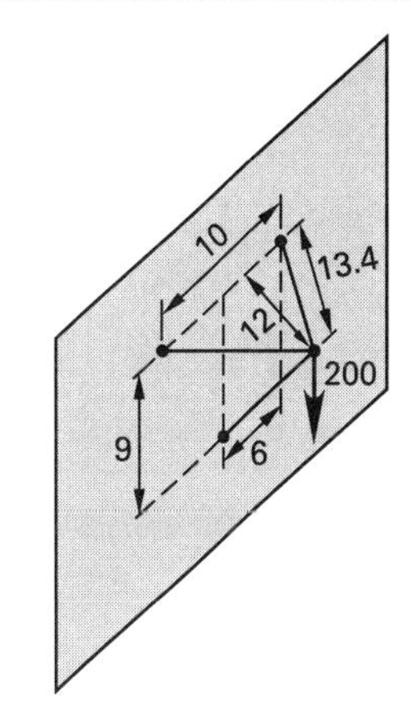

(A) 120

(B) 170

(C) 250

(D) 330

3. The ideal truss shown is supported by a pinned connection at point D and a roller connection at point C. Loads are applied at points A and F. What is most nearly the force in member DE?

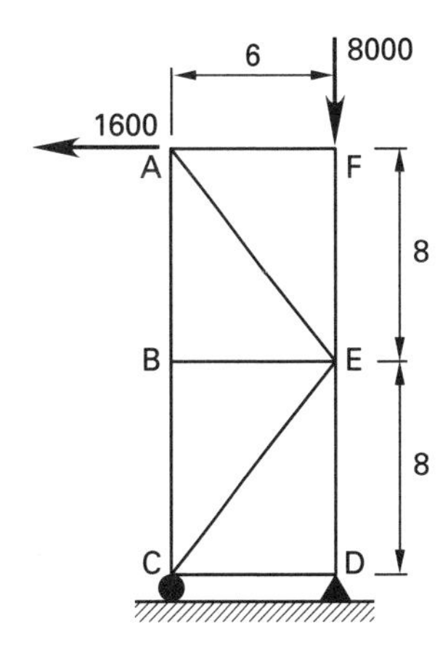

(A) 1200

(B) 2700

(C) 3300

(D) 3700

Statics

4. A pin-connected tripod is loaded at the apex by a horizontal force of 1200, as shown. What is most nearly the magnitude of the force in member AD?

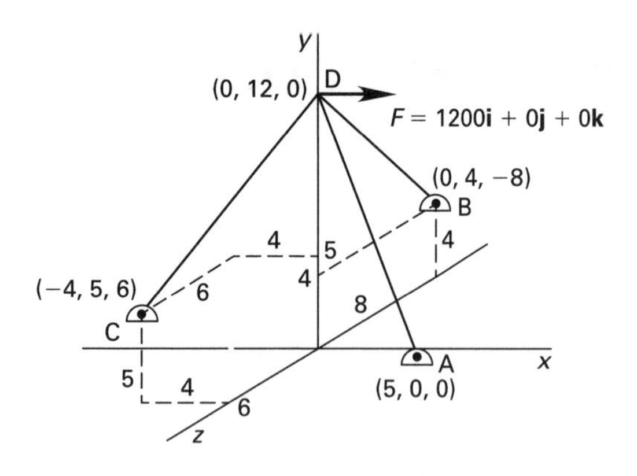

(A) 1100

(B) 1300

(C) 1800

(D) 2500

5. A truss is loaded by forces of 4000 at each upper connection point and forces of 60,000 at each lower connection point, as shown.

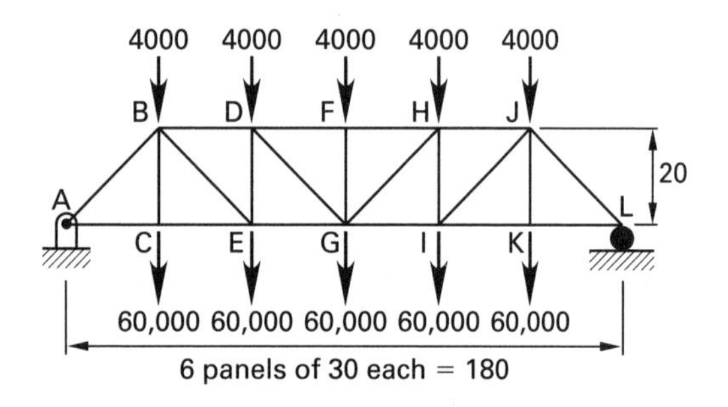

(a) The force in member DE is most nearly

(A) 36,000

(B) 45,000

(C) 60,000

(D) 160,000

(b) The force in member HJ is most nearly

(A) 24,000

(B) 60,000

(C) 160,000

(D) 380,000

6. The rigid rod AO is supported by guy wires BO and CO, as shown. (Points A, B, and C are all in the same vertical plane. Points A, O, and C are all in the same horizontal plane. Points A, O, and B are all in the same vertical plane.) Vertical and horizontal forces are 12,000 and 6000, respectively, as carried at the end of the rod. What are most nearly the x-, y-, and z-components of the reactions at point C?

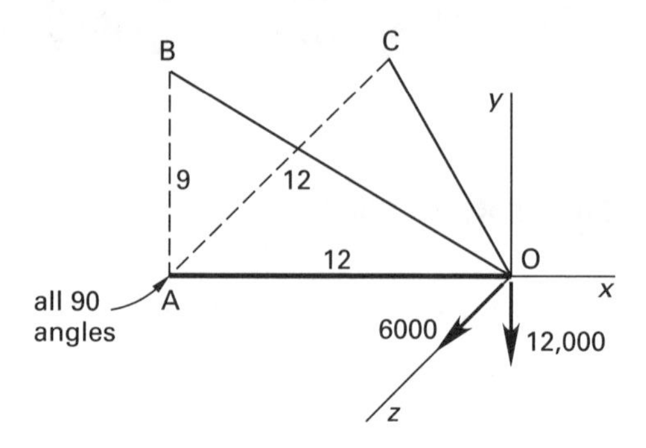

(A) $(C_x, C_y, C_z) = (0, 6000, 0)$

(B) $(C_x, C_y, C_z) = (6000, 0, 6000)$

(C) $(C_x, C_y, C_z) = (6000, 0, 4200)$

(D) $(C_x, C_y, C_z) = (4200, 0, 8500)$

7. When the temperature is 70°F (21.11°C), sections of steel railroad rail are welded end to end to form a continuous, horizontal track that is exactly 1 mi (1.6 km) long. Both ends of the track are constrained by preexisting installed sections of rail. Before the 1 mi section of track can be nailed to the ties, however, the sun warms it to a uniform temperature of 99.14°F (37.30°C). Laborers watch in amazement as the rail pops up in the middle and takes on a parabolic shape. The laborers prop the rail up (so that it does not buckle over) while they take souvenir pictures. Approximately how high off the ground is the midpoint of the hot rail?

(A) 0.8 ft (0.24 m)

(B) 2.1 ft (0.64 m)

(C) 17 ft (5.1 m)

(D) 45 ft (14 m)

8. (a) What is most nearly the force in member FC?

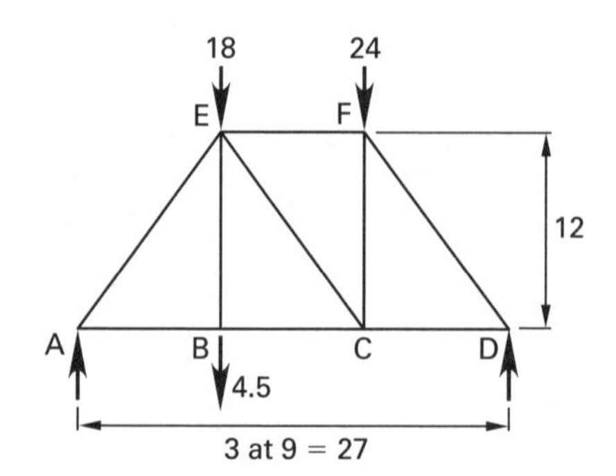

Statics

(A) 0.5

(B) 17

(C) 23

(D) 29

(b) The force in member CE is most nearly

(A) 0.50

(B) 0.63

(C) 11

(D) 17

9. The lever shown is pinned at and free to rotate about point O. The lever is acted upon by forces F_1 and F_2, but not by gravity. What is the algebraic expression for force F_2 in terms of F_1 that will keep the lever stationary?

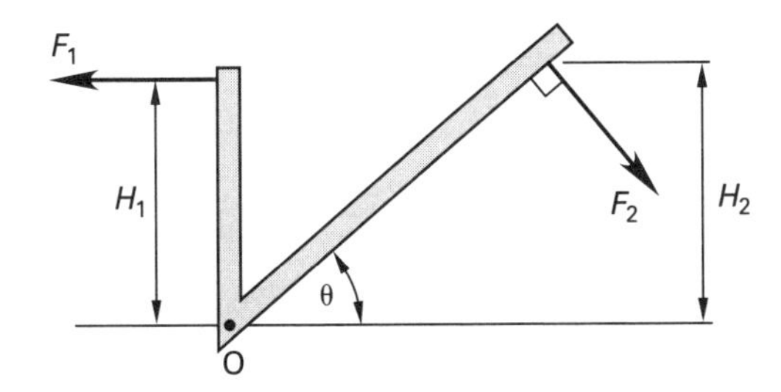

(A) $F_2 = \dfrac{F_1 H_1 \sin\theta}{H_2}$

(B) $F_2 = \dfrac{F_1 H_1}{H_2 \sin\theta}$

(C) $F_2 = \dfrac{F_1 H_1}{H_2}$

(D) $F_2 = \dfrac{F_1 H_1 \cos\theta}{H_2}$

10. What name best describes the type of crane illustrated?

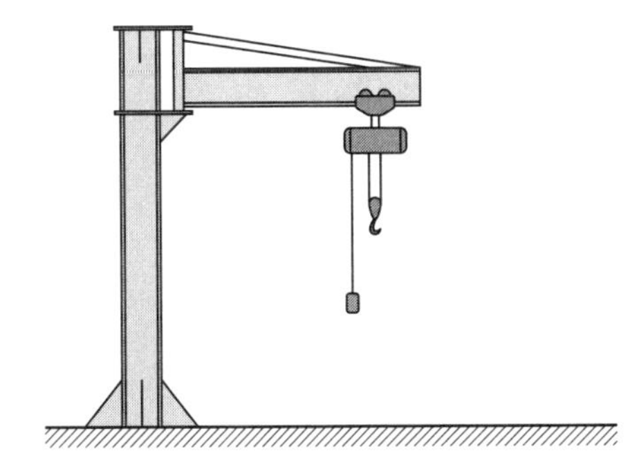

(A) cherry picker crane

(B) jib crane

(C) luffing crane

(D) overhead crane

11. A power pole loaded horizontally by electrical wires T_1 and T_2 is braced by a guy wire, as shown. The guy wire is 25 ft long, is attached 20 ft up the pole, and has an internal tensile force of 12,500 lbf. The loads are perpendicular to the vertical power pole, and the bases of the guy wire and power pole are both even and at ground level. All forces are in equilibrium. What are most nearly the magnitudes of the horizontal loads T_1 and T_2, respectively?

(A) $T_1 = 3500$ lbf; $T_2 = 5800$ lbf

(B) $T_1 = 3800$ lbf; $T_2 = 6500$ lbf

(C) $T_1 = 3900$ lbf; $T_2 = 6800$ lbf

(D) $T_1 = 6300$ lbf; $T_2 = 6300$ lbf

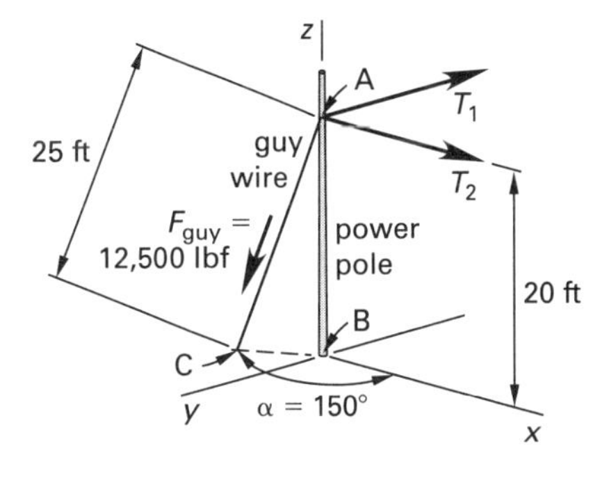

12. A 30 ft × 10 ft box-shaped 9000 lbf piece of machinery is lifted using a three-cable hoist ring, as shown. Two cables of the hoist are connected to opposite corners of the 10 ft side, and the other cable is connected to the center of the other 10 ft side. The hoist ring is 20 ft directly above the center of the machinery. What are most nearly the forces in hoist cables C_1, C_2, and C_3, respectively?

(A) 2300 lbf, 2300 lbf, and 6800 lbf

(B) 2500 lbf, 2700 lbf, and 3900 lbf

(C) 2900 lbf, 2900 lbf, and 5600 lbf

(D) 3000 lbf, 3000 lbf, and 3000 lbf

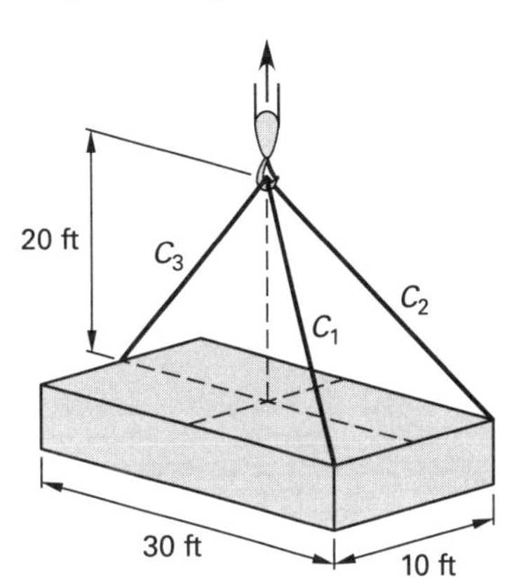

Statics

SOLUTIONS

1. *Customary U.S. Solution*

(a) Use Eq. 45.68 to relate the midpoint sag, S, to the constant, c.

$$S = c\left(\cosh\left(\frac{a}{c}\right) - 1\right)$$

$$10 \text{ ft} = c\left(\cosh\left(\frac{50 \text{ ft}}{c}\right) - 1\right)$$

Solve by trial and error.

$$c = 126.6 \text{ ft}$$

Use Eq. 45.52(b) and Eq. 45.70 to find the midpoint tension.

$$H = wc = m\left(\frac{g}{g_c}\right)c$$

$$= \left(2 \ \frac{\text{lbm}}{\text{ft}}\right)\left(\frac{32.2 \ \frac{\text{ft}}{\text{sec}^2}}{32.2 \ \frac{\text{lbm-ft}}{\text{lbf-sec}^2}}\right)(126.6 \text{ ft})$$

$$= \boxed{253.2 \text{ lbf} \quad (250 \text{ lbf})}$$

The answer is (B).

(b) Use Eq. 45.72 to find the maximum tension.

$$T = wy = w(c+S) = m\left(\frac{g}{g_c}\right)(c+S)$$

$$= \left(2 \ \frac{\text{lbm}}{\text{ft}}\right)\left(\frac{32.2 \ \frac{\text{ft}}{\text{sec}^2}}{32.2 \ \frac{\text{lbm-ft}}{\text{lbf-sec}^2}}\right)(126.6 \text{ ft} + 10 \text{ ft})$$

$$= \boxed{273.2 \text{ lbf} \quad (270 \text{ lbf})}$$

The answer is (C).

(c) From $T = wy$,

$$y = \frac{T}{w} = \frac{500 \text{ lbf}}{2 \ \frac{\text{lbf}}{\text{ft}}} = 250 \text{ ft} \quad [\text{at right support}]$$

$$250 \text{ ft} = c\left(\cosh \frac{50 \text{ ft}}{c}\right)$$

By trial and error, $c = 245$ ft.

Substitute into Eq. 45.68.

$$S = c\left(\cosh\left(\frac{a}{c}\right) - 1\right) = (245 \text{ ft})\left(\cosh\left(\frac{50 \text{ ft}}{245 \text{ ft}}\right) - 1\right)$$

$$= \boxed{5.1 \text{ ft}}$$

The answer is (D).

SI Solution

(a) Use Eq. 45.68 to relate the midpoint sag, S, to the constant, c.

$$S = c\left(\cosh\left(\frac{a}{c}\right) - 1\right)$$

$$3 \text{ m} = c\left(\cosh\left(\frac{15 \text{ m}}{c}\right) - 1\right)$$

Solve by trial and error.

$$c = 38.0 \text{ m}$$

Use Eq. 45.52(a) and Eq. 45.70 to find the midpoint tension

$$H = wc = mgc = \left(3 \ \frac{\text{kg}}{\text{m}}\right)\left(9.81 \ \frac{\text{m}}{\text{s}^2}\right)(38.0 \text{ m})$$

$$= \boxed{1118.3 \text{ N} \quad (1.1 \text{ kN})}$$

The answer is (B).

(b) Use Eq. 45.72 to find the maximum tension.

$$T = wy = w(c+S) = mg(c+S)$$

$$= \left(3 \ \frac{\text{kg}}{\text{m}}\right)\left(9.81 \ \frac{\text{m}}{\text{s}^2}\right)(38.0 \text{ m} + 3.0 \text{ m})$$

$$= \boxed{1206.6 \text{ N} \quad (1.2 \text{ kN})}$$

The answer is (C).

(c) From $T = wy$,

$$y = \frac{T}{w} = \frac{2200 \text{ N}}{\left(3 \ \frac{\text{kg}}{\text{m}}\right)\left(9.81 \ \frac{\text{m}}{\text{s}^2}\right)}$$

$$= 74.75 \text{ m} \quad [\text{at right support}]$$

$$74.75 \text{ m} = c\left(\cosh \frac{15 \text{ m}}{c}\right)$$

By trial and error, $c = 73.2$ m.

Substitute into Eq. 45.68.

$$S = c\left(\cosh\left(\frac{a}{c}\right) - 1\right)$$

$$= (73.2 \text{ m})\left(\cosh\left(\frac{15 \text{ m}}{73.2 \text{ m}}\right) - 1\right)$$

$$= \boxed{1.54 \text{ m} \quad (1.5 \text{ m})}$$

The answer is (D).

2. *step 1:* Draw the tripod with the origin at the apex.

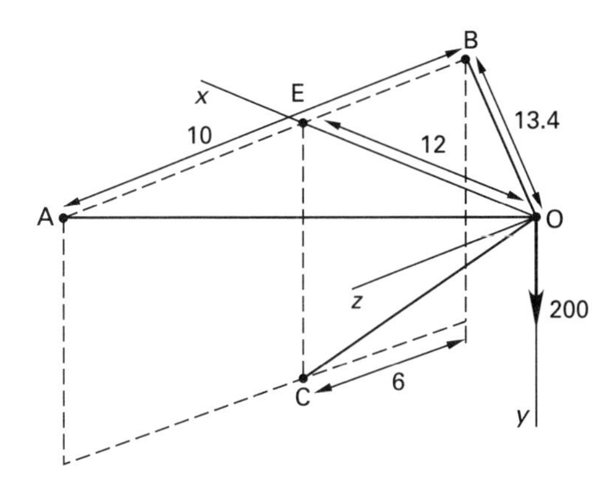

step 2: By inspection, the force components are $F_x = 0$, $F_y = 200$, and $F_z = 0$.

step 3: First, from triangle EBO, length BE is

$$\begin{aligned} \text{BE} &= \sqrt{(13.4 \text{ units})^2 - (12 \text{ units})^2} \\ &= 5.96 \text{ units} \quad [\text{use 6 units}] \end{aligned}$$

The (x, y, z) coordinates of the three support points are

point A: (12, 0, 4)
point B: (12, 0, −6)
point C: (12, 9, 0)

step 4: Find the lengths of the legs.

$$\begin{aligned} \text{AO} &= \sqrt{(x_\text{A} - x_\text{O})^2 + (y_\text{A} - y_\text{O})^2 + (z_\text{A} - z_\text{O})^2} \\ &= \sqrt{(12 \text{ units})^2 + (0 \text{ units})^2 + (4 \text{ units})^2} \\ &= 12.65 \text{ units} \end{aligned}$$

$$\begin{aligned} \text{BO} &= \sqrt{(x_\text{B} - x_\text{O})^2 + (y_\text{B} - y_\text{O})^2 + (z_\text{B} - z_\text{O})^2} \\ &= \sqrt{(12 \text{ units})^2 + (0 \text{ units})^2 + (-6 \text{ units})^2} \\ &= 13.4 \text{ units} \end{aligned}$$

$$\begin{aligned} \text{CO} &= \sqrt{(x_\text{C} - x_\text{O})^2 + (y_\text{C} - y_\text{O})^2 + (z_\text{C} - z_\text{O})^2} \\ &= \sqrt{(12 \text{ units})^2 + (9 \text{ units})^2 + (0 \text{ units})^2} \\ &= 15.0 \text{ units} \end{aligned}$$

step 5: Use Eq. 45.76, Eq. 45.77, and Eq. 45.78 to find the direction cosines for each leg.

For leg AO,

$$\cos\theta_{\text{A},x} = \frac{x_\text{A}}{\text{AO}} = \frac{12 \text{ units}}{12.65 \text{ units}} = 0.949$$

$$\cos\theta_{\text{A},y} = \frac{y_\text{A}}{\text{AO}} = \frac{0 \text{ units}}{12.65 \text{ units}} = 0$$

$$\cos\theta_{\text{A},z} = \frac{z_\text{A}}{\text{AO}} = \frac{4 \text{ units}}{12.65 \text{ units}} = 0.316$$

For leg BO,

$$\cos\theta_{\text{B},x} = \frac{x_\text{B}}{\text{BO}} = \frac{12 \text{ units}}{13.4 \text{ units}} = 0.896$$

$$\cos\theta_{\text{B},y} = \frac{y_\text{B}}{\text{BO}} = \frac{0 \text{ units}}{13.4 \text{ units}} = 0$$

$$\cos\theta_{\text{B},z} = \frac{z_\text{B}}{\text{BO}} = \frac{-6 \text{ units}}{13.4 \text{ units}} = -0.448$$

For leg CO,

$$\cos\theta_{\text{C},x} = \frac{x_\text{C}}{\text{CO}} = \frac{12 \text{ units}}{15.0 \text{ units}} = 0.80$$

$$\cos\theta_{\text{C},y} = \frac{y_\text{C}}{\text{CO}} = \frac{9 \text{ units}}{15.0 \text{ units}} = 0.60$$

$$\cos\theta_{\text{C},z} = \frac{z_\text{C}}{\text{CO}} = \frac{0 \text{ units}}{15.0 \text{ units}} = 0$$

steps 6 and 7: Substitute Eq. 45.79, Eq. 45.80, and Eq. 45.81 into equilibrium Eq. 45.82, Eq. 45.83, and Eq. 45.84.

$$\begin{aligned} F_\text{A}\cos\theta_{\text{A},x} + F_\text{B}\cos\theta_{\text{B},x} + F_\text{C}\cos\theta_{\text{C},x} + F_x &= 0 \\ F_\text{A}\cos\theta_{\text{A},y} + F_\text{B}\cos\theta_{\text{B},y} + F_\text{C}\cos\theta_{\text{C},y} + F_y &= 0 \\ F_\text{A}\cos\theta_{\text{A},z} + F_\text{B}\cos\theta_{\text{B},z} + F_\text{C}\cos\theta_{\text{C},z} + F_z &= 0 \\ 0.949F_\text{A} + 0.896F_\text{B} + 0.80F_\text{C} + 0 &= 0 \\ 0F_\text{A} + 0F_\text{B} + 0.60F_\text{C} + 200 &= 0 \\ 0.316F_\text{A} - 0.448F_\text{B} + 0F_\text{C} + 0 &= 0 \end{aligned}$$

Solve the three equations simultaneously.

$$F_\text{A} = 168.6 \quad \text{(T)}$$

$$F_\text{B} = 118.9 \quad \text{(T)}$$

$$F_\text{C} = \boxed{-333.3 \quad \text{(C)} \quad (330)}$$

The answer is (D).

Statics

3. First, find the vertical reaction at point D.

$$\sum M_C = (CD)D_y - (AF)(8000) + (AC)(1600) = 0$$
$$6D_y - (6)(8000) + (16)(1600) = 0$$

Solve for $D_y = 3733.3$.

The free-body diagram of pin D is as follows.

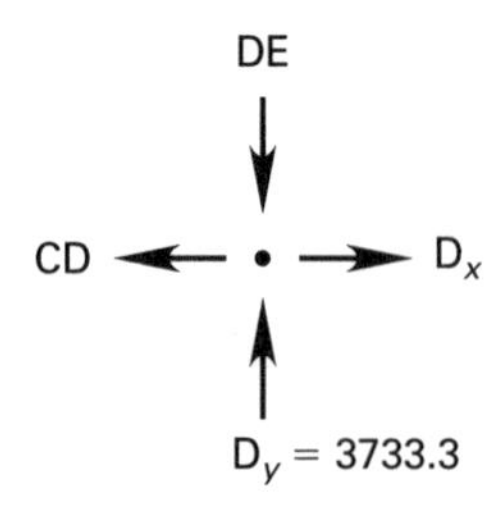

$$\sum F_y = D_y - DE = 0$$

Therefore,

$$DE = D_y = \boxed{3733.3\ (C)\quad (3700)}$$

The answer is (D).

4. *step 1:* Move the origin to the apex of the tripod. Call this point O.

step 2: By inspection, the force components are $F_x = 1200$, $F_y = 0$, and $F_z = 0$.

step 3: The (x, y, z) coordinates of the three support points are

point A: (5, −12, 0)
point B: (0, −8, −8)
point C: (−4, −7, 6)

step 4: Find the lengths of the legs.

$$\begin{aligned} AO &= \sqrt{(x_A - x_O)^2 + (y_A - y_O)^2 + (z_A - z_O)^2} \\ &= \sqrt{(5)^2 + (-12)^2 + (0)^2} \\ &= 13.0 \end{aligned}$$

$$\begin{aligned} BO &= \sqrt{(x_B - x_O)^2 + (y_B - y_O)^2 + (z_B - z_O)^2} \\ &= \sqrt{(0)^2 + (-8)^2 + (-8)^2} \\ &= 11.31 \end{aligned}$$

$$\begin{aligned} CO &= \sqrt{(x_C - x_O)^2 + (y_C - y_O)^2 + (z_C - z_O)^2} \\ &= \sqrt{(-4)^2 + (-7)^2 + (6)^2} \\ &= 10.05 \end{aligned}$$

step 5: Use Eq. 45.76, Eq. 45.77, and Eq. 45.78 to find the direction cosines for each leg.

For leg AO,

$$\cos\theta_{A,x} = \frac{x_A}{AO} = \frac{5}{13.0} = 0.385$$
$$\cos\theta_{A,y} = \frac{y_A}{AO} = \frac{-12}{13.0} = -0.923$$
$$\cos\theta_{A,z} = \frac{z_A}{AO} = \frac{0}{13.0} = 0$$

For leg BO,

$$\cos\theta_{B,x} = \frac{x_B}{BO} = \frac{0}{11.31} = 0$$
$$\cos\theta_{B,y} = \frac{y_B}{BO} = \frac{-8}{11.31} = -0.707$$
$$\cos\theta_{B,z} = \frac{z_B}{BO} = \frac{-8}{11.31} = 0.707$$

For leg CO,

$$\cos\theta_{C,x} = \frac{x_C}{CO} = \frac{-4}{10.05} = -0.398$$
$$\cos\theta_{C,y} = \frac{y_C}{CO} = \frac{-7}{10.05} = -0.697$$
$$\cos\theta_{C,z} = \frac{z_C}{CO} = \frac{6}{10.05} = 0.597$$

steps 6 and 7: Substitute Eq. 45.79, Eq. 45.80, and Eq. 45.81 into equilibrium Eq. 45.82, Eq. 45.83, and Eq. 45.84.

$$F_A\cos\theta_{A,x} + F_B\cos\theta_{B,x} + F_C\cos\theta_{C,x} + F_x = 0$$
$$F_A\cos\theta_{A,y} + F_B\cos\theta_{B,y} + F_C\cos\theta_{C,y} + F_y = 0$$
$$F_A\cos\theta_{A,z} + F_B\cos\theta_{B,z} + F_C\cos\theta_{C,z} + F_z = 0$$
$$0.385F_A + 0F_B - 0.398F_C + 1200 = 0$$
$$-0.923F_A - 0.707F_B - 0.697F_C + 0 = 0$$
$$0F_A - 0.707F_B + 0.597F_C + 0 = 0$$

Solve the three equations simultaneously.

$$\boxed{F_A = -1794\quad (C)\quad (1800)}$$
$$F_B = 1081\quad (T)$$
$$F_C = 1280\quad (T)$$

The answer is (C).

5. First, find the vertical reactions.

$$\sum F_y = A_y + L_y - (5)(4000) - (5)(60{,}000) = 0$$

Statics

By symmetry, $A_y = L_y$.

$$2A_y = (5)(4000) + (5)(60{,}000)$$
$$A_y = 160{,}000$$
$$L_y = 160{,}000$$

(a) For DE, make a cut in members BD, DE, and EG.

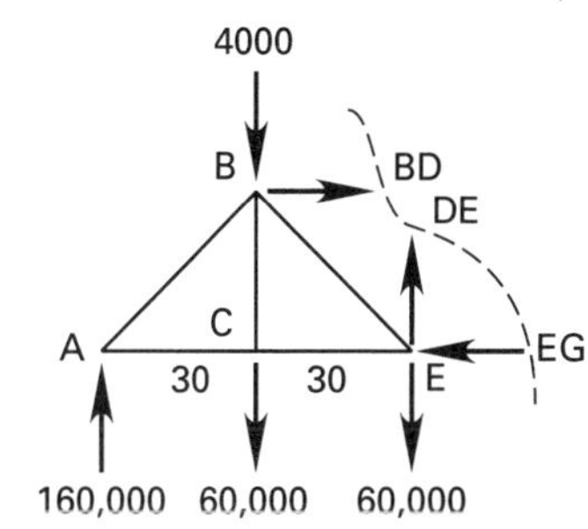

$$\sum F_y = 160{,}000 - 60{,}000 - 60{,}000 + \text{DE} - 4000 = 0$$

$$\text{DE} = \boxed{-36{,}000 \quad (\text{C}) \quad (36{,}000)}$$

The answer is (A).

(b) For HJ, make a cut in members HJ, HI, and GI.

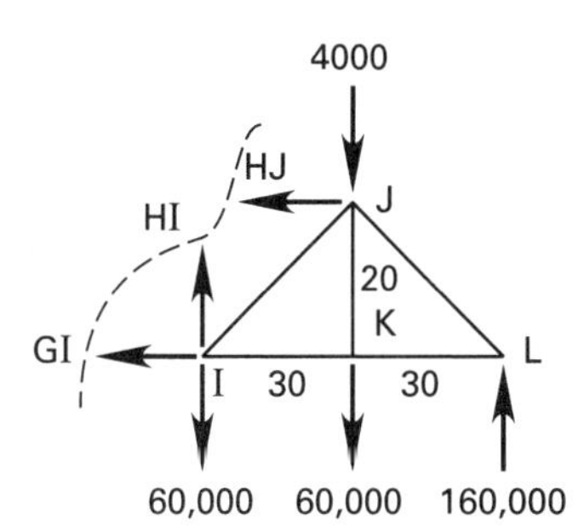

$$\sum M_\text{I} = (160{,}000)(60) - (60{,}000)(30) - (4000)(30) + \text{HJ}(20) = 0$$

$$\text{HJ} = \boxed{-384{,}000 \quad (\text{C}) \quad (380{,}000)}$$

The answer is (D).

6. First, consider a free-body diagram at point O in the vertical (x-y) plane.

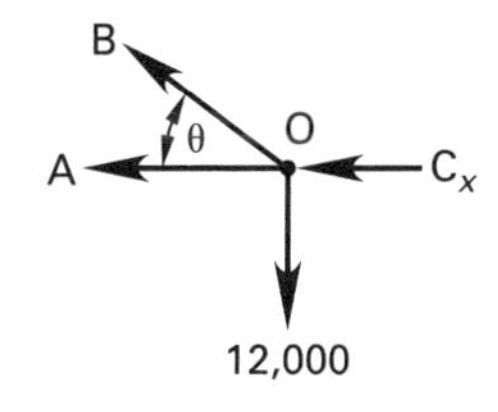

θ is obtained from triangle AOB.

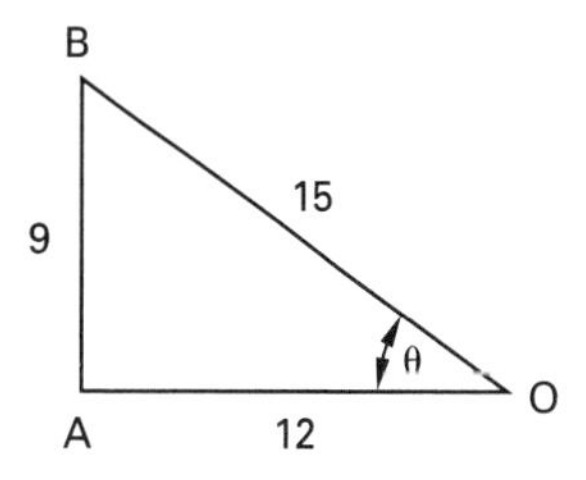

C_x is obtained from triangle AOC.

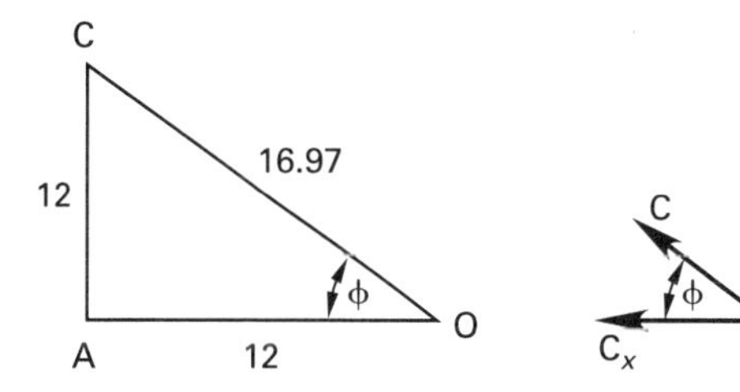

Equilibrium in the vertical (x-y) plane at point O requires $\sum F_y = 0$.

$$\text{B}\sin\theta = 12{,}000$$
$$\text{B} = \frac{12{,}000}{\sin\theta} = \frac{12{,}000}{\frac{9}{15}} = 20{,}000$$
$$\text{B}_x = \text{B}\cos\theta = (20{,}000)\left(\frac{12}{15}\right) = 16{,}000$$
$$\text{B}_y = \text{B}\sin\theta = (20{,}000)\left(\frac{9}{15}\right) = 12{,}000$$

Since BO is in the x-y plane, $\text{B}_z = 0$.

Next, consider a free-body diagram at point O in the horizontal (x-z) plane.

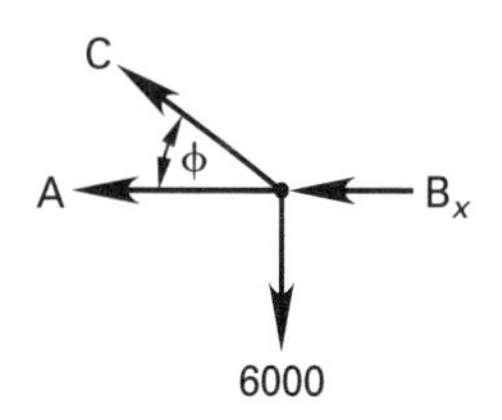

$\sum F_x = 0$:

$$\text{A} + \text{B}_x + \text{C}\cos\phi = 0$$
$$\text{A} + \text{C}\cos\phi = -\text{B}_x = -\text{B}\cos\theta = (-20{,}000)\left(\frac{12}{15}\right) = -16{,}000$$

Statics

$\sum F_y = 0$:

$$\begin{aligned} \mathrm{C}\sin\phi &= 6000 \\ \mathrm{C} &= \frac{6000}{\sin\phi} \\ &= \frac{6000}{\frac{12}{16.97}} \\ &= 8485 \end{aligned}$$

Therefore,

$$\begin{aligned} \mathrm{A} &= -\mathrm{C}\cos\phi - 16{,}000 \\ &= (-8485)\left(\frac{12}{16.97}\right) - 16{,}000 \\ &= -22{,}000 \end{aligned}$$

Since AO is on the x-axis,

$$\begin{aligned} \mathrm{A}_x &= -22{,}000 \\ \mathrm{A}_y &= 0 \\ \mathrm{A}_z &= 0 \end{aligned}$$

Solve for C reactions.

$$\begin{aligned} \mathrm{C}_x &= \mathrm{C}\cos\phi = (8485)\left(\frac{12}{16.97}\right) \\ &= \boxed{6000} \\ \mathrm{C}_z &= \mathrm{C}\sin\phi = (8485)\left(\frac{12}{16.97}\right) \\ &= \boxed{6000} \end{aligned}$$

Since CO is in the horizontal (x-z) plane, $C_y = \boxed{0.}$

The answer is (B).

7. *Customary U.S. Solution*

First, find the amount of thermal expansion. From Table 51.2, the coefficient of thermal expansion for steel is 6.5×10^{-6} 1/°F. Use Eq. 51.9.

$$\begin{aligned} \Delta L &= \alpha L(T_2 - T_1) \\ &= \left(6.5 \times 10^{-6}\ \frac{1}{^\circ\mathrm{F}}\right)(1)\left(5280\ \frac{\mathrm{ft}}{\mathrm{mi}}\right) \\ &\quad \times (99.14^\circ\mathrm{F} - 70^\circ\mathrm{F}) \\ &= 1.000085\ \mathrm{ft} \quad (1\ \mathrm{ft}) \end{aligned}$$

Assume the distributed load is uniform along the length of the rail. This resembles the case of a cable under its own weight. From the parabolic cable figure, as shown in Fig. 45.18, when distance S is small relative to distance a, the problem can be solved by using the parabolic equation, Eq. 45.58.

$$L \approx a\left(1 + \tfrac{2}{3}\left(\frac{S}{a}\right)^2 - \tfrac{2}{5}\left(\frac{S}{a}\right)^4\right)$$

$$\frac{5280\ \mathrm{ft} + 1\ \mathrm{ft}}{2} \approx (2640\ \mathrm{ft})\left(\begin{array}{c} 1 + \frac{2}{3}\left(\frac{S}{2640\ \mathrm{ft}}\right)^2 \\ -\frac{2}{5}\left(\frac{S}{2640\ \mathrm{ft}}\right)^4 \end{array}\right)$$

Using trial and error, $S \approx \boxed{44.5\ \mathrm{ft}\ (45\ \mathrm{ft}).}$

The answer is (D).

SI Solution

First, find the amount of thermal expansion. From Table 51.2, the coefficient of thermal expansion for steel is 11.7×10^{-6} 1/°C. Use Eq. 51.9.

$$\begin{aligned} \Delta L &= \alpha L_o(T_2 - T_1) \\ &= \left(11.7 \times 10^{-6}\ \frac{1}{^\circ\mathrm{C}}\right)(1.6\ \mathrm{km})\left(1000\ \frac{\mathrm{m}}{\mathrm{km}}\right) \\ &\quad \times (37.30^\circ\mathrm{C} - 21.11^\circ\mathrm{C}) \\ &= 0.30308\ \mathrm{m} \quad (0.30\ \mathrm{m}) \end{aligned}$$

Assume the distributed load is uniform along the length of the rail. This resembles the case of a cable under its own weight. From the parabolic cable figure, as shown in Fig. 45.18, when distance S is small relative to distance a, the problem can be solved by using the parabolic equation, Eq. 45.58.

$$L \approx a\left(1 + \tfrac{2}{3}\left(\frac{S}{a}\right)^2 - \tfrac{2}{5}\left(\frac{S}{a}\right)^4\right)$$

$$\frac{1600\ \mathrm{m} + 0.3\ \mathrm{m}}{2} \approx (800\ \mathrm{m})\left(\begin{array}{c} 1 + \frac{2}{3}\left(\frac{S}{800\ \mathrm{m}}\right)^2 \\ -\frac{2}{5}\left(\frac{S}{800\ \mathrm{m}}\right)^4 \end{array}\right)$$

Using trial and error, $S \approx \boxed{13.6\ \mathrm{m}\ (14\ \mathrm{m}).}$

The answer is (D).

8. (a) First, find the reactions. Take clockwise moments about A as positive.

$$\begin{aligned} \sum M_\mathrm{A} &= (27)(-\mathrm{D}_y) + (18)(24) + (9)(22.5) = 0 \\ \mathrm{D}_y &= 23.5 \\ \mathrm{A}_y &= 18 + 24 + 4.5 - 23.5 = 23 \end{aligned}$$

Either the method of sections (easiest) or a member-by-member analysis can be used.

The general force triangle is

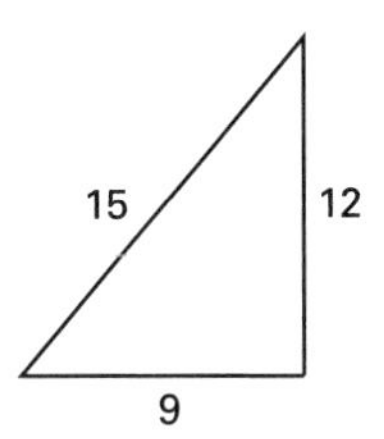

At pin A,

$$\text{AE}_y = 23$$

$$\text{AE}_x = \left(\frac{9}{12}\right)(23) = 17.25$$

$$\text{AE} = \left(\frac{15}{12}\right)(23) = 28.75 \quad \text{(C)}$$

$$\text{AB} = \text{AE}_x = 17.25 \quad \text{(T)}$$

At pin B,

$$\text{BE} = 4.5 \quad \text{(T)}$$

$$\text{BC} = \text{AB} = 17.25 \quad \text{(T)}$$

At pin D,

$$\text{DF}_y = 23.5$$

$$\text{DF}_x = \left(\frac{9}{12}\right)(23.5) = 17.63$$

$$\text{DF} = \left(\frac{15}{12}\right)(23.5) = 29.38 \quad \text{(C)}$$

$$\text{DC} = \text{DF}_x = 17.63 \quad \text{(T)}$$

At pin F,

$$\text{FE} = \text{DF}_x = 17.63 \quad \text{(C)}$$

$$\text{FC} = 24 - \text{DF}_y = 24 - 23.5 = \boxed{0.5 \quad \text{(C)}}$$

The answer is (A).

(b) At pin C,

$$\text{CE}_y = \text{FC} = 0.5$$

$$\text{CE} = \left(\frac{15}{12}\right)(0.5) = \boxed{0.63 \quad \text{(T)}}$$

The answer is (B).

9. For the lever to be stationary, the sum of the moments about point O must be zero.

$$\sum M_\text{O} = F_1H_1 - \frac{F_2H_2}{\sin\theta} = 0$$

Solve for F_2.

$$\boxed{F_2 = \frac{F_1H_1\sin\theta}{H_2}}$$

The answer is (A).

10. This is clearly not a cherry picker (basket) crane used to transport a passenger. (Such a crane consists of an aerial work platform with supporting basket-like platform large enough to carry a worker.) Nor is it an overhead crane that runs on two parallel tracks. A luffing crane has a boom that can move vertically up and down as well as another special apparatus that keeps the hook stationary. This is a jib crane (occasionally, "jib pole"), a small overhead block and tackle mechanism, usually mounted on a motorized transport carried by a swinging horizontal cantilever-arm. Jib cranes are primarily used in moving loads within a small area. (Many types of cranes have booms, and the term "jib" is used interchangeably for "boom." However, "jib crane" is used to describe only the type of lifting mechanism shown.)

The answer is (B).

11. *step 1:* Establish point A as the origin.

step 2: By inspection, length AC = 25 ft and length AB = 20 ft. Use the Pythagorean theorem to solve for length BC.

$$\begin{aligned}\text{BC} &= \sqrt{(\text{AC})^2 - (\text{AB})^2} = \sqrt{(25\text{ ft})^2 - (20\text{ ft})^2} \\ &= 15\text{ ft}\end{aligned}$$

Because loads T_1 and T_2 are horizontal (i.e., in the x-y plane), neglect all z components. Use trigonometry to find the x- and y-coordinates for point C.

$$x_\text{C} = \text{BC}\cos\alpha = (15\text{ ft})\cos 150° = -13\text{ ft}$$

$$y_\text{C} = \text{BC}\sin\alpha = (15\text{ ft})\sin 150° = 7.5\text{ ft}$$

step 3: Use Eq. 7.25 and Eq. 7.26 to find the x- and y-direction cosines for the guy wire.

$$\cos\theta_{\text{guy},x} = \frac{d_x}{\text{AC}} = \frac{-13\text{ ft} - 0\text{ ft}}{25\text{ ft}} = -0.52\text{ ft}$$

$$\cos\theta_{\text{guy},y} = \frac{d_y}{\text{AC}} = \frac{7.5\text{ ft} - 0\text{ ft}}{25\text{ ft}} = 0.3\text{ ft}$$

The x- and y-components of the guy wire's force are

$$F_{\text{guy},x} = -0.52F_\text{guy}$$

$$F_{\text{guy},y} = 0.3F_\text{guy}$$

Statics

steps 4 and 5: Use Eq. 45.35, Eq. 45.39, and Eq. 45.40 to write the equilibrium equations for the forces in the vertical direction.

$$F_{\text{guy},y} - T_1 = 0$$
$$F_{\text{guy},x} + T_2 = 0$$

Substitute and solve for the magnitudes of T_1 and T_2.

$$\begin{aligned} 0.3F_{\text{guy}} - T_1 &= 0 \\ (0.3)(12{,}500 \text{ lbf}) - T_1 &= 0 \\ T_1 &= \boxed{3750 \text{ lbf} \quad (3800 \text{ lbf})} \end{aligned}$$

$$\begin{aligned} -0.52F_{\text{guy}} + T_2 &= 0 \\ (-0.52)(12{,}500 \text{ lbf}) + T_2 &= 0 \\ T_2 &= \boxed{6500 \text{ lbf}} \end{aligned}$$

The answer is (B).

12. Follow the procedure outlined in Sec. 45.46.

steps 1 and 2: Establish the apex as the origin, O.

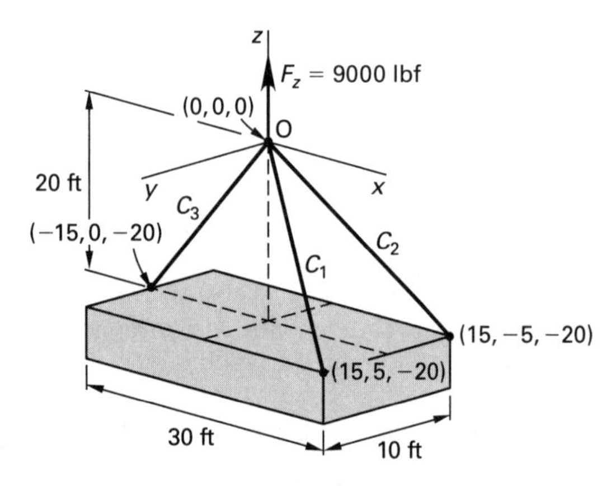

step 3: Determine the x-, y-, and z-components for cables C_1, C_2, and C_3. By inspection, the coordinates are $C_1=(15 \text{ ft}, 5 \text{ ft}, -20 \text{ ft})$, $C_2=(15 \text{ ft}, -5 \text{ ft}, -20 \text{ ft})$, and $C_3=(-15 \text{ ft}, 0 \text{ ft}, -20 \text{ ft})$.

step 4: Use Eq. 45.75 to find the cable lengths, L_{C_i}. By inspection, $L_{C_1} = L_{C_2}$.

$$\begin{aligned} L_{C_1} = L_{C_2} &= \sqrt{x^2 + y^2 + z^2} \\ &= \sqrt{(x_O - x_{C_1})^2 + (y_O - y_{C_1})^2 + (z_O - z_{C_1})^2} \\ &= \sqrt{\begin{matrix}(0 \text{ ft} - 15 \text{ ft})^2 + (0 \text{ ft} - 5 \text{ ft})^2 \\ + \left(0 \text{ ft} - (-20 \text{ ft})\right)^2\end{matrix}} \\ &= 25.5 \text{ ft} \end{aligned}$$

$$\begin{aligned} L_{C_3} &= \sqrt{x^2 + y^2 + z^2} \\ &= \sqrt{(x_O - x_{C_3})^2 + (y_O - y_{C_3})^2 + (z_O - z_{C_3})^2} \\ &= \sqrt{\begin{matrix}\left(0 \text{ ft} - (-15 \text{ ft})\right)^2 + (0 \text{ ft} - 0 \text{ ft})^2 \\ + \left(0 \text{ ft} - (-20 \text{ ft})\right)^2\end{matrix}} \\ &= 25 \text{ ft} \end{aligned}$$

steps 5 and 6: Use Eq. 7.25, Eq. 7.26, and Eq. 7.27 to find the direction cosine for each cable.

For cable C_1,

$$\cos\theta_{C_1,x} = \frac{x_{C_1}}{L_{C_1}} = \frac{0 \text{ ft} - 15 \text{ ft}}{25.5 \text{ ft}} = -0.588$$
$$\cos\theta_{C_1,y} = \frac{y_{C_1}}{L_{C_1}} = \frac{0 \text{ ft} - 5 \text{ ft}}{25.5 \text{ ft}} = -0.196$$
$$\cos\theta_{C_1,z} = \frac{z_{C_1}}{L_{C_1}} = \frac{0 \text{ ft} - (-20 \text{ ft})}{25.5 \text{ ft}} = 0.784$$

The direction cosines and x-, y-, and z-components of the force in cable C_1 are

$$\begin{aligned} F_{C_1x} &= -0.588F_{C_1} \\ F_{C_1,y} &= -0.196F_{C_1} \\ F_{C_1,z} &= 0.784F_{C_1} \end{aligned}$$

For cable C_2,

$$\cos\theta_{C_2,x} = \frac{x_{C_2}}{L_{C_2}} = \frac{0 \text{ ft} - 15 \text{ ft}}{25.5 \text{ ft}} = -0.588$$
$$\cos\theta_{C_2,y} = \frac{y_{C_2}}{L_{C_2}} = \frac{0 \text{ ft} - (-5 \text{ ft})}{25.5 \text{ ft}} = 0.196$$
$$\cos\theta_{C_2,z} = \frac{z_{C_2}}{L_{C_2}} = \frac{0 \text{ ft} - (-20 \text{ ft})}{25.5 \text{ ft}} = 0.784$$

The direction cosines and x-, y-, and z-components of the force in cable C_2 are

$$\begin{aligned} F_{C_2,x} &= -0.588F_{C_2} \\ F_{C_2,y} &= 0.196F_{C_2} \\ F_{C_2,z} &= 0.784F_{C_2} \end{aligned}$$

For cable C_3,

$$\cos\theta_{C_3,x} = \frac{x_{C_3}}{L_{C_3}} = \frac{0 \text{ ft} - (-15 \text{ ft})}{25 \text{ ft}} = 0.6$$
$$\cos\theta_{C_3,y} = \frac{y_{C_3}}{L_{C_3}} = \frac{0 \text{ ft} - 0 \text{ ft}}{25 \text{ ft}} = 0$$
$$\cos\theta_{C_3,z} = \frac{z_{C_3}}{L_{C_3}} = \frac{0 \text{ ft} - (-20 \text{ ft})}{25 \text{ ft}} = 0.8$$

The direction cosines and x-, y-, and z-components of the force in cable C_3 are

$$F_{C_3,x} = 0.6F_{C_3}$$
$$F_{C_3,y} = 0F_{C_3}$$
$$F_{C_3,z} = 0.8F_{C_3}$$

step 7: Write the three sum-of-forces equilibrium equations for the apex using Eq. 45.82, Eq. 45.83, and Eq. 45.74.

$$-0.588F_{C_1} - 0.588F_{C_2} + 0.6F_{C_3} = 0$$
$$-0.196F_{C_1} + 0.196F_{C_2} + 0F_{C_3} = 0$$
$$0.784F_{C_1} + 0.784F_{C_2} + 0.8F_{C_3} = -9000$$

The solution to these simultaneous equations is

$$F_{C_1} = \boxed{-2870 \text{ lbf} \quad (2900 \text{ lbf})}$$
$$F_{C_2} = \boxed{-2870 \text{ lbf} \quad (2900 \text{ lbf})}$$
$$F_{C_3} = \boxed{-5625 \text{ lbf} \quad (5600 \text{ lbf})}$$

The answer is (C).

Statics

46 Indeterminate Statics

PRACTICE PROBLEMS

Degree of Indeterminacy

1. What is the degree of indeterminacy of the structure shown?

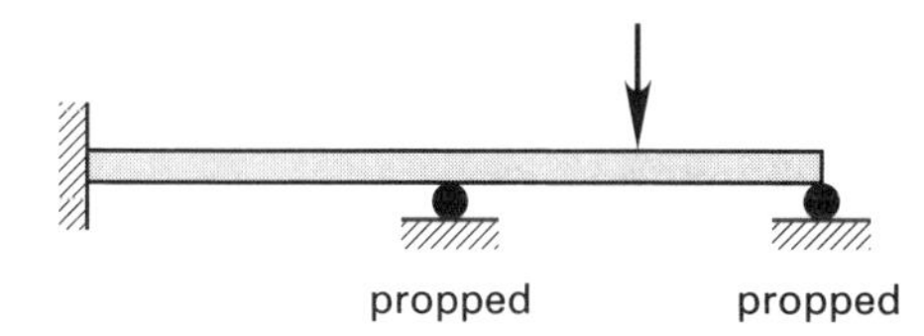

(A) 1

(B) 2

(C) 3

(D) 4

2. What is the degree of indeterminacy of the truss shown?

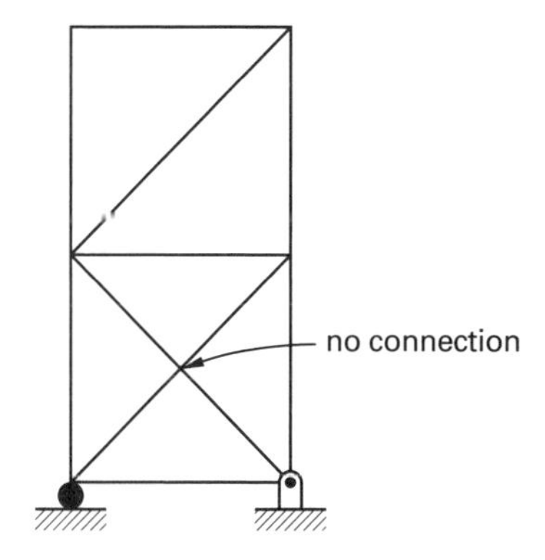

(A) 1

(B) 2

(C) 3

(D) 4

Elastic Deformation

3. A 1 in × 2 in × 10 in (2 cm × 5 cm × 30 cm) copper tie rod experiences a 55°F (30°C) increase from the no-stress temperature. The rod has a modulus of elasticity of 18×10^6 lbf/in^2(12×10^4 MPa). The compressive load is most nearly

(A) 8000 lbf (26 kN)

(B) 12,000 lbf (38 kN)

(C) 15,000 lbf (48 kN)

(D) 18,000 lbf (58 kN)

4. A 2 in (5 cm) diameter and 15 in (40 cm) long steel rod supports a 2250 lbf (10 000 N) compressive load. The rod has a modulus of elasticity of 29×10^6 lbf/in^2 (20×10^4 MPa).

(a) The stress is most nearly

(A) 570 lbf/in^2 (4.0 MPa)

(B) 720 lbf/in^2 (5.1 MPa)

(C) 1100 lbf/in^2 (7.8 MPa)

(D) 1400 lbf/in^2 (9.9 MPa)

(b) The decrease in length is most nearly

(A) 8.7×10^{-5} in (2.3×10^{-6} m)

(B) 1.1×10^{-4} in (3.0×10^{-6} m)

(C) 3.7×10^{-4} in (1.0×10^{-5} m)

(D) 9.3×10^{-4} in (2.6×10^{-5} m)

Consistent Deformation Method

5. The following questions refer to the structure shown.

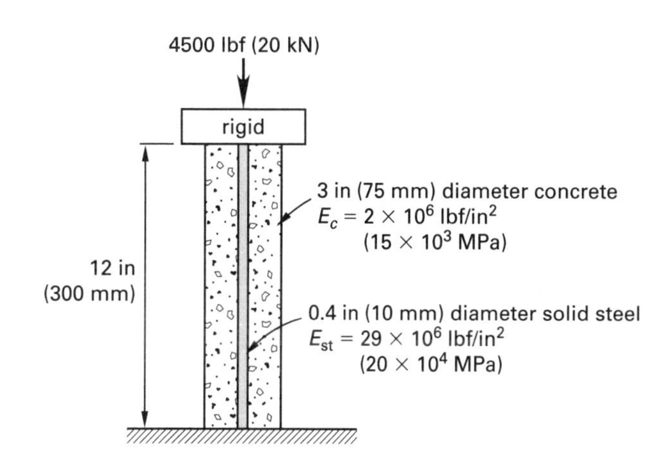

(a) What is most nearly the force in the concrete and steel members, respectively?

(A) 3570 lbf, 940 lbf (16.10 kN, 3.90 kN)

(B) 3580 lbf, 920 lbf (16.20 kN, 3.80 kN)

(C) 4140 lbf, 360 lbf (18.70 kN, 1.50 kN)

(D) 4420 lbf, 80 lbf (20 kN, 0.30 kN)

(b) What is most nearly the stress in the concrete and steel members, respectively?

(A) 500 lbf/in^2, 350 lbf/in^2 (3.60 MPa, 2.30 MPa)

(B) 510 lbf/in^2, 7440 lbf/in^2 (3.70 MPa, 50.0 MPa)

(C) 597 lbf/in^2, 2840 lbf/in^2 (4.30 MPa, 20.0 MPa)

(D) 636 lbf/in^2, 636 lbf/in^2 (4.60 MPa, 4.20 MPa)

(c) What is most nearly the total deflection?

(A) 2.6×10^{-4} in (62 μm)

(B) 1.2×10^{-3} in (28 μm)

(C) 2.9×10^{-3} in (69 μm)

(D) 3.1×10^{-3} in (74 μm)

6. The concentric pipe assembly shown is rigidly attached at both ends. The rigid ends are unconstrained.

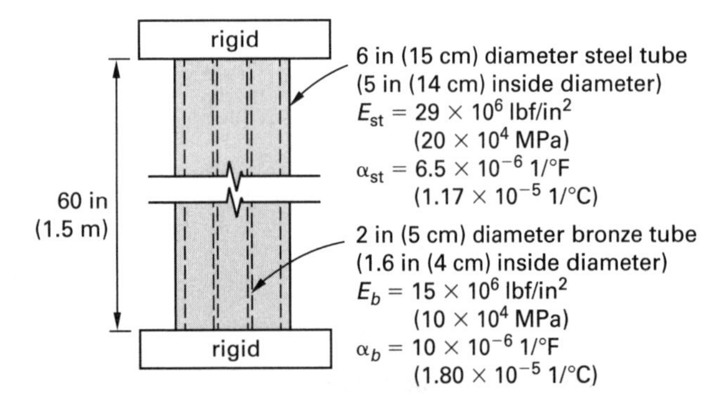

What is most nearly the stress generated in the steel and bronze pipes, respectively, if the temperature of the assembly rises 200°F (110°C)?

(A) 1100 lbf/in^2, 9930 lbf/in^2 (16 MPa, 61 MPa)

(B) 1200 lbf/in^2, 9990 lbf/in^2 (17 MPa, 60 MPa)

(C) 1290 lbf/in^2, 9830 lbf/in^2 (19 MPa, 60 MPa)

(D) 2990 lbf/in^2, 8660 lbf/in^2 (43 MPa, 55 MPa)

7. For the structure shown, the beam is made of steel, with an area moment of inertia of 10 in^4 (4.17×10^6 mm^4). The cable cross section is 0.0124 in^2 (8 mm^2). Before the 270 lbf load is applied, the cable is taut but carries no load.

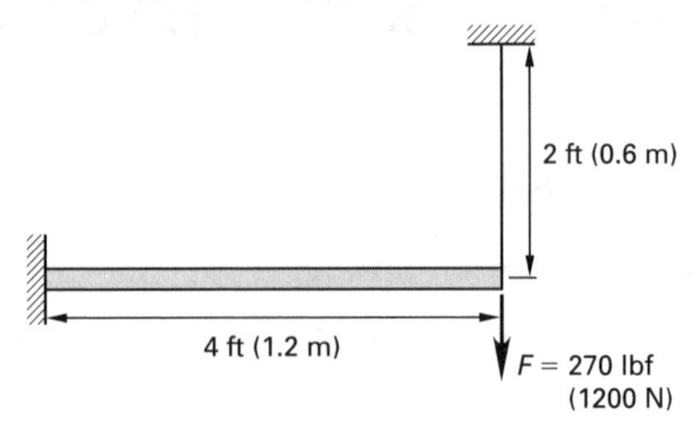

What is most nearly the load carried by the steel cable?

(A) 180 lbf (780 N)

(B) 200 lbf (870 N)

(C) 220 lbf (950 N)

(D) 240 lbf (1000 N)

8. A beam is simply supported at its ends and by a column of area A at its center, as shown. The beam and column are of the same material. The beam is subject to a uniform load, w.

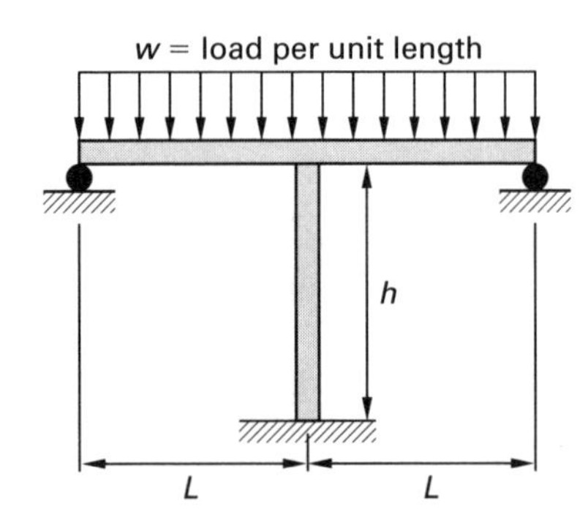

What are each of the left and right support reactions?

(A) $R = wL + \dfrac{5AwL^4}{48hI + 8AL^3}$

(B) $R = wL\left(\dfrac{48hI + 3AL^3}{48hI + 8AL^3}\right)$

(C) $R = wL + wL\left(\dfrac{5AL^3}{48hI + 8AL^3}\right)$

(D) $R = wL + \dfrac{10AwL^4}{24hI + 4AL^3}$

9. A rigid bar is supported by a hinge and two aluminum rods as shown. Deformations are small, and angular geometry changes are negligible.

$a = 6$ ft (2 m)

$b = 3$ ft (1 m)

$c = 3$ ft (1 m)

$d = 12$ ft (4 m)

$P = 4500$ lbf (20 kN)

$A_{\text{rod}} = 0.124\ \text{in}^2$ ($80\ \text{mm}^2$)

$E_{\text{rod}} = 10 \times 10^6\ \text{lbf/in}^2$ (70×10^3 MPa)

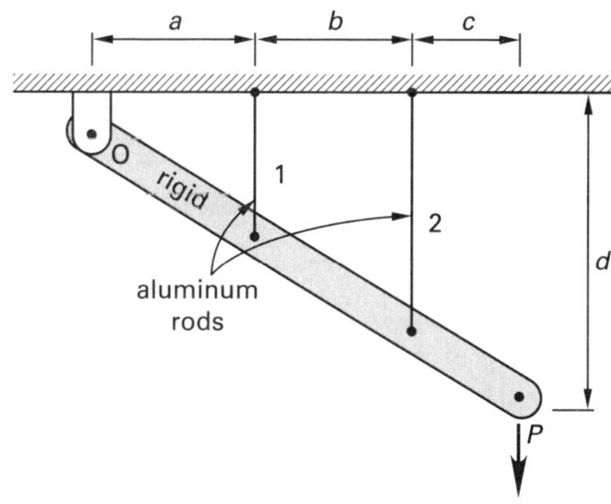

(a) The force in each aluminum rod is most nearly

(A) 3200 lbf (14 kN)

(B) 3600 lbf (16 kN)

(C) 4000 lbf (18 kN)

(D) 4300 lbf (19 kN)

(b) The elongations of the two aluminum rods are most nearly

(A) 0.13 in and 0.26 in (3.3 mm and 6.6 mm)

(B) 0.21 in and 0.31 in (5.7 mm and 8.6 mm)

(C) 0.21 in and 0.47 in (5.7 mm and 12 mm)

(D) 0.26 in and 0.62 in (6.6 mm and 16 mm)

10. The beams shown are identical in length, cross-sectional shape and area, and material. What is most nearly the reaction at each support?

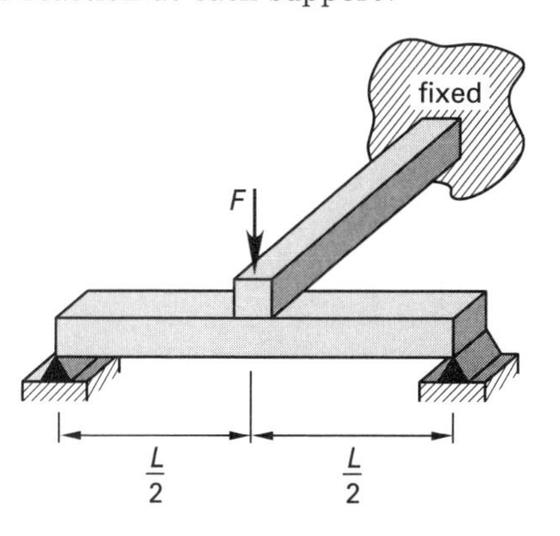

(A) $F/13$

(B) $4F/21$

(C) $8F/17$

(D) $2F/3$

Superposition Method

11. Using the superposition method, what is most nearly the intermediate support reaction, R_2, for the structure?

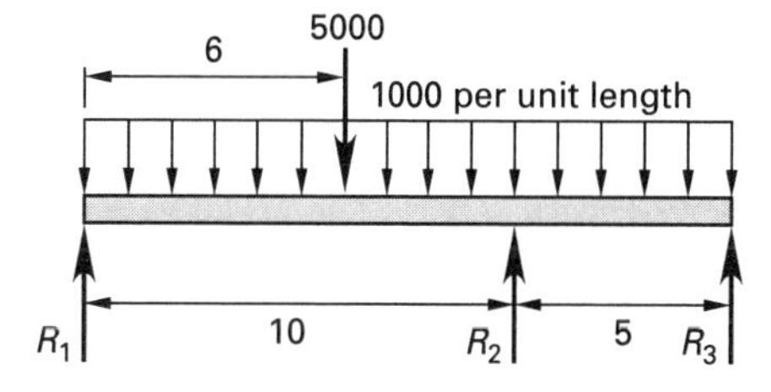

(A) 9000

(B) 11,000

(C) 15,000

(D) 18,000

12. Using the superposition method, what is most nearly the reaction, R, at the prop?

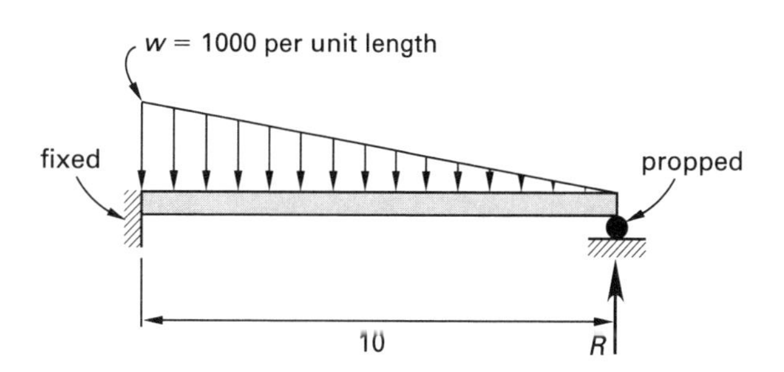

(A) 800

(B) 1000

(C) 1200

(D) 1400

Statics

Three-Moment Equation

13. What are most nearly the reactions R_1, R_2, and R_3, respectively, in the structure shown?

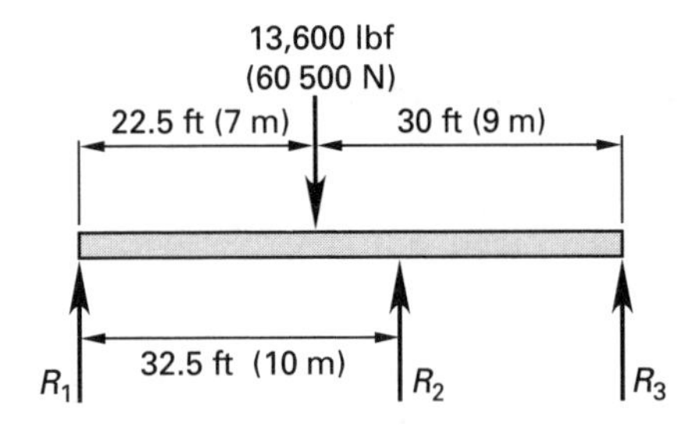

(A) 2000 lbf; 15,000 lbf; −3600 lbf (8500 N; 68 000 N; −16 000 N)

(B) 2700 lbf; 13,000 lbf; −2500 lbf (11 000 N; 60 000 N; −11 000 N)

(C) 3900 lbf; 10,000 lbf; −540 lbf (16 000 N; 46 000 N; −2500 N)

(D) 4700 lbf; 8200 lbf; 770 lbf (20 000 N; 37 000 N; −3500 N)

14. What are most nearly the reactions R_1, R_2, and R_3, respectively, in the structure shown?

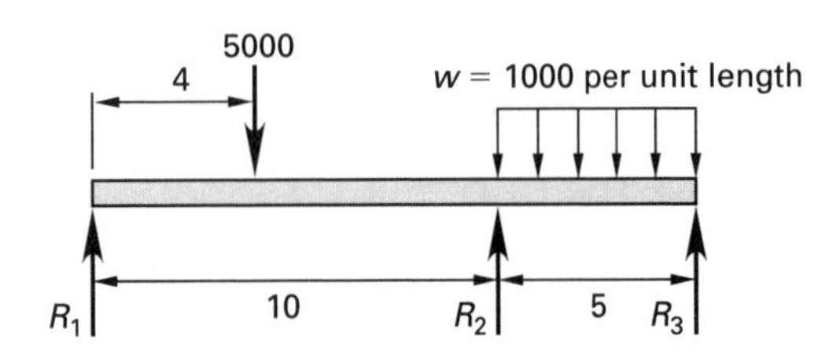

(A) 2230, 6810, 960

(B) 2240, 6780, 980

(C) 2340, 6490, 1170

(D) 2450, 6680, 870

Fixed-End Moments

15. What are most nearly the vertical reactions, R_2 and R_1, respectively, at the ends of the structure?

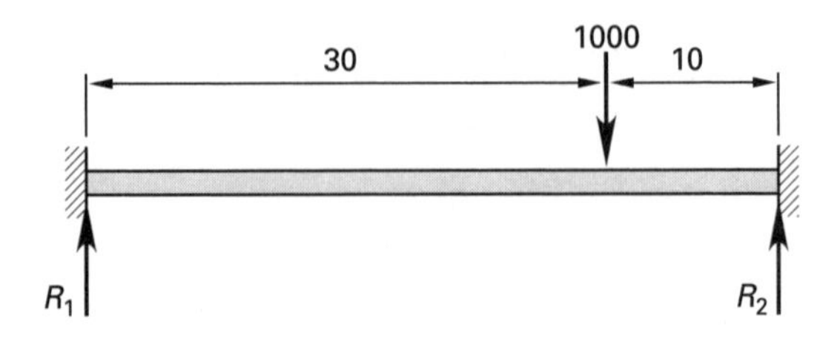

(A) 710, 290

(B) 840, 160

(C) 880, 130

(D) 890, 110

16. The truss shown carries a moving uniform live load of 2 kips/ft and a moving concentrated live load of 15 kips.

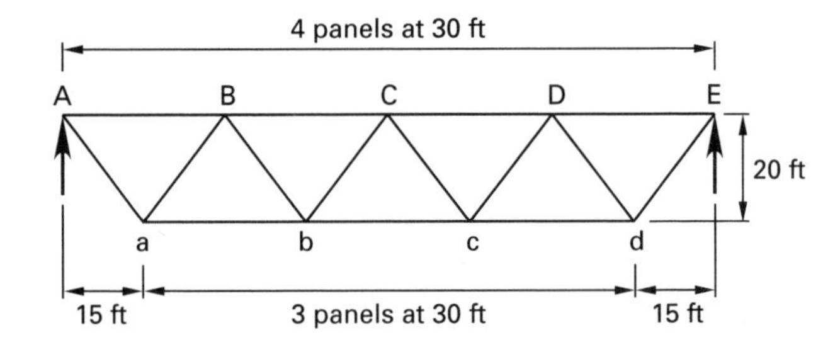

(a) The maximum force in member Bb is most nearly

(A) 40 kips

(B) 51 kips

(C) 59 kips

(D) 73 kips

(b) The maximum force in member BC is most nearly

(A) 15 kips

(B) 80 kips

(C) 95 kips

(D) 170 kips

17. The truss shown carries the group of four live loads shown along its bottom chords. What is most nearly the maximum force in member CD?

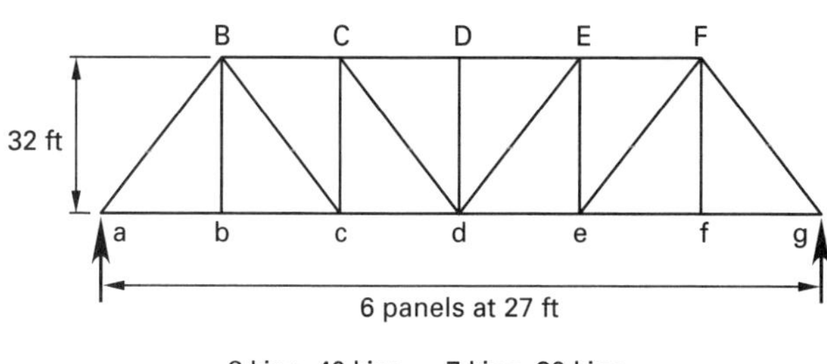

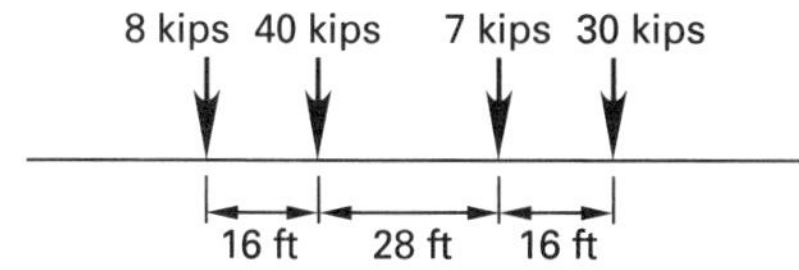

(A) 34 kips

(B) 82 kips

(C) 180 kips

(D) 220 kips

18. A moving load, consisting of two 30 kip forces separated by a constant 6 ft, travels over a two-span bridge as shown. The bridge has an interior expansion joint that can be considered to be an ideal hinge.

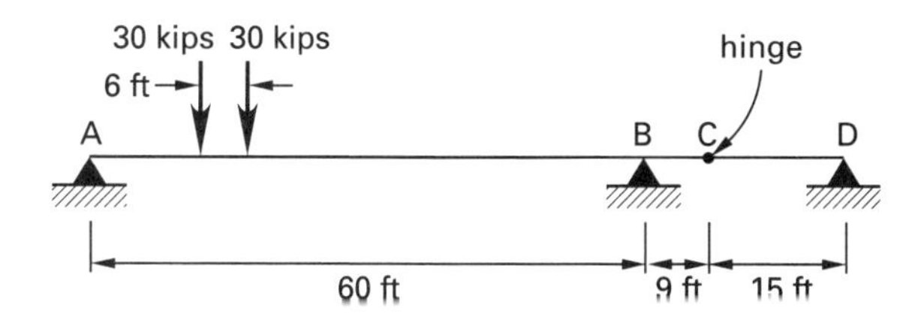

(a) The maximum moment at point B is most nearly

(A) 250 ft-kips

(B) 310 ft-kips

(C) 380 ft-kips

(D) 430 ft-kips

(b) The maximum reaction at point B is most nearly

(A) 30 kips

(B) 60 kips

(C) 70 kips

(D) 90 kips

SOLUTIONS

1. The degree of indeterminacy is $\boxed{2.}$ Remove the two props (vertical reactions) in order to make the structure statically determinate.

The answer is (B).

2. From Eq. 46.1,

$$I = r + m - 2j = 3 + 10 - (2)(6)$$
$$= \boxed{1}$$

The answer is (A).

3. *Customary U.S. Solution*

From Table 51.2, the coefficient of linear thermal expansion for copper is 8.9×10^{-6} 1/°F. The thermal strain is

$$\epsilon_{\text{th}} = \alpha\Delta T = \left(8.9 \times 10^{-6}\ \frac{1}{^\circ\text{F}}\right)(55^\circ\text{F}) = 0.00049$$

The strain must be counteracted to maintain the rod in its original position.

$$\sigma = E\epsilon_{\text{th}} = \left(18 \times 10^6\ \frac{\text{lbf}}{\text{in}^2}\right)(0.00049) = 8820\ \text{lbf/in}^2$$
$$F = \sigma A = \left(8820\ \frac{\text{lbf}}{\text{in}^2}\right)(1\ \text{in})(2\ \text{in})$$
$$= \boxed{17{,}640\ \text{lbf}\quad (18{,}000\ \text{lbf})}$$

The answer is (D).

SI Solution

From Table 51.2, the coefficient of linear thermal expansion for copper is 16.0×10^{-6} 1/°C. The thermal strain is

$$\epsilon_{\text{th}} = \alpha\Delta T = \left(16.0 \times 10^{-6}\ \frac{1}{^\circ\text{C}}\right)(30^\circ\text{C}) = 0.00048$$

The strain must be counteracted to maintain the rod in its original position.

$$\sigma = E\epsilon_{\text{th}} = (12 \times 10^4\ \text{MPa})(0.00048)$$
$$= 57.6\ \text{MPa}$$
$$F = \sigma A = (57.6 \times 10^6\ \text{Pa})(0.02\ \text{m})(0.05\ \text{m})$$
$$= \boxed{57\,600\ \text{N}\quad (58\ \text{kN})}$$

The answer is (D).

Statics

4. *Customary U.S. Solution*

(a) The rod's stress is

$$\sigma = \frac{F}{A} = \frac{2250 \text{ lbf}}{\frac{\pi(2 \text{ in})^2}{4}} = \boxed{716.2 \text{ lbf/in}^2 \quad (720 \text{ lbf/in}^2)}$$

The answer is (B).

(b) The decrease in length can be found from Eq. 51.2 and Eq. 51.4.

$$\epsilon = \frac{\sigma}{E} = \frac{\delta}{L_o}$$

$$\delta = L_o\left(\frac{\sigma}{E}\right) = (15 \text{ in})\left(\frac{716.2 \ \frac{\text{lbf}}{\text{in}^2}}{29 \times 10^6 \ \frac{\text{lbf}}{\text{in}^2}}\right)$$

$$= \boxed{3.7 \times 10^{-4} \text{ in}}$$

The answer is (C).

SI Solution

(a) The rod's stress is

$$\sigma = \frac{F}{A} = \frac{10\,000 \text{ N}}{\frac{\pi(0.05 \text{ m})^2}{4}} = \boxed{5.093 \times 10^6 \text{ Pa} \quad (5.1 \text{ MPa})}$$

The answer is (B).

(b) The decrease in length can be found from Eq. 51.4 and Eq. 51.15.

$$\epsilon = \frac{\sigma}{E} = \frac{\delta}{L_o}$$

$$\delta = L_o\left(\frac{\sigma}{E}\right) = (0.4 \text{ m})\left(\frac{5.093 \text{ MPa}}{20 \times 10^4 \text{ MPa}}\right)$$

$$= \boxed{1.02 \times 10^{-5} \text{ m} \quad (1.0 \times 10^{-5} \text{ m})}$$

The answer is (C).

5. *Customary U.S. Solution*

(a) Let F_c and F_{st} be the loads carried by the concrete and steel, respectively.

$$F = F_c + F_{st} = 4500 \text{ lbf} \quad \text{[Eq. I]}$$

The deformation of the steel is

$$\delta_{st} = \frac{F_{st}L}{A_{st}E_{st}}$$

The deformation of the concrete is

$$\delta_c = \frac{F_cL}{A_cE_c}$$

The geometric constraint is $\delta_{st} = \delta_c$, or

$$\frac{F_{st}L}{A_{st}E_{st}} = \frac{F_cL}{A_cE_c} \quad \text{[Eq. II]}$$

Solving Eq. I and Eq. II,

$$F_c = \frac{F}{1 + \frac{A_{st}E_{st}}{A_cE_c}}$$

$$= \frac{4500 \text{ lbf}}{1 + \frac{\left(\frac{\pi(0.4 \text{ in})^2}{4}\right)\left(29 \times 10^6 \ \frac{\text{lbf}}{\text{in}^2}\right)}{\left(\frac{\pi\left((3.0 \text{ in})^2 - (0.4 \text{ in})^2\right)}{4}\right)\left(2 \times 10^6 \ \frac{\text{lbf}}{\text{in}^2}\right)}}$$

$$= \boxed{3565 \text{ lbf} \quad (3570 \text{ lbf})}$$

For the steel,

$$F_{st} = 4500 \text{ lbf} - F_c = 4500 \text{ lbf} - 3565 \text{ lbf}$$

$$= \boxed{935 \text{ lbf} \quad (940 \text{ lbf})}$$

The answer is (A).

(b) Let σ_c and σ_{st} be the stress carried by the concrete and steel, respectively.

$$\sigma_c = \frac{F_c}{A_c} = \frac{3565 \text{ lbf}}{\frac{\pi\left((3 \text{ in})^2 - (0.4 \text{ in})^2\right)}{4}}$$

$$= \boxed{513 \text{ lbf/in}^2 \quad (510 \text{ lbf/in}^2)}$$

$$\sigma_{st} = \frac{F_{st}}{A_{st}} = \frac{935 \text{ lbf}}{\frac{\pi(0.4 \text{ in})^2}{4}} = \boxed{7440 \text{ lbf/in}^2}$$

The answer is (B).

(c) The total deflection is

$$\delta = \frac{F_{st}L}{A_{st}E_{st}} = \frac{(936 \text{ lbf})(12 \text{ in})}{\left(\frac{\pi(0.4 \text{ in})^2}{4}\right)\left(29 \times 10^6 \ \frac{\text{lbf}}{\text{in}^2}\right)}$$

$$= \boxed{3.08 \times 10^{-3} \text{ in} \quad (3.1 \times 10^{-3} \text{ in})}$$

The answer is (D).

SI Solution

(a) Let F_c and F_{st} be the loads carried by the concrete and steel, respectively.

$$F = F_c + F_{st} = 20 \text{ kN} \qquad \text{[Eq. I]}$$

The deformation of the steel is

$$\delta_{st} = \frac{F_{st}L}{A_{st}E_{st}}$$

The deformation of the concrete is

$$\delta_c = \frac{F_cL}{A_cE_c}$$

The geometric constraint is $\delta_{st} = \delta_c$, or

$$\frac{F_{st}L}{A_{st}E_{st}} = \frac{F_cL}{A_cE_c} \qquad \text{[Eq. II]}$$

Solving Eq. I and Eq. II,

$$\begin{aligned} F_c &= \frac{F}{1 + \dfrac{A_{st}E_{st}}{A_cE_c}} \\ &= \frac{20 \text{ kN}}{1 + \dfrac{\left(\dfrac{\pi(0.01 \text{ m})^2}{4}\right)(20 \times 10^4 \text{ MPa})}{\left(\dfrac{\pi\left((0.075 \text{ m})^2 - (0.01 \text{ m})^2\right)}{4}\right)(15 \times 10^3 \text{ MPa})}} \\ &= \boxed{16.11 \text{ kN} \quad (16.10 \text{ kN})} \end{aligned}$$

For the steel,

$$\begin{aligned} F_{st} &= 20 \text{ kN} - F_c = 20 \text{ kN} - 16.11 \text{ kN} \\ &= \boxed{3.89 \text{ kN} \quad (3.90 \text{ kN})} \end{aligned}$$

The answer is (A).

(b) Let σ_c and σ_{st} be the stress carried by the concrete and steel, respectively.

$$\begin{aligned} \sigma_c &= \frac{F_c}{A_c} = \frac{16.11 \times 10^{-3} \text{ MN}}{\dfrac{\pi\left((0.075 \text{ m})^2 - (0.01 \text{ m})^2\right)}{4}} \\ &= \boxed{3.71 \text{ MPa} \quad (3.70 \text{ MPa})} \\ \sigma_{st} &= \frac{F_{st}}{A_{st}} = \frac{3.89 \times 10^{-3} \text{ MN}}{\dfrac{\pi(0.01 \text{ m})^2}{4}} = \boxed{49.5 \text{ MPa} \quad (50 \text{ MPa})} \end{aligned}$$

The answer is (B).

(c) The total deflection is

$$\begin{aligned} \delta &= \frac{F_{st}L}{A_{st}E_{st}} \\ &= \frac{(3.89 \times 10^3 \text{ N})(0.3 \text{ m})}{\left(\dfrac{\pi(0.01 \text{ m})^2}{4}\right)(20 \times 10^{10} \text{ Pa})} \\ &= \boxed{7.43 \times 10^{-5} \text{ m} \quad (74 \ \mu\text{m})} \end{aligned}$$

The answer is (D).

6. *Customary U.S. Solution*

Let subscript st refer to steel and subscript b refer to bronze.

$$A_{st} = \frac{\pi\left((6 \text{ in})^2 - (5 \text{ in})^2\right)}{4} = 8.639 \text{ in}^2$$

$$A_b = \frac{\pi\left((2 \text{ in})^2 - (1.6 \text{ in})^2\right)}{4} = 1.131 \text{ in}^2$$

From Table 51.2, the coefficients of thermal expansion for brass and steel are 10×10^{-6} 1/°F and 6.5×10^{-6} 1/°F, respectively. The thermal deformation that would occur if the pipes were free is

$$\begin{aligned} \delta_b &= \alpha_b L \Delta T = \left(10 \times 10^{-6} \ \frac{1}{^\circ\text{F}}\right)(60 \text{ in})(200^\circ\text{F}) \\ &= 0.120 \text{ in} \\ \delta_{st} &= \alpha_{st} L \Delta T = \left(6.5 \times 10^{-6} \ \frac{1}{^\circ\text{F}}\right)(60 \text{ in})(200^\circ\text{F}) \\ &= 0.078 \text{ in} \end{aligned}$$

The two pipes expand by the same amount, δ, so that $\delta_{st} < \delta < \delta_b$.

$$\delta - \delta_{st} = \frac{F_{st}L}{A_{st}E_{st}} \qquad \text{[Eq. I]}$$

$$\delta_b - \delta = \frac{F_bL}{A_bE_b} \qquad \text{[Eq. II]}$$

F_{st} is a tensile force, and F_b is a compressive force. Since there is no external force,

$$F_{st} = F_b = F$$

Statics

Adding Eq. I to Eq. II,

$$\delta_b - \delta_{st} = \frac{FL}{A_{st}E_{st}} + \frac{FL}{A_bE_b} = FL\left(\frac{1}{A_{st}E_{st}} + \frac{1}{A_bE_b}\right)$$

$$F = \frac{\dfrac{\delta_b - \delta_{st}}{L}}{\dfrac{1}{A_{st}E_{st}} + \dfrac{1}{A_bE_b}}$$

$$= \frac{\dfrac{0.120\text{ in} - 0.078\text{ in}}{60\text{ in}}}{\dfrac{1}{(8.639\text{ in}^2)\left(29\times10^6\ \dfrac{\text{lbf}}{\text{in}^2}\right)} + \dfrac{1}{(1.131\text{ in}^2)\left(15\times10^6\ \dfrac{\text{lbf}}{\text{in}^2}\right)}}$$

$$= 11{,}122\text{ lbf}$$

$$\sigma_{st} = \frac{F}{A_{st}} = \frac{11{,}122\text{ lbf}}{8.639\text{ in}^2}$$

$$= \boxed{1287\text{ lbf/in}^2\quad(1290\text{ lbf/in}^2)}$$

$$\sigma_b = \frac{F}{A_b} = \frac{11{,}122\text{ lbf}}{1.131\text{ in}^2}$$

$$= \boxed{9834\text{ lbf/in}^2\quad(9830\text{ lbf/in}^2)}$$

The answer is (C).

SI Solution

Let subscript st refer to steel and subscript b refer to bronze.

$$A_{st} = \frac{\pi\left((15\text{ cm})^2 - (14\text{ cm})^2\right)}{4} = 22.78\text{ cm}^2$$

$$= 0.002\,278\text{ m}^2$$

$$A_b = \frac{\pi\left((5\text{ cm})^2 - (4\text{ cm})^2\right)}{4} = 7.069\text{ cm}^2$$

$$= 0.000\,706\,9\text{ m}^2$$

From Table 51.2, the coefficients of linear thermal expansion for brass and steel are 18×10^{-6} 1/°C and 11.7×10^{-6} 1/°C, respectively. The thermal deformation that would occur if the pipes were free is

$$\delta_b = \alpha_b L\Delta T = \left(18\times10^{-6}\ \frac{1}{^\circ\text{C}}\right)(1.5\text{ m})(110^\circ\text{C})$$

$$= 0.002\,97\text{ m}$$

$$\delta_{st} = \alpha_{st} L\Delta T = \left(11.7\times10^{-6}\ \frac{1}{^\circ\text{C}}\right)(1.5\text{ m})(110^\circ\text{C})$$

$$= 0.001\,93\text{ m}$$

The two pipes expand by the same amount, δ, so that $\delta_{st} < \delta < \delta_b$.

$$\delta - \delta_{st} = \frac{F_{st}L}{A_{st}E_{st}}\quad\text{[Eq. I]}$$

$$\delta_b - \delta = \frac{F_bL}{A_bE_b}\quad\text{[Eq. II]}$$

F_{st} is a tensile force, and F_b is a compressive force. Since there is no external force,

$$F_{st} = F_b = F$$

Adding Eq. I to Eq. II,

$$\delta_b - \delta_{st} = \frac{FL}{A_{st}E_{st}} + \frac{FL}{A_bE_b} = FL\left(\frac{1}{A_{st}E_{st}} + \frac{1}{A_bE_b}\right)$$

$$F = \frac{\dfrac{\delta_b - \delta_{st}}{L}}{\dfrac{1}{A_{st}E_{st}} + \dfrac{1}{A_bE_b}}$$

$$= \frac{\dfrac{0.002\,97\text{ m} - 0.001\,93\text{ m}}{1.5\text{ m}}}{\dfrac{1}{(0.002\,278\text{ m}^2)(20\times10^{10}\text{ Pa})} + \dfrac{1}{(0.000\,706\,9\text{ m}^2)(10\times10^{10}\text{ Pa})}}$$

$$= 4.24\times10^4\text{ N}$$

$$\sigma_{st} = \frac{F}{A_{st}} = \frac{4.24\times10^4\text{ N}}{0.002\,278\text{ m}^2}$$

$$= \boxed{1.86\times10^7\text{ Pa}\quad(19\text{ MPa})}$$

$$\sigma_b = \frac{F}{A_b} = \frac{4.24\times10^4\text{ N}}{0.000\,706\,9\text{ m}^2}$$

$$= \boxed{6\times10^7\text{ Pa}\quad(60\text{ MPa})}$$

The answer is (C).

7. *Customary U.S. Solution*

The deflection of the beam is $\delta_b = P_bL^3/3EI$, where P_b is the net load at the beam tip. If P_c is the tension in the cable, the elongation of the cable is

$$\delta_c = \frac{P_cL_c}{AE}$$

$\delta_c = \delta_b$ is the constraint on the deformation. Therefore,

$$\frac{P_bL_b^3}{3EI} = \frac{P_cL_c}{AE}\quad\text{[Eq. I]}$$

Statics

Another equation is the equilibrium equation.

$$F - P_c = P_b \quad \text{[Eq. II]}$$

Solving Eq. 1 and Eq. II simultaneously,

$$P_c = \frac{\frac{FL_b^3}{3I}}{\frac{L_c}{A} + \frac{L_b^3}{3I}}$$

$$= \frac{\frac{(270 \text{ lbf})\left((4 \text{ ft})\left(12 \frac{\text{in}}{\text{ft}}\right)\right)^3}{(3)(10 \text{ in}^4)}}{\frac{(2 \text{ ft})\left(12 \frac{\text{in}}{\text{ft}}\right)}{0.0124 \text{ in}^2} + \frac{\left((4 \text{ ft})\left(12 \frac{\text{in}}{\text{ft}}\right)\right)^3}{(3)(10 \text{ in}^4)}}$$

$$= \boxed{177 \text{ lbf} \quad (180 \text{ lbf})}$$

The answer is (A).

SI Solution

The deflection of the beam is $\delta_b = P_b L^3/3EI$, where P_b is the net load at the beam tip. If P_c is the tension in the cable, the elongation of the cable is

$$\delta_c = \frac{P_c L_c}{AE}$$

$\delta_c = \delta_b$ is the constraint on the deformation. Therefore,

$$\frac{P_b L_b^3}{3EI} = \frac{P_c L_c}{AE} \quad \text{[Eq. I]}$$

Another equation is the equilibrium equation.

$$F - P_c = P_b \quad \text{[Eq. II]}$$

Solving Eq. I and Eq. II simultaneously,

$$P_c = \frac{\frac{FL_b^3}{3I}}{\frac{L_c}{A} + \frac{L_b^3}{3I}} = \frac{\frac{(1200 \text{ N})(1.2 \text{ m})^3\left(1000 \frac{\text{mm}}{\text{m}}\right)^4}{(3)(4.17 \times 10^6 \text{ mm}^4)}}{\frac{(0.6 \text{ m})\left(1000 \frac{\text{mm}}{\text{m}}\right)^2}{8 \text{ mm}^2} + \frac{(1.2 \text{ m})^3\left(1000 \frac{\text{mm}}{\text{m}}\right)^4}{(3)(4.17 \times 10^6 \text{ mm}^4)}}$$

$$= \boxed{778 \text{ N} \quad (780 \text{ N})}$$

The answer is (A).

8. Let deflection downward be positive.

The deflection at the center of the beam is

$$\delta_b = \frac{5w(2L)^4}{384EI} - \frac{F(2L)^3}{48EI}$$

F is the force applied by the column at the beam center. The beam deflection is equal to the shortening of the column.

$$\delta_c = \frac{Fh}{EA}$$

Since $\delta_b = \delta_c$,

$$\frac{Fh}{EA} = \frac{5w(2L)^4}{384EI} - \frac{F(2L)^3}{48EI}$$

$$F = \boxed{\frac{5AwL^4}{24hI + 4AL^3}} \quad \text{[Eq. I]}$$

Another equation is the equilibrium equation. Let R_1 and R_2 be the left and right support reactions, respectively, on the beam. By symmetry,

$$R_1 = R_2 = R$$

$$2R + F - 2wL = 0 \quad \text{[Eq. II]}$$

Solving Eq. I and Eq. II simultaneously,

$$R = \frac{2wL - F}{2}$$

$$= \boxed{wL\left(\frac{48hI + 3AL^3}{48hI + 8AL^3}\right)}$$

The answer is (B).

9. *Customary U.S. Solution*

(a) Let F_1 and F_2 and δ_1 and δ_2 be the tensions and the deformations in the rods, respectively. The moment equilibrium equation is taken at the hinge of the rigid bar and is

$$\sum M_o = aF_1 + (a+b)F_2 - (a+b+c)P = 0 \quad \text{[Eq. I]}$$

Since the bar is rigid and angle changes are negligible, the elongations of the two aluminum rods are proportional to distances from the hinge at point O. The relationship between the elongations is

$$\frac{\delta_1}{a} = \frac{\delta_2}{a+b}$$

This can be rewritten as

$$\frac{F_1 L_1}{AEa} = \frac{F_2 L_2}{AE(a+b)}$$

Since $L_1/a = L_2/(a+b)$,

$$F_1 = F_2 \qquad \text{[Eq. II]}$$

Solving Eq. I and Eq. II,

$$aF + (a+b)F - (a+b+c)P = 0$$

$$\begin{aligned} F = F_1 = F_2 &= \left(\frac{a+b+c}{2a+b}\right)P \\ &= \left(\frac{6 \text{ ft} + 3 \text{ ft} + 3 \text{ ft}}{(2)(6 \text{ ft}) + 3 \text{ ft}}\right)(4500 \text{ lbf}) \\ &= \boxed{3600 \text{ lbf}} \end{aligned}$$

The answer is (B).

(b) Take downward as a positive deflection. The slope of the rigid member is

$$\frac{d}{a+b+c}$$

$$\begin{aligned} \delta_1 &= \left(\frac{F_1}{AE}\right)L_1 = \left(\frac{F_1}{AE}\right)\left(\frac{d}{a+b+c}\right)a \\ &= \left(\frac{3600 \text{ lbf}}{(0.124 \text{ in}^2)\left(10 \times 10^6 \ \frac{\text{lbf}}{\text{in}^2}\right)}\right) \\ &\quad \times \left(\frac{12 \text{ ft}}{6 \text{ ft} + 3 \text{ ft} + 3 \text{ ft}}\right)(6 \text{ ft})\left(12 \ \frac{\text{in}}{\text{ft}}\right) \\ &= \boxed{0.21 \text{ in}} \end{aligned}$$

$$\begin{aligned} \delta_2 &= \left(\frac{F_2}{AE}\right)L_2 = \left(\frac{F_2}{AE}\right)\left(\frac{d}{a+b+c}\right)(a+b) \\ &= \left(\frac{3600 \text{ lbf}}{(0.124 \text{ in}^2)\left(10 \times 10^6 \ \frac{\text{lbf}}{\text{in}^2}\right)}\right) \\ &\quad \times \left(\frac{12 \text{ ft}}{6 \text{ ft} + 3 \text{ ft} + 3 \text{ ft}}\right)(6 \text{ ft} + 3 \text{ ft})\left(12 \ \frac{\text{in}}{\text{ft}}\right) \\ &= \boxed{0.31 \text{ in}} \end{aligned}$$

The answer is (B).

SI Solution

(a) Let F_1 and F_2 and δ_1 and δ_2 be the tensions and the deformations in the rods, respectively. The moment equilibrium equation is taken at the hinge of the rigid bar and is

$$\sum M_o = aF_1 + (a+b)F_2 - (a+b+c)P = 0 \qquad \text{[Eq. I]}$$

Since the bar is rigid and angle changes are negligible, the elongations of the two aluminum rods are proportional to distances from the hinge at point O. The relationship between the elongations is

$$\frac{\delta_1}{a} = \frac{\delta_2}{a+b}$$

This can be rewritten as

$$\frac{F_1 L_1}{AEa} = \frac{F_2 L_2}{AE(a+b)}$$

Since $L_1/a = L_2/(a+b)$,

$$F_1 = F_2 \qquad \text{[Eq. II]}$$

Solving Eq. I and Eq. II,

$$\begin{aligned} aF + (a+b)F - (a+b+c)P &= 0 \\ F = F_1 = F_2 &= \left(\frac{a+b+c}{2a+b}\right)P \\ &= \left(\frac{2 \text{ m} + 1 \text{ m} + 1 \text{ m}}{(2)(2 \text{ m}) + 1 \text{ m}}\right)(20 \text{ kN}) \\ &= \boxed{16 \text{ kN}} \end{aligned}$$

The answer is (B).

(b) Take downward as a positive deflection. The slope of the rigid member is

$$\frac{d}{a+b+c}$$

$$\begin{aligned} \delta_1 &= \left(\frac{F_1}{AE}\right)L_1 = \left(\frac{F_1}{AE}\right)\left(\frac{d}{a+b+c}\right)a \\ &= \left(\frac{(16\,000 \text{ N})\left(1000 \ \frac{\text{mm}}{\text{m}}\right)^2}{(80 \text{ mm}^2)(70 \times 10^9 \text{ Pa})}\right) \\ &\quad \times \left(\frac{4 \text{ m}}{2 \text{ m} + 1 \text{ m} + 1 \text{ m}}\right)(2 \text{ m})\left(1000 \ \frac{\text{mm}}{\text{m}}\right) \\ &= \boxed{5.7 \text{ mm}} \end{aligned}$$

Statics

$$\delta_2 = \left(\frac{F_2}{AE}\right)L_2 = \left(\frac{F_2}{AE}\right)\left(\frac{d}{a+b+c}\right)(a+b)$$

$$= \left(\frac{(16\,000\text{ N})\left(1000\ \frac{\text{mm}}{\text{m}}\right)^2}{(80\text{ mm}^2)(70\times10^9\text{ Pa})}\right)$$

$$\times\left(\frac{4\text{ m}}{2\text{ m}+1\text{ m}+1\text{ m}}\right)(2\text{ m}+1\text{ m})\left(1000\ \frac{\text{mm}}{\text{m}}\right)$$

$$= \boxed{8.6\text{ mm}}$$

The answer is (B).

10. Let subscript s refer to the supported beam and subscript c refer to the cantilever beam. The deflections are equal.

$$\delta_s = \delta_c$$

$$\frac{P_sL^3}{48EI} = \frac{P_cL^3}{3EI}$$

$$P_s = 16P_c \quad \text{[Eq. I]}$$

P_s and P_c are the net loads exerted on the supported beam center and the cantilever beam tip, respectively. The equilibrium equation for the supported beam is

$$\sum F_{s,y} = 2R - P_s = 0 \quad \text{[Eq. II]}$$

The equilibrium equation for the cantilever beam is

$$\sum F_{c,y} = P_c = F - P_s \quad \text{[Eq. III]}$$

Solving Eq. I and Eq. III simultaneously,

$$P_c = \frac{F}{17}$$

$$P_s = \tfrac{16}{17}F$$

From Eq. II,

$$R = \boxed{\tfrac{8}{17}F}$$

The answer is (C).

11. Assume deflection downward is positive.

step 1: Remove support 2 to make the structure statically determinate.

step 2: The deflection at the location of (removed) support 2 is the sum of the deflections induced by the discrete load and the distributed load.

Using App. 51.A, case 7, the deflection induced by the discrete load is

$$\delta_{\text{discrete}} = \frac{Pb}{6EIL}\left(\left(\frac{L}{b}\right)(x-a)^3 + (L^2-b^2)x - x^3\right)$$

$$(P = 5000, L = 15, a = 6, b = 9, x = 10)$$

Using App. 51.A, case 9, the deflection induced by the distributed load is

$$\delta_{\text{distributed}} = \frac{-w}{24EI}(L^3x - 2Lx^3 + x^4)$$

step 3: The deflection induced by R_2 alone considered as a load is

$$\delta_{R_2} = \frac{-R_2a^2b^2}{3EIL}$$

step 4: The total deflection at the location of support 2 is zero.

$$\delta_{\text{discrete}} + \delta_{\text{distributed}} + \delta_{R_2} = 0$$

$$\left(\frac{(5000)(9)}{(6)(15)}\right) \times\left(\left(\frac{15}{9}\right)(10-6)^3 + \left((15)^2-(9)^2\right)(10) - (10)^3\right)$$

$$+\left(\frac{1000}{24}\right)\left((15)^3(10) - (2)(15)(10)^3 + (10)^4\right)$$

$$+\frac{-R_2(10)^2(5)^2}{(3)(15)} = 0$$

$$R_2 = \boxed{15{,}233\quad(15{,}000)}$$

The answer is (C).

12. First, remove the prop to make the structure statically determinate. The deflection induced by the distributed load at the tip is

$$\delta_{\text{distributed}} = \frac{-wL^4}{30EI} \quad \text{[down]}$$

The deflection caused by load R is

$$\delta_R = \frac{RL^3}{3EI} \quad \text{[up]}$$

Since the deflection is actually zero at the tip,

$$\delta_{\text{distributed}} + \delta_R = 0$$

$$\frac{-wL^4}{30EI} + \frac{RL^3}{3EI} = 0$$

$$R = \frac{wL}{10} = \frac{(1000)(10)}{10} = \boxed{1000}$$

The answer is (B).

13. *Customary U.S. Solution*

Use the three-moment method. The first moment of the area is

$$\begin{aligned} A_1 a &= \tfrac{1}{6}Fc(L^2 - c^2) \\ &= \left(\tfrac{1}{6}\right)(13{,}600 \text{ lbf})(22.5 \text{ ft})\left((32.5 \text{ ft})^2 - (22.5 \text{ ft})^2\right) \\ &= 28{,}050{,}000 \text{ ft}^3\text{-lbf} \end{aligned}$$

Since there is no force between R_2 and R_3, $A_2 b = 0$.

The left and right ends of the beam are simply supported; M_1 and M_3 are zero. Therefore, the three-moment equation becomes

$$\begin{aligned} 2M_2(32.5 \text{ ft} + 20 \text{ ft}) &= (-6)\left(\frac{28{,}050{,}000 \text{ ft}^3\text{-lbf}}{32.5 \text{ ft}}\right) \\ M_2 &= -49{,}318.7 \text{ ft-lbf} \end{aligned}$$

M_2 can be written in terms of the load and reactions to the left of support 2.

$$\begin{aligned} M_2 &= (-13{,}600 \text{ lbf})(10 \text{ ft}) + (32.5 \text{ ft})R_1 \\ &= -49{,}318.7 \text{ ft-lbf} \\ R_1 &= \boxed{2667.1 \text{ lbf} \quad (2700 \text{ lbf})} \end{aligned}$$

Now that R_1 is known, moments can be taken about support 3 to the left.

$$\begin{aligned} \sum M_3 &= (2667.1 \text{ lbf})(52.5 \text{ ft}) - (13{,}600 \text{ lbf})(30 \text{ ft}) \\ &\quad + R_2(20 \text{ ft}) = 0 \\ R_2 &= \boxed{13{,}398.9 \text{ lbf} \quad (13{,}000 \text{ lbf})} \end{aligned}$$

R_3 can be obtained by taking moments about support 1.

$$\begin{aligned} \sum M_1 &= (22.5 \text{ ft})(13{,}600 \text{ lbf}) - (32.5 \text{ ft})(13{,}398.9 \text{ lbf}) \\ &\quad - (52.5)R_3 = 0 \\ R_3 &= \boxed{-2466.0 \text{ lbf} \quad (-2500 \text{ lbf})} \end{aligned}$$

The answer is (B).

SI Solution

Use the three-moment method. The first moment of the area is

$$\begin{aligned} A_1 a &= \tfrac{1}{6}Fc(L^2 - c^2) \\ &= \left(\tfrac{1}{6}\right)(60\,500 \text{ N})(7 \text{ m})\left((10 \text{ m})^2 - (7 \text{ m})^2\right) \\ &= 3.60 \times 10^6 \text{ N·m}^3 \end{aligned}$$

Since there is no force between R_2 and R_3, $A_2 b = 0$.

The left and right ends of the beam are simply supported; M_1 and M_3 are zero. Therefore, the three-moment equation becomes

$$\begin{aligned} 2M_2(10 \text{ m} + 6 \text{ m}) &= (-6)\left(\frac{3.6 \times 10^6 \text{ N·m}^3}{10 \text{ m}}\right) \\ M_2 &= -67\,500 \text{ N·m} \end{aligned}$$

M_2 can be written in terms of the load and reactions to the left of support 2.

$$\begin{aligned} M_2 &= (-60\,500 \text{ N})(10 \text{ m} - 7 \text{ m}) + (10 \text{ m})R_1 \\ &= -67\,500 \text{ N·m} \\ R_1 &= \boxed{11\,400 \text{ N} \quad (11\,000 \text{ N})} \end{aligned}$$

Now that R_1 is known, moments can be taken about support 3 to the left.

$$\begin{aligned} \sum M_3 &= (11\,400 \text{ N})(16 \text{ m}) - (60\,500 \text{ N})(9 \text{ m}) \\ &\quad + R_2(6 \text{ m}) = 0 \\ R_2 &= \boxed{60\,350 \text{ N} \quad (60\,000 \text{ N})} \end{aligned}$$

R_3 can be obtained by taking moments about support 1.

$$\begin{aligned} \sum M_1 &= (7 \text{ m})(60\,500 \text{ N}) - (10 \text{ m})(60\,350 \text{ N}) \\ &\quad - (16 \text{ m})R_3 = 0 \\ R_3 &= \boxed{-11\,250 \text{ N} \quad (-11\,000 \text{ N})} \end{aligned}$$

The answer is (B).

14. Use the three-moment method. The first moments of the areas are

$$\begin{aligned} A_1 a &= \tfrac{1}{6}Fc(L_1^2 - c^2) = \left(\tfrac{1}{6}\right)(5000)(4)\left((10)^2 - (4)^2\right) \\ &= 280{,}000 \\ A_2 b &= \frac{wL_2^4}{24} = \frac{(1000)(5)^4}{24} = 26{,}042 \end{aligned}$$

The two ends of the beam are simply supported; M_1 and M_3 are zero. Therefore, the three moment equation becomes

$$\begin{aligned} 2M_2(10 + 5) &= (-6)\left(\frac{280{,}000}{10} + \frac{26{,}042}{5}\right) \\ M_2 &= -6642 \end{aligned}$$

M_2 can also be written in terms of the load and reactions to the left of support 2.

$$\begin{aligned} M_2 &= (-5000)(6) + 10R_1 = -6642 \\ R_1 &= \boxed{2336 \quad (2340)} \end{aligned}$$

Sum the moments around support 3.

$$\sum M_3 = (2336)(15) - (5000)(11) + 5R_2 - (5000)(2.5) = 0$$

$$R_2 = \boxed{6492 \quad (6490)}$$

Sum the moments around support 1.

$$\sum M_1 = (5000)(4) - (6492)(10) + (5000)(12.5) - 15R_3 = 0$$

$$R_3 = \boxed{1172 \quad (1170)}$$

The answer is (C).

15. The equilibrium requirement is

$$R_1 + R_2 = 1000$$

The moment equation at support 1 is

$$\sum M_{R_1} = M_1 + M_2 + (1000)(30) - 40R_2 = 0$$

From a table of fixed-end moments,

$$M_1 = \frac{-Fb^2a}{L^2} = \frac{-(1000)(10)^2(30)}{(40)^2} = -1875$$

$$M_2 = \frac{Fa^2b}{L^2} = \frac{(1000)(30)^2(10)}{(40)^2} = 5625$$

$$R_2 = \boxed{843.75 \quad (840)}$$

The moment equation at support 2 is

$$\sum M_{R_2} = M_1 + M_2 - (1000)(10) + 40R_1 = 0$$

$$R_1 = \boxed{156.25 \quad (160)}$$

The equilibrium requirement is

$$R_1 + R_2 = 1000$$

$$156.25 + 843.75 = 1000 \quad \text{[check]}$$

The answer is (B).

16. (a) The force in member Bb depends on the shear, V, across the cut shown.

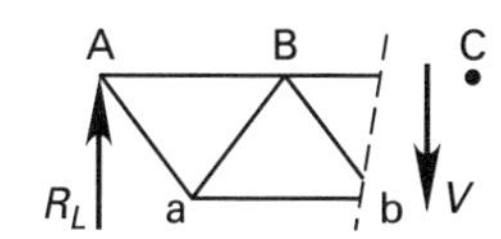

- Influence diagram for shear across panel Bb:

If the unit load is to the right of point C, the reaction, R_L, will be

$$R_L = \frac{x}{120 \text{ ft}}$$

(x is the distance from the right reaction to the unit load.)

$$V_L = R_L = \frac{x}{120 \text{ ft}}$$

If the unit load is to the left of point B,

$$R_L = \frac{x}{120 \text{ ft}}$$

$$V = R_L - 1 = \frac{x}{120 \text{ ft}} - 1$$

At points B and C,

$$V_B = \frac{90 \text{ ft}}{120 \text{ ft}} - 1 = -0.25$$

$$V_C = \frac{60 \text{ ft}}{120 \text{ ft}} = 0.5$$

The influence diagram for shear in member Bb is

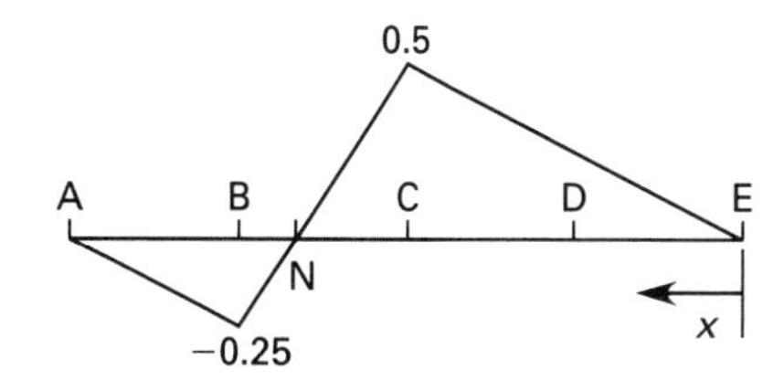

The neutral point, N, is located at

$$x = (2)(30 \text{ ft}) + \left(\frac{0.5}{0.5 + 0.25}\right)(30 \text{ ft}) = 80 \text{ ft}$$

- Maximum shear due to moving uniform load:

The moving load, perhaps representing a stream of cars, is allowed to be over any part or all of the bridge deck. The shear will be maximum in member Bb if the load is distributed from N to E.

The area under the influence line from N to E is

$$\left(\tfrac{1}{2}\right)\Big((20 \text{ ft})(0.5) + (2)(30 \text{ ft})(0.5)\Big) = 20 \text{ ft}$$

The maximum shear, V, is

$$(20 \text{ ft})\left(2 \ \frac{\text{kips}}{\text{ft}}\right) = 40 \text{ kips}$$

- Maximum shear due to moving concentrated load:

From the influence diagram, maximum shear will occur when the concentrated load is at point C. The shear in panel Bb is

$$(0.5)(15 \text{ kips}) = 7.5 \text{ kips}$$

Statics

- Tension in member Bb:

The force triangle is

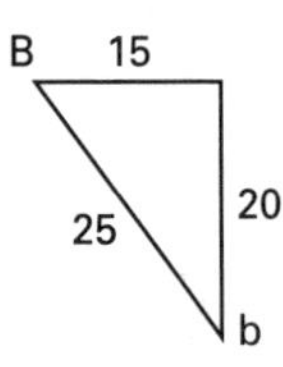

The total maximum shear across panel Bb is

$$40 \text{ kips} + 7.5 \text{ kips} = 47.5 \text{ kips}$$

The total maximum tension in member Bb is

$$(47.5 \text{ ft})\left(\frac{25 \text{ ft}}{20 \text{ ft}}\right) = \boxed{59.375 \text{ kips} \quad (59 \text{ kips})}$$

The answer is (C).

(b) • Influence diagram for moment at point b:

Since the horizontal member BC cannot resist vertical shear, the shear influence diagram previously used will not work.

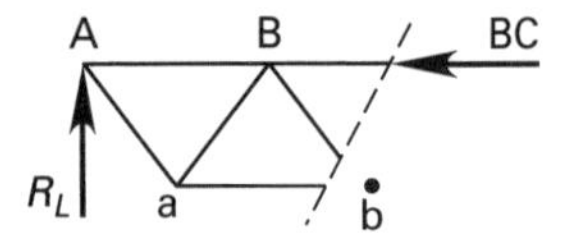

With no loads between A and C, the force in member BC can be found by summing the moments about point b. Taking clockwise moments as positive,

$$\begin{aligned}\sum M_b &= (45 \text{ ft})R_L - (20 \text{ ft})(\text{BC}) \\ &= 0 \\ \text{BC} &= \frac{45R_L}{20 \text{ ft}}\end{aligned}$$

$45R_L$ is the moment that the moment from force BC opposes. In general,

$$\text{BC} = \frac{M_b}{20 \text{ ft}}$$

If the load is between C and E,

$$R_L = \frac{x}{120 \text{ ft}} \quad [x \text{ is measured from E}]$$

The moment caused by R_L is

$$\begin{aligned}M_b &= (45 \text{ ft})\left(\frac{x}{120 \text{ ft}}\right) \\ &= 0.375x\end{aligned}$$

If the load is between A and B, the reaction is

$$R_L = \frac{x}{120 \text{ ft}}$$

The moment at b is also affected by the load between A and B.

$$\begin{aligned}M_b &= 0.375x - (1)(x - 75 \text{ ft}) \\ &= 75 \text{ ft} - 0.625x\end{aligned}$$

Plotting these values versus x,

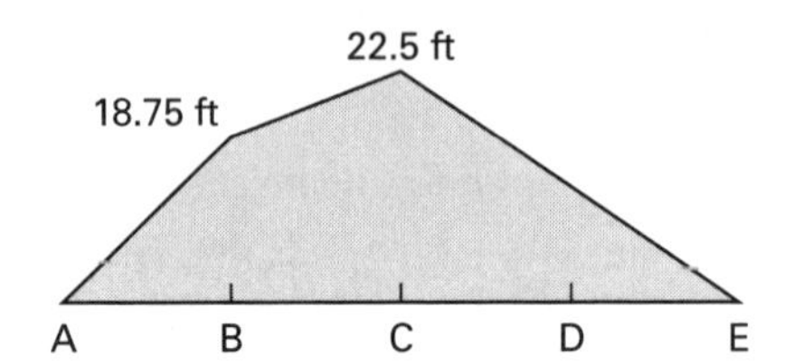

- Maximum moment due to uniform load:

The moment at b is maximum when the entire truss is loaded from A to E. The area under the curve is

$$\begin{aligned}&\left(\tfrac{1}{2}\right)(30 \text{ ft})(18.75 \text{ ft}) + \left(\tfrac{1}{2}\right)(60 \text{ ft})(22.5 \text{ ft}) \\ &\quad + (30 \text{ ft})(18.75 \text{ ft}) \\ &\quad + \left(\tfrac{1}{2}\right)(30 \text{ ft})(22.5 \text{ ft} - 18.75 \text{ ft}) \\ &\quad = 1575 \text{ ft}^2\end{aligned}$$

The maximum moment is

$$M_b = \left(2 \ \frac{\text{kips}}{\text{ft}}\right)(1575 \text{ ft}^2) = 3150 \text{ ft-kips}$$

- Maximum moment due to concentrated load:

Maximum moment will occur when the load is at C.

$$\begin{aligned}M_b &= (15 \text{ ft})(22.5 \text{ kips}) \\ &= 337.5 \text{ ft-kips}\end{aligned}$$

- Total maximum moment:

$$\begin{aligned}M_b &= 337.5 \text{ ft-kips} + 3150 \text{ ft-kips} \\ &= 3487.5 \text{ ft-kips}\end{aligned}$$

- Compression in BC:

$$\begin{aligned}\text{BC} &= \frac{M_b}{20 \text{ ft}} = \frac{3487.5 \text{ ft-kips}}{20 \text{ ft}} \\ &= \boxed{174.4 \text{ kips} \quad (170 \text{ kips})}\end{aligned}$$

The answer is (D).

17. The force in member CD is a function of the moment at point d.

If the unit load is between d and g, the left reaction is

$$R_L = \frac{x}{162 \text{ ft}}$$

The moment at point d is

$$M_d = (3)(27 \text{ ft})\left(\frac{x}{162 \text{ ft}}\right) = 0.5x \quad \text{[d to g]}$$

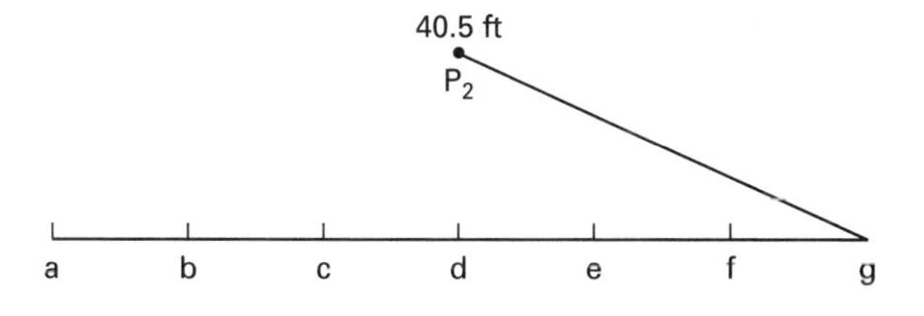

If the unit load is between a and c, the moment at point d is

$$\begin{aligned} M_d &= 0.5x - (1)(x - (3)(27 \text{ ft})) \\ &= 81 \text{ ft} - 0.5x \quad \text{[a to c]} \end{aligned}$$

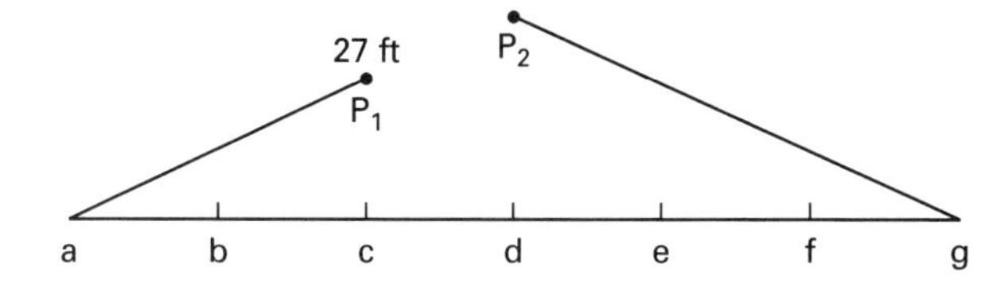

Complete the influence diagram by joining points P_1 and P_2. Observe that the slope of P_1 is the same as that of P_2. This is because point d is at the center of the truss.

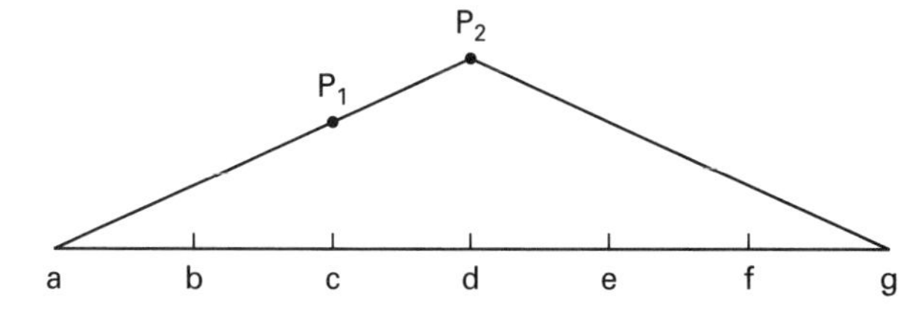

The resultant of the load group and its location are

$$8 \text{ kips} + 40 \text{ kips} + 7 \text{ kips} + 30 \text{ kips} = 85 \text{ kips}$$

$$\frac{\begin{array}{c}(40 \text{ kips})(16 \text{ ft}) + (7 \text{ kips})(16 \text{ ft} + 28 \text{ ft}) \\ + (30 \text{ kips})(16 \text{ ft} + 28 \text{ ft} + 16 \text{ ft})\end{array}}{85 \text{ kips}} = 32.33 \text{ ft}$$

The resultant is located 32.33 ft to the right of the 8 kip load.

Assume the load group moves from right to left.

- Case 1:

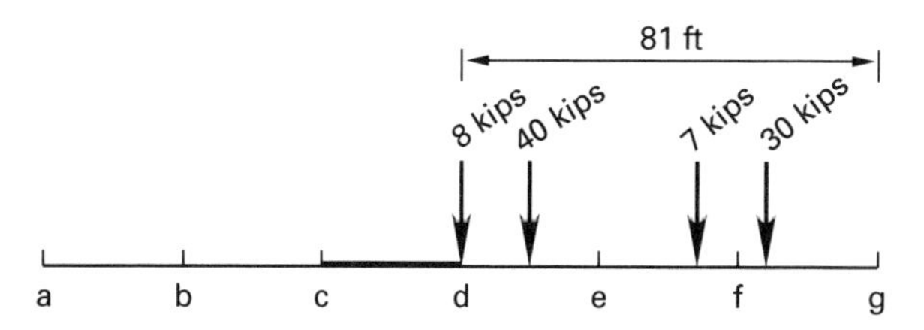

Taking counterclockwise moments as positive,

$$\begin{aligned} \sum M_g &= (85 \text{ kips})(81 \text{ ft} - 32.33 \text{ ft}) - (162 \text{ ft})R_L \\ R_L &= 25.54 \text{ kips} \end{aligned}$$

The shear does not change sign under panel cd.

- Case 2:

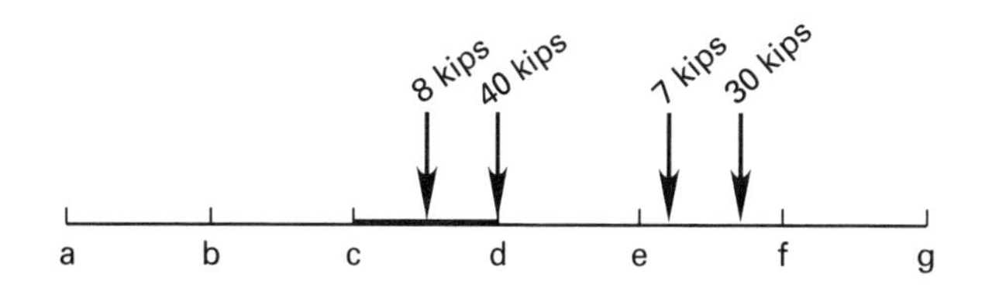

$$\begin{aligned} R_L &= 33.9 \text{ kips} \\ V_{CD} &= 33.9 \text{ kips} - 8 \text{ kips} - 40 \text{ kips} \\ &= -14.1 \text{ kips} \end{aligned}$$

Since the shear changes sign (goes through zero), the moment is maximum when the 40 kip load is at point d.

load	x	influence diagram height	moment
8 kips	97 ft	32.5 ft	(8)(32.5) = 260 ft-kips
40 kips	81 ft	40.5 ft	(40)(40.5) = 1620 ft-kips
7 kips	53 ft	26.5 ft	(7)(26.5) = 185.5 ft-kips
30 kips	37 ft	18.5 ft	(30)(18.5) = 555 ft-kips
total			2620.5 ft-kips

The compression in CD is

$$CD = \frac{2620.5 \text{ ft-kips}}{32 \text{ ft}} = \boxed{81.9 \text{ kips}}$$

Since $7 + 30 < 8 + 40$, if the load moves from left to right it must reach the same position as case 2 for the moment to be maximum. Therefore, the left-to-right analysis is not needed.

The answer is (B).

Statics

- Alternate solution:

Having determined that the 40 kip load should be at point d, find the left reaction.

$$\begin{aligned}\sum M_{R_R} &= (162 \text{ ft})R_L - (8 \text{ kips})(97 \text{ ft}) \\ &\quad - (40 \text{ kips})(81 \text{ ft}) - (7 \text{ kips})(53 \text{ ft}) \\ &\quad - (30 \text{ kips})(37 \text{ ft}) = 0 \\ R_L &= 33.93 \text{ kips}\end{aligned}$$

Sum the moments about point d and use the method of sections.

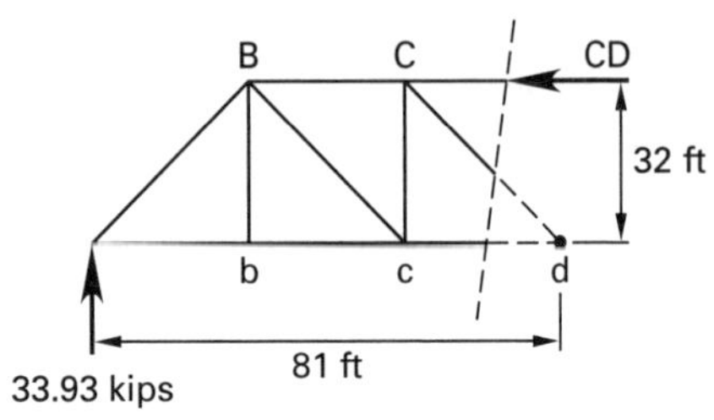

Taking clockwise moments as positive,

$$\begin{aligned}\sum M_{\text{d}} &= (33.93 \text{ kips})(81 \text{ ft}) - (32 \text{ ft})(\text{CD}) \\ &\quad - (8 \text{ kips})(16 \text{ ft}) = 0 \\ \text{CD} &= \boxed{81.9 \text{ kips} \quad (82 \text{ kips})}\end{aligned}$$

The answer is (B).

18. (a) Moment:

Put a hinge at point B and rotate.

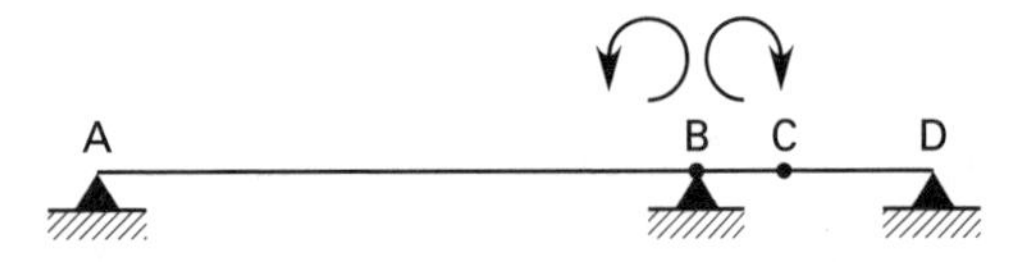

The moment influence diagram is

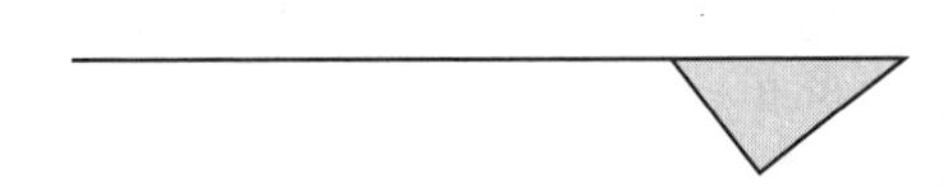

One of the loads should be at point C.

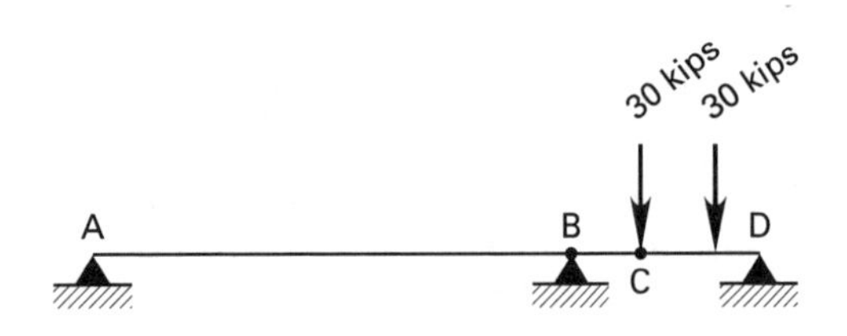

Since the slope of the influence line is less between C and D, the ordinate 6 ft to the right of point C will be larger than the ordinate 6 ft to the left of point C. The reaction at the hinge due to the 30 kip load 6 ft to the right of point C is

$$R = (30 \text{ kips})\left(\frac{15 \text{ ft} - 6 \text{ ft}}{15 \text{ ft}}\right) = 18 \text{ kips}$$

The moment at point B is

$$\begin{aligned}M_{\text{B}} &= (9 \text{ ft})(30 \text{ kips} + 18 \text{ kips}) \\ &= \boxed{432 \text{ ft-kips} \quad (430 \text{ ft-kips})}\end{aligned}$$

The answer is (D).

(b) Shear:

Use the method of virtual displacement to draw the shear influence diagram. Since the point is a reaction point, lift the point a distance of 1. The shear diagram is

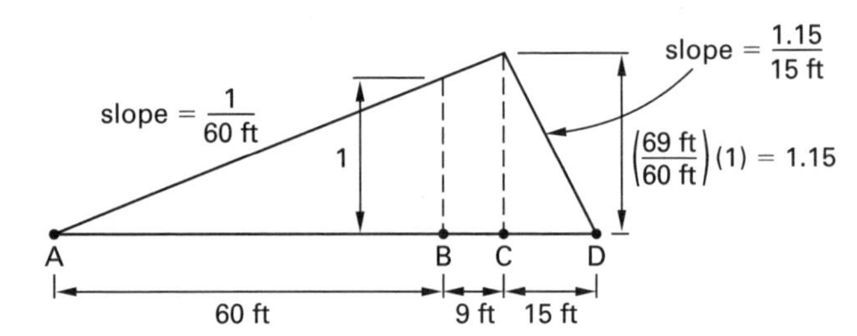

For maximum shear, one load or the other must be at point C.

By inspection, the effect of having both loads to the left of C is greater than having them to the right.

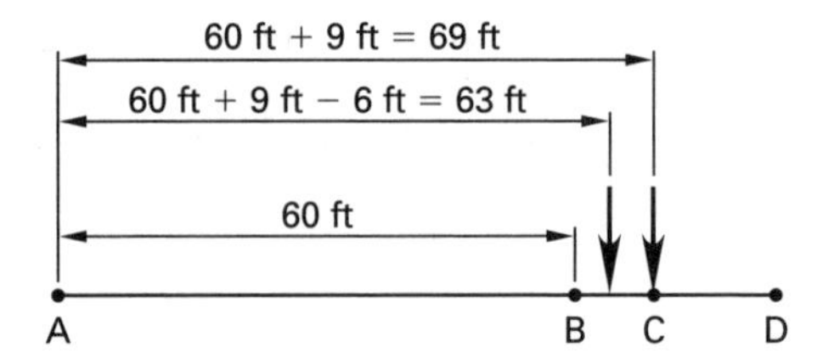

The maximum shear is

$$V_{\text{max}} = (30 \text{ kips})\left(\frac{63 \text{ ft}}{60 \text{ ft}} + \frac{69 \text{ ft}}{60 \text{ ft}}\right) = \boxed{66 \text{ kips} \quad (70 \text{ kips})}$$

The answer is (C).

47 Engineering Materials

PRACTICE PROBLEMS

1. Approximately how much (in molecules of HCl per gram of PVC) HCl should be used as an initiator in PVC if the efficiency is 20% and an average molecular weight of 7000 g/mol is desired? The final polymer has the structure shown.

```
   H  H  H  H          H  H  H  H
   |  |  |  |          |  |  |  |
H- C -C -C -C -•••- C -C -C -C -Cl
   |  |  |  |          |  |  |  |
   H  Cl H  Cl         H  Cl H  Cl
```

(A) 7.8×10^{18} molecules HCl per gram PVC

(B) 4.3×10^{20} molecules HCl per gram PVC

(C) 9.5×10^{21} molecules HCl per gram PVC

(D) 3.6×10^{23} molecules HCl per gram PVC

2. 10 ml of a 0.2% solution (by weight) of hydrogen peroxide is added to 12 g of ethylene to stabilize the polymer. Hydrogen peroxide breaks down according to

$$H_2O_2 \rightarrow 2(OH^-) + \cdots$$

Assume that the stabilized polymer has the structure shown.

```
     H  H  H          H  H  H
     |  |  |          |  |  |
OH - C -C -C -•••- C -C -C - OH
     |  |  |          |  |  |
     H  H  H          H  H  H
```

What is most nearly the average degree of polymerization if the hydrogen peroxide is completely utilized?

(A) 180

(B) 730

(C) 910

(D) 1200

SOLUTIONS

1. The vinyl chloride mer is

```
H   H
|   |
C = C
|   |
H   Cl
```

With 20% efficiency, 5 molecules of HCl per PVC molecule are required to supply each end Cl atom. This is the same as 5 mol HCl per mole PVC. Using Avogadro's number of 6.022×10^{23} molecules per mole, the number of molecules of HCl per gram of PVC is

$$\frac{\left(5 \ \frac{\text{mol HCl}}{\text{mol PVC}}\right)\left(6.022 \times 10^{23} \ \frac{\text{molecules}}{\text{mol}}\right)}{7000 \ \frac{\text{g}}{\text{mol}}} = \boxed{4.3 \times 10^{20} \text{ molecules HCl/gram PVC}}$$

The answer is (B).

2. The molecular weight of hydrogen peroxide, H_2O_2, is

$$(2)\left(1 \ \frac{\text{g}}{\text{mol}}\right) + (2)\left(16 \ \frac{\text{g}}{\text{mol}}\right) = 34 \text{ g/mol}$$

The weight of H_2O_2 is

$$(10 \text{ mL})\left(1 \ \frac{\text{g}}{\text{mL}}\right) = 10 \text{ g}$$

The number of H_2O_2 molecules in a 0.2% solution is

$$\frac{(10 \text{ g})\left(\frac{0.2\%}{100\%}\right)\left(6.022 \times 10^{23} \ \frac{\text{molecules}}{\text{mol}}\right)}{34 \ \frac{\text{g}}{\text{mol}}} = 3.54 \times 10^{20} \text{ molecules}$$

The molecular weight of ethylene, C_2H_4, is

$$(2)\left(12 \ \frac{\text{g}}{\text{mol}}\right) + (4)\left(1 \ \frac{\text{g}}{\text{mol}}\right) = 28 \text{ g/mol}$$

The number of ethylene molecules is

$$\frac{(12\text{ g})\left(6.022\times 10^{23}\ \frac{\text{molecules}}{\text{mol}}\right)}{28\ \frac{\text{g}}{\text{mol}}} = 2.58\times 10^{23}\text{ molecules}$$

Since it takes one H_2O_2 molecule (i.e., $2OH^-$ radicals) to stabilize a polyethylene molecule, there are 3.54×10^{20} polymers.

The degree of polymerization is

$$\text{DP} = \frac{2.58\times 10^{23}\ C_2H_4\text{ molecules}}{3.54\times 10^{20}\text{ polymers}} = \boxed{729\quad(730)}$$

The answer is (B).

Materials

48 Material Properties and Testing

PRACTICE PROBLEMS

1. The engineering stress and engineering strain for a copper specimen are 20,000 lbf/in^2 (140 MPa) and 0.0200 in/in (0.0200 mm/mm), respectively. Poisson's ratio for the specimen is 0.3.

(a) The true stress is most nearly

(A) 14,000 lbf/in^2 (98 MPa)

(B) 18,000 lbf/in^2 (130 MPa)

(C) 20,000 lbf/in^2 (140 MPa)

(D) 22,000 lbf/in^2 (160 MPa)

(b) The true strain is most nearly

(A) 0.0182 in/in (0.0182 mm/mm)

(B) 0.0189 in/in (0.0189 mm/mm)

(C) 0.0194 in/in (0.0194 mm/mm)

(D) 0.0198 in/in (0.0198 mm/mm)

2. A graph of engineering stress-strain is shown. Poisson's ratio for the material is 0.3.

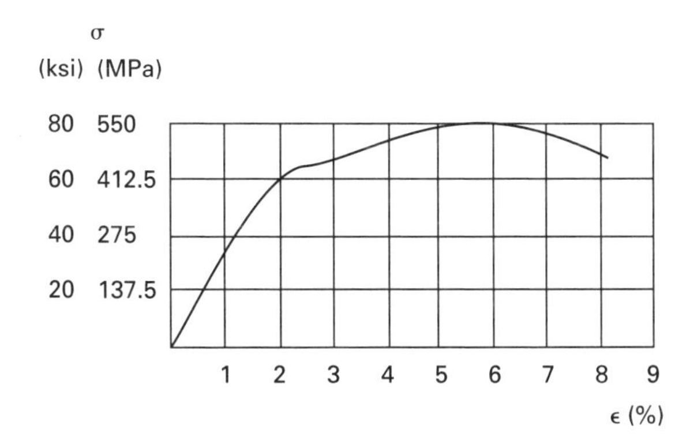

(a) The 0.5% yield strength is most nearly

(A) 70,000 lbf/in^2 (480 MPa)

(B) 76,000 lbf/in^2 (530 MPa)

(C) 84,000 lbf/in^2 (590 MPa)

(D) 98,000 lbf/in^2 (690 MPa)

(b) The elastic modulus is most nearly

(A) 2.4×10^6 lbf/in^2 (17 GPa)

(B) 2.7×10^6 lbf/in^2 (19 GPa)

(C) 2.9×10^6 lbf/in^2 (20 GPa)

(D) 3.0×10^6 lbf/in^2 (21 GPa)

(c) The ultimate strength is most nearly

(A) 72,000 lbf/in^2 (500 MPa)

(B) 76,000 lbf/in^2 (530 MPa)

(C) 80,000 lbf/in^2 (550 MPa)

(D) 84,000 lbf/in^2 (590 MPa)

(d) The fracture strength is most nearly

(A) 62,000 lbf/in^2 (430 MPa)

(B) 70,000 lbf/in^2 (480 MPa)

(C) 76,000 lbf/in^2 (530 MPa)

(D) 84,000 lbf/in^2 (590 MPa)

(e) The percentage of elongation at fracture is most nearly

(A) 6%

(B) 8%

(C) 10%

(D) 12%

(f) The shear modulus is most nearly

(A) 0.9×10^6 lbf/in^2 (6.3 GPa)

(B) 1.1×10^6 lbf/in^2 (7.7 GPa)

(C) 1.2×10^6 lbf/in^2 (7.9 GPa)

(D) 1.5×10^6 lbf/in^2 (11 GPa)

(g) The toughness is most nearly

(A) 5100 in-lbf/in^3 (35 MJ/m^3)

(B) 5700 in-lbf/in^3 (38 MJ/m^3)

(C) 6300 in-lbf/in^3 (42 MJ/m^3)

(D) 8900 in-lbf/in^3 (60 MJ/m^3)

3. A specimen with an unstressed cross-sectional area of 4 in^2 (25 cm^2) necks down to 3.42 in^2 (22 cm^2) before breaking in a standard tensile test. The percentage reduction in area of the material is most nearly

(A) 9.4% (8.9%)

(B) 10% (9.4%)

(C) 13% (10%)

(D) 15% (12%)

4. A constant 15,000 lbf/in^2 (100 MPa) tensile stress is applied to a specimen. The stress is known to be less than the material's yield strength. The strain is measured at various times. What is most nearly the steady-state creep rate for the material?

time (hr)	strain (in/in)
5	0.018
10	0.022
20	0.026
30	0.031
40	0.035
50	0.040
60	0.046
70	0.058

(A) 0.00037 hr^{-1}

(B) 0.00041 hr^{-1}

(C) 0.00046 hr^{-1}

(D) 0.00049 hr^{-1}

5. At a particular instant during a tensile test, a 0.5 in diameter steel cylinder 2 in long experienced a length of 2.6 in at a load of 4200 lbf. The cylinder was prepared according to ASTM specifications. The true stress in the steel cylinder at that instant was most nearly

(A) 28,000 lbf/in^2

(B) 32,000 lbf/in^2

(C) 36,000 lbf/in^2

(D) 40,000 lbf/in^2

6. True stress and true strain from a tensile test performed on a metal bar are plotted on a log-log graph in accordance with ASTM E646. If the slope of the curve is between 0.5 and 1.0, which conclusion can be drawn about the metal?

(A) The metal is brittle.

(B) The metal is hard.

(C) The metal is highly malleable and ductile.

(D) The metal has a high strain-hardening capacity.

7. What property describes the variance of a metal's crystalline structure depending on temperature?

(A) allotropism

(B) isotropism

(C) polymorphism

(D) isogamy

8. What kind of material has properties that depend on direction?

(A) anisometropic

(B) anisotropic

(C) anisometric

(D) anisophyllic

9. According to ASTM standards, which test is utilized to test the hardness of metals?

(A) Bauschinger

(B) Charpy

(C) rupture

(D) scleroscopic

SOLUTIONS

1. *Customary U.S. Solution*

(a) The fractional reduction in diameter is

$$\nu e = (0.3)\left(0.020\ \frac{\text{in}}{\text{in}}\right) = 0.006$$

The true stress is given by Eq. 48.6.

$$\sigma = \frac{s}{(1-\nu e)^2} = \frac{20{,}000\ \frac{\text{lbf}}{\text{in}^2}}{(1-0.006)^2} = \boxed{20{,}242\ \text{lbf/in}^2\quad (20{,}000\ \text{lbf/ft}^2)}$$

The answer is (C).

(b) The true strain is given by Eq. 48.7.

$$\epsilon = \ln(1+e) = \ln(1+0.020) = \boxed{0.0198\ \text{in/in}}$$

The answer is (D).

SI Solution

(a) The fractional reduction in diameter is

$$\nu e = (0.3)\left(0.020\ \frac{\text{mm}}{\text{mm}}\right) = 0.006$$

The true stress is given by Eq. 48.6.

$$\sigma = \frac{s}{(1-\nu e)^2} = \frac{140\ \text{MPa}}{(1-0.006)^2} = \boxed{141.7\ \text{MPa}\quad (140\ \text{MPa})}$$

The answer is (C).

(b) The true strain is given by Eq. 48.7.

$$\epsilon = \ln(1+e) = \ln(1+0.020) = \boxed{0.0198\ \text{mm/mm}}$$

The answer is (D).

2. *Customary U.S. Solution*

(a) Extend a line from the 0.5% offset strain value parallel to the linear portion of the curve.

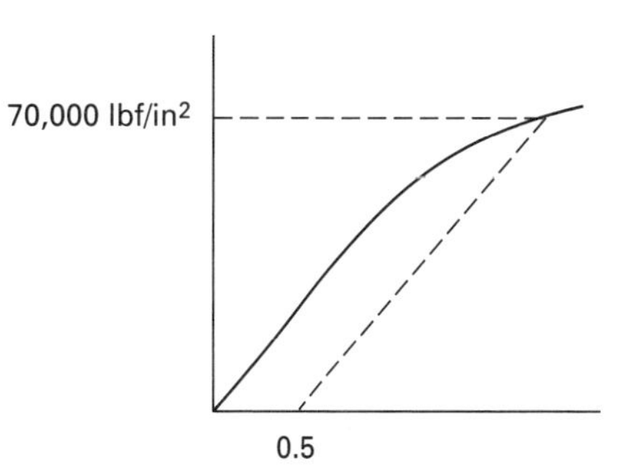

The 0.5% yield strength is $\boxed{70{,}000\ \text{lbf/in}^2.}$

The answer is (A).

(b) At the elastic limit, the stress is 60,000 lbf/in^2, and the percent strain is 2. The elastic modulus is

$$E = \frac{\text{stress}}{\text{strain}} = \frac{60{,}000\ \frac{\text{lbf}}{\text{in}^2}}{0.02} = \boxed{3.0\times 10^6\ \text{lbf/in}^2}$$

The answer is (D).

(c) The ultimate strength is the highest point of the curve. This value is $\boxed{80{,}000\ \text{lbf/in}^2.}$

The answer is (C).

(d) The fracture strength is at the end of the curve. This value is $\boxed{70{,}000\ \text{lbf/in}^2.}$

The answer is (B).

(e) The percent elongation at fracture is determined by extending a straight line parallel to the initial strain line from the fracture point. This gives an approximate value of $\boxed{6\%.}$

The answer is (A).

(f) The shear modulus is given by Eq. 48.22.

$$G = \frac{E}{2(1+\nu)} = \frac{3\times 10^6\ \frac{\text{lbf}}{\text{in}^2}}{(2)(1+0.3)} = \boxed{1.15\times 10^6\ \text{lbf/in}^2\quad (1.2\times 10^6\ \text{lbf/in}^2)}$$

The answer is (C).

Materials

(g) The toughness is the area under the stress-strain curve. Divide the area into squares of 20 ksi × 1%. There are about 25.5 squares covered.

$$(25.5)\left(20{,}000\ \frac{\text{lbf}}{\text{in}^2}\right)\left(0.01\ \frac{\text{in}}{\text{in}}\right) = \boxed{5100\ \text{in-lbf/in}^3}$$

(Using Eq. 48.18 would be inappropriate for a material as elastic as this one.)

The answer is (A).

SI Solution

(a) Extend a line from the 0.5% offset strain value parallel to the linear portion of the curve.

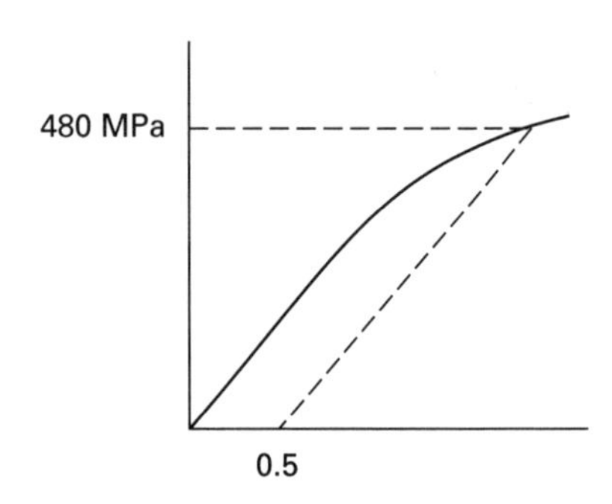

The 0.5% yield strength is $\boxed{480\ \text{MPa.}}$

The answer is (A).

(b) At the elastic limit, the stress is 410 MPa, and the percent strain is 2. The elastic modulus is

$$E = \frac{\text{stress}}{\text{strain}} = \frac{410\ \text{MPa}}{(0.02)\left(1000\ \frac{\text{MPa}}{\text{GPa}}\right)}$$
$$= \boxed{20.5\ \text{GPa}\quad(21\ \text{GPa})}$$

The answer is (D).

(c) The ultimate strength is the highest point of the curve. This value is $\boxed{550\ \text{MPa.}}$

The answer is (C).

(d) The fracture strength is at the end of the curve. This value is $\boxed{480\ \text{MPa.}}$

The answer is (B).

(e) The percent elongation at fracture is determined by extending a straight line parallel to the initial strain line from the fracture point. This gives an approximate value of $\boxed{6\%.}$

The answer is (A).

(f) The shear modulus is given by Eq. 48.22.

$$G = \frac{E}{2(1+\nu)} = \frac{20.5\ \text{GPa}}{(2)(1+0.3)}$$
$$= \boxed{7.88\ \text{GPa}\quad(7.9\ \text{GPa})}$$

The answer is (C).

(g) The toughness is the area under the stress-strain curve. Divide the area into squares of 137.5 MPa × 1%. There are about 25.5 squares covered.

$$(25.5)(137.5\ \text{MPa})\left(0.01\ \frac{\text{m}}{\text{m}}\right) = \boxed{35\ \text{MJ/m}^3}$$

The answer is (A).

3. *Customary U.S. Solution*

Use Eq. 48.13.

$$q_f = \frac{A_o - A_f}{A_o} \times 100\%$$
$$= \frac{4.0\ \text{in}^2 - 3.42\ \text{in}^2}{4.0\ \text{in}^2} \times 100\%$$
$$= \boxed{14.5\%\quad(15\%)}$$

The answer is (D).

SI Solution

Use Eq. 48.13.

$$q_f = \frac{A_o - A_f}{A_o} \times 100\%$$
$$= \frac{25\ \text{cm}^2 - 22\ \text{cm}^2}{25\ \text{cm}^2} \times 100\%$$
$$= \boxed{12\%}$$

The answer is (D).

4. Plot the data and draw a straight line. Disregard the first and last data points, as these represent primary and tertiary creep, respectively.

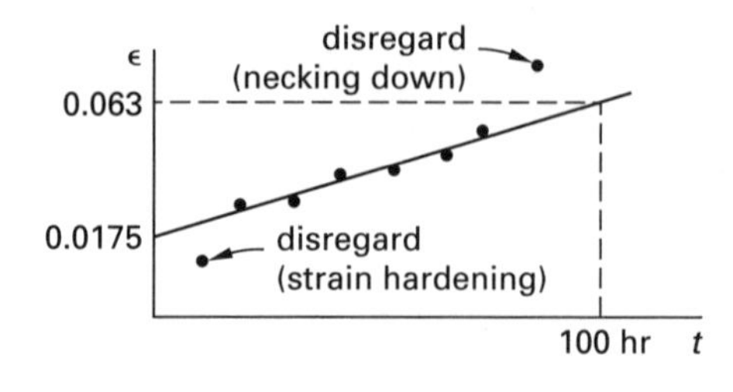

The creep rate is the slope of the line.

$$\text{creep rate} = \frac{\Delta\epsilon}{\Delta t} = \frac{0.063\ \frac{\text{in}}{\text{in}} - 0.0175\ \frac{\text{in}}{\text{in}}}{100\ \text{hr}}$$
$$= \boxed{0.000455\ 1/\text{hr}\quad (0.00046\ 1/\text{hr})}$$

The answer is (C).

5. As volume is constant, use Eq. 48.8 to calculate instantaneous area, A.

$$AL = A_oL_o$$
$$A = \frac{A_oL_o}{L} = \frac{\pi(0.25\ \text{in})^2(2\ \text{in})}{2.6\ \text{in}} = 0.15\ \text{in}^2$$

Use Eq. 48.5 to solve for true stress, σ.

$$\sigma = \frac{F}{A} = \frac{4200\ \text{lbf}}{0.15\ \text{in}^2} = \boxed{28{,}000\ \text{lbf/in}^2}$$

The answer is (A).

6. (See Eq. 48.10.) The true stress-true strain curve of many metals in the region of uniform plastic deformation can be expressed by the power curve relationship $\sigma_{\text{true}} = K\epsilon_{\text{true}}^n$. When data following this equation is plotted in log-log format, the result will be a straight line with linear slope n. The slope is known as the strain-hardening exponent. The strain-hardening exponent may have values from 0 (perfectly plastic solid) to 1 (perfectly elastic solid). For most metals, n is between 0.10 and 0.50. Only copper based alloys have values in the vicinity of 0.50. Annealed steels, for example, have a strain-hardening exponent between 0.10 and 0.25. A strain-hardening exponent between 0.5 and 1.0 would indicate that the material becomes stronger and harder as it is strained. That is, it has a high-strain hardening capacity.

The answer is (D).

7. Different crystalline structures require different amounts of energy to form; the energy from a temperature change can cause the crystalline structures of some metals to change. The property that describes the variance of a metal's crystalline structure depending on temperature is allotropism.

The answer is (A).

8. An anisotropic material looks and behaves differently in some or all directions.

The answer is (B).

9. In order to measure a metal's hardness, a scleroscope, which has a diamond-tipped hammer, is dropped on the metal from a standard height. The height of the scleroscope's rebound determines the hardness of the test specimen. Due to the portable nature of the unit, a scleroscopic test is used for measuring the hardness of large objects. (A Charpy test is used to measure fracture energy and (indirectly) transition temperature.)

The answer is (D).

Materials

49 Thermal Treatment of Metals

PRACTICE PROBLEMS

1. Refer to the following equilibrium diagram for an alloy of elements A and B.

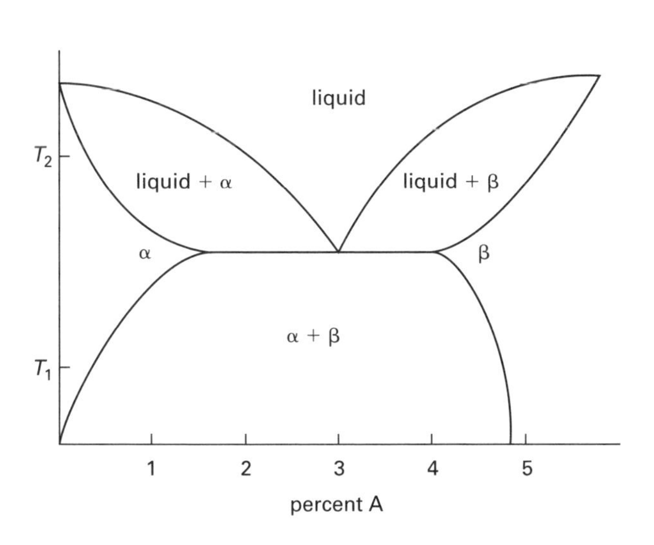

(a) For a 4%A alloy at temperature T_1, what are the compositions of solids α and β?

(b) For a 4%A alloy at temperature T_1, what is the percentage of α and β?

(c) For a 1%A alloy at temperature T_2, how much liquid and how much solid are present?

2. Write the procedures used with 2011 aluminum for (a) annealing and (b) precipitation hardening.

SOLUTIONS

1. (a) For temperature T_1, the equilibrium diagram is

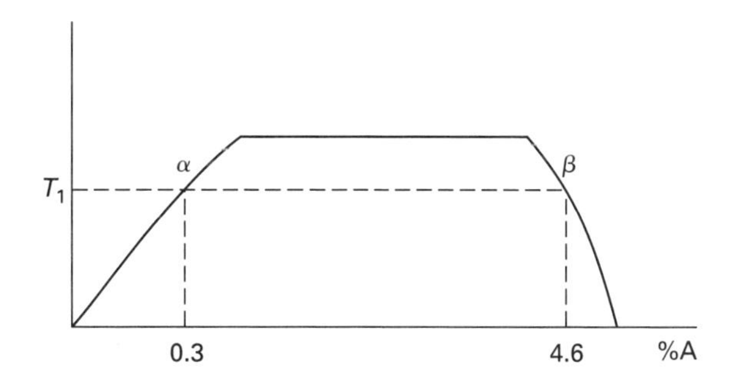

From the diagram, solid α is $\boxed{0.3\%\text{A.}}$

The %B for solid α is

$$\%\text{B} = 100\% - 0.3\% = \boxed{99.7\%}$$

For solid β,

$$\%\text{A} = \boxed{4.6\%}$$

$$\%\text{B} = 100\% - 4.6\% = \boxed{95.4\%}$$

(b) The percentages of α and β are

$$\begin{aligned}\%\alpha &= \frac{4.6\% - 4\%}{4.6\% - 0.3\%} \\ &= \boxed{0.14 \quad (14.0\%)} \\ \%\beta &= 100\% - \%\alpha \\ &= 100\% - 14.0\% \\ &= \boxed{86.0\%}\end{aligned}$$

(c) For temperature T_2, the equilibrium diagram is

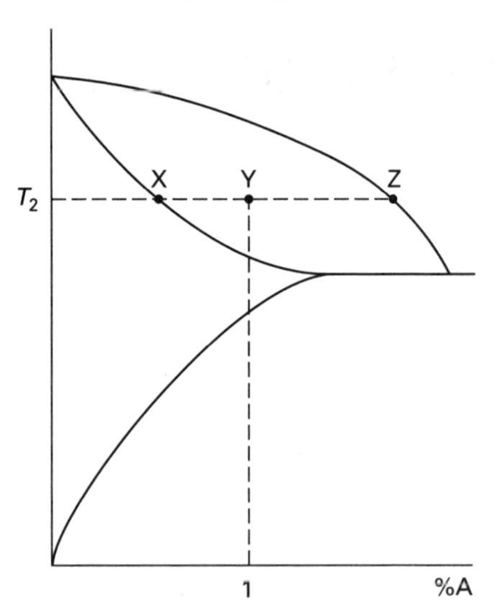

From the diagram, the following distances are measured.

1. The distance from point X to point Y is 9 mm.
2. The distance from point X to point Z is 21 mm.

From Eq. 49.6, the percent liquid is

$$\left(\frac{\text{XY}}{\text{XZ}}\right)(100\%) = \left(\frac{9 \text{ mm}}{21 \text{ mm}}\right)(100\%) = \boxed{42.9\%}$$

From Eq. 49.5, the percent solid is

$$100\% - \text{percent liquid} = 100\% - 42.9\% = \boxed{57.1\%}$$

2. (a) Since 2011 aluminum is a nonferrous substance, it does not readily form allotropes and thus its properties cannot be changed by the controlled cooling in an annealing process.

(b) The procedures for precipitation hardening of 2011 aluminum are

1. precipitation
2. rapid quenching
3. artificial aging

Materials

50 Properties of Areas

PRACTICE PROBLEMS

1. Where is the x-coordinate of the centroid of the area most nearly located?

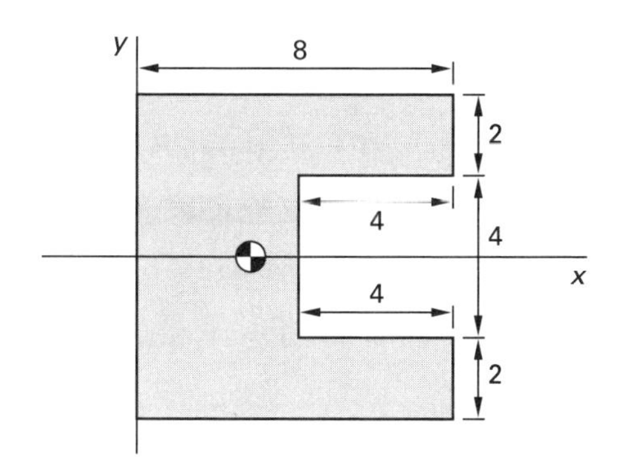

(A) 2.7 units

(B) 2.9 units

(C) 3.1 units

(D) 3.3 units

2. Replace the distributed load with three concentrated loads, and indicate the points of application.

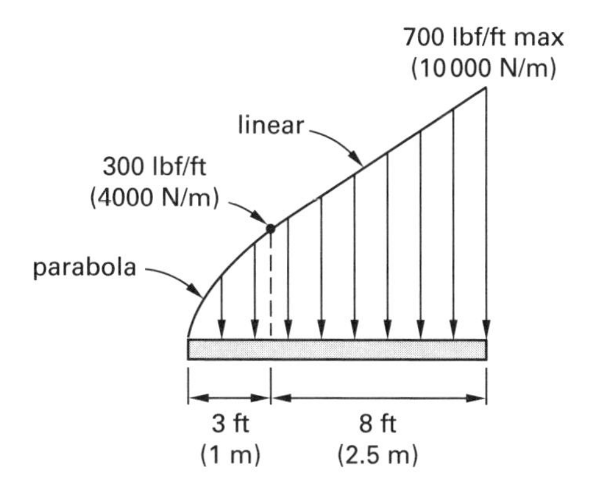

3. Most nearly, what is the centroidal moment of inertia about an axis parallel to the x-axis?

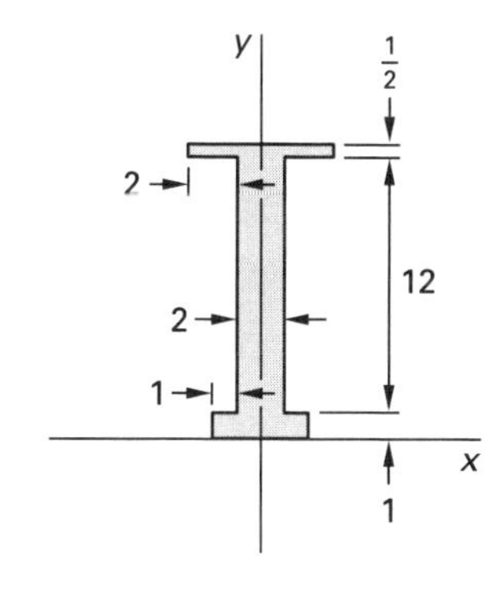

(A) 160 units^4

(B) 290 units^4

(C) 570 units^4

(D) 740 units^4

4. A rectangular 4 in × 10 in area has a 2 in diameter hole in its geometric center. What is most nearly the moment of inertia about the y-axis?

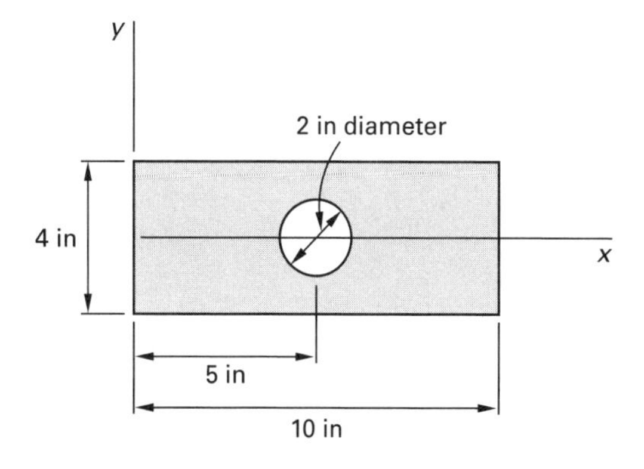

(A) 1030 in^4

(B) 1150 in^4

(C) 1250 in^4

(D) 1370 in^4

5. What is most nearly the moment of inertia about the x-axis of the area OAB?

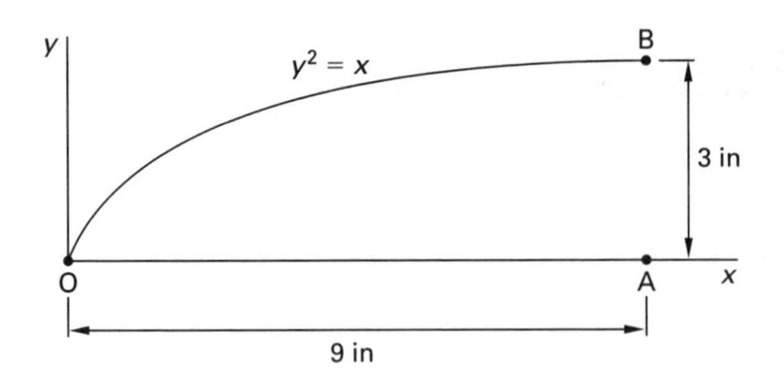

(A) 32 in^4

(B) 67 in^4

(C) 70 in^4

(D) 76 in^4

6. An annular flat ring has an outer diameter of 4 in and an inner diameter of 2 in. What is most nearly the radius of gyration of the ring about the y-axis?

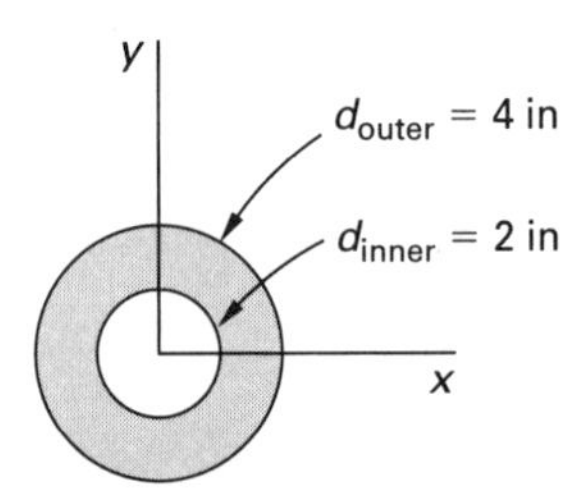

(A) 0.89 in

(B) 1.1 in

(C) 1.3 in

(D) 2.2 in

SOLUTIONS

1. The area is divided into three basic shapes.

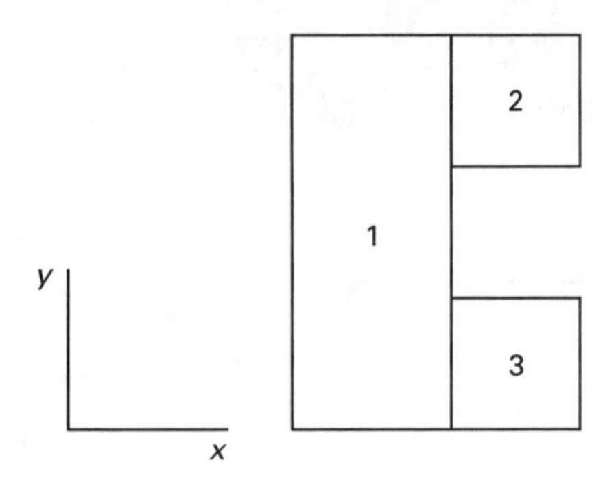

First, calculate the areas of the basic shapes.

$$A_1 = (4)(8) = 32 \text{ units}^2$$
$$A_2 = (4)(2) = 8 \text{ units}^2$$
$$A_3 = (4)(2) = 8 \text{ units}^2$$

Next, find the x-components of the centroids of the basic shapes.

$$x_{c,1} = 2 \text{ units}$$
$$x_{c,2} = 6 \text{ units}$$
$$x_{c,3} = 6 \text{ units}$$

Finally, use Eq. 50.5.

$$x_c = \frac{\sum A_i x_{c,i}}{\sum A_i}$$

$$= \frac{(32 \text{ units}^2)(2 \text{ units}) + (8 \text{ units}^2)(6 \text{ units}) + (8 \text{ units}^2)(6 \text{ units})}{32 \text{ units}^2 + 8 \text{ units}^2 + 8 \text{ units}^2}$$

$$= \boxed{3.33 \text{ units} \quad (3.3 \text{ units})}$$

The answer is (D).

2. *Customary U.S. Solution*

The parabolic shape is

$$f(x) = \left(300 \ \frac{\text{lbf}}{\text{ft}}\right)\sqrt{\frac{x}{3}}$$

Materials

First, use Eq. 50.3 to find the concentrated load given by the area.

$$A = \int f(x)\,dx = \int_{0\text{ ft}}^{3\text{ ft}} 300\sqrt{\frac{x}{3}}\,dx$$

$$= \left(\frac{300\ \frac{\text{lbf}}{\text{ft}}}{\sqrt{3}}\right)\left(\frac{x^{3/2}}{\frac{3}{2}}\Bigg|_{0\text{ ft}}^{3\text{ ft}}\right)$$

$$= \left(\frac{300\ \frac{\text{lbf}}{\text{ft}}}{\frac{3}{2}\sqrt{3}}\right)(3^{3/2} - 0^{3/2})$$

$$= \boxed{600\text{ lbf}}\quad \text{[first concentrated load]}$$

Next, from Eq. 50.4,

$$dA = f(x)\,dx = 300\sqrt{\frac{x}{3}}$$

Finally, use Eq. 50.1 to find the location, x_c, of the concentrated load from the left end.

$$x_c = \frac{\int x\,dA}{A} = \frac{1}{600\text{ lbf}}\int_{0\text{ ft}}^{3\text{ ft}} 300x\sqrt{\frac{x}{3}}\,dx$$

$$= \frac{300}{600\sqrt{3}}\int_{0\text{ ft}}^{3\text{ ft}} x^{3/2}\,dx$$

$$= \frac{300x^{5/2}}{600\sqrt{3}\left(\frac{5}{2}\right)}\Bigg|_{0\text{ ft}}^{3\text{ ft}}$$

$$= \left(\frac{300\ \frac{\text{lbf}}{\text{ft}}}{(600\text{ lbf})\sqrt{3}\left(\frac{5}{2}\right)}\right)(3^{5/2} - 0^{5/2})$$

$$= \boxed{1.8\text{ ft}}\quad \text{[location]}$$

Alternative solution for the parabola:

Use App. 50.A.

$$A = \frac{2bh}{3} = \frac{(2)\left(300\ \frac{\text{lbf}}{\text{ft}}\right)(3\text{ ft})}{3} = \boxed{600\text{ lbf}}$$

The centroid is located at a distance from the left end of

$$\frac{3h}{5} = \frac{(3)(3\text{ ft})}{5} = \boxed{1.8\text{ ft}}$$

The concentrated load for the triangular shape is the area from App. 50.A.

$$A = \frac{bh}{2} = \frac{\left(700\ \frac{\text{lbf}}{\text{ft}} - 300\ \frac{\text{lbf}}{\text{ft}}\right)(8\text{ ft})}{2}$$

$$= \boxed{1600\text{ lbf}}\quad \text{[second concentrated load]}$$

From App. 50.A, the location of the concentrated load from the right end is

$$\frac{h}{3} = \frac{8\text{ ft}}{3} = \boxed{2.67\text{ ft}}$$

The concentrated load for the rectangular shape is the area from App. 50.A.

$$A = bh = \left(300\ \frac{\text{lbf}}{\text{ft}}\right)(8\text{ ft})$$

$$= \boxed{2400\text{ lbf}}\quad \text{[third concentrated load]}$$

From App. 50.A, the location of the concentrated load from the right end is

$$\frac{h}{2} = \frac{8\text{ ft}}{2} = \boxed{4\text{ ft}}$$

SI Solution

The parabolic shape is

$$f(x) = \left(4000\ \frac{\text{N}}{\text{m}}\right)\sqrt{x}$$

First, use Eq. 50.3 to find the concentrated load given by the area.

$$A = \int f(x)\,dx = \int_{0\text{ m}}^{1\text{ m}} 4000\sqrt{x}\,dx = \frac{4000x^{3/2}}{\frac{3}{2}}\Bigg|_{0\text{ m}}^{1\text{ m}}$$

$$= \left(\frac{4000\ \frac{\text{N}}{\text{m}}}{\frac{3}{2}}\right)(1^{3/2} - 0^{3/2})$$

$$= \boxed{2666.7\text{ N}}\quad \text{[first concentrated load]}$$

Next, from Eq. 50.4,

$$dA = f(x)\,dx = 4000\sqrt{x}\,dx$$

Finally, use Eq. 50.1 to find the location, x_c, of the concentrated load from the left end.

$$\begin{aligned} x_c &= \frac{\int x\,dA}{A} = \frac{1}{2666.7\ \text{N}} \int_{0\ \text{m}}^{1\ \text{m}} 4000x\sqrt{x}\,dx \\ &= \frac{4000}{2666.7} \int_{0\ \text{m}}^{1\ \text{m}} x^{3/2}\,dx \\ &= \frac{4000x^{5/2}}{(2666.7)\left(\frac{5}{2}\right)} \Bigg|_{0\ \text{m}}^{1\ \text{m}} \\ &= \left(\frac{4000\ \frac{\text{N}}{\text{m}}}{(2666.7\ \text{N})\left(\frac{5}{2}\right)} \right) \left(1^{5/2} - 0^{5/2}\right) \\ &= \boxed{0.60\ \text{m}} \quad \text{[location]} \end{aligned}$$

Alternative solution for the parabola:

Use App. 50.A.

$$\begin{aligned} A &= \frac{2bh}{3} = \frac{(2)\left(4000\ \frac{\text{N}}{\text{m}}\right)(1\ \text{m})}{3} \\ &= \boxed{2666.7\ \text{N}} \end{aligned}$$

The centroid is located at a distance from the left end of

$$\frac{3h}{5} = \frac{(3)(1\ \text{m})}{5} = \boxed{0.6\ \text{m}}$$

The concentrated load for the triangular shape is the area from App. 50.A.

$$\begin{aligned} A &= \frac{bh}{2} = \frac{\left(10\,000\ \frac{\text{N}}{\text{m}} - 4000\ \frac{\text{N}}{\text{m}}\right)(2.5\ \text{m})}{2} \\ &= \boxed{7500\ \text{N}} \quad \text{[second concentrated load]} \end{aligned}$$

From App. 50.A, the location of the concentrated load from the right end is

$$\frac{h}{3} = \frac{2.5\ \text{m}}{3} = \boxed{0.83\ \text{m}}$$

The concentrated load for the rectangular shape is the area from App. 50.A.

$$\begin{aligned} A &= bh = \left(4000\ \frac{\text{N}}{\text{m}}\right)(2.5\ \text{m}) \\ &= \boxed{10\,000\ \text{N}} \quad \text{[third concentrated load]} \end{aligned}$$

From App. 50.A, the location of the concentrated load from the right end is

$$\frac{h}{2} = \frac{2.5\ \text{m}}{2} = \boxed{1.25\ \text{m}}$$

3. The area is divided into three basic shapes.

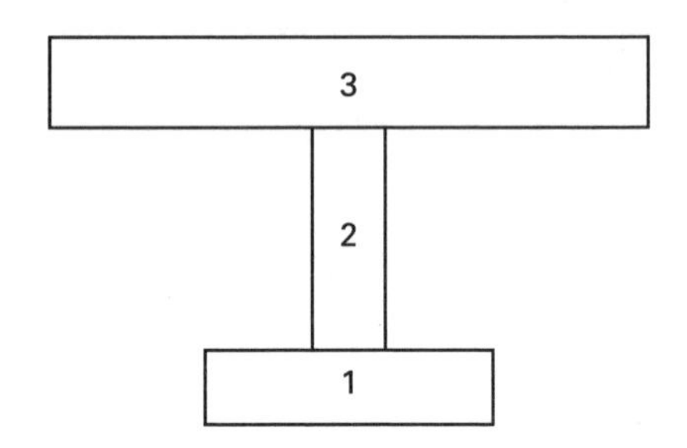

First, calculate the areas of the basic shapes.

$$\begin{aligned} A_1 &= (4)(1) = 4\ \text{units}^2 \\ A_2 &= (2)(12) = 24\ \text{units}^2 \\ A_3 &= (6)(0.5) = 3\ \text{units}^2 \end{aligned}$$

Next, find the y-components of the centroids of the basic shapes.

$$\begin{aligned} y_{c,1} &= 0.5\ \text{units} \\ y_{c,2} &= 7\ \text{units} \\ y_{c,3} &= 13.25\ \text{units} \end{aligned}$$

From Eq. 50.6, the centroid of the area is

$$\begin{aligned} y_c &= \frac{\sum A_i y_{c,i}}{\sum A_i} = \frac{(4)(0.5) + (24)(7) + (3)(13.25)}{4 + 24 + 3} \\ &= 6.77\ \text{units} \end{aligned}$$

From App. 50.A, the moment of inertia of basic shape 1 about its own centroid is

$$I_{cx,1} = \frac{bh^3}{12} = \frac{(4)(1)^3}{12} = 0.33\ \text{units}^4$$

The moment of inertia of basic shape 2 about its own centroid is

$$I_{cx,2} = \frac{bh^3}{12} = \frac{(2)(12)^3}{12} = 288\ \text{units}^4$$

The moment of inertia of basic shape 3 about its own centroid is

$$I_{cx,3} = \frac{bh^3}{12} = \frac{(6)(0.5)^3}{12} = 0.063 \text{ units}^4$$

From the parallel axis theorem, Eq. 50.20, the moment of inertia of basic shape 1 about the centroidal axis of the section is

$$\begin{aligned} I_{x,1} &= I_{cx,1} + A_1 d_1^2 = 0.33 + (4)(6.77 - 0.5)^2 \\ &= 157.6 \text{ units}^4 \end{aligned}$$

The moment of inertia of basic shape 2 about the centroidal axis of the section is

$$\begin{aligned} I_{x,2} &= I_{cx,2} + A_2 d_2^2 = 288 + (24)(7.0 - 6.77)^2 \\ &= 289.3 \text{ units}^4 \end{aligned}$$

The moment of inertia of basic shape 3 about the centroidal axis of the section is

$$\begin{aligned} I_{x,3} &= I_{cx,3} + A_3 d_3^2 = 0.063 + (3)(13.25 - 6.77)^2 \\ &= 126.0 \text{ units}^4 \end{aligned}$$

The total moment of inertia about the centroidal axis of the section is

$$\begin{aligned} I_x &= I_{x,1} + I_{x,2} + I_{x,3} \\ &= 157.6 \text{ units}^4 + 289.3 \text{ units}^4 + 126.0 \text{ units}^4 \\ &= \boxed{572.9 \text{ units}^4 \quad (570 \text{ units}^4)} \end{aligned}$$

The answer is (C).

4. This is a composite area. Divide the composite area into two basic shapes, a rectangle and a circle.

From App. 50.A, the moment of inertia of the rectangle with respect to an edge (in this case, the y-axis) is

$$I_{y,\text{rectangle}} = \frac{bh^3}{3} = \frac{(4 \text{ in})(10 \text{ in})^3}{3} = 1333.33 \text{ in}^4$$

From App. 50.A, the centroidal moment of inertia of the circle is

$$I_{c,\text{circle}} = \frac{\pi r^4}{4} = \frac{\pi(1 \text{ in})^4}{4} = 0.785 \text{ in}^4$$

From the parallel axis theorem, the moment of inertia of the circle with respect to the y-axis is

$$\begin{aligned} I_{y,\text{circle}} &= I_{c,\text{circle}} + d_2 A = 0.785 \text{ in}^4 + (5 \text{ in})^2 \pi (1 \text{ in})^2 \\ &= 79.285 \text{ in}^4 \end{aligned}$$

The moment of inertia about the y-axis is

$$\begin{aligned} I_{y,\text{composite area}} &= I_{y,\text{rectangle}} - I_{y,\text{circle}} \\ &= 1333.33 \text{ in}^4 - 79.285 \text{ in}^4 \\ &= \boxed{1254.05 \text{ in}^4 \quad (1250 \text{ in}^4)} \end{aligned}$$

The answer is (C).

5. Determine dA, which is the shaded area within the curve parallel to the x-axis.

$$dA = (9 - x)dy$$

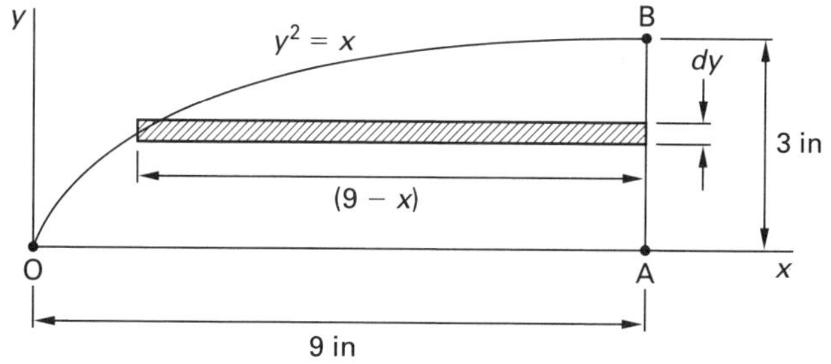

Using Eq. 50.17, calculate the moment of inertia from $y = 0$ in to $y = 3$ in.

$$\begin{aligned} I_x &= \int y^2 dA = \int_{0 \text{ in}}^{3 \text{ in}} y^2(9 - x)dy \\ &= \int_{0 \text{ in}}^{3 \text{ in}} y^2(9 - y^2)dy \\ &= 9\int_{0 \text{ in}}^{3 \text{ in}} y^2 dy - \int_{0 \text{ in}}^{3 \text{ in}} y^4 dy \\ &= (9)\left(\frac{y^3}{3}\Bigg|_{0 \text{ in}}^{3 \text{ in}}\right) - \frac{y^5}{5}\Bigg|_{0 \text{ in}}^{3 \text{ in}} \\ &= \boxed{32.4 \text{ in}^4 \quad (32 \text{ in}^4)} \end{aligned}$$

The answer is (A).

Materials

6. Due to symmetry, the moment of inertia and radius of gyration about the y-axis will be the same as for the x-axis. From App. 50.A, the moment of inertia of the annular ring is

$$I = I_{\text{outer}} - I_{\text{inner}} = \frac{\pi r_{\text{outer}}^4}{4} - \frac{\pi r_{\text{inner}}^4}{4}$$

$$= \frac{\pi\left(\frac{4\ \text{in}}{2}\right)^4}{4} - \frac{\pi\left(\frac{2\ \text{in}}{2}\right)^4}{4}$$

$$= 11.78\ \text{in}^4$$

The area of the composite area a is

$$A = \pi r_{\text{outer}}^2 - \pi r_{\text{inner}}^2 = \pi\left(\frac{4\ \text{in}}{2}\right)^2 - \pi\left(\frac{2\ \text{in}}{2}\right)^2$$

$$= 9.42\ \text{in}^2$$

From Eq. 50.25, the radius of gyration of the annular ring about the y-axis is

$$k = \sqrt{\frac{I}{A}} = \sqrt{\frac{11.78\ \text{in}^4}{9.42\ \text{in}^2}}$$

$$= \boxed{1.12\ \text{in}\quad (1.1\ \text{in})}$$

The answer is (B).

Materials

51 Strength of Materials

PRACTICE PROBLEMS

Shear and Moment Diagrams

1. A beam 14 ft (4.2 m) long is simply supported at the left end and 2 ft (0.6 m) from the right end. The beam has a mass of 20 lbm/ft (30 kg/m). A 100 lbf (450 N) load is applied 2 ft (0.6 m) from the left end. An 80 lbf (350 N) load is applied at the right end.

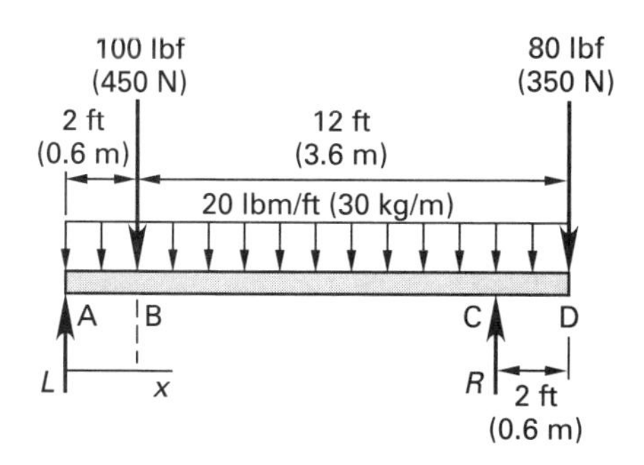

(a) The maximum moment is most nearly

(A) 150 ft-lbf (200 N·m)

(B) 250 ft-lbf (340 N·m)

(C) 390 ft-lbf (520 N·m)

(D) 830 ft-lbf (1100 N·m)

(b) The maximum shear is most nearly

(A) 80 lbf (360 N)

(B) 120 lbf (530 N)

(C) 150 lbf (650 N)

(D) 190 lbf (830 N)

Properties of Cross Sections

2. A rod with a circular cross section is used to replace a rod with a solid square cross section with side length s. The section modulus of the circular cross section must be at least 12 times that of the square-shaped cross section. The radius of the circular cross section is

(A) $\dfrac{\sqrt[3]{\pi}}{2s}$

(B) $\dfrac{2s}{\sqrt[3]{\pi}}$

(C) $\dfrac{12s^2}{\sqrt[3]{\pi}}$

(D) $\dfrac{12\sqrt[3]{\pi}}{s^2}$

Beam Deflections

3. A cantilever beam is 6 ft (1.8 m) in length. The cross section is 6 in wide by 4 in high (150 mm wide by 100 mm high). The modulus of elasticity is 1.5×10^6 psi (10 GPa). The beam is loaded by two concentrated forces: 200 lbf (900 N) located 1 ft (0.3 m) from the free end and 120 lbf (530 N) located 2 ft (0.6 m) from the free end. The tip deflection is most nearly

(A) 0.29 in (0.0084 m)

(B) 0.31 in (0.0090 m)

(C) 0.47 in (0.014 m)

(D) 0.55 in (0.015 m)

Thermal Deformation

4. A straight steel beam 200 ft (60 m) long is installed when the temperature is 40°F (4°C). It is supported in such a manner as to allow only 0.5 in (12 mm) longitudinal expansion. Lateral support is provided to prevent buckling. If the temperature increases to 110°F (43°C), the compressive stress in the member will be most nearly

(A) 3900 lbf/in^2 (28 MPa)

(B) 7200 lbf/in^2 (51 MPa)

(C) 9200 lbf/in^2 (67 MPa)

(D) 12,000 lbf/in^2 (88 MPa)

5. A 0.25% carbon steel supporting strap in an oven furnace carries a constant tensile load while the oven temperature is increased from 400°F to 800°F. The modulus of elasticity of steel at 400°F is 27,000,000 lbf/in^2. The modulus of elasticity of steel at 800°F is 24,200,000 lbf/in^2. Compared to the strain at 400°F, what is most nearly the strain at 800°F?

(A) $1.1\epsilon_{400°F}$

(B) $1.3\epsilon_{400°F}$

(C) $1.4\epsilon_{400°F}$

(D) $2.6\epsilon_{400°F}$

6. A 1 in thick steel plate and 1 in thick aluminum plate are held together with a $^3/_4$ in no. 10 UNC bolt and nut. The bolt is snug, with no initial preload. The modulus of elasticity of the steel and aluminum are 30×10^6 lbf/in^2 and 10×10^6 lbf/in^2, respectively. The coefficients of linear thermal expansion of steel and aluminum are 6.5×10^{-6} 1/°F and 12.8×10^{-6} 1/°F, respectively. If the temperature is increased 250°F, the tensile force in the bolt will be most nearly

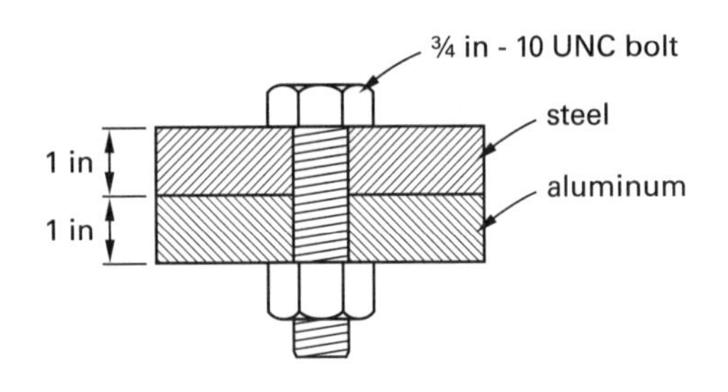

(A) 1200 lbf

(B) 6500 lbf

(C) 7900 lbf

(D) 24,000 lbf

Elastic Deformation

7. A 1 in (25 mm) diameter soft steel rod carries a tensile load of 15,000 lbf (67 kN). The elongation is 0.158 in (4 mm). The modulus of elasticity is 2.9×10^7 lbf/in^2 (200 GPa). All known parameters are precise to at least five significant digits. The total length of the rod is most nearly

(A) 239.46 in (5.853 m)

(B) 239.93 in (5.857 m)

(C) 240.03 in (5.861 m)

(D) 240.07 in (5.865 m)

Stress Analysis

8. A 3 in (75 mm) diameter horizontal shaft carries a 32 in (80 cm) diameter, 600 lbm (270 kg) pulley on an overhung (cantilever) end. The pulley is 8 in (200 mm) from the face of the outboard bearing. The pulley belt approaches and leaves horizontally. The belt carries upper and lower tensions of 1500 lbf (6.7 kN) and 350 lbf (1.6 kN), respectively. The maximum stress in the shaft is most nearly

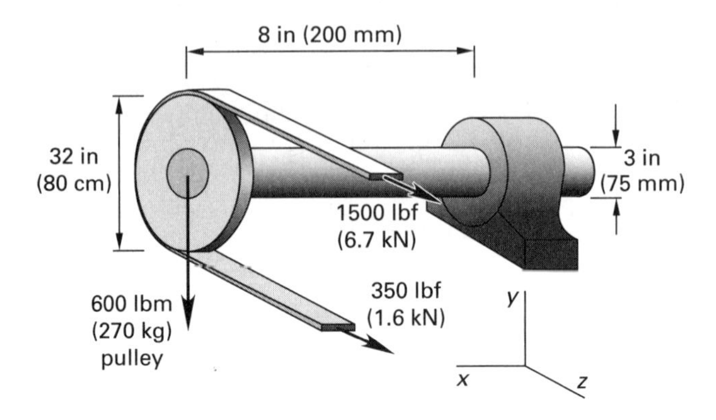

(A) 4500 lbf/in^2 (32 MPa)

(B) 5500 lbf/in^2 (39 MPa)

(C) 6500 lbf/in^2 (46 MPa)

(D) 7500 lbf/in^2 (53 MPa)

9. A 1.0 in (25 mm) diameter solid rod is held firmly in a chuck. A wrench with a 12 in (300 mm) moment arm applies 60 lbf (270 N) of force 8 in (200 mm) up from the chuck.

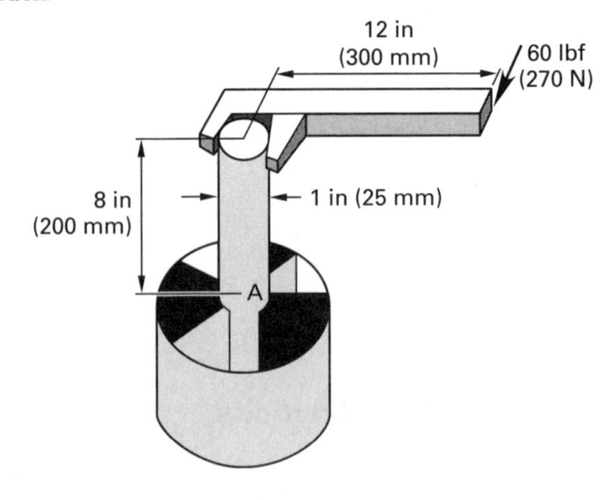

(a) The maximum shear stress in the rod is most nearly

(A) 2900 lbf/in^2 (20 MPa)

(B) 3700 lbf/in^2 (26 MPa)

(C) 4400 lbf/in^2 (32 MPa)

(D) 6800 lbf/in^2 (48 MPa)

(b) The maximum normal stress in the rod is most nearly

(A) 5400 lbf/in^2 (38 MPa)

(B) 6900 lbf/in^2 (49 MPa)

(C) 7500 lbf/in^2 (54 MPa)

(D) 11,000 lbf/in^2 (77 MPa)

10. A 0.5 in × 0.5 in × 3 in (13 mm × 13 mm × 75 mm) horizontal square bar with a face area of 1.5 in^2 (9.7 cm^2) is acted upon by an 18,000 lbf (80 kN) compressive load evenly distributed over each face. The shear stress in the horizontal direction on a cross section at a particular point is 4000 lbf/in^2 (28 MPa).

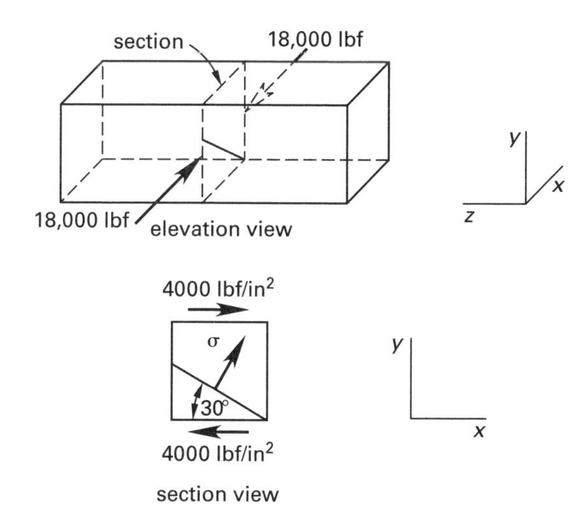

(a) The normal stress on a plane inclined +30° from the horizontal is most nearly

(A) 460 lbf/in^2 (3.6 MPa)

(B) 930 lbf/in^2 (6.5 MPa)

(C) 1700 lbf/in^2 (12 MPa)

(D) 2400 lbf/in^2 (17 MPa)

(b) The shear stress on a plane inclined +30° from the horizontal is most nearly

(A) 1700 lbf/in^2 (12 MPa)

(B) 2800 lbf/in^2 (20 MPa)

(C) 3200 lbf/in^2 (22 MPa)

(D) 4100 lbf/in^2 (29 MPa)

11. The offset wrench handle shown is constructed from a 5/8 in (16 mm) diameter round bar. The bar's modulus of elasticity is 29.6×10^6 lbf/in^2 (204 GPa). The 3 in (75 mm) rise is in the plane of the socket.

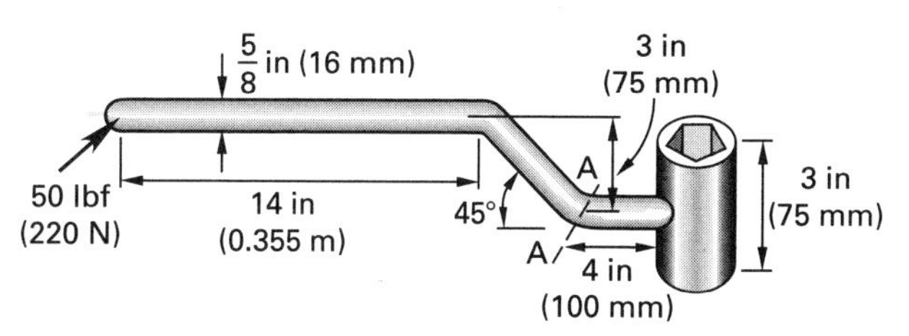

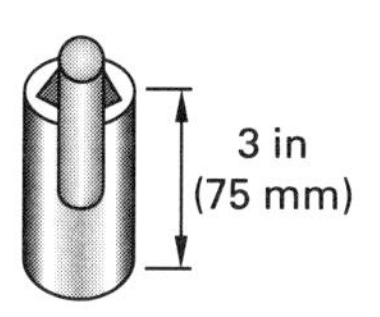

(a) The moment of inertia is most nearly

(A) 0.0030 in^4 (1.0×10^{-9} m^4)

(B) 0.0075 in^4 (3.0×10^{-9} m^4)

(C) 0.010 in^4 (4.0×10^{-8} m^4)

(D) 0.12 in^4 (5.0×10^{-8} m^4)

(b) The polar moment of inertia is most nearly

(A) 0.015 in^4 (6.4×10^{-9} m^4)

(B) 0.030 in^4 (1.3×10^{-8} m^4)

(C) 0.24 in^4 (1.0×10^{-7} m^4)

(D) 0.49 in^4 (0.013 m^4)

(c) The moment at section A-A is most nearly

(A) 150 in-lbf (17 N·m)

(B) 500 in-lbf (56 N·m)

(C) 700 in-lbf (78 N·m)

(D) 850 in-lbf (95 N·m)

(d) The maximum bending stress at the extreme fiber at section A-A is most nearly

(A) 30,000 lbf/in^2 (190 MPa)

(B) 35,000 lbf/in^2 (240 MPa)

(C) 58,000 lbf/in^2 (390 MPa)

(D) 71,000 lbf/in^2 (470 MPa)

(e) The maximum torsional shear stress at the extreme fiber at section A-A is most nearly

(A) 1000 lbf/in^2 (6.9 MPa)

(B) 2500 lbf/in^2 (16 MPa)

(C) 3100 lbf/in^2 (20 MPa)

(D) 6300 lbf/in^2 (41 MPa)

(f) The principal stresses at section A-A are most nearly

(A) 36,000 lbf/in^2, −270 lbf/in^2 (240 MPa, −1.8 MPa)

(B) 46,000 lbf/in^2, 700 lbf/in^2 (280 MPa, 4.5 MPa)

(C) 53,000 lbf/in^2, 17,000 lbf/in^2 (320 MPa, 110 MPa)

(D) 54,000 lbf/in^2, −18,000 lbf/in^2 (330 MPa, 120 MPa)

(g) The maximum shear at section A-A is most nearly

(A) 18,000 lbf/in^2 (120 MPa)

(B) 36,000 lbf/in^2 (240 MPa)

(C) 47,000 lbf/in^2 (310 MPa)

(D) 54,000 lbf/in^2 (360 MPa)

12. A brass tube 6 ft (1.8 m) long, with a 2.0 in (50 mm) outside diameter, a 1.0 in (25 mm) inside diameter, and a modulus of elasticity of 1.5×10^7 lbf/in^2 (100 GPa) is used as a cantilever beam. When a concentrated load of 50 lbf (220 N) is applied at the free end, the tip deflection is found to be excessive. To reduce the deflection, it is suggested that a tight-fitting 1 in (25 mm) outside diameter soft steel rod with a modulus of elasticity of 2.9×10^7 lbf/in^2 (200 GPa) be inserted into the entire length of the brass tube.

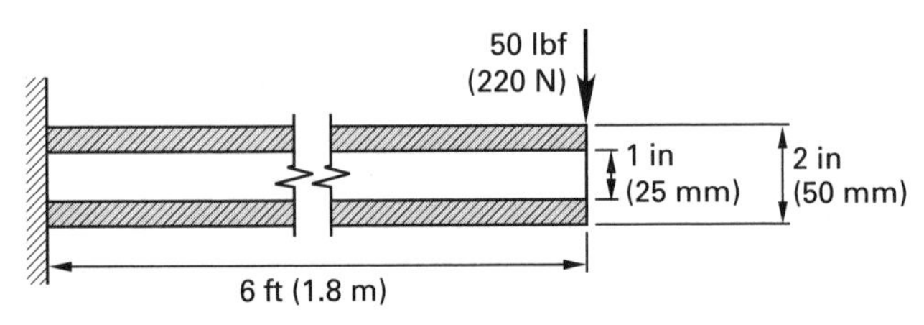

(a) The moment of inertia of the brass tube annular cross section is most nearly

(A) 0.050 in^4 (1.9×10^{-8} m^4)

(B) 0.70 in^4 (2.9×10^{-7} m^4)

(C) 0.80 in^4 (3.1×10^{-7} m^4)

(D) 0.90 in^4 (3.7×10^{-7} m^4)

(b) The moment of inertia of the steel rod insert circular cross section is most nearly

(A) 0.050 in^4 (2.0×10^{-8} m^4)

(B) 0.060 in^4 (0.30×10^{-7} m^4)

(C) 0.70 in^4 (3.0×10^{-7} m^4)

(D) 0.80 in^4 (4.0×10^{-7} m^4)

(c) The product EI for the brass tube is most nearly

(A) 7.4×10^5 lbf-in^2 (1900 N·m^2)

(B) 1.1×10^7 lbf-in^2 (29 000 N·m^2)

(C) 1.2×10^7 lbf-in^2 (30 000 N·m^2)

(D) 1.4×10^7 lbf-in^2 (36 000 N·m^2)

(d) The product EI for the steel rod insert is most nearly

(A) 0.14×10^7 lbf-in^2 (3800 N·m^2)

(B) 0.18×10^7 lbf-in^2 (4900 N·m^2)

(C) 1.9×10^7 lbf-in^2 (50 000 N·m^2)

(D) 2.3×10^7 lbf-in^2 (62 000 N·m^2)

(e) The total EI for the composite is most nearly

(A) 1.2×10^7 lbf-in^2 (33 000 N·m^2)

(B) 1.9×10^7 lbf-in^2 (48 000 N·m^2)

(C) 2.4×10^7 lbf-in^2 (62 000 N·m^2)

(D) 3.7×10^7 lbf-in^2 (96 000 N·m^2)

(f) Does the suggestion have merit?

(A) Yes; the insert decreases the EI product.

(B) Yes; the deflection is inversely proportional to the EI.

(C) Yes; the insert increases the EI product.

(D) No; the deflection is directly proportional to the EI.

(g) Neglecting self-weights, the percentage change in the tip deflection is most nearly

(A) 5.7%

(B) 7.1%

(C) 9.6%

(D) 11%

13. A 2500 lbm (1100 kg) stationary flywheel is mounted with a 4 in (100 mm) overhang on a stepped shaft as shown. The step fillet has a 5/16 in (8 mm) radius. Disregarding yielding in the vicinity of the fillet, what is most nearly the maximum bending stress in the overhanging section of the shaft?

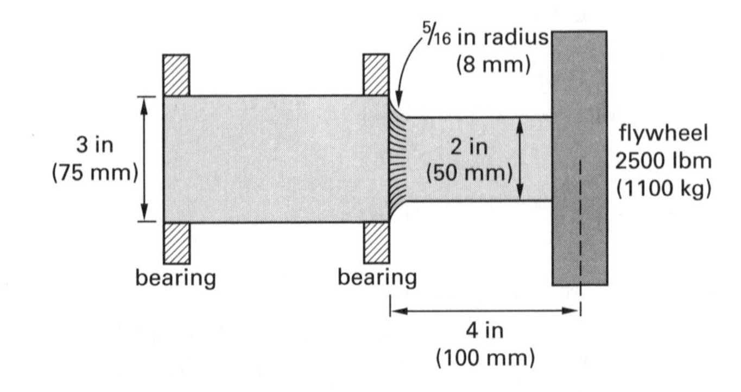

(A) 11,000 lbf/in^2 (77 MPa)

(B) 13,000 lbf/in^2 (91 MPa)

(C) 16,000 lbf/in^2 (110 MPa)

(D) 19,000 lbf/in^2 (130 MPa)

14. The discontinuity stress at an abrupt change in geometry is the

(A) total stress

(B) incremental stress

(C) residual stress after load relaxation

(D) membrane stress

Composite Structures

15. A bimetallic spring is constructed of an aluminum strip with a $^1/_8$ in by $1^1/_2$ in (3.2 mm by 38 mm) cross section bonded on top of a $^1/_{16}$ in by $1^1/_2$ in (1.6 mm by 38 mm) steel strip. The modulus of elasticity of the aluminum is 10×10^6 lbf/in^2 (70 GPa). The modulus of elasticity of the steel is 30×10^6 lbf/in^2 (200 GPa). What is most nearly the equivalent, all-aluminum centroidal area moment of inertia of the cross section?

(A) 2.4×10^{-4} in^4 (190 mm^4)

(B) 5.2×10^{-4} in^4 (390 mm^4)

(C) 1.3×10^{-3} in^4 (550 mm^4)

(D) 2.6×10^{-3} in^4 (1100 mm^4)

16. A 500,000 lbf tensile load is jointly supported by two members in axial tension: a steel bar and a reinforced concrete member. The load carried by the steel bar is 400,000 lbf. Both ends of the steel and concrete members share rigid end caps. The concrete member has a length of 10 ft and cross-sectional area of 12 in^2. The modulus of elasticity for the concrete is 5,000,000 lbf/in^2. The steel member has a length of 10 ft. The modulus of elasticity for the steel bar is 29,000,000 lbf/in^2. The elongation experienced by the steel bar is most nearly

(A) 0.15 in

(B) 0.20 in

(C) 0.25 in

(D) 0.27 in

SOLUTIONS

1. *Customary U.S. Solution*

First, determine the reactions. The uniform load can be assumed to be concentrated at the center of the beam.

Sum the moments about A.

$$(100 \text{ lbf})(2 \text{ ft}) + (80 \text{ lbf})(14 \text{ ft}) + \left(20 \ \frac{\text{lbm}}{\text{ft}}\right)\left(\frac{32.2 \ \frac{\text{ft}}{\text{sec}^2}}{32.2 \ \frac{\text{ft-lbm}}{\text{lbf-sec}^2}}\right) \times (14 \text{ ft})(7 \text{ ft}) - R(12 \text{ ft}) = 0$$

$$R = 273.3 \text{ lbf}$$

Sum the forces in the vertical direction.

$$L + 273.3 \text{ lbf} = 100 \text{ lbf} + 80 \text{ lbf} + \left(20 \ \frac{\text{lbm}}{\text{ft}}\right)\left(\frac{32.2 \ \frac{\text{ft}}{\text{sec}^2}}{32.2 \ \frac{\text{ft-lbm}}{\text{lbf-sec}^2}}\right)(14 \text{ ft})$$

$$L = 186.7 \text{ lbf}$$

The shear diagram starts at +186.7 lbf at the left reaction and decreases linearly at a rate of 20 lbf/ft to 146.7 lbf at point B. The concentrated load reduces the shear to 46.7 lbf. The shear then decreases linearly at a rate of 20 lbf/ft to point C. Measuring x from the left, the shear line goes through zero at

$$x = 2 \text{ ft} + \frac{46.7 \text{ lbf}}{20 \ \frac{\text{lbf}}{\text{ft}}} = 4.3 \text{ ft}$$

The shear at the right of the beam at point D is 80 lbf and increases linearly at a rate of 20 lbf/ft to 120 lbf at point C. The reaction, R, at point C decreases the shear to −153.3 lbf. This is sufficient to draw the shear diagram.

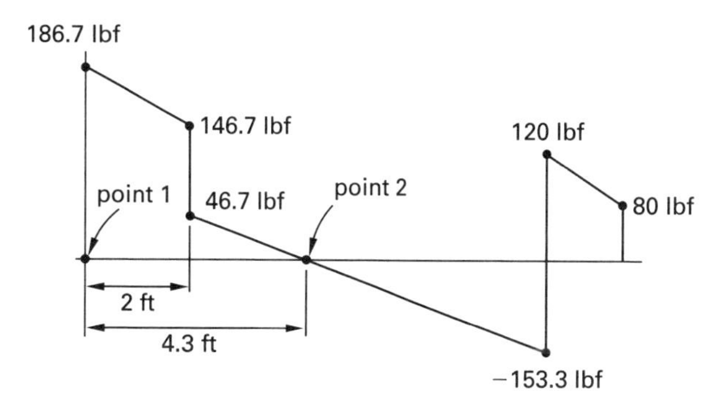

Materials

(a) From the shear diagram, the maximum moment occurs when the shear is zero. Call this point 2. The moment at the left reaction is zero. Call this point 1. Use Eq. 51.27.

$$M_2 = M_1 + \int_{x_1}^{x_2} V\,dx$$

The integral is the area under the curve from $x_1 = 0$ to $x_2 = 4.3$ ft.

$$\begin{aligned} M_2 &= 0 + (146.7 \text{ lbf})(2 \text{ ft}) \\ &\quad + \left(\tfrac{1}{2}\right)(186.7 \text{ lbf} - 146.7 \text{ lbf})(2 \text{ ft}) \\ &\quad + \left(\tfrac{1}{2}\right)(46.7 \text{ lbf})(4.3 \text{ ft} - 2 \text{ ft}) \\ &= \boxed{387.1 \text{ ft-lbf} \quad (390 \text{ ft-lbf})} \end{aligned}$$

The answer is (C).

(b) From the shear diagram, the maximum shear is $\boxed{186.7 \text{ lbf } (190 \text{ lbf}).}$

The answer is (D).

SI Solution

First, determine the reactions. The uniform load can be assumed to be concentrated at the center of the beam.

Sum the moments about A.

$$\begin{aligned} &(450 \text{ N})(0.6 \text{ m}) + (350 \text{ N})(4.2 \text{ m}) \\ &\quad + \left(30 \ \frac{\text{kg}}{\text{m}}\right)(4.2 \text{ m})\left(9.81 \ \frac{\text{N}}{\text{kg}}\right)(2.1 \text{ m}) \\ &\qquad - R(3.6 \text{ m}) = 0 \\ &\qquad R = 1204.4 \text{ N} \end{aligned}$$

Sum the forces in the vertical direction.

$$\begin{aligned} L + 1204.4 \text{ N} &= 450 \text{ N} + 350 \text{ N} \\ &\quad + \left(30 \ \frac{\text{kg}}{\text{m}}\right)(4.2 \text{ m})\left(9.81 \ \frac{\text{N}}{\text{kg}}\right) \\ L &= 831.7 \text{ N} \end{aligned}$$

The shear diagram starts at +831.7 N at the left reaction and decreases linearly at a rate of 294.3 N/m to 655.1 N at point B. The concentrated load reduces the shear to 205.1 N. The shear then decreases linearly at a rate of 294.3 N/m to point C. Measuring x from the left, the shear line goes through zero at

$$x = 0.6 \text{ m} + \frac{205.1 \text{ N}}{294.3 \ \frac{\text{N}}{\text{m}}} = 1.3 \text{ m}$$

The shear at the right of the beam at point D is 350 N and increases linearly at a rate of 294.3 N/m to 526.3 N at point C. The reaction, R, at point C decreases the shear to −677.8 N. This is sufficient to draw the shear diagram.

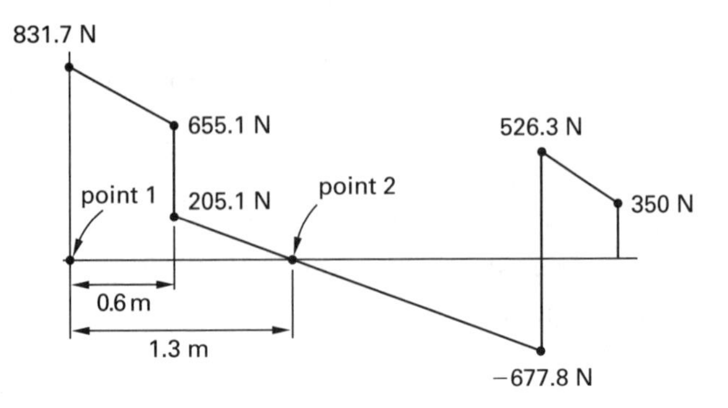

(a) From the shear diagram, the maximum moment occurs when the shear is zero. Call this point 2. The moment at the left reaction is zero. Call this point 1. Use Eq. 51.27.

$$M_2 = M_1 + \int_{x_1}^{x_2} V\,dx$$

The integral is the area under the curve from $x_1 = 0$ to $x_2 = 1.3$ m.

$$\begin{aligned} M_2 &= 0 + (655.1 \text{ N})(0.6 \text{ m}) \\ &\quad + \left(\tfrac{1}{2}\right)(831.7 \text{ N} - 655.1 \text{ N})(0.6 \text{ m}) \\ &\quad + \left(\tfrac{1}{2}\right)(205.1 \text{ N})(1.3 \text{ m} - 0.6 \text{ m}) \\ &= \boxed{517.8 \text{ N·m} \quad (520 \text{ N·m})} \end{aligned}$$

The answer is (C).

(b) From the shear diagram, the maximum shear is $\boxed{831.7 \text{ N } (830 \text{ N}).}$

The answer is (D).

2. From Eq. 51.40, the section modulus of a rectangular cross section is

$$Z_{\text{rectangular}} = \frac{bh^2}{6} = \frac{s^3}{6}$$

From Eq. 51.39 and App. 50.A, the section modulus of a circular cross section with radius r is

$$Z_{\text{circular}} = \frac{I_{c,\text{circular}}}{c_{\text{circular}}} = \frac{\frac{\pi r^4}{4}}{r} = \frac{\pi r^3}{4}$$

Materials

Solve for the radius of the circular cross section.

$$Z_{\text{circular}} = 12Z_{\text{rectangular}}$$

$$\frac{\pi r^3}{4} = (12)\left(\frac{s^3}{6}\right)$$

$$r^3 = \frac{8s^3}{\pi}$$

$$r = \boxed{\frac{2s}{\sqrt[3]{\pi}}}$$

The answer is (B).

3. *Customary U.S. Solution*

First, the moment of inertia of the beam cross section is

$$I - \frac{bh^3}{12} = \frac{(6\text{ in})(4\text{ in})^3}{12} = 32\text{ in}^4$$

From case 1 in App. 51.A, the deflection of the 200 lbf load is

$$y_1 = \frac{PL^3}{3EI} = \frac{(200\text{ lbf})(72\text{ in} - 12\text{ in})^3}{(3)\left(1.5 \times 10^6\ \frac{\text{lbf}}{\text{in}^2}\right)(32\text{ in}^4)} = 0.30\text{ in}$$

The slope at the 200 lbf load is

$$\theta = \frac{PL^2}{2EI} = \frac{(200\text{ lbf})(72\text{ in} - 12\text{ in})^2}{(2)\left(1.5 \times 10^6\ \frac{\text{lbf}}{\text{in}^2}\right)(32\text{ in}^4)} = 0.0075\text{ in/in}$$

The additional deflection at the tip of the beam is

$$y_{1'} = \left(0.0075\ \frac{\text{in}}{\text{in}}\right)(12\text{ in}) = 0.09\text{ in}$$

From case 1 in App. 51.A, the deflection at the 120 lbf load is

$$y_2 = \frac{PL^3}{3EI} = \frac{(120\text{ lbf})(72\text{ in} - 24\text{ in})^3}{(3)\left(1.5 \times 10^6\ \frac{\text{lbf}}{\text{in}^2}\right)(32\text{ in}^4)} = 0.0922\text{ in}$$

The slope at the 120 lbf load is

$$\theta_2 = \frac{PL^2}{2EI} = \frac{(120\text{ lbf})(72\text{ in} - 24\text{ in})^2}{(2)\left(1.5 \times 10^6\ \frac{\text{lbf}}{\text{in}^2}\right)(32\text{ in}^4)}$$
$$= 0.00288\text{ in/in}$$

The additional deflection at the tip of the beam is

$$y_{2'} = \left(0.00288\ \frac{\text{in}}{\text{in}}\right)(24\text{ in}) = 0.0691\text{ in}$$

The total deflection is the sum of the preceding four parts.

$$\begin{aligned} y_{\text{tip}} &= y_1 + y_{1'} + y_2 + y_{2'} \\ &= 0.30\text{ in} + 0.09\text{ in} + 0.0922\text{ in} + 0.0691\text{ in} \\ &= \boxed{0.5513\text{ in}\quad(0.55\text{ in})} \end{aligned}$$

The answer is (D).

SI Solution

First, the moment of inertia of the beam cross section is

$$I = \frac{bh^3}{12} = \frac{(0.15\text{ m})(0.10\text{ m})^3}{12} = 1.25 \times 10^{-5}\text{ m}^4$$

From case 1 in App. 51.A, the deflection at the 900 N load is

$$\begin{aligned} y_1 &= \frac{PL^3}{3EI} = \frac{(900\text{ N})(1.8\text{ m} - 0.3\text{ m})^3}{(3)(10 \times 10^9\text{ Pa})(1.25 \times 10^{-5}\text{ m}^4)} \\ &= 0.0081\text{ m} \end{aligned}$$

The slope at the 900 N load is

$$\begin{aligned} \theta_1 &= \frac{PL^2}{2EI} = \frac{(900\text{ N})(1.8\text{ m} - 0.3\text{ m})^2}{(2)(10 \times 10^9\text{ Pa})(1.25 \times 10^{-5}\text{ m}^4)} \\ &= 0.0081\text{ m/m} \end{aligned}$$

The additional deflection at the tip of the beam is

$$y_{1'} = \left(0.0081\ \frac{\text{m}}{\text{m}}\right)(0.3\text{ m}) = 0.00243\text{ m}$$

From case 1 in App. 51.A, the deflection at the 530 N load is

$$\begin{aligned} y_2 &= \frac{PL^3}{3EI} = \frac{(530\text{ N})(1.8\text{ m} - 0.6\text{ m})^3}{(3)(10 \times 10^9\text{ Pa})(1.25 \times 10^{-5}\text{ m}^4)} \\ &= 0.00244\text{ m} \end{aligned}$$

The slope at the 530 N load is

$$\begin{aligned} \theta_2 &= \frac{PL^2}{2EI} = \frac{(530\text{ N})(1.8\text{ m} - 0.6\text{ m})^2}{(2)(10 \times 10^9\text{ Pa})(1.25 \times 10^{-5}\text{ m}^4)} \\ &= 0.00305\text{ m/m} \end{aligned}$$

The additional deflection at the tip of the beam is

$$y_{2'} = \left(0.00305\ \frac{\text{m}}{\text{m}}\right)(0.6\ \text{m}) = 0.00183\ \text{m}$$

The total deflection is the sum of the preceding four parts.

$$\begin{aligned} y_{\text{tip}} &= y_1 + y_{1'} + y_2 + y_{2'} \\ &= 0.0081\ \text{m} + 0.00243\ \text{m} + 0.00244\ \text{m} + 0.00183\ \text{m} \\ &= \boxed{0.0148\ \text{m}\quad (0.015\ \text{m})} \end{aligned}$$

The answer is (D).

4. *Customary U.S. Solution*

Use Eq. 51.9 to find the amount of expansion for an unconstrained beam. Use Table 51.2.

$$\begin{aligned} \Delta L &= \alpha L_o(T_2 - T_1) \\ &= \left(6.5 \times 10^{-6}\ \frac{1}{^\circ\text{F}}\right)(200\ \text{ft})\left(12\ \frac{\text{in}}{\text{ft}}\right)(110^\circ\text{F} - 40^\circ\text{F}) \\ &= 1.092\ \text{in} \end{aligned}$$

The constrained length is

$$\Delta L_c = 1.092\ \text{in} - 0.5\ \text{in} = 0.592\ \text{in}$$

From Eq. 51.14, the thermal strain is

$$\begin{aligned} \epsilon_{\text{th}} &= \frac{\Delta L_c}{L_o + 0.5\ \text{in}} = \frac{0.592\ \text{in}}{2400\ \text{in} + 0.5\ \text{in}} \\ &= 2.466 \times 10^{-4}\ \text{in/in} \end{aligned}$$

From Eq. 51.15, the compressive thermal stress is

$$\begin{aligned} \sigma_{\text{th}} = E\epsilon_{\text{th}} &= \left(29 \times 10^6\ \frac{\text{lbf}}{\text{in}^2}\right)(2.466 \times 10^{-4}) \\ &= \boxed{7151\ \text{lbf/in}^2\quad (7200\ \text{lbf/in}^2)} \end{aligned}$$

The answer is (B).

SI Solution

Use Eq. 51.9 to find the amount of expansion for an unconstrained beam. Use Table 51.2.

$$\begin{aligned} \Delta L &= \alpha L_o(T_2 - T_1) \\ &= \left(11.7 \times 10^{-6}\ \frac{1}{^\circ\text{C}}\right)(60\ \text{m})(43^\circ\text{C} - 4^\circ\text{C}) \\ &= 0.02738\ \text{m} \end{aligned}$$

The constrained length is

$$\Delta L_c = 0.02738\ \text{m} - 0.012\ \text{m} = 0.01538\ \text{m}$$

From Eq. 51.14, the thermal strain is

$$\begin{aligned} \epsilon_{\text{th}} &= \frac{\Delta L_c}{L_o + 0.012\ \text{m}} = \frac{0.01538\ \text{m}}{60\ \text{m} + 0.012\ \text{m}} \\ &= 0.000256\ \text{m/m} \end{aligned}$$

From Eq. 51.15, the compressive thermal stress is

$$\begin{aligned} \sigma_{\text{th}} = E\epsilon_{\text{th}} &= (20 \times 10^{10}\ \text{Pa})\left(0.000256\ \frac{\text{m}}{\text{m}}\right) \\ &= \boxed{5.12 \times 10^7\ \text{Pa}\quad (51\ \text{MPa})} \end{aligned}$$

The answer is (B).

5. The strain at 400°F is

$$\epsilon_{400^\circ\text{F}} = \frac{\sigma}{E_{400^\circ\text{F}}}$$

The strain at 800°F is

$$\epsilon_{800^\circ\text{F}} = \frac{\sigma}{E_{800^\circ\text{F}}}$$

Since the stress is constant, the strain at 800°F is

$$\begin{aligned} \epsilon_{800^\circ\text{F}} &= \left(\frac{E_{400^\circ\text{F}}}{E_{800^\circ\text{F}}}\right)\epsilon_{400^\circ\text{F}} = \left(\frac{27{,}000{,}000\ \frac{\text{lbf}}{\text{in}^2}}{24{,}200{,}000\ \frac{\text{lbf}}{\text{in}^2}}\right)\epsilon_{400^\circ\text{F}} \\ &= \boxed{1.12\epsilon_{400^\circ\text{F}}\quad (1.1\epsilon_{400^\circ\text{F}})} \end{aligned}$$

The answer is (A).

6. From Eq. 51.9, the unconstrained thermal deformations are

$$\begin{aligned} \Delta L_{\text{bolt}} &= \alpha L_o(T_2 - T_1) \\ &= \left(6.5 \times 10^{-6}\ \frac{1}{^\circ\text{F}}\right)(2\ \text{in})(250^\circ\text{F}) \\ &= 0.00325\ \text{in} \\ \Delta L_{\text{steel}} &= \alpha L_o(T_2 - T_1) \\ &= \left(6.5 \times 10^{-6}\ \frac{1}{^\circ\text{F}}\right)(1\ \text{in})(250^\circ\text{F}) \\ &= 0.001625\ \text{in} \\ \Delta L_{\text{aluminum}} &= \alpha L_o(T_2 - T_1) \\ &= \left(12.8 \times 10^{-6}\ \frac{1}{^\circ\text{F}}\right)(1\ \text{in})(250^\circ\text{F}) \\ &= 0.0032\ \text{in} \end{aligned}$$

The unrealized elongation of the bolt is

$$\begin{aligned} \Delta L &= \Delta L_{\text{aluminum}} + \Delta L_{\text{steel}} - \Delta L_{\text{bolt}} \\ &= 0.0032\ \text{in} + 0.001625\ \text{in} - 0.00325\ \text{in} \\ &= 0.001575\ \text{in} \end{aligned}$$

The resulting thermal strain in the bolt is

$$\epsilon_{\text{bolt}} = \frac{\Delta L}{L_o} = \frac{0.001575 \text{ in}}{2 \text{ in}} = 0.0007875$$

The resulting tensile stress in the bolt is

$$\begin{aligned}\sigma_{\text{bolt}} &= E_{\text{steel}}\epsilon_{\text{bolt}} = \left(30 \times 10^6 \ \frac{\text{lbf}}{\text{in}^2}\right)(0.0007875) \\ &= 23{,}625 \text{ lbf/in}^2\end{aligned}$$

From Table 53.5, the tensile area of the bolt is 0.334 in^2. The tensile force in the bolt is

$$\begin{aligned}F_{\text{bolt}} &= \sigma_{\text{bolt}} A_t = \left(23{,}625 \ \frac{\text{lbf}}{\text{in}^2}\right)(0.334 \text{ in}^2) \\ &= \boxed{7891 \text{ lbf} \quad (7900 \text{ lbf})}\end{aligned}$$

The answer is (C).

7. *Customary U.S. Solution*

Since the answer choices have five significant digits, use at least six significant digits in the interim calculations.

First, the cross-sectional area of the rod is

$$A = \frac{\pi d^2}{4} = \frac{\pi (1 \text{ in})^2}{4} = \pi/4 \text{ in}^2$$

From Eq. 51.4, the unstretched length of the rod is

$$\begin{aligned}L_o &= \frac{\delta EA}{F} = \frac{(0.158 \text{ in})\left(2.9 \times 10^7 \ \frac{\text{lbf}}{\text{in}^2}\right)\left(\frac{\pi}{4} \text{ in}^2\right)}{15{,}000 \text{ lbf}} \\ &= 239.913 \text{ in}\end{aligned}$$

The total length of the rod is given by Eq. 51.5.

$$\begin{aligned}L &= L_o + \delta = 239.913 \text{ in} + 0.158 \text{ in} \\ &= \boxed{240.071 \text{ in} \quad (240.07 \text{ in})}\end{aligned}$$

The answer is (D).

SI Solution

Since the answer choices have four significant digits, use at least five significant digits in the interim calculations.

First, the cross-sectional area of the rod is

$$A = \frac{\pi d^2}{4} = \frac{\pi (0.025 \text{ m})^2}{4} = 4.9087 \times 10^{-4} \text{ m}^2$$

From Eq. 51.4, the unstretched length of the rod is

$$\begin{aligned}L_o &= \frac{\delta EA}{F} \\ &= \frac{(0.004 \text{ m})\left(20 \times 10^{10} \ \frac{\text{N}}{\text{m}^2}\right)(4.9087 \times 10^{-4} \text{ m}^2)}{67 \times 10^3 \text{ N}} \\ &= 5.8611 \text{ m}\end{aligned}$$

The total length of the rod is given by Eq. 51.5.

$$\begin{aligned}L &= L_o + \delta = 5.8611 \text{ m} + 0.004 \text{ m} \\ &= \boxed{5.8651 \text{ m} \quad (5.865 \text{ m})}\end{aligned}$$

The answer is (D).

8. *Customary U.S. Solution*

First, find the properties of the shaft cross section. The area is

$$A = \frac{\pi d^2}{4} = \frac{\pi (3 \text{ in})^2}{4} = 7.07 \text{ in}^2$$

The moment of inertia is

$$I_c = \frac{\pi r^4}{4} = \frac{\pi \left(\frac{3 \text{ in}}{2}\right)^4}{4} = 3.98 \text{ in}^4$$

The polar moment of inertia is

$$J = \frac{\pi r^4}{2} = \frac{\pi \left(\frac{3 \text{ in}}{2}\right)^4}{2} = 7.95 \text{ in}^4$$

The moment at the bearing face due to the pulley weight is

$$M_y = (600 \text{ lbm})\left(\frac{32.2 \ \frac{\text{ft}}{\text{sec}^2}}{32.2 \ \frac{\text{lbm-ft}}{\text{lbm-sec}^2}}\right)(8 \text{ in}) = 4800 \text{ in-lbf}$$

From Eq. 51.37, the maximum bending stress at the extreme fiber on the y-axis is

$$\sigma_y = \frac{M_y c}{I_c} = \frac{(4800 \text{ in-lbf})\left(\frac{3 \text{ in}}{2}\right)}{3.98 \text{ in}^4} = 1809 \text{ lbf/in}^2$$

σ_y occurs at the upper and lower surfaces ("6 and 12 o'clock") at the bearing face.

The moment at the bearing face due to the belt tensions is

$$M_z = (1500 \text{ lbf} + 350 \text{ lbf})(8 \text{ in}) = 14{,}800 \text{ in-lbf}$$

From Eq. 51.37, the maximum bending stress at the extreme fiber on the z-axis is

$$\sigma_z = \frac{M_z c}{I_c} = \frac{(14{,}800 \text{ in-lbf})\left(\frac{3 \text{ in}}{2}\right)}{3.98 \text{ in}^4} = 5578 \text{ lbf/in}^2$$

σ_z occurs at the near and far sides of the shaft ("3 and 9 o'clock") at the bearing face.

σ_y and σ_z combine in tension in the fourth quadrant (i.e., between "9 and 12 o'clock"), and they combine in compression in the second quadrant (i.e., between "3 and 6 o'clock"). The magnitude of the combination is

$$\sigma_c = \sqrt{\sigma_y^2 + \sigma_z^2} = \sqrt{\left(1809\ \frac{\text{lbf}}{\text{in}^2}\right)^2 + \left(5578\ \frac{\text{lbf}}{\text{in}^2}\right)^2}$$
$$= 5864\ \text{lbf/in}^2$$

The net torque from the belt tensions is

$$T = (1500\ \text{lbf} - 350\ \text{lbf})(16\ \text{in}) = 18{,}400\ \text{in-lbf}$$

The maximum torsional shear stress at the extreme fiber is

$$\tau = \frac{Tr}{J} = \frac{(18{,}400\ \text{in-lbf})\left(\frac{3\ \text{in}}{2}\right)}{7.95\ \text{in}^4} = 3472\ \text{lbf/in}^2$$

The direct shear stress is zero at the surface of the shaft.

Use Eq. 51.19 and Eq. 51.20 to find the maximum stress in the shaft.

$$\sigma_1 = \frac{\sigma_c}{2} + \tfrac{1}{2}\sqrt{\sigma_c^2 + (2\tau)^2}$$
$$= \frac{5864\ \frac{\text{lbf}}{\text{in}^2}}{2} + \tfrac{1}{2}\sqrt{\left(5864\ \frac{\text{lbf}}{\text{in}^2}\right)^2 + \left((2)\left(3472\ \frac{\text{lbf}}{\text{in}^2}\right)\right)^2}$$
$$= \boxed{7476\ \text{lbf/in}^2 \quad (7500\ \text{lbf/in}^2)}$$

The answer is (D).

SI Solution

First, find the properties of the shaft cross section. The area is

$$A = \frac{\pi d^2}{4} = \frac{\pi(0.075\ \text{m})^2}{4} = 4.42 \times 10^{-3}\ \text{m}^2$$

The moment of inertia is

$$I_c = \frac{\pi r^4}{4} = \frac{\pi\left(\frac{0.075\ \text{m}}{2}\right)^4}{4} = 1.55 \times 10^{-6}\ \text{m}^4$$

The polar moment of inertia is

$$J = \frac{\pi r^4}{2} = \frac{\pi\left(\frac{0.075\ \text{m}}{2}\right)^4}{2} = 3.11 \times 10^{-6}\ \text{m}^4$$

The moment at the bearing face due to the pulley weight is

$$M_y = (270\ \text{kg})\left(9.81\ \frac{\text{m}}{\text{s}^2}\right)(0.2\ \text{m}) = 529.74\ \text{N·m}$$

From Eq. 51.37, the maximum bending stress at the extreme fiber on the y-axis is

$$\sigma_y = \frac{M_y c}{I_c} = \frac{(529.74\ \text{N·m})\left(\frac{0.075\ \text{m}}{2}\right)}{1.55 \times 10^{-6}\ \text{m}^4} = 1.282 \times 10^7\ \text{Pa}$$

σ_y occurs at the upper and lower surfaces ("6 and 12 o'clock") at the bearing face.

The moment at the bearing face due to the belt tensions is

$$M_z = (6.7 \times 10^3\ \text{N} + 1.6 \times 10^3\ \text{N})(0.2\ \text{m}) = 1660\ \text{N·m}$$

From Eq. 51.37, the maximum bending stress at the extreme fiber on the z-axis is

$$\sigma_z = \frac{M_z c}{I_c} = \frac{(1660\ \text{N·m})\left(\frac{0.075\ \text{m}}{2}\right)}{1.55 \times 10^{-6}\ \text{m}^4} = 4.016 \times 10^7\ \text{Pa}$$

σ_z occurs at the near and far sides of the shaft ("3 and 9 o'clock") at the bearing face.

σ_y and σ_z combine in tension in the fourth quadrant (i.e., between "9 and 12 o'clock"), and they combine in compression in the second quadrant (i.e., between "3 and 6 o'clock"). The magnitude of the combination is

$$\sigma_c = \sqrt{\sigma_y^2 + \sigma_z^2}$$
$$= \sqrt{(1.282 \times 10^7\ \text{Pa})^2 + (4.016 \times 10^7\ \text{Pa})^2}$$
$$= 4.216 \times 10^7\ \text{Pa}$$

The net torque from the belt tensions is

$$T = (6.7 \times 10^3\ \text{N} - 1.6 \times 10^3\ \text{N})(0.40\ \text{m}) = 2040\ \text{N·m}$$

The maximum torsional shear stress at the extreme fiber is

$$\tau = \frac{Tr}{J} = \frac{(2040\ \text{N·m})\left(\frac{0.075\ \text{m}}{2}\right)}{3.11 \times 10^{-6}\ \text{m}^4} = 2.460 \times 10^7\ \text{Pa}$$

The direct shear stress is zero at the surface of the shaft.

Use Eq. 51.19 and Eq. 51.20 to find the maximum stress in the shaft.

$$\begin{aligned}\sigma_1 &= \frac{\sigma_c}{2} + \tfrac{1}{2}\sqrt{\sigma_c^2 + (2\tau)^2} \\ &= \frac{4.216\times 10^7\ \text{Pa}}{2} \\ &\quad + \tfrac{1}{2}\sqrt{\begin{array}{l}(4.216\times 10^7\ \text{Pa})^2 \\ \quad + \Big((2)(2.460\times 10^7\ \text{Pa})\Big)^2\end{array}} \\ &= \boxed{5.347\times 10^7\ \text{Pa}\quad (53\ \text{MPa})}\end{aligned}$$

The answer is (D).

9. *Customary U.S. Solution*

First, find the properties of the rod cross section. The area is

$$A = \frac{\pi d^2}{4} = \frac{\pi(1\ \text{in})^2}{4} = 0.7854\ \text{in}^2$$

The moment of inertia is

$$I_c = \frac{\pi r^4}{4} = \frac{\pi\left(\frac{1.0\ \text{in}}{2}\right)^4}{4} = 0.04909\ \text{in}^4$$

The polar moment of inertia is

$$J = \frac{\pi r^4}{2} = \frac{\pi\left(\frac{1\ \text{in}}{2}\right)^4}{2} = 0.09817\ \text{in}^4$$

The moment at the chuck is

$$M = (60\ \text{lbf})(8\ \text{in}) = 480\ \text{in-lbf}$$

From Eq. 51.37, the maximum bending stress at the extreme fiber of the rod is

$$\sigma = \frac{Mc}{I_c} = \frac{(480\ \text{in-lbf})\left(\frac{1.0\ \text{in}}{2}\right)}{0.04909\ \text{in}^4} = 4889\ \text{lbf/in}^2$$

The torque applied to the rod is

$$T = (60\ \text{lbf})(12\ \text{in}) = 720\ \text{in-lbf}$$

The maximum torsional shear stress at the extreme fiber of the rod is

$$\tau = \frac{Tr}{J} = \frac{(720\ \text{in-lbf})\left(\frac{1.0\ \text{in}}{2}\right)}{0.09817\ \text{in}^4} = 3667\ \text{lbf/in}^2$$

The direct shear stress is zero at the surface of the rod.

(a) Use Eq. 51.20 to find the maximum shear stress in the rod.

$$\begin{aligned}\tau_1 &= \tfrac{1}{2}\sqrt{\sigma^2 + (2\tau)^2} \\ &= \tfrac{1}{2}\sqrt{\left(4889\ \frac{\text{lbf}}{\text{in}^2}\right)^2 + \left((2)\left(3667\ \frac{\text{lbf}}{\text{in}^2}\right)\right)^2} \\ &= \boxed{4407\ \text{lbf/in}^2\quad (4400\ \text{lbf/in}^2)}\end{aligned}$$

The answer is (C).

(b) Use Eq. 51.19 to find the maximum normal stress in the rod.

$$\begin{aligned}\sigma_1 &= \frac{\sigma}{2} + \tau_1 = \frac{4889\ \frac{\text{lbf}}{\text{in}^2}}{2} + 4407\ \frac{\text{lbf}}{\text{in}^2} \\ &= \boxed{6852\ \text{lbf/in}^2\quad (6900\ \text{lbf/in}^2)}\end{aligned}$$

The answer is (B).

SI Solution

First, find the properties of the rod cross section. The area is

$$A = \frac{\pi d^2}{4} = \frac{\pi(0.025\ \text{m})^2}{4} = 4.909\times 10^{-4}\ \text{m}^2$$

The moment of inertia is

$$I_c = \frac{\pi r^4}{4} = \frac{\pi\left(\frac{0.025\ \text{m}}{2}\right)^4}{4} = 1.917\times 10^{-8}\ \text{m}^4$$

The polar moment of inertia is

$$J = \frac{\pi r^4}{2} = \frac{\pi\left(\frac{0.025\ \text{m}}{2}\right)^4}{2} = 3.835\times 10^{-8}\ \text{m}^4$$

The moment at the chuck is

$$M = (270\ \text{N})(0.2\ \text{m}) = 54\ \text{N·m}$$

From Eq. 51.37, the maximum bending stress at the extreme fiber of the rod is

$$\begin{aligned}\sigma &= \frac{Mc}{I_c} = \frac{(54\ \text{N·m})\left(\frac{0.025\ \text{m}}{2}\right)}{1.917\times 10^{-8}\ \text{m}^4} \\ &= 3.52\times 10^7\ \text{Pa}\quad (35.2\ \text{MPa})\end{aligned}$$

The torque applied to the rod is

$$T = (270\ \text{N})(0.3\ \text{m}) = 81\ \text{N·m}$$

The maximum torsional shear stress at the extreme fiber of the rod is

$$\begin{aligned}\tau = \frac{Tr}{J} &= \frac{(81\ \text{N·m})\left(\frac{0.025\ \text{m}}{2}\right)}{3.835\times 10^{-8}\ \text{m}^4}\\ &= 2.64\times 10^{7}\ \text{Pa}\quad (26.4\ \text{MPa})\end{aligned}$$

The direct shear stress is zero at the surface of the rod.

(a) Use Eq. 51.20 to find the maximum shear stress in the rod.

$$\begin{aligned}\tau_1 &= \tfrac{1}{2}\sqrt{\sigma^2 + (2\tau)^2}\\ &= \tfrac{1}{2}\sqrt{(35.2\ \text{MPa})^2 + \big((2)(26.4\ \text{MPa})\big)^2}\\ &= \boxed{31.7\ \text{MPa}\quad (32\ \text{MPa})}\end{aligned}$$

The answer is (C).

(b) Use Eq. 51.19 to find the maximum normal stress in the rod.

$$\begin{aligned}\sigma_1 = \frac{\sigma}{2} + \tau_1 &= \frac{35.2\ \text{MPa}}{2} + 31.7\ \text{MPa}\\ &= \boxed{49.3\ \text{MPa}\quad (49\ \text{MPa})}\end{aligned}$$

The answer is (B).

10. *Customary U.S. Solution*

The normal stress in the bar is compressive (i.e., negative).

$$\sigma_x = \frac{F}{A} = \frac{-18{,}000\ \text{lbf}}{1.5\ \text{in}^2} = -12{,}000\ \text{lbf/in}^2$$

The shear stress in the bar, τ, is 4000 lbf/in^2.

From Fig. 51.7, a plane inclined +30° from the horizontal gives an angle θ of 60°. $2\theta = (2)(60°) = 120°$.

(a) From Eq. 51.17, the normal stress on the plane is

$$\begin{aligned}\sigma_\theta &= \frac{\sigma_x}{2} + \left(\frac{\sigma_x}{2}\right)(\cos 2\theta) + \tau\sin 2\theta\\ &= \frac{-12{,}000\ \frac{\text{lbf}}{\text{in}^2}}{2} + \frac{\left(-12{,}000\ \frac{\text{lbf}}{\text{in}^2}\right)(\cos 120°)}{2}\\ &\quad + \left(4000\ \frac{\text{lbf}}{\text{in}^2}\right)(\sin 120°)\\ &= \boxed{464\ \text{lbf/in}^2\quad (460\ \text{lbf/in}^2)}\end{aligned}$$

The answer is (A).

(b) From Eq. 51.18, the shear stress on the plane is

$$\begin{aligned}\tau_\theta &= -\frac{\sigma_x\sin 2\theta}{2} + \tau\cos 2\theta\\ &= -\frac{\left(-12{,}000\ \frac{\text{lbf}}{\text{in}^2}\right)(\sin 120°)}{2}\\ &\quad + \left(4000\ \frac{\text{lbf}}{\text{in}^2}\right)(\cos 120°)\\ &= \boxed{3196\ \text{lbf/in}^2\quad (3200\ \text{lbf/in}^2)}\end{aligned}$$

The answer is (C).

SI Solution

The normal stress in the bar is compressive (i.e., negative).

$$\begin{aligned}\sigma_x = \frac{F}{A} &= \frac{(-80\times 10^3\ \text{N})\left(100\ \frac{\text{cm}}{\text{m}}\right)^2}{9.7\ \text{cm}^2}\\ &= -8.247\times 10^{7}\ \text{Pa}\quad (-82.5\ \text{MPa})\end{aligned}$$

The shear stress in the bar, τ, is 28 MPa. From Fig. 51.17, a plane inclined +30° from the horizontal gives an angle θ of 60°. $2\theta = (2)(60°) = 120°$.

(a) From Eq. 51.17, the normal stress on the plane is

$$\begin{aligned}\sigma_\theta &= \frac{\sigma_x}{2} + \left(\frac{\sigma_x}{2}\right)(\cos 2\theta) + \tau\sin 2\theta\\ &= \frac{-82.5\ \text{MPa}}{2} + \frac{(-82.5\ \text{MPa})(\cos 120°)}{2}\\ &\quad + (28\ \text{MPa})(\sin 120°)\\ &= \boxed{3.62\ \text{MPa}\quad (3.6\ \text{MPa})}\end{aligned}$$

The answer is (A).

(b) From Eq. 51.18, the shear stress on the plane is

$$\begin{aligned}\tau_\theta &= -\frac{\sigma_x\sin 2\theta}{2} + \tau\cos 2\theta\\ &= -\frac{(-82.4\ \text{MPa})(\sin 120°)}{2} + (28\ \text{MPa})(\cos 120°)\\ &= \boxed{21.7\ \text{MPa}\quad (22\ \text{MPa})}\end{aligned}$$

The answer is (C).

11. *Customary U.S. Solution*

(a) Find the properties of the handle cross section. The moment of inertia is

$$\begin{aligned}I_c = \frac{\pi r^4}{4} &= \frac{\pi\left(\frac{0.625\ \text{in}}{2}\right)^4}{4}\\ &= \boxed{0.00749\ \text{in}^4\quad (0.0075\ \text{in}^4)}\end{aligned}$$

The answer is (B).

Materials

(b) The polar moment of inertia is

$$J = \frac{\pi r^4}{2} = \frac{\pi\left(\frac{0.625 \text{ in}}{2}\right)^4}{2}$$
$$= \boxed{0.01498 \text{ in}^4 \quad (0.015 \text{ in}^4)}$$

The answer is (A).

(c) The moment at section A-A is

$$M = (50 \text{ lbf})(14 \text{ in} + 3 \text{ in})$$
$$= \boxed{850 \text{ in-lbf}}$$

The answer is (D).

(d) From Eq. 51.37, the maximum bending stress at the extreme fiber at section A-A is

$$\sigma = \frac{Mc}{I_c} = \frac{(850 \text{ in-lbf})\left(\frac{0.625 \text{ in}}{2}\right)}{0.00749 \text{ in}^4}$$
$$= \boxed{35{,}464 \text{ lbf/in}^2 \quad (35{,}000 \text{ lbf/in}^2)}$$

The answer is (B).

(e) The torque at section A-A is

$$T = (50 \text{ lbf})(3 \text{ in}) = 150 \text{ in-lbf}$$

The maximum torsional shear stress at the extreme fiber at section A-A is

$$\tau = \frac{Tr}{J} = \frac{(150 \text{ in-lbf})\left(\frac{0.625 \text{ in}}{2}\right)}{0.01498 \text{ in}^4}$$
$$= \boxed{3129 \text{ lbf/in}^2 \quad (3100 \text{ lbf/in}^2)}$$

The answer is (C).

(f) The direct shear at the extreme fiber is assumed to be small enough to be neglected.

Use Eq. 51.20 to find the maximum shear stress at section A-A.

$$\tau_1 = \tfrac{1}{2}\sqrt{\sigma^2 + (2\tau)^2}$$
$$= \tfrac{1}{2}\sqrt{\left(35{,}464 \ \frac{\text{lbf}}{\text{in}^2}\right)^2 + \left((2)\left(3129 \ \frac{\text{lbf}}{\text{in}^2}\right)\right)^2}$$
$$= 18{,}006 \text{ lbf/in}^2$$

Use Eq. 51.19 to find the principal stresses.

$$\sigma_1 = \frac{\sigma}{2} + \tau_1 = \frac{35{,}464 \ \frac{\text{lbf}}{\text{in}^2}}{2} + 18{,}006 \ \frac{\text{lbf}}{\text{in}^2}$$
$$= \boxed{35{,}738 \text{ lbf/in}^2 \quad (36{,}000 \text{ lbf/in}^2)}$$

$$\sigma_2 = \frac{\sigma}{2} - \tau_1 = \frac{35{,}464 \ \frac{\text{lbf}}{\text{in}^2}}{2} - 18{,}006 \ \frac{\text{lbf}}{\text{in}^2}$$
$$= \boxed{-274 \text{ lbf/in}^2 \quad (-270 \text{ lbf/in}^2)}$$

The answer is (A).

(g) From part (f), the maximum shear at section A-A is $\tau_1 = \boxed{18{,}006 \text{ lbf/in}^2 \ (18{,}000 \text{ lbf/in}^2).}$

The answer is (A).

SI Solution

(a) Find the properties of the handle cross section. The moment of inertia is

$$I_c = \frac{\pi r^4}{4} = \frac{\pi\left(\frac{0.016 \text{ m}}{2}\right)^4}{4}$$
$$= \boxed{3.22 \times 10^{-9} \text{ m}^4 \quad (3.0 \times 10^{-9} \text{ m}^4)}$$

The answer is (B).

(b) The polar moment of inertia is

$$J = \frac{\pi r^4}{2} = \frac{\pi\left(\frac{0.016 \text{ m}}{2}\right)^4}{2}$$
$$= \boxed{6.43 \times 10^{-9} \text{ m}^4 \quad (6.4 \times 10^{-9} \text{ m}^4)}$$

The answer is (A).

(c) The moment at section A-A is

$$M = (220 \text{ N})(0.355 \text{ m} + 0.075 \text{ m})$$
$$= \boxed{94.6 \text{ N·m} \quad (95 \text{ N·m})}$$

The answer is (D).

(d) From Eq. 51.37, the maximum bending stress at the extreme fiber at section A-A is

$$\sigma = \frac{Mc}{I_c} = \frac{(94.6 \text{ N·m})\left(\frac{0.016 \text{ m}}{2}\right)}{3.22 \times 10^{-9} \text{ m}^4}$$
$$= \boxed{2.35 \times 10^8 \text{ Pa} \quad (240 \text{ MPa})}$$

The answer is (B).

(e) The torque at section A-A is

$$T = (220\ \text{N})(0.075\ \text{m}) = 16.5\ \text{N·m}$$

The maximum torsional shear stress at the extreme fiber at section A-A is

$$\tau = \frac{Tr}{J} = \frac{(16.5\ \text{N·m})\left(\frac{0.016\ \text{m}}{2}\right)}{6.43\times 10^{-9}\ \text{m}^4} = \boxed{2.05\times 10^7\ \text{Pa}\quad (20\ \text{MPa})}$$

The answer is (C).

(f) The direct shear at the extreme fiber is assumed to be small enough to be neglected.

Use Eq. 51.20 to find the maximum shear stress at section A-A.

$$\begin{aligned}\tau_1 &= \tfrac{1}{2}\sqrt{\sigma^2 + (2\tau)^2}\\ &= \tfrac{1}{2}\sqrt{(235\ \text{MPa})^2 + \left((2)(20.5\ \text{MPa})^2\right)}\\ &= 119.3\ \text{MPa}\end{aligned}$$

Use Eq. 51.19 to find the principal stresses.

$$\begin{aligned}\sigma_1 &= \frac{\sigma}{2} + \tau_1 = \frac{235\ \text{MPa}}{2} + 119.3\ \text{MPa}\\ &= \boxed{237\ \text{MPa}\quad (240\ \text{MPa})}\\ \sigma_2 &= \frac{\sigma}{2} - \tau_1 = \frac{235\ \text{MPa}}{2} - 119.3\ \text{MPa}\\ &= \boxed{-1.8\ \text{MPa}}\end{aligned}$$

The answer is (A).

(g) From part (f), the maximum shear at section A-A is $\tau_1 = \boxed{119.3\ \text{MPa}\ (120\ \text{MPa}).}$

The answer is (A).

12. *Customary U.S. Solution*

(a) The moment of inertia of the brass tube annular cross section is

$$\begin{aligned}I_{\text{brass}} &= \frac{\pi}{4}(r_o^4 - r_i^4) = \left(\frac{\pi}{4}\right)\left(\left(\frac{2.0\ \text{in}}{2}\right)^4 - \left(\frac{1.0\ \text{in}}{2}\right)^4\right)\\ &= \boxed{0.736\ \text{in}^4\quad (0.70\ \text{in}^4)}\end{aligned}$$

The answer is (B).

(b) The moment of inertia of the steel rod insert circular cross section is

$$I_{\text{steel}} = \frac{\pi r^4}{4} = \frac{\pi\left(\frac{1.0\ \text{in}}{2}\right)^4}{4} = \boxed{0.0491\ \text{in}^4\quad (0.050\ \text{in}^4)}$$

The answer is (A).

(c) The product EI for the brass tube is

$$\begin{aligned}E_{\text{brass}}I_{\text{brass}} &= \left(1.5\times 10^7\ \frac{\text{lbf}}{\text{in}^2}\right)(0.736\ \text{in}^4)\\ &= \boxed{1.104\times 10^7\ \text{lbf-in}^2\quad (1.1\times 10^7\ \text{lbf-in}^2)}\end{aligned}$$

The answer is (B).

(d) The product EI for the steel rod insert is

$$\begin{aligned}E_{\text{steel}}I_{\text{steel}} &= \left(2.9\times 10^7\ \frac{\text{lbf}}{\text{in}^2}\right)(0.0491\ \text{in}^4)\\ &= \boxed{0.142\times 10^7\ \text{lbf-in}^2\quad (0.14\times 10^7\ \text{lbf-in}^2)}\end{aligned}$$

The answer is (A).

(e) The total EI for the composite is

$$\begin{aligned}E_cI_c &= E_{\text{brass}}I_{\text{brass}} + E_{\text{steel}}I_{\text{steel}}\\ &= 1.104\times 10^7\ \text{lbf-in}^2 + 0.142\times 10^7\ \text{lbf-in}^2\\ &= \boxed{1.246\times 10^7\ \text{lbf-in}^2\quad (1.2\times 10^7\ \text{lbf-in}^2)}\end{aligned}$$

The answer is (A).

(f) From case 1 in App. 51.A, the tip deflection of the tube is

$$y_{\text{tip}} = \frac{PL^3}{3EI}$$

Since the deflection is inversely proportional to the EI, the suggestion of inserting a steel rod does have merit because the insert increases the EI product.

The answer is (C).

(g) The percent change in tip deflection is

$$\begin{aligned}\text{percent} &= \frac{y_{\text{tip}_{\text{brass}}} - y_{\text{tip}_{\text{brass+steel}}}}{y_{\text{tip}_{\text{brass}}}}\times 100\%\\ &= \frac{\dfrac{PL^3}{3E_{\text{brass}}I_{\text{brass}}} - \dfrac{PL^3}{3E_cI_c}}{\dfrac{PL^3}{3E_{\text{brass}}I_{\text{brass}}}}\times 100\%\end{aligned}$$

Simplify.

$$\text{percent} = \frac{\frac{1}{E_{\text{brass}}I_{\text{brass}}} - \frac{1}{E_cI_c}}{\frac{1}{E_{\text{brass}}I_{\text{brass}}}} \times 100\%$$

$$= \frac{\frac{1}{1.104\times 10^7\ \text{lbf-in}^2} - \frac{1}{1.246\times 10^7\ \text{lbf-in}^2}}{\frac{1}{1.104\times 10^7\ \text{lbf-in}^2}} \times 100\%$$

$$= \boxed{11.4\%\quad (11\%)}$$

The answer is (D).

SI Solution

(a) The moment of inertia of the brass tube annular cross section is

$$I_{\text{brass}} = \frac{\pi}{4}(r_o^4 - r_i^4)$$

$$= \left(\frac{\pi}{4}\right)\left(\left(\frac{0.050\ \text{m}}{2}\right)^4 - \left(\frac{0.025\ \text{m}}{2}\right)^4\right)$$

$$= \boxed{2.876\times 10^{-7}\ \text{m}^4\quad (2.9\times 10^{-7}\ \text{m}^4)}$$

The answer is (B).

(b) The moment of inertia of the steel rod insert circular cross section is

$$I_{\text{steel}} = \frac{\pi r^4}{4} = \frac{\pi\left(\frac{0.025\ \text{m}}{2}\right)^4}{4}$$

$$= \boxed{1.917\times 10^{-8}\ \text{m}^4\quad (2.0\times 10^{-8}\ \text{m}^4)}$$

The answer is (A).

(c) The product EI for the brass tube is

$$E_{\text{brass}}I_{\text{brass}} = (100\times 10^9\ \text{Pa})(2.876\times 10^{-7}\ \text{m}^4)$$

$$= \boxed{28\,760\ \text{N·m}^2\quad (29\,000\ \text{N·m}^2)}$$

The answer is (B).

(d) The product EI for the steel rod insert is

$$E_{\text{steel}}I_{\text{steel}} = (200\times 10^9\ \text{Pa})(1.917\times 10^{-8}\ \text{m}^4)$$

$$= \boxed{3834\ \text{N·m}^2\quad (3800\ \text{N·m}^2)}$$

The answer is (A).

(e) The total EI for the composite is

$$E_cI_c = E_{\text{brass}}I_{\text{brass}} + E_{\text{steel}}I_{\text{steel}}$$

$$= 28\,760\ \text{N·m}^2 + 3834\ \text{N·m}^2$$

$$= \boxed{32\,594\ \text{N·m}^2\quad (33\,000\ \text{N·m}^2)}$$

The answer is (A).

(f) From case 1 in App. 51.A, the tip deflection of the tube is

$$y_{\text{tip}} = \frac{PL^3}{3EI}$$

Since the deflection is inversely proportional to the EI, the suggestion of inserting a steel rod does have merit because the insert increases the EI product.

The answer is (C).

(g) The percent change in tip deflection is

$$\text{percent} = \frac{y_{\text{tip}_{\text{brass}}} - y_{\text{tip}_{\text{brass+steel}}}}{y_{\text{tip}_{\text{brass}}}} \times 100\%$$

$$= \frac{\frac{PL^3}{3E_{\text{brass}}I_{\text{brass}}} - \frac{PL^3}{3E_cI_c}}{\frac{PL^3}{3E_{\text{brass}}I_{\text{brass}}}} \times 100\%$$

Simplify.

$$\text{percent} = \frac{\frac{1}{E_{\text{brass}}I_{\text{brass}}} - \frac{1}{E_cI_c}}{\frac{1}{E_{\text{brass}}I_{\text{brass}}}} \times 100\%$$

$$= \frac{\frac{1}{28\,760\ \text{N·m}^2} - \frac{1}{32\,594\ \text{N·m}^2}}{\frac{1}{28\,760\ \text{N·m}^2}} \times 100\%$$

$$= \boxed{11.8\%\quad (11\%)}$$

The answer is (D).

13. *Customary U.S. Solution*

The centroidal moment of inertia of the shaft is

$$I = \frac{\pi r^4}{4} = \frac{\pi\left(\frac{2\ \text{in}}{2}\right)^4}{4} = 0.7854\ \text{in}^4$$

Although the stress is probably greatest at the fillet toe, it is difficult to specify exactly where the greatest stress occurs. By common convention, it is assumed to occur at the shoulder.

The moment at the shoulder is

$$M = Fd = m\left(\frac{g}{g_c}\right)d = (2500\ \text{lbm})\left(\frac{32.2\ \frac{\text{ft}}{\text{sec}^2}}{32.2\ \frac{\text{ft-lbm}}{\text{lbf-sec}^2}}\right)(4\ \text{in})$$
$$= 10{,}000\ \text{in-lbf}$$

Use App. 51.B, Fig. (c).

$$\frac{r}{d} = \frac{\frac{5}{16}\ \text{in}}{2\ \text{in}} = 0.156$$
$$\frac{D}{d} = \frac{3\ \text{in}}{2\ \text{in}} = 1.5$$

From App. 51.B, the stress concentration factor is approximately 1.5.

The bending stress is

$$\sigma = K\left(\frac{Mc}{I}\right) = \frac{(1.5)(10{,}000\ \text{in-lbf})\left(\frac{2\ \text{in}}{2}\right)}{0.7854\ \text{in}^4}$$
$$= \boxed{19{,}099\ \text{lbf/in}^2\quad (19{,}000\ \text{lbf/in}^2)}$$

(Note that shear stress is zero where bending stress is maximum.)

The answer is (D).

SI Solution

The centroidal moment of inertia of the shaft is

$$I = \frac{\pi r^4}{4} = \frac{\pi\left(\frac{50\ \text{mm}}{(2)\left(1000\ \frac{\text{mm}}{\text{m}}\right)}\right)^4}{4} = 3.068\times 10^{-7}\ \text{m}^4$$

Although the stress is probably greatest at the fillet toe, it is difficult to specify exactly where the greatest stress occurs. By common convention, it is assumed to occur at the shoulder.

The moment at the shoulder is

$$M = Fd = mgd = (1100\ \text{kg})\left(9.81\ \frac{\text{m}}{\text{s}^2}\right)\left(\frac{100\ \text{mm}}{1000\ \frac{\text{mm}}{\text{m}}}\right)$$
$$= 1079\ \text{N·m}$$

Use App. 51.B, Fig. (c).

$$\frac{r}{d} = \frac{8\ \text{mm}}{50\ \text{mm}} = 0.16$$
$$\frac{D}{d} = \frac{75\ \text{mm}}{50\ \text{mm}} = 1.5$$

From App. 51.B, the stress concentration factor is approximately 1.5.

The bending stress is

$$\sigma = K\left(\frac{Mc}{I}\right)$$
$$= \frac{(1.5)(1079\ \text{N·m})\left(\frac{50\ \text{mm}}{(2)\left(1000\ \frac{\text{mm}}{\text{m}}\right)}\right)}{3.068\times 10^{-7}\ \text{m}^4}$$
$$= \boxed{1.319\times 10^8\ \text{Pa}\quad (130\ \text{MPa})}$$

The answer is (D).

14. A discontinuity stress is an additional localized stress (i.e., an incremental stress) due to an abrupt change in geometry (thickness, diameter, direction) or material. The term is primarily encountered in pressure vessel work where the total stress is equal to the (primary) membrane stress plus the (secondary) discontinuity stress.

The answer is (B).

15. *Customary U.S. Solution*

First, determine an equivalent aluminum area for the steel. The ratio of equivalent aluminum area to steel area is the modular ratio.

$$n = \frac{30\times 10^6\ \frac{\text{lbf}}{\text{in}^2}}{10\times 10^6\ \frac{\text{lbf}}{\text{in}^2}} = 3$$

The equivalent aluminum width to replace the steel is

$$(1.5\ \text{in})(3) = 4.5\ \text{in}$$

The equivalent all-aluminum cross section is

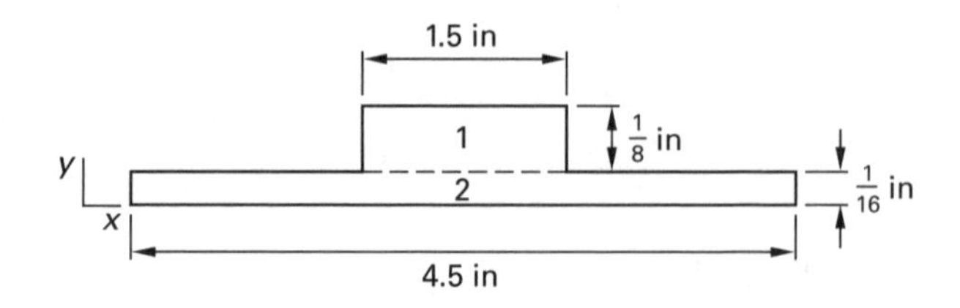

Materials

To find the centroid of the section, first calculate the areas of the basic shapes.

$$A_1 = (1.5 \text{ in})\left(\tfrac{1}{8} \text{ in}\right) = 0.1875 \text{ in}^2$$
$$A_2 = (4.5 \text{ in})\left(\tfrac{1}{16} \text{ in}\right) = 0.28125 \text{ in}^2$$

Next, find the y-components of the centroids of the basic shapes.

$$y_{c1} = \tfrac{1}{16} \text{ in} + \left(\tfrac{1}{2}\right)\left(\tfrac{1}{8} \text{ in}\right) = 0.125 \text{ in}$$
$$y_{c2} = \frac{\tfrac{1}{16} \text{ in}}{2} = 0.03125 \text{ in}$$

The centroid of the section is

$$\begin{aligned} y_c &= \frac{\sum_i A_i y_{ci}}{\sum A_i} \\ &= \frac{(0.1875 \text{ in}^2)(0.125 \text{ in}) + (0.28125 \text{ in}^2)(0.03125 \text{ in})}{0.1875 \text{ in}^2 + 0.28125 \text{ in}^2} \\ &= 0.06875 \text{ in} \end{aligned}$$

The moment of inertia of basic shape 1 about its own centroid is

$$I_{cy1} = \frac{bh^3}{12} = \frac{(1.5 \text{ in})(0.125 \text{ in})^3}{12} = 2.441 \times 10^{-4} \text{ in}^4$$

The moment of inertia of basic shape 2 about its own centroid is

$$I_{cy2} = \frac{bh^3}{12} = \frac{(4.5 \text{ in})(0.0625 \text{ in})^3}{12} = 9.155 \times 10^{-5} \text{ in}^4$$

The distance from the centroid of basic shape 1 to the section centroid is

$$d_1 = y_{c1} - y_c = 0.125 \text{ in} - 0.06875 \text{ in} = 0.05625 \text{ in}$$

The distance from the centroid of basic shape 2 to the section centroid is

$$d_2 = y_c - y_{c2} = 0.06875 \text{ in} - 0.03125 \text{ in} = 0.0375 \text{ in}$$

From the parallel axis theorem (use Eq. 50.29), the moment of inertia of basic shape 1 about the centroid of the section is

$$\begin{aligned} I_{y1} &= I_{yc1} + A_1 d_1^2 \\ &= 2.411 \times 10^{-4} \text{ in}^4 + (0.1875 \text{ in}^2)(0.05625 \text{ in})^2 \\ &= 8.344 \times 10^{-4} \text{ in}^4 \end{aligned}$$

The moment of inertia of basic shape 2 about the centroid of the section is

$$\begin{aligned} I_{y2} &= I_{yc2} + A_2 d_2^2 \\ &= 9.155 \times 10^{-5} \text{ in}^2 + (0.28125 \text{ in}^2)(0.0375 \text{ in})^2 \\ &= 4.871 \times 10^{-4} \text{ in}^2 \end{aligned}$$

The total moment of inertia of the section about the centroid of the section is

$$\begin{aligned} I_y &= I_{y1} + I_{y2} = 8.344 \times 10^{-4} \text{ in}^4 + 4.871 \times 10^{-4} \text{ in}^4 \\ &= \boxed{1.322 \times 10^{-3} \text{ in}^4 \quad (1.3 \times 10^{-3} \text{ in}^4)} \end{aligned}$$

The answer is (C).

SI Solution

First, determine an equivalent aluminum area for the steel. The ratio of equivalent aluminum area to steel area is the modular ratio.

$$n = \frac{20 \times 10^4 \text{ MPa}}{70 \times 10^3 \text{ MPa}} = 2.86$$

The equivalent aluminum width to replace the steel is

$$(38 \text{ mm})(2.86) = 108.7 \text{ mm}$$

The equivalent all-aluminum cross section is

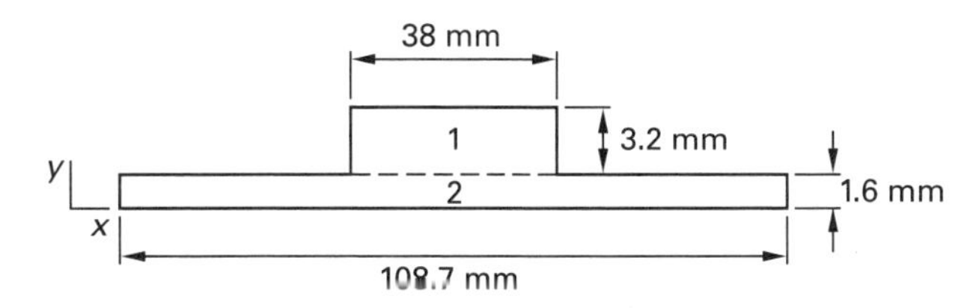

To find the centroid of the section, first calculate the areas of the basic shapes.

$$A_1 = (38 \text{ mm})(3.2 \text{ mm}) = 121.6 \text{ mm}^2$$
$$A_2 = (108.7 \text{ mm})(1.6 \text{ mm}) = 173.9 \text{ mm}^2$$

Next, find the y-components of the centroids of the basic shapes.

$$y_{c1} = 1.6 \text{ mm} + \left(\tfrac{1}{2}\right)(3.2 \text{ mm}) = 3.2 \text{ mm}$$
$$y_{c2} = \frac{1.6 \text{ mm}}{2} = 0.8 \text{ mm}$$

Materials

The centroid of the section is

$$\begin{aligned} y_c &= \frac{\sum_i A_i y_{ci}}{\sum A_i} \\ &= \frac{(121.6\ \text{mm}^2)(3.2\ \text{mm}) + (173.9\ \text{mm}^2)(0.8\ \text{mm})}{121.6\ \text{mm}^2 + 173.9\ \text{mm}^2} \\ &= 1.79\ \text{mm} \end{aligned}$$

The moment of inertia of basic shape 1 about its own centroid is

$$I_{cy1} = \frac{bh^3}{12} = \frac{(38\ \text{mm})(3.2\ \text{mm})^3}{12} = 103.77\ \text{mm}^4$$

The moment of inertia of basic shape 2 about its own centroid is

$$I_{cy2} = \frac{bh^3}{12} = \frac{(108.7\ \text{mm})(1.6\ \text{mm})^3}{12} = 37.10\ \text{mm}^4$$

The distance from the centroid of basic shape 1 to the section centroid is

$$d_1 = y_{c1} - y_c = 3.2\ \text{mm} - 1.79\ \text{mm} = 1.41\ \text{mm}$$

The distance from the centroid of basic shape 2 to the section centroid is

$$d_2 = y_c - y_{c2} = 1.79\ \text{mm} - 0.8\ \text{mm} = 0.99\ \text{mm}$$

From the parallel axis theorem (use Eq. 50.20), the moment of inertia of basic shape 1 about the centroid of the section is

$$\begin{aligned} I_{y1} &= I_{yc1} + A_1 d_1^2 \\ &= 103.77\ \text{mm}^4 + (121.6\ \text{mm}^2)(1.41\ \text{mm})^2 \\ &= 345.52\ \text{mm}^4 \end{aligned}$$

The moment of inertia of basic shape 2 about the centroid of the section is

$$\begin{aligned} I_{y2} &= I_{yc2} + A_2 d_2^2 \\ &= 37.10\ \text{mm}^4 + (173.9\ \text{mm}^2)(0.99\ \text{mm})^2 \\ &= 207.54\ \text{mm}^4 \end{aligned}$$

The total moment of inertia of the section about the centroid of the section is

$$\begin{aligned} I_y &= I_{y1} + I_{y2} = 345.52\ \text{mm}^4 + 207.54\ \text{mm}^4 \\ &= \boxed{553.06\ \text{mm}^4 \quad (550\ \text{mm}^4)} \end{aligned}$$

The answer is (C).

16. The axial load supported by the concrete member is

$$\begin{aligned} P_{\text{concrete}} &= P_t - P_{\text{steel}} = 500{,}000\ \text{lbf} - 400{,}000\ \text{lbf} \\ &= 100{,}000\ \text{lbf} \end{aligned}$$

Since the concrete and the steel bar are rigidly connected, the elongation of the steel bar is the same as the elongation of the concrete. The elongation is

$$\begin{aligned} \delta_{\text{concrete}} &= \frac{P_{\text{concrete}} L}{E_{\text{concrete}} A_{\text{concrete}}} = \frac{(100{,}000\ \text{lbf})(120\ \text{in})}{\left(5{,}000{,}000\ \frac{\text{lbf}}{\text{in}^2}\right)(12\ \text{in}^2)} \\ &= \boxed{0.20\ \text{in}} \end{aligned}$$

The answer is (B).

Materials

52 Failure Theories

PRACTICE PROBLEMS

1. A shaft with a 1.125 in (28.6 mm) diameter receives 400 in-lbf (45 N·m) of torque through a pinned sleeve. The pin is manufactured from steel with a tensile yield strength of 73.9 ksi (510 MPa). Using a factor of safety of 2.5, what is most nearly the pin diameter needed?

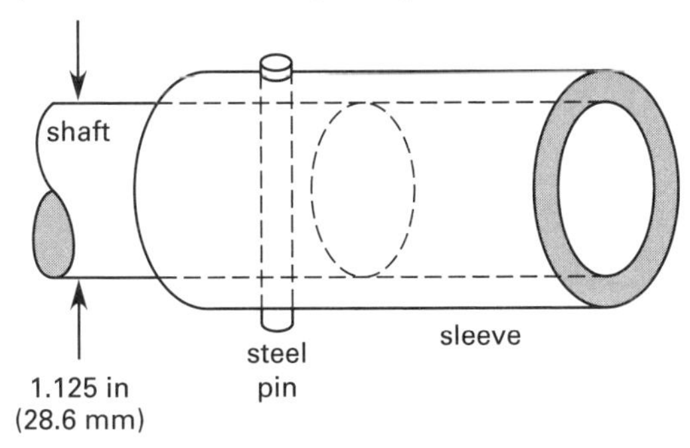

(A) 0.16 in (0.0041 m)

(B) 0.19 in (0.0048 m)

(C) 0.26 in (0.0066 m)

(D) 0.37 in (0.0094 m)

2. A 5/8 in UNC bolt (with rolled threads) and nut carry a simple tensile load that fluctuates between 1000 lbf and 8000 lbf (4400 N and 35,000 N). Both the bolt and nut are made from medium carbon steel with a yield strength of 57,000 lbf/in^2 (390 MPa) and an endurance limit of 30,000 lbf/in^2 (205 MPa). Using Soderberg theory, and including a stress concentration factor for the threads, the factor of safety for the bolt is most nearly

(A) 0.5

(B) 0.7

(C) 1.2

(D) 2.4

3. The spool in the spool valve shown is constructed of aluminum with a yield strength of 19,000 lbf/in^2 (130 MPa). A 500 psig (3.5 MPa) pressure differential exists across the spool shaft. The shaft has a 1.0 in (25 mm) outside diameter and a 0.050 in (1.3 mm) wall thickness. The end disks have a 2.0 in (50 mm) outside diameter and a 0.375 in (9.5 mm) thickness. The end disks are rigidly attached to the tube. Disregard spool collapse.

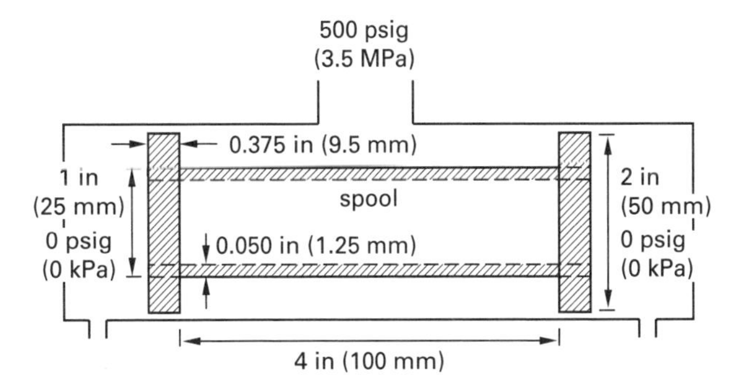

(a) The end disc area exposed to the 500 psig (3.5 MPa) pressure is most nearly

(A) 0.80 in^2 (5.0×10^{-4} m^2)

(B) 1.6 in^2 (10×10^{-4} m^2)

(C) 2.4 in^2 (1.5×10^{-3} m^2)

(D) 3.1 in^2 (2.0×10^{-3} m^2)

(b) The longitudinal force produced by the 500 psig (3.5 MPa) pressure is most nearly

(A) 390 lbf (1700 N)

(B) 790 lbf (3400 N)

(C) 1200 lbf (5200 N)

(D) 1600 lbf (6900 N)

(c) The annular area of the spool tube is most nearly

(A) 0.64 in^2 (4.0×10^{-5} m^2)

(B) 0.15 in^2 (9.3×10^{-5} m^2)

(C) 0.19 in^2 (1.2×10^{-4} m^2)

(D) 0.23 in^2 (1.4×10^{-4} m^2)

(d) The longitudinal stress in the spool tube is most nearly

(A) 5300 lbf/in^2 (37×10^6 Pa)

(B) 6200 lbf/in^2 (43×10^6 Pa)

(C) 7900 lbf/in^2 (55×10^6 Pa)

(D) 11,000 lbf/in^2 (74×10^6 Pa)

(e) The compressive circumferential stress in the tube is most nearly

(A) -5300 lbf/in^2 (-37×10^6 Pa)

(B) -840 lbf/in^2 (-60×10^5 Pa)

(C) 1300 lbf/in^2 (-87×10^5 Pa)

(D) 2500 lbf/in^2 (-18×10^6 Pa)

(f) What are most nearly the principal stresses, σ_1, and σ_2, respectively?

(A) 5300 lbf/in^2, -840 lbf/in^2 (37 MPa, -60 MPa)

(B) 6200 lbf/in^2, 1300 lbf/in^2 (43 MPa, 87 MPa)

(C) 7900 lbf/in^2, -5300 lbf/in^2 (55 MPa, -37 MPa)

(D) 11,000 lbf/in^2, 2500 lbf/in^2 (74 MPa, 18 MPa)

(g) Using the distortion energy theory, the von Mises stress is most nearly

(A) 5600 lbf/in^2 (38 MPa)

(B) 5700 lbf/in^2 (40 MPa)

(C) 9500 lbf/in^2 (67 MPa)

(D) 11,000 lbf/in^2 (80 MPa)

(h) Using the von Mises stress calculated in part (g), the factor of safety for the spool is most nearly

(A) 0.8

(B) 1.1

(C) 1.6

(D) 2.6

4. The pressure in a small pressure vessel operating at room temperature, 70°F (21°C), varies continually between the extremes of 50 psig and 350 psig (340 kPa and 2400 kPa). The vessel is closed by a $^1/_2$ in (12 mm) plate. The plate is attached to the vessel flange with six $^3/_8$-24 UNF bolts evenly spaced around a $9^1/_2$ in (240 mm) circle. Each bolt is tightened to an initial preload of 3700 lbf (16.4 kN). The bolts and nuts are constructed of cold-rolled steel with a 90 ksi (620 MPa) yield strength and 110 ksi (760 MPa) ultimate strength. The plate, flange, and vessel are constructed of steel with a 30 ksi (205 MPa) yield strength and a 50 ksi (345 MPa) ultimate strength. The vessel is intended to be used indefinitely. Neglect the effects of the gasket (not shown). Neglect bending of the plate.

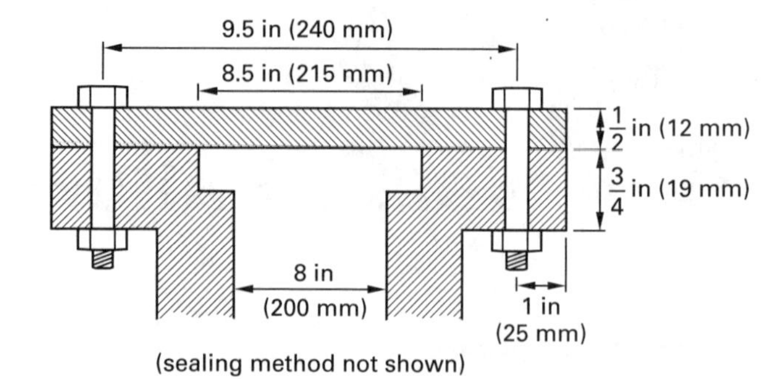

(a) The total bolt stiffness is most nearly

(A) 2.7×10^6 lbf/in (4.6×10^8 N/m)

(B) 1.6×10^7 lbf/in (2.7×10^9 N/m)

(C) 3.2×10^7 lbf/in (5.5×10^9 N/m)

(D) 4.1×10^7 lbf/in (7.1×10^9 N/m)

(b) The plate/vessel stiffness is most nearly

(A) 1.1×10^9 lbf/in (1.8×10^{11} N/m)

(B) 5.8×10^9 lbf/in (10×10^{11} N/m)

(C) 6.4×10^{10} lbf/in (1.1×10^{12} N/m)

(D) 8.4×10^{10} lbf/in (1.4×10^{13} N/m)

(c) The end plate area exposed to pressure is most nearly

(A) 57 in^2 (0.036 m^2)

(B) 77 in^2 (0.048 m^2)

(C) 114 in^2 (0.073 m^2)

(D) 227 in^2 (0.14 m^2)

(d) What are most nearly the maximum and minimum forces on the end plate, respectively?

(A) 2800 lbf, 3800 lbf (12×10^3 N, 16×10^3 N)

(B) 5700 lbf, 5700 lbf (25×10^3 N, 24×10^3 N)

(C) 20,000 lbf, 2800 lbf (87×10^3 N, 12×10^3 N)

(D) 26,000 lbf, 11,000 lbf (120×10^3 N, 50×10^3 N)

(e) The maximum stress in each bolt is most nearly

(A) 42,200 lbf/in^2 (2.900×10^8 Pa)

(B) 42,300 lbf/in^2 (2.905×10^8 Pa)

(C) 42,700 lbf/in^2 (2.930×10^8 Pa)

(D) 42,900 lbf/in^2 (2.940×10^8 Pa)

(f) The minimum stress in each bolt is most nearly

(A) 42,220 lbf/in^2 (2.899×10^8 Pa)

(B) 42,240 lbf/in^2 (2.900×10^8 Pa)

(C) 42,300 lbf/in^2 (2.904×10^8 Pa)

(D) 42,450 lbf/in^2 (2.915×10^8 Pa)

(g) The mean bolt stress is most nearly

(A) 42,220 lbf/in^2 (280 MPa)

(B) 42,300 lbf/in^2 (290 MPa)

(C) 42,500 lbf/in^2 (291 MPa)

(D) 42,700 lbf/in^2 (293 MPa)

(h) The ideal endurance strength of the bolt is most nearly

(A) 40,000 lbf/in^2 (280 MPa)

(B) 45,000 lbf/in^2 (310 MPa)

(C) 50,000 lbf/in^2 (345 MPa)

(D) 55,000 lbf/in^2 (380 MPa)

(i) Using a thread stress concentration factor of 2 and appropriate endurance strength derating factors for a 95% reliability, disregarding miscellaneous effects such as residual stresses and corrosion, the factor of safety is most nearly

(A) 1.2

(B) 1.8

(C) 2.1

(D) 2.5

5. A structural member with the S-N curve shown is subjected to repeated loadings. 10% of the time, the member experiences cycles at 117% of the endurance strength; 15% of the time, the cycles are at 110% of the endurance strength; and 20% of the time, the cycles are at 105% of the endurance strength. The rest of the time, the stress is below the endurance limit. How many cycles can the member experience before failure?

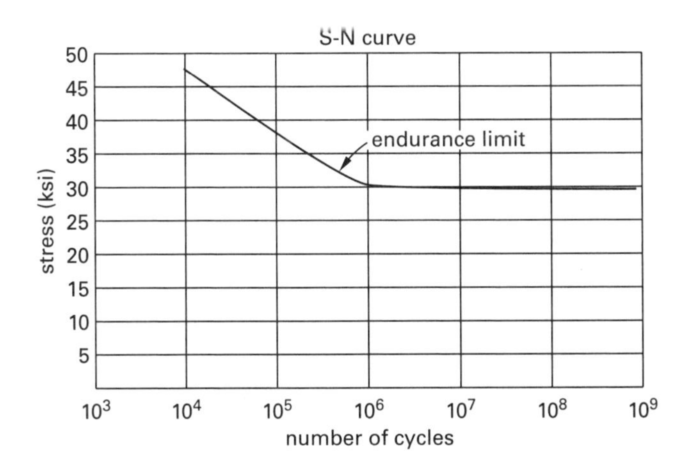

(A) 390,000

(B) 450,000

(C) 620,000

(D) 730,000

SOLUTIONS

1. *Customary U.S. Solution*

Use the distortion energy failure theory. From Eq. 52.27, the yield strength in shear is

$$S_{ys} = 0.577S_{yt} = (0.577)\left(73.9 \times 10^3 \ \frac{\text{lbf}}{\text{in}^2}\right) = 42.64 \times 10^3 \ \text{lbf/in}^2$$

For a safety factor of 2.5, the maximum allowable shear stress is

$$\tau_{\text{max}} = \frac{S_{ys}}{\text{FS}} = \frac{42.64 \times 10^3 \ \frac{\text{lbf}}{\text{in}^2}}{2.5} = 17.06 \times 10^3 \ \text{lbf/in}^2$$

The total shear at the pin is

$$V = \frac{\text{shaft torque}}{\text{shaft radius}} = \frac{400 \ \text{in-lbf}}{\frac{1.125 \ \text{in}}{2}} = 711.1 \ \text{lbf}$$

This is not a case of biaxial stress. There is no bending. The direct shear stress in the pin is

$$\tau_{\text{ave}} = \frac{V}{A}$$

Solve for the required total pin area.

$$A = \frac{V}{\tau_{\text{ave}}} = \frac{711.1 \ \text{lbf}}{17.06 \times 10^3 \ \frac{\text{lbf}}{\text{in}^2}} = 0.04168 \ \text{in}^2$$

Since two surfaces of the pin resist the shear,

$$A = (2)\left(\frac{\pi d^2}{4}\right) = \frac{\pi d^2}{2}$$

Solve for the pin diameter.

$$d = \sqrt{\frac{2A}{\pi}} = \sqrt{\frac{(2)(0.04168 \ \text{in}^2)}{\pi}} = \boxed{0.163 \ \text{in} \quad (0.16 \ \text{in})}$$

The answer is (A).

SI Solution

Use the distortion energy failure theory. From Eq. 52.27, the yield strength in shear is

$$S_{ys} = 0.577S_{yt} = (0.577)(510 \times 10^6 \ \text{Pa}) = 294.3 \times 10^6 \ \text{Pa}$$

Materials

For a safety factor of 2.5, the maximum allowable shear stress is

$$\tau_{max} = \frac{S_{ys}}{FS} = \frac{294.3 \times 10^6 \text{ Pa}}{2.5} = 117.7 \times 10^6 \text{ Pa}$$

The total shear at the pin is

$$V = \frac{\text{shaft torque}}{\text{shaft radius}} = \frac{45 \text{ N·m}}{\frac{0.0286 \text{ m}}{2}} = 3146.9 \text{ N}$$

This is not a case of biaxial stress. There is no bending. The direct shear stress in the pin is

$$\tau_{ave} = \frac{V}{A}$$

Solve for the required total pin area.

$$A = \frac{V}{\tau_{ave}} = \frac{3146.9 \text{ N}}{117.7 \times 10^6 \text{ Pa}} = 2.674 \times 10^{-5} \text{ m}^2$$

Since two surfaces of the pin resist the shear, the total pin area is

$$A = (2)\left(\frac{\pi d^2}{4}\right) = \frac{\pi d^2}{2}$$

Solve for the pin diameter.

$$d = \sqrt{\frac{2A}{\pi}} = \sqrt{\frac{(2)(2.674 \times 10^{-5} \text{ m}^2)}{\pi}}$$
$$= \boxed{0.00413 \text{ m} \quad (0.0041 \text{ m})}$$

The answer is (A).

2. *Customary U.S. Solution*

From Table 53.5, the tensile stress area for a 5/8 in UNC bolt is $A = 0.226 \text{ in}^2$.

The maximum stress is

$$\sigma_{max} = \frac{F_{max}}{A} = \frac{8000 \text{ lbf}}{0.226 \text{ in}^2} = 35{,}398 \text{ lbf/in}^2$$

The minimum stress is

$$\sigma_{min} = \frac{F_{min}}{A} = \frac{1000 \text{ lbf}}{0.226 \text{ in}^2} = 4425 \text{ lbf/in}^2$$

From Eq. 52.28, the mean stress is

$$\sigma_m = \frac{\sigma_{max} + \sigma_{min}}{2} = \frac{35{,}398 \frac{\text{lbf}}{\text{in}^2} + 4425 \frac{\text{lbf}}{\text{in}^2}}{2}$$
$$= 19{,}912 \text{ lbf/in}^2$$

From Eq. 52.30, the alternating stress is

$$\sigma_{alt} = \tfrac{1}{2}(\sigma_{max} - \sigma_{min}) = \left(\tfrac{1}{2}\right)\left(35{,}398 \frac{\text{lbf}}{\text{in}^2} - 4425 \frac{\text{lbf}}{\text{in}^2}\right)$$
$$= 15{,}487 \text{ lbf/in}^2$$

From Sec. 53.13, the stress concentration factor for rolled threads is 2.2. Thus, the alternating stress is

$$\sigma_{alt} = \left(15{,}487 \frac{\text{lbf}}{\text{in}^2}\right)(2.2) = 34{,}071 \text{ lbf/in}^2$$

Draw the Soderberg line and locate the (σ_m, σ_{alt}) point.

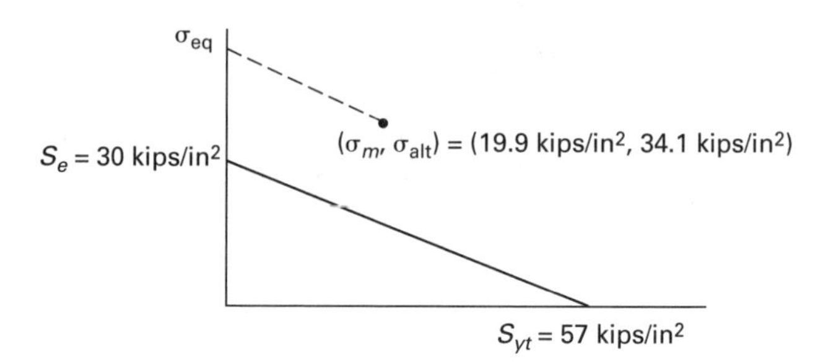

The Soderberg equivalent stress is

$$\sigma_{eq} = \sigma_{alt} + \left(\frac{S_e}{S_{yt}}\right)\sigma_m$$
$$= 34{,}071 \frac{\text{lbf}}{\text{in}^2} + \left(\frac{30{,}000 \frac{\text{lbf}}{\text{in}^2}}{57{,}000 \frac{\text{lbf}}{\text{in}^2}}\right)\left(19{,}912 \frac{\text{lbf}}{\text{in}^2}\right)$$
$$= 44{,}551 \text{ lbf/in}^2$$

The factor of safety is

$$FS = \frac{S_e}{\sigma_{eq}} = \frac{30{,}000 \frac{\text{lbf}}{\text{in}^2}}{44{,}551 \frac{\text{lbf}}{\text{in}^2}} = \boxed{0.673 \quad (0.7)}$$

The answer is (B).

SI Solution

From Table 53.5, the tensile stress area for a 5/8 in UNC bolt is

$$A = \frac{0.226 \text{ in}^2}{\left(39.36 \frac{\text{in}}{\text{m}}\right)^2} = 1.459 \times 10^{-4} \text{ m}^2$$

The maximum stress is

$$\sigma_{max} = \frac{F_{max}}{A} = \frac{35\,000 \text{ N}}{1.459 \times 10^{-4} \text{ m}^2} = 239.89 \times 10^6 \text{ Pa}$$

The minimum stress is

$$\sigma_{min} = \frac{F_{min}}{A} = \frac{4400 \text{ N}}{1.459 \times 10^{-4} \text{ m}^2} = 30.16 \times 10^6 \text{ Pa}$$

From Eq. 52.28, the mean stress is

$$\sigma_m = \frac{\sigma_{\max} + \sigma_{\min}}{2} = \frac{239.89 \times 10^6\ \text{Pa} + 30.16 \times 10^6\ \text{Pa}}{2}$$
$$= 135.03 \times 10^6\ \text{Pa} \quad (135.03\ \text{MPa})$$

From Eq. 52.30, the alternating stress is

$$\sigma_{\text{alt}} = \tfrac{1}{2}(\sigma_{\max} - \sigma_{\min})$$
$$= \left(\tfrac{1}{2}\right)(239.89 \times 10^6\ \text{Pa} - 30.16 \times 10^6\ \text{Pa})$$
$$= 104.87 \times 10^6\ \text{Pa} \quad (104.87\ \text{MPa})$$

From Sec. 53.13, the stress concentration factor for rolled threads is 2.2. Thus, the alternating stress is

$$\sigma_{\text{alt}} = (104.87\ \text{MPa})(2.2) = 230.71\ \text{MPa}$$

Draw the Soderberg line and locate the $(\sigma_m, \sigma_{\text{alt}})$ point.

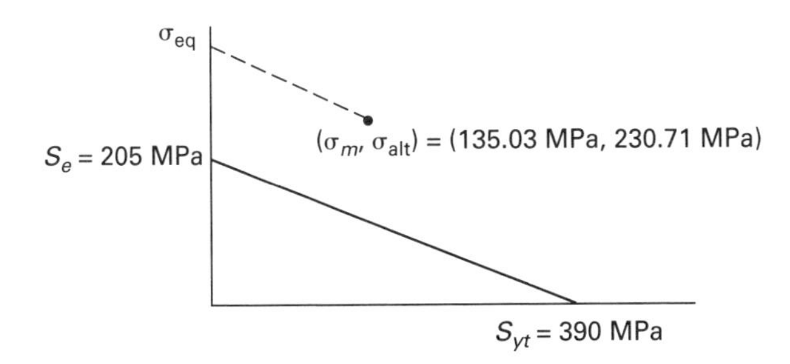

The Soderberg equivalent stress is

$$\sigma_{\text{eq}} = \sigma_{\text{alt}} + \left(\frac{S_e}{S_{yt}}\right)\sigma_m$$
$$= 230.71\ \text{MPa} + \left(\frac{205\ \text{MPa}}{390\ \text{MPa}}\right)(135.03\ \text{MPa})$$
$$= 301.7\ \text{MPa}$$

The factor of safety is

$$\text{FS} = \frac{S_e}{\sigma_{\text{eq}}} = \frac{205\ \text{MPa}}{301.7\ \text{MPa}} = \boxed{0.679 \quad (0.7)}$$

The answer is (B).

3. *Customary U.S. Solution*

(a) The spool appears to be a thin-walled tube. Only the circumferential stress and the longitudinal stress are significant.

The end disc area exposed to the 500 psi pressure is

$$A_e = \left(\frac{\pi}{4}\right)\left((2.0\ \text{in})^2 - (1.0\ \text{in})^2\right)$$
$$= \boxed{2.356\ \text{in}^2 \quad (2.4\ \text{in}^2)}$$

The answer is (C).

(b) The longitudinal force produced by the 500 psig pressure is

$$F = pA_e = \left(500\ \frac{\text{lbf}}{\text{in}^2}\right)(2.356\ \text{in}^2)$$
$$= \boxed{1178\ \text{lbf} \quad (1200\ \text{lbf})}$$

The answer is (C).

(c) The annular area of the spool tube is

$$A = \frac{\pi}{4}(d_o^2 - d_i^2)$$
$$= \left(\frac{\pi}{4}\right)\left((1.0\ \text{in})^2 - \left(1\ \text{in} - (2)(0.05\ \text{in})\right)^2\right)$$
$$= \boxed{0.149\ \text{in}^2 \quad (0.15\ \text{in}^2)}$$

The answer is (B).

(d) The longitudinal stress in the spool tube is

$$\sigma_{\text{long}} = \frac{F}{A} = \frac{1178\ \text{lbf}}{0.149\ \text{in}^2} = \boxed{7906\ \text{lbf/in}^2 \quad (7900\ \text{lbf/in}^2)}$$

The answer is (C).

(e) Since $t/d = 0.05\ \text{in}/1\ \text{in} = 0.05 < 0.10$, this qualifies as a thin-walled cylinder. However, since the tube is exposed to external pressure (not internal), use Lamé's equation for a thick-walled cylinder. This is in the range for Lamé's solution.

The compressive circumferential stress in the tube is given in Table 53.2.

$$r_o = \frac{1\ \text{in}}{2} = 0.5\ \text{in}$$
$$r_i = 0.5\ \text{in} - 0.050\ \text{in} = 0.45\ \text{in}$$
$$\sigma_{\text{ci}} = \frac{-2r_o^2 p}{r_o^2 - r_i^2} = \frac{(-2)(0.5\ \text{in})^2\left(500\ \frac{\text{lbf}}{\text{in}^2}\right)}{(0.5\ \text{in})^2 - (0.45\ \text{in})^2}$$
$$= \boxed{-5263\ \text{lbf/in}^2 \quad (-5300\ \text{lbf/in}^2)}$$

[negative because compression]

The answer is (A).

(f) The principal stresses are

$$\sigma_1 = \sigma_{\text{long}} = \boxed{7906\ \text{lbf/in}^2 \quad (7900\ \text{lbf/in}^2)}$$
$$\sigma_2 = \sigma_{\text{ci}} = \boxed{-5263\ \text{lbf/in}^2 \quad (-5300\ \text{lbf/in}^2)}$$

The answer is (C).

(g) For the distortion energy theory, use Eq. 52.23 to find the von Mises stress.

$$\sigma' = \sqrt{\sigma_1^2 + \sigma_2^2 - \sigma_1\sigma_2}$$
$$= \sqrt{\begin{array}{l}\left(7906\ \frac{\text{lbf}}{\text{in}^2}\right)^2 + \left(-5263\ \frac{\text{lbf}}{\text{in}^2}\right)^2 \\ \quad - \left(7906\ \frac{\text{lbf}}{\text{in}^2}\right)\left(-5263\ \frac{\text{lbf}}{\text{in}^2}\right)\end{array}}$$
$$= \boxed{11{,}481\ \text{lbf/in}^2 \quad (11{,}000\ \text{lbf/in}^2)}$$

The answer is (D).

(h) From Eq. 52.26, the factor of safety for the spool is

$$\text{FS} = \frac{S_{yt}}{\sigma'} = \frac{19{,}000\ \frac{\text{lbf}}{\text{in}^2}}{11{,}481\ \frac{\text{lbf}}{\text{in}^2}} = \boxed{1.65 \quad (1.6)}$$

The answer is (C).

SI Solution

(a) The spool appears to be a thin-walled tube. Only the circumferential stress and the longitudinal stress are significant.

The end disc area exposed to the 3.5 MPa pressure is

$$A_e = \left(\frac{\pi}{4}\right)\left((0.050\ \text{m})^2 - (0.025\ \text{m})^2\right)$$
$$= \boxed{0.001473\ \text{m}^2 \quad (1.5\times 10^{-3}\ \text{m}^2)}$$

The answer is (C).

(b) The longitudinal force produced by the 3.5 MPa pressure is

$$F = pA_e = (3.5\times 10^6\ \text{Pa})(0.001473\ \text{m}^2)$$
$$= \boxed{5155.5\ \text{N} \quad (5200\ \text{N})}$$

The answer is (C).

(c) The annular area of the spool tube is

$$A = \frac{\pi}{4}(d_o^2 - d_i^2)$$
$$= \left(\frac{\pi}{4}\right)\left((0.025\ \text{m})^2 - \left(0.025\ \text{m} - (2)(0.00125\ \text{m})\right)^2\right)$$
$$= \boxed{9.327\times 10^{-5}\ \text{m}^2 \quad (9.3\times 10^{-5}\ \text{m}^2)}$$

The answer is (B).

(d) The longitudinal stress in the spool tube is

$$\sigma_{\text{long}} = \frac{F}{A} = \frac{5155.5\ \text{N}}{9.327\times 10^{-5}\ \text{m}^2}$$
$$= \boxed{55.28\times 10^6\ \text{Pa} \quad (55\times 10^6\ \text{Pa})}$$

The answer is (C).

(e) Since $d/t = 1.3\ \text{mm}/2.5\ \text{mm} = 0.05 < 0.10$, this qualifies as a thin-walled cylinder. However, since the tube is exposed to external pressure (not internal), use Lamé's equation for a thick-walled cylinder. This is in the range for Lamé's solution.

The compressive circumferential stress in the tube is given in Table 53.2.

$$r_o = \frac{25\ \text{mm}}{2} = 12.5\ \text{mm}$$
$$r_i = 12.5\ \text{mm} - 1.25\ \text{mm} = 11.25\ \text{mm}$$
$$\sigma_{\text{ci}} = \frac{-2r_o^2 p}{r_o^2 - r_1^2}$$
$$= \frac{(-2)(12.5\ \text{mm})^2(3.5\ \text{MPa})\left(10^6\ \frac{\text{Pa}}{\text{MPa}}\right)}{(12.5\ \text{mm})^2 - (11.25\ \text{mm})^2}$$
$$= \boxed{-36.84\times 10^6\ \text{Pa} \quad (-37\times 10^6\ \text{Pa})}$$

[negative because compression]

The answer is (A).

(f) The principal stresses are

$$\sigma_1 = \sigma_{\text{long}} = \boxed{55.28\times 10^6\ \text{Pa} \quad (55\ \text{MPa})}$$
$$\sigma_2 = \sigma_{\text{ci}} = \boxed{-36.84\times 10^6\ \text{Pa} \quad (-37\ \text{MPa})}$$

The answer is (C).

(g) For the distortion energy theory, use Eq. 52.23 to find the von Mises stress.

$$\sigma' = \sqrt{\sigma_1^2 + \sigma_2^2 - \sigma_1\sigma_2}$$
$$= \sqrt{\begin{array}{l}(55.28\ \text{MPa})^2 + (-36.84\ \text{MPa})^2 \\ \quad - (55.28\ \text{MPa})(-36.84\ \text{MPa})\end{array}}$$
$$= \boxed{80.31\ \text{MPa} \quad (80\ \text{MPa})}$$

The answer is (D).

(h) From Eq. 52.26, the factor of safety for the spool is

$$\text{FS} = \frac{S_{yt}}{\sigma'} = \frac{130\ \text{MPa}}{80.31\ \text{MPa}} = \boxed{1.62 \quad (1.6)}$$

The answer is (C).

4. *Customary U.S. Solution*

(a) First, find the portion of the pressure load, p, taken by the bolts.

The length of the bolt in tension is

$$L = 0.75 \text{ in} + 0.50 \text{ in} = 1.25 \text{ in}$$

The modulus of elasticity of cold-rolled bolts is $E = 30 \times 10^6 \text{ lbf/in}^2$.

The total bolt stiffness is

$$k_{\text{bolt}} = (6)\left(\frac{AE}{L}\right) = (6)\left(\frac{\left(\frac{\pi}{4}\right)(0.375 \text{ in})^2\left(30 \times 10^6 \ \frac{\text{lbf}}{\text{in}^2}\right)}{1.25 \text{ in}}\right)$$
$$= \boxed{1.590 \times 10^7 \text{ lbf/in} \quad (1.6 \times 10^7 \text{ lbf/in})}$$

The answer is (B).

(b) The plate/vessel contact area is an annulus with an 8.5 in inside diameter and an 11.5 in outside diameter.

$$A = \left(\frac{\pi}{4}\right)\left((11.5 \text{ in})^2 - (8.5 \text{ in})^2\right) = 47.12 \text{ in}^2$$

The modulus of elasticity of the plate/vessel material is $E = 29 \times 10^6 \text{ lbf/in}^2$.

The plate/vessel stiffness is

$$k_{\text{plate/vessel}} = \frac{AE}{L} = \frac{(47.12 \text{ in}^2)\left(29 \times 10^6 \ \frac{\text{lbf}}{\text{in}^2}\right)}{1.25 \text{ in}}$$
$$= \boxed{1.093 \times 10^9 \text{ lbf/in} \quad (1.1 \times 10^9 \text{ lbf/in})}$$

The answer is (A).

(c) Let x be the decimal portion of the pressure load, p, taken by the bolts. The increase in bolt length is

$$\delta_{\text{bolt}} = \frac{xp}{k_{\text{bolt}}}$$

The plate/vessel deformation is

$$\delta_{\text{plate/vessel}} = \frac{(1-x)p}{k_{\text{plate/vessel}}}$$

Set $\delta_{\text{bolt}} = \delta_{\text{plate/vessel}}$.

$$\frac{xp}{k_{\text{bolt}}} = \frac{(1-x)p}{k_{\text{plate/vessel}}}$$

Canceling p gives

$$\frac{x}{1.590 \times 10^7 \ \frac{\text{lbf}}{\text{in}^2}} = \frac{1-x}{1.093 \times 10^9 \ \frac{\text{lbf}}{\text{in}}}$$
$$x = 0.0143$$

Next, find the stresses in the bolts.

The end plate area exposed to pressure is

$$A_{\text{plate}} = \left(\frac{\pi}{4}\right)(8.5 \text{ in})^2 = \boxed{56.75 \text{ in}^2 \quad (57 \text{ in}^2)}$$

The answer is (A).

(d) The maximum and minimum forces on the end plate are

$$F_{\text{max}} = p_{\text{max}}A_{\text{plate}} = \left(350 \ \frac{\text{lbf}}{\text{in}^2}\right)(56.75 \text{ in}^2)$$
$$= \boxed{19{,}863 \text{ lbf} \quad (20{,}000 \text{ lbf})}$$
$$F_{\text{min}} = p_{\text{min}}A_{\text{plate}} = \left(50 \ \frac{\text{lbf}}{\text{in}^2}\right)(56.75 \text{ in}^2)$$
$$= \boxed{2838 \text{ lbf} \quad (2800 \text{ lbf})}$$

The answer is (C).

(e) From Table 53.5, the stress area of a 3/8-24 UNF bolt is $A = 0.0878 \text{ in}^2$.

For a preload, F_o, the maximum stress in each bolt is

$$\sigma_{\text{max}} = \frac{F_o + \frac{xF_{\text{max}}}{6}}{A} = \frac{3700 \text{ lbf} + \frac{(0.0143)(19{,}863 \text{ lbf})}{6}}{0.0878 \text{ in}^2}$$
$$= \boxed{42{,}680 \text{ lbf/in}^2 \quad (42{,}700 \text{ lbf/in}^2)}$$

The answer is (C).

(f) The minimum stress in each bolt is

$$\sigma_{\text{min}} = \frac{F_o + \frac{xF_{\text{min}}}{6}}{A} = \frac{3700 \text{ lbf} + \frac{(0.0143)(2838 \text{ lbf})}{6}}{0.0878 \text{ in}^2}$$
$$= \boxed{42{,}218 \text{ lbf/in}^2 \quad (42{,}220 \text{ lbf/in}^2)}$$

The answer is (A).

(g) From Eq. 52.28, the mean stress is

$$\sigma_m = \frac{\sigma_{\text{max}} + \sigma_{\text{min}}}{2} = \frac{42{,}680 \ \frac{\text{lbf}}{\text{in}^2} + 42{,}218 \ \frac{\text{lbf}}{\text{in}^2}}{2}$$
$$= \boxed{42{,}449 \text{ lbf/in}^2 \quad (42{,}500 \text{ lbf/in}^2)}$$

The answer is (C).

(h) For a thread stress concentration factor of 2, use Eq. 52.30 to find the alternating stress.

$$\begin{aligned}\sigma_{\text{alt}} &= (2)\left(\frac{\sigma_{\max} - \sigma_{\min}}{2}\right) = \sigma_{\max} - \sigma_{\min} \\ &= 42{,}680\ \frac{\text{lbf}}{\text{in}^2} - 42{,}218\ \frac{\text{lbf}}{\text{in}^2} \\ &= 462\ \text{lbf/in}^2\end{aligned}$$

The ideal endurance strength of the bolt is considered to be one-half the ultimate strength.

$$S'_e = 0.5S_{ut} = (0.5)\left(110{,}000\ \frac{\text{lbf}}{\text{in}^2}\right) = \boxed{55{,}000\ \text{lbf/in}^2}$$

The answer is (D).

(i) Use the endurance strength derating factors (from Shigley and Mischke, *Mechanical Engineering Design*). (Note: The third edition of Shigley and Mischke gives $K_b = 0.85$ without explanation. By the fifth edition, a quantitative method based on Kugel's theory yields 0.96.)

- surface finish: $K_a = 0.72$
- size: $K_b = 0.85$
- reliability: $K_c = 0.90$

The derated endurance strength of the bolt is

$$\begin{aligned}S_e &= K_aK_bK_cS'_e = (0.72)(0.85)(0.90)\left(55{,}000\ \frac{\text{lbf}}{\text{in}^2}\right) \\ &= 30{,}294\ \text{lbf/in}^2\end{aligned}$$

An even more sophisticated method of estimating the derated endurance limit, S_e, from the experimental value derived from a polished specimen in a rotary beam method is presented in the eighth edition of *Mechanical Engineering Design*. Methodology and values of the derating factors are as presented in the eighth edition.

$$S_e = k_ak_bk_ck_dk_ek_fS'_e$$

- surface factor:

$$\begin{aligned}k_a &= aS_{ut}^b = (2.70)(110\ \text{ksi})^{-0.265} \\ &= 0.777\quad [\text{cold rolled bolt}]\end{aligned}$$

- size modification factor:

$$k_b = 1.00\quad [\text{axially loaded bolt}]$$

- loading factor: $k_c = 0.85$ [axially loaded bolt]
- temperature factor: $k_d = 1.00$ [room temperature]
- reliability factor: $k_e = 0.868$ [95% reliability]
- miscellaneous effects factor:

$$k_f = 1.00\quad [\text{to be disregarded}]$$

The derated endurance strength of the bolt is

$$\begin{aligned}S_e &= k_ak_bk_ck_dk_ek_fS'_e \\ &= (0.777)(1.00)(0.85)(1.00)(0.868)(1.00) \\ &\quad\times\left(55{,}000\ \frac{\text{lbf}}{\text{in}^2}\right) \\ &= 31{,}530\ \text{lbf/in}^2\end{aligned}$$

Draw the Goodman line and locate the $(\sigma_m, \sigma_{\text{alt}})$ point.

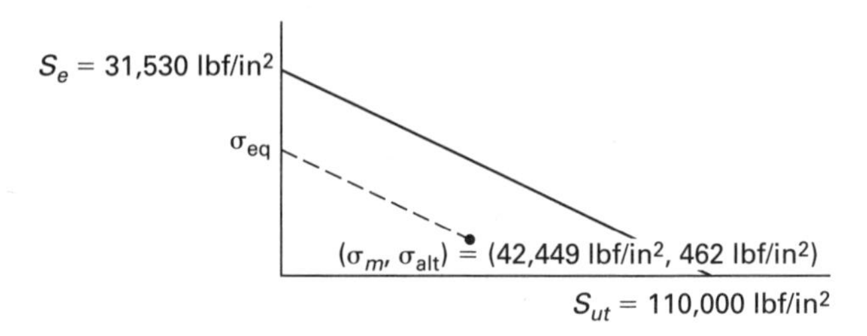

The Goodman equivalent stress is

$$\begin{aligned}\sigma_{\text{eq}} &= \sigma_{\text{alt}} + \left(\frac{S_e}{S_{ut}}\right)\sigma_m \\ &= 462\ \frac{\text{lbf}}{\text{in}^2} + \left(\frac{31{,}530\ \frac{\text{lbf}}{\text{in}^2}}{110{,}000\ \frac{\text{lbf}}{\text{in}^2}}\right)\left(42{,}449\ \frac{\text{lbf}}{\text{in}^2}\right) \\ &= 12{,}629\ \text{lbf/in}^2\end{aligned}$$

From Eq. 52.32, the factor of safety is

$$\text{FS} = \frac{S_e}{\sigma_{\text{eq}}} = \frac{31{,}530\ \frac{\text{lbf}}{\text{in}^2}}{12{,}629\ \frac{\text{lbf}}{\text{in}^2}} = \boxed{2.5}$$

The answer is (D).

SI Solution

(a) First, find the portion of the pressure load, p, taken by the bolts.

The length of the bolt in tension is

$$L = \frac{19\ \text{mm} + 12\ \text{mm}}{1000\ \frac{\text{mm}}{\text{m}}} = 0.031\ \text{m}$$

The modulus of elasticity of cold-rolled bolts is $E = 20 \times 10^4$ MPa.

The total bolt stiffness is

$$\begin{aligned}k_{\text{bolt}} &= (6)\left(\frac{AE}{L}\right) = (6)\left(\frac{\left(\frac{\pi}{4}\right)(0.0095\ \text{m})^2(20 \times 10^{10}\ \text{Pa})}{0.031\ \text{m}}\right) \\ &= \boxed{2.744 \times 10^9\ \text{N/m}\quad (2.7 \times 10^9\ \text{N/m})}\end{aligned}$$

The answer is (B).

(b) The plate/vessel contact area is an annulus with a 215 mm inside diameter and a 290 mm outside diameter.

$$A = \left(\frac{\pi}{4}\right)\left((0.290 \text{ m})^2 - (0.215 \text{ m})^2\right) = 0.02975 \text{ m}^2$$

The modulus of elasticity of the plate/vessel material is $E = 19 \times 10^4$ MPa.

The plate/vessel stiffness is

$$k_{\text{plate/vessel}} = \frac{AE}{L} = \frac{(0.02975 \text{ m}^2)(19 \times 10^{10} \text{ Pa})}{0.031 \text{ m}}$$
$$= \boxed{1.823 \times 10^{11} \text{ N/m} \quad (1.8 \times 10^{11} \text{ N/m})}$$

The answer is (A).

(c) Let x be the decimal portion of the pressure load, p, taken by the bolts. The increase in bolt length is

$$\delta_{\text{bolt}} = \frac{xp}{k_{\text{bolt}}}$$

The plate/vessel deformation is

$$\delta_{\text{plate/vessel}} = \frac{(1-x)p}{k_{\text{plate/vessel}}}$$

Set $\delta_{\text{bolt}} = \delta_{\text{plate/vessel}}$.

$$\frac{xp}{k_{\text{bolt}}} = \frac{(1-x)p}{k_{\text{plate/vessel}}}$$

Canceling p gives

$$\frac{x}{2.744 \times 10^9 \ \frac{\text{N}}{\text{m}}} = \frac{1-x}{1.823 \times 10^{11} \ \frac{\text{N}}{\text{m}}}$$
$$x = 0.0148$$

Next, find the stresses in the bolts.

The end plate area exposed to pressure is

$$A_{\text{plate}} = \left(\frac{\pi}{4}\right)(0.215 \text{ m})^2 = \boxed{0.0363 \text{ m}^2 \quad (0.036 \text{ m}^2)}$$

The answer is (A).

(d) The maximum and minimum forces on the end plate are

$$F_{\max} = p_{\max} A_{\text{plate}} = (2400 \times 10^3 \text{ Pa})(0.0363 \text{ m}^2)$$
$$= \boxed{87.12 \times 10^3 \text{ N} \quad (87 \times 10^3 \text{ N})}$$
$$F_{\min} = p_{\min} A_{\text{plate}} = (340 \times 10^3 \text{ Pa})(0.0363 \text{ m}^2)$$
$$= \boxed{12.34 \times 10^3 \text{ N} \quad (12 \times 10^3 \text{ N})}$$

The answer is (C).

(e) From Table 53.5, the stress area of a 3/8-24 UNF bolt is

$$A = \frac{0.0878 \text{ in}^2}{\left(39.36 \ \frac{\text{m}}{\text{in}}\right)^2} = 5.667 \times 10^{-5} \text{ m}^2$$

For a preload F_o, the maximum stress in each bolt is

$$\sigma_{\max} = \frac{F_o + \frac{xF_{\max}}{6}}{A}$$
$$= \frac{16.4 \times 10^3 \text{ N} + \frac{(0.0148)(87.12 \times 10^3 \text{ N})}{6}}{5.667 \times 10^{-5} \text{ m}^2}$$
$$= \boxed{2.932 \times 10^8 \text{ Pa} \quad (2.930 \times 10^8 \text{ Pa})}$$

The answer is (C).

(f) The minimum stress in each bolt is

$$\sigma_{\min} = \frac{F_o + \frac{xF_{\min}}{6}}{A}$$
$$= \frac{16.4 \times 10^3 \text{ N} + \frac{(0.0148)(12.34 \times 10^3 \text{ N})}{6}}{5.667 \times 10^{-5} \text{ m}^2}$$
$$= \boxed{2.899 \times 10^8 \text{ Pa}}$$

The answer is (A).

(g) From Eq. 52.28, the mean stress is

$$\sigma_m = \frac{\sigma_{\max} + \sigma_{\min}}{2} = \frac{2.932 \times 10^8 \text{ Pa} + 2.899 \times 10^8 \text{ Pa}}{2}$$
$$= \boxed{2.916 \times 10^8 \text{ Pa} \quad (291 \text{ MPa})}$$

The answer is (C).

(h) For a thread stress concentration factor of 2, use Eq. 52.30 to find the alternating stress.

$$\sigma_{\text{alt}} = (2)\left(\frac{\sigma_{\max} - \sigma_{\min}}{2}\right) = \sigma_{\max} - \sigma_{\min}$$
$$= 2.932 \times 10^8 \text{ Pa} - 2.899 \times 10^8 \text{ Pa}$$
$$= 0.033 \times 10^8 \text{ Pa} \quad (3.5 \text{ MPa})$$

The ideal endurance strength of the bolt is considered to be one-half the ultimate strength.

$$S'_e = 0.5 S_{ut} = (0.5)(760 \text{ MPa}) = \boxed{380 \text{ MPa}}$$

The answer is (D).

(i) Use the endurance strength derating factors (from Shigley and Mischke, *Mechanical Engineering Design*). (Note: The third edition of Shigley and Mischke gives $K_b=0.85$ without explanation. By the fifth edition, a quantitative method based on Kugel's theory yields 0.96.)

- surface finish: $K_a=0.72$
- size: $K_b=0.85$
- reliability: $K_c=0.90$

The derated endurance strength of the bolt is

$$\begin{aligned} S_e &= K_a K_b K_c S'_e = (0.72)(0.85)(0.90)(380\ \text{MPa}) \\ &= 209.3\ \text{MPa} \end{aligned}$$

An even more sophisticated method of estimating the derated endurance limit, S_e, from the experimental value derived from a polished specimen in a rotary beam method is presented in the eighth edition of *Mechanical Engineering Design*. Methodology and values of the derating factors are as presented in the eighth edition.

$$S_e = k_a k_b k_c k_d k_e k_f S'_e$$

- surface factor:

$$\begin{aligned} k_a &= aS_{ut}^b = (4.51)(760\ \text{MPa})^{-0.265} \\ &= 0.778 \quad [\text{cold rolled bolt}] \end{aligned}$$

- size modification factor:

$$k_b = 1.00 \quad [\text{axially loaded bolt}]$$

- loading factor: $k_c = 0.85$ [axially loaded bolt]
- temperature factor: $k_d = 1.00$ [room temperature]
- reliability factor: $k_e = 0.868$ [95% reliability]
- miscellaneous effects factor:

$$k_f = 1.00 \quad [\text{to be disregarded}]$$

The derated endurance strength of the bolt is

$$\begin{aligned} S_e &= k_a k_b k_c k_d k_e k_f S'_e \\ &= (0.778)(1.00)(0.85)(1.00)(0.868)(1.00) \\ &\quad \times (380\ \text{MPa}) \\ &= 218.1\ \text{MPa} \end{aligned}$$

Draw the Goodman line and locate the (σ_m, σ_{alt}) point.

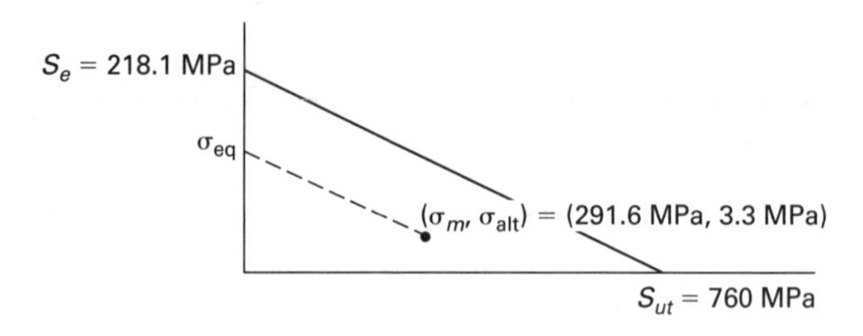

The Goodman equivalent stress is

$$\begin{aligned} \sigma_{eq} &= \sigma_{alt} + \left(\frac{S_e}{S_{ut}}\right)\sigma_m \\ &= 3.3\ \text{MPa} + \left(\frac{218.1\ \text{MPa}}{760\ \text{MPa}}\right)(291.6\ \text{MPa}) \\ &= 87.0\ \text{MPa} \end{aligned}$$

From Eq. 52.32, the factor of safety is

$$\text{FS} = \frac{S_e}{\sigma_{eq}} = \frac{218.1\ \text{MPa}}{87.0\ \text{MPa}} = \boxed{2.5}$$

The answer is (D).

5. The endurance limit is approximately 30 ksi. Determine the fatigue life for each stress level.

For a stress level of 117% of S_e, for example,

$$\sigma = 1.17S_e = (1.17)\left(30\ \frac{\text{kips}}{\text{in}^2}\right) = 35\ \text{ksi}$$

The following table is generated similarly.

amount of time (%)	stress level (% of S_e)	stress (ksi)	fatigue life, N_i (cycles)
10%	117%	35	1.2×10^5
15%	110%	33	3.3×10^5
20%	105%	31.5	6.0×10^5

Use the Palmgren-Miner rule, Eq. 52.35. Let N^* represent the number of cycles at failure. Then, the number of cycles experienced at the 117% stress level is $1.1N^*$, and so on. From Eq. 52.35,

$$\begin{aligned} \sum \frac{n_i}{N_i} &= \frac{0.1N^*}{1.2 \times 10^5} + \frac{0.15N^*}{3.3 \times 10^5} + \frac{0.2N^*}{6.0 \times 10^5} = 1 \\ N^* &= \boxed{616{,}822 \quad (620{,}000)} \end{aligned}$$

The answer is (C).

Materials

53 Basic Machine Design

PRACTICE PROBLEMS

Note: Unless instructed otherwise in a problem, use the following properties:

steel: $E = 30 \times 10^6$ lbf/in^2 (20×10^4 MPa)
$G = 11.5 \times 10^6$ lbf/in^2 (8.0×10^4 MPa)
$\alpha = 6.5 \times 10^{-6}$ 1/°F (1.2×10^{-5} 1/°C)
$\nu = 0.3$
aluminum: $E = 10 \times 10^6$ lbf/in^2 (70×10^3 MPa)
copper: $E = 17.5 \times 10^6$ lbf/in^2 (12×10^4 MPa)

1. The yield strength of a structural steel member is 36,000 lbf/in^2 (250 MPa). The tensile stress is 8240 lbf/in^2 (57 MPa). The factor of safety in tension is most nearly

(A) 2.5

(B) 3.1

(C) 3.6

(D) 4.4

2. A structural steel member 50 ft (15 m) long is used as a long column to support 75,000 lbf (330 kN). Both ends are built-in, and there are no intermediate supports. A factor of safety of 2.5 is used. The required moment of inertia is most nearly

(A) 48 in^4 (2.0×10^{-5} m^4)

(B) 72 in^4 (3.0×10^{-5} m^4)

(C) 96 in^4 (4.0×10^{-5} m^4)

(D) 130 in^4 (5.5×10^{-5} m^4)

3. A long steel column has pinned ends and a yield strength of 36,000 lbf/in^2 (250 MPa). The column is 25 ft (7.5 m) long and has a cross-sectional area of 25.6 in^2 (165 cm^2), a centroidal moment of inertia of 350 in^4 (14 600 cm^4), and a distance from the neutral axis to the extreme fiber of 7 in (180 mm). It carries an axial concentric load of 100,000 lbf (440 kN) and an eccentric load of 150,000 lbf (660 kN) located 3.33 in (80 mm) from the longitudinal axial axis. Using the secant formula, the stress factor of safety is most nearly

(A) 1.6

(B) 2.1

(C) 2.4

(D) 3.0

4. A rectangular tank with dimensions of 2 ft by 2 ft by 2 ft (60 cm by 60 cm by 60 cm) is pressurized to 2 lbf/in^2 gage (14 kPa). The steel plate used to construct a tank has a thickness of 0.25 in (6.3 mm) and a yield stress of 36 ksi (250 MPa). Neglecting stress concentration factors at the corners and edges, the factor of safety is most nearly

(A) 3.1

(B) 3.7

(C) 4.2

(D) 6.3

5. A short section of pipe with an inside diameter of 1.750 in (44.5 mm) is to be produced by turning and boring bar stock. The steel has an allowable normal stress of 20,000 lbf/in^2 (140 MPa). The pipe will be pressurized internally to 2000 lbf/in^2 gage (14 MPa). Disregard longitudinal stress and other end effects. Using a thick-walled analysis, the required outside diameter is most nearly

(A) 1.9 in (49 mm)

(B) 2.2 in (56 mm)

(C) 2.6 in (66 mm)

(D) 3.0 in (00 mm)

6. A pressurized cylinder has an inside diameter of 0.742 in (18.8 mm) and an outside diameter of 1.486 in (37.7 mm). The cylinder is subjected to an external pressure of 400 lbf/in^2 gage (2.8 MPa). The maximum stress developed in the cylinder is most nearly

(A) −3600 lbf/in^2 (−28 MPa)

(B) −2500 lbf/in^2 (−18 MPa)

(C) −1600 lbf/in^2 (−12 MPa)

(D) −1100 lbf/in^2 (−7.5 MPa)

7. A shell with an outside diameter of 16 in (406 mm) and a wall thickness of 0.10 in (2.54 mm) is subjected to a 40,000 lbf/in^2 (280 MPa) tensile load and a 400,000 in-lbf torque (45 kN·m).

(a) What are the approximate principal stresses?

(b) What is the approximate maximum shear stress?

8. A projectile launcher is formed by shrinking a jacket over a tube of the same length. The assembly is made with the maximum allowable diametral interference. The tube inside and outside diameters are 4.7 in and 7.75 in (119 mm and 197 mm), respectively. The outside diameter of the jacket is 12 in (300 mm). The jacket and tube are both steel with a Poisson's ratio of 0.3 and modulus of elasticity of 3.0×10^7 lbf/in^2 (204 GPa). The maximum allowable stress at the interface is 18,000 lbf/in^2 (124 MPa).

(a) The diametral interference is most nearly

(A) 0.0064 in (0.16 mm)

(B) 0.0088 in (0.22 mm)

(C) 0.012 in (0.30 mm)

(D) 0.022 in (0.56 mm)

(b) The maximum circumferential stress in the jacket is most nearly

(A) 12,000 lbf/in^2 (81.6 MPa)

(B) 14,000 lbf/in^2 (95.2 MPa)

(C) 16,000 lbf/in^2 (106 MPa)

(D) 18,000 lbf/in^2 (124 MPa)

(c) The minimum circumferential stress in the tube is most nearly

(A) −18,000 lbf/in^2 (−124 MPa)

(B) −16,000 lbf/in^2 (−106 MPa)

(C) −14,000 lbf/in^2 (−95.2 MPa)

(D) −12,000 lbf/in^2 (−81.6 MPa)

9. Two hollow cylinders of the same length are press-fitted together with a diametral interference of 0.012 in (0.3 mm). The outer cylinder has a wall thickness of 0.079 in (2 mm) and an outside diameter of 4.7 in (120 mm). The inner cylinder has a wall thickness of 0.12 in (3 mm). Both cylinders are the same high-strength material with a Poisson's ratio of 0.3 and modulus of elasticity of 3×10^7 lbf/in^2 (207 GPa).

(a) At the interface between the two cylinders, the circumferential stress for the inner cylinder is most nearly

(A) −130,000 lbf/in^2 (−900 MPa)

(B) −71,000 lbf/in^2 (−490 MPa)

(C) −31,000 lbf/in^2 (−210 MPa)

(D) −26,000 lbf/in^2 (−180 MPa)

(b) At the interface between the two cylinders, the circumferential stress for the outer cylinder is most nearly

(A) 48,000 lbf/in^2 (330 MPa)

(B) 53,000 lbf/in^2 (370 MPa)

(C) 148,000 lbf/in^2 (1020 MPa)

(D) 210,000 lbf/in^2 (1400 MPa)

10. A $^3/_4$-16 UNF steel bolt is used without washers to clamp two rigid steel plates, each 2 in (50 mm) thick. 0.75 in (19 mm) of the threaded section of the bolt remains under the nut. The nut has six threads. Assume half of the bolt in the nut contributes to elongation. The nut is tightened until the stress in the bolt body is 40,000 lbf/in^2 (280 MPa). The bolt's modulus of elasticity is 2.9×10^7 lbf/in^2 (200 GPa). Approximately how much does the bolt stretch?

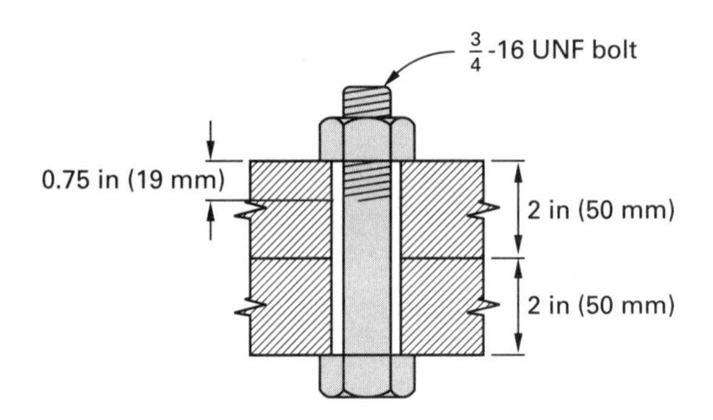

(A) 0.006 in (0.15 mm)

(B) 0.012 in (0.30 mm)

(C) 0.024 in (0.60 mm)

(D) 0.048 in (1.2 mm)

11. A class 30 gray cast-iron hub with an o1utside diameter of 12 in (300 mm) is pressed onto a soft steel solid shaft with an outside diameter of 6 in (150 mm). Poisson's ratio and modulus of elasticity for the cast iron are 0.27 and 1.45×10^7 lbf/in^2 (100 GPa), respectively. Poisson's ratio and modulus of elasticity for the steel are 0.30 and 2.9×10^7 lbf/in^2 (200 GPa), respectively. The stress-strain curve for gray cast iron becomes nonlinear at low stresses. The maximum radial interference such that the stress-strain relationship remains linear is most nearly

(A) 0.0014 in (0.036 mm)

(B) 0.0060 in (0.15 mm)

(C) 0.0087 in (0.22 mm)

(D) 0.011 in (0.28 mm)

12. The bracket shown is attached to a column with three 0.75 in (19 mm) bolts arranged in an equilateral triangular layout. A force is applied with a moment arm of 20 in (500 mm) measured to the centroid of the bolt group. The maximum shear stress in the bolts is limited to 15,000 lbf/in^2 (100 MPa). Perform an elastic analysis to determine the approximate maximum force that the connection can support.

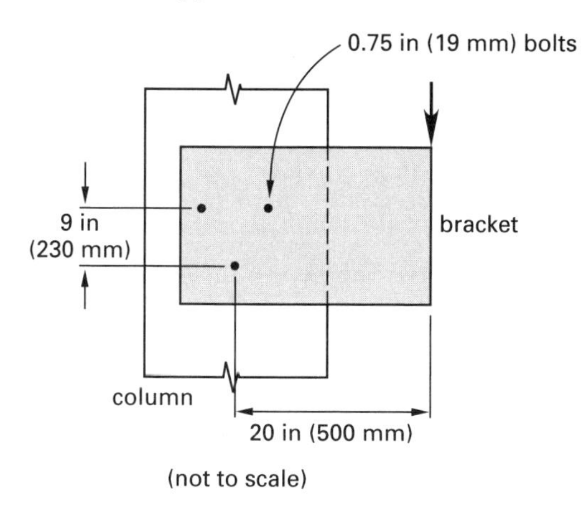

(A) 700 lbf (3.1 kN)

(B) 1200 lbf (5.3 kN)

(C) 2400 lbf (11 kN)

(D) 4700 lbf (20 kN)

13. A fillet weld is used to secure a steel bracket to a column. The bracket supports a 10,000 lbf (44 kN) force applied 12 in (300 mm) from the face of the column, as shown. The maximum shear stress in the weld material is 8000 lbf/in^2 (55 MPa). Perform an elastic analysis to determine the approximate size fillet weld required.

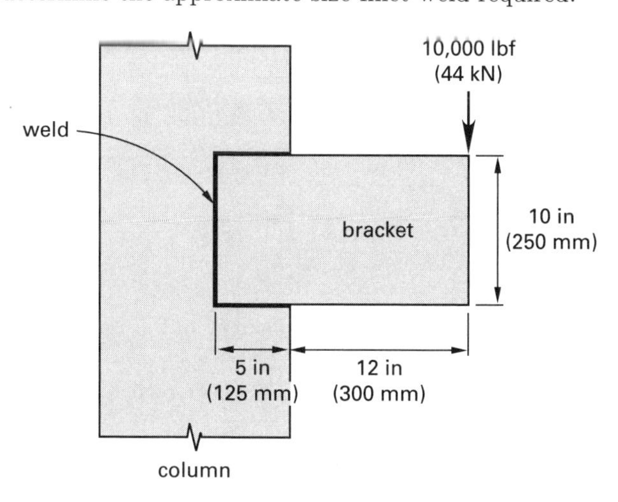

(A) $^3/_8$ in (9.5 mm)

(B) $^1/_2$ in (13 mm)

(C) $^5/_8$ in (16 mm)

(D) $^3/_4$ in (19 mm)

14. A 2 in (50 mm) diameter steel solid shaft rotates at 3500 rpm. A 16 in (406 mm) outside diameter, 2 in (50 mm) thick steel disk flywheel is press fitted to the shaft. The contact pressure between the shaft and flywheel is 1250 lbf/in^2 (8.6 MPa) when the shaft is turning. Centrifugal expansion of the shaft diameter is negligible. For both the shaft and the flywheel, Poisson's ratio is 0.3, modulus of elasticity is 2.9×10^7 lbf/in^2 (20 GPa), and mass density is 0.283 lbm/in^3 (7830 kg/m^3). The radial interference required when the shaft is not turning is most nearly

(A) 1.0×10^{-4} in (2.5×10^{-3} mm)

(B) 1.5×10^{-4} in (3.7×10^{-3} mm)

(C) 2.1×10^{-4} in (5.2×10^{-3} mm)

(D) 2.7×10^{-4} in (6.7×10^{-3} mm)

15. Two brackets are connected by two bolts with individual cross-sectional areas of A_{bolt}. Each bracket carries a load of $2F$.

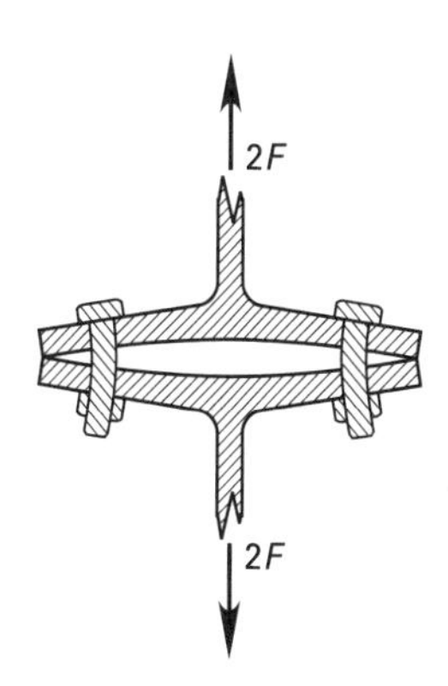

(a) What best describes the bolt configuration shown?

(A) prying action

(B) complex double shear

(C) complex double tension

(D) plastic bending

(b) Neglecting secondary effects, what is the tensile stress in each bolt?

(A) $F/2A_{\text{bolt}}$

(B) F/A_{bolt}

(C) $2F/A_{\text{bolt}}$

(D) $4F/A_{\text{bolt}}$

16. An American Unified Standard Threaded bolt is preloaded with a tensile force of 80,000 lbf. The bolt's modulus of elasticity is 20×10^6 lbf/in^2, the original length is 4 in, and the maximum allowable elongation due to preloading is 0.05 in. The minimum appropriate coarse series nominal size is

(A) $^1/_4$ in

(B) $^3/_8$ in

(C) $^7/_{16}$ in

(D) $^3/_4$ in

17. An SAE bolt is loaded by a tensile proof load. The sum of the bolt's root diameter and pitch diameter is 1 in. Under a proof stress of 70,000 lbf/in^2, the bolt elongates 0.0147 in. The spring constant for the bolt is most nearly

(A) 6.2×10^5 lbf/in

(B) 9.4×10^5 lbf/in

(C) 17×10^5 lbf/in

(D) 22×10^5 lbf/in

18. A bolt is preloaded with a force of 500,000 lbf. The bolt has a cross-sectional area of 1 in^2, a spring constant of 10×10^6 lbf/in, and a modulus of elasticity of 20×10^6 lbf/in^2. The elongation of the bolt as a percentage of the original length is most nearly

(A) 1.0%

(B) 1.5%

(C) 1.7%

(D) 2.5%

SOLUTIONS

1. *Customary U.S. Solution*

The factor of safety is

$$\text{FS} = \frac{S_{yt}}{\sigma} = \frac{36{,}000\ \frac{\text{lbf}}{\text{in}^2}}{8240\ \frac{\text{lbf}}{\text{in}^2}} = \boxed{4.37 \quad (4.4)}$$

The answer is (D).

SI Solution

The factor of safety is

$$\text{FS} = \frac{S_{yt}}{\sigma} = \frac{250\ \text{MPa}}{57\ \text{MPa}} = \boxed{4.39 \quad (4.4)}$$

The answer is (D).

2. *Customary U.S. Solution*

The design load for a factor of safety of 2.5 is

$$F = (75{,}000\ \text{lbf})(2.5) = 187{,}500\ \text{lbf}$$

From Table 53.1, the theoretical end restraint coefficient for built-in ends is $C=0.5$, and the minimum design value is 0.65.

From Eq. 53.9, the effective length of the column is

$$L' = CL = (0.65)(50\ \text{ft})\left(12\ \frac{\text{in}}{\text{ft}}\right) = 390\ \text{in}$$

Set the design load equal to the Euler load F_e and use Eq. 53.7 to find the required moment of inertia.

$$I = \frac{F_e(L')^2}{\pi^2 E} = \frac{(187{,}500\ \text{lbf})(390\ \text{in})^2}{\pi^2\left(30 \times 10^6\ \frac{\text{lbf}}{\text{in}^2}\right)} = \boxed{96.32\ \text{in}^4 \quad (96\ \text{in}^4)}$$

The answer is (C).

SI Solution

The design load for a factor of safety of 2.5 is

$$F = (330 \times 10^3\ \text{N})(2.5) = 825 \times 10^3\ \text{N}$$

From Table 53.1, the theoretical end restraint coefficient for built-in ends is $C=0.5$, and the minimum design value is 0.65.

From Eq. 53.9, the effective length of the column is

$$L' = CL = (0.65)(15\ \text{m}) = 9.75\ \text{m}$$

Set the design load equal to the Euler load F_e and use Eq. 53.1 to find the required moment of inertia.

$$\begin{aligned} I &= \frac{F_e(L')^2}{\pi^2 E} = \frac{(825 \times 10^3\ \text{N})(9.75\ \text{m})^2}{\pi^2\left(20 \times 10^{10}\ \frac{\text{N}}{\text{m}^2}\right)} \\ &= \boxed{3.973 \times 10^{-5}\ \text{m}^4 \quad (4.0 \times 10^{-5}\ \text{m}^4)} \end{aligned}$$

The answer is (C).

3. *Customary U.S. Solution*

The radius of gyration is

$$r = \sqrt{\frac{I}{A}} = \sqrt{\frac{350\ \text{in}^4}{25.6\ \text{in}^2}} = 3.70\ \text{in}$$

The slenderness ratio is

$$\frac{L}{r} = \frac{(25\ \text{ft})\left(12\ \frac{\text{in}}{\text{ft}}\right)}{3.70\ \text{in}} = 81.08$$

The total buckling load is

$$F = 100{,}000\ \text{lbf} + 150{,}000\ \text{lbf} = 250{,}000\ \text{lbf}$$

From Eq. 53.15,

$$\begin{aligned} \phi &= \tfrac{1}{2}\left(\frac{L}{r}\right)\sqrt{\frac{F}{AE}} \\ &= \left(\tfrac{1}{2}\right)(81.08)\sqrt{\frac{250{,}000\ \text{lbf}}{(25.6\ \text{in}^2)\left(29 \times 10^6\ \frac{\text{lbf}}{\text{in}^2}\right)}} \\ &= 0.744\ \text{rad} \end{aligned}$$

The effective eccentricity is

$$\begin{aligned} e &= \frac{M}{F} = \frac{F'e'}{F} = \frac{(150{,}000\ \text{lbf})(3.33\ \text{in})}{250{,}000\ \text{lbf}} \\ &= 2.0\ \text{in} \end{aligned}$$

From Eq. 53.14, the critical column stress is

$$\begin{aligned} \sigma_{\text{max}} &= \left(\frac{F}{A}\right)\left(1 + \left(\frac{ec}{r^2}\right)\sec\phi\right) \\ &= \left(\frac{250{,}000\ \text{lbf}}{25.6\ \text{in}^2}\right) \\ &\quad \times \left(1 + \left(\frac{(2.0\ \text{in})(7\ \text{in})}{(3.70\ \text{in})^2}\right)\sec(0.744\ \text{rad})\right) \\ &= 23{,}339\ \text{lbf/in}^2 \end{aligned}$$

The stress factor of safety for the column is

$$\begin{aligned} \text{FS} &= \frac{S_y}{\sigma_{\text{max}}} = \frac{36{,}000\ \frac{\text{lbf}}{\text{in}^2}}{23{,}339\ \frac{\text{lbf}}{\text{in}^2}} \\ &= \boxed{1.54 \quad (1.6)} \end{aligned}$$

The answer is (A).

SI Solution

The radius of gyration is

$$r = \sqrt{\frac{I}{A}} = \sqrt{\frac{14\,600\ \text{cm}^4}{165\ \text{cm}^2}} = 9.41\ \text{cm}$$

The slenderness ratio is

$$\frac{L}{r} = \frac{(7.5\ \text{m})\left(100\ \frac{\text{cm}}{\text{m}}\right)}{9.41\ \text{cm}} = 79.70$$

The total buckling load is

$$\begin{aligned} F &= (440\ \text{kN} + 660\ \text{kN})\left(1000\ \frac{\text{N}}{\text{kN}}\right) \\ &= 1.1 \times 10^6\ \text{N} \end{aligned}$$

From Eq. 53.15,

$$\begin{aligned} \phi &= \tfrac{1}{2}\left(\frac{L}{r}\right)\sqrt{\frac{F}{AE}} \\ &= \left(\tfrac{1}{2}\right)(79.70)\sqrt{\frac{1.1 \times 10^6\ \text{N}}{\left(\frac{165\ \text{cm}^2}{\left(100\ \frac{\text{cm}}{\text{m}}\right)^2}\right)(20 \times 10^4\ \text{MPa}) \times \left(10^6\ \frac{\text{Pa}}{\text{MPa}}\right)}} \\ &= 0.728\ \text{rad} \end{aligned}$$

The eccentricity is

$$\begin{aligned} e &= \frac{M}{F} = \frac{F'e'}{F} = \frac{(660\ \text{kN})(80\ \text{mm})}{1100\ \text{kN}} \\ &= 48\ \text{mm} \end{aligned}$$

From Eq. 53.14, the critical column stress is

$$\sigma_{max} = \left(\frac{F}{A}\right)\left(1 + \left(\frac{ec}{r^2}\right)\sec\phi\right)$$

$$= \left(\frac{\dfrac{1.1\times10^6\ \text{N}}{165\ \text{cm}^2}}{\left(100\ \dfrac{\text{cm}}{\text{m}}\right)^2}\right) \times \left(1 + \left(\frac{(48\ \text{mm})(180\ \text{mm})}{(9.41\ \text{cm})^2\left(10\ \dfrac{\text{mm}}{\text{cm}}\right)^2}\right) \times (\sec(0.728\ \text{rad}))\right) \times \left(\frac{1\ \text{MPa}}{1\times10^6\ \text{Pa}}\right)$$

$$= 153.8\ \text{MPa}$$

The stress factor of safety for the column is

$$\text{FS} = \frac{S_y}{\sigma_{max}} = \frac{250\ \text{MPa}}{153.8\ \text{MPa}} = \boxed{1.63\quad(1.6)}$$

The answer is (A).

4. *Customary U.S. Solution*

The dimensions of the face are

$$a = (2\ \text{ft})\left(12\ \frac{\text{in}}{\text{ft}}\right) = 24\ \text{in}$$

$$b = (2\ \text{ft})\left(12\ \frac{\text{in}}{\text{ft}}\right) = 24\ \text{in}$$

The ratio of the side dimensions is

$$\frac{a}{b} = \frac{24\ \text{in}}{24\ \text{in}} = 1$$

The plate is considered to have built-in edges. From Table 53.7, the maximum stress is

$$\sigma_{max} = \frac{C_3 p b^2}{t^2} = \frac{(0.308)\left(2\ \dfrac{\text{lbf}}{\text{in}^2}\right)(24\ \text{in})^2}{(0.25\ \text{in})^2}$$

$$= 5677\ \text{lbf/in}^2$$

The factor of safety is

$$\text{FS} = \frac{S_y}{\sigma_{max}} = \frac{36{,}000\ \dfrac{\text{lbf}}{\text{in}^2}}{5677\ \dfrac{\text{lbf}}{\text{in}^2}} = \boxed{6.34\quad(6.3)}$$

The answer is (D).

SI Solution

The dimensions of the face are $a = 60$ cm and $b = 60$ cm.

The ratio of the side dimensions is

$$\frac{a}{b} = \frac{60\ \text{cm}}{60\ \text{cm}} = 1$$

The plate is considered to have built-in edges. From Table 53.7, the maximum stress is

$$\sigma_{max} = \frac{C_3 p b^2}{t^2}$$

$$= \frac{(0.308)(14\ \text{kPa})\left(\dfrac{60\ \text{cm}}{100\ \dfrac{\text{cm}}{\text{m}}}\right)^2}{\left(\dfrac{6.3\ \text{mm}}{1000\ \dfrac{\text{mm}}{\text{m}}}\right)^2\left(1000\ \dfrac{\text{kPa}}{\text{MPa}}\right)}$$

$$= 39.11\ \text{MPa}$$

The factor of safety is

$$\text{FS} = \frac{S_y}{\sigma_{max}} = \frac{250\ \text{MPa}}{39.11\ \text{MPa}} = \boxed{6.39\quad(6.3)}$$

The answer is (D).

5. *Customary U.S. Solution*

The inside radius is

$$r_i = \frac{d_i}{2} = \frac{1.750\ \text{in}}{2} = 0.875\ \text{in}$$

From Table 53.2, the maximum stress is the circumferential stress at the inside radius. This is a normal stress.

$$\sigma_{c,i} = \frac{(r_o^2 + r_i^2)p}{r_o^2 - r_i^2}$$

$$20{,}000\ \frac{\text{lbf}}{\text{in}^2} = \frac{\left(r_o^2 + (0.875\ \text{in})^2\right)\left(2000\ \dfrac{\text{lbf}}{\text{in}^2}\right)}{r_o^2 - (0.875\ \text{in})^2}$$

Simplify.

$$(10)\left(r_o^2 - (0.875\ \text{in})^2\right) = r_o^2 + (0.875\ \text{in})^2$$

$$r_o^2 = \frac{(11)(0.875\ \text{in})^2}{9}$$

$$r_o = 0.9673\ \text{in}$$

The required outside diameter is

$$d_o = 2r_o = (2)(0.9673\ \text{in}) = \boxed{1.935\ \text{in}\quad(1.9\ \text{in})}$$

The answer is (A).

SI Solution

The inside radius is

$$r_i = \frac{d_i}{2} = \frac{44.5 \text{ mm}}{2} = 22.25 \text{ mm}$$

From Table 53.2, the maximum stress is the circumferential stress at the inside radius. This is a normal stress.

$$\sigma_{c,i} = \frac{(r_o^2 + r_i^2)p}{r_o^2 - r_i^2}$$

$$140 \text{ MPa} = \frac{\left(r_o^2 + (22.25 \text{ mm})^2\right)(14 \text{ MPa})}{r_o^2 - (22.25 \text{ mm})^2}$$

Simplify.

$$(10)\left(r_o^2 - (22.25 \text{ mm})^2\right) = r_o^2 + (22.25 \text{ mm})^2$$

$$r_o^2 = \frac{(11)(22.25 \text{ mm})^2}{9}$$

$$r_o = 24.598 \text{ mm}$$

The required outside diameter is

$$d_o = 2r_o = (2)(24.598 \text{ mm}) = \boxed{49.20 \text{ mm} \quad (49 \text{ mm})}$$

The answer is (A).

6. *Customary U.S. Solution*

The inside radius is

$$r_i = \frac{0.742 \text{ in}}{2} = 0.371 \text{ in}$$

The outside radius is

$$r_o = \frac{1.486 \text{ in}}{2} = 0.743 \text{ in}$$

From Table 53.2, the maximum stress developed in the cylinder is the circumferential stress at the inside radius.

$$\sigma_{c,i} = \frac{-2r_o^2 p}{r_o^2 - r_i^2} = \frac{(-2)(0.743 \text{ in})^2\left(400 \ \frac{\text{lbf}}{\text{in}^2}\right)}{(0.743 \text{ in})^2 - (0.371 \text{ in})^2}$$

$$= \boxed{-1066 \text{ lbf/in}^2 \text{ (C)} \quad (-1100 \text{ lbf/in}^2)}$$

The answer is (D).

SI Solution

The inside radius is

$$r_i = \frac{18.8 \text{ mm}}{2} = 9.4 \text{ mm}$$

The outside radius is

$$r_o = \frac{37.7 \text{ mm}}{2} = 18.85 \text{ mm}$$

From Table 53.2, the maximum stress developed in the cylinder is the circumferential stress at the inside radius.

$$\sigma_{c,i} = \frac{-2r_o^2 p}{r_o^2 - r_i^2} = \frac{(-2)(18.85 \text{ mm})^2(2.8 \text{ MPa})}{(18.85 \text{ mm})^2 - (9.4 \text{ mm})^2}$$

$$= \boxed{-7.45 \text{ MPa (C)} \quad (-7.5 \text{ MPa})}$$

The answer is (D).

7. *Customary U.S. Solution*

The inside diameter of the shell is

$$d_i = d_o - 2t = 16 \text{ in} - (2)(0.10 \text{ in}) = 15.8 \text{ in}$$

The polar moment of inertia is

$$J = \left(\frac{\pi}{32}\right)(d_o^4 - d_i^4) = \left(\frac{\pi}{32}\right)\left((16 \text{ in})^4 - (15.8 \text{ in})^4\right)$$

$$= 315.7 \text{ in}^4$$

The shear stress is

$$\tau = \frac{Tc}{J} = \frac{(400{,}000 \text{ in-lbf})(8 \text{ in})}{315.7 \text{ in}^4} = 10{,}136 \text{ lbf/in}^2$$

(b) The maximum shear stress is

$$\tau_{\text{max}} = \sqrt{\left(\frac{\sigma}{2}\right)^2 + \tau^2}$$

$$= \sqrt{\left(\frac{40{,}000 \ \frac{\text{lbf}}{\text{in}^2}}{2}\right)^2 + \left(10{,}136 \ \frac{\text{lbf}}{\text{in}^2}\right)^2}$$

$$= \boxed{22{,}422 \text{ lbf/in}^2}$$

(a) The principal stresses are

$$\sigma_1 = \frac{\sigma}{2} + \tau_{\text{max}} = \frac{40{,}000 \ \frac{\text{lbf}}{\text{in}^2}}{2} + 22{,}422 \ \frac{\text{lbf}}{\text{in}^2}$$

$$= \boxed{42{,}422 \text{ lbf/in}^2}$$

$$\sigma_2 = \frac{\sigma}{2} - \tau_{\text{max}} = \frac{40{,}000 \ \frac{\text{lbf}}{\text{in}^2}}{2} - 22{,}422 \ \frac{\text{lbf}}{\text{in}^2}$$

$$= \boxed{-2422 \text{ lbf/in}^2}$$

SI Solution

The inside diameter of the shell is

$$d_i = d_o - 2t = 406 \text{ mm} - (2)(2.54 \text{ mm}) = 400.92 \text{ mm}$$

The polar moment of inertia is

$$\begin{aligned} J &= \left(\frac{\pi}{32}\right)(d_o^4 - d_i^4) \\ &= \frac{\left(\frac{\pi}{32}\right)\left((406 \text{ mm})^4 - (400.92 \text{ mm})^4\right)}{\left(1000 \ \frac{\text{mm}}{\text{m}}\right)^4} \\ &= 0.000131 \text{ m}^4 \end{aligned}$$

The shear stress is

$$\begin{aligned} \tau &= \frac{Tc}{J} = \frac{(45 \text{ kN·m})(0.203 \text{ m})\left(1000 \ \frac{\text{N}}{\text{kN}}\right)}{0.000131 \text{ m}^4} \\ &= \boxed{69.73 \times 10^6 \text{ Pa} \quad (69.73 \text{ MPa})} \end{aligned}$$

(b) The maximum shear stress is

$$\begin{aligned} \tau_{\max} &= \sqrt{\frac{\sigma^2}{2} + \tau^2} \\ &= \sqrt{\left(\frac{280 \text{ MPa}}{2}\right)^2 + (69.73 \text{ MPa})^2} \\ &= \boxed{156.4 \text{ MPa}} \end{aligned}$$

(a) The principal stresses are

$$\sigma_1 = \frac{\sigma}{2} + \tau_{\max} = \frac{280 \text{ MPa}}{2} + 156.4 \text{ MPa} = \boxed{296.4 \text{ MPa}}$$

$$\sigma_2 = \frac{\sigma}{2} - \tau_{\max} = \frac{280 \text{ MPa}}{2} - 156.4 \text{ MPa} = \boxed{-16.4 \text{ MPa}}$$

8. *Customary U.S. Solution*

For the jacket,

$$\begin{aligned} r_o &= \frac{12 \text{ in}}{2} = 6 \text{ in} \\ r_i &= \frac{7.75 \text{ in}}{2} = 3.875 \text{ in} \end{aligned}$$

From Table 53.2, the highest stress is the circumferential stress at the inside surface.

$$\sigma_{c,i} = \frac{(r_o^2 + r_i^2)p}{r_o^2 - r_i^2}$$

Note that the maximum allowable stress is given for the interface, not for the tube. The stress is higher than 18,000 lbf/in^2 at the inside of the tube. Note, also, that radial and circumferential stresses are principal stresses, and do not combine.

Set $\sigma_{c,i}$ equal to the maximum allowable stress.

$$18{,}000 \ \frac{\text{lbf}}{\text{in}^2} = \frac{\left((6 \text{ in})^2 + (3.875 \text{ in})^2\right)p}{(6 \text{ in})^2 - (3.875 \text{ in})^2}$$

$$p = 7404 \text{ lbf/in}^2$$

The radial stress at the inside surface is

$$\sigma_{r,i} = -p = -7404 \text{ lbf/in}^2$$

From Eq. 53.24, the diametral strain is

$$\begin{aligned} \epsilon &= \frac{\sigma_c - \nu(\sigma_r + \sigma_l)}{E} \\ &= \frac{18{,}000 \ \frac{\text{lbf}}{\text{in}^2} - (0.3)\left(-7404 \ \frac{\text{lbf}}{\text{in}^2} + 0 \text{ MPa}\right)}{3 \times 10^7 \ \frac{\text{lbf}}{\text{in}^2}} \\ &= 6.7404 \times 10^{-4} \end{aligned}$$

The diametral deflection is

$$\Delta d = \epsilon d_i = (6.7404 \times 10^{-4})(7.75 \text{ in}) = 0.00522 \text{ in}$$

For the tube,

$$\begin{aligned} r_o &= \frac{7.75 \text{ in}}{2} = 3.875 \text{ in} \\ r_i &= \frac{4.7 \text{ in}}{2} = 2.35 \text{ in} \end{aligned}$$

The external pressure acting on the tube is the same as the radial stress at the inside surface of the jacket. Thus, $p = 7404$ lbf/in^2.

From Table 53.2, the circumferential stress at the outside surface is

$$\begin{aligned} \sigma_{c,o} &= \frac{-(r_o^2 + r_i^2)p}{r_o^2 + r_i^2} \\ &= \frac{-\left((3.875 \text{ in})^2 + (2.35 \text{ in})^2\right)\left(7404 \ \frac{\text{lbf}}{\text{in}^2}\right)}{(3.875 \text{ in})^2 - (2.35 \text{ in})^2} \\ &= -16{,}018 \text{ lbf/in}^2 \end{aligned}$$

The radial stress at the outside surface is

$$\sigma_{r,o} = -p = -7404 \text{ lbf/in}^2$$

From Eq. 53.24, the diametral strain is

$$\begin{aligned}\epsilon &= \frac{\sigma_c - \nu(\sigma_r + \sigma_l)}{E} \\ &= \frac{-16{,}018\ \frac{\text{lbf}}{\text{in}^2} - (0.3)\left(-7404\ \frac{\text{lbf}}{\text{in}^2} + 0\ \frac{\text{lbf}}{\text{in}^2}\right)}{3 \times 10^7\ \frac{\text{lbf}}{\text{in}^2}} \\ &= -4.599 \times 10^{-4}\end{aligned}$$

The diametral deflection is

$$\Delta d = \epsilon d_o = (-4.599 \times 10^{-4})(7.75\ \text{in}) = -0.00356\ \text{in}$$

(a) The diametral interference is the sum of the magnitudes of the two deflections. From Eq. 53.26,

$$\begin{aligned}I_{\text{diametral}} &= |\Delta d_{\text{jacket}}| + |\Delta d_{\text{tube}}| \\ &= |0.00522\ \text{in}| + |-0.00356\ \text{in}| \\ &= \boxed{0.00878\ \text{in}\quad (0.0088\ \text{in})}\end{aligned}$$

The answer is (B).

(b) The maximum circumferential stress in the jacket is at the inside surface.

$$\sigma_{\max} = \sigma_{c,i} = \boxed{18{,}000\ \text{lbf/in}^2}$$

The answer is (D).

(c) The minimum circumferential stress in the tube is at the outside surface.

$$\sigma_{\min} = \sigma_{c,o} = \boxed{-16{,}018\ \text{lbf/in}^2\quad (-16{,}000\ \text{lbf/in}^2)}$$

The answer is (B).

SI Solution

For the jacket,

$$\begin{aligned}r_o &= \frac{300\ \text{mm}}{2} = 150\ \text{mm} \\ r_i &= \frac{197\ \text{mm}}{2} = 98.5\ \text{mm}\end{aligned}$$

From Table 53.2, the highest stress is the circumferential stress at the inside surface.

$$\sigma_{c,i} = \frac{(r_o^2 + r_i^2)p}{r_o^2 - r_i^2}$$

Note that the maximum allowable stress is given for the interface, not for the tube. The stress is higher than 124 MPa at the inside of the tube. Note, also, that radial and circumferential stresses are principal stresses, and do not combine.

Set $\sigma_{c,i}$ equal to the maximum allowable stress.

$$\begin{aligned}124\ \text{MPa} &= \frac{\left((150\ \text{mm})^2 + (98.5\ \text{mm})^2\right)p}{(150\ \text{mm})^2 - (98.5\ \text{mm})^2} \\ p &= 49.28\ \text{MPa}\end{aligned}$$

The radial stress at the inside surface is

$$\sigma_{r,i} = -p = -49.28\ \text{MPa}$$

From Eq. 53.24, the diametral strain is

$$\begin{aligned}\epsilon &= \frac{\sigma_c - \nu(\sigma_r + \sigma_l)}{E} \\ &= \frac{124\ \text{MPa} - (0.3)(-49.28\ \text{MPa} + 0\ \text{MPa})}{(204\ \text{GPa})\left(1000\ \frac{\text{MPa}}{\text{GPa}}\right)} \\ &= 6.803 \times 10^{-4}\end{aligned}$$

The diametral deflection is

$$\Delta d = \epsilon d_i = (6.803 \times 10^{-4})(197\ \text{mm}) = 0.1340\ \text{mm}$$

For the tube,

$$\begin{aligned}r_o &= \frac{197\ \text{mm}}{2} = 98.5\ \text{mm} \\ r_i &= \frac{119\ \text{mm}}{2} = 59.5\ \text{mm}\end{aligned}$$

The external pressure acting on the tube is the same as the radial stress at the inside surface of the jacket. Thus, $p = 49.28$ MPa.

From Table 53.2, the circumferential stress at the outside surface is

$$\begin{aligned}\sigma_{c,o} &= \frac{-(r_o^2 + r_i^2)p}{r_o^2 + r_i^2} \\ &= \frac{-\left((98.5\ \text{mm})^2 + (59.5\ \text{mm})^2\right)(49.28\ \text{MPa})}{(98.5\ \text{m})^2 - (59.5\ \text{m})^2} \\ &= -105.91\ \text{MPa}\end{aligned}$$

The radial stress at the outside surface is

$$\sigma_{r,o} = -p = -49.28\ \text{MPa}$$

From Eq. 53.24, the diametral strain is

$$\begin{aligned}\epsilon &= \frac{\sigma_c - \nu(\sigma_r + \sigma_l)}{E} \\ &= \frac{-105.91\ \text{MPa} - (0.3)(-49.28\ \text{MPa} - 0\ \text{MPa})}{(204\ \text{GPa})\left(1000\ \frac{\text{MPa}}{\text{GPa}}\right)} \\ &= -4.467 \times 10^{-4}\end{aligned}$$

The diametral deflection is

$$\Delta d = \epsilon d_o = (-4.467\times 10^{-4})(197\text{ mm}) = -0.0880\text{ mm}$$

(a) The diametral interference is the sum of the magnitudes of the two deflections. From Eq. 53.26,

$$\begin{aligned} I_{\text{diametral}} &= |\Delta d_{\text{jacket}}| + |\Delta d_{\text{tube}}| \\ &= |0.1340\text{ mm}| + |-0.0880\text{ mm}| \\ &= \boxed{0.222\text{ mm}\quad(0.22\text{ mm})} \end{aligned}$$

The answer is (B).

(b) The maximum circumferential stress in the jacket is at the inside surface.

$$\sigma_{\max} = \sigma_{c,i} = \boxed{124\text{ MPa}}$$

The answer is (D).

(c) The minimum circumferential stress in the tube is at the outside surface.

$$\sigma_{\min} = \sigma_{c,o} = \boxed{-105.91\text{ MPa}\quad(-106\text{ MPa})}$$

The answer is (B).

9. *Customary U.S. Solution*

For the inner cylinder,

$$r_o = \frac{4.7\text{ in} - (2)(0.079\text{ in})}{2} = 2.27\text{ in}$$

Give all of the interference to the outside of the inner cylinder, since it is easier to cut the outside diameter than to ream an inside diameter.

$$r_i = 2.27\text{ in} - 0.12\text{ in} = 2.15\text{ in}$$

From Table 53.2, the circumferential stress at the outside surface due to external pressure, p, is

$$\begin{aligned} \sigma_{c,o} &= \frac{-(r_o^2 + r_i^2)p}{r_o^2 + r_i^2} \\ &= \frac{-\left((2.27\text{ in})^2 + (2.15\text{ in})^2\right)p}{(2.27\text{ in})^2 - (2.15\text{ in})^2} \\ &= -18.43p\text{ in lbf/in}^2 \end{aligned}$$

The radial stress in lbf/in^2 is $\sigma_{r,o} = -p$.

From Eq. 53.24, the diametral deflection is

$$\begin{aligned} \Delta d_{\text{inner}} &= \left(\frac{d}{E}\right)\left(\sigma_c - \nu(\sigma_r + \sigma_l)\right) \\ &= \frac{(4.54\text{ in})\left(-18.43p - (0.3)(-p+0)\right)}{3\times 10^7\ \frac{\text{lbf}}{\text{in}^2}} \\ &= -27.437\times 10^{-7}p\quad[\text{in in}] \end{aligned}$$

For the outer cylinder,

$$\begin{aligned} r_o &= \frac{4.7\text{ in}}{2} = 2.35\text{ in} \\ r_i &= 2.35\text{ in} - 0.079\text{ in} = 2.27\text{ in} \end{aligned}$$

From Table 53.2, the circumferential stress at the inside surface due to internal pressure, p, is

$$\begin{aligned} \sigma_{c,i} &= \frac{(r_o^2 + r_i^2)p}{r_o^2 - r_i^2} = \frac{\left((2.35\text{ in})^2 + (2.27\text{ in})^2\right)p}{(2.35\text{ in})^2 - (2.27\text{ in})^2} \\ &= 28.88p\quad[\text{in lbf/in}^2] \end{aligned}$$

The radial stress in lbf/in^2 is $\sigma_{r,i} = -p$.

From Eq. 53.24, the diametral deflection is

$$\begin{aligned} \Delta d_{\text{outer}} &= \left(\frac{d}{E}\right)\left(\sigma_e - \nu(\sigma_r + \sigma_o)\right) \\ &= \frac{(4.54\text{ in})\left(28.88p - (0.3)(-p+0)\right)}{3\times 10^7\ \frac{\text{lbf}}{\text{in}^2}} \\ &= 44.159\times 10^{-7}p\quad[\text{in in}] \end{aligned}$$

From Eq. 53.26, the diametral interference is

$$\begin{aligned} I_{\text{diametral}} &= |\Delta d_{\text{inner}}| + |\Delta d_{\text{outer}}| \\ 0.012\text{ in} &= |27.437\times 10^{-7}p| + |44.159\times 10^{-7}p| \\ p &= 1676\text{ lbf/in}^2 \end{aligned}$$

The circumferential stress at the interface between the two cylinders is as follows.

(a) For the inner cylinder,

$$\begin{aligned} \sigma_{c,o} &= -18.43p = (-18.43)\left(1676\ \frac{\text{lbf}}{\text{in}^2}\right) \\ &= \boxed{-30{,}889\text{ lbf/in}^2\quad(-31{,}000\text{ lbf/in}^2)} \end{aligned}$$

The answer is (C).

(b) For the outer cylinder,

$$\begin{aligned} \sigma_{c,i} &= 28.88p = (28.88)\left(1676\ \frac{\text{lbf}}{\text{in}^2}\right) \\ &= \boxed{48{,}403\text{ lbf/in}^2\quad(48{,}000\text{ lbf/in}^2)} \end{aligned}$$

The answer is (A).

SI Solution

For the inner cylinder,

$$r_o = \frac{120 \text{ mm} - (2)(2 \text{ mm})}{2} = 58 \text{ mm}$$

Give all of the interference to the outside of the inner cylinder, since it is easier to cut the outside diameter than to ream an inside diameter.

$$r_i = 58 \text{ mm} - 3 \text{ mm} = 55 \text{ mm}$$

From Table 53.2, the circumferential stress at the outside surface due to external pressure, p, is

$$\begin{aligned} \sigma_{c,o} &= \frac{-(r_o^2 + r_i^2)p}{r_o^2 - r_i^2} \\ &= \frac{-\left((58 \text{ mm})^2 + (55 \text{ mm})^2\right)p}{(58 \text{ mm})^2 - (55 \text{ mm})^2} \\ &= -18.85p \quad \text{[in MPa]} \end{aligned}$$

The radial stress in MPa is $\sigma_{r,o} = -p$.

From Eq. 53.24, the diametral deflection is

$$\begin{aligned} \Delta d_{\text{inner}} &= \left(\frac{d}{E}\right)\left(\sigma_c - \nu(\sigma_r + \sigma_l)\right) \\ &= \frac{(116 \text{ mm})\left(-18.85p - (0.3)(-p + 0)\right)}{(207 \text{ GPa})\left(1000 \frac{\text{MPa}}{\text{GPa}}\right)} \\ &= -1.040 \times 10^{-2}p \quad \text{[in mm]} \end{aligned}$$

For the outer cylinder,

$$\begin{aligned} r_o &= \frac{120 \text{ mm}}{2} = 60 \text{ mm} \\ r_i &= 60 \text{ mm} - 2 \text{ mm} = 58 \text{ mm} \end{aligned}$$

From Table 53.2, the circumferential stress at the inside surface due to internal pressure, p, is

$$\begin{aligned} \sigma_{c,i} &= \frac{(r_o^2 + r_i^2)p}{r_o^2 - r_i^2} = \frac{\left((60 \text{ mm})^2 + (58 \text{ mm})^2\right)p}{(60 \text{ mm})^2 - (58 \text{ mm})^2} \\ &= 29.51p \quad \text{[in MPa]} \end{aligned}$$

The radial stress in MPa is $\sigma_{r,i} = -p$.

From Eq. 53.24, the diametral deflection is

$$\begin{aligned} \Delta d_{\text{outer}} &= \left(\frac{d}{E}\right)\left(\sigma_e - \nu(\sigma_r - \sigma_l)\right) \\ &= \frac{(116 \text{ mm})\left(29.51p - (0.3)(-p + 0)\right)}{(207 \text{ GPa})\left(1000 \frac{\text{MPa}}{\text{GPa}}\right)} \\ &= 1.671 \times 10^{-2}p \quad \text{[in mm]} \end{aligned}$$

From Eq. 53.26, the diametral interference is

$$\begin{aligned} I_{\text{diametral}} &= |\Delta d_{\text{inner}}| + |\Delta d_{\text{outer}}| \\ 0.3 \text{ mm} &= |1.040 \times 10^{-2}p| + |1.671 \times 10^{-2}p| \\ p &= 11.07 \text{ MPa} \end{aligned}$$

The circumferential stress at the interface between the two cylinders is as follows.

(a) For the inner cylinder,

$$\begin{aligned} \sigma_{c,o} &= -18.85p = (-18.85)(11.07 \text{ MPa}) \\ &= \boxed{-209 \text{ MPa} \quad (-210 \text{ MPa})} \end{aligned}$$

The answer is (C).

(b) For the outer cylinder,

$$\begin{aligned} \sigma_{c,i} &= 29.51p = (29.51)(11.07 \text{ MPa}) \\ &= \boxed{327 \text{ MPa} \quad (330 \text{ MPa})} \end{aligned}$$

The answer is (A).

10. *Customary U.S. Solution*

Since the plates are rigid, they do not deform.

The area of the bolt body is

$$A_1 = \frac{\pi d^2}{4} = \frac{\pi(0.75 \text{ in})^2}{4} = 0.4418 \text{ in}^2$$

The tensile force in the bolt is

$$\begin{aligned} F &= \sigma_2 A_2 = \left(40{,}000 \ \frac{\text{lbf}}{\text{in}^2}\right)(0.4418 \text{ in}^2) \\ &= 17{,}672 \text{ lbf} \end{aligned}$$

From Table 53.5, the stress area (in the threaded section) for a $^3\!/_4$-16 UNF bolt is $A_2 = 0.373 \text{ in}^2$.

The lengths of the unthreaded and threaded sections are

$$\begin{aligned} L_1 &= 2 \text{ in} + 2 \text{ in} - 0.75 \text{ in} = 3.25 \text{ in} \\ L_2 &= 0.75 \text{ in} + \left(\frac{3 \text{ threads}}{16 \ \frac{\text{threads}}{\text{in}}}\right) = 0.9375 \text{ in} \end{aligned}$$

The stiffnesses of the unthreaded and threaded sections are

$$k_1 = \frac{A_1 E}{L_1} = (0.4418 \text{ in}^2)\left(\frac{2.9 \times 10^7 \ \frac{\text{lbf}}{\text{in}^2}}{3.25 \text{ in}}\right)$$

$$= 3.94 \times 10^6 \text{ lbf/in}$$

$$k_2 = \frac{A_2 E}{L_2} = (0.373 \text{ in}^2)\left(\frac{2.9 \times 10^7 \ \frac{\text{lbf}}{\text{in}^2}}{0.9375 \text{ in}}\right)$$

$$= 11.5 \times 10^6 \text{ lbf/in}$$

The two parts of the bolt act like two springs in series to clamp the plates together. From Eq. 54.5, the equivalent spring constant is

$$\frac{1}{k_{\text{eq}}} = \frac{1}{k_1} + \frac{1}{k_2}$$

$$k_{\text{eq}} = \frac{k_1 k_2}{k_1 + k_2}$$

$$= \frac{\left(3.94 \times 10^6 \ \frac{\text{lbf}}{\text{in}}\right)\left(11.5 \times 10^6 \ \frac{\text{lbf}}{\text{in}}\right)}{3.94 \times 10^6 \ \frac{\text{lbf}}{\text{in}} + 11.5 \times 10^6 \ \frac{\text{lbf}}{\text{in}}}$$

$$= 2.93 \times 10^6 \text{ lbf/in}$$

The total elongation is

$$\delta = \frac{F}{k_{\text{eq}}} = \frac{17{,}672 \text{ lbf}}{2.93 \times 10^6 \ \frac{\text{lbf}}{\text{in}}}$$

$$= \boxed{0.006 \text{ in}}$$

The answer is (A).

SI Solution

Since the plates are rigid, they do not deform.

The area of the bolt body is

$$A_1 = \frac{\pi d^2}{4} = \frac{\pi (0.75 \text{ in})^2 \left(25.4 \ \frac{\text{mm}}{\text{in}}\right)^2}{4} = 285.0 \text{ mm}^2$$

The tensile force in the bolt is

$$F = \sigma_2 A_2 = \frac{(280 \text{ MPa})\left(10^6 \ \frac{\text{Pa}}{\text{MPa}}\right)(285.0 \text{ mm}^2)}{\left(1000 \ \frac{\text{mm}}{\text{m}}\right)^2}$$

$$= 79\,800 \text{ N}$$

From Table 53.5, the stress area (in the threaded section) for a 3/4-16 UNF bolt is $A_2 = 0.373 \text{ in}^2$.

$$A_2 = (0.373 \text{ in}^2)\left(25.4 \ \frac{\text{mm}}{\text{in}}\right)^2 = 240.6 \text{ mm}^2$$

The lengths of the unthreaded and threaded sections are

$$L_1 = 50 \text{ mm} + 50 \text{ mm} - 19 \text{ mm} = 81 \text{ mm}$$

$$L_2 = 19 \text{ mm} + \left(\frac{3 \text{ threads}}{16 \ \frac{\text{threads}}{\text{in}}}\right)\left(25.4 \ \frac{\text{mm}}{\text{in}}\right) = 23.76 \text{ mm}$$

The stiffnesses of the unthreaded and threaded sections are

$$k_1 = \frac{A_1 E}{L_1} = \frac{(285.0 \text{ mm}^2)(200 \text{ GPa})\left(10^9 \ \frac{\text{Pa}}{\text{GPa}}\right)}{(81 \text{ mm})\left(1000 \ \frac{\text{mm}}{\text{m}}\right)^2}$$

$$= 7.04 \times 10^5 \text{ N/mm}$$

$$k_2 = \frac{A_2 E}{L_2} = \frac{(240.6 \text{ mm}^2)(200 \text{ GPa})\left(10^9 \ \frac{\text{Pa}}{\text{GPa}}\right)}{(23.76 \text{ mm})\left(1000 \ \frac{\text{mm}}{\text{m}}\right)^2}$$

$$= 20.25 \times 10^5 \text{ N/mm}$$

The two parts of the bolt act like two springs in series to clamp the plates together. From Eq. 54.5, the equivalent spring constant is

$$\frac{1}{k_{\text{eq}}} = \frac{1}{k_1} + \frac{1}{k_2}$$

$$k_{\text{eq}} = \frac{k_1 k_2}{k_1 + k_2}$$

$$= \frac{\left(7.04 \times 10^5 \ \frac{\text{N}}{\text{mm}}\right)\left(20.25 \times 10^5 \ \frac{\text{N}}{\text{mm}}\right)}{7.04 \times 10^5 \ \frac{\text{N}}{\text{mm}} + 20.25 \times 10^5 \ \frac{\text{N}}{\text{mm}}}$$

$$= 5.22 \times 10^5 \text{ N/mm}$$

The total elongation is

$$\delta = \frac{F}{k_{\text{eq}}} = \frac{79\,800 \text{ N}}{5.22 \times 10^5 \ \frac{\text{N}}{\text{mm}}} = \boxed{0.15 \text{ mm}}$$

The answer is (A).

11. *Customary U.S. Solution*

For the cast-iron hub,

$$r_o = \frac{12 \text{ in}}{2} = 6 \text{ in}$$

$$r_i = \frac{6 \text{ in}}{2} = 3 \text{ in}$$

From Table 53.2, the circumferential stress at the inside radius due to internal pressure, p, is

$$\sigma_{c,i} = \frac{(r_o^2 + r_i^2)p}{r_o^2 - r_i^2}$$

The ultimate strength of cast iron is $S_{ut} = 30{,}000 \text{ lbf/in}^2$. For cast iron, nonlinearity begins at approximately $\frac{1}{6}S_{ut}$.

$$\sigma_{c,i} = \frac{S_{ut}}{\text{FS}} = \frac{30{,}000 \ \frac{\text{lbf}}{\text{in}^2}}{6} = 5000 \text{ lbf/in}^2$$

The $\sigma_{c,i}$ relationship is

$$5000 \ \frac{\text{lbf}}{\text{in}^2} = \frac{\left((6 \text{ in})^2 + (3 \text{ in})^2\right)p}{(6 \text{ in})^2 - (3 \text{ in})^2}$$

$$p = 3000 \text{ lbf/in}^2$$

The radial stress is

$$\sigma_{r,i} = -p = -3000 \text{ lbf/in}^2$$

From Eq. 53.24, the radial deflection at the inside of the cast iron hub is

$$\begin{aligned}\Delta r_{\text{inner}} &= \frac{r\left(\sigma_c - v(\sigma_r + \sigma_l)\right)}{E} \\ &= \frac{(3 \text{ in})\left(5000 \ \frac{\text{lbf}}{\text{in}^2} - (0.27) \times \left(-3000 \ \frac{\text{lbf}}{\text{in}^2} + 0 \ \frac{\text{lbf}}{\text{in}^2}\right)\right)}{1.45 \times 10^7 \ \frac{\text{lbf}}{\text{in}^2}} \\ &= 1.202 \times 10^{-3} \text{ in}\end{aligned}$$

For the steel shaft,

$$r_o = \frac{6 \text{ in}}{2} = 3 \text{ in}$$

$$r_i = 0$$

From Table 53.2, the circumferential stress at the outside radius due to external pressure, p, is

$$\begin{aligned}\sigma_{c,o} &= \frac{-(r_o^2 + r_i^2)p}{r_o^2 - r_i^2} \\ &= \frac{-\left((3 \text{ in})^2 + (0 \text{ in})^2\right)\left(3000 \ \frac{\text{lbf}}{\text{in}^2}\right)}{(3 \text{ in})^2 - (0 \text{ in})^2} \\ &= -3000 \text{ lbf/in}^2\end{aligned}$$

The radial stress is

$$\sigma_{r,o} = -p = -3000 \text{ lbf/in}^2$$

From Eq. 53.24, the radial deflection at the outside of the steel shaft is

$$\begin{aligned}\Delta r_{\text{outer}} &= \frac{r\left(\sigma_c - v(\sigma_r + \sigma_l)\right)}{E} \\ &= \frac{(3 \text{ in})\left(-3000 \ \frac{\text{lbf}}{\text{in}^2} - (0.30)\left(-3000 \ \frac{\text{lbf}}{\text{in}^2} + 0 \ \frac{\text{lbf}}{\text{in}^2}\right)\right)}{2.9 \times 10^7 \ \frac{\text{lbf}}{\text{in}^2}} \\ &= 2.172 \times 10^{-4} \text{ in}\end{aligned}$$

From Eq. 53.26, the radial interference is

$$\begin{aligned}I_{\text{radial}} &= |\Delta r_{\text{outer}}| + |\Delta r_{\text{inner}}| \\ &= |-2.172 \times 10^{-4} \text{ in}| + |1.202 \times 10^{-3} \text{ in}| \\ &= \boxed{1.419 \times 10^{-3} \text{ in} \quad (0.0014 \text{ in})}\end{aligned}$$

The answer is (A).

SI Solution

For the cast-iron hub,

$$r_o = \frac{300 \text{ mm}}{2} = 150 \text{ mm}$$

$$r_i = \frac{150 \text{ mm}}{2} = 75 \text{ mm}$$

From Table 53.2, the circumferential stress at the inside radius due to internal pressure, p, is

$$\sigma_{c,i} = \frac{(r_o^2 + r_i^2)p}{r_o^2 - r_i^2}$$

The ultimate strength of cast iron is $S_{ut} = 210$ MPa. For cast iron, nonlinearity begins at approximately $1/6 S_{ut}$.

$$\sigma_{c,i} = \frac{S_{ut}}{FS} = \frac{210\ \text{MPa}}{6} = 35\ \text{MPa}$$

The $\sigma_{c,i}$ relationship is

$$35\ \text{MPa} = \frac{\left((150\ \text{mm})^2 + (75\ \text{mm})^2\right)p}{(150\ \text{mm})^2 - (75\ \text{mm})^2}$$
$$p = 21\ \text{MPa}$$

The radial stress is

$$\sigma_{r,i} = -p = -21\ \text{MPa}$$

From Eq. 53.24, the radial deflection at the inside of the cast iron hub is

$$\begin{aligned}\Delta r_{\text{inner}} &= \frac{r\left(\sigma_c - \nu(\sigma_r + \sigma_l)\right)}{E}\\ &= \frac{(75\ \text{mm}) \times \left(35\ \text{MPa} - (0.27)(-21\ \text{MPa} + 0\ \text{MPa})\right)}{(100\ \text{GPa})\left(1000\ \frac{\text{MPa}}{\text{GPa}}\right)}\\ &= 0.03050\ \text{mm}\end{aligned}$$

For the steel shaft,

$$r_o = \frac{150\ \text{mm}}{2} = 75\ \text{mm}$$
$$r_i = 0$$

From Table 53.2, the circumferential stress at the outside radius due to external pressure, p, is

$$\begin{aligned}\sigma_{c,o} &= \frac{-(r_o^2 + r_i^2)p}{r_o^2 - r_i^2}\\ &= \frac{-\left((75\ \text{mm})^2 + (0\ \text{mm})^2\right)(21\ \text{MPa})}{(75\ \text{mm})^2 - (0\ \text{mm})^2}\\ &= -21\ \text{MPa}\end{aligned}$$

The radial stress is

$$\sigma_{r,o} = -p = -21\ \text{MPa}$$

From Eq. 53.24, the radial deflection at the outside of the steel shaft is

$$\begin{aligned}\Delta r_{\text{outer}} &= \frac{r\left(\sigma_c - \nu(\sigma_r + \sigma_l)\right)}{E}\\ &= \frac{(75\ \text{mm}) \times \left(\begin{array}{l}-21\ \text{MPa}\\ \quad - (0.30)(-21\ \text{MPa} + 0\ \text{MPa})\end{array}\right)}{(200\ \text{GPa})\left(1000\ \frac{\text{MPa}}{\text{GPa}}\right)}\\ &= -0.00551\ \text{mm}\end{aligned}$$

From Eq. 53.26, the radial interference is

$$\begin{aligned}I_{\text{radial}} &= |\Delta r_{\text{outer}}| + |\Delta r_{\text{inner}}|\\ &= |-0.00551\ \text{mm}| + |0.03050\ \text{mm}|\\ &= \boxed{0.03601\ \text{mm} \quad (0.036\ \text{mm})}\end{aligned}$$

The answer is (A).

12. *Customary U.S. Solution*

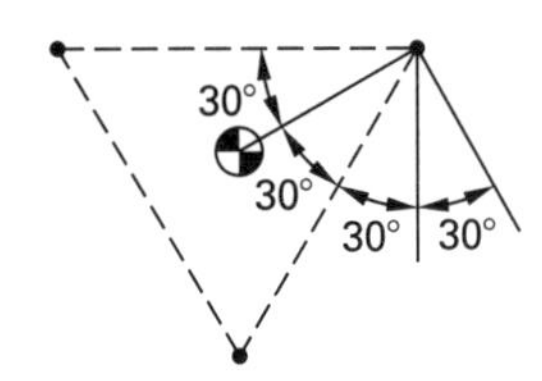

Find the properties of the bolt area. For an equilateral triangular layout, the distance from the centroid of the bolt group to the center of each bolt is

$$r = \left(\tfrac{2}{3}\right)(9\ \text{in}) = 6\ \text{in}$$

The area of each bolt is

$$A = \frac{\pi d^2}{4} = \frac{\pi(0.75\ \text{in})^2}{4} = 0.442\ \text{in}^2$$

The polar moment of inertia of a bolt about the centroid is

$$J = Ar^2 = (0.442\ \text{in}^2)(6\ \text{in})^2 = 15.91\ \text{in}^4$$

The vertical shear load at each bolt is

$$F_v = \frac{F}{3} = 0.333F \quad [\text{in lbf}]$$

The moment applied to each bolt is

$$M = \frac{(20\ \text{in})F}{3} = 6.67F \quad [\text{in in-lbf}]$$

The vertical shear stress in each bolt is

$$\tau_v = \frac{F_v}{A} = \frac{0.333F}{0.442 \text{ in}^2} = 0.753F \quad [\text{in lbf/in}^2]$$

The torsional shear stress in each bolt is

$$\tau = \frac{Mr}{J} = \frac{(6.67F)(6 \text{ in})}{15.91 \text{ in}^4} = 2.52F \quad [\text{in lbf/in}^2]$$

The most highly stressed bolt is the right-most bolt. The shear stress configuration is

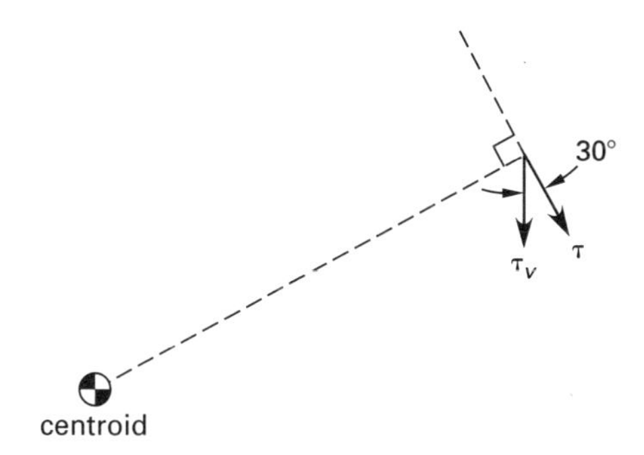

The stresses are combined to find the maximum stress.

$$\tau_{\max} = \sqrt{(\tau \sin 30°)^2 + (\tau_v + \tau \cos 30°)^2}$$

$$15{,}000 \ \frac{\text{lbf}}{\text{in}^2} = \sqrt{\begin{array}{l}((2.52F)(\sin 30°))^2 \\ \quad + (0.753F + (2.52F)(\cos 30°))^2\end{array}}$$

$$= 3.19F$$

$$F = \boxed{4702 \text{ lbf} \quad (4700 \text{ lbf})}$$

The answer is (D).

SI Solution

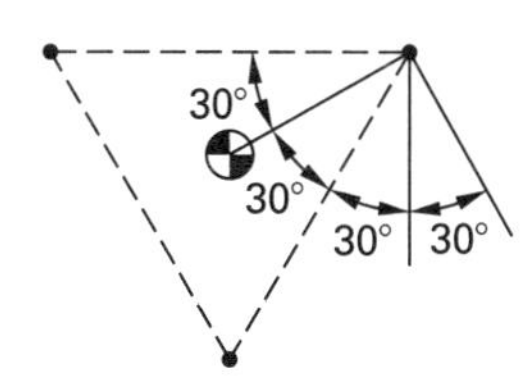

Find the properties of the bolt area. For an equilateral triangular layout, the distance from the centroid of the bolt group to each bolt is

$$r = \left(\tfrac{2}{3}\right)(230 \text{ mm}) = 153 \text{ mm}$$

The area of each bolt is

$$A = \frac{\pi d^2}{4} = \frac{\pi (19 \text{ mm})^2}{(4)\left(1000 \ \frac{\text{mm}}{\text{m}}\right)^2} = 2.835 \times 10^{-4} \text{ m}^2$$

The polar moment of inertia of a bolt about the centroid is

$$J = Ar^2 = (2.835 \times 10^{-4} \text{ m}^2)(0.153 \text{ m})^2 = 6.636 \times 10^{-6} \text{ m}^4$$

The vertical shear load at each bolt is

$$F_v = \frac{F}{3} = 0.333F \quad [\text{in N}]$$

The moment applied to each bolt is

$$M = \frac{(0.5 \text{ m})F}{3} = 0.167F \quad [\text{in N·m}]$$

The vertical shear stress in each bolt is

$$\tau_v = \frac{F_v}{A} = \frac{0.333F}{2.835 \times 10^{-4} \text{ m}^2} = 1175F \quad [\text{in Pa}]$$

The torsional shear stress in each bolt is

$$\tau = \frac{Mr}{J} = \frac{(0.167F)(0.153 \text{ m})}{6.636 \times 10^{-6} \text{ m}^4} = 3850F \quad [\text{in Pa}]$$

The most highly stressed bolt is the right-most bolt. The shear stress configuration is

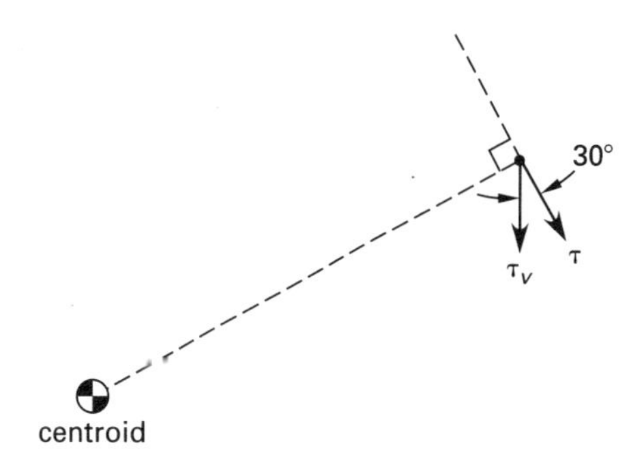

The stresses are combined to find the maximum stress.

$$\tau_{\max} = \sqrt{(\tau \sin 30°)^2 + (\tau_v + \tau \cos 30°)^2}$$

$$(100 \text{ MPa}) \times \left(1000 \ \frac{\text{kPa}}{\text{MPa}}\right) = \sqrt{\begin{array}{l}\left((3850F)(\sin 30°)\right)^2 \\ +\left((1175F) + (3850F)(\cos 30°)\right)^2\end{array}}$$

$$= 4903F \quad [\text{in Pa}]$$

$$F = \boxed{20.4 \text{ kN} \quad (20 \text{ kN})}$$

The answer is (D).

13. *Customary U.S. Solution*

The fillet weld consists of three basic shapes, each with a throat size of t.

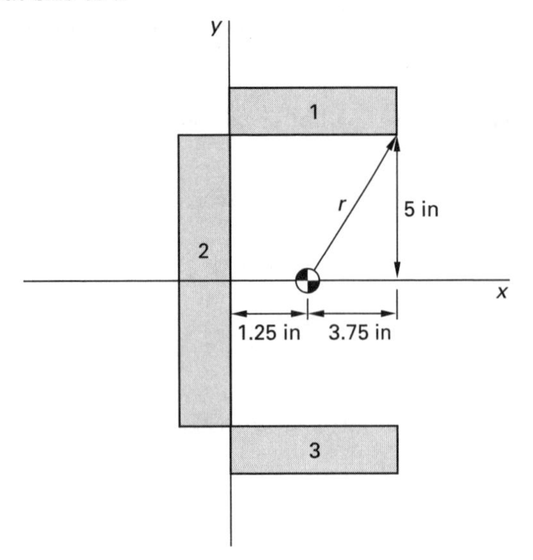

First, find the centroid.

By inspection, $y_c = 0$.

The areas of the basic shapes are

$$A_1 = 5t \quad [\text{in in}^2]$$
$$A_2 = 10t \quad [\text{in in}^2]$$
$$A_3 = 5t \quad [\text{in in}^2]$$

The x-components of the centroids of the basic shapes are

$$x_{c1} = \frac{5 \text{ in}}{2} = 2.5 \text{ in}$$
$$x_{c2} = -\frac{t}{2} = 0 \text{ in} \quad [\text{for small } t]$$
$$x_{c3} = \frac{5 \text{ in}}{2} = 2.5 \text{ in}$$
$$\begin{aligned} x_c &= \frac{\sum\limits_i A_i x_{ci}}{\sum A_i} \\ &= \frac{(5t)(2.5 \text{ in}) + (10t)(0 \text{ in}) + (5t)(2.5 \text{ in})}{5t + 10t + 5t} \\ &= 1.25 \text{ in} \end{aligned}$$

Next, find the centroidal moments of inertia.

The moments of inertia of basic shape 1 about its own centroid are

$$I_{cx1} = \frac{bh^3}{12} = \frac{(5 \text{ in})t^3}{12} = 0.417t^3 \quad [\text{in in}^4]$$
$$I_{cy1} = \frac{bh^3}{12} = \frac{t(5 \text{ in})^3}{12} = 10.417t \quad [\text{in in}^4]$$

The moments of inertia of basic shape 2 about its own centroid are

$$I_{cx2} = \frac{bh^3}{12} = \frac{t(10 \text{ in})^3}{12} = 83.333t \quad [\text{in in}^4]$$
$$I_{cy2} = \frac{bh^3}{12} = \frac{(10 \text{ in})t^3}{12} = 0.833t^3 \quad [\text{in in}^4]$$

The moments of inertia of basic shape 3 about its own centroid are the same as for basic shape 1.

$$I_{cx3} = I_{cx1} = 0.417t^3 \quad [\text{in in}^4]$$
$$I_{cy3} = I_{cy1} = 10.417t \quad [\text{in in}^4]$$

From the parallel axis theorem, the moments of inertia of basic shape 1 about the centroidal axis of the section are

$$\begin{aligned} I_{x1} - I_{cx1} + A_1 d_1^2 - 0.417t^3 + (5t)(5 \text{ in})^2 \\ &= 0.417t^3 + 125t \\ I_{y1} = I_{cy1} + A_1 d_1^2 \\ &= 10.417t + 5t\left(\frac{5 \text{ in}}{2} - 1.25 \text{ in}\right)^2 \\ &= 18.23t \quad [\text{in in}^4] \end{aligned}$$

The moments of inertia of basic shape 2 about the centroidal axis of the section are

$$\begin{aligned} I_{x2} &= I_{cx2} + A_2 d_2^2 = 83.333t + (10t)(0 \text{ in}) \\ &= 83.333t \quad [\text{in in}^4] \\ I_{y2} &= I_{cy2} + A_2 d_2^2 = 0.833t^3 + (10t)(1.25 \text{ in})^2 \\ &= 0.833t^3 + 15.625t \end{aligned}$$

The moment of inertia of basic shape 3 about the centroidal axis of the section are the same as for basic shape 1.

$$I_{x3} = I_{x1} = 0.417t^3 + 125t$$
$$I_{y3} = I_{y1} = 18.23t \quad [\text{in in}^4]$$

The total moments of inertia about the centroidal axis of the section are

$$\begin{aligned} I_x &= I_{x1} + I_{x2} + I_{x3} \\ &= 0.417t^3 + 125t + 83.333t + 0.417t^3 + 125t \\ &= 0.834t^3 + 333.33t \\ I_y &= I_{y1} + I_{y2} + I_{y3} \\ &= 18.23t + 0.833t^3 + 15.625t + 18.23t \\ &= 0.833t^3 + 52.09t \end{aligned}$$

Since t will be small and since the coefficient of the t^3 term is smaller than the coefficient of the t term, the higher-order t^3 term may be neglected.

$$I_x = 333.33t \quad [\text{in in}^4]$$
$$I_y = 52.09t \quad [\text{in in}^4]$$

The polar moment of inertia for the section is

$$\begin{aligned} J &= I_x + I_y = 333.33t + 52.09t \\ &= 385.42t \quad [\text{in in}^4] \end{aligned}$$

The maximum shear stress will occur at the right-most point of basic shape 1 since this point is farthest from the centroid of the section.

This distance is

$$\begin{aligned} r &= \sqrt{(5\text{ in} - 1.25\text{ in})^2 + \left(\frac{10\text{ in}}{2}\right)^2} \\ &= 6.25\text{ in} \end{aligned}$$

The applied moment is

$$\begin{aligned} M &= Fe = (10{,}000\text{ lbf})(12\text{ in} + 5\text{ in} - 1.25\text{ in}) \\ &= 157{,}500\text{ in-lbf} \end{aligned}$$

The torsional shear stress is

$$\begin{aligned} \tau &= \frac{Mr}{J} = \frac{(157{,}500\text{ in-lbf})(6.25\text{ in})}{385.42t} \\ &= 2554.0/t \quad [\text{in lbf/in}^2] \end{aligned}$$

The shear stress is perpendicular to the line from the point to the centroid. The x- and y-components are shown in the following figure.

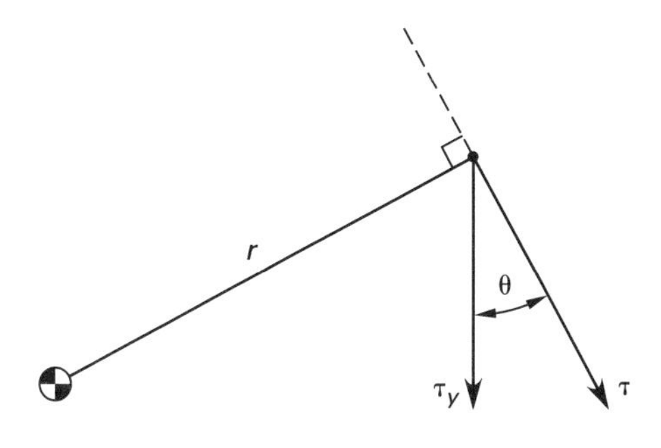

The angle θ is

$$\theta = 90° - \tan^{-1}\left(\frac{5\text{ in} - 1.25\text{ in}}{\frac{10\text{ in}}{2}}\right) = 53.1°$$

$$\begin{aligned} \tau_x &= \tau\sin\theta = \left(\frac{2554.0\ \frac{\text{lbf}}{\text{in}^2}}{t}\right)(\sin 53.1°) \\ &= 2042.4/t \quad [\text{in lbf/in}^2] \end{aligned}$$

$$\begin{aligned} \tau_y &= \tau\cos\theta = \left(\frac{2554.0\ \frac{\text{lbf}}{\text{in}^2}}{t}\right)(\cos 53.1°) \\ &= 1533.5/t \quad [\text{in lbf/in}^2] \end{aligned}$$

The vertical shear stress due to the load is

$$\begin{aligned} \tau_{vy} &= \frac{F}{\sum A_i} = \frac{10{,}000\text{ lbf}}{5t + 10t + 5t} \\ &= 500/t \quad [\text{in lbf/in}^2] \end{aligned}$$

The resultant shear stress is

$$\begin{aligned} \tau &= \sqrt{\tau_x^2 + (\tau_y + \tau_{vy})^2} \\ &= \sqrt{\left(\frac{2042.4}{t}\right)^2 + \left(\frac{1533.5}{t} + \frac{500}{t}\right)^2} \\ &= 2882.1/t \quad [\text{in lbf/in}^2] \end{aligned}$$

For a maximum shear stress of 8000 lbf/in^2,

$$8000\ \frac{\text{lbf}}{\text{in}^2} = \frac{2882.1\ \frac{\text{lbf}}{\text{in}^2}}{t}$$
$$t = 0.360\text{ in}$$

From Eq. 53.48, the required weld size is

$$\begin{aligned} y &= \frac{t}{\frac{\sqrt{2}}{2}} = \frac{0.360\text{ in}}{0.707} \\ &= \boxed{0.509\text{ in} \quad (1/2\text{ in})} \end{aligned}$$

The answer is (B).

SI Solution

The fillet weld consists of three basic shapes, each with a throat size of t.

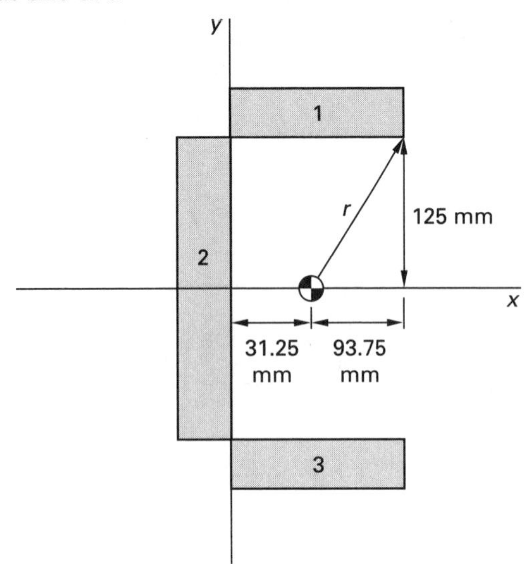

First, find the centroid.

By inspection, $y_c = 0$.

The areas of the basic shapes are

$$A_1 = 125t \quad \text{[in mm}^2\text{]}$$
$$A_2 = 250t \quad \text{[in mm}^2\text{]}$$
$$A_3 = 125t \quad \text{[in mm}^2\text{]}$$

The x-components of the centroids of the basic shapes are

$$x_{c1} = \frac{125 \text{ mm}}{2} = 62.5 \text{ mm}$$
$$x_{c2} = -\frac{t}{2} = 0 \text{ mm} \quad \text{[for small } t\text{]}$$
$$x_{c3} = \frac{125 \text{ mm}}{2} = 62.5 \text{ mm}$$
$$\begin{aligned} x_c &= \frac{\sum_i A_i x_{ci}}{\sum A_i} \\ &= \frac{\begin{array}{c}(125t)(62.5 \text{ mm}) + (250t)(0 \text{ mm}) \\ + (125t)(6.25 \text{ mm})\end{array}}{125t + 250t + 125t} \\ &= 31.25 \text{ mm} \end{aligned}$$

Next, find the centroidal moments of inertia.

The moments of inertia of basic shape 1 about its own centroid are

$$I_{cx1} = \frac{bh^3}{12} = \frac{(125 \text{ mm})t^3}{12} = 10.42t^3 \quad \text{[in mm}^4\text{]}$$
$$I_{cy1} = \frac{bh^3}{12} = \frac{t(125 \text{ mm})^3}{12} = 162\,760t \quad \text{[in mm}^4\text{]}$$

The moments of inertia of basic shape 2 about its own centroid are

$$I_{cx2} = \frac{bh^3}{12} = \frac{t(250 \text{ mm})^3}{12} = 1\,302\,083t \quad \text{[in mm}^4\text{]}$$
$$I_{cy2} = \frac{bh^3}{12} = \frac{(250 \text{ mm})t^3}{12} = 20.83t^3 \quad \text{[in mm}^4\text{]}$$

The moments of inertia of basic shape 3 about its own centroid are the same as for basic shape 1.

$$I_{cx3} = I_{cx1} = 10.42t^3 \quad \text{[in mm}^4\text{]}$$
$$I_{cy3} = I_{cy1} = 162\,760t \quad \text{[in mm}^4\text{]}$$

From the parallel axis theorem, the moments of inertia of basic shape 1 about the centroidal axis of the section are

$$\begin{aligned} I_{x1} &= I_{cx1} + A_1 d_1^2 = 10.42t^3 + (125t)(125 \text{ mm})^2 \\ &= 10.42t^3 + 1\,953\,125t \\ I_{y1} &= I_{cy1} + A_1 d_1^2 \\ &= 162\,760t + (125t)\left(\frac{125 \text{ mm}}{2} - 31.25 \text{ mm}\right)^2 \\ &= 284\,830t \quad \text{[in mm}^4\text{]} \end{aligned}$$

The moments of inertia of basic shape 2 about the centroidal axis of the section are

$$\begin{aligned} I_{x2} &= I_{cx2} + A_2 d_2^2 = 1\,302\,083t + (250t)(0 \text{ mm})^2 \\ &= 1\,302\,083t \quad \text{[in mm}^4\text{]} \\ I_{y2} &= I_{cy2} + A_2 d_2^2 = 20.83t^3 + (250t)(31.25 \text{ mm})^2 \\ &= 20.83t^3 + 244\,141t \end{aligned}$$

The moments of inertia of basic shape 3 about the centroidal axis of the section are the same as for basic shape 1.

$$I_{x3} = I_{x1} = 10.42t^3 + 1\,953\,125t$$
$$I_{y3} = I_{y1} = 284\,830t \quad \text{[in mm}^4\text{]}$$

The total moments of inertia about the centroidal axis of the section are

$$\begin{aligned} I_x &= I_{x1} + I_{x2} + I_{x3} \\ &= 10.42t^3 + 1\,953\,125t + 1\,302\,083t + 10.42t^3 \\ &\quad + 1\,953\,125t \\ &= 20.84t^3 + 5\,208\,333t \\ I_y &= I_{y1} + I_{y2} + I_{y3} \\ &= 284\,830t + 20.83t^3 + 244\,140t + 284\,830t \\ &= 20.83t^3 + 813\,800t \end{aligned}$$

Since t will be small and since the coefficient of the t^3 term is smaller than the coefficient of the t term, the t^3 term may be neglected.

$$I_x = 5\,208\,333t \quad [\text{in mm}^4]$$
$$I_y = 813\,800t \quad [\text{in mm}^4]$$

The polar moment of inertia for the section is

$$\begin{aligned} J &= I_x + I_y = 5\,208\,333t + 813\,800t \\ &= 6\,022\,133t \quad [\text{in mm}^4] \end{aligned}$$

The maximum shear stress will occur at the right-most point of basic shape 1 since this point is farthest from the centroid of the section.

The distance is

$$\begin{aligned} r &= \sqrt{(125\text{ mm} - 31.25\text{ mm})^2 + \left(\frac{250\text{ mm}}{2}\right)^2} \\ &= 156.25\text{ mm} \end{aligned}$$

The applied moment is

$$\begin{aligned} M = Fe &= (44\text{ kN})(300\text{ mm} + 125\text{ mm} - 31.25\text{ mm}) \\ &= 17\,325\text{ kN·mm} \end{aligned}$$

The torsional shear stress is

$$\begin{aligned} \tau &= \frac{Mr}{J} \\ &= \frac{\left(\frac{(17\,325\text{ kN·mm})(156.25\text{ mm})}{6\,022\,133t}\right)\left(1000\ \frac{\text{mm}}{\text{m}}\right)^2}{1000\ \frac{\text{kPa}}{\text{MPa}}} \\ &= 449.5/t \quad [\text{in MPa}] \end{aligned}$$

The shear stress is perpendicular to the line from the point to the centroid. The x- and y-components are shown in the following figure.

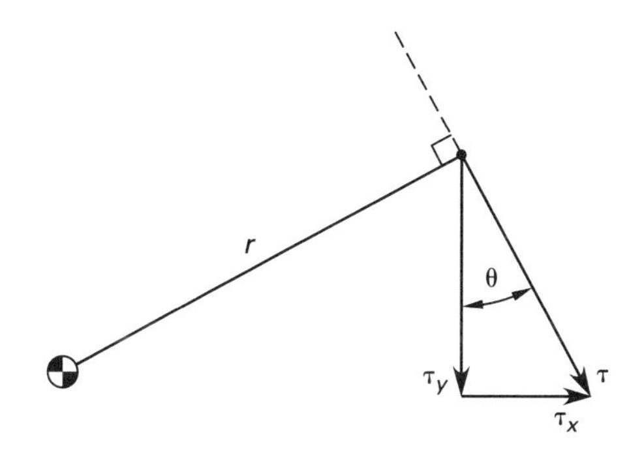

The angle θ is

$$\theta = 90° - \tan^{-1}\left(\frac{125\text{ mm} - 31.25\text{ mm}}{\frac{250\text{ mm}}{2}}\right) = 53.1°$$

$$\begin{aligned} \tau_x &= \tau\sin\theta = \left(\frac{449.5\text{ MPa}}{t}\right)(\sin 53.1°) \\ &= 359.5/t \quad [\text{in MPa}] \\ \tau_y &= \tau\cos\theta = \left(\frac{449.5\text{ MPa}}{t}\right)(\cos 53.1°) \\ &= 269.9/t \quad [\text{in MPa}] \end{aligned}$$

The vertical shear stress due to the load is

$$\begin{aligned} \tau_{vy} &= \frac{F}{\sum A_i} \\ &= \frac{(44\text{ kN})\left(1000\ \frac{\text{mm}}{\text{m}}\right)^2}{(125t + 250t + 125t)\left(1000\ \frac{\text{kPa}}{\text{MPa}}\right)} \\ &= 88.0/t \quad [\text{in MPa}] \end{aligned}$$

The resultant shear stress is

$$\begin{aligned} \tau &= \sqrt{\tau_x^2 + (\tau_y + \tau_{vy})^2} \\ &= \sqrt{\left(\frac{359.5}{t}\right)^2 + \left(\frac{269.9}{t} + \frac{88.0}{t}\right)^2} \\ &= 507.3/t \quad [\text{in MPa}] \end{aligned}$$

For a maximum shear stress of 55 MPa,

$$\begin{aligned} 55\text{ MPa} &= \frac{507.3\text{ MPa}}{t} \\ t &= 9.22\text{ mm} \end{aligned}$$

From Eq. 53.48, the required weld size is

$$y = \frac{t}{\frac{\sqrt{2}}{2}} = \frac{9.22\text{ mm}}{0.707} = \boxed{13.0\text{ mm}}$$

The answer is (B).

14. *Customary U.S. Solution*

The rotational speed is

$$\omega = \frac{\left(3500\ \frac{\text{rev}}{\text{min}}\right)\left(2\pi\ \frac{\text{rad}}{\text{rev}}\right)}{60\ \frac{\text{sec}}{\text{min}}} = 366.5\ \text{rad/sec}$$

From Eq. 59.19(b), the tangential stress at the inside radius of the flywheel is

$$\begin{aligned}\sigma &= \left(\frac{\rho\omega^2}{4g_c}\right)\left((3+\nu)r_0^2 + (1-\nu)r_i^2\right)\\ &= \left(\frac{\left(0.283\ \frac{\text{lbm}}{\text{in}^3}\right)\left(366.5\ \frac{\text{rad}}{\text{sec}}\right)^2}{(4)\left(386\ \frac{\text{in-lbm}}{\text{lbf-sec}^2}\right)}\right)\\ &\quad\times\left((3+0.3)(8\ \text{in})^2 + (1-0.3)(1\ \text{in})^2\right)\\ &= 5217\ \text{lbf/in}^2\end{aligned}$$

The circumferential (tangential) strain is

$$\epsilon = \frac{\sigma}{E} = \frac{5217\ \frac{\text{lbf}}{\text{in}^2}}{2.9\times10^7\ \frac{\text{lbf}}{\text{in}^2}} = 1.799\times10^{-4}$$

The circumferential, diametral, and radial strains are all equal.

The change in the inner radius is

$$\Delta r = r\epsilon = (1\ \text{in})(1.799\times10^{-4}) = 1.799\times10^{-4}\ \text{in}$$

Since the shaft and hub are both steel, the radial interference is obtained from Eq. 53.28.

$$\begin{aligned}I_{\text{radial}} &= \frac{I_{\text{diametral}}}{2} = \left(\frac{2pr_{\text{shaft}}}{E}\right)\left(\frac{1}{1-\left(\frac{r_{\text{shaft}}}{r_{o,\text{hub}}}\right)^2}\right)\\ &= \left(\frac{(2)\left(1250\ \frac{\text{lbf}}{\text{in}^2}\right)(1\ \text{in})}{2.9\times10^7\ \frac{\text{lbf}}{\text{in}^2}}\right)\left(\frac{1}{1-\left(\frac{1\ \text{in}}{8\ \text{in}}\right)^2}\right)\\ &= 8.758\times10^{-5}\ \text{in}\end{aligned}$$

The required initial interference is

$$\begin{aligned}I &= \Delta r + I_{\text{radial}}\\ &= 1.799\times10^{-4}\ \text{in} + 8.758\times10^{-5}\ \text{in}\\ &= \boxed{2.67\times10^{-4}\ \text{in}\quad(2.7\times10^{-4}\ \text{in})}\end{aligned}$$

The answer is (D).

SI Solution

The rotational speed is

$$\omega = \frac{\left(3500\ \frac{\text{rev}}{\text{min}}\right)\left(2\pi\ \frac{\text{rad}}{\text{rev}}\right)}{60\ \frac{\text{s}}{\text{min}}} = 366.5\ \text{rad/s}$$

From Eq. 59.19(a), the tangential stress at the inside radius of the flywheel is

$$\begin{aligned}\sigma &= \left(\frac{\rho\omega^2}{4}\right)\left((3+\nu)r_o^2 + (1-\nu)r_i^2\right)\\ &= \left(\frac{\left(7830\ \frac{\text{kg}}{\text{m}^3}\right)\left(366.5\ \frac{\text{rad}}{\text{s}}\right)^2}{4}\right)\\ &\quad\times\left((3+0.3)(0.203\ \text{m})^2 + (1-0.3)(0.025\ \text{m})^2\right)\\ &= 35.87\times10^6\ \text{Pa}\quad(35.87\ \text{MPa})\end{aligned}$$

The circumferential (tangential) strain is

$$\epsilon = \frac{\sigma}{E} = \frac{35.87\ \text{MPa}}{(200\ \text{GPa})\left(1000\ \frac{\text{MPa}}{\text{GPa}}\right)} = 1.794\times10^{-4}$$

The circumferential, diametral, and radial strains are all equal.

The change in inner radius is

$$\begin{aligned}\Delta r &= r\epsilon = (25\ \text{mm})(1.794\times10^{-4})\\ &= 44.85\times10^{-4}\ \text{mm}\end{aligned}$$

Since the shaft and hub are both steel, the radial interference is obtained from Eq. 53.28.

$$\begin{aligned}I_{\text{radial}} &= \frac{I_{\text{diametral}}}{2} = \left(\frac{2pr_{\text{shaft}}}{E}\right)\left(\frac{1}{1-\left(\frac{r_{\text{shaft}}}{r_{o,\text{hub}}}\right)^2}\right)\\ &= \left(\frac{(2)(8.6\ \text{MPa})(25\ \text{mm})}{(200\ \text{GPa})\left(1000\ \frac{\text{MPa}}{\text{GPa}}\right)}\right)\left(\frac{1}{1-\left(\frac{25\ \text{mm}}{203\ \text{mm}}\right)^2}\right)\\ &= 21.83\times10^{-4}\ \text{mm}\end{aligned}$$

The required initial interference is

$$\begin{aligned}I &= \Delta r + I_{\text{radial}}\\ &= 44.85\times10^{-4}\ \text{mm} + 21.83\times10^{-4}\ \text{mm}\\ &= \boxed{6.67\times10^{-3}\ \text{mm}\quad(6.7\times10^{-3}\ \text{mm})}\end{aligned}$$

The answer is (D).

15. (a) The bolts are experiencing $\boxed{\text{prying action.}}$ The deformation may or may not be plastic.

The answer is (A).

(b) Either load can be considered the "applied" load, while the other load can be considered the "resisting" load. The two loads are not additive. The bending (secondary effects) are disregarded.

$$\sigma_t = \frac{P_{\text{total}}}{A_{\text{total}}} = \frac{2F}{2A_{\text{bolt}}} = \boxed{\frac{F}{A_{\text{bolt}}}}$$

The answer is (B).

16. Since the elongation can be calculated as PL/AE, the cross-sectional area of the major diameter is

$$A = \frac{PL}{\delta E} = \frac{(80{,}000\ \text{lbf})(4\ \text{in})}{(0.05\ \text{in})\left(20{,}000{,}000\ \frac{\text{lbf}}{\text{in}^2}\right)} = 0.32\ \text{in}^2$$

From Table 53.5, using the tensile stress area column, the nominal size is $\boxed{{}^3/_4\ \text{in coarse series.}}$

The answer is (D).

17. The tensile stress area is calculated from the average of the major and minor diameters.

$$d_{\text{ave}} = \frac{d_{\text{root}} + d_{\text{pitch}}}{2} = \frac{1\ \text{in}}{2} = 0.5\ \text{in}$$

$$A_t = \left(\frac{\pi}{4}\right) d_{\text{ave}}^2 = \left(\frac{\pi}{4}\right)(0.5\ \text{in})^2 = 0.1963\ \text{in}^2$$

The proof load is

$$\begin{aligned} P_{\text{proof}} &= S_{\text{proof}} A_{\text{tensile}} = \left(70{,}000\ \frac{\text{lbf}}{\text{in}^2}\right)(0.1963\ \text{in}^2) \\ &= 13{,}741\ \text{lbf} \end{aligned}$$

The spring constant is

$$\begin{aligned} k &= \frac{P_{\text{proof}}}{\delta} = \frac{13{,}741\ \text{lbf}}{0.0147\ \text{in}} \\ &= \boxed{9.35 \times 10^5\ \text{lbf/in} \quad (9.4 \times 10^5\ \text{lbf/in})} \end{aligned}$$

The answer is (B).

18. The original bolt length is

$$L = \frac{AE}{k} = \frac{(1\ \text{in}^2)\left(20{,}000{,}000\ \frac{\text{lbf}}{\text{in}^2}\right)}{10{,}000{,}000\ \frac{\text{lbf}}{\text{in}}} = 2.0\ \text{in}$$

The elongation is

$$\delta = \frac{P}{k} = \frac{500{,}000\ \text{lbf}}{10{,}000{,}000\ \frac{\text{lbf}}{\text{in}}} = 0.05\ \text{in}$$

The elongation as a percentage of original bolt length is

$$\frac{0.05\ \text{in}}{2.0\ \text{in}} \times 100\% = \boxed{2.5\%}$$

Alternative Solution

$$\begin{aligned} \epsilon &= \frac{\sigma}{E} = \frac{F}{AE} = \frac{500{,}000\ \text{lbf}}{(1\ \text{in}^2)\left(20{,}000{,}000\ \frac{\text{lbf}}{\text{in}^2}\right)} \\ &= 0.025 \quad (2.5\%) \end{aligned}$$

The answer is (D).

54 Advanced Machine Design

PRACTICE PROBLEMS

Keyway Stresses

1. A shaft with a keyway transmits 300 hp at 1200 rev/min. The shaft is 2 ft long with a diameter of 3 in. The keyway has a width of 1 in and a length of 3 in. The shear stress of the keyway is most nearly

(A) 440 lbf/in^2

(B) 1200 lbf/in^2

(C) 1800 lbf/in^2

(D) 3500 lbf/in^2

Spring Design

2. A spring with 12 active coils and a spring index of 9 supports a static load of 50 lbf (220 N) with a deflection of 0.5 in (12 mm). The shear modulus of the spring material is 1.2×10^7 lbf/in^2 (83 GPa).

(a) The theoretical wire diameter is most nearly

(A) 0.58 in (15 mm)

(B) 0.63 in (16 mm)

(C) 0.69 in (18 mm)

(D) 0.74 in (19 mm)

(b) The mean spring diameter is most nearly

(A) 5.2 in (140 mm)

(B) 6.4 in (160 mm)

(C) 7.1 in (180 mm)

(D) 7.9 in (200 mm)

3. A severe service valve spring is to be manufactured from unpeened ASTM A230 steel wire in standard W&M sizes operating continuously between 20 lbf and 30 lbf (100 N and 150 N). The valve lift is 0.3 in (8 mm). The spring index is 10. The ultimate tensile strength of the wire material is 205,000 lbf/in^2 (1414.5 MPa). The factor of safety is 1.5. The minimum design fatigue service life is 10,000,000 cycles.

(a) The wire diameter is most nearly

(A) 0.12 in (3.0 mm)

(B) 0.15 in (4.1 mm)

(C) 0.21 in (53 mm)

(D) 0.28 in (71 mm)

(b) The spring constant is most nearly

(A) 10 lbf/in (1.9 kN/m)

(B) 21 lbf/in (3.9 kN/m)

(C) 33 lbf/in (6.3 kN/m)

(D) 48 lbf/in (92 kN/m)

(c) If the ends are squared and ground, the minimum number of active coils is most nearly

(A) 4

(B) 5

(C) 7

(D) 9

(d) The minimum total number of coils is most nearly

(A) 6

(B) 7

(C) 8

(D) 9

(e) The solid height is most nearly

(A) 1.2 in (35 mm)

(B) 1.7 in (43 mm)

(C) 2.3 in (58 mm)

(D) 2.8 in (71 mm)

(f) To prevent the spring from experiencing destructive stress during its lifetime, the spring force at solid height is most nearly

(A) 32 lbf (10 N)

(B) 46 lbf (250 N)

(C) 53 lbf (290 N)

(D) 62 lbf (330 N)

(g) The deflection at solid height is most nearly

(A) 1.4 in (39 mm)

(B) 1.7 in (43 mm)

(C) 2.1 in (53 mm)

(D) 2.4 in (61 mm)

(h) The minimum free height is most nearly

(A) 2.1 in (53 mm)

(B) 2.6 in (75 mm)

(C) 3.2 in (81 mm)

(D) 4.5 in (110 mm)

4. The material in a spring wire has a shear modulus of 1.2×10^7 lbf/in^2 (83 GPa). The maximum allowable stress under design conditions is 50,000 lbf/in^2 (350 MPa). The spring index is 7. Assume the maximum stress is experienced when a 700 lbm (320 kg) object falls from a height of 46 in (120 cm) above the tip of the spring, impacts squarely on the spring, and deflects the spring 10 in (26 cm).

(a) The minimum wire diameter is most nearly

(A) 0.58 in (15 mm)

(B) 0.88 in (23 mm)

(C) 1.4 in (36 mm)

(D) 1.8 in (47 mm)

(b) The mean coil diameter is most nearly

(A) 8.6 in (220 mm)

(B) 10 in (250 mm)

(C) 13 in (330 mm)

(D) 16 in (410 mm)

(c) The number of active coils is most nearly

(A) 8

(B) 11

(C) 12

(D) 14

5. A 6 in (150 mm) wide, 24 in (610 mm) long cantilever steel spring supports an 800 lbf (3.5 kN) load at its tip. The deflection is to be less than 1 in (25 mm). The bending stress is limited to 50,000 lbf/in^2 (345 MPa).

(a) The minimum thickness as limited by deflection alone is most nearly

(A) 0.37 in (9.4 mm)

(B) 0.45 in (11 mm)

(C) 0.63 in (16 mm)

(D) 0.77 in (20 mm)

(b) The minimum thickness as limited by bending stress is most nearly

(A) 0.62 in (16 mm)

(B) 0.66 in (17 mm)

(C) 0.72 in (18 mm)

(D) 0.78 in (20 mm)

6. Two concentric springs are constructed with squared-and-ground ends from oil-hardened steel. The ultimate tensile strength for the steel is 204,000 lbf/in^2 (1.3 GPa). The steel's yield strength is to be estimated as 75% of the ultimate tensile strength. The shear modulus for the steel is 11.5×10^6 lbf/in^2 (79 GPa). The springs support a static force of 150 lbf (660 N). The spring dimensions and properties are as follows.

inner spring		
wire diameter:	0.177 in	(4.5 mm)
mean coil diameter:	1.5 in	(38 mm)
free length:	4.5 in	(115 mm)
total number of coils:	12.75	
outer spring		
wire diameter:	0.2253 in	(5.723 mm)
mean coil diameter:	2.0 in	(51 mm)
free length:	3.75 in	(95.3 mm)
total number of coils:	10.25	

(a) The deflection of the inner spring is most nearly

(A) 2.0 in (48 mm)

(B) 2.4 in (60 mm)

(C) 2.9 in (73 mm)

(D) 3.1 in (78 mm)

(b) The maximum force exerted by the inner spring is most nearly

(A) 57 lbf (240 N)

(B) 67 lbf (280 N)

(C) 79 lbf (330 N)

(D) 86 lbf (360 N)

(c) The maximum shear stress in the inner spring is most nearly

(A) 47 kips/in^2 (320 MPa)

(B) 52 kips/in^2 (350 MPa)

(C) 57 kips/in^2 (390 MPa)

(D) 64 kips/in^2 (410 MPa)

(d) The factor of safety in shear for the inner spring is most nearly

(A) 1.2

(B) 1.5

(C) 1.8

(D) 2.2

(e) Specify the winding helix direction for each spring.

(A) parallel

(B) right hand

(C) clockwise

(D) reverse

7. A bathroom scale design makes use of four cantilever flat steel springs (beams) located at the corners of a rectangular load plate. Each beam is made from high strength 0-gage standard plate and is $1^3/_4$ in wide. The four beams are equally loaded. Strain gauges are mounted on the upper surface of each of the four beams. A digital readout is scaled to convert the resulting voltage due to load to a person's weight. All pieces other than the beam springs are rigid. The modulus of elasticity, E, is 30×10^6 lbf/in^2.

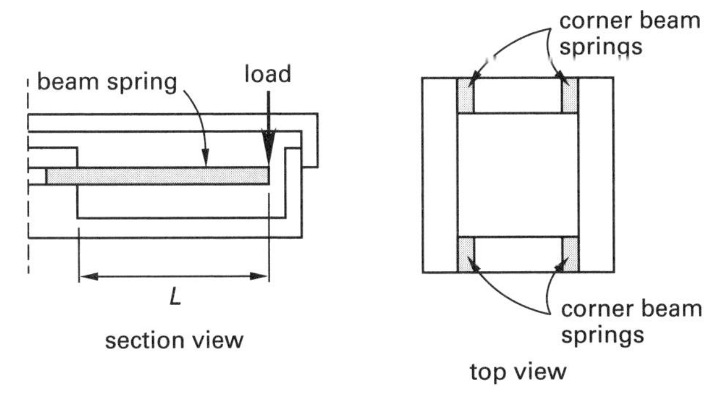

Given a spring constant of 25,758 lbf/in for each beam spring, the user weight that would produce an engineering strain of 1.2×10^{-3} in/in on the upper surface of one of the beams is most nearly

(A) 77 lbf

(B) 150 lbf

(C) 410 lbf

(D) 1600 lbf

8. A 2 in NPS schedule-80 steel pipe is rigidly attached to a wall at one end and is connected to a solid 2 in × 2 in square aluminum rod at the other end, as shown. The effective length of each piece is 12 in. The assembly is part of a torque-limiting system and experiences torsion.

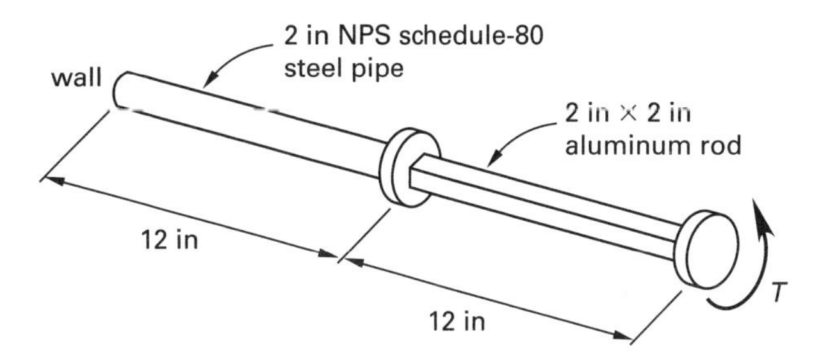

Neglecting the connection points, the equivalent torsional spring constant for a torque applied to the end of the aluminum rod is most nearly

(A) 5.0×10^5 in-lbf/rad

(B) 7.0×10^5 in-lbf/rad

(C) 17×10^5 in-lbf/rad

(D) 24×10^5 in-lbf/rad

General Gear Design

9. A gear train is to have a speed reduction of 600:1. The gears used can have no fewer than 12 teeth and no more than 96 teeth. The pinion gears in the first three stages have the same number of teeth. The number of stages needed is most nearly

(A) 2

(B) 3

(C) 4

(D) 5

10. The power transmission system shown consists of helical gears and AISI 1045 cold-drawn steel shafting. The gears have a 25° helix angle, a 20° normal pressure angle, and a diametral pitch of 5 (module of 5 mm). The yield strength of the 1045 steel is 69,000 lbf/in^2 (480 MPa). The mesh efficiency of each gear set is 98%. Loading is slow and steady. Use a factor of safety of 2.

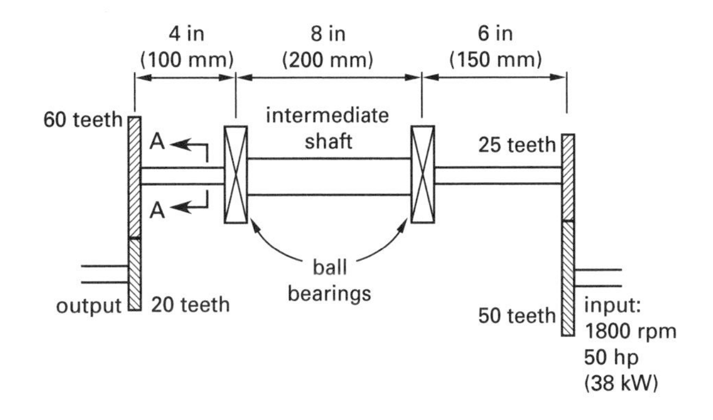

(a) The speed of the output shaft is most nearly

(A) 7000 rev/min

(B) 9000 rev/min

(C) 11,000 rev/min

(D) 12,000 rev/min

(b) The torque output is most nearly

(A) 230 in-lbf (26 N·m)

(B) 280 in-lbf (32 N·m)

(C) 350 in-lbf (35 N·m)

(D) 550 in-lbf (61 N·m)

(c) The pitch circle velocity is most nearly

(A) 11,000 ft/min (57 m/s)

(B) 12,000 ft/min (60 m/s)

(C) 16,000 ft/min (80 m/s)

(D) 18,000 ft/min (90 m/s)

(d) What is the transmitted load?

(A) 130 lbf (600 N)

(B) 140 lbf (660 N)

(C) 150 lbf (700 N)

(D) 160 lbf (750 N)

(e) Assume static loading and a worst-case scenario where all moments on the shaft are additive. The minimum shaft diameter at section A-A is most nearly

(A) 0.51 in (13 mm)

(B) 0.57 in (15 mm)

(C) 0.70 in (18 mm)

(D) 0.80 in (20 mm)

11. The diametral pitch of an American standard involute gear is π in^{-1}. The base pitch of a basic rack for the involute gear shown is most nearly

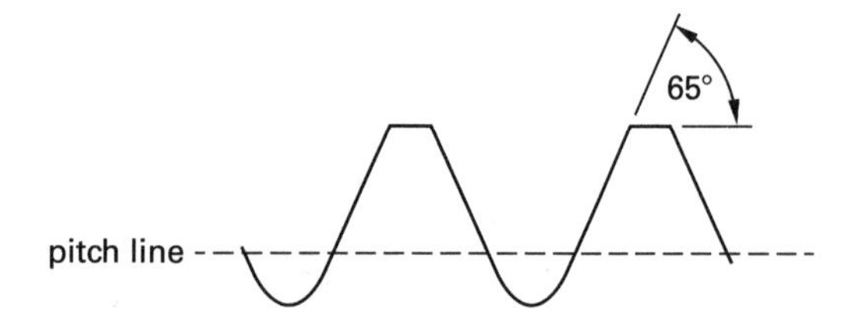

(A) 0.55 in

(B) 0.75 in

(C) 0.82 in

(D) 0.91 in

12. A basic rack for an American standard full-depth involute 20° gear is shown. The circular pitch is 1 in, and the pitch circle diameter is 8 in. The clearance, *C*, is most nearly

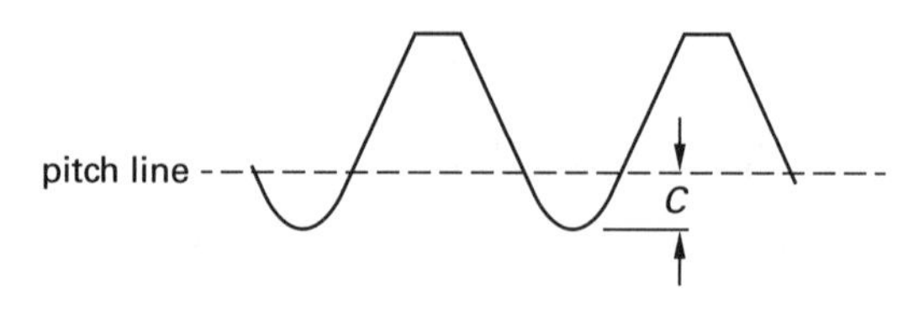

(A) 0.04 in

(B) 0.05 in

(C) 0.06 in

(D) 0.10 in

13. An American standard involute gear system consisting of one pinion and one gear has a gear ratio of 4. The pinion gear has a pressure angle of 20° and a pitch diameter of 20 in. The gear has a diametral pitch of 1 in^{-1} and a pitch diameter of 80 in. The diametral pitch of the pinion gear is most nearly

(A) 0.5 in^{-1}

(B) 0.7 in^{-1}

(C) 0.8 in^{-1}

(D) 1 in^{-1}

14. A basic rack of an American standard full-depth involute 14.5° gear system has a base circular pitch of 1 in. The tooth thickness, *t*, on the pitch line shown is most nearly

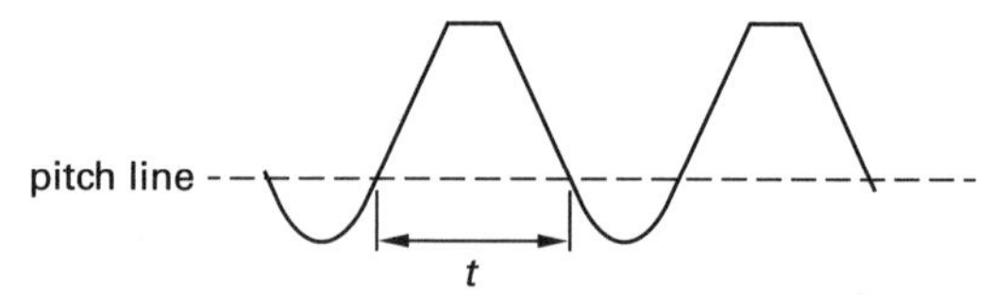

(A) 0.43 in

(B) 0.52 in

(C) 0.56 in

(D) 0.65 in

15. A pinion-driven gear has a circular pitch of 1 in and 50 teeth. The pinion rotates at 1000 rev/min with a gear ratio of 4. The pitch (line) velocity of the pinion gear is most nearly

(A) 1000 ft/min

(B) 4200 ft/min

(C) 12,000 ft/min

(D) 17,000 ft/min

16. The helical gear shown has a normal rack pitch of π in. The number of helical gear teeth is 50. The helical gear's transverse diametral pitch is most nearly

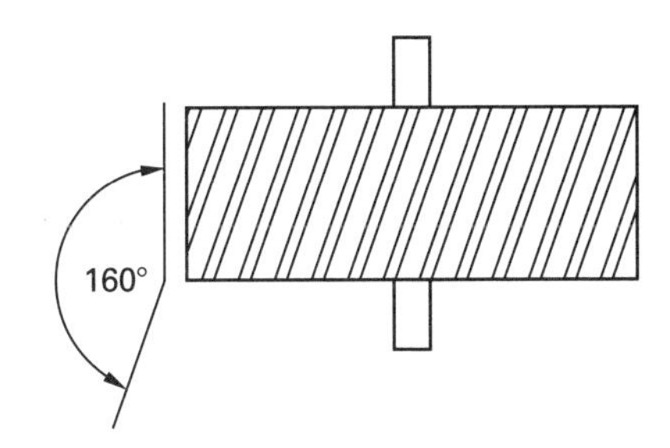

(A) 0.55 1/in

(B) 0.75 1/in

(C) 0.86 1/in

(D) 0.94 1/in

17. The common differential gear of an automobile

(A) equalizes the traction of both wheels and permits one wheel to turn faster than the other, as needed on curves

(B) transmits engine torque to the torque converter

(C) provides better traction when one wheel is slipping

(D) maintains the final gear ratio of the transmission

18. The two identical bevel gears in an automobile differential gear serve the same function as which gear(s) in an epicyclic gear set?

(A) ring

(B) sun

(C) both ring and sun

(D) the planet gears

19. Which kind of bevel gears connect shafts at angles other than 90°?

(A) angular

(B) miter (mitre)

(C) Coniflex

(D) Gleason

20. Spiral bevel gears are used for applications involving

(A) speeds in excess of 1000 rev/min

(B) high loads

(C) quiet operations

(D) all of the above

21. Which of the following angles is the angle between mating bevel-gear axes?

(A) shaft angle

(B) root angle

(C) spiral angle

(D) pitch angle

22. A helical gear has a tangential pressure angle of 22.5° and a normal pressure angle of 20°. The gear is exposed to an axial force of 1000 lbf. The total force on the gear is most nearly

(A) 1200 lbf

(B) 2200 lbf

(C) 4400 lbf

(D) 5500 lbf

23. A helical gear with a tangential pitch of 1 in carries an axial force of 1000 lbf and a tangential force of 1500 lbf. The normal circular pitch is most nearly

(A) 0.15 in

(B) 0.21 in

(C) 0.25 in

(D) 0.83 in

Gear Tooth Strength

24. A mechanism is driven by a 150 hp (110 kW) motor. The motor turns at 1200 rev/min, but a pair of $14^1/_2°$ gears on 15 in (380 mm) centers is to reduce the speed of the mechanism to 270 rev/min. The pinion is to be SAE 1045 steel with an endurance strength of 90,000 lbf/in^2 (620 MPa). The pinion width is 6 in (150 mm). The gear is to be cast steel with an endurance strength of 50,000 lbf/in^2 (345 MPa). A safety factor of 3 is used in the design. Use the Lewis beam strength theory. Do not check for undercutting.

(a) What are most nearly the pitch diameters? (Assume d_1 = gear and d_2 = pinion.)

(A) $d_1 = 16$ in, $d_2 = 14$ in ($d_1 = 400$ mm, $d_2 = 370$ mm)

(B) $d_1 = 18$ in, $d_2 = 12$ in ($d_1 = 450$ mm, $d_2 = 310$ mm)

(C) $d_1 = 21$ in, $d_2 = 9$ in ($d_1 = 530$ mm, $d_2 = 230$ mm)

(D) $d_1 = 24$ in, $d_2 = 6$ in ($d_1 = 620$ mm, $d_2 = 140$ mm)

(b) Given a pinion face width of 6 in (150 mm), the diametral pitch (gear module) is most nearly

(A) 4 1/in (5 mm)

(B) 6 1/in (4 mm)

(C) 8 1/in (3 mm)

(D) 10 1/in (2 mm)

(c) Given a diametral pitch of 4 in (a module of 6 mm), and disregarding stress concentrations, the face width for the gear is most nearly

(A) 4.5 in (110 mm)

(B) 5.0 in (130 mm)

(C) 6.0 in (150 mm)

(D) 7.5 in (190 mm)

25. Two 20° involute spur gears are mounted such that their centers are 15 in (380 mm) apart. The pinion is untreated steel with an allowable stress of 30,000 lbf/in^2 (210 MPa). The pinion width is 6 in (150 mm). The gear set reduces the speed of a 250 hp (190 kW) motor with an integral speed reducer from 250 rev/min to 83$^1/_3$ rev/min. The gear is cast steel with a maximum strength of 50,000 lbf/in^2 (345 MPa). A factor of safety of 3 is used. Use the Lewis beam strength theory.

(a) What are most nearly the pitch diameters? (Assume d_1 = gear and d_2 = pinion.)

(A) $d_1 = 7.0$ in, $d_2 = 23$ in ($d_1 = 190$ mm, $d_2 = 570$ mm)

(B) $d_1 = 13$ in, $d_2 = 17$ in ($d_1 = 320$ mm, $d_2 = 440$ mm)

(C) $d_1 = 17$ in, $d_2 = 13$ in ($d_1 = 440$ mm, $d_2 = 320$ mm)

(D) $d_1 = 23$ in, $d_2 = 8.0$ in ($d_1 = 570$ mm, $d_2 = 190$ mm)

(b) The diametral pitch is most nearly

(A) 0.75 1/in (0.029 1/mm)

(B) 1.5 1/in (0.058 1/mm)

(C) 1.8 1/in (0.068 1/mm)

(D) 3.5 1/in (0.14 1/mm)

(c) Given a diametral pitch of 2 in^{-1} (0.0787 mm^{-1}), approximately how many teeth are on each gear?

(A) $N_{pinion} = 15$, $N_{gear} = 45$

(B) $N_{pinion} = 22$, $N_{gear} = 30$

(C) $N_{pinion} = 30$, $N_{gear} = 22$

(D) $N_{pinion} = 39$, $N_{gear} = 13$

(d) Given a diametral pitch of 2 in^{-1} (0.0787 mm^{-1}) and a pinion with 13 teeth, the minimum face width of the pinion is most nearly

(A) 8.0 in (200 mm)

(B) 9.0 in (230 mm)

(C) 10 in (270 mm)

(D) 14 in (350 mm)

(e) Given a diametral pitch of 2 in^{-1} (0.0787 mm^{-1}) and a gear with 39 teeth, the minimum face width of the gear is most nearly

(A) 8.4 in (220 mm)

(B) 10 in (250 mm)

(C) 14 in (340 mm)

(D) 18 in (430 mm)

AGMA Gear Design

26. An AGMA spur gear has 48 teeth and a circular pitch of 0.785 in. Based on AGMA Standard 2000 and using an AGMA quality number of 10, the pitch tolerance is most nearly

(A) 0.00035 in

(B) 0.00046 in

(C) 0.00052 in

(D) 0.00059 in

27. A spur gear in an AGMA 20° spur gear system has 24 teeth and a pitch diameter of 15 in. The load is applied at the tip of the tooth. The tooth strength of the spur gear is most nearly

(A) 120,000 lbf/in^2

(B) 240,000 lbf/in^2

(C) 310,000 lbf/in^2

(D) 640,000 lbf/in^2

28. An AGMA gear with an A-1 classification is designed for 10,000,000 load cycles. The gear has a surface hardness of 180 BHN. The hardness ratio factor, reliability factor, and temperature factor are 1.05, 1.50, and 1.00, respectively. Using AGMA standards for pitting resistance, the maximum permissible calculated contact stress for the gear is most nearly

(A) 55,000 lbf/in^2

(B) 65,000 lbf/in^2

(C) 67,000 lbf/in^2

(D) 95,000 lbf/in^2

29. An AGMA spur gear has a circular pitch of 0.785 in and 24 teeth. Based on AGMA standards and an AGMA quality number of 11, the runout tolerance is most nearly

(A) 0.0008 in

(B) 0.0012 in

(C) 0.0017 in

(D) 0.0027 in

Clutches

30. A wet, steel-backed asbestos clutch with hardened steel plates is being designed to transmit 300 in-lbf (33 N·m). The coefficient of friction is 0.12. Slip will occur at 300% of the rated torque. The maximum and minimum friction surface diameters are 4.5 in and 2.5 in (115 mm and 65 mm), respectively. The contact pressure is 100 psi (700 kPa). Uniform wear is expected.

(a) The number of plates needed is most nearly

(A) 2

(B) 3

(C) 4

(D) 5

(b) The number of disks needed is most nearly

(A) 2

(B) 3

(C) 4

(D) 5

Belt Drives

31. A piece of equipment is driven by a 15 hp (11.2 kW) motor through a standard B90 V-belt. The motor runs at 1750 rev/min. The nominal speed of the equipment must be 800 rev/min. The motor's sheave diameter is 10 in (254 mm).

(a) The pitch length of the belt is most nearly

(A) 77 in (1960 mm)

(B) 81 in (2060 mm)

(C) 90 in (2290 mm)

(D) 92 in (2330 mm)

(b) The pitch diameter of the sheave on the equipment drive is most nearly

(A) 20 in (520 mm)

(B) 22 in (560 mm)

(C) 24 in (610 mm)

(D) 26 in (660 mm)

Journal Bearings

32. A 3 in (76 mm) diameter shaft turns at 1200 rev/min in a journal bearing. The bearing has an axial length of 3.5 in (90 mm) and a radial clearance ratio (c_r/r) of 0.001. The transverse load allocated to the bearing is 880 lbf (4 kN). The lubricating oil has a temperature of 165°F (75°C) and a viscosity of 1.184×10^{-6} lbf-sec/in^2 (8.16 cP).

(a) The bearing characteristic number is most nearly

(A) 0.28

(B) 0.31

(C) 0.34

(D) 0.40

(b) The axial length-to-diameter ratio is most nearly

(A) 0.86

(B) 1.0

(C) 1.2

(D) 1.3

(c) The minimum film thickness is most nearly

(A) 0.00032 in (0.0080 mm)

(B) 0.00064 in (0.016 mm)

(C) 0.00096 in (0.024 mm)

(D) 0.0013 in (0.032 mm)

(d) The friction torque is most nearly

(A) 5.1 in-lbf (0.59 N·m)

(B) 7.9 in-lbf (0.91 N·m)

(C) 10 in-lbf (1.2 N·m)

(D) 15 in-lbf (1.7 N·m)

(e) The power lost to friction is most nearly

(A) 0.12 hp (0.088 kW)

(B) 0.15 hp (0.11 kW)

(C) 0.25 hp (0.18 kW)

(D) 0.44 hp (0.32 kW)

(f) Is this bearing operating within or outside its optimum capacity?

(A) It is within the optimum capacity; minimum film thickness is to the left of the maximum load-carrying ability curve.

(B) It is within the optimum capacity; minimum film thickness is to the right of the maximum load-carrying ability curve.

(C) It is outside the optimum capacity; minimum film thickness is to the left of the maximum load-carrying ability curve.

(D) It is outside the optimum capacity; minimum film thickness is to the right of the maximum load-carrying ability curve.

33. A 1 in pump journal bearing is exposed to a 2500 lbf lateral shaft load. The pressure on the journal bearing is most nearly

(A) 1050 lbf/in^2

(B) 1250 lbf/in^2

(C) 1310 lbf/in^2

(D) 1550 lbf/in^2

34. A plain cylindrical journal bearing has an operating characteristic value of 1.0 and a radial clearance of 0.01 in. The angle between the 5500 lbf bearing load and the entering edge of the oil film is 30°, resulting in an eccentricity ratio of 0.6. The minimum film thickness of the journal bearing is most nearly

(A) 0.004 in

(B) 0.005 in

(C) 0.006 in

(D) 0.007 in

Rolling Element Bearings

35. A ball bearing with a single row of balls has a basic design capacity of 5100 lbf.

(a) Based on a 90% reliability, approximately how many revolutions can it be expected to turn if it is subjected to an equivalent loading of 2500 lbf?

(A) 0.12×10^6 rev

(B) 0.50×10^6 rev

(C) 1.3×10^6 rev

(D) 8.5×10^6 rev

(b) If the shaft turns at 1800 rev/min while being loaded at 2500 lbf, approximately how many hours will the bearing last based on L_{10} life?

(A) 80 hr

(B) 190 hr

(C) 240 hr

(D) 610 hr

36. A ball bearing is intended to carry a radial load of 5000 lbf. The ball bearing is a double-row 200 series with a 20 mm bore. The basic load rating of the bearing is 3480 lbf. The L_{10} rated life in revolutions of the bearing is most nearly

(A) 340,000 rev

(B) 360,000 rev

(C) 450,000 rev

(D) 930,000 rev

37. A 200 series double-row ball bearing is to carry a radial load of 1000 lbf and a thrust load of 1500 lbf. The radial and thrust load factors are 0.63 and 1.25, respectively. The basic load rating is 2000 lbf. The rated life, L_{10}, is most nearly

(A) 400,000 rev

(B) 500,000 rev

(C) 600,000 rev

(D) 700,000 rev

38. A 200 series double-row, deep-row bearing has a basic load rating of 2000 lbf and a permissible equivalent load of 1820 lbf. The equivalent load factor is most nearly

(A) 1.1

(B) 1.3

(C) 2.1

(D) 2.2

39. A roller bearing is exposed to an equivalent radial load of 1000 lbf. The basic load rating of the bearing is 3480 lbf. The rotational speed is 200 rev/min. The rated life, L_{10}, of the bearing is most nearly

(A) 3200 hr

(B) 5300 hr

(C) 5400 hr

(D) 5500 hr

40. A ball bearing is to carry an equivalent radial load of 1000 lbf. The rated life, L_{10}, is 500,000,000 revolutions. The basic load rating is most nearly

(A) 6500 lbf

(B) 7100 lbf

(C) 7900 lbf

(D) 8000 lbf

SOLUTIONS

1. Rearranging Eq. 54.33(b), the torque of the shaft, T, is

$$T = \frac{63{,}025P}{n} = \frac{(63{,}025)(300\text{ hp})}{1200\ \frac{\text{rev}}{\text{min}}} = 15{,}756\text{ in-lbf}$$

The keyway's shear stress, τ, is

$$\begin{aligned}\tau &= \frac{2T}{dwL} = \frac{(2)(15{,}756\text{ in-lbf})}{(3\text{ in})(1\text{ in})(3\text{ in})} \\ &= \boxed{3501\text{ lbf/in}^2\quad(3500\text{ lbf/in}^2)}\end{aligned}$$

The answer is (D).

2. *Customary U.S. Solution*

Select a spring with a spring index of $C=9$.

(a) From Eq. 54.10, the minimum wire diameter is

$$\begin{aligned}d &= \frac{8FC^3n_a}{G\delta} = \frac{(8)(50\text{ lbf})(9)^3(12)}{\left(1.2\times10^7\ \frac{\text{lbf}}{\text{in}^2}\right)(0.5\text{ in})} \\ &= \boxed{0.583\text{ in}\quad(0.58\text{ in})}\end{aligned}$$

The answer is (A).

(b) From Eq. 54.8, the mean spring diameter is

$$D = Cd = (9)(0.583\text{ in}) = \boxed{5.247\text{ in}\quad(5.2\text{ in})}$$

The answer is (A).

SI Solution

Select a spring with a spring index of $C=9$.

(a) From Eq. 54.10, the minimum wire diameter is

$$\begin{aligned}d &= \frac{8FC^3n_a}{G\delta} = \left(\frac{(8)(220\text{ N})(9)^3(12)}{(83\times10^9\text{ Pa})(0.012\text{ m})}\right)\left(1000\ \frac{\text{mm}}{\text{m}}\right) \\ &= \boxed{15.46\text{ mm}\quad(15\text{ mm})}\end{aligned}$$

The answer is (A).

(b) From Eq. 54.8, the mean spring diameter is

$$D = Cd = (9)(15.46\text{ mm}) = \boxed{139.1\text{ mm}\quad(140\text{ mm})}$$

The answer is (A).

3. *Customary U.S. Solution*

From Table 54.3, the fatigue loading shear stress factor is 0.30. From Eq. 54.7, the maximum allowable shear stress is

$$\begin{aligned}\tau_{\text{max}} &= \frac{(\text{factor})S_{ut}}{\text{FS}} = \frac{(0.3)\left(205{,}000\ \frac{\text{lbf}}{\text{in}^2}\right)}{1.5} \\ &= 41{,}000\text{ lbf/in}^2\end{aligned}$$

From Eq. 54.13, the Wahl correction factor is

$$\begin{aligned}W &= \frac{4C-1}{4C-4} + \frac{0.615}{C} = \frac{(4)(10)-1}{(4)(10)-4} + \frac{0.615}{10} \\ &= 1.145\end{aligned}$$

(a) From Eq. 54.14, the required wire diameter is

$$\begin{aligned}d &= \sqrt{\frac{8FCW}{\pi\tau_{\text{max}}}} = \sqrt{\frac{(8)(30\text{ lbf})(10)(1.145)}{\pi\left(41{,}000\ \frac{\text{lbf}}{\text{in}^2}\right)}} \\ &= 0.146\text{ in}\end{aligned}$$

From App. 54.A, use W&M wire no. 9 with $d=$ $\boxed{0.1483\text{ in }(0.15\text{ in}).}$

The answer is (B).

(b) From Eq. 54.2, the spring constant is

$$\begin{aligned}k &= \frac{F_1 - F_2}{\delta_1 - \delta_2} = \frac{30\text{ lbf} - 20\text{ lbf}}{0.3\text{ in}} \\ &= \boxed{33.33\text{ lbf/in}\quad(33\text{ lbf/in})}\end{aligned}$$

The answer is (C).

(c) From Table 54.1, the shear modulus of ASTM A230 steel wire is $G=11.5\times10^6$ lbf/in^2.

From Eq. 54.12, and rounding up, the number of active coils is

$$\begin{aligned}n_a &= \frac{Gd}{8kC^3} = \frac{\left(11.5\times10^6\ \frac{\text{lbf}}{\text{in}^2}\right)(0.1483\text{ in})}{(8)\left(33.33\ \frac{\text{lbf}}{\text{in}}\right)(10)^3} \\ &= \boxed{6.40\quad(7)}\end{aligned}$$

The answer is (C).

(d) From Table 54.4, and rounding up, the total number of coils for squared and ground ends is

$$n_t = n_a + 2 = 6.40 + 2 = \boxed{8.40\quad(9)}$$

The answer is (D).

(e) From Table 54.4, the solid height is

$$h_s = dn_t = (0.1483 \text{ in})(8.40) = \boxed{1.25 \text{ in} \quad (1.2 \text{ in})}$$

The answer is (A).

(f) The spring should not experience destructive stress if it is accidentally compressed solid. At solid height, the maximum shear stress is

$$\tau_{\max} = (\text{factor})S_{ut} = (0.3)\left(205{,}000 \ \frac{\text{lbf}}{\text{in}^2}\right) = 61{,}500 \text{ lbf/in}^2$$

From Eq. 54.14, the force at solid height is

$$F_s = \frac{\tau_{\max}\pi d^2}{8CW} = \frac{\left(61{,}500 \ \frac{\text{lbf}}{\text{in}^2}\right)\pi(0.1483 \text{ in})^2}{(8)(10)(1.145)} = \boxed{46.39 \text{ lbf} \quad (46 \text{ lbf})}$$

The answer is (B).

(g) From Eq. 54.10, the deflection at solid height is

$$\delta_s = \frac{F_s}{k} = \frac{46.39 \text{ lbf}}{33.33 \ \frac{\text{lbf}}{\text{in}}} = \boxed{1.39 \text{ in} \quad (1.4 \text{ in})}$$

The answer is (A).

(h) From Eq. 54.15, the minimum free height is

$$h_f = h_s + \delta_s = 1.25 \text{ in} + 1.39 \text{ in} = \boxed{2.64 \text{ in} \quad (2.6 \text{ in})}$$

The answer is (B).

SI Solution

From Table 54.3, the fatigue loading shear stress factor is 0.30. From Eq. 54.7, the maximum allowable shear stress is

$$\tau_{\max} = \frac{(\text{factor})S_{ut}}{\text{FS}} = \frac{(0.3)(1414.5 \text{ MPa})}{1.5} = 282.9 \text{ MPa}$$

From Eq. 54.13, the Wahl correction factor is

$$W = \frac{4C-1}{4C-4} + \frac{0.615}{C} = \frac{(4)(10)-1}{(4)(10)-4} + \frac{0.615}{10} = 1.145$$

(a) From Eq. 54.14, the required wire diameter is

$$d = \sqrt{\frac{8FCW}{\pi\tau_{\max}}} = \sqrt{\frac{(8)(150 \text{ N})(10)(1.145)\left(1000 \ \frac{\text{mm}}{\text{m}}\right)^2}{\pi(282.9 \times 10^6 \text{ Pa})}} = 3.93 \text{ mm}$$

From App. 54.A, use W&M wire no. 8 with a diameter of

$$d = (0.162 \text{ in})\left(25.4 \ \frac{\text{mm}}{\text{in}}\right) = \boxed{4.115 \text{ mm} \quad (4.1 \text{ mm})}$$

The answer is (B).

(b) From Eq. 54.2, the spring constant is

$$k = \frac{F_1 - F_2}{\delta_1 - \delta_2} = \frac{(150 \text{ N} - 100 \text{ N})\left(1000 \ \frac{\text{mm}}{\text{m}}\right)}{8 \text{ mm}} = \boxed{6.25 \times 10^3 \text{ N/m} \quad (6.3 \text{ kN/m})}$$

The answer is (C).

(c) From Table 54.1, the shear modulus of ASTM A230 steel wire is

$$G = \left(11.5 \times 10^6 \ \frac{\text{lbf}}{\text{in}^2}\right)\left(6.89 \times 10^3 \ \frac{\text{Pa}}{\frac{\text{lbf}}{\text{in}^2}}\right) = 79.2 \times 10^9 \text{ Pa}$$

From Eq. 54.12, and rounding up, the number of active coils is

$$n_a = \frac{Gd}{8kC^3} = \frac{(79.2 \times 10^9 \text{ Pa})(4.115 \text{ mm})}{(8)\left(6.25 \times 10^3 \ \frac{\text{N}}{\text{m}}\right)(10)^3\left(1000 \ \frac{\text{mm}}{\text{m}}\right)} = \boxed{6.52 \quad (7)}$$

The answer is (C).

(d) From Table 54.4, and rounding up, the total number of coils for squared and ground ends is

$$n_t = n_a + 2 = 6.52 + 2 = \boxed{8.52 \quad (9)}$$

The answer is (D).

(e) From Table 54.4, the solid height is

$$h_s = dn_t = (4.115 \text{ mm})(8.52) = \boxed{35.06 \text{ mm} \quad (35 \text{ mm})}$$

The answer is (A).

(f) The spring should not experience destructive stress if it is accidentally compressed solid. At solid height, the maximum shear stress is

$$\tau_{max} = (\text{factor})S_{ut} = (0.3)(1414.5\ \text{MPa}) = 424.4\ \text{MPa}$$

From Eq. 54.14, the force at solid height is

$$F_s = \frac{\tau_{max}\pi d^2}{8CW} = \frac{(424.4\ \text{MPa})\pi(4.115\ \text{mm})^2 \times \left(10^6\ \frac{\text{Pa}}{\text{MPa}}\right)}{(8)(10)(1.145)\left(1000\ \frac{\text{mm}}{\text{m}}\right)^2}$$
$$= \boxed{246.5\ \text{N}\quad(250\ \text{N})}$$

The answer is (B).

(g) From Eq. 54.10, the deflection at solid height is

$$\delta_s = \frac{F_s}{k} = \left(\frac{246.5\ \text{N}}{6.25\times 10^3\ \frac{\text{N}}{\text{m}}}\right)\left(1000\ \frac{\text{mm}}{\text{m}}\right)$$
$$= \boxed{39.44\ \text{mm}\quad(39\ \text{mm})}$$

The answer is (A).

(h) From Eq. 54.15, the minimum free height is

$$h_f = h_s + \delta_s = 35.06\ \text{mm} + 39.44\ \text{mm}$$
$$= \boxed{74.5\ \text{mm}\quad(75\ \text{mm})}$$

The answer is (B).

4. *Customary U.S. Solution*

The potential energy absorbed is

$$E_p = \frac{mg\Delta h}{g_c} = \frac{(700\ \text{lbm})\left(32.2\ \frac{\text{ft}}{\text{sec}^2}\right)(46\ \text{in} + 10\ \text{in})}{32.2\ \frac{\text{lbm-ft}}{\text{lbf-sec}^2}}$$
$$= 39{,}200\ \text{in-lbf}$$

The work done by the spring is equal to the potential energy.

$$W = \tfrac{1}{2}k\delta^2 = E_p$$
$$k = \frac{2E_p}{\delta^2} = \frac{(2)(39{,}200\ \text{in-lbf})}{(10\ \text{in})^2} = 784\ \text{lbf/in}$$

The equivalent spring force is

$$F = k\delta = \left(784\ \frac{\text{lbf}}{\text{in}}\right)(10\ \text{in}) = 7840\ \text{lbf}$$

From Eq. 54.13, the Wahl correction factor is

$$W = \frac{4C-1}{4C-4} + \frac{0.615}{C} = \frac{(4)(7)-1}{(4)(7)-4} + \frac{0.615}{7} = 1.213$$

(a) From Eq. 54.14, the wire diameter is

$$d = \sqrt{\frac{8FCW}{\pi\tau_{allowable}}} = \sqrt{\frac{(8)(7840\ \text{lbf})(7)(1.213)}{\pi\left(50{,}000\ \frac{\text{lbf}}{\text{in}^2}\right)}}$$
$$= \boxed{1.84\ \text{in}\quad(1.8\ \text{in})}$$

The answer is (D).

(b) From Eq. 54.8, the mean coil diameter is

$$D = Cd = (7)(1.84\ \text{in}) = \boxed{12.88\ \text{in}\quad(13\ \text{in})}$$

The answer is (C).

(c) From Eq. 54.12, the number of active coils is

$$n_a = \frac{Gd}{8kC^3} = \frac{\left(1.2\times 10^7\ \frac{\text{lbf}}{\text{in}^2}\right)(1.84\ \text{in})}{(8)\left(784\ \frac{\text{lbf}}{\text{in}}\right)(7)^3}$$
$$= \boxed{10.3\quad(11)}$$

The answer is (B).

SI Solution

The potential energy absorbed is

$$E_p = mg\Delta h = (320\ \text{kg})\left(9.81\ \frac{\text{m}}{\text{s}^2}\right)(1.2\ \text{m} + 0.26\ \text{m})$$
$$= 4583\ \text{J}$$

The work done by the spring is equal to the potential energy.

$$W_k = \tfrac{1}{2}k\delta^2 = E_p$$
$$k = \frac{2E_p}{\delta^2} = \frac{(2)(4583\ \text{J})}{(0.26\ \text{m})^2} = 135\,592\ \text{N/m}$$

The equivalent spring force is

$$F = k\delta = \left(135\,592\ \frac{\text{N}}{\text{m}}\right)(0.26\ \text{m}) = 35\,254\ \text{N}$$

From Eq. 54.13, the Wahl correction factor is

$$W = \frac{4C-1}{4C-4} + \frac{0.615}{C} = \frac{(4)(7)-1}{(4)(7)-4} + \frac{0.615}{7} = 1.213$$

(a) From Eq. 54.14, the wire diameter is

$$d = \sqrt{\frac{8FCW}{\pi\tau_{\text{allowable}}}}$$

$$= \sqrt{\frac{(8)(35\,254\text{ N})(7)(1.213)\left(10^3\ \frac{\text{mm}}{\text{m}}\right)^2}{\pi(350\times10^6\text{ Pa})}}$$

$$= \boxed{46.7\text{ mm}\quad(47\text{ mm})}$$

The answer is (D).

(b) From Eq. 54.8, the mean coil diameter is

$$D = Cd = (7)(46.7\text{ mm}) = \boxed{327\text{ mm}\quad(330\text{ mm})}$$

The answer is (C).

(c) From Eq. 54.12, the number of active coils is

$$n_a = \frac{Gd}{8kC^3} = \frac{(83\times10^9\text{ Pa})(46.7\text{ mm})}{(8)\left(135\,592\ \frac{\text{N}}{\text{m}}\right)(7)^3\left(10^3\ \frac{\text{mm}}{\text{m}}\right)}$$

$$= \boxed{10.4\quad(11)}$$

The answer is (B).

5. *Customary U.S. Solution*

The moment of inertia of the beam cross section is

$$I = \frac{bh^3}{12} = \frac{(6\text{ in})h^3}{12} = 0.5h^3\quad[\text{in}^4]$$

The section modulus is

$$\frac{I}{c} = \frac{0.5h^3}{\frac{h}{2}} = h^2\quad[\text{in}^3]$$

From Table 54.1, the modulus of elasticity for a steel spring is $E = 30\times10^6$ lbf/in^2.

(a) The tip deflection is

$$y = \frac{FL^3}{3EI} = \frac{FL^3}{3E(0.5)h^3}$$

$$h = \left(\frac{FL^3}{1.5yE}\right)^{1/3} = \left(\frac{(800\text{ lbf})(24\text{ in})^3}{(1.5\text{ in})(1.0\text{ in})\left(30\times10^6\ \frac{\text{lbf}}{\text{in}^2}\right)}\right)^{1/3}$$

$$= \boxed{0.626\text{ in}\quad(0.63\text{ in})}$$

The width-thickness ratio is

$$\frac{w}{t} = \frac{6.0\text{ in}}{0.626\text{ in}} \approx 9.6$$

Since $w/t \approx 10$, this could be considered a wide beam.

The answer is (C).

(b) The moment at the fixed end is

$$M = FL = (800\text{ lbf})(24\text{ in}) = 19{,}200\text{ in-lbf}$$

The bending stress is

$$\sigma = \frac{M}{\frac{I}{c}} = \frac{M}{h^2}$$

$$h = \sqrt{\frac{M}{\sigma}} = \sqrt{\frac{19{,}200\text{ in-lbf}}{(1.0\text{ in})\left(50{,}000\ \frac{\text{lbf}}{\text{in}^2}\right)}} = \boxed{0.62\text{ in}}$$

The answer is (A).

SI Solution

The moment of inertia of the beam cross section is

$$I = \frac{bh^3}{12} = \frac{(150\text{ mm})h^3}{12} = 12.5h^3\quad[\text{mm}^4]$$

The section modulus is

$$\frac{I}{c} = \frac{12.5h^3}{\frac{h}{2}} = 25h^2\quad[\text{mm}^3]$$

From Table 54.1, the modulus of elasticity for a steel spring is

$$E = \left(30\times10^6\ \frac{\text{lbf}}{\text{in}^2}\right)\left(6.89\times10^3\ \frac{\text{Pa}}{\frac{\text{lbf}}{\text{in}^2}}\right)$$

$$= 206.7\times10^9\text{ Pa}$$

(a) The tip deflection is

$$y = \frac{FL^3}{3EI} = \frac{FL^3}{3E(12.5)h^3}$$

$$h = \left(\frac{FL^3}{37.5yE}\right)^{1/3}$$

$$= \left(\frac{(3.5 \text{ kN})\left(1000 \ \frac{\text{N}}{\text{kN}}\right) \times (610 \text{ mm})^3\left(1000 \ \frac{\text{mm}}{\text{m}}\right)^2}{(37.5)(25 \text{ mm})(206.7 \times 10^9 \text{ Pa})}\right)^{1/3}$$

$$= \boxed{16 \text{ mm}}$$

The width-thickness ratio is

$$\frac{w}{t} = \frac{150 \text{ mm}}{16 \text{ mm}} \approx 9.4$$

Since $w/t \approx 10$, this could be considered a wide beam.

The answer is (C).

(b) The moment at the fixed end is

$$M = FL = \frac{(3.5 \text{ kN})\left(1000 \ \frac{\text{N}}{\text{kN}}\right)(610 \text{ mm})}{1000 \ \frac{\text{mm}}{\text{m}}}$$

$$= 2.135 \times 10^3 \text{ N·m}$$

The bending stress is

$$\sigma = \frac{M}{\frac{I}{c}} = \frac{M}{25h^2}$$

$$h = \sqrt{\frac{M}{25\sigma}}$$

$$= \sqrt{\frac{(2.135 \times 10^3 \text{ N·m})\left(1000 \ \frac{\text{mm}}{\text{m}}\right)}{(25 \text{ mm})(345 \text{ MPa})\left(10^6 \ \frac{\text{Pa}}{\text{MPa}}\right)}}$$

$$= \boxed{15.7 \text{ mm} \quad (16 \text{ mm})}$$

The answer is (A).

6. *Customary U.S. Solution*

From Table 54.4, the numbers of active coils are as follows.

For the inner spring,

$$n_a = n_t - 2 = 12.75 - 2 = 10.75$$

For the outer spring,

$$n_a = n_t - 2 = 10.25 - 2 = 8.25$$

From Eq. 54.8, the spring indices are as follows.

For the inner spring,

$$C = \frac{D}{d} = \frac{1.5 \text{ in}}{0.177 \text{ in}} = 8.47$$

For the outer spring,

$$C = \frac{D}{d} = \frac{2.0 \text{ in}}{0.2253 \text{ in}} = 8.88$$

From Eq. 54.12, the spring constants are as follows.

For the inner spring,

$$k_i = \frac{Gd}{8C^3n_a} = \frac{\left(11.5 \times 10^6 \ \frac{\text{lbf}}{\text{in}^2}\right)(0.177 \text{ in})}{(8)(8.47)^3(10.75)}$$

$$= 38.95 \text{ lbf/in}$$

For the outer spring,

$$k_o = \frac{Gd}{8C^3n_a} = \frac{\left(11.5 \times 10^6 \ \frac{\text{lbf}}{\text{in}^2}\right)(0.2253 \text{ in})}{(8)(8.88)^3(8.25)}$$

$$= 56.06 \text{ lbf/in}$$

The inner spring is longer by $\delta_i = 4.5 \text{ in} - 3.75 \text{ in} = 0.75 \text{ in}$, so the force absorbed by the inner spring over this distance is

$$F_i = k_i\delta_i = \left(38.95 \ \frac{\text{lbf}}{\text{in}}\right)(0.75 \text{ in}) = 29.2 \text{ lbf}$$

The force shared by both springs is

$$F = F_s - F_i = 150 \text{ lbf} - 29.2 \text{ lbf} = 120.8 \text{ lbf}$$

The composite spring constant for both springs is

$$k = k_i + k_o = 38.95 \ \frac{\text{lbf}}{\text{in}} + 56.06 \ \frac{\text{lbf}}{\text{in}} = 95.0 \text{ lbf/in}$$

The deflection of the composite spring is

$$\delta_c = \frac{F}{k} = \frac{120.8 \text{ lbf}}{95.0 \ \frac{\text{lbf}}{\text{in}}} = 1.27 \text{ in}$$

(a) The total deflection of the inner spring is

$$\delta_{\text{total}} = \delta_i + \delta_c = 0.75 \text{ in} + 1.27 \text{ in} = \boxed{2.02 \text{ in} \quad (2.0 \text{ in})}$$

The answer is (A).

(b) The maximum force exerted by the inner spring is

$$F = k_i\delta_{\text{total}} = \left(38.95 \ \frac{\text{lbf}}{\text{in}}\right)(2.02 \text{ in}) = \boxed{78.7 \text{ lbf} \quad (79 \text{ lbf})}$$

The answer is (C).

(c) From Eq. 54.13, the Wahl correction factor for the inner spring is

$$W = \frac{4C-1}{4C-4} + \frac{0.615}{C} = \frac{(4)(8.47)-1}{(4)(8.47)-4} + \frac{0.615}{8.47} = 1.173$$

From Eq. 54.14, the maximum shear stress in the inner spring is

$$\tau_{\text{max}} = \frac{8FCW}{\pi d^2} = \frac{(8)(78.7 \text{ lbf})(8.47)(1.173)}{\pi(0.177 \text{ in})^2} = \boxed{63{,}555 \text{ lbf/in}^2 \quad (64 \text{ kips/in}^2)}$$

The answer is (D).

(d) The yield strength is 75% of the ultimate strength.

$$S_{yt} = 0.75S_{ut} = (0.75)\left(204{,}000 \ \frac{\text{lbf}}{\text{in}^2}\right) = 153{,}000 \text{ lbf/in}^2$$

According to the maximum shear stress failure theory, the maximum allowable shear stress is

$$\tau_{\text{max},a} = 0.5S_{yt} = (0.5)\left(153{,}000 \ \frac{\text{lbf}}{\text{in}^2}\right) = 76{,}500 \text{ lbf/in}^2$$

The factor of safety in shear for the inner spring is

$$\text{FS} = \frac{\tau_{\text{max},a}}{\tau_{\text{max}}} = \frac{76{,}500 \ \frac{\text{lbf}}{\text{in}^2}}{63{,}555 \ \frac{\text{lbf}}{\text{in}^2}} = \boxed{1.2}$$

The answer is (A).

(e) The inner and outer springs should be wound with opposite (i.e., $\boxed{\text{reverse}}$) direction helixes. This configuration will minimize resonance and prevent coils from one spring from entering the other spring's gaps.

The answer is (D).

SI Solution

From Table 54.4, the numbers of active coils are as follows.

For the inner spring,

$$n_a = n_t - 2 = 12.75 - 2 = 10.75$$

For the outer spring,

$$n_1 = n_t - 2 = 10.25 - 2 = 8.25$$

From Eq. 54.8, the spring indices are as follows.

For the inner spring,

$$C = \frac{D}{d} = \frac{38 \text{ mm}}{4.5 \text{ mm}} = 8.44$$

For the outer spring,

$$C = \frac{D}{d} = \frac{51 \text{ mm}}{5.723 \text{ mm}} = 8.91$$

From Eq. 54.12, the spring constants are as follows.

For the inner spring,

$$k_i = \frac{Gd}{8C^3n_a} = \frac{(79 \times 10^9 \text{ Pa})(4.5 \text{ mm})}{(8)(8.44)^3(10.75)\left(1000 \ \frac{\text{mm}}{\text{m}}\right)} = 6876 \text{ N/m}$$

For the outer spring,

$$k_o = \frac{Gd}{8C^3n_a} = \frac{(79 \times 10^9 \text{ Pa})(5.723 \text{ mm})}{(8)(8.91)^3(8.25)\left(1000 \ \frac{\text{mm}}{\text{m}}\right)} = 9684 \text{ N/m}$$

The inner spring is longer by $\delta_i = 115 \text{ mm} - 95.3 \text{ mm} = 19.7 \text{ mm}$, so the force absorbed by the inner spring over this distance is

$$F_i = k_i\delta_i = \frac{\left(6876 \ \frac{\text{N}}{\text{m}}\right)(19.7 \text{ mm})}{1000 \ \frac{\text{mm}}{\text{m}}} = 135.5 \text{ N}$$

The force shared by both springs is

$$F = F_s - F_i = 600\ \text{N} - 135.5\ \text{N} = 464.5\ \text{N}$$

The composite spring constant for both springs is

$$k = k_i + k_o = 6876\ \frac{\text{N}}{\text{m}} + 9684\ \frac{\text{N}}{\text{m}} = 16\,560\ \text{N/m}$$

The deflection of the composite spring is

$$\delta_c = \frac{F}{k} = \left(\frac{(464.5\ \text{N})\left(1000\ \frac{\text{mm}}{\text{m}}\right)}{16\,560\ \frac{\text{N}}{\text{m}}}\right) = 28.0\ \text{mm}$$

(a) The total deflection of the inner spring is

$$\begin{aligned}\delta_{\text{total}} &= \delta_i + \delta_c = 19.7\ \text{mm} + 28.0\ \text{mm} \\ &= \boxed{47.7\ \text{mm} \quad (48\ \text{mm})}\end{aligned}$$

The answer is (A).

(b) The maximum force exerted by the inner spring is

$$\begin{aligned}F = k_i\delta_{\text{total}} &= \frac{\left(6876\ \frac{\text{N}}{\text{m}}\right)(47.7\ \text{mm})}{1000\ \frac{\text{mm}}{\text{m}}} \\ &= \boxed{328.0\ \text{N} \quad (330\ \text{N})}\end{aligned}$$

The answer is (C).

(c) From Eq. 54.13, the Wahl correction factor for the inner spring is

$$\begin{aligned}W &= \frac{4C-1}{4C-4} + \frac{0.615}{C} = \frac{(4)(8.44)-1}{(4)(8.44)-4} + \frac{0.615}{8.44} \\ &= 1.174\end{aligned}$$

From Eq. 54.14, the maximum shear stress in the inner spring is

$$\begin{aligned}\tau_{\max} = \frac{8FCW}{\pi d^2} &= \frac{(8)(328.0\ \text{N})(8.44) \times (1.174)\left(1000\ \frac{\text{mm}}{\text{m}}\right)^2}{\pi(4.5\ \text{mm})^2} \\ &= \boxed{4.087 \times 10^8\ \text{Pa} \quad (410\ \text{MPa})}\end{aligned}$$

The answer is (D).

(d) The yield strength is 75% of the ultimate strength.

$$\begin{aligned}S_{yt} &= 0.75S_{ut} = (0.75)(1.3 \times 10^9\ \text{Pa}) \\ &= 9.75 \times 10^8\ \text{Pa}\end{aligned}$$

According to the maximum shear stress failure theory, the maximum allowable shear stress is

$$\begin{aligned}\tau_{\max,a} &= 0.5S_{yt} = (0.5)(9.75 \times 10^8\ \text{Pa}) \\ &= 4.875 \times 10^8\ \text{Pa} \quad (487.5\ \text{MPa})\end{aligned}$$

The factor of safety in shear for the inner spring is

$$\text{FS} = \frac{\tau_{\max,a}}{\tau_{\max}} = \frac{487.5\ \text{MPa}}{408.7\ \text{MPa}} = \boxed{1.19 \quad (1.2)}$$

The answer is (A).

(e) The inner and outer springs should be wound with opposite (i.e., $\boxed{\text{reverse}}$) direction helixes. This configuration will minimize resonance and prevent coils from the spring entering from the other spring's gaps.

The answer is (D).

7. From App. 54.A, the thickness of 0-gage standard plate is $h = 0.313$ in.

The moment of inertia, I, for one beam is

$$\begin{aligned}I &= \tfrac{1}{12}bh^3 = \left(\tfrac{1}{12}\right)(1.75\ \text{in})(0.313\ \text{in})^3 \\ &= 4.47 \times 10^{-3}\ \text{in}^4\end{aligned}$$

From case 1 of App. 51.A, the deflection for an end load is

$$\delta = \frac{PL^3}{3EI}$$

From Hooke's law, the resulting spring constant is

$$k = \frac{P}{\delta} = \frac{3EI}{L^3}$$

Solving this equation for length,

$$\begin{aligned}L &= \sqrt[3]{\frac{3EI}{k}} = \sqrt[3]{\frac{(3)\left(30 \times 10^6\ \frac{\text{lbf}}{\text{in}^2}\right)(4.47 \times 10^{-3}\ \text{in}^4)}{25{,}758\ \frac{\text{lbf}}{\text{in}}}} \\ &= 2.5\ \text{in}\end{aligned}$$

The longitudinal strain at the surface is related to the stress, moment, and force.

$$\sigma = Ee = \frac{Mc}{I} = \frac{PLc}{I}$$

Solve for the tip force, P.

$$\begin{aligned} P &= \frac{EeI}{Lc} = \frac{2EeI}{Lh} \\ &= \frac{(2)\left(30 \times 10^6 \ \frac{\text{lbf}}{\text{in}^2}\right)\left(1.2 \times 10^{-3} \ \frac{\text{in}}{\text{in}}\right) \times (4.47 \times 10^{-3} \ \text{in}^4)}{(2.5 \ \text{in})(0.313 \ \text{in})} \\ &= 411.3 \ \text{lbf} \end{aligned}$$

With the beams equally loaded, the user weight, W, is the load per beam multiplied by 4.

$$\begin{aligned} W &= 4P \\ &= (4)(411.3 \ \text{lbf}) \\ &= \boxed{1645.2 \ \text{lbf} \quad (1600 \ \text{lbf})} \end{aligned}$$

The answer is (D).

8. The angle of twist, γ, of a shaft due to an applied torque is given by Eq. 53.54.

$$\gamma = \frac{TL}{GJ}$$

From Eq. 53.54 and Hooke's law, the torsional spring constant, k, is

$$k = \frac{T}{\gamma} = \frac{GJ}{L}$$

The shear modulus values are found from Table 48.5.

$$G_{\text{steel}} = 11.5 \times 10^6 \ \text{lbf/in}^2$$

$$G_{\text{aluminum}} = 3.8 \times 10^6 \ \text{lbf/in}^2$$

The inner and outer diameters for 2 in NPS schedule-80 pipe are found from App. 16.B.

$$d_i = 1.939 \ \text{in}$$

$$d_o = 2.375 \ \text{in}$$

The polar moment of inertia for the steel circular shaft, J_{steel}, is

$$\begin{aligned} J_{\text{steel}} &= \frac{\pi}{32}(d_o^4 - d_i^4) \\ &= \left(\frac{\pi}{32}\right)\left((2.375 \ \text{in})^4 - (1.939 \ \text{in})^4\right) \\ &= 1.736 \ \text{in}^4 \end{aligned}$$

The polar moment of inertia for the aluminum solid section, J_{aluminum}, is found from Table 53.6.

$$\begin{aligned} J_{\text{aluminum}} &= 0.1406a^4 \\ &= (0.1406)(2 \ \text{in})^4 \\ &= 2.250 \ \text{in}^4 \end{aligned}$$

The torsional spring constants for the steel and aluminum sections are

$$\begin{aligned} k_{\text{steel}} &= \frac{G_{\text{steel}} J_{\text{steel}}}{L_{\text{steel}}} \\ &= \frac{\left(11.5 \times 10^6 \ \frac{\text{lbf}}{\text{in}^2}\right)(1.736 \ \text{in}^4)}{12 \ \text{in}} \\ &= 16.6 \times 10^5 \ \text{in-lbf/rad} \end{aligned}$$

$$\begin{aligned} k_{\text{aluminum}} &= \frac{G_{\text{aluminum}} J_{\text{aluminum}}}{L_{\text{aluminum}}} \\ &= \frac{\left(3.8 \times 10^6 \ \frac{\text{lbf}}{\text{in}^2}\right)(2.250 \ \text{in}^4)}{12 \ \text{in}} \\ &= 7.13 \times 10^5 \ \text{in-lbf/rad} \end{aligned}$$

The equivalent torsional spring constant, k_{eq}, is found from Eq. 54.5 for springs in series.

$$\begin{aligned} k_{\text{eq}} &= \left(\frac{1}{k_{\text{steel}}} + \frac{1}{k_{\text{aluminum}}}\right)^{-1} \\ &= \left(\frac{1}{16.6 \times 10^5 \ \frac{\text{in-lbf}}{\text{rad}}} + \frac{1}{7.13 \times 10^5 \ \frac{\text{in-lbf}}{\text{rad}}}\right)^{-1} \\ &= \boxed{4.99 \times 10^5 \ \text{in-lbf/rad} \quad (5.0 \times 10^5 \ \text{in-lbf/rad})} \end{aligned}$$

The answer is (A).

9. The maximum speed ratio is

$$\frac{N_{\text{max}}}{N_{\text{min}}} = \frac{96 \ \text{teeth}}{12 \ \text{teeth}} = 8$$

For three stages,

$$(8)^3 = 512$$

Since this is less than 600, $\boxed{\text{four stages are required.}}$

$$\text{check: } (8)^4 = 4096$$

The answer is (C).

10. *Customary U.S. Solution*

(a) From Eq. 54.40, the speed of the intermediate shaft is

$$\begin{aligned} n_{\text{int}} &= n_{\text{input}}\left(\frac{N_{\text{input}}}{N_{\text{int at input}}}\right) \\ &= \left(1800\ \frac{\text{rev}}{\text{min}}\right)\left(\frac{50\ \text{teeth}}{25\ \text{teeth}}\right) \\ &= 3600\ \text{rev/min} \end{aligned}$$

The speed of the output shaft is

$$\begin{aligned} n_o &= n_{\text{int}}\left(\frac{N_{\text{int at input}}}{N_o}\right) \\ &= \left(3600\ \frac{\text{rev}}{\text{min}}\right)\left(\frac{60\ \text{teeth}}{20\ \text{teeth}}\right) \\ &= \boxed{10{,}800\ \text{rev/min} \quad (11{,}000\ \text{rev/min})} \end{aligned}$$

The answer is (C).

(b) From Eq. 54.33(b), with two gear sets, the torque output is

$$\begin{aligned} T_o &= \frac{P_o\left(63{,}025\ \frac{\text{in-lbf}}{\text{hp-min}}\right)}{n_o} = \frac{\eta_{\text{mesh}}^2 P_i\left(63{,}025\ \frac{\text{in-lbf}}{\text{hp-min}}\right)}{n_o} \\ &= \frac{(0.98)^2(50\ \text{hp})\left(63{,}025\ \frac{\text{in-lbf}}{\text{hp-min}}\right)}{10{,}800\ \frac{\text{rev}}{\text{min}}} \\ &= \boxed{280.2\ \text{in-lbf} \quad (280\ \text{in-lbf})} \end{aligned}$$

The answer is (B).

(c) For the gear at section A-A, the pitch diameter is

$$d = \frac{N}{P} = \frac{60\ \text{teeth}}{5\ \frac{\text{teeth}}{\text{in}}} = 12\ \text{in}$$

From Eq. 54.32, the pitch circle velocity is

$$\begin{aligned} \text{v}_t &= \pi d n_{\text{rpm}} \\ &= \frac{\pi(12\ \text{in})\left(3600\ \frac{\text{rev}}{\text{min}}\right)}{12\ \frac{\text{in}}{\text{ft}}} \\ &= \boxed{11{,}310\ \text{ft/min} \quad (11{,}000\ \text{ft/min})} \end{aligned}$$

The answer is (A).

(d) From Eq. 54.46, the horsepower transmitted is

$$\begin{aligned} P &= \eta_{\text{mesh}} P_i \\ &= (0.98)(50\ \text{hp}) \\ &= 49\ \text{hp} \end{aligned}$$

From Eq. 54.49(b), the transmitted load is

$$\begin{aligned} F_t &= \frac{P_{\text{hp}}\left(33{,}000\ \frac{\text{ft-lbf}}{\text{hp-min}}\right)}{\text{v}_t} \\ &= \frac{(49\ \text{hp})\left(33{,}000\ \frac{\text{ft-lbf}}{\text{hp-min}}\right)}{11{,}310\ \frac{\text{ft}}{\text{min}}} \\ &= \boxed{143.0\ \text{lbf} \quad (140\ \text{lbf})} \end{aligned}$$

The answer is (B).

(e) The radial thrust (separation) force for a helical gear is

$$\begin{aligned} F_r &= F_t \tan\phi_t \\ &= \frac{F_t \tan\phi_n}{\cos\psi} \\ &= \frac{(143\ \text{lbf})\tan 20^\circ}{\cos 25^\circ} \\ &= 57.4\ \text{lbf} \end{aligned}$$

The bending moment on the shaft due to the thrust force is

$$\begin{aligned} M_r &= F_r L \\ &= (57.4\ \text{lbf})(4\ \text{in}) \\ &= 230\ \text{in-lbf} \end{aligned}$$

The bending moment on the shaft due to the tangential force is

$$\begin{aligned} M_t &= F_t r \\ &= (143\ \text{lbf})(4\ \text{in}) \\ &= 572\ \text{in-lbf} \end{aligned}$$

The moments are additive, so the total moment in the vertical plane is

$$\begin{aligned} M &= M_r + M_t \\ &= 230 \text{ in-lbf} + 572 \text{ in-lbf} \\ &= 802 \text{ in-lbf} \end{aligned}$$

The torsional moment on the shaft is

$$\begin{aligned} T &= F_t\left(\frac{d}{2}\right) \\ &= (143.0 \text{ lbf})\left(\frac{12 \text{ in}}{2}\right) \\ &= 858 \text{ in-lbf} \end{aligned}$$

According to the maximum shear stress failure theory, the shear stress is limited to

$$\begin{aligned} \tau_{\max} &= 0.5S_{yt} \\ &= (0.5)\left(69{,}000 \ \frac{\text{lbf}}{\text{in}^2}\right) \\ &= 34{,}500 \text{ lbf/in}^2 \end{aligned}$$

With a factor of safety of 2, the allowable shear stress is

$$\begin{aligned} \tau_a &= \frac{\tau_{\max}}{2} \\ &= \frac{34{,}500 \ \frac{\text{lbf}}{\text{in}^2}}{2} \\ &= 17{,}250 \text{ lbf/in}^2 \end{aligned}$$

Disregarding the axial loading, from Eq. 53.58, the minimum shaft diameter is

$$\begin{aligned} d_s &= \left(\frac{16\sqrt{M^2 + T^2}}{\pi\tau_{\max}}\right)^{1/3} \\ &= \left(\frac{16\sqrt{(802 \text{ in-lbf})^2 + (858 \text{ in-lbf})^2}}{\pi\left(17{,}250 \ \frac{\text{lbf}}{\text{in}^2}\right)}\right)^{1/3} \\ &= 0.70 \text{ in} \end{aligned}$$

Check the normal stress criterion. The maximum allowable normal stress is

$$\sigma_{\max} = \frac{S_{yt}}{\text{FS}} = \frac{69{,}000 \ \frac{\text{lbf}}{\text{in}^2}}{2} = 34{,}500 \text{ lbf/in}^2$$

From Eq. 53.59, the minimum diameter is

$$\begin{aligned} d &= \left(\frac{16}{\pi\sigma_{\max}}\sqrt{4M^2 + 3T^2}\right)^{1/3} \\ &= \left(\frac{16}{\pi\left(34{,}500 \ \frac{\text{lbf}}{\text{in}^2}\right)}\sqrt{\begin{array}{c}(4)(802 \text{ in-lbf})^2 \\ + (3)(858 \text{ in-lbf})^2\end{array}}\right)^{1/3} \\ &= 0.69 \text{ in} \end{aligned}$$

This is smaller than the diameter calculated from the shear stress criterion, so $d = \boxed{0.70 \text{ in.}}$

Check the axial stress.

From Eq. 54.57, the axial force is

$$F_a = F_t \tan\psi = (143.0 \text{ lbf})(\tan 25°) = 66.7 \text{ lbf}$$

The cross-sectional area of the shaft is

$$A = \frac{\pi}{4}d_s^2 = \left(\frac{\pi}{4}\right)(0.70 \text{ in})^2 = 0.385 \text{ in}^2$$

The axial stress is

$$\sigma_{\text{axial}} = \frac{F_a}{A} = \frac{66.7 \text{ lbf}}{0.385 \text{ in}^2} = 173 \text{ lbf/in}^2$$

Since σ_{axial} is much less than the yield strength, it is insignificant in this problem.

The answer is (C).

SI Solution

(a) From Eq. 54.40, the speed of the intermediate shaft is

$$\begin{aligned} n_{\text{int}} &= n_{\text{input}}\left(\frac{N_{\text{input}}}{N_{\text{int at output}}}\right) = \left(1800 \ \frac{\text{rev}}{\text{min}}\right)\left(\frac{50 \text{ teeth}}{25 \text{ teeth}}\right) \\ &= 3600 \text{ rev/min} \end{aligned}$$

The speed of the output shaft is

$$\begin{aligned} n_o &= n_{\text{int}}\left(\frac{N_{\text{int at output}}}{N_o}\right) = \left(3600 \ \frac{\text{rev}}{\text{min}}\right)\left(\frac{60 \text{ teeth}}{20 \text{ teeth}}\right) \\ &= \boxed{10{,}800 \text{ rev/min} \quad (11{,}000 \text{ rev/min})} \end{aligned}$$

The answer is (C).

(b) From Eq. 54.33(a), with two gear sets, the torque output is

$$T_o = \frac{P_o\left(9549\ \frac{\text{N·m}}{\text{kW·min}}\right)}{n_o} = \frac{\eta_{\text{mesh}}^2 P_i\left(9549\ \frac{\text{N·m}}{\text{kW·min}}\right)}{n_o}$$

$$= \frac{(0.98)^2(38\ \text{kW})\left(9549\ \frac{\text{N·m}}{\text{kW·min}}\right)}{10{,}800\ \frac{\text{rev}}{\text{min}}}$$

$$= \boxed{32.27\ \text{N·m}\quad(32\ \text{N·m})}$$

The answer is (B).

(c) For the gear at section A-A, the pitch diameter is

$$d = Nm = (60\ \text{teeth})\left(5\ \frac{\text{mm}}{\text{tooth}}\right) = 300\ \text{mm}$$

From Eq. 54.32, the pitch circle velocity is

$$\text{v}_t = \pi d n_{\text{rpm}} = \frac{\pi(300\ \text{mm})\left(3600\ \frac{\text{rev}}{\text{min}}\right)}{\left(60\ \frac{\text{sec}}{\text{min}}\right)\left(1000\ \frac{\text{mm}}{\text{m}}\right)}$$

$$= \boxed{56.55\ \text{m/s}\quad(57\ \text{m/s})}$$

The answer is (A).

(d) From Eq. 54.46, the power transmitted is

$$P = n_{\text{mesh}} P_i = (0.98)(38\ \text{kW}) = 37.24\ \text{kW}$$

From Eq. 54.49(a), the transmitted load is

$$F_t = \frac{P_{\text{kW}}\left(1000\ \frac{\text{W}}{\text{kW}}\right)}{\text{v}_t} = \frac{(37.24\ \text{kW})\left(1000\ \frac{\text{W}}{\text{kW}}\right)}{56.55\ \frac{\text{m}}{\text{s}}}$$

$$= \boxed{658.5\ \text{N}\quad(660\ \text{N})}$$

The answer is (B).

(e) The radial thrust (separation) force for a helical gear is

$$F_r = F_t \tan\phi_t = \frac{F_t \tan\phi_n}{\cos\psi} = \frac{(658.5\ \text{N})\tan 20°}{\cos 25°}$$

$$= 264.5\ \text{N}$$

The bending moment on the shaft due to the thrust force is

$$M_r = F_r L = \frac{(264.5\ \text{N})(100\ \text{mm})}{1000\ \frac{\text{mm}}{\text{m}}} = 26.45\ \text{N·m}$$

The bending moment on the shaft due to the tangential force is

$$M_t = F_t r = (658.5\ \text{N})\left(\frac{300\ \text{mm}}{(2)\left(1000\ \frac{\text{mm}}{\text{m}}\right)}\right)$$

$$= 98.78\ \text{N·m}$$

The moments are additive, so the total moment in the vertical plane is

$$M = M_r + M_t = 26.45\ \text{N·m} + 98.78\ \text{N·m}$$

$$= 125.23\ \text{N·m}$$

The torsional moment on the shaft is

$$T = F_t\left(\frac{d}{2}\right) = \frac{(658.5\ \text{N})\left(\frac{300\ \text{mm}}{2}\right)}{1000\ \frac{\text{mm}}{\text{m}}} = 98.78\ \text{N·m}$$

According to the maximum shear stress failure theory, the shear stress is limited to

$$\tau_{\max} = 0.5 S_{yt} = (0.5)(480\ \text{MPa}) = 240\ \text{MPa}$$

With a factor of safety of 2, the allowable shear stress is

$$\tau_a = \frac{\tau_{\max}}{2} = \frac{240\ \text{MPa}}{2} = 120\ \text{MPa}$$

Disregarding the axial loading, from Eq. 53.58, the minimum shaft diameter is

$$d_s = \left(\frac{16\sqrt{M^2 + T^2}}{\pi\tau_{\max}}\right)^{1/3}$$

$$= \left(\frac{16\sqrt{(125.23\ \text{N·m})^2 + (98.78\ \text{N·m})^2}}{\pi(120\times 10^6\ \text{Pa})}\right)^{1/3}$$

$$= 0.0189\ \text{m}$$

Check the normal stress criterion. The maximum allowable normal stress is

$$\sigma_{\max} = \frac{S_{yt}}{\text{FS}} = \frac{480\ \text{MPa}}{2} = 240\ \text{MPa}$$

From Eq. 53.59, the minimum diameter is

$$d = \left(\frac{16}{\pi\tau_{\max}}\sqrt{4M^2 + 3T^2}\right)^{1/3}$$

$$= \left(\frac{16}{\pi(240 \times 10^6 \text{ Pa})}\sqrt{\begin{array}{l}(4)(125.23 \text{ N·m})^2 \\ + (3)(98.78 \text{ N·m})^2\end{array}}\right)^{1/3}$$

$$= 0.0186 \text{ m}$$

This is smaller than the diameter calculated from the shear stress criterion, so $d = \boxed{0.0189 \text{ m (19 mm)}}.$

Check the axial stress.

From Eq. 54.57, the axial force is

$$F_a = F_t \tan\psi = (658.5 \text{ N})(\tan 25°) = 307.1 \text{ N}$$

The cross-sectional area of the shaft is

$$A = \frac{\pi}{4}d_s^2 = \left(\frac{\pi}{4}\right)(0.0189 \text{ m})^2 = 2.81 \times 10^{-4} \text{ m}$$

The axial stress is

$$\sigma_{\text{axial}} = \frac{F_a}{A} = \frac{307.1 \text{ N}}{2.81 \times 10^{-4} \text{ m}} = 1.09 \times 10^6 \text{ Pa}$$

Since σ_{axial} is much less than the yield strength, it is insignificant in this problem.

The answer is (C).

11. The circular pitch, p, depends on the diametral pitch, P.

$$p = \pi\frac{1}{P} = \pi\left(\frac{1}{\pi \text{ in}^{-1}}\right) = 1 \text{ in}$$

The base pitch, p_b, is a function of the circular pitch and pressure angle.

$$p_b = p\cos\phi$$

The pressure angle is

$$\phi = 90° - 65° = 25°$$

The base pitch is

$$p_b = p\cos\phi = (1 \text{ in})(\cos 25°) = \boxed{0.906 \text{ in} \quad (0.91 \text{ in})}$$

The answer is (D).

12. The clearance, C, depends on the diametral pitch, P.

$$P = \frac{\pi}{p} = \frac{\pi}{1 \text{ in}} = \pi \text{ in}^{-1}$$

The minimum clearance, C, is

$$C = \frac{0.157}{P} = \frac{0.157}{\pi \text{ in}^{-1}} = \boxed{0.049 \text{ in} \quad (0.05 \text{ in})}$$

The answer is (B).

13. The number of teeth on the gear, N_G, is

$$N_G = d_G P_G = (80 \text{ in})(1 \text{ in}^{-1}) = 80$$

The velocity ratio, VR, is equivalent to the gear ratio.

$$\text{VR} = 4$$

The number of pinion teeth, N_P, is

$$N_P = \frac{N_G}{\text{VR}} = \frac{80}{4} = 20$$

The pinion gear's diametral pitch, P_P, is

$$P_P = \frac{N_P}{d_P} = \frac{20}{20 \text{ in}} = \boxed{1 \text{ in}^{-1}}$$

The answer is (D).

14. The circular pitch, p, is

$$p = \frac{p_b}{\cos\phi} = \frac{1 \text{ in}}{\cos 14.5°} = 1.03 \text{ in}$$

The diametral pitch, P, is

$$P = \frac{\pi}{p} = \frac{\pi}{1.03 \text{ in}} = 3.05 \text{ in}^{-1}$$

The thickness, t, on the pitch line of a full-depth involute gear system is

$$t = \frac{1.5708}{P} = \frac{1.5708}{3.05 \text{ in}^{-1}} = \boxed{0.515 \text{ in} \quad (0.52 \text{ in})}$$

The answer is (B).

15. The gear's diametral pitch, P, is

$$P = \frac{\pi}{p} = \frac{\pi}{1 \text{ in}} = \pi \text{ in}^{-1}$$

The gear's pitch diameter, d_G, is

$$d_G = \frac{N}{P} = \frac{50}{\pi \text{ in}^{-1}} = 15.9 \text{ in}$$

The velocity ratio, VR, is the same as the gear ratio. The velocity ratio is required to calculate the rotational speed of the gear, n_G.

$$n_G = \frac{n_P}{\text{VR}} = \frac{1000\ \frac{\text{rev}}{\text{min}}}{4} = 250\ \text{rev/min}$$

The tangential pitch velocity, v_t, of the pinion is the same as the gear's pitch velocity.

$$\text{v}_t = \pi d_G n_G = \frac{\left(\pi \frac{1}{\text{rev}}\right)(15.9\ \text{in})\left(250\ \frac{\text{rev}}{\text{min}}\right)}{12\ \frac{\text{in}}{\text{ft}}}$$

$$= \boxed{1041\ \text{ft/min}\quad (1000\ \text{ft/min})}$$

The answer is (A).

16. The basic rack helix angle, ψ_r, is

$$\psi_r = 180° - 160° = 20°$$

The transverse rack pitch, p_{tr}, is

$$p_{\text{tr}} = \frac{p_{\text{nr}}}{\cos\psi_r} = \frac{\pi\ \text{in}}{\cos 20°} = 3.34\ \text{in}$$

The transverse diametral pitch, P_t, is a function of transverse rack pitch.

$$P_t = \frac{\pi}{p_{\text{tr}}} = \frac{\pi}{3.34\ \text{in}} = \boxed{0.94\ \text{in}^{-1}\quad (0.94\ \text{in})}$$

The answer is (D).

17. The differential gear's singular function is to equalize traction of both wheels and permit one wheel to turn faster than the other, as needed, on curves.

The answer is (A).

18. The two identical bevel gears in an automobile differential gear act as idler gears and separate the two identical gears that drive the axle shafts. Therefore, the bevel gears serve the same function as the ring and sun gears in an epicyclic gear set.

The answer is (C).

19. Angular bevel gears connect shafts at angles other than 90°.

The answer is (A).

20. The three types of bevel gears are spiral, zerol, and straight. Spiral bevel gears are used for speeds in excess of 1000 rev/min, high loads, and quiet operations. Zerol and straight bevel gears are used for speeds less than 1000 rev/min and light loads.

The answer is (D).

21. The spiral angle is the angle between the tooth trace and an element of the pitch cone. The root angle is the angle formed between a tooth root element and the axis of the bevel gear. The shaft angle is the angle between the mating bevel-gear axes; it is also the sum of the two pitch angles.

The answer is (A).

22. The helix angle, ψ, is a function of the tangential, ϕ_t, and normal pressure angles, ϕ_n.

$$\cos\psi = \frac{\tan\phi_n}{\tan\phi_t} = \frac{\tan 20°}{\tan 22.5°} = 0.879$$

$$\arccos\psi = \arccos 0.879$$

$$\psi = 28.5°$$

The total force, F_t, is

$$F_t = \frac{F_a}{\cos\phi_n \sin\psi} = \frac{1000\ \text{lbf}}{(\cos 20°)(\sin 28.5°)}$$

$$= \boxed{2230\ \text{lbf}\quad (2200\ \text{lbf})}$$

The answer is (B).

23. The helix angle, ψ, is a function of the axial, F_a, and tangential, F_t, forces.

$$\tan\psi = \frac{F_a}{F_t} = \frac{1000\ \text{lbf}}{1500\ \text{lbf}} = 0.667$$

$$\arctan\psi = \arctan 0.667$$

$$\psi = 33.7°$$

The normal circular pitch, p_n, is

$$p_n = p_t\cos\psi = (1\ \text{in})(\cos 33.7°) = \boxed{0.832\ \text{in}\quad (0.83\ \text{in})}$$

The answer is (D).

24. *Customary U.S. Solution*

(a) Let gear 1 be the gear and gear 2 be the pinion.

$$\frac{d_1}{2} + \frac{d_2}{2} = \text{center distance} = 15\ \text{in}$$

From Eq. 54.40,

$$\frac{r_1}{r_2} = \frac{d_1}{d_2} = \frac{n_2}{n_1} = \frac{1200 \ \frac{\text{rev}}{\text{min}}}{270 \ \frac{\text{rev}}{\text{min}}}$$

$$= 4.44$$

Solving these two equations simultaneously gives

$$d_1 = \boxed{24.49 \text{ in} \quad (24 \text{ in})}$$

$$d_2 = \boxed{5.51 \text{ in} \quad (6 \text{ in})}$$

The answer is (D).

(b) From Eq. 54.32, the pitch circle velocity is

$$\text{v}_t = \pi d_1 n_1 = \frac{\pi(24.49 \text{ in})\left(270 \ \frac{\text{rev}}{\text{min}}\right)}{12 \ \frac{\text{in}}{\text{ft}}}$$

$$= 1731 \text{ ft/min}$$

From Eq. 54.49(b), the transmitted load is

$$F_t = \frac{P_{\text{hp}}\left(33{,}000 \ \frac{\text{ft-lbf}}{\text{hp-min}}\right)}{\text{v}_t} = \frac{(150 \text{ hp})\left(33{,}000 \ \frac{\text{ft-lbf}}{\text{hp-min}}\right)}{1731 \ \frac{\text{ft}}{\text{min}}}$$

$$= 2860 \text{ lbf}$$

The allowable bending stress for the pinion is

$$\sigma_a = \frac{S_e}{\text{FS}} = \frac{90{,}000 \ \frac{\text{lbf}}{\text{in}^2}}{3} = 30{,}000 \text{ lbf/in}^2$$

From Eq. 54.65, the Barth speed factor is

$$k_d = \frac{a}{a + \text{v}_t} = \frac{600 \ \frac{\text{ft}}{\text{min}}}{600 \ \frac{\text{ft}}{\text{min}} + 1731 \ \frac{\text{ft}}{\text{min}}} = 0.257$$

Use an approximate form factor of $Y = 0.3$.

From Eq. 54.64, the diametral pitch is

$$P = \frac{k_d \sigma_a w Y}{F_t} = \frac{(0.257)\left(30{,}000 \ \frac{\text{lbf}}{\text{in}^2}\right)(6 \text{ in})(0.3)}{2860 \text{ lbf}}$$

$$= 4.85 \text{ 1/in} \quad (4.0 \text{ 1/in})$$

Use a standard diametral pitch of $\boxed{4.0 \text{ 1/in},}$ rounding down so as not to exceed the allowable stress.

The answer is (A).

(c) The number of teeth on the pinion is

$$N_2 = Pd_2 = \left(4 \ \frac{1}{\text{in}}\right)(5.51 \text{ in}) = 22.04 \text{ teeth} \quad (22 \text{ teeth})$$

The number of teeth on the gear is

$$N_1 = Pd_1 = \left(4 \ \frac{1}{\text{in}}\right)(24.49 \text{ in})$$

$$= 97.96 \text{ teeth} \quad (98 \text{ teeth})$$

For a $14^1/_2°$, 98-tooth gear, the form factor is $Y = 0.37$. The allowable bending stress for the gear is

$$\sigma_a = \frac{S_e}{\text{FS}} = \frac{50{,}000 \ \frac{\text{lbf}}{\text{in}^2}}{3} = 16{,}667 \text{ lbf/in}^2$$

From Eq. 54.64, the face width of the gear is

$$w = \frac{PF_t}{k_d \sigma_a Y} = \frac{\left(4 \ \frac{1}{\text{in}}\right)(2860 \text{ lbf})}{(0.257)\left(16{,}667 \ \frac{\text{lbf}}{\text{in}^2}\right)(0.37)}$$

$$= \boxed{7.22 \text{ in} \quad (7.5 \text{ in})}$$

For a $14^1/_2°$, 22-tooth gear, the form factor is $Y \approx 0.29$. The allowable bending stress for the pinion is

$$\sigma_a = \frac{90{,}000 \ \frac{\text{lbf}}{\text{in}^2}}{3} = 30{,}000 \text{ lbf/in}^2$$

From Eq. 54.64, the face width of the pinion is

$$w = \frac{PF_t}{k_d \sigma_a Y} = \frac{(4)(2860 \text{ lbf})}{(0.257)\left(30{,}000 \ \frac{\text{lbf}}{\text{in}^2}\right)(0.29)}$$

$$= 5.11 \text{ in}$$

Each gear would be 7.5 in wide or larger.

The answer is (D).

SI Solution

(a) Let gear 1 be the gear and gear 2 be the pinion.

$$\frac{d_1}{2} + \frac{d_2}{2} = \text{center distance} = 380 \text{ mm}$$

From Eq. 54.40,

$$\frac{r_1}{r_2} = \frac{d_1}{d_2} = \frac{n_2}{n_1} = \frac{1200 \ \frac{\text{rev}}{\text{min}}}{270 \ \frac{\text{rev}}{\text{min}}}$$

$$= 4.44$$

Solving these two equations simultaneously gives

$$d_1 = \boxed{620.4 \text{ mm} \quad (620 \text{ mm})}$$

$$d_2 = \boxed{139.6 \text{ mm} \quad (140 \text{ mm})}$$

The answer is (D).

(b) From Eq. 54.32, the pitch circle velocity is

$$\mathrm{v}_t = \pi d_1 n_1 = \frac{\pi(0.6204\ \text{m})\left(270\ \frac{\text{rev}}{\text{min}}\right)}{60\ \frac{\text{sec}}{\text{min}}} = 8.77\ \text{m/s}$$

From Eq. 54.49(a), the transmitted load is

$$\begin{aligned} F_t &= \frac{P_{\text{kW}}}{\mathrm{v}_t} = \frac{110\ \text{kW}}{8.77\ \frac{\text{m}}{\text{s}}} \\ &= 12.54\ \text{kN} \end{aligned}$$

The allowable bending stress for the pinion is

$$\sigma_a = \frac{S_e}{\text{FS}} = \frac{620\ \text{MPa}}{3} = 206.7\ \text{MPa}$$

From Eq. 54.65, the Barth speed factor is

$$\begin{aligned} k_d &= \frac{a}{a + \mathrm{v}_t} = \frac{600\ \frac{\text{ft}}{\text{min}}}{600\ \frac{\text{ft}}{\text{min}} + \left(8.77\ \frac{\text{m}}{\text{s}}\right)\left(196.8\ \frac{\frac{\text{ft}}{\text{min}}}{\frac{\text{m}}{\text{s}}}\right)} \\ &= 0.258 \end{aligned}$$

Use an approximate form factor of $Y = 0.3$.

From Eq. 54.64, the diametral pitch is

$$\begin{aligned} P &= \frac{k_d \sigma_a w Y}{F_t} \\ &= \frac{(0.258)(206.7\ \text{MPa})\left(1000\ \frac{\text{kPa}}{\text{MPa}}\right) \times (150\ \text{mm})(0.3)}{(12.54\ \text{kN})\left(1000\ \frac{\text{mm}}{\text{m}}\right)^2} \\ &= 0.191\ 1/\text{mm} \end{aligned}$$

The module is

$$m = \frac{1}{p} = \frac{1}{0.191\ \frac{1}{\text{mm}}} = 5.2\ \text{mm}$$

Use a standard module of $\boxed{5\ \text{mm}}$, rounding down so as not to exceed the allowable stress.

The answer is (A).

(c) The number of teeth on the pinion is

$$N_2 = \frac{d_2}{m} = \frac{139.6\ \text{mm}}{6\ \text{mm}} = 23.3\ \text{teeth} \quad (23\ \text{teeth})$$

The number of teeth on the gear is

$$N_1 = \frac{d_1}{m} = \frac{620.4\ \text{mm}}{6\ \text{mm}} = 103.4\ \text{teeth} \quad (103\ \text{teeth})$$

For a $14^1/_2°$, 103-tooth gear, the form factor is $Y = 0.37$.

The allowable bending stress for the gear is

$$\begin{aligned} \sigma_a &= \frac{S_e}{\text{FS}} = \frac{345\ \text{MPa}}{3} \\ &= 115\ \text{MPa} \end{aligned}$$

From Eq. 54.64, the face width of the gear is

$$\begin{aligned} w &= \frac{PF_t}{k_d \sigma_a Y} = \frac{F_t}{m k_d \sigma_a Y} \\ &= \frac{12.54\ \text{kN}}{(6\ \text{mm})(0.258)(115\ \text{MPa})(0.37)} \times \left(\frac{1000\ \frac{\text{kPa}}{\text{MPa}}}{\left(1000\ \frac{\text{mm}}{\text{m}}\right)^2}\right) \\ &= \boxed{190.4\ \text{mm} \quad (190\ \text{mm})} \end{aligned}$$

For a $14^1/_2°$, 23-tooth gear, the form factor is $Y \approx 0.29$. The allowable bending stress for the pinion is

$$\sigma_a = \frac{620\ \text{MPa}}{3} = 206.7\ \text{MPa}$$

From Eq. 54.64, the face width of the pinion is

$$\begin{aligned} w &= \frac{PF_t}{k_d \sigma_a Y} = \frac{F_t}{m k_d \sigma_a Y} \\ &= \frac{12.54\ \text{kN}}{(6\ \text{mm})(0.258)(206.7\ \text{MPa})(0.29)} \times \left(\frac{1000\ \frac{\text{kPa}}{\text{MPa}}}{\left(1000\ \frac{\text{mm}}{\text{m}}\right)^2}\right) \\ &= 135.1\ \text{mm} \quad (135\ \text{mm}) \end{aligned}$$

Each gear would be 190 mm wide or larger.

The answer is (D).

25. *Customary U.S. Solution*

(a) Let gear 1 be the gear and gear 2 be the pinion. The center distance is

$$\frac{d_1}{2} + \frac{d_2}{2} = 15\ \text{in}$$

From Eq. 54.40,

$$\frac{d_1}{d_2} = \frac{n_2}{n_1} = \frac{250\ \frac{\text{rev}}{\text{min}}}{83.33\ \frac{\text{rev}}{\text{min}}} = 3.0$$

$$d_1 = \boxed{22.5\ \text{in}\quad (23\ \text{in})}$$

$$d_2 = \boxed{7.5\ \text{in}\quad (8.0\ \text{in})}$$

The answer is (D).

(b) From Eq. 54.32, the pitch circle velocity is

$$\text{v}_t = \pi d_1 n_1 = \frac{\pi(22.5\ \text{in})\left(83.33\ \frac{\text{rev}}{\text{min}}\right)}{12\ \frac{\text{in}}{\text{ft}}} = 490.9\ \text{ft/min}$$

From Eq. 54.49(b), the transmitted load is

$$\begin{aligned} F_t &= \frac{P_{\text{hp}}\left(33{,}000\ \frac{\text{ft-lbf}}{\text{hp-min}}\right)}{\text{v}_t} \\ &= \frac{(250\ \text{hp})\left(33{,}000\ \frac{\text{ft-lbf}}{\text{hp-min}}\right)}{490.9\ \frac{\text{ft}}{\text{min}}} \\ &= 16{,}806\ \text{lbf} \end{aligned}$$

From Eq. 54.65, the Barth speed factor is

$$k_d = \frac{a}{a + \text{v}_t} = \frac{600\ \frac{\text{ft}}{\text{min}}}{600\ \frac{\text{ft}}{\text{min}} + 490.9\ \frac{\text{ft}}{\text{min}}} = 0.55$$

Select an approximate form factor of $Y = 0.3$. This value can be refined once the number of teeth is known.

The allowable stress is given so the factor of safety is not used.

From Eq. 54.64, the diametral pitch is

$$\begin{aligned} P &= \frac{k_d \sigma_a w Y}{F_t} = \frac{(0.55)\left(30{,}000\ \frac{\text{lbf}}{\text{in}^2}\right)(6\ \text{in})(0.3)}{16{,}806\ \text{lbf}} \\ &= 1.77\ 1/\text{in}\quad (1.8\ 1/\text{in}) \\ &\boxed{\text{Select } P = 1.8\ 1/\text{in as a standard size.}} \end{aligned}$$

The answer is (C).

(c) The number of teeth on the pinion is

$$\begin{aligned} N_{\text{pinion}} &= P d_2 = \left(2\ \frac{1}{\text{in}}\right)(7.5\ \text{in}) \\ &= \boxed{15\ \text{teeth}} \end{aligned}$$

The number of teeth on the gear is

$$\begin{aligned} N_{\text{gear}} &= P d_1 = \left(2\ \frac{1}{\text{in}}\right)(22.5\ \text{in}) \\ &= \boxed{45\ \text{teeth}} \end{aligned}$$

The answer is (A).

(d) For a 20°, 13-tooth pinion, the form factor is $Y = 0.26$.

From Eq. 54.64, the minimum face width of the pinion is

$$\begin{aligned} w &= \frac{P F_t}{k_d \sigma_a Y} = \frac{\left(2\ \frac{1}{\text{in}}\right)(16{,}806\ \text{lbf})}{(0.55)\left(30{,}000\ \frac{\text{lbf}}{\text{in}^2}\right)(0.26)} \\ &= \boxed{7.835\ \text{in}\quad (8.0\ \text{in})} \end{aligned}$$

The answer is (A).

(e) For a 20°, 39-tooth gear, the form factor is $Y = 0.38$.

The allowable bending stress for the gear is

$$\sigma_a = \frac{\sigma_{\max}}{\text{FS}} = \frac{50{,}000\ \frac{\text{lbf}}{\text{in}^2}}{3} = 16{,}667\ \text{lbf/in}^2$$

From Eq. 54.64, the minimum face width of the gear is

$$\begin{aligned} w &= \frac{P F_t}{k_d \sigma_a Y} = \frac{\left(2\ \frac{1}{\text{in}}\right)(16{,}806\ \text{lbf})}{(0.55)\left(16{,}667\ \frac{\text{lbf}}{\text{in}^2}\right)(0.38)} \\ &= \boxed{9.65\ \text{in}\quad (10\ \text{in})} \end{aligned}$$

The answer is (B).

SI Solution

(a) Let gear 1 be the gear and gear 2 be the pinion. The center distance is

$$\frac{d_1}{2} + \frac{d_2}{2} = 380\ \text{mm}$$

From Eq. 54.40,

$$\frac{d_1}{d_2} = \frac{n_2}{n_1} = \frac{250\ \frac{\text{rev}}{\text{min}}}{83.33\ \frac{\text{rev}}{\text{min}}} = 3.0$$

$$d_1 = \boxed{570\ \text{mm}}$$

$$d_2 = \boxed{190\ \text{mm}}$$

The answer is (D).

(b) From Eq. 54.32, the pitch circle velocity is

$$v_t = \pi d_1 n_1 = \frac{\pi(0.57\ \text{m})\left(83.33\ \frac{\text{rev}}{\text{min}}\right)}{60\ \frac{\text{sec}}{\text{min}}} = 2.49\ \text{m/s}$$

From Eq. 54.49(a), the transmitted load is

$$\begin{aligned} F_t &= \frac{P_{\text{kW}}}{v_t} \\ &= \frac{190\ \text{kW}}{2.49\ \frac{\text{m}}{\text{s}}} \\ &= 76.3\ \text{kN} \end{aligned}$$

From Eq. 54.65, the Barth speed factor is

$$\begin{aligned} k_d &= \frac{a}{a + v_t} \\ &= \frac{600\ \frac{\text{ft}}{\text{min}}}{600\ \frac{\text{ft}}{\text{min}} + \left(2.49\ \frac{\text{m}}{\text{s}}\right)\left(196.8\ \frac{\frac{\text{ft}}{\text{min}}}{\frac{\text{m}}{\text{s}}}\right)} \\ &= 0.55 \end{aligned}$$

Select an approximate form factor of $Y = 0.3$. This value can be refined once the number of teeth is known. The allowable stress is given so the factor of safety is not used.

From Eq. 54.64, the diametral pitch is

$$\begin{aligned} P &= \frac{k_d \sigma_a w Y}{F_t} \\ &= \frac{(0.55)(210\ \text{MPa})\left(1000\ \frac{\text{kPa}}{\text{MPa}}\right) \times (150\ \text{mm})(0.3)}{(76.3\ \text{kN})\left(1000\ \frac{\text{mm}}{\text{m}}\right)^2} \\ &= \boxed{0.068\ 1/\text{mm}} \end{aligned}$$

The answer is (C).

(c) The number of teeth on the pinion is

$$\begin{aligned} N_{\text{pinion}} &= P d_2 = \left(0.0787\ \frac{1}{\text{mm}}\right)(190\ \text{mm}) \\ &= \boxed{14.95\ \text{teeth}\quad (15\ \text{teeth})} \end{aligned}$$

The number of teeth on the gear is

$$\begin{aligned} N_{\text{gear}} &= P d_1 = \left(0.0787\ \frac{1}{\text{mm}}\right)(570\ \text{mm}) \\ &= \boxed{44.9\ \text{teeth}\quad (45\ \text{teeth})} \end{aligned}$$

The answer is (A).

(d) For a 20°, 13-tooth pinion, the form factor is $Y = 0.26$.

From Eq. 54.64, the minimum face width of the pinion is

$$\begin{aligned} w &= \frac{P F_t}{k_d \sigma_a Y} \\ &= \frac{\left(0.0787\ \frac{1}{\text{mm}}\right)(76.3\ \text{kN})\left(1000\ \frac{\text{mm}}{\text{m}}\right)^2}{(0.55)(210\ \text{MPa})\left(1000\ \frac{\text{kPa}}{\text{MPa}}\right)(0.26)} \\ &= \boxed{199\ \text{mm}\quad (200\ \text{mm})} \end{aligned}$$

The answer is (A).

(e) For a 20°, 39-tooth gear, the form factor is $Y = 0.38$.

The allowable bending stress for the gear is

$$\sigma_a = \frac{\sigma_{\text{max}}}{\text{FS}} = \frac{345\ \text{MPa}}{3} = 115\ \text{MPa}$$

From Eq. 54.64, the minimum face width of the gear is

$$\begin{aligned} w &= \frac{P F_t}{k_d \sigma_a Y} = \frac{\left(0.0787\ \frac{1}{\text{mm}}\right)(76.3\ \text{kN})\left(1000\ \frac{\text{mm}}{\text{m}}\right)^2}{(0.55)(115\ \text{MPa})\left(1000\ \frac{\text{kPa}}{\text{MPa}}\right)(0.38)} \\ &= \boxed{249.83\ \text{mm}\quad (250\ \text{mm})} \end{aligned}$$

The answer is (B).

26. To determine the pitch tolerance from AGMA Standard 2000, the diametral pitch and pitch diameter are needed.

The diametral pitch, P, is

$$P = \frac{\pi}{p} = \frac{\pi}{0.785\ \text{in}} = 4.00\ \text{in}^{-1}$$

The pitch diameter, d, is

$$d = \frac{N}{P} = \frac{48}{4.00\ \text{in}^{-1}} = 12.00\ \text{in}$$

Using these values, from AGMA Standard 2000, the pitch tolerance is $\boxed{0.00059\ \text{in.}}$

The answer is (D).

27. The diametral pitch, P, is

$$P = \frac{N}{d} = \frac{24}{15\ \text{in}} = 1.6\ \text{in}^{-1}$$

The bending geometry factor, J, is determined from AGMA Standard 2001. For a spur gear having 24 teeth,

$$J = 0.25$$

The tooth strength, S_t, is

$$S_t = \left(100{,}000\ \frac{\text{lbf}}{\text{in}}\right)\left(\frac{P}{J}\right) = \left(100{,}000\ \frac{\text{lbf}}{\text{in}}\right)\left(\frac{1.6\ \text{in}^{-1}}{0.25}\right)$$

$$= \boxed{640{,}000\ \text{lbf/in}^2}$$

The answer is (D).

28. From the allowable contact stress, s_{ac}, tables in AGMA Standard 2001, the range of allowable contact stress is approximately 85,000 lbf/in^2 to 95,000 lbf/in^2. The largest allowable contact stress value, 95,000 lbf/in^2, is selected to calculate the maximum permissible calculated contact stress.

From AGMA Standard 2001, the pitting life factor, C_L, is 1.00 for 10,000,000 load cycles.

The maximum permissible calculated contact stress, s_c, is

$$s_c = s_{\text{ac}}\left(\frac{C_L C_H}{C_T C_R}\right) = \left(95{,}000\ \frac{\text{lbf}}{\text{in}^2}\right)\left(\frac{(1.00)(1.05)}{(1.00)(1.50)}\right)$$

$$= \boxed{66{,}500\ \text{lbf/in}^2 \quad (67{,}000\ \text{lbf/in}^2)}$$

The answer is (C).

29. To determine the runout tolerance from AGMA Standard 2000, the diametral pitch and pitch diameter must be calculated. The diametral pitch, P, is

$$P = \frac{\pi}{p} = \frac{\pi}{0.785\ \text{in}} = 4.00\ \text{in}^{-1}$$

The pitch diameter, d, is

$$d = \frac{N}{P} = \frac{24}{4.00\ \text{in}^{-1}} = 6.00\ \text{in}$$

From AGMA Standard 2000, the runout tolerance is $\boxed{0.00166\ \text{in}\ (0.0017\ \text{in.})}$

The answer is (C).

30. *Customary U.S. Solution*

From Eq. 54.103, the torque per contact surface is

$$T = \pi f p_{\max} r_i (r_o^2 - r_i^2)$$

$$= \pi(0.12)\left(100\ \frac{\text{lbf}}{\text{in}^2}\right)\left(\frac{2.5\ \text{in}}{2}\right)\left(\left(\frac{4.5\ \text{in}}{2}\right)^2 - \left(\frac{2.5\ \text{in}}{2}\right)^2\right)$$

$$= 164.9\ \text{in-lbf}$$

The slipping torque is

$$T_{\text{slip}} = 3T_{\text{rated}} = (3)(300\ \text{in-lbf}) = 900\ \text{in-lbf}$$

The slipping torque is equal to the torque per contact surface.

$$T_{\text{slip}} = T$$

$$900\ \text{in-lbf} = 164.9M \quad [\text{in in-lbf}]$$

$$M = 5.5$$

Use six contact surfaces. The arrangement is

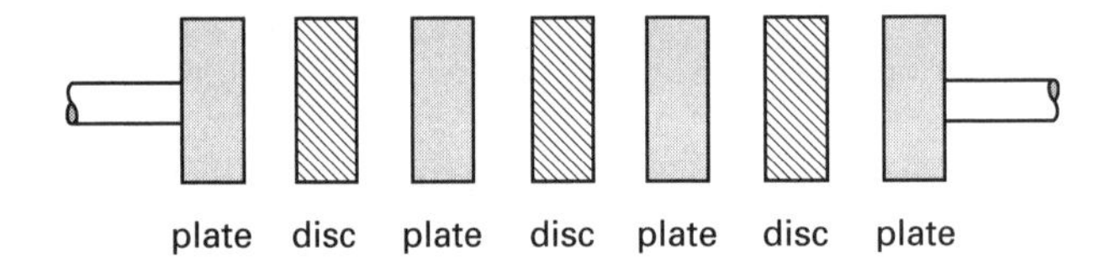

(a) $\boxed{\text{Use four plates.}}$

The answer is (C).

(b) $\boxed{\text{Use three discs.}}$

The answer is (B).

SI Solution

Assume uniform wear. From Eq. 54.103, the torque per contact surface is

$$T = \pi f p_{\max} r_i (r_o^2 - r_i^2)^2$$

$$= \frac{\pi(0.12)(700\ \text{kPa})\left(1000\ \frac{\text{Pa}}{\text{kPa}}\right)\left(\frac{65\ \text{mm}}{2}\right) \times\left(\left(\frac{115\ \text{mm}}{2}\right)^2 - \left(\frac{65\ \text{mm}}{2}\right)^2\right)}{\left(1000\ \frac{\text{mm}}{\text{m}}\right)^3}$$

$$= 19.29\ \text{N·m}$$

The slipping torque is

$$T_{slip} = 3T_{rated} = (3)(33\ \text{N·m}) = 99\ \text{N·m}$$

The slipping torque is equal to the torque per contact surface.

$$T_{slip} = T$$
$$99\ \text{N·m} = 19.30M \quad [\text{in N·m}]$$
$$M = 5.1$$

Use six contact surfaces. The arrangement is

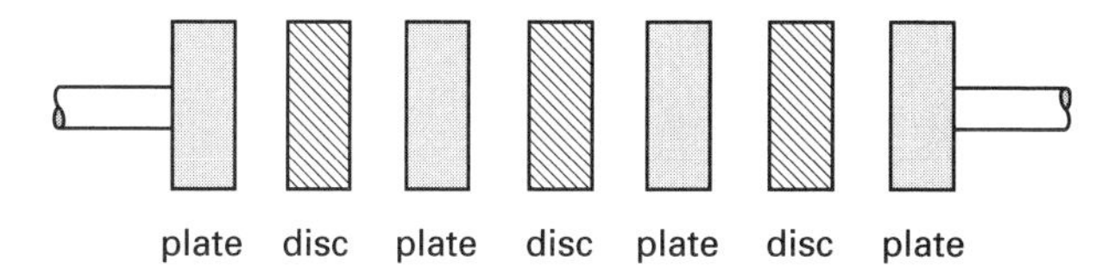

(Equation 54.105 could also have been used.)

(a) $\boxed{\text{Use four plates.}}$

The answer is (C).

(b) $\boxed{\text{Use three discs.}}$

The answer is (B).

31. *Customary U.S. Solution*

(a) A B90 belt has an inside length of 90 in. For a B-type v-belt, the belt length correction is 1.8 in. The pitch length is

$$\begin{aligned} L_p &= L_{inside} + \text{correction} = 90\ \text{in} + 1.8\ \text{in} \\ &= \boxed{91.8\ \text{in} \quad (92\ \text{in})} \end{aligned}$$

The answer is (D).

(b) The ratio of speeds determines the sheave sizes.

$$\begin{aligned} d_{equipment} &= d_{motor}\left(\frac{n_{motor}}{n_{equipment}}\right) = (10\ \text{in})\left(\frac{1750\ \frac{\text{rev}}{\text{min}}}{800\ \frac{\text{rev}}{\text{min}}}\right) \\ &= \boxed{21.875\ \text{in} \quad (22\ \text{in})} \end{aligned}$$

The answer is (B).

SI Solution

(a) A B90 belt has an inside length of 90 in. For a B-type v-belt, the belt length correction is 1.8 in. The pitch length is

$$\begin{aligned} L_p &= L_{inside} + \text{correction} = (90\ \text{in} + 1.8\ \text{in})\left(25.4\ \frac{\text{mm}}{\text{in}}\right) \\ &= \boxed{2332\ \text{mm} \quad (2330\ \text{mm})} \end{aligned}$$

The answer is (D).

(b) The ratio of speeds determines the sheave sizes.

$$\begin{aligned} d_{equipment} &= d_{motor}\left(\frac{n_{motor}}{n_{equipment}}\right) = (254\ \text{mm})\left(\frac{1750\ \frac{\text{rev}}{\text{min}}}{800\ \frac{\text{rev}}{\text{min}}}\right) \\ &= \boxed{555.625\ \text{mm} \quad (560\ \text{mm})} \end{aligned}$$

The answer is (B).

32. *Customary U.S. Solution*

(a) From Eq. 54.131, the bearing pressure is

$$p = \frac{\text{lateral shaft load}}{Ld} = \frac{880\ \text{lbf}}{(3.5\ \text{in})(3\ \text{in})} = 83.81\ \text{lbf/in}^2$$

From Eq. 54.130, the ratio c_d/d is

$$\frac{c_d}{d} = \frac{c_r}{r} = 0.001$$

This is the radial clearance ratio.

From Eq. 54.137, the bearing characteristic number is

$$\begin{aligned} S &= \left(\frac{d}{c_d}\right)^2\left(\frac{\mu n_{rps}}{p}\right) = \left(\frac{r}{c_r}\right)^2\left(\frac{\mu n_{rps}}{p}\right) \\ &= \left(\frac{1}{0.001}\right)^2\left(\frac{\left(1.184\times 10^{-6}\ \frac{\text{lbf-sec}}{\text{in}^2}\right)\times\left(1200\ \frac{\text{rev}}{\text{min}}\right)}{\left(83.81\ \frac{\text{lbf}}{\text{in}^2}\right)\left(60\ \frac{\text{sec}}{\text{min}}\right)}\right) \\ &= \boxed{0.283 \quad (0.28)} \end{aligned}$$

The answer is (A).

(b) The axial length-to-diameter ratio is

$$\frac{L}{d} = \frac{3.5\ \text{in}}{3\ \text{in}} = \boxed{1.17 \quad (1.2)}$$

The answer is (C).

(c) From App. 54.B, the minimum film thickness variable is approximately 0.64. From Eq. 54.139, the minimum film thickness is

$$\begin{aligned} h_0 &= r\left(\frac{c_d}{d}\right)(\text{minimum film thickness variable}) \\ &= \left(\frac{3\ \text{in}}{2}\right)(0.001)(0.64) \\ &= \boxed{0.00096\ \text{in}} \end{aligned}$$

The answer is (C).

(d) From App. 54.B, the coefficient of friction variable is approximately 6.

From Eq. 54.138, the coefficient of friction is

$$f = \left(\frac{c_d}{d}\right)(\text{coefficient of friction variable}) = (0.001)(6) = 0.006$$

The friction torque is

$$T = fF_r = (0.006)(880 \text{ lbf})\left(\frac{3 \text{ in}}{2}\right) = \boxed{7.92 \text{ in-lbf} \quad (7.9 \text{ in-lbf})}$$

The answer is (B).

(e) From Eq. 54.33(b), the power lost to friction is

$$P_{hp} = \frac{T_{\text{in-lbf}}n_{\text{rpm}}}{63{,}025 \ \frac{\text{in-lbf}}{\text{hp-min}}} = \frac{(7.92 \text{ in-lbf})\left(1200 \ \frac{\text{rev}}{\text{min}}\right)}{63{,}025 \ \frac{\text{in-lbf}}{\text{hp-min}}} = \boxed{0.151 \text{ hp} \quad (0.15 \text{ hp})}$$

The answer is (B).

(f) Since the minimum film thickness variable is out of the shaded area of the figure in App. 54.B and $\boxed{\text{to the right}}$ of the maximum load-carrying ability curve, the bearing is operating $\boxed{\text{below}}$ its optimum capacity.

The answer is (D).

SI Solution

(a) From Eq. 54.131, the bearing pressure is

$$p = \frac{\text{lateral shaft load}}{Ld} = \frac{(4000 \text{ N})\left(1000 \ \frac{\text{mm}}{\text{m}}\right)^2}{(90 \text{ mm})(76 \text{ mm})} = 584\,795 \text{ Pa}$$

From Eq. 54.130, the ratio c_d/d is

$$\frac{c_d}{d} = \frac{c_r}{r} = 0.001$$

This is the radial clearance ratio.

From Eq. 54.137, the bearing characteristic number is

$$S = \left(\frac{d}{c_d}\right)^2\left(\frac{\mu n_{\text{rps}}}{p}\right) = \left(\frac{r}{c_r}\right)^2\left(\frac{\mu n_{\text{rps}}}{p}\right)$$

$$= \left(\frac{1}{0.001}\right)^2 \times \left(\frac{(8.16 \text{ cP})\left(1200 \ \frac{\text{rev}}{\text{min}}\right)}{(584\,795 \text{ Pa})\left(1000 \ \frac{\text{cP}}{\frac{\text{N·s}}{\text{m}^2}}\right)\left(60 \ \frac{\text{s}}{\text{min}}\right)}\right)$$

$$= \boxed{0.279 \quad (0.28)}$$

The answer is (A).

(b) The axial length-to-diameter ratio is

$$\frac{L}{d} = \frac{90 \text{ mm}}{76 \text{ mm}} = \boxed{1.18 \quad (1.2)}$$

The answer is (C).

(c) From App. 54.B, the minimum film thickness variable is approximately 0.64.

From Eq. 54.139, the minimum film thickness is

$$h_0 = r\left(\frac{c_d}{d}\right)(\text{minimum film thickness variable}) = \left(\frac{76 \text{ mm}}{2}\right)(0.001)(0.64) = \boxed{0.0243 \text{ mm} \quad (0.024 \text{ mm})}$$

The answer is (C).

(d) From App. 54.B, the coefficient of friction variable is approximately 6. From Eq. 54.138, the coefficient of friction is

$$f = \left(\frac{c_d}{d}\right)(\text{coefficient of friction variable}) = (0.001)(6) = 0.006$$

The friction torque is

$$T = fF_r = \frac{(0.006)(4 \text{ kN})\left(1000 \ \frac{\text{N}}{\text{kN}}\right)\left(\frac{76 \text{ mm}}{2}\right)}{1000 \ \frac{\text{mm}}{\text{m}}} = \boxed{0.912 \text{ N·m} \quad (0.91 \text{ N·m})}$$

The answer is (B).

(e) From Eq. 54.33(a), the power lost to friction is

$$P_{\text{kW}} = \frac{T_{\text{N·m}} n_{\text{rpm}}}{9549 \ \frac{\text{N·m}}{\text{kW·min}}} = \frac{(0.912 \ \text{N·m})\left(1200 \ \frac{\text{rev}}{\text{min}}\right)}{9549 \ \frac{\text{N·m}}{\text{kW·min}}}$$
$$= \boxed{0.1146 \ \text{kW} \quad (0.11 \ \text{kW})}$$

The answer is (B).

(f) Since the minimum film thickness variable is out of the shaded area of the figure in App. 54.B and $\boxed{\text{to the right}}$ of the maximum load-carrying ability curve, the bearing is operating $\boxed{\text{below}}$ its optimum capacity.

The answer is (D).

33. The length-to-diameter ratio for a journal bearing is

$$\frac{L}{D} = 2$$

The projected area, A, of the journal bearing that is exposed to the shaft load is

$$A = \frac{L}{D} D^2 = (2)(1 \ \text{in})^2 = 2 \ \text{in}^2$$

The pressure, p, on the journal bearing is a function of load and projected area.

$$p = \frac{\text{lateral shaft load}}{A} = \frac{2500 \ \text{lbf}}{2 \ \text{in}^2} = \boxed{1250 \ \text{lbf/in}^2}$$

The answer is (B).

34. The minimum film thickness, h_0, is

$$h_0 = c_r(1 - \epsilon) = (0.01 \ \text{in})(1 - 0.6) = \boxed{0.004 \ \text{in}}$$

The answer is (A).

35. (a) A 90% reliability indicates that the L_{10} life is applicable. From Eq. 54.117, the life is

$$L_{10} = \left(\frac{C_r}{P_r}\right)^3 \times 10^6$$
$$= \left(\frac{5100 \ \text{lbf}}{2500 \ \text{lbf}}\right)^3 \times 10^6$$
$$= 8.49 \times 10^6 \quad (8.5 \times 10^6)$$

The bearing has an L_{10} life of

$$\boxed{8.5 \ \text{million rev} \quad (8.5 \times 10^6 \ \text{rev})}$$

The answer is (D).

(b) From Eq. 54.119, the lifetime for the ball bearing is

$$L_{10} = \left(\frac{10^6}{n_{\text{rpm}}\left(60 \ \frac{\text{min}}{\text{hr}}\right)}\right)\left(\frac{C_r}{P_r}\right)^3$$
$$= \left(\frac{10^6}{\left(1800 \ \frac{\text{rev}}{\text{min}}\right)\left(60 \ \frac{\text{min}}{\text{hr}}\right)}\right)\left(\frac{5100 \ \text{lbf}}{2500 \ \text{lbf}}\right)^3$$
$$= \boxed{78.6 \ \text{hr} \quad (80 \ \text{hr})}$$

The answer is (A).

36. The L_{10} rated life is a function of the basic load rating, C_r, equivalent radial load, P_r, and an exponent, p. From Eq. 54.117, for ball bearings,

$$L_{10} = \left(\frac{C_r}{P_r}\right)^3 \times 10^6$$
$$= \left(\frac{3480 \ \text{lbf}}{5000 \ \text{lbf}}\right)^3 \times 10^6$$
$$= \boxed{337{,}154 \ \text{rev} \quad (340{,}000 \ \text{rev})}$$

The answer is (A).

37. Rated life, L_{10}, is a function of the basic load rating, C_r, and the equivalent radial load, $P_{r,\text{eq}}$. The basic load rating, C_r, is 2000 lbf.

The equivalent radial load, $P_{r,\text{eq}}$, is a function of the actual radial load, F_r, radial thrust factor, X, thrust load, F_a, and thrust factor, Y. From Eq. 54.112,

$$P_{r,\text{eq}} = XF_r + YF_a$$
$$= (0.63)(1000 \ \text{lbf}) + (1.25)(1500 \ \text{lbf})$$
$$= 2505 \ \text{lbf}$$

From Eq. 54.117, the rated life, in revolutions, is

$$L_{10} = \left(\frac{C_r}{P_r}\right)^3 \times 10^6 = \left(\frac{2000 \ \text{lbf}}{2505 \ \text{lbf}}\right)^3 \times 10^6$$
$$= \boxed{508{,}940 \ \text{rev} \quad (500{,}000 \ \text{rev})}$$

The answer is (B).

38. The equivalent load factor, LF, is

$$\mathrm{LF} = \frac{2000\ \mathrm{lbf}}{1820\ \mathrm{lbf}} = \boxed{1.1}$$

The answer is (A).

39. Rated life, L_{10}, is a function of the basic load rating, C_r, the equivalent radial load, P_r, the rotational speed, n, and the bearing exponent, p. From Eq. 54.120,

$$\begin{aligned} L_{10} &= \left(\frac{10^6}{n_{\mathrm{rpm}} \left(60\ \frac{\mathrm{min}}{\mathrm{hr}} \right)} \right) \left(\frac{C_r}{P_r} \right)^{10/3} \\ &= \left(\frac{10^6}{\left(200\ \frac{\mathrm{rev}}{\mathrm{min}} \right) \left(60\ \frac{\mathrm{min}}{\mathrm{hr}} \right)} \right) \left(\frac{3480\ \mathrm{lbf}}{1000\ \mathrm{lbf}} \right)^{10/3} \\ &= \boxed{5322\ \mathrm{hr}\quad (5300\ \mathrm{hr})} \end{aligned}$$

The answer is (B).

40. Rated life, L_{10}, is a function of the basic load rating, C_r, and the equivalent radial load, P_r. Solving for C_r, the basic load rating is

$$L_{10} = \left(\frac{C_r}{P_r} \right)^3 \times 10^6$$

$$\begin{aligned} C_r &= P_r \sqrt[3]{\frac{L_{10}}{10^6}} \\ &= 1000\ \mathrm{lbf} \sqrt[3]{\frac{500{,}000{,}000\ \mathrm{rev}}{10^6}} \\ &= \boxed{7937\ \mathrm{lbf}\quad (7900\ \mathrm{lbf})} \end{aligned}$$

The answer is (C).

55 Pressure Vessels

PRACTICE PROBLEMS

Codes and Basic Concepts

1. A cylindrical structure with a 24 in outside diameter has design parameters of 400 psi and 300°F. The structure is constructed of SA-516 grade 70 steel, and all welds are double-weld butt joints. No radiography has been performed. The structure is an integral part of a system used to transport process fluid from one location to another. Which standard should be referenced to determine the minimum required wall thickness?

(A) API 510

(B) ASME BPVC Sec. VIII

(C) ASME B31.3

(D) API 570

2. Which of the following symptoms or conditions is NOT one of the main causes of pressure vessel failure?

(A) embrittlement

(B) chattering

(C) corrosion

(D) erosion

3. In regard to an ASME "code" vessel, what is the definition of "lethal service"?

(A) installation within 25 ft (8.1 m) of occupied space

(B) processing of oxygen-deficient gases

(C) remote internal inspection only

(D) containment of poisons

4. In regard to a pressure vessel originally designed and manufactured according to the ASME *Boiler and Pressure Vessel Code* (BPVC), which activities may a shop with an R stamp issued by the National Board of Boiler and Pressure Vessel Inspectors perform?

I. modify the pressure vessel

II. repair the pressure vessel

III. file an "R-form" with the National Board of Boiler and Pressure Vessel Inspectors

(A) I only

(B) II only

(C) II and III only

(D) I, II, and III

5. What is the major distinction between pressure vessels marked with "U" and "UM" stamps?

(A) UM-stamped pressure vessels may be used in corrosive environments.

(B) UM-stamped pressure vessels are not intended for installation in European Commonwealth countries.

(C) UM stamps are for pressure vessels intended for U.S. military installations.

(D) UM stamps are for miniature pressure vessels.

6. A U-stamped pressurized heating boiler is to be designed and manufactured by a shop with an "H" certificate. Which of the following requirements for the design and manufacture of the boiler is NOT correctly stated?

(A) A shop engineer must confirm that the design is in accordance with the ASME code.

(B) An independent authorized inspector must establish hold points when he/she wants to see the vessel in production.

(C) Materials with complete traceability must be used.

(D) The nameplate must be stamped with both ASME and National Board stamps.

7. The purpose of the RT vessel stamping system is to

(A) confirm the presence of an authorized inspector during manufacturing

(B) indicate the extent of radiographic examination

(C) identify pressure vessels intended to operate at room temperature

(D) identify pressure vessels allowed to operate without downtime (100% duty cycle)

8. Which RT mark indicates that the entire vessel satisfies the spot radiography requirements of ASME *Boiler and Pressure Vessel Code* (BPVC) Sec. VIII, Div. 1, UW-52?

(A) RT-1

(B) RT-2

(C) RT-3

(D) RT-4

9. Which pressure range of pressure vessels is NOT covered by Section VIII of the ASME *Boiler and Pressure Vessel Code* (BPVC)?

(A) less than 3000 psig

(B) 3000 psig to 10,000 psig

(C) greater than 10,000 psig

(D) none of the above

10. A "code" pressure vessel is designed to operate at room temperature under a maximum allowable working pressure of 250 psig. The minimum hydrostatic test pressure required is most nearly

(A) 250 psig

(B) 330 psig

(C) 380 psig

(D) 500 psig

11. An explosion caused by the rupture of a vessel that has reached a temperature well above its atmospheric boiling point is known as a(n)

(A) accelerated exposure escape of boiling vapor (AEEBV)

(B) boiling liquid expanding vapor explosion (BLEVE)

(C) vapor escape reaction of boiling liquid (VERBL)

(D) liquid expansion vapor explosive relief (LEVER)

12. A 60 in diameter horizontal pressure vessel has a hemispherical head, two 72 in shell courses, and three 12 in nozzles located on the shell at the top of the vessel. The design pressure for the vessel is 200 psi, and the temperature is 300°F. Adding a new 14 in nozzle to the shell and operating the vessel at 400°F would be classified as a(n)

(A) alteration and rerating

(B) rerating and repair

(C) alteration and repair

(D) derating and repair

Corrosion

13. A 48 in diameter vertical pressure vessel was built 14 years ago with one 96 in shell course made from $^3/_8$ in plate material. The current shell thickness is 0.207 in. The minimum required shell thickness is 0.154 in. The remaining life of the vessel is most nearly

(A) 1 yr

(B) 3 yr

(C) 4 yr

(D) 6 yr

Wall Thickness

14. A 72 in inside diameter cylindrical pressure vessel has seamless torispherical heads. Each head has an inside crown radius of 72 in and a knuckle radius of 4.5 in. The vessel is constructed of SA-515 grade 70 steel, and the heads are attached with spot radiographed double-welded butt joints. The design parameters are 150 psi at 600°F. Exclusive of a corrosion allowance, the required thickness of the heads is most nearly

(A) 0.49 in

(B) 0.53 in

(C) 0.58 in

(D) 0.70 in

15. A 10 in nominal diameter schedule-80 pipe is used as the basis for constructing a nozzle on the shell of a 48 in inside diameter cylindrical pressure vessel. The vessel operates at 400 psi and 300°F. The vessel and nozzle are constructed of SA-516 grade 70 and SA-106 grade B steel, respectively. All welds are double-welded butt joints. No radiography will be performed. Exclusive of corrosion allowances, the required nozzle neck thickness is most nearly

(A) 0.2 in

(B) 0.3 in

(C) 0.4 in

(D) 0.5 in

16. A seamless pressure vessel has an inside diameter of 60 in. The vessel is made from 0.75 in SA-515, grade 60 plate and is designed for service at 200 psig at 750°F. A seamless 10 in diameter nozzle neck made from 0.5 in SA-105 is attached. The nozzle neck is flush with the vessel walls. The corrosion allowance is $^1/_{16}$ in. Full radiography is used.

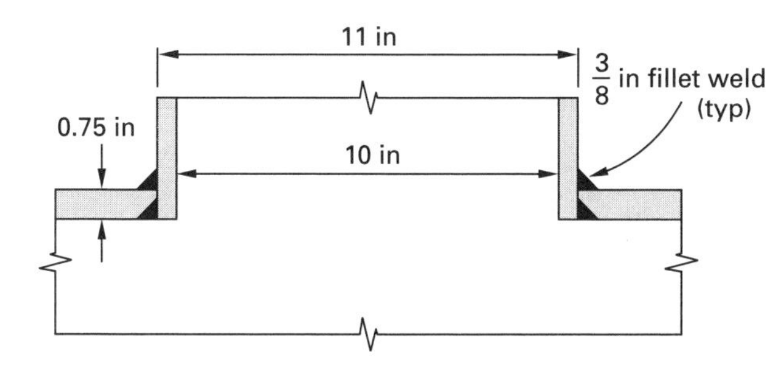

(a) Are the specified thicknesses for the shell and nozzle neck adequate?

(A) shell, inadequate; nozzle neck, inadequate

(B) shell, inadequate; nozzle neck, adequate

(C) shell, adequate; nozzle neck, inadequate

(D) shell, adequate; nozzle neck, adequate

(b) Assume the required shell thickness is 0.47 in, and the required nozzle neck thickness is 0.069 in. Evaluate the required and available areas, and determine whether a reinforcing pad is necessary.

(A) $A_{\text{required}} > A_{\text{available}}$; reinforcement required

(B) $A_{\text{required}} < A_{\text{available}}$; reinforcement required

(C) $A_{\text{required}} > A_{\text{available}}$; reinforcement not required

(D) $A_{\text{required}} < A_{\text{available}}$; reinforcement not required

17. The pressure vessel shown is intended for general service. No radiography is performed. The corrosion allowance is $^1/_8$ in. The vessel is seamless, operates at 350 psig and 500°F, and is made from SA-106 steel. The shell has a 48 in outside diameter and a nominal thickness of 0.75 in. The sump is a seamless pipe with a 16 in outside diameter and 0.625 in thickness.

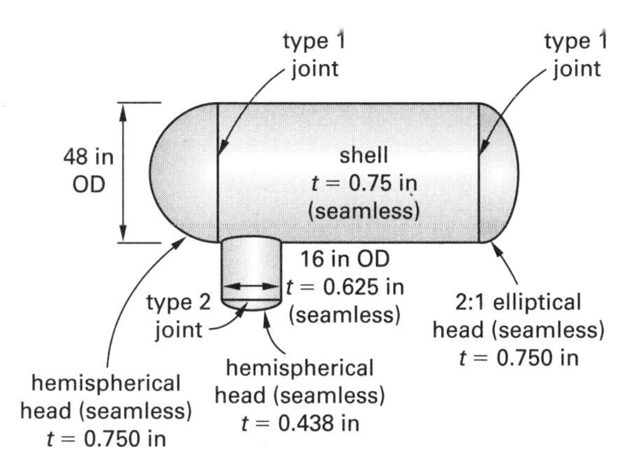

(a) Is the thickness of the shell adequate?

(A) 0.34 in required; adequate

(B) 0.40 in required; adequate

(C) 0.57 in required; adequate

(D) 0.70 in required; adequate

(b) Are the thicknesses of the elliptical and hemispherical heads, respectively, adequate?

(A) elliptical, inadequate; hemispherical, inadequate

(B) elliptical, inadequate; hemispherical, adequate

(C) elliptical, adequate; hemispherical, inadequate

(D) elliptical, adequate; hemispherical, adequate

(c) Are the thicknesses of the sump shell and head, respectively, adequate?

(A) shell, inadequate; head, inadequate

(B) shell, inadequate; head, adequate

(C) shell, adequate; head, inadequate

(D) shell, adequate; head, adequate

18. The pressure vessel shown weighs 21,000 lbf empty. The contents weigh 40,000 lbf, and the liquid head is 46 ft when the vessel is full. The MAWP at the top of the tank is 150 psig. The tank is supported by a 0.25 in thick skirt with the same outside diameter as the tank. The allowable tensile and compressive stresses in the supporting skirt are 14,000 lbf/in^2 and 17,000 lbf/in^2, respectively. The vessel is subjected to a uniform wind load. The tower anchorage is sufficient.

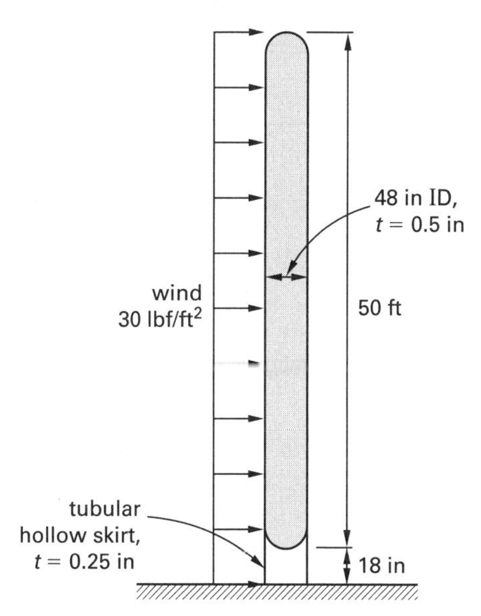

(a) Evaluate the maximum tensile and compressive stresses, and determine whether a skirt thickness of 0.25 in is inadequate.

(A) $S_{\text{tensile}} = 500$ lbf/in^2, $S_{\text{compressive}} = 3600$ lbf/in^2; adequate

(B) $S_{\text{tensile}} = 2600$ lbf/in^2, $S_{\text{compressive}} = 5800$ lbf/in^2; adequate

(C) $S_{\text{tensile}} = 15{,}000$ lbf/in^2, $S_{\text{compressive}} = 18{,}000$ lbf/in^2; inadequate

(D) $S_{\text{tensile}} = 20{,}000$ lbf/in^2, $S_{\text{compressive}} = 17{,}000$ lbf/in^2; inadequate

(b) The maximum stress that should be considered in determining the adequacy of the pressure vessel wall thickness is most nearly

(A) 3600 psi

(B) 5000 psi

(C) 6600 psi

(D) 7900 psi

19. A pressure vessel previously used in a refinery has been out of service for a number of years. The vessel is known to be constructed from 0.625 in SA-516, grade 65 steel plate, but no other documentation is available. The longitudinal seam joint in the vessel is type 1, and the heads are attached with type 2 joints. The vessel is spot radiographed and operates at 850°F. The corrosion allowance is $^1/_{16}$ in.

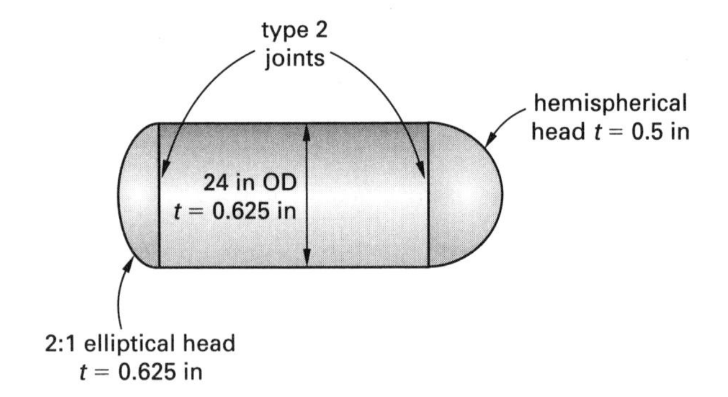

(a) The maximum design pressure is most nearly

(A) 300 psig

(B) 350 psig

(C) 430 psig

(D) 530 psig

(b) If the vessel is tested at room temperature, the hydrostatic test pressure is most nearly

(A) 600 psig

(B) 800 psig

(C) 900 psig

(D) 1000 psig

20. A pressure vessel with a 48 in inside diameter is fabricated from $^3/_4$ in thick SA-105 plate. The vessel is constructed with a type 1 longitudinal seam weld. A 12 in inside diameter, $^1/_2$ in thick, seamless nozzle made from SA-515, grade 70 steel is attached. The nozzle does not pass through the seam weld. The operating temperature of the vessel is 750°F at 400 psig. The vessel is fully radiographed. An annular reinforcing pad 0.5 in thick, made from SA-105 steel, abuts the nozzle neck and is attached to the vessel as shown. Disregard corrosion allowance.

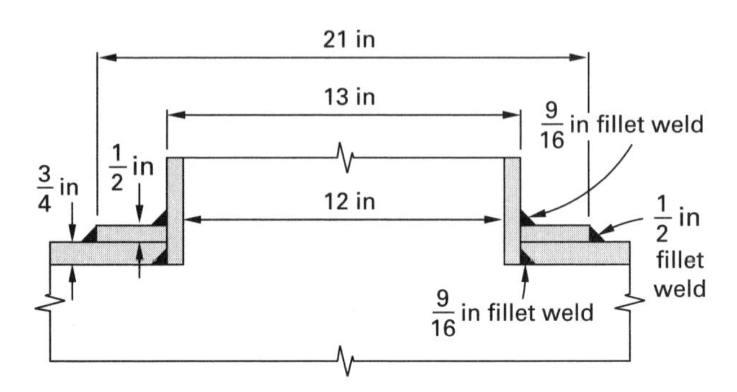

(a) According to Section VIII, Division 1, determine whether the thickness requirements for the shell and nozzle neck, respectively, are adequate or inadequate.

(A) shell, inadequate; nozzle neck, adequate

(B) shell, inadequate; nozzle neck, inadequate

(C) shell, adequate; nozzle neck, adequate

(D) shell, adequate; nozzle neck, inadequate

(b) Determine whether the inner and outer nozzle welds have been properly sized. (Do not perform a stress analysis on the welds.)

(A) inner nozzle weld, inadequate; outer nozzle weld, adequate

(B) inner nozzle weld, adequate; outer nozzle weld, adequate

(C) inner nozzle weld, adequate; outer nozzle weld, inadequate

(D) inner nozzle weld, inadequate; outer nozzle weld, inadequate

(c) Does the vessel meet requirements for the required area?

(A) $A_{\text{available}} = 6.1\ \text{in}^2$; no

(B) $A_{\text{available}} = 7.9\ \text{in}^2$; no

(C) $A_{\text{available}} = 8.7\ \text{in}^2$; yes

(D) $A_{\text{available}} = 9.5\ \text{in}^2$; yes

Flat Heads

21. Which of the flat unstayed heads shown is/are not permitted by the ASME *Boiler and Pressure Vessel Code* (BPVC)?

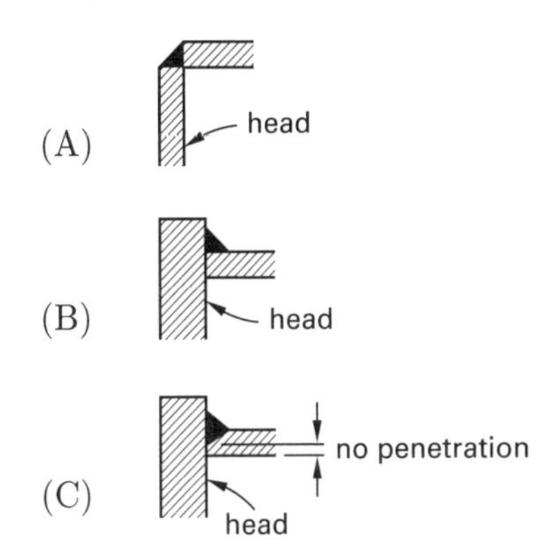

(D) all of the above

22. A large pressure vessel with a circular flat unstayed integral head is to operate at 20 psig. The shell and head are constructed from material with a maximum allowable stress of 20,000 psi. The inside diameter of the head is 147 in. The flange attachment factor is 0.2. Spot radiography has been used on the head attachment welds. Exclusive of any corrosion allowance, what is most nearly the minimum head thickness?

(A) 2.1 in

(B) 2.3 in

(C) 2.5 in

(D) 2.7 in

23. A pressure vessel with a circular flat unstayed head is to operate at 650 psig. Flanges are constructed from 6 in NPS schedule-80 pipe and are capped with welded flat unstayed plates with a maximum allowable stress of 14,400 psi. The inside diameter of the flange is 5.761 in. The flange attachment factor is 0.33. Exclusive of any corrosion allowance, what is most nearly the minimum head thickness?

(A) 0.70 in

(B) 0.87 in

(C) 0.96 in

(D) 1.2 in

Pressure Relief Components

24. Normally, how should a pressure relief valve (PRV) and rupture discs (RD) be used together?

(A) The PRV and RD should be installed side-by-side, both exposed to the same vessel pressure.

(B) Both should be installed in the same relief line, with the PRV closest to the pressure vessel.

(C) Both should be installed in the same relief line, with the RD closest to the pressure vessel.

(D) Two RDs should be installed in the relief line, one RD closest to the pressure vessel and before the PRV, and the other RD after the PRV.

25. When a pressure relief valve (PRV) and rupture disc (RD) are used together, how should the RD rated burst pressure compare to the PRV setting?

(A) RD burst pressure should be at least 20% below the valve setting.

(B) RD burst pressure should be equal to the valve setting.

(C) RD burst pressure should be within ±10% of the valve setting.

(D) RD burst pressure should be at least 20% above the valve setting.

26. When may a safety relief valve that is fully open at 116% of the maximum allowable working pressure (MAWP) be installed?

(A) When the safety valve is the sole safety device used.

(B) When another safety valve is fully open by 110% of the MAWP.

(C) When another safety valve is fully open by 120% of the MAWP.

(D) When another safety valve is fully open by 130% of the MAWP.

SOLUTIONS

1. *Pressure Vessel Inspection Code: Maintenance Inspection, Rating, Repair, and Alteration* (API 510) covers the maintenance, inspection, repair, alteration, and rerating procedures for pressure vessels used by the petroleum and chemical process industries.

The ASME *Boiler and Pressure Vessel Code* (BPVC) covers the design, fabrication, and inspection of boilers and pressure vessels. However, Sec. VIII does not cover structures whose primary function is the transport of fluids from one location to another within a system of which it is an integral part, such as piping systems.

Process Piping (ASME B31.3) covers the design of piping for all fluids (with some exclusions for fluid service and high pressure applications). Excluded are pressure vessels and piping systems with internal pressures below 15 psig (103 kPa).

Piping Inspection Code: In-Service Inspection, Repair, and Alteration of Piping Systems(API 570) covers metallic piping systems that have been in service; pressure vessels are excluded. Minimum thickness requirement calculations are not included. API 570 references ASME B31.3 for thickness requirement calculations.

In this case, the structure is an integral part of a system used to transport process fluid from one location to another. With no reference to heads, shell, or use as a pressure vessel, the structure is considered a piping system. While API 570 can assist in the inspection of the structure, minimum thickness requirements are addressed in ASME B31.3.

The answer is (C).

2. Embrittlement, corrosion, and erosion all compromise material integrity. Chattering is the rapid opening and closing of a pressure relief valve due to some installation or operational valve defect. The resultant vibration can result in failure of the valve and/or associated piping, but valve chattering probably isn't a cause for pressure vessel failure.

The answer is (B).

3. ASME BPVC Sec. VIII, Div. 1, UW-2 defines lethal substances as poisonous gases, vapors, or liquids of such a nature that very small amounts mixed or unmixed with air are dangerous to life when inhaled. It does not identify which substances are lethal. It only specifies manufacturing requirements, such as requiring butt-welded joints in vessels that contain lethal substances to be fully radiographed. The definition of lethal substances is found in federal, state, and local regulations. For example, class A poisons (e.g., hydrogen cyanide and phosgene gas) are defined in the United States by the Code of Federal Regulations (CFR), Title 49.

The answer is (D).

4. A shop with an R stamp issued by the National Board of Boiler and Pressure Vessel Inspectors may modify and/or repair the pressure vessel. The shop is required to file an R-form with the National Board.

The answer is (D).

5. "UM" stamps are for miniature pressure vessels (inside diameters no greater than 16 in, working pressures no greater than 100 psig, and internal volumes no greater than 5 ft^3).

The answer is (D).

6. The ASME "H" certificate allows the shop to design and construct heating boilers. The design must be reviewed by an outside Authorized Inspector (AI), licensed by the National Board of Boiler and Pressure Vessel Inspectors. The AI verifies that the design meets the ASME code. The AI also sets the hold points in the manufacturing process, observes the manufacturing at those points, and observes the hydrostatic testing.

The answer is (A).

7. An RT mark (stamp) indicates the use and extent of radiographic examination of a vessel's welds.

The answer is (B).

8. "RT-1" means 100% of all longitudinal and circumferential seams have been radiographed. It also indicates 100% of nozzle welds over 10 in (25 mm) diameter have been radiographed. This level is "full radiography" and yields a 1.0 joint efficiency on all welds. RT-1 is mandatory for head/shell thicknesses greater than 1.25 in (32 mm). (Vessels designed to ASME BPVC Sec. I and certain nozzles require 100% RT on circumferential and longitudinal seams.)

"RT-2" means 100% of longitudinal weld seams have been radiographed and spot RT has been done on circumferential seams. This level is also considered "full radiography" and yields a 1.0 joint efficiency for thickness calculations. No radiographic testing is done on nozzle welds for this level.

"RT-3" means, with only a few exceptions, spot radiographic inspection has been performed on all longitudinal and circumferential seams. This yields a 0.85 joint efficiency. No nozzle connection welds have been radiographed.

"RT-4" means some radiographic examination took place, but one cannot describe the amount with the RT numbering system. This level yields a 0.70 joint efficiency.

The answer is (C).

9. BPVC Sec. VIII is divided into three divisions. Division 1 covers pressure vessels that operate at less than 3000 psig; Division 2 covers the pressure range of 3000 psig to 10,000 psig; and Division 3 covers pressures greater than 10,000 psig.

The answer is (D).

10. From ASME *Boiler and Pressure Vessel Code* (BPVC) Sec. VIII, Div. 1, UG-99(b), *Standard Hydrostatic Test*, "...vessels designed for internal pressure shall be subjected to a hydrostatic test pressure which at every point in the vessel is at least equal to 1.3 times the maximum allowable working pressure to be marked on the vessel multiplied by the lowest ratio (for the materials of which the vessel is constructed) of the stress value S for the test temperature on the vessel to the stress value S for the design temperature..."

$$p_{\text{test}} = (1.3)\left(250\ \frac{\text{lbf}}{\text{in}^2}\right) = \boxed{325\text{ psig}\quad(330\text{ psig})}$$

The answer is (B).

11. An explosion caused by the rupture of a vessel that has reached a temperature well above its atmospheric boiling point is known as boiling liquid expanding vapor explosion (BLEVE).

The answer is (B).

12. According to API 510, *Pressure Vessel Inspection Code: Maintenance Inspection, Rating, Repair, and Alteration*, an *alteration* is a physical change in any component or a rerating that has design implications affecting the pressure-containing capability of a pressure vessel beyond the scope of the items described in existing data reports. The following are not considered alterations: (1) comparable or duplicate replacements, (2) additions of reinforced nozzles less than or equal to the size of existing reinforced nozzles, and (3) additions of nozzles not requiring reinforcement.

A *repair* is work necessary to restore a vessel to a condition suitable for safe operation at the design conditions. Repairs include additions or replacements of pressure or nonpressure parts that do not change the rating of the vessel.

A *rerating* is a change in either the temperature ratings or the maximum allowable working pressure rating of a vessel, or a change in both. A *derating* below the original design conditions is sometimes used to provide for additional corrosion allowance and extend useful life.

Adding a new 14 in nozzle and increasing the design temperature would be considered an alteration and a rerating.

The answer is (A).

13. Find the corrosion rate, CR. The original nominal thickness, t_{previous}, was 3/8 in (0.375 in), the current thickness, t_{current}, is 0.207 in, and the vessel age, N, is 14 years.

$$\begin{aligned}\text{CR} &= \frac{t_{\text{previous}} - t_{\text{current}}}{N} = \frac{0.375\text{ in} - 0.207\text{ in}}{14\text{ yr}} \\ &= 0.012\text{ in/yr}\end{aligned}$$

The remaining life, RL, is

$$\begin{aligned}\text{RL} &= \frac{t_{\text{current}} - t_{\text{required}}}{\text{CR}} = \frac{0.207\text{ in} - 0.154\text{ in}}{0.012\ \frac{\text{in}}{\text{yr}}} \\ &= \boxed{4.42\text{ yr}\quad(4\text{ yr})}\end{aligned}$$

The answer is (C).

14. From Table 55.3, the allowable stress for SA-515 grade 70 steel at 600°F is 19,400 psi.

From Table 55.8, for torispherical heads, the joint efficiency of spot radiographed double-butt welded joints is 1.00. The inside crown radius, L, is 72 in. The knuckle radius, r_k, is 4.5 in. From Table 55.10,

$$M = \left(\tfrac{1}{4}\right)\left(3 + \sqrt{\frac{L}{r_k}}\right) = \left(\tfrac{1}{4}\right)\left(3 + \sqrt{\frac{72\text{ in}}{4.5\text{ in}}}\right) = 1.75$$

From Table 55.10, the minimum required thickness, t, is

$$\begin{aligned}t &= \frac{pLM}{2SE - 0.2p} \\ &= \frac{\left(150\ \frac{\text{lbf}}{\text{in}^2}\right)(72\text{ in})(1.75)}{(2)\left(19{,}400\ \frac{\text{lbf}}{\text{in}^2}\right)(1.00) - (0.2)\left(150\ \frac{\text{lbf}}{\text{in}^2}\right)} \\ &= \boxed{0.487\text{ in}\quad(0.49\text{ in})}\end{aligned}$$

The answer is (A).

15. The required minimum thickness of a nozzle neck depends on three values.

(1) *nozzle neck thickness due to internal pressure*

From App. 16.B, the inside diameter for a 10 in schedule-80 pipe is 9.564 in. The inside radius of the nozzle is

$$R = \frac{D_i}{2} = \frac{9.564\text{ in}}{2} = 4.782\text{ in}$$

From Table 55.3, the allowable stress for SA-106 grade B at 300°F is 17,100 psi.

No radiography was performed, so the joint efficiency is 0.70. From Eq. 55.5, the nozzle neck thickness is

$$t_a = \frac{pR}{SE - 0.6p} = \frac{\left(400\ \frac{\text{lbf}}{\text{in}^2}\right)(4.782\ \text{in})}{\left(17{,}100\ \frac{\text{lbf}}{\text{in}^2}\right)(0.70) - (0.6)\left(400\ \frac{\text{lbf}}{\text{in}^2}\right)} = 0.163\ \text{in}$$

(2) *required vessel wall thickness due to internal pressure (based on E= 1.0)*

The inside radius of the vessel is

$$R = \frac{D_i}{2} = \frac{48\ \text{in}}{2} = 24\ \text{in}$$

From Table 55.3, the allowable stress for SA-516 grade 70 steel at 300°F is 20,000 psi.

From Eq. 55.5, the nozzle neck thickness is

$$t_{b1} = \frac{pR}{SE - 0.6p} = \frac{\left(400\ \frac{\text{lbf}}{\text{in}^2}\right)(24\ \text{in})}{\left(20{,}000\ \frac{\text{lbf}}{\text{in}^2}\right)(1.0) - (0.6)\left(400\ \frac{\text{lbf}}{\text{in}^2}\right)} = 0.486\ \text{in} \geq \tfrac{1}{16}\ \text{in} \quad [\text{OK}]$$

(3) *minimum thickness of standard pipe with a manufacturing tolerance of 12.5%*

From App. 16.B, schedule-40 is standard for 10 in diameter pipe ($t_{\text{standard}} = 0.365$ in).

$$t_{b3} = t_{\text{standard}}(1 - 0.125) = (0.365\ \text{in})(1 - 0.125) = 0.319\ \text{in}$$

$$t_{\text{UG-45}} = \max\begin{cases} t_a \\ t_b = \min\begin{cases} t_{b1} \geq \frac{1}{16}\ \text{in} \\ t_{b3} \end{cases}\end{cases} = \max\begin{cases} 0.163\ \text{in} \\ \min\begin{cases} 0.486\ \text{in} \\ 0.319\ \text{in} \end{cases}\end{cases} = \max\begin{cases} 0.163\ \text{in} \\ 0.319\ \text{in} \end{cases} = \boxed{0.319\ \text{in}\quad (0.3\ \text{in})}$$

The answer is (B).

16. (a) To determine if the specified thicknesses are adequate, calculate the minimum required wall thicknesses for the shell and nozzle.

Shell:

$$R = 30\ \text{in} + 0.0625\ \text{in} = 30.0625\ \text{in}$$

[correction for corrosion allowance]

From Table 55.3, $S = 13{,}000$ lbf/in^2.

$$E = 1.0 \quad [\text{seamless vessel}]$$

$$t_{r,\text{shell}} = \frac{pR}{SE - 0.6p} + c = \frac{\left(200\ \frac{\text{lbf}}{\text{in}^2}\right)(30.0625\ \text{in})}{\left(13{,}000\ \frac{\text{lbf}}{\text{in}^2}\right)(1.0) - (0.6)\left(200\ \frac{\text{lbf}}{\text{in}^2}\right)} + \tfrac{1}{16}\ \text{in} = 0.4668\ \text{in} + 0.0625\ \text{in} = 0.5293\ \text{in}$$

Since 0.75 in > 0.5293 in, the specified shell thickness is $\boxed{\text{adequate.}}$

Nozzle neck:

From Table 55.3, $S = 14{,}800$ lbf/in^2.

$$R = 5\ \text{in} + 0.0625\ \text{in} = 5.0625\ \text{in}$$

$$t_a = \frac{pR_n}{SE - 0.6p} + c = \frac{\left(200\ \frac{\text{lbf}}{\text{in}^2}\right)(5.0625\ \text{in})}{\left(14{,}800\ \frac{\text{lbf}}{\text{in}^2}\right)(1.0) - (0.6)\left(200\ \frac{\text{lbf}}{\text{in}^2}\right)} + \tfrac{1}{16}\ \text{in} = 0.0697\ \text{in} + 0.0625\ \text{in} = 0.1315\ \text{in}$$

$$t_{b1} = t_{r,\text{shell},E=1} = 0.5293\ \text{in} > \tfrac{1}{16}\ \text{in} \quad [\text{OK}]$$

The nozzle neck outside diameter is 11 in. The nearest (larger) schedule-40 pipe size is 12 in, with a wall thickness of 0.406 in.

$$t_{b3} = (0.406\ \text{in})(1 - 0.125) + \tfrac{1}{16}\ \text{in} = 0.4178\ \text{in}$$

$$t_{\text{UG-45}} = \max\begin{cases} t_a \\ t_b = \min\begin{cases} t_{b1} \\ t_{b3} \end{cases}\end{cases} = \max\begin{cases} 0.1315\ \text{in} \\ t_b = \min\begin{cases} 0.5293\ \text{in} \\ 0.4178\ \text{in} \end{cases}\end{cases} = 0.4178\ \text{in}$$

Since 0.5 in > 0.4178 in, the nozzle neck thickness is adequate.

The answer is (D).

(b) To determine whether reinforcement is required, compare the area required with the available area.

$$f_{r1} = f_{r2} = \frac{S_n}{S_v} = \frac{14{,}800\ \frac{\text{lbf}}{\text{in}^2}}{13{,}000\ \frac{\text{lbf}}{\text{in}^2}} > 1.0$$

Therefore,

$$f_{r1} = f_{r2} = 1.0$$

$$D = 10\ \text{in} + 0.125\ \text{in} = 10.125\ \text{in} \quad \text{[correction for corrosion allowance]}$$

The required area is

$$\begin{aligned} A_r &= Dt_r + 2t_nt_r(1 - f_{r1}) = (10.125\ \text{in})(0.47\ \text{in}) + 0 \\ &= 4.76\ \text{in}^2 \end{aligned}$$

The area available is $A = A_1 + A_2 + A_{41}$.

Calculation of A_1 (area available in shell):

$$t = 0.75\ \text{in} - 0.0625\ \text{in} = 0.6875\ \text{in}$$

$$t_n = 0.5\ \text{in} - 0.0625\ \text{in} = 0.4375\ \text{in}$$

A_1 is the larger of

$$\begin{aligned} &D(Et - t_r) - 2t_n(Et - t_r)(1 - f_{r1}) \\ &\quad = (10.125\ \text{in})\big((1.0)(0.6875\ \text{in}) - 0.47\ \text{in}\big) - 0 \\ &\quad = 2.20\ \text{in}^2 \end{aligned}$$

And,

$$\begin{aligned} &2(t + t_n)(Et - t_r) - 2t_n(Et - t_r)(1 - f_{r1}) \\ &\quad = (2)(0.6875\ \text{in} + 0.4375\ \text{in}) \\ &\qquad \times \big((1)(0.6875\ \text{in}) - 0.47\ \text{in}\big) - 0 \\ &\quad = 0.49\ \text{in}^2 \end{aligned}$$

$A_1 = 2.20\ \text{in}^2$.

Calculation of A_2 (area available from nozzle):

A_2 is the smaller of

$$\begin{aligned} &5(t_n - t_{rn})f_{r2}t \\ &\quad = (5)(0.4375\ \text{in} - 0.069\ \text{in})(1.0)(0.6875\ \text{in}) \\ &\quad = 1.27\ \text{in}^2 \end{aligned}$$

And,

$$\begin{aligned} &5(t_n - t_{rn})f_{r2}t_n \\ &\quad = (5)(0.4375\ \text{in} - 0.069\ \text{in})(1.0)(0.4375\ \text{in}) \\ &\quad = 0.81\ \text{in}^2 \end{aligned}$$

$A_2 = 0.81\ \text{in}^2$.

Calculation of A_{41} (area available from welds):

$$f_{r3} = \frac{S_n}{S_v} = \frac{14{,}800\ \frac{\text{lbf}}{\text{in}^2}}{13{,}000\ \frac{\text{lbf}}{\text{in}^2}} > 1.0 \quad [\text{use } f_{r3} = 1.0]$$

$$\begin{aligned} A_{41} &= (\text{leg})^2 f_{r3} = (0.375\ \text{in})^2(1.0) \\ &= 0.14\ \text{in}^2 \end{aligned}$$

$$\begin{aligned} \text{total available area} &= A_1 + A_2 + A_{41} \\ &= 2.20\ \text{in}^2 + 0.81\ \text{in}^2 + 0.14\ \text{in}^2 \\ &= 3.15\ \text{in}^2 \end{aligned}$$

Since $4.76\ \text{in}^2 > 3.15\ \text{in}^2$ (i.e., $A_{\text{required}} > A_{\text{available}}$), additional reinforcement is required.

This can be accomplished by adding an annular reinforcing pad around the nozzle neck.

$$f_{r4} = \frac{\text{allowable stress of pad}}{\text{allowable stress of shell}} = 1.0$$

For a 15 in OD pad that is 0.5 in thick, the additional area added is

$$\begin{aligned} A_5 &= (D_p - D - 2t_n)t_pf_{r4} \\ &= \big(15\ \text{in} - 10\ \text{in} - (2)(0.5\ \text{in})\big)(0.5\ \text{in}) \\ &= 2.0\ \text{in}^2 \end{aligned}$$

This will satisfy the area requirements.

The answer is (A).

17. From Table 55.3, at 500°F, the allowable stress for SA-106 is 17,100 psi.

(a) *Shell:*

Circumferential stress (seamless):

$$E = 0.85 \quad \text{[seamless shell, no radiography]}$$

$$t = 0.75\ \text{in}$$

Check this design value against the minimum required thickness to determine whether it is adequate.

$$R = 24.000 \text{ in} - 0.625 \text{ in} = 23.375 \text{ in}$$

$$\begin{aligned} t_{\text{circumferential}} &= \frac{pR}{SE - 0.6p} + c \\ &= \frac{\left(350 \ \frac{\text{lbf}}{\text{in}^2}\right)(23.375 \text{ in})}{\left(17{,}100 \ \frac{\text{lbf}}{\text{in}^2}\right)(0.85) - (0.6)\left(350 \ \frac{\text{lbf}}{\text{in}^2}\right)} + \tfrac{1}{8} \text{ in} \\ &= \boxed{0.696 \text{ in} \quad (0.70 \text{ in})} \end{aligned}$$

Longitudinal stress:

$$E = 0.70 \quad \text{[no radiography, type 1 joint]}$$

$$\begin{aligned} t_{\text{longitudinal}} &= \frac{pR}{2SE + 0.4p} + c \\ &= \frac{\left(350 \ \frac{\text{lbf}}{\text{in}^2}\right)(23.375 \text{ in})}{(2)\left(17{,}100 \ \frac{\text{lbf}}{\text{in}^2}\right)(0.70) + (0.4)\left(350 \ \frac{\text{lbf}}{\text{in}^2}\right)} + \tfrac{1}{8} \text{ in} \\ &= 0.465 \text{ in} \end{aligned}$$

Since 0.75 in > 0.696 in, the specified shell thickness is $\boxed{\text{adequate.}}$

The answer is (D).

(b) The basic head weld efficiency, E, is 0.85, unless full radiography is used (in which case, the efficiency is 1.00), or unless the head is hemispherical (in which case, the efficiency may be less than 0.85). See Table 55.8.

Elliptical head (seamless):

$$E = 0.85 \quad \text{[seamless head, no radiography]}$$

$$t = 0.75 \text{ in}$$

$$D = 48 \text{ in} - (2)(0.625 \text{ in}) = 46.75 \text{ in}$$

$$\begin{aligned} K &= \left(\tfrac{1}{6}\right)\left(2 + \left(\frac{D}{2h}\right)^2\right) = \left(\tfrac{1}{6}\right)\left(2 + (2)^2\right) \\ &= 1 \end{aligned}$$

$$\begin{aligned} t_{\text{elliptical}} &= \frac{pDK}{2SE - 0.2p} \\ &= \frac{\left(350 \ \frac{\text{lbf}}{\text{in}^2}\right)(46.75 \text{ in})(1)}{(2)\left(17{,}100 \ \frac{\text{lbf}}{\text{in}^2}\right)(0.85) - (0.2)\left(350 \ \frac{\text{lbf}}{\text{in}^2}\right)} + \tfrac{1}{8} \text{ in} \\ &= 0.689 \text{ in} \end{aligned}$$

Since 0.75 in > 0.689 in, the head thickness is $\boxed{\text{adequate.}}$

Hemispherical head (seamless):

$$E = 0.70 \quad \left[\begin{array}{l}\text{seamless hemispherical head,} \\ \text{type 1 joint, no radiography}\end{array}\right]$$

$$R = 24 \text{ in} - 0.625 \text{ in} = 23.375 \text{ in}$$

$$\begin{aligned} t_{\text{hemispherical}} &= \frac{pR}{2SE - 0.2p} + c \\ &= \frac{\left(350 \ \frac{\text{lbf}}{\text{in}^2}\right)(23.375 \text{ in})}{(2)\left(17{,}100 \ \frac{\text{lbf}}{\text{in}^2}\right)(0.70) - (0.2)\left(350 \ \frac{\text{lbf}}{\text{in}^2}\right)} + \tfrac{1}{8} \text{ in} \\ &= 0.468 \text{ in} \end{aligned}$$

Since 0.75 in > 0.468 in, the head thickness is $\boxed{\text{adequate.}}$

The answer is (D).

(c) *Sump shell (seamless, no radiography):*

Circumferential stress:

$$E = 0.85 \quad \text{[seamless shell, no radiography]}$$

$$t = 0.625 \text{ in}$$

$$R = 8 \text{ in} - 0.500 \text{ in} = 7.50 \text{ in}$$

$$\begin{aligned} t_{\text{circumferential}} &= \frac{pR}{SE - 0.6p} + c \\ &= \frac{\left(350 \ \frac{\text{lbf}}{\text{in}^2}\right)(7.50 \text{ in})}{\left(17{,}100 \ \frac{\text{lbf}}{\text{in}^2}\right)(0.85) - (0.6)\left(350 \ \frac{\text{lbf}}{\text{in}^2}\right)} + \tfrac{1}{8} \text{ in} \\ &= 0.308 \text{ in} \end{aligned}$$

Longitudinal stress:

$$E = 0.60 \quad \text{[type 3 joint]}$$

$$\begin{aligned} t_{\text{longitudinal}} &= \frac{pR}{2SE + 0.4p} + c \\ &= \frac{\left(350 \ \frac{\text{lbf}}{\text{in}^2}\right)(7.50 \text{ in})}{(2)\left(17{,}100 \ \frac{\text{lbf}}{\text{in}^2}\right)(0.60) + (0.4)\left(350 \ \frac{\text{lbf}}{\text{in}^2}\right)} + \tfrac{1}{8} \text{ in} \\ &= 0.252 \text{ in} \end{aligned}$$

Since 0.625 in > 0.308 in and 0.252 in, the sump shell thickness is $\boxed{\text{adequate.}}$

Hemispherical sump head (type 2 joint):

$$E = 0.65$$

$$t = 0.438 \text{ in}$$

$$R = 8.000 \text{ in} - 0.313 \text{ in} = 7.687 \text{ in}$$

$$t_{\text{hemispherical}} = \frac{pR}{2SE - 0.2p} + c$$

$$= \frac{\left(350 \frac{\text{lbf}}{\text{in}^2}\right)(7.687 \text{ in})}{(2)\left(17{,}100 \frac{\text{lbf}}{\text{in}^2}\right)(0.65) - (0.2)\left(350 \frac{\text{lbf}}{\text{in}^2}\right)} + \tfrac{1}{8} \text{ in}$$

$$= 0.246 \text{ in}$$

Since 0.438 in > 0.246 in, the sump head thickness is adequate.

The answer is (D).

18. (a) Calculate the stress due to the weight of the vessel and its contents and compare with the allowable stress in the skirt.

Stress on skirt

= axial stress due to vessel and content weight

+ bending stress due to wind load

Axial stress due to the weight of the vessel and its contents:

The skirt dimensions are

$$D_o = 48 \text{ in} + (2)(0.5 \text{ in}) = 49 \text{ in}$$

$$D_i = 49 \text{ in} - (2)(0.25 \text{ in}) = 48.5 \text{ in}$$

$$\text{weight} = 21{,}000 \text{ lbf} + 40{,}000 \text{ lbf} = 61{,}000 \text{ lbf}$$

$$\text{skirt area} = \left(\frac{\pi}{4}\right)(D_o^2 - D_i^2)$$

$$= \left(\frac{\pi}{4}\right)\left((49 \text{ in})^2 - (48.5 \text{ in})^2\right)$$

$$= 38.29 \text{ in}^2$$

$$\text{axial stress} = \frac{\text{weight}}{\text{skirt area}} = \frac{61{,}000 \text{ lbf}}{38.29 \text{ in}^2}$$

$$= 1593 \text{ lbf/in}^2$$

Bending stress due to wind load:

For thin, circular cross sections, the moment at the bottom of the skirt due to a uniform wind load is

$$M = Fy = pAy = pD_o h\left(\frac{h}{2}\right)$$

$$= \left(30 \frac{\text{lbf}}{\text{ft}^2}\right)\left(\frac{49 \text{ in}}{12 \frac{\text{in}}{\text{ft}}}\right)(51.5 \text{ ft})\left(\frac{51.5 \text{ ft}}{2}\right)$$

$$= 162{,}450 \text{ ft-lbf}$$

$$R_o = \frac{D_o}{2} = \frac{49 \text{ in}}{2} = 24.5 \text{ in}$$

$$R_i = \frac{D_i}{2} = \frac{48.5 \text{ in}}{2} = 24.25 \text{ in}$$

$$I = I_{\text{circle},R_o} - I_{\text{circle},R_i} = \left(\frac{\pi}{4}\right)(R_o^4 - R_i^4)$$

$$= \left(\frac{\pi}{4}\right)\left((24.5 \text{ in})^4 - (24.25 \text{ in})^4\right)$$

$$= 11{,}375 \text{ in}^4$$

The maximum bending stress is

$$\sigma = \frac{Mc}{I} = \frac{(162{,}450 \text{ ft-lbf})\left(12 \frac{\text{in}}{\text{ft}}\right)(24.5 \text{ in})}{11{,}375 \text{ in}^4}$$

$$= 4199 \text{ lbf/in}^2$$

The maximum tensile stress is

$$S_{\text{tensile}} = 4199 \frac{\text{lbf}}{\text{in}^2} - 1593 \frac{\text{lbf}}{\text{in}^2}$$

$$= \boxed{2606 \text{ lbf/in}^2 \quad (2600 \text{ lbf/in}^2)}$$

The maximum compressive stress is

$$S_{\text{compressive}} = 4199 \frac{\text{lbf}}{\text{in}^2} + 1593 \frac{\text{lbf}}{\text{in}^2}$$

$$= \boxed{5792 \text{ lbf/in}^2 \quad (5800 \text{ lbf/in}^2)}$$

The compressive and tensile stresses are below the allowable stresses, so a thickness of 0.25 in is adequate.

The answer is (B).

(b) From Sec. VIII, Div. 2, by inspection, Table 5.3, *Load Case Combinations and Allowable Membrane Stresses for an Elastic Analysis*, Case 5 results in the maximum stress. (Case 4 is only for an overturning analysis. Since the anchorage was specified as being adequate, overturning is not considered.)

$$0.9P + P_s + D + (W \text{ or } 0.7E) \leq S$$

From Sec. VIII, Div. 2, Table 5.2, P is determined from the MAWP. From Eq. 55.1, the longitudinal stress is

$$P = \frac{(\text{MAWP})R_i}{2t} = \frac{\left(150 \frac{\text{lbf}}{\text{in}^2}\right)(24 \text{ in})}{(2)(0.5 \text{ in})}$$

$$= 3600 \text{ psi} \quad \text{[tensile]}$$

P_s is the longitudinal stress due to the static head of the vessel contents. ($P + P_s$ is essentially the static pressure at the bottom of the tank.)

$$p_s = \gamma h = \frac{\text{contents weight}}{\text{bottom area}} \approx \frac{40{,}000 \text{ lbf}}{\frac{\pi}{4}(48 \text{ in})^2}$$
$$= 22.1 \text{ psi}$$

This pressure acts against the vessel walls as well as vertically downward.

$$P_s = \frac{p_s R_i}{2t} = \frac{\left(22.1 \ \frac{\text{lbf}}{\text{in}^2}\right)(24 \text{ in})}{(2)(0.5 \text{ in})} = 530 \text{ psi} \quad \text{[tensile]}$$

D is the stress from the dead load (i.e., the tank and contents weight). Since the vessel and skirt axial compressive stresses are essentially inversely proportional to their wall thicknesses,

$$D_{\text{vessel}} \approx \left(\frac{t_{\text{skirt}}}{t_{\text{vessel}}}\right) D_{\text{skirt}}$$
$$= \left(\frac{0.25 \text{ in}}{0.5 \text{ in}}\right)\left(1593 \ \frac{\text{lbf}}{\text{in}^2}\right)$$
$$= 797 \text{ psi} \quad \text{[compressive]}$$

Using the concept of radius of gyration ($I = Ar^2$), the area moment of inertia of the annular vessel cross section is also proportional to the thickness, so the vessel and skirt wind moment stresses are also inversely proportional to the wall thicknesses.

The wind-induced stress is

$$W_{\text{vessel}} \approx \left(\frac{t_{\text{skirt}}}{t_{\text{vessel}}}\right) W_{\text{skirt}}$$
$$= \left(\frac{0.25 \text{ in}}{0.5 \text{ in}}\right)\left(4029 \ \frac{\text{lbf}}{\text{in}^2}\right)$$
$$= 2015 \text{ psi} \quad \text{[tensile and compressive]}$$

Use the load group from Sec. VIII, Div. 2, Table 5.3, Case 5. Considering tensile stresses as positive, the stress that should be used to determine the adequacy of the pressure vessel is

$$\begin{aligned} &0.9P + P_s + D \\ &\quad + (W \text{ or } 0.7E) = (0.9)\left(3600 \ \frac{\text{lbf}}{\text{in}^2}\right) + 530 \ \frac{\text{lbf}}{\text{in}^2} \\ &\qquad - 797 \ \frac{\text{lbf}}{\text{in}^2} + 2015 \ \frac{\text{lbf}}{\text{in}^2} \\ &\qquad = \boxed{4988 \text{ psi} \quad (5000 \text{ psi})} \end{aligned}$$

(An argument could be made that conservatism requires that the dead load be positive, or at least, the load combination should not be reduced by the dead load in order to account for an empty tank. In the case of an empty tank, however, the static pressure term would also be absent, resulting in essentially the same answer. Other assumptions could be made.)

The answer is (B).

19. (a) Check the maximum allowable pressure for the shell and heads.

Shell:

Longitudinal stress (circumferential joint):

From Table 55.3, $S = 8700 \text{ lbf/in}^2$.

$$E = 0.80 \quad \text{[spot radiography, type 2 joint]}$$
$$t_{\text{corroded}} = 0.625 \text{ in} - 0.0625 \text{ in} = 0.563 \text{ in}$$
$$R = 12.000 \text{ in} - 0.563 \text{ in} = 11.437 \text{ in}$$
$$p = \frac{2SEt}{R - 0.4t} = \frac{(2)\left(8700 \ \frac{\text{lbf}}{\text{in}^2}\right)(0.80)(0.563 \text{ in})}{11.437 \text{ in} - (0.4)(0.563 \text{ in})}$$
$$= 699 \text{ lbf/in}^2$$

Longitudinal joint:

$$E = 0.85 \quad \text{[spot radiography, type 1 joint]}$$
$$p = \frac{SEt}{R + 0.6t} = \frac{\left(8700 \ \frac{\text{lbf}}{\text{in}^2}\right)(0.85)(0.563 \text{ in})}{11.437 \text{ in} + (0.6)(0.563 \text{ in})}$$
$$= 354 \text{ lbf/in}^2$$

Hemispherical head (based on inside radius):

$$E = 0.8 \quad \text{[spot radiography, type 2 joint]}$$
$$t = 0.500 \text{ in} - 0.0625 \text{ in} = 0.438 \text{ in}$$
$$R = 12.000 \text{ in} - 0.438 \text{ in} = 11.562 \text{ in}$$
$$p = \frac{2SEt}{R + 0.2t} = \frac{(2)\left(8700 \ \frac{\text{lbf}}{\text{in}^2}\right)(0.8)(0.438 \text{ in})}{11.562 \text{ in} + (0.2)(0.438 \text{ in})}$$
$$= 523 \text{ lbf/in}^2$$

Elliptical head (based on inside diameter):

$$E = 1.0$$

For a 2:1 ellipsoidal head, $h = D/4$.

$$K = \left(\tfrac{1}{6}\right)\left(2 + \left(\frac{D}{2h}\right)^2\right)$$
$$= \left(\tfrac{1}{6}\right)\left(2 + \left(\frac{D}{(2)\left(\frac{D}{4}\right)}\right)^2\right)$$
$$= 1$$
$$t = 0.625 \text{ in} - 0.0625 \text{ in} = 0.563 \text{ in}$$
$$D = 24.000 \text{ in} - (2)(0.563 \text{ in}) = 22.874 \text{ in}$$
$$p = \frac{2SEt}{KD + 0.2t} = \frac{(2)\left(8700 \ \frac{\text{lbf}}{\text{in}^2}\right)(1.0)(0.563 \text{ in})}{(1)(22.874 \text{ in}) + (0.2)(0.563 \text{ in})}$$
$$= 426 \text{ lbf/in}^2$$

The design pressure is the smallest of the calculated pressures, which is 354 psig (350 psig). This will be the MAWP for the vessel.

The answer is (B).

(b) $S_{\text{allowable at room temp}} = 18{,}600 \text{ lbf/in}^2$

$S_{\text{allowable at operating temp}} = 8700 \text{ lbf/in}^2$

The hydrostatic pressure at room temperature is

$$\frac{1.3P_{\max}S_{\text{allowable at room temp}}}{S_{\text{allowable at operating temp}}} = \frac{(1.3)\left(350 \dfrac{\text{lbf}}{\text{in}^2}\right)\left(18{,}600 \dfrac{\text{lbf}}{\text{in}^2}\right)}{8700 \dfrac{\text{lbf}}{\text{in}^2}} = \boxed{973 \text{ lbf/in}^2 \quad (1000 \text{ psig})}$$

The answer is (D).

20. (a) Calculate shell and nozzle neck thicknesses.

Determine the allowable stresses in the shell and nozzle neck from Table 55.3.

$$S_{\text{shell}} = S_{\text{nozzle}} = 14{,}800 \text{ lbf/in}^2$$

The circumferential stress will determine shell and nozzle neck thicknesses.

Shell:

$p = 400$ psig

$R = 24$ in

$E = 1.0$ [type 1 joint, full radiography]

$$t_r = \frac{pR}{SE - 0.6p} = \frac{\left(400 \dfrac{\text{lbf}}{\text{in}^2}\right)(24 \text{ in})}{\left(14{,}800 \dfrac{\text{lbf}}{\text{in}^2}\right)(1.0) - (0.6)\left(400 \dfrac{\text{lbf}}{\text{in}^2}\right)} = 0.659 \text{ in}$$

Since 0.75 in > 0.659 in, the specified shell thickness is adequate.

Nozzle neck:

$R = 6$ in

$E = 1.0$ [seamless pipe, full radiography]

$$t_a = \frac{pR}{SE - 0.6p} = \frac{\left(400 \dfrac{\text{lbf}}{\text{in}^2}\right)(6 \text{ in})}{\left(14{,}800 \dfrac{\text{lbf}}{\text{in}^2}\right)(1.0) - (0.6)\left(400 \dfrac{\text{lbf}}{\text{in}^2}\right)} = 0.165 \text{ in}$$

$$t_{b1} = t_{r,\text{shell},E=1} = 0.659 \text{ in}$$

For a 14 in schedule-40 pipe, $t = 0.438$ in.

$$t_{b3} = (0.438 \text{ in})(1 - 0.125) = 0.3833 \text{ in}$$

$$t_{\text{UG-45}} = \max\begin{cases} t_a \\ t_b = \min\begin{cases} t_{b1} \\ t_{b3} \end{cases} \end{cases} = \max\begin{cases} 0.165 \text{ in} \\ t_b = \min\begin{cases} 0.659 \text{ in} \\ 0.3833 \text{ in} \end{cases} \end{cases} = 0.3833 \text{ in}$$

Since 0.5 in > 0.3833 in, the specified thickness is adequate for the working pressure.

The answer is (C).

(b) The decimal equivalent of $^9/_{16}$ in is 0.563 in.

Weld sizes are specified on plans by their leg dimensions. The weld throat dimension is $\sqrt{2}/2$ (approximately 0.7) times the weld leg dimension.

This problem does not require a stress analysis, so only the weld sizes need to be validated. The requirements for fillet weld sizing are specified in UW-16 of Part IX of the ASME BPVC. To be properly sized, a weld must be greater than or equal to the minimum size required. The minimum fillet weld size depends on the thickness of the thinnest piece being welded, designated as $t_{\min}$, which does not have to be greater than 0.75 in.

The BPVC specifies the minimum weld size by both the weld throat size and the weld leg size. The throat size is designated in UW-16 as t_w, which does not mean the weld (leg) size. Generally, the smallest weld throat size is $^1/_4$ in. Depending on the location of the weld, the minimum weld throat size may be $0.5t_{\min}$ or $0.7t_{\min}$. If the weld leg size is greater than or equal to $t_{\min}$, the weld throat size will be greater than or equal to $0.7t_{\min}$. This condition satisfies both minimum requirements.

The maximum fillet weld size is generally $1.4t_{\min}$.

Inner shell-to-nozzle weld:

The shell thickness is 0.75 in, and the nozzle thickness is 0.5 in. Therefore, $t_{\min}$ is 0.5 in.

For this location, UW-16 (case j) specifies the minimum leg size, t_2, as $0.7t_{\min}$. Since the weld leg size is specified as $^9/_{16}$ in, the minimum weld leg size requirement is satisfied.

Outer pad-to-nozzle weld:

The pad (reinforcement ring) thickness is 0.5 in, and the nozzle thickness is also 0.5 in. Therefore, $t_{\min}$ is 0.5 in.

For this location, UW-16 (case h, j, or similar) specifies the minimum leg size, t_1 or t_w, as $0.7t_{\min}$. Since the weld leg size is specified as $^9/_{16}$ in, the minimum weld leg size requirement is satisfied.

The inner nozzle weld is adequate.

Outer shell-to-pad weld:

The shell thickness is 0.75 in, and the pad thickness is 0.5 in. Therefore, t_{min} is 0.5 in.

For this location, UW-16 (case h) specifies the minimum throat size as $0.5t_{min}$. Since the weld leg size is specified as 1/2 in, the minimum weld leg size requirement is satisfied.

The outer nozzle weld size is $\boxed{\text{adequate.}}$

The answer is (B).

(c) To determine whether reinforcement is adequate, compare the area required with the available area.

$$f_{r1} = f_{r2} = \frac{S_n}{S_v} = \frac{14{,}800\ \frac{\text{lbf}}{\text{in}^2}}{14{,}800\ \frac{\text{lbf}}{\text{in}^2}} = 1.0$$

The area required is

$$\begin{aligned}A_r &= Dt_r + 2t_nt_r(1-f_{r1}) = (12\text{ in})(0.659\text{ in})\\ &= 7.91\text{ in}^2\end{aligned}$$

Calculation of A_1 (area available from shell):

A_1 = the larger of

$$\begin{aligned}&D(Et-t_r) - 2t_n(Et-t_r)(1-f_{r1})\\ &\quad = (12\text{ in})\Big((1)(0.75\text{ in}) - 0.659\text{ in}\Big) - 0\\ &\quad = 1.09\text{ in}^2\end{aligned}$$

And,

$$\begin{aligned}&2(t+t_n)(Et-t_r) - 2t_n(Et-t_r)(1-f_{r1})\\ &\quad = (2)(0.75\text{ in} + 0.5\text{ in})\\ &\qquad \times \Big((1)(0.75\text{ in}) - 0.659\text{ in}\Big) - 0\\ &\quad = 0.23\text{ in}^2\end{aligned}$$

$A_1 = 1.09\text{ in}^2$.

Calculation of A_2 (area available from nozzle):

The equations for A_2 with a reinforcing pad are used.

A_2 = the smaller of

$$\begin{aligned}5(t_n - t_{rn})tf_{r2} &= (5)(0.5\text{ in} - 0.3833\text{ in})(0.75\text{ in})(1.0)\\ &= 0.438\text{ in}^2\end{aligned}$$

And,

$$\begin{aligned}&2(t_n - t_{rn})(2.5t_n + t_p)f_{r2}\\ &\quad = (2)(0.5\text{ in} - 0.3833\text{ in})\\ &\qquad \times \big((2.5)(0.5\text{ in}) + 0.5\text{ in}\big)(1.0)\\ &\quad = 0.408\text{ in}^2\end{aligned}$$

$A_2 = 0.408\text{ in}^2$.

Calculation of A_4 (area available from welds):

$$\begin{aligned}f_{r3} &= f_{r4} = 1.0\\ A_4 &= A_{41} + A_{42} = (\text{leg})^2 f_{r3} + (\text{leg})^2 f_{r4}\\ &= (0.563\text{ in})^2(1.0) + (0.563\text{ in})^2(1.0)\\ &= 0.634\text{ in}^2\end{aligned}$$

Calculation of A_5 (area available from pad):

$$\begin{aligned}A_5 &= (D_p - D - 2t_n)t_pf_{r4}\\ &= \big(21\text{ in} - 12\text{ in} - (2)(0.5\text{ in})\big)(0.5\text{ in})(1.0)\\ &= 4.0\text{ in}^2\end{aligned}$$

The available area is

$$\begin{aligned}A &= A_1 + A_2 + A_4 + A_5\\ &= 1.09\text{ in}^2 + 0.408\text{ in}^2 + 0.634\text{ in}^2 + 4.0\text{ in}^2\\ &= \boxed{6.13\text{ in}^2 \quad (6.1\text{ in}^2)}\end{aligned}$$

$\boxed{\text{No,}}$ the vessel does not meet requirements. Since $6.13\text{ in}^2 < 7.91\text{ in}^2$, the pad is not acceptable as designed. An additional 1.78 in^2 of reinforcement is required. An acceptable design can be achieved by increasing the outside diameter of the pad by 4.0 in.

$$\begin{aligned}D_p &= 21\text{ in} + 4.0\text{ in} = 25.0\text{ in}\\ A_5 &= (D_p - D - 2t_n)t_pf_{r4}\\ &= \big(25.0\text{ in} - 12\text{ in} - (2)(0.5\text{ in})\big)(0.5\text{ in})(1.0)\\ &= 6.0\text{ in}^2\end{aligned}$$

The available area is

$$\begin{aligned}A &= A_1 + A_2 + A_4 + A_5\\ &= 1.09\text{ in}^2 + 0.408\text{ in}^2 + 0.634\text{ in}^2 + 6.0\text{ in}^2\\ &= 8.13\text{ in}^2\end{aligned}$$

The answer is (A).

21. In all three situations shown, the weld's eccentricity relative to the shell induces a moment that tends to open up the joint. Rather than being a self-limiting effect relieving local peak stresses, the moment-induced stress is increased by migrating failure. All of these joint types are subject to premature failure under cyclic loading.

The answer is (D).

22. Butt welds are not mentioned, which would affect the joint efficiency. From Table 55.8, the weld efficiency for spot-radiographed "others" is 1.00.

$$t = d\sqrt{\frac{Cp}{SE}}$$

$$= 147 \text{ in}\sqrt{\frac{(0.2)\left(20\ \frac{\text{lbf}}{\text{in}^2}\right)}{\left(20{,}000\ \frac{\text{lbf}}{\text{in}^2}\right)(1.0)}}$$

$$= \boxed{2.08 \text{ in} \quad (2.1 \text{ in})}$$

The answer is (A).

23. From Eq. 55.9, the minimum head thickness for a circular welded plate is

$$t = d\sqrt{\frac{Cp}{SE}}$$

$$= 5.761 \text{ in}\sqrt{\frac{(0.33)\left(650\ \frac{\text{lbf}}{\text{in}^2}\right)}{\left(14{,}400\ \frac{\text{lbf}}{\text{in}^2}\right)(1.00)}}$$

$$= \boxed{0.703 \text{ in} \quad (0.70 \text{ in})}$$

The answer is (A).

24. Rupture discs should be installed in the relief line closest to the pressure vessel. In this manner, they can protect the pressure relief valve from corrosive pressure vessel contents and prevent fugitive emissions.

The answer is (C).

25. From ASME *Boiler and Pressure Vessel Code* (BPVC) Sec. VIII, Div. 1, UG-127 footnote 52: "...result in opening of the valve coincident with the bursting of the rupture disc." From ASME Sec. VIII, Div. 1, UG-132(a)(4)(a): "The marked burst pressure shall be between 90% and 100% of the marked set pressure of the valve."

The answer is (B).

26. Per ASME *Boiler and Pressure Vessel Code* (BPVC) UG-134(a), "The safety-relief valve shall be set at or below MAWP." Therefore, a single safety relief valve must be set at the MAWP. The allowable accumulation pressure (the difference between the MAWP and actual fully open pressure) for a single valve is 3 psi or 110% of MAWP, whichever is greater, per UG-125(c). In a fire contingency, the maximum accumulation pressure is 21% (i.e., the vessel pressure is permitted to reach 121% of MAWP.) Therefore, only if it is a second safety valve used for a fire contingency can it reach full capacity at 116% of the MAWP, in which case the first valve must be fully open by 110% of MAWP.

The answer is (B).

56 Properties of Solid Bodies

PRACTICE PROBLEMS

1. A spoked flywheel has an outside diameter of 60 in (1500 mm). The rim thickness is 6 in (150 mm) and the width is 12 in (300 mm). The cylindrical hub has an outside diameter of 12 in (300 mm), a thickness of 3 in (75 mm), and a width of 12 in (300 mm). The rim and hub are connected by six equally spaced cylindrical radial spokes, each having a diameter of 4.25 in (110 mm). All parts of the flywheel are ductile cast iron with a density of 0.256 lbm/in^3 (7080 kg/m^3). What is the rotational mass moment of inertia of the entire flywheel?

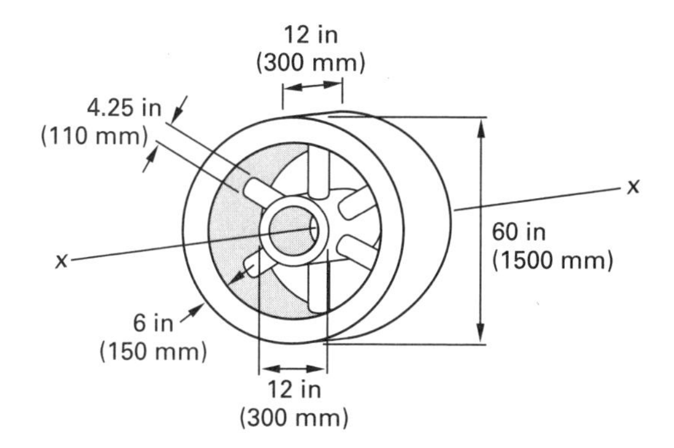

(A) 16,000 $lbm\text{-}ft^2$ (625 $kg\cdot m^2$)

(B) 16,200 $lbm\text{-}ft^2$ (631 $kg\cdot m^2$)

(C) 16,800 $lbm\text{-}ft^2$ (654 $kg\cdot m^2$)

(D) 193,000 $lbm\text{-}ft^2$ (7500 $kg\cdot m^2$)

2. Which option best describes the difference between the mass moment of inertia for a solid body and the area moment of inertia for an area?

(A) The mass moment of inertia characterizes the object's resistance to changes in rotational speed, and the area moment of inertia characterizes the object's resistance to bending.

(B) The mass moment of inertia characterizes the object's resistance to bending, and the area moment of inertia characterizes the object's resistance changes in rotational speed.

(C) The mass moment of inertia characterizes the object's resistance to torsion, and the area moment of inertia characterizes the object's resistance changes in bending.

(D) The mass moment of inertia characterizes the object's resistance to bending, and the area moment of inertia characterizes the object's resistance changes in torsion.

3. Which option best describes the difference between the moment of inertia for a mass and the polar moment of inertia for an area?

(A) The mass polar moment of inertia characterizes the object's resistance to bending, and the area polar moment of inertia characterizes the object's resistance to changes in rotational speed.

(B) The mass polar moment of inertia characterizes the object's resistance to torsion, and the area polar moment of inertia characterizes the object's resistance to changes in rotational speed.

(C) The mass polar moment of inertia characterizes the object's resistance to changes in rotational speed, and the area polar moment of inertia characterizes the object's resistance to torsion.

(D) The mass moment of inertia characterizes the object's resistance to torsion, and the area moment of inertia characterizes the object's resistance changes in bending.

4. Which of the following is/are NOT correct units for mass moment of inertia?

I. slug-ft^2

II. poundal-ft^2

III. lbf-sec^2-ft

(A) II only

(B) III only

(C) II and III only

(D) I, II, and III

SOLUTIONS

1. *Customary U.S. Solution*

From the hollow circular cylinder in App. 56.A, the mass moment of inertia of the rim about the x-axis (i.e., the rotational axis) is

$$\begin{aligned} I_{x,1} &= \left(\frac{\pi\rho L}{2}\right)(r_o^4 - r_i^4) \\ &= \frac{\left(\frac{\pi}{2}\right)\left(0.256\ \frac{\text{lbm}}{\text{in}^3}\right)(12\ \text{in})}{\left(12\ \frac{\text{in}}{\text{ft}}\right)^2} \\ &\quad \times \left(\left(\frac{60\ \text{in}}{2}\right)^4 - \left(\frac{60\ \text{in} - 12\ \text{in}}{2}\right)^4\right) \\ &= 16{,}025\ \text{lbm-ft}^2 \end{aligned}$$

From the hollow circular cylinder in App. 56.A, the mass moment of inertia of the hub is

$$\begin{aligned} I_{x,2} &= \left(\frac{\pi\rho L}{2}\right)(r_o^4 - r_i^4) \\ &= \frac{\left(\frac{\pi}{2}\right)\left(0.256\ \frac{\text{lbm}}{\text{in}^3}\right)(12\ \text{in})}{\left(12\ \frac{\text{in}}{\text{ft}}\right)^2} \\ &\quad \times \left(\left(\frac{12\ \text{in}}{2}\right)^4 - \left(\frac{12\ \text{in} - 6\ \text{in}}{2}\right)^4\right) \\ &= 41\ \text{lbm-ft}^2 \end{aligned}$$

The length of a cylindrical spoke is

$$L = 24\ \text{in} - 6\ \text{in} = 18\ \text{in}$$

The mass of a spoke is

$$\begin{aligned} m &= \rho A L = \rho\pi r^2 L \\ &= \left(0.256\ \frac{\text{lbm}}{\text{in}^3}\right)\pi\left(\frac{4.25\ \text{in}}{2}\right)^2(18\ \text{in}) \\ &= 65.37\ \text{lbm} \end{aligned}$$

From the solid circular cylinder in App. 56.A, the mass moment of inertia of a spoke about its own centroidal axis is

$$\begin{aligned} I_{c,x} &= \frac{m(3r^2 + L^2)}{12} \\ &= \frac{(65.37\ \text{lbm})\left((3)\left(\frac{4.25\ \text{in}}{2}\right)^2 + (18\ \text{in})^2\right)}{(12)\left(12\ \frac{\text{in}}{\text{ft}}\right)^2} \\ &= 13\ \text{lbm-ft}^2 \end{aligned}$$

Use the parallel axis theorem, Eq. 56.12, to find the mass moment of inertia of a spoke about the axis of the flywheel.

$$d = \frac{12 \text{ in}}{2} + \frac{18 \text{ in}}{2} = 15 \text{ in}$$

$$\begin{aligned} I_{x,3\,\text{per spoke}} &= I_{c,x} + md^2 \\ &= 13 \text{ lbm-ft}^2 + (65.37 \text{ lbm})\left(\frac{15 \text{ in}}{12 \frac{\text{in}}{\text{ft}}}\right)^2 \\ &= 115 \text{ lbm-ft}^2 \end{aligned}$$

The total for six spokes is

$$\begin{aligned} I_{x,3} &= 6I_{x,3\,\text{per spoke}} \\ &= (6)(115 \text{ lbm-ft}^2) \\ &= 690 \text{ lbm-ft}^2 \end{aligned}$$

Finally, the total rotational mass moment of inertia of the flywheel is

$$\begin{aligned} I &= I_{x,1} + I_{x,2} + I_{x,3} \\ &= 16{,}025 \text{ lbm-ft}^2 + 41 \text{ lbm-ft}^2 + 690 \text{ lbm-ft}^2 \\ &= \boxed{16{,}756 \text{ lbm-ft}^2 \quad (16{,}800 \text{ lbm-ft}^2)} \end{aligned}$$

The answer is (C).

SI Solution

From the hollow circular cylinder in App. 56.A, the mass moment of inertia of the rim about the x-axis (i.e., the rotational axis) is

$$\begin{aligned} I_{x,1} &= \left(\frac{\pi\rho L}{2}\right)(r_o^4 - r_i^4) \\ &= \left(\frac{\pi\left(7080 \frac{\text{kg}}{\text{m}^3}\right)(0.3 \text{ m})}{2}\right) \\ &\quad \times \left(\left(\frac{1.5 \text{ m}}{2}\right)^4 - \left(\frac{1.5 \text{ m} - 0.30 \text{ m}}{2}\right)^4\right) \\ &= 623.3 \text{ kg·m}^2 \end{aligned}$$

From the hollow circular cylinder in App. 56.A, the mass moment of inertia of the hub is

$$\begin{aligned} I_{x,2} &= \left(\frac{\pi\rho L}{2}\right)(r_o^4 - r_i^4) \\ &= \left(\frac{\pi\left(7080 \frac{\text{kg}}{\text{m}^3}\right)(0.3 \text{ m})}{2}\right) \\ &\quad \times \left(\left(\frac{0.3 \text{ m}}{2}\right)^4 - \left(\frac{0.3 \text{ m} - 0.15 \text{ m}}{2}\right)^4\right) \\ &= 1.6 \text{ kg·m}^2 \end{aligned}$$

The length of a cylindrical spoke is

$$L = 0.6 \text{ m} - 0.15 \text{ m} = 0.45 \text{ m}$$

The mass of a spoke is

$$\begin{aligned} m &= \rho AL = \rho\pi r^2 L \\ &= \left(7080 \frac{\text{kg}}{\text{m}^3}\right)\pi\left(\frac{0.11 \text{ m}}{2}\right)^2(0.45 \text{ m}) \\ &= 30.3 \text{ kg} \end{aligned}$$

From the solid circular cylinder in App. 56.A, the mass moment of inertia of a spoke about its own centroidal axis is

$$\begin{aligned} I_{c,x} &= \frac{m(3r^2 + L^2)}{12} \\ &= \frac{(30.3 \text{ kg})\left((3)\left(\frac{0.11 \text{ m}}{2}\right)^2 + (0.45 \text{ m})^2\right)}{12} \\ &= 0.53 \text{ kg·m}^2 \end{aligned}$$

Use the parallel axis theorem, Eq. 56.12, to find the mass moment of inertia of a spoke about the axis of the flywheel.

$$d = \frac{0.30 \text{ m}}{2} + \frac{0.45 \text{ m}}{2} = 0.375 \text{ m}$$

$$\begin{aligned} I_{x,3\,\text{per spoke}} &= I_{c,x} + md^2 \\ &= 0.53 \text{ kg·m}^2 + (30.3 \text{ kg})(0.375 \text{ m})^2 \\ &= 4.8 \text{ kg·m}^2 \end{aligned}$$

The total for six spokes is

$$\begin{aligned} I_{x,3} &= 6I_{x,3\,\text{per spoke}} = (6)(4.8\ \text{kg·m}^2) \\ &= 28.8\ \text{kg·m}^2 \end{aligned}$$

The total rotational mass moment of inertia of the flywheel is

$$\begin{aligned} I &= I_{x,1} + I_{x,2} + I_{x,3} \\ &= 623.3\ \text{kg·m}^2 + 1.6\ \text{kg·m}^2 + 28.8\ \text{kg·m}^2 \\ &= \boxed{653.7\ \text{kg·m}^2 \quad (654\ \text{kg·m}^2)} \end{aligned}$$

The answer is (C).

2. The mass moment of inertia is the limit of the product of the mass of the solid body and the square of the distance from a given axis. It measures the object's resistance to changes in rotational speed. Typical use is in the equation $T = I\alpha$. A solid body with a small amount of inertia will be less likely to resist a change in rotational speed.

The area moment of inertia, I, is the limit of the product of the area of the solid body and the square of the distance from a given axis. It measures the object's resistance to bending. Typical use is in the equation $\sigma = Mc/I$. A solid body with a small amount of inertia will bend more than a solid body with greater inertia.

The answer is (A).

3. The mass moment of inertia is the limit of the product of the mass of the solid body and the square of the distance from a given axis. It measures the object's resistance to changes in rotational speed. Typical use is in the equation $T = I\alpha$. A solid body with a small amount of inertia will be less likely to resist a change in rotational speed.

The area polar moment of inertia is measured about an axis perpendicular to the plane of the area. It characterizes the object's resistance to torsion. Typical use is in the equation $\theta = TL/JG$. A solid body with a small area polar moment inertia will twist more than a solid body with greater inertia.

The answer is (C).

4. Mass moment of inertia has units of mass × length^2. A slug is a unit of mass whose basic units can be derived from $m = F/a$.

$$\text{units of } \frac{F}{a} = \frac{\text{lbf}}{\frac{\text{ft}}{\text{sec}^2}} = \frac{\text{lbf-sec}^2}{\text{ft}}$$

The units of slug-ft^2 resolve into $\boxed{\text{lbf-sec}^2\text{-ft.}}$ A poundal is an obscure and seldom used unit of force, not mass.

The answer is (A).

57 Kinematics

PRACTICE PROBLEMS

Linear Particle Motion

1. A particle moves horizontally according to the formula $s = 2t^2 - 8t + 3$.

(a) When $t = 2$, what are most nearly the position, velocity, and acceleration, respectively?

(A) $s = -5$, $v = 0$, $a = 4$

(B) $s = 0$, $v = 0$, $a = 4$

(C) $s = 3$, $v = -4$, $a = 4$

(D) $s = 3$, $v = -8$, $a = 4$

(b) What are most nearly the linear displacement and total distance traveled between $t = 1$ and $t = 3$, respectively?

(A) 0, 0

(B) 0, 4

(C) 6, 6

(D) 6, 10

Uniform Acceleration

2. An aircraft starts from rest and acquires a speed of 180 mph (290 km/h) in 60 sec. Its acceleration is most nearly

(A) 4.4 ft/sec^2 (1.3 m/s^2)

(B) 6.3 ft/sec^2 (1.9 m/s^2)

(C) 8.9 ft/sec^2 (2.7 m/s^2)

(D) 12 ft/sec^2 (3.6 m/s^2)

3. The acceleration of a train that increases its speed from 5 ft/sec to 20 ft/sec (1.5 m/s to 6 m/s) in 2 min is most nearly

(A) 0.13 ft/sec^2 (0.038 m/s^2)

(B) 0.39 ft/sec^2 (0.12 m/s^2)

(C) 0.82 ft/sec^2 (0.25 m/s^2)

(D) 1.3 ft/sec^2 (0.39 m/s^2)

4. A car traveling at 60 mph (100 km/h) applies its brakes, decelerates uniformly, and stops in 5 sec. What are most nearly its acceleration and distance traveled, respectively, before stopping?

(A) $a = -88$ ft/sec^2, $s = 220$ ft
($a = -27.8$ m/s^2, $s = 220$ m)

(B) $a = -29$ ft/sec^2, $s = 363$ ft
($a = -8.8$ m/s^2, $s = 110$ m)

(C) $a = -18$ ft/sec^2, $s = 220$ ft
($a = -5.7$ m/s^2, $s = 70$ m)

(D) $a = -11$ ft/sec^2, $s = 33$ ft
($a = -3.3$ m/s^2, $s = 10$ m)

Projectile Motion

5. A projectile is fired at 45° from the horizontal with an initial velocity of 2700 ft/sec (820 m/s). Neglecting air friction, what are most nearly the maximum altitude and range, respectively?

(A) $H = 57{,}000$ ft, $R = 230{,}000$ ft
($H = 17\,000$ m, $R = 69\,000$ m)

(B) $H = 82{,}000$ ft, $R = 202{,}000$ ft
($H = 25\,000$ m, $R = 62\,000$ m)

(C) $H = 57{,}000$ ft, $R = 113{,}000$ ft
($H = 17\,000$ m, $R = 50\,000$ m)

(D) $H = 82{,}000$ ft, $R = 164{,}000$ ft
($H = 25\,000$ m, $R = 50\,000$ m)

6. A projectile is launched with an initial velocity of 900 ft/sec (270 m/s). The target is 12,000 ft (3600 m) away and 2000 ft (600 m) higher than the launch point. Air friction is to be neglected. Most nearly, at what angle should the projectile be launched?

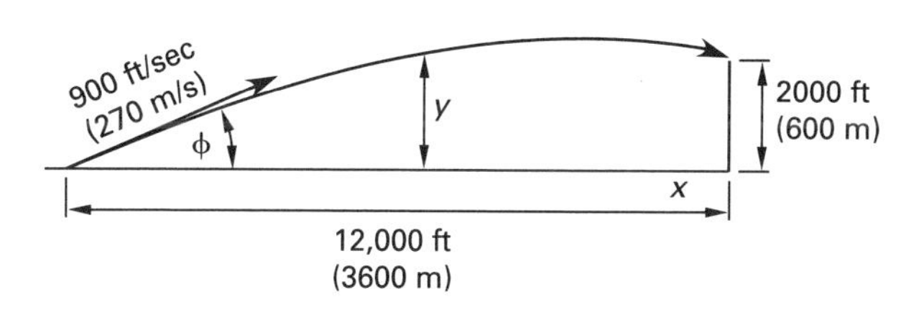

(A) 75°, 15°

(B) 75°, 24°

(C) 1.3°, 0.25°

(D) 1.3°, 0.40°

7. A baseball is hit at 60 ft/sec (20 m/s) and 36.87° from the horizontal. It strikes a fence 72 ft (22 m) away. What are most nearly the velocity components and the elevation, respectively, above the origin at impact?

(A) $v_x = 16$ ft/sec, $v_y = -36$ ft/sec, $y = -18$ ft ($v_x = 5$ m/s, $v_y = -11$ m/s, $y = 6$ m)

(B) $v_x = 48$ ft/sec, $v_y = -12$ ft/sec, $y = 18$ ft ($v_x = 16$ m/s, $v_y = -1.5$ m/s, $y = 7$ m)

(C) $v_x = 41$ ft/sec, $v_y = -93$ ft/sec, $y = -103$ ft ($v_x = 12$ m/s, $v_y = -28$ m/s, $y = -31$ m)

(D) $v_x = 130$ ft/sec, $v_y = -190$ ft/sec, $y = 230$ ft ($v_x = 41$ m/s, $v_y = -57$ m/s, $y = 70$ m)

8. A bomb is dropped from a plane that is climbing at 30° and 600 ft/sec (180 m/s) while passing through 12,000 ft (3600 m) altitude.

(a) The bomb's maximum altitude is most nearly

(A) 12,500 ft (3700 m)

(B) 13,000 ft (3900 m)

(C) 13,400 ft (4000 m)

(D) 14,200 ft (4300 m)

(b) Approximately how long will it take for the bomb to reach the ground from the release point?

(A) 24 sec

(B) 38 sec

(C) 43 sec

(D) 55 sec

Rotational Particle Motion

9. A point starts from rest and travels in a circle according to $\omega = 6t^2 - 10t$, where t is time measured in seconds, and ω is measured in radians per second. At $t = 2$, the direction of motion is clockwise.

(a) The angular velocity at $t = 2$ is most nearly

(A) 4

(B) 6

(C) 8

(D) 12

(b) The maximum displacement between the point's positions at $t = 1$ and $t = 3$ is most nearly

(A) 0.57 rad

(B) 0.82 rad

(C) 12 rad

(D) 15 rad

(c) Assume $\theta(0) = 0$. The total angle turned through between $t = 1$ and $t = 3$ is most nearly

(A) 6 rad

(B) 8 rad

(C) 12 rad

(D) 15 rad

10. The linear speed of a point on the edge of a 14 in diameter (36 cm diameter) disk turning at 40 rpm is most nearly

(A) 2.4 ft/sec (0.75 m/s)

(B) 4.5 ft/sec (1.4 m/s)

(C) 7.8 ft/sec (2.3 m/s)

(D) 9.2 ft/sec (2.8 m/s)

11. The angular acceleration that is required to increase an electric motor's speed from 1200 rpm to 3000 rpm in 10 sec is most nearly

(A) 10 rad/sec^2

(B) 14 rad/sec^2

(C) 18 rad/sec^2

(D) 25 rad/sec^2

12. An apparatus for determining the speed of a bullet consists of 2 paper disks mounted 5 ft (1.5 m) apart on a single horizontal shaft that is turning at 1750 rpm. A bullet pierces both disks at radius 6 in (15 cm) and an angle of 18° exists between each hole. The bullet velocity is most nearly

(A) 1700 ft/sec (510 m/s)

(B) 2100 ft/sec (630 m/s)

(C) 2400 ft/sec (720 m/s)

(D) 2900 ft/sec (880 m/s)

13. Disks B and C are in contact and rotate without slipping. A and B are splined together and rotate counterclockwise. Angular velocity and acceleration of disk C are 2 rad/sec (2 rad/s) and 6 rad/sec^2 (6 rad/s^2), respectively. What are most nearly the velocity and acceleration, respectively, of point D?

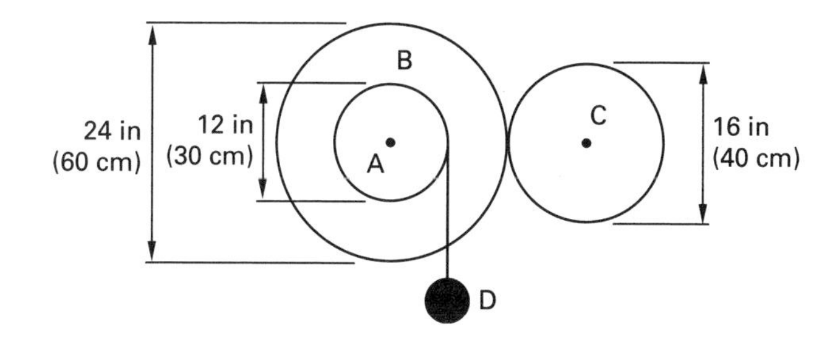

(A) $v_D = 0.7$ ft/sec, $a_D = 2$ ft/sec^2
($v_D = 0.2$ m/s, $a_D = 0.6$ m/s^2)

(B) $v_D = 1.5$ ft/sec, $a_D = 4.5$ ft/sec^2
($v_D = 0.5$ m/s, $a_D = 1.4$ m/s^2)

(C) $v_D = 2$ ft/sec, $a_D = 0.67$ ft/sec^2
($v_D = 0.6$ m/s, $a_D = 0.2$ m/s^2)

(D) $v_D = 4.5$ ft/sec, $a_D = 1.5$ ft/sec^2
($v_D = 1.4$ m/s, $a_D = 0.5$ m/s^2)

Relative Motion

14. The center of a wheel with an outer diameter of 24 in (610 mm) is moving at 28 mph (12.5 m/s). There is no slippage between the wheel and surface. A valve stem is mounted 6 in (150 mm) from the center. What are most nearly the velocity and the direction, respectively, of the valve stem when it is 45° from the horizontal?

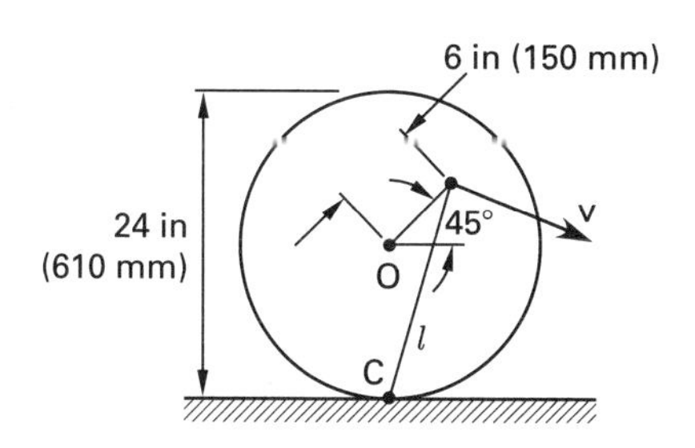

(A) v = 30 ft/sec, perpendicular to l
(v = 9 m/s, perpendicular to l)

(B) v = 57 ft/sec, perpendicular to l
(v = 17 m/s, perpendicular to l)

(C) v = 61 ft/sec, perpendicular to l
(v = 19 m/s, perpendicular to l)

(D) v = 97 ft/sec, perpendicular to l
(v = 30 m/s, perpendicular to l)

15. A balloon is 200 ft (60 m) above the ground and is rising at a constant 15 ft/sec (4.5 m/s). An automobile passes under it, traveling along a straight and level road at 45 mph (72 km/h). Approximately how fast is the distance between them changing 1 sec later?

(A) 13 ft/sec (4.0 m/s)

(B) 34 ft/sec (10 m/s)

(C) 37 ft/sec (11 m/s)

(D) 51 ft/sec (15 m/s)

Rotation About a Fixed Axis

16. The axle that the wheel is attached to moves to the right at 10 ft/sec (3 m/s). What are most nearly the velocities of points A and B, respectively, with respect to point O if the wheel rolls without slipping?

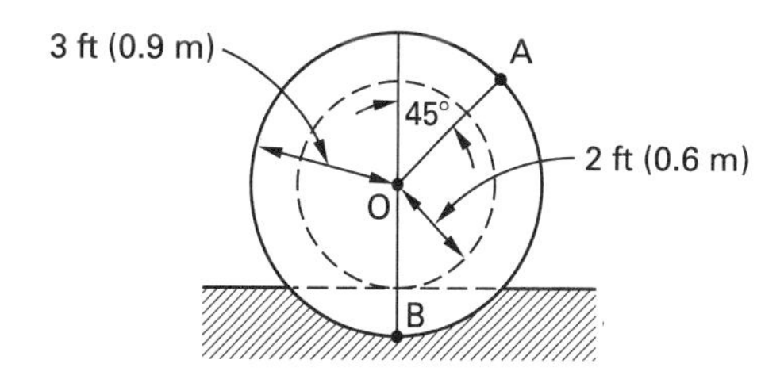

(A) $v_{A/O} = 5$ ft/sec, 45° below horizontal, to the right; $v_{B/O} = 5$ ft/sec, horizontal, to the left
($v_{A/O} = 1.5$ m/s, 45° below horizontal, to the right; $v_{B/O} = 1.5$ m/s, horizontal, to the left)

(B) $v_{A/O} = 6.7$ ft/sec, 45° below horizontal, to the right; $v_{B/O} = 6.7$ ft/sec, horizontal, to the left
($v_{A/O} = 2.0$ m/s, 45° below horizontal, to the right; $v_{B/O} = 2.0$ m/s, horizontal, to the left)

(C) $v_{A/O} = 10$ ft/sec, 45° below horizontal, to the right; $v_{B/O} = 10$ ft/sec, horizontal, to the left
($v_{A/O} = 3.3$ m/s, 45° below horizontal, to the right; $v_{B/O} = 3.3$ m/s, horizontal, to the left)

(D) $v_{A/O} = 15$ ft/sec, 45° below horizontal, to the right; $v_{B/O} = 15$ ft/sec, horizontal, to the left
($v_{A/O} = 4.5$ m/s, 45° below horizontal, to the right; $v_{B/O} = 4.5$ m/s, horizontal, to the left)

17. The velocity of point B with respect to point A in Prob. 16 is most nearly

(A) 28 ft/sec, 23° above horizontal, to the left
(8.3 m/s, 23° above horizontal, to the left)

(B) 30.0 ft/sec, 22.5° above horizontal, to the left
(9.1 m/s, 22.5° above horizontal, to the left)

(C) 41.3 ft/sec, 22.5° above horizontal, to the left
(12.5 m/s, 22.5° above horizontal, to the left)

(D) 42.3 ft/sec, 22.5° above horizontal, to the left
(12.8 m/s, 22.5° above horizontal, to the left)

SOLUTIONS

1. (a) From the problem statement,

$$s(t) = 2t^2 - 8t + 3$$

Using Eq. 57.3 and Eq. 57.4,

$$\mathrm{v}(t) = \frac{ds(t)}{dt} = 4t - 8$$

$$a(t) = \frac{d\mathrm{v}(t)}{dt} = 4$$

At $t = 2$,

$$s = (2)(2)^2 - (8)(2) + 3 = \boxed{-5}$$

$$\mathrm{v} = (4)(2) - 8 = \boxed{0}$$

$$a = \boxed{4}$$

The answer is (A).

(b) At $t = 1$,

$$s = (2)(1)^2 - (8)(1) + 3 = -3$$

At $t = 3$,

$$s = (2)(3)^2 - (8)(3) + 3 = -3$$

From Eq. 57.7,

$$\text{displacement} = s(t_2) - s(t_1) = -3 - (-3) = \boxed{0}$$

The total distance traveled from $t = 1$ to $t = 3$ is

$$\begin{aligned} s &= \int_1^3 |\mathrm{v}(t)|dt = \int_1^3 |4t - 8|dt \\ &= \int_1^2 (8 - 4t)dt + \int_2^3 (4t - 8)dt \\ &= 2 + 2 \\ &= \boxed{4} \end{aligned}$$

The answer is (B).

2. *Customary U.S. Solution*

$$a = \frac{\Delta \mathrm{v}}{\Delta t} = \frac{\left(180\ \frac{\text{mi}}{\text{hr}}\right)\left(5280\ \frac{\text{ft}}{\text{mi}}\right)}{(60\ \text{sec})\left(60\ \frac{\text{sec}}{\text{min}}\right)\left(60\ \frac{\text{min}}{\text{hr}}\right)} = \boxed{4.4\ \text{ft/sec}^2}$$

The answer is (A).

SI Solution

$$a = \frac{\Delta \mathrm{v}}{\Delta t} = \frac{\left(290\ \frac{\text{km}}{\text{h}}\right)\left(1000\ \frac{\text{m}}{\text{km}}\right)}{(60\ \text{s})\left(60\ \frac{\text{s}}{\text{min}}\right)\left(60\ \frac{\text{min}}{\text{h}}\right)} = \boxed{1.3\ \text{m/s}^2}$$

The answer is (A).

3. *Customary U.S. Solution*

Use Table 57.1.

$$\begin{aligned} a = \frac{\mathrm{v} - \mathrm{v}_0}{t} &= \frac{20\ \frac{\text{ft}}{\text{sec}} - 5\ \frac{\text{ft}}{\text{sec}}}{(2\ \text{min})\left(60\ \frac{\text{sec}}{\text{min}}\right)} \\ &= \boxed{0.125\ \text{ft/sec}^2 \quad (0.13\ \text{ft/sec}^2)} \end{aligned}$$

The answer is (A).

SI Solution

Use Table 57.1.

$$\begin{aligned} a = \frac{\mathrm{v} - \mathrm{v}_0}{t} &= \frac{6\ \frac{\text{m}}{\text{s}} - 1.5\ \frac{\text{m}}{\text{s}}}{(2\ \text{min})\left(60\ \frac{\text{s}}{\text{min}}\right)} \\ &= \boxed{0.0375\ \text{m/s}^2 \quad (0.038\ \text{m/s}^2)} \end{aligned}$$

The answer is (A).

4. *Customary U.S. Solution*

The initial velocity is

$$\mathrm{v}_0 = \frac{\left(60\ \frac{\text{mi}}{\text{hr}}\right)\left(5280\ \frac{\text{ft}}{\text{mi}}\right)}{\left(60\ \frac{\text{sec}}{\text{min}}\right)\left(60\ \frac{\text{min}}{\text{hr}}\right)} = 88\ \text{ft/sec}$$

$$a = \frac{v - v_0}{t} = \frac{0\ \frac{\text{ft}}{\text{sec}} - 88\ \frac{\text{ft}}{\text{sec}}}{5\ \text{sec}}$$
$$= \boxed{-17.6\ \text{ft/sec}^2 \quad (-18\ \text{ft/sec}^2)}$$
$$s = \tfrac{1}{2}t(v_0 + v) = \tfrac{1}{2}tv_0 = \left(\tfrac{1}{2}\right)(5\ \text{sec})\left(88\ \frac{\text{ft}}{\text{sec}}\right)$$
$$= \boxed{220\ \text{ft}}$$

The answer is (C).

SI Solution

The initial velocity is

$$v_0 = \frac{\left(100\ \frac{\text{km}}{\text{h}}\right)\left(1000\ \frac{\text{m}}{\text{km}}\right)}{\left(60\ \frac{\text{s}}{\text{min}}\right)\left(60\ \frac{\text{min}}{\text{h}}\right)} = 27.78\ \text{m/s}$$
$$a = \frac{v - v_0}{t} = \frac{0\ \frac{\text{m}}{\text{s}} - 27.78\ \frac{\text{m}}{\text{s}}}{5\ \text{s}}$$
$$= \boxed{-5.56\ \text{m/s}^2 \quad (-5.7\ \text{m/s}^2)}$$
$$s = \tfrac{1}{2}t(v_0 + v) = \tfrac{1}{2}tv_0 = \left(\tfrac{1}{2}\right)(5\ \text{s})\left(27.78\ \frac{\text{m}}{\text{s}}\right)$$
$$= \boxed{69.5\ \text{m} \quad (70\ \text{m})}$$

The answer is (C).

5. *Customary U.S. Solution*

Use Table 57.2.

$$H = \frac{v_0^2 \sin^2 \phi}{2g} = \frac{\left(2700\ \frac{\text{ft}}{\text{sec}}\right)^2 (\sin 45°)^2}{(2)\left(32.2\ \frac{\text{ft}}{\text{sec}^2}\right)}$$
$$= \boxed{56{,}599\ \text{ft} \quad (57{,}000\ \text{ft})}$$
$$R = \frac{v_0^2 \sin 2\phi}{g} = \frac{\left(2700\ \frac{\text{ft}}{\text{sec}}\right)^2 (\sin(2)(45°))}{32.2\ \frac{\text{ft}}{\text{sec}^2}}$$
$$= \boxed{226{,}398\ \text{ft} \quad (230{,}000\ \text{ft})}$$

The answer is (A).

SI Solution

Use Table 57.2.

$$H = \frac{v_0^2 \sin^2 \phi}{2g} = \frac{\left(820\ \frac{\text{m}}{\text{s}}\right)^2 (\sin 45°)^2}{(2)\left(9.81\ \frac{\text{m}}{\text{s}^2}\right)}$$
$$= \boxed{17\,136\ \text{m} \quad (17\,000\ \text{m})}$$
$$R = \frac{v_0^2 \sin 2\phi}{g} = \frac{\left(820\ \frac{\text{m}}{\text{s}}\right)^2 (\sin(2)(45°))}{9.81\ \frac{\text{m}}{\text{s}^2}}$$
$$= \boxed{68\,542\ \text{m} \quad (69\,000\ \text{m})}$$

The answer is (A).

6. *Customary U.S. Solution*

From Table 57.2, the x-distance is

$$x = (v_0 \cos \phi)t$$

Solve for t.

$$t = \frac{x}{v_0 \cos \phi} = \frac{12{,}000\ \text{ft}}{\left(900\ \frac{\text{ft}}{\text{sec}}\right)(\cos \phi)} = \frac{13.33}{\cos \phi}$$

From Table 57.2, the y-distance is

$$y = (v_0 \sin \phi)t - \tfrac{1}{2}gt^2$$

Substitute t and the given value of y.

$$2000\ \text{ft} = \left(900\ \frac{\text{ft}}{\text{sec}}\right)(\sin \phi)\left(\frac{13.33}{\cos \phi}\right) - \left(\tfrac{1}{2}\right)\left(32.2\ \frac{\text{ft}}{\text{sec}^2}\right)\left(\frac{13.33}{\cos \phi}\right)^2$$

Simplify.

$$1 = 6.0 \tan \phi - \frac{1.43}{\cos^2 \phi}$$

Use the following identity.

$$\frac{1}{\cos^2 \phi} = 1 + \tan^2 \phi$$
$$1 = 6.0 \tan \phi - 1.43 - 1.43 \tan^2 \phi$$

Simplify.

$$\tan^2\phi - 4.20\tan\phi + 1.70 = 0$$

Use the quadratic formula.

$$\begin{aligned}\tan\phi &= \frac{4.20 \pm \sqrt{(4.20)^2 - (4)(1)(1.70)}}{(2)(1)}\\ &= 3.75,\ 0.454 \quad \text{[radians]}\\ \phi &= \tan^{-1}3.75,\ \tan^{-1}0.454\\ &= \boxed{75.1°,\ 24.4° \quad (75°,\ 24°)}\end{aligned}$$

The answer is (B).

SI Solution

From Table 57.1, the x-distance is

$$x = (\mathrm{v}_0\cos\phi)t$$

Solve for t.

$$t = \frac{x}{\mathrm{v}_0\cos\phi} = \frac{3600\ \text{m}}{\left(270\ \frac{\text{m}}{\text{s}}\right)(\cos\phi)} = \frac{13.33}{\cos\phi}$$

From Table 57.1, the y-distance is

$$y = (\mathrm{v}_0\sin\phi)t - \tfrac{1}{2}gt^2$$

Substitute t and the given value of y.

$$\begin{aligned}600\ \text{m} = &\left(270\ \frac{\text{m}}{\text{s}}\right)(\sin\phi)\left(\frac{13.33}{\cos\phi}\right)\\ &- \left(\tfrac{1}{2}\right)\left(9.81\ \frac{\text{m}}{\text{s}^2}\right)\left(\frac{13.33}{\cos\phi}\right)^2\end{aligned}$$

Simplify.

$$1 = 6.0\tan\phi - \frac{1.45}{\cos^2\phi}$$

Use the following identity.

$$\frac{1}{\cos^2\phi} = 1 + \tan^2\phi$$

$$1 = 6.0\tan\phi - 1.45 - 1.45\tan^2\phi$$

Simplify.

$$\tan^2\phi - 4.14\tan\phi + 1.69 = 0$$

Use the quadratic formula.

$$\begin{aligned}\tan\phi &= \frac{4.14 \pm \sqrt{(4.14)^2 - (4)(1)(1.69)}}{(2)(1)}\\ &= 3.68,\ 0.459 \quad \text{[radians]}\\ \phi &= \tan^{-1}3.68,\ \tan^{-1}0.459\\ &= \boxed{74.8°,\ 24.7° \quad (75°,\ 24°)}\end{aligned}$$

The answer is (B).

7. *Customary U.S. Solution*

Neglect air friction. From Table 57.1,

$$\begin{aligned}\mathrm{v}_x(t) &= \mathrm{v}_0\cos\phi = \left(60\ \frac{\text{ft}}{\text{sec}}\right)(\cos 36.87°)\\ &= \boxed{48\ \text{ft/sec} \quad \text{[constant]}}\end{aligned}$$

$$t = \frac{s}{\mathrm{v}_x} = \frac{72\ \text{ft}}{48\ \frac{\text{ft}}{\text{sec}}} = 1.5\ \text{sec}$$

$$\begin{aligned}\mathrm{v}_y(t) &= \mathrm{v}_0\sin\phi - gt\\ &= \left(60\ \frac{\text{ft}}{\text{sec}}\right)(\sin 36.87°) - \left(32.2\ \frac{\text{ft}}{\text{sec}^2}\right)(1.5\ \text{sec})\\ &= \boxed{-12.3\ \text{ft/sec} \quad (-12\ \text{ft/sec})}\end{aligned}$$

$$\begin{aligned}y &= (\mathrm{v}_0\sin\phi)t - \tfrac{1}{2}gt^2\\ &= \left(60\ \frac{\text{ft}}{\text{sec}}\right)(\sin 36.87°)(1.5\ \text{sec})\\ &\quad - \left(\tfrac{1}{2}\right)\left(32.2\ \frac{\text{ft}}{\text{sec}^2}\right)(1.5\ \text{sec})^2\\ &= \boxed{17.78\ \text{ft} \quad (18\ \text{ft})}\end{aligned}$$

The answer is (B).

SI Solution

Neglect air friction. From Table 57.1,

$$\begin{aligned}\mathrm{v}_x(t) &= \mathrm{v}_0\cos\phi = \left(20\ \frac{\text{m}}{\text{s}}\right)(\cos 36.87°)\\ &= \boxed{16\ \text{m/s} \quad \text{[constant]}}\end{aligned}$$

$$t = \frac{s}{\mathrm{v}_x} = \frac{22\ \text{m}}{16\ \frac{\text{m}}{\text{s}}} = 1.375\ \text{s}$$

$$\begin{aligned} v_y(t) &= v_0 \sin\phi - gt \\ &= \left(20\ \frac{\text{m}}{\text{s}}\right)(\sin 36.87^\circ) - \left(9.81\ \frac{\text{m}}{\text{s}^2}\right)(1.375\ \text{s}) \\ &= \boxed{-1.49\ \text{m/s}\quad(-1.5\ \text{m/s})} \end{aligned}$$

$$\begin{aligned} y &= (v_0 \sin\phi)t - \tfrac{1}{2}gt^2 \\ &= \left(20\ \frac{\text{m}}{\text{s}}\right)(\sin 36.87^\circ)(1.375\ \text{s}) \\ &\quad - \left(\tfrac{1}{2}\right)\left(9.81\ \frac{\text{m}}{\text{s}^2}\right)(1.375\ \text{s})^2 \\ &= \boxed{7.227\ \text{m}\quad(7\ \text{m})} \end{aligned}$$

The answer is (B).

8. *Customary U.S. Solution*

(a) Using Table 57.1, the bomb's maximum altitude is

$$\begin{aligned} H &= z + \frac{v_0^2 \sin^2\phi}{2g} \\ &= 12{,}000\ \text{ft} + \frac{\left(600\ \frac{\text{ft}}{\text{sec}}\right)^2 (\sin 30^\circ)^2}{(2)\left(32.2\ \frac{\text{ft}}{\text{sec}^2}\right)} \\ &= \boxed{13{,}398\ \text{ft}\quad(13{,}400\ \text{ft})} \end{aligned}$$

The answer is (C).

(b) Let t_1 be the time the bomb takes to reach the maximum altitude.

$$t_1 = \tfrac{1}{2}\left(\frac{2v_0 \sin\phi}{g}\right) = \frac{\left(600\ \frac{\text{ft}}{\text{sec}}\right)(\sin 30^\circ)}{32.2\ \frac{\text{ft}}{\text{sec}^2}} = 9.32\ \text{sec}$$

Let t_2 be the time the bomb takes to fall from H.

$$\begin{aligned} t_2 &= \sqrt{\frac{2H}{g}} = \sqrt{\frac{(2)(13{,}398\ \text{ft})}{32.2\ \frac{\text{ft}}{\text{sec}^2}}} = 28.85\ \text{sec} \\ t &= t_1 + t_2 = 9.32\ \text{sec} + 28.85\ \text{sec} \\ &= \boxed{38.17\ \text{sec}\quad(38\ \text{sec})} \end{aligned}$$

The answer is (B).

SI Solution

(a) Using Table 57.1, the bomb's maximum altitude is

$$\begin{aligned} H &= z + \frac{v_0^2 \sin^2\phi}{2g} \\ &= 3600\ \text{m} + \frac{\left(180\ \frac{\text{m}}{\text{s}}\right)^2 (\sin 30^\circ)^2}{(2)\left(9.81\ \frac{\text{m}}{\text{s}^2}\right)} \\ &= \boxed{4012.8\ \text{m}\quad(4000\ \text{m})} \end{aligned}$$

The answer is (C).

(b) Let t_1 be the time the bomb takes to reach the maximum altitude.

$$t_1 = \tfrac{1}{2}\left(\frac{2v_0 \sin\phi}{g}\right) = \frac{\left(180\ \frac{\text{m}}{\text{s}}\right)(\sin 30^\circ)}{9.81\ \frac{\text{m}}{\text{s}^2}} = 9.17\ \text{s}$$

Let t_2 be the time the bomb takes to fall from H.

$$\begin{aligned} t_2 &= \sqrt{\frac{2H}{g}} = \sqrt{\frac{(2)(4012.8\ \text{m})}{9.81\ \frac{\text{m}}{\text{s}^2}}} = 28.60\ \text{s} \\ t &= t_1 + t_2 = 9.17\ \text{s} + 28.60\ \text{s} \\ &= \boxed{37.77\ \text{s}\quad(38\ \text{s})} \end{aligned}$$

The answer is (B).

9. (a) The angular velocity is

$$\begin{aligned} \omega(t) &= 6t^2 - 10t \\ \omega(2) &= (6)(2)^2 - (10)(2) = \boxed{4} \end{aligned}$$

The answer is (A).

(b) The displacement is

$$\begin{aligned} \theta &= \theta(3) - \theta(1) = \int_1^3 \omega(t)\,dt \\ &= \int_1^3 (6t^2 - 10t)\,dt = 2t^3 - 5t^2 + C\Big|_1^3 \\ &\qquad [C = 0 \text{ since } \theta(0) = 0] \\ &= (2)(3)^3 - (5)(3)^2 - (2)(1)^3 + (5)(1)^2 \\ &= 12\ \text{rad} \end{aligned}$$

The point's position at time, t, is

$$\theta(t) = \int \omega(t)dt = \int (6t^2 - 10t)dt = 2t^3 - 5t^2 + C$$

Since the point starts from rest at an arbitrary starting position, let $C = 0$.

The point's position at $t = 1$ is

$$\theta(1) = (2)(1)^3 - (5)(1)^2 = -3$$

Since the point travels in a circle, and since a circle has 2π radians, the position can be expressed as

$$\theta(1) = -3 + 2\pi = 3.28\ \text{rad}$$

The point's position at $t = 3$ is

$$\theta(3) = (2)(3)^3 - (5)(3)^2 = 9$$

Since the point travels in a circle, it returns to its starting position after 2π radians. The position can be expressed as

$$\theta(3) = 9 - 2\pi = 2.72 \text{ rad}$$

The maximum displacement (shortest distance between the particles) is

$$\begin{aligned}\delta &= |\theta_2 - \theta_1| = |2.72 \text{ rad} - 3.28 \text{ rad}| \\ &= \boxed{0.566 \text{ rad} \quad (0.57 \text{ rad})}\end{aligned}$$

The answer is (A).

(c) To find the total distance traveled, check for sign reversals in $\omega(t)$ over the interval $t = 1$ to $t = 3$.

$$6t^2 - 10t = 0$$

$$\text{sign reversal at } t = \tfrac{5}{3}$$

The total angle turned is

$$\begin{aligned}\int_1^3 |\omega(t)|\,dt &= \int_1^{5/3} (10t - 6t^2)\,dt + \int_{5/3}^3 (6t^2 - 10t)\,dt \\ &= \boxed{15.26 \text{ rad} \quad (15 \text{ rad})}\end{aligned}$$

The answer is (D).

10. *Customary U.S. Solution*

Use Eq. 57.19.

$$\begin{aligned}\text{v}(t) &= \omega r = (2\pi f)r = d\pi f \\ &= \frac{(14 \text{ in})\pi\left(40 \ \frac{\text{rev}}{\text{min}}\right)}{\left(12 \ \frac{\text{in}}{\text{ft}}\right)\left(60 \ \frac{\text{sec}}{\text{min}}\right)} \\ &= \boxed{2.44 \text{ ft/sec} \quad (2.4 \text{ ft/sec})}\end{aligned}$$

The answer is (A).

SI Solution

Use Eq. 57.19.

$$\begin{aligned}\text{v}(t) &= \omega r = (2\pi f)r = d\pi f \\ &= \frac{(0.36 \text{ m})\pi\left(40 \ \frac{\text{rev}}{\text{min}}\right)}{60 \ \frac{\text{s}}{\text{min}}} \\ &= \boxed{0.754 \text{ m/s} \quad (0.75 \text{ m/s})}\end{aligned}$$

The answer is (A).

11. Convert 1200 rpm and 3000 rpm into radians per second. There are 2π radians per complete revolution.

$$\begin{aligned}\omega_1 = 2\pi f_1 &= \frac{\left(2\pi \ \frac{\text{rad}}{\text{rev}}\right)\left(1200 \ \frac{\text{rev}}{\text{min}}\right)}{60 \ \frac{\text{sec}}{\text{min}}} \\ &= 125.66 \text{ rad/sec}\end{aligned}$$

$$\begin{aligned}\omega_2 = 2\pi f_2 &= \frac{\left(2\pi \ \frac{\text{rad}}{\text{rev}}\right)\left(3000 \ \frac{\text{rev}}{\text{min}}\right)}{60 \ \frac{\text{sec}}{\text{min}}} \\ &= 314.16 \text{ rad/sec}\end{aligned}$$

$$\begin{aligned}\alpha = \frac{\omega_2 - \omega_1}{\Delta t} &= \frac{314.16 \ \frac{\text{rad}}{\text{sec}} - 125.66 \ \frac{\text{rad}}{\text{sec}}}{10 \text{ sec}} \\ &= \boxed{18.85 \text{ rad/sec}^2 \quad (18 \text{ rad/sec}^2)}\end{aligned}$$

The answer is (C).

12. *Customary U.S. Solution*

$$f = \frac{1750 \ \frac{\text{rev}}{\text{min}}}{60 \ \frac{\text{sec}}{\text{min}}} = 29.167 \text{ rev/sec}$$

$$\theta = (18^\circ)\left(\frac{1 \text{ rev}}{360^\circ}\right) = 0.05 \text{ rev}$$

$$t = \frac{\theta}{f} = \frac{0.05 \text{ rev}}{29.167 \ \frac{\text{rev}}{\text{sec}}} = 0.001714 \text{ sec}$$

$$\text{v} = \frac{s}{t} = \frac{5 \text{ ft}}{0.001714 \text{ sec}} = \boxed{2916.7 \text{ ft/sec} \quad (2900 \text{ ft/sec})}$$

The answer is (D).

SI Solution

$$\omega = \frac{\left(1750 \ \frac{\text{rev}}{\text{min}}\right)\left(2\pi \ \frac{\text{rad}}{\text{rev}}\right)}{60 \ \frac{\text{s}}{\text{min}}} = 183.26 \text{ rad/s}$$

$$\theta = (18^\circ)\left(\frac{2\pi \text{ rad}}{360^\circ}\right) = 0.314 \text{ rad}$$

$$t = \frac{\theta}{\omega} = \frac{0.314 \text{ rad}}{183.26 \ \frac{\text{rad}}{\text{s}}} = 0.001713 \text{ s}$$

$$\text{v} = \frac{s}{t} = \frac{1.5 \text{ m}}{0.001713 \text{ s}} = \boxed{875.66 \text{ m/s} \quad (880 \text{ m/s})}$$

The answer is (D).

13. *Customary U.S. Solution*

$$\omega_B = \left(\frac{16 \text{ in}}{24 \text{ in}}\right)\omega_C = \left(\frac{16 \text{ in}}{24 \text{ in}}\right)\left(2 \ \frac{\text{rad}}{\text{sec}}\right) = 1.333 \text{ rad/sec}$$

$$\alpha_B = \left(\frac{16 \text{ in}}{24 \text{ in}}\right)\alpha_C = \left(\frac{16 \text{ in}}{24 \text{ in}}\right)\left(6 \ \frac{\text{rad}}{\text{sec}^2}\right) = 4 \text{ rad/sec}^2$$

Since A and B are splined together,

$$\omega_A = \omega_B$$

$$\alpha_A = \alpha_B$$

$$\begin{aligned} v_D = r_A\omega_A &= (0.5 \text{ ft})\left(1.333 \ \frac{\text{rad}}{\text{sec}}\right) \\ &= \boxed{0.6665 \text{ ft/sec} \quad (0.7 \text{ ft/sec})} \end{aligned}$$

$$a_D = r_A\alpha_A = (0.5 \text{ ft})\left(4 \ \frac{\text{rad}}{\text{scc}^2}\right) = \boxed{2 \text{ ft/sec}^2}$$

The answer is (A).

SI Solution

$$\omega_B = \left(\frac{40 \text{ cm}}{60 \text{ cm}}\right)\omega_C = \left(\frac{40 \text{ cm}}{60 \text{ cm}}\right)\left(2 \ \frac{\text{rad}}{\text{s}}\right) = 1.333 \text{ rad/s}$$

$$\alpha_B = \left(\frac{40 \text{ cm}}{60 \text{ cm}}\right)\alpha_C = \left(\frac{40 \text{ cm}}{60 \text{ cm}}\right)\left(6 \ \frac{\text{rad}}{\text{s}^2}\right) = 4 \text{ rad/s}^2$$

Since A and B are splined together,

$$\omega_A = \omega_B$$

$$\alpha_A = \alpha_B$$

$$v_D = r_A\omega_A = (0.15 \text{ m})\left(1.333 \ \frac{\text{rad}}{\text{s}}\right) = \boxed{0.2 \text{ m/s}}$$

$$a_D = r_A\alpha_A = (0.15 \text{ m})\left(4 \ \frac{\text{rad}}{\text{s}^2}\right) = \boxed{0.6 \text{ m/s}^2}$$

The answer is (A).

14. *Customary U.S. Solution*

The angular velocity of the wheel is

$$\begin{aligned} \omega = \frac{v_C}{r} &= \frac{\left(28 \ \frac{\text{mi}}{\text{hr}}\right)\left(5280 \ \frac{\text{ft}}{\text{mi}}\right)\left(12 \ \frac{\text{in}}{\text{ft}}\right)}{(12 \text{ in})\left(60 \ \frac{\text{scc}}{\text{min}}\right)\left(60 \ \frac{\text{min}}{\text{hr}}\right)} \\ &= 41.07 \text{ rad/sec} \end{aligned}$$

The distance from the valve stem to the instant center of the point of contact of the wheel and the surface is determined from the law of cosines for the triangle defined by the valve stem, instant center, and center of wheel.

$$\begin{aligned} l^2 &= r_{\text{wheel}}^2 + r_{\text{stem}}^2 - 2r_{\text{wheel}}r_{\text{stem}}\cos\phi \\ &= (12 \text{ in})^2 + (6 \text{ in})^2 - (2)(12 \text{ in})(6 \text{ in})(\cos 135^\circ) \\ l &= 16.79 \text{ in} \end{aligned}$$

Use Eq. 57.52.

$$\begin{aligned} v = l\omega &= \frac{(16.79 \text{ in})\left(41.07 \ \frac{\text{rad}}{\text{sec}}\right)}{12 \ \frac{\text{in}}{\text{ft}}} \\ &= \boxed{57.46 \text{ ft/sec} \quad (57 \text{ ft/sec})} \end{aligned}$$

The direction is $\boxed{\text{perpendicular}}$ to the direction of l.

The answer is (B).

SI Solution

The angular velocity of the wheel is

$$\omega = \frac{v_C}{r} = \frac{12.5 \ \frac{\text{m}}{\text{s}}}{0.305 \text{ m}} = 40.98 \ \text{rad/s}$$

The distance from the valve stem to the instant center at the point of contact of the wheel and the surface is determined from the law of cosines for the triangle defined by the valve stem, instant center, and center of wheel.

$$\begin{aligned} l^2 &= r_{\text{wheel}}^2 + r_{\text{stem}}^2 - 2r_{\text{wheel}}r_{\text{stem}}\cos\phi \\ &= (0.305 \text{ m})^2 + (0.150 \text{ m})^2 \\ &\quad - (2)(0.305 \text{ m})(0.150 \text{ m})(\cos 135^\circ) \\ l &= 0.425 \text{ m} \end{aligned}$$

Use Eq. 57.52.

$$\begin{aligned} v &= l\omega \\ &= (0.425 \text{ m})\left(40.98 \ \frac{\text{rad}}{\text{s}}\right) \\ &= \boxed{17.4 \text{ m/s} \quad (17 \text{ m/s})} \end{aligned}$$

The direction is $\boxed{\text{perpendicular}}$ to the direction of l.

The answer is (B).

15. *Customary U.S. Solution*

The velocity of the car is

$$v_{car} = \frac{\left(45\ \frac{mi}{hr}\right)\left(5280\ \frac{ft}{mi}\right)}{\left(60\ \frac{sec}{min}\right)\left(60\ \frac{min}{hr}\right)} = 66\ ft/sec$$

The separation distance after 1 sec is

$$s(1) = \sqrt{\begin{array}{c}\left(200\ ft + \left(15\ \frac{ft}{sec}\right)(1\ sec)\right)^2 \\ + \left(\left(66\ \frac{ft}{sec}\right)(1\ sec)\right)^2\end{array}} = 224.9\ ft$$

The separation velocity is the difference in components of the car's and balloon's velocities along a mutually parallel line. Use the separation vector as this line.

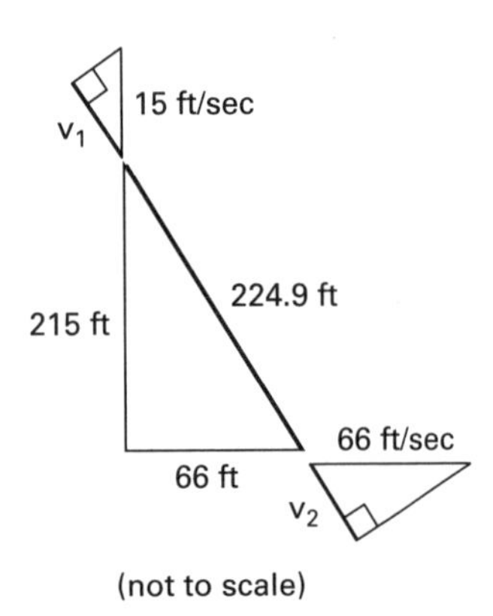

(not to scale)

$$v_1 = \left(15\ \frac{ft}{sec}\right)\left(\frac{215\ ft}{224.9\ ft}\right) = 14.34\ ft/sec$$

$$v_2 = \left(66\ \frac{ft}{sec}\right)\left(\frac{66\ ft}{224.9\ ft}\right) = 19.37\ ft/sec$$

$$\Delta v = v_1 + v_2 = 14.34\ \frac{ft}{sec} + 19.37\ \frac{ft}{sec}$$

$$= \boxed{33.71\ ft/sec\quad (34\ ft/sec)}$$

The answer is (B).

SI Solution

The velocity of the car is

$$v_{car} = \frac{\left(72\ \frac{km}{h}\right)\left(1000\ \frac{m}{km}\right)}{\left(60\ \frac{s}{min}\right)\left(60\ \frac{min}{h}\right)} = 20\ m/s$$

The separation distance after 1 second is

$$s(1) = \sqrt{\left(60\ m + \left(4.5\ \frac{m}{s}\right)(1\ s)\right)^2 + \left(\left(20\ \frac{m}{s}\right)(1\ s)\right)^2}$$
$$= 67.53\ m$$

The separation velocity is the difference in components of the car's and balloon's velocities along a mutually parallel line. Use the separation vector as this line.

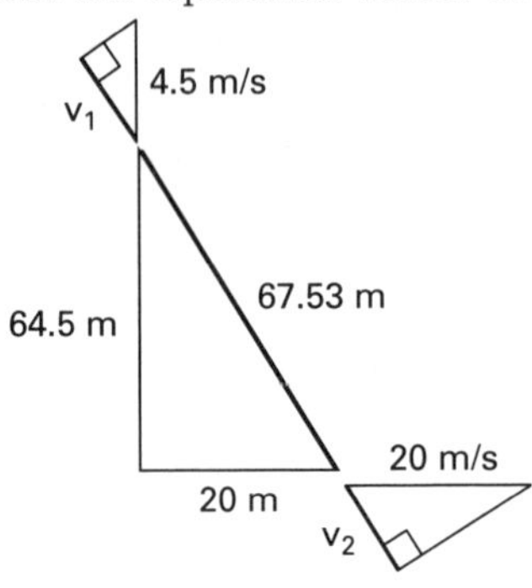

(not to scale)

$$v_1 = \left(4.5\ \frac{m}{s}\right)\left(\frac{64.5\ m}{67.53\ m}\right) = 4.298\ m/s$$

$$v_2 = \left(20\ \frac{m}{s}\right)\left(\frac{20\ m}{67.53\ m}\right) = 5.923\ m/s$$

$$\Delta v = v_1 + v_2 = 4.298\ \frac{m}{s} + 5.923\ \frac{m}{s}$$

$$= \boxed{10.22\ m/s\quad (10\ m/s)}$$

The answer is (B).

16. *Customary U.S. Solution*

From Eq. 57.52,

$$\omega = \frac{v_0}{r_{inner}} = \frac{10\ \frac{ft}{sec}}{2\ \frac{ft}{rad}} = 5\ rad/sec$$

$$v_{A/O} = r_{outer}\omega = \left(3\ \frac{ft}{rad}\right)\left(5\ \frac{rad}{sec}\right)$$

$$= \boxed{\begin{array}{l}15\ ft/sec,\ 45^\circ\ below\ horizontal, \\ to\ the\ right\end{array}}$$

$$v_{B/O} = r_{outer}\,\omega = \boxed{15\ ft/sec,\ horizontal,\ to\ the\ left}$$

The answer is (D).

SI Solution

From Eq. 57.52,

$$\omega = \frac{\mathrm{v}_0}{r_{\text{inner}}} = \frac{3\ \frac{\text{m}}{\text{s}}}{0.6\ \frac{\text{m}}{\text{rad}}} = 5\ \text{rad/s}$$

$$\begin{aligned}\mathrm{v}_{\text{A/O}} &= r_{\text{outer}}\omega = \left(0.9\ \frac{\text{m}}{\text{rad}}\right)\left(5\ \frac{\text{rad}}{\text{s}}\right) \\ &= \boxed{\begin{array}{l}4.5\ \text{m/s},\ 45°\ \text{below horizontal,} \\ \text{to the right}\end{array}}\end{aligned}$$

$$\mathrm{v}_{\text{B/O}} = r_{\text{outer}}\omega = \boxed{4.5\ \text{m/s, horizontal, to the left}}$$

The answer is (D).

17. *Customary U.S. Solution*

From the law of cosines,

$$\begin{aligned}|\text{AB}|^2 &= r_\text{A}^2 + r_\text{B}^2 - 2r_\text{A}r_\text{B}\cos\phi \\ &= (3\ \text{ft})^2 + (3\ \text{ft})^2 - (2)(3\ \text{ft})^2(\cos 135°) \\ &= 30.728\ \text{ft}^2 \\ |\text{AB}| &= 5.543\ \text{ft}\end{aligned}$$

$$\begin{aligned}\mathrm{v}_{\text{B/A}} &= |\text{AB}|\omega = (5.543\ \text{ft})\left(5\ \frac{\text{rad}}{\text{sec}}\right) \\ &= \boxed{\begin{array}{l}27.72\ \text{ft/sec}\ (28\ \text{ft/sec}), 22.5°\ (23°) \\ \text{above horizontal, to the left}\end{array}}\end{aligned}$$

The answer is (A).

SI Solution

From the law of cosines,

$$\begin{aligned}|\text{AB}|^2 &= r_\text{A}^2 + r_\text{B}^2 - 2r_\text{A}r_\text{B}\cos\phi \\ &= (0.9\ \text{m})^2 + (0.9\ \text{m})^2 - (2)(0.9\ \text{m})^2(\cos 135°) \\ &- 2.7655\ \text{m}^2 \\ |\text{AB}| &= 1.663\ \text{m}\end{aligned}$$

$$\begin{aligned}\mathrm{v}_{\text{B/A}} &= |\text{AB}|\omega = (1.663\ \text{m})\left(5\ \frac{\text{rad}}{\text{s}}\right) \\ &= \boxed{\begin{array}{l}8.315\ \text{m/s}\ (8.3\ \text{m/s}),\ 22.5°\ (23°) \\ \text{above horizontal, to the left}\end{array}}\end{aligned}$$

The answer is (A).

58 Kinetics

PRACTICE PROBLEMS

Centripetal and Centrifugal Forces

1. The superelevation (in percent) necessary on a curve with a 6000 ft (1800 m) radius so that at 60 mph (100 km/h) cars will not have to rely on friction to stay on the roadway is most nearly

(A) 4.0%

(B) 7.0%

(C) 11%

(D) 14%

2. The 8.05 lbm (3.65 kg) object is rotating at 20 ft/sec (6 m/s). What is most nearly the angle between the pole and the wire if the radius of the path is 4 ft (1.2 m)?

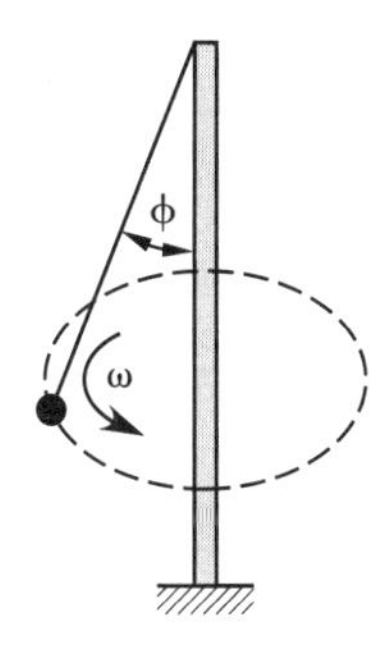

(A) 46°

(B) 58°

(C) 72°

(D) 79°

3. A 10 lbm (5 kg) mass is tied to a 2 ft (50 cm) string and whirled at 5 rev/sec horizontally to the ground.

(a) The centripetal acceleration is most nearly

(A) 1400 ft/sec^2 (350 m/s^2)

(B) 1600 ft/sec^2 (400 m/s^2)

(C) 1800 ft/sec^2 (450 m/s^2)

(D) 2000 ft/sec^2 (490 m/s^2)

(b) The centrifugal force is most nearly

(A) 490 lbf (1900 N)

(B) 610 lbf (2500 N)

(C) 750 lbf (2900 N)

(D) 820 lbf (3100 N)

(c) The centripetal force is most nearly

(A) 490 lbf (1900 N)

(B) 610 lbf (2500 N)

(C) 750 lbf (2900 N)

(D) 820 lbf (3100 N)

(d) The angular momentum is most nearly

(A) 39 ft-lbf-sec (39 J·s)

(B) 44 ft-lbf-sec (44 J·s)

(C) 53 ft-lbf-sec (53 J·s)

(D) 71 ft-lbf-sec (71 J·s)

Friction

4. A 100 lbm (50 kg) body has an initial velocity of 12.88 ft/sec (3.93 m/s) while moving on a plane with a coefficient of friction of 0.2. The distance it will travel before coming to rest is most nearly

(A) 13 ft (3.9 m)

(B) 15 ft (4.5 m)

(C) 18 ft (5.4 m)

(D) 24 ft (7.2 m)

5. A box is dropped onto a conveyor belt moving at 10 ft/sec (3 m/s). The coefficient of friction between the box and the belt is 0.333. Approximately how long will it take before the box stops slipping on the belt?

(A) 0.48 sec

(B) 0.67 sec

(C) 0.93 sec

(D) 1.2 sec

6. A motorcycle and rider weigh 400 lbm (200 kg). They travel horizontally around the inside of a hollow, right-angle cylinder of 100 ft (30 m) inside diameter. What is most nearly the coefficient of friction that will allow a speed of 40 mph (60 km/h)?

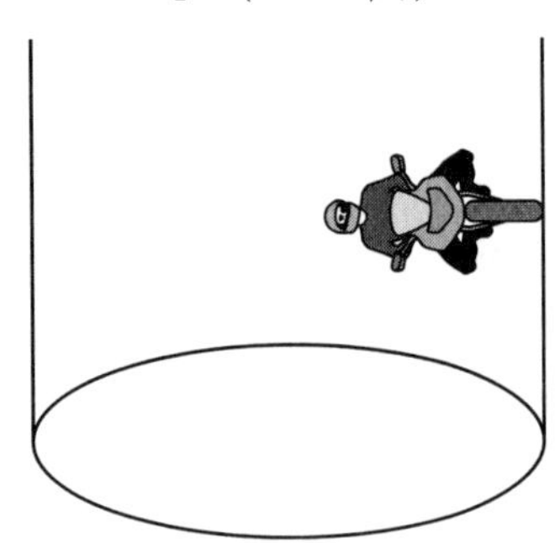

(A) 0.3

(B) 0.4

(C) 0.5

(D) 0.6

7. When a force acts on a 10 lbm (5 kg) body initially at rest, a speed of 12 ft/sec (3.6 m/s) is attained in 36 ft (11 m). If the coefficient of friction is 0.25 and the acceleration is constant, the force is most nearly

(A) 1.8 lbf (8.6 N)

(B) 2.4 lbf (13 N)

(C) 3.1 lbf (15 N)

(D) 4.6 lbf (22 N)

8. A 130 lbm (60 kg) block slides up a 22.62° incline with a coefficient of friction of 0.1 and an initial velocity of 30 ft/sec (9 m/s). Approximately how far will the block slide up the incline before coming to rest?

(A) 18 ft (5.4 m)

(B) 29 ft (8.7 m)

(C) 42 ft (13 m)

(D) 57 ft (17 m)

9. A 100 lbm (50 kg) block is acted upon by a 100 lbf (400 N) force while resting on a horizontal surface with a coefficient of friction of 0.2. A 50 lbm (25 kg) block sits on top of the 100 lbm (50 kg) block. The minimum coefficient of friction between the 50 lbm and 100 lbm (25 kg and 50 kg) blocks for there to be no slipping is most nearly

(A) 0.010

(B) 0.34

(C) 0.47

(D) 0.53

10. A flat package with a mass of 1 lbm is placed on a horizontal, flat conveyor belt. The coefficient of static friction between the conveyor and the package is 0.15. The package enters a curve on the conveyor where the radius is 5 ft.

(a) What is most nearly the maximum tangential velocity such that the package does not slide outward on the curve?

(A) 3.7 ft/sec

(B) 4.9 ft/sec

(C) 6.4 ft/sec

(D) 24 ft/sec

(b) What is most nearly the maximum centrifugal force that the package can experience without sliding?

(A) 0.03 lbf

(B) 0.15 lbf

(C) 0.30 lbf

(D) 4.8 lbf

(c) The maximum experienced frictional force between the package and the conveyer is most nearly

(A) 0.15 lbf

(B) 0.28 lbf

(C) 0.42 lbf

(D) 1.3 lbf

11. A curved, tubular conveyance chute allows material to fall 10 ft vertically while changing the material's direction of travel. The friction factor between the conveyed material and chute is 0.10. At one moment in time, a 2 lbm slug of material enters the chute and achieves an instantaneous speed of 5 ft/sec in the curve. The material "rides up" the curve until it is superelevated 20°. The radius of curvature at that point is most nearly

(A) 0.43 ft

(B) 0.85 ft

(C) 1.6 ft

(D) 2.4 ft

Rigid Body Motion

12. A constant force of 20 lbf (90 N) is applied to a 100 lbm (50 kg) door supported on rollers at A and B.

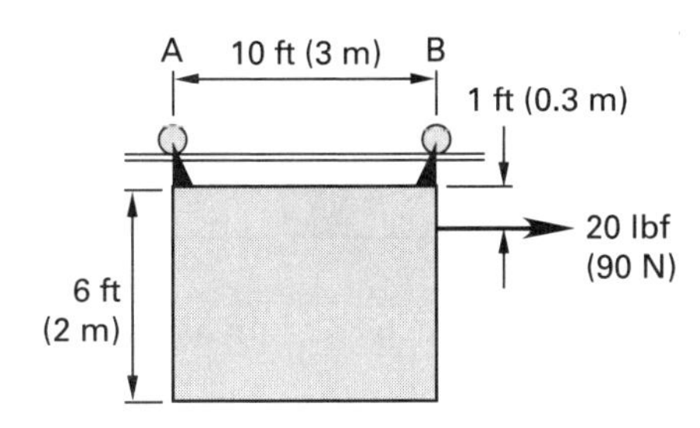

(a) The acceleration of the door is most nearly

(A) 2.0 ft/sec^2 (0.56 m/s^2)

(B) 3.1 ft/sec^2 (0.87 m/s^2)

(C) 5.3 ft/sec^2 (1.5 m/s^2)

(D) 6.4 ft/sec^2 (1.8 m/s^2)

(b) The reactions at B and A, respectively, are most nearly

(A) 34 lbf, 66 lbf (170 N, 322 N)

(B) 38 lbf, 62 lbf (190 N, 300 N)

(C) 42 lbf, 58 lbf (210 N, 280 N)

(D) 54 lbf, 46 lbf (270 N, 220 N)

(c) Where would the 20 lbf (90 N) force have to be applied for the reactions at A and B to be equal?

(A) in line with the center of gravity

(B) at roller A

(C) between roller A and roller B

(D) 6 ft (2m) below roller B

Constrained Motion

13. A solid sphere rolls without slipping down a 30° incline, starting from rest. Its speed after 2 sec is most nearly

(A) 17 ft/sec (5.1 m/s)

(B) 23 ft/sec (7.0 m/s)

(C) 34 ft/sec (10 m/s)

(D) 86 ft/sec (26 m/s)

14. Object A weighs 10 lbm (5 kg) and rests on a frictionless plane with a 36.87° slope (from the horizontal). Object B weighs 20 lbm (10 kg). What is most nearly the velocity of object B 3 sec after release?

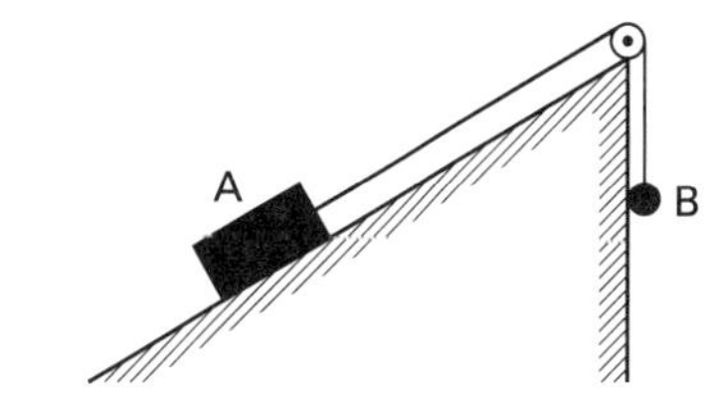

(A) 14 ft/sec (4.2 m/s)

(B) 22 ft/sec (6.6 m/s)

(C) 38 ft/sec (11 m/s)

(D) 45 ft/sec (14 m/s)

Impulse and Momentum

15. Sand is dropping at the rate of 560 lbm/min (250 kg/min) onto a conveyor belt moving with a velocity of 3.2 ft/sec (0.98 m/s). The force required to keep the belt moving is most nearly

(A) 0.93 lbf (4.1 N)

(B) 1.2 lbf (5.3 N)

(C) 2.3 lbf (10 N)

(D) 4.8 lbf (14 N)

16. The impulse imparted to a 0.4 lbm (0.2 kg) baseball that approaches the batter at 90 ft/sec (30 m/s) and leaves at 130 ft/sec (40 m/s) is most nearly

(A) 1.8 lbf-sec (9.4 N·s)

(B) 2.7 lbf-sec (14 N·s)

(C) 4.1 lbf-sec (21 N·s)

(D) 8.9 lbf-sec (46 N·s)

17. At what approximate velocity will a 1000 lbm (500 kg) gun mounted on wheels recoil if a 2.6 lbm (1.2 kg) projectile is propelled to 2100 ft/sec (650 m/s)?

(A) 3.7 ft/sec (1.1 m/s)

(B) 4.8 ft/sec (1.4 m/s)

(C) 5.5 ft/sec (1.6 m/s)

(D) 15 ft/sec (4.5 m/s)

18. A 0.15 lbm (60 g) bullet traveling 2300 ft/sec (700 m/s) embeds itself in a 9 lbm (4.5 kg) wooden block initially at rest. The block's velocity immediately after impact is most nearly

(A) 24 ft/sec (7.2 m/s)

(B) 38 ft/sec (9.2 m/s)

(C) 66 ft/sec (20 m/s)

(D) 110 ft/sec (33 m/s)

19. A nozzle discharges 40 gal/min (0.15 m^3/min) of water at 60 ft/sec (20 m/s). There is no splashback. The total force required to hold a flat vertical plate in front of the nozzle is most nearly

(A) 10 lbf (50 N)

(B) 20 lbf (100 N)

(C) 40 lbf (200 N)

(D) 80 lbf (400 N)

20. In a water turbine, 100 gal/sec (0.4 m^3/s) impinge on a stationary blade. The water is turned through an angle of 160° and exits at 57 ft/sec (17 m/s). The water impinges at 60 ft/sec (20 m/s).

(a) The force exerted by the blade on the stream is most nearly

(A) 1600 lbf (7800 N)

(B) 2300 lbf (11 000 N)

(C) 3000 lbf (15 000 N)

(D) 3500 lbf (17 000 N)

(b) The angle from the horizontal of the force is most nearly

(A) 18° (clockwise from horizontal)

(B) 72° (counterclockwise from horizontal)

(C) 120° (clockwise from horizontal)

(D) 170° (counterclockwise from horizontal)

Impacts

21. Two 2 lbm billiard balls collide as shown. The coefficient of restitution is 0.8. What are most nearly the velocities after impact?

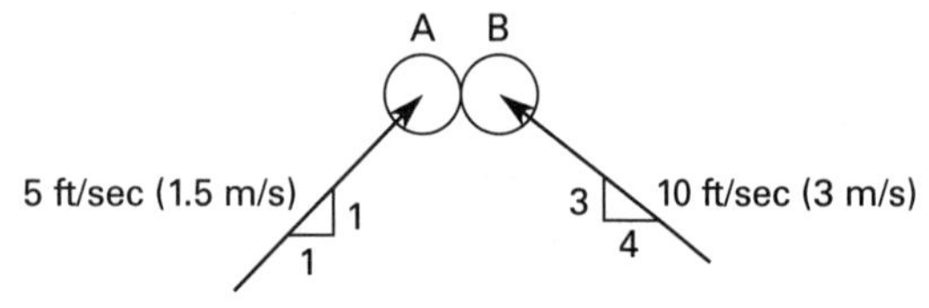

(a) The velocities of balls A and B, respectively, after impact are most nearly

(A) 7.0 ft/sec, 8.8 ft/sec (2.1 m/s, 2.6 m/s)

(B) 7.7 ft/sec, 6.5 ft/sec (2.3 m/s, 1.9 m/s)

(C) 8.3 ft/sec, 7.2 ft/sec (2.5 m/s, 2.1 m/s)

(D) 9.4 ft/sec, 4.9 ft/sec (2.8 m/s, 1.4 m/s)

(b) The angles of balls A and B, respectively, are most nearly

(A) 30°, 24°

(B) 60°, 130°

(C) 120°, 100°

(D) 150°, 68°

22. A 10 lbm (5 kg) pendulum is released from rest and strikes a 50 lbm (25 kg) block. The coefficient of restitution is 0.7. The block slides on a frictionless surface.

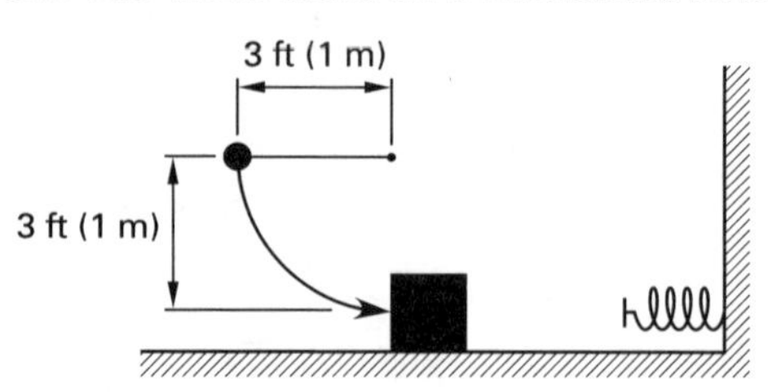

(a) The velocity of the pendulum at impact is most nearly

(A) 14 ft/sec (4.4 m/s)

(B) 26 ft/sec (7.8 m/s)

(C) 38 ft/sec (11 m/s)

(D) 51 ft/sec (15 m/s)

(b) The tension in the cord at impact is most nearly

(A) 10 lbf (50 N)

(B) 20 lbf (100 N)

(C) 30 lbf (150 N)

(D) 40 lbf (200 N)

(c) The block's velocity immediately after impact is most nearly

(A) 2.2 ft/sec (0.66 m/s)

(B) 3.9 ft/sec (1.3 m/s)

(C) 4.6 ft/sec (1.4 m/s)

(D) 5.8 ft/sec (1.7 m/s)

(d) The required spring constant to stop the block with less than 6 in (15 cm) deflection is most nearly

(A) 54 lbf/ft (1200 N/m)

(B) 62 lbf/ft (1400 N/m)

(C) 75 lbf/ft (1600 N/m)

(D) 96 lbf/ft (1800 N/m)

23. Two freight cars weighing 5 tons (5000 kg) each roll toward each other and couple. The left car has a velocity of 5 ft/sec (1.5 m/s) and the right car has a velocity of 4 ft/sec (1.2 m/s) prior to the impact. The velocity of the two cars coupled together after the impact is most nearly

(A) 0 ft/sec (0 m/s)

(B) 0.5 ft/sec (0.15 m/s)

(C) 1 ft/sec (0.30 m/s)

(D) 2 ft/sec (0.60 m/s)

24. An electron (0.0005486 atomic mass units) collides with a hydrogen atom (1.007277 atomic mass units) initially at rest. The electron's initial velocity is 1500 ft/sec (500 m/s). Its final velocity is 65 ft/sec (20.2 m/s) with a path 30° from its original path. Find the approximate velocity of the hydrogen atom in the x-direction if it recoils 1.2° from the original path of the electron.

(A) 0.043 ft/sec (0.013 m/s)

(B) 0.79 ft/sec (0.26 m/s)

(C) 4.6 ft/sec (1.4 m/s)

(D) 53 ft/sec (16 m/s)

25. A contractor needs to drag a heavy object horizontally across the floor and closer to the east wall. To do so, the contractor connects the object through snatch blocks and a dead lead to a wire rope "come-along." The object requires a net force of 2000 lbf to maintain motion once the object starts sliding. What is most nearly the initial tension in the wire rope needed to start the object moving?

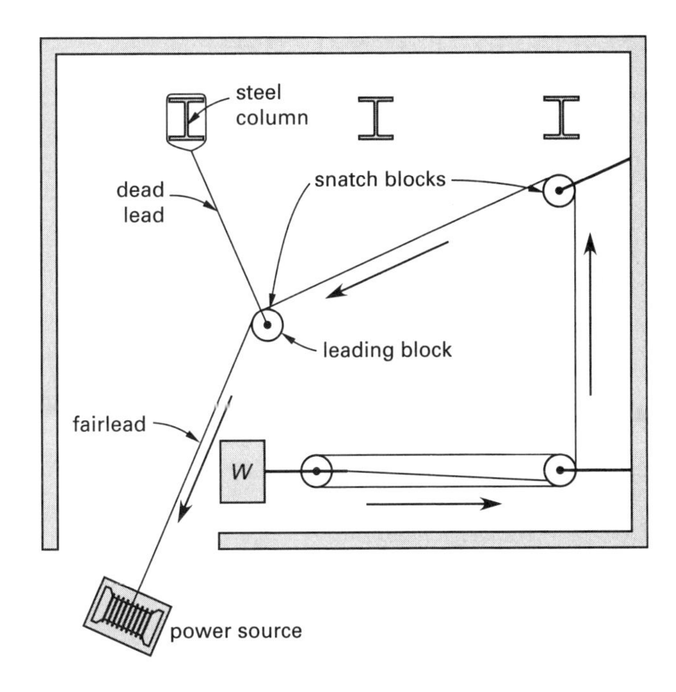

(A) 200 lbf

(B) 250 lbf

(C) 560 lbf

(D) 680 lbf

SOLUTIONS

1. *Customary U.S. Solution*

The tangential velocity is

$$v_t = \frac{\left(60\ \frac{\text{mi}}{\text{hr}}\right)\left(5280\ \frac{\text{ft}}{\text{mi}}\right)}{\left(60\ \frac{\text{min}}{\text{hr}}\right)\left(60\ \frac{\text{sec}}{\text{min}}\right)} = 88\ \text{ft/sec}$$

From Eq. 58.29, the superelevation is

$$\tan\phi = \frac{v_t^2}{gr} = \frac{\left(88\ \frac{\text{ft}}{\text{sec}}\right)^2}{\left(32.2\ \frac{\text{ft}}{\text{sec}^2}\right)(6000\ \text{ft})} = \boxed{0.04\quad(4.0\%)}$$

The answer is (A).

SI Solution

The tangential velocity is

$$v_t = \frac{\left(100\ \frac{\text{km}}{\text{h}}\right)\left(1000\ \frac{\text{m}}{\text{km}}\right)}{\left(60\ \frac{\text{min}}{\text{h}}\right)\left(60\ \frac{\text{s}}{\text{min}}\right)} = 27.78\ \text{m/s}$$

From Eq. 58.29, the superelevation is

$$\tan\phi = \frac{v_t^2}{gr} = \frac{\left(27.78\ \frac{\text{m}}{\text{s}}\right)^2}{\left(9.81\ \frac{\text{m}}{\text{s}^2}\right)(1800\ \text{m})} = \boxed{0.0437\quad(4.0\%)}$$

The answer is (A).

2. *Customary U.S. Solution*

The net acceleration vector is directed along the string. Using Eq. 58.29,

$$\phi = \arctan\frac{a_c}{a_n} = \arctan\frac{v_t^2}{gr}$$

$$= \arctan\frac{\left(20\ \frac{\text{ft}}{\text{sec}}\right)^2}{\left(32.2\ \frac{\text{ft}}{\text{sec}^2}\right)(4\ \text{ft})}$$

$$= \boxed{72.15^\circ\quad(72^\circ)\quad\left[\begin{array}{c}\text{independent of}\\ \text{object's mass}\end{array}\right]}$$

The answer is (C).

SI Solution

The net acceleration vector is directed radially along the string. Using Eq. 58.29,

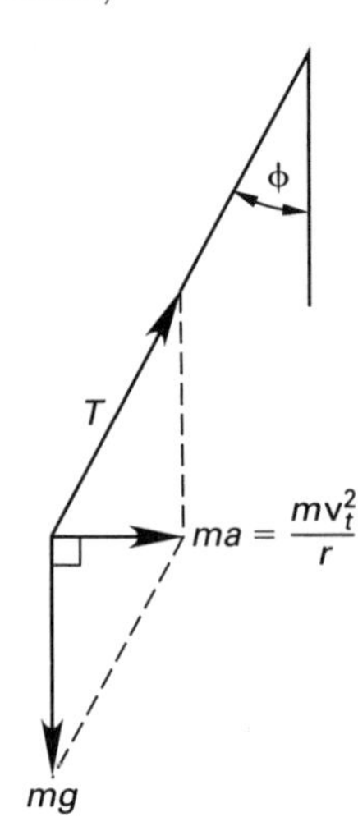

$$\frac{m\mathrm{v}_t^2}{r}\cos\phi = mg\sin\phi$$

$$\tan\phi = \frac{\mathrm{v}_t^2}{rg}$$

$$\phi = \arctan\frac{\mathrm{v}_t^2}{gr}$$

$$= \arctan\frac{\left(6\ \frac{\mathrm{m}}{\mathrm{s}}\right)^2}{\left(9.81\ \frac{\mathrm{m}}{\mathrm{s}^2}\right)(1.2\ \mathrm{m})}$$

$$= \boxed{71.89^\circ\quad(72^\circ)\quad\left[\begin{array}{c}\text{independent of}\\ \text{object's mass}\end{array}\right]}$$

The answer is (C).

3. *Customary U.S. Solution*

(a) The tangential velocity is

$$\mathrm{v}_t = \omega r = 2\pi f r$$
$$= \left(2\pi\ \frac{\mathrm{rad}}{\mathrm{rev}}\right)\left(5\ \frac{\mathrm{rev}}{\mathrm{sec}}\right)(2\ \mathrm{ft})$$
$$= 62.83\ \mathrm{ft/sec}$$

The centripetal acceleration (i.e., the acceleration normal to the path of motion) is

$$a_n = \frac{\mathrm{v}_t^2}{r} = \frac{\left(62.83\ \frac{\mathrm{ft}}{\mathrm{sec}}\right)^2}{2\ \mathrm{ft}}$$
$$= \boxed{1974\ \mathrm{ft/sec^2}\quad(2000\ \mathrm{ft/sec^2})}$$

The answer is (D).

(b) From Eq. 58.12, the centrifugal force is

$$F_{\mathrm{centrifugal}} = \frac{ma_n}{g_c} = \frac{(10\ \mathrm{lbm})\left(1974\ \frac{\mathrm{ft}}{\mathrm{sec}^2}\right)}{32.2\ \frac{\mathrm{lbm\text{-}ft}}{\mathrm{lbf\text{-}sec^2}}}$$
$$= \boxed{613.0\ \mathrm{lbf}\quad(610\ \mathrm{lbf})\quad[\text{directed outward}]}$$

The answer is (B).

(c) The centripetal force is

$$|\mathbf{F}_{\mathrm{centripetal}}| = |\mathbf{F}_{\mathrm{centrifugal}}|$$
$$F_{\mathrm{centripetal}} = \boxed{613.0\ \mathrm{lbf}\quad(610\ \mathrm{lbf})\quad[\text{directed inward}]}$$

The answer is (B).

(d) From Eq. 58.8, the angular momentum is

$$\mathbf{h} = \frac{\mathbf{r}\times m\mathbf{v}_t}{g_c}$$
$$h = \frac{rm\mathrm{v}_t}{g_c}\quad[\text{since } \mathbf{r}\perp\mathbf{v}_t]$$
$$= \frac{(2\ \mathrm{ft})(10\ \mathrm{lbm})\left(62.83\ \frac{\mathrm{ft}}{\mathrm{sec}}\right)}{32.2\ \frac{\mathrm{lbm\text{-}ft}}{\mathrm{lbf\text{-}sec^2}}}$$
$$= \boxed{39.02\ \mathrm{ft\text{-}lbf\text{-}sec}\quad(39\ \mathrm{ft\text{-}lbf\text{-}sec})}$$

The answer is (A).

SI Solution

(a) The tangential velocity is

$$\mathrm{v}_t = \omega r = 2\pi f r$$
$$= \left(2\pi\ \frac{\mathrm{rad}}{\mathrm{rev}}\right)\left(5\ \frac{\mathrm{rev}}{\mathrm{s}}\right)(0.5\ \mathrm{m})$$
$$= 15.708\ \mathrm{m/s}$$

The centripetal acceleration (i.e., the acceleration normal to the path of motion) is

$$a_n = \frac{\mathrm{v}_t^2}{r} = \frac{\left(15.708\ \frac{\mathrm{m}}{\mathrm{s}}\right)^2}{0.5\ \mathrm{m}} = \boxed{493.5\ \mathrm{m/s^2}\quad(490\ \mathrm{m/s^2})}$$

The answer is (D).

(b) From Eq. 58.12, the centrifugal force is

$$F_{\mathrm{centrifugal}} = ma_n = (5\ \mathrm{kg})\left(493.5\ \frac{\mathrm{m}}{\mathrm{s}^2}\right)$$
$$= \boxed{2467.5\ \mathrm{N}\quad(2500\ \mathrm{N})\quad[\text{directed outward}]}$$

The answer is (B).

(c) The centripetal force is

$$|\mathbf{F}_{\text{centripetal}}| = |\mathbf{F}_{\text{centrifugal}}|$$

$$F_{\text{centripetal}} = \boxed{2467.5 \text{ N} \quad (2500 \text{ N}) \quad [\text{directed inward}]}$$

The answer is (B).

(d) From Eq. 58.8, the angular momentum is

$$\mathbf{h} = \mathbf{r} \times m\mathbf{v}_t$$

$$h = rm\text{v}_t \quad [\text{since } \mathbf{r} \perp \mathbf{v}_t]$$

$$= (0.5 \text{ m})(5 \text{ kg})\left(15.708 \ \frac{\text{m}}{\text{s}}\right)$$

$$= \boxed{39.27 \text{ J·s} \quad (39 \text{ J·s})}$$

The answer is (A).

4. *Customary U.S. Solution*

The deceleration is

$$a = \frac{F_f}{\frac{m}{g_c}} = \frac{fN}{\frac{m}{g_c}} = \frac{\frac{fmg}{g_c}}{\frac{m}{g_c}}$$

$$= fg$$

$$= (0.2)\left(32.2 \ \frac{\text{ft}}{\text{sec}^2}\right)$$

$$= 6.44 \text{ ft/sec}^2$$

From Table 57.1, the skidding distance is

$$s_{\text{skidding}} = \frac{\text{v}^2}{2a} = \frac{\left(12.88 \ \frac{\text{ft}}{\text{sec}}\right)^2}{(2)\left(6.44 \ \frac{\text{ft}}{\text{sec}^2}\right)} = \boxed{12.88 \text{ ft} \quad (13 \text{ ft})}$$

The answer is (A).

SI Solution

The deceleration is

$$a = \frac{F_f}{m} = \frac{fN}{m} = \frac{fmg}{m}$$

$$= fg$$

$$= (0.2)\left(9.81 \ \frac{\text{m}}{\text{s}^2}\right)$$

$$= 1.962 \text{ m/s}^2$$

From Table 57.1, the skidding distance is

$$s_{\text{skidding}} = \frac{\text{v}^2}{2a} = \frac{\left(3.93 \ \frac{\text{m}}{\text{s}}\right)^2}{(2)\left(1.962 \ \frac{\text{m}}{\text{s}^2}\right)} = \boxed{3.936 \text{ m} \quad (3.9 \text{ m})}$$

The answer is (A).

5. *Customary U.S. Solution*

The deceleration is

$$a = \frac{F_f}{\frac{m}{g_c}} = \frac{fN}{\frac{m}{g_c}} = \frac{\frac{fmg}{g_c}}{\frac{m}{g_c}} = fg$$

$$= (0.333)\left(32.2 \ \frac{\text{ft}}{\text{sec}^2}\right)$$

$$= 10.72 \text{ ft/sec}^2$$

From Table 57.1,

$$\Delta t = \frac{\text{v}}{a} = \frac{10 \ \frac{\text{ft}}{\text{sec}}}{10.72 \ \frac{\text{ft}}{\text{sec}^2}} = \boxed{0.93 \text{ sec}}$$

The answer is (C).

SI Solution

The deceleration is

$$a = \frac{F_f}{m} = \frac{fN}{m} = \frac{fmg}{m} = fg$$

$$= (0.333)\left(9.81 \ \frac{\text{m}}{\text{s}^2}\right)$$

$$= 3.267 \text{ m/s}^2$$

From Table 57.1,

$$\Delta t = \frac{\text{v}}{a} = \frac{3 \ \frac{\text{m}}{\text{s}}}{3.267 \ \frac{\text{m}}{\text{s}^2}} = \boxed{0.918 \text{ s} \quad (0.93 \text{ s})}$$

The answer is (C).

6. *Customary U.S. Solution*

The tangential velocity is

$$\text{v}_t = \frac{\left(40 \ \frac{\text{mi}}{\text{hr}}\right)\left(5280 \ \frac{\text{ft}}{\text{mi}}\right)}{\left(60 \ \frac{\text{min}}{\text{hr}}\right)\left(60 \ \frac{\text{sec}}{\text{min}}\right)}$$

$$= 58.67 \text{ ft/sec}$$

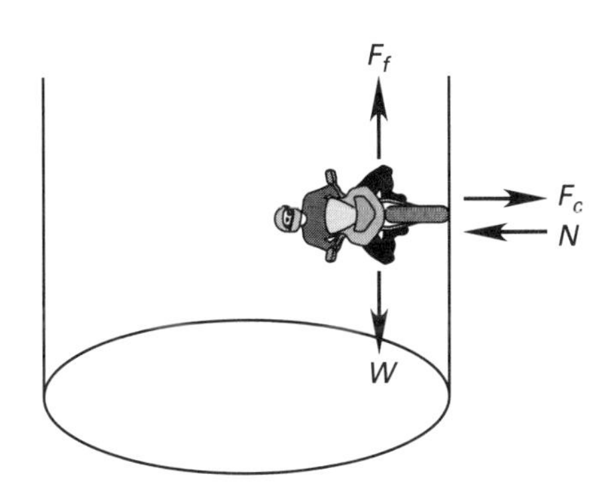

The normal force is calculated from Eq. 58.15.

$$N = F_c = ma_n = \frac{m\mathrm{v}_t^2}{g_c r}$$

The work, W, is equal to the frictional force, given by Eq. 58.21.

$$W = F_f = fN$$

So, from Eq. 58.21, the coefficient of friction is

$$f = \frac{W}{N} = \frac{\frac{mg}{g_c}}{\frac{m\mathrm{v}_t^2}{g_c r}} = \frac{gr}{\mathrm{v}_t^2} = \frac{\left(32.2\ \frac{\text{ft}}{\text{sec}^2}\right)(50\ \text{ft})}{\left(58.67\ \frac{\text{ft}}{\text{sec}}\right)^2} = \boxed{0.468\quad(0.5)}$$

The answer is (C).

SI Solution

The tangential force is

$$\mathrm{v}_t = \frac{\left(60\ \frac{\text{km}}{\text{h}}\right)\left(1000\ \frac{\text{m}}{\text{km}}\right)}{\left(60\ \frac{\text{min}}{\text{h}}\right)\left(60\ \frac{\text{s}}{\text{min}}\right)} = 16.67\ \text{m/s}$$

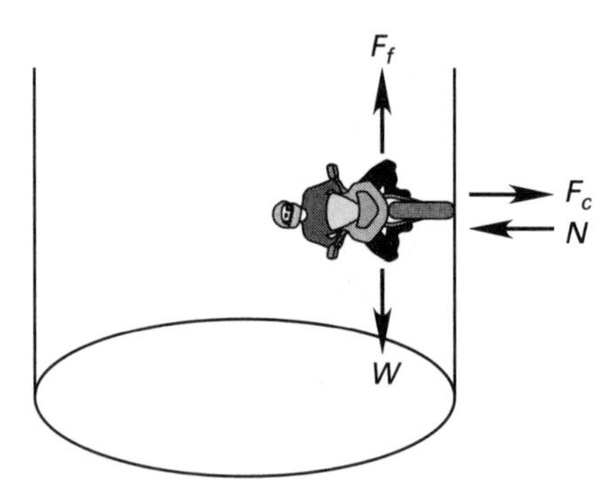

The normal force is calculated from Eq. 58.15.

$$N = F_c = ma_n = \frac{m\mathrm{v}_t^2}{r}$$

The work, W, is equal to the frictional force, given by Eq. 58.21.

$$W = F_f = fN$$

So, from Eq. 58.21, the coefficient of friction is

$$f = \frac{W}{N} = \frac{mg}{\frac{m\mathrm{v}_t^2}{r}} = \frac{gr}{\mathrm{v}_t^2} = \frac{\left(9.81\ \frac{\text{m}}{\text{s}^2}\right)(15\ \text{m})}{\left(16.67\ \frac{\text{m}}{\text{s}}\right)^2} = \boxed{0.53\quad(0.5)}$$

The answer is (C).

7. *Customary U.S. Solution*

From Table 57.1, the uniform acceleration is

$$a = \frac{\mathrm{v}^2}{2s} = \frac{\left(12\ \frac{\text{ft}}{\text{sec}}\right)^2}{(2)(36\ \text{ft})} = 2\ \text{ft/sec}^2$$

Using Eq. 58.21, the frictional force is

$$F_f = fN = fW = (0.25)(10\ \text{lbf}) = 2.5\ \text{lbf}$$

Therefore, the force is

$$F = F_f + \frac{ma}{g_c} = 2.5\ \text{lbf} + \frac{(10\ \text{lbm})\left(2\ \frac{\text{ft}}{\text{sec}^2}\right)}{32.2\ \frac{\text{lbm-ft}}{\text{lbf-sec}^2}} = \boxed{3.12\ \text{lbf}\quad(3.1\ \text{lbf})}$$

The answer is (C).

SI Solution

From Table 57.1, the uniform acceleration is

$$a = \frac{\mathrm{v}^2}{2s} = \frac{\left(3.6\ \frac{\text{m}}{\text{s}}\right)^2}{(2)(11\ \text{m})} = 0.589\ \text{m/s}^2$$

Using Eq. 58.21, the frictional force is

$$F_f = fN = fW = (0.25)(5\ \text{kg})\left(9.81\ \frac{\text{m}}{\text{s}^2}\right) = 12.26\ \text{N}$$

Therefore, the force is

$$F = F_f + ma = 12.26\ \text{N} + (5\ \text{kg})\left(0.589\ \frac{\text{m}}{\text{s}^2}\right) = \boxed{15.21\ \text{N}\quad(15\ \text{N})}$$

The answer is (C).

8. *Customary U.S. Solution*

The acceleration is

$$\begin{aligned} a &= g\sin\theta + fg\cos\theta \\ &= \left(32.2\ \frac{\text{ft}}{\text{sec}^2}\right)(\sin 22.62^\circ) \\ &\quad + (0.1)\left(32.2\ \frac{\text{ft}}{\text{sec}^2}\right)(\cos 22.62^\circ) \\ &= 15.36\ \text{ft/sec}^2 \end{aligned}$$

From Table 57.1, the sliding distance is

$$\begin{aligned} s &= \frac{\text{v}_0^2}{2a} = \frac{\left(30\ \frac{\text{ft}}{\text{sec}}\right)^2}{(2)\left(15.36\ \frac{\text{ft}}{\text{sec}^2}\right)} \\ &= \boxed{29.30\ \text{ft}\quad (29\ \text{ft})} \end{aligned}$$

The answer is (B).

SI Solution

The acceleration is

$$\begin{aligned} a &= g\sin\theta + fg\cos\theta \\ &= \left(9.81\ \frac{\text{m}}{\text{s}^2}\right)(\sin 22.62^\circ) + (0.1)\left(9.81\ \frac{\text{m}}{\text{s}^2}\right)(\cos 22.62^\circ) \\ &= 4.679\ \text{m/s}^2 \end{aligned}$$

From Table 57.1, the sliding distance is

$$\begin{aligned} s &= \frac{\text{v}_0^2}{2a} = \frac{\left(9\ \frac{\text{m}}{\text{s}}\right)^2}{(2)\left(4.679\ \frac{\text{m}}{\text{s}^2}\right)} \\ &= \boxed{8.656\ \text{m}\quad (8.7\ \text{m})} \end{aligned}$$

The answer is (B).

9. *Customary U.S. Solution*

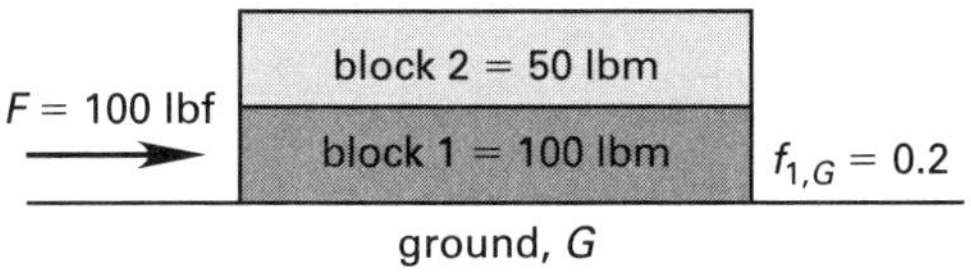

The friction force is given by Eq. 58.21.

$$\begin{aligned} F_{f(1,G)} &= (W_1 + W_2)f_{1,G} = (100\ \text{lbf} + 50\ \text{lbf})(0.2) \\ &= 30\ \text{lbf}\quad \left[\begin{array}{l} < F;\ \text{blocks move together at accel-} \\ \text{eration } a,\ \text{assuming no slipping} \end{array}\right] \end{aligned}$$

The acceleration is

$$\begin{aligned} a &= \frac{\left(F - F_{f(1,G)}\right)g_c}{m_1 + m_2} \\ &= \frac{(100\ \text{lbf} - 30\ \text{lbf})\left(32.2\ \frac{\text{lbm-ft}}{\text{lbf-sec}^2}\right)}{100\ \text{lbm} + 50\ \text{lbm}} \\ &= 15.03\ \text{ft/sec}^2 \end{aligned}$$

If block 2 is not slipping,

$$F_{f(1,2)} = f_{1,2}W_2 = \frac{m_2 a}{g_c}$$

Therefore, the coefficient of friction between blocks 1 and 2 will be

$$f_{1,2} = \frac{a}{g} = \frac{15.03\ \frac{\text{ft}}{\text{sec}^2}}{32.2\ \frac{\text{ft}}{\text{sec}^2}} = \boxed{0.467\quad (0.47)}$$

The answer is (C).

SI Solution

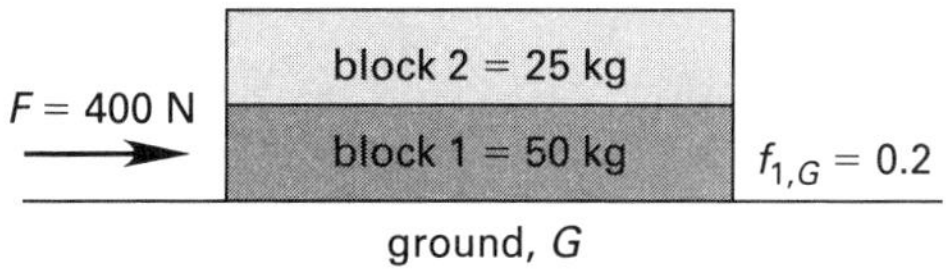

The frictional force is given by Eq. 58.21.

$$\begin{aligned} F_{f(1,G)} &= (W_1 + W_2)f_{1,G} \\ &= (50\ \text{kg} + 25\ \text{kg})\left(9.81\ \frac{\text{m}}{\text{s}^2}\right)(0.2) \\ &= 147.2\ \text{N}\quad \left[\begin{array}{l} < F;\ \text{blocks move together at accel-} \\ \text{eration } a,\ \text{assuming no slipping} \end{array}\right] \end{aligned}$$

The acceleration is

$$\begin{aligned} a &= \frac{F - F_{f(1,G)}}{m_1 + m_2} = \frac{400\ \text{N} - 147.2\ \text{N}}{50\ \text{kg} + 25\ \text{kg}} \\ &= 3.371\ \text{m/s}^2 \end{aligned}$$

If block 2 is not slipping,

$$F_{f(1,2)} = f_{1,2}W_2 = m_2 a$$

Therefore, the coefficient of friction between blocks 1 and 2 will be

$$f_{1,2} = \frac{a}{g} = \frac{3.371\ \frac{\text{m}}{\text{s}^2}}{9.81\ \frac{\text{m}}{\text{s}^2}} = \boxed{0.344\quad(0.34)}$$

The answer is (B).

10. (a) Use Eq. 58.31. Since the conveyor belt is flat and horizontal, $\phi = 0$.

$$\tan\phi = \frac{\text{v}_t^2 - fgr}{gr + f\text{v}_t^2}$$

$$0 = \frac{\text{v}_t^2 - (0.15)\left(32.2\ \frac{\text{ft}}{\text{sec}^2}\right)(5\text{ ft})}{\left(32.2\ \frac{\text{ft}}{\text{sec}^2}\right)(5\text{ ft}) + (0.15)\text{v}_t^2}$$

$$\text{v}_t^2 = 24.15\ \frac{\text{ft}^2}{\text{sec}^2}$$

$$\text{v}_t = \sqrt{24.15\ \frac{\text{ft}^2}{\text{sec}^2}} = \boxed{4.914\text{ ft/sec}\quad(4.9\text{ ft/sec})}$$

The answer is (B).

(b) The maximum centrifugal force is

$$F_c = \frac{m\text{v}_t^2}{g_c r} = \frac{(1\text{ lbm})\left(4.914\ \frac{\text{ft}}{\text{sec}}\right)^2}{\left(32.2\ \frac{\text{lbm-ft}}{\text{lbf-sec}^2}\right)(5\text{ ft})} = \boxed{0.15\text{ lbf}}$$

The answer is (B).

(c) The normal force on a 1 lbm mass is

$$N = \frac{mg}{g_c} = \frac{(1\text{ lbm})\left(32.2\ \frac{\text{ft}}{\text{sec}^2}\right)}{32.2\ \frac{\text{lbm-ft}}{\text{lbf-sec}^2}} = 1\text{ lbf}$$

The limiting frictional force is

$$F_f = fN = (0.15)(1\text{ lbf}) = \boxed{0.15\text{ lbf}}$$

The answer is (A).

11. From Eq. 58.31, the radius is

$$r = \frac{\text{v}_t^2(1 - f\tan\phi)}{g(\tan\phi + f)} = \frac{\left(5\ \frac{\text{ft}}{\text{sec}}\right)^2\left(1 - (0.10)\tan 20^\circ\right)}{\left(32.2\ \frac{\text{ft}}{\text{sec}^2}\right)(\tan 20^\circ + 0.10)} = \boxed{1.61\text{ ft}\quad(1.6\text{ ft})}$$

The answer is (C).

12. *Customary U.S. Solution*

(a)

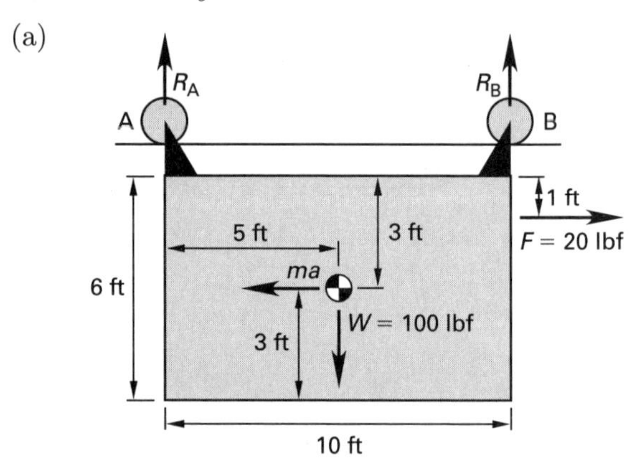

From Eq. 58.12, the acceleration is

$$a_x = \frac{F_x g_c}{m} = \frac{(20\text{ lbf})\left(32.2\ \frac{\text{lbm-ft}}{\text{lbf-sec}^2}\right)}{100\text{ lbm}} = \boxed{6.44\text{ ft/sec}^2\quad(6.4\text{ ft/sec}^2)}$$

The answer is (D).

(b) The inertial force, ma, is also 20 lbf.

$$\sum M_\text{A} = -(5\text{ ft})(100\text{ lbf}) + (10\text{ ft})R_\text{B} + (1\text{ ft})(20\text{ lbf}) - (3\text{ ft})(20\text{ lbf}) = 0$$

$$R_\text{B} = \boxed{54\text{ lbf}\quad[\text{upward}]}$$

$$\sum F_y = R_\text{A} + 54\text{ lbf} - 100\text{ lbf} = 0$$

$$R_\text{A} = \boxed{46\text{ lbf}\quad[\text{upward}]}$$

The answer is (D).

(c) Let y be the distance (positive upwards) from the center of gravity to F's line of action for $R_\text{A} = R_\text{B} = R$.

$$\sum M_\text{CG} = W(0) - (5\text{ ft})R + (5\text{ ft})R + y(20\text{ lbf}) = 0$$

$$y = 0$$

$$\boxed{\text{Apply the force in line with the center of gravity.}}$$

The answer is (A).

SI Solution

(a)

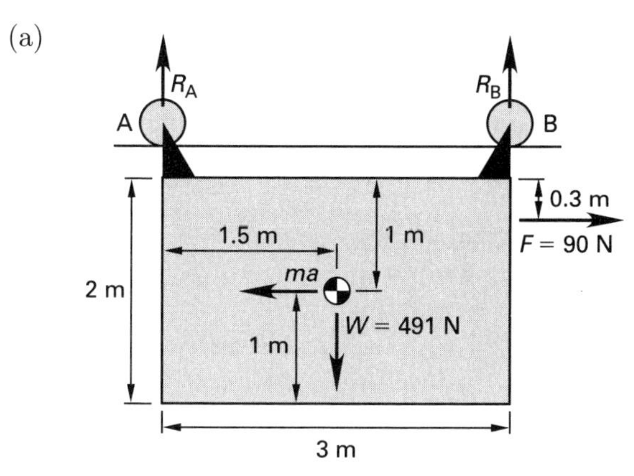

From Eq. 58.12, the acceleration is

$$a_x = \frac{F_x}{m} = \frac{90 \text{ N}}{50 \text{ kg}} = \boxed{1.8 \text{ m/s}^2}$$

The answer is (D).

(b) The inertial force is also 90 N.

$$\sum M_A = -(1.5 \text{ m})(491 \text{ N}) + (3 \text{ m})R_B + (0.3 \text{ m})(90 \text{ N}) - (1 \text{ m})(90 \text{ N}) = 0$$

$$R_B = \boxed{266.5 \text{ N} \quad (270 \text{ N}) \quad [\text{upward}]}$$

$$\sum F_y = R_A + 266.5 \text{ N} - 491 \text{ N} = 0$$

$$R_A = \boxed{224.5 \text{ N} \quad (220 \text{ N}) \quad [\text{upward}]}$$

The answer is (D).

(c) Let y be the distance (positive upward) from the center of gravity to F's line of action for $R_A = R_B = R$.

$$\sum M_{CG} = W(0) - (1.5 \text{ m})R + (1.5 \text{ m})R + y(90 \text{ N}) = 0$$

$$y = 0$$

$$\boxed{\text{Apply the force in line with the center of gravity.}}$$

The answer is (A).

13. *Customary U.S. Solution*

$$F_f = \frac{mg}{g_c}\sin\phi - \frac{ma}{g_c} \qquad \text{[Eq. I]}$$

For constrained motion,

$$F_f r = I_0\left(\frac{\alpha}{g_c}\right) = \left(\tfrac{2}{5}mr^2\right)\left(\frac{a}{r}\right)\left(\frac{1}{g_c}\right) = \tfrac{2}{5}\left(\frac{mar}{g_c}\right)$$

$$F_f = \tfrac{2}{5}\left(\frac{ma}{g_c}\right) \qquad \text{[Eq. II]}$$

From Eq. I and Eq. II, $mg\sin\phi - ma = (2/5)ma$. The sphere's acceleration will be

$$a = \tfrac{5}{7}g\sin\phi = \left(\tfrac{5}{7}\right)\left(32.2 \ \frac{\text{ft}}{\text{sec}^2}\right)(\sin 30^\circ) = 11.5 \text{ ft/sec}^2$$

Therefore, the sphere speed is

$$\text{v} = at = \left(11.5 \ \frac{\text{ft}}{\text{sec}^2}\right)(2 \text{ sec}) = \boxed{23 \text{ ft/sec}}$$

The answer is (B).

SI Solution

$$F_f = mg\sin\phi - ma \qquad \text{[Eq. I]}$$

For constrained motion,

$$F_f r = I_0\alpha = \left(\tfrac{2}{5}mr^2\right)\left(\frac{a}{r}\right) = \tfrac{2}{5}mar$$

$$F_f = \tfrac{2}{5}ma \qquad \text{[Eq. II]}$$

From Eq. I and Eq. II, $mg\sin\phi - ma = (2/5)ma$. The sphere's acceleration will be

$$a = \tfrac{5}{7}g\sin\phi = \left(\tfrac{5}{7}\right)\left(9.81 \ \frac{\text{m}}{\text{s}^2}\right)(\sin 30^\circ) = 3.504 \text{ m/s}^2$$

Therefore, the sphere speed is

$$\text{v} = at = \left(3.504 \ \frac{\text{m}}{\text{s}^2}\right)(2 \text{ s}) = \boxed{7.008 \text{ m/s} \quad (7.0 \text{ m/s})}$$

The answer is (B).

14. *Customary U.S. Solution*

Let B's acceleration and velocity be a and v, respectively, positive upwards. Then, A's acceleration and velocity are $-a$ and $-\text{v}$, respectively.

For B:

$$T - \frac{m_B g}{g_c} = \frac{m_B a}{g_c}$$

$$T = \frac{m_B a + m_B g}{g_c}$$

For A:

$$T - \frac{m_A g\sin\phi}{g_c} = \frac{m_A(-a)}{g_c}$$

$$T = \frac{m_A g\sin\phi - m_A a}{g_c}$$

Find the acceleration.

$$a = \frac{m_{\mathrm{A}} g \sin\phi - m_{\mathrm{B}} g}{m_{\mathrm{B}} + m_{\mathrm{A}}}$$

$$= \frac{(10 \text{ lbm})\left(32.2 \ \frac{\text{ft}}{\text{sec}^2}\right)(\sin 36.87°) - (20 \text{ lbm})\left(32.2 \ \frac{\text{ft}}{\text{sec}^2}\right)}{20 \text{ lbm} + 10 \text{ lbm}}$$

$$= -15.03 \text{ ft/sec}^2$$

Therefore, the velocity of object B is

$$\text{v} = at = \left(-15.03 \ \frac{\text{ft}}{\text{sec}^2}\right)(3 \text{ sec})$$

$$= \boxed{-45.09 \text{ ft/sec} \quad (45 \text{ ft/sec}) \quad [\text{downward}]}$$

The answer is (D).

SI Solution

Let B's acceleration and velocity be a and v, respectively, positive upwards. Then, A's acceleration and velocity are $-a$ and $-\text{v}$, respectively.

For B: $T - m_{\mathrm{B}} g = m_{\mathrm{B}} a$

$$T = m_{\mathrm{B}} a + m_{\mathrm{B}} g$$

For A: $T - m_{\mathrm{A}} g \sin\phi = m_{\mathrm{A}}(-a)$

$$T = m_{\mathrm{A}} g \sin\phi - m_{\mathrm{A}} a$$

Find the acceleration.

$$a = \frac{m_{\mathrm{A}} g \sin\phi - m_{\mathrm{B}} g}{m_{\mathrm{A}} + m_{\mathrm{B}}}$$

$$= \frac{(5 \text{ kg})\left(9.81 \ \frac{\text{m}}{\text{s}^2}\right)(\sin 36.87°) - (10 \text{ kg})\left(9.81 \ \frac{\text{m}}{\text{s}^2}\right)}{5 \text{ kg} + 10 \text{ kg}}$$

$$= -4.578 \text{ m/s}^2$$

Therefore, the velocity of object B is

$$\text{v} = at = \left(-4.578 \ \frac{\text{m}}{\text{s}^2}\right)(3 \text{ s})$$

$$= \boxed{-13.7 \text{ m/s} \quad (14 \text{ m/s}) \quad [\text{downward}]}$$

The answer is (D).

15. *Customary U.S. Solution*

From Eq. 58.57(b), the force is

$$F = \frac{\dot{m}\Delta\text{v}}{g_c} = \frac{\left(560 \ \frac{\text{lbm}}{\text{min}}\right)\left(3.2 \ \frac{\text{ft}}{\text{sec}}\right)}{\left(32.2 \ \frac{\text{lbm-ft}}{\text{lbf-sec}^2}\right)\left(60 \ \frac{\text{sec}}{\text{min}}\right)}$$

$$= \boxed{0.9275 \text{ lbf} \quad (0.93 \text{ lbf})}$$

The answer is (A).

SI Solution

From Eq. 58.57(a), the force is

$$F = \dot{m}\Delta\text{v} = \frac{\left(250 \ \frac{\text{kg}}{\text{min}}\right)\left(0.98 \ \frac{\text{m}}{\text{s}}\right)}{60 \ \frac{\text{s}}{\text{min}}}$$

$$= \boxed{4.083 \text{ N} \quad (4.1 \text{ N})}$$

The answer is (A).

16. *Customary U.S. Solution*

From Eq. 58.54, the impulse is

$$|\mathbf{Imp}| = \frac{|\Delta\mathbf{p}|}{g_c} = \frac{|\mathbf{p}_2 - \mathbf{p}_1|}{g_c} = \frac{m\text{v}_2 - (-m\text{v}_1)}{g_c}$$

$$= \frac{(0.4 \text{ lbm})\left(90 \ \frac{\text{ft}}{\text{sec}} + 130 \ \frac{\text{ft}}{\text{sec}}\right)}{32.2 \ \frac{\text{lbm-ft}}{\text{lbf-sec}^2}}$$

$$= \boxed{2.73 \text{ lbf-sec} \quad (2.7 \text{ lbf-sec})}$$

The answer is (B).

SI Solution

From Eq. 58.54, the impulse is

$$|\mathbf{Imp}| = |\Delta\mathbf{p}| = |\mathbf{p}_2 - \mathbf{p}_1| = m\text{v}_2 - (-m\text{v}_1)$$

$$= (0.2 \text{ kg})\left(40 \ \frac{\text{m}}{\text{s}} + 30 \ \frac{\text{m}}{\text{s}}\right)$$

$$= \boxed{14 \text{ N·s}}$$

The answer is (B).

17. *Customary U.S. Solution*

Momentum is always conserved.

$$\Delta p_{\text{gun}} = -\Delta p_{\text{proj}}$$

$$\frac{m_{\text{gun}} \text{v}_{\text{gun}}}{g_c} = \frac{-m_{\text{proj}} \text{v}_{\text{proj}}}{g_c}$$

$$\text{v}_{\text{gun}} = -\frac{m_{\text{proj}} \text{v}_{\text{proj}}}{m_{\text{gun}}} = \frac{-(2.6 \text{ lbm})\left(2100 \ \frac{\text{ft}}{\text{sec}}\right)}{1000 \text{ lbm}}$$

$$= \boxed{-5.46 \text{ ft/sec} \quad (5.5 \text{ ft/sec}) \quad \begin{bmatrix}\text{opposite projectile} \\ \text{direction}\end{bmatrix}}$$

The answer is (C).

SI Solution

Momentum is always conserved.

$$\Delta p_{\text{gun}} = -\Delta p_{\text{proj}}$$

$$m_{\text{gun}} \text{v}_{\text{gun}} = -m_{\text{proj}} \text{v}_{\text{proj}}$$

$$\text{v}_{\text{gun}} = -\frac{m_{\text{proj}} \text{v}_{\text{proj}}}{m_{\text{gun}}} = \frac{-(1.2 \text{ kg})\left(650 \ \frac{\text{m}}{\text{s}}\right)}{500 \text{ kg}}$$

$$= \boxed{-1.56 \text{ m/s} \quad (1.6 \text{ m/s}) \quad \begin{bmatrix}\text{opposite projectile} \\ \text{direction}\end{bmatrix}}$$

The answer is (C).

18. *Customary U.S. Solution*

In absence of external forces, $\mathbf{\Delta p} = 0$.

$$\frac{m_{\text{bullet}} \text{v}_{\text{bullet}}}{g_c} = \frac{m_{(\text{bullet+block})} \text{v}_{(\text{bullet+block})}}{g_c}$$

$$\text{v}_{(\text{bullet+block})} = \frac{m_{\text{bullet}} \text{v}_{\text{bullet}}}{m_{(\text{bullet+block})}}$$

$$= \frac{(0.15 \text{ lbm})\left(2300 \ \frac{\text{ft}}{\text{sec}}\right)}{0.15 \text{ lbm} + 9 \text{ lbm}}$$

$$= \boxed{37.7 \text{ ft/sec} \quad (38 \text{ ft/sec})}$$

The answer is (B).

SI Solution

In absence of external forces, $\mathbf{\Delta p} = 0$.

$$m_{\text{bullet}} \text{v}_{\text{bullet}} = m_{(\text{bullet+block})} \text{v}_{(\text{bullet+block})}$$

$$\text{v}_{(\text{bullet+block})} = \frac{m_{\text{bullet}} \text{v}_{\text{bullet}}}{m_{(\text{bullet+block})}} = \frac{(0.06 \text{ kg})\left(700 \ \frac{\text{m}}{\text{s}}\right)}{4.5 \text{ kg} + 0.06 \text{ kg}}$$

$$= \boxed{9.21 \text{ m/s} \quad (9.2 \text{ m/s})}$$

The answer is (B).

19. *Customary U.S. Solution*

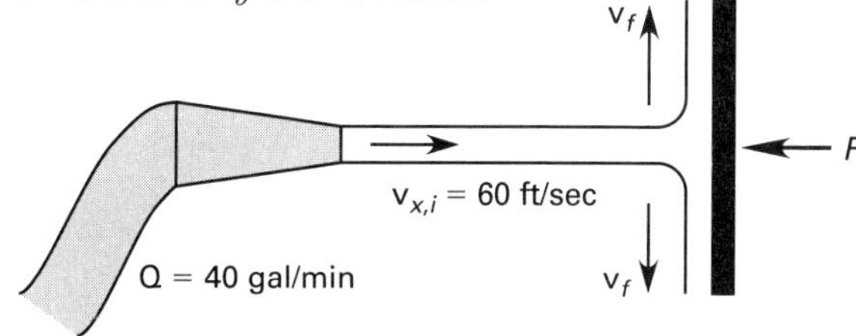

The mass flow rate is

$$\dot{m} = \rho Q = \frac{\left(62.4 \ \frac{\text{lbm}}{\text{ft}^3}\right)\left(40 \ \frac{\text{gal}}{\text{min}}\right)}{\left(60 \ \frac{\text{sec}}{\text{min}}\right)\left(7.48 \ \frac{\text{gal}}{\text{ft}^3}\right)}$$

$$= 5.56 \text{ lbm/sec}$$

Neglecting gravity, water is turned through an angle of 90° in equal portions in all directions. For every direction other than along the x-axis, equal amounts of water are directed in opposite senses; no net force is applied in these directions. From Eq. 58.57,

$$F_x = \frac{\dot{m}\Delta \text{v}_x}{g_c} = \frac{-\dot{m}\text{v}_{x,i}}{g_c} = \frac{-\left(5.56 \ \frac{\text{lbm}}{\text{sec}}\right)\left(60 \ \frac{\text{ft}}{\text{sec}}\right)}{32.2 \ \frac{\text{lbm-ft}}{\text{lbf-sec}^2}}$$

$$= \boxed{-10.36 \text{ lbf} \quad (10 \text{ lbf}) \quad \begin{bmatrix}\text{opposite water} \\ \text{direction}\end{bmatrix}}$$

The answer is (A).

SI Solution

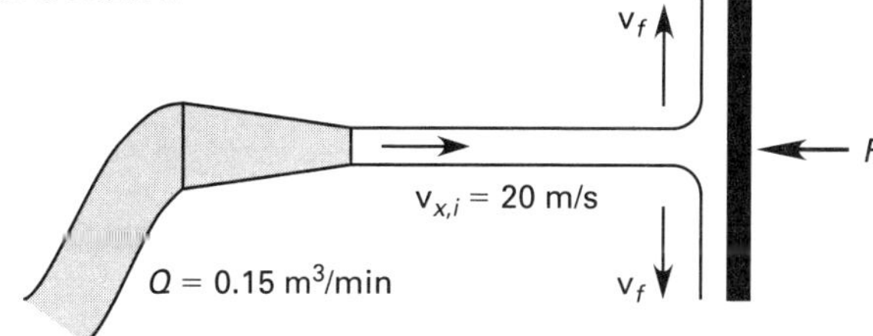

The mass flow rate is

$$\dot{m} = \rho Q$$

$$= \frac{\left(1000 \ \frac{\text{kg}}{\text{m}^3}\right)\left(0.15 \ \frac{\text{m}^3}{\text{min}}\right)}{60 \ \frac{\text{s}}{\text{min}}} = 2.5 \text{ kg/s}$$

Neglecting gravity, water is turned through an angle of 90° in equal portions in all directions. For every direction other than along the x-axis, equal amounts of water are directed in opposite senses; no net force is applied in these directions. From Eq. 58.57,

$$F_x = \dot{m}\Delta \text{v}_x = -\dot{m}\text{v}_{x,i} = -\left(2.5 \ \frac{\text{kg}}{\text{s}}\right)\left(20 \ \frac{\text{m}}{\text{s}}\right)$$

$$= \boxed{-50 \text{ N} \quad (50 \text{ N}) \quad [\text{opposite water direction}]}$$

The answer is (A).

20. *Customary U.S. Solution*

(a) The mass flow rate is

$$\dot{m} = \rho Q = \frac{\left(62.4\ \frac{\text{lbm}}{\text{ft}^3}\right)\left(100\ \frac{\text{gal}}{\text{sec}}\right)}{7.48\ \frac{\text{gal}}{\text{ft}^3}} = 834.2\ \text{lbm/sec}$$

Let inward directions be positive.

The outward momentum flow rates are

$$\begin{aligned}\dot{p}_{\text{out},x} &= \dot{m}\text{v}_{\text{out}}\cos\phi \\ &= \left(834.2\ \frac{\text{lbm}}{\text{sec}}\right)\left(57\ \frac{\text{ft}}{\text{sec}}\right)(\cos 160^\circ) \\ &= -44{,}682\ \text{lbm-ft/sec}^2\end{aligned}$$

$$\begin{aligned}\dot{p}_{\text{out},y} &= \dot{m}\text{v}_{\text{out}}\sin\phi \\ &= \left(834.2\ \frac{\text{lbm}}{\text{sec}}\right)\left(57\ \frac{\text{ft}}{\text{sec}}\right)(\sin 160^\circ) \\ &= 16{,}263\ \text{lbm-ft/sec}^2\end{aligned}$$

The inward momentum flow rate is

$$\dot{p}_{\text{in},x} = \dot{m}\text{v}_{\text{in}} = \left(834.2\ \frac{\text{lbm}}{\text{sec}}\right)\left(60\ \frac{\text{ft}}{\text{sec}}\right) = 50{,}052\ \text{lbm-ft/sec}^2$$

The forces are

$$\begin{aligned}F_x &= \frac{\Delta\dot{p}_x}{g_c} = \frac{\dot{p}_{\text{out},x} - \dot{p}_{\text{in},x}}{g_c} \\ &= \frac{-44{,}682\ \frac{\text{lbm-ft}}{\text{sec}^2} - 50{,}052\ \frac{\text{lbm-ft}}{\text{sec}^2}}{32.2\ \frac{\text{lbm-ft}}{\text{lbf-sec}^2}} \\ &= -2942\ \text{lbf}\end{aligned}$$

$$F_y = \frac{\Delta\dot{p}_y}{g_c} = \frac{16{,}263\ \frac{\text{lbm-ft}}{\text{sec}^2} - 0\ \frac{\text{lbm-ft}}{\text{sec}^2}}{32.2\ \frac{\text{lbm-ft}}{\text{lbf-sec}^2}} = 505.1\ \text{lbf}$$

$$\begin{aligned}F_{\text{blade}} &= \sqrt{F_x^2 + F_y^2} = \sqrt{(-2942\ \text{lbf})^2 + (505.1\ \text{lbf})^2} \\ &= \boxed{2985\ \text{lbf}\quad (3000\ \text{lbf})}\end{aligned}$$

The answer is (C).

(b) From the horizontal component of the force, the angle is

$$\begin{aligned}\phi &= \arctan\frac{F_y}{F_x} = \arctan\frac{505.1\ \text{lbf}}{-2942\ \text{lbf}} \\ &= -9.74^\circ\quad [\text{second quadrant}] \\ &= \boxed{-170.26^\circ\quad (170^\circ)\quad \left[\begin{array}{c}\text{counterclockwise}\\ \text{from horizontal}\end{array}\right]}\end{aligned}$$

The answer is (D).

SI Solution

(a) The mass flow rate is

$$\begin{aligned}\dot{m} = \rho Q &= \left(1000\ \frac{\text{kg}}{\text{m}^3}\right)\left(0.4\ \frac{\text{m}^3}{\text{s}}\right) \\ &= 400\ \text{kg/s}\end{aligned}$$

Let inward directions be positive.

The outward momentum flow rates are

$$\begin{aligned}\dot{p}_{\text{out},x} &= \dot{m}\text{v}_{\text{out}}\cos\phi \\ &= \left(400\ \frac{\text{kg}}{\text{s}}\right)\left(17\ \frac{\text{m}}{\text{s}}\right)(\cos 160^\circ) \\ &= -6390\ \text{N}\end{aligned}$$

$$\begin{aligned}\dot{p}_{\text{out},y} &= \dot{m}\text{v}_{\text{out}}\sin\phi \\ &= \left(400\ \frac{\text{kg}}{\text{s}}\right)\left(17\ \frac{\text{m}}{\text{s}}\right)(\sin 160^\circ) \\ &= 2326\ \text{N}\end{aligned}$$

The inward momentum flow rate is

$$\begin{aligned}\dot{p}_{\text{in},x} = \dot{m}\text{v}_{\text{in}} &= \left(400\ \frac{\text{kg}}{\text{s}}\right)\left(20\ \frac{\text{m}}{\text{s}}\right) \\ &= 8000\ \text{N}\end{aligned}$$

The forces are

$$\begin{aligned}F_x &= \Delta\dot{p}_x = \dot{p}_{\text{out},x} - \dot{p}_{\text{in},x} = -6390\ \text{N} - 8000\ \text{N} \\ &= -14\,390\ \text{N}\end{aligned}$$

$$F_y = \Delta\dot{p}_y = 2326\ \text{N} - 0\ \text{N} = 2326\ \text{N}$$

$$\begin{aligned}F_{\text{blade}} &= \sqrt{F_x^2 + F_y^2} = \sqrt{(-14\,390\ \text{N})^2 + (2326\ \text{N})^2} \\ &= \boxed{14\,577\ \text{N}\quad (15\,000\ \text{N})}\end{aligned}$$

The answer is (C).

(b) From the horizontal component of the force, the angle is

$$\begin{aligned}\phi &= \arctan\frac{F_y}{F_x} = \arctan\frac{2326\ \text{N}}{-14\,390\ \text{N}} \\ &= -9.18^\circ\quad [\text{second quadrant}] \\ &= \boxed{-170.82^\circ\quad (170^\circ)\quad \left[\begin{array}{c}\text{counterclockwise}\\ \text{from horizontal}\end{array}\right]}\end{aligned}$$

The answer is (D).

21. *Customary U.S. Solution*

(a)

$$v_{A,y} = \left(5\ \frac{ft}{sec}\right)(\sin 45°) = 3.536\ ft/sec$$

$$v_{A,x} = \left(5\ \frac{ft}{sec}\right)(\cos 45°) = 3.536\ ft/sec$$

$$v_{B,y} = \left(10\ \frac{ft}{sec}\right)\left(\frac{3}{5}\right) = 6\ ft/sec$$

$$v_{B,x} = \left(10\ \frac{ft}{sec}\right)\left(-\frac{4}{5}\right) = -8\ ft/sec$$

The force of impact is in the x-direction only.

$$v'_{A,y} = v_{A,y}$$

$$v'_{B,y} = v_{B,y}$$

In the x-direction,

$$e = \frac{v'_{A,x} - v'_{B,x}}{v_{B,x} - v_{A,x}} = \frac{v'_{A,x} - v'_{B,x}}{-8\ \frac{ft}{sec} - 3.536\ \frac{ft}{sec}}$$

$$\begin{aligned} v'_{A,x} - v'_{B,x} &= (0.8)\left(-8\ \frac{ft}{sec} - 3.536\ \frac{ft}{sec}\right) \\ &= -9.229\ ft/sec \qquad \text{[Eq. I]} \end{aligned}$$

$$m_A v_{A,x} + m_B v_{B,x} = m_A v'_{A,x} + m_B v'_{B,x}$$

Since $m_A = m_B$,

$$v'_{A,x} + v'_{B,x} = 3.536\ \frac{ft}{sec} + \left(-8\ \frac{ft}{sec}\right) = -4.464\ ft/sec \qquad \text{[Eq. II]}$$

Solving Eq. I and Eq. II simultaneously,

$$v'_{A,x} = -6.846\ ft/sec$$

$$v'_{B,x} = 2.382\ ft/sec$$

$$\begin{aligned} v'_A &= \sqrt{(v'_{A,x})^2 + (v'_{A,y})^2} \\ &= \sqrt{\left(-6.846\ \frac{ft}{sec}\right)^2 + \left(3.536\ \frac{ft}{sec}\right)^2} \\ &= \boxed{7.705\ ft/sec \quad (7.7\ ft/sec)} \end{aligned}$$

$$\begin{aligned} v'_B &= \sqrt{(v'_{B,x})^2 + (v'_{B,y})^2} \\ &= \sqrt{\left(2.382\ \frac{ft}{sec}\right)^2 + \left(6\ \frac{ft}{sec}\right)^2} \\ &= \boxed{6.456\ ft/sec \quad (6.5\ ft/sec)} \end{aligned}$$

The answer is (B).

(b)

$$\phi_A = \arctan \frac{3.536\ \frac{ft}{sec}}{-6.846\ \frac{ft}{sec}} = \boxed{152.7° \quad (150°)}$$

$$\phi_B = \arctan \frac{6\ \frac{ft}{sec}}{2.382\ \frac{ft}{sec}} = \boxed{68.3° \quad (68°)}$$

The answer is (D).

SI Solution

(a)

$$v_{A,y} = \left(1.5\ \frac{m}{s}\right)(\sin 45°) = 1.061\ m/s$$

$$v_{A,x} = \left(1.5\ \frac{m}{s}\right)(\cos 45°) = 1.061\ m/s$$

$$v_{B,y} = \left(3\ \frac{m}{s}\right)\left(\frac{3}{5}\right) = 1.8\ m/s$$

$$v_{B,x} = \left(3\ \frac{m}{s}\right)\left(-\frac{4}{5}\right) = -2.4\ m/s$$

The force of impact is in the x-direction only.

$$v'_{A,y} = v_{A,y}$$

$$v'_{B,y} = v_{B,y}$$

In the x-direction,

$$e = \frac{v'_{A,x} - v'_{B,x}}{v_{B,x} - v_{A,x}} = \frac{v'_{A,x} - v'_{B,x}}{-2.4\ \frac{m}{s} - 1.061\ \frac{m}{s}}$$

$$\begin{aligned} v'_{A,x} - v'_{B,x} &= (0.8)\left(-2.4\ \frac{m}{s} - 1.061\ \frac{m}{s}\right) \\ &= -2.769\ m/s \qquad \text{[Eq. I]} \end{aligned}$$

$$m_A v_{A,x} + m_B v_{B,x} = m_A v'_{A,x} + m_B v'_{B,x}$$

Since $m_A = m_B$,

$$v'_{A,x} + v'_{B,x} = 1.061\ \frac{m}{s} + \left(-2.4\ \frac{m}{s}\right) = -1.339\ m/s \qquad \text{[Eq. II]}$$

Solving Eq. I and Eq. II simultaneously,

$$v'_{A,x} = -2.054\ m/s$$

$$v'_{B,x} = 0.715\ m/s$$

$$\begin{aligned} v'_A &= \sqrt{(v'_{A,x})^2 + (v'_{A,y})^2} \\ &= \sqrt{\left(-2.054\ \frac{m}{s}\right)^2 + \left(1.061\ \frac{m}{s}\right)^2} \\ &= \boxed{2.312\ m/s\quad (2.3\ m/s)} \end{aligned}$$

$$\begin{aligned} v'_B &= \sqrt{(v'_{B,x})^2 + (v'_{B,y})^2} \\ &= \sqrt{\left(0.715\ \frac{m}{s}\right)^2 + \left(1.8\ \frac{m}{s}\right)^2} \\ &= \boxed{1.937\ m/s\quad (1.9\ m/s)} \end{aligned}$$

(b)

$$\phi_A = \arctan\frac{1.061\ \frac{m}{s}}{-2.054\ \frac{m}{s}} = \boxed{152.7^\circ\quad (150^\circ)}$$

$$\phi_B = \arctan\frac{1.8\ \frac{m}{s}}{0.715\ \frac{m}{s}} = \boxed{68.34^\circ\quad (68^\circ)}$$

The answer is (D).

22. *Customary U.S. Solution*

(a) Energy is conserved.

$$mgh = \tfrac{1}{2}mv^2$$

$$\begin{aligned} v &= \sqrt{2gh} = \sqrt{(2)\left(32.2\ \frac{ft}{sec^2}\right)(3\ ft)} \\ &= \boxed{13.9\ ft/sec\quad (14\ ft/sec)} \end{aligned}$$

The answer is (A).

(b) The tension in the cord resists the pendulum's weight and the centrifugal force.

$$F_c = \frac{mv^2}{g_c r} = \frac{(10\ lbm)\left(13.9\ \frac{ft}{sec}\right)^2}{\left(32.2\ \frac{lbm\text{-}ft}{lbf\text{-}sec^2}\right)(3\ ft)} = 20\ lbf$$

$$T = W + F_c = 10\ lbf + 20\ lbf = \boxed{30\ lbf}$$

The answer is (C).

(c) The block's velocity immediately after impact is found as follows.

$$e = \frac{v'_1 - v'_2}{v_2 - v_1} = \frac{v'_1 - v'_2}{0\ \frac{ft}{sec} - 13.9\ \frac{ft}{sec}}$$

$$v'_1 - v'_2 = (0.7)\left(-13.9\ \frac{ft}{sec}\right) = -9.73\ ft/sec \qquad \text{[Eq. I]}$$

$$m_1v_1 + m_2v_2 = m_1v'_1 + m_2v'_2$$

$$(10\ lbm)v'_1 + (50\ lbm)v'_2 = 139\ ft/sec \qquad \text{[Eq. II]}$$

Solving Eq. I and Eq. II simultaneously,

$$\begin{aligned} v'_1 &= -5.79\ ft/sec \\ v'_2 &= \boxed{3.94\ ft/sec\quad (3.9\ ft/sec)} \end{aligned}$$

The answer is (B).

(d) The required spring constant, k, is

$$\frac{mv^2}{2g_c} = \frac{kx^2}{2}$$

$$\begin{aligned} k &= \frac{mv^2}{g_c x^2} = \frac{(50\ lbm)\left(3.94\ \frac{ft}{sec}\right)^2}{\left(32.2\ \frac{lbm\text{-}ft}{lbf\text{-}sec^2}\right)(0.5\ ft)^2} \\ &= \boxed{96.4\ lbf/ft\quad (96\ lbf/ft)} \end{aligned}$$

The answer is (D).

SI Solution

(a) Energy is conserved.

$$mgh = \frac{mv^2}{2}$$

$$\begin{aligned} v &= \sqrt{2gh} = \sqrt{(2)\left(9.81\ \frac{m}{s^2}\right)(1\ m)} \\ &= \boxed{4.43\ m/s\quad (4.4\ m/s)} \end{aligned}$$

The answer is (A).

(b) The tension in the cord resists the pendulum's weight and the centrifugal force.

$$F_c = \frac{mv^2}{r} = \frac{(5\ kg)\left(4.43\ \frac{m}{s}\right)^2}{1\ m} = 98.1\ N$$

$$\begin{aligned} T &= W + F_c = (5\ kg)\left(9.81\ \frac{m}{s^2}\right) + 98.1\ N \\ &= \boxed{147\ N\quad (150\ N)} \end{aligned}$$

The answer is (C).

(c) The block's velocity immediately after impact is found as follows.

$$e = \frac{v'_1 - v'_2}{v_2 - v_1} = \frac{v'_1 - v'_2}{0\ \frac{m}{s} - 4.43\ \frac{m}{s}}$$

$$v'_1 - v'_2 = (0.7)\left(-4.43\ \frac{m}{s}\right) = -3.1\ m/s \qquad \text{[Eq. I]}$$

$$m_1v_1 + m_2v_2 = m_1v'_1 + m_2v'_2$$

$$(5\ kg)v'_1 + (25\ kg)v'_2 = 22.15\ m/s \qquad \text{[Eq. II]}$$

Solving Eq. I and Eq. II simultaneously,

$$v_1' = -1.845\ \text{m/s}$$
$$v_2' = \boxed{1.255\ \text{m/s}\quad (1.3\ \text{m/s})}$$

The answer is (B).

(d) The required spring constant, k, is

$$\frac{mv^2}{2} = \frac{kx^2}{2}$$

$$k = \frac{mv^2}{x^2} = \frac{(25\ \text{kg})\left(1.255\ \frac{\text{m}}{\text{s}}\right)^2}{(0.15\ \text{m})^2}$$
$$= \boxed{1750\ \text{N/m}\quad (1800\ \text{N/m})}$$

The answer is (D).

23. *Customary U.S. Solution*

$$m_{\text{left}}v_{\text{left}} + m_{\text{right}}v_{\text{right}} = m_{\text{couple}}v_{\text{couple}}$$

$$v_{\text{couple}} = \frac{(5\ \text{tons})\left(5\ \frac{\text{ft}}{\text{sec}}\right) + (5\ \text{tons})\left(-4\ \frac{\text{ft}}{\text{sec}}\right)}{10\ \text{tons}}$$
$$= \boxed{0.5\ \text{ft/sec}\quad [\text{to the right}]}$$

The answer is (B).

SI Solution

$$m_{\text{left}}v_{\text{left}} + m_{\text{right}}v_{\text{right}} = m_{\text{couple}}v_{\text{couple}}$$

$$v_{\text{couple}} = \frac{(5000\ \text{kg})\left(1.5\ \frac{\text{m}}{\text{s}}\right) + (5000\ \text{kg})\left(-1.2\ \frac{\text{m}}{\text{s}}\right)}{10\,000\ \text{kg}}$$
$$= \boxed{0.15\ \text{m/s}\quad [\text{to the right}]}$$

The answer is (B).

24. *Customary U.S. Solution*

Since the electron is deflected from its original path, the collision is oblique. The ratio of the masses is

$$\frac{m_{\text{H}}}{m_e} = \frac{1.007277u}{0.0005486u} = 1836$$

Kinetic energy may or may not be conserved in this collision; there is insufficient information to make the determination. Momentum is always conserved, regardless of the axis along which it is evaluated. Consider the original path of the electron to be parallel to the x-axis. Then, by conservation of momentum in the x-direction,

$$m_e v_{e,x} + m_{\text{H}} v_{\text{H},x} = m_e v'_{e,x} + m_{\text{H}} v'_{\text{H},x}$$

Recognizing that the x-component of v'_{H} was requested and $v_{\text{H},x} = 0$, substitute the ratio of masses.

$$v_{e,x} = v'_e \cos\theta_e + 1836 v'_{\text{H},x}$$
$$1500\ \frac{\text{ft}}{\text{sec}} = \left(65\ \frac{\text{ft}}{\text{sec}}\right)(\cos 30^\circ) + 1836 v'_{\text{H},x}$$
$$v'_{\text{H},x} = \boxed{0.786\ \text{ft/sec}\quad (0.79\ \text{ft/sec})}$$

The answer is (B).

SI Solution

Since the electron is deflected from its original path, the collision is oblique. The ratio of the masses is

$$\frac{m_{\text{H}}}{m_e} = \frac{1.007277u}{0.0005486u} = 1836$$

Kinetic energy may or may not be conserved in this collision; there is insufficient information to make the determination. Momentum is always conserved, regardless of the axis along which it is evaluated. Consider the original path of the electron to be parallel to the x-axis. Then, by conservation of momentum in the x-direction,

$$m_e v_{e,x} + m_{\text{H}} v_{\text{H},x} = m_e v'_{e,x} + m_{\text{H}} v'_{\text{H},x}$$

Recognizing that the x-component of v'_{H} was requested and $v_{\text{H},x} = 0$, substitute the ratio of masses.

$$v_{e,x} = v'_e \cos\theta_e + 1836 v'_{\text{H},x}$$
$$500\ \frac{\text{m}}{\text{s}} = \left(20.2\ \frac{\text{m}}{\text{s}}\right)(\cos 30^\circ) + 1836 v'_{\text{H},x}$$
$$v'_{\text{H},x} = \boxed{0.26\ \text{m/s}}$$

The answer is (B).

25. The running block (the block attached to the object) has 3 falls (cables) running from it. Therefore, the mechanical advantage of this arrangement is 3:1. Once the block is moving, the tension in the cable is $(1/3)(2000\ \text{lbf}) = 667\ \text{lbf}$. However, since the coefficient of static friction is higher than the coefficient of dynamic friction, the initial cable tension will be slightly higher. Of the options, only $\boxed{680\ \text{lbf}}$ is higher than 667.

The answer is (D).

59 Mechanisms and Power Transmission Systems

PRACTICE PROBLEMS

Position, Velocity, and Acceleration

1. The lever shown rotates about point O. Arms OA and OB are fixed in relation to one another. The length of OA is 3 ft. The angle that arm OA makes with the horizontal, ϕ, is 20°. About point O, the angular velocity of the lever, ω, is 10 rad/sec, and the angular acceleration of the lever, α, is –20 rad/sec^2. What is most nearly the magnitude of the total acceleration of point A?

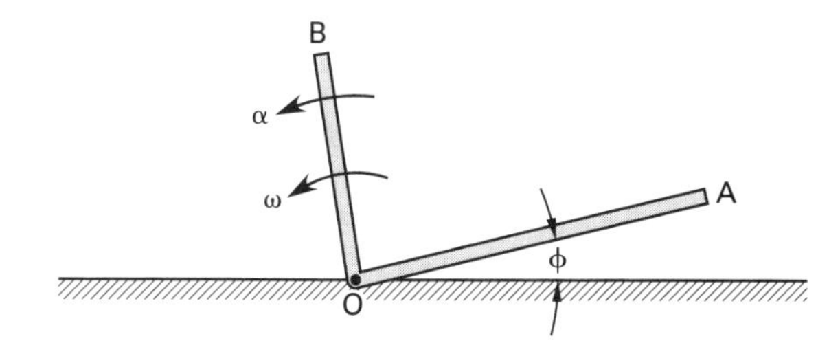

(A) 60 ft/sec^2

(B) 310 ft/sec^2

(C) 350 ft/sec^2

(D) 410 ft/sec^2

2. The lever shown rotates freely about point O. Arms OA and OB are fixed in relation to one another. The length of OA is 4 ft. The angular velocity of the lever, ω, is 100 rad/sec, and the angular acceleration of the lever, α, is constant at –5 rad/sec^2. What is most nearly the distance traveled by point A before the rotational speed is reduced to zero?

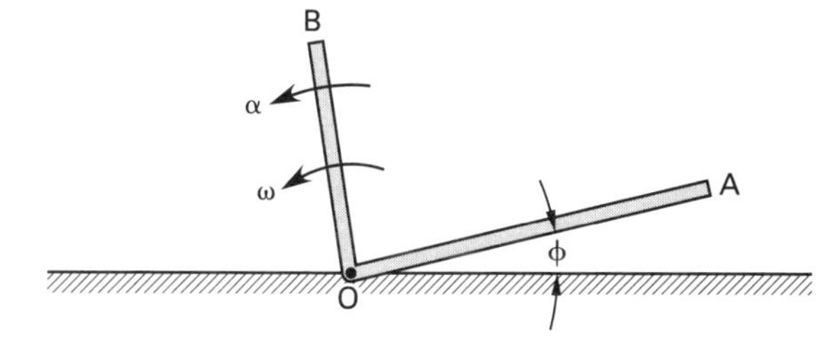

(A) 2000 ft

(B) 3000 ft

(C) 4000 ft

(D) 5000 ft

3. The lever shown is free to rotate about point O. Arms OA and OB are fixed in relation to one another. The length of OA is 1.5 ft, and the length of OB is 2.5 ft. At a particular moment, the velocity of point A, v_A, is $-70.5\mathbf{i} - 25.5\mathbf{j}$ ft/sec. Considering clockwise rotation as positive, what is most nearly the velocity of point B at that moment?

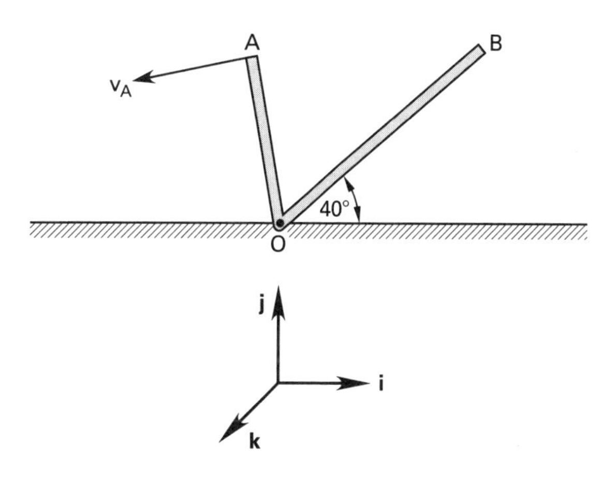

(A) $-77.4\mathbf{i} - 111\mathbf{j}$ ft/sec

(B) $-80.5\mathbf{i} + 96.0\mathbf{j}$ ft/sec

(C) $-98.4\mathbf{i} - 42.5\mathbf{j}$ ft/sec

(D) $-105\mathbf{i} + 48.5\mathbf{j}$ ft/sec

Flywheels

4. A flywheel is designed as a solid disk, 2 in (51 mm) thick and 20 in (510 mm) in diameter, with a concentric 4 in (102 mm) diameter mounting hole. It is manufactured from cast iron with an ultimate tensile strength of 30,000 lbf/in^2 (207 MPa) and a Poisson's ratio of 0.27. The density of the cast iron is 0.26 lbm/in^3 (7200 kg/m^3).

Using a factor of safety of 10, the maximum safe speed for the flywheel is most nearly

(A) 1700 rpm

(B) 2200 rpm

(C) 3500 rpm

(D) 8700 rpm

Epicyclic Gearsets

5. A simple epicyclic gearbox with one planet has gears with 24, 40, and 104 teeth on the sun, planet, and internal ring gears, respectively. The sun rotates clockwise at 50 rpm. The ring gear is fixed. The rotational velocity of the planet carrier is most nearly

(A) 9.4 rpm

(B) 12 rpm

(C) 17 rpm

(D) 23 rpm

6. Refer to the epicyclic gear set illustrated. Gear A rotates counterclockwise on a fixed center at 100 rpm. The ring gear rotates. The planet carrier rotates clockwise at 60 rpm. Each gear has the number of teeth indicated. What is most nearly the rotational speed of the sun gear?

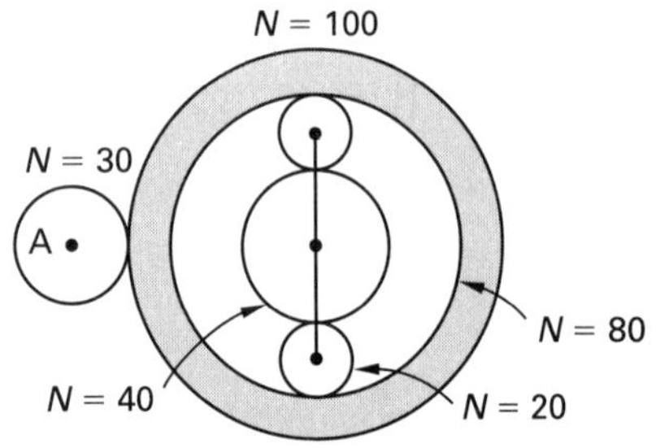

(A) 30 rpm

(B) 60 rpm

(C) 90 rpm

(D) 120 rpm

7. An epicyclic gear train consists of a ring gear, three planets, and a fixed sun gear. 15 hp (11 kW) are transmitted through the input ring gear, which turns clockwise at 1500 rpm. The diametral pitch is 10 per inch with a 20° pressure angle. The pitch diameters are at 5 in, $2^1/_2$ in, and 10 in (127 mm, 63.5 mm, 254 mm) for the sun, planets, and ring gear, respectively.

(a) Approximately how many teeth are on the sun, planet, and ring gears, respectively?

(A) 25, 50, 50

(B) 50, 50, 100

(C) 50, 25, 100

(D) 100, 25, 50

(b) What are most nearly the velocities of the sun, ring, and carrier gears, respectively?

(A) 0 rpm, 1500 rpm, 1000 rpm

(B) 0 rpm, 1000 rpm, 1500 rpm

(C) 1000 rpm, 1500 rpm, 1500 rpm

(D) 1500 rpm, 0 rpm, 1500 rpm

(c) In what directions do the ring, carrier, and sun gears turn?

(A) The ring is fixed, the carrier turns clockwise, and the sun turns counterclockwise.

(B) The ring and carrier turn clockwise, but the sun is fixed.

(C) The ring turns clockwise, and the carrier and sun are fixed.

(D) The ring turns counterclockwise, the carrier is fixed, and the sun turns clockwise.

(d) Approximately what torques are on the input and output shafts, respectively?

(A) 50 ft-lbf (70 N·m); 80 ft-lbf (110 N·m)

(B) 60 ft-lbf (80 N·m); 50 ft-lbf (70 N·m)

(C) 80 ft-lbf (110 N·m); 60 ft-lbf (80 N·m)

(D) 90 ft-lbf (120 N·m); 80 ft-lbf (110 N·m)

8. The epicyclic gear set shown has an overall speed reduction of 3:1. The planet carrier is the driven element. The driving gear is the sun gear, which turns clockwise at 1000 rpm. The planet has 20 teeth and a diametral pitch of 10.

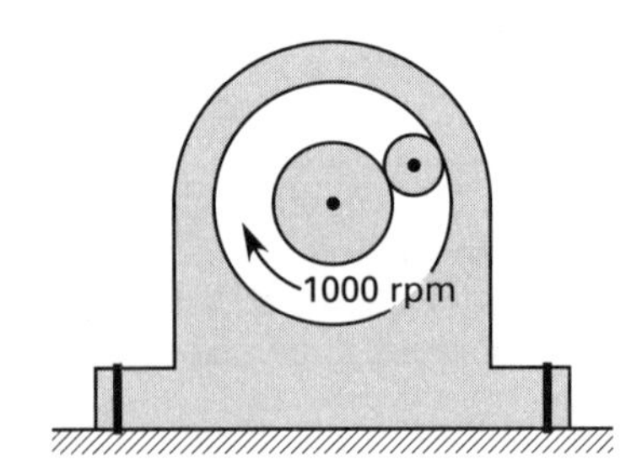

(a) The train value is most nearly

(A) −2

(B) −1

(C) 1

(D) 2

(b) The ratio of the numbers of teeth on the ring and sun gears is most nearly

(A) 1

(B) 2

(C) 3

(D) 4

(c) Disregarding direction of rotation, the angular velocity of the planet with respect to the carrier is most nearly

(A) 330 rpm

(B) 670 rpm

(C) 1000 rpm

(D) 1200 rpm

(d) Disregarding direction of rotation, the angular velocity of the planet with respect to the sun gear's axis of rotation is most nearly

(A) 330 rpm

(B) 670 rpm

(C) 1000 rpm

(D) 1300 rpm

9. A gear train is constructed as shown. The output turns at 250 rpm counterclockwise.

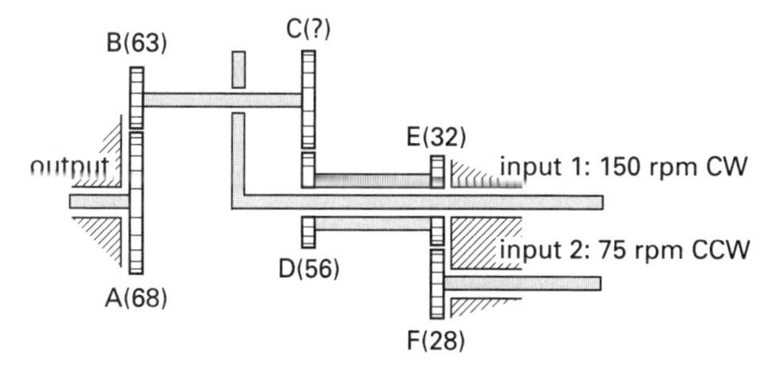

(a) At what approximate speed will input no. 2 cause gear D–E to rotate?

(A) 55 rpm

(B) 65 rpm

(C) 75 rpm

(D) 85 rpm

(b) Approximately how many teeth does gear C have?

(A) 11 teeth

(B) 14 teeth

(C) 21 teeth

(D) 38 teeth

Differentials

10. An automobile differential gear set is arranged as shown. Gear C is turned by the automobile driveshaft. The axle shafts are driven by gears A and B.

gear A: 30 teeth, 50 rpm counterclockwise, as viewed from the gear looking up the shaft
gear B: 30 teeth
gear C: 18 teeth, 600 rpm counterclockwise, as viewed from the gear looking up the shaft
gear D: 54 teeth
gear E: 15 teeth
gear F: 15 teeth

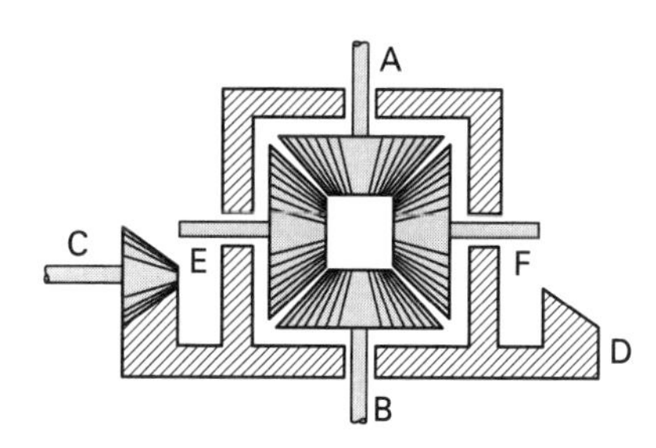

(a) Relative to gear C, the algebraic speed of rotation of gear D is most nearly

(A) −600 rpm

(B) −200 rpm

(C) 50 rpm

(D) 200 rpm

(b) The train value is most nearly

(A) −1

(B) −0.5

(C) 1

(D) 2

(c) The speed of rotation for gear B is most nearly

(A) −500 rpm

(B) −350 rpm

(C) −100 rpm

(D) 130 rpm

(d) What is the direction of rotation for gear B?

(A) clockwise and in the opposite direction of gear C

(B) clockwise and in the same direction as gear C

(C) counterclockwise and in the same direction as gear C

(D) counterclockwise and in the opposite direction of gear C

Cams

11. A cam is turning at a constant speed of 120 rpm. A radial follower starting from rest rises 0.5 in (12 mm) with constant acceleration as the cam turns 60°. The follower returns to rest with constant acceleration during the next 90° of cam movement.

(a) The magnitude of the acceleration during the first 60° is most nearly

(A) 60 in/sec^2 (1.4 m/s^2)

(B) 92 in/sec^2 (2.2 m/s^2)

(C) 120 in/sec^2 (2.9 m/s^2)

(D) 140 in/sec^2 (3.5 m/s^2)

(b) The magnitude of the deceleration during the last 90° is most nearly

(A) 80 in/sec^2 (1.9 m/s^2)

(B) 100 in/sec^2 (2.3 m/s^2)

(C) 120 in/sec^2 (2.9 m/s^2)

(D) 150 in/sec^2 (3.6 m/s^2)

(c) Approximately how far does the follower move during the 150° of rotation?

(A) 0.5 in (12 mm)

(B) 0.8 in (19 mm)

(C) 1.0 in (25 mm)

(D) 1.3 in (30 mm)

Linkages

12. Bars r_0, r_1, r_2, and r_3 constitute a four-bar linkage as shown. With the origin, O, at the base of bar r_3, and positive directions upward and to the right, what are the coordinates of the rotating end of r_3 (point B) when θ_1 is 45°?

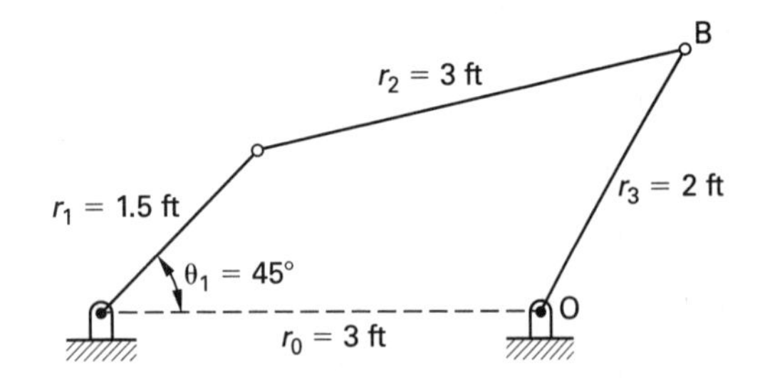

(A) $x = 1.0$ ft; $y = 1.7$ ft

(B) $x = 0.88$ ft; $y = 1.5$ ft

(C) $x = 0.67$ ft; $y = 1.5$ ft

(D) $x = 0.64$ ft; $y = 1.9$ ft

13. Bars r_0, r_1, r_2, and r_3 constitute a four-bar linkage as shown. At a particular moment, θ_1 is 45°, θ_2 is 13.1°, ω_1 is 150 rad/sec, and ω_2 is –27.2 rad/sec. Using the shown unit vectors **i**, **j**, and **k**, what is most nearly the velocity of point B at that moment?

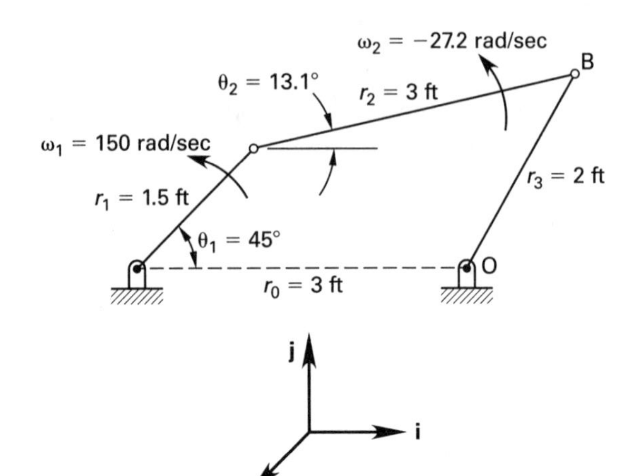

(A) 110 ft/sec∠140°

(B) 130 ft/sec∠150°

(C) 150 ft/sec∠150°

(D) 160 ft/sec∠150°

SOLUTIONS

1. The magnitude of the total acceleration is found by summing the components of normal acceleration, a_n, and tangential acceleration, a_t.

$$a_n = \omega^2 L = \left(10\ \frac{\text{rad}}{\text{sec}}\right)^2 (3\ \text{ft}) = 300\ \text{ft/sec}^2$$

$$a_t = \alpha L = \left(-20\ \frac{\text{rad}}{\text{sec}^2}\right)(3\ \text{ft}) = -60\ \text{ft/sec}^2$$

$$a = \sqrt{a_n^2 + a_t^2} = \sqrt{\left(300\ \frac{\text{ft}}{\text{sec}^2}\right)^2 + \left(-60\ \frac{\text{ft}}{\text{sec}^2}\right)^2}$$
$$= \boxed{306\ \text{ft/sec}^2 \quad (310\ \text{ft/sec}^2)}$$

The answer is (B).

2. Use the equation for the angular velocity of the lever, ω. When the rotational speed is zero, $\omega = 0$. Rearranging the equation to solve for time, t,

$$\omega = \omega_0 + \alpha t$$

$$t = \frac{-\omega_0}{\alpha} = \frac{-100\ \frac{\text{rad}}{\text{sec}}}{-5\ \frac{\text{rad}}{\text{sec}^2}} = 20\ \text{sec}$$

The angular distance traveled by the lever, θ, is

$$\theta = \theta_0 + \omega_0 t + \frac{\alpha t^2}{2}$$
$$= 0 + \left(100\ \frac{\text{rad}}{\text{sec}}\right)(20\ \text{sec}) + \frac{\left(-5\ \frac{\text{rad}}{\text{sec}^2}\right)(20\ \text{sec})^2}{2}$$
$$= 1000\ \text{rad}$$

The distance traveled by point A, L_A, is

$$L_A = \theta L = (1000\ \text{rad})(4\ \text{ft}) = \boxed{4000\ \text{ft}}$$

The answer is (C).

3. Use the traditional definitions of unit vectors **i**, **j**, and **k**. The rotation is counterclockwise since both components of the velocity are negative. The magnitude of the angular velocity of the lever, ω, is found by dividing the magnitude of the velocity by the length of the link.

$$\omega = \frac{\text{v}}{L} = \frac{\sqrt{\text{v}_x^2 + \text{v}_y^2}}{L} = \frac{\sqrt{\left(-70.5\ \frac{\text{ft}}{\text{sec}}\right)^2 + \left(-25.5\ \frac{\text{ft}}{\text{sec}}\right)^2}}{1.5\ \text{ft}}$$
$$= 50\ \text{rad/sec}$$

The position coordinates of OB_x and OB_y are

$$\text{OB}_x = \text{OB}\cos\theta = (2.5\ \text{ft})\cos 40° = 1.92\ \text{ft}$$
$$\text{OB}_y = \text{OB}\sin\theta = (2.5\ \text{ft})\sin 40° = 1.61\ \text{ft}$$

Unit vector **k** is oriented along the rotational axis. The velocity of point B, $\mathbf{v}_B$, is found from the cross product of the angular velocity and the link length.

$$\mathbf{v}_B = \omega \times \mathbf{OB}$$

Calculate the vector cross product.

$$\mathbf{v}_B = \begin{vmatrix} \mathbf{i} & \mathbf{j} & \mathbf{k} \\ 0 & 0 & 50\ \text{rad/sec} \\ 1.92\ \text{ft} & 1.61\ \text{ft} & 0 \end{vmatrix}$$
$$= \boxed{-80.5\mathbf{i} + 96.0\mathbf{j}\ \text{ft/sec}}$$

The answer is (B).

4. *Customary U.S. Solution*

Treat the flywheel as a rotating hub. First, consider the maximum tangential stress.

For a factor of safety of 10,

$$\sigma_{t,\text{max}} = \frac{S_{ut}}{\text{FS}} = \frac{30{,}000\ \frac{\text{lbf}}{\text{in}^2}}{10} = 3000\ \text{lbf/in}^2$$

Use Eq. 59.19 to solve for ω.

$$\omega = \sqrt{\frac{4 g_c \sigma_{t,\text{max}}}{\rho\left((3+\nu)r_o^2 + (1-\nu)r_i^2\right)}}$$
$$= \sqrt{\frac{(4)\left(32.2\ \frac{\text{lbm-ft}}{\text{lbf-sec}^2}\right)\left(12\ \frac{\text{in}}{\text{ft}}\right)\left(3000\ \frac{\text{lbf}}{\text{in}^2}\right)}{\left(0.26\ \frac{\text{lbm}}{\text{in}^3}\right)\left((3+0.27)\left(\frac{20\ \text{in}}{2}\right)^2 + (1-0.27)\left(\frac{4\ \text{in}}{2}\right)^2\right)}}$$
$$= 232.5\ \text{rad/sec}$$

$$n = \frac{\omega}{2\pi} = \frac{\left(232.5\ \frac{\text{rad}}{\text{sec}}\right)\left(60\ \frac{\text{sec}}{\text{min}}\right)}{2\pi\ \frac{\text{rad}}{\text{rev}}}$$
$$= 2220\ \text{rev/min}$$

Dynamics and Vibrations

Next, consider the maximum radial stress. Use Eq. 59.20 to solve for ω. Use the same maximum stress.

$$\omega = \sqrt{\frac{8g_c\sigma_{r,\max}}{\rho(3+\nu)(r_o-r_i)^2}}$$

$$= \sqrt{\frac{(8)\left(32.2\ \frac{\text{lbm-ft}}{\text{lbf-sec}^2}\right)\left(12\ \frac{\text{in}}{\text{ft}}\right)\left(3000\ \frac{\text{lbf}}{\text{in}^2}\right)}{\left(0.26\ \frac{\text{lbm}}{\text{in}^3}\right)(3+0.27)\left(\frac{20\text{ in}}{2}-\frac{4\text{ in}}{2}\right)^2}}$$

$$= 412.8\text{ rad/sec}$$

$$n = \frac{\omega}{2\pi} = \frac{\left(412.8\ \frac{\text{rad}}{\text{sec}}\right)\left(60\ \frac{\text{sec}}{\text{min}}\right)}{2\pi\ \frac{\text{rad}}{\text{rev}}}$$

$$= 3942\text{ rev/min}$$

The flywheel's maximum safe speed is limited by tangential stress and is $\boxed{2220\text{ rpm (2200 rpm)}}$.

The von Mises stress is

$$\sigma' = \sqrt{\sigma_r^2+\sigma_t^2-\sigma_r\sigma_t}$$

Using the von Mises criterion requires knowing the radial and tangential stresses at the same point. The stresses obtained from Eq. 59.19 and Eq. 59.20 are not at the same point.

The answer is (B).

SI Solution

Treat the flywheel as a rotating hub. First, consider the maximum tangential stress.

For a factor of safety of 10,

$$\sigma_{t,\max} = \frac{S_{ut}}{\text{FS}} = \frac{207\times10^6\text{ Pa}}{10} = 20.7\times10^6\text{ Pa}$$

Use Eq. 59.19 to solve for ω.

$$\omega = \sqrt{\frac{4\sigma_{t,\max}}{\rho\left((3+\nu)r_o^2+(1-\nu)r_i^2\right)}}$$

$$= \sqrt{\frac{(4)(20.7\times10^6\text{ Pa})}{\left(7200\ \frac{\text{kg}}{\text{m}^3}\right)\left((3+0.27)\left(\frac{0.51\text{ m}}{2}\right)^2 + (1-0.27)\left(\frac{0.102\text{ m}}{2}\right)^2\right)}}$$

$$= 231.5\text{ rad/s}$$

$$n = \frac{\omega}{2\pi} = \frac{\left(231.5\ \frac{\text{rad}}{\text{s}}\right)\left(60\ \frac{\text{s}}{\text{min}}\right)}{2\pi\ \frac{\text{rad}}{\text{rev}}}$$

$$= 2211\text{ rev/min}$$

Next, consider the maximum radial stress. Use Eq. 59.20 to solve for ω. Use the same maximum stress.

$$\omega = \sqrt{\frac{8\sigma_{r,\max}}{\rho(3+\nu)(r_o-r_i)^2}}$$

$$= \sqrt{\frac{(8)(20.7\times10^6\text{ Pa})}{\left(7200\ \frac{\text{kg}}{\text{m}^3}\right)(3+0.27)\left(\frac{0.51\text{ m}}{2}-\frac{0.102\text{ m}}{2}\right)^2}}$$

$$= 411.1\text{ rad/s}$$

$$n = \frac{\omega}{2\pi} = \frac{\left(411.1\ \frac{\text{rad}}{\text{s}}\right)\left(60\ \frac{\text{s}}{\text{min}}\right)}{2\pi\ \frac{\text{rad}}{\text{rev}}}$$

$$= 3926\text{ rev/min}$$

The flywheel's maximum safe speed is limited by tangential stress and is $\boxed{2211\text{ rpm (2200 rpm)}}$.

The von Mises stress is

$$\sigma' = \sqrt{\sigma_r^2+\sigma_t^2-\sigma_r\sigma_t}$$

Using the von Mises criterion requires knowing the radial and tangential stresses at the same point. The stresses obtained from Eq. 59.19 and Eq. 59.20 are not at the same point.

The answer is (B).

5. If the arm was locked and the ring was free to rotate, the sun and ring gears would rotate in different directions. Therefore, the train value is negative.

From Eq. 59.34,

$$\text{TV} = \frac{N_{\text{ring}}}{N_{\text{sun}}} = -\frac{104\text{ teeth}}{24\text{ teeth}} = -4.333$$

From Eq. 59.35, the rotational velocity of the sun is

$$\omega_{\text{sun}} = (\text{TV})\omega_{\text{ring}} + (1-\text{TV})\omega_{\text{carrier}}$$

Since the ring gear is fixed, $\omega_{\text{ring}} = 0$, and the rotational velocity of the carrier is

$$\omega_{\text{carrier}} = \frac{\omega_{\text{sun}}}{1-\text{TV}} = \frac{50\ \frac{\text{rev}}{\text{min}}}{1-(-4.333)}$$

$$= \boxed{9.38\text{ rpm}\quad(9.4\text{ rpm})}$$

The answer is (A).

6. Since the ring gear rotates in a different direction from gear A, the rotational velocity of the ring gear is

$$\omega_{\text{ring}} = \omega_{\text{A}}\left(-\frac{N_{\text{A}}}{N_{\text{ring}}}\right) = \left(-100\ \frac{\text{rev}}{\text{min}}\right)\left(-\frac{30\ \text{teeth}}{100\ \text{teeth}}\right)$$
$$= 30\ \text{rev/min}\quad [\text{clockwise}]$$

Since the ring and sun gears rotate in different directions, the train value is negative. From Eq. 59.34,

$$\text{TV} = -\frac{N_{\text{ring}}}{N_{\text{sun}}} = -\frac{80\ \text{teeth}}{40\ \text{teeth}} = -2$$

From Eq. 59.35, the rotational velocity of the sun gear is

$$\omega_{\text{sun}} = (\text{TV})\omega_{\text{ring}} + (1-\text{TV})\omega_{\text{carrier}}$$
$$= (-2)\left(30\ \frac{\text{rev}}{\text{min}}\right) + (1-(-2))\left(60\ \frac{\text{rev}}{\text{min}}\right)$$
$$= \boxed{120\ \text{rpm}}\quad [\text{clockwise}]$$

The answer is (D).

7. *Customary U.S. Solution*

(a) From Eq. 54.35, the number of teeth are as follows.

For the sun gear,

$$N_{\text{sun}} = Pd = \left(10\ \frac{\text{teeth}}{\text{in}}\right)(5\ \text{in}) = \boxed{50\ \text{teeth}}$$

For the planet gears,

$$N_{\text{planet}} = Pd = \left(10\ \frac{\text{teeth}}{\text{in}}\right)(2.5\ \text{in}) = \boxed{25\ \text{teeth}}$$

For the ring gear,

$$N_{\text{ring}} = Pd = \left(10\ \frac{\text{teeth}}{\text{in}}\right)(10\ \text{in}) = \boxed{100\ \text{teeth}}$$

The answer is (C).

(b) Since the sun gear is fixed, its rotational speed is zero $\boxed{(\omega_{\text{sun}} = 0\ \text{rpm}).}$

The rotational speed of the ring gear is $\omega_{\text{ring}} = \boxed{1500\ \text{rpm}.}$

Since the ring and sun gears rotate in different directions, the train value is negative. From Eq. 59.34,

$$\text{TV} = -\frac{N_{\text{ring}}}{N_{\text{sun}}} = -\frac{100\ \text{teeth}}{50\ \text{teeth}} = -2$$

From Eq. 59.35, the rotational speed of the sun gear is

$$\omega_{\text{sun}} = (\text{TV})\omega_{\text{ring}} + (1-\text{TV})\omega_{\text{carrier}}$$
$$0\ \frac{\text{rev}}{\text{min}} = (-2)\left(1500\ \frac{\text{rev}}{\text{min}}\right) + (1-(-2))\omega_{\text{carrier}}$$

Solve for the rotational speed of the carrier gear.

$$\omega_{\text{carrier}} = \frac{(2)\left(1500\ \frac{\text{rev}}{\text{min}}\right)}{3} = \boxed{1000\ \text{rpm}}$$

The answer is (A).

(c) From part (b), the direction of rotation of the ring and carrier gears is $\boxed{\text{clockwise.}}$ The sun gear is $\boxed{\text{fixed.}}$

The answer is (B).

(d) From Eq. 59.12(b), the torque on the input shaft is

$$T_{\text{in,ft-lbf}} = \frac{P_{\text{hp}}\left(63{,}025\ \frac{\text{in-lbf}}{\text{hp-min}}\right)}{n_{\text{rpm,ring}}}$$
$$= \frac{(15\ \text{hp})\left(63{,}025\ \frac{\text{in-lbf}}{\text{hp-min}}\right)}{\left(1500\ \frac{\text{rev}}{\text{min}}\right)\left(12\ \frac{\text{in}}{\text{ft}}\right)}$$
$$= \boxed{52.52\ \text{ft-lbf}\quad (50\ \text{ft-lbf})}$$

The torque on the output shaft is

$$T_{\text{out,ft-lbf}} = \frac{P_{\text{hp}}\left(63{,}025\ \frac{\text{in-lbf}}{\text{hp-min}}\right)}{n_{\text{carrier}}}$$
$$= \frac{(15\ \text{hp})\left(63{,}025\ \frac{\text{in-lbf}}{\text{hp-min}}\right)}{\left(1000\ \frac{\text{rev}}{\text{min}}\right)\left(12\ \frac{\text{in}}{\text{ft}}\right)}$$
$$= \boxed{78.78\ \text{ft-lbf}\quad (80\ \text{ft-lbf})}$$

The answer is (A).

SI Solution

(a) From Eq. 54.35, the number of teeth are as follows.

For the sun gear,

$$N_{\text{sun}} = Pd = \frac{\left(10\ \frac{\text{teeth}}{\text{in}}\right)(127\ \text{mm})}{25.4\ \frac{\text{mm}}{\text{in}}} = \boxed{50\ \text{teeth}}$$

For the planet gears,

$$N_{\text{planet}} = Pd = \frac{\left(10\ \frac{\text{teeth}}{\text{in}}\right)(63.5\ \text{mm})}{25.4\ \frac{\text{mm}}{\text{in}}} = \boxed{25\ \text{teeth}}$$

Dynamics and Vibrations

For the ring gear,

$$N_{\text{ring}} = Pd = \frac{\left(10\ \frac{\text{teeth}}{\text{in}}\right)(254\ \text{mm})}{25.4\ \frac{\text{mm}}{\text{in}}} = \boxed{100\ \text{teeth}}$$

The answer is (C).

(b) Since the sun gear is fixed, its rotational velocity is zero $\boxed{(\omega_{\text{sun}} = 0\ \text{rpm}).}$

The rotational speed of the ring gear is $\omega_{\text{ring}} = \boxed{1500\ \text{rpm}.}$

Since the ring and sun gears rotate in different directions, the train value is negative. From Eq. 59.34,

$$\begin{aligned}\text{TV} &= -\frac{N_{\text{ring}}}{N_{\text{sun}}} = -\frac{100\ \text{teeth}}{50\ \text{teeth}}\\ &= -2\end{aligned}$$

From Eq. 59.35, the rotational velocity of the sun gear is

$$\omega_{\text{sun}} = (\text{TV})\omega_{\text{ring}} + (1 - \text{TV})\omega_{\text{carrier}}$$

$$0\ \frac{\text{rev}}{\text{min}} = (-2)\left(1500\ \frac{\text{rev}}{\text{min}}\right) + \big(1 - (-2)\big)\omega_{\text{carrier}}$$

Solve for the rotational velocity of the carrier gear.

$$\omega_{\text{carrier}} = \frac{(2)\left(1500\ \frac{\text{rev}}{\text{min}}\right)}{3} = \boxed{1000\ \text{rpm}}$$

The answer is (A).

(c) From part (b), the direction of rotation of the ring and carrier gears is $\boxed{\text{clockwise.}}$ The sun gear is $\boxed{\text{fixed.}}$

The answer is (B).

(d) From Eq. 59.12(a), the torque on the input shaft is

$$\begin{aligned}T_{\text{in,N·m}} &= \frac{P_{\text{kW}}\left(9549\ \frac{\text{N·m}}{\text{kW·min}}\right)}{n_{\text{rpm,ring}}}\\ &= \frac{(11\ \text{kW})\left(9549\ \frac{\text{N·m}}{\text{kW·min}}\right)}{1500\ \frac{\text{rev}}{\text{min}}}\\ &= \boxed{70.03\ \text{N·m}\quad (70\ \text{N·m})}\end{aligned}$$

The torque on the output shaft is

$$\begin{aligned}T_{\text{out,N·m}} &= \frac{P_{\text{kW}}\left(9549\ \frac{\text{N·m}}{\text{kW·min}}\right)}{n_{\text{carrier}}}\\ &= \frac{(11\ \text{kW})\left(9549\ \frac{\text{N·m}}{\text{kW·min}}\right)}{1000\ \frac{\text{rev}}{\text{min}}}\\ &= \boxed{105\ \text{N·m}\quad (110\ \text{N·m})}\end{aligned}$$

The answer is (A).

8. (a) From the overall speed reduction of 3:1, the speed of the carrier is

$$\begin{aligned}\omega_{\text{carrier}} &= \frac{\omega_{\text{sun}}}{3} = \frac{1000\ \frac{\text{rev}}{\text{min}}}{3}\\ &= 333.3\ \text{rev/min}\end{aligned}$$

The speed of the ring gear is zero. From Eq. 59.34,

$$\begin{aligned}\text{TV} &= \frac{\omega_{\text{sun}} - \omega_{\text{carrier}}}{\omega_{\text{ring}} - \omega_{\text{carrier}}} = \frac{1000\ \frac{\text{rev}}{\text{min}} - 333.3\ \frac{\text{rev}}{\text{min}}}{0\ \frac{\text{rev}}{\text{min}} - 333.3\ \frac{\text{rev}}{\text{min}}}\\ &= \boxed{-2}\end{aligned}$$

The answer is (A).

(b) From Eq. 59.34, the ratio of the number of teeth on the ring and sun gears is

$$\frac{N_{\text{ring}}}{N_{\text{sun}}} = \text{TV} = \boxed{2}\quad [\text{negative sign is not required}]$$

The answer is (B).

(c) Assume $N_{\text{sun}} = 40$ teeth and $N_{\text{ring}} = 80$ teeth to satisfy part (a).

Then, from Eq. 54.35,

$$d_{\text{sun}} = \frac{N_{\text{sun}}}{P} = \frac{40\ \text{teeth}}{10\ \frac{\text{teeth}}{\text{in}}} = 4\ \text{in}$$

$$d_{\text{ring}} = \frac{N_{\text{ring}}}{P} = \frac{80\ \text{teeth}}{10\ \frac{\text{teeth}}{\text{in}}} = 8\ \text{in}$$

$$d_{\text{planet}} = \frac{N_{\text{planet}}}{P} = \frac{20\ \text{teeth}}{10\ \frac{\text{teeth}}{\text{in}}} = 2\ \text{in}$$

Check the sum of the gear diameters.

$$d_{\text{ring}} = d_{\text{sun}} + 2d_{\text{planet}}$$
$$8 \text{ in} = 4 \text{ in} + (2)(2 \text{ in}) = 8 \text{ in}$$

Since the gears fit, $N_{\text{sun}} = 40$ teeth and $N_{\text{ring}} = 80$ teeth is a valid solution.

The speed of the planet with respect to the carrier is the speed of the planet with a stopped carrier, the same as the speed of the planet with respect to its own axis of rotation.

Use Eq. 59.36 to solve for ω_{planet}.

$$\begin{aligned}\omega_{\text{planet}} &= \omega_{\text{carrier}} - \left(\frac{N_{\text{sun}}}{N_{\text{planet}}}\right)(\omega_{\text{sun}} - \omega_{\text{carrier}}) \\ &= 333.3\ \frac{\text{rev}}{\text{min}} \\ &\quad - \left(\frac{40 \text{ teeth}}{20 \text{ teeth}}\right)\left(1000\ \frac{\text{rev}}{\text{min}} - 333.3\ \frac{\text{rev}}{\text{min}}\right) \\ &= \boxed{-1000 \text{ rpm}}\end{aligned}$$

The negative sign indicates that the planet rotates in an opposite direction to the sun.

The answer is (C).

(d) The speed of the planet with respect to the sun gear's rotational axis is

$$\begin{aligned}\omega_{\text{planet}|\text{sun}} &= \omega_{\text{planet}} + \omega_{\text{sun carrier}} \\ &= -1000\ \frac{\text{rev}}{\text{min}} + 333.3\ \frac{\text{rev}}{\text{min}} \\ &= \boxed{-666.7 \text{ rpm} \quad (670 \text{ rpm})}\end{aligned}$$

The negative sign indicates that the planet rotates in an opposite direction to the sun.

The answer is (B).

9. (a) First, simplify the problem. Input no. 2 causes gear D–E to rotate at

$$\begin{aligned}\omega_{\text{D}} &= \left(-75\ \frac{\text{rev}}{\text{min}}\right)\left(-\frac{N_{\text{F}}}{N_{\text{E}}}\right) \\ &= \left(-75\ \frac{\text{rev}}{\text{min}}\right)\left(-\frac{28 \text{ teeth}}{32 \text{ teeth}}\right) \\ &= \boxed{65.625 \text{ rpm} \quad (65 \text{ rpm})}\end{aligned}$$

The answer is (B).

(b) Use the eight-step procedure from Sec. 59.15.

step 1: Gears A and D have the same center of rotation as the arm.

	A	D
row 1		
row 2		
row 3		

step 2: Write ω_{carrier} in the first row.

	A	D
row 1	ω_{carrier}	ω_{carrier}
row 2		
row 3		

step 3: Arbitrarily select gear D as the gear with unknown speed.

	A	D
row 1	ω_{carrier}	ω_{carrier}
row 2		
row 3		ω_{D}

step 4:

	A	D
row 1	ω_{carrier}	ω_{carrier}
row 2		$\omega_{\text{D}} - \omega_{\text{carrier}}$
row 3		ω_{D}

step 5: The translational path D to A is a compound mesh.

From Fig. 59.10,

$$\begin{aligned}\omega_{\text{A}} &= \omega_{\text{D}}\left(\frac{N_{\text{B}}N_{\text{D}}}{N_{\text{A}}N_{\text{C}}}\right) = \omega_{\text{D}}\left(\frac{(63 \text{ teeth})(56 \text{ teeth})}{(68 \text{ teeth})N_{\text{C}}}\right) \\ &= 51.8823\omega_{\text{D}}/N_{\text{C}}\end{aligned}$$

step 6: With respect to gear D, the speed ratio of gear A is $51.8823/N_{\text{C}}$.

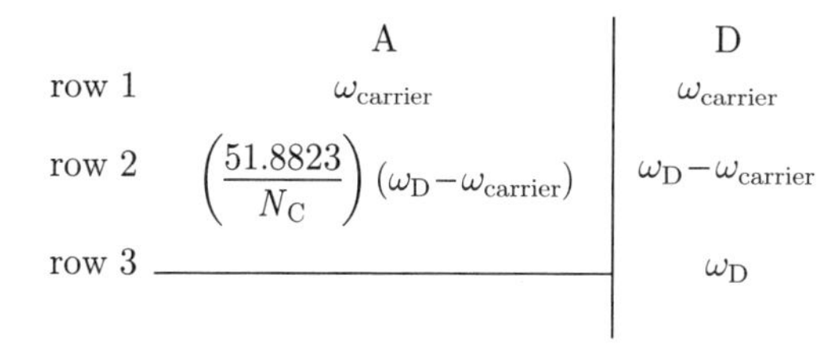

	A	D
row 1	ω_{carrier}	ω_{carrier}
row 2	$\left(\frac{51.8823}{N_{\text{C}}}\right)(\omega_{\text{D}} - \omega_{\text{carrier}})$	$\omega_{\text{D}} - \omega_{\text{carrier}}$
row 3		ω_{D}

step 7: Insert the known values for ω_{carrier} and ω_D.

	A	D
row 1	150	150
row 2	$\left(\frac{51.8823}{N_C}\right)(65.625-150)$	$65.625-150$
row 3	-250	65.625

step 8: From Eq. 59.37, the characteristic equation for column 1 is

$$\text{row 1} + \text{row 2} = \text{row 3}$$

$$150\ \frac{\text{rev}}{\text{min}} + \left(\frac{51.8823}{N_C}\right) \times \left(65.525\ \frac{\text{rev}}{\text{min}} - 150\ \frac{\text{rev}}{\text{min}}\right) = -250\ \frac{\text{rev}}{\text{min}}$$

$$N_C = \boxed{10.94 \quad (11 \text{ teeth})}$$

The answer is (A).

10. (This problem can be solved with Eq. 59.38.) The automobile differential is equivalent to the imaginary gear set shown.

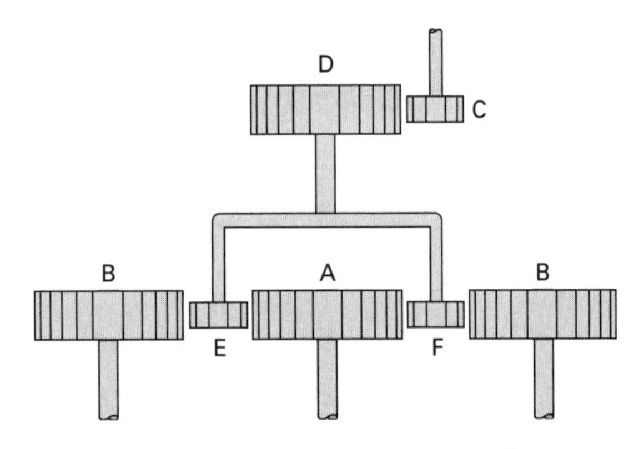

This is the same as a conventional epicyclic gear train except that the ring gear is replaced with two external gears B.

(a) Start with $\omega_C = +600$ rev/min.

Gear D rotates in the opposite direction of gear C.

$$\omega_D = -\left(\frac{N_C}{N_D}\right)\omega_C = -\left(\frac{18 \text{ teeth}}{54 \text{ teeth}}\right)\left(600\ \frac{\text{rev}}{\text{min}}\right) = \boxed{-200 \text{ rpm}}$$

The answer is (B).

(b) Gear A rotates in the same direction as gear D.

$$\omega_A = -50 \text{ rev/min}$$

From the actual gear set, turn gear A. Then gear B moves in the opposite direction of gear A. Thus, the train value is negative. From Eq. 59.34,

$$\text{TV} = -\frac{N_B}{N_A} = -\frac{30 \text{ teeth}}{30 \text{ teeth}} = \boxed{-1}$$

The answer is (A).

(c) From Eq. 59.35,

$$\omega_A = (\text{TV})\omega_B + (1 - \text{TV})\omega_D$$

The speed of rotation for gear B is most nearly

$$\omega_B = \frac{\omega_A - (1 - \text{TV})\omega_D}{\text{TV}} = \frac{-50\ \frac{\text{rev}}{\text{min}} - (1-(-1.0))\left(-200\ \frac{\text{rev}}{\text{min}}\right)}{-1.0} = \boxed{-350 \text{ rpm}}$$

The answer is (B).

(d) Gear B rotates in a direction $\boxed{\text{opposite}}$ to gear C; that is, it rotates $\boxed{\text{clockwise.}}$

The answer is (A).

11. *Customary U.S. Solution*

(a) The angular velocity is

$$\omega = 2\pi n_{\text{rps}} = \frac{(2\pi)\left(120\ \frac{\text{rev}}{\text{min}}\right)}{60\ \frac{\text{sec}}{\text{min}}} = 12.57 \text{ rad/sec}$$

The time to turn 60° is

$$t = \frac{\theta}{\omega} = \frac{(60^\circ)\left(\frac{\pi \text{ rad}}{180^\circ}\right)}{12.57\ \frac{\text{rad}}{\text{sec}}} = 0.08331 \text{ sec}$$

Using the uniform acceleration formula $x = \frac{1}{2}at^2$, the constant acceleration during the first 60° is

$$a = \frac{2x}{t^2} = \frac{(2)(0.5 \text{ in})}{(0.08331 \text{ sec})^2} = \boxed{144.08 \text{ in/sec}^2 \quad (140 \text{ in/sec}^2)}$$

The answer is (D).

(b) The velocity at the end of the first 60° of rotation is

$$\mathrm{v} = at = \left(144.08\ \frac{\text{in}}{\text{sec}^2}\right)(0.08331\ \text{sec}) = 12.0\ \text{in/sec}$$

The time between $\theta = 60°$ and $\theta = 150°$ is

$$\Delta t = \frac{\Delta\theta}{\omega} = \frac{(150° - 60°)\left(\frac{\pi\ \text{rad}}{180°}\right)}{12.57\ \frac{\text{rad}}{\text{sec}}} = 0.1250\ \text{sec}$$

Since the cam follower returns to rest at $\theta = 150°$, its velocity is zero there. The constant deceleration during the last 90° is

$$a = \frac{\Delta \mathrm{v}}{\Delta t} = \frac{0\ \frac{\text{in}}{\text{sec}} - 12.0\ \frac{\text{in}}{\text{sec}}}{0.1250\ \text{sec}} = \boxed{-96\ \text{in/sec}^2 \quad (100\ \text{in/scc}^2)}$$

The answer is (B).

(c) The distance moved by the cam follower between $\theta = 60°$ and $\theta = 150°$ is

$$x = \tfrac{1}{2}at^2 = \left(\tfrac{1}{2}\right)\left(95.92\ \frac{\text{in}}{\text{sec}^2}\right)(0.1250\ \text{sec})^2 = 0.75\ \text{in}$$

The total follower movement for 150° of rotation is

$$x_{\text{total}} = 0.50\ \text{in} + 0.75\ \text{in} = \boxed{1.25\ \text{in} \quad (1.3\ \text{in})}$$

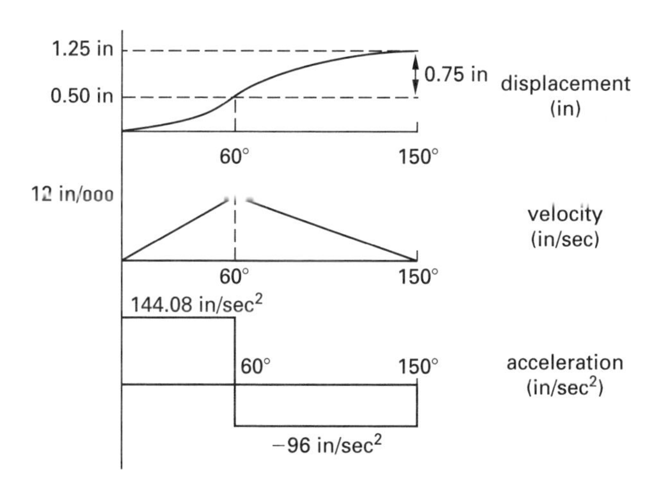

The answer is (D).

SI Solution

(a) The angular velocity is

$$\omega = 2\pi n_{\text{rps}} = \frac{(2\pi)\left(120\ \frac{\text{rev}}{\text{min}}\right)}{60\ \frac{\text{s}}{\text{min}}} = 12.57\ \text{rad/s}$$

The time to turn 60° is

$$t = \frac{\theta}{\omega} = \frac{(60°)\left(\frac{\pi\ \text{rad}}{180°}\right)}{12.57\ \frac{\text{rad}}{\text{s}}} = 0.08331\ \text{s}$$

Using the uniform acccleration formula $x = \frac{1}{2}at^2$, the constant acceleration during the first 60° is

$$a = \frac{2x}{t^2} = \frac{(2)(0.012\ \text{m})}{(0.08331\ \text{s})^2} = \boxed{3.458\ \text{m/s}^2 \quad (3.5\ \text{m/s}^2)}$$

The answer is (D).

(b) The velocity during the first 60° is

$$\mathrm{v} = at = \left(3.458\ \frac{\text{m}}{\text{s}^2}\right)(0.08331\ \text{s}) = 0.288\ \text{m/s}$$

The time between $\theta = 60°$ and $\theta = 150°$ is

$$\Delta t = \frac{\Delta\theta}{\omega} = \frac{(150° - 60°)\left(\frac{\pi\ \text{rad}}{180°}\right)}{12.57\ \frac{\text{rad}}{\text{s}}} = 0.1250\ \text{s}$$

Since the cam follower returns to rest at $\theta = 150°$, its velocity is zero there. The constant deceleration during the last 90° is

$$a = \frac{\Delta \mathrm{v}}{\Delta t} = \frac{0\ \frac{\text{m}}{\text{s}} - 0.288\ \frac{\text{m}}{\text{s}}}{0.1250\ \text{s}} = \boxed{-2.3\ \text{m/s}^2}$$

The answer is (B).

(c) The distance moved by the cam follower between $\theta = 60°$ and $\theta = 150°$ is

$$x = \tfrac{1}{2}at^2 = \left(\tfrac{1}{2}\right)\left(2.30\ \frac{\text{m}}{\text{s}^2}\right)(0.1250\ \text{s})^2\left(1000\ \frac{\text{mm}}{\text{m}}\right) = 18.0\ \text{mm}$$

The total follower movement for 150° of rotation is

$$x_{\text{total}} = 12.0\ \text{mm} + 18.0\ \text{mm} = \boxed{30\ \text{mm}}$$

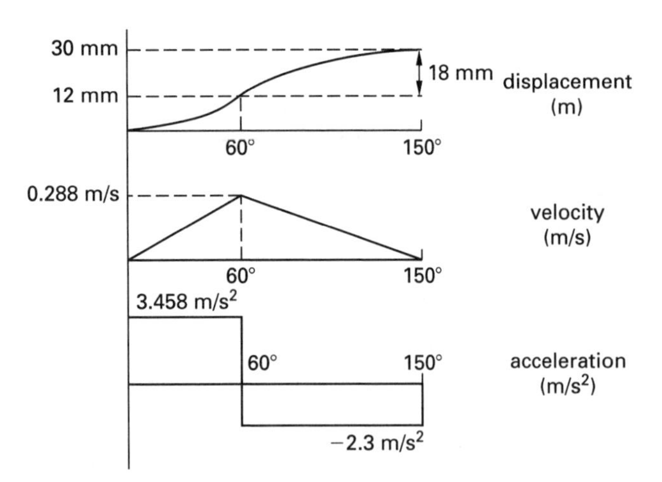

The answer is (D).

12. Use a diagonal line to divide the four-bar linkage into two triangles as shown.

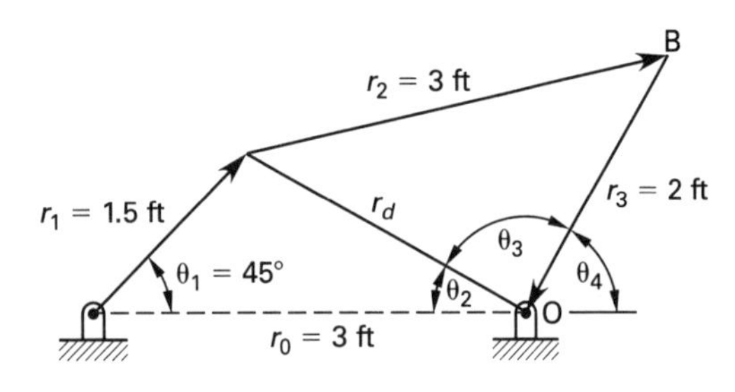

From the law of cosines, the length of the diagonal, r_d, is

$$\begin{aligned} r_d &= \sqrt{r_0^2 + r_1^2 - 2r_0r_1\cos\theta_1} \\ &= \sqrt{(3\text{ ft})^2 + (1.5\text{ ft})^2 - (2)(3\text{ ft})(1.5\text{ ft})\cos 45^\circ} \\ &= 2.2\text{ ft} \end{aligned}$$

Use the law of sines to solve for θ_2.

$$\begin{aligned} \frac{\sin\theta_2}{r_1} &= \frac{\sin\theta_1}{r_d} \\ \sin\theta_2 &= \frac{r_1\sin\theta_1}{r_d} = \frac{(1.5\text{ ft})\sin 45^\circ}{2.2\text{ ft}} = 0.482 \\ \theta_2 &= 28.82^\circ \end{aligned}$$

Use the law of cosines to solve for θ_3.

$$\begin{aligned} \cos\theta_3 &= \frac{r_2^2 - r_d^2 - r_3^2}{-2r_dr_3} = \frac{(3\text{ ft})^2 - (2.2\text{ ft})^2 - (2\text{ ft})^2}{(-2)(2.2\text{ ft})(3\text{ ft})} \\ &= -0.012 \\ \theta_3 &= 90.69^\circ \end{aligned}$$

Angle θ_4 is

$$\begin{aligned} \theta_4 &= 180^\circ - \theta_3 - \theta_2 = 180^\circ - 90.69^\circ - 28.82^\circ \\ &= 60.5^\circ \end{aligned}$$

With the origin $(0,0)$ at the base of bar r_3, the coordinates for the rotating end of bar r_3 are

$$x = r_3\cos\theta_4 = (2\text{ ft})\cos 60.5^\circ = \boxed{0.99\text{ ft}\ \ (1.0\text{ ft})}$$

$$y = r_3\sin\theta_4 = (2\text{ ft})\sin 60.5^\circ = \boxed{1.74\text{ ft}\ \ (1.7\text{ ft})}$$

The answer is (A).

13. With respect to the base of bar r_1, the coordinates of point A are

$$\begin{aligned} x &= r_1\cos\theta_1 = (1.5\text{ ft})\cos 45^\circ = 1.06\text{ ft} \\ y &= r_1\sin\theta_1 = (1.5\text{ ft})\sin 45^\circ = 1.06\text{ ft} \end{aligned}$$

With respect to the base of bar r_2, the coordinates of point B are

$$\begin{aligned} x &= r_2\cos\theta_2 = (3\text{ ft})\cos 13.1^\circ = 2.92\text{ ft} \\ y &= r_2\cos\theta_2 = (3\text{ ft})\sin 13.1^\circ = 0.68\text{ ft} \end{aligned}$$

Expressing both velocities in terms of the unit vectors, the velocity of point B, $\mathbf{v}_B$, is the simple sum of the two vector velocities.

$$\mathbf{v}_B = \mathbf{v}_1 + \mathbf{v}_2 = (\boldsymbol{\omega}_1 \times \mathbf{r}_1) + (\boldsymbol{\omega}_2 \times \mathbf{r}_2)$$

Take the vector cross products.

$$\begin{aligned} \mathbf{v}_B &= \begin{vmatrix} \mathbf{i} & \mathbf{j} & \mathbf{k} \\ 0 & 0 & 150\text{ rad/sec} \\ 1.06\text{ ft} & 1.06\text{ ft} & 0 \end{vmatrix} \\ &\quad + \begin{vmatrix} \mathbf{i} & \mathbf{j} & \mathbf{k} \\ 0 & 0 & -27.2\text{ rad/sec} \\ 2.92\text{ ft} & 0.68\text{ ft} & 0 \end{vmatrix} \\ &= -140.5\mathbf{i}\text{ ft/sec} + 79.6\mathbf{j}\text{ ft/sec} \\ &= \boxed{161\text{ ft/sec}\angle 150.5^\circ\ \ (160\text{ ft/sec}\angle 150^\circ)} \end{aligned}$$

The answer is (D).

60 Vibrating Systems

PRACTICE PROBLEMS

1. A 2 in (50 mm) steel shaft 40 in (1020 mm) long is supported on frictionless bearings at its two ends. The shaft carries a 100 lbm (45 kg) disk 15 in (380 mm) from the left bearing, and a 75 lbm (34 kg) disk 25 in (640 mm) from the left bearing. The shaft weight is negligible. There is no damping. The critical speed of the shaft is most nearly

(A) 1500 rpm

(B) 1700 rpm

(C) 2000 rpm

(D) 2400 rpm

2. An 800 lbm (360 kg), single-cylinder vertical compressor operates at 1200 rpm. The compressor has an intrinsic, rotational out-of-balance condition. The compressor is to be mounted to the floor on four identical, equally loaded springs at its corners. It is desired to reduce the maximum force transmitted to the mounting base from an applied force of 25 lbf to a transmitted force of 3 lbf (110 N to 13 N). Damping is negligible. The new maximum oscillation is most nearly

(A) 0.49×10^{-4} in (1.2×10^{-5} m)

(B) 0.63×10^{-4} in (1.6×10^{-5} m)

(C) 0.72×10^{-4} in (1.8×10^{-5} m)

(D) 8.6×10^{-4} in (2.2×10^{-5} m)

3. When an 8 lbm (3.6 kg) mass is attached on the end of a spring, the spring stretches 5.9 in (150 mm). A dashpot with a damping coefficient of 0.50 lbf-sec/ft (7.3 N·s/m) opposes movement of the mass. A forcing function of $4\cos 2t$ lbf ($18\cos 2t$ N) is applied to the mass. The mass is initially at rest.

(a) The natural frequency is most nearly

(A) 0.4 rad/sec

(B) 2.0 rad/sec

(C) 8.0 rad/sec

(D) 65 rad/sec

(b) The damping ratio is most nearly

(A) 0.02

(B) 0.08

(C) 0.10

(D) 0.40

(c) The maximum excursion is most nearly

(A) 0.1 in (0.003 m)

(B) 2.0 in (0.05 m)

(C) 3.0 in (0.08 m)

(D) 9.0 in (0.20 m)

(d) The response of the system is most nearly

(A) $0.17\sin 4t - 0.55\cos 4t - (0.17) \times (e^{-t}\cos 2t + 0.26e^{-t}\sin 8t)$

$\big(0.0533\sin 4t + 0.173\cos 4t - (0.0533)(e^{-t}\cos 8t + 0.26e^{-t}\sin 8t)\big)$

(B) $0.26\cos 2t + 0.017\sin 2t - (0.26) \times (e^{-t}\cos 8t + 0.14e^{-t}\sin 8t)$

$\big(0.0816\cos 4t + 0.00535\sin 2t - (0.0816)(e^{-t}\cos 8t + 0.14e^{-t}\sin 8t)\big)$

(C) $0.37\sin 4t - 0.09\cos 8t - (0.37) \times (e^{-t}\cos 4t + 0.17e^{-t}\sin 4t)$

$\big(0.1161\sin 4t + 0.0283\cos 8t - (0.1161)(e^{-t}\cos 8t + 0.17e^{-t}\sin 8t)\big)$

(D) $0.55\cos 4t - 0.34\sin 4t - (0.55) \times (e^{-t}\cos 2t + 0.12e^{-t}\sin 2t)$

$\big(0.1726\cos 4t + 0.107\sin 4t - (0.1726)(e^{-t}\cos 8t + 0.12e^{-t}\sin 8t)\big)$

4. A uniform bar with a mass of 5 lbm (2.3 kg) carries a concentrated mass of 3 lbm (1.4 kg) at its free end. The bar is hinged at one end and supported by an outboard spring as shown. The deflection of the spring from its unstressed position is 0.55 in (1.4 cm). There is no damping. What is most nearly the natural frequency of vibration?

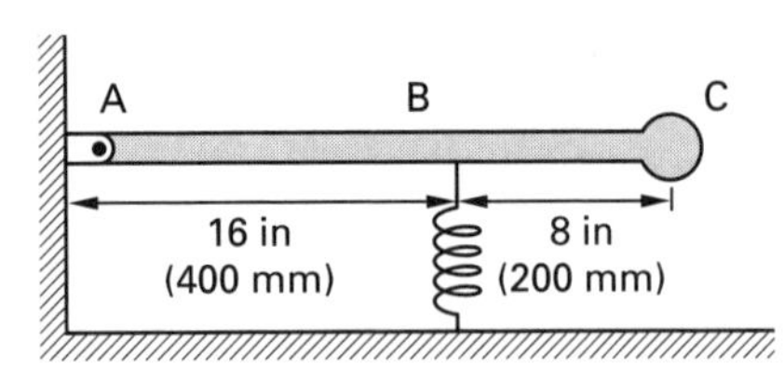

(A) 4 Hz

(B) 7 Hz

(C) 20 Hz

(D) 50 Hz

5. A 300 lbm (140 kg) electromagnet at the end of a cable holds 200 lbm (90 kg) of scrap metal. The total equivalent stiffness of the cable and crane boom is 1000 lbf/in (175 kN/m). The current to the electromagnet is cut off suddenly, and the scrap falls away. Neglect damping.

(a) The frequency of oscillation of the electromagnet is most nearly

(A) 5.7 Hz

(B) 10 Hz

(C) 23 Hz

(D) 84 Hz

(b) The minimum cable tension is most nearly

(A) 100 lbf (490 N)

(B) 200 lbf (980 N)

(C) 300 lbf (1500 N)

(D) 400 lbf (2000 N)

6. A 50 lbm (23 kg) motor is supported by four identical, equally loaded springs, each with a spring constant of 1000 lbf/in (175 kN/m). When the motor is turning at 800 rpm, the rotor imbalance is equivalent to a 1 oz (30 g) mass located 5 in (130 mm) from the shaft's longitudinal axis. The damping factor is $^1/_8$ of critical damping. The maximum vertical vibration amplitude is most nearly

(A) 0.0012 in (3.0×10^{-5} m)

(B) 0.0018 in (5.0×10^{-5} m)

(C) 0.0046 in (1.2×10^{-4} m)

(D) 0.0099 in (2.5×10^{-4} m)

7. Eight horizontal high-strength steel plates are supported on rigid rollers and support a 20,000 lbm (9100 kg) load, as shown. Each of the plates is 40 in × 30 in × $^1/_2$ in (1020 mm × 760 mm × 12 mm). The modulus of elasticity of the steel is 2.9×10^7 lbf/in^2 (200 GPa) and Poisson's ratio of 0.3. The masses of the plates and rollers are insignificant compared to the supported load. The yield point of the steel is not exceeded.

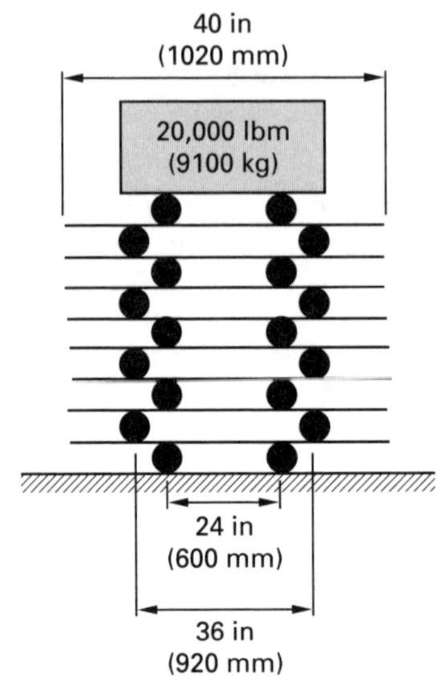

(a) The moment of inertia of the cross section of the plate is most nearly

(A) 0.30 in^4 (1.0×10^{-7} m^4)

(B) 0.40 in^4 (1.5×10^{-7} m^4)

(C) 0.60 in^4 (2.0×10^{-7} m^4)

(D) 0.80 in^4 (3.0×10^{-7} m^4)

(b) The total vertical static deflection of the load is most nearly

(A) 4.0 in (0.13 m)

(B) 4.4 in (0.15 m)

(C) 6.2 in (0.16 m)

(D) 6.9 in (0.17 m)

(c) The maximum stress in the plates is most nearly

(A) 48 kips/in^2 (390 MPa)

(B) 60 kips/in^2 (480 MPa)

(C) 96 kips/in^2 (780 MPa)

(D) 120 kips/in^2 (970 MPa)

(d) The natural frequency of oscillation in the vertical direction is most nearly

(A) 1.6 Hz

(B) 2.7 Hz

(C) 4.2 Hz

(D) 7.6 Hz

8. A 175 lbm (80 kg), single-cylinder air compressor is mounted on four identical, equally loaded corner springs. The motor turns at 1200 rpm. During each cycle, a disturbing force is generated by a 3.6 lbm (1.6 kg) imbalance acting at a radius of 3 in (75 mm). Damping is insignificant. It is desired that the dynamic force transmitted to the base be limited to 5% of the disturbing force.

(a) The out-of-balance force caused by the rotating imbalance is most nearly

(A) 40 (190 N)

(B) 160 (690 N)

(C) 400 (1700 N)

(D) 440 (1900 N)

(b) The required natural frequency is most nearly

(A) 10 rad/sec

(B) 15 rad/sec

(C) 27 rad/sec

(D) 32 rad/sec

(c) The individual spring stiffness required is most nearly

(A) 44 lbf/in (7.5 kN/m)

(B) 68 lbf/in (12 kN/m)

(C) 85 lbf/in (15 kN/m)

(D) 170 lbf/in (29 kN/m)

(d) The amplitude of vibration is most nearly

(A) 0.018 in (0.00045 m)

(B) 0.033 in (0.00083 m)

(C) 0.065 in (0.0016 m)

(D) 0.097 in (0.0024 m)

9. A centrifugal fan has 8 driving blades and 64 fan blades. The fan turns at 600 rpm and is driven by a 1725 rpm, 60 Hz, 4 pole motor. The fan and motor pulley are $11^1/_2$ in (290 mm) and 4 in (100 mm) in diameter, respectively. The drive belt has a total length of 72 in (1.83 m).

(a) The frequency of vibration produced from the fan rotation is most nearly

(A) 10 Hz

(B) 60 Hz

(C) 64 Hz

(D) 600 Hz

(b) The frequency of vibration produced from the driving blades is most nearly

(A) 10 Hz

(B) 60 Hz

(C) 80 Hz

(D) 160 Hz

(c) The frequency vibration produced from the fan blades is most nearly

(A) 10 Hz

(B) 64 Hz

(C) 600 Hz

(D) 640 Hz

(d) The frequency vibration produced from the motor rotation is most nearly

(A) 30 Hz

(B) 80 Hz

(C) 140 Hz

(D) 210 Hz

(e) The frequency vibration produced from the poles is most nearly

(A) 30 Hz

(B) 60 Hz

(C) 120 Hz

(D) 170 Hz

(f) The frequency of sound produced from the electrical hum is most nearly

(A) 10 Hz

(B) 60 Hz

(C) 80 Hz

(D) 100 Hz

(g) The frequencies of vibration produced from the motor and the fan pulleys are most nearly

(A) motor = 10 Hz; fan = 80 Hz

(B) motor = 29 Hz; fan = 10 Hz

(C) motor = 60 Hz; fan = 120 Hz

(D) motor = 80 Hz; fan = 100 Hz

(h) The frequency of vibration produced from the belt is most nearly

(A) 2.0 Hz

(B) 2.5 Hz

(C) 5.0 Hz

(D) 8.0 Hz

10. When not running, a hydraulic oil pump compresses a cork mounting pad 0.02 in (0.5 mm). The pump is turned at 1725 rpm. The transmissibility of the pad is most nearly increased by

(A) 25%

(B) 50%

(C) 75%

(D) 100%

11. A length of square structural steel tubing (modulus of elasticity 29×10^6 lbf/in^2) is used as a flagpole. The tube measures 4 in × 4 in externally, and the wall thickness is 0.125 in. The flagpole is fixed at the bottom and extends vertically 22 ft. The total flagpole mass is 150 lbm. The flagpole has a damping ratio of 0.119.

(a) The undamped fundamental natural frequency of the flagpole is most nearly

(A) 1.3 Hz

(B) 2.0 Hz

(C) 5.1 Hz

(D) 7.1 Hz

(b) The damped fundamental natural frequency of the flagpole is most nearly

(A) 1.2 Hz

(B) 2.0 Hz

(C) 5.0 Hz

(D) 7.0 Hz

SOLUTIONS

1. *Customary U.S. Solution*

The moment of inertia of the circular cross section is

$$I = \frac{\pi r^4}{4} = \frac{\pi\left(\frac{2.0\text{ in}}{2}\right)^4}{4} = 0.7854\text{ in}^4$$

Use App. 51.A. The deflection due to the 100 lbm disk at the 100 lbm disk is

$$\begin{aligned}\delta &= \frac{Fa^2b^2}{3EIL}\\ &= \frac{(100\text{ lbm})\left(\dfrac{32.2\ \frac{\text{ft}}{\text{sec}^2}}{32.2\ \frac{\text{lbm-ft}}{\text{lbf-sec}^2}}\right)(15\text{ in})^2(25\text{ in})^2}{(3)\left(30\times 10^6\ \frac{\text{lbf}}{\text{in}^2}\right)(0.7854\text{ in}^4)(40\text{ in})}\\ &= 0.00497\text{ in}\end{aligned}$$

This installation is symmetrical with respect to load positioning.

Find the deflection due to the 100 lbm disk at the 75 lbm disk. By looking at the shaft from the back, the formula used is

$$\begin{aligned}\delta &= \left(\frac{Fbx}{6EIL}\right)(L^2 - b^2 - x^2)\\ &= \frac{(100\text{ lbm})\left(\dfrac{32.2\ \frac{\text{ft}}{\text{sec}^2}}{32.2\ \frac{\text{lbm-ft}}{\text{lbf-sec}^2}}\right)(15\text{ in})(15\text{ in})\times\left((40\text{ in})^2 - (15\text{ in})^2 - (15\text{ in})^2\right)}{(6)\left(30\times 10^6\ \frac{\text{lbf}}{\text{in}^2}\right)(0.7854\text{ in}^4)(40\text{ in})}\\ &= 0.00458\text{ in}\end{aligned}$$

The deflection due to the 75 lbm disk at the 75 lbm disk is

$$\begin{aligned}\delta &= \frac{Fa^2b^2}{3EIL}\\ &= \frac{(75\text{ lbm})\left(\dfrac{32.2\ \frac{\text{ft}}{\text{sec}^2}}{32.2\ \frac{\text{lbm-ft}}{\text{lbf-sec}^2}}\right)(25\text{ in})^2(15\text{ in})^2}{(3)\left(30\times 10^6\ \frac{\text{lbf}}{\text{in}^2}\right)(0.7854\text{ in}^4)(40\text{ in})}\\ &= 0.00373\text{ in}\end{aligned}$$

The deflection due to the 75 lbm disk at the 100 lbm disk is

$$\delta = \left(\frac{Fbx}{6EIL}\right)(L^2 - b^2 - x^2)$$

$$= \frac{(75\ \text{lbm})\left(\dfrac{32.2\ \frac{\text{ft}}{\text{sec}^2}}{32.2\ \frac{\text{lbm-ft}}{\text{lbf-sec}^2}}\right)(15\ \text{in})(15\ \text{in}) \times \left((40\ \text{in})^2 - (15\ \text{in})^2 - (15\ \text{in})^2\right)}{(6)\left(30\times 10^6\ \frac{\text{lbf}}{\text{in}^2}\right)(0.7854\ \text{in}^4)(40\ \text{in})}$$

$$= 0.00343\ \text{in}$$

The total deflection at the 100 lbm disk is

$$\delta_{\text{st},1} = 0.00497\ \text{in} + 0.00343\ \text{in} = 0.00840\ \text{in}$$

The total deflection at the 75 lbm disk is

$$\delta_{\text{st},2} = 0.00458\ \text{in} + 0.00373\ \text{in} = 0.00831\ \text{in}$$

Use Eq. 60.60 to find the natural frequency.

$$f = \left(\frac{1}{2\pi}\right)\sqrt{\frac{g\sum m_i\delta_{\text{st},i}}{\sum m_i\delta^2_{\text{st},i}}}$$

$$= \left(\frac{1}{2\pi}\right)\sqrt{\frac{\left(386.4\ \frac{\text{in}}{\text{sec}^2}\right)\left((100\ \text{lbm})(0.00840\ \text{in}) + (75\ \text{lbm})(0.00831\ \text{in})\right)}{(100\ \text{lbm})(0.00840\ \text{in})^2 + (75\ \text{lbm})(0.00831\ \text{in})^2}}$$

$$= 34.21\ \text{Hz}$$

The critical speed of the shaft is

$$n = (34.21\ \text{Hz})\left(60\ \frac{\text{sec}}{\text{min}}\right) = \boxed{2053\ \text{rpm}\quad(2000\ \text{rpm})}$$

The answer is (C).

SI Solution

The moment of inertia of the circular cross section is

$$I = \frac{\pi r^4}{4} = \frac{\pi\left(\frac{0.05\ \text{m}}{2}\right)^4}{4} = 3.068\times 10^{-7}\ \text{m}^4$$

The deflection due to the 45 kg disk at the 45 kg disk is

$$\delta = \frac{Fa^2b^2}{3EIL}$$

$$= \frac{(45\ \text{kg})\left(9.81\ \frac{\text{m}}{\text{s}^2}\right)(0.38\ \text{m})^2(0.64\ \text{m})^2}{(3)(200\times 10^9\ \text{Pa})(3.068\times 10^{-7}\ \text{m}^4)(1.02\ \text{m})}$$

$$= 0.000139\ \text{m}$$

The deflection due to the 45 kg disk at the 34 kg disk is

$$\delta = \left(\frac{Fbx}{6EIL}\right)(L^2 - b^2 - x^2)$$

$$= \frac{(45\ \text{kg})\left(9.81\ \frac{\text{m}}{\text{s}^2}\right)(0.38\ \text{m})(0.38\ \text{m}) \times \left((1.02\ \text{m})^2 - (0.38\ \text{m})^2 - (0.38\ \text{m})^2\right)}{(6)(200\times 10^9\ \text{Pa})(3.068\times 10^{-7}\ \text{m}^4)(1.02\ \text{m})}$$

$$= 0.000128\ \text{m}$$

The deflection due to the 34 kg disk at the 34 kg disk is

$$\delta = \frac{Fa^2b^2}{3EIL}$$

$$= \frac{(34\ \text{kg})\left(9.81\ \frac{\text{m}}{\text{s}^2}\right)(0.64\ \text{m})^2(0.38\ \text{m})^2}{(3)(200\times 10^9\ \text{Pa})(3.068\times 10^{-7}\ \text{m}^4)(1.02\ \text{m})}$$

$$= 0.000105\ \text{m}$$

The deflection due to the 34 kg disk at the 45 kg disk is

$$\delta = \left(\frac{Fbx}{6EIL}\right)(L^2 - b^2 - x^2)$$

$$= \frac{(34\ \text{kg})\left(9.81\ \frac{\text{m}}{\text{s}^2}\right)(0.38\ \text{m})(0.38\ \text{m}) \times \left((1.02\ \text{m})^2 - (0.38\ \text{m})^2 - (0.38\ \text{m})^2\right)}{(6)(200\times 10^9\ \text{Pa})(3.068\times 10^{-7}\ \text{m}^4)(1.02\ \text{m})}$$

$$= 0.000096\ \text{m}$$

The total deflection at the 45 kg disk is

$$\delta_{\text{st},1} = 0.000139\ \text{m} + 0.000096\ \text{m} = 0.000235\ \text{m}$$

The total deflection at the 34 kg disk is

$$\delta_{\text{st},2} = 0.000128\ \text{m} + 0.000105\ \text{m} = 0.000233\ \text{m}$$

Use Eq. 60.60 to find the natural frequency.

$$f=\left(\frac{1}{2\pi}\right)\sqrt{\frac{g\sum m_i\delta_{st,i}}{\sum m_i\delta_{st,i}^2}}$$

$$=\left(\frac{1}{2\pi}\right)\sqrt{\frac{\left(9.81\ \frac{\text{m}}{\text{s}^2}\right)\left(\begin{array}{c}(45\text{ kg})(0.000235\text{ m})\\ +\ (34\text{ kg})(0.000233\text{ m})\end{array}\right)}{\begin{array}{c}(45\text{ kg})(0.000235\text{ m})^2\\ +\ (34\text{ kg})(0.000233\text{ m})^2\end{array}}}$$

$$=32.6\text{ Hz}$$

The critical speed of the shaft is

$$n=(32.6\text{ Hz})\left(60\ \frac{\text{sec}}{\text{min}}\right)=\boxed{1956\text{ rpm}\quad(2000\text{ rpm})}$$

The answer is (C).

2. *Customary U.S. Solution*

The transmissibility is

$$\text{TR}=\frac{|F_{\text{transmitted}}|}{F_{\text{applied}}}=\frac{3\text{ lbf}}{25\text{ lbf}}=0.12$$

The angular forcing frequency is

$$\omega_f=\frac{\left(1200\ \frac{\text{rev}}{\text{min}}\right)\left(2\pi\ \frac{\text{rad}}{\text{rev}}\right)}{60\ \frac{\text{sec}}{\text{min}}}=125.7\text{ rad/sec}$$

For negligible damping, Eq. 60.55 and Eq. 60.56 can be simplified.

$$\text{TR}=\frac{1}{|1-r^2|}=\frac{1}{\left|1-\left(\frac{\omega_f}{\omega}\right)^2\right|}$$

$$=\frac{1}{\left(\frac{\omega_f}{\omega}\right)^2-1}\quad[\text{for }\omega_f>\omega]$$

$$\frac{\omega_f}{\omega}=\sqrt{\frac{1}{\text{TR}}+1}=\sqrt{\frac{1}{0.12}+1}=3.055$$

The required natural frequency is

$$\omega=\frac{\omega_f}{3.055}=\frac{125.7\ \frac{\text{rad}}{\text{sec}}}{3.055}=41.15\text{ rad/sec}$$

From Eq. 60.3(b), the equivalent stiffness of the springs is

$$k=\frac{m\omega^2}{g_c}=\frac{(800\text{ lbm})\left(41.15\ \frac{\text{rad}}{\text{sec}}\right)^2}{32.2\ \frac{\text{lbm-ft}}{\text{lbf-sec}^2}}$$

$$=42{,}070\text{ lbf/ft}$$

The reduced pseudo-static deflection is

$$\frac{F_0}{k}=\frac{25\text{ lbf}}{42{,}070\ \frac{\text{lbf}}{\text{ft}}}=5.94\times10^{-4}\text{ ft}$$

From Eq. 60.50, the magnification factor (transmissibility) is

$$\beta=\frac{D}{\frac{F_0}{k}}=\left|\frac{1}{1-\left(\frac{\omega_f}{\omega}\right)^2}\right|=\text{TR}=0.12$$

The new maximum oscillation is

$$D=\left(\frac{F_0}{k}\right)\beta=(5.94\times10^{-4}\text{ ft})\left(12\ \frac{\text{in}}{\text{ft}}\right)(0.12)$$

$$=\boxed{8.56\times10^{-4}\text{ in}\quad(8.6\times10^{-4}\text{ in})}$$

The answer is (D).

SI Solution

The transmissibility is

$$\text{TR}=\frac{|F_{\text{transmitted}}|}{F_{\text{applied}}}=\frac{13\text{ N}}{110\text{ N}}=0.118$$

The angular forcing frequency is

$$\omega_f=\frac{\left(1200\ \frac{\text{rev}}{\text{min}}\right)\left(2\pi\ \frac{\text{rad}}{\text{rev}}\right)}{60\ \frac{\text{s}}{\text{min}}}=125.7\text{ rad/s}$$

For negligible damping, Eq. 60.55 and Eq. 60.56 can be simplified.

$$\text{TR}=\frac{1}{|1-r^2|}=\frac{1}{\left|1-\left(\frac{\omega_f}{\omega}\right)^2\right|}$$

$$=\frac{1}{\left(\frac{\omega_f}{\omega}\right)^2-1}\quad[\text{for }\omega_f>\omega]$$

$$\frac{\omega_f}{\omega}=\sqrt{\frac{1}{\text{TR}}+1}=\sqrt{\frac{1}{0.118}+1}=3.078$$

The required natural frequency is

$$\omega = \frac{\omega_f}{3.078} = \frac{125.7\ \frac{\text{rad}}{\text{s}}}{3.078} = 40.84\ \text{rad/s}$$

From Eq. 60.3(a), the equivalent stiffness of the springs is

$$k = m\omega^2 = (360\ \text{kg})\left(40.84\ \frac{\text{rad}}{\text{s}}\right)^2 = 600\,446\ \text{N/m}$$

The pseudo-static deflection is

$$\frac{F_0}{k} = \frac{110\ \text{N}}{600\,446\ \frac{\text{N}}{\text{m}}} = 1.83 \times 10^{-4}\ \text{m}$$

From Eq. 60.50, the magnification factor (transmissibility) is

$$\beta = \frac{D}{\frac{F_0}{k}} = \left|\frac{1}{1 - \left(\frac{\omega_f}{\omega}\right)^2}\right| = \text{TR} = 0.118$$

The new maximum oscillation is

$$\begin{aligned} D &= \left(\frac{F_0}{k}\right)\beta = (1.83 \times 10^{-4}\ \text{m})(0.118) \\ &= \boxed{2.16 \times 10^{-5}\ \text{m} \quad (2.2 \times 10^{-5}\ \text{m})} \end{aligned}$$

The answer is (D).

3. *Customary U.S. Solution*

The static deflection is

$$\begin{aligned} \delta_{\text{st}} &= \frac{\text{weight}}{k} = \frac{m\left(\frac{g}{g_c}\right)}{k} \\ k &= \frac{m\left(\frac{g}{g_c}\right)}{\delta_{\text{st}}} = \left(\frac{8\ \text{lbm}}{5.9\ \text{in}}\right)\left(\frac{32.2\ \frac{\text{ft}}{\text{sec}^2}}{32.2\ \frac{\text{lbm-ft}}{\text{lbf-sec}^2}}\right) \\ &= 1.356\ \text{lbf/in} \end{aligned}$$

(a) From Eq. 60.3(b) and Eq. 60.7, the natural frequency is

$$\begin{aligned} \omega &= \sqrt{\frac{kg_c}{m}} = \sqrt{\frac{g}{\delta_{\text{st}}}} = \sqrt{\frac{386.4\ \frac{\text{in}}{\text{sec}^2}}{5.9\ \text{in}}} \\ &= \boxed{8.09\ \text{rad/sec} \quad (8.0\ \text{rad/sec})} \end{aligned}$$

The answer is (C).

(b) From Eq. 60.40(b), the damping ratio is

$$\begin{aligned} \zeta &= \frac{C}{2\sqrt{\frac{mk}{g_c}}} \\ &= \frac{0.50\ \frac{\text{lbf-sec}}{\text{ft}}}{(2)\sqrt{\frac{(8\ \text{lbm})\left(1.356\ \frac{\text{lbf}}{\text{in}}\right)\left(12\ \frac{\text{in}}{\text{ft}}\right)}{32.2\ \frac{\text{lbm-ft}}{\text{lbf-sec}^2}}}} \\ &= \boxed{0.124 \quad (0.10)} \end{aligned}$$

The answer is (C).

(c) The forcing frequency is $\omega_f = 2$ rad/sec.

The pseudo-static deflection is

$$\frac{F_0}{k} = \frac{4\ \text{lbf}}{1.356\ \frac{\text{lbf}}{\text{in}}} = 2.95\ \text{in}$$

The ratio of frequencies is

$$\begin{aligned} r &= \frac{\omega_f}{\omega} = \frac{2\ \frac{\text{rad}}{\text{sec}}}{8.09\ \frac{\text{rad}}{\text{sec}}} \\ &= 0.247 \end{aligned}$$

From Eq. 60.53, the magnification factor is

$$\begin{aligned} \beta &= \frac{D}{\frac{F_0}{k}} = \left|\frac{1}{\sqrt{(1 - r^2)^2 + (2\zeta r)^2}}\right| \\ &= \left|\frac{1}{\sqrt{(1 - (0.247)^2)^2 + ((2)(0.124)(0.247))^2}}\right| \\ &= 1.063 \end{aligned}$$

The maximum excursion of the system is

$$\begin{aligned} D &= \beta\left(\frac{F_0}{k}\right) = (1.063)(2.95\ \text{in}) \\ &= \boxed{3.14\ \text{in} \quad (3.0\ \text{in})} \end{aligned}$$

The answer is (C).

Dynamics and Vibrations

(d) From Eq. 60.52(b), the differential equation of motion is

$$\frac{m}{g_c}\frac{d^2x}{dt^2} = -kx - C\frac{dx}{dt} + F(t)$$

$$\left(\frac{8\ \text{lbm}}{32.2\ \frac{\text{lbm-ft}}{\text{lbm-sec}^2}}\right)x'' = \frac{-(8\ \text{lbm})\left(\frac{32.2\ \frac{\text{ft}}{\text{sec}^2}}{32.2\ \frac{\text{lbm-ft}}{\text{lbm-sec}^2}}\right)\times\left(12\ \frac{\text{in}}{\text{ft}}\right)x}{5.9\ \text{in}} - \left(0.50\ \frac{\text{lbf-sec}}{\text{ft}}\right)x' + 4\cos 2t$$

$$0.25x'' + 0.50x' + 16.27x = 4\cos 2t$$

$$x'' + 2x' + 65x = 16\cos 2t$$

[coefficients rounded for convenience]

Initial conditions are

$$x_0 = 0$$
$$x_0' = 0$$

There are a variety of methods to solve this differential equation. Use Laplace transforms.

Taking the Laplace transform of both sides,

$$\mathcal{L}(x'') + \mathcal{L}(2x') + \mathcal{L}(65x) = \mathcal{L}(16\cos 2t)$$

$$s^2\mathcal{L}(x) - sx_0 - x_0' + 2s\mathcal{L}(x) - 2x_0 + 65\mathcal{L}(x) = (16)\left(\frac{s}{s^2+4}\right)$$

$$\mathcal{L}(x) = \frac{16s}{(s^2+4)(s^2+2s+65)}$$

Use partial fractions.

$$\mathcal{L}(x) = \frac{16s}{(s^2+4)(s^2+2s+65)} = \frac{As+B}{s^2+4} + \frac{Cs+D}{s^2+2s+65}$$

$$16s = As^2 + Bs^2 + 2As^2 + 2Bs + 65As + 65B + Cs^3 + Ds^2 + 4Cs + 4D$$

Then,

$$A + C = 0 \quad C = -A = -\frac{61}{8}B$$

$$B + 2A + D = 0 \quad B + 2A - \frac{65}{4}B = 0 \rightarrow A = \frac{61}{8}B$$

$$65B + 4D = 0 \quad D = -\frac{65}{4}B$$

$$2B + 65A + 4C = 16$$

$$2B + (65)\left(\frac{61}{8}B\right) + (4)\left(-\frac{61}{8}B\right) = 16$$

$$B = 0.0342521$$
$$A = 0.2611721$$
$$C = -0.2611721$$
$$D = -0.5565962$$

$$\mathcal{L}(x) = \frac{0.2611721s + 0.0342521}{s^2+4} - \frac{0.2611721s + 0.5566}{(s+1)^2 + (8)^2}$$

$$= (0.26)\left(\frac{s}{s^2+(2)^2}\right) + (0.017)\left(\frac{2}{s^2+(2)^2}\right) - (0.26)\left(\frac{s-(-1)}{\left(s-(-1)\right)^2+(8)^2} + \frac{1.1311472}{\left(s-(-1)\right)^2+(8)^2}\right)$$

Take the inverse transform. The response is

$$\boxed{\begin{aligned} x(t) &= 0.26\cos 2t + 0.017\sin 2t \\ &\quad - (0.26)(e^{-t}\cos 8t + 0.14e^{-t}\sin 8t) \end{aligned}}$$

Compare the values here with those obtained previously.

$$D = (0.26\ \text{ft})\left(12\ \frac{\text{in}}{\text{ft}}\right) = 3.12\ \text{in}$$

This checks with part (c).

The natural frequency is 8 rad/sec, which corresponds to ω from part (a). (The coefficient of 0.25 in the differential equation was rounded from 0.248, which accounts for the difference.)

The answer is (B).

SI Solution

The static deflection is

$$\delta_{\text{st}} = \frac{\text{weight}}{k} = \frac{mg}{k}$$

$$k = \frac{mg}{\delta_{\text{st}}} = \frac{(3.6\ \text{kg})\left(9.81\ \frac{\text{m}}{\text{s}^2}\right)}{0.15\ \text{m}} = 235.4\ \text{N/m}$$

(a) From Eq. 60.3(a) and Eq. 60.7, the natural frequency is

$$\omega = \sqrt{\frac{k}{m}} = \sqrt{\frac{g}{\delta_{\text{st}}}} = \sqrt{\frac{9.81 \ \frac{\text{m}}{\text{s}^2}}{0.15 \text{ m}}}$$

$$= \boxed{8.09 \text{ rad/s} \quad (8.0 \text{ rad/s})}$$

The answer is (C).

(b) From Eq. 60.40(a), the damping ratio is

$$\zeta = \frac{C}{2\sqrt{mk}} = \frac{7.3 \ \frac{\text{N}\cdot\text{s}}{\text{m}}}{2\sqrt{(3.6 \text{ kg})\left(235.4 \ \frac{\text{N}}{\text{m}}\right)}}$$

$$= \boxed{0.125 \quad (0.10)}$$

The answer is (C).

(c) The forcing frequency is $\omega_f = 2$ rad/s.

The pseudo-static deflection is

$$\frac{F_0}{k} = \frac{18 \text{ N}}{235.4 \ \frac{\text{N}}{\text{m}}} = 0.0765 \text{ m}$$

The ratio of frequencies is

$$r = \frac{\omega_f}{\omega} = \frac{2 \ \frac{\text{rad}}{\text{s}}}{8.09 \ \frac{\text{rad}}{\text{s}}} = 0.247$$

From Eq. 60.53, the magnification factor is

$$\beta = \frac{D}{\frac{F_0}{k}} = \left|\frac{1}{\sqrt{(1-r^2)^2 + (2\zeta r)^2}}\right|$$

$$= \left|\frac{1}{\sqrt{\left(1-(0.247)^2\right)^2 + \left((2)(0.125)(0.247)\right)^2}}\right|$$

$$= 1.063$$

The response of the system is

$$D = \beta\left(\frac{F_0}{k}\right) = (1.063)(0.0765 \text{ m})$$

$$= \boxed{0.0813 \text{ m} \quad (0.08 \text{ m})}$$

(d) From Eq. 60.52(a), the differential equation of motion is

$$m\frac{d^2x}{dt^2} = -kx - C\frac{dx}{dt} + F(t)$$

$$(3.6 \text{ kg})x'' = \left(\frac{(-3.6 \text{ kg})\left(9.81 \ \frac{\text{m}}{\text{s}^2}\right)}{0.15 \text{ m}}\right)x$$

$$-\left(7.3 \ \frac{\text{N}\cdot\text{m}}{\text{s}}\right)x' + 18\cos 2t$$

$$3.6x'' + 7.3x' + 235.4x = 18\cos 2t$$

$$x'' + 2x' + 65x = 5\cos 2t$$

[coefficients rounded for convenience]

Initial conditions are

$$x_0 = 0$$

$$x'_0 = 0$$

There are a variety of methods to solve this differential equation. Use Laplace transforms.

Take the Laplace transform of both sides.

$$\mathcal{L}(x'') + \mathcal{L}(2x') + \mathcal{L}(65x) = \mathcal{L}(5\cos 2t)$$

$$s^2\mathcal{L}(x) - sx_0 - x'_0 + 2s\mathcal{L}(x) - 2x_0 + 65\mathcal{L}(x)$$

$$= (5)\left(\frac{s}{s^2+4}\right)$$

$$\mathcal{L}(x) = \frac{5s}{(s^2+4)(s^2+2s+65)}$$

Use partial fractions.

$$\mathcal{L}(x) = \frac{5s}{(s^2+4)(s^2+2s+65)} = \frac{As+B}{s^2+4} + \frac{Cs+D}{s^2+2s+65}$$

$$5s = As^2 + Bs^2 + 2As^2 + 2Bs + 65As + 65B$$

$$+ Cs^3 + Ds^2 + 4Cs + 4D$$

Then,

$$A + C = 0 \quad C = -A = -\frac{61}{8}B$$

$$B + 2A + D = 0 \quad B + 2A - \frac{65}{4}B = 0 \rightarrow A = \frac{61}{8}B$$

$$65B + 4D = 0 \quad D = -\frac{65}{4}B$$

$$2B + 65A + 4C = 5$$

$$2B + (65)\left(\frac{61}{8}B\right) + (4)\left(-\frac{61}{8}B\right) = 5$$

$$B = 0.0107$$
$$A = 0.0816$$
$$C = -0.0816$$
$$D = -0.1739$$

$$\mathcal{L}(x) = \frac{0.0816s + 0.0107}{s^2 + 4} - \frac{0.0816 + 0.1739}{(s+1)^2 + (8)^2}$$
$$= (0.0816)\left(\frac{s}{s^2 + (2)^2}\right) + (0.00535)\left(\frac{2}{s^2 + (2)^2}\right) - (0.0816)\left(\frac{s-(-1)}{\left(s-(-1)\right)^2 + (8)^2} + \frac{1.1311}{\left(s-(-1)\right)^2 + (8)^2}\right)$$

Take the inverse transform. The response is

$$\boxed{x(t) = 0.0816\cos 2t + 0.00535\sin 2t - (0.0816)(e^{-t}\cos 8t + 0.14e^{-t}\sin 8t)}$$

Compare the values here with those obtained previously.

$$D = 0.0816 \text{ m}$$

This checks with part (c).

The natural frequency is 8 rad/sec, which corresponds to ω from part (a).

The answer is (B).

4. *Customary U.S. Solution*

First, consider static equilibrium. The bar mass is considered concentrated at 12 in from the hinge.

$$\sum M_A = 0$$

$$(3 \text{ lbm})\left(\frac{32.2 \ \frac{\text{ft}}{\text{sec}^2}}{32.2 \ \frac{\text{lbm-ft}}{\text{lbf-sec}^2}}\right)(24 \text{ in}) + (5 \text{ lbm})\left(\frac{32.2 \ \frac{\text{ft}}{\text{sec}^2}}{32.2 \ \frac{\text{lbm-ft}}{\text{lbf-sec}^2}}\right)(12 \text{ in}) - M_{\text{spring}} = 0$$

$$M_{\text{spring}} = 132 \text{ in-lbf}$$

The angle of rotation is

$$\theta = \frac{\delta}{L} = \frac{0.55 \text{ in}}{16 \text{ in}} = 0.0344 \text{ rad}$$

The equivalent torsional spring constant is

$$k_r = \frac{M_{\text{spring}}}{\theta} = \frac{132 \text{ in-lbf}}{0.0344 \text{ rad}} = 3837 \text{ in-lbf}$$

From App. 56.A, the mass moment of inertia of the bar rotating about its end is

$$I_{\text{bar}} = \tfrac{1}{3}mL^2 = \left(\tfrac{1}{3}\right)(5 \text{ lbm})(24 \text{ in})^2 = 960 \text{ lbm-in}^2$$

The mass moment of inertia of the concentrated mass is

$$I_{\text{mass}} = mL^2 = (3 \text{ lbm})(24 \text{ in})^2 = 1728 \text{ lbm-in}^2$$

The total mass moment of inertia of the system is

$$I = I_{\text{bar}} + I_{\text{mass}} = 960 \text{ lbm-in}^2 + 1728 \text{ lbm-in}^2 = 2688 \text{ lbm-in}^2$$

From Eq. 60.22(b), the natural frequency is

$$\omega = \sqrt{\frac{k_r g_c}{I}} = \sqrt{\frac{(3837 \text{ in-lbf})\left(386.4 \ \frac{\text{lbm-in}}{\text{lbf-sec}^2}\right)}{2688 \text{ lbm-in}^2}} = 23.49 \text{ rad/sec}$$

From Eq. 60.4 (Table 60.1 could also be used),

$$f = \frac{\omega}{2\pi} = \frac{23.49 \ \frac{\text{rad}}{\text{sec}}}{2\pi \ \frac{\text{rad}}{\text{rev}}} = \boxed{3.74 \text{ Hz} \quad (4 \text{ Hz})}$$

The answer is (A).

SI Solution

First, consider static equilibrium. The bar mass is considered concentrated at 0.30 m from the hinge.

$$\sum M_A = 0$$

$$(1.4 \text{ kg})\left(9.81 \ \frac{\text{m}}{\text{s}^2}\right)(0.6 \text{ m}) + (2.3 \text{ kg})\left(9.81 \ \frac{\text{m}}{\text{s}^2}\right)(0.3 \text{ m}) - M_{\text{spring}} = 0$$

$$M_{\text{spring}} = 15.01 \text{ N·m}$$

The angle of rotation is

$$\theta = \frac{\delta}{L} = \frac{0.014\text{ m}}{0.40\text{ m}} = 0.035\text{ rad}$$

The equivalent torsional spring constant is

$$k_r = \frac{M_{\text{spring}}}{\theta} = \frac{15.01\text{ N·m}}{0.035\text{ rad}} = 428.9\text{ N·m}$$

From App. 56.A, the mass moment of inertia of the bar rotating about its end is

$$\begin{aligned} I_{\text{bar}} &= \tfrac{1}{3}mL^2 = \left(\tfrac{1}{3}\right)(2.3\text{ kg})(0.60\text{ m})^2 \\ &= 0.276\text{ kg·m}^2 \end{aligned}$$

The mass moment of inertia of the concentrated mass is

$$I_{\text{mass}} = mL^2 = (1.4\text{ kg})(0.60\text{ m})^2 = 0.504\text{ kg·m}^2$$

The total mass moment of inertia of the system is

$$\begin{aligned} I &= I_{\text{bar}} + I_{\text{mass}} = 0.276\text{ kg·m}^2 + 0.504\text{ kg·m}^2 \\ &= 0.78\text{ kg·m}^2 \end{aligned}$$

From Eq. 60.22(a), the natural frequency is

$$\omega = \sqrt{\frac{k_r}{I}} = \sqrt{\frac{428.9\text{ N·m}}{0.78\text{ kg·m}^2}} = 23.45\text{ rad/s}$$

From Eq. 60.4 (Table 60.1 could also be used),

$$f = \frac{\omega}{2\pi} = \frac{23.45\ \frac{\text{rad}}{\text{sec}}}{2\pi\ \frac{\text{rad}}{\text{rev}}} = \boxed{3.73\text{ Hz}\quad(4\text{ Hz})}$$

The answer is (A).

5. *Customary U.S. Solution*

The static deflection caused by the electromagnet is

$$\begin{aligned} \delta_{\text{st}} &= \frac{\text{weight}}{k} = \frac{m\left(\frac{g}{g_c}\right)}{k} \\ &= \left(\frac{300\text{ lbm}}{1000\ \frac{\text{lbf}}{\text{in}}}\right)\left(\frac{32.2\ \frac{\text{ft}}{\text{sec}^2}}{32.2\ \frac{\text{lbm-ft}}{\text{lbf-sec}^2}}\right) \\ &= 0.3\text{ in} \end{aligned}$$

(a) From Eq. 60.4 and Eq. 60.7, the natural frequency is

$$\begin{aligned} f &= \frac{\omega}{2\pi} = \left(\frac{1}{2\pi}\right)\sqrt{\frac{g}{\delta_{\text{st}}}} = \left(\frac{1}{2\pi}\right)\sqrt{\frac{386.4\ \frac{\text{in}}{\text{sec}^2}}{0.3\text{ in}}} \\ &= \boxed{5.71\text{ Hz}\quad(5.7\text{ Hz})} \end{aligned}$$

The answer is (A).

(b) The minimum tension occurs at the upper limit of travel. The decrease in tension at that point is the same as the increase in tension at the lower limit caused by the scrap.

$$\begin{aligned} F_{\text{min}} &= (300\text{ lbm} - 200\text{ lbm})\left(\frac{32.2\ \frac{\text{ft}}{\text{sec}^2}}{32.2\ \frac{\text{lbm-ft}}{\text{lbf-sec}^2}}\right) \\ &= \boxed{100\text{ lbf}} \end{aligned}$$

The answer is (A).

SI Solution

The static deflection caused by the electromagnet is

$$\begin{aligned} \delta_{\text{st}} &= \frac{mg}{k} = \frac{(140\text{ kg})\left(9.81\ \frac{\text{m}}{\text{s}^2}\right)}{\left(175\ \frac{\text{kN}}{\text{m}}\right)\left(1000\ \frac{\text{N}}{\text{kN}}\right)} \\ &= 0.00785\text{ m} \end{aligned}$$

(a) From Eq. 60.4 and Eq. 60.7, the natural frequency is

$$\begin{aligned} f &= \frac{\omega}{2\pi} = \left(\frac{1}{2\pi}\right)\sqrt{\frac{g}{\delta_{\text{st}}}} = \left(\frac{1}{2\pi}\right)\sqrt{\frac{9.81\ \frac{\text{m}}{\text{s}^2}}{0.00785\text{ m}}} \\ &= \boxed{5.63\text{ Hz}\quad(5.7\text{ Hz})} \end{aligned}$$

The answer is (A).

(b) The minimum tension occurs at the upper limit of travel. The decrease in tension at that point is the same as the increase in tension at the lower limit caused by the scrap.

$$\begin{aligned} F_{\text{min}} &= (140\text{ kg} - 90\text{ kg})\left(9.81\ \frac{\text{m}}{\text{s}^2}\right) \\ &= \boxed{490.5\text{ N}\quad(490\text{ N})} \end{aligned}$$

The answer is (A).

6. *Customary U.S. Solution*

The equivalent spring constant is

$$k_{eq} = (4)\left(1000\ \frac{\text{lbf}}{\text{in}}\right) = 4000\ \text{lbf/in}$$

From Eq. 60.3(b), the natural frequency of the system is

$$\omega = \sqrt{\frac{k_{eq}g_c}{m}} = \sqrt{\frac{\left(4000\ \frac{\text{lbf}}{\text{in}}\right)\left(386.4\ \frac{\text{in-lbm}}{\text{lbf-sec}^2}\right)}{50\ \text{lbm}}}$$
$$= 175.8\ \text{rad/sec}$$

The forcing frequency is

$$\omega_f = \frac{\left(800\ \frac{\text{rev}}{\text{min}}\right)\left(2\pi\ \frac{\text{rad}}{\text{rev}}\right)}{60\ \frac{\text{sec}}{\text{min}}} = 83.77\ \text{rad/sec}$$

The out-of-balance force caused by the rotating eccentric mass is

$$F_0 = \frac{m_0\omega^2 e}{g_c} = \frac{\left(\frac{1\ \text{oz}}{16\ \frac{\text{oz}}{\text{lbm}}}\right)\left(83.77\ \frac{\text{rad}}{\text{sec}}\right)^2(5\ \text{in})}{386.4\ \frac{\text{in-lbm}}{\text{lbf-sec}^2}}$$
$$= 5.675\ \text{lbf}$$

The pseudo-static deflection is

$$\frac{F_0}{k_{eq}} = \frac{5.675\ \text{lbf}}{4000\ \frac{\text{lbf}}{\text{in}}} = 0.00142\ \text{in}$$

The ratio of frequencies is

$$r = \frac{\omega_f}{\omega} = \frac{83.77\ \frac{\text{rad}}{\text{sec}}}{175.8\ \frac{\text{rad}}{\text{sec}}} = 0.477$$

From Eq. 60.53, the magnification factor is

$$\beta = \frac{D}{\frac{F_0}{k}} = \left|\frac{1}{\sqrt{(1-r^2)^2 + (2\zeta r)^2}}\right|$$
$$= \left|\frac{1}{\sqrt{\left(1-(0.477)^2\right)^2 + \left((2)(0.125)(0.477)\right)^2}}\right|$$
$$= 1.28$$

The maximum vertical displacement is

$$D = \beta\left(\frac{F_0}{k_{eq}}\right) = (1.28)(0.00142\ \text{in})$$
$$= \boxed{0.00182\ \text{in}\quad (0.0018\ \text{in})}$$

The answer is (B).

SI Solution

The equivalent spring constant is

$$k_{eq} = (4)\left(175\ \frac{\text{kN}}{\text{m}}\right)\left(1000\ \frac{\text{N}}{\text{kN}}\right) = 700\,000\ \text{N/m}$$

From Eq. 60.3(a), the natural frequency of the system is

$$\omega = \sqrt{\frac{k_{eq}}{m}} = \sqrt{\frac{700\,000\ \frac{\text{N}}{\text{m}}}{23\ \text{kg}}} = 174.5\ \text{rad/s}$$

The forcing frequency is

$$\omega_f = \frac{\left(800\ \frac{\text{rev}}{\text{min}}\right)\left(2\pi\ \frac{\text{rad}}{\text{rev}}\right)}{60\ \frac{\text{s}}{\text{min}}} = 83.77\ \text{rad/s}$$

The out-of-balance force caused by the rotating eccentric mass is

$$F_0 = m_0\omega^2 e = \frac{(30\ \text{g})\left(83.77\ \frac{\text{rad}}{\text{s}}\right)^2(0.130\ \text{m})}{1000\ \frac{\text{g}}{\text{kg}}}$$
$$= 27.37\ \text{N}$$

The pseudo-static deflection is

$$\frac{F_0}{k_{eq}} = \frac{27.37\ \text{N}}{700\,000\ \frac{\text{N}}{\text{m}}} = 3.91\times 10^{-5}\ \text{m}$$

The ratio of frequencies is

$$r = \frac{\omega_f}{\omega} = \frac{83.77\ \frac{\text{rad}}{\text{s}}}{174.5\ \frac{\text{rad}}{\text{s}}}$$
$$= 0.48$$

From Eq. 60.53, the magnification factor is

$$\beta = \frac{D}{\frac{F_0}{k}} = \left| \frac{1}{\sqrt{(1-r^2)^2 + (2\zeta r)^2}} \right|$$

$$= \left| \frac{1}{\sqrt{\left(1-(0.48)^2\right)^2 + \left((2)(0.125)(0.48)\right)^2}} \right|$$

$$= 1.28$$

The maximum vertical displacement is

$$D = \beta\left(\frac{F_0}{k_{\text{eq}}}\right) = (1.28)(3.91 \times 10^{-5}\ \text{m})$$

$$= \boxed{5.0 \times 10^{-5}\ \text{m} \quad (5.0 \times 10^{-5}\ \text{m})}$$

The answer is (B).

7. *Customary U.S Solution*

(a) The first plate under the load is a simple beam with two concentrated forces and may be modeled as in case 8 of App. 51.A.

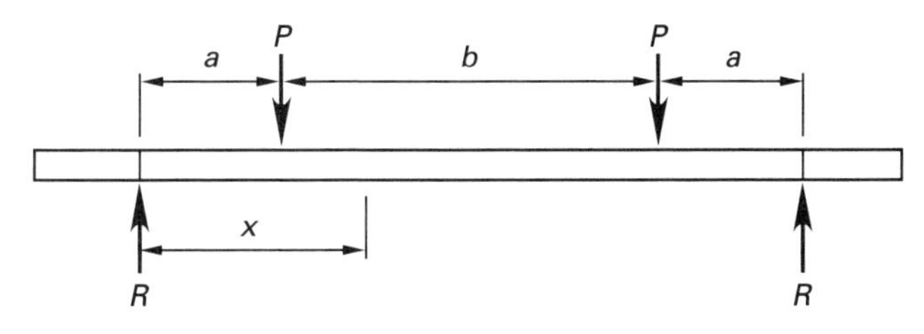

The load, P, is

$$P = \frac{20{,}000\ \text{lbf}}{2} = 10{,}000\ \text{lbf}$$

The locations of P are

$$a = \frac{36\ \text{in} - 24\ \text{in}}{2} = 6\ \text{in}$$

$$b = 24\ \text{in} \quad \text{[given]}$$

The moment of inertia of the cross section of the plate is

$$I = \frac{bh^3}{12} = \frac{(30\ \text{in})(0.5\ \text{in})^3}{12}$$

$$= \boxed{0.3125\ \text{in}^4 \quad (0.30\ \text{in}^4)}$$

The answer is (A).

(b) The overhangs do not contribute to the rigidity of the plate. The length of the beam model is

$$L = 2a + b = (2)(6\ \text{in}) + 24\ \text{in} = 36\ \text{in}$$

The deflection at the load $P(x = a)$ is

$$y = \left(\frac{P}{6EI}\right)(3Lax - 3a^2x - x^3)$$

$$= \left(\frac{Px}{6EI}\right)\left((3a)(L-a) - x^2\right)$$

$$= \frac{(10{,}000\ \text{lbf})(6\ \text{in})\left((3)(6\ \text{in})(36\ \text{in} - 6\ \text{in}) - (6\ \text{in})^2\right)}{(6)\left(2.9 \times 10^7\ \frac{\text{lbf}}{\text{in}^2}\right)(0.3125\ \text{in}^4)}$$

$$= 0.556\ \text{in}$$

The wide-beam correction is needed here. The deflection is multiplied by

$$1 - \nu^2 = 1 - (0.3)^2 = 0.91$$

$$y = (0.91)(0.556\ \text{in}) = 0.506\ \text{in}$$

The second plate is loaded exactly the same as the first plate, only upside down. Therefore, its deflection is also 0.556 in at the load. Since plates 3, 5, and 7 are loaded the same as plate 1 and since plates 4, 6, and 8 are loaded the same as plate 2, the total static deflection for the eight plates is

$$y_{\text{total}} = (4)(0.506\ \text{in}) + (4)(0.506\ \text{in})$$

$$= \boxed{4.048\ \text{in} \quad (4.0\ \text{in})}$$

The answer is (A).

(c) The maximum moment in the plates is at the load, P, and is

$$M_{\text{max}} = Pa$$

$$= (10{,}000\ \text{lbf})(6\ \text{in})$$

The maximum stress in the plates is at the extreme fiber and is

$$\sigma_{\text{max}} = \frac{M_{\text{max}}c}{I} = \frac{(60{,}000\ \text{in-lbf})\left(\frac{0.50\ \text{in}}{2}\right)}{0.3125\ \text{in}^4}$$

$$= \boxed{48{,}000\ \text{lbf/in}^2 \quad (48\ \text{kips/in}^2)}$$

The answer is (A).

(d) From Eq. 60.4 and Eq. 60.7, the natural frequency of oscillation is

$$f = \left(\frac{1}{2\pi}\right)\sqrt{\frac{g}{\delta_{st}}} = \left(\frac{1}{2\pi}\right)\sqrt{\frac{g}{y_{total}}} = \left(\frac{1}{2\pi}\right)\sqrt{\frac{386.4\ \frac{\text{in}}{\text{sec}^2}}{4.048\ \text{in}}} = \boxed{1.55\ \text{Hz}\quad (1.6\ \text{Hz})}$$

The answer is (A).

SI Solution

(a) The first plate under the load is a simple beam with two concentrated forces and may be modeled as in case 8 of App. 51.A.

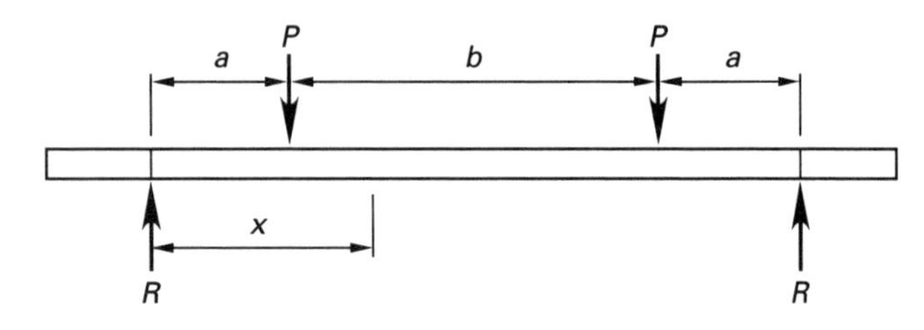

The load, P, is

$$P = \frac{(9100\ \text{kg})\left(9.81\ \frac{\text{m}}{\text{s}^2}\right)}{2} = 44\,636\ \text{N}$$

The locations of P are

$$a = \frac{0.92\ \text{m} - 0.60\ \text{m}}{2} = 0.16\ \text{m}$$

$$b = 0.60\ \text{m}\ \ [\text{given}]$$

The moment of inertia of the cross section of the plate is

$$I = \frac{bh^3}{12} = \frac{(0.76\ \text{m})(0.012\ \text{m})^3}{12} = \boxed{1.094\times 10^{-7}\ \text{m}^4\quad (1.0\times 10^{-7}\ \text{in}^4)}$$

The answer is (A).

(b) The overhangs do not contribute to the rigidity of the plate. The length of the beam model is

$$L = 2a + b = (2)(0.16\ \text{m}) + 0.60\ \text{m} = 0.92\ \text{m}$$

The deflection at the load $P(x = a)$ is

$$y = \left(\frac{P}{6EI}\right)(3Lax - 3a^2x - x^3) = \left(\frac{Px}{6EI}\right)\left((3a)(L-a) - x^2\right)$$

$$= \frac{(44\,636\ \text{N})(0.16\ \text{m}) \times \left(\begin{array}{c}(3)(0.16\ \text{m})(0.92\ \text{m} - 0.16\ \text{m}) \\ -\ (0.16\ \text{m})^2\end{array}\right)}{(6)(200\times 10^9\ \text{Pa})(1.094\times 10^{-7}\ \text{m}^4)} = 0.01845\ \text{m}$$

The wide-beam correction is needed here. The deflection is multiplied by

$$1 - \nu^2 - 1 - (0.3)^2 = 0.91$$

$$y = (0.91)(0.01845\ \text{m}) = 0.01679\ \text{m}$$

The second plate is loaded exactly the same as the first plate, only upside down. Therefore, its deflection is also 0.01845 m at the load. Since plates 3, 5, and 7 are loaded the same as plate 1 and since plates 4, 6, and 8 are loaded the same as plate 2, the total static deflection for the eight plates is

$$y_{total} = (4)(0.01679\ \text{m}) + (4)(0.01679\ \text{m}) = \boxed{0.1343\ \text{m}\quad (0.13\ \text{m})}$$

The answer is (A).

(c) The maximum moment in the plates is at the load, P, and is

$$M_{max} = Pa = (44\,636\ \text{N})(0.16\ \text{m}) = 7142\ \text{N·m}$$

The maximum stress in the plates is at the extreme fiber and is

$$\sigma_{max} = \frac{M_{max}c}{I} = \frac{(7142\ \text{N·m})\left(\frac{0.012\ \text{m}}{2}\right)}{1.094\times 10^{-7}\ \text{m}^4} = \boxed{3.92\times 10^8\ \text{Pa}\quad (390\ \text{MPa})}$$

The answer is (A).

(d) From Eq. 60.4 and Eq. 60.7, the natural frequency of oscillation is

$$f = \left(\frac{1}{2\pi}\right)\sqrt{\frac{g}{\delta_{st}}} = \left(\frac{1}{2\pi}\right)\sqrt{\frac{g}{y_{total}}} = \left(\frac{1}{2\pi}\right)\sqrt{\frac{9.81\ \frac{\text{m}}{\text{s}^2}}{0.1343\ \text{m}}} = \boxed{1.36\ \text{Hz}\quad (1.6\ \text{Hz})}$$

The answer is (A).

8. *Customary U.S. Solution*

(a) The forcing frequency is

$$\omega_f = \frac{\left(1200\ \frac{\text{rev}}{\text{min}}\right)\left(2\pi\ \frac{\text{rad}}{\text{rev}}\right)}{60\ \frac{\text{sec}}{\text{min}}} = 125.7\ \text{rad/sec}$$

The out-of-balance force caused by the rotating imbalance is

$$\begin{aligned} F_0 &= \frac{m_0\omega^2 e}{g_c} = \frac{(3.6\ \text{lbm})\left(125.7\ \frac{\text{rad}}{\text{sec}}\right)^2(3\ \text{in})}{386.4\ \frac{\text{in-lbm}}{\text{lbf-sec}^2}} \\ &= \boxed{441.6\ \text{lbf}\quad(440\ \text{lbf})} \end{aligned}$$

The answer is (D).

(b) The transmissibility is

$$\text{TR} = \frac{|F_{\text{transmitted}}|}{F_{\text{applied}}} = 0.05$$

From Eq. 60.55 and Eq. 60.56, the transmissibility for negligible damping and a value of TR < 1 is

$$\text{TR} = \frac{1}{\left(\frac{\omega_f}{\omega}\right)^2 - 1}$$

$$\frac{\omega_f}{\omega} = \sqrt{\frac{1}{\text{TR}} + 1} = \sqrt{\frac{1}{0.05} + 1} = 4.5826$$

The required natural frequency is

$$\begin{aligned} \omega &= \frac{\omega_f}{4.5826} = \frac{125.7\ \frac{\text{rad}}{\text{sec}}}{4.5826} \\ &= \boxed{27.43\ \text{rad/sec}\quad(27\ \text{rad/sec})} \end{aligned}$$

The answer is (C).

(c) From Eq. 60.3(a), the required stiffness of the system is

$$k_{\text{eq}} = \frac{m\omega^2}{g_c} = \frac{(175\ \text{lbm})\left(27.43\ \frac{\text{rad}}{\text{sec}}\right)^2}{386.4\ \frac{\text{in-lbm}}{\text{lbf-sec}^2}} = 340.8\ \text{lbf/in}$$

For four identical springs in parallel, the required stiffness for an individual spring is

$$k = \frac{k_{\text{eq}}}{4} = \frac{340.8\ \frac{\text{lbf}}{\text{in}}}{4} = \boxed{85.2\ \text{lbf/in}\quad(85\ \text{lbf/in})}$$

The answer is (C).

(d) The pseudo-static deflection is

$$\frac{F_0}{k_{\text{eq}}} = \frac{441.6\ \text{lbf}}{340.8\ \frac{\text{lbf}}{\text{in}}} = 1.30\ \text{in}$$

From Eq. 60.50, the amplitude of vibration is

$$\begin{aligned} D &= \left(\frac{F_0}{k_{\text{eq}}}\right)\left|\frac{1}{1 - \left(\frac{\omega_f}{\omega}\right)^2}\right| = \left(\frac{F_0}{k_{\text{eq}}}\right)(\text{TR}) \\ &= (1.30\ \text{in})(0.05) \\ &= \boxed{0.065\ \text{in}\quad(0.065\ \text{in})} \end{aligned}$$

The answer is (C).

SI Solution

(a) The forcing frequency is

$$\omega_f = \frac{\left(1200\ \frac{\text{rev}}{\text{min}}\right)\left(2\pi\ \frac{\text{rad}}{\text{rev}}\right)}{60\ \frac{\text{s}}{\text{min}}} = 125.7\ \text{rad/s}$$

The out-of-balance force caused by the rotating imbalance is

$$\begin{aligned} F_0 &= m_0\omega^2 e = (1.6\ \text{kg})\left(125.7\ \frac{\text{rad}}{\text{s}}\right)^2(0.075\ \text{m}) \\ &= \boxed{1896\ \text{N}\quad(1900\ \text{N})} \end{aligned}$$

The answer is (D).

(b) The transmissibility is

$$\text{TR} = \frac{|F_{\text{transmitted}}|}{F_{\text{applied}}} = 0.05$$

From Eq. 60.55 and Eq. 60.56, the transmissibility for negligible damping and a value of TR < 1 is

$$\text{TR} = \frac{1}{\left(\frac{\omega_f}{\omega}\right)^2 + 1}$$

$$\frac{\omega_f}{\omega} = \sqrt{\frac{1}{\text{TR}} + 1} = \sqrt{\frac{1}{0.05} + 1} = 4.5826$$

The required natural frequency is

$$\begin{aligned} \omega &= \frac{\omega_f}{4.5826} = \frac{125.7\ \frac{\text{rad}}{\text{s}}}{4.5826} \\ &= \boxed{27.43\ \text{rad/s}\quad(27\ \text{rad/s})} \end{aligned}$$

The answer is (C).

(c) From Eq. 60.3(a), the required stiffness of the system is

$$k_{eq} = m\omega^2 = (80\ \text{kg})\left(27.43\ \frac{\text{rad}}{\text{s}}\right)^2 = 60\,192\ \text{N/m}$$

For four identical springs in parallel, the required stiffness for an individual spring is

$$k = \frac{k_{eq}}{4} = \frac{60\,192\ \frac{\text{N}}{\text{m}}}{4} = \boxed{15\,048\ \text{N/m}\quad (15\ \text{kN/m})}$$

The answer is (C).

(d) The pseudo-static deflection is

$$\frac{F_0}{k_{eq}} = \frac{1896\ \text{N}}{60\,192\ \frac{\text{N}}{\text{m}}} = 0.0315\ \text{m}$$

From Eq. 60.50, the amplitude of vibration is

$$\begin{aligned} D &= \left(\frac{F_0}{k_{eq}}\right)\left|\frac{1}{1-\left(\frac{\omega_f}{\omega}\right)^2}\right| = \left(\frac{F_0}{k_{eq}}\right)(\text{TR}) \\ &= (0.0315\ \text{m})(0.05) \\ &= \boxed{0.00157\ \text{m}\quad (0.0016\ \text{m})} \end{aligned}$$

The answer is (C).

9. *Customary U.S. Solution*

(a) From the fan rotation,

$$\frac{600\ \frac{\text{rev}}{\text{min}}}{60\ \frac{\text{sec}}{\text{min}}} = \boxed{10\ \text{Hz}}$$

The answer is (A).

(b) From the driving blades,

$$\frac{\left(600\ \frac{\text{rev}}{\text{min}}\right)(8)}{60\ \frac{\text{sec}}{\text{min}}} = \boxed{80\ \text{Hz}}$$

The answer is (C).

(c) From the fan blades,

$$\frac{\left(600\ \frac{\text{rev}}{\text{min}}\right)(64)}{60\ \frac{\text{sec}}{\text{min}}} = \boxed{640\ \text{Hz}}$$

The answer is (D).

(d) From the motor rotation,

$$\frac{1725\ \frac{\text{rev}}{\text{min}}}{60\ \frac{\text{sec}}{\text{min}}} = \boxed{28.75\ \text{Hz}\quad (30\ \text{Hz})}$$

The answer is (A).

(e) From the poles,

$$\frac{\left(1725\ \frac{\text{rev}}{\text{min}}\right)(4)}{60\ \frac{\text{sec}}{\text{min}}} = \boxed{115\ \text{Hz}\quad (120\ \text{Hz})}$$

The answer is (C).

(f) The electrical hum is $\boxed{60\ \text{Hz.}}$

The answer is (B).

(g) The pulleys are

$$\begin{aligned} \text{motor pulley (same as motor)} &= \boxed{29\ \text{Hz}} \\ \text{fan pulley (same as fan)} &= \boxed{10\ \text{Hz}} \end{aligned}$$

The answer is (B).

(h) For the belt,

$$\begin{aligned} \text{belt speed} = \pi D n &= \pi(4\ \text{in})\left(\frac{1725\ \frac{\text{rev}}{\text{min}}}{60\ \frac{\text{sec}}{\text{min}}}\right) \\ &= 361.3\ \text{in/sec} \end{aligned}$$

The frequency is

$$f = \frac{361.3\ \frac{\text{in}}{\text{sec}}}{72\ \text{in}} = \boxed{5.02\ \text{Hz}\quad (5.0\ \text{Hz})}$$

The answer is (C).

SI Solution

(a) From the fan rotation,

$$\frac{600\ \frac{\text{rev}}{\text{min}}}{60\ \frac{\text{s}}{\text{min}}} = \boxed{10\ \text{Hz}}$$

The answer is (A).

(b) From the driving blades,

$$\frac{\left(600\ \frac{\text{rev}}{\text{min}}\right)(8)}{60\ \frac{\text{s}}{\text{min}}} = \boxed{80\ \text{Hz}}$$

The answer is (C).

(c) From the fan blades,

$$\frac{\left(600\ \frac{\text{rev}}{\text{min}}\right)(64)}{60\ \frac{\text{s}}{\text{min}}} = \boxed{640\ \text{Hz}}$$

The answer is (D).

(d) From the motor rotation,

$$\frac{1725\ \frac{\text{rev}}{\text{min}}}{60\ \frac{\text{s}}{\text{min}}} = \boxed{28.75\ \text{Hz}\quad (30\ \text{Hz})}$$

The answer is (A).

(e) From the poles,

$$\frac{\left(1725\ \frac{\text{rev}}{\text{min}}\right)(4)}{60\ \frac{\text{sec}}{\text{min}}} = \boxed{115\ \text{Hz}\quad (120\ \text{Hz})}$$

The answer is (C).

(f) The electrical hum is $\boxed{60\ \text{Hz.}}$

The answer is (B).

(g) The pulleys are

$$\text{motor pulley (same as motor)} = \boxed{29\ \text{Hz}}$$
$$\text{fan pulley (same as fan)} = \boxed{10\ \text{Hz}}$$

The answer is (B).

(h) For the belt,

$$\text{belt speed} = \frac{(\pi)(100\ \text{mm})\left(\dfrac{1725\ \frac{\text{rev}}{\text{min}}}{60\ \frac{\text{s}}{\text{min}}}\right)}{1000\ \frac{\text{mm}}{\text{m}}} = 9.032\ \text{m/s}$$

The frequency is

$$f = \frac{9.032\ \frac{\text{m}}{\text{s}}}{1.83\ \text{m}} = \boxed{4.94\ \text{Hz}\quad (5.0\ \text{Hz})}$$

The answer is (C).

10. *Customary U.S Solution*

The forcing frequency is

$$f_f = \frac{1725\ \frac{\text{rev}}{\text{min}}}{60\ \frac{\text{sec}}{\text{min}}} = 28.75\ \text{Hz}$$

The natural frequency is

$$f = \left(\frac{1}{2\pi}\right)\sqrt{\frac{g}{\delta_{\text{st}}}} = \left(\frac{1}{2\pi}\right)\sqrt{\frac{386.4\ \frac{\text{in}}{\text{sec}^2}}{0.02\ \text{in}}} = 22.12\ \text{Hz}$$

From Eq. 60.56 with negligible damping,

$$\text{TR} = \frac{1}{\sqrt{(1-r^2)^2}} = \frac{1}{\sqrt{\left(1-\left(\frac{f_f}{f}\right)^2\right)^2}} = \frac{1}{\sqrt{\left(1-\left(\frac{28.75\ \text{Hz}}{22.12\ \text{Hz}}\right)^2\right)^2}} = 1.451$$

$\boxed{\text{This is a 45.1\% (50\%) increase in force.}}$

The answer is (B).

SI Solution

The forcing frequency is

$$f_f = \frac{1725\ \frac{\text{rev}}{\text{min}}}{60\ \frac{\text{sec}}{\text{min}}} = 28.75\ \text{Hz}$$

The natural frequency is

$$f = \left(\frac{1}{2\pi}\right)\sqrt{\frac{g}{\delta_{\text{st}}}} = \left(\frac{1}{2\pi}\right)\sqrt{\frac{\left(9.81\ \frac{\text{m}}{\text{s}^2}\right)\left(1000\ \frac{\text{mm}}{\text{m}}\right)}{0.5\ \text{mm}}} = 22.29\ \text{Hz}$$

From Eq. 60.56 with negligible damping,

$$\text{TR} = \frac{1}{\sqrt{(1-r^2)^2}} = \frac{1}{\sqrt{\left(1-\left(\frac{f_f}{f}\right)^2\right)^2}} = \frac{1}{\sqrt{\left(1-\left(\frac{28.75\ \text{Hz}}{22.29\ \text{Hz}}\right)^2\right)^2}} = 1.507$$

$\boxed{\text{This is a 50.7\% (50\%) increase in force.}}$

The answer is (B).

Dynamics and Vibrations

11. (a) The moment of inertia of the flagpole is

$$
\begin{aligned}
I &= \tfrac{1}{12}(b_o h_o^3 - b_i h_i^3) \\
&= \left(\tfrac{1}{12}\right)\left(\begin{array}{l}(4\text{ in})(4\text{ in})^3 - \big(4\text{ in} - (2)(0.125\text{ in})\big) \\ \quad \times \big(4\text{ in} - (2)(0.125\text{ in})\big)^3\end{array}\right) \\
&= 4.854\text{ in}^4
\end{aligned}
$$

For a simple cantilever with distributed weight, $w = W/L$, the stiffness can be calculated as W/δ. From App. 51.A, case 2, the stiffness is

$$
\begin{aligned}
k &= \frac{W}{\delta} = \frac{W}{\dfrac{wL^4}{8EI}} = \frac{W}{\dfrac{WL^3}{8EI}} \\
&= \frac{8EI}{L^3} \\
&= \frac{(8)\left(29\times 10^6\ \frac{\text{lbf}}{\text{in}^2}\right)(4.854\text{ in}^4)}{\left((22\text{ ft})\left(12\ \frac{\text{in}}{\text{ft}}\right)\right)^3} \\
&= 61.20\text{ lbf/in}
\end{aligned}
$$

Only the linear frequency, f, has units of Hz (not rad/sec). From Eq. 60.3(b) and Eq. 60.4, the undamped natural (linear) frequency is

$$
\begin{aligned}
f &= \frac{\omega}{2\pi} = \frac{1}{2\pi}\sqrt{\frac{kg_c}{m}} \\
&= \frac{1}{2\pi}\sqrt{\frac{\left(61.20\ \frac{\text{lbf}}{\text{in}}\right)\left(32.2\ \frac{\text{lbm-ft}}{\text{lbf-sec}^2}\right)\left(12\ \frac{\text{in}}{\text{ft}}\right)}{150\text{ lbm}}} \\
&= \boxed{1.998\text{ Hz}\quad(2.0\text{ Hz})}
\end{aligned}
$$

The answer is (B)

(b) Based on Eq. 60.41, the damped natural (linear) frequency is

$$
\begin{aligned}
\omega_d &= \omega\sqrt{1-\zeta^2} \\
2\pi f_d &= 2\pi f\sqrt{1-\zeta^2} \\
f_d &= f\sqrt{1-\zeta^2} = 1.998\text{ Hz}\sqrt{1-(0.119)^2} \\
&= \boxed{1.984\text{ Hz}\quad(2.0\text{ Hz})}
\end{aligned}
$$

The answer is (B).

61 Modeling of Engineering Systems

PRACTICE PROBLEMS

1. For the system of ideal elements shown, (a) draw the system diagram. (b) What is the differential equation for node 1 and node 2, respectively?

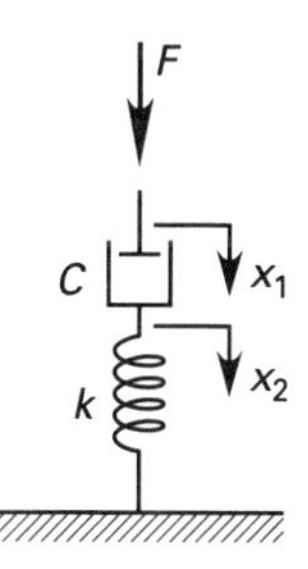

(A) $F = (x_1' - x_2')$; $F = kx_1$

(B) $F = (x_1 - x_2)$; $F = kx_1 + kx_2$

(C) $F = C(x_1' - x_2')$; $F = kx_2$

(D) $F = C(x_2' - x_1')$; $F = kx_2 - kx_1$

2. For the system of ideal elements shown, (a) draw the system diagram. (b) What is the differential equation for node 1 and node 2, respectively?

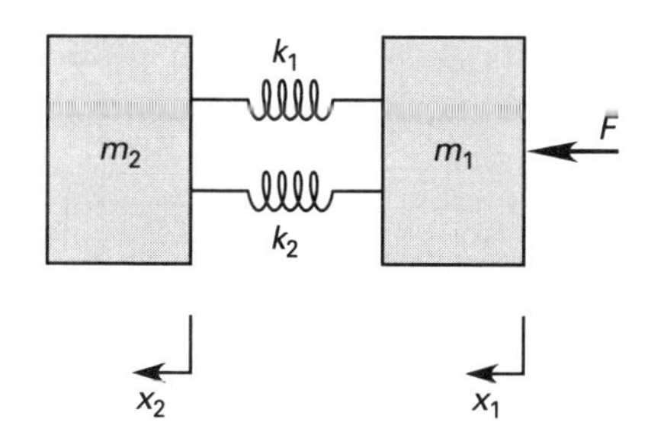

(A) $F = m_1x_1'' + (k_1 + k_2)(x_1 - x_2)$;
$0 = m_2x_2'' + (k_1 + k_2)(x_2 - x_1)$

(B) $F = m_2x_1'' + (k_1 - k_2)(x_2 - x_1)$;
$0 = m_2x_2'' + (k_1 - k_2)(x_2 - x_1)$

(C) $F = m_1x_1'' + (k_1 + k_2)(x_1 + x_2)$;
$0 = m_2x_2'' + (k_1 - k_2)(x_1 + x_2)$

(D) $F = m_2x_2'' + (k_1 - k_2)(x_1 - x_2)$;
$0 = m_2x_2'' + (k_2 + k_1)(x_1 - x_2)$

3. For the system of ideal elements shown, (a) draw the system diagram. (b) What is the differential equation for the system?

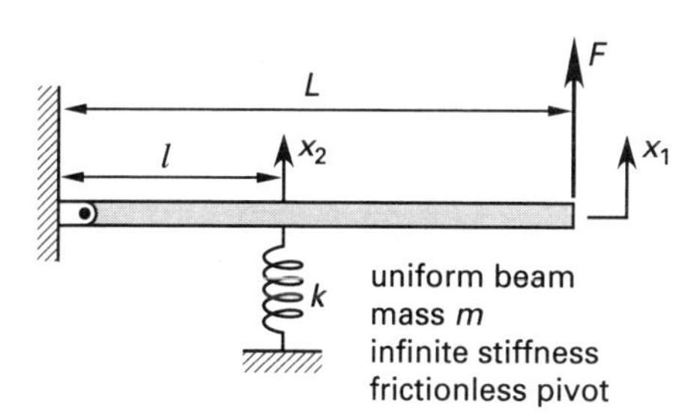

(A) $FL = (\frac{1}{5}mL^2)\theta - kl^2\theta$

(B) $FL = (\frac{1}{5}mL^2)\theta'' + kl^2\theta$

(C) $FL = (\frac{1}{3}mL^2)\theta'' + kl^2\theta$

(D) $FL = (\frac{1}{3}mL^2)\theta' - kl^2\theta$

4. For the system of ideal elements shown, (a) draw the system diagram. (b) What are the transformer equations for the system?

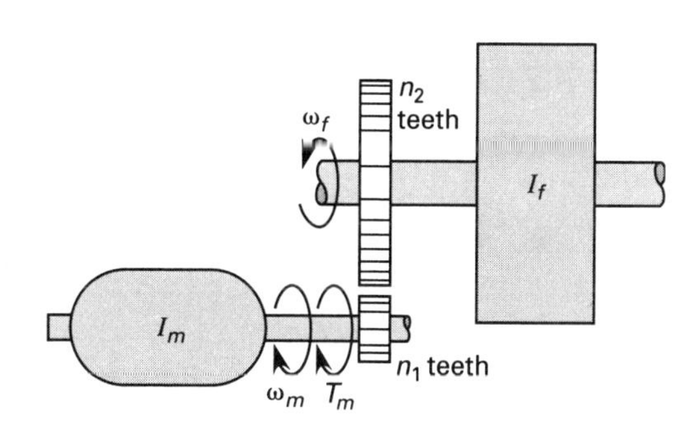

(A) $T_2 = \left(\frac{n_2}{n_1}\right)T_1$; $\theta_m = \left(\frac{n_2}{n_1}\right)\theta_f$

(B) $T_2 = \left(\frac{n_2}{n_1}\right)T_f$; $\theta_m = \left(\frac{n_2}{n_1}\right)\theta_1$

(C) $T_2 = \left(\frac{n_1}{n_2}\right)T_m$; $\theta_m = \left(\frac{n_1}{n_2}\right)\theta_1$

(D) $T_2 = \left(\frac{n_1}{n_2}\right)T_1$; $\theta_m = \left(\frac{n_1}{n_2}\right)\theta_2$

5. For the system of ideal elements shown, (a) draw the system diagram. (b) What is the differential equation for node 1 and node 2, respectively?

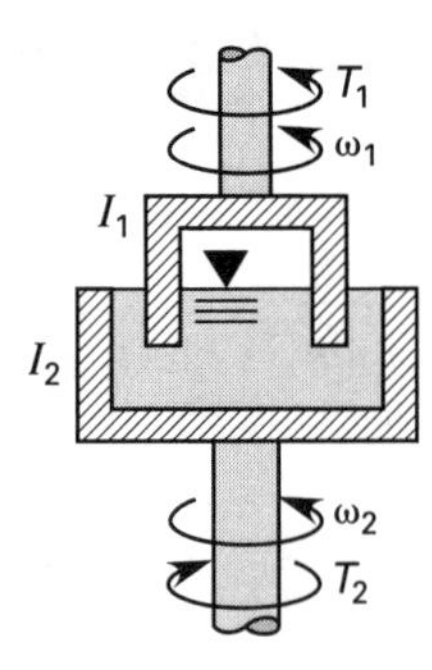

(A) $T_1 = \theta_1'' + C_1(\theta_1'' - \theta_2'')$;
$-T_2 = I_2\theta_2' - C_r(\theta_2'' - \theta_1'')$

(B) $T_1 = I_1\theta_1'' + C_r(\theta_1' - \theta_2')$;
$-T_2 = I_2\theta_2'' + C_r(\theta_2' - \theta_1')$

(C) $T_1 = I_1\theta_1' - C_r(\theta_1'' - \theta_2')$;
$-T_2 = I_2\theta_2' + C_r(\theta_1' - \theta_2')$

(D) $T_1 = I_1\theta_1'' + C_1(\theta_2' - \theta_1')$;
$-T_2 = I_2\theta_2'' - C_r(\theta_2'' - \theta_1'')$

6. The coupling of a railroad car is modeled as the mechanical system shown. Assume all elements are linear. What are the system equations that describe the positions x_1 and x_2 as functions of time?

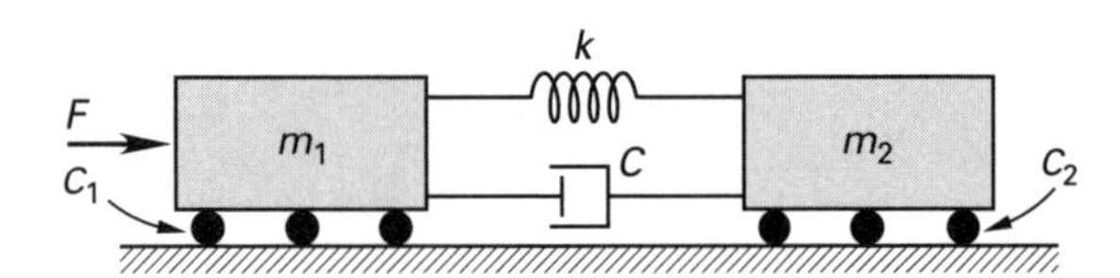

(A) $F = m_1x_1' + C_1x_1' + C(x_1' + x_2') + k(x_1 - x_2)$;
$0 = C_2x_2' + m_2x_2'' + C(x_1' - x_2') + k(x_2 + x_1)$

(B) $F = m_1x_1'' + C_1x_1'' + C(x_2' + x_1') + k(x_1 + x_2)$;
$0 = C_2x_2' + m_2x_2'' + C(x_1' - x_2') + k(x_2 + x_1)$

(C) $F = m_1x_1'' + C_1x_1'' + C(x_1' - x_2') + k(x_1 - x_2)$;
$0 = C_2x_2'' + m_2x_2' + C(x_1' + x_2') + k(x_2 - x_1)$

(D) $F = m_1x_1'' + C_1x_1' + C(x_1' - x_2') + k(x_1 - x_2)$;
$0 = C_2x_2' + m_2x_2'' + C(x_2' - x_1') + k(x_2 - x_1)$

7. Water is discharged freely at a constant rate into an open tank. Water flows out of the tank through a drain with a resistance to flow.

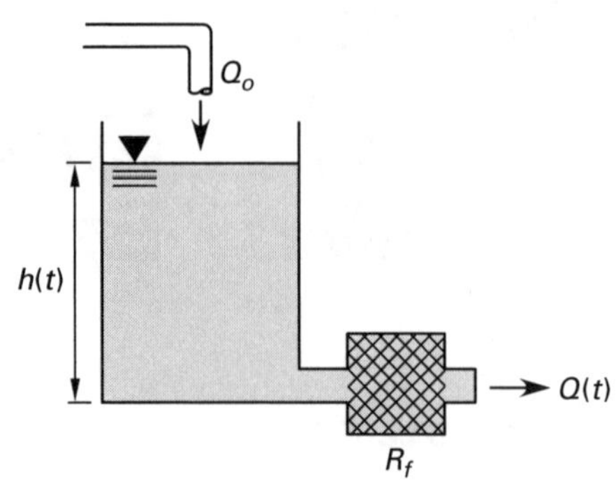

(a) Draw the system diagram using idealized elements. (b) What is the differential equation for node 1 and node 2, respectively, that describes the response of the system?

(A) $Q_1 = C_{f_1}\left(\frac{dp_1}{dt}\right) + \left(\frac{1}{R_f}\right)(p_1 - p_2)$;
$Q_2 = \left(\frac{1}{R_f}\right)(p_1 - p_2)$

(B) $Q_1 = C_{f_1}\left(\frac{dp_1}{dt}\right) - \left(\frac{1}{R_f}\right)(p_2 + p_1)$;
$Q_2 = \left(\frac{1}{C_{f_1}}\right)(p_2 + p_1)$

(C) $Q_1 = C_{f_1}\left(\frac{dp_2}{dt}\right) + \left(\frac{1}{R_f}\right)(p_2 - p_1)$;
$Q_2 = \left(\frac{1}{R_f}\right)(p_1 + p_2)$

(D) $Q_1 = C_{f_1}\left(\frac{dp_1}{dt}\right) - \left(\frac{1}{R_f}\right)(p_1 - p_2)$;
$Q_2 = \left(\frac{1}{C_{f_1}}\right)(p_1 - p_2)$

8. Water is pumped into the bottom of an open tank.

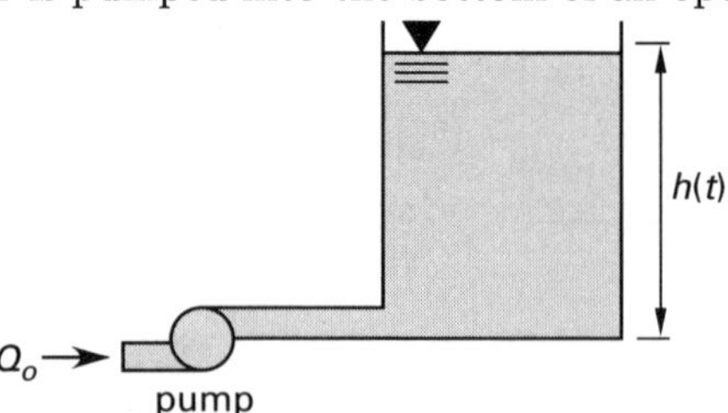

For the system of ideal elements shown, (a) draw the system diagram. (b) What is the differential equation for the resistor and the capacitor, respectively, that describes the response of the system?

(A) $Q = \frac{p_1 + p_2}{C_f}$; $Q = R_f\left(\frac{dp_2 - dp_1}{dt}\right)$

(B) $Q = \frac{p_2 - p_1}{R_f}$; $Q = C_f\left(\frac{dp_1 + dp_2}{dt}\right)$

(C) $Q = \frac{p_1 - p_2}{C_f}$; $Q = C_f\left(\frac{dp_1 + dp_2}{dt}\right)$

(D) $Q = \frac{p_1 - p_2}{R_f}$; $Q = C_f\left(\frac{dp_2}{dt}\right)$

SOLUTIONS

1. (a) The velocity of the plunger is v_1. The velocity of the body of the damper is the same as the upper part of the spring, v_2. By Rule 2, the other end of the force and the spring is attached to the stationary wall at $v=0$. The system diagram is

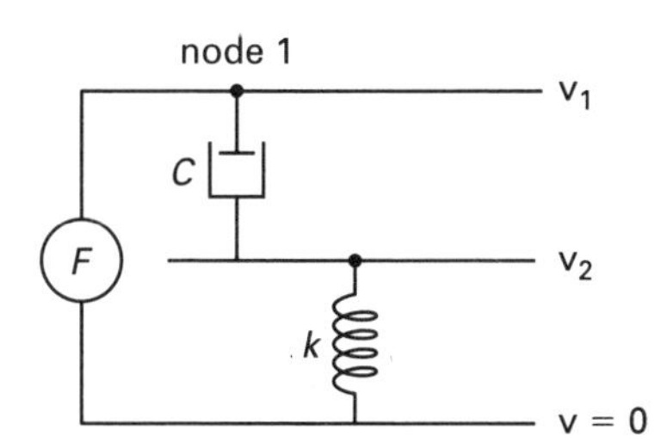

(b) By Rule 3, the force from the source is the same force experienced by the dashpot. One of the system equations is based on node 1. Using Rule 4 and expanding with Eq. 61.5,

$$\boxed{F = F_C = C(v_1 - v_2) = C(x_1' - x_2')}$$

By Rule 3, the force from the source is the same force experienced by the spring. A second system equation is based on node 2. Using Rule 4 and expanding with Eq. 61.4,

$$\boxed{F = F_k = k(x_2 - 0) = kx_2}$$

The answer is (C).

2. (a) The velocity of the ends of the springs connected to m_i is v_1. The velocity of the ends of the springs connected to m_2 is v_2. By Rule 1, the other end of each mass connects to $v=0$. The system diagram is

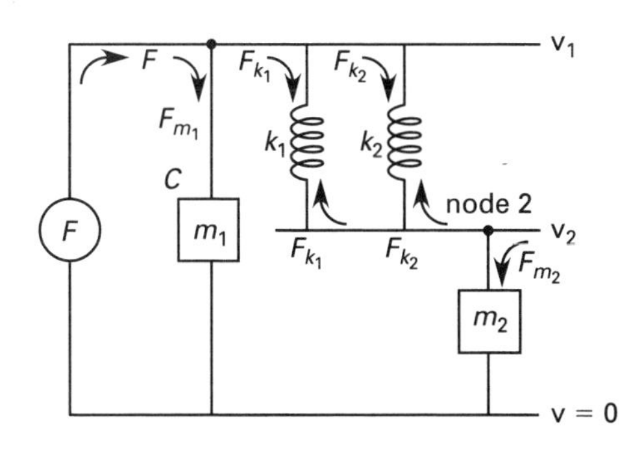

(b) The force leaving the source splits: some of it goes through m_1, some of it goes through k_1, and some of it goes through k_2. One of the system equations is based on node 1.

$$F = F_{m_1} + F_{k_1} + F_{k_2}$$

Using Rule 4 and expanding with Eq. 61.3 and Eq. 61.4,

$$\boxed{\begin{aligned} F &= m_1 a_1 + k_1(x_1 - x_2) + k_2(x_1 - x_2) \\ &= m_1 x_1'' + (k_1 + k_2)(x_1 - x_2) \end{aligned}}$$

A second system equation is based on node 2. The conservation law is written to conserve force in the v_2 line.

$$0 = F_{m_2} + F_{k_2} + F_{k_1}$$

Using Rule 4 and expanding with Eq. 61.3 and Eq. 61.4,

$$\boxed{\begin{aligned} 0 &= m_2 a_2 + k_2(x_2 - x_1) + k_1(x_2 - x_1) \\ &= m_2 x_2'' + (k_1 + k_2)(x_2 - x_1) \end{aligned}}$$

The answer is (A).

3. (a) Treat this as a rotational system. The applied rotational torque is

$$T = FL$$

The equivalent torsional spring constant is

$$k_r = \frac{M_{\text{resisting}}}{\theta} = \frac{F_k l}{\theta} = \frac{k x_2 l}{\theta}$$

However, $x_2 = l \sin\theta$ and $\theta \approx \sin\theta$ for small angles.

$$k_r = kl^2$$

The moment of inertia of the beam about the hinge point is

$$I = \tfrac{1}{3}mL^2$$

The equivalent rotational system is

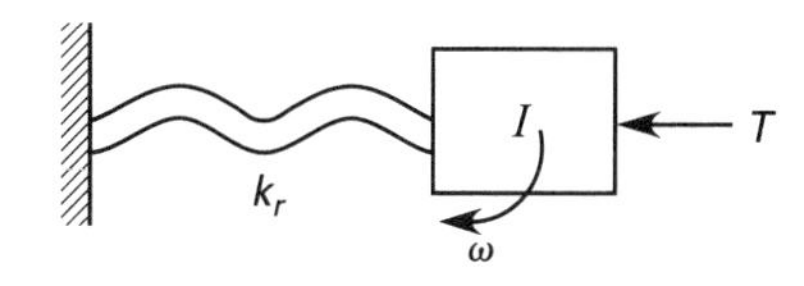

Control Systems

The angular velocity of the end of the spring connected to the inertial element is ω. By Rule 2, the other end of the spring is attached to the stationary wall at $\omega = 0$. The system diagram is

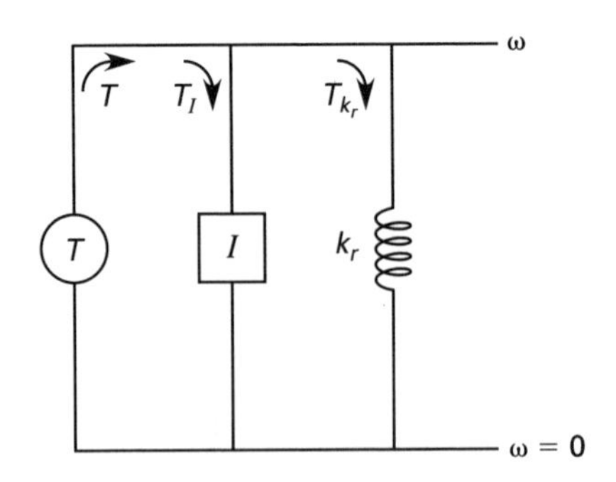

(b) The torque leaving the source splits: some of it goes through I and some of it goes through k_r. The conservation law is written to conserve torque in the ω line.

$$T = T_I + T_{k_r}$$

Using Rule 4 and expanding with Eq. 61.6 and Eq. 61.7,

$$T = I\alpha + k_r(\theta - 0)$$

$$\boxed{FL = (\tfrac{1}{3}mL^2)\theta'' + kl^2\theta}$$

The answer is (C).

4. (a) The angular velocity of the small gear is ω_m, and the angular velocity of the large gear is ω_f. By Rule 2, the other end of each inertia connects to $\omega = 0$. The gearing transforms the torque and angular displacement from gear 1 to gear 2. The system diagram is

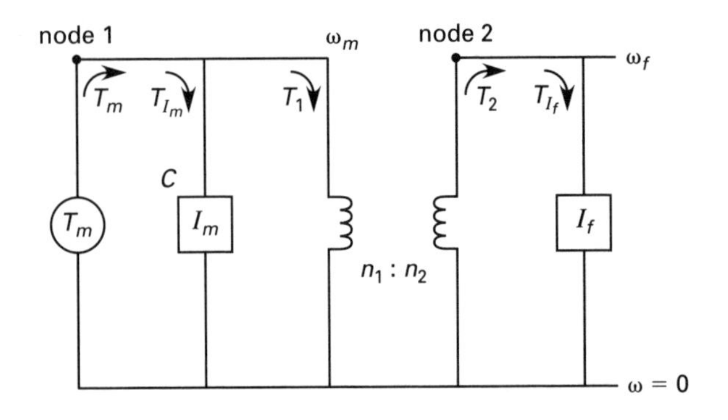

(b) The conservation law based on node 1 is written to conserve torque in the ω_m line.

$$T_m = T_{I_m} + T_1 = I_m\alpha_m + T_1 = I_m\theta_m'' + T_1$$

The same conservation principle based on node 2 is used to conserve torque in the ω_f line.

$$T_2 = T_{I_f} = I_f\alpha_f = I_f\theta_f''$$

The transformer equations are

$$\boxed{\begin{aligned} T_2 &= \left(\frac{n_2}{n_1}\right)T_1 \\ \theta_m &= \left(\frac{n_2}{n_1}\right)\theta_f \end{aligned}}$$

The answer is (A).

5. (a) Consider the fluid to act as a damper with coefficient C_r. The plunger is connected to velocity ω_1, and the body is connected to velocity ω_2. By Rule 2, the ends of the inertia elements are connected to $\omega = 0$. The system diagram is

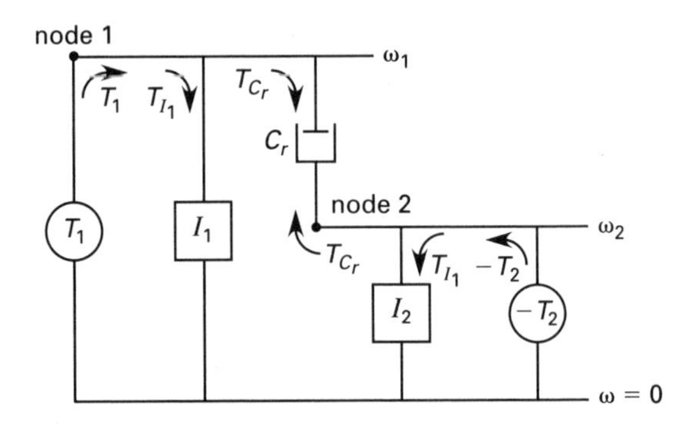

(b) One of the system equations is based on conservation of torque at node 1.

$$T_1 = T_{I_1} + T_{C_r}$$

Using Rule 4 and expanding with Eq. 61.6 and Eq. 61.8,

$$\boxed{T_1 = I_1\alpha_1 + C_r(\omega_1 - \omega_2) = I_1\theta_1'' + C_r(\theta_1' - \theta_2')}$$

The second system equation is based on conservation of torque at node 2. T_2 is negative because it is acting in the opposite direction of T_1, ω_1, and ω_2.

$$\boxed{\begin{aligned} -T_2 &= T_{I_2} + T_{C_r} = I_2\alpha_2 + C_r(\omega_2 - \omega_1) \\ &= I_2\theta_2'' + C_r(\theta_2' - \theta_1') \end{aligned}}$$

The answer is (B).

6. The velocity of the end of the spring connected to m_1 is v_1. This is also the velocity of the plunger and the velocity of the viscous damper, C_1. The velocity of the end of the spring connected to m_2 is v_2. This is also the velocity of the body of the damper and the velocity of the viscous damper, C_2. By Rule 2, the other end of each mass connects to $v = 0$. The system diagram is

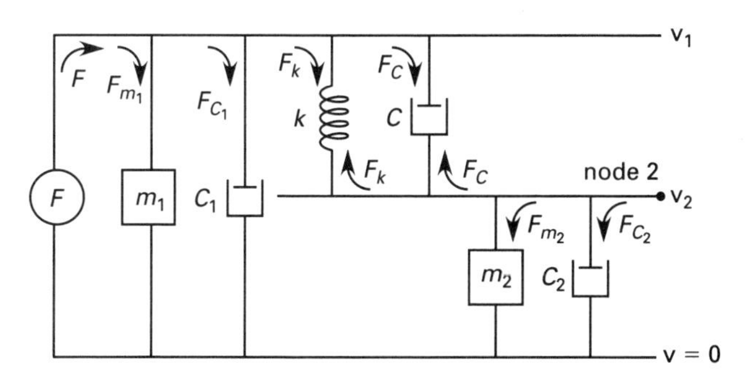

The force leaving the source splits: some of it goes through m_1, C_1, k, and C. The conservation law is written to conserve force in the v_1 line. The equation is based on node 1.

$$F = F_{m_1} + F_{C_1} + F_k + F_C$$

Using Rule 4 and expanding with Eq. 61.3, Eq. 61.4, and Eq. 61.5,

$$\begin{aligned} F &= m_1 a_1 + C_1(v_1 - 0) + C(v_1 - v_2) + k(x_1 - x_2) \\ &= m_1 x_1'' + C_1 x_1' + C(x_1' - x_2') + k(x_1 - x_2) \end{aligned}$$

The same conservation principle based on node 2 is used to conserve force in the v_2 line. Using Rule 4 and expanding with Eq. 61.3, Eq. 61.4, and Eq. 61.5,

$$\begin{aligned} 0 &= F_{C_2} + F_{m_2} + F_C + F_k \\ &= C_2(v_2 - 0) + m_2 a_2 + C(v_2 - v_1) + k(x_2 - x_1) \\ &= C_2 x_2' + m_2 x_2'' + C(x_2' - x_1') + k(x_2 - x_1) \end{aligned}$$

The answer is (D).

7. (a) The fluid capacitance of the water in the tank is C_f. From Rule 2, one end of each of the two energy sources, Q_1 and Q_2, connects to $p=0$. The system diagram is

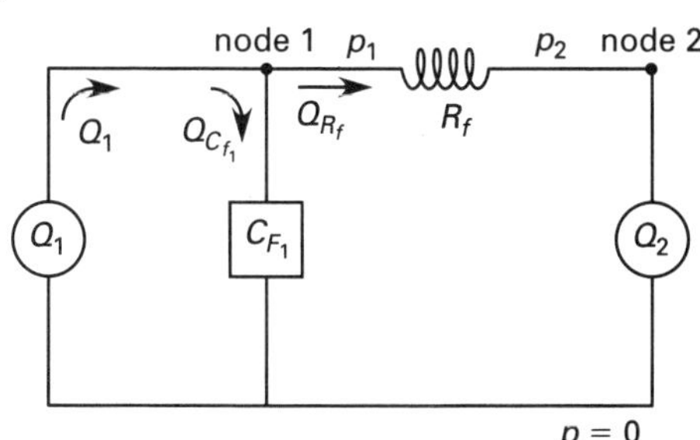

(b) From Eq. 61.10, the flow through the capacitor is

$$Q_{C_{f_1}} = C_{f_1}\left(\frac{dp_1}{dt}\right)$$

From Eq. 61.12, the flow through the resistor is

$$Q_{R_f} = \frac{p_1 - p_2}{R_f}$$

One of the system equations is based on conservation of flow at node 1.

$$\begin{aligned} Q_1 &= Q_{C_{f_1}} + Q_{R_f} \\ &= C_{f_1}\left(\frac{dp_1}{dt}\right) + \left(\frac{1}{R_f}\right)(p_1 - p_2) \end{aligned}$$

The second system equation is based on conservation of flow at node 2.

$$Q_2 = \left(\frac{1}{R_f}\right)(p_1 - p_2)$$

The answer is (A).

8. (a) The fluid resistance in the entrance pipe is R_f. The pressure at the entrance is p_1, and the pressure in the tank is p_2. The fluid capacitance of the water is C_f. The pressure at the open top of the tank is $p=0$. The system diagram is

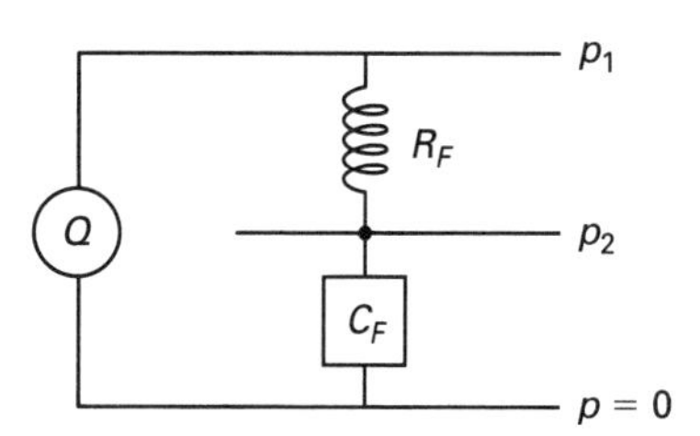

(b) By Rule 3, the source flow, Q, is the same flow through the resistor and the capacitor. Use Eq. 61.12 for the resistor.

$$Q = \frac{p_1 - p_2}{R_f}$$

Use Eq. 61.10 for the capacitor.

$$Q = C_f\left(\frac{dp_2}{dt}\right)$$

The answer is (D).

Control Systems

62 Analysis of Engineering Systems

PRACTICE PROBLEMS

1. Simplify the following block diagrams and determine the overall system gain.

(a)

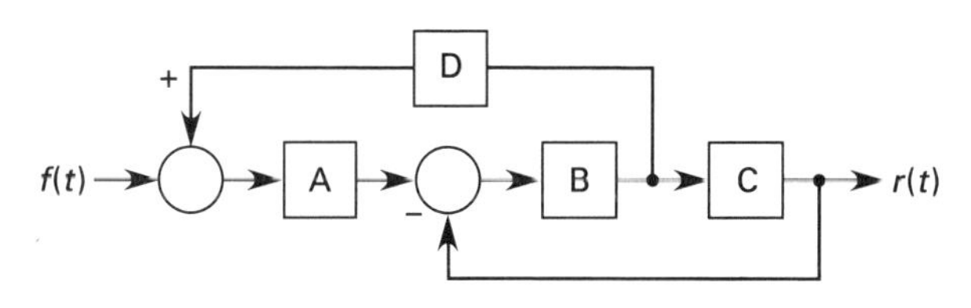

(A) $\dfrac{ABC}{1+BC-ABD}$

(B) $\dfrac{AB}{1-BC-AD}$

(C) $\dfrac{BC}{1+AC+ABD}$

(D) $\dfrac{ABD}{1-ABD}$

(b)

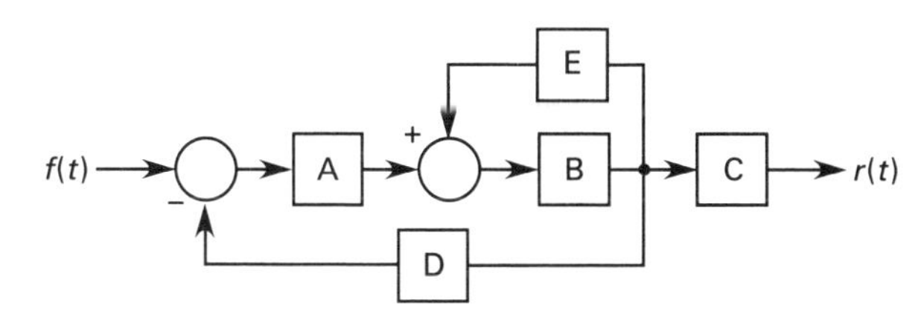

(A) $\dfrac{ABC}{BC-ABD}$

(B) $\dfrac{AEC}{1+AD-ABD}$

(C) $\dfrac{BC}{1+BC-ABD}$

(D) $\dfrac{ABC}{1-BE+ABD}$

2. A mass of 100 lbm (45 kg) is supported uniformly by a spring system. The spring system has a combined stiffness of 1200 lbf/ft (17.5 kN/m). A dashpot with a damping coefficient of 60 lbf-sec/ft (880 N·s/m) has been installed.

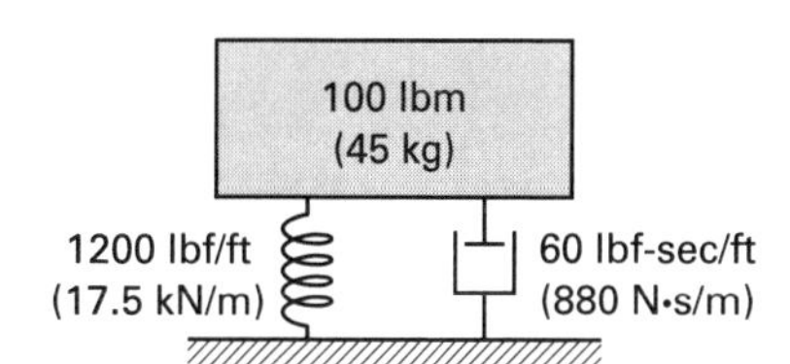

(a) The undamped natural frequency is most nearly

(A) 3.0 rad/sec

(B) 5.0 rad/sec

(C) 13 rad/sec

(D) 20 rad/sec

(b) The damping ratio is most nearly

(A) 0.24

(B) 0.50

(C) 0.76

(D) 0.97

(c) Sketch the magnitude and phase characteristics of the frequency response.

(d) Sketch the response to a unit step input.

3. A constant-speed motor/magnetic clutch drive train is monitored and controlled by a speed-sensing tachometer. The entire system is modeled as a control system block diagram, as shown. (The lowercase letters represent small-signal increments from the reference values.) When the control system is operating, the desired motor speed, n (in rpm), is set with a speed-setting potentiometer. The setting is compared to the tachometer output. The comparator output error (in volts), controls the clutch. A current, i (in amps), passes through the clutch coil. The external load torque, t_Lm (in in-lbf), is seen by the clutch and is countered by the clutch output torque, t (in in-lbf).

(a) Plot the open-loop frequency response.

(b) The open-loop steady-state gain is most nearly

(A) $T_1(0) = 25$, $T_2(0) = 50$

(B) $T_1(0) = 50$, $T_2(0) = 500$

(C) $T_1(0) = 100$, $T_2(0) = 200$

(D) $T_1(0) = 120$, $T_2(0) = 75$

(c) Plot the unity feedback closed-loop frequency response.

(d) The closed-loop steady-state gain is most nearly

(A) $T_1(0) = 0.1$, $T_2(0) = 2$

(B) $T_1(0) = 0.3$, $T_2(0) = 5$

(C) $T_1(0) = 0.6$, $T_2(0) = 12$

(D) $T_1(0) = 1.0$, $T_2(0) = 9.8$

(e) Plot the system sensitivity.

(f) Describe the closed-loop response to a step change in the desired output angular velocity. Is it damped or oscillatory? Is there a steady-state error? Why or why not?

(g) Describe the closed-loop response to a step change in the load torque. Is the response damped or oscillatory? Is there a steady-state error? Why or why not?

(h) Assume that you have to select the comparator gain and that it doesn't have to be 0.1. Using the root-locus method or either the Routh or Nyquist stability criterion, the limits of the comparator gain that cause the closed-loop system to be unstable are most nearly

(A) $1 + 250K > 0$, $K > -0.004$

(B) $1 + 250K < 1$, $K > -0.002$

(C) $1 + 500K > 0$, $K > -0.002$

(D) $1 + 500K < 0$, $K > -0.004$

(i) What kind of control is necessary to improve the steady-state response of the closed-loop system to constant disturbances in the load torque, t_L?

(A) integral

(B) logic

(C) on-off

(D) proportional

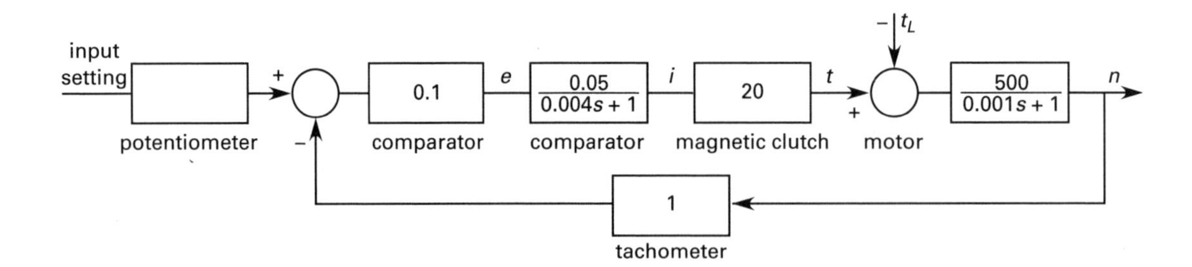

SOLUTIONS

1. (a) Draw the first block diagram.

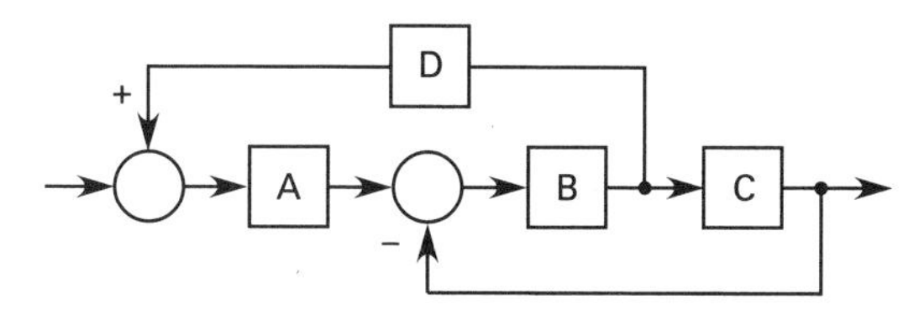

From the rules of simplifying block diagrams, use case 7 to move the extreme right pick-off point to the left of C.

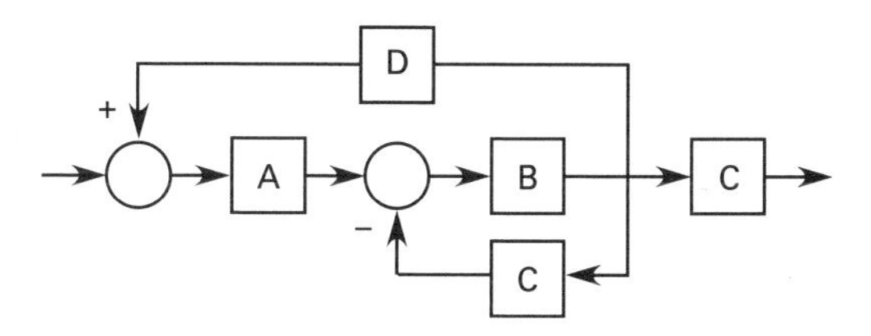

Use case 6 to combine the two summing points on the left.

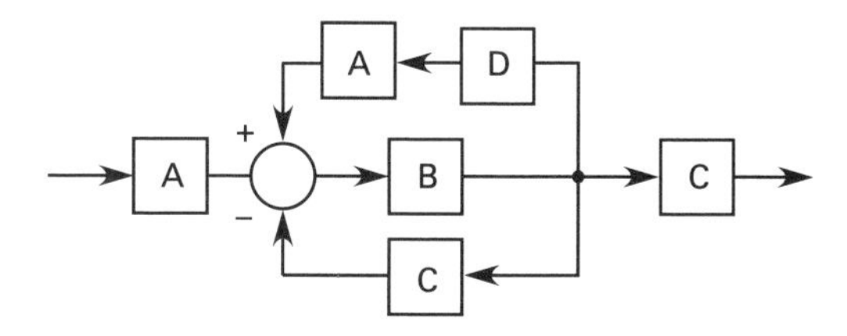

Use case 1 to combine boxes in series in the upper feedback loop.

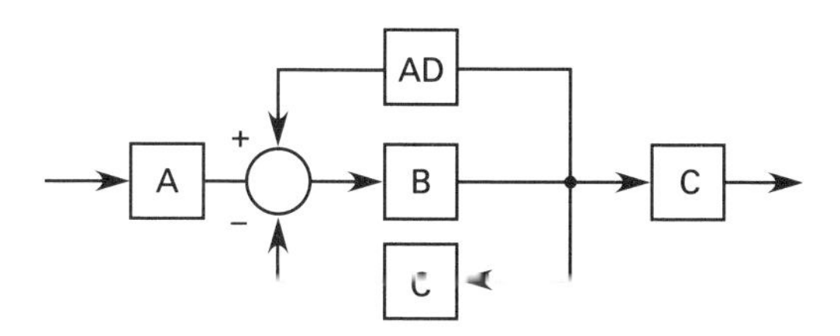

Use case 2 to combine the two feedback loops.

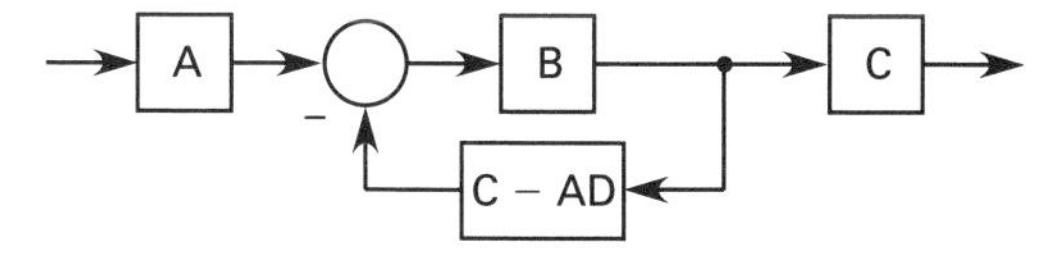

Use case 3 to simplify the remaining feedback loop.

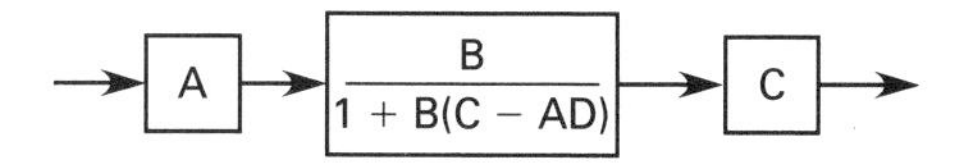

Use case 1 to combine boxes in series to determine the system gain.

$$G_{\text{loop}} = \boxed{\frac{\text{ABC}}{1 + \text{BC} - \text{ABD}}}$$

The answer is (A).

(b) Draw the second block diagram.

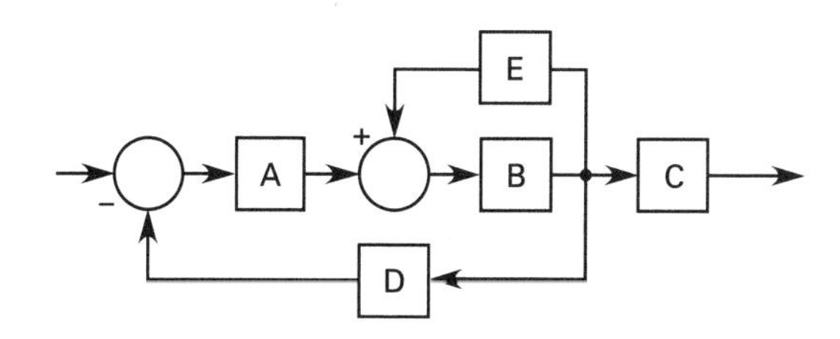

From the rules of simplifying block diagrams, use case 6 to combine the two summing points on the left.

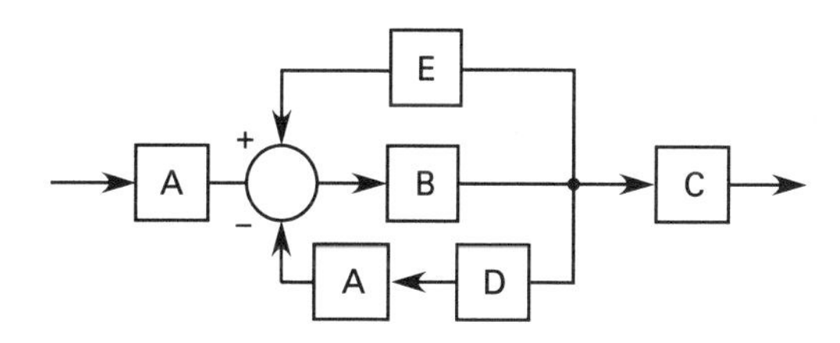

Use case 1 to combine boxes in series in the lower feedback loop.

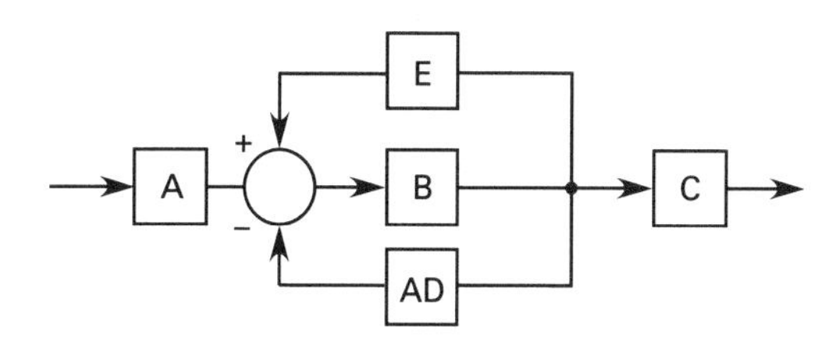

Use case 2 to combine the two feedback loops.

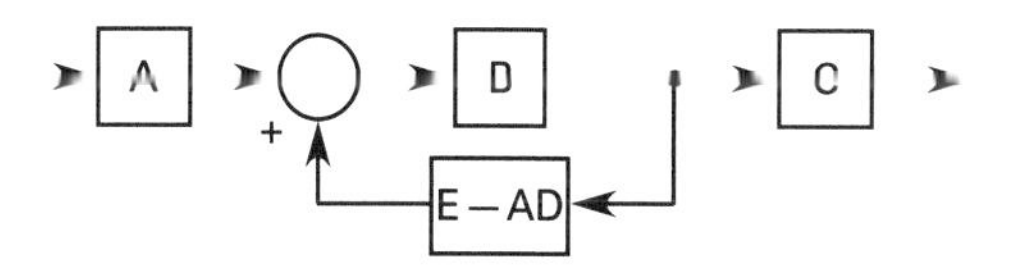

Use case 3 to simplify the remaining feedback loop.

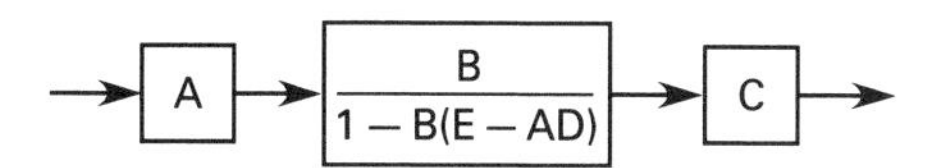

Use case 1 to combine boxes in series to determine the system gain.

$$G_{\text{loop}} = \boxed{\frac{\text{ABC}}{1 - \text{BE} + \text{ABD}}}$$

The answer is (D).

2. *Customary U.S. Solution*

The system differential equation for a force input, f, is

$$mx'' + Bx' + kx = f$$

Divide by m to write the equation in terms of natural frequency and damping factor.

$$x'' + \left(\frac{B}{m}\right)x' + \left(\frac{k}{m}\right)x = \frac{f}{m}$$

$$x'' + 2\zeta\omega_n x' + \omega_n^2 x = \frac{f}{\frac{k}{\omega_n^2}} = \omega_n^2\left(\frac{f}{k}\right)$$

Define the forcing function as $h = f/k$.

The equation is the same as Eq. 62.3.

$$x'' + 2\zeta\omega_n x' + \omega_n^2 x = \omega_n^2 h$$

(a) The undamped natural frequency is

$$\omega_n = \sqrt{\frac{kg_c}{m}} = \sqrt{\frac{\left(1200\ \frac{\text{lbf}}{\text{ft}}\right)\left(32.2\ \frac{\text{lbm-ft}}{\text{lbf-sec}^2}\right)}{100\ \text{lbm}}}$$
$$= \boxed{19.66\ \text{rad/sec}\quad (20\ \text{rad/sec})}$$

The answer is (D).

(b) The damping ratio is

$$\zeta = \frac{\frac{B}{m}}{2\omega_n} = \frac{B}{2\omega_n m} = \frac{\left(60\ \frac{\text{lbf-sec}}{\text{ft}}\right)\left(32.2\ \frac{\text{lbm-ft}}{\text{lbf-sec}^2}\right)}{(2)\left(19.66\ \frac{\text{rad}}{\text{sec}}\right)(100\ \text{lbm})}$$
$$= \boxed{0.491\quad (0.50)}$$

The answer is (B).

(c) Take the Laplace transform of Eq. 62.3. Consider zero initial conditions.

$$s^2x(s) + 2\zeta\omega_n s x(s) + \omega_n^2 x(s) = \omega_n^2 H(s)$$

Determine the transfer function.

$$T(s) = \frac{x(s)}{H(s)} = \frac{\omega_n^2}{s^2 + 2\zeta\omega_n s + \omega_n^2}$$

The frequency response is obtained by letting $s = j\omega$.

$$T(j\omega) = \frac{\omega_n^2}{(\omega_n^2 - \omega^2) + j2\zeta\omega_n\omega}$$

Write the equation in polar form.

$$T(j\omega) = |T(j\omega)|e^{j\phi(\omega)}$$

The magnitude is

$$|T(j\omega)| = \frac{\omega_n^2}{\sqrt{(\omega_n^2 - \omega^2)^2 + (2\zeta\omega_n\omega)^2}}$$

At $\omega = 0, |T(j\omega)| = 1$.

At $\omega = \omega_n$,

$$|T(j\omega)| = \frac{1}{2\zeta} = \frac{1}{(2)(0.491)} \approx 1.0$$

At $\omega \to \infty, |T(j\omega)| \to 0$.

A peak occurs near $\omega = \omega_n$. To obtain the location ω_p, set the derivative of $|T(j\omega)|$ equal to zero.

$$\frac{d|T(j\omega)|}{d\omega} = -\tfrac{1}{2}\omega_n^2\left((\omega_n^2 - \omega^2) + (2\zeta\omega_n\omega)^2\right)^{-3/2}$$
$$\times\left((2)(\omega_n^2 - \omega^2)(-2\omega) + (2)(2\zeta\omega_n\omega)(2\zeta\omega_n)\right)$$
$$= 0$$

$$\omega_p = \omega_n\sqrt{1 - 2\zeta^2} = \left(19.66\ \frac{\text{rad}}{\text{sec}}\right)\sqrt{1 - (2)(0.491)^2}$$
$$= 14.15\ \text{rad/sec}$$

At $\omega = \omega_p$,

$$|T(j\omega)| = \frac{\omega_n^2}{\sqrt{(\omega_n^2 - \omega_p^2)^2 + (2\zeta\omega_n\omega_p)^2}}$$

$$= \frac{\left(19.66\ \frac{\text{rad}}{\text{sec}}\right)^2}{\sqrt{\begin{array}{l}\left(\left(19.66\ \frac{\text{rad}}{\text{sec}}\right)^2 - \left(14.15\ \frac{\text{rad}}{\text{sec}}\right)^2\right)^2 \\ + \left(\begin{array}{c}(2)(0.491)\left(19.66\ \frac{\text{rad}}{\text{sec}}\right) \\ \times\left(14.15\ \frac{\text{rad}}{\text{sec}}\right)\end{array}\right)^2\end{array}}}$$
$$= 1.17$$

A sketch of the frequency response magnitude is

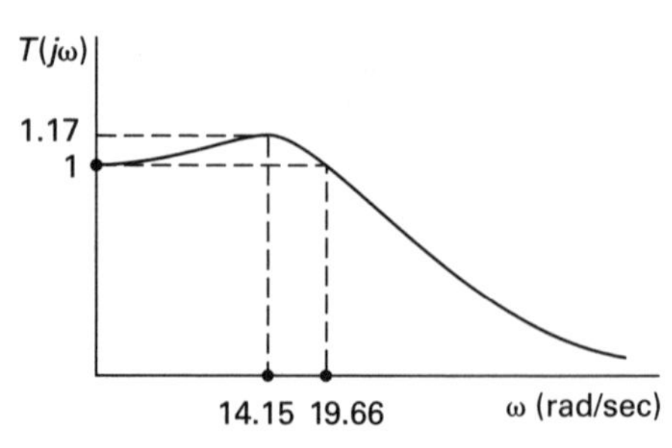

The phase is

$$\phi(\omega) = \tan^{-1}\left(\frac{-2\zeta\omega_n\omega}{\omega_n^2 - \omega^2}\right)$$

At $\omega = 0, \phi(\omega) = 0$.

At $\omega = \omega_n, \phi(\omega) = -\pi/2$.

At $\omega \to \infty, \phi(\omega) = -\pi$.

A sketch of the phase is

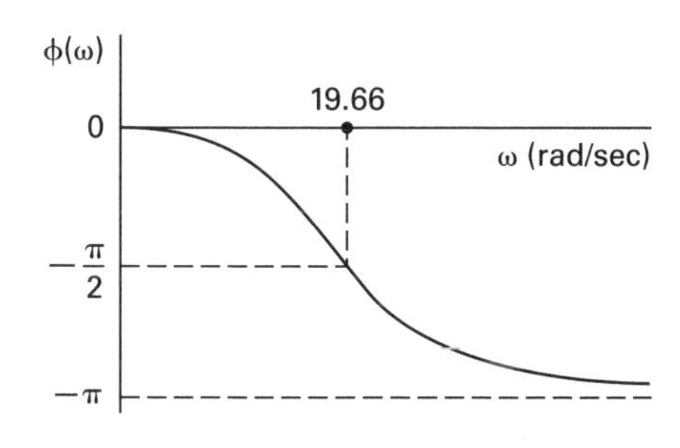

(d) The final value of the step response is obtained from the final value theorem.

$$\begin{aligned} x_{\text{final}} &= \lim_{s\to 0} sx(s) = \lim_{s\to 0} sT(s)H(s) \\ &= \lim_{s\to 0} sT(s)\left(\frac{1\ \text{ft}}{s}\right) = \left(\frac{\omega_n^2}{\omega_n^2}\right)(1\ \text{ft}) \\ &= 1.0\ \text{ft} \end{aligned}$$

From Eq. 62.7, the fraction of overshoot is

$$\begin{aligned} M_p &= \exp\left(\frac{-\pi\zeta}{\sqrt{1-\zeta^2}}\right) = \exp\left(\frac{-\pi(0.491)}{\sqrt{1-(0.491)^2}}\right) \\ &= 0.17 \end{aligned}$$

The peak is $x_{\text{max}} = 1.17$ ft.

The damped natural frequency is

$$\begin{aligned} \omega_d &= \omega_n\sqrt{1-\zeta^2} = \left(19.66\ \frac{\text{rad}}{\text{sec}}\right)\sqrt{1-(0.491)^2} \\ &= 17.1\ \text{rad/sec} \end{aligned}$$

The peak time from Eq. 62.6 is

$$t_p = \frac{\pi}{\omega_d} = \frac{\pi}{17.1\ \frac{\text{rad}}{\text{sec}}} = 0.18\ \text{sec}$$

The 5% criterion settling time from Eq. 62.9 is

$$t_s = \frac{3.00}{\zeta\omega_n} = \frac{3.00}{(0.491)\left(19.66\ \frac{\text{rad}}{\text{sec}}\right)} = 0.31\ \text{sec}$$

The sketch of the response to a unit step input is obtained from Fig. 62.2 and the preceding calculations.

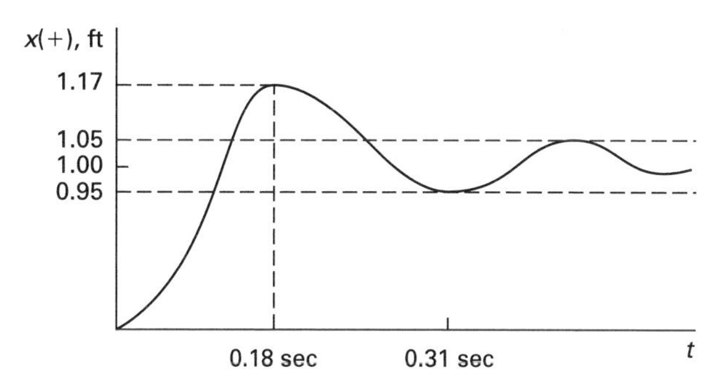

SI Solution

The system differential equation for a force input, f, is

$$mx'' + Bx' + kx = f$$

Divide by m to write the equation in terms of natural frequency and damping factor.

$$x'' + \left(\frac{B}{m}\right)x' + \left(\frac{k}{m}\right)x = \frac{f}{m}$$

$$x'' + 2\zeta\omega_n x' + \omega_n^2 x = \frac{\frac{f}{k}}{\omega_n^2} = \omega_n^2\left(\frac{f}{k}\right)$$

Define the forcing function as $h = f/k$.

The equation is the same as Eq. 62.3.

$$x'' + 2\zeta\omega_n x' + \omega_n^2 x = \omega_n^2 h$$

(a) The undamped natural frequency is

$$\omega_n = \sqrt{\frac{k}{m}} = \sqrt{\frac{17\,500\ \frac{\text{N}}{\text{m}}}{45\ \text{kg}}} = \boxed{19.7\ \text{rad/s}\quad (20\ \text{rad/s})}$$

The answer is (D).

(b) The damping ratio is

$$\zeta = \frac{\frac{B}{m}}{2\omega_n} = \frac{\frac{880\ \frac{\text{N·s}}{\text{m}}}{45\ \text{kg}}}{(2)\left(19.7\ \frac{\text{rad}}{\text{s}}\right)} = \boxed{0.50}$$

The answer is (B).

(c) Take the Laplace transform of Eq. 62.3. Consider zero initial conditions.

$$s^2x(s) + 2\zeta\omega_n s x(s) + \omega_n^2 x(s) = \omega_n^2 H(s)$$

Determine the transfer function.

$$T(s) = \frac{x(s)}{H(s)} = \frac{\omega_n^2}{s^2 + 2\zeta\omega_n s + \omega_n^2}$$

The frequency response is obtained by letting $s = j\omega$.

$$T(j\omega) = \frac{\omega_n^2}{(\omega_n^2 - \omega^2) + j2\zeta\omega_n\omega}$$

Write the equation in polar form.

$$T(j\omega) = |T(j\omega)|e^{j\phi(\omega)}$$

The magnitude is

$$|T(j\omega)| = \frac{\omega_n^2}{\sqrt{(\omega_n^2 - \omega^2)^2 + (2\zeta\omega_n\omega)^2}}$$

At $\omega = 0, |T(j\omega)| = 1$.

At $\omega = \omega_n$,

$$|T(j\omega)| = \frac{1}{2s} = \frac{1}{(2)(0.50)} = 1.0$$

At $\omega \to \infty, |T(j\omega)| \to 0$.

A peak occurs near $\omega = \omega_n$. To obtain the location ω_p, set the derivative of $|T(j\omega)|$ equal to zero.

$$\begin{aligned}\frac{d|T(j\omega)|}{d\omega} &= -\tfrac{3}{2}\omega_n^2\left((\omega_n^2 - \omega^2) + (2\zeta\omega_n\omega)^2\right)^{-3/2} \\ &\quad \times \left(2(\omega_n^2 - \omega^2)(-2\omega) + 2(2\zeta\omega_n\omega)2\zeta\omega_n\right) \\ &= 0 \\ \omega_p &= \omega_n\sqrt{1 - 2\zeta^2} = \left(19.7\ \frac{\text{rad}}{\text{s}}\right)\sqrt{1 - (2)(0.50)^2} \\ &= 13.9\ \text{rad/s}\end{aligned}$$

At $\omega = \omega_p$,

$$\begin{aligned}|T(j\omega)| &= \frac{\omega_n^2}{\sqrt{(\omega_n^2 - \omega_p^2)^2 + (2\zeta\omega_n\omega_p)^2}} \\ &= \frac{\left(19.7\ \frac{\text{rad}}{\text{s}}\right)^2}{\sqrt{\left(\left(19.7\ \frac{\text{rad}}{\text{s}}\right)^2 - \left(13.9\ \frac{\text{rad}}{\text{s}}\right)^2\right)^2 + \left((2)(0.50)\left(19.7\ \frac{\text{rad}}{\text{s}}\right) \times \left(13.9\ \frac{\text{rad}}{\text{s}}\right)\right)^2}} \\ &= 1.16\end{aligned}$$

A sketch of the frequency response magnitude is

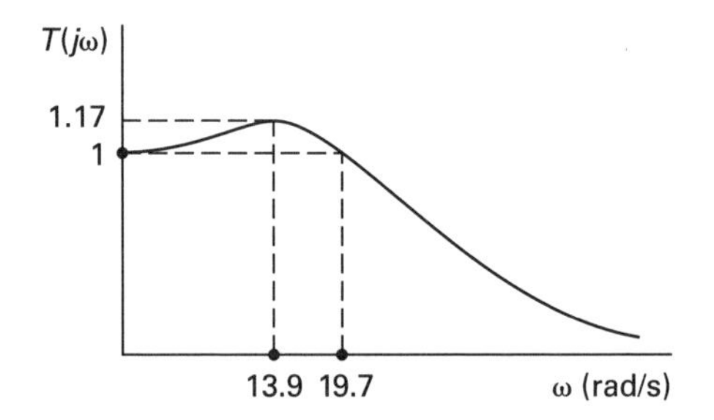

The phase is

$$\phi(\omega) = \tan^{-1}\left(\frac{-2\zeta\omega_n\omega}{\omega_n^2 - \omega^2}\right)$$

At $\omega = 0,\ \phi(\omega) = 0$.

At $\omega = \omega_n,\ \phi(\omega) = -\pi/2$.

At $\omega \to \infty,\ \phi(\omega) = -\pi$.

A sketch of the phase is

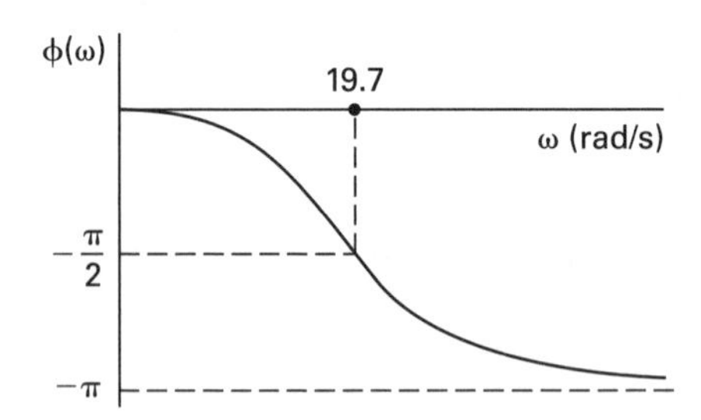

(d) The final value of the step response is obtained from the final value theorem.

$$\begin{aligned}x_{\text{final}} &= \lim_{s\to 0} s x(s) = \lim_{s\to 0} sT(s)H(s) \\ &= \lim_{s\to 0} sT(s)\left(\frac{1\ \text{m}}{s}\right) = \left(\frac{\omega_n^2}{\omega_n^2}\right)(1\ \text{m}) \\ &= 1.0\ \text{m}\end{aligned}$$

From Eq. 62.7, the fraction of overshoot is

$$M_p = \exp\left(\frac{-\pi\zeta}{\sqrt{1-\zeta^2}}\right) = \exp\left(\frac{-\pi(0.50)}{\sqrt{1-(0.50)^2}}\right)$$
$$= 0.16$$

The peak is $x_{\max} = 1.16$ m.

The damped natural frequency is

$$\omega_d = \omega_n\sqrt{1-\zeta^2} = \left(19.7\ \frac{\text{rad}}{\text{s}}\right)\sqrt{1-(0.50)^2}$$
$$= 17.1 \text{ rad/s}$$

The peak time from Eq. 62.6 is

$$t_p = \frac{\pi}{\omega_d} = \frac{\pi}{17.1\ \frac{\text{rad}}{\text{s}}} = 0.18 \text{ s}$$

The 5% criterion settling time from Eq. 62.9 is

$$t_s = \frac{3.00}{\zeta\omega_n} = \frac{3.00}{(0.50)\left(19.7\ \frac{\text{rad}}{\text{s}}\right)} = 0.30 \text{ s}$$

The sketch of the response to a unit step input is obtained from Fig. 62.2 and the preceding calculations.

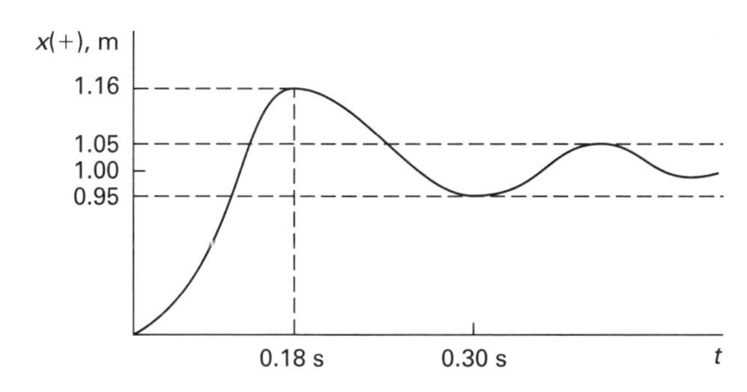

3. First, redraw the system in more traditional form.

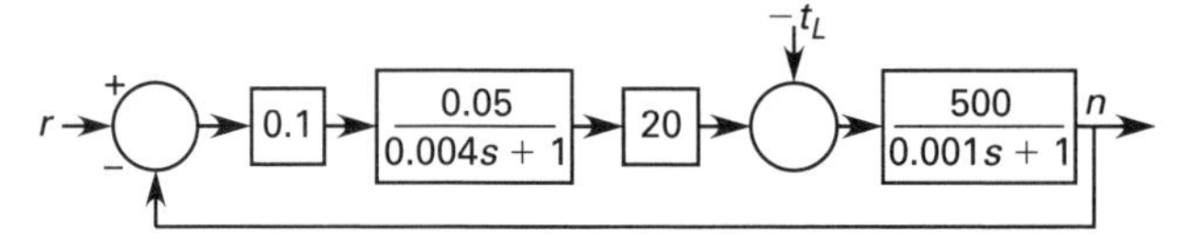

From Fig. 62.5, use case 1 to combine boxes in series.

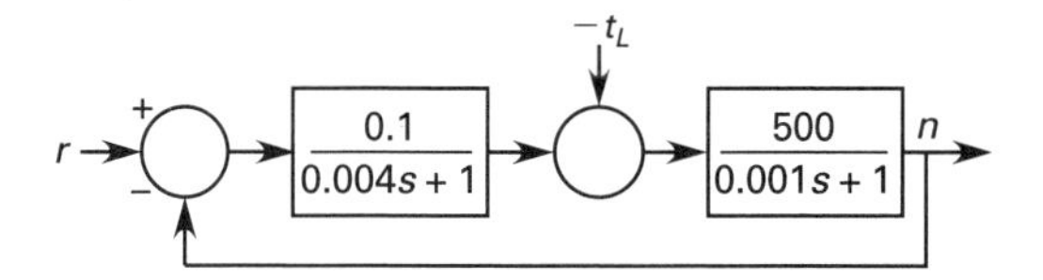

Use case 5 to combine the two summing points.

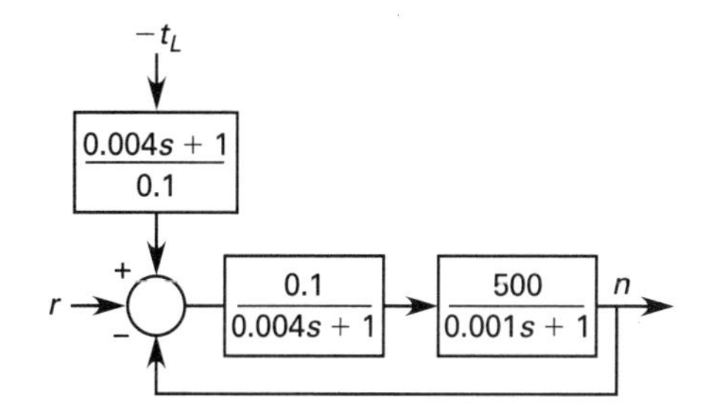

Use case 1 to combine boxes in series.

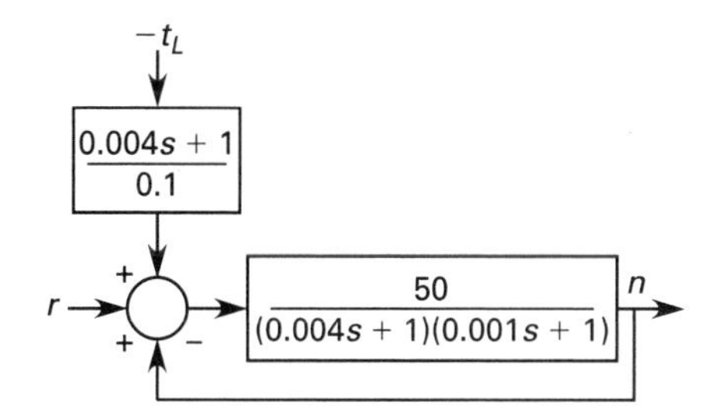

(a) Ignoring the feedback loop, the open-loop transfer function from r to n is

$$T_1(s) = \frac{N(s)}{R(s)} = \frac{50}{(0.004s+1)(0.001s+1)}$$

The open-loop transfer function from $-t_L$ to n is

$$T_2(s) = \frac{N(s)}{-T_L(s)}$$
$$= \left(\frac{0.004s+1}{0.1}\right)\left(\frac{50}{(0.004s+1)(0.001s+1)}\right)$$
$$= \frac{500}{0.001s+1}$$

The open-loop frequency response for $T_1(s)$ is

$$T_1(j\omega) = |T_1(j\omega)|e^{j\phi_1(\omega)}$$

From Eq. 62.34, the gain is $20\log|T_1(j\omega)|$ (in dB).

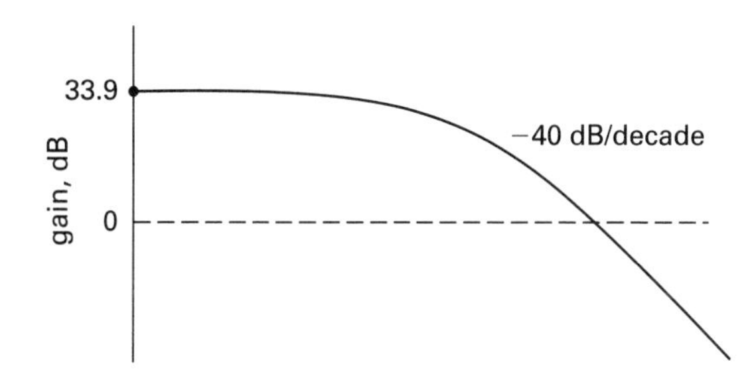

The phase is $\phi_1(\omega)$.

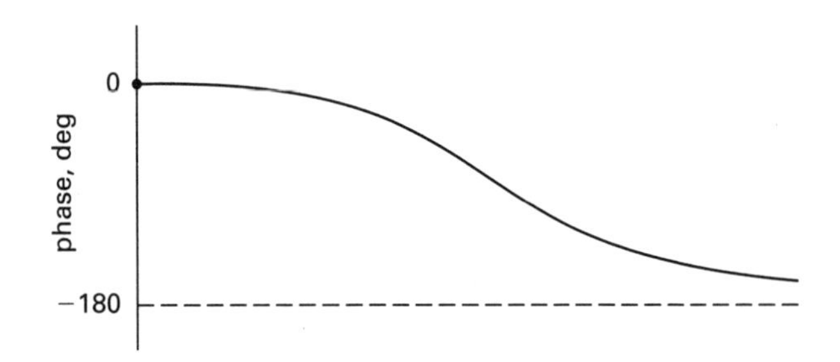

The open-loop frequency response for $T_2(s)$ is

$$T_2(j\omega) = |T_2(j\omega)|e^{j\phi_2(\omega)}$$

From Eq. 62.34, the gain is $20\log|T_2(j\omega)|$ (in dB).

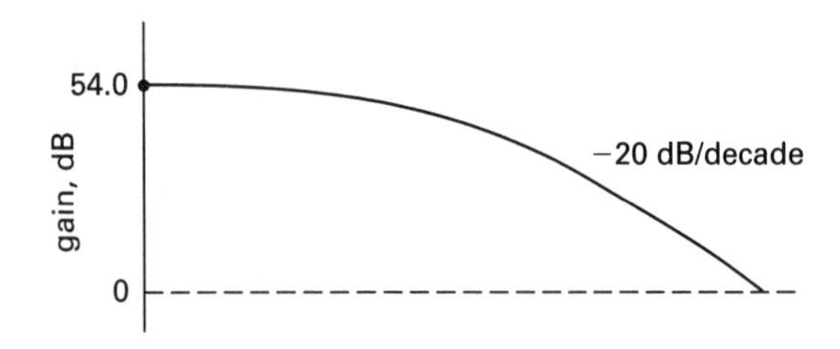

The phase is $\phi_2(\omega)$.

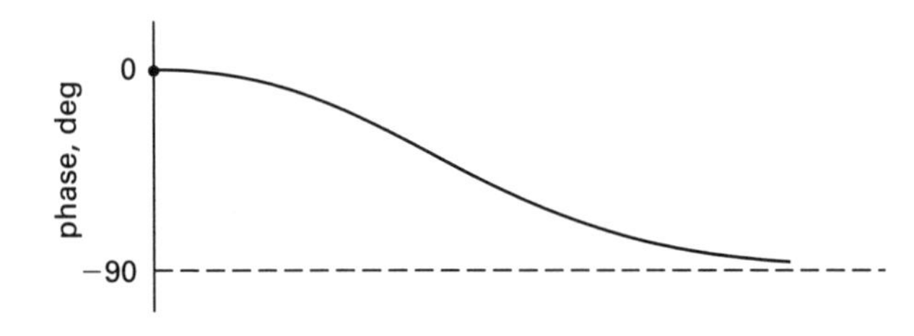

(b) The open-loop steady-state gain for $T_1(s)$ is

$$T_1(0) = \frac{50}{\Big((0.004)(0)+1\Big)\Big((0.001)(0)+1\Big)} = \boxed{50}$$

The open-loop steady-state gain for $T_2(s)$ is

$$T_2(0) = \frac{500}{(0.001)(0)+1} = \boxed{500}$$

The answer is (B).

(c) From Fig. 62.5, use case 3 to simplify the feedback loop.

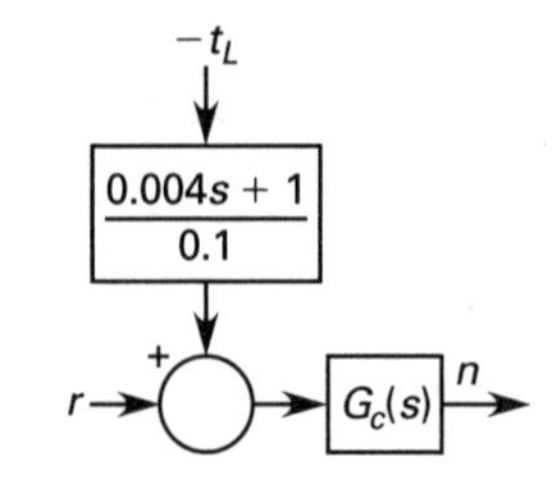

$$\begin{aligned} G_c(s) &= \frac{\dfrac{50}{(0.004s+1)(0.001s+1)}}{1+\dfrac{50}{(0.004s+1)(0.001s+1)}} \\ &= \frac{50}{(0.004s+1)(0.001s+1)+50} \\ &= \frac{50}{(4\times10^{-6})s^2+0.005s+51} \end{aligned}$$

The closed-loop transfer function from r to n is

$$\begin{aligned} T_1(s) &= \frac{N(s)}{R(s)} = G_c(s) \\ &= \frac{50}{(4\times10^{-6})s^2+0.005s+51} \end{aligned}$$

The closed-loop transfer function from $-t_L$ to n is

$$\begin{aligned} T_2(s) &= \frac{N(s)}{-T_L(s)} = \left(\frac{0.004s+1}{0.1}\right)G_c(s) \\ &= \frac{(500)(0.004s+1)}{(4\times10^{-6})s^2+0.005s+51} \end{aligned}$$

The gain for the closed-loop frequency response of $T_1(s)$ is

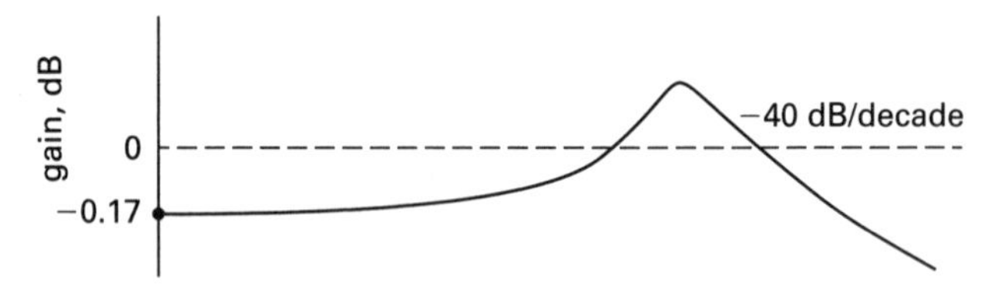

The phase for the closed-loop frequency response of $T_1(s)$ is

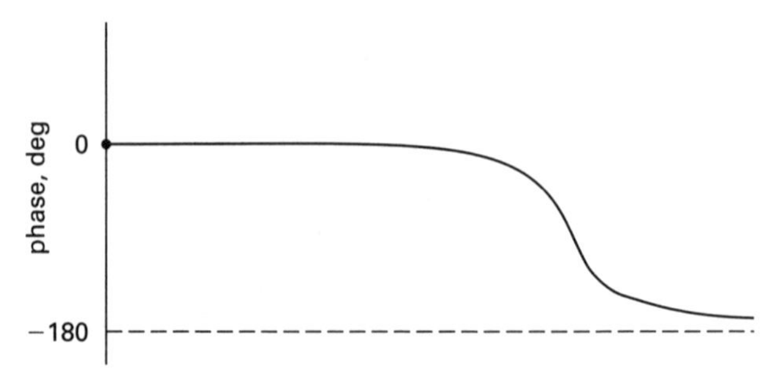

The gain for the closed-loop frequency response of $T_2(s)$ is

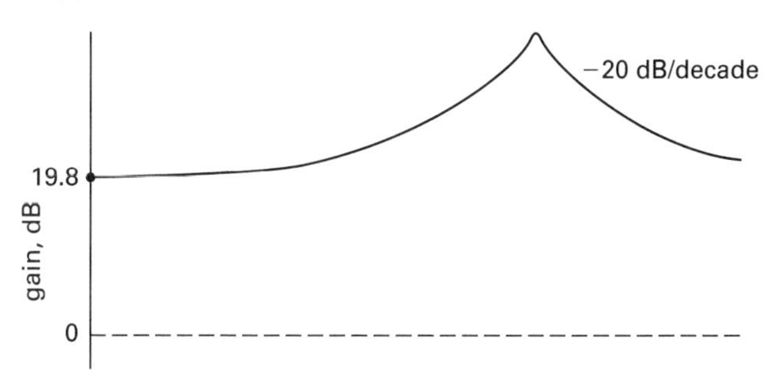

The phase for the closed-loop frequency response of $T_2(s)$ is

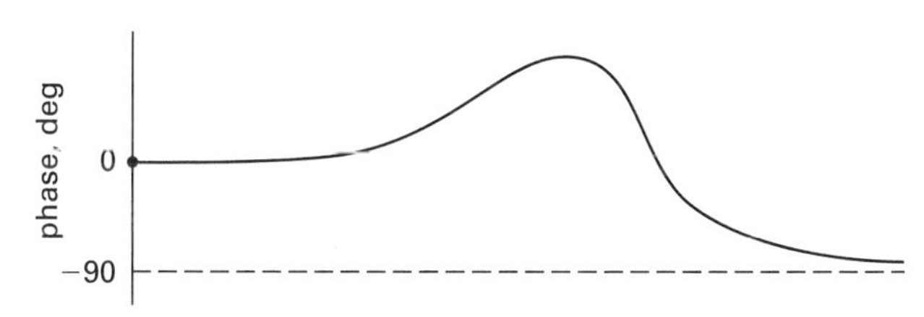

(d) The closed-loop steady-state gain for $T_1(s)$ is

$$\begin{aligned} T_1(0) &= \frac{50}{(4\times 10^{-6})(0)^2 + (0.005)(0) + 51} \\ &= \boxed{0.98 \quad (1.0)} \end{aligned}$$

The closed-loop steady-state gain for $T_2(s)$ is

$$T_2(0) = \frac{(500)\Big((0.004)(0)+1\Big)}{(4\times 10^{-6})(0)^2 + (0.005)(0) + 51} = \boxed{9.8}$$

The answer is (D).

(e) From Fig. 62.4, use Eq. 62.22 to find the system sensitivity.

$$\begin{aligned} S &= \frac{1}{1+GH} = \frac{1}{1+\left(\dfrac{50}{(0.004s+1)(0.001s+1)}\right)(1)} \\ &= \frac{(0.004s+1)(0.001s+1)}{50+(0.004s+1)(0.001s+1)} \\ &= \frac{(0.004s+1)(0.001s+1)}{(4\times 10^{-6})s^2 + 0.005s + 51} \end{aligned}$$

The frequency response for S is

$$S(j\omega) = |S(j\omega)|e^{j\phi(\omega)}$$

From Eq. 62.34, the gain is $20\log|S(j\omega)|$ (in dB).

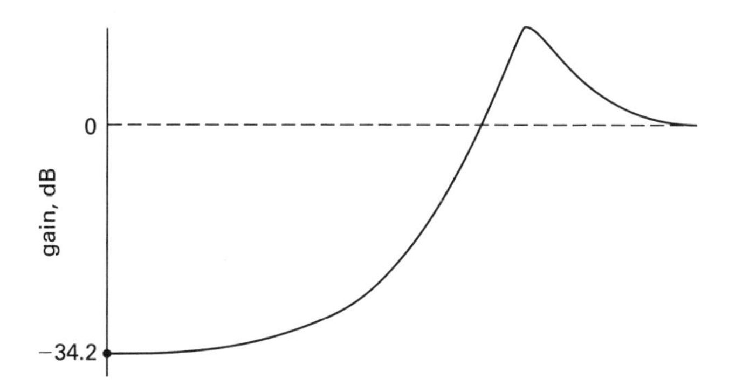

The phase is $\phi(\omega)$.

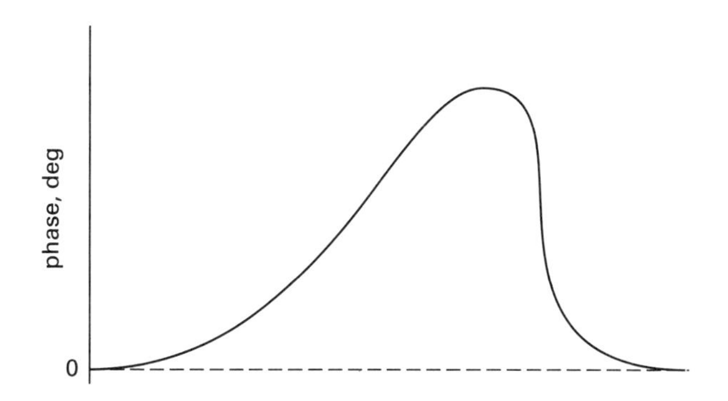

(f) From part (c), the response to a step change in input r is

$$\begin{aligned} N(s) &= R(s)G_c(s) \\ &= \left(\frac{R}{s}\right)\left(\frac{50}{(4\times 10^{-6})s^2 + 0.005s + 51}\right) \\ &= \frac{(1.25\times 10^7)\left(\dfrac{R}{s}\right)}{s^2 + 1250s + (1.275\times 10^7)} \end{aligned}$$

Equate the denominator of $N(s)$ to the standard second-order form $s^2 + 2\varsigma\omega_n s + \omega_n^2$.

$$\omega_n = \sqrt{1.275\times 10^7 \ \frac{\text{rad}^2}{\text{sec}^2}} = 3570.7 \text{ rad/sec}$$

Set $2\zeta\omega_n$ equal to 1250 and solve for ζ.

$$\zeta = \frac{1250 \ \dfrac{\text{rad}}{\text{sec}}}{2\omega_n} = \frac{1250 \ \dfrac{\text{rad}}{\text{sec}}}{(2)\left(3570.7 \ \dfrac{\text{rad}}{\text{sec}}\right)} = 0.175$$

Since ζ is less than one, the response is oscillatory.

Use Eq. 62.7 to find the fraction overshoot.

$$M_p = \exp\left(\frac{-\pi\zeta}{\sqrt{1-\zeta^2}}\right) = \exp\left(\frac{-\pi(0.175)}{\sqrt{1-(0.175)^2}}\right)$$
$$= 0.57$$

Use Eq. 62.5 to find the 90% rise time.

$$t_r = \frac{\pi - \arccos\zeta}{\omega_d} = \frac{\pi - \arccos\zeta}{\omega_n\sqrt{1-\zeta^2}}$$
$$= \frac{\pi - \arccos 0.175}{\left(3570.7\ \frac{\text{rad}}{\text{sec}}\right)\sqrt{1-(0.175)^2}}$$
$$= 4.97 \times 10^{-4}\ \text{sec}$$

The response is very fast but highly oscillatory.

Use the final value theorem, Eq. 62.27, to find the steady-state value.

$$n = \lim_{s\to 0} sN(s) = \lim_{s\to 0}\left(\frac{s(1.25\times 10^7)\left(\frac{R}{s}\right)}{s^2 + 1250s + 1.257 + 10^7}\right)$$
$$= 0.98R$$

The steady-state error is

$$e = R - n = R - 0.98R = 0.02R$$

There is a steady-state error proportional to the desired output angular velocity, R.

The plot for $n(t)$ is

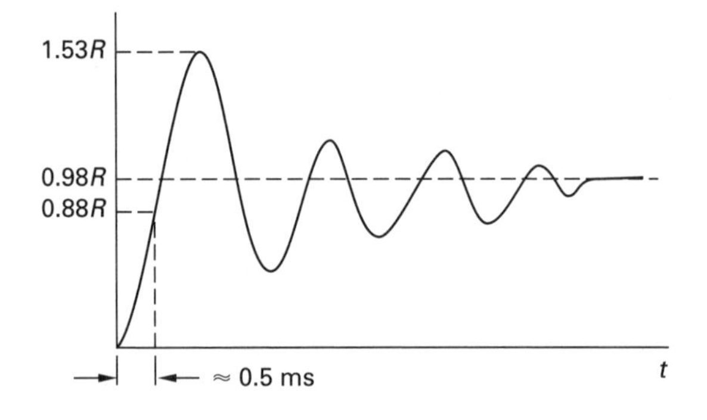

(g) From part (c), the response to a step change in input t_L is

$$N(s) = -T_L(s)\left(\frac{0.004s+1}{0.1}\right)G_c(s)$$
$$= \left(\frac{-T_L}{s}\right)\left(\frac{(500)(0.004s+1)}{(4\times 10^{-6})s^2 + 0.005s + 51}\right)$$
$$= \frac{-(1.25\times 10^8)\left(\frac{T_L}{s}\right)(0.004s+1)}{s^2 + 1250s + (1.275\times 10^7)}$$

Since the denominator is the same as that for $N(s)$ in Prob. 3(f), the response to a step change in t_L will be similar to a step change in r. However, the numerator, $0.004s+1$, will cause the response to deviate from second order. The response will still be oscillatory and very fast.

Use the final value theorem, Eq. 62.27, to find the steady-state error.

$$n = \lim_{s\to 0} sN(s)$$
$$= \lim_{s\to 0}\left(\frac{-s(1.25\times 10^8)\left(\frac{T_L}{s}\right)(0.004s+1)}{s^2 + 1250s + (1.27\times 10^7)}\right)$$
$$= -9.84T_L$$

The new steady-state error for both r and t_L inputs is

$$e = R - n = R - (0.98R - 9.84T_L) = 0.02R + 9.84T_L$$

Thus, the load torque, $-t_L$, contributes to the steady-state error.

The response is oscillatory. The plot of the response due to a step change in $-t_L$ is

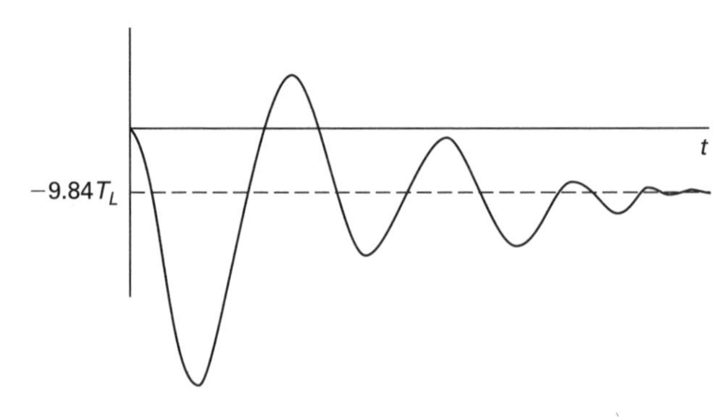

(h) Replace the comparator gain of 0.1 with K. The reduced system is

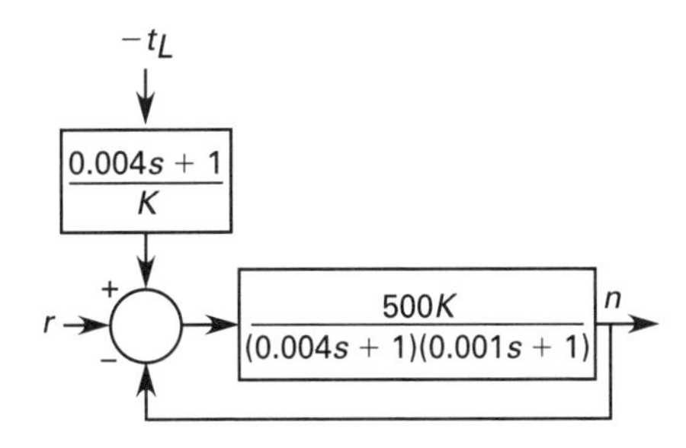

The closed-loop transfer function from r to n is

$$T_1(s) = \frac{N(s)}{R(s)} = \frac{\dfrac{500K}{(0.004s+1)(0.001s+1)}}{1 + \dfrac{500K}{(0.004s+1)(0.001s+1)}}$$

$$= \frac{500K}{(4 \times 10^{-6})s^2 + 0.005s + 1 + 500K}$$

The Routh-Hurwitz table is

$$\begin{bmatrix} a_0 & a_2 \\ a_1 & a_3 \\ b_1 & b_2 \end{bmatrix} = \begin{bmatrix} 4 \times 10^{-6} & 1 + 500K \\ 0.005 & 0 \\ b_1 & b_2 \end{bmatrix}$$

Use Eq. 62.40.

$$b_1 = \frac{a_1 a_2 - a_0 a_3}{a_1}$$
$$= \frac{(0.005)(1 + 500K) - (4 \times 10^{-6})(0)}{0.005}$$
$$= 1 + 500K$$

For a stable system, there can be no sign changes in the first column of the table.

Thus,

$$\boxed{\begin{aligned} 1 + 500K &> 0 \\ K &> -0.002 \end{aligned}}$$

The answer is (C).

(i) The closed-loop system steady-state response can be improved by adding integral control. This will effectively compensate for any steady-state disturbances due to t_L and will provide a zero steady-state error for a step input for r. This addition has a side effect of reducing the stability margin of the system. However, if properly designed, the system will still be stable.

The answer is (A).

Control Systems

63 Management Science

PRACTICE PROBLEMS

1. (*Time limit: one hour*) Printed circuit boards are manufactured in four consecutive departmental operations. Units move sequentially through departments 1, 2, 3, and 4. Each operation occurs at a station, each of which has a single employee overseeing a specific machine. Employees in all departments work from 8:00 a.m. to 5:00 p.m. and have one hour total for lunch and personal breaks. Defects created within a department are found before the units are passed to the subsequent department. Defective units are discarded. There are 52 work weeks per year, each of which has 5 work days. No units are produced during set-up, downtime, maintenance, or record-keeping periods. Units left incomplete at the end of one day are completed on the following day.

	department			
	1	2	3	4
production time (sec/unit)	6	10	11	45
set-up time (min/day)	16	8	20	5
downtime (min/day)	12	10	15	0
maintenance time (min/day)	8	12	8	0
record-keeping (min/day)	6	6	6	30
percentage defects	4%	6%	3%	2%

(a) How many units can be produced in one year if department 4 has a single station (machine) and an infinite inventory of units completed by departments 1, 2, and 3?

(A) 72,000 units/yr

(B) 110,000 units/yr

(C) 140,000 units/yr

(D) 150,000 units/yr

(b) What is the maximum number of units that can be produced in one year if each department has only a single station (machine)?

(A) 72,000 units/yr

(B) 110,000 units/yr

(C) 140,000 units/yr

(D) 150,000 units/yr

(c) If the production goal is 900,000 completed units per year, what is the minimum number (total) of machines needed in departments 1, 2, 3, and 4, respectively?

(A) 1, 2, 2, 6

(B) 1, 2, 3, 7

(C) 2, 2, 2, 6

(D) 2, 2, 3, 7

2. (*Time limit: one hour*) Four workers perform operations 1, 2, 3, and 4 in sequence on a manual assembly line. Each station performs its operation only once on the product before sending the product on to the next operation. Operation times at the stations are as given. (Travel times are included in the operation times.)

station	time (min)
1	0.6
2	0.6
3	0.9
4	0.8

A fifth "floating" station has the ability to assist any of the four stations. The fifth station works with the same efficiencies and times as the four stations. There is no fixed assignment for this fifth station. The fifth station is allowed to help any station that needs it.

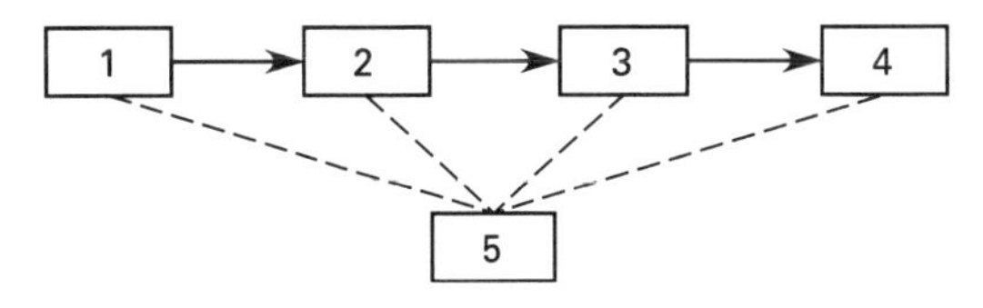

The operators of all five stations are permitted a 10 minute break each hour.

(a) What is the fraction of station 5's time allocated to each operation?

(b) What is the maximum number of products that can be produced assuming that the fifth station is assigned to work optimally? Neglect the initial (transient) performance.

Plant Engineering

3. The activities that constitute a project are listed. The project starts at $t=0$.

activity	predecessors	successors	duration
start	–	A	0
A	start	B, C, D	7
B	A	G	6
C	A	E, F	5
D	A	G	2
E	C	H	13
F	C	H, I	4
G	D, B	I	18
H	E, F	finish	7
I	F, G	finish	5
finish	H, I	–	0

(a) Draw the critical path network.

(b) What is the critical path?

(A) A-B-G-I

(B) A-C-F-I

(C) A-C-E-H

(D) A-D-G-I

(c) The earliest finish is most nearly

(A) 25

(B) 29

(C) 32

(D) 36

(d) The latest finish is most nearly

(A) 25

(B) 29

(C) 32

(D) 36

(e) The slack along the critical path is most nearly

(A) 0

(B) 1

(C) 4

(D) 5

(f) The float along the critical path is most nearly

(A) 0

(B) 1

(C) 4

(D) 5

4. PERT activities constituting a short project are listed with their characteristic completion times. If the project starts at $t=15$, what is most nearly the probability that the project will be completed on or before $t=42$?

activity	predecessors	successors	t_{min}	$t_{most\ likely}$	t_{max}
start	–	A	0	0	0
A	start	B, D	1	2	5
B	A	C	7	9	20
C	B	D	5	12	18
D	A, C	finish	2	4	7
finish	D	–	0	0	0

(A) 18%

(B) 29%

(C) 56%

(D) 71%

5. (*Time limit: one hour*) Your manufacturing facility produces two models of municipal transit buses, designated B-1 and B-2. You can sell as many of either model as you produce. The per-bus profits are $800,000 for model B-1 and $650,000 for model B-2. You would like to maximize your company's profit by determining the number of each model buses to produce. However, it is not a simple matter of producing only B-1 models because certain common parts are in limited supply.

part	number available
gage sending units	2000
wheel housing flares	1800
intake grilles	3600

Due to differences in design, the quantity of each common part varies between the two models.

	number in model	
part	B-1	B-2
gage sending units	8	10
wheel housing flares	6	4
intake grilles	3	2

6. A small company makes two chemicals, designated C-1 and C-2. The process for both includes fermentation and purification. Labor limits all fermentation operations to 300 hr per month, and purification is limited to 120 hr per month. Each unit of product C-1 requires 10 hr for fermentation and 8 hr for purification. Each unit of product C-2 requires 20 hr for fermentation and 3 hr for purification. The profit per unit of product C-1 is $3000; the profit per unit of product C-2 is $5000.

Plant Engineering

(a) Approximately how much should the company make of each chemical per month if partial units are permitted?

(A) $x_1 = 0, x_2 = 0$

(B) $x_1 = 0, x_2 = 15$

(C) $x_1 = 12, x_2 = 9.0$

(D) $x_1 = 15, x_2 = 0$

(b) Approximately how much should the company make of each chemical per month if only whole units are permitted?

(A) $x_1 = 9.0, x_2 = 11$

(B) $x_1 = 10, x_2 = 10$

(C) $x_1 = 11, x_2 = 9.0$

(D) $x_1 = 11, x_2 = 11$

7. A linear wage incentive program provides for 50% participation and a bonus that begins at a productivity level of 66.7% of standard. A worker produces 1900 units in 8 hr. The standard time for each unit is 0.004 hr. What is most nearly the worker's relative earnings?

(A) 14%

(B) 21%

(C) 120%

(D) 160%

SOLUTIONS

1. The number of work days per year is

$$\text{WDY} = \left(5\ \frac{\text{days}}{\text{wk}}\right)\left(52\ \frac{\text{wk}}{\text{yr}}\right) = 260 \text{ days/yr}$$

Each employee works 8 hours per day, so the number of minutes worked per employee is

$$\left(8\ \frac{\text{hr}}{\text{day}}\right)\left(60\ \frac{\text{min}}{\text{hr}}\right) = 480 \text{ min/day}$$

In general, the productive time available per machine is

$$M = 480\ \frac{\text{min}}{\text{day}} - \left(\begin{array}{r}\dfrac{\text{set-up}}{\text{day}} + \dfrac{\text{downtime}}{\text{day}} \\ + \dfrac{\text{maintenance}}{\text{day}} \\ + \dfrac{\text{record-keeping}}{\text{day}}\end{array}\right)$$

The production from a single department is

$$D_{\text{out}} = \frac{(1 - \text{defect fraction})(\text{time available})}{\text{production time}}$$

(a) The productive time available in department 4 is

$$\begin{aligned} M_4 &= 480\ \frac{\text{min}}{\text{day}} - \left(5\ \frac{\text{min}}{\text{day}} + 0\ \frac{\text{min}}{\text{day}} + 0\ \frac{\text{min}}{\text{day}} + 30\ \frac{\text{min}}{\text{day}}\right) \\ &= 445 \text{ min/day} \end{aligned}$$

The annual production rate of defect-free units from a single machine is

$$\begin{aligned} D_{4,\text{out}} &= \frac{(1 - 0.02)\left(445\ \frac{\text{min}}{\text{day}}\right)\left(60\ \frac{\text{sec}}{\text{min}}\right)\left(260\ \frac{\text{days}}{\text{yr}}\right)}{45\ \frac{\text{sec}}{\text{unit}}} \\ &= \boxed{151{,}181 \text{ units/yr} \quad (150{,}000 \text{ units/yr})} \end{aligned}$$

The answer is (D).

(b) Calculate the annual production rate for each of the departments. Assume each department has an infinite backlog of units processed by the previous department.

Department 1:

The productive time available is

$$\begin{aligned} M_1 &= 480\ \frac{\text{min}}{\text{day}} - \left(16\ \frac{\text{min}}{\text{day}} + 12\ \frac{\text{min}}{\text{day}} + 8\ \frac{\text{min}}{\text{day}} + 6\ \frac{\text{min}}{\text{day}}\right) \\ &= 438 \text{ min/day} \end{aligned}$$

Plant Engineering

The annual production rate of defect-free units from a single machine is

$$D_{1,\text{out}} = \frac{(1-0.04)\left(438\ \frac{\text{min}}{\text{day}}\right)\left(60\ \frac{\text{sec}}{\text{min}}\right)\left(260\ \frac{\text{days}}{\text{yr}}\right)}{6\ \frac{\text{sec}}{\text{unit}}}$$
$$= 1{,}093{,}248\ \text{units/yr}$$

Department 2:

The productive time available is

$$M_2 = 480\ \frac{\text{min}}{\text{day}} - \left(8\ \frac{\text{min}}{\text{day}} + 10\ \frac{\text{min}}{\text{day}} + 12\ \frac{\text{min}}{\text{day}} + 6\ \frac{\text{min}}{\text{day}}\right)$$
$$= 444\ \text{min/day}$$

The annual production rate of defect-free units from a single machine is

$$D_{2,\text{out}} = \frac{(1-0.06)\left(444\ \frac{\text{min}}{\text{day}}\right)\left(60\ \frac{\text{sec}}{\text{min}}\right)\left(260\ \frac{\text{days}}{\text{yr}}\right)}{10\ \frac{\text{sec}}{\text{unit}}}$$
$$= 651{,}082\ \text{units/yr}$$

Department 3:

The productive time available is

$$M_3 = 480\ \frac{\text{min}}{\text{day}} - \left(20\ \frac{\text{min}}{\text{day}} + 15\ \frac{\text{min}}{\text{day}} + 8\ \frac{\text{min}}{\text{day}} + 6\ \frac{\text{min}}{\text{day}}\right)$$
$$= 431\ \text{min/day}$$

The annual production rate of defect-free units from a single machine is

$$D_{3,\text{out}} = \frac{(1-0.03)\left(431\ \frac{\text{min}}{\text{day}}\right)\left(60\ \frac{\text{sec}}{\text{min}}\right)\left(260\ \frac{\text{days}}{\text{yr}}\right)}{11\ \frac{\text{sec}}{\text{unit}}}$$
$$= 592{,}899\ \text{units/yr}$$

Department 4:

From part (a), the productive time available is

$$M_4 = 445\ \text{min/day}$$

The annual production rate of defect-free units from a single machine is

$$D_{4,\text{out}} = 151{,}181\ \text{units/yr}$$

Department 4 is the bottleneck. The annual production rate is limited to $\boxed{151{,}181\ \text{units/yr}\ (150{,}000\ \text{units/yr}).}$

The answer is (D).

(c) Work backward from the desired output.

Department 4:

Each machine in department 4 produces 151,181 defect-free units per year. In order to produce 900,000 units per year, the number of machines needed is

$$n_4 = \frac{900{,}000\ \frac{\text{units}}{\text{yr}}}{151{,}181\ \frac{\text{units}}{\text{machine-yr}}}$$
$$= \boxed{5.95\ \text{machines}\quad (6\ \text{machines})}$$

Department 3:

In order to produce 900,000 units per year and supply the extra that will become defective in the subsequent processing, the number of machines needed is

$$n_3 = \frac{900{,}000\ \frac{\text{units}}{\text{yr}}}{(1-0.02)\left(592{,}899\ \frac{\text{units}}{\text{machine-yr}}\right)}$$
$$= \boxed{1.55\ \text{machines}\quad (2\ \text{machines})}$$

Department 2:

In order to produce 900,000 units per year and supply the extra that will become defective in the subsequent processing, the number of machines needed is

$$n_2 = \frac{900{,}000\ \frac{\text{units}}{\text{yr}}}{(1-0.03)(1-0.02)\left(651{,}082\ \frac{\text{units}}{\text{machine-yr}}\right)}$$
$$= \boxed{1.45\ \text{machines}\quad (2\ \text{machines})}$$

Department 1:

In order to produce 900,000 units per year and supply the extra that will become defective in the subsequent processing, the number of machines needed is

$$n_4 = \frac{900{,}000\ \frac{\text{units}}{\text{yr}}}{(1-0.06)(1-0.03)(1-0.02) \times \left(1{,}093{,}248\ \frac{\text{units}}{\text{machine-yr}}\right)}$$
$$= \boxed{0.92\ \text{machine}\quad (1\ \text{machine})}$$

The answer is (A).

Plant Engineering

2. (a) For the time being, disregard the 10 min per hour shift break since this break reduces the capacity of all stations by the same percentage.

Determine the maximum output per hour for each station.

station	output
1	$\dfrac{60\ \frac{\text{min}}{\text{hr}}}{0.6\ \frac{\text{min}}{\text{unit}}} = 100\text{ units/hr}$
2	$\dfrac{60\ \frac{\text{min}}{\text{hr}}}{0.6\ \frac{\text{min}}{\text{unit}}} = 100\text{ units/hr}$
3	$\dfrac{60\ \frac{\text{min}}{\text{hr}}}{0.9\ \frac{\text{min}}{\text{unit}}} = 66.67\text{ units/hr}$
4	$\dfrac{60\ \frac{\text{min}}{\text{hr}}}{0.8\ \frac{\text{min}}{\text{unit}}} = 75\text{ units/hr}$

Stations 3 and 4 are the bottleneck operations. Intuitively, operation 3 needs help the most, followed by operation 4.

Start by allocating station 5 capacity to the slowest operation, operation 3. Try to bring station 3 up to the same capacity as stations 1 and 2. To do so requires station 5 to produce $100 - 66.67 = 33.33$ units per hour. Since station 5 works at the same speed as station 3, the fraction of time station 5 needs to assist station 3 is $33.33/66.67 = 0.5$ (50%). This leaves 50% of station 5's time available to assist other stations.

Next, allocate the remaining station 5 time to station 4. To bring station 4 up to 100 units per hour will require station 5 to produce $100 - 75 = 25$ units per hour. Since station 5 works at the same rate as station 4, the fraction of time station 5 needs to assist station 4 is $25/75 = 0.3333$ (33.33%).

So, all of the stations have been brought up to 100 units per hour. Station 5 still has some remaining capacity: $100\% - 50\% - 33.33\% = 16.67\%$. This remaining capacity needs to be allocated to all of the remaining stations to bring them all up to the same output rate.

Suppose raising the output of the assembly line by 1 unit (i.e., from 100 to 101 units/hr) is desired. How much time would this take? It would take 0.6 min for operation 1, 0.6 min for operation 2, 0.9 min for operation 3, and 0.8 min for operation 4. Suppose raising the output by 2 units/hr is desired. That would take 1.2 min for operation 1, 1.2 min for operation 2, 1.8 min for operation 3, and 1.6 min for operation 4. Notice that the ratios of times between stations remain the same. The additional time for operation 2, for example, is always the same as for operation 1.

All of the extra time must come from the remaining capacity of station 5, since all other stations are working at their individual capacities. Station 5 has 17% of its time left, and it must allocate its time in the same fraction (ratio) as the assembly times.

operation	time	fraction of total	ratio × 17%
1	0.6	0.2069	3.52%
2	0.6	0.2069	3.52%
3	0.9	0.3103	5.27%
4	0.8	0.2759	4.69%
totals	2.9	1.0000	17.00%

Therefore, 3.52% (0.04) of station 5's time will be given to operations 1 and 2. Operation 3 will receive $50\% + 5.27\% = 55.27\%$ (0.60) of station 5's time. Operation 4 will receive $33.33\% + 4.69\% = 38.02\%$ (0.40).

(b) The production rates of each of the operations are as follows.

Operations 1 and 2:

$$\frac{(1.0352)\left(60\ \frac{\text{min}}{\text{hr}}\right)}{0.6\ \frac{\text{min}}{\text{unit}}} = 103.5\text{ units/hr}$$

Operation 3:

$$\frac{(1.5527)\left(60\ \frac{\text{min}}{\text{hr}}\right)}{0.9\ \frac{\text{min}}{\text{unit}}} = 103.5\text{ units/hr}$$

Operation 4:

$$\frac{(1.3802)\left(60\ \frac{\text{min}}{\text{hr}}\right)}{0.8\ \frac{\text{min}}{\text{unit}}} = 103.5\text{ units/hr}$$

So, the capacity of the assembly line is 103 units per hour.

(Check: The total time required per product is 2.9 minutes. With all five stations working optimally, the total available time per hour is (5 stations)(60 min/station) = 300 min. The optimal production rate would be (300 min/hr)/(2.9 min/unit) = 103.5 units/hr. This checks.)

However, everybody takes a 10 min break per hour, and the stations only work 50 min per hour. So, the overall capacity is reduced proportionally.

$$\begin{aligned}\text{capacity} &= \left(103.5\ \frac{\text{units}}{\text{hr}}\right)\left(\frac{50\text{ min}}{60\text{ min}}\right)\\ &= 86.3\text{ units/hr}\end{aligned}$$

Plant Engineering

3. (a) The critical path network diagram is as follows.

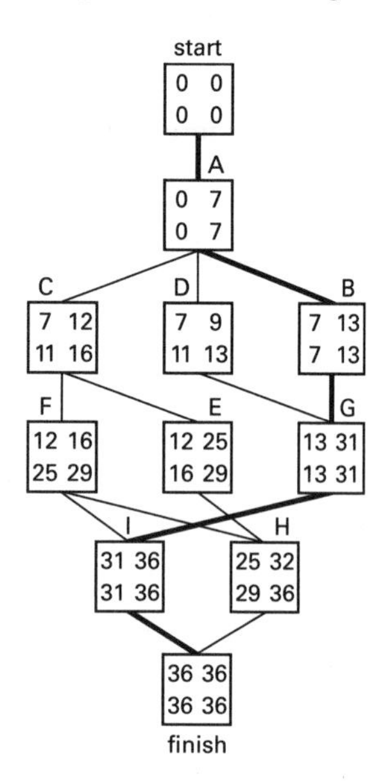

ES (earliest start) Rule: The earliest start time for an activity leaving a particular node is equal to the largest of the earliest finish times for all activities entering the node.

LF (latest finish) Rule: The latest finish time for an activity entering a particular node is equal to the smallest of the latest start times for all activities leaving the node.

The activity is critical if the earliest start equals the latest start.

(b) The critical path is A-B-G-I.

The answer is (A).

(c) The earliest finish is 36.

The answer is (D).

(d) The latest finish is 36.

The answer is (D).

(e) The slack along the critical path is 0.

The answer is (A).

(f) The float along the critical path is 0.

The answer is (A).

4.

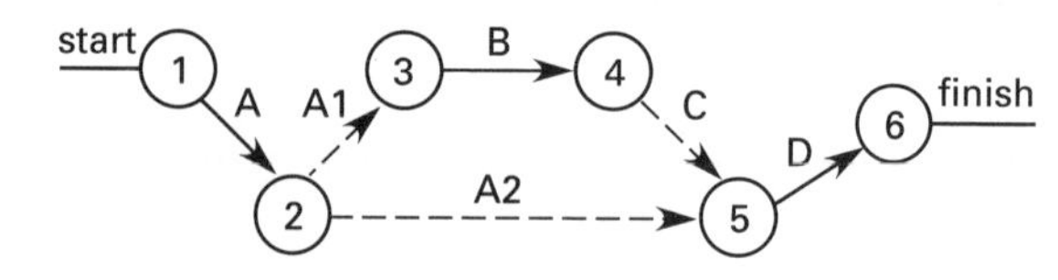

The critical path is A-A1-B-C-D.

Perform a PERT analysis. (See *Table for Sol. 4.*)

The following probability calculations assume that all activities are independent. Use the following theorems for the sum of independent random variables and use the normal distribution for T (project time).

From Eq. 63.1, the mean time is

$$\mu = \tfrac{1}{6}(t_{\text{min}} + 4t_{\text{most likely}} + t_{\text{max}})$$

From Eq. 63.2, squaring both sides, the variance is

$$\sigma^2 = \left(\tfrac{1}{6}(t_{\text{max}} - t_{\text{min}})\right)^2$$

$$\mu_{\text{total}} = t_{\text{A}} + t_{\text{B}} + t_{\text{C}} + t_{\text{D}}$$

$$\sigma^2_{\text{total}} = \sigma^2_{\text{A}} + \sigma^2_{\text{B}} + \sigma^2_{\text{C}} + \sigma^2_{\text{D}}$$

The variance is 10.52778 and the standard deviation is 3.244654.

$$\mu_{\text{total}} = 43.83333$$

$$\sigma^2_{\text{total}} = 10.52778$$

$$\sigma_{\text{total}} = 3.244654$$

$$z = \frac{t - \mu_{\text{total}}}{\sigma} = \left|\frac{42 - 43.83333}{3.244654}\right| = 0.565$$

The probability of finishing for $T \le 42$ is 0.286037 (29%).

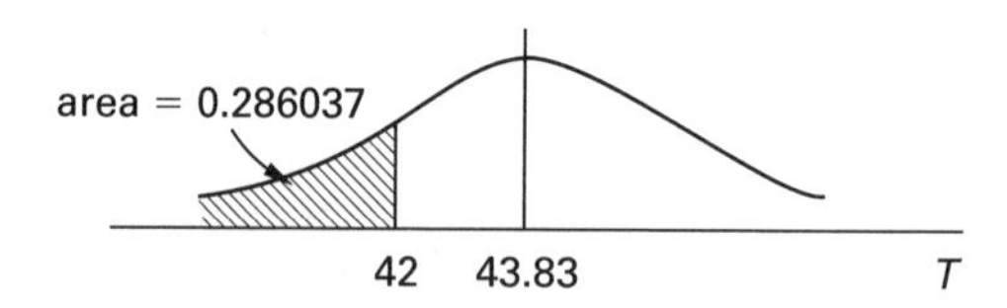

The answer is (B).

Table for Sol. 4

activity								
no.	name	exp. time	variance	earliest start	latest start	earliest finish	latest finish	slack LS-ES
1	A	+2.33333	+0.44444	15	15	+17.3333	+17.3333	0
2	A1	0	0	+17.3333	+17.3333	+17.3333	+17.3333	0
3	A2	0	0	+17.3333	+39.6667	+17.3333	+39.6667	+22.3333
4	B	+10.5000	+4.69444	+17.3333	+17.3333	+27.8333	+27.8333	0
5	C	+11.8333	+4.69444	+27.8333	+27.8333	+39.6667	+39.6667	0
6	D	+4.16667	+0.69444	+39.6667	+39.6667	+43.8333	+43.8333	0

expected completion time = 43.83333

5. This is a two-dimensional linear programming problem.

$$x_1 = \text{no. of B-1 buses produced}$$
$$x_2 = \text{no. of B-2 buses produced}$$
$$Z = \text{total profit}$$

The objective function is

$$\text{maximize } Z = 800{,}000x_1 + 650{,}000x_2$$

The constraints are

$$8x_1 + 10x_2 \leq 2000$$
$$6x_1 + 4x_2 \leq 1800$$
$$3x_1 + 2x_2 \leq 3600$$
$$x_1 \geq 0, x_2 \geq 0$$

The simplex theory states that the optimal solution will be a feasible corner point.

feasible corner points (x_1, x_2)	Z
(0, 0)	0
(0, 200)	130,000,000
optimal solution → (250, 0)	200,000,000

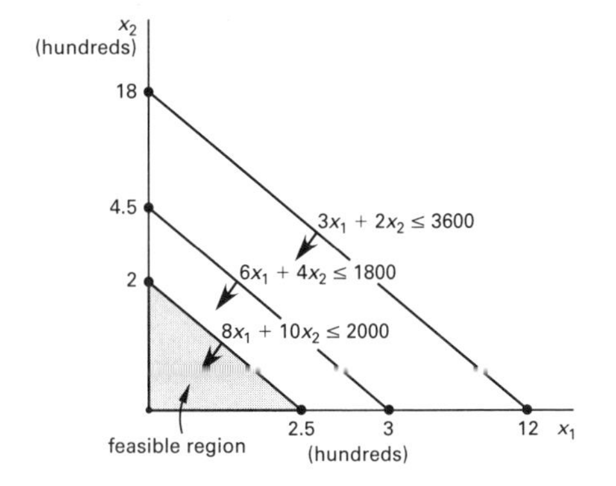

6. (a) Solve as a linear programming problem, since partial units are permitted.

$$x_1 = \text{no. of units of C-1 produced/month}$$
$$x_2 = \text{no. of units of C-2 produced/month}$$
$$Z = \text{total profit}$$

The objective function is

$$\text{maximize } Z = 3000x_1 + 5000x_2$$

The constraints are

$$10x_1 + 20x_2 \leq 300$$
$$8x_1 + 3x_2 \leq 120$$
$$x_1 \geq 0, x_2 \geq 0$$

The simplex theory states that the optimal solution will be a feasible corner point.

feasible corner points (x_1, x_2)	Z
(0, 0)	0
(0, 15)	75,000
optimal solution → $\left(\frac{150}{13}, \frac{120}{13}\right)$	80,769.23
(15, 0)	45,000

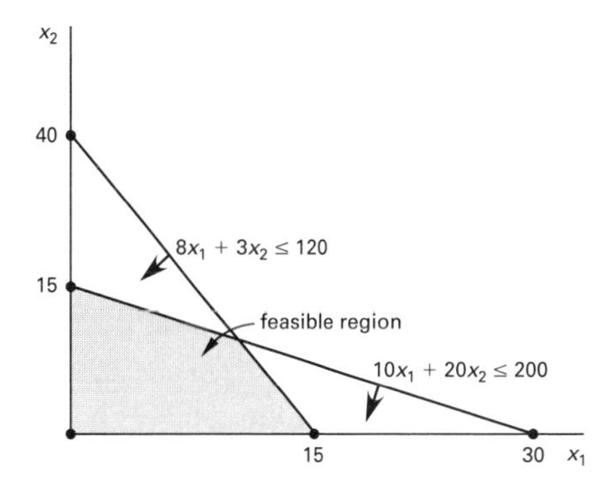

At the intersection point of the two constraints,

$$65x_2 = 600$$
$$x_2 = \frac{600}{65} = 120/13$$

Substituting x_2 into the first constraint yields

$$10x_1 + (20)\left(\frac{120}{13}\right) \leq 300$$
$$x_1 = 150/13$$
$$(x_1, x_2) = \left(\frac{150}{13}, \frac{120}{13}\right)$$
$$= \boxed{11.5, 9.23 \quad (12, 9.0)}$$

The answer is (C).

(b) Integer programming methods are required when variables are constrained to integer values. Merely deleting the fractional parts in the solution doesn't always work. In this case, deleting the fractional parts yields

$$x_1 = \text{INT}\left(\frac{150}{13}\right) = 11$$
$$x_2 = \text{INT}\left(\frac{120}{13}\right) = 9$$
$$Z = 3000x_1 + 5000x_2 = 78{,}000$$

However, $x_1 = x_2 = \boxed{10}$ also satisfies the constraints and yields the optimum $Z = 80{,}000$. In the absence of integer programming tools, trial and error in the vicinity of the feasible corners is required.

The answer is (B).

7.

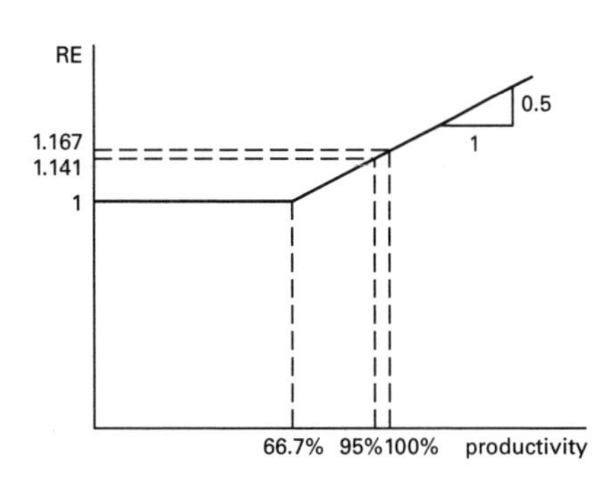

This is a linear gain sharing plan because the participation is 50% (i.e., less than 100%).

$$\text{participation} = 0.5$$

The productivity is the number of units produced in an hour divided by the number of standard units.

$$\begin{aligned}\text{productivity} &= \frac{\text{actual production}}{\text{ideal production}} = \frac{\frac{1900 \text{ units}}{8 \text{ hr}}}{0.004 \ \frac{\text{units}}{\text{hr}}} \\ &= 0.95\end{aligned}$$

In this plan, relative earnings are 1.0 until productivity is 0.667 (i.e., the threshold is 66.7%). Since the productivity of 95% is greater than 66.7%, the employee will be paid a bonus.

The bonus factor is determined from Eq. 63.75.

$$\begin{aligned}\text{bonus factor} &= 1 - \text{threshold} \\ &= 1 - 0.667 \\ &= 0.333\end{aligned}$$

The equation that describes the relative earnings, RE, is

$$\begin{aligned}\text{RE} &= 1 + \text{participation}(\text{productivity} + \text{bonus factor} - 1) \\ &= 1 + (0.5)(0.95 + 0.333 - 1) \\ &= \boxed{1.141 \quad (14\%)}\end{aligned}$$

The answer is (A).

64 Instrumentation and Measurements

PRACTICE PROBLEMS

Indicator Diagrams

1. Diesel engines in a waste-to-energy plant are powered by a gaseous mixture of methane and other digestion gases. The combustion conditions in each cylinder are continuously monitored by pressure and temperature transducers. A particular cylinder's indicator diagram is shown. The scale selected in the monitoring software is set to 111 kPa/mm. Most nearly, what is the mean effective pressure in the cylinder?

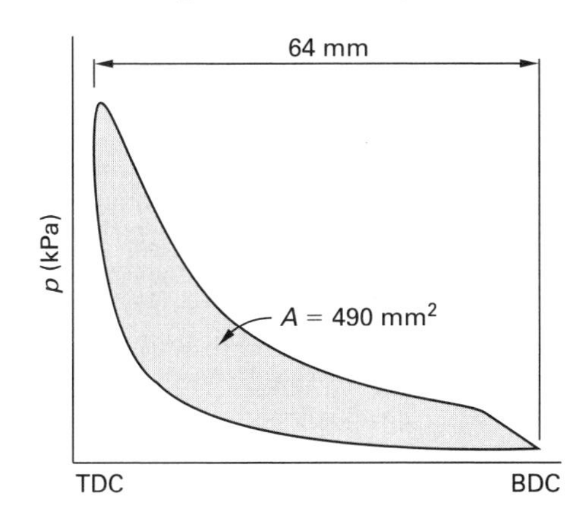

(A) 7.7 kPa

(B) 15 kPa

(C) 360 kPa

(D) 850 kPa

Variable Inductance Transformers

2. Which of the statements regarding the linear variable differential transformer (LVDT) shown is true?

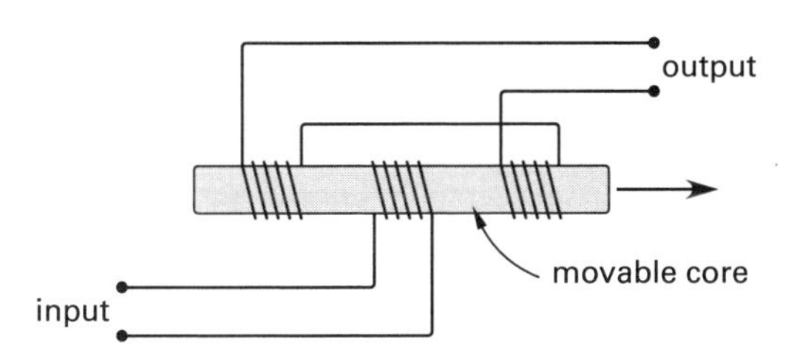

(A) The LVDT is a variable reluctance transducer, and its output voltage is directly proportional to its core movement.

(B) The LVDT is a variable reluctance transducer, and its output current is directly proportional to its core movement.

(C) The LVDT is a variable capacitance transducer, and its output current is directly proportional to core movement.

(D) The LVDT is a variable inductance transducer, and its output voltage is directly proportional to core movement.

Resistance Temperature Detectors

3. A resistance temperature detector (RTD) measures temperature in a bridge circuit. The resistances of resistors 1, 2, and 3 are 100 Ω, 200 Ω, and 50 Ω, respectively. The power supply voltage is 10 V. The reference resistance of the RTD is 100 Ω at 0°C. The alpha value for the RTD is 0.00392 1/°C. The variation of RTD resistance with temperature is linear. What is most nearly the temperature of the RTD when the voltmeter reads 4.167 V?

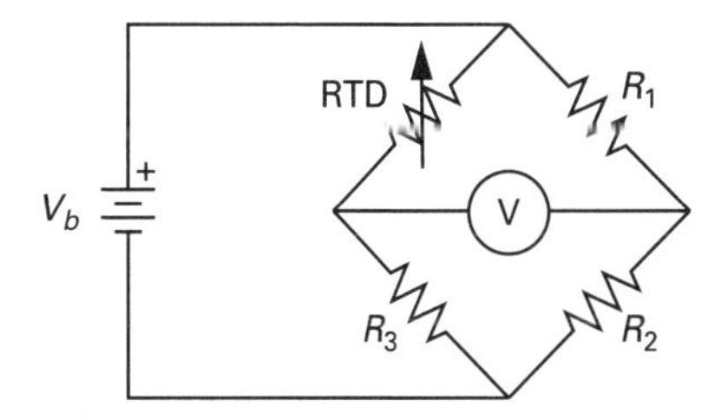

(A) 130°F

(B) 210°F

(C) 260°F

(D) 330°F

4. The temperature of a steel girder during a fire test is measured with a type 404 platinum RTD and a simple two-wire bridge, as shown. The characteristics of the RTD are: resistance, R, 100 Ω at 0°C; temperature coefficient, α, 0.00385 1/°C. The values of the bridge resistances are $R_1 = 1000\ \Omega$, and $R_3 = 1000\ \Omega$. When the meter is nulled out during a test, $R_2 = 376\ \Omega$. The lead resistances are each 100 Ω. Most nearly, what is the temperature of the beam during the test?

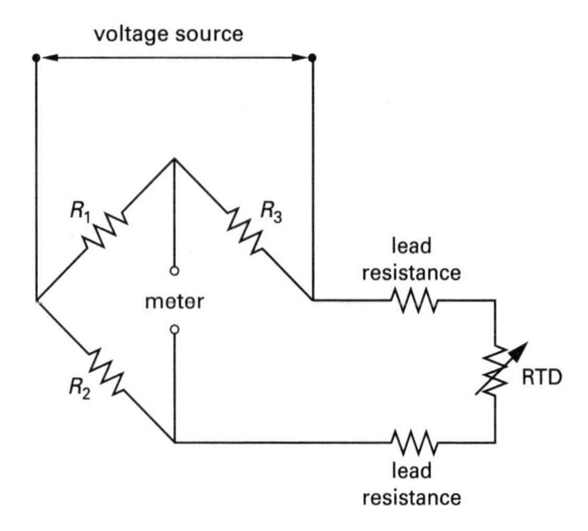

(A) 130°C

(B) 175°C

(C) 200°C

(D) 450°C

Piezoelectric Transducers

5. A compression-style crystal piezoelectric accelerometer has a sensitivity of 2.0 mV/g. It falls to the floor from a height of 3 ft and comes to rest 0.02 sec after contact with the floor. What is most nearly the average voltage output of the accelerometer during impact?

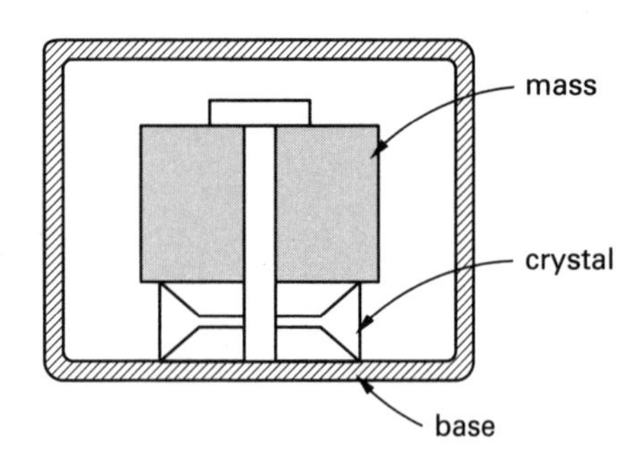

(A) 21 mV

(B) 32 mV

(C) 43 mV

(D) 57 mV

6. A piezoelectric transducer is connected to a charge amplifier as shown. What is the purpose of capacitor C_t?

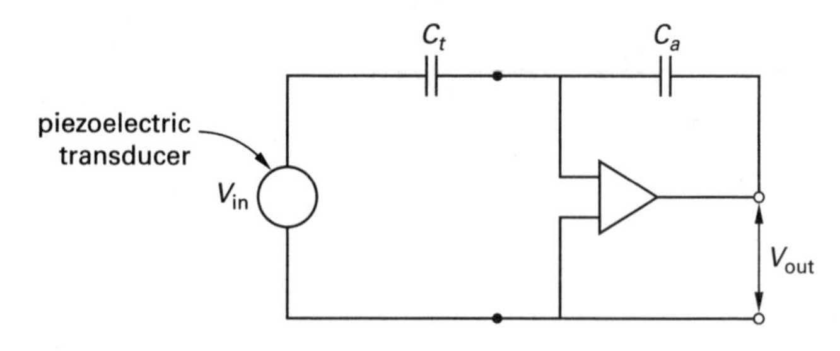

(A) C_t biases the transducer output.

(B) C_t isolates the transducer from inadvertent amplifier transients.

(C) C_t rectifies the transducer output.

(D) C_t prevents DC voltages from being passed to the amplifier.

Photosensitive Conductors

7. A velocity indicating system attached to a machine consists of a light source, a light sensor, a disk that rotates with the machine's rotation, and a counting circuit. There are 12 reflective strips around the periphery of the disk. As a machine starts up, the disk spins, and the sensor detects the pulses reflected back by the reflective strips. Given the numbers of counts per minute recorded by the sensor, what is most nearly the average angular acceleration of the disk between 2 sec and 4 sec?

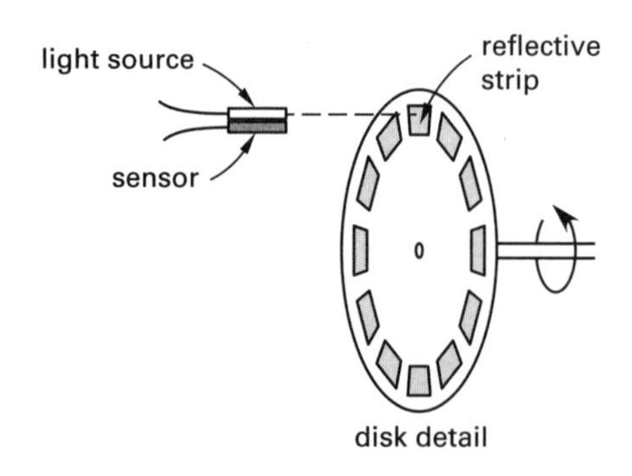

startup counts

time (sec)	counts (min)
0	0
1	150
2	1200
4	10,800

(A) 3.5 rad/sec^2

(B) 21 rad/sec^2

(C) 42 rad/sec^2

(D) 84 rad/sec^2

Hydraulic Rams

8. A 2 ft diameter (outer dimension) oil-filled hydraulic scale has 2 in thick sidewalls. When the scale is unloaded, the pressure gauge reads zero. When the pressure gauge reads 50 psig, the weight of the object is most nearly

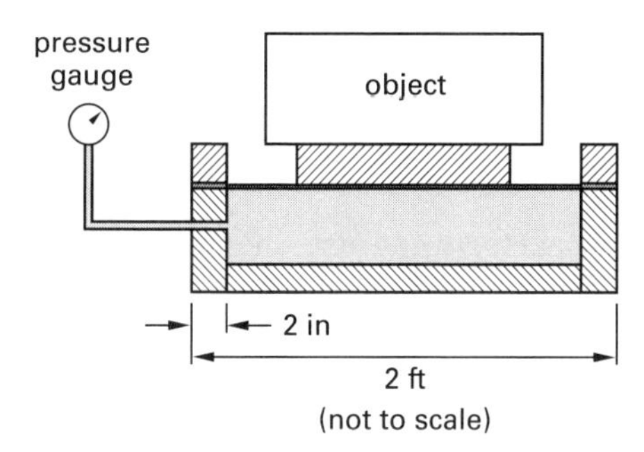

(A) 8000 lbf

(B) 16,000 lbf

(C) 23,000 lbf

(D) 63,000 lbf

Strain Gauges

9. A load cell measures tensile force with a steel bar and a bonded strain gauge. The modulus of elasticity of the steel is 30×10^6 lbf/in^2. The bar's tensile area is 1.0 in^2. The gage factor of the strain gauge is 6.0. The unloaded strain gauge resistance is 100 Ω, and the loaded resistance is 100.1 Ω. The applied force is most nearly

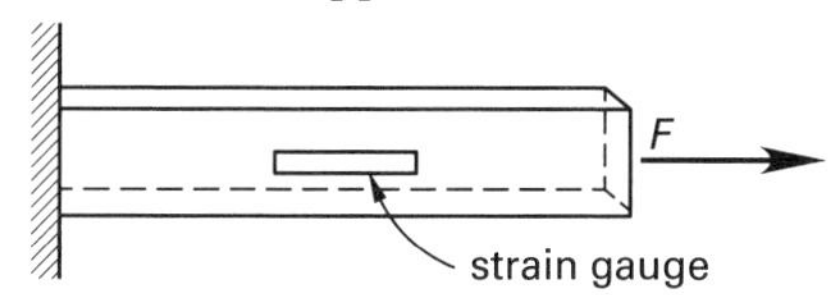

(A) 2500 lbf

(B) 5000 lbf

(C) 17,000 lbf

(D) 20,000 lbf

10. A cylindrical, stainless steel tank rests on the ground. The tank has a diameter of 1 ft, is 4 ft long, and has a wall thickness of 0.35 in. The steel has a modulus of elasticity of 30×10^6 lbf/in^2 and a Poisson's ratio of 0.3. A strain gauge (gage factor of 4; unstrained resistance of 200 Ω) is bonded to the surface and aligned to measure longitudinal strain. When pressurized, the change in strain gauge resistance is 0.12 Ω. What is most nearly the pressure in the tank?

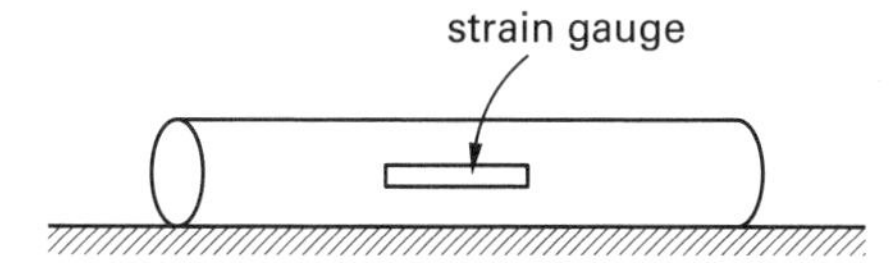

(A) 1100 psig

(B) 1400 psig

(C) 1700 psig

(D) 2100 psig

11. A strain gauge is bonded to each of the four vertical bars of a grain hopper scale. The hopper has a tare weight of 1500 lbf. The support bars have a modulus of elasticity of 20×10^6 lbf/in^2, and each bar has a cross-sectional area of 2 in^2. Each strain gauge has a gage factor of 3 and had an initial resistance of 300 Ω when originally applied to the unstressed bars. After the grain is loaded into the hopper, the strain gauge resistances are 300.05 Ω, 300.09 Ω, 300.11 Ω, and 300.03 Ω, respectively. What is most nearly the weight of the grain loaded into the hopper?

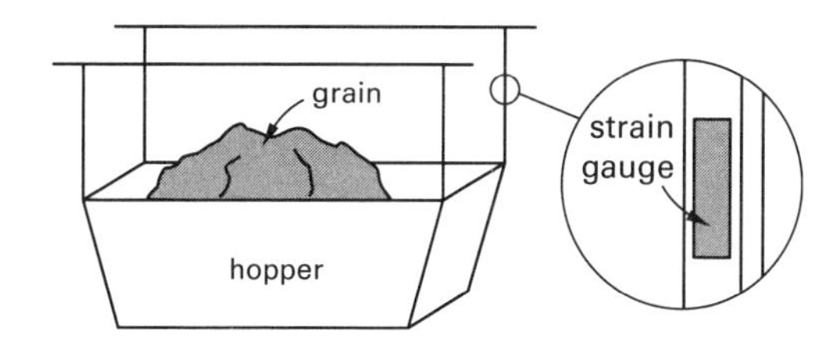

(A) 9500 lbf

(B) 11,000 lbf

(C) 12,000 lbf

(D) 14,000 lbf

12. A strain gauge with a gage factor of 2.0 and resistance of 120 Ω experiences a microstrain of 500. Most nearly, what is the percentage change in resistance?

(A) 0.00001%

(B) 0.001%

(C) 0.1%

(D) 1000%

13. A strain gauge has a gage factor of 3.0 and a nominal (strain-free) resistance of 350 Ω. The gauge is bonded to the top of a rectangular aluminum bar 12 cm long, 2 cm high, and 4 cm wide. The aluminum has a modulus of elasticity of 73 GPa. The bar is loaded as a cantilever beam. The distance from the applied load to the center of the strain gauge is 10 cm. The instrumentation consists of a simple Wheatstone bridge consisting of two 1000 Ω resistors, a variable resistor, and the strain gauge. Lead resistance is negligible. Most nearly, what is the change in resistance of the strain gauge when the free end of the aluminum bar is loaded vertically by a 1 kg mass?

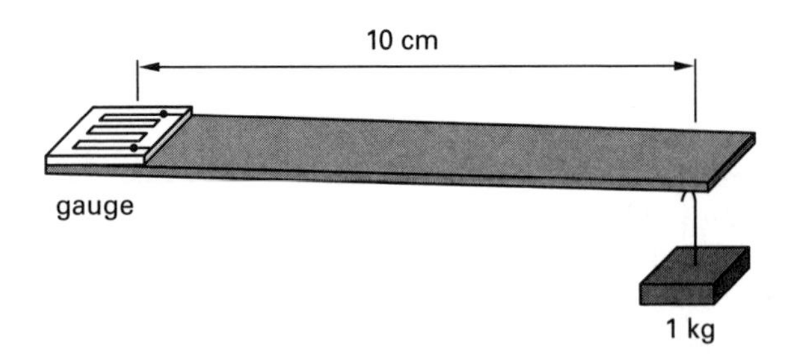

(A) 0.00054 Ω

(B) 0.0053 Ω

(C) 0.048 Ω

(D) 0.50 Ω

14. A cylindrical 2 in diameter solid shaft used to power an aerator blade is mounted as a stationary horizontal cantilever and instrumented with a strain gauge, R_1, bonded to its upper surface. The strain gauge is connected to an uncompensated bridge circuit where R_2 is 160 Ω and R_3 is 2500 Ω. R_4 is the adjustable resistance with a resistance of 2505 Ω when the shaft is unloaded, and 2511.7 Ω when the shaft tip 24 in from the midpoint of the strain gauge is loaded with vertical force of 1100 lbf. The gage factor of the strain gauge is 2.10. Disregard slip ring and lead resistances. Most nearly, what is the modulus of elasticity of the shaft?

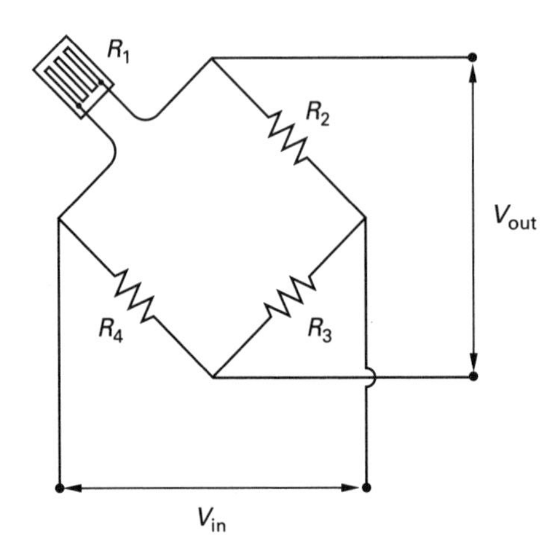

(A) 12×10^6 lbf/in^2

(B) 18×10^6 lbf/in^2

(C) 23×10^6 lbf/in^2

(D) 26×10^6 lbf/in^2

15. A 45° (rectangular) strain gauge rosette is bonded to the web of a large A36 steel built-up girder beam. The web's modulus of elasticity is 200 GPa, and Poisson's ratio is 0.3. The microstrains are $\epsilon_A = 650$, $\epsilon_B = -300$, and $\epsilon_C = 480$.

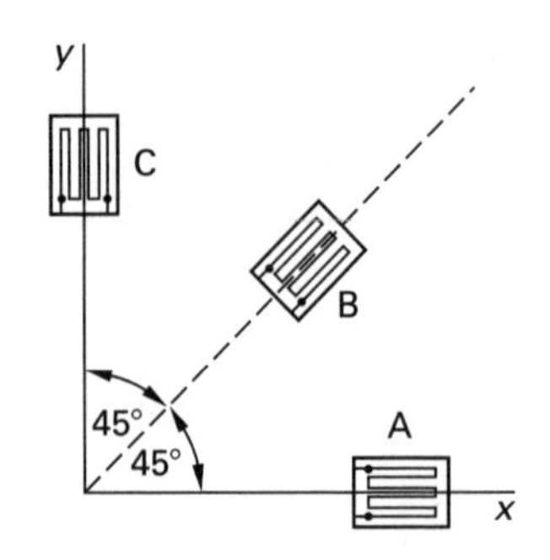

(a) Most nearly, what are the principal strains in the web?

(A) 840, 290

(B) 1400, −300

(C) 2300, 1200

(D) 2900, −610

(b) Most nearly, what are the principal stresses in the web?

(A) 250 MPa, 17 MPa

(B) 290 MPa, 57 MPa

(C) 300 MPa, 28 MPa

(D) 330 MPa, 57 MPa

Controllers and Control Logic

16. The schematic for a diesel generator's starter motor is shown. The table describes the conditions that affect the starting circuit's individual contacts.

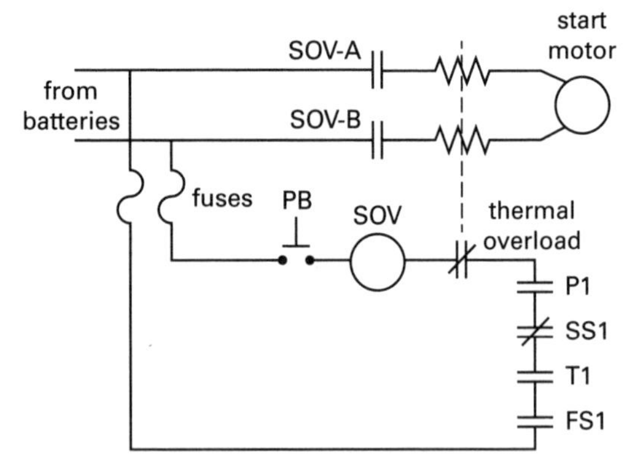

SOV = shutoff valve
PB = push button
P = pressure (switch)
SS = speed (switch)
T = temperature (switch)
FS = flow (switch)

contact	action
thermal overload	opens upon thermal overload
P1	closes when diesel oil pressure is greater than 3 psig
SS1	opens when diesel speed is greater than 100 rev/min
T1	closes when diesel crankcase temperature is greater than 45°F
FS1	closes when diesel cooling water flow is adequate

Which of the following conditions is NOT required for the diesel motor to run?

I. fuses must not be blown

II. push button must be pressed and held

III. no thermal overload on the starter motor

IV. diesel oil pressure must be greater than 3 psig

V. diesel speed must be above 100 rev/min

VI. diesel crankcase temperature must be greater than 45°F

VII. diesel standby cooling water pump must be running

(A) V only

(B) VI only

(C) III and IV only

(D) IV and V only

17. An operator presses push button PB on the test circuit shown. Which statement best describes the condition of the circuit nine seconds after button PB is pressed?

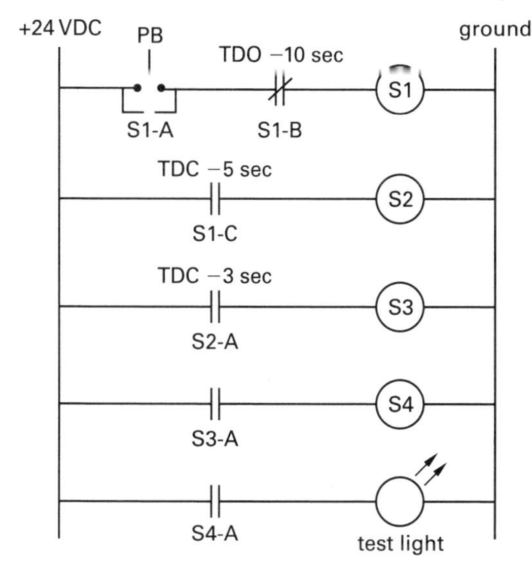

(A) Nothing is energized, and the test light is not lit.

(B) S1, S2, and S3 are energized, and the test light is not lit.

(C) S1 and S2 are deenergized, S3 and S4 are energized, and the test light is lit.

(D) S1, S2, S3, and S4 are energized, and the test light is lit.

18. A room's temperature is held constant by using a proportional controller. The thermostat has a temperature range of 60–80°F, and the controller has a proportional range of 0–100%. When the thermostat and controller gain are set to 70°F and 50%, respectively, the temperature of the room is constant at 66°F. Which option will keep the temperature of the room at the setpoint?

(A) The gain on the proportional controller should be increased.

(B) The gain on the proportional controller should be decreased.

(C) The controller should be changed to a proportional-integral controller.

(D) The controller should be changed to a proportional-derivative controller.

19. Which statement about a proportional-integral-derivative (PID) controller is correct?

(A) The integral term ensures the controller controls at setpoint, and the derivative term increases controller damping and enhances controller stability.

(B) The derivative term ensures the PID controls at setpoint. The integral term increases controller damping and enhances controller stability.

(C) Adjusting the controller gain to the optimum setting will force the controller to control at setpoint.

(D) Controller stability is independent of the controller gain.

Thermocouples

20. The voltage across a copper-constantan thermocouple is 4.108 mV when it is placed in a 250°F oil bath. The temperature of the thermocouple reference junction is most nearly

(A) 32°F

(B) 65°F

(C) 80°F

(D) 85°F

21. Two iron-constantan thermocouples are wired in series to monitor the cylinder head temperature in an aircraft engine. The voltage across the thermocouples is read in the cockpit as 16.52 mV. The cockpit temperature is 70°F. What is most nearly the average temperature of the cylinder head?

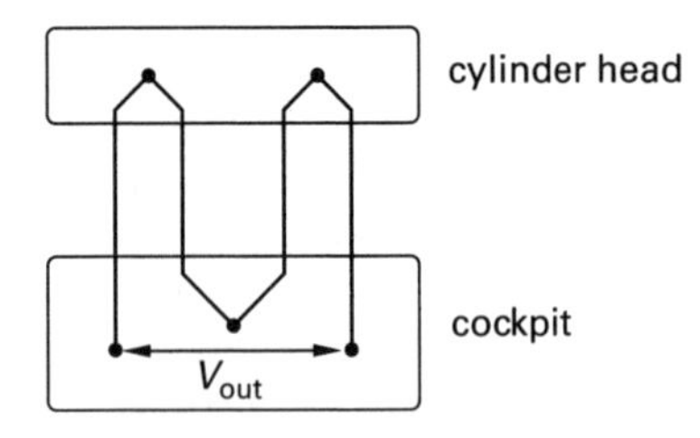

(A) 290°F

(B) 330°F

(C) 350°F

(D) 380°F

22. A copper-constantan thermocouple is used with an ice bath to measure the temperature of waste sludge during digestion. The voltage is proportional to the difference in temperature between the junctions. When the thermocouple is calibrated, it generates a voltage of 5.07×10^{-3} V when the temperature of the hot junction is 108.9°C and the reference temperature is at 0°C. Most nearly, what is the temperature of waste sludge when the voltage is 1.83×10^{-3} V?

(A) 40°C

(B) 45°C

(C) 50°C

(D) 55°C

SOLUTIONS

1. The average height of the pressure plot is

$$\bar{h} = \frac{A}{w} = \frac{490\ \text{mm}^2}{64\ \text{mm}} = 7.656\ \text{mm}$$

The average pressure is

$$\begin{aligned}\bar{p} &= k\bar{h} = \left(111\ \frac{\text{kPa}}{\text{mm}}\right)(7.656\ \text{mm}) \\ &= \boxed{849.8\ \text{kPa}\quad (850\ \text{kPa})}\end{aligned}$$

The answer is (D).

2. A coil of wire, particularly a coil with an iron core, behaves as an inductor in an AC circuit. In an LVDT, the movable transformer core changes the inductance between the primary and secondary windings. The transformer is constructed so that the output voltage of the LVDT is directly proportional to the movable core position. A linear variable differential transformer (LVDT) is a variable inductance transducer.

The answer is (D).

3. Since the voltage is not zero, the bridge is not balanced. Rearrange Eq. 64.16 to find the RTD resistance.

$$\begin{aligned}R_{\text{RTD}} &= \frac{V_b R_1 R_3 + V(R_1 R_3 + R_2 R_3)}{V_b R_2 - V(R_1 + R_2)} \\ &= \frac{\begin{array}{c}(10\ \text{V})(100\ \Omega)(50\ \Omega) \\ + (4.167\ \text{V})\left(\begin{array}{c}(100\ \Omega)(50\ \Omega) \\ + (200\ \Omega)(50\ \Omega)\end{array}\right)\end{array}}{\begin{array}{c}(10\ \text{V})(200\ \Omega) - (4.167\ \text{V}) \\ \times (100\ \Omega + 200\ \Omega)\end{array}} \\ &= 150.0\ \Omega\end{aligned}$$

Rearrange Eq. 64.3 to find the temperature. Since temperature varies linearly with resistance, only the alpha-value, α, is required.

$$\begin{aligned}T &= \frac{R_{\text{ref}}(\alpha T_{\text{ref}} - 1) + R_{\text{RTD}}}{R_{\text{ref}}\alpha} \\ &= \frac{(100\ \Omega)\left(\left(0.00392\ \frac{1}{^\circ\text{C}}\right)(0^\circ\text{C}) - 1\right) + 150.0\ \Omega}{(100\ \Omega)\left(0.00392\ \frac{1}{^\circ\text{C}}\right)} \\ &= 127.6^\circ\text{C}\end{aligned}$$

Convert to the Fahrenheit scale.

$$T_{°F} = 32° + \tfrac{9}{5}T_{°C} = 32°F + \left(\tfrac{9}{5}\right)(127.6°C) = \boxed{261.7°F \quad (260°F)}$$

The answer is (C).

4. When the meter is nulled out, the resistance of the leg containing the RTD is

$$R_4 = R_{RTD} + 2R_{lead} = \frac{R_2R_3}{R_1} = \frac{(376\ \Omega)(1000\ \Omega)}{1000\ \Omega} = 376\ \Omega$$

The resistance of the RTD is

$$R_{RTD} = R_4 - 2R_{lead} = 376\ \Omega - (2)(100\ \Omega) = 176\ \Omega$$

The relationship between temperature and resistance for an RTD is

$$R_T = R_0(1 + AT + BT^2)$$

For a platinum RTD with $\alpha = 0.00385\ 1/°C$, the resistance is

$$\begin{aligned} R_T &= R_0\left(\begin{array}{l} 1 + (3.9083 \times 10^{-3})T \\ \quad + (-5.775 \times 10^{-7})T^2 \end{array}\right) \\ &\approx R_0\left(1 + (3.9083 \times 10^{-3})T\right) \\ 176\ \Omega &\approx (100\ \Omega)\left(1 + (3.9083 \times 10^{-3})T\right) \\ T &\approx \boxed{194.5°C \quad (200°C)} \end{aligned}$$

(If the squared term is kept, the temperature is 200.3°C.)

The answer is (C).

5. The displacement of the falling accelerometer is

$$x = x_0 + \mathrm{v}_0 t - \tfrac{1}{2}gt^2$$

The initial displacement and the initial velocity are both zero. The time required to fall a distance of x is

$$t = \sqrt{\frac{2x}{g}}$$

The impact velocity is

$$\begin{aligned} \mathrm{v} &= gt = g\sqrt{\frac{2x}{g}} = \sqrt{2xg} \\ &= \sqrt{(2)(3\ \text{ft})\left(32.2\ \frac{\text{ft}}{\text{sec}^2}\right)} \\ &= 13.9\ \text{ft/sec} \end{aligned}$$

The average acceleration is

$$\overline{a} = \frac{\mathrm{v}}{\Delta t} = \frac{13.9\ \frac{\text{ft}}{\text{sec}}}{(0.02\ \text{sec})\left(32.2\ \frac{\text{ft}}{\text{sec}^2\text{-g}}\right)} = 21.58\ \text{g}$$

The average voltage output is

$$\overline{V} = k\overline{a} = \left(2.0\ \frac{\text{mV}}{\text{g}}\right)(21.58\ \text{g}) = \boxed{43.16\ \text{mV} \quad (43\ \text{mV})}$$

The answer is (C).

6. The transducer capacitance, C_t, prevents DC voltages from being passed to the amplifier. The value of C_t does not influence the amplifier gain.

The answer is (D).

7. Each revolution produces 12 counts. Using Eq. 57.18, the average angular acceleration is

$$\begin{aligned} \alpha_{ave} &= \frac{\omega_2 - \omega_1}{t_2 - t_1} \\ &= \frac{\left(2\pi\ \frac{\text{rad}}{\text{rev}}\right)\left(10{,}800\ \frac{\text{counts}}{\text{min}} - 1200\ \frac{\text{counts}}{\text{min}}\right)}{(4\ \text{sec} - 2\ \text{sec})\left(12\ \frac{\text{counts}}{\text{rev}}\right)\left(60\ \frac{\text{sec}}{\text{min}}\right)} \\ &= \boxed{41.89\ \text{rad/sec}^2 \quad (42\ \text{rad/sec}^2)} \end{aligned}$$

The answer is (C).

8. The weight of the object is equal to the force on the scale.

$$\begin{aligned} F &= pA = p\left(\frac{\pi}{4}\right)(D - 2t)^2 \\ &= \left(50\ \frac{\text{lbf}}{\text{in}^2}\right)\left(\frac{\pi}{4}\right)\left((2\ \text{ft})\left(12\ \frac{\text{in}}{\text{ft}}\right) - (2)(2\ \text{in})\right)^2 \\ &= \boxed{15{,}708\ \text{lbf} \quad (16{,}000\ \text{lbf})} \end{aligned}$$

The answer is (B).

9. From Eq. 64.11, the strain is

$$\epsilon = \frac{\Delta R_g}{(\text{GF})R_g} = \frac{100.1\ \Omega - 100\ \Omega}{(6.0)(100\ \Omega)} = 1.67 \times 10^{-4}$$

From Eq. 64.24, the stress is

$$\sigma = E\epsilon = \left(30 \times 10^6\ \frac{\text{lbf}}{\text{in}^2}\right)(1.67 \times 10^{-4}) = 5.0 \times 10^3\ \text{lbf/in}^2$$

Plant Engineering

Rearranging Eq. 64.33, the applied force is

$$F = \sigma A = \left(5.0 \times 10^3 \ \frac{\text{lbf}}{\text{in}^2}\right)(1.0 \ \text{in}^2) = \boxed{5.0 \times 10^3 \ \text{lbf} \quad (5000 \ \text{lbf})}$$

The answer is (B).

10. From Eq. 64.11, the strain is

$$\epsilon_{\text{in/in}} = \frac{\Delta R}{(\text{GF})R_g} = \frac{0.12 \ \Omega}{(4)(200 \ \Omega)} = 0.00015 \ \text{in/in}$$

Since the wall thickness-to-radius ratio of the tank is less than 0.1, the tank is thin-walled.

From Eq. 53.17 and Eq. 53.18, the hoop stress, σ_h, and the longitudinal stress, σ_l, are

$$\sigma_h = \frac{pr}{t}$$
$$\sigma_l = \frac{pr}{2t}$$

The longitudinal strain, ϵ, is given by Eq. 53.19.

$$\epsilon = \frac{\sigma_l - \nu\sigma_h}{E} = \frac{\frac{pr}{2t} - \nu\frac{pr}{t}}{E} = \frac{\frac{pr}{t}\left(\frac{1}{2} - \nu\right)}{E}$$

Since the tank is thin-walled, either the inside radius or the outside radius may be used. Use the inside radius. The pressure in the tank is

$$p = \frac{\epsilon E}{\frac{r_i}{t}\left(\frac{1}{2} - \nu\right)} = \frac{\left(0.00015 \ \frac{\text{in}}{\text{in}}\right)\left(30 \times 10^6 \ \frac{\text{lbf}}{\text{in}^2}\right)}{\left(\frac{\left(\frac{1 \ \text{ft}}{2}\right)\left(12 \ \frac{\text{in}}{\text{ft}}\right) - 0.35 \ \text{in}}{0.35 \ \text{in}}\right)\left(\frac{1}{2} - 0.3\right)} = \boxed{1393.8 \ \text{lbf/in}^2 \quad (1400 \ \text{psig})}$$

The answer is (B).

11. Work with average values. The average strain gauge resistance is

$$R_{\text{ave}} = \frac{300.05 \ \Omega + 300.09 \ \Omega + 300.11 \ \Omega + 300.03 \ \Omega}{4} = 300.07 \ \Omega$$

From Eq. 64.11, the average strain is

$$\epsilon_{\text{ave}} = \frac{\Delta R_g}{(\text{GF})R_g} = \frac{300.07 \ \Omega - 300 \ \Omega}{(3)(300 \ \Omega)} = 7.777 \times 10^{-5} \ \text{in/in}$$

From Eq. 64.24, the average stress is

$$\sigma_{\text{ave}} = E\epsilon_{\text{ave}} = \left(20 \times 10^6 \ \frac{\text{lbf}}{\text{in}^2}\right)\left(7.777 \times 10^{-5} \ \frac{\text{in}}{\text{in}}\right) = 1555 \ \text{lbf/in}^2$$

Rearranging Eq. 64.33, the average force is

$$F_{\text{ave}} = \sigma_{\text{ave}} A = \left(1555 \ \frac{\text{lbf}}{\text{in}^2}\right)(2 \ \text{in}^2) = 3110 \ \text{lbf}$$

The total weight of the hopper and grain is

$$W_t = 4F_{\text{ave}} = (4)(3110 \ \text{lbf}) = 12{,}440 \ \text{lbf}$$

Therefore, the weight of the grain is

$$W_g = W_t - W_h = 12{,}440 \ \text{lbf} - 1500 \ \text{lbf} = \boxed{10{,}940 \ \text{lbf} \quad (11{,}000 \ \text{lbf})}$$

The answer is (B).

12. Use Eq. 64.11.

$$\frac{\Delta R_g}{R_g} = (\text{GF})\epsilon = (2.0)(500 \times 10^{-6}) = \boxed{0.001 \quad (0.1\%)}$$

The answer is (C).

13. The moment of inertia of the bar in bending is

$$I = \frac{bh^3}{12} = \frac{(4 \ \text{cm})(2 \ \text{cm})^3}{12} = 2.667 \ \text{cm}^4$$

The stress in the bar is

$$\sigma = \frac{Mc}{I} = \frac{mglc}{I} = \frac{(1 \ \text{kg})\left(9.81 \ \frac{\text{m}}{\text{s}^2}\right)(10 \ \text{cm})\left(\frac{2 \ \text{cm}}{2}\right)\left(100 \ \frac{\text{cm}}{\text{m}}\right)^2}{2.667 \ \text{cm}^4} = 367{,}829 \ \text{Pa}$$

The strain is

$$\epsilon = \frac{\sigma}{E} = \frac{367{,}829 \text{ Pa}}{(73 \text{ GPa})\left(10^9 \ \frac{\text{Pa}}{\text{GPa}}\right)} = 5.04 \times 10^{-6}$$

Use Eq. 64.11.

$$\begin{aligned}\Delta R_g &= \epsilon(\text{GF})R_g = (5.04 \times 10^{-6})(3.0)(350 \ \Omega) \\ &= \boxed{0.00529 \ \Omega \quad (0.0053 \ \Omega)}\end{aligned}$$

The answer is (B).

14. First, use the information from the unloaded case to determine the unstressed resistance of the strain gauge. Since the bridge numbering docs not correspond to Fig. 64.6, rewrite Eq. 64.16 using the ratio of the vertical legs, top to bottom.

$$\frac{R_1}{R_4} = \frac{R_2}{R_3}$$

The unstressed resistance of the strain gauge is

$$R_1 = \frac{R_4 R_2}{R_3} = \frac{(2505 \ \Omega)(160 \ \Omega)}{2500 \ \Omega} = 160.32 \ \Omega$$

The resistance of the strain gauge when the shaft is loaded is

$$R_1 = \frac{R_4 R_2}{R_3} = \frac{(2511.7 \ \Omega)(160 \ \Omega)}{2500 \ \Omega} = 160.7488 \ \Omega$$

Now, use Eq. 64.11 to determine the strain when the shaft is loaded.

$$\begin{aligned}\epsilon &= \frac{\Delta R_g}{(\text{GF})R_g} = \frac{160.7488 \ \Omega - 160.32 \ \Omega}{(2.10)(160.32 \ \Omega)} \\ &= 0.00127364\end{aligned}$$

The shaft's moment of inertia as a beam is

$$I = \frac{\pi r^4}{4} = \frac{\pi\left(\frac{2 \text{ in}}{2}\right)^4}{4} = 0.7854 \text{ in}^4$$

The bending stress in the shaft is

$$\begin{aligned}\sigma &= \frac{Mc}{I} = \frac{Flc}{I} = \frac{(1100 \text{ lbf})(24 \text{ in})\left(\frac{2 \text{ in}}{2}\right)}{0.7854 \text{ in}^4} \\ &= 33{,}613 \text{ lbf/in}^2\end{aligned}$$

The modulus of elasticity is

$$\begin{aligned}E &= \frac{\sigma}{\epsilon} = \frac{33{,}613 \ \frac{\text{lbf}}{\text{in}^2}}{0.00127364} \\ &= \boxed{26.39 \times 10^6 \text{ lbf/in}^2 \quad (26 \times 10^6 \text{ lbf/in}^2)}\end{aligned}$$

The answer is (D).

15. (a) Use Table 64.6. (The strain gauge designations are changed.) Work in microstrains (μm/m). The principal strains are

$$\begin{aligned}\epsilon_p, \epsilon_q &= \tfrac{1}{2}\left(\epsilon_A + \epsilon_C \pm \sqrt{2(\epsilon_A - \epsilon_B)^2 + 2(\epsilon_B - \epsilon_C)^2}\right) \\ &= \left(\tfrac{1}{2}\right)\left(650 + 480 \pm \sqrt{\begin{aligned}&(2)\big(650 - (-300)\big)^2 \\ &\quad + (2)(-300 - 480)^2\end{aligned}}\right) \\ &= \left(\tfrac{1}{2}\right)(1130 \pm 1738) \\ &= \boxed{1434, -304 \quad (1400, -300)}\end{aligned}$$

The answer is (B).

(b) Use Table 64.6. Work in microstrains (μm/m). The principal stresses are

$$\begin{aligned}\sigma_p, \sigma_q &= \frac{E}{2}\left(\begin{aligned}&\frac{\epsilon_A + \epsilon_C}{1 - \nu} \pm \frac{1}{1 + \nu} \\ &\quad \times \sqrt{2(\epsilon_A - \epsilon_B)^2 + 2(\epsilon_B - \epsilon_C)^2}\end{aligned}\right) \\ &= \left(\frac{200 \text{ GPa}}{2}\right)\left(10^9 \ \frac{\text{Pa}}{\text{GPa}}\right) \\ &\quad \times \left(\begin{aligned}&\frac{650 + 480}{1 - 0.3} \pm \left(\frac{1}{1 + 0.3}\right) \\ &\quad \times \sqrt{\begin{aligned}&(2)\big(650 - (-300)\big)^2 \\ &\quad + (2)(-300 - 480)^2\end{aligned}}\end{aligned}\right) \\ &\quad \times \left(10^{-6} \ \frac{\text{m}}{\mu\text{m}}\right) \\ &= (1614 \pm 1337) \times 10^5 \text{ Pa} \\ &= 295 \times 10^6 \text{ Pa}, 27.7 \times 10^6 \text{ Pa}\end{aligned}$$

$$\boxed{\text{The principal stresses are 295 MPa and 27.7 MPa (300 MPa and 28 MPa).}}$$

The answer is (C).

Plant Engineering

16. In order for the starter motor to run, fuses must not be blown (condition I); the push button must be pressed and held (condition II); there can be no thermal overload on the starter motor (condition III); the diesel oil pressure must be greater than 3 psig (condition IV); the diesel speed must be less than 100 rev/min (condition V); and the diesel standby cooling water pump must be running (condition VII). The starter motor disengages as soon as the engine speed increases above 100 rev/min. Condition V is incorrect.

The answer is (A).

17. This is a ladder diagram, not a circuit diagram. All of the S1 elements are part of the same switch. Pressing the push button energizes S1 and causes S1-A to close to maintain power to S1. After a 5 sec delay, S1-C closes to energize S2. After an additional 3 sec delay, S2-A closes to energize S3. This causes S3 to close and immediately energize S4. S4-A then closes to illuminate the test light. The total time for this process is the sum of the delay times, which is 8 sec. At 10 sec, S1-B opens to de-energize the circuit.

Therefore, at 9 sec, S1, S2, S3, and S4 are energized, and the test light is lit.

The answer is (D).

18. The signal from a proportional-only controller is proportional to the difference between the actual temperature and the setpoint. The signal decreases as the temperature approaches the setpoint, so the temperature will approach, but never actually reach, the setpoint. Adjusting the gain on a proportional controller will not change this behavior and may make the controller unstable by causing the signal (and, hence, the temperature) to overshoot. A proportional-derivative (PD) controller responds to changing temperature (at any temperature), enhances stability, and increases damping of the controller. However, since the derivative of a constant is zero, a PD controller does not respond to a stable temperature and, therefore, cannot maintain the temperature at the setpoint. A proportional-integral (PI) controller eliminates offset and is able to control the temperature at the setpoint. Therefore, the controller should be changed to a PI controller.

The answer is (C).

19. Controller stability is affected by controller gain. If the gain is too high, the controller may become unstable; if the gain is too low, the controller response time may be too slow. The controller gain cannot be forced to operate at setpoint by adjusting the controller gain to an optimum setting. The integral term compensates for offset and ensures the controller controls at setpoint. The derivative term increases controller damping and enhances controller stability.

The answer is (A).

20. From App. 64.A, the voltage output from a copper-constantan thermocouple at 250°F is 5.280 mV. (Read horizontally on the 200°F line to the 50°F column.) The voltage corresponding to the reference junction temperature, V_{ref}, is

$$\begin{aligned} V_{ref} &= V_{250°F} - V = 5.280\text{ mV} - 4.108\text{ mV} \\ &= 1.172\text{ mV} \end{aligned}$$

By interpolation from App. 64.A, this corresponds to a reference junction temperature of $\boxed{85°F.}$

The answer is (D).

21. The average voltage across each thermocouple, V_{ave}, is

$$\begin{aligned} V_{ave} &= \frac{V}{2} = \frac{16.52\text{ mV}}{2} \\ &= 8.26\text{ mV} \end{aligned}$$

The reference junction is at 70°F in the cockpit. From App. 64.A, the thermoelectric constant for an iron-constantan thermocouple at 70°F is 1.07 mV. This value, added to V_{ave}, represents the voltage that would be obtained if the cockpit were at the reference temperature of 32°F.

$$\begin{aligned} V &= V_{ave} + 1.07\text{ mV} = 8.26\text{ mV} + 1.07\text{ mV} \\ &= 9.33\text{ mV} \end{aligned}$$

Interpolating from App. 64.A, the voltage yields an average cylinder head temperature of $\boxed{345°F\ (350°F).}$

The answer is (C).

22. During calibration, the temperature difference is 108.9°C − 0°C = 108.9°C. The thermoelectric constant is

$$k_T = \frac{V}{\Delta T} = \frac{5.07 \times 10^{-3}\text{ V}}{108.9°\text{C}} = 4.656 \times 10^{-5}\text{ V/°C}$$

Use Eq. 64.9. The temperature of the waste sludge is

$$\begin{aligned} T &= T_{ref} + \frac{V}{k_T} \\ &= 0°\text{C} + \frac{1.83 \times 10^{-3}\text{ V}}{4.656 \times 10^{-5}\ \frac{\text{V}}{°\text{C}}} \\ &= \boxed{39.3°\text{C}\quad(40°\text{C})} \end{aligned}$$

The answer is (A).

65 Manufacturing Processes

PRACTICE PROBLEMS

Nondestructive Testing

1. An example of destructive weld testing is

(A) penetrant testing

(B) hardness testing

(C) ultrasonic inspection

(D) radiographic inspection

Joining and Fastening

2. A 10 mil (250 μm) adhesive with a shear strength of 1500 lbf/in^2 (10 MPa) is used in a lap joint between two 0.20 in (5 mm) aluminum sheets. The aluminum has a yield strength of 15,000 lbf/in^2 (100 MPa). Assume the adhesive is loaded in pure shear, but use a stress concentration factor of 2. If the joint is to be as strong as the aluminum, the width (overlap) of adhesive joint that is required is most nearly

(A) 2.3 in (0.058 m)

(B) 4.0 in (0.10 m)

(C) 5.3 in (0.13 m)

(D) 6.9 in (0.17 m)

3. The welding symbol shown calls for a

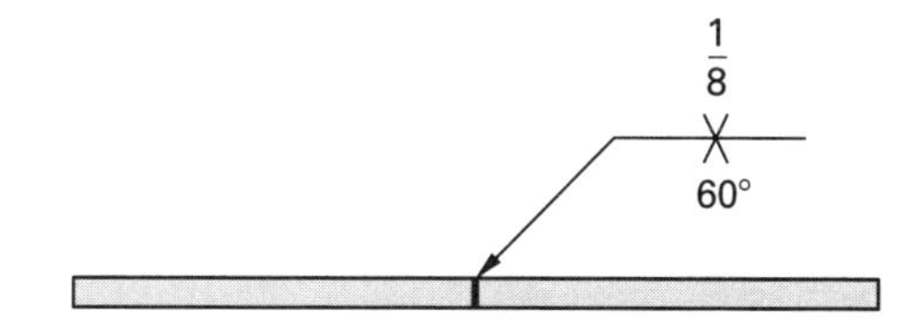

(A) double-welded butt joint

(B) single-welded butt joint

(C) single-full fillet lap joint

(D) double-welded fillet lap joint

Metal Cutting Operations

4. The tool life equation for tool bit A is $v_A T_A^{0.3} = 605$, and the equation for tool bit B is $v_B T_B^{0.15} = 386$. v is the cutting speed in feet per minute, and T is the tool life in minutes. At a particular cutting speed, tool bit B will last twice as long as tool bit A before developing wear. This cutting speed is most nearly

(A) 120 ft/min

(B) 200 ft/min

(C) 340 ft/min

(D) 480 ft/min

5. A hole is to be drilled into a 2 in thick aluminum workpiece using a 0.5 in drill bit. The average material removal rate for the drilling operation is 0.6 in^3/min. The time required to drill the hole is most nearly

(A) 39 sec

(B) 42 sec

(C) 51 sec

(D) 58 sec

6. For an orthogonal metal cutting operation, the cutting speed is 200 ft/min, the thrust force is 60 lbf, and the resultant force is 100 lbf. The amount of power input needed for the cutting operation is most nearly

(A) 0.25 hp

(B) 0.36 hp

(C) 0.48 hp

(D) 0.63 hp

7. A square opening in a 2 in thick aluminum alloy sheet has a side length of 0.5 in. The aluminum has an ultimate tensile strength of 120,000 lbf/in^2. The initial force required to punch the opening is most nearly

(A) 48 tons

(B) 62 tons

(C) 95 tons

(D) 140 tons

Metal Forming Operations

8. A cylindrical shell is being drawn from a 0.125 in thick flat steel blank. The blank is circular and has a diameter of 7.25 in. A 3.5 in diameter punch is used. The tensile strength of the steel is 65,000 psi. The drawing factor (correction to the drawing ratio) to account for friction and bending is 0.6. The angle of radial stress at die entry is 90°. Neglect clearance. The drawing force needed is most nearly

(A) 45 tons

(B) 53 tons

(C) 63 tons

(D) 72 tons

9. A round 6 in diameter steel billet is extruded at 1500°F down to a diameter of 2.5 in. The billet has an extrusion constant of $k = 30{,}000$ psi. The extrusion force needed is most nearly

(A) 520 tons

(B) 570 tons

(C) 610 tons

(D) 740 tons

SOLUTIONS

1. Of the four kinds of testing listed, hardness testing is the only one that plastically deforms or scratches the weld. Penetrant testing, ultrasonic inspection, and radiographic inspection do not permanently alter or destroy the object being tested.

The answer is (B).

2.

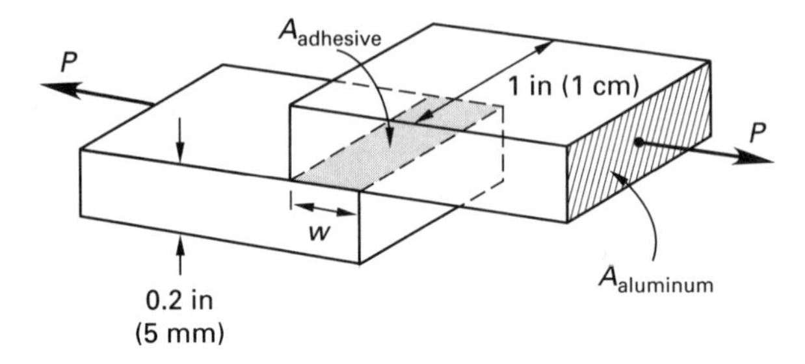

Customary U.S. Solution

step 1: Assume the unit length ($L = 1$ in) for the joint.

$$A_{adhesive} = wL = w_{in}(1\text{ in}) = w$$
$$A_{aluminum} = Lt = (1\text{ in})(0.2\text{ in}) = 0.2\text{ in}^2$$

step 2: Find the tensile force, P, as follows.

The yield strength is

$$\begin{aligned} P = S_y A_{aluminum} &= \left(15{,}000\ \frac{\text{lbf}}{\text{in}^2}\right)(0.2\text{ in}^2) \\ &= 3000\text{ lbf} \end{aligned}$$

step 3: Find the required width, w, with a stress concentration factor of 2.

The shear stress is

$$\tau = \frac{2P}{A_{adhesive}}$$
$$1500\ \frac{\text{lbf}}{\text{in}^2} = \frac{(2)(3000\text{ lbf})}{w}$$
$$w = \boxed{4.0\text{ in}}$$

The answer is (B).

SI Solution

step 1: Assume the unit length ($L = 1$ cm) for the joint.

$$A_{adhesive} = wL = w_{in}(0.01\text{ m}) = 0.01w$$
$$\begin{aligned} A_{aluminum} = Lt &= (0.01\text{ m})(0.005\text{ m}) \\ &= 0.00005\text{ m}^2 \end{aligned}$$

step 2: Find the tensile force, P, as follows.

The yield strength is

$$P = S_y A_{\text{aluminum}} = (100 \times 10^6 \text{ Pa})(0.00005 \text{ m}^2) = 5000 \text{ N}$$

step 3: Find the required width, w, with a stress concentration factor of 2.

The shear stress is

$$\tau = \frac{2P}{A_{\text{adhesive}}}$$

$$10 \times 10^6 \text{ Pa} = \frac{(2)(5000 \text{ N})}{0.01w}$$

$$w = \boxed{0.10 \text{ m}}$$

The answer is (B).

3. The two pieces are joined end to end, so this is a butt joint. The V shapes on both sides of the arrow indicate welding on both sides of the joint. The symbol indicates a $\boxed{\text{double-welded butt joint.}}$

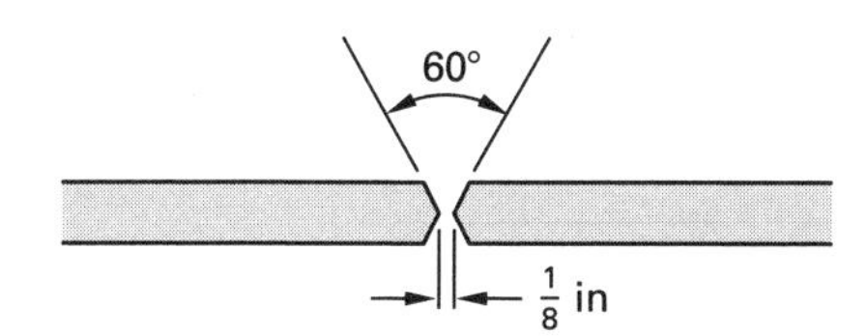

The answer is (A).

4. Rearrange the two tool life equations.

$$\text{v}_\text{A} = \frac{605}{T_\text{A}^{0.3}}$$

$$\text{v}_\text{B} = \frac{386}{T_\text{B}^{0.15}}$$

If the same cutting speed is used for both tool bits,

$$\frac{605}{T_\text{A}^{0.3}} = \frac{386}{T_\text{B}^{0.15}}$$

Since $T_\text{B} = 2T_\text{A}$, substitute $2T_\text{A}$ for T_B and solve for T_A.

$$\frac{605}{T_\text{A}^{0.3}} = \frac{386}{(2T_\text{A})^{0.15}}$$

$$\frac{T_\text{A}^{0.3}}{T_\text{A}^{0.15}} = T_\text{A}^{0.15} = \frac{(605)(2)^{0.15}}{386} = 1.739$$

$$T_\text{A} = 39.99$$

The cutting speed is

$$\text{v}_\text{A} = \frac{605}{T_\text{A}^{0.3}} = \frac{605}{(39.99 \text{ min})^{0.3}} = \boxed{200.06 \text{ ft/min} \quad (200 \text{ ft/min})}$$

The answer is (B).

5. The volume of material removed is

$$V = AL = \left(\frac{\pi}{4}\right)d^2L = \left(\frac{\pi}{4}\right)(0.5 \text{ in})^2(2 \text{ in}) = 0.3927 \text{ in}^3$$

The time required to remove all of the material is

$$t = \frac{V}{\dot{V}} = \frac{(0.3927 \text{ in}^3)\left(60 \frac{\text{sec}}{\text{min}}\right)}{0.6 \frac{\text{in}^3}{\text{min}}} = \boxed{39.27 \text{ sec} \quad (39 \text{ sec})}$$

The answer is (A).

6. The cutting force is

$$F_c = \sqrt{R^2 - F_t^2} = \sqrt{(100 \text{ lbf})^2 - (60 \text{ lbf})^2} = 80 \text{ lbf}$$

The power needed for the orthogonal cutting operation is

$$P = F_c\text{v} = \frac{(80 \text{ lbf})\left(200 \frac{\text{ft}}{\text{min}}\right)}{33{,}000 \frac{\text{ft-lbf}}{\text{hp-min}}} = \boxed{0.4848 \text{ hp} \quad (0.48 \text{ hp})}$$

The answer is (C).

7. There is insufficient information to use a more sophisticated shear force model, so use a punch-force model. The total sheared length of the perimeter of the square is

$$L = 4s = (4)(0.5 \text{ in}) = 2 \text{ in}$$

The initial area in shear is

$$A_{\text{shear}} = Lt = (2 \text{ in})(2 \text{ in}) = 4 \text{ in}^2$$

The ultimate shear strength can be estimated from the ultimate tensile strength. Various correlations produce a range of ratios from 0.5 to 0.65.

$$\begin{aligned} S_{us} &= 0.6S_{ut} = (0.6)\left(120{,}000\ \frac{\text{lbf}}{\text{in}^2}\right) \\ &= 72{,}000\ \text{lbf/in}^2 \end{aligned}$$

The initial force required is

$$\begin{aligned} F = S_{us}A &= \frac{\left(72{,}000\ \frac{\text{lbf}}{\text{in}^2}\right)(4\ \text{in}^2)}{2000\ \frac{\text{lbf}}{\text{ton}}} \\ &= \boxed{144\ \text{tons}\quad (140\ \text{tons})} \end{aligned}$$

The answer is (D).

8. Without knowing the clearance or final wall thickness, the mean shell diameter, d_1, can't be found, so use the punch diameter.

The drawing ratio is

$$\text{DR} = \frac{D_{\text{blank}}}{d_1} = \frac{7.25\ \text{in}}{3.5\ \text{in}} = 2.07$$

Generally, the drawing ratio is no larger than 2.0, so use DR = 2.0.

The flow stress is the stress that must be applied to cause a material to deform at a constant strain rate in its plastic range. Most materials will harden under such conditions, so the flow stress is a function of the amount of plastic strain. Although there are relationships that can be used to calculate flow stress, there is insufficient information in this problem to apply them. So, use the tensile strength as given.

Using the standard metal-drawing model, the force required to draw cylindrical shells from flat stock is

$$\begin{aligned} F &= \pi d_1 t S_{ut}(\text{DR} - C)\sin\alpha \\ &= \frac{\pi(3.5\ \text{in})(0.125\ \text{in})\left(65{,}000\ \frac{\text{lbf}}{\text{in}^2}\right) \times (2.0 - 0.6)\sin 90^\circ}{2000\ \frac{\text{lbf}}{\text{ton}}} \\ &= \boxed{62.54\ \text{tons}\quad (63\ \text{tons})} \end{aligned}$$

The answer is (C).

9. The original cross-sectional area is

$$A_o = \frac{\pi d^2}{4} = \frac{\pi(6\ \text{in})^2}{4} = 28.27\ \text{in}^2$$

The final cross-sectional area is

$$A_f = \frac{\pi d^2}{4} = \frac{\pi(2.5\ \text{in})^2}{4} = 4.91\ \text{in}^2$$

The extrusion force is

$$\begin{aligned} F &= A_o k \ln\frac{A_o}{A_f} \\ &= \frac{(28.27\ \text{in}^2)\left(30{,}000\ \frac{\text{lbf}}{\text{in}^2}\right)\ln\frac{28.27\ \text{in}^2}{4.91\ \text{in}^2}}{2000\ \frac{\text{lbf}}{\text{ton}}} \\ &= \boxed{742.3\ \text{tons}\quad (740\ \text{tons})} \end{aligned}$$

The answer is (D).

66 Materials Handling and Processing

PRACTICE PROBLEMS

Elevators

1. The work performed in raising a 500 kg elevator 30 m is most nearly

(A) 75 kJ

(B) 150 kJ

(C) 15 000 kJ

(D) 150 000 kJ

2. A 700 kg elevator starts from rest. It accelerates uniformly for 3 s, at which time it reaches a speed of 1.8 m/s.

(a) What is most nearly the average power delivered by the elevator motor during the acceleration?

(A) 0.4 kW

(B) 6.2 kW

(C) 6.6 kW

(D) 20 kW

(b) What is most nearly the instantaneous power at $t = 3$ s?

(A) 0.8 kW

(B) 12 kW

(C) 13 kW

(D) 18 kW

3. An elevator with an empty mass of 1200 kg has a passenger capacity of 900 kg. The guide rail system imparts a constant frictional force of 4200 N.

(a) What is most nearly the minimum motor power required to lift the full elevator at a constant speed of 4.5 m/s?

(A) 89 kW

(B) 110 kW

(C) 120 kW

(D) 130 kW

(b) What is most nearly the instantaneous power developed when the elevator accelerates from 4.5 m/s at 1.75 m/s^2?

(A) 89 kW

(B) 110 kW

(C) 120 kW

(D) 130 kW

Bulk Storage Piles

4. The amount of corn (angle of internal friction = 21°) that can be stored in a triangular windrow pile 50 ft wide × 100 ft long is most nearly

(A) 15,000 bu

(B) 18,000 bu

(C) 23,000 bu

(D) 29,000 bu

Bin and Hoppers

5. A baghouse collector captures and stores chips, dust, and shavings from a woodworking plant. The particulate matter gathers in the funnel-shaped base of the collector and then falls through an orifice at the bottom. If the particulate matter is cohesive enough, it may form an arched cap over this orifice and block the flow. What is this behavior called?

(A) bridging

(B) choking

(C) corking

(D) piping

6. A large flat-bottom steel silo has a diameter of 12 ft and a height of 60 ft. The silo is filled with polyethylene pellets (bulk specific weight of 35 lbf/ft^3 and angle of internal friction of 20°). Disregard friction between the pellets and silo wall. What is most nearly the pressure on the vertical wall at the bottom of the silo?

(A) 280 lbf/ft^2

(B) 340 lbf/ft^2

(C) 390 lbf/ft^2

(D) 570 lbf/ft^2

7. A rail cargo hopper has a capacity of 42,000 lbm. Polyethylene pellets (bulk density = 35 lbm/ft^3) enter the hopper through a 6 in schedule-40 opening from a mass flow silo whose walls are inclined 65° from the horizontal. The time required for the compartment to fill is most nearly

(A) 15 min

(B) 19 min

(C) 26 min

(D) 34 min

Belt Conveyors

8. A continuous conveyor belt operates at 120 ft/min using a 5 hp drive. All of the conveyor is inclined 10° from the horizontal and raises goods a vertical distance of 10 ft. The maximum loading that the conveyor can transport is most nearly

(A) 320,000 lbm/hr

(B) 690,000 lbm/hr

(C) 770,000 lbm/hr

(D) 890,000 lbm/hr

9. A continuous rubber belt conveyor operates at 120 ft/min. The coefficient of friction between the belt and its pulleys is 0.35. The belt wrap angle on the driving pulley is 240°. The maximum belt tension is 750 lbf. The power required to drive the belt is most nearly

(A) 0.6 hp

(B) 1.4 hp

(C) 2.1 hp

(D) 2.9 hp

10. A continuous conveyor belt operates at 120 ft/min at temperatures between 30°F and 100°F. Belt tension and pulley friction are adequate at both temperature extremes. The belt is 0.25 in thick and 24 in wide. The belt material has a thermal coefficient of 1.8×10^{-6} 1/°F and a Young's modulus of 1.0×10^6 lbf/in^2. What is most nearly the decrease in motor power required when the temperature increases from 30°F to 100°F?

(A) 0.84 hp

(B) 1.5 hp

(C) 2.7 hp

(D) 4.7 hp

Cyclone Separators

11. A particular dust is modeled as spherical particles with a diameter of 5 μm and a solid density of 2.0 g/cm^3. The dust enters a conventional 2D2D air cyclone in standard air at 20 m/s. The cyclone diameter is 1.0 m, the entrance width is 0.25 m, and the entrance height is 0.5 m. What is most nearly the removal efficiency?

(A) 28%

(B) 32%

(C) 39%

(D) 43%

12. 68°F air tangentially enters a 2 ft wide × 5 ft high rectangular entrance of a high-efficiency gas cyclone at 50 ft/sec. The clean air leaves through a 5 ft diameter exit. What is most nearly the pressure drop?

(A) 19 lbf/ft^2

(B) 24 lbf/ft^2

(C) 35 lbf/ft^2

(D) 60 lbf/ft^2

Venturi Feeders

13. 1800 lbm/hr of sawdust with a bulk density of 11 lbm/ft^3 is conveyed by air at standard conditions through a 6 in schedule-40 pipe. The material loading ratio is 1:2. What is most nearly the air velocity?

(A) 1700 ft/min

(B) 2200 ft/min

(C) 2600 ft/min

(D) 4000 ft/min

SOLUTIONS

1. The work is

$$W = Fd = mgd = \frac{(500\ \text{kg})\left(9.81\ \frac{\text{m}}{\text{s}^2}\right)(30\ \text{m})}{1000\ \frac{\text{J}}{\text{kJ}}}$$
$$= \boxed{147.15\ \text{kJ}\quad (150\ \text{kJ})}$$

The answer is (B).

2. (a) The acceleration is

$$a = \frac{\text{v}}{t} = \frac{1.8\ \frac{\text{m}}{\text{s}}}{3\ \text{s}} = 0.6\ \text{m/s}^2$$

The distance traveled in 3 s is

$$\Delta y = \tfrac{1}{2}at^2 = \left(\tfrac{1}{2}\right)\left(0.6\ \frac{\text{m}}{\text{s}^2}\right)(3\ \text{s})^2 = 2.7\ \text{m}$$

The tension in the cable is

$$\begin{aligned} T &= F_{\text{acceleration}} + W = ma + mg \\ &= (700\ \text{kg})\left(0.6\ \frac{\text{m}}{\text{s}^2}\right) + (700\ \text{kg})\left(9.81\ \frac{\text{m}}{\text{s}^2}\right) \\ &= 7287\ \text{N} \end{aligned}$$

The work performed by the motor is

$$W = T\Delta y = \frac{(7287\ \text{N})(2.7\ \text{m})}{1000\ \frac{\text{J}}{\text{kJ}}} = 19.675\ \text{kJ}$$

The average power delivered by the motor is

$$P = \frac{W}{t} = \frac{19.675\ \text{kJ}}{3\ \text{s}} = \boxed{6.558\ \text{kW}\quad (6.6\ \text{kW})}$$

The answer is (C).

(b) The instantaneous power at $t = 3$ s is

$$P = T\text{v} = \frac{(7287\ \text{N})\left(1.8\ \frac{\text{m}}{\text{s}}\right)}{1000\ \frac{\text{W}}{\text{kW}}} = \boxed{13.117\ \text{kW}\quad (13\ \text{kW})}$$

The answer is (C).

3. (a) The total mass of the loaded elevator is

$$m = 1200\ \text{kg} + 900\ \text{kg} = 2100\ \text{kg}$$

Since there is no acceleration, the tension in the cable includes only weight and frictional components.

$$T = mg + F_f = (2100\ \text{kg})\left(9.81\ \frac{\text{m}}{\text{s}^2}\right) + 4200\ \text{N} = 24\,801\ \text{N}$$

The power required is

$$P = T\text{v} = \frac{(24\,801\ \text{N})\left(4.5\ \frac{\text{m}}{\text{s}}\right)}{1000\ \frac{\text{W}}{\text{kW}}} = \boxed{111.6\ \text{kW}\ \ (110\ \text{kW})}$$

The answer is (B).

(b) When accelerating, the tension in the cable is

$$\begin{aligned} T &= mg + ma + F_f \\ &= (2100\ \text{kg})\left(9.81\ \frac{\text{m}}{\text{s}^2}\right) + (2100\ \text{kg})\left(1.75\ \frac{\text{m}}{\text{s}^2}\right) + 4200\ \text{N} \\ &= 28\,476\ \text{N} \end{aligned}$$

The instantaneous power is

$$P = T\text{v} = \frac{(28\,476\ \text{N})\left(4.5\ \frac{\text{m}}{\text{s}}\right)}{1000\ \frac{\text{W}}{\text{kW}}} = \boxed{128.1\ \text{kW}\ \ (130\ \text{kW})}$$

The answer is (D).

4. Windrows of grain naturally form piles that can be divided into three sections: a half cone on each of the two ends, and a section between the conical ends that is triangular in cross section. The two ends together make a full cone. The diameter of the conical ends is 50 ft. The angle of repose of the corn is 21°. The height of the cone is

$$h = \frac{b}{2}\tan\phi = \left(\frac{50\ \text{ft}}{2}\right)\tan 21^\circ = 9.60\ \text{ft}$$

The total volume of the two ends is

$$\begin{aligned} V_{\text{cone}} &= 2\left(\frac{V_{\text{cone}}}{2}\right) = \frac{\pi r^2 h}{3} = \frac{\pi\left(\frac{50\ \text{ft}}{2}\right)^2(9.60\ \text{ft})}{3} \\ &= 6283\ \text{ft}^3 \end{aligned}$$

The length of the pile includes the semicircular ends. The length of the triangular wedge is

$$L = 100\ \text{ft} - (2)(25\ \text{ft}) = 50\ \text{ft}$$

Plant Engineering

The volume of the triangular wedge is

$$\begin{aligned} V_{\text{wedge}} &= AL = \tfrac{1}{2}bhL = \left(\tfrac{1}{2}\right)(50 \text{ ft})(9.60 \text{ ft})(50 \text{ ft}) \\ &= 12{,}000 \text{ ft}^3 \end{aligned}$$

The total volume is

$$\begin{aligned} V_t &= V_{\text{cone}} + V_{\text{wedge}} \\ &= (6283 \text{ ft}^3 + 12{,}000 \text{ ft}^3)\left(0.8036 \ \frac{\text{bu}}{\text{ft}^3}\right) \\ &= \boxed{14{,}692.22 \text{ bu} \quad (15{,}000 \text{ bu})} \end{aligned}$$

The answer is (A).

5. Bridging, also called *arching*, occurs when the contents of a bin or hopper settle into an arched cap that covers the opening. Bridging can be eliminated by reducing cohesion in the material or by enlarging the opening.

The answer is (A).

6. Since friction between the pellets and silo wall can be disregarded, the pressure ratio is consistent with the Rankine model of earth pressure. From Eq. 66.16,

$$k = \tan^2\left(45° - \frac{\phi}{2}\right) = \tan^2\left(45° - \frac{20°}{2}\right) = 0.490$$

Use the Janssen equation, Eq. 66.17, and the relationship given in Eq. 66.15 to calculate the vertical pressure on the silo's flat bottom.

$$\begin{aligned} p_v &= \frac{\rho g D}{4 g_c \mu k}\left(1 - \exp\left(\frac{-4h\mu k}{D}\right)\right) \\ &= \left(\frac{\left(35 \ \frac{\text{lbf}}{\text{ft}^3}\right)\left(32.2 \ \frac{\text{ft}}{\text{sec}^2}\right)(12 \text{ ft})}{(4)\left(32.2 \ \frac{\text{ft-lbm}}{\text{lbf-sec}^2}\right)\tan 20°(0.490)}\right) \\ &\quad \times \left(1 - \exp\left(\frac{-(4)(60 \text{ ft})\tan 20°(0.490)}{12 \text{ ft}}\right)\right) \\ &= 572.1 \text{ lbf/ft}^2 \end{aligned}$$

Using Eq. 66.18, the lateral pressure is

$$\begin{aligned} p_h &= kp_v = (0.490)\left(572.1 \ \frac{\text{lbf}}{\text{ft}^2}\right) \\ &= \boxed{280.3 \text{ lbf/ft}^2 \quad (280 \text{ lbf/ft}^2)} \end{aligned}$$

The answer is (A).

7. Since the silo is mass flow, the Johanson equation is applicable. From App. 16.B, the internal area of a 6 in schedule-40 pipe is 0.2006 ft^2; the internal diameter is 6.065 in. From Eq. 66.19, the flow rate is

$$\begin{aligned} \dot{m} &= \rho_b A\sqrt{\frac{gd}{4\tan\theta}} \\ &= \left(35 \ \frac{\text{lbm}}{\text{ft}^3}\right)(0.2006 \text{ ft}^2)\sqrt{\frac{\left(32.2 \ \frac{\text{ft}}{\text{sec}^2}\right)\left(\frac{6.065 \text{ in}}{12 \ \frac{\text{in}}{\text{ft}}}\right)}{4\tan(90° - 65°)}} \\ &= 20.74 \text{ lbm/sec} \end{aligned}$$

The filling time is

$$\begin{aligned} t &= \frac{m}{\dot{m}} = \frac{42{,}000 \text{ lbm}}{\left(20.74 \ \frac{\text{lbm}}{\text{sec}}\right)\left(60 \ \frac{\text{sec}}{\text{min}}\right)} \\ &= \boxed{33.75 \text{ min} \quad (34 \text{ min})} \end{aligned}$$

The answer is (D).

8. Calculate the length of the belt.

$$L = \frac{10 \text{ ft}}{\sin 10°} = 57.59 \text{ ft}$$

The power needed for a conveyor belt can be approximated from Eq. 66.24. This is an empirical formula and not dimensionally consistent. Rearranging, the maximum mass flow rate for a 5 hp drive is

$$P_{\text{hp}} = 0.4 + \left(\frac{\dot{m}_{\text{lbm/hr}}}{2000}\right)\left(\frac{0.00325L_{\text{ft}}}{100} + \frac{0.01h_{\text{ft}}}{10}\right)$$

$$\begin{aligned} \dot{m}_{\text{lbm/hr}} &= \frac{2000(P_{\text{hp}} - 0.4)}{\frac{0.00325L_{\text{ft}}}{100} + \frac{0.01h_{\text{ft}}}{10}} \\ &= \frac{(2000)(5 \text{ hp} - 0.4)}{\frac{(0.00325)(57.59 \text{ ft})}{100} + \frac{(0.01)(10 \text{ ft})}{10}} \\ &= \boxed{774{,}954 \text{ lbm/hr} \quad (770{,}000 \text{ lbm/hr})} \end{aligned}$$

The answer is (C).

9. The belt wrap angle is

$$\theta = (240°)\left(\frac{2\pi \text{ rad}}{360°}\right) = \frac{4\pi}{3} \text{ rad}$$

Plant Engineering

The relationship among the tight-side and loose-side tensions, T_1 and T_2, the coefficient of friction, f, and the belt wrap angle (in radians), θ, is given by Eq. 66.22. Rearranging, the loose-side tension is

$$\frac{T_1}{T_2} = e^{f\theta}$$

$$T_2 = \frac{T_1}{e^{f\theta}} = \frac{750 \text{ lbf}}{e^{(0.35)(4\pi/3 \text{ rad})}} = 173.1 \text{ lbf}$$

From Eq. 66.21, the power required is

$$P = (T_1 - T_2)\text{v} = \frac{(750 \text{ lbf} - 173.1 \text{ lbf})\left(120 \frac{\text{ft}}{\text{min}}\right)}{33{,}000 \frac{\text{ft-lbf}}{\text{hp-min}}} = \boxed{2.098 \text{ hp} \quad (2.1 \text{ hp})}$$

The answer is (C).

10. The temperature increase causes an elongation of the belt slack and tight sides. From Eq. 51.9, the increase in belt length is

$$\Delta L = \alpha L_o(T_2 - T_1)$$

The length of the belt affects the belt tension. From Eq. 51.4, the elongation of the belt due to the change in belt tension is

$$\Delta\delta_{\text{tight}} = \frac{L_o(F_{\text{tight},1} - F_{\text{tight},2})}{EA}$$

$$\Delta\delta_{\text{loose}} = \frac{L_o(F_{\text{loose},1} - F_{\text{loose},2})}{EA}$$

The changes in belt length due to the temperature increase and due to the change in tension are equal.

$$\Delta L = \Delta\delta$$

$$\alpha L_o(T_2 - T_1) = \frac{L_o(F_{\text{tight},1} - F_{\text{tight},2})}{EA} = \frac{L_o(F_{\text{loose},1} - F_{\text{loose},2})}{EA}$$

$$\alpha EA(T_2 - T_1) = F_{\text{tight},1} - F_{\text{tight},2} = F_{\text{loose},1} - F_{\text{loose},2}$$

Therefore,

$$F_{\text{tight},1} - F_{\text{tight},2} = F_{\text{loose},1} - F_{\text{loose},2} = \Delta F$$

Since the same change in tension occurs in the loose and tight sides, the change in net tension in the belt is

$$\Delta F_{\text{belt,net}} = F_{\text{tight}} - F_{\text{loose}} = \alpha EA(T_2 - T_1)$$

The cross-sectional area of the belt is

$$A = tw = (0.25 \text{ in})(24 \text{ in}) = 6 \text{ in}^2$$

From Eq. 66.21, the change in the power needed to drive the belt is

$$\Delta P = \Delta F_{\text{belt,net}}\text{v} = \alpha EA(T_2 - T_1)\text{v}$$

$$= \frac{\left(1.8 \times 10^{-6} \frac{1}{{}^\circ\text{F}}\right)(6 \text{ in}^2)\left(1.0 \times 10^6 \frac{\text{lbf}}{\text{in}^2}\right) \times (100^\circ\text{F} - 30^\circ\text{F})\left(120 \frac{\text{ft}}{\text{min}}\right)}{\left(550 \frac{\text{ft-lbf}}{\text{hp-sec}}\right)\left(60 \frac{\text{sec}}{\text{min}}\right)}$$

$$= \boxed{2.749 \text{ hp} \quad (2.7 \text{ hp})}$$

The answer is (C).

11. The length of the barrel, L, and the convergent section, Z, are both 2D.

$$L = Z = 2D = (2)(1.0 \text{ m}) = 2 \text{ m}$$

Use Eq. 66.33 to calculate the effective number of turns.

$$N_e = \frac{1}{H}\left(L + \frac{Z}{2}\right) = \left(\frac{1}{0.5 \text{ m}}\right)\left(2 \text{ m} + \frac{2 \text{ m}}{2}\right) = 6 \text{ turns}$$

Use Eq. 66.40. From App. 14.E, the absolute viscosity, μ, of standard air (0°C) is 1.723×10^{-5} Pa·s. The collection efficiency is

$$\eta_{d_p} = 1 - \exp\left(\frac{-\pi d_p^2(\rho_s - \rho_g)N_e\text{v}_i}{9\mu g_c B}\right)$$

$$= 1 - \exp\left(\frac{-\pi(5 \times 10^{-6} \text{ m})^2 \times \left(\left(2.0 \frac{\text{g}}{\text{cm}^3}\right)\left(1000 \frac{\text{kg}\cdot\text{cm}^3}{\text{m}^3\cdot\text{g}}\right) - 1.29 \frac{\text{kg}}{\text{m}^3}\right) \times (6 \text{ turns})\left(20 \frac{\text{m}}{\text{s}}\right)}{(9)(1.723 \times 10^{-5} \text{ Pa·s})(0.25 \text{ m})}\right)$$

$$= 0.385 \quad (39\%)$$

The answer is (C).

Plant Engineering

12. From App. 14.D, the density of 68°F air is 0.0752 lbm/ft^3. Use Eq. 66.43. $K = 16$ for tangential inlets.

$$\Delta p = \frac{K\rho_g H B \text{v}_i^2}{2g_c D_e^2}$$

$$= \frac{(16)\left(0.0752\ \frac{\text{lbm}}{\text{ft}^3}\right)(5\ \text{ft})(2\ \text{ft})\left(50\ \frac{\text{ft}}{\text{sec}}\right)^2}{(2)\left(32.2\ \frac{\text{ft-lbm}}{\text{lbf-sec}^2}\right)(5\ \text{ft})^2}$$

$$= \boxed{18.68\ \text{lbf/ft}^2 \quad (19\ \text{lbf/ft}^2)}$$

The answer is (A).

13. The rate of material conveyance is

$$\dot{m}_{\text{sawdust}} = \frac{1800\ \frac{\text{lbm}}{\text{hr}}}{60\ \frac{\text{min}}{\text{hr}}} = 30\ \text{lbm/min}$$

The material-air ratio is 1:2, and the density of dry air at standard conditions is 0.075 lbm/ft^3. The mass flow rate of air is

$$Q_{\text{air}} = \frac{\dot{m}_{\text{air}}}{\rho_{\text{air}}} = \frac{(2)\left(30\ \frac{\text{lbm}}{\text{min}}\right)}{0.075\ \frac{\text{lbm}}{\text{ft}^3}} = 800\ \text{ft}^3/\text{min}$$

The internal area of a 6 in schedule-40 pipe is 0.2006 ft^2. The air velocity is

$$\text{v}_{\text{air}} = \frac{Q_{\text{air}}}{A_{\text{pipe}}} = \frac{800\ \frac{\text{ft}^3}{\text{min}}}{0.2006\ \text{ft}^2}$$

$$= \boxed{3988\ \text{ft/min} \quad (4000\ \text{ft/min})}$$

The answer is (D).

67 Fire Protection Sprinkler Systems

PRACTICE PROBLEMS

Hazardous Materials

1. The MSDS for a liquid substance lists the following properties.

flash point (closed cap)	–4–160°F
flammable range (in air)	0.8–16%
VOC (lbm/gal)	0.0–7.5

How should this material be classified?

(A) flammable

(B) combustible

(C) explosive

(D) not flammable, combustible, or explosive

2. A gas stream contains a mixture of acetone at 1000 ppm, benzene at 2000 ppm, and toluene at 500 ppm. The lower explosive limits (LELs) are

acetone	25,000 ppm
benzene	12,000 ppm
toluene	11,000 ppm

What is most nearly the LEL of the mixture?

(A) 14,000 ppm

(B) 16,000 ppm

(C) 17,000 ppm

(D) 48,000 ppm

Hazard Classification

3. The main floor of a public library includes a circulation area, a reference area, and some office space. The basement contains an extensive stack room with floor-to-ceiling shelves of books. According to NFPA 13, which of these spaces are classified as light hazard (LH) occupancies for the purpose of fire sprinkler design?

(A) circulation area and office space

(B) reference area and stack room

(C) circulation area, reference area, and office space

(D) circulation area, reference area, and stack room

4. An elementary school includes classrooms, office space, and an auditorium with a performance stage. According to NFPA 13, which of these spaces is/are classified as ordinary hazard group 2 (OH-2) occupancies for the purpose of fire sprinkler design?

(A) only the stage

(B) the office space and the auditorium, not including the stage

(C) the classrooms and the auditorium, including the stage

(D) all spaces in the school

Hydrants and Nozzle Flow

5. A jet of water flows from an open hydrant hose port at a rate of 900 gal/min. The centerline of the port is 30 in above level grade, and the inside diameter of the port is 2.5 in. Wind and air resistance are negligible. Approximately how far does the center of the water jet travel before hitting the ground?

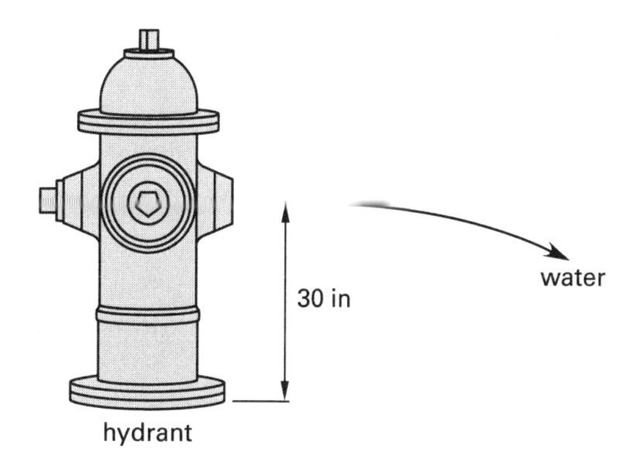

(A) 15 ft

(B) 23 ft

(C) 32 ft

(D) 76 ft

6. Firefighters can get no nearer than 50 ft from a burning building. They want to direct water from a fire hose into the center of an open window on an upper floor. The effective inside diameter of the hose nozzle is 1.5 in, and water flows from the hose at a rate of 550 gal/min. The window is 4 ft tall, and its center is 30 ft above the fire hose nozzle. Wind and air resistance are negligible.

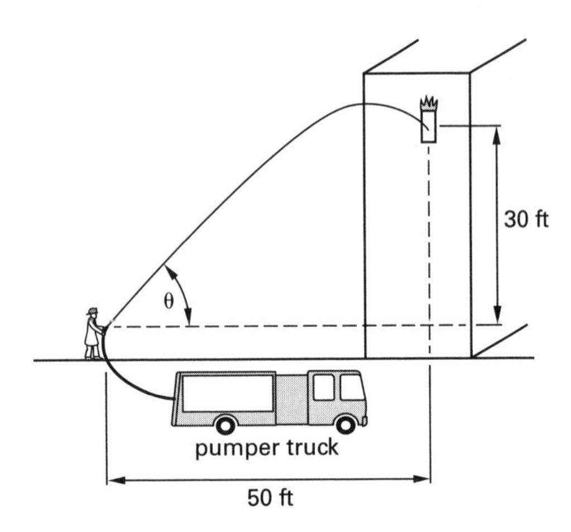

The angle at which the firefighters need to aim the fire hose nozzle is most nearly

(A) 29°

(B) 32°

(C) 36°

(D) 39°

7. Two hydrants are located adjacent to each other next to a building. The hydrants are fed by the same main as a fire sprinkler system in the building. The static pressure in the main is 85 psig. The sprinkler system has a demand of 750 gal/min. During a fire hydrant flow test, when 600 gal/min of water flow out of one hydrant, a residual pressure of 65 psig is measured at the second hydrant. The expected residual pressure at the second hydrant at a flow of 750 gal/min is most nearly

(A) 30 psig

(B) 48 psig

(C) 55 psig

(D) 62 psig

Design of Sprinkler Systems

8. The post indicator valve (PIV) in a fire sprinkler system makes it possible to

I. restart the system in the event of failure

II. stop the water supply from outside the building

III. adjust and optimize flow to the sprinklers

IV. prevent freezing in winter

V. maintain and repair the system

(A) I and III only

(B) I and V only

(C) II and IV only

(D) II and V only

9. An unmonitored fire suppression system is being planned for a combined horse barn (moderate amounts of hay), stable, and attached garage/workshop. The building has 11 ft ceilings and a protected area of 2500 ft^2. Water for the system will come from a dedicated tank on the property. A 20% allowance will be added to the sprinkler demand to account for friction loss and multiple fire heads operating simultaneously. What is most nearly the minimum tank size required for the system?

(A) 49,000 gal

(B) 63,000 gal

(C) 72,000 gal

(D) 76,000 gal

10. A wet pipe sprinkler system includes an alarm check valve. The alarm check valve is equipped with a retard chamber. The main purpose of the retard chamber is to minimize

(A) noise and vibrations from water hammer

(B) the number of spurious alarms

(C) cavitation

(D) damage to the piping from pressure transients

11. Pipe intended for a water main is marked "PE3408/PE4710 - PE100." What pipe characteristics are indicated by the designation "PE4710"?

(A) conventional polyethylene pipe with a hydrostatic design basis of 1000 psi

(B) polypropylene pipe with a slow crack growth resistance of more than 7 hr

(C) plastic pipe with a sandbox crush resistance performance evaluation of at least 47 psi

(D) high-performance polyethylene pipe with a hydrostatic design stress of 1000 psi

12. A hydraulically calculated sprinkler design uses a total of 20 identical sprinklers arranged in five identical four-sprinkler runs. The design is based on a minimum water pressure of 15 psi in the supply line. The sprinkler at the end of each run has a discharge coefficient of 6.2 (customary U.S. units) for pure water (SG = 1). According to NFPA 13, what is most nearly the water flow rate that must be supplied for the design to function properly?

(A) 96 gpm

(B) 190 gpm

(C) 240 gpm

(D) 480 gpm

13. A wet pipe sprinkler system is installed in a building classified as an ordinary hazard group 2 (OH-2) occupancy. The system was originally designed for pure water, but since the interior of the building is expected to experience sustained temperatures of −20°F, a decision has been made to charge the system with a 50% propylene glycol-water solution. The design density for the system is 0.2 gpm/ft^2. A loop in the system has an internal volume of 45 gal. Sprinklers with a K-factor of 8.0 have been selected, and the sprinklers are spaced so each covers an area of 130 ft^2. The specific gravity of pure water is 1, and the specific gravity of the propylene glycol-water solution is 1.085. What is most nearly each sprinkler's expected flow rate with the propylene glycol-water solution?

(A) 24 gpm

(B) 25 gpm

(C) 26 gpm

(D) 27 gpm

14. A section of sprinkler system piping spans a distance of 100 ft. The pipe run is level and includes two 45° elbows (K = 0.3). The pressure at the entrance is 90 psia. The minimum residual pressure head at the end of the run is 4 ft wg. The pipe is standard weight, 2 in B36.10 sprinkler system piping, which has a specific roughness of 0.000005 ft. What is most nearly the maximum flow rate of 60°F water through the pipe run?

(A) 210 gpm

(B) 410 gpm

(C) 620 gpm

(D) 860 gpm

Pumps and Pumping

15. When selecting a pump for a fire protection system, which two of the following statements are true?

I. Churn flow head must be no greater than 140% of the rated head.

II. Churn flow head must be at least 140% of the rated head.

III. The capacity at 65% of the rated head must be no greater than 150% of total capacity at the design point.

IV. The capacity at 65% of the rated head must be at least 150% of total capacity at the design point.

(A) I and III only

(B) I and IV only

(C) II and III only

(D) II and IV only

16. A new pump is being selected for a fire protection sprinkler system. Local codes require the pump be selected on the basis of the design point. The sprinkler system has a design point of 600 gpm at 55 ft total head. The pump curves for two pumps that meet this design point are shown.

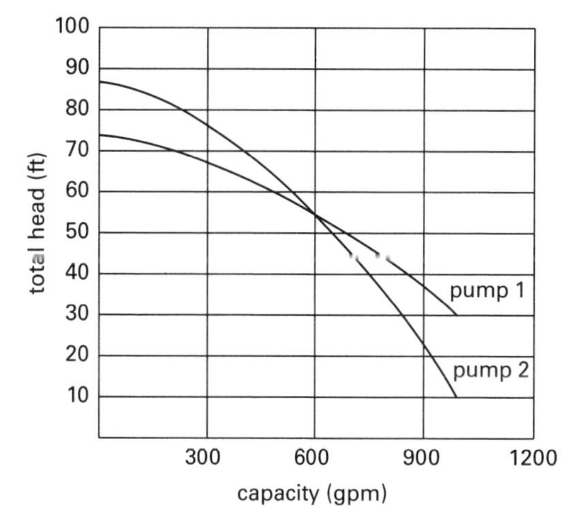

Which pump(s) meet(s) the criteria for fire sprinkler design given by NFPA 20?

(A) pump 1 only

(B) pump 2 only

(C) both pumps

(D) neither pump

17. A pump curve and a sprinkler system characteristic curve are shown. Per NFPA 13, the sprinkler system curve includes 100 gpm for an outside hose stream. The design point for the system is 675 gpm at 80 ft of total head. What effect would closing the valve to the hose have on flow rate and total head, and how would the curves change?

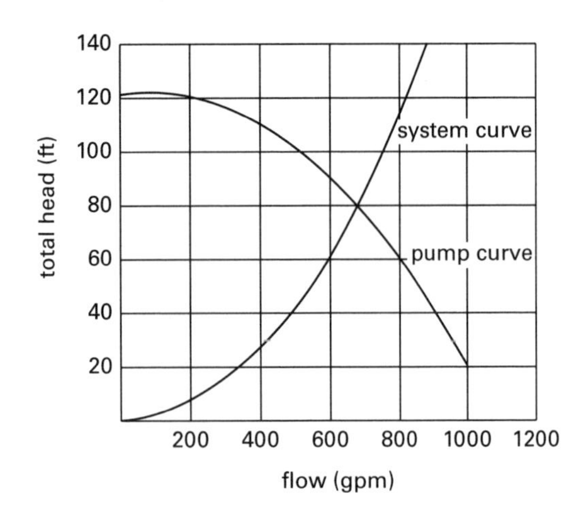

(A) The system would flow 575 gpm at a higher total head. The system curve would become steeper. The pump curve would become steeper.

(B) The system would flow less than 675 gpm at a higher total head. The system curve would become steeper, but the pump curve would remain unchanged. The operating point would shift to the left on the pump curve.

(C) System flow would drop, and total head would remain the same. The system curve would remain unchanged. The pump curve would shift lower on the graph.

(D) System flow would drop, and total head would increase. The system curve would become steeper. The pump curve would become flatter.

18. A pumper truck draws water from an open reservoir below. The pumper truck and suction piping are 20 ft above the surface of the reservoir. The equivalent length of the suction piping is 45 ft, the diameter of the suction piping is 5 in, and the Hazen-Williams coefficient (C-value) for the suction piping is 90. The truck draws water at a rate of 500 gpm. The atmospheric pressure is 14.7 psia. The temperature in the reservoir is 65°F.

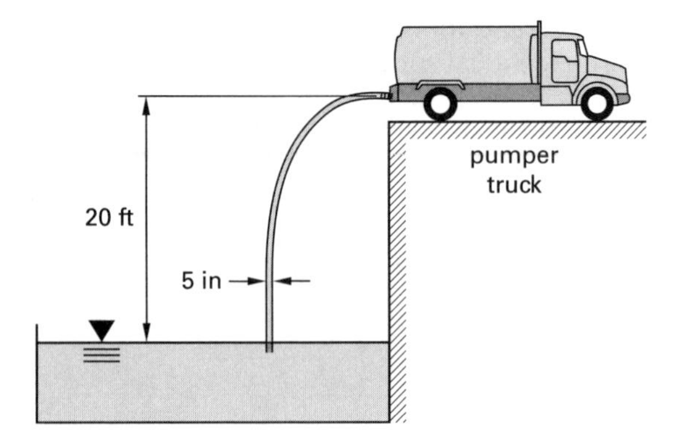

The net positive suction head available (NPSHA) to the pump is most nearly

(A) 5.3 ft

(B) 7.1 ft

(C) 8.8 ft

(D) 9.7 ft

19. A pumper truck draws water from an open reservoir below. The pump suction piping is 14 ft above the surface of the reservoir. The diameter of the suction piping is 4 in, and the Hazen-Williams coefficient (C-value) for the suction piping is 85. The truck draws water from the reservoir at a rate of 600 gpm. The net positive suction head required (NPSHR) for the pump to draw at this rate is 7 ft. The atmospheric pressure is 14.7 psia, and the temperature in the reservoir is 75°F.

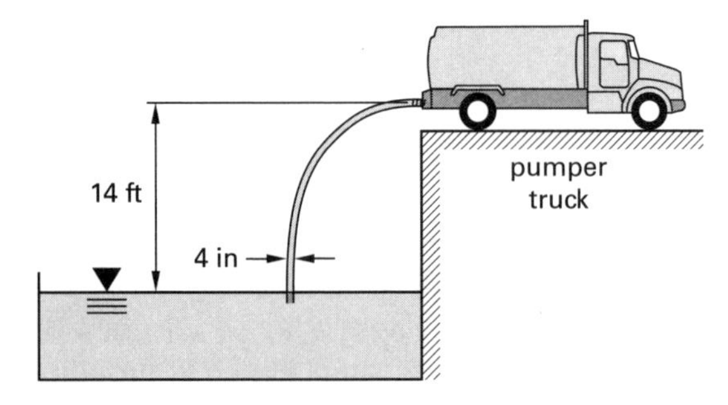

The maximum permissible equivalent length of the suction piping is most nearly

(A) 26 ft

(B) 57 ft

(C) 73 ft

(D) 96 ft

SOLUTIONS

1. The lowest flash point is lower than 100°F, so the substance is $\boxed{\text{flammable.}}$ An explosive material does not need an oxidizer. Gasoline requires the oxygen in atmospheric air to burn; therefore, gasoline is not explosive. Flammable range and VOC content are not used in classifying the liquid.

The answer is (A).

2. Using the concentrations in ppm, calculate the volumetric percentage of the total volume for each compound.

$$V = 1000 \text{ ppm} + 2000 \text{ ppm} + 500 \text{ ppm} = 3500 \text{ ppm}$$

$$\begin{aligned} C_{V,\text{acetone},\%} &= \frac{1000 \text{ ppm}}{3500 \text{ ppm}} \times 100\% \\ &= 28.6\% \\ C_{V,\text{benzene},\%} &= \frac{2000 \text{ ppm}}{3500 \text{ ppm}} \times 100\% \\ &= 57.1\% \\ C_{V,\text{toluene},\%} &= \frac{500 \text{ ppm}}{3500 \text{ ppm}} \times 100\% \\ &= 14.3\% \end{aligned}$$

Use Eq. 67.14.

$$\begin{aligned} \text{LEL}_{\text{mixture,ppm}} &= \frac{100\%}{\sum \dfrac{C_{V,i\%}}{\text{LEL}_{i,\text{ppm}}}} \\ &= \frac{100\%}{\dfrac{28.6\%}{25{,}000 \text{ ppm}} + \dfrac{57.1\%}{12{,}000 \text{ ppm}} + \dfrac{14.3\%}{11{,}000 \text{ ppm}}} \\ &= \boxed{13{,}884 \text{ ppm} \quad (14{,}000 \text{ ppm})} \end{aligned}$$

The answer is (A).

3. According to NFPA 13 Sec. A-5.2, all areas in a library are light hazard (LH) occupancies except for large stack rooms. According to NFPA 13 Sec. A-5.3.2, large stack rooms are ordinary hazard group 2 (OH-2) occupancies. Therefore, $\boxed{\text{the circulation area, the reference area, and the office space}}$ would be classified as LH occupancies.

The answer is (C).

4. According to NFPA 13 Sec. A-2.1.1, light hazard (LH) occupancies include $\boxed{\text{educational spaces, office spaces, and auditoriums.}}$ Per NFPA 13 Sec. A-2.1.2.2, stages are OH-2 occupancies.

The answer is (A).

5. Use the projectile motion equations. From Table 57.2, the horizontal and vertical distances traveled by the center of the water jet starting at $(0,0)$ in time t are

$$\begin{aligned} x(t) &= \text{v}_0 t \\ y(t) &= -\tfrac{1}{2}gt^2 \end{aligned}$$

When the water hits the ground, the vertical distance traveled, $y(t)$, is -30 in. Determine the time taken to travel this distance.

$$\begin{aligned} y(t) &= -\tfrac{1}{2}gt^2 \\ t &= \sqrt{\frac{-2y(t)}{g}} \\ &= \sqrt{\frac{(-2)(-30 \text{ in})}{\left(32.2 \ \frac{\text{ft}}{\text{sec}^2}\right)\left(12 \ \frac{\text{in}}{\text{ft}}\right)}} \\ &= 0.394 \text{ sec} \end{aligned}$$

The flow rate of the water is

$$Q_0 = \frac{900 \ \frac{\text{gal}}{\text{min}}}{\left(7.48 \ \frac{\text{gal}}{\text{ft}^3}\right)\left(60 \ \frac{\text{sec}}{\text{min}}\right)} = 2.005 \text{ ft}^3/\text{sec}$$

The cross-sectional area of the port is

$$\begin{aligned} A &= \frac{\pi d^2}{4} = \frac{\pi\left(\dfrac{2.5 \text{ in}}{12 \ \frac{\text{in}}{\text{ft}}}\right)^2}{4} \\ &= 0.034 \text{ ft}^2 \end{aligned}$$

The average velocity is

$$\begin{aligned} \text{v}_0 &= \frac{Q_0}{A} = \frac{2.005 \ \frac{\text{ft}^3}{\text{sec}}}{0.034 \text{ ft}^2} \\ &= 58.97 \text{ ft/sec} \end{aligned}$$

The horizontal distance traveled by the water before it hits the ground is

$$\begin{aligned} x(t) &= \text{v}_0 t \\ &= \left(58.97 \ \frac{\text{ft}}{\text{sec}}\right)(0.394 \text{ sec}) \\ &= \boxed{23.23 \text{ ft} \quad (23 \text{ ft})} \end{aligned}$$

The answer is (B).

6. The cross-sectional area of the nozzle is

$$A = \frac{\pi d^2}{4} = \frac{\pi\left(\frac{1.5 \text{ in}}{12 \frac{\text{in}}{\text{ft}}}\right)^2}{4} = 0.01227 \text{ ft}^2$$

The flow rate is

$$Q_0 = \frac{550 \frac{\text{gal}}{\text{min}}}{\left(7.48 \frac{\text{gal}}{\text{ft}^3}\right)\left(60 \frac{\text{sec}}{\text{min}}\right)} = 1.225 \text{ ft}^3/\text{sec}$$

The velocity of the water discharged from the nozzle is

$$\text{v}_0 = \frac{Q_0}{A} = \frac{1.225 \frac{\text{ft}^3}{\text{sec}}}{0.01227 \text{ ft}^2} = 99.84 \text{ ft/sec}$$

The height of the water when it reaches the window is

$$y(t) = (\text{v}_0 \sin\phi)t - \tfrac{1}{2}gt^2 = 30 \text{ ft}$$

Both the nozzle angle and the water's travel time are unknown. There is no way to solve directly for the angle, so the angle must be found by trial and error. Test each of the answer options.

If the water traveled in a straight line, the needed angle would be

$$\phi = \arctan\frac{30 \text{ ft}}{50 \text{ ft}} = 30.96°$$

To allow for the effect of gravity on the water, the actual angle must be greater than this, so option A can be eliminated.

From Table 57.2, the time it will take the water to reach the building is

$$t = \frac{R}{\text{v}_0 \cos\phi}$$

For an initial angle of 32° (option B),

$$t = \frac{R}{\text{v}_0 \cos\phi} = \frac{50 \text{ ft}}{\left(99.84 \frac{\text{ft}}{\text{sec}}\right)\cos 32°} = 0.591 \text{ sec}$$

$$\begin{aligned} y(t) &= (\text{v}_0 \sin\phi)t - \tfrac{1}{2}gt^2 \\ &= \left(\left(99.84 \frac{\text{ft}}{\text{sec}}\right)\sin 32°\right)(0.591 \text{ sec}) \\ &\quad - \left(\tfrac{1}{2}\right)\left(32.2 \frac{\text{ft}}{\text{sec}^2}\right)(0.591 \text{ sec})^2 \\ &= 25.64 \text{ ft} \end{aligned}$$

For an initial angle of 36° (option C),

$$t = \frac{R}{\text{v}_0 \cos\phi} = \frac{50 \text{ ft}}{\left(99.84 \frac{\text{ft}}{\text{sec}}\right)\cos 36°} = 0.619 \text{ sec}$$

$$\begin{aligned} y(t) &= (\text{v}_0 \sin\phi)t - \tfrac{1}{2}gt^2 \\ &= \left(\left(99.84 \frac{\text{ft}}{\text{sec}}\right)\sin 36°\right)(0.619 \text{ sec}) \\ &\quad - \left(\tfrac{1}{2}\right)\left(32.2 \frac{\text{ft}}{\text{sec}^2}\right)(0.619 \text{ sec})^2 \\ &= 30.16 \text{ ft} \end{aligned}$$

For an initial angle of 39° (option D),

$$t = \frac{R}{\text{v}_0 \cos\phi} = \frac{50 \text{ ft}}{\left(99.84 \frac{\text{ft}}{\text{sec}}\right)\cos 39°} = 0.644 \text{ sec}$$

$$\begin{aligned} y(t) &= (\text{v}_0 \sin\phi)t - \tfrac{1}{2}gt^2 \\ &= \left(\left(99.84 \frac{\text{ft}}{\text{sec}}\right)\sin 39°\right)(0.644 \text{ sec}) \\ &\quad - \left(\tfrac{1}{2}\right)\left(32.2 \frac{\text{ft}}{\text{sec}^2}\right)(0.644 \text{ sec})^2 \\ &= 33.79 \text{ ft} \end{aligned}$$

An initial angle of $\boxed{36°}$ is closest.

The answer is (C).

7. Pressure loss in a hydraulically designed sprinkler system is calculated using the Hazen-Williams equation, Eq. 67.10(b).

$$P_{f,\text{psi}} = \frac{4.52 L_{\text{ft}} Q_{\text{gpm}}^{1.85}}{C^{1.85} d_{\text{in}}^{4.87}}$$

The Hazen-Williams coefficient, C, is constant for a particular pipe system and does not vary with the velocity of flow. The pipe length, L, and diameter, d, also do not vary with flow. Therefore, within a particular system, pressure loss varies in proportion with $Q^{1.85}$.

$$\frac{P_{f,1}}{P_{f,2}} = \frac{Q_1^{1.85}}{Q_2^{1.85}}$$

Plant Engineering

The pressure loss at the second hydrant at a flow rate of 600 gal/min is the difference between the static pressure and the residual pressure, or 85 psig − 65 psig = 20 psig. Solve for the pressure loss at 750 gal/min.

$$\frac{P_{f,750\text{ gpm}}}{P_{f,600\text{ gpm}}} = \frac{Q^{1.85}_{750\text{ gpm}}}{Q^{1.85}_{600\text{ gpm}}}$$

$$\begin{aligned} P_{f,750\text{ gpm}} &= P_{f,600\text{ gpm}}\left(\frac{Q_{750\text{ gpm}}}{Q_{600\text{ gpm}}}\right)^{1.85} \\ &= \left(20\ \frac{\text{lbf}}{\text{in}^2}\right)\left(\frac{750\ \frac{\text{gal}}{\text{min}}}{600\ \frac{\text{gal}}{\text{min}}}\right)^{1.85} \\ &= 30.22\text{ lbf/in}^2 \end{aligned}$$

The residual pressure at 750 gal/min will be the static pressure minus the pressure loss, or 85 psig − 30.22 psig = 54.78 psig (55 psig).

The answer is (C).

8. The post indicator valve (PIV) cannot be used to restart the system in the event of failure, so I is incorrect. The PIV makes it possible to stop the water supply from outside the building in case an emergency situation makes it dangerous to enter, so II is correct. The PIV can only be fully open or fully closed, so adjustment of flow is not possible, making III incorrect. Supply piping for a sprinkler system is buried deep enough to prevent freezing in the winter, so the PIV plays no role in this, and IV is incorrect. The PIV permits the system to be disconnected from the water supply for maintenance and repair, so V is correct.

The answer is (D).

9. Determine the hazard classification for the building. A hazard classification of ordinary hazard group 2 (OH-2) is standard for garages and workshops. Moderate amounts of hay would not change this classification. From Fig. 67.10, the required design density, ρ_S, for an OH-2 occupancy with a protected area, A_c, of 2500 ft^2 is 0.18 gpm/ft^2. The sprinkler demand is

$$\begin{aligned} Q_{\text{sprinkler}} &= \rho_S A_c \\ &= \left(0.18\ \frac{\frac{\text{gal}}{\text{min}}}{\text{ft}^2}\right)(2500\text{ ft}^2) \\ &= 450\text{ gpm} \end{aligned}$$

From Table 67.5, the sprinkler flow for an unmonitored system in an OH occupancy must be sustained for 90 minutes. The water volume required to support the sprinkler demand, including the 20% allowance, is

$$\begin{aligned} V_{\text{sprinkler}} &= 1.2Q_{\text{sprinkler}}t = (1.2)\left(450\ \frac{\text{gal}}{\text{min}}\right)(90\text{ min}) \\ &= 48{,}600\text{ gal} \end{aligned}$$

From Table 67.5, the total combined inside and outside hose stream allowance is 250 gpm. The water volume required is

$$V_{\text{hose}} = Q_{\text{hose}}t = \left(250\ \frac{\text{gal}}{\text{min}}\right)(90\text{ min}) = 22{,}500\text{ gal}$$

The minimum tank size is

$$\begin{aligned} V_{\text{tank}} &= V_{\text{sprinkler}} + V_{\text{hose}} = 48{,}600\text{ gal} + 22{,}500\text{ gal} \\ &= 71{,}100\text{ gal} \end{aligned}$$

Of the options listed, only a tank size of 76,000 gal will hold 71,100 gal.

The answer is (D).

10. A retard chamber absorbs fluctuations in the supply system pressure.

This reduces the number of spurious alarms.

The answer is (B).

11. It is standard to mark plastic pipe with a designation indicating the type of plastic, followed by four numbers that describe its key properties. From ASTM D3350, "PE" refers to polyethylene; "4" refers to high-performance, high-density polyethylene with a density cell class of 4 (0.947–0.955 g/cc); "7" refers to slow crack growth (SCG) resistance cell class 7 (Pennsylvania notch test (PENT) value > 500 hr); and "10" refers to a 1000 psi hydrostatic design stress with water at 73°F.

The answer is (D).

12. Use Eq. 67.4(b). The system flow rate is

$$\begin{aligned} Q_{\text{total}} &= NQ_{\text{single sprinkler}} = N\left(K\sqrt{\frac{p}{\text{SG}}}\right) \\ &= (20)\left(6.2\sqrt{\frac{15\ \frac{\text{lbf}}{\text{in}^2}}{1}}\right) \\ &= 480\text{ gpm} \end{aligned}$$

The answer is (D).

13. Using Eq. 67.7, calculate the sprinkler flow rate for pure water using the design density.

$$Q = \rho_S A_c = \left(0.2\ \frac{\frac{\text{gal}}{\text{min}}}{\text{ft}^2}\right)(130\ \text{ft}^2) = 26\ \text{gal/min}$$

Solve Eq. 67.4(b) for the required pressure for pure water.

$$Q = K\sqrt{\frac{p}{\text{SG}}}$$

$$\begin{aligned} p &= \text{SG}\left(\frac{Q}{K}\right)^2 = (1)\left(\frac{26\ \frac{\text{gal}}{\text{min}}}{8.0}\right)^2 \\ &= 10.56\ \text{lbf/in}^2 \end{aligned}$$

Calculate the actual flow rate with the propylene glycol-water solution. The initial hydraulic pressure must be adjusted higher to provide the required density with the higher-density propylene glycol-water mixture, so round the result up.

$$\begin{aligned} Q &= K\sqrt{\frac{p}{\text{SG}}} = 8\sqrt{\frac{10.56\ \frac{\text{lbf}}{\text{in}^2}}{1.085}} \\ &= \boxed{24.96\ \text{gpm}\quad (25\ \text{gpm})} \end{aligned}$$

The answer is (B).

14. The specific roughness is given. However, sprinkler calculations most commonly use the Hazen-Williams equation, which requires a value for the Hazen-Williams coefficient. An exact conversion from specific roughness to Hazen-Williams coefficient is only possible if the Reynolds number is known. Otherwise, either an iterative solution or an estimate based on sound logic is required. Apparently, the specific roughness is given to help select the Hazen-Williams coefficient.

In this case, per Table 17.2, the specific roughness corresponds to a smooth pipe. In actual practice, this degree of smoothness isn't achievable by steel pipe, so the pipe will either be made of copper or plastic. For smooth copper and plastic pipes, a Hazen-Williams coefficient of 150 is commonly used.

Find the total equivalent length of the pipe. Appendix 67.A lists the equivalent length of a 2 in 45° elbow as 2 ft. Since App. 67.A is based on a Hazen-Williams coefficient of 120, use the modifying factor of 1.51 from Table 67.8.

The total equivalent pipe length of the pipe and two elbows is

$$L_e = L_{\text{pipe}} + L_{2\ \text{elbows}} = 100\ \text{ft} + \frac{(2)(2\ \text{ft})}{1.51} = 102.6\ \text{ft}$$

The residual pressure head at the pipe end cannot be less than 4 ft of water, which is equivalent to a pressure of

$$p = \gamma h = \frac{\left(62.4\ \frac{\text{lbf}}{\text{ft}^3}\right)(4\ \text{ft})}{\left(12\ \frac{\text{in}}{\text{ft}}\right)^2} = 1.73\ \text{lbf/in}^2$$

The maximum friction loss due to the pipe run and elbows is

$$\begin{aligned} \Delta P_{f,\max} &= p_1 - p_2 = 90\ \frac{\text{lbf}}{\text{in}^2} - 1.73\ \frac{\text{lbf}}{\text{in}^2} \\ &= 88.27\ \text{lbf/in}^2 \end{aligned}$$

From App. 16.B, the inside diameter, d, of 2 in B36.10 pipe is 2.067 in. Rearrange Eq. 67.10(b) to solve for the flow rate.

$$P_{f,\text{psi}} = \frac{4.52 L_{\text{ft}} Q_{\text{gpm}}^{1.85}}{C^{1.85} d_{\text{in}}^{4.87}}$$

$$\begin{aligned} Q &= \left(\frac{P_{f,\text{psi}} C^{1.85} d_{\text{in}}^{4.87}}{4.52 L_{\text{ft}}}\right)^{1/1.85} \\ &= \left(\frac{\left(88.27\ \frac{\text{lbf}}{\text{in}^2}\right)(150)^{1.85}(2.067\ \text{in})^{4.87}}{(4.52)(102.6\ \text{ft})}\right)^{1/1.85} \\ &= \boxed{413.8\ \text{gpm}\quad (410\ \text{gpm})} \end{aligned}$$

The answer is (B).

15. According to NFPA 20, the head at churn (or shutoff) flow must be no greater than 140% of the rated head. The capacity at 65% of rated head must be no less than 150% of total capacity at the design point.

The answer is (B).

16. According to NFPA 20, pumps used in fire sprinkler systems must meet the following criteria.

- Head at churn (or shutoff) flow must be no greater than 140% of the rated head.
- The capacity at 65% of the rated head must be no less than 150% of total capacity at the design point.

Total head at the design point is 55 ft, so the head at churn flow (zero capacity) must be no greater than (1.4)(55 ft) = 77 ft. Pump 1 meets this requirement, but for pump 2, the head at churn flow is over 80 ft.

Total capacity at the design point is 600 gpm, so the capacity at 65% of the rated head, or (0.65)(55 ft) = 35.75 ft, must be no less than (1.5)(600 gpm) = 900 gpm. Pump 1 meets this requirement, but for pump 2, the capacity at 35.75 ft of head is only about 800 gpm.

$$\boxed{\text{Pump 1 is adequate for the design, but pump 2 is not.}}$$

The answer is (A).

17. Closing a valve in the system affects the system curve, but not the pump curve. The system will "ride the pump curve," and the new operating point will be at a flow less than the 675 gpm design flow but at a higher total head. Because of the higher head, the flow will be greater than 575 gpm. The new flow will be between 575 gpm and 675 gpm at a higher total head.

The answer is (B).

18. From Eq. 18.4(b), the atmospheric head is

$$h_{\text{atm}} = \frac{p_{\text{atm}}}{\gamma_{\text{water}}} = \frac{\left(14.7\ \frac{\text{lbf}}{\text{in}^2}\right)\left(12\ \frac{\text{in}}{\text{ft}}\right)^2}{62.4\ \frac{\text{lbf}}{\text{ft}^3}} = 33.92\ \text{ft}$$

The static suction lift, $h_{z(s)}$, is the distance from the pump inlet to the surface of the reservoir, −20 ft. From the Hazen-Williams equation, Eq. 17.29, the friction loss in feet is

$$\begin{aligned} h_{f,\text{ft}} &= \frac{10.44 L_{\text{ft}} Q_{\text{gpm}}^{1.85}}{C^{1.85} d_{\text{in}}^{4.87}} = \frac{(10.44)(45\ \text{ft})\left(500\ \frac{\text{gal}}{\text{min}}\right)^{1.85}}{(90)^{1.85}(5\ \text{in})^{4.87}} \\ &= 4.42\ \text{ft} \end{aligned}$$

(The Hazen-Williams equation is an empirical formula and not dimensionally consistent.)

By interpolation from App. 23.A, the vapor pressure of water at 65°F is 0.3060 lbf/in^2, so the vapor pressure head is

$$\begin{aligned} h_{\text{vp}} &= \frac{p_{\text{vapor}}}{\gamma_{\text{water}}} = \frac{\left(0.3060\ \frac{\text{lbf}}{\text{in}^2}\right)\left(12\ \frac{\text{in}}{\text{ft}}\right)^2}{62.4\ \frac{\text{lbf}}{\text{ft}^3}} \\ &= 0.7062\ \text{ft} \end{aligned}$$

From Eq. 18.30, the NPSHA is

$$\begin{aligned} \text{NPSHA} &= h_{\text{atm}} + h_{z(s)} - h_{f(s)} - h_{\text{vp}} \\ &= 33.92\ \text{ft} + (-20\ \text{ft}) - 4.42\ \text{ft} - 0.7062\ \text{ft} \\ &= \boxed{8.79\ \text{ft}\quad (8.8\ \text{ft})} \end{aligned}$$

The answer is (C).

19. From Eq. 18.4(b), the atmospheric head is

$$h_{\text{atm}} = \frac{p_{\text{atm}}}{\gamma_{\text{water}}} = \frac{\left(14.7\ \frac{\text{lbf}}{\text{in}^2}\right)\left(12\ \frac{\text{in}}{\text{ft}}\right)^2}{62.4\ \frac{\text{lbf}}{\text{ft}^3}} = 33.92\ \text{ft}$$

The static suction lift, $h_{z(s)}$, is the distance from the pump inlet to the surface of the reservoir, −14 ft.

By interpolation from App. 23.A, the vapor pressure of water at 75°F is 0.4304 lbf/in^2, so the vapor pressure head is

$$h_{\text{vp}} = \frac{p_{\text{vapor}}}{\gamma_{\text{water}}} = \frac{\left(0.4304\ \frac{\text{lbf}}{\text{in}^2}\right)\left(12\ \frac{\text{in}}{\text{ft}}\right)^2}{62.4\ \frac{\text{lbf}}{\text{ft}^3}} = 0.9932\ \text{ft}$$

The net positive suction head available (NPSHA) must be greater than or equal to NPSHR. Solving Eq. 18.30 for the suction piping loss, and substituting NPSHR for NPSHA, the maximum allowable suction piping loss is

$$\begin{aligned} \text{NPSHA} &= h_{\text{atm}} + h_{z(s)} - h_{f(s)} - h_{\text{vp}} \\ h_{f(s)} &= h_{\text{atm}} + h_{z(s)} - h_{\text{vp}} - \text{NPSHR} \\ &= 33.92\ \text{ft} + (-14\ \text{ft}) - 0.9932\ \text{ft} - 7\ \text{ft} \\ &= 11.93\ \text{ft} \end{aligned}$$

Rearrange the Hazen-Williams equation, Eq. 17.29, to find the allowable equivalent length of suction piping. (The Hazen-Williams equation is an empirical formula and not dimensionally consistent.)

$$\begin{aligned} h_{f,\text{ft}} &= \frac{10.44 L_{\text{ft}} Q_{\text{gpm}}^{1.85}}{C^{1.85} d_{\text{in}}^{4.87}} \\ L_{\text{ft}} &= \frac{h_{f,\text{ft}} C^{1.85} d_{\text{in}}^{4.87}}{10.44 Q_{\text{gpm}}^{1.85}} = \frac{(11.93\ \text{ft})(85)^{1.85}(4\ \text{in})^{4.87}}{(10.44)\left(600\ \frac{\text{gal}}{\text{min}}\right)^{1.85}} \\ &= \boxed{26.29\ \text{ft}\quad (26\ \text{ft})} \end{aligned}$$

The answer is (A).

68 Pollutants in the Environment

PRACTICE PROBLEMS

Dust and Particles

1. A PM-10 sample is collected with a high-volume sampler over a sampling interval of 24 hours and 10 minutes. The sampler draws an average flow of 41 ft^3/min and collects particles on a glass fiber filter. The initial and final masses of the filter are 4.4546 g and 4.4979 g, respectively. The PM-10 concentration is most nearly

(A) 11 $\mu g/m^3$

(B) 26 $\mu g/m^3$

(C) 42 $\mu g/m^3$

(D) 55 $\mu g/m^3$

2. At one setting of rotational speed and air flow rate, 1.0 μm diameter particles entering a centrifuge particulate sampler travel 11 in from the inlet to the inside of the outer collection wall. A second setting increases the rotational speed of the sampling duct by 30% and reduces the gas volumetric flow rate by 40%. The gas viscosity, instrument calibration constant, collection efficiency, and particle density remain the same. What is most nearly the diameter of the particles that travel 17 in from the inlet to the wall?

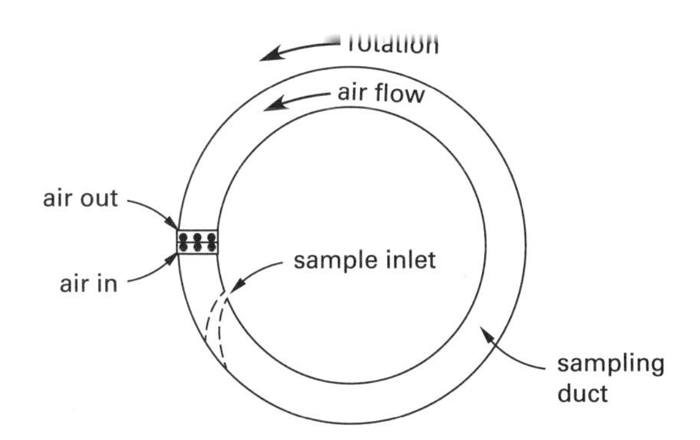

(A) 0.48 μm

(B) 0.72 μm

(C) 0.80 μm

(D) 0.93 μm

3. An optical discriminator uses light attenuation to size particles. When a particle flows into the sampling chamber, the brightness of a light source at the photodiode is diminished, decreasing the voltage in the detection circuit. When a spherical particle of diameter d enters the sampling chamber, the voltage drop is ΔE_1. Subsequently, when two spherical particles, one with the same diameter, d, and the other with diameter $0.5d$, enter the sampling chamber, the voltage drop is ΔE_2. Most nearly, what is the ratio $\Delta E_2/\Delta E_1$?

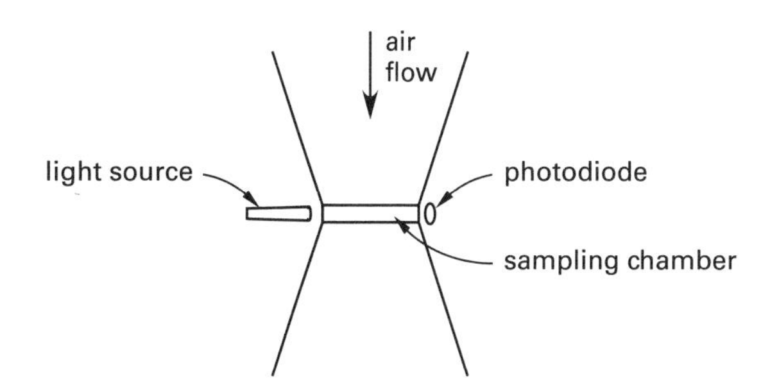

(A) 1.1

(B) 1.3

(C) 1.5

(D) 1.8

Smog

4. Which statement related to smog is INCORRECT?

(A) Smog is produced when ozone precursors such as nitrogen dioxide, hydrocarbons, and volatile organic compounds (VOCs) react with sunlight.

(B) Peroxyacyl nitrates contribute to the formation of smog.

(C) Volatile organic compounds (VOCs) contributing to smog are the products of incomplete combustion reactions in automobiles, refineries, and industrial boilers.

(D) Ground-level ozone is a secondary pollutant that contributes to the formation of smog.

5. Smog generation generally increases with all of the following conditions EXCEPT

(A) temperature inversions

(B) absence of wind

(C) increased commuter traffic

(D) early morning clouds and fog

6. According to air pollutant criteria defined in the Clean Air Act, smog is a

(A) hazardous air pollutant (HAP)

(B) criteria air pollutant

(C) volatile organic compound (VOC)

(D) primary air toxic

7. The concentration, C, of nitrogen oxides in a city over a 24 hour period is shown.

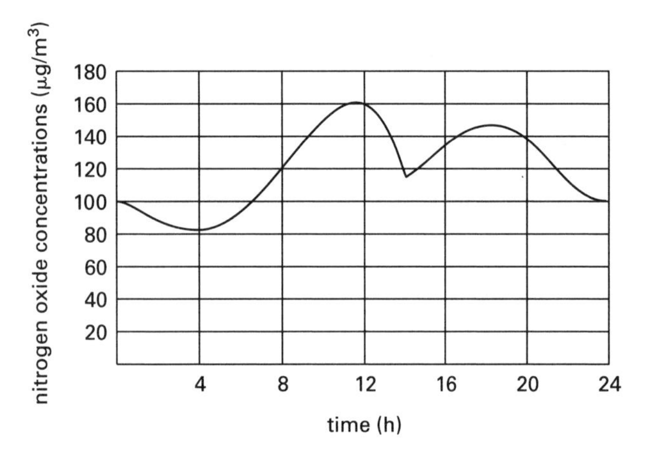

The concentration from $t=0$ h to $t=14$ h is described by

$$C_{\mu g/m^3} = 100 - 3.7848t - 2.2609t^2 + 0.6381t^3 - 0.0324t^4$$

The concentration from $t=14$ h to $t=24$ h is described by

$$C_{\mu g/m^3} = 5.1885 \times 10^3 - 1.1899 \times 10^3 t + 101.8740t^2 - 3.7638t^3 + 0.0507t^4$$

What is most nearly the average nitrogen oxide concentration over the 24 hour period?

(A) 118 μg/m^3

(B) 121 μg/m^3

(C) 123 μg/m^3

(D) 125 μg/m^3

SOLUTIONS

1. The duration of sampling is

$$t = (24 \text{ hr})\left(60 \ \frac{\text{min}}{\text{hr}}\right) + 10 \text{ min} = 1450 \text{ min}$$

The volume of air sampled is

$$V = Q_{\text{ave}}t = \left(\frac{41 \ \frac{\text{ft}^3}{\text{min}}}{\left(3.281 \ \frac{\text{ft}}{\text{m}}\right)^3}\right)(1450 \text{ min}) = 1683 \text{ m}^3$$

The PM-10 concentration is

$$C_{\text{PM-10}} = \frac{m_{\text{particles}}}{V} = \frac{m_{\text{final}} - m_{\text{initial}}}{V} = \frac{(4.4979 \text{ g} - 4.4546 \text{ g})\left(10^6 \ \frac{\mu\text{g}}{\text{g}}\right)}{1683 \text{ m}^3} = \boxed{25.7 \ \mu\text{g/m}^3 \quad (26 \ \mu\text{g/m}^3)}$$

The answer is (B).

2. Simplify Eq. 70.9. The constant K accumulates fixed dimensions, air properties, and various constants and fixed terms.

$$d_p = \sqrt{\frac{KQ}{\omega^2 L}}$$

$$\frac{d_{p,2}}{d_{p,1}} = \frac{\sqrt{\frac{KQ_2}{\omega_2^2 L_2}}}{\sqrt{\frac{KQ_1}{\omega_1^2 L_1}}}$$

$$d_{p,2} = d_{p,1}\frac{\sqrt{\frac{KQ_2}{\omega_2^2 L_2}}}{\sqrt{\frac{KQ_1}{\omega_1^2 L_1}}} = d_{p,1}\sqrt{\frac{\omega_1^2 Q_2 L_1}{\omega_2^2 Q_1 L_2}}$$

$$= d_{p,1}\sqrt{\frac{\omega_1^2(0.6Q_1)L_1}{(1.3\omega_1)^2 Q_1 L_2}}$$

$$= 1.0 \ \mu\text{m}\sqrt{\frac{(0.6)(11 \text{ in})}{(1.3)^2(17 \text{ in})}}$$

$$= \boxed{0.48 \ \mu\text{m}}$$

The answer is (A).

3. From Eq. 70.14, the voltage drop for a single particle is

$$\Delta E = \frac{A_{\text{shadow}} E_0}{A_{\text{detector}}}$$

The ratio of the voltage drop for two particles, E_{2P}, to the voltage drop with the single particle, E_{1P}, is

$$\frac{\Delta E_{2P}}{\Delta E_{1P}} = \frac{\dfrac{A_{2P} E_0}{A_{\text{detector}}}}{\dfrac{A_{1P} E_0}{A_{\text{detector}}}} = \frac{\dfrac{\left(\pi\left(\dfrac{d}{2}\right)^2 + \pi\left(\dfrac{\frac{d}{2}}{2}\right)^2\right) E_0}{A_{\text{detector}}}}{\dfrac{\pi\left(\dfrac{d}{2}\right)^2 E_0}{A_{\text{detector}}}}$$

$$= \frac{\left(\frac{1}{2}\right)^2 + \left(\frac{1}{4}\right)^2}{\left(\frac{1}{2}\right)^2}$$

$$= \boxed{1.25 \quad (1.3)}$$

The answer is (B).

4. Volatile organic compounds (VOCs) are emitted by manufacturing and refining processes, dry cleaners, gasoline stations, print shops, painting operations, and municipal wastewater treatment plants, not by combustion sources.

The answer is (C).

5. Temperature inversions are conducive to smog generation because they keep warm air near the ground. An absence of wind causes smog to increase because smog precursors are not dispersed. Smog also increases with traffic, as automobile emissions contain nitrogen oxides and hydrocarbons, which are precursors to smog. Smog is primarily a sunlight-induced reaction, and smog production usually peaks in early afternoon when there is the most sunlight.

The answer is (D).

6. The Environmental Protection Agency (EPA) has identified six air pollutants common in the United States. These common air pollutants, known as *criteria air pollutants*, include ground-level ozone, carbon monoxide, sulfur oxides, nitrogen oxides, particle pollution, and lead. Since smog is composed primarily of ground-level ozone, it would be classified by the EPA as a $\boxed{\text{criteria air pollutant.}}$

The answer is (B).

7. Use integration to calculate the average, C_{ave}.

$$C_{\text{ave},\mu\text{g/m}^3} = \frac{\displaystyle\int_{0\text{ hr}}^{14\text{ hr}} \begin{pmatrix} 100 - 3.7848t - 2.2609t^2 \\ + 0.6381t^3 - 0.0324t^4 \end{pmatrix} dt}{24 \text{ hr}} + \frac{\displaystyle\int_{14\text{ hr}}^{24\text{ hr}} \begin{pmatrix} 5.1885 \times 10^3 \\ - 1.1899 \times 10^3 t \\ + 101.8740t^2 \\ - 3.7638t^3 \\ + 0.0507t^4 \end{pmatrix} dt}{24 \text{ hr}}$$

$$= \frac{\begin{pmatrix} 100t - 1.8924t^2 \\ - 7.5363 \times 10^{-1} t^3 \\ - 0.1595t^4 + 6.4800 \times 10^{-3} t^5 \end{pmatrix}\Bigg|_{0\text{ hr}}^{14\text{ hr}}}{24 \text{ hr}} + \frac{\begin{pmatrix} 5.1885 \times 10^3 t \\ - 5.9495 \times 10^2 t^2 \\ + 33.9580t^3 - 0.9410t^4 \\ + 1.0140 \times 10^{-2} t^5 \end{pmatrix}\Bigg|_{14\text{ hr}}^{24\text{ hr}}}{24 \text{ hr}}$$

$$= \boxed{121 \ \mu\text{g/m}^3}$$

The answer is (B).

Plant Engineering

69 Storage and Disposition of Hazardous Materials

PRACTICE PROBLEMS

1. The operator of an underground oil storage tank wishes to permanently convert to above-ground storage. To satisfy federal regulations, what must the operator do?

(A) Drain the existing tank.

(B) Drain, clean, and pressurize the existing tank to 150% of atmospheric pressure.

(C) Drain, clean, and fill the existing tank with sand.

(D) Remove the existing tank, remove and replace soil equal to 100% of the tank volume, and install test wells as per code.

2. According to the Environmental Protection Agency (EPA), leaking tanks and pipes may be repaired as long as the repairs meet "industry codes and standards." Which organization publishes such codes and standards?

(A) American Welding Society (AWS)

(B) National Institute of Standards and Technology (NIST)

(C) International Code Council (ICC)

(D) National Fire Protection Association (NFPA)

3. Contaminants migrate through soils and aquifers at a velocity equal to the

(A) pore velocity

(B) effective velocity

(C) superficial velocity

(D) Darcy velocity

4. Common paint solvents accumulated by a painting contractor are processed in a properly licensed and operating hazardous waste incineration facility. A small amount of mineral ash remains after incineration. Which statement regarding the mineral ash is true?

(A) The ash is considered nonhazardous solid waste and may be disposed of in a municipal solid waste facility that accepts ashes.

(B) The ash is considered an ignitablc solid waste and must be buried in a buried (RCRA) municipal solid waste facility.

(C) The waste is considered a "derived from" (D) hazardous waste and must be disposed of in a RCRA hazardous waste facility.

(D) The ash is considered a "K" waste and must be disposed of in a RCRA hazardous waste facility.

5. A large section of town contains vacant factory buildings that, over the years, manufactured numerous unknown products. A developer now wishes to purchase the property from the current owners and convert it to noncommercial use. Public concern centers on possible site contamination that the developer cannot afford to clean up. The property can be referred to as a

(A) Superfund site

(B) designated protection zone

(C) brownfield

(D) undesignated watch site

SOLUTIONS

1. When an underground tank is decommissioned, (1) the regulatory authority must be notified at least 30 days before closing; (2) any contamination must be remediated; (3) the tank must be drained and cleaned by removing all liquids, dangerous vapor levels, and accumulated sludge; and (4) the tank must be removed from the ground or filled with a harmless, chemically inactive solid, such as sand.

The answer is (C).

2. The EPA identifies the following organizations as having relevant codes and standards: API (American Petroleum Institute); ASTM International (formerly American Society for Testing and Materials); KWA (Ken Wilcox Associates, Inc.); NACE International (formerly the National Association of Corrosion Engineers); NFPA (National Fire Protection Association); NLPA (National Leak Prevention Association); PEI (Petroleum Equipment Institute); STI (Steel Tank Institute); and UL (Underwriters Laboratories Inc.).

The answer is (D).

3. Contaminants migrate at the pore velocity, also known as seepage velocity or flow front velocity. Darcy velocity (also known as effective velocity and superficial velocity) does not take the porosity into consideration.

The answer is (A).

4. Although the source material was ignitable, the residual ash is not itself a "specially listed" waste due to its ignitability, corrosivity, reactivity, and toxicity characteristics. Painting is not designated as a specific source industry (such as wood preserving, petroleum refining, and organic chemical manufacturing), so the ash is not a "K" waste. The ash is derived from a hazardous waste, so it is a "derived from" (D) hazardous waste. It must be disposed of in a RCRA hazardous waste facility.

The answer is (C).

5. The current owner, not the developer, would be required to clean up any contamination found. Brownfields are abandoned, idled, or underused industrial and commercial properties where expansion or redevelopment is complicated by actual or suspected environmental contamination.

The answer is (C).

70 Testing and Sampling

PRACTICE PROBLEMS

There are no problems in this book corresponding to Chap. 70 of the *Mechanical Engineering Reference Manual.*

71 Environmental Remediation

PRACTICE PROBLEMS

Oil Spill Remediation

1. Which statement about petroleum spills and spill remediation is INCORRECT?

(A) As long as the roots of vegetation are kept moist by soil moisture, in situ burning of surface petroleum contaminants has limited long-term environmental effects.

(B) When properly applied in large bodies of water, oil dispersants have little environmental impact.

(C) Washing oil-contaminated sands, soils, and rocks with steam and hot water also removes organisms and nutrients that would otherwise contribute to bioremediation.

(D) Magnetic particle technology helps recover spilled petroleum products for reuse.

2. The treatment of oil-contaminated soil in a large plastic-covered tank is known as

(A) bioremediation

(B) biofiltration

(C) bioreaction

(D) bioventing

Baghouses

3. The performance of a pilot study baghouse follows the filter drag model, with $K_e = 0.6$ in wg-min/ft and $K_s = 0.07$ in wg-ft^2-min/gr. After 20 minutes of operation of the scaled-up baghouse, the dust loading is 27 gr/ft^3, and the air-to-cloth ratio is 3.0 ft/min. Most nearly, what is the pressure drop for the full scale baghouse after 20 minutes?

(A) 2.5 in wg

(B) 4.1 in wg

(C) 5.2 in wg

(D) 7.5 in wg

4. 2500 ft^3/min of air flow into a baghouse. The air has a particulate concentration of 0.8 g/m^3. The instantaneous collection efficiency of the baghouse increases with time as its pores fill with particles, as shown. The collection efficiency is consistent with a rate constant of -1.26 1/hr. Most nearly, what mass of particulate matter is removed in the first five hours of operation?

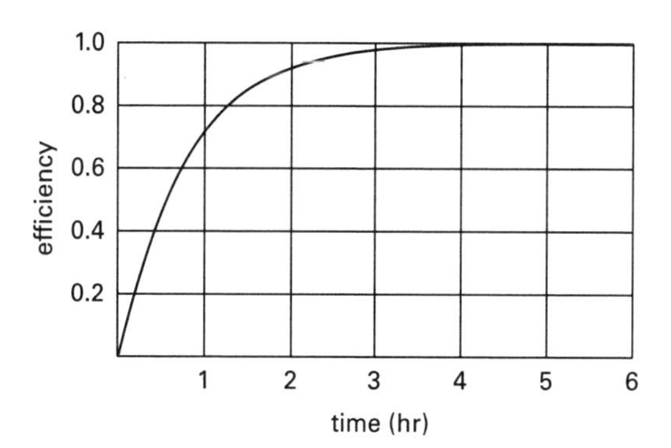

(A) 32 lbm

(B) 36 lbm

(C) 42 lbm

(D) 63 lbm

Incineration of Municipal Solid Waste

5. Which statement about incineration of municipal solid waste material is INCORRECT?

(A) The primary source of lead and cadmium in the fly ash is the combustion of plastics.

(B) Fly ash typically contains lead in concentrations 5000 times higher than natural soils and cadmium in concentrations 20,000 times higher than natural soils.

(C) The level of carcinogenic metals found in the fly ash is typically high enough to classify the fly ash as extremely hazardous waste under state and federal regulations.

(D) Incineration releases metals that are mobilized in air emissions or concentrated in ash residues, causing them to be highly bioavailable.

Incineration of Hazardous Waste

6. Which statement concerning hazardous waste incinerators is INCORRECT?

(A) The majority of incinerated radioactive waste is high-level waste (HLW) from nuclear power plants.

(B) Open pit incineration results in excessive smoke and high particulate emission.

(C) Single chamber incinerators generally do not meet federal air emission standards.

(D) Multiple chamber incinerators have low particulate emissions and generally meet federal air emission standards without additional air pollution control equipment.

Air Scrubbers

7. A 6 ft high spray tower has a collection efficiency of 80% when removing 10 μm diameter particles. If the tower's height is increased to 8 ft, and if all other characteristics are unchanged, the collection efficiency when removing 7 μm diameter particles will be most nearly

(A) 85%

(B) 91%

(C) 95%

(D) 98%

Electrostatic Precipitators

8. An electrostatic precipitator is being designed for 110,000 ACFM of 400°F flue gas containing incinerator fly ash at a concentration of 3.2 gr/SCF. For this installation, the effective drift velocity is 0.31 ft/sec. The required collection efficiency is 99.9%. The maximum pressure loss is 5 in wg. Most nearly, what should the total collection area be?

(A) 26,000 ft^2

(B) 31,000 ft^2

(C) 34,000 ft^2

(D) 41,000 ft^2

9. An electrostatic precipitator (ESP) has an efficiency of 95%. By increasing the collection area, the ESP's efficiency processing the same gas is increased to 99%. The ratio of the new collection area to the original collection area is most nearly

(A) 1.5

(B) 1.7

(C) 2.1

(D) 2.5

10. A two-chamber electrostatic precipitator (ESP) has an overall efficiency of 98% when the flow is split equally between the two chambers. If 80% of the same total flow goes through one chamber and 20% goes through the other, what is most nearly the new efficiency of the ESP?

(A) 84%

(B) 89%

(C) 92%

(D) 93%

SOLUTIONS

1. Dispersants are ultimately damaging to the environment. Rather than removing the oil, dispersants break up and distribute the oil, essentially hiding it and making it impossible to remove it from the environment.

The answer is (B).

2. *Bioremediation* is the use of microorganisms to remove pollutants. *Biofiltration* is the use of composting and soil beds to remove pollutants. *Bioreactors* are open or closed tanks that contain dozens or hundreds of slowly rotating disks covered with a biological film of microorganisms used to remove pollutants. *Bioventing* is the treatment of contaminated soil in a large plastic-covered tank. Clean air, water, and nutrients are continuously supplied to the tank while off-gases are suctioned off. The off-gases are cleaned with activated carbon adsorption or with thermal or catalytic oxidation prior to discharge.

The answer is (D).

3. The filter drag is

$$\begin{aligned} S &= K_e + K_s W \\ &= 0.6\ \frac{\text{in wg-min}}{\text{ft}} + \left(0.07\ \frac{\text{in wg-ft}^2\text{-min}}{\text{gr}}\right)\left(27\ \frac{\text{gr}}{\text{ft}^3}\right) \\ &= 2.49\ \text{in wg-min/ft} \end{aligned}$$

The air-to-cloth ratio is the same as the filtering velocity. The pressure drop is

$$\begin{aligned} \Delta p &= \text{v}_{\text{filtering}} S = \left(3.0\ \frac{\text{ft}}{\text{min}}\right)\left(2.49\ \frac{\text{in wg-min}}{\text{ft}}\right) \\ &= \boxed{7.47\ \text{in wg}\quad (7.5\ \text{in wg})} \end{aligned}$$

The answer is (D).

4. Instantaneous efficiency (given by the graph) is not average efficiency over time. From Eq. 71.13, the average efficiency of the baghouse, η, over five hours is

$$\begin{aligned} \eta_{\text{ave}} &= \frac{\int_{0\ \text{hr}}^{5\ \text{hr}} (1 - e^{-1.26t})\,dt}{5\ \text{hr}} = \frac{\left(t + \frac{e^{-1.26t}}{1.26}\right)\Big|_{0\ \text{hr}}^{5\ \text{hr}}}{5\ \text{hr}} \\ &= \frac{\left(5\ \text{hr} + \frac{e^{\left(-1.26\ \frac{1}{\text{hr}}\right)(5\ \text{hr})}}{1.26}\right) - \left(0\ \text{hr} + \frac{e^{\left(-1.26\ \frac{1}{\text{hr}}\right)(0\ \text{hr})}}{1.26}\right)}{5\ \text{hr}} \\ &= 0.8416 \end{aligned}$$

The total mass of particulate matter entering the baghouse in five hours is

$$\begin{aligned} m_{\text{in}} &= cQt \\ &= \frac{\left(0.8\ \frac{\text{g}}{\text{m}^3}\right)\left(2500\ \frac{\text{ft}^3}{\text{min}}\right)(5\ \text{hr})\left(60\ \frac{\text{min}}{\text{hr}}\right)}{\left(453.6\ \frac{\text{g}}{\text{lbm}}\right)\left(3.281\ \frac{\text{ft}}{\text{m}}\right)^3} \\ &= 37.45\ \text{lbm} \end{aligned}$$

The mass of particulate matter removed by the filters, m_r, is

$$\begin{aligned} m_r &= \eta_{\text{ave}} m_{\text{in}} = (0.8416)(37.45\ \text{lbm}) \\ &= \boxed{31.52\ \text{lbm}\quad (32\ \text{lbm})} \end{aligned}$$

The answer is (A).

5. The primary sources of lead and cadmium in incinerator fly ash are lead acid batteries and nickel cadmium batteries, respectively.

The answer is (A).

6. The majority of incinerated radioactive waste is low-level waste (LLW).

The answer is (A).

7. Use Eq. 71.40 with the initial conditions to calculate the scrubber constant.

$$\begin{aligned} K' &= \frac{\ln(1-\eta)d_p}{L} \\ &= \frac{\ln(1-0.8)(10\ \mu\text{m})\left(10^{-6}\ \frac{\text{m}}{\mu\text{m}}\right)\left(3.281\ \frac{\text{ft}}{\text{m}}\right)}{6\ \text{ft}} \\ &= -8.80\times 10^{-6} \end{aligned}$$

Using the constant, the scrubber efficiency under the new conditions is

$$\begin{aligned} \eta &= 1 - \exp\left(\frac{K'L}{d_p}\right) \\ &= 1 - \exp\left(\frac{(-8.80\times 10^{-6})(8\ \text{ft})}{(7\ \mu\text{m})\left(10^{-6}\ \frac{\text{m}}{\mu\text{m}}\right)\left(3.281\ \frac{\text{ft}}{\text{m}}\right)}\right) \\ &= \boxed{0.953\quad (95\%)} \end{aligned}$$

The answer is (C).

Plant Engineering

8. Solve Eq. 71.26 for the specific collection area.

$$\eta = 1 - \exp\left(-0.06 w_{e,\text{ft/sec}} \text{SCA}_{\text{ft}^2/1000\text{ ACFM}}\right)$$

$$\begin{aligned} \text{SCA} &= \frac{-\ln(1-\eta)}{0.06 w_e} \\ &= \frac{-\ln(1-0.999)}{(0.06)\left(0.31 \ \frac{\text{ft}}{\text{sec}}\right)} \\ &= 371.4 \text{ ft}^2/1000 \text{ ACFM} \end{aligned}$$

Since the actual volumetric flow rate was given, it does not have to be calculated from the fly ash concentration. The total required plate area is

$$\begin{aligned} \text{SCA} &= \frac{A_p}{Q} \\ A_p &= Q(\text{SCA}) \\ &= \left(\frac{110{,}000 \ \frac{\text{ft}^3}{\text{min}}}{1000 \ \frac{\text{ft}^3}{1000 \text{ ft}^3}}\right)\left(371.4 \ \frac{\text{ft}^2}{1000 \ \frac{\text{ft}^3}{\text{min}}}\right) \\ &= \boxed{40{,}854 \text{ ft}^2 \quad (41{,}000 \text{ ft}^2)} \end{aligned}$$

The answer is (D).

9. Rearrange Eq. 71.21.

$$A = \frac{-\ln(1-\eta)Q}{w}$$

The flow and drift velocity are unchanged. The ratio of the final area to the original area is

$$\begin{aligned} \frac{A_2}{A_1} &= \frac{-\ln(1-\eta_2)}{-\ln(1-\eta_1)} \\ &= \frac{-\ln(1-0.99)}{-\ln(1-0.95)} \\ &= \boxed{1.5} \end{aligned}$$

The answer is (A).

10. Use Eq. 71.21. Let $Q_t = Q_1 + Q_2$. Use x_1 as the fraction of flow through chamber 1 and x_2 as the fraction of flow through chamber 2.

$$\eta = x_1\left(1 - e^{\frac{-wA_p}{Q_1}}\right) + x_2\left(1 - e^{\frac{-wA_p}{Q_2}}\right)$$

The efficiency is 0.98 when x_1 and x_2 are both 0.5. Rearranging and combining,

$$\begin{aligned} \eta_{50\text{-}50} &= x_1\left(1 - e^{\frac{-wA_p}{Q_1}}\right) + x_2\left(1 - e^{\frac{-wA_p}{Q_2}}\right) \\ &= (0.50)\left(1 - e^{\frac{-wA_p}{0.5Q_t}}\right) + (0.50)\left(1 - e^{\frac{-wA_p}{0.5Q_t}}\right) \\ &= (0.50)\left(2 - 2e^{\frac{-wA_p}{0.5Q_t}}\right) \\ &= 1 - e^{\frac{-wA_p}{0.5Q_t}} \end{aligned}$$

$$\begin{aligned} wA_p &= -\ln(1-\eta_{50\text{-}50})(0.5)Q_t \\ &= -\ln(1-0.98)(0.5)Q_t \\ &= 1.956 Q_t \end{aligned}$$

Let $Q_t = Q_1 + Q_2$. With an 80%-20% split, $x_1 = 0.80$, $x_2 = 0.20$, $Q_1 = 0.8Q_t$, and $Q_2 = 0.2Q_t$. The total efficiency is

$$\begin{aligned} \eta_{80\text{-}20} &= x_1\left(1 - e^{\frac{-wA_p}{Q_1}}\right) + x_2\left(1 - e^{\frac{-wA_p}{Q_2}}\right) \\ &= (0.80)\left(1 - e^{-\frac{1.956Q_t}{0.8Q_t}}\right) + (0.20)\left(1 - e^{-\frac{1.956Q_t}{0.2Q_t}}\right) \\ &= (0.80)(1 - e^{-2.445}) + (0.20)(1 - e^{-9.780}) \\ &= \boxed{0.9306 \quad (93\%)} \end{aligned}$$

The answer is (D).

72 Electricity and Electrical Equipment

PRACTICE PROBLEMS

Circuit Analysis

1. Most nearly, how much power is dissipated by the circuit shown?

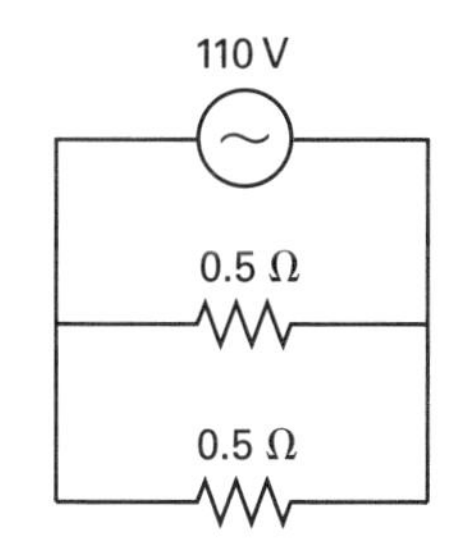

(A) 1.0 kW

(B) 3.0 kW

(C) 4.0 kW

(D) 48 kW

Transformers

2. An ideal step-down transformer has 200 primary coil turns and 50 secondary coil turns. 440 V are applied across the primary side. The resistance of an external secondary load is 5 Ω. The amount of heat dissipated in the resistance is most nearly

(A) 1200 W

(B) 2400 W

(C) 2600 W

(D) 3300 W

Induction Motors

3. The speed (in rpm) at which an induction motor rotates is

(A) $\dfrac{120(\text{frequency of the electric source, Hz})}{\text{number of poles}}$

(B) $\dfrac{120(\text{number of poles})}{\text{frequency of the electric source, Hz}}$

(C) $\dfrac{120(\text{voltage of the electric source, V})}{\text{frequency of the electric source, Hz}}$

(D) none of the above

4. The full-load phase current drawn by a 440 V (rms) 60 hp (total) three-phase induction motor having a full-load efficiency of 86% and a full-load power factor of 76% is most nearly

(A) 52 A

(B) 78 A

(C) 160 A

(D) 700 A

5. A 200 hp, three-phase, four-pole, 60 Hz, 440 V (rms) squirrel-cage induction motor operates at full load with an efficiency of 85%, power factor of 91%, and 3% slip.

(a) The speed is most nearly

(A) 1500 rpm

(B) 1700 rpm

(C) 2300 rpm

(D) 5800 rpm

(b) The torque developed is most nearly

(A) 180 ft-lbf

(B) 450 ft-lbf

(C) 600 ft-lbf

(D) 690 ft-lbf

(c) The line current is most nearly

(A) 34 A

(B) 150 A

(C) 250 A

(D) 280 A

6. A factory's induction motor load draws 550 kW at 82% power factor. What is most nearly the size of an additional synchronous motor required to produce 250 hp and raise the power factor to 95%? The line voltage is 220 V (rms).

(A) 140 kVA

(B) 230 kVA

(C) 240 kVA

(D) 330 kVA

7. The nameplate of an induction motor lists 960 rpm as the full-load speed. The frequency the motor was designed for is most nearly

(A) 24 Hz

(B) 34 Hz

(C) 48 Hz

(D) 50 Hz

Synchronous Motors

8. The speed (in rpm) at which a synchronous motor rotates is

(A) $\dfrac{120(\text{frequency of the electric source, Hz})}{\text{number of poles}}$

(B) $\dfrac{120(\text{number of poles})}{\text{frequency of the electric source, Hz}}$

(C) $\dfrac{120(\text{voltage of the electric source, V})}{\text{frequency of the electric source, Hz}}$

(D) none of the above

Three-Phase Power

9. The power triangle shown represents the total of all phases of a three-phase generator operating at 22 kV (rms). What is most nearly the rms line current for one phase?

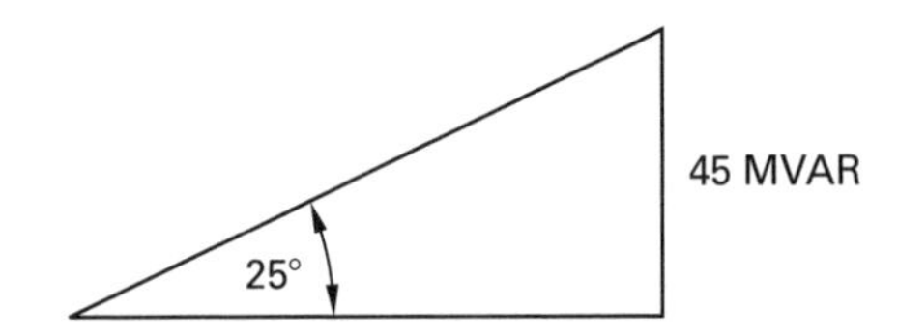

(A) 1.6 kA

(B) 2.2 kA

(C) 2.8 kA

(D) 4.8 kA

10. A 5000 ft^3/min insulated air handler is equipped with a resistive 40 kW heater. The heater is three-phase, 460 V (rms). The temperature entering the air handler is 60°F.

(a) Since the air handler is insulated, operation is adiabatic. The temperature of the air leaving the air handler is most nearly

(A) 67°F

(B) 73°F

(C) 79°F

(D) 85°F

(b) The rms line current drawn by the heater is most nearly

(A) 29 A

(B) 50 A

(C) 87 A

(D) 150 A

11. A 25 hp, three-phase motor draws 28 A (rms) at 480 V (rms). The motor efficiency is 92%. The total reactive load drawn by the motor is most nearly

(A) 8.6 kVAR

(B) 11 kVAR

(C) 16 kVAR

(D) 17 kVAR

Power Factor Correction

12. A machine shop that operates 14 hours per day, 349 days per year is considering some upgrades to decrease its energy usage. The shop currently uses electric motors generating a total of 600 hp. The upgrade would increase motor electrical-to-mechanical efficiencies from 86% to 95%. The shop is also considering replacing its 50 kW of incandescent lighting with 50 kW of fluorescent lighting. After the upgrade, the overall power factor of the shop will increase from 0.85 to 0.89. The cost for the upgrade is $160,000, and the electricity rate is $0.065/kVA-hr, based on apparent power. The simple payback period for this upgrade is most nearly

(A) 4.3 yr

(B) 5.9 yr

(C) 6.7 yr

(D) 8.1 yr

13. A manufacturing plant uses several large, 4160 V (rms), three-phase induction motors in its manufacturing process. The total motor power generated is 3960 hp, and the apparent drawn power is 3750 kVA. The motor electrical-to-mechanical efficiency is 90%, and the electricity cost is $0.045/kVA-hr, based on apparent power. The plant operates for 24 hours per day, seven days a week, but is shut down for maintenance two weeks per year. Several induction motors are replaced with synchronous motors to increase the plant power factor to 0.95. The synchronous motor efficiency is the same as the induction motor efficiency. Total plant motor power remains the same. The annual decrease in operating costs is most nearly

(A) $45,000

(B) $76,000

(C) $110,000

(D) $120,000

SOLUTIONS

1. The two resistors are in parallel. From Eq. 72.22, the equivalent resistance for the circuit is

$$\frac{1}{R_e} = \frac{1}{R_1} + \frac{1}{R_2} = \frac{1}{0.5\ \Omega} + \frac{1}{0.5\ \Omega}$$
$$R_e = 0.25\ \Omega$$

Although the voltage is AC, the circuit is purely resistive. The power dissipation is real power. The electrical power dissipated is

$$P = \frac{V^2}{R_e} = \frac{(110\ \text{V})^2}{(0.25\ \Omega)\left(1000\ \frac{\text{W}}{\text{kW}}\right)} = \boxed{48.4\ \text{kW}\quad(48\ \text{kW})}$$

The answer is (D).

2. From Eq. 72.43, the transformer turns ratio is

$$a = \frac{N_p}{N_s} = \frac{200}{50} = 4$$

The secondary voltage is

$$V_s = \frac{V_p}{a} = \frac{440\ \text{V}}{4} = 110\ \text{V}$$

The secondary current is

$$I_s = \frac{V_s}{R_s} = \frac{110\ \text{V}}{5\ \Omega} = 22\ \text{A}$$

The heat dissipated in the resistance is

$$P_s = I_s^2 R_s = (22\ \text{A})^2(5\ \Omega) = \boxed{2420\ \text{W}\quad(2400\ \text{W})}$$

The answer is (B).

3. An induction motor is drawn forward (accelerated) by an induced current that decreases as the motor approaches the synchronous speed at which a synchronous motor would turn, option A. For this reason, an induction motor always turns a few percent slower than the synchronous speed. None of the options mention this "slip."

The answer is (D).

Plant Engineering

4. From Eq. 72.67, the full-load phase current is

$$I_p = \frac{P_p}{\eta V_p \cos\phi} = \frac{\left(\frac{60 \text{ hp}}{3}\right)\left(745.7 \ \frac{\text{W}}{\text{hp}}\right)}{(0.86)(440 \text{ V})(0.76)} = \boxed{51.86 \text{ A} \quad (52 \text{ A})}$$

The answer is (A).

5. (a) From Eq. 72.76 and Eq. 72.78,

$$n_r = n_{\text{synchronous}}(1-s) = \left(\frac{120f}{p}\right)(1-s) = \left(\frac{(2)\left(60 \ \frac{\text{sec}}{\text{min}}\right)(60 \text{ Hz})}{4}\right)(1-0.03) = \boxed{1746 \text{ rpm} \quad (1700 \text{ rpm})}$$

The answer is (B).

(b) From Eq. 72.71(b), the torque developed is

$$T = \frac{5252P}{n_r} = \frac{\left(5252 \ \frac{\text{ft-lbf}}{\text{hp-sec}}\right)(200 \text{ hp})}{1746 \ \frac{\text{rev}}{\text{min}}} = \boxed{602 \text{ ft-lbf} \quad (600 \text{ ft-lbf})}$$

The answer is (C).

(c) From Eq. 72.67, the line current is

$$I_l = \frac{P_t}{\sqrt{3}\eta V_l \cos\phi} = \frac{(200 \text{ hp})\left(745.7 \ \frac{\text{W}}{\text{hp}}\right)}{(\sqrt{3})(0.85)(440 \text{ V})(0.91)} = \boxed{253 \text{ A} \quad (250 \text{ A})}$$

The answer is (C).

6. Draw the power triangle.

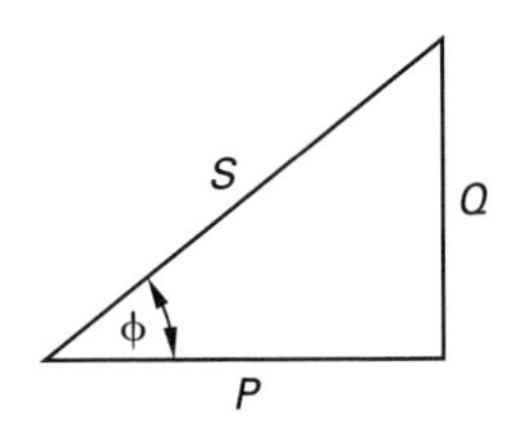

The original power angle is

$$\phi_i = \arccos 0.82 = 34.92°$$
$$P_1 = 550 \text{ kW}$$

From Eq. 72.56,

$$Q_i = P_1 \tan\phi_i = (550 \text{ kW})(\tan 34.92°) = 384.0 \text{ kVAR}$$

The new conditions are

$$P_2 = (250 \text{ hp})\left(0.7457 \ \frac{\text{kW}}{\text{hp}}\right) = 186.4 \text{ kW}$$
$$\phi_f = \arccos 0.95 = 18.19°$$

Since both motors perform real work,

$$P_f = P_1 + P_2 = 550 \text{ kW} + 186.4 \text{ kW} = 736.4 \text{ kW}$$

The new reactive power is

$$Q_f = P_f \tan\phi_f = (736.4 \text{ kW})(\tan 18.19°) = 242.0 \text{ kVAR}$$

The change in reactive power is

$$\Delta Q = 384.0 \text{ kVAR} - 242.0 \text{ kVAR} = 142 \text{ kVAR}$$

Synchronous motors used for power factor correction are rated by apparent power.

$$S = \sqrt{(\Delta P)^2 + (\Delta Q)^2} = \sqrt{(186.4 \text{ kW})^2 + (142 \text{ kVAR})^2} = \boxed{234.3 \text{ kVA} \quad (230 \text{ kVA})}$$

The answer is (B).

7. From Eq. 72.76 and Eq. 72.78, the frequency is

$$f = \frac{pn_{\text{synchronous}}}{120} = \frac{pn_{\text{actual}}}{(2)\left(60 \ \frac{\text{sec}}{\text{min}}\right)(1-s)}$$

The slip and number of poles are unknown. Assume $s = 0$ and $p = 4$.

$$f = \frac{(4)\left(960 \ \frac{\text{rev}}{\text{min}}\right)}{(2)\left(60 \ \frac{\text{sec}}{\text{min}}\right)(1-0)} = 32 \text{ Hz}$$

Plant Engineering

32 Hz is not close to any frequency in commercial use. Try $p = 6$.

$$f = \frac{(6)\left(960\ \frac{\text{rev}}{\text{min}}\right)}{(2)\left(60\ \frac{\text{sec}}{\text{min}}\right)(1-0)} = 48\ \text{Hz}$$

With a 4% slip, $f = 50$ Hz.

$$\boxed{50\ \text{Hz (European)}}$$

The answer is (D).

8. A synchronous motor's speed is dependent on the speed of the stator rotating field. The motor can rotate at the speed of the applied alternating current or submultiples thereof. From Eq. 72.76,

$$n_{\text{synchronous}} = \frac{120f}{p}$$

The synchronous speed is 120 times the frequency of the electric source divided by the number of poles.

The answer is (A).

9. The total real power is

$$\begin{aligned} P_t &= \frac{Q_t}{\tan\phi} \\ &= \frac{(45\ \text{MVAR})\left(1000\ \frac{\text{kVAR}}{\text{MVAR}}\right)}{\tan 25^\circ} \\ &= 96{,}502\ \text{kVAR} \end{aligned}$$

From Eq. 72.67, the line current is

$$\begin{aligned} I_l &= \frac{P_t}{\sqrt{3}V_l\cos\phi} \\ &= \frac{96{,}502\ \text{kVAR}}{(\sqrt{3})(22\ \text{kV})\left(1000\ \frac{\text{A}}{\text{kA}}\right)\cos 25^\circ} \\ &= \boxed{2.79\ \text{kA}\quad(2.8\ \text{kA})} \end{aligned}$$

The answer is (C).

10. (a) Use Eq. 36.6(c) to solve for the temperature leaving the air handler. The constant 1.08 is the product of an air density of 0.075 lbm/ft^3, a specific heat of 0.24 Btu/lbm-°F, and a conversion of 60 min/hr.

$$\begin{aligned} T_{\text{out}} &= \frac{\dot{q}}{\left(1.08\ \frac{\text{Btu-min}}{\text{ft}^3\text{-hr-}^\circ\text{F}}\right)\dot{V}} + T_{\text{in}} \\ &= \frac{(40\ \text{kW})\left(1000\ \frac{\text{W}}{\text{kW}}\right)\left(3.412\ \frac{\text{Btu}}{\text{W-hr}}\right)}{\left(1.08\ \frac{\text{Btu-min}}{\text{ft}^3\text{-hr-}^\circ\text{F}}\right)\left(5000\ \frac{\text{ft}^3}{\text{min}}\right)} + 60^\circ\text{F} \\ &= \boxed{85.27^\circ\text{F}\quad(85^\circ\text{F})} \end{aligned}$$

The answer is (D).

(b) Since the load is purely resistive, the power factor is 1.0. The load is three-phase. From Eq. 72.68, the line current is

$$I_l = \frac{P}{\sqrt{3}V_l} = \frac{(40\ \text{kW})\left(1000\ \frac{\text{W}}{\text{kW}}\right)}{(\sqrt{3})(460\ \text{V})} = \boxed{50.2\ \text{A}\quad(50\ \text{A})}$$

The answer is (B).

11. From Eq. 72.72, the total real electrical power drawn from the line by the motor is

$$\begin{aligned} P_{\text{electrical}} &= \frac{P_{\text{rated}}}{\eta_m} = \frac{(25\ \text{hp})\left(745.7\ \frac{\text{W}}{\text{hp}}\right)}{(0.92)\left(1000\ \frac{\text{W}}{\text{kW}}\right)} \\ &= 20.26\ \text{kW} \end{aligned}$$

Rearrange Eq. 72.79 for the power factor.

$$\begin{aligned} \text{pf} &= \frac{P_{\text{electrical}}}{\sqrt{3}V_lI_l} = \frac{(20.26\ \text{kW})\left(1000\ \frac{\text{W}}{\text{kW}}\right)}{(\sqrt{3})(480\ \text{V})(28\ \text{A})} \\ &= 0.870 \end{aligned}$$

The power factor angle is

$$\begin{aligned} \phi &= \arccos \text{pf} = \arccos 0.870 \\ &= 29.5^\circ \end{aligned}$$

From Eq. 72.68, the total apparent power is

$$\begin{aligned} S_t &= \sqrt{3}V_lI_l \\ &= \frac{(\sqrt{3})(480\ \text{V})(28\ \text{A})}{1000\ \frac{\text{VA}}{\text{kVA}}} \\ &= 23.28\ \text{kVA} \end{aligned}$$

Therefore, from Eq. 72.55, the total reactive load is

$$\begin{aligned} Q_t &= S_t \sin\phi = (23.28 \text{ kVA})\sin 29.5° \\ &= \boxed{11.46 \text{ kVAR} \quad (11 \text{ kVAR})} \end{aligned}$$

The answer is (B).

12. The lighting changes do not decrease the lighting power draw, but they do affect the apparent power. The real and apparent powers before the motor upgrade are

$$\begin{aligned} P_{\text{before}} &= \frac{P_m}{\eta_{\text{before}}} = \left(\frac{600 \text{ hp}}{0.86}\right)\left(0.7457 \ \frac{\text{kW}}{\text{hp}}\right) + 50 \text{ kW} \\ &= 570.3 \text{ kW} \end{aligned}$$

$$\begin{aligned} S_{\text{before}} &= \frac{P_{\text{before}}}{\cos\phi} = \frac{P_{\text{before}}}{\text{pf}} = \frac{570.3 \text{ kW}}{0.85} \\ &= 670.9 \text{ kVA} \end{aligned}$$

The real and apparent powers after the motor upgrade are

$$\begin{aligned} P_{\text{after}} &= \frac{P_m}{\eta_{\text{after}}} = \left(\frac{600 \text{ hp}}{0.95}\right)\left(0.7457 \ \frac{\text{kW}}{\text{hp}}\right) + 50 \text{ kW} \\ &= 521.0 \text{ kW} \\ S_{\text{after}} &= \frac{P_{\text{after}}}{\cos\phi} = \frac{P_{\text{after}}}{\text{pf}} = \frac{521.0 \text{ kW}}{0.89} \\ &= 585.4 \text{ kVA} \end{aligned}$$

The annual savings in cost from the upgrade, C_A, is

$$\begin{aligned} \Delta A &= (S_{\text{before}} - S_{\text{after}})\left(\frac{\text{hr of operation}}{\text{yr}}\right) C_{\text{electricity}} \\ &= (671.2 \text{ kVA} - 585.4 \text{ kVA}) \\ &\quad \times \left(14 \ \frac{\text{hr}}{\text{day}}\right)\left(349 \ \frac{\text{day}}{\text{yr}}\right)\left(0.065 \ \frac{\$}{\text{kVA-hr}}\right) \\ &= \$27{,}249/\text{yr} \end{aligned}$$

The simple payback period is

$$\begin{aligned} \text{payback} &= \frac{C_{\text{upgrade}}}{\Delta A} = \frac{\$160{,}000}{27{,}249 \ \frac{\$}{\text{yr}}} \\ &= \boxed{5.87 \text{ yr} \quad (5.9 \text{ yr})} \end{aligned}$$

The answer is (B).

13. From Eq. 72.72, the total real electrical power drawn by the motors from the line is

$$\begin{aligned} P_{\text{electrical}} &= \frac{P_{\text{rated}}}{\eta_m} = \frac{3960 \text{ hp}}{0.90} \\ &= 4400 \text{ hp} \end{aligned}$$

Rearranging Eq. 72.57, the new total apparent power after the synchronous motors have been installed is

$$\begin{aligned} S_{\text{after}} &= \frac{P_{\text{electrical}}}{\text{pf}} = \frac{(4400 \text{ hp})\left(745.7 \ \frac{\text{W}}{\text{hp}}\right)}{(0.95)\left(1000 \ \frac{\text{VA}}{\text{kVA}}\right)} \\ &= 3454 \text{ kVA} \end{aligned}$$

The annual decrease in operating cost is

$$\begin{aligned} \Delta A &= (S_{\text{before}} - S_{\text{after}})\left(\frac{\text{hr of operation}}{\text{yr}}\right) C_{\text{electricity}} \\ &= (3750 \text{ kVA} - 3454 \text{ kVA}) \\ &\quad \times \left(24 \ \frac{\text{hr}}{\text{day}}\right)\left(365 \ \frac{\text{day}}{\text{yr}} - 14 \ \frac{\text{day}}{\text{yr}}\right) \\ &\quad \times \left(0.045 \ \frac{\$}{\text{kVA-hr}}\right) \\ &= \boxed{\$112{,}207/\text{yr} \quad (\$110{,}000)} \end{aligned}$$

The answer is (C).

Plant Engineering

73 Illumination and Sound

PRACTICE PROBLEMS

Illumination

1. A 60 m × 24 m product assembly area is illuminated by a gridded arrangement of pendant lamps, each producing 18 000 lm. The minimum illumination required at the work surface level is 200 lux (lm/m^2). The lamps are well maintained and have a maintenance factor of 0.80. The utilization factor is 0.4. Approximately how many lamps are required?

(A) 34

(B) 42

(C) 50

(D) 66

Noise and Sound

2. A pipe is covered with acoustic pipe insulation to reduce noise. What is the best predictor of the perceived sound reduction?

(A) insertion loss (IL)

(B) transmission loss (TL)

(C) noise reduction rating (NRR)

(D) noise reduction coefficient (NRC)

3. A gear-driven electric power generation system has a prime mover noise level of 88 dBA, a gear system noise level of 82 dBA, and a generator noise level of 95 dBA. The overall noise level is most nearly

(A) 75 dBA

(B) 86 dBA

(C) 96 dBA

(D) 99 dBA

4. A 40 dB source is placed adjacent to a 35 dB source. The combined sound pressure level is most nearly

(A) 17 dB

(B) 28 dB

(C) 37 dB

(D) 41 dB

5. With no machinery operating, the background noise in a room has a sound pressure level of 43 dB. With the machinery operating, the sound pressure level is 45 dB. The sound pressure level due to the machinery alone is most nearly

(A) 2.0 dB

(B) 41 dB

(C) 47 dB

(D) 49 dB

6. An unenclosed source produces a sound pressure level of 100 dB. An enclosure is constructed from a material having a transmission loss of 30 dB. The sound pressure level inside the enclosure from the enclosed source increases to 110 dB. The reduction in sound pressure level outside the enclosure is most nearly

(A) 20 dB

(B) 30 dB

(C) 80 dB

(D) 90 dB

7. 4 ft (1.2 m) from an isotropic sound source, the sound pressure level is 92 dB. The sound pressure level 12 ft (3.6 m) from the source is most nearly

(A) 62 dB

(B) 73 dB

(C) 83 dB

(D) 87 dB

8. Octave band measurements of a noise source were made. The measurements were 85 dB, 90 dB, 92 dB, 87 dB, 82 dB, 78 dB, 65 dB, and 54 dB at frequencies of 63 Hz, 125 Hz, 250 Hz, 500 Hz, 1000 Hz, 2000 Hz, 4000 Hz, and 8000 Hz, respectively. The overall A-weighted sound pressure level is most nearly

(A) 83 dBA

(B) 86 dBA

(C) 89 dBA

(D) 93 dBA

9. If the number of sabins is 50% of the total room area, the maximum possible reduction in sound pressure level is most nearly a

(A) 1.2 dB decrease

(B) 3.0 dB decrease

(C) 4.6 dB decrease

(D) 6.1 dB decrease

10. A storage room has dimensions of 100 ft × 400 ft × 20 ft (30 m × 120 m × 6 m). All surfaces are plain concrete. 40% of the walls are treated acoustically with a material having a sound absorption coefficient of 0.8. The reduction in sound pressure level is most nearly a

(A) 1.2 dB decrease

(B) 3.0 dB decrease

(C) 4.6 dB decrease

(D) 6.1 dB decrease

11. A meeting room has dimensions of 20 ft × 50 ft × 10 ft (6 m × 15 m × 3 m). The floor is covered with roll vinyl. The walls and ceiling are sheetrock. 20% of the walls are glass windows. There are 15 seats, lightly upholstered, with 15 occupants, and 5 miscellaneous sabins. After complaints from the occupants, the ceiling is treated with a sound absorbing material with a sound absorption coefficient of 0.7. The reduction in sound level is most nearly a

(A) 3.5 dB decrease

(B) 5.7 dB decrease

(C) 6.3 dB decrease

(D) 12 dB decrease

12. A room has dimensions of 15 ft × 20 ft × 10 ft (4.5 m × 6 m × 3 m). The sound absorption coefficients are 0.03, 0.5, and 0.06 for the floor, ceiling, and walls, respectively. A machine with a sound power level of 65 dB is located at the intersection of the floor and wall, 7.5 ft (2.25 m) from the perpendicular walls. The ambient sound pressure level is 50 dB everywhere in the room. The sound pressure level 5 ft (1.5 m) from the machine is most nearly

(A) 45 dBA

(B) 61 dBA

(C) 73 dBA

(D) 81 dBA

SOLUTIONS

1. Use Eq. 73.8. The number of lamps required is

$$N = \frac{EA}{\Phi(\text{CU})(\text{LLF})} = \frac{\left(200\ \frac{\text{lm}}{\text{m}^2}\right)(60\ \text{m})(24\ \text{m})}{(18\,000\ \text{lm})(0.4)(0.8)} = \boxed{50}$$

The answer is (C).

2. The noise reduction coefficient (NRC) relates to sound absorption by features and furnishings within a space. The noise reduction rating (NRR) relates to hearing protection devices. The transmission loss (TL) is the amount of sound energy lost when sound passes through a barrier, such as acoustic pipe insulation. While the insulation may be quite dense, it is in direct contact with the pipe, and a high degree of structural vibration (i.e., noises) will transfer directly to the outer surface of the insulation. The insertion loss (IL) is the difference in sound pressure levels before and after insulation is applied. Therefore, the $\boxed{\text{insertion loss}}$ is the best predictor of the perceived sound reduction.

The answer is (A).

3. From Eq. 73.22, the governing equation for combining multiple noise sources is

$$L_p = 10\log\sum 10^{L_i/10} = 10\log\begin{pmatrix} 10^{88\ \text{dBA}/10} \\ +\,10^{82\ \text{dBA}/10} \\ +\,10^{95\ \text{dBA}/10} \end{pmatrix} = \boxed{96\ \text{dBA}}$$

The answer is (C).

4. Use Eq. 73.22.

$$L_p = 10\log\sum 10^{L_i/10} = 10\log\left(10^{40\ \text{dB}/10} + 10^{35\ \text{dB}/10}\right) = \boxed{41.2\ \text{dB}\quad(41\ \text{dB})}$$

The answer is (D).

5. Use Eq. 73.22.

$$L_p = 10\log\sum 10^{L_i/10} = 10\log\left(10^{43\ \text{dB}/10} + 10^{L/10}\right) = 45\ \text{dB}$$

Plant Engineering

Solve for the unknown machinery sound pressure level, L.

$$10^{L/10} = 10^{45\text{ dB}/10} - 10^{43\text{ dB}/10}$$
$$L = 10\log\left(10^{45\text{ dB}/10} - 10^{43\text{ dB}/10}\right)$$
$$= \boxed{40.7\text{ dB}\quad(41\text{ dB})}$$

The answer is (B).

6. Define $L_{p,1}$ as the sound pressure level inside the enclosure, and define $L_{p,2}$ as the sound pressure level outside the enclosure. Use Eq. 73.29.

$$L_{p,2} = L_{p,1} - \text{TL} = 110\text{ dB} - 30\text{ dB} = 80\text{ dB}$$

Define $L_{p,1}$ as the sound pressure level for the unenclosed source, and define $L_{p,2}$ as the sound pressure level for the enclosed source.

Use Eq. 73.28 to solve for the insertion loss.

$$\text{IL} = L_{p,1} - L_{p,2} = 100\text{ dB} - 80\text{ dB} = \boxed{20\text{ dB}}$$

The answer is (A).

7. The free-field sound pressure is inversely proportional to the distance from the source.

$$\frac{p_2}{p_1} = \frac{r_1}{r_2}$$

From Eq. 73.19,

$$L_{p,2} = L_{p,1} + 10\log\left(\frac{p_1}{p_2}\right)^2 = L_{p,1} + 20\log\frac{p_1}{p_2}$$
$$= L_{p,1} + 20\log\frac{r_1}{r_2}$$

Customary U.S. Solution

$$L_{p,2} = L_{p,1} + 20\log\frac{r_1}{r_2} = 92\text{ dB} + 20\log\frac{4\text{ ft}}{12\text{ ft}}$$
$$= 92\text{ dB} - 9.5\text{ dB}$$
$$= \boxed{82.5\text{ dB}\quad(83\text{ dB})}$$

The answer is (C).

SI Solution

$$L_{p,2} = L_{p,1} + 20\log\frac{r_1}{r_2} = 92\text{ dB} + 20\log\frac{1.2\text{ m}}{3.6\text{ m}}$$
$$= 92\text{ dB} - 9.5\text{ dB}$$
$$= \boxed{82.5\text{ dB}\quad(83\text{ dB})}$$

The answer is (C).

8. Add the corrections from Table 73.5 to the measurements.

frequency (Hz)	measurement (dB)	correction (dB)	corrected value (dB)
63	85	–26.2	58.8
125	90	–16.1	73.9
250	92	–8.6	83.4
500	87	–3.2	83.8
1000	82	0	82.0
2000	78	+1.2	79.2
4000	65	+1.0	66.0
8000	54	–1.1	52.9

Use Eq. 73.22.

$$L_p = 10\log\sum 10^{L_i/10}$$
$$= 10\log\begin{pmatrix} 10^{58.8\text{ dB}/10} + 10^{73.9\text{ dB}/10} \\ + 10^{83.4\text{ dB}/10} + 10^{83.8\text{ dB}/10} \\ + 10^{82.0\text{ dB}/10} + 10^{79.2\text{ dB}/10} \\ + 10^{66.0\text{ dB}/10} + 10^{52.9\text{ dB}/10} \end{pmatrix}$$
$$= \boxed{88.6\text{ dB}\quad(89\text{ dBA})}$$

The answer is (C).

9. Define A as the total room area.

$$\sum S_1 = 0.50A$$

The maximum number of sabins is equal to the room area.

$$\sum S_2 = A$$

Use Eq. 73.27.

$$\text{NR} = 10\log\frac{\sum S_1}{\sum S_2} = 10\log\frac{0.50A}{A} = 10\log 0.50$$
$$= \boxed{-3.0\text{ dB}\quad(3.0\text{ dB decrease})}$$

The answer is (B).

Plant Engineering

10. *Customary U.S. Solution*

The surface area of the room walls is

$$A_1 = \big((2)(100\ \text{ft}) + (2)(400\ \text{ft})\big)(20\ \text{ft}) = 20{,}000\ \text{ft}^2$$

The surface area of the room floor and ceiling is

$$A_2 = (2)(100\ \text{ft})(400\ \text{ft}) = 80{,}000\ \text{ft}^2$$

The sound absorption coefficient of concrete is the NRC value of 0.02 from App. 73.A.

Define the sound absorption coefficient of precast concrete as α_{concrete}.

The sabin area of the room with all precast concrete is

$$\begin{aligned}\sum S_2 &= \alpha_{\text{concrete}}(A_1 + A_2)\\ &= (0.02)(20{,}000\ \text{ft}^2 + 80{,}000\ \text{ft}^2)\\ &= 2000\ \text{ft}^2\end{aligned}$$

Define the sound absorption coefficient of the wall acoustical treatment as α_{wall}.

The sabin area of the room with 40% of the walls treated with $\alpha_{\text{wall}} = 0.8$ is

$$\begin{aligned}\sum S_1 &= \alpha_{\text{concrete}}A_2 + \alpha_{\text{concrete}}(0.6A_1) + \alpha_{\text{wall}}(0.4A_1)\\ &= (0.02)(80{,}000\ \text{ft}^2) + (0.02)(0.6)(20{,}000\ \text{ft}^2)\\ &\quad + (0.8)(0.4)(20{,}000\ \text{ft}^2)\\ &= 8240\ \text{ft}^2\end{aligned}$$

Use Eq. 73.27.

$$\begin{aligned}\text{NR} &= 10\log\frac{\sum S_1}{\sum S_2} = 10\log\frac{8240\ \text{ft}^2}{2000\ \text{ft}^2}\\ &= \boxed{6.1\ \text{dB decrease}}\end{aligned}$$

The answer is (D).

SI Solution

The surface area of the room walls is

$$A_1 = \big((2)(30\ \text{m}) + (2)(120\ \text{m})\big)(6\ \text{m}) = 1800\ \text{m}^2$$

The surface area of the room floor and ceiling is

$$A_2 = (2)(30\ \text{m})(120\ \text{m}) = 7200\ \text{m}^2$$

The sound absorption coefficient of precast concrete is the NRC value of 0.02 from App. 73.A.

Define the sound absorption coefficient of precast concrete as α_{concrete}.

The sabin area of the room with all precast concrete is

$$\begin{aligned}\sum S_2 &= \alpha_{\text{concrete}}(A_1 + A_2)\\ &= (0.02)(1800\ \text{m}^2 + 7200\ \text{m}^2)\\ &= 180\ \text{m}^2\end{aligned}$$

Define the sound absorption coefficient of the wall acoustical treatment as α_{wall}.

The sabin area of the room with 40% of the walls treated with $\alpha_{\text{wall}} = 0.8$ is

$$\begin{aligned}\sum S_1 &= \alpha_{\text{concrete}}A_2 + \alpha_{\text{concrete}}(0.6A_1) + \alpha_{\text{wall}}(0.4A_1)\\ &= (0.02)(7200\ \text{m}^2) + (0.02)(0.6)(1800\ \text{m}^2)\\ &\quad + (0.8)(0.4)(1800\ \text{m}^2)\\ &= 741.6\ \text{m}^2\end{aligned}$$

Use Eq. 73.27.

$$\begin{aligned}\text{NR} &= 10\log\frac{\sum S_1}{\sum S_2} = 10\log\frac{741.6\ \text{m}^2}{180\ \text{m}^2}\\ &= \boxed{6.1\ \text{dB decrease}}\end{aligned}$$

The answer is (D).

11. *Customary U.S. Solution*

The gross area of the walls is

$$\begin{aligned}A_1 &= \big((2)(20\ \text{ft}) + (2)(50\ \text{ft})\big)(10\ \text{ft}) = 1400\ \text{ft}^2\\ A_{\text{walls}} &= (1 - 0.20)(1400\ \text{ft}^2) = 1120\ \text{ft}^2\\ A_{\text{glass}} &= (0.20)(1400\ \text{ft}^2) = 280\ \text{ft}^2\end{aligned}$$

From App. 73.A, the sound absorption coefficient of glass is $\alpha_1 = 0.03$.

The area of the floor is

$$A_2 = (20\ \text{ft})(50\ \text{ft}) = 1000\ \text{ft}^2$$

From App. 73.A, the sound absorption coefficient of roll vinyl is $\alpha_2 = 0.03$.

The area of the ceiling is

$$A_3 = (20\ \text{ft})(50\ \text{ft}) = 1000\ \text{ft}^2$$

The sound absorption coefficient of the sheetrock is $\alpha_3 = 0.05$.

From App. 73.A, the seats have approximately 1.5 sabins each, and the occupants have approximately 5.0 sabins each.

The total sabin area of the untreated room is

$$\begin{aligned}\sum S_2 &= \alpha_{\text{glass}} A_{\text{glass}} + \alpha_{\text{walls}} A_{\text{walls}} + \alpha_{\text{floor}} A_{\text{floor}} \\ &\quad + \alpha_{\text{ceiling}} A_{\text{ceiling}} + \text{seats} + \text{occupants} \\ &\quad + \text{miscellaneous} \\ &= (0.03)(280\ \text{ft}^2) + (0.05)(1120\ \text{ft}^2) \\ &\quad + (0.03)(1000\ \text{ft}^2) + (0.05)(1000\ \text{ft}^2) \\ &\quad + (15)(1.5\ \text{ft}^2) + (15)(5.0\ \text{ft}^2) + 5.0\ \text{ft}^2 \\ &= 246.9\ \text{ft}^2\end{aligned}$$

The total sabin area excluding the ceiling is

$$\begin{aligned}246.9\ \text{ft}^2 - \alpha_3 A_3 &= 246.9\ \text{ft}^2 - (0.03)(1000\ \text{ft}^2) \\ &= 216.9\ \text{ft}^2\end{aligned}$$

The total sabin area of the room with the ceiling treated with $\alpha_3 = 0.7$ sound absorption is

$$\begin{aligned}\sum S_1 &= 216.9\ \text{ft}^2 + \alpha_3 A_3 \\ &= 216.9\ \text{ft}^2 + (0.7)(1000\ \text{ft}^2) \\ &= 916.9\ \text{ft}^2\end{aligned}$$

Use Eq. 73.27.

$$\begin{aligned}\text{NR} &= 10 \log \frac{\sum S_1}{\sum S_2} = 10 \log \frac{916.9\ \text{ft}^2}{246.9\ \text{ft}^2} \\ &= \boxed{5.7\ \text{dB decrease}}\end{aligned}$$

The answer is (B).

SI Solution

The gross area of the walls is

$$A_1 = \Big((2)(6\ \text{m}) + (2)(15\ \text{m})\Big)(3\ \text{m}) = 126\ \text{m}^2$$

$$A_{\text{walls}} = (1 - 0.20)(126\ \text{m}^2) = 100.8\ \text{m}^2$$

$$A_{\text{glass}} = (0.20)(126\ \text{m}^2) = 25.2\ \text{m}^2$$

From App. 73.A, the sound absorption coefficient of glass is $\alpha_1 = 0.03$.

The area of the floor is

$$A_2 = (6\ \text{m})(15\ \text{m}) = 90\ \text{m}^2$$

From App. 73.A, the sound absorption coefficient of roll vinyl is $\alpha_2 = 0.03$.

The area of the ceiling is

$$A_3 = (6\ \text{m})(15\ \text{m}) = 90\ \text{m}^2$$

The sound absorption coefficient of the sheetrock is $\alpha_3 = 0.05$.

From App. 73.A, the seats have approximately 1.5 sabins each, and the occupants have approximately 5.0 sabins each.

The total sabin area of the untreated room is

$$\begin{aligned}\sum S_2 &= \alpha_{\text{glass}} A_{\text{glass}} + \alpha_{\text{walls}} A_{\text{walls}} + \alpha_{\text{floor}} A_{\text{floor}} \\ &\quad + \alpha_{\text{ceiling}} A_{\text{ceiling}} + \text{seats} + \text{occupants} \\ &\quad + \text{miscellaneous} \\ &= (0.03)(25.2\ \text{m}^2) + (0.05)(100.8\ \text{m}^2) \\ &\quad + (0.03)(90\ \text{m}^2) + (0.05)(90\ \text{m}^2) \\ &\quad + (15)(1.5\ \text{ft}^2)\left(0.3048\ \frac{\text{m}}{\text{ft}}\right)^2 + (15)(5.0\ \text{ft}^2) \\ &\quad \times \left(0.3048\ \frac{\text{m}}{\text{ft}}\right)^2 + (5.0\ \text{ft}^2)\left(0.3048\ \frac{\text{m}}{\text{ft}}\right)^2 \\ &= 22.5\ \text{m}^2\end{aligned}$$

The total sabin area excluding the ceiling is

$$\begin{aligned}22.5\ \text{m}^2 - \alpha_3 A_3 &= 22.5\ \text{m}^2 - (0.03)(90\ \text{m}^2) \\ &= 19.8\ \text{m}^2\end{aligned}$$

The total sabin area of the room with the ceiling treated with $\alpha_3 = 0.7$ sound absorption is

$$\begin{aligned}\sum S_1 &= 19.8\ \text{m}^2 + \alpha_3 A_3 = 19.8\ \text{m}^2 + (0.7)(90\ \text{m}^2) \\ &= 82.8\ \text{m}^2\end{aligned}$$

Use Eq. 73.27.

$$\begin{aligned}\text{NR} &= 10 \log \frac{\sum S_1}{\sum S_2} = 10 \log \frac{82.8\ \text{m}^2}{22.5\ \text{m}^2} \\ &= \boxed{5.7\ \text{dB decrease}}\end{aligned}$$

The answer is (B).

12. *Customary U.S. Solution*

Define the sound absorption coefficients of the floor, ceiling, and walls as α_1, α_2, and α_3, respectively.

The floor area is

$$A_1 = (15\ \text{ft})(20\ \text{ft}) = 300\ \text{ft}^2$$

The ceiling area is

$$A_2 = (15\text{ ft})(20\text{ ft}) = 300\text{ ft}^2$$

The area of the walls is

$$A_3 = \Big((2)(15\text{ ft}) + (2)(20\text{ ft})\Big)(10\text{ ft}) = 700\text{ ft}^2$$

The total surface area of the room is

$$\begin{aligned} A &= \sum A_i = A_1 + A_2 + A_3 = 300\text{ ft}^2 + 300\text{ ft}^2 + 700\text{ ft}^2 \\ &= 1300\text{ ft}^2 \end{aligned}$$

From Eq. 73.24, the average sound absorption coefficient of the room is

$$\begin{aligned} \overline{\alpha} &= \frac{\sum S_i}{\sum A_i} = \frac{\alpha_1 A_1 + \alpha_2 A_2 + \alpha_3 A_3}{A} \\ &= \frac{(0.03)(300\text{ ft}^2) + (0.5)(300\text{ ft}^2) + (0.06)(700\text{ ft}^2)}{1300\text{ ft}^2} \\ &= 0.155 \end{aligned}$$

From Eq. 73.25, the room constant is

$$R = \frac{\overline{\alpha}A}{1-\overline{\alpha}} = \frac{(0.155)(1300\text{ ft}^2)}{1-0.155} = 238.5\text{ ft}^2$$

From Eq. 73.21(b), the sound pressure level due to the machine is

$$\begin{aligned} L_{p,\text{dBA}} &= 10.5 + L_W + 10\log\left(\frac{Q}{4\pi r^2} + \frac{4}{R}\right) \\ &= 10.5\text{ dB} + 65\text{ dB} \\ &\quad + 10\log\left(\frac{4}{(4\pi)(5\text{ ft})^2} + \frac{4}{238.5\text{ ft}^2}\right) \\ &= 60.2\text{ dBA} \end{aligned}$$

Use Eq. 73.22 to combine the machine sound pressure level with the ambient sound pressure level.

$$\begin{aligned} L_p &= 10\log\sum 10^{L_i/10} \\ &= 10\log\left(10^{60.2\text{ dBA}/10} + 10^{50\text{ dBA}/10}\right) \\ &= \boxed{60.6\text{ dBA}\quad(61\text{ dBA})} \end{aligned}$$

The answer is (B).

SI Solution

Define the sound absorption coefficients of the floor, ceiling, and walls as α_1, α_2, and α_3, respectively.

The floor area is

$$A_1 = (4.5\text{ m})(6\text{ m}) = 27\text{ m}^2$$

The ceiling area is

$$A_2 = (4.5\text{ m})(6\text{ m}) = 27\text{ m}^2$$

The area of the walls is

$$A_3 = \Big((2)(4.5\text{ m}) + (2)(6\text{ m})\Big)(3\text{ m}) = 63\text{ m}^2$$

The total surface area of the room is

$$\begin{aligned} A &= \sum A_i = A_1 + A_2 + A_3 = 27\text{ m}^2 + 27\text{ m}^2 + 63\text{ m}^2 \\ &= 117\text{ m}^2 \end{aligned}$$

From Eq. 73.24, the average sound absorption coefficient of the room is

$$\begin{aligned} \overline{\alpha} &= \frac{\sum S_i}{\sum A_i} = \frac{\alpha_1 A_1 + \alpha_2 A_2 + \alpha_3 A_3}{A} \\ &= \frac{(0.03)(27\text{ m}^2) + (0.5)(27\text{ m}^2) + (0.06)(63\text{ m}^2)}{117\text{ m}^2} \\ &= 0.155 \end{aligned}$$

From Eq. 73.25, the room constant is

$$R = \frac{\overline{\alpha}A}{1-\overline{\alpha}} = \frac{(0.155)(117\text{ m}^2)}{1-0.155} = 21.5\text{ m}^2$$

From Eq. 73.21(a), the sound pressure level due to the machine is

$$\begin{aligned} L_{p,\text{dBA}} &= 0.2 + L_W + 10\log\left(\frac{Q}{4\pi r^2} + \frac{4}{R}\right) \\ &= 0.2\text{ dB} + 65\text{ dB} + 10\log\left(\frac{4}{(4\pi)(1.5\text{ m})^2} + \frac{4}{21.5\text{ m}^2}\right) \\ &= 60.4\text{ dBA} \end{aligned}$$

Use Eq. 73.22 to combine the machine sound pressure level with the ambient sound pressure level.

$$\begin{aligned} L_p &= 10\log\sum 10^{L_i/10} \\ &= 10\log\left(10^{60.4\text{ dBA}/10} + 10^{50\text{ dBA}/10}\right) \\ &= \boxed{60.8\text{ dBA}\quad(61\text{ dBA})} \end{aligned}$$

The answer is (B).

74 Engineering Economic Analysis

PRACTICE PROBLEMS

1. At 6% effective annual interest, approximately how much will be accumulated if $1000 is invested for ten years?

(A) $560

(B) $790

(C) $1600

(D) $1800

2. At 6% effective annual interest, the present worth of $2000 that becomes available in four years is most nearly

(A) $520

(B) $580

(C) $1600

(D) $2500

3. At 6% effective annual interest, approximately how much should be invested to accumulate $2000 in 20 years?

(A) $620

(B) $1400

(C) $4400

(D) $6400

4. At 6% effective annual interest, the year-end annual amount deposited over seven years that is equivalent to $500 invested now is most nearly

(A) $90

(B) $210

(C) $300

(D) $710

5. At 6% effective annual interest, the accumulated amount at the end of ten years if $50 is invested at the end of each year for ten years is most nearly

(A) $90

(B) $370

(C) $660

(D) $900

6. At 6% effective annual interest, approximately how much should be deposited at the start of each year for ten years (a total of 10 deposits) in order to empty the fund by drawing out $200 at the end of each year for ten years (a total of 10 withdrawals)?

(A) $190

(B) $210

(C) $220

(D) $250

7. At 6% effective annual interest, approximately how much should be deposited at the start of each year for five years to accumulate $2000 on the date of the last deposit?

(A) $350

(B) $470

(C) $510

(D) $680

8. At 6% effective annual interest, approximately how much will be accumulated in ten years if three payments of $100 are deposited every other year for four years, with the first payment occurring at $t = 0$?

(A) $180

(B) $480

(C) $510

(D) $540

9. $500 is compounded monthly at a 6% nominal annual interest rate. Approximately how much will have accumulated in five years?

(A) $515

(B) $530

(C) $675

(D) $690

10. The effective annual rate of return on an $80 investment that pays back $120 in seven years is most nearly

(A) 4.5%

(B) 5.0%

(C) 5.5%

(D) 6.0%

11. A new machine will cost $17,000 and will have a resale value of $14,000 after five years. Special tooling will cost $5000. The tooling will have a resale value of $2500 after five years. Maintenance will be $2000 per year. The effective annual interest rate is 6%. The average annual cost of ownership during the next five years will be most nearly

(A) $2000

(B) $2300

(C) $4300

(D) $5500

12. An old covered wooden bridge can be strengthened at a cost of $9000, or it can be replaced for $40,000. The present salvage value of the old bridge is $13,000. It is estimated that the reinforced bridge will last for 20 years, will have an annual cost of $500, and will have a salvage value of $10,000 at the end of 20 years. The estimated salvage value of the new bridge after 25 years is $15,000. Maintenance for the new bridge would cost $100 annually. The effective annual interest rate is 8%. Which is the best alternative?

(A) Strengthen the old bridge.

(B) Build the new bridge.

(C) Both options are economically identical.

(D) Not enough information is given.

13. A firm expects to receive $32,000 each year for 15 years from sales of a product. An initial investment of $150,000 will be required to manufacture the product. Expenses will run $7530 per year. Salvage value is zero, and straight-line depreciation is used. The income tax rate is 48%. The after-tax rate of return is most nearly

(A) 8.0%

(B) 9.0%

(C) 10%

(D) 11%

14. A public works project has initial costs of $1,000,000, benefits of $1,500,000, and disbenefits of $300,000.

(a) The benefit/cost ratio is most nearly

(A) 0.20

(B) 0.47

(C) 1.2

(D) 1.7

(b) The excess of benefits over costs is most nearly

(A) $200,000

(B) $500,000

(C) $700,000

(D) $800,000

15. A speculator in land pays $14,000 for property that he expects to hold for ten years. $1000 is spent in renovation, and a monthly rent of $75 is collected from the tenants. (Use the year-end convention.) Taxes are $150 per year, and maintenance costs are $250 per year. In ten years, the sale price needed to realize a 10% rate of return is most nearly

(A) $26,000

(B) $31,000

(C) $34,000

(D) $36,000

16. The effective annual interest rate for a payment plan of 30 equal payments of $89.30 per month when a lump sum payment of $2000 would have been an outright purchase is most nearly

(A) 27%

(B) 35%

(C) 43%

(D) 51%

17. A depreciable item is purchased for $500,000. The salvage value at the end of 25 years is estimated at $100,000.

(a) The depreciation in each of the first three years using the straight line method is most nearly

(A) $4000

(B) $16,000

(C) $20,000

(D) $24,000

(b) The depreciation in each of the first three years using the sum-of-the-years' digits method is most nearly

(A) $16,000; $16,000; $16,000

(B) $30,000; $28,000; $27,000

(C) $31,000; $30,000; $28,000

(D) $32,000; $31,000; $30,000

(c) The depreciation in each of the first three years using the double-declining balance method is most nearly

(A) $16,000; $16,000; $16,000

(B) $31,000; $30,000; $28,000

(C) $37,000; $34,000; $31,000

(D) $40,000; $37,000; $34,000

18. Equipment that is purchased for $12,000 now is expected to be sold after ten years for $2000. The estimated maintenance is $1000 for the first year, but it is expected to increase $200 each year thereafter. The effective annual interest rate is 10%.

(a) The present worth is most nearly

(A) $16,000

(B) $17,000

(C) $21,000

(D) $22,000

(b) The annual cost is most nearly

(A) $1100

(B) $2200

(C) $3600

(D) $3700

19. A new grain combine with a 20-year life can remove seven pounds of rocks from its harvest per hour. Any rocks left in its output hopper will cause $25,000 damage in subsequent processes. Several investments are available to increase the rock-removal capacity, as listed in the table. The effective annual interest rate is 10%. What should be done?

rock removal rate	annual probability of exceeding rock removal rate	required investment to achieve removal rate
7	0.15	0
8	0.10	$15,000
9	0.07	$20,000
10	0.03	$30,000

(A) Do nothing.

(B) Invest $15,000.

(C) Invest $20,000.

(D) Invest $30,000.

20. A mechanism that costs $10,000 has operating costs and salvage values as given. An effective annual interest rate of 20% is to be used.

year	operating cost	salvage value
1	$2000	$8000
2	$3000	$7000
3	$4000	$6000
4	$5000	$5000
5	$6000	$4000

(a) The cost of owning and operating the mechanism in year two is most nearly

(A) $3800

(B) $4700

(C) $5800

(D) $6000

(b) The cost of owning and operating the mechanism in year five is most nearly

(A) $5600

(B) $6500

(C) $7000

(D) $7700

(c) The economic life of the mechanism is most nearly

(A) one year

(B) two years

(C) three years

(D) five years

(d) Assuming that the mechanism has been owned and operated for four years already, the cost of owning and operating the mechanism for one more year is most nearly

(A) $6400

(B) $7200

(C) $8000

(D) $8200

21. (*Time limit: one hour*) A salesperson intends to purchase a car for $50,000 for personal use, driving 15,000 miles per year. Insurance for personal use costs $2000 per year, and maintenance costs $1500 per year. The car gets 15 miles per gallon, and gasoline costs $1.50 per gallon. The resale value after five years will be $10,000. The salesperson's employer has asked that the car be used for business driving of 50,000 miles per year and has offered a reimbursement of $0.30 per mile.

Using the car for business would increase the insurance cost to $3000 per year and maintenance to $2000 per year. The salvage value after five years would be reduced to $5000. If the employer purchased a car for the salesperson to use, the initial cost would be the same, but insurance, maintenance, and salvage would be $2500, $2000, and $8000, respectively. The salesperson's effective annual interest rate is 10%.

(a) Is the reimbursement offer adequate?

(b) With a reimbursement of $0.30 per mile, approximately how many miles must the car be driven per year to justify the employer buying the car for the salesperson to use?

(A) 20,000 mi

(B) 55,000 mi

(C) 82,000 mi

(D) 150,000 mi

22. Alternatives A and B are being evaluated. The effective annual interest rate is 10%.

	alternative A	alternative B
first cost	$80,000	$35,000
life	20 years	10 years
salvage value	$7000	0
annual costs		
years 1–5	$1000	$3000
years 6–10	$1500	$4000
years 11–20	$2000	0
additional cost		
year 10	$5000	0

(a) The present worth for alternative A is most nearly

(A) $91,000

(B) $93,000

(C) $100,000

(D) $120,000

(b) The equivalent uniform annual cost (EUAC) for alternative A is most nearly

(A) $10,000

(B) $11,000

(C) $12,000

(D) $14,000

(c) The present worth for alternative B is most nearly

(A) $56,000

(B) $62,000

(C) $70,000

(D) $78,000

(d) The EUAC for alternative B is most nearly

(A) $9100

(B) $10,000

(C) $11,000

(D) $13,000

(e) Which alternative is economically superior?

(A) Alternative A is economically superior.

(B) Alternative B is economically superior.

(C) Alternatives A and B are economically equivalent.

(D) Not enough information is provided.

23. A car is needed for three years. Plans A and B for acquiring the car are being evaluated. An effective annual interest rate of 10% is to be used.

Plan A: lease the car for $0.25/mile (all inclusive)

Plan B: purchase the car for $30,000; keep the car for three years; sell the car after three years for $7200; pay $0.14 per mile for oil and gas; pay other costs of $500 per year

(a) What is the annual mileage of plan A?

(b) What is the annual mileage of plan B?

(c) For an equal annual cost, what is the annual mileage?

(d) Which plan is economically superior?

(A) Plan A is economically superior.

(B) Plan B is economically superior.

(C) Plans A and B are economically equivalent.

(D) Not enough information is provided.

24. Two methods are being considered to meet strict air pollution control requirements over the next ten years. Method A uses equipment with a life of ten years. Method B uses equipment with a life of five years that will be replaced with new equipment with an additional life of five years. Capacities of the two methods are different, but operating costs do not depend on the throughput. Operation is 24 hours per day, 365 days per year. The effective annual interest rate for this evaluation is 7%.

	method A	method B	
	years 1–10	years 1–5	years 6–10
installation cost	$13,000	$6000	$7000
equipment cost	$10,000	$2000	$2200
operating cost per hour	$10.50	$8.00	$8.00
salvage value	$5000	$2000	$2000
capacity (tons/yr)	50	20	20
life	10 years	5 years	5 years

Economics

(a) The uniform annual cost per ton for method A is most nearly

(A) $1800

(B) $1900

(C) $2100

(D) $2200

(b) The uniform annual cost per ton for method B is most nearly

(A) $3500

(B) $3600

(C) $4200

(D) $4300

(c) Over what range of throughput (in units of tons/yr) does each method have the minimum cost?

25. A transit district has asked for assistance in determining the proper fare for its bus system. An effective annual interest rate of 7% is to be used. The following additional information was compiled.

cost per bus	$60,000
bus life	20 years
salvage value	$10,000
miles driven per year	37,440
number of passengers per year	80,000
operating cost	$1.00 per mile in the first year, increasing $0.10 per mile each year thereafter

(a) If the fare is to remain constant for the next 20 years, the break-even fare per passenger is most nearly

(A) $0.51/passenger

(B) $0.61/passenger

(C) $0.84/passenger

(D) $0.88/passenger

(b) If the transit district decides to set the per-passenger fare at $0.35 for the first year, approximately how much should the passenger fare go up each year thereafter such that the district can break even in 20 years?

(A) $0.022 increase per year

(B) $0.036 increase per year

(C) $0.067 increase per year

(D) $0.072 increase per year

(c) If the transit district decides to set the per-passenger fare at $0.35 for the first year and the per-passenger fare goes up $0.05 each year thereafter, the additional government subsidy (per passenger) needed for the district to break even in 20 years is most nearly

(A) $0.11

(B) $0.12

(C) $0.16

(D) $0.21

26. Make a recommendation to your client to accept one of the following alternatives. Use the present worth comparison method. (Initial costs are the same.)

Alternative A: a 25 year annuity paying $4800 at the end of each year, where the interest rate is a nominal 12% per annum

Alternative B: a 25 year annuity paying $1200 every quarter at 12% nominal annual interest

(A) Alternative A is economically superior.

(B) Alternative B is economically superior.

(C) Alternatives A and B are economically equivalent.

(D) Not enough information is provided.

27. A firm has two alternatives for improvement of its existing production line. The data are as follows.

	alternative A	alternative B
initial installment cost	$1500	$2500
annual operating cost	$800	$650
service life	5 years	8 years
salvage value	0	0

Determine the best alternative using an interest rate of 15%.

(A) Alternative A is economically superior.

(B) Alternative B is economically superior.

(C) Alternatives A and B are economically equivalent.

(D) Not enough information is provided.

28. Two mutually exclusive alternatives requiring different investments are being considered. The life of both alternatives is estimated at 20 years with no salvage values. The minimum rate of return that is considered acceptable is 4%. Which alternative is best?

	alternative A	alternative B
investment required	$70,000	$40,000
net income per year	$5620	$4075
rate of return on total investment	5%	8%

(A) Alternative A is economically superior.

(B) Alternative B is economically superior.

(C) Alternatives A and B are economically equivalent.

(D) Not enough information is provided.

29. Compare the costs of two plant renovation schemes, A and B. Assume equal lives of 25 years, no salvage values, and interest at 25%.

	alternative A	alternative B
first cost	$20,000	$25,000
annual expenditure	$3000	$2500

(a) Determine the best alternative using the present worth method.

(A) Alternative A is economically superior.

(B) Alternative B is economically superior.

(C) Alternatives A and B are economically equivalent.

(D) Not enough information is provided.

(b) Determine the best alternative using the capitalized cost comparison.

(A) Alternative A is economically superior.

(B) Alternative B is economically superior.

(C) Alternatives A and B are economically equivalent.

(D) Not enough information is provided.

(c) Determine the best alternative using the annual cost comparison.

(A) Alternative A is economically superior.

(B) Alternative B is economically superior.

(C) Alternatives A and B are economically equivalent.

(D) Not enough information is provided.

30. A machine costs $18,000 and has a salvage value of $2000. It has a useful life of 8 years. The interest rate is 8%.

(a) Using straight line depreciation, its book value at the end of 5 years is most nearly

(A) $2000

(B) $3000

(C) $6000

(D) $8000

(b) Using the sinking fund method, the depreciation in the third year is most nearly

(A) $1500

(B) $1600

(C) $1800

(D) $2000

(c) Repeat part (a) using double declining balance depreciation. The balance value at the fifth year is most nearly

(A) $2000

(B) $4000

(C) $6000

(D) $8000

31. A chemical pump motor unit is purchased for $14,000. The estimated life is 8 years, after which it will be sold for $1800. Find the depreciation in the first two years by the sum-of-the-years' digits method. The after-tax depreciation recovery using 15% interest with 52% income tax is most nearly

(A) $3600

(B) $3900

(C) $4100

(D) $6300

32. A soda ash plant has the water effluent from processing equipment treated in a large settling basin. The settling basin eventually discharges into a river that runs alongside the basin. Recently enacted environmental regulations require all rainfall on the plant to be diverted and treated in the settling basin. A heavy rainfall will cause the entire basin to overflow. An uncontrolled overflow will cause environmental damage and heavy fines. The construction of additional height on the existing basic walls is under consideration.

Data on the costs of construction and expected costs for environmental cleanup and fines are shown. Data on 50 typical winters have been collected. The soda ash plant management considers 12% to be their minimum rate of return, and it is felt that after 15 years the plant will be closed. The company wants to select the alternative that minimizes its total expected costs.

additional basin height (ft)	number of winters with basin overflow	expense for environmental clean up per year	construction cost
0	24	$550,000	0
5	14	$600,000	$600,000
10	8	$650,000	$710,000
15	3	$700,000	$900,000
20	1	$800,000	$1,000,000
	50		

The additional height the basin should be built to is most nearly

(A) 5.0 ft

(B) 10 ft

(C) 15 ft

(D) 20 ft

Economics

33. A wood processing plant installed a waste gas scrubber at a cost of $30,000 to remove pollutants from the exhaust discharged into the atmosphere. The scrubber has no salvage value and will cost $18,700 to operate next year, with operating costs expected to increase at the rate of $1200 per year thereafter. Money can be borrowed at 12%. Approximately when should the company consider replacing the scrubber?

(A) 3 yr

(B) 6 yr

(C) 8 yr

(D) 10 yr

34. Two alternative piping schemes are being considered by a water treatment facility. Head and horsepower are reflected in the hourly cost of operation. On the basis of a 10-year life and an interest rate of 12%, what is most nearly the number of hours of operation for which the two installations will be equivalent?

	alternative A	alternative B
pipe diameter	4 in	6 in
head loss for required flow	48 ft	26 ft
size motor required	20 hp	7 hp
energy cost per hour of operation	$0.30	$0.10
cost of motor installed	$3600	$2800
cost of pipes and fittings	$3050	$5010
salvage value at end of 10 years	$200	$280

(A) 1000 hr

(B) 3000 hr

(C) 5000 hr

(D) 6000 hr

35. An 88% learning curve is used with an item whose first production time was 6 weeks.

(a) Approximately how long will it take to produce the fourth item?

(A) 4.5 wk

(B) 5.0 wk

(C) 5.5 wk

(D) 6.0 wk

(b) Approximately how long will it take to produce the sixth through fourteenth items?

(A) 35 wk

(B) 40 wk

(C) 45 wk

(D) 50 wk

36. A company is considering two alternatives, only one of which can be selected.

alternative	initial investment	salvage value	annual net profit	life
A	$120,000	$15,000	$57,000	5 yr
B	$170,000	$20,000	$67,000	5 yr

The net profit is after operating and maintenance costs, but before taxes. The company pays 45% of its year-end profit as income taxes. Use straight line depreciation. Do not use investment tax credit.

(a) Determine whether each alternative has an ROR greater than the MARR.

(A) Alternative A has ROR > MARR.

(B) Alternative B has ROR > MARR.

(C) Both alternatives have ROR > MARR.

(D) Neither alternative has ROR > MARR.

(b) Find the best alternative if the company's minimum attractive rate of return is 15%.

(A) Alternative A is economically superior.

(B) Alternative B is economically superior.

(C) Alternatives A and B are economically equivalent.

(D) Not enough information is provided.

37. A company is considering the purchase of equipment to expand its capacity. The equipment cost is $300,000. The equipment is needed for 5 years, after which it will be sold for $50,000. The company's before-tax cash flow will be improved $90,000 annually by the purchase of the asset. The corporate tax rate is 48%, and straight line depreciation will be used. The company will take an investment tax credit of 6.67%. What is the after-tax rate of return associated with this equipment purchase?

(A) 10.9%

(B) 11.8%

(C) 12.2%

(D) 13.2%

38. A 120-room hotel is purchased for $2,500,000. A 25-year loan is available for 12%. The year-end convention applies to loan payments. A study was conducted to determine the various occupancy rates.

occupancy	probability
65% full	0.40
70%	0.30
75%	0.20
80%	0.10

Economics

The operating costs of the hotel are as follows.

taxes and insurance	\$20,000 annually
maintenance	\$50,000 annually
operating	\$200,000 annually

The life of the hotel is figured to be 25 years when operating 365 days per year. The salvage value after 25 years is \$500,000. The new hotel owners want to receive an annual rate of return of 15% on their investment. Neglect tax credit and income taxes.

(a) The distributed profit is most nearly

(A) \$300,000

(B) \$320,000

(C) \$340,000

(D) \$380,000

(b) The annual daily receipts are most nearly

(A) \$2300

(B) \$2400

(C) \$2500

(D) \$2600

(c) The average occupancy is most nearly

(A) 0.65

(B) 0.70

(C) 0.75

(D) 0.80

(d) The average rate that should be charged per room per night is most nearly

(A) \$27

(B) \$29

(C) \$30

(D) \$31

39. A company is insured for \$3,500,000 against fire and the insurance rate is \$0.69/\$1000. The insurance company will decrease the rate to \$0.47/\$1000 if fire sprinklers are installed. The initial cost of the sprinklers is \$7500. Annual costs are \$200; additional taxes are \$100 annually. The system life is 25 years.

(a) The annual savings is most nearly

(A) \$300

(B) \$470

(C) \$570

(D) \$770

(b) The rate of return is most nearly

(A) 3.8%

(B) 5.0%

(C) 13%

(D) 16%

40. Heat losses through the walls in an existing building cost a company \$1,300,000 per year. This amount is considered excessive, and two alternatives are being evaluated. Neither of the alternatives will increase the life of the existing building beyond the current expected life of 6 years, and neither of the alternatives will produce a salvage value. Improvements can be depreciated.

Alternative A: Do nothing, and continue with current losses.

Alternative B: Spend \$2,000,000 immediately to upgrade the building and reduce the loss by 80%. Annual maintenance will cost \$150,000.

Alternative C: Spend \$1,200,000 immediately. Then, repeat the \$1,200,000 expenditure 3 years from now. Heat loss the first year will be reduced 80%. Due to deterioration, the reduction will be 55% and 20% in the second and third years. (The pattern is repeated starting after the second expenditure.) There are no maintenance costs.

All energy and maintenance costs are considered expenses for tax purposes. The company's tax rate is 48%, and straight line depreciation is used. 15% is regarded as the effective annual interest rate. Evaluate each alternative on an after-tax basis.

(a) The present worth of alternative A is most nearly

(A) −\$5.9 million

(B) −\$4.9 million

(C) −\$2.6 million

(D) −\$2.4 million

(b) The present worth of alternative B is most nearly

(A) −\$3.4 million

(B) −\$2.8 million

(C) −\$2.6 million

(D) −\$2.2 million

(c) The present worth of alternative C is most nearly

(A) −\$3.2 million

(B) −\$3.1 million

(C) −\$2.4 million

(D) −\$2.0 million

(d) Which alternative should be recommended?

(A) alternative A

(B) alternative B

(C) alternative C

(D) not enough information

41. You have been asked to determine if a 7-year-old machine should be replaced. Give a full explanation for your recommendation. Base your decision on a before-tax interest rate of 15%.

The existing machine is presumed to have a 10-year life. It has been depreciated on a straight line basis from its original value of $1,250,000 to a current book value of $620,000. Its ultimate salvage value was assumed to be $350,000 for purposes of depreciation. Its present salvage value is estimated at $400,000, and this is not expected to change over the next 3 years. The current operating costs are not expected to change from $200,000 per year.

A new machine costs $800,000, with operating costs of $40,000 the first year, and increasing by $30,000 each year thereafter. The new machine has an expected life of 10 years. The salvage value depends on the year the new machine is retired.

year retired	salvage
1	$600,000
2	$500,000
3	$450,000
4	$400,000
5	$350,000
6	$300,000
7	$250,000
8	$200,000
9	$150,000
10	$100,000

42. A company estimates that the demand for its product will be 500,000 units per year. The product incorporates three identical valves. The acquisition cost per valve is $6.50, and the cost of keeping a valve in storage per year is 40% of the acquisition cost per valve per year. It costs the company an average of $49.50 to process an order. The company receives no quantity discounts and does not use inventory safety stocks. The most economical quantity of valves to order is most nearly

(A) 2700 valves

(B) 4400 valves

(C) 4800 valves

(D) 7600 valves

43. As production facilities move toward just-in-time manufacturing, it is important to minimize

(A) demand rate

(B) production rate

(C) inventory carrying cost

(D) setup cost

SOLUTIONS

1.

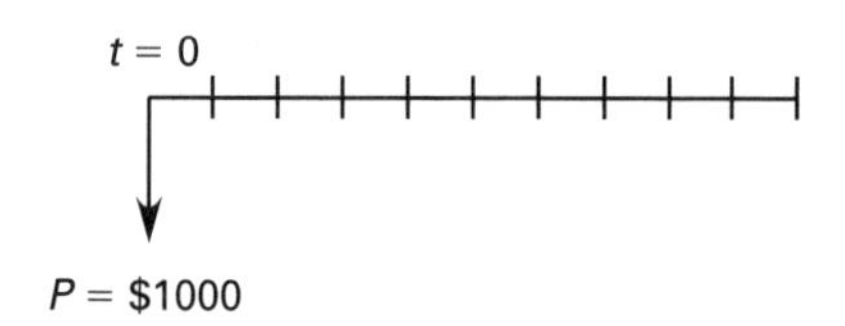

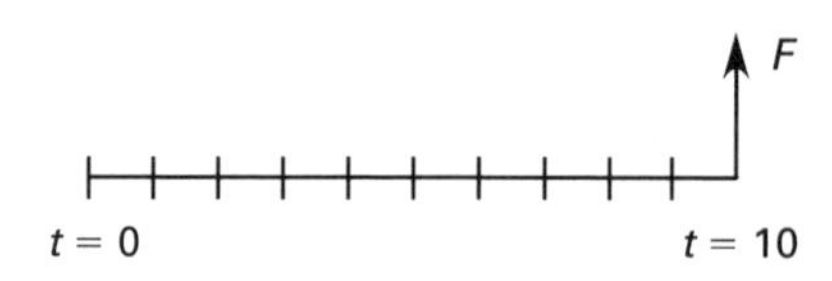

i = 6% a year

Using the formula from Table 74.1,

$$F = P(1+i)^n = (\$1000)(1+0.06)^{10}$$
$$= \boxed{\$1790.85 \quad (\$1800)}$$

Using the factor from App. 74.B, $(F/P,\ i,\ n) = 1.7908$ for $i = 6\%$ a year and $n = 10$ years.

$$F = P(F/P, 6\%, 10) = (\$1000)(1.7908)$$
$$= \boxed{\$1790.80 \quad (\$1800)}$$

The answer is (D).

2.

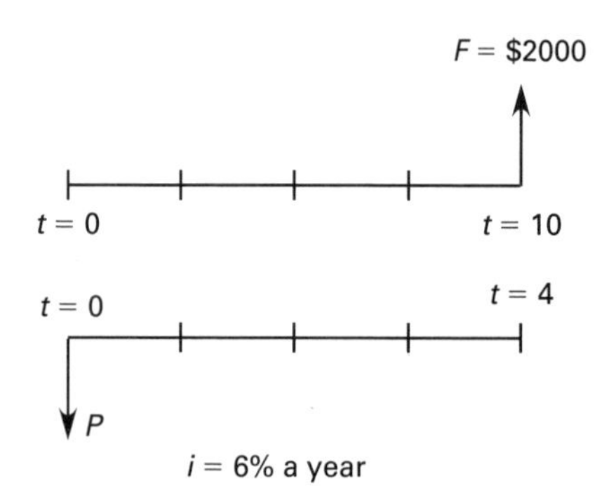

Using the formula from Table 74.1,

$$P = \frac{F}{(1+i)^n} = \frac{\$2000}{(1+0.06)^4}$$
$$= \boxed{\$1584.19 \quad (\$1600)}$$

Using the factor from App. 74.B, $(P/F,\ i,\ n) = 0.7921$ for $i = 6\%$ a year and $n = 4$ years.

$$P = F(P/F, 6\%, 4) = (\$2000)(0.7921)$$
$$= \boxed{\$1584.20 \quad (\$1600)}$$

The answer is (C).

3.

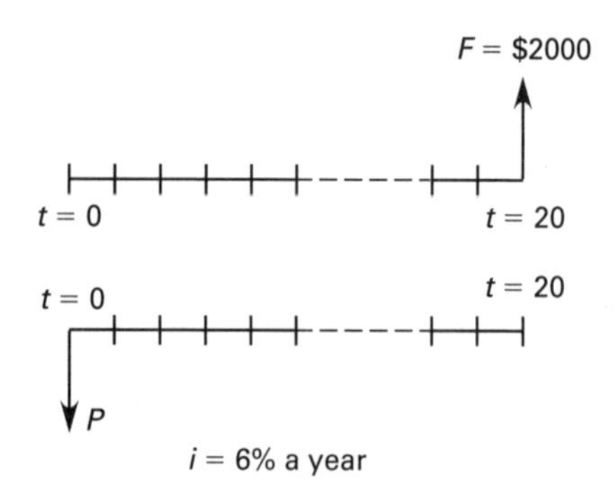

Using the formula from Table 74.1,

$$P = \frac{F}{(1+i)^n} = \frac{\$2000}{(1+0.06)^{20}}$$
$$= \boxed{\$623.61 \quad (\$620)}$$

Using the factor from App. 74.B, $(P/F,\ i,\ n) = 0.3118$ for $i = 6\%$ a year and $n = 20$ years.

$$P = F(P/F, 6\%, 20) = (\$2000)(0.3118)$$
$$= \boxed{\$623.60 \quad (\$620)}$$

The answer is (A).

4.

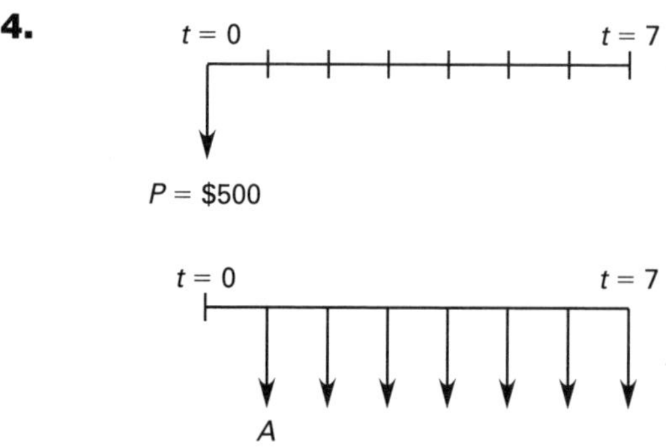

Using the formula from Table 74.1,

$$A = P\left(\frac{i(1+i)^n}{(1+i)^n - 1}\right) = (\$500)\left(\frac{(0.06)(1+0.06)^7}{(1+0.06)^7 - 1}\right)$$
$$= \boxed{\$89.57 \quad (\$90)}$$

Using the factor from App. 74.B, $(A/P,\ i,\ n) = 0.1791$ for $i = 6\%$ a year and $n = 7$ years.

$$A = P(A/P, 6\%, 7) = (\$500)(0.17914)$$
$$= \boxed{\$89.55 \quad (\$90)}$$

The answer is (A).

5.

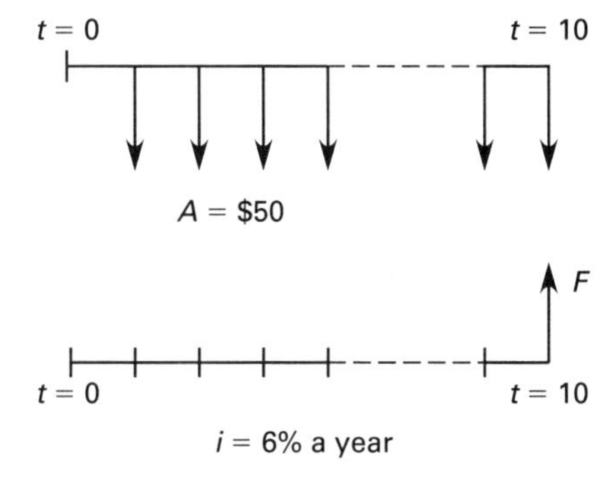

Using the formula from Table 74.1,

$$\begin{aligned} F &= A\left(\frac{(1+i)^n - 1}{i}\right) \\ &= (\$50)\left(\frac{(1+0.06)^{10} - 1}{0.06}\right) \\ &= \boxed{\$659.04 \quad (\$660)} \end{aligned}$$

Using the factor from App. 74.B, $(F/A,\ i,\ n) = 13.1808$ for $i = 6\%$ a year and $n = 10$ years.

$$\begin{aligned} F &= A(F/A, 6\%, 10) \\ &= (\$50)(13.1808) \\ &= \boxed{\$659.04 \quad (\$660)} \end{aligned}$$

The answer is (C).

6.

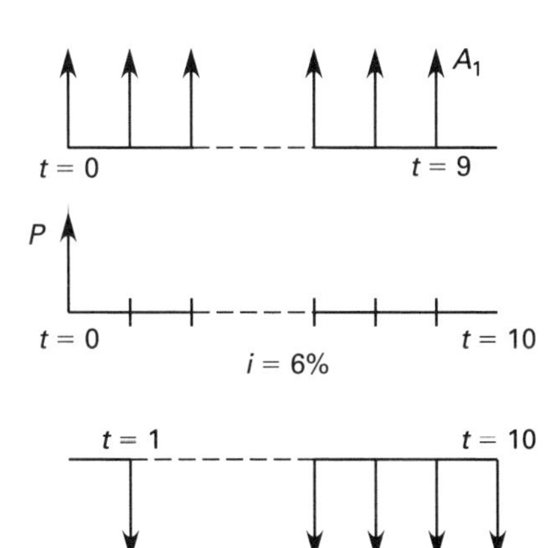

From Table 74.1, for each cash flow diagram,

$$\begin{aligned} P &= A_1 + A_1\left(\frac{(1+0.06)^9 - 1}{(0.06)(1+0.06)^9}\right) \\ &= A_2\left(\frac{(1+0.06)^{10} - 1}{(0.06)(1+0.06)^{10}}\right) \end{aligned}$$

Therefore, for $A_2 = \$200$,

$$\begin{aligned} A_1 + A_1\left(\frac{(1+0.06)^9 - 1}{(0.06)(1+0.06)^9}\right) &= (\$200)\left(\frac{(1+0.06)^{10} - 1}{(0.06)(1+0.06)^{10}}\right) \\ 7.80A_1 &= \$1472.02 \\ A_1 &= \boxed{\$188.72 \quad (\$190)} \end{aligned}$$

Using factors from App. 74.B,

$$\begin{aligned} (P/A, 6\%, 9) &= 6.8017 \\ (P/A, 6\%, 10) &= 7.3601 \\ A_1 + A_1(6.8017) &= (\$200)(7.3601) \\ 7.8017A_1 &= \$1472.02 \\ A_1 &= \frac{\$1472.02}{7.8017} \\ &= \boxed{\$188.68 \quad (\$190)} \end{aligned}$$

The answer is (A).

7.

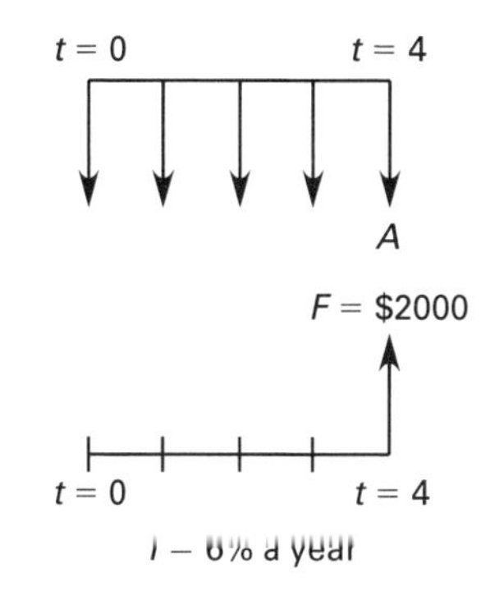

From Table 74.1,

$$F = A\left(\frac{(1+i)^n - 1}{i}\right)$$

Since the deposits start at the beginning of each year, five deposits are made that contribute to the final amount. This is equivalent to a cash flow that starts at $t = -1$ without a deposit and has a duration (starting at $t = -1$) of 5 years.

$$\begin{aligned} F &= A\left(\frac{(1+i)^n - 1}{i}\right) = \$2000 \\ &= A\left(\frac{(1+0.06)^5 - 1}{0.06}\right) \end{aligned}$$

Economics

$$\$2000 = 5.6371A$$
$$A = \frac{\$2000}{5.6371}$$
$$= \boxed{\$354.79 \quad (\$350)}$$

Using factors from App. 74.B,

$$F = A\big((F/P, 6\%, 4) + (F/A, 6\%, 4)\big)$$
$$\$2000 = A(1.2625 + 4.3746)$$
$$A = \boxed{\$354.79 \quad (\$350)}$$

The answer is (A).

8.

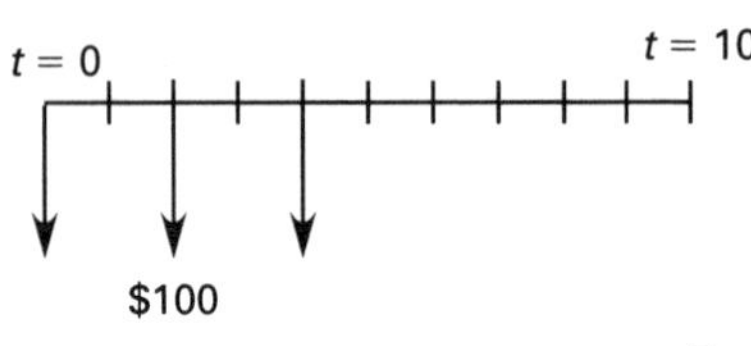

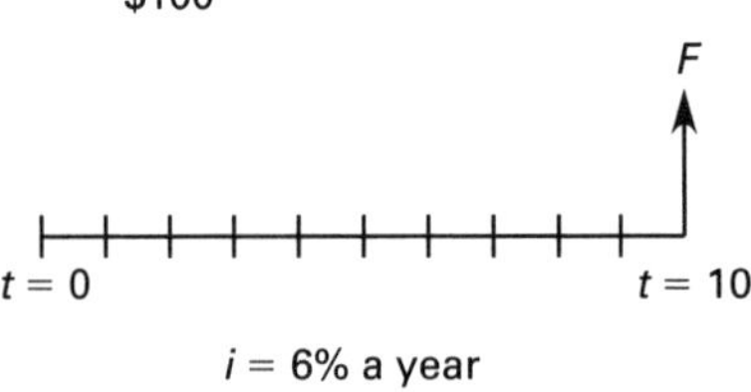

From Table 74.1, $F = P(1 + i)^n$. If each deposit is considered as P, each will accumulate interest for periods of 10, 8, and 6 years.

Therefore,

$$F = (\$100)(1 + 0.06)^{10} + (\$100)(1 + 0.06)^{8} + (\$100)(1 + 0.06)^{6}$$
$$= (\$100)(1.7908 + 1.5938 + 1.4185)$$
$$= \boxed{\$480.31 \quad (\$480)}$$

Using App. 74.B,

$$(F/P, i, n) = 1.7908 \text{ for } i = 6\% \text{ and } n = 10$$
$$= 1.5938 \text{ for } i = 6\% \text{ and } n = 8$$
$$= 1.4185 \text{ for } i = 6\% \text{ and } n = 6$$

By summation,

$$F = (\$100)(1.7908 + 1.5938 + 1.4185)$$
$$= \boxed{\$480.31 \quad (\$480)}$$

The answer is (B).

9.

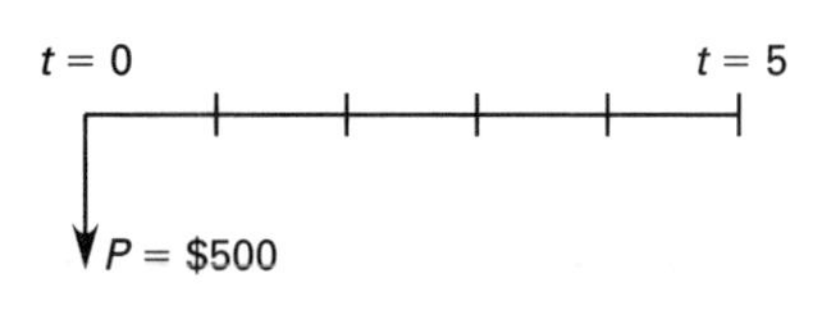

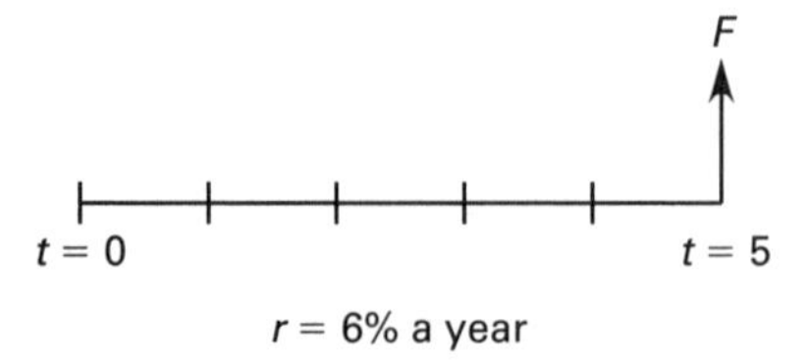

Since the deposit is compounded monthly, the effective interest rate should be calculated from Eq. 74.51.

$$i = \left(1 + \frac{r}{k}\right)^k - 1 = \left(1 + \frac{0.06}{12}\right)^{12} - 1$$
$$= 0.061678 \quad (6.1678\%)$$

From Table 74.1,

$$F = P(1 + i)^n = (\$500)(1 + 0.061678)^5$$
$$= \boxed{\$674.43 \quad (\$680)}$$

To use App. 74.B, interpolation is required.

$i\%$	factor F/P
6	1.3382
6.1678	desired
7	1.4026

$$\Delta(F/P) = \left(\frac{6.1678\% - 6\%}{7\% - 6\%}\right)(1.4026 - 1.3382)$$
$$= 0.0108$$

Therefore,

$$F/P = 1.3382 + 0.0108 = 1.3490$$
$$F = P(F/P, 6.1677\%, 5) = (\$500)(1.3490)$$
$$= \boxed{\$674.50 \quad (\$675)}$$

The answer is (C).

Economics

10.

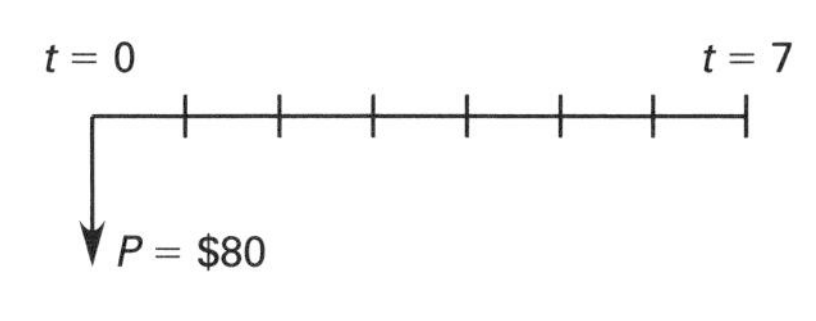

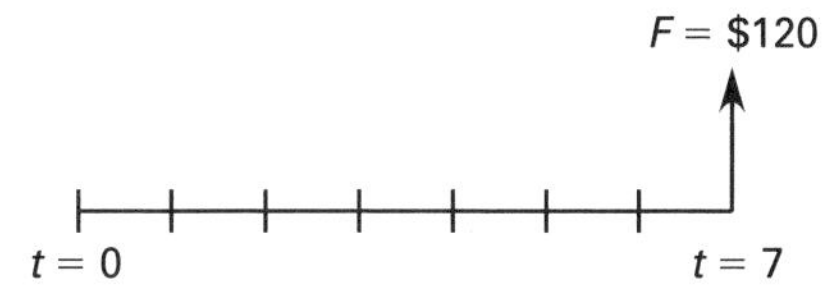

From Table 74.1,

$$F = P(1+i)^n$$

Therefore,

$$(1+i)^n = F/P$$

$$i = (F/P)^{1/n} - 1 = \left(\frac{\$120}{\$80}\right)^{1/7} - 1$$

$$= 0.0596 \approx \boxed{6.0\%}$$

From App. 74.B,

$$F = P(F/P, i\%, 7)$$

$$(F/P, i\%, 7) = F/P = \frac{\$120}{\$80} = 1.5$$

Searching App. 74.B,

$$\begin{aligned}(F/P, i\%, 7) &= 1.4071 \text{ for } i = 5\% \\ &= 1.5036 \text{ for } i = 6\% \\ &= 1.6058 \text{ for } i = 7\%\end{aligned}$$

Therefore, $i \approx \boxed{6.0\%}$

The answer is (D).

11.

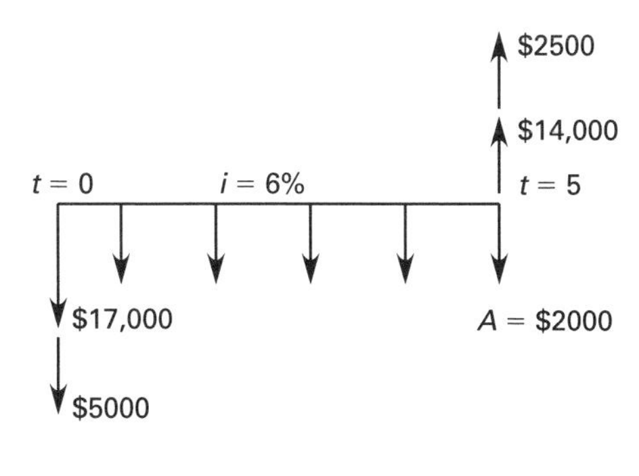

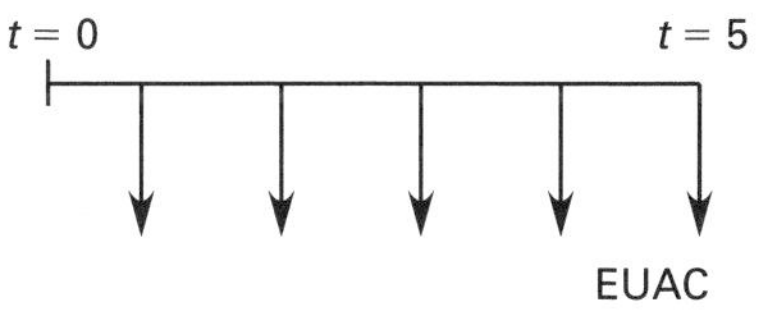

Annual cost of ownership, EUAC, can be obtained by the factors converting P to A and F to A.

$$\begin{aligned}P &= \$17{,}000 + \$5000 \\ &= \$22{,}000 \\ F &= \$14{,}000 + \$2500 \\ &= \$16{,}500 \\ \text{EUAC} &= A + P(A/P, 6\%, 5) - F(A/F, 6\%, 5) \\ (A/P, 6\%, 5) &= 0.2374 \\ (A/F, 6\%, 5) &= 0.1774 \\ \text{EUAC} &= \$2000 + (\$22{,}000)(0.2374) \\ &\quad - (\$16{,}500)(0.1774) \\ &= \boxed{\$4295.70 \quad (\$4300)}\end{aligned}$$

The answer is (C).

12.

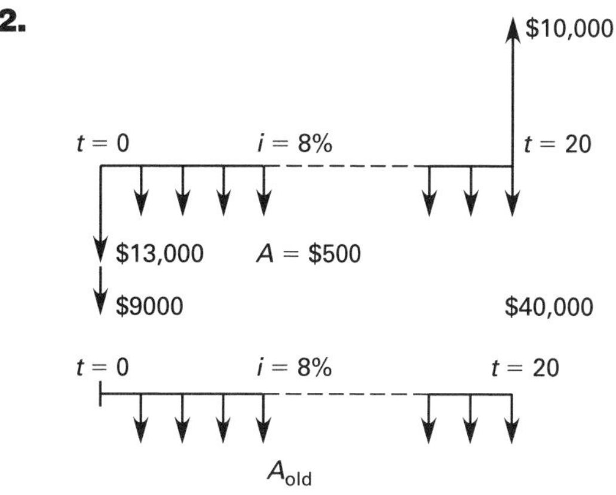

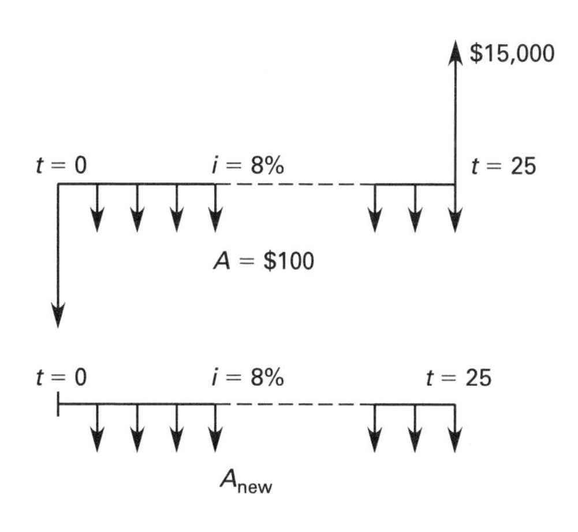

Consider the salvage value as a benefit lost (cost).

$$\begin{aligned}\text{EUAC}_{\text{old}} &= \$500 + (\$9000 + \$13{,}000)(A/P, 8\%, 20) \\ &\quad - (\$10{,}000)(A/F, 8\%, 20)\end{aligned}$$

$$(A/P, 8\%, 20) = 0.1019$$

$$(A/F, 8\%, 20) = 0.0219$$

$$\begin{aligned}\text{EUAC}_{\text{old}} &= \$500 + (\$22{,}000)(0.1019) \\ &\quad - (\$10{,}000)(0.0219) \\ &= \$2522.80\end{aligned}$$

Similarly,

$$\begin{aligned}\text{EUAC}_{\text{new}} &= \$100 + (\$40{,}000)(A/P, 8\%, 25) \\ &\quad - (\$15{,}000)(A/F, 8\%, 25)\end{aligned}$$

$$(A/P, 8\%, 25) = 0.0937$$

$$(A/F, 8\%, 25) = 0.0137$$

$$\begin{aligned}\text{EUAC}_{\text{new}} &= \$100 + (\$40{,}000)(0.0937) \\ &\quad - (\$15{,}000)(0.0137) \\ &= \$3642.50\end{aligned}$$

Therefore, the new bridge is more costly.

$$\boxed{\text{The best alternative is to strengthen the old bridge.}}$$

The answer is (A).

13.

$32,000

t = 0

t = 15

$7530

$150,000

The annual depreciation is

$$D = \frac{C - S_n}{n} = \frac{\$150{,}000}{15} = \$10{,}000/\text{year}$$

The taxable income is

$$\$32{,}000 - \$7530 - \$10{,}000 = \$14{,}470/\text{year}$$

Taxes paid are

$$(\$14{,}470)(0.48) = \$6945.60/\text{year}$$

The after-tax cash flow is

$$\$32{,}000 - \$7530 - \$6945.60 = \$17{,}524.40$$

The present worth of the alternate is zero when evaluated at its ROR.

$$0 = -\$150{,}000 + (\$17{,}524.40)(P/A, i\%, 15)$$

Therefore,

$$(P/A, i\%, 15) = \frac{\$150{,}000}{\$17{,}524.40} = 8.55949$$

Searching App. 74.B, this factor matches $i = 8\%$.

$$\boxed{\text{ROR} = 8.0\%}$$

The answer is (A).

14. (a) The conventional benefit/cost ratio is

$$B/C = \frac{B - D}{D}$$

The benefit/cost ratio will be

$$B/C = \frac{\$1{,}500{,}000 - \$300{,}000}{\$1{,}000{,}000} = \boxed{1.2}$$

The answer is (C).

(b) The excess of benefits over cost is $\boxed{\$200{,}000.}$

The answer is (A).

15. The annual rent is

$$(\$75)\left(12\ \frac{\text{months}}{\text{year}}\right) = \$900$$

$$P = P_1 + P_2 = \$15{,}000$$

$$A_1 = -\$900$$

$$A_2 = \$250 + \$150 = \$400$$

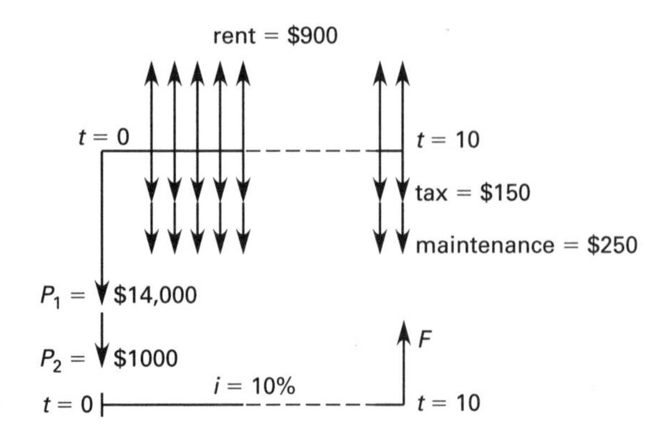

Use App. 74.B.

$$F = (\$15{,}000)(F/P, 10\%, 10) + (\$400)(F/A, 10\%, 10) - (\$900)(F/A, 10\%, 10)$$

$$(F/P, 10\%, 10) = 2.5937$$

$$(F/A, 10\%, 10) = 15.9374$$

$$F = (\$15{,}000)(2.5937) + (\$400)(15.9374) - (\$900)(15.9374) = \boxed{\$30{,}937 \quad (\$31{,}000)}$$

The answer is (B).

16.

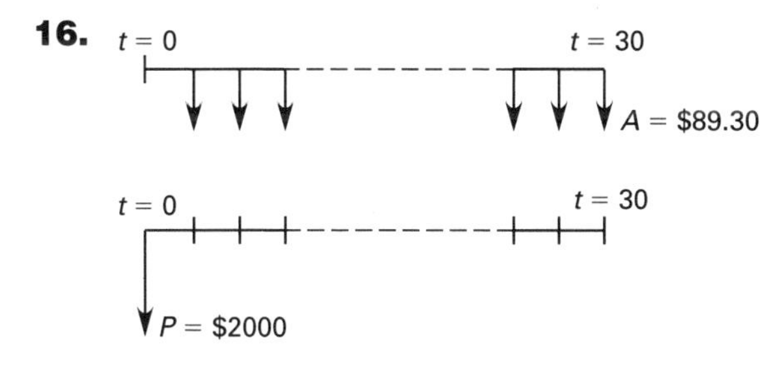

From Table 74.1,

$$P = A\left(\frac{(1+i)^n - 1}{i(1+i)^n}\right)$$

$$\frac{(1+i)^{30} - 1}{i(1+i)^{30}} = \frac{\$2000}{\$89.30} = 22.40$$

By trial and error,

$i\%$	$(1+i)^{30}$	$\frac{(1+i)^{30}-1}{i(1+i)^{30}}$
10	17.45	9.43
6	5.74	13.76
4	3.24	17.29
2	1.81	22.40

2% per month is close.

$$i = (1 + 0.02)^{12} - 1 = \boxed{0.2682 \quad (27\%)}$$

The answer is (A).

17. (a) Use the straight line method, Eq. 74.25.

$$D = \frac{C - S_n}{n}$$

Each year depreciation will be the same.

$$D = \frac{\$500{,}000 - \$100{,}000}{25} = \boxed{\$16{,}000}$$

The answer is (B).

(b) Use Eq. 74.27.

$$T = \tfrac{1}{2}n(n+1) = \left(\tfrac{1}{2}\right)(25)(25+1) = 325$$

Sum-of-the-years' digits (SOYD) depreciation can be calculated from Eq. 74.28.

$$D_j = \frac{(C - S_n)(n - j + 1)}{T}$$

$$D_1 = \frac{(\$500{,}000 - \$100{,}000)(25 - 1 + 1)}{325} = \boxed{\$30{,}769 \quad (\$31{,}000)}$$

$$D_2 = \frac{(\$500{,}000 - \$100{,}000)(25 - 2 + 1)}{325} = \boxed{\$29{,}538 \quad (\$30{,}000)}$$

$$D_3 = \frac{(\$500{,}000 - \$100{,}000)(25 - 3 + 1)}{325} = \boxed{\$28{,}308 \quad (\$28{,}000)}$$

The answer is (C).

(c) The double declining balance (DDB) method can be used. By Eq. 74.32,

$$D_j = dC(1 - d)^{j-1}$$

Use Eq. 74.31.

$$d = \frac{2}{n} = \frac{2}{25}$$

$$D_1 = \left(\tfrac{2}{25}\right)(\$500{,}000)\left(1 - \tfrac{2}{25}\right)^0 = \boxed{\$40{,}000}$$

$$D_2 = \left(\tfrac{2}{25}\right)(\$500{,}000)\left(1 - \tfrac{2}{25}\right)^1 = \boxed{\$36{,}800 \quad (\$37{,}000)}$$

$$D_3 = \left(\tfrac{2}{25}\right)(\$500{,}000)\left(1 - \tfrac{2}{25}\right)^2 = \boxed{\$33{,}856 \quad (\$34{,}000)}$$

The answer is (D).

18.

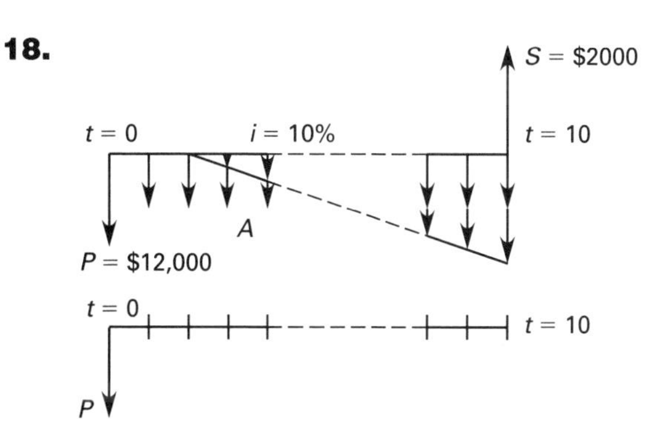

(a) A is \$1000, and G is \$200 for $t = n - 1 = 9$ years. With $F = S = \$2000$, the present worth is

$$\begin{aligned} P &= \$12{,}000 + A(P/A, 10\%, 10) + G(P/G, 10\%, 10) \\ &\quad - F(P/F, 10\%, 10) \\ &= \$12{,}000 + (\$1000)(6.1446) + (\$200)(22.8913) \\ &\quad - (\$2000)(0.3855) \\ &= \boxed{\$21{,}952 \quad (\$22{,}000)} \end{aligned}$$

t = 0
t = 10
A

The answer is (D).

(b) The annual cost is

$$\begin{aligned} A &= (\$12{,}000)(A/P, 10\%, 10) + \$1000 \\ &\quad + (\$200)(A/G, 10\%, 10) \\ &\quad - (\$2000)(A/F, 10\%, 10) \\ &= (\$12{,}000)(0.1627) + \$1000 + (\$200)(3.7255) \\ &\quad - (\$2000)(0.0627) \\ &= \boxed{\$3572.10 \quad (\$3600)} \end{aligned}$$

The answer is (C).

19. An increase in rock removal capacity can be achieved by a 20-year loan (investment). Different cases available can be compared by equivalent uniform annual cost (EUAC).

$$\begin{aligned} \text{EUAC} &= \text{annual loan cost} \\ &\quad + \text{expected annual damage} \\ &= \text{cost}\,(A/P, 10\%, 20) \\ &\quad + (\$25{,}000)(\text{probability}) \end{aligned}$$

$$(A/P, 10\%, 20) = 0.1175$$

A table can be prepared for different cases.

rock removal rate	cost (\$)	annual loan cost (\$)	expected annual damage (\$)	EUAC (\$)
7	0	0	3750	3750.00
8	15,000	1761.90	2500	4261.90
9	20,000	2349.20	1750	4099.20
10	30,000	3523.80	750	4273.80

$\boxed{\text{It is cheapest to do nothing.}}$

The answer is (A).

20. (a) Calculate the cost of owning and operating for years one and two.

$$\begin{aligned} A_1 &= (\$10{,}000)(A/P, 20\%, 1) + \$2000 \\ &\quad - (\$8000)(A/F, 20\%, 1) \end{aligned}$$

$$(A/P, 20\%, 1) = 1.2$$
$$(A/F, 20\%, 1) = 1.0$$

$$\begin{aligned} A_1 &= (\$10{,}000)(1.2) + \$2000 - (\$8000)(1.0) \\ &= \$6000 \end{aligned}$$

$$\begin{aligned} A_2 &= (\$10{,}000)(A/P, 20\%, 2) + \$2000 \\ &\quad + (\$1000)(A/G, 20\%, 2) \\ &\quad - (\$7000)(A/F, 20\%, 2) \end{aligned}$$

$$(A/P, 20\%, 2) = 0.6545$$
$$(A/G, 20\%, 2) = 0.4545$$
$$(A/F, 20\%, 2) = 0.4545$$

$$\begin{aligned} A_2 &= (\$10{,}000)(0.6545) + \$2000 \\ &\quad + (\$1000)(0.4545) - (\$7000)(0.4545) \\ &= \boxed{\$5818 \quad (\$5800)} \end{aligned}$$

The answer is (C).

(b) Calculate the cost of owning and operating for years three through five.

$$\begin{aligned} A_3 &= (\$10{,}000)(A/P, 20\%, 3) + \$2000 \\ &\quad + (\$1000)(A/G, 20\%, 3) \\ &\quad - (\$6000)(A/F, 20\%, 3) \end{aligned}$$

$$(A/P, 20\%, 3) = 0.4747$$
$$(A/G, 20\%, 3) = 0.8791$$
$$(A/F, 20\%, 3) = 0.2747$$

$$\begin{aligned} A_3 &= (\$10{,}000)(0.4747) + \$2000 \\ &\quad + (\$1000)(0.8791) \\ &\quad - (\$6000)(0.2747) \\ &= \$5977.90 \end{aligned}$$

$$\begin{aligned} A_4 &= (\$10{,}000)(A/P, 20\%, 4) \\ &\quad + \$2000 + (\$1000)(A/G, 20\%, 4) \\ &\quad - (\$5000)(A/F, 20\%, 4) \end{aligned}$$

$$(A/P, 20\%, 4) = 0.3863$$
$$(A/G, 20\%, 4) = 1.2742$$
$$(A/F, 20\%, 4) = 0.1863$$

$$\begin{aligned} A_4 &= (\$10{,}000)(0.3863) + \$2000 \\ &\quad + (\$1000)(1.2742) - (\$5000)(0.1863) \\ &= \$6205.70 \end{aligned}$$

Economics

$$A_5 = (\$10{,}000)(A/P, 20\%, 5) + \$2000 + (\$1000)(A/G, 20\%, 5) - (\$4000)(A/F, 20\%, 5)$$

$$(A/P, 20\%, 5) = 0.3344$$
$$(A/G, 20\%, 5) = 1.6405$$
$$(A/F, 20\%, 5) = 0.1344$$

$$A_5 = (\$10{,}000)(0.3344) + \$2000 + (\$1000)(1.6405) - (\$4000)(0.1344) = \boxed{\$6446.90 \quad (\$6500)}$$

The answer is (B).

(c) Since the annual owning and operating cost is smallest after two years of operation, it is advantageous to sell the mechanism after the second year.

$$\boxed{\text{The economic life is two years.}}$$

The answer is (B).

(d) After four years of operation, the owning and operating cost of the mechanism for one more year will be

$$A = \$6000 + (\$5000)(1 + i) - \$4000$$
$$i = 0.2 \quad (20\%)$$
$$A = \$6000 + (\$5000)(1.2) - \$4000 = \boxed{\$8000}$$

The answer is (C).

21. (a) To find out if the reimbursement is adequate, calculate the business-related expense.

Charge the company for business travel.

insurance:	$3000 – $2000 = $1000
maintenance:	$2000 – $1500 = $500
drop in salvage value:	$10,000 – $5000 = $5000

The annual portion of the drop in salvage value is

$$A = (\$5000)(A/F, 10\%, 5)$$
$$(A/F, 10\%, 5) = 0.1638$$
$$A = (\$5000)(0.1638) = \$819/\text{yr}$$

The annual cost of gas is

$$\left(\frac{50{,}000 \text{ mi}}{15 \ \frac{\text{mi}}{\text{gal}}}\right)\left(\frac{\$1.50}{\text{gal}}\right) = \$5000$$

$$\text{EUAC per mile} = \frac{\$1000 + \$500 + \$819 + \$5000}{50{,}000 \text{ mi}} = \$0.14638/\text{mi}$$

Since the reimbursement per mile was $0.30 and since $\$0.30 > \$0.14638$, the reimbursement is $\boxed{\text{adequate.}}$

(b) Determine (with reimbursement) how many miles the car must be driven to break even.

If the car is driven M miles per year,

$$\left(\frac{\$0.30}{1 \text{ mi}}\right)M = (\$50{,}000)(A/P, 10\%, 5) + \$2500 + \$2000 - (\$8000)(A/F, 10\%, 5) + \left(\frac{M}{15 \ \frac{\text{mi}}{\text{gal}}}\right)(\$1.50)$$

$$(A/P, 10\%, 5) = 0.2638$$
$$(A/F, 10\%, 5) = 0.1638$$

$$0.3M = (\$50{,}000)(0.2638) + \$2500 + \$2000 - (\$8000)(0.1638) + 0.1M$$
$$0.2M = \$16{,}379.60$$
$$M = \frac{\$16{,}379.60}{\frac{\$0.20}{1 \text{ mi}}} = \boxed{81{,}898 \text{ mi} \quad (82{,}000 \text{ mi})}$$

The answer is (C).

22. The present worth of alternative A is

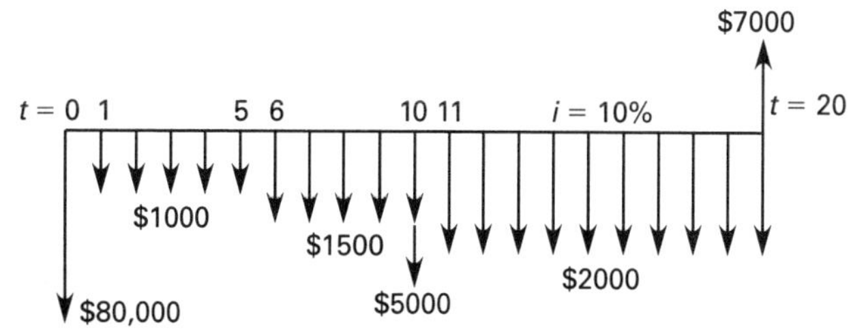

$$P_A = \$80{,}000 + (\$1000)(P/A, 10\%, 5) + (\$1500)(P/A, 10\%, 5)(P/F, 10\%, 5) + (\$2000)(P/A, 10\%, 10)(P/F, 10\%, 10) + (\$5000)(P/F, 10\%, 10) - (\$7000)(P/F, 10\%, 20)$$

$$(P/A,10\%,5) = 3.7908$$
$$(P/F,10\%,5) = 0.6209$$
$$(P/A,10\%,10) = 6.1446$$
$$(P/F,10\%,10) = 0.3855$$
$$(P/F,10\%,20) = 0.1486$$

$$\begin{aligned} P_{\text{A}} &= \$80{,}000 + (\$1000)(3.7908) \\ &\quad + (\$1500)(3.7908)(0.6209) \\ &\quad + (\$2000)(6.1446)(0.3855) \\ &\quad + (\$5000)(0.3855) \\ &\quad - (\$7000)(0.1486) \\ &= \boxed{\$92{,}946.15 \quad (\$93{,}000)} \end{aligned}$$

The answer is (B).

(b) Since the lives are different, compare by EUAC.

$$\begin{aligned} \text{EUAC(A)} &= (\$92{,}946.14)(A/P,10\%,20) \\ &= (\$92{,}946.14)(0.1175) \\ &= \boxed{\$10{,}921 \quad (\$11{,}000)} \end{aligned}$$

The answer is (B).

(c) Evaluate alternative B.

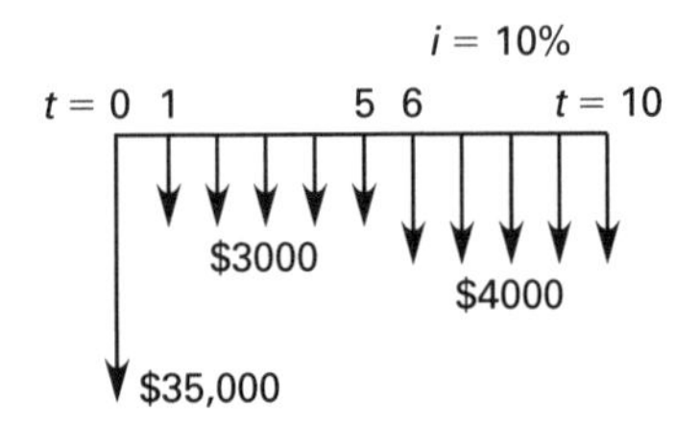

$$\begin{aligned} P_{\text{B}} &= \$35{,}000 + (\$3000)(P/A,10\%,5) \\ &\quad + (\$4000)(P/A,10\%,5)(P/F,10\%,5) \end{aligned}$$
$$(P/A,10\%,5) = 3.7908$$
$$(P/F,10\%,5) = 0.6209$$
$$\begin{aligned} P_{\text{B}} &= \$35{,}000 + (\$3000)(3.7908) \\ &\quad + (\$4000)(3.7908)(0.6209) \\ &= \boxed{\$55{,}787.23 \quad (\$56{,}000)} \end{aligned}$$

The answer is (A).

(d) Since the lives are different, compare by EUAC.

$$\begin{aligned} \text{EUAC(B)} &= (\$55{,}787.23)(A/P,10\%,10) \\ &= (\$55{,}787.23)(0.1627) \\ &= \boxed{\$9077 \quad (\$9100)} \end{aligned}$$

The answer is (A).

(e) Since EUAC(B) < EUAC(A),

$$\boxed{\text{Alternative B is economically superior.}}$$

The answer is (B).

23. (a) If the annual cost is compared with a total annual mileage of M, for plan A,

$$A_{\text{A}} = \boxed{\$0.25M}$$

(b) For plan B,

$$\begin{aligned} A_{\text{B}} &= (\$30{,}000)(A/P,10\%,3) + \$0.14M \\ &\quad + \$500 - (\$7200)(A/F,10\%,3) \end{aligned}$$
$$(A/P,10\%,3) = 0.4021$$
$$(A/F,10\%,3) = 0.3021$$
$$\begin{aligned} A_{\text{B}} &= (\$30{,}000)(0.4021) + \$0.14M + \$500 \\ &\quad - (\$7200)(0.3021) \\ &= \boxed{\$10{,}387.88 + \$0.14M} \end{aligned}$$

(c) For an equal annual cost $A_{\text{A}} = A_{\text{B}}$,

$$\$0.25M = \$10{,}387.88 + \$0.14M$$
$$\$0.11M = \$10{,}387.88$$
$$M = 94{,}435 \quad (94{,}000)$$

An annual mileage would be $\boxed{M = 94{,}000 \text{ mi.}}$

(d) For an annual mileage less than that, $A_{\text{A}} < A_{\text{B}}$.

$$\boxed{\text{Plan A is economically superior until 94,000 mi is exceeded.}}$$

The answer is (A).

24. (a) Method A:

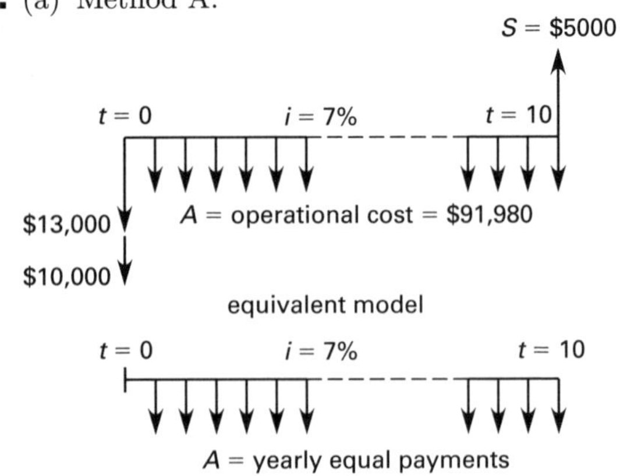

$$24\text{ hr/day}$$
$$365\text{ days/yr}$$
$$\text{total of }(24)(365) = 8760\text{ hr/yr}$$
$$\$10.50\text{ operational cost/hr}$$
$$\text{total of }(8760)(\$10.50) = \$91{,}980\text{ operational cost/yr}$$

$$\begin{aligned} A &= \$91{,}980 + (\$23{,}000)(A/P,7\%,10) \\ &\quad - (\$5000)(A/F,7\%,10) \end{aligned}$$

$$(A/P,7\%,10) = 0.1424$$
$$(A/F,7\%,10) = 0.0724$$

$$\begin{aligned} A &= \$91{,}980 + (\$23{,}000)(0.1424) \\ &\quad - (\$5000)(0.0724) \\ &= \$94{,}893.20/\text{yr} \end{aligned}$$

Therefore, the uniform annual cost per ton each year will be

$$\frac{\$94{,}893.20}{50\text{ ton}} = \boxed{\$1897.86 \quad (\$1900)}$$

The answer is (B).

(b) Method B:

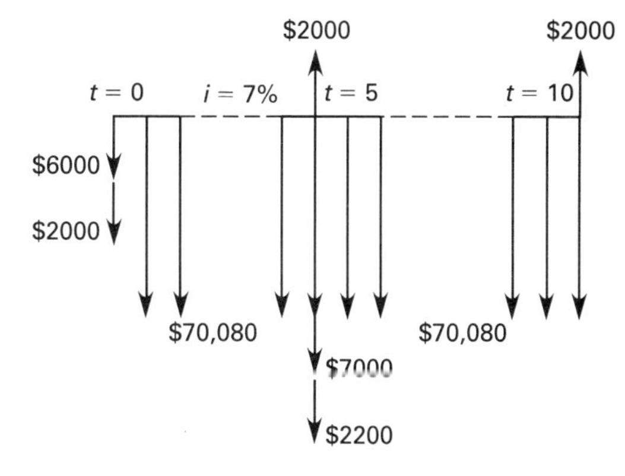

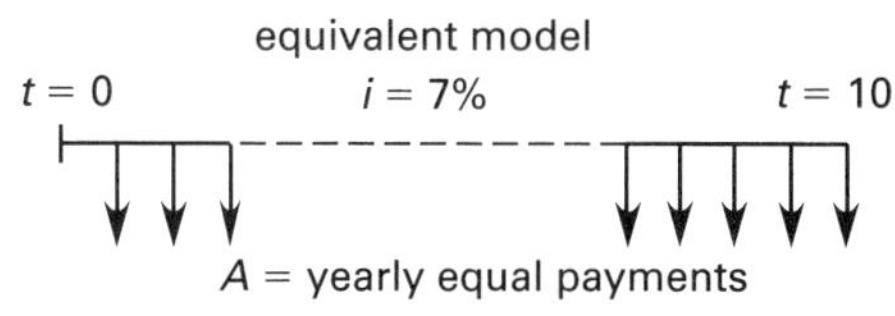

$$8760\text{ hr/yr}$$
$$\$8\text{ operational cost/hr}$$
$$\text{total of }\$70{,}080\text{ operational cost/yr}$$

$$\begin{aligned} A &= \$70{,}080 + (\$6000 + \$2000) \\ &\quad \times (A/P,7\%,10) \\ &\quad + (\$7000 + \$2200 - \$2000) \\ &\quad \times (P/F,7\%,5)(A/P,7\%,10) \\ &\quad - (\$2000)(A/F,7\%,10) \end{aligned}$$

$$(A/P,7\%,10) = 0.1424$$
$$(A/F,7\%,10) = 0.0724$$
$$(P/F,7\%,5) = 0.7130$$

$$\begin{aligned} A &= \$70{,}080 + (\$8000)(0.1424) \\ &\quad + (\$7200)(0.7130)(0.1424) \\ &\quad - (\$2000)(0.0724) \\ &= \$71{,}805.42/\text{yr} \end{aligned}$$

Therefore, the uniform annual cost per ton each year will be

$$\frac{\$71{,}805.42}{20\text{ ton}} = \boxed{\$3590.27 \quad (\$3600)}$$

The answer is (B).

(c) The following table can be used to determine which alternative is least expensive for each throughput range.

tons/yr	cost of using A	cost of using B	cheapest
0–20	\$94,893 (1×)	\$71,805 (1×)	B
20–40	\$94,893 (1×)	\$143,610 (2×)	A
40–50	\$94,893 (1×)	\$215,415 (3×)	A
50–60	\$189,786 (2×)	\$215,415 (3×)	A
60–80	\$189,786 (2×)	\$287,220 (4×)	A

25.

$$\begin{aligned} A_e &= (\$60{,}000)(A/P,7\%,20) + A \\ &\quad + G(P/G,7\%,20)(A/P,7\%,20) \\ &\quad - (\$10{,}000)(A/F,7\%,20) \end{aligned}$$

$$(A/P,7\%,20) = 0.0944$$

$$A = (37{,}440\text{ mi})\left(\frac{\$1.0}{1\text{ mi}}\right) = \$37{,}440$$

$$G = 0.1A = (0.1)(\$37{,}440) = \$3744$$

$$(P/G,7\%,20) = 77.5091$$
$$(A/F,7\%,20) = 0.0244$$

$$\begin{aligned} A_e &= (\$60{,}000)(0.0944) + \$37{,}440 \\ &\quad + (\$3744)(77.5091)(0.0944) \\ &\quad - (\$10{,}000)(0.0244) \\ &= \$70{,}254.32 \end{aligned}$$

S = \$10,000
t = 0
i = 7%
t = 20
G = 0.1A = \$3744
A = \$37,440
P = \$60,000
equivalent model
A_e
t = 0
t = 20

(a) With 80,000 passengers a year, the break-even fare per passenger would be

$$\text{fare} = \frac{A_e}{80{,}000} = \frac{\$70{,}254.32}{80{,}000}$$
$$= \boxed{\$0.878/\text{passenger} \quad (\$0.88/\text{passenger})}$$

The answer is (D).

(b) The passenger fare should go up each year by

$$\$0.878 = \$0.35 + G(A/G, 7\%, 20)$$
$$G = \frac{\$0.878 - \$0.35}{7.3163}$$
$$= \boxed{\$0.072 \text{ increase per year}}$$

The answer is (D).

(c) As in part (b), the subsidy should be

$$\text{subsidy} = \text{cost} - \text{revenue}$$
$$P = \$0.878 - \big(\$0.35 + (\$0.05)(A/G, 7\%, 20)\big)$$
$$= \$0.878 - \big(\$0.35 + (\$0.05)(7.3163)\big)$$
$$= \boxed{\$0.162 \quad (\$0.16)}$$

The answer is (C).

26. Use the present worth comparison method.

$$P(\text{A}) = (\$4800)(P/A, 12\%, 25)$$
$$= (\$4800)(7.8431)$$
$$= \$37{,}646.88$$

(4 quarters)(25 years) = 100 compounding periods

$$P(\text{B}) = (\$1200)(P/A, 3\%, 100)$$
$$= (\$1200)(31.5989)$$
$$= \$37{,}918.68$$

$$\boxed{\text{Alternative B is economically superior.}}$$

The answer is (B).

27. Use the equivalent uniform annual cost method.

$$\text{EUAC(A)} = (\$1500)(A/P, 15\%, 5) + \$800$$
$$= (\$1500)(0.2983) + \$800$$
$$= \$1247.45$$
$$\text{EUAC(B)} = (\$2500)(A/P, 15\%, 8) + \$650$$
$$= (\$2500)(0.2229) + \$650$$
$$= \$1207.25$$

$$\boxed{\text{Alternative B is economically superior.}}$$

The answer is (B).

28. The data given imply that both investments return 4% or more. However, the increased investment of \$30,000 may not be cost effective. Do an incremental analysis.

$$\text{incremental cost} = \$70{,}000 - \$40{,}000 = \$30{,}000$$
$$\text{incremental income} = \$5620 - \$4075 = \$1545$$
$$0 = -\$30{,}000 + (\$1545)(P/A, i\%, 20)$$
$$(P/A, i\%, 20) = 19.417$$
$$i \approx 0.25\% < 4\%$$

$$\boxed{\text{Alternative B is economically superior.}}$$

(The same conclusion could be reached by taking the present worths of both alternatives at 4%.)

The answer is (B).

29. (a) The present worth comparison is

$$P(\text{A}) = (-\$3000)(P/A, 25\%, 25) - \$20{,}000$$
$$= (-\$3000)(3.9849) - \$20{,}000$$
$$= -\$31{,}954.70$$
$$P(\text{B}) = (-\$2500)(3.9849) - \$25{,}000$$
$$= -\$34{,}962.25$$

$$\boxed{\text{Alternative A is economically superior.}}$$

The answer is (A).

(b) The capitalized cost comparison is

$$\text{CC(A)} = \$20{,}000 + \frac{\$3000}{0.25} = \$32{,}000$$
$$\text{CC(B)} = \$25{,}000 + \frac{\$2500}{0.25} = \$35{,}000$$

$$\boxed{\text{Alternative A is economically superior.}}$$

The answer is (A).

(c) The annual cost comparison is

$$\text{EUAC(A)} = (\$20{,}000)(A/P, 25\%, 25) + \$3000$$
$$= (\$20{,}000)(0.2509) + \$3000$$
$$= \$8018.00$$
$$\text{EUAC(B)} = (\$25{,}000)(0.2509) + \$2500$$
$$= \$8772.50$$

$$\boxed{\text{Alternative A is economically superior.}}$$

The answer is (A).

30. (a) The depreciation in the first 5 years is

$$\text{BV} = \$18{,}000 - (5)\left(\frac{\$18{,}000 - \$2000}{8}\right) = \boxed{\$8000}$$

The answer is (D).

(b) With the sinking fund method, the basis is

$$\begin{aligned}(\$18{,}000 - \$2000)(A/F, 8\%, 8) &= (\$18{,}000 - \$2000)(0.0940)\\ &= \$1504\end{aligned}$$

$$D_1 = (\$1504)(1.000) = \$1504$$

$$D_2 = (\$1504)(1.0800) = \$1624$$

$$D_3 = (\$1504)(1.0800)^2 = \boxed{\$1754 \quad (\$1800)}$$

The answer is (C).

(c) Using double declining balance depreciation, the first five years' depreciation is

$$D_1 = \left(\tfrac{2}{8}\right)(\$18{,}000) = \$4500$$

$$D_2 = \left(\tfrac{2}{8}\right)(\$18{,}000 - \$4500) = \$3375$$

$$D_3 = \left(\tfrac{2}{8}\right)(\$18{,}000 - \$4500 - \$3375) = \$2531$$

$$D_4 = \left(\tfrac{2}{8}\right)(\$18{,}000 - \$4500 - \$3375 - \$2531) = \$1898$$

$$D_5 = \left(\tfrac{2}{8}\right)(\$18{,}000 - \$4500 - \$3375 - \$2531 - \$1898) = \$1424$$

The balance value at the fifth year is

$$\begin{aligned}\text{BV} &= \$18{,}000 - \$4500 - \$3375 - \$2531 - \$1898 - \$1424\\ &= \boxed{\$4272 \quad (\$4000)}\end{aligned}$$

The answer is (B).

31. The after-tax depreciation recovery is

$$T = \left(\tfrac{1}{2}\right)(8)(9) = 36$$

$$D_1 = \left(\tfrac{8}{36}\right)(\$14{,}000 - \$1800) = \$2711$$

$$\Delta D = \left(\tfrac{1}{36}\right)(\$14{,}000 - \$1800) = \$339$$

$$D_2 = \$2711 - \$339 = \$2372$$

$$\begin{aligned}\text{DR} &= (0.52)(\$2711)(P/A, 15\%, 8) - (0.52)(\$339)(P/G, 15\%, 8)\\ &= (0.52)(\$2711)(4.4873) - (0.52)(\$339)(12.4807)\\ &= \boxed{\$4125.74 \quad (\$4100)}\end{aligned}$$

The answer is (C).

32. Use the equivalent uniform annual cost method to find the best alternative.

$$(A/P, 12\%, 15) = 0.1468$$

$$\begin{aligned}\text{EUAC}_{5\,\text{ft}} &= (\$600{,}000)(0.1468) + \left(\tfrac{14}{50}\right)(\$600{,}000) + \left(\tfrac{8}{50}\right)(\$650{,}000)\\ &\quad + \left(\tfrac{3}{50}\right)(\$700{,}000) + \left(\tfrac{1}{50}\right)(\$800{,}000)\\ &= \$418{,}080\end{aligned}$$

$$\begin{aligned}\text{EUAC}_{10\,\text{ft}} &= (\$710{,}000)(0.1468) + \left(\tfrac{8}{50}\right)(\$650{,}000) + \left(\tfrac{3}{50}\right)(\$700{,}000)\\ &\quad + \left(\tfrac{1}{50}\right)(\$800{,}000)\\ &= \$266{,}228\end{aligned}$$

$$\begin{aligned}\text{EUAC}_{15\,\text{ft}} &= (\$900{,}000)(0.1468) + \left(\tfrac{3}{50}\right)(\$700{,}000) + \left(\tfrac{1}{50}\right)(\$800{,}000)\\ &= \$190{,}120\end{aligned}$$

$$\begin{aligned}\text{EUAC}_{20\,\text{ft}} &= (\$1{,}000{,}000)(0.1468) + \left(\tfrac{1}{50}\right)(\$800{,}000)\\ &= \$162{,}800\end{aligned}$$

$$\boxed{\text{Build to 20 ft.}}$$

The answer is (D).

33. Assume replacement after 1 year.

$$\begin{aligned}\text{EUAC}(1) &= (\$30{,}000)(A/P, 12\%, 1) + \$18{,}700\\ &= (\$30{,}000)(1.12) + \$18{,}700\\ &= \$52{,}300\end{aligned}$$

Assume replacement after 2 years.

$$\begin{aligned}\text{EUAC}(2) &= (\$30{,}000)(A/P, 12\%, 2) + \$18{,}700 + (\$1200)(A/G, 12\%, 2)\\ &= (\$30{,}000)(0.5917) + \$18{,}700 + (\$1200)(0.4717)\\ &= \$37{,}017\end{aligned}$$

Economics

Assume replacement after 3 years.

$$\begin{aligned}\text{EUAC}(3) &= (\$30{,}000)(A/P, 12\%, 3) \\ &\quad + \$18{,}700 + (\$1200)(A/G, 12\%, 3) \\ &= (\$30{,}000)(0.4163) + \$18{,}700 \\ &\quad + (\$1200)(0.9246) \\ &= \$32{,}299\end{aligned}$$

Similarly, calculate to obtain the numbers in the following table.

years in service	EUAC
1	\$52,300
2	\$37,017
3	\$32,299
4	\$30,207
5	\$29,152
6	\$28,602
7	\$28,335
8	\$28,234
9	\$28,240
10	\$28,312

$$\boxed{\text{Replace after 8 yr.}}$$

The answer is (C).

34. Since the head and horsepower data are already reflected in the hourly operating costs, there is no need to work with head and horsepower.

Let $N =$ no. of hours operated each year.

$$\begin{aligned}\text{EUAC(A)} &= (\$3600 + \$3050)(A/P, 12\%, 10) \\ &\quad - (\$200)(A/F, 12\%, 10) + 0.30N \\ &= (\$6650)(0.1770) - (\$200)(0.0570) + 0.30N \\ &= 1165.65 + 0.30N \\ \text{EUAC(B)} &= (\$2800 + \$5010)(A/P, 12\%, 10) \\ &\quad + (\$280)(A/F, 12\%, 10) + 0.10N \\ &= (\$7810)(0.1770) - (\$280)(0.0570) + 0.10N \\ &= 1366.41 + 0.10N\end{aligned}$$

$$\begin{aligned}\text{EUAC(A)} &= \text{EUAC(B)} \\ 1165.65 + 0.30N &= 1366.41 + 0.10N \\ N &= \boxed{1003.8 \text{ hr} \quad (1000 \text{ hr})}\end{aligned}$$

The answer is (A).

35. (a) From Eq. 74.88,

$$\begin{aligned}\frac{T_2}{T_1} &= 0.88 = 2^{-b} \\ \log 0.88 &= -b \log 2 \\ -0.0555 &= -(0.3010)b \\ b &= 0.1843\end{aligned}$$

$$T_4 = (6)(4)^{-0.1843} = \boxed{4.65 \text{ wk} \quad (4.5 \text{ wk})}$$

The answer is (A).

(b) From Eq. 74.89,

$$\begin{aligned}T_{6-14} &= \left(\frac{T_1}{1-b}\right)\left(\left(n_2 + \tfrac{1}{2}\right)^{1-b} - \left(n_1 - \tfrac{1}{2}\right)^{1-b}\right) \\ &= \left(\frac{6}{1 - 0.1843}\right) \\ &\quad \times \left(\left(14 + \tfrac{1}{2}\right)^{1-0.1843} - \left(6 - \tfrac{1}{2}\right)^{1-0.1843}\right) \\ &= \left(\frac{6}{0.8157}\right)(8.857 - 4.017) \\ &= \boxed{35.6 \text{ wk} \quad (35 \text{ wk})}\end{aligned}$$

The answer is (A).

36. (a) First check that both alternatives have an ROR greater than the MARR. Work in thousands of dollars. Evaluate alternative A.

$$\begin{aligned}P(\text{A}) &= -\$120 + (\$15)(P/F, i\%, 5) \\ &\quad + (\$57)(P/A, i\%, 5)(1 - 0.45) \\ &\quad + \left(\frac{\$120 - \$15}{5}\right)(P/A, i\%, 5)(0.45) \\ &= -\$120 + (\$15)(P/F, i\%, 5) \\ &\quad + (\$40.8)(P/A, i\%, 5)\end{aligned}$$

Try 15%.

$$\begin{aligned}P(\text{A}) &= -\$120 + (\$15)(0.4972) + (\$40.8)(3.3522) \\ &= \$24.23\end{aligned}$$

Try 25%.

$$\begin{aligned}P(\text{A}) &= -\$120 + (\$15)(0.3277) + (\$40.8)(2.6893) \\ &= -\$5.36\end{aligned}$$

Since $P(\text{A})$ goes through 0,

$$(\text{ROR})_{\text{A}} > \text{MARR} = 15\%$$

Next, evaluate alternative B.

$$\begin{aligned} P(\text{B}) &= -\$170 + (\$20)(P/F, i\%, 5) \\ &\quad + (\$67)(P/A, i\%, 5)(1 - 0.45) \\ &\quad + \left(\frac{\$170 - \$20}{5}\right)(P/A, i\%, 5)(0.45) \\ &= -\$170 + (\$20)(P/F, i\%, 5) \\ &\quad + (\$50.35)(P/A, i\%, 5) \end{aligned}$$

Try 15%.

$$\begin{aligned} P(\text{B}) &= -\$170 + (\$20)(0.4972) + (\$50.35)(3.3522) \\ &= \$8.73 \end{aligned}$$

Since $P(\text{B}) > 0$ and will decrease as i increases,

$$(\text{ROR})_\text{B} > 15\%$$

$$\boxed{\text{ROR} > \text{MARR for both alternatives.}}$$

The answer is (C).

(b) Do an incremental analysis to see if it is worthwhile to invest the extra $170 − $120 = $50.

$$\begin{aligned} P(\text{B} - \text{A}) &= -\$50 + (\$20 - \$15)(P/F, i\%, 5) \\ &\quad + (\$50.35 - \$40.8)(P/A, i\%, 5) \end{aligned}$$

Try 15%.

$$\begin{aligned} P(\text{B} - \text{A}) &= -\$50 + (\$5)(0.4972) \\ &\quad + (\$9.55)(3.3522) \\ &= -\$15.50 \end{aligned}$$

Since $P(\text{B} - \text{A}) < 0$ and would become more negative as i increases, the ROR of the added investment is greater than 15%.

$$\boxed{\text{Alternative A is superior.}}$$

The answer is (A).

37. Use the year-end convention with the tax credit. The purchase is made at $t = 0$. However, the credit is received at $t = 1$ and must be multiplied by $(P/F, i\%, 1)$.

$$\begin{aligned} P &= -\$300{,}000 + (0.0667)(\$300{,}000)(P/F, i\%, 1) \\ &\quad + (\$90{,}000)(P/A, i\%, 5)(1 - 0.48) \\ &\quad + \left(\frac{\$300{,}000 - \$50{,}000}{5}\right)(P/A, i\%, 5)(0.48) \\ &\quad + (\$50{,}000)(P/F, i\%, 5) \\ &= -\$300{,}000 + (\$20{,}010)(P/F, i\%, 1) \\ &\quad + (\$46{,}800)(P/A, i\%, 5) \\ &\quad + (\$24{,}000)(P/A, i\%, 5) \\ &\quad + (\$50{,}000)(P/F, i\%, 5) \end{aligned}$$

By trial and error,

i	P
10%	$17,625
15%	−$20,409
12%	$1456
13%	−$6134
$12^1/_4$%	−$472

$$\boxed{i \text{ is between } 12\% \text{ and } 12^1/_4\%.}$$

The answer is (C).

38. (a) The distributed profit is

$$\begin{aligned} \text{distributed profit} &= (0.15)(\$2{,}500{,}000) \\ &= \boxed{\$375{,}000 \quad (\$380{,}000)} \end{aligned}$$

The answer is (D).

(b) Find the annual loan payment.

$$\begin{aligned} \text{payment} &= (\$2{,}500{,}000)(A/P, 12\%, 25) \\ &= (\$2{,}500{,}000)(0.1275) \\ &= \$318{,}750 \end{aligned}$$

After paying all expenses and distributing the 15% profit, the remainder should be 0.

$$\begin{aligned} 0 &= \text{EUAC} \\ &= \$20{,}000 + \$50{,}000 + \$200{,}000 \\ &\quad + \$375{,}000 + \$318{,}750 - \text{annual receipts} \\ &\quad - (\$500{,}000)(A/F, 15\%, 25) \\ &= \$963{,}750 - \text{annual receipts} \\ &\quad - (\$500{,}000)(0.0047) \end{aligned}$$

Economics

This calculation assumes $i = 15\%$, which equals the desired return. However, this assumption only affects the salvage calculation, and since the number is so small, the analysis is not sensitive to the assumption.

$$\text{annual receipts} = \$961{,}400$$

The average daily receipts are

$$\frac{\$961{,}400}{365} = \boxed{\$2634 \quad (\$2600)}$$

The answer is (D).

(c) Use the expected value approach. The average occupancy is

$$\begin{aligned}(0.40)(0.65) &+ (0.30)(0.70) + (0.20)(0.75)\\ &+ (0.10)(0.80) = \boxed{0.70}\end{aligned}$$

The answer is (B).

(d) The average number of rooms occupied each night is

$$(0.70)(120 \text{ rooms}) = 84 \text{ rooms}$$

The minimum required average daily rate per room is

$$\frac{\$2634}{84} = \boxed{\$31.36 \quad (\$31)}$$

The answer is (D).

39. (a) The annual savings are

$$\begin{array}{c}\text{annual}\\ \text{savings}\end{array} = \left(\frac{0.69 - 0.47}{1000}\right)(\$3{,}500{,}000) = \boxed{\$770}$$

The answer is (D).

(b)

$$\begin{aligned}P &= -\$7500 + (\$770 - \$200 - \$100)\\ &\quad \times (P/A, i\%, 25) = 0\end{aligned}$$

$$(P/A, i\%, 25) = 15.957$$

Searching the tables and interpolating, the rate of return is

$$i \approx \boxed{3.75\% \quad (3.8\%)}$$

The answer is (A).

40. (a) Evaluate alternative A, working in millions of dollars.

$$\begin{aligned}P(\text{A}) &= -(\$1.3)(1 - 0.48)(P/A, 15\%, 6)\\ &= -(\$1.3)(0.52)(3.7845)\\ &= \boxed{-\$2.56 \quad (-\$2.6) \quad [\text{millions}]}\end{aligned}$$

The answer is (C).

(b) Use straight line depreciation to evaluate alternative B.

$$D_j = \frac{\$2}{6} = \$0.333$$

$$\begin{aligned}P(\text{B}) &= -\$2 - (0.20)(\$1.3)(1 - 0.48)(P/A, 15\%, 6)\\ &\quad - (\$0.15)(1 - 0.48)(P/A, 15\%, 6)\\ &\quad + (\$0.333)(0.48)(P/A, 15\%, 6)\\ &= -\$2 - (0.20)(\$1.3)(0.52)(3.7845)\\ &\quad - (\$0.15)(0.52)(3.7845)\\ &\quad + (\$0.333)(0.48)(3.7845)\\ &= \boxed{-\$2.202 \quad (-\$2.2) \quad [\text{millions}]}\end{aligned}$$

The answer is (D).

(c) Evaluate alternative C.

$$D_j = \frac{1.2}{3} = 0.4$$

$$\begin{aligned}P(\text{C}) &= -(\$1.2)\big(1 + (P/F, 15\%, 3)\big)\\ &\quad - (\$0.20)(\$1.3)(1 - 0.48)\\ &\quad \times \big((P/F, 15\%, 1) + (P/F, 15\%, 4)\big)\\ &\quad - (\$0.45)(\$1.3)(1 - 0.48)\\ &\quad \times \big((P/F, 15\%, 2) + (P/F, 15\%, 5)\big)\\ &\quad - (\$0.80)(\$1.3)(1 - 0.48)\\ &\quad \times \big((P/F, 15\%, 3) + (P/F, 15\%, 6)\big)\\ &\quad + (\$0.4)(\$0.48)(P/A, 15\%, 6)\\ &= -(\$1.2)(1.6575)\\ &\quad - (\$0.20)(\$1.3)(0.52)(0.8696 + 0.5718)\\ &\quad - (\$0.45)(\$1.3)(0.52)(0.7561 + 0.4972)\\ &\quad - (\$0.80)(\$1.3)(0.52)(0.6575 + 0.4323)\\ &\quad + (\$0.4)(0.48)(3.7845)\\ &= \boxed{-\$2.428 \quad (-\$2.4) \quad [\text{millions}]}\end{aligned}$$

The answer is (C).

(d) From parts (a) through (c), $\boxed{\text{alternative B is superior.}}$

The answer is (B).

41. This is a replacement study. Since production capacity and efficiency are not a problem with the defender, the only question is when to bring in the challenger.

Since this is a before-tax problem, depreciation is not a factor, nor is book value.

The cost of keeping the defender one more year is

$$\begin{aligned} \text{EUAC(defender)} &= \$200{,}000 + (0.15)(\$400{,}000) \\ &= \$260{,}000 \end{aligned}$$

For the challenger,

$$\begin{aligned} &\text{EUAC(challenger)} \\ &\quad = (\$800{,}000)(A/P, 15\%, 10) + \$40{,}000 \\ &\qquad + (\$30{,}000)(A/G, 15\%, 10) \\ &\qquad - (\$100{,}000)(A/F, 15\%, 10) \\ &\quad = (\$800{,}000)(0.1993) + \$40{,}000 \\ &\qquad + (\$30{,}000)(3.3832) \\ &\qquad - (\$100{,}000)(0.0493) \\ &\quad = \$296{,}006 \end{aligned}$$

Since the defender is cheaper, keep it. The same analysis next year will give identical answers. Therefore, keep the defender for the next 3 years, at which time the decision to buy the challenger will be automatic.

Having determined that it is less expensive to keep the defender than to maintain the challenger for 10 years, determine whether the challenger is less expensive if retired before 10 years.

If retired in 9 years,

$$\begin{aligned} \text{EUAC(challenger)} &= (\$800{,}000)(A/P, 15\%, 9) + \$40{,}000 \\ &\quad + (\$30{,}000)(A/G, 15\%, 9) \\ &\quad - (\$150{,}000)(A/F, 15\%, 9) \\ &= (\$800{,}000)(0.2096) \\ &\quad + \$40{,}000 + (\$30{,}000)(3.0922) \\ &\quad - (\$150{,}000)(0.0596) \\ &= \$291{,}506 \end{aligned}$$

Similar calculations yield the following results for all the retirement dates.

n	EUAC
10	\$296,000
9	\$291,506
8	\$287,179
7	\$283,214
6	\$280,016
5	\$278,419
4	\$279,909
3	\$288,013
2	\$313,483
1	\$360,000

Since none of these equivalent uniform annual costs are less than that of the defender, it is not economical to buy and keep the challenger for any length of time.

$$\boxed{\text{Keep the defender.}}$$

42. The annual demand for valves is

$$\begin{aligned} D &= (500{,}000 \text{ products})\left(3\ \frac{\text{valves}}{\text{product}}\right) \\ &= 1{,}500{,}000 \text{ valves} \end{aligned}$$

The cost of holding (storing) a valve for one year is

$$h = \left(0.4\ \frac{1}{\text{yr}}\right)(\$6.50) = \$2.60\ 1/\text{yr}$$

The most economical quantity of valves to order is

$$\begin{aligned} Q^* &= \sqrt{\frac{2aK}{h}} = \sqrt{\frac{(2)\left(1{,}500{,}000\ \frac{1}{\text{yr}}\right)(\$49.50)}{\$2.60\ \frac{1}{\text{yr}}}} \\ &= \boxed{7557 \text{ valves} \quad (7600 \text{ valves})} \end{aligned}$$

The answer is (D).

43. With just-in-time manufacturing, production is one-at-a-time, according to demand. When one is needed, one is made. In order to make the EOQ approach zero, the $\boxed{\text{setup cost}}$ must approach zero.

The answer is (D).

75 Professional Services, Contracts, and Engineering Law

PRACTICE PROBLEMS

1. List the different forms of company ownership. What are the advantages and disadvantages of each?

2. Define the requirements for a contract to be enforceable.

3. What standard features should a written contract include?

4. Describe the ways a consulting fee can be structured.

5. What is a retainer fee?

6. Which of the following organizations is NOT a contributor to the standard design and construction contract documents developed by the Engineers Joint Contract Documents Committee (EJCDC)?

(A) National Society of Professional Engineers

(B) Construction Specifications Institute

(C) Associated General Contractors of America

(D) American Institute of Architects

7. To be affected by the Fair Labor Standards Act (FLSA) and be required to pay minimum wage, construction firms working on bridges and highways must generally have

(A) 1 or more employees

(B) 2 or more employees and annual gross billings of $500,000

(C) 10 or more employees and be working on federally funded projects

(D) 50 or more employees and have been in business for longer than 6 months

8. A "double-breasted" design firm

(A) has errors and omissions as well as general liability insurance coverage

(B) is licensed to practice in both engineering and architecture

(C) serves both union and nonunion clients

(D) performs post-construction certification for projects it did not design

9. The phrase "without expressed authority" means which of the following when used in regards to partnerships of design professionals?

(A) Each full member of a partnership is a general agent of the partnership and has complete authority to make binding commitments, enter into contracts, and otherwise act for the partners within the scope of the business.

(B) The partnership may act in a manner that it considers best for the client, even though the client has not been consulted.

(C) Only plans, specifications, and documents that have been signed and stamped (sealed) by the authority of the licensed engineer may be relied upon.

(D) Only officers to the partnership may obligate the partnership.

10. A limited partnership has 1 managing general partner, 2 general partners, 1 silent partner, and 3 limited partners. If all partners cast a single vote when deciding on an issue, how many votes will be cast?

(A) 1

(B) 3

(C) 4

(D) 7

11. Which of the following are characteristics of a limited liability company (LLC)?

I. limited liability for all members

II. no taxation as an entity (no double taxation)

III. more than one class of stock

IV. limited to fewer than 25 members

V. fairly easy to establish

VI. no "continuity of life" like regular corporation

(A) I, II, and IV

(B) I, II, IV, and VI

(C) I, II, III, IV, and V

(D) I, II, III, V, and VI

12. Which of the following statements is FALSE in regard to joint ventures?

(A) Members of a joint venture may be any combination of sole proprietorships, partnerships, and corporations.

(B) A joint venture is a business entity separate from its members.

(C) A joint venture spreads risk and rewards, and it pools expertise, experience, and resources. However, bonding capacity is not aggregated.

(D) A joint venture usually dissolves after the completion of a specific project.

13. Which of the following construction business types can have unlimited shareholders?

I. S corporation

II. LLC

III. corporation

IV. sole proprietorship

(A) II and III only

(B) I, II, and III

(C) I, III, and IV

(D) I, II, III, and IV

14. The phrase "or approved equal" allows a contractor to

(A) substitute one connection design for another

(B) substitute a more expensive feature for another

(C) replace an open-shop subcontractor with a union subcontractor

(D) install a product whose brand name and model number are not listed in the specifications

15. Cities, other municipalities, and departments of transportation often have standard specifications, in addition to the specifications issued as part of the construction document set, that cover such items as

(A) safety requirements

(B) environmental requirements

(C) concrete, fire hydrant, manhole structures, and curb requirements

(D) procurement and accounting requirements

16. What is intended to prevent a contractor from bidding on a project and subsequently backing out after being selected for the project?

(A) publically recorded bid

(B) property lien

(C) surety bond

(D) proposal bond

17. Which of the following is illegal, in addition to being unethical?

(A) bid shopping

(B) bid peddling

(C) bid rigging

(D) bid unbalancing

18. Which of the following is NOT normally part of a construction contract?

I. invitation to bid

II. instructions to bidders

III. general conditions

IV. supplementary conditions

V. liability insurance policy

VI. technical specifications

VII. drawings

VIII. addenda

IX. proposals

X. bid bond

XI. agreement

XII. performance bond

XIII. labor and material payment bond

XIV. nondisclosure agreement

(A) I

(B) II

(C) V

(D) XIV

19. Once a contract has been signed by the owner and contractor, changes to the contract

(A) cannot be made

(B) can be made by the owner, but not by the contractor

(C) can be made by the contractor, but not by the owner

(D) can be made by both the owner and the contractor

20. A constructive change is a change to the contract that can legally be construed to have been made, even though the owner did not issue a specific, written change order. Which of the following situations is normally a constructive change?

(A) request by the engineer-architect to install OSHA-compliant safety features

(B) delay caused by the owner's failure to provide access

(C) rework mandated by the building official

(D) expense and delay due to adverse weather

21. A bid for foundation construction is based on owner supplied soil borings showing sandy clay to a depth of 12 ft. However, after the contract has been assigned and during construction, the backhoe encounters large pieces of concrete buried throughout the construction site. This situation would normally be referred to as

(A) concealed conditions

(B) unexplained features

(C) unexpected characteristics

(D) hidden detriment

22. If a contract has a value engineering clause and a contractor suggests to the owner that a feature or method be used to reduce the annual maintenance cost of the finished project, what will be the most likely outcome?

(A) The contractor will be able to share one time in the owner's expected cost savings.

(B) The contractor will be paid a fixed amount (specified by the contract) for making the suggestion, but only if the suggestion is accepted.

(C) The contract amount will be increased by some amount specified in the contract.

(D) The contractor will receive an annuity payment over some time period specified in the contract.

23. A contract has a value engineering clause that allows the parties to share in improvements that reduce cost. The contractor had originally planned to transport concrete on site for a small pour with motorized wheelbarrows. On the day of the pour, however, a concrete pump is available and is used, substantially reducing the contractor's labor cost for the day. This is an example of

(A) value engineering whose benefit will be shared by both contractor and owner

(B) efficient methodology whose benefit is to the contractor only

(C) value engineering whose benefit is to the owner only

(D) cost reduction whose benefit will be shared by both contractor and laborers

24. A material breach of contract occurs when the

(A) contractor uses material not approved by the contract to use

(B) contractor's material order arrives late

(C) owner becomes insolvent

(D) contractor installs a feature incorrectly

25. When an engineer stops work on a job site after noticing unsafe conditions, the engineer is acting as a(n)

I. agent

II. local official

III. OSHA safety inspector

IV. competent person

(A) I only

(B) I and IV

(C) II and III

(D) III and IV

26. While performing duties pursuant to a contract for professional services with an owner/developer, a professional engineer gives erroneous instruction to a subcontractor hired by the prime contractor. To whom would the subcontractor look for financial relief?

I. professional engineer

II. owner/developer

III. subcontractor's bonding company

IV. prime contractor

(A) III only

(B) IV only

(C) I and II only

(D) I, II, III, and IV

27. A professional engineer employed by a large, multinational corporation is the lead engineer in designing and producing a consumer product that proves injurious to some purchasers. What can the engineer expect in the future?

(A) legal action against him/her from consumers

(B) termination and legal action against him/her from his/her employer

(C) legal action against him/her from the U.S. Consumer Protection Agency

(D) thorough review of his/her work, legal support from the employer, and possible termination

28. A professional engineer in private consulting practice makes a calculation error that causes the collapse of a structure during the construction process. What can the engineer expect in the future?

(A) cancellation of his/her Errors & Omissions insurance

(B) criminal prosecution

(C) loss of his/her professional engineering license

(D) incarceration

29. A professional engineer is hired by a homeowner to design a septic tank and leach field. The septic tank fails after 18 years of operation. What sentence best describes what comes next?

(A) The homeowner could pursue a tort action claiming a septic tank should retain functionality longer than 18 years.

(B) The engineer will be protected by a statute of limitations law.

(C) The homeowner may file a claim with the engineer's original bonding company.

(D) The engineer is ethically bound to provide remediation services to the homeowner.

SOLUTIONS

1. The three different forms of company ownership are the (1) sole proprietorship, (2) partnership, and (3) corporation.

A *sole proprietor* is his or her own boss. This satisfies the proprietor's ego and facilitates quick decisions, but unless the proprietor is trained in business, the company will usually operate without the benefit of expert or mitigating advice. The sole proprietor also personally assumes all the debts and liabilities of the company. A sole proprietorship is terminated upon the death of the proprietor.

A *partnership* increases the capitalization and the knowledge base beyond that of a proprietorship, but offers little else in the way of improvement. In fact, the partnership creates an additional disadvantage of one partner's possible irresponsible actions creating debts and liabilities for the remaining partners.

A *corporation* has sizable capitalization (provided by the stockholders) and a vast knowledge base (provided by the board of directors). It keeps the company and owner liability separate. It also survives the death of any employee, officer, or director. Its major disadvantage is the administrative work required to establish and maintain the corporate structure.

2. To be legal, a contract must contain an *offer*, some form of *consideration* (which does not have to be equitable), and an *acceptance* by both parties. To be enforceable, the contract must be voluntarily entered into, both parties must be competent and of legal age, and the contract cannot be for illegal activities.

3. A written contract will identify both parties, state the purpose of the contract and the obligations of the parties, give specific details of the obligations (including relevant dates and deadlines), specify the consideration, state the boilerplate clauses to clarify the contract terms, and leave places for signatures.

4. A consultant will either charge a fixed fee, a variable fee, or some combination of the two. A one-time fixed fee is known as a *lump-sum fee*. In a *cost plus fixed fee* contract, the consultant will also pass on certain costs to the client. Some charges to the client may depend on other factors, such as the salary of the consultant's staff, the number of days the consultant works, or the eventual cost or value of an item being designed by the consultant.

5. A *retainer* is a (usually) nonreturnable advance paid by the client to the consultant. While the retainer may be intended to cover the consultant's initial expenses until the first big billing is sent out, there does not need to be any rational basis for the retainer. Often, a small retainer is used by the consultant to qualify the client (i.e., to make sure the client is not just shopping around and getting free initial consultations) and as a security deposit (to make sure the client does not change consultants after work begins).

6. The Engineers Joint Contract Documents Committee (EJCDC) consists of the National Society of Professional Engineers, the American Council of Engineering Companies (formerly the American Consulting Engineers Council), the American Society of Civil Engineers, Construction Specifications Institute, and the Associated General Contractors of America. The American Institute of Architects is not a member, and it has its own standardized contract documents.

The answer is (D).

7. A business in the construction industry must have two or more employees and a minimum annual gross sales volume of $500,000 to be subject to the Fair Labor Standards Act (FLSA). Individual coverage also applies to employees whose work regularly involves them in commerce between states (i.e., interstate commerce). Any person who works on, or otherwise handles, goods moving in interstate commerce, or who works on the expansion of existing facilities of commerce, is individually subject to the protection of the FLSA and the current minimum wage and overtime pay requirements, regardless of the sales volume of the employer.

The answer is (B).

8. A *double-breasted* design firm serves both union and nonunion clients. When union-affiliated companies find themselves uncompetitive in bidding on nonunion projects, the company owners may decide to form and operate a second company that is *open shop* (i.e., employees are not required to join a union as a condition of employment). Although there are some restrictions requiring independence of operation, the common ownership of two related firms is legal.

The answer is (C).

9. *Without expressed authority* means each member of a partnership has full authority to obligate the partnership (and the other partners).

The answer is (A).

10. Only general partners can vote in a partnership. Both silent and limited partners share in the profit and benefits of the partnership, but they only contribute financing and do not participate in the management. The identities of silent partners are often known only to a few, whereas limited partners are known to all.

The answer is (B).

11. Limited liability companies have limited liability for all members, no double taxation (unless electing to be taxed as a coproration), more than one type of stock, and no continuity of life. They are not limited in members, and they are comparatively fairly easy to establish.

The answer is (D).

12. One of the reasons for forming joint ventures is that the bonding capacity is aggregated. Even if each contractor cannot individually meet the minimum bond requirements, the total of the bonding capacities may be sufficient.

The answer is (C).

13. Normal corporations, S corporations, and limited liability companies (LLCs) can have unlimited shareholders. Sole proprietorships are for individuals.

The answer is (B).

14. When the specifications include a nonstructural, brand-named article and the accompanying phrase "or approved equal," the contractor can substitute something with the same functionality, even though it is not the brand-named article. "Or approved equal" would not be used with a structural detail such as a connection design.

The answer is (D).

15. Municipalities that experience frequent construction projects within their boundaries have standard specifications that are included by reference in every project's construction document set. This document set would cover items such as concrete, fire hydrants, manhole structures, and curb requirements.

The answer is (C).

16. *Proposal bonds*, also known as *bid bonds*, are insurance policies payable to the owner in the event that the contractor backs out after submitting a qualified bid.

The answer is (D).

17. *Bid rigging*, also known as *price fixing*, is an illegal arrangement between contractors to control the bid prices of a construction project or to divide up customers or market areas. *Bid shopping* before or after the bid letting is where the general contractor tries to secure better subcontract proposals by negotiating with the subcontractors. *Bid peddling* is done by the subcontractor to try to lower its proposal below the lowest proposal. *Bid unbalancing* is where a contractor pushes the payment for some expense items to prior construction phases in order to improve cash flow.

The answer is (C).

18. The construction contract includes many items, some explicit only by reference. The contractor may be required to carry liability insurance, but the policy itself is between the contractor and its insurance company, and is not normally part of the construction contract.

The answer is (C).

19. Changes to the contract can be made by either the owner or the contractor. The method for making such changes is indicated in the contract. Almost always, the change must be agreed to by both parties in writing.

The answer is (D).

20. A *constructive change* to the contract is the result of an action or lack of action of the owner or its agent. If the project is delayed by the owner's failure to provide access, the owner has effectively changed the contract.

The answer is (B).

21. *Concealed conditions* are also known as *changed conditions* and *differing site conditions.* Most, but not all, contracts have provisions dealing with changed conditions. Some place the responsibility to confirm the site conditions before bidding on the contractor. Others detail the extent of changed conditions that will trigger a review of reimbursable expenses.

The answer is (A).

22. Changes to a structure's performance, safety, appearance, or maintenance that benefit the owner in the long run will be covered by the value engineering clause of a contract. Normally, the contractor is able to share in cost savings in some manner by receiving a payment or credit to the contract.

The answer is (A).

23. The problem gives an example of efficient methodology, where the benefit is to the contractor only. It is not an example of value engineering, as the change affects the contractor, not the owner. Performance, safety, appearance, and maintenance are unaffected.

The answer is (B).

24. *A material breach of the contract* is a significant event that is grounds for cancelling the contract entirely. Typical triggering events include failure of the owner to make payments, the owner causing delays, the owner declaring bankruptcy, the contractor abandoning the job, or the contractor getting substantially off schedule.

The answer is (C).

25. An engineer with the authority to stop work gets his/her authority from the agency clause of the contract for professional services with the owner/developer. It is unlikely that the local building department or OSHA would have authorized the engineer to act on their behalves. It is also possible that the engineer may have been designated as the jobsite's "competent person" (as required by OSHA) for one or more aspects of the job, with authority to implement corrective action.

The answer is (B).

26. In the absence of other information, the subcontractor only has a contractual relationship with the prime contractor, so the request for financial relief would initially go to the prime contractor. The prime contractor would ask for relief from the owner/developer, who in turn may want relief from the professional engineer. The subcontractor's bonding company's role is to guarantee the subcontractor's work product against subcontractor errors, not engineering errors.

The answer is (B).

27. Manufacturers of consumer products are generally held responsible for the safety of their products. Individual team members are seldom held personally accountable, and when they are, there is evidence of intentional wrongdoing and/or gross incompetence. The manufacturer will organize and absorb the costs of the legal defense. If the engineer is fired, it is likely this will occur after the case is closed.

The answer is (D).

28. A professional engineer is not held to the standard of perfection, so making a calculation error is neither a criminal act nor sufficient evidence of incompetence to result in loss of license. Without criminal prosecution for fraud or other wrongdoing, it is unlikely that the engineer would ever see the inside of a jail. However, it is likely that the engineer's bonding company will cancel his/her policy (after defending the case).

The answer is (A).

29. Satisfactory operation for 18 years is proof that the design was partially, if not completely, adequate. There is no ethical obligation to remediate normal wear-and-tear, particularly when the engineer was not involved in how the septic system was used or maintained. While the homeowner can threaten legal action and even file a complaint, in the absence of a warranty to the contrary, the engineer is probably well-protected by a statute of limitations.

The answer is (B).

76 Engineering Ethics

PRACTICE PROBLEMS

(Each problem has two parts. Determine whether the situation is (or can be) permitted legally. Then, determine whether the situation is permitted ethically.)

1. (a) Was it legal and/or ethical for an engineer to sign and seal plans that were not prepared by him or prepared under his responsible direction, supervision, or control?

(b) Was it legal and/or ethical for an engineer to sign and seal plans that were not prepared by him but were prepared under his responsible direction, supervision, and control?

2. Under what conditions would it be legal and/or ethical for an engineer to rely on the information (e.g., elevations and amounts of cuts and fills) furnished by a grading contractor?

3. Was it legal and/or ethical for an engineer to alter the soils report prepared by another engineer for his client?

4. Under what conditions would it be legal and/or ethical for an engineer to assign work called for in his contract to another engineer?

5. A licensed professional engineer was convicted of a felony totally unrelated to his consulting engineering practice.

(a) What actions would you recommend be taken by the state registration board?

(b) What actions would you recommend be taken by the professional or technical society (e.g., ASCE, ASME, IEEE, NSPE, and so on)?

6. An engineer came across some work of a predecessor. After verifying the validity and correctness of all assumptions and calculations, the engineer used the work. Under what conditions would such use be legal and/or ethical?

7. A building contractor made it a policy to provide cell phones to the engineers of projects he was working on. Under what conditions could the engineers accept the phones?

8. An engineer designed a tilt-up slab building for a client. The design engineer sent the design out to another engineer for checking. The checking engineer sent the plans to a concrete contractor for review. The concrete contractor made suggestions that were incorporated into the recommendations of the checking contractor. These recommendations were subsequently incorporated into the plans by the original design engineer. What steps must be taken to keep the design process legal and/or ethical?

9. A consulting engineer registered his corporation as "John Williams, P.E. and Associates, Inc." even though he had no associates. Under what conditions would this name be legal and/or ethical?

10. When it became known that a chemical plant was planning on producing a toxic product, an engineer employed by the plant wrote to the local newspaper condemning the chemical plant's action. Under what conditions would the engineer's action be legal and/or ethical?

11. An engineer signed a contract with a client. The fee the client agreed to pay was based on the engineer's estimate of time required. The engineer was able to complete the contract satisfactorily in half the time he expected. Under what conditions would it be legal and/or ethical for the engineer to keep the full fee?

12. After working on a project for a client, the engineer was asked by a competitor of the client to perform design services. Under what conditions would it be legal and/or ethical for the engineer to work for the competitor?

13. Two engineers submitted bids to a prospective client for a design project. The client told engineer A how much engineer B had bid and invited engineer A to beat the amount. Under what conditions could engineer A legally/ethically submit a lower bid?

14. A registered civil engineer specializing in well-drilling, irrigation pipelines, and farmhouse sanitary systems took a booth at a county fair located in a farming town. By a random drawing, the engineer's booth was located next to a hog-breeder's booth, complete with live (prize) hogs. The engineer gave away helium balloons with his name and phone number to all visitors to the booth. Did the engineer violate any laws/ethical guidelines?

15. While in a developing country supervising construction of a project an engineer designed, the engineer discovered the client's project manager was treating local workers in an unsafe and inhumane (but, for that country, legal) manner. When the engineer objected, the client told the engineer to mind his own business. Later, the local workers asked the engineer to participate in a walkout and strike with them.

(a) What legal/ethical positions should the engineer take?

(b) Should it have made any difference if the engineer had or had not yet accepted any money from the client?

16. While working for a client, an engineer learns confidential knowledge of a proprietary production process being used by the client's chemical plant. The process is clearly destructive to the environment, but the client will not listen to the objections of the engineer. To inform the proper authorities will require the engineer to release information that was gained in confidence. Is it legal and/or ethical for the engineer to expose the client?

17. While working for an engineering design firm, an engineer was moonlighting as a soils engineer. At night, the engineer used the employer's facilities to analyze and plot the results of soils tests. He then used his employer's computers to write his reports. The equipment and computers would otherwise be unused. Under what conditions could the engineer's actions be considered legal and/or ethical?

18. Ethical codes and state legislation forbidding competitive bidding by design engineers are

(A) enforceable in some states

(B) not enforceable on public (nonfederal) projects

(C) enforceable for projects costing less than $5 million

(D) not enforceable

SOLUTIONS

Introduction to the Solutions

Case studies in law and ethics can be interpreted in many ways. The problems presented are simple thumbnail outlines. In most real cases, there will be more facts to influence a determination than are presented in the case scenarios. In some cases, a state may have specific laws affecting the determination; in other cases, prior case law will have been established.

The determination of whether an action is legal can be made in two ways. The obvious interpretation of an illegal action is one that violates a specific law or statute. An action can also be *found to be illegal* if it is judged in court to be a breach of a written, verbal, or implied contract. Both of these approaches are used in the following solutions.

These answers have been developed to teach legal and ethical principles. While being realistic, they are not necessarily based on actual incidents or prior case law.

1. (a) Stamping plans for someone else is illegal. The registration laws of all states permit a registered engineer to stamp/sign/seal only plans that were prepared by him personally or were prepared under his direction, supervision, or control. This is sometimes called being in *responsible charge.* The stamping/signing/sealing, for a fee or gratis, of plans produced by another person, whether that person is registered or not and whether that person is an engineer or not, is illegal.

An illegal act, being a concealed act, is intrinsically unethical. In addition, stamping/signing/sealing plans that have not been checked violates the rule contained in all ethical codes that requires an engineer to protect the public.

(b) This is both ethical and legal. Consulting engineering firms typically operate in this manner, with a senior engineer being in responsible charge for the work of subordinate engineers.

2. Unless the engineer and contractor worked together such that the engineer had personal knowledge that the information was correct, accepting the contractor's information is illegal. Not only would using unverified data violate the state's registration law (for the same reason that stamping/signing/sealing unverified plans in Prob. 1 was illegal), but the engineer's contract clause dealing with assignment of work to others would probably be violated.

The act is unethical. An illegal act, being a concealed act, is intrinsically unethical. In addition, using unverified data violates the rule contained in all ethical codes that requires an engineer to protect the client.

3. It is illegal to alter a report to bring it "more into line" with what the client wants unless the alterations represent actual, verified changed conditions. Even when the alterations are warranted, however, use of the unverified remainder of the report is a violation of the state registration law requiring an engineer only to stamp/sign/seal plans developed by or under him. Furthermore, this would be a case of fraudulent misrepresentation unless the originating engineer's name was removed from the report.

Unless the engineer who wrote the original report has given permission for the modification, altering the report would be unethical.

4. Assignment of engineering work is legal (1) if the engineer's contract permitted assignment, (2) all prerequisites (e.g., notifying the client) were met, and (3) the work was performed under the direction of another licensed engineer.

Assignment of work is ethical (1) if it is not illegal, (2) if it is done with the awareness of the client, and (3) if the assignor has determined that the assignee is competent in the area of the assignment.

5. (a) The registration laws of many states require a hearing to be held when a licensee is found guilty of unrelated, but nevertheless unforgivable, felonies (e.g., moral turpitude). The specific action (e.g., suspension, revocation of license, public censure, and so on) taken depends on the customs of the state's registration board.

(b) By convention, it is not the responsibility of technical and professional organizations to monitor or judge the personal actions of their members. Such organizations do not have the authority to discipline members (other than to revoke membership), nor are they immune from possible retaliatory libel/slander lawsuits if they publicly censure a member.

6. The action is legal because, by verifying all the assumptions and checking all the calculations, the engineer effectively does the work. Very few engineering procedures are truly original; the fact that someone else's effort guided the analysis does not make the action illegal.

The action is probably ethical, particularly if the client and the predecessor are aware of what has happened (although it is not necessary for the predecessor to be told). It is unclear to what extent (if at all) the predecessor should be credited. There could be other extenuating circumstances that would make referring to the original work unethical.

7. Gifts, per se, are not illegal. Unless accepting the phones violates some public policy or other law, or is in some way an illegal bribe to induce the engineer to favor the contractor, it is probably legal to accept the phones.

Ethical acceptance of the phones requires (among other considerations) that (1) the phones be required for the job, (2) the phones be used for business only, (3) the phones are returned to the contractor at the end of the job, and (4) the contractor's and engineer's clients know and approve of the transaction.

8. There are two issues: (1) the assignment and (2) the incorporation of work done by another. To avoid a breach, the contracts of both the design and checking engineers must permit the assignments. To avoid a violation of the state registration law requiring engineers to be in responsible charge of the work they stamp/sign/seal, both the design and checking engineers must verify the validity of the changes.

To be ethical, the actions must be legal and all parties (including the design engineer's client) must be aware that the assignments have occurred and that the changes have been made.

9. The name is probably legal. If the name was accepted by the state's corporation registrar, it is a legally formatted name. However, some states have engineering registration laws that restrict what an engineering corporation may be named. For example, all individuals listed in the name (e.g., "Cooper, Williams, and Somerset—Consulting Engineers") may need to be registered. Whether having "Associates" in the name is legal depends on the state.

Using the name is unethical. It misleads the public and represents unfair competition with other engineers running one-person offices.

10. Unless the engineer's accusation is known to be false or exaggerated, or the engineer has signed an agreement (confidentiality, nondisclosure, and so on) with his employer forbidding the disclosure, the letter to the newspaper is probably not illegal.

The action is probably unethical. (If the letter to the newspaper is unsigned it is a concealed action and is definitely unethical.) While whistle-blowing to protect the public is implicitly an ethical procedure, unless the engineer is reasonably certain that manufacture of the toxic product represents a hazard to the public, he has a responsibility to the employer. Even then, the engineer should exhaust all possible remedies to render the manufacture nonhazardous before blowing the whistle. Of course, the engineer may quit working for the chemical plant and be as critical as the law allows without violating engineer-employer ethical considerations.

11. Unless the engineer's payment was explicitly linked in the contract to the amount of time spent on the job, taking the full fee would not be illegal or a breach of the contract.

An engineer has an obligation to be fair in estimates of cost, particularly when the engineer knows no one else is providing a competitive bid. Taking the full fee would be ethical if the original estimate was arrived at logically and was not meant to deceive or take advantage of the client. An engineer is permitted to take advantage of economies of scale, state-of-the-art techniques, and break-through methods. (Similarly, when a job costs more than the estimate, the engineer may be ethically bound to stick with the original estimate.)

12. In the absence of a nondisclosure or noncompetition agreement or similar contract clause, working for the competitor is probably legal.

Working for both clients is unethical. Even if both clients know and approve, it is difficult for the engineer not to "cross-pollinate" his work and improve one client's position with knowledge and insights gained at the expense of the other client. Furthermore, the mere appearance of a conflict of interest of this type is a violation of most ethical codes.

13. In the absence of a sealed-bid provision mandated by a public agency and requiring all bids to be opened simultaneously (and the award going to the lowest bidder), the action is probably legal.

It is unethical for an engineer to undercut the price of another engineer. Not only does this violate a standard of behavior expected of professionals, it unfairly benefits one engineer because a similar chance is not given to the other engineer. Even if both engineers are bidding openly against each other (in an auction format), the client must understand that a lower price means reduced service. Each reduction in price is an incentive to the engineer to reduce the quality or quantity of service.

14. It is generally legal for an engineer to advertise his services. Unless the state has relevant laws, the engineer probably did not engage in illegal actions.

Most ethical codes prohibit unprofessional advertising. The unfortunate location due to a random drawing might be excusable, but the engineer should probably refuse to participate. In any case, the balloons are a form of unprofessional advertising, and as such, are unethical.

15. (a) As stated in the scenario statement, the client's actions are legal for that country. The fact that the actions might be illegal in another country is irrelevant. Whether or not the strike is legal depends on the industry and the laws of the land. Some or all occupations (e.g., police and medical personnel) may be forbidden to strike. Assuming the engineer's contract does not prohibit participation, the engineer should determine the legality of the strike before making a decision to participate.

If the client's actions are inhumane, the engineer has an ethical obligation to withdraw from the project. Not doing so associates the profession of engineering with human misery.

(b) The engineer has a contract to complete the project for the client. (It is assumed that the contract between the engineer and client was negotiated in good faith, that the engineer had no knowledge of the work conditions prior to signing, and that the client did not falsely induce the engineer to sign.) Regardless of the reason for withdrawing, the engineer is breaching his contract. In the absence of proof of illegal actions by the client, withdrawal by the engineer requires a return of all fees received. Even if no fees have been received, withdrawal exposes the engineer to other delay-related claims by the client.

16. A contract for an illegal action cannot be enforced. Therefore, any confidentiality or nondisclosure agreement that the engineer has signed is unenforceable if the production process is illegal, uses illegal chemicals, or violates laws protecting the environment. If the production process is not illegal, it is not legal for the engineer to expose the client.

Society and the public are at the top of the hierarchy of an engineer's responsibilities. Obligations to the public take precedence over the client. If the production process is illegal, it would be ethical to expose the client.

17. It is probably legal for the engineer to use the facilities, particularly if the employer is aware of the use. (The question of whether the engineer is trespassing or violating a company policy cannot be answered without additional information.)

Moonlighting, in general, is not ethical. Most ethical codes prohibit running an engineering consulting business while receiving a salary from another employer. The rationale is that the moonlighting engineer is able to offer services at a much lower price, placing other consulting engineers at a competitive disadvantage. The use of someone else's equipment compounds the problem since the engineer does not have to pay for using the equipment, and so does not have to charge any clients for it. This places the engineer at an unfair competitive advantage compared to other consultants who have invested heavily in equipment.

18. Ethical bans on competitive bidding are not enforceable. The National Society of Professional Engineers' (NSPE) ethical ban on competitive bidding was struck down by the U.S. Supreme Court in 1978 as a violation of the Sherman Antitrust Act of 1890.

The answer is (D).